ADIRONDACK
DISCARDED
LIBRARY
BAY ROAD

*Mothers and Daughters of Invention*

ANIMAL & COMMUNITY
DISPOSAL
BAY ROAD

# Mothers and Daughters of Invention

## NOTES FOR A REVISED HISTORY OF TECHNOLOGY

*by Autumn Stanley*

**RUTGERS UNIVERSITY PRESS**
*New Brunswick, New Jersey*

First published in hardcover by The Scarecrow Press, Inc., 1993

First published in paperback by Rutgers University Press, 1995

**Library of Congress Cataloging-in-Publication Data**

Stanley, Autumn, 1933–
    Mothers and daughters of invention : notes for a revised history of technology / by Autumn Stanley.
      p.     cm.
    Includes bibliographical references and index.
    ISBN 0-8135-2197-1
    1. Women in technology—History.   2. Women inventors—History.
  I. Title.
  T36.S73     1995
  604′.82—dc20                                  94-46973

British Library Cataloguing-in-Publication data available

Copyright © 1993 by Autumn Stanley
All rights reserved.
Manufactured in the United States of America
Printed on acid-free paper

*To Becky Schroeder,*
 *who started it all by getting a patent at age 12;*

*to Charlotte Smith,*
 *who wanted to write this book a hundred years ago;*

*and, most of all, to my mother, Donna Stanley,*
 *who saw to it that I got a college education.*

# Contents

# *Preface*

The question most often asked when people heard I was writing a book on women inventors was, "How did you ever get interested in that?" In retrospect, it seems the topic came looking for me. In the 1970s I was a senior developmental editor, specializing in the sciences, for a large publisher of college texts. We had nearly finished a big new biology text when we realized it did not mention a single woman biologist. I called the authors, who all but said, "There aren't any." Appalled, I started a crash course of research into women's major achievements in the biological sciences, interviewed feminist scientists, and managed belatedly to get some important women mentioned.

Determined never to face such a situation again, I expanded my research in women's achievement to other disciplines, creating an archive of women's contributions in each field for our editors and authors to draw upon. Our authors were far more willing to insert women in appropriate chapters if I had already done the necessary work and could present them with the scientific and biographical information ready to use.

In the midst of that crash research program, I chanced to read a newspaper story about **Becky Schroeder,** an Ohio girl who patented a luminous-lined writing board at twelve. Perhaps partly because I'm from Ohio, I was struck by her story, and suddenly wondered why I had never heard of any other female inventors. Women are so obviously creative, and I had just been studying women's scientific achievements in depth, and yet I couldn't name a single one. None of my friends, all college graduates, could name any, either. I started looking for material on the subject and found absolutely nothing.

Right then, I knew I had found my topic. For the same reason that the first thing I ever knitted was a pair of argyll socks in four colors, that my freshman term paper was on Lucky Luciano, and that I had just done an asparagus cookbook, I now had to write about women inventors. Things that range from unlikely to impossible, things that have never been done before—those are the things I like to do. Thus it was that I willingly stepped up and took a tiger by the tail, a tiger that has not let me rest for more than ten years.

It was not just the difficulty of the thing—the late Stacy Jones, long-time patents columnist for the *New York Times,* told me he'd considered the idea and decided it was too hard—or the fact that no one had ever done it before that appealed to me. The project was the ultimate in detective work—a little like trying to solve a murder that happened hundreds or thousands of years ago, one that the authorities do not really want solved and most people deny ever happened. I always wanted to be Nancy Drew.

Perhaps, too, I always wanted to be an inventor. Certainly my mind is always inventing, improving, solving problems, throwing out ideas. Some of my ideas have been patented (by others) during the course of my research, and I felt I had some instinctive understanding of the inventor's psyche.

When I began, I intended to write an article. At the end, the manuscript could have run 3,000 pages. In the course of researching and writing the book as it finally took shape, I have re-educated myself in several sciences; studied anthropology, archaeology, and history of technology; created a substantial private archive on women's achievement in science and technology, among other fields; found material and ideas for several more books—and discovered my true work. I met Charlotte Smith through this research (see

Dedication) and look forward to the small biography I will write on her. I now know the shape of what I hope will be my masterpiece.

One of the reasons the project has taken so long is that I wanted to write so that if I ever got a rejection letter from a publisher, it would read something like the following letter from an apocryphal Chinese publisher:

> We have read your manuscript with boundless delight. If we were to publish your paper, it would be impossible for us to publish any work of a lower standard. And as it is unthinkable that, in the next 1,000 years, we shall see its equal, we are, to our regret, compelled to return your divine composition and beg you a thousand times to overlook our short sight and timidity.

The other reasons for the length of the project are too familiar in women's history to need rehearsing here. There were, however, some special difficulties of a pioneering work—no close models, no ready-made core bibliography, virtually no reliable secondary sources, no mentors or fully informed evaluators to give advice along the way (only experts in various fields to comment on parts). And it is only as a pioneering work that the resulting book deserves any special indulgence. Otherwise, though I can't be sure that the Chinese publisher would respond as above, I send it forth with some pride and much joy, as well as with extreme relief at letting go the tiger's tail.

Autumn Stanley
Portola Valley, CA
1993

# *Acknowledgments*

In the course of a ten-year project, a scholar gets help from more people than can easily be imagined. I only hope that I have recorded the great majority of the names—and that those few who fall through the cracks will add forgiveness to their other virtues.

First must be mentioned those women who encouraged me, in the thin early days of the project, to believe that it was possible. Here I refer not to my own ability to do the task—of which I had perhaps too little doubt—but to whether other scholars would value the work, and whether the academic establishment would eventually recognize it. Martha Trescott, Joan Rothschild, and Nancy Martin provided me forums for my work, gave me material on promising women, guided me to background reading, and read early drafts of the first writing to come out of the project. Joan published both my first article on women and agricultural technology (in a special edition of *Women's Studies International Quarterly,* later separately published by Pergamon) and my most important work until the appearance of this book, "Women Hold Up Two-Thirds of the Sky," in *Machina Ex Dea* (Pergamon, 1983), and invited me to and joined me on panels at scholarly meetings both here and abroad. Joan's book, *Teaching Technology from a Feminist Perspective,* is available from Pergamon, and her book on reproductive technologies is forthcoming.

Martha lent me a copy of an important dissertation on women engineers, corresponded with me, and shared her work and her faith in me over the years. Specifically, she pointed me to important information on **Lillian Gilbreth, Bertha Lammé,** and **Julia Hall;** most recently, she commented on a late draft of the "Machines" chapter. Happily, her own book on women engineers will soon appear.

Elizabeth Reed, retired geneticist and active advocate for women in science, answered my appeal for information in *Ms.* magazine in the 1970s. She sent me names and references over the years, recently steering me to an important inventor of soybean products. Her book on early American women scientists is in press at the Women's Press.

Others who sent me information early on are Dorothy Stephenson, then President of the Inventors' Association of New England; Neil Daniel (a Texas college teacher) who alerted me to **Bette Graham,** inventor of Liquid Paper®; and the late Bette Graham herself, who answered queries about her invention and her life, sending me the manuscript of a biographical chapter on her in a book on self-made women. Eleanor Maass sent me material on women inventors from *New Century for Women.* Claire Radford of the Pio Pico Museum in Whittier, CA, searched southern California archives on **Harriet R. Strong,** sending me data far beyond what was available in *Notable American Women.*

Deborah Warner, Assistant Curator of Physical Sciences at the Smithsonian, sent me two of the three or four most important existing sources for my research: the U.S. Patent Office's list of patents granted to women between 1809 and February 28, 1895; and the first issue of Charlotte Smith's 1891 periodical *The Woman Inventor.* Author of the most informative and carefully researched article on women inventors up to its date ("Women Inventors at the Centennial" in Martha Trescott, ed., *Dynamos and Virgins Revisited* [1979]), she has been an inspiration for my work.

Since encouragement and validation are both so crucial and so difficult for independent scholars, a special vote of thanks goes to the Center for Research on Women at Stanford University (now the Institute for Research on Women and Gender) for appointing me an affiliated scholar for two terms (1984–8). The Institute gave me a forum for my work, opened professional doors, and introduced me to many marvelous women. Most important to me—doing exciting work of their own, but never too busy to encourage a fellow scholar—are Annette Lawson, author of an important study of adultery; Edith Gelles, whose biography of Abigail Adams is in press; and the fine playwright Elizabeth Roden. Historian Karen Offen helped me formulate my application, and gave incisive advice on presented work.

Staff members of inventors' associations and institutes for invention and innovation, whether private or connected with universities, were often quite helpful. Noteworthy here is Maggie Weisberg, then of Inventors' Workshop International in the Los Angeles area, now of M&M Associates, Chehalis, WA, who sent me names, brief biographies, achievements, and current addresses of many inventors from her organization, and who continued to respond to further inquiries over the entire decade.

Later in the project, Fred Amram of the Communications Department at the University of Minnesota told me about **Patsy Sherman,** co-inventor of Scotchgard®, and sent me copies of some of his articles. In 1988 he offered kind hospitality in Minneapolis after the NWSA sessions, and sent me information about his exhibit on women inventors at the Goldstein Gallery in Minneapolis ("Her Works Praise Her").

Many other scholars were generous with their time, their research, and their books. Cynthia Cockburn, a London sociologist, sent me copies of two of her books, *Brothers* and *Machinery of Dominance,* spent hours with me discussing our work, and read a draft of one of my chapters. She has been unfailingly encouraging. Anthropologist Ruby Rohrlich, a fount of ideas, was admirably patient with a novice's queries, and sent me an important paper. Francesca Bray of the East Asian History of Science Library in Cambridge, England, talked with me in Cambridge in the early 1980s, answered questions about women's participation in Asian agriculture by letter, and sent me a copy of her very fine *Agriculture* volume (Part II, vol. 6) in Joseph Needham's series, *Science and Civilisation in China.* Kathleen Ochs of the Colorado School of Mines, early contributor to Garland's bibliography on women and technology (Bindocci), read an early version of Chapter 3. Catherine Callaghan of Ohio State's Linguistics Department answered linguistic questions, sent me her articles, and enthusiastically discussed the book with me. Londa Schiebinger, a scholar and academic in history of science, now at Pennsylvania State University, invited me to lecture in her VTSS class at Stanford, and commented on the Health/Medicine chapter.

Helen Irvin, formerly on the faculty at my undergraduate alma mater, Transylvania College, sent me a vital reference about **Mary Walton**'s significant environmental invention (a means of deadening the sound of elevated railroads), as well as considerable information on Shaker women's inventions. Her book, *Women in Kentucky,* appeared in 1979.

Virginia Beauchamp of the University of Maryland pointed me to such highly significant Maryland women as the inventor of buffered aspirin (**Mrs. Shipley** of Annapolis), a black woman engineer (**Mary James**) who designed some components of the Stanford Linear Accelerator, and **Mary Shorb,** inventor of an important assay leading to a treatment for pernicious anemia. She also housed me on a research trip to her area, and intelligently discussed my ongoing work. Her book, *A Private War: Letters and Diaries of Madge Preston, 1862–1867,* appeared in 1987.

B.M.E. O'Mahoney of Wetherden, Suffolk, England, showed me his work on **Henrietta Vansittart,** arguably Britain's first woman engineer. The paper later appeared in *The Woman Engineer,* and his longer manuscript is now in the Fawcett Library. He also invited me down to the country for a lovely lunch with him and Mrs. O'Mahoney, and an afternoon's good discussion. North Carolina historian Elizabeth Reid Murray sent me vital background information on **Beulah Louise Henry,** the "Lady Edison."

Sandra Myres, University of Texas (Arlington) historian and author of *Westering Women* (1982) ransacked her files for women inventors and innovators, and raised my spirits with her letters over the years.

A San Francisco genealogical researcher—Wilber Leeds—freely shared with me his research on the Dietz family, in response to a query about the **Dietz sisters** who invented the Cyclone Snowplow in the 1880s; and retired engineer Guenter Holzer shared with me his notes and unpublished work on **Bertha Lammé.** Emily Stothart of Coushatta, LA, gave me detailed information about her grandmother, early-20th-century calculator inventor **Emily Duncan,** and sent me draft and published copies of her books on Emily and her husband (both inventors).

Diane Swendeman of NASA/Ames located several women patentees at NASA, and led me to good source people at other NASA facilities. Marva Jett did the same at the Department of Agriculture. Also helpful at the DOA were G. Terry Sharrer and Vivian Wiser.

Friends and acquaintances, people I met at parties, on buses and trains or at scholarly meetings—all kinds of people—got interested in my book and gave me tips about inventors they knew or knew of. Thanks are especially due to Jan Butin Sloan of the Kansas City Art Institute, who told me about **Fannie Hesse;** to Joanne Verplanck, who led me to **Beulah Henry;** to Matthias Urigoyem, whose mother, **Anne Gackle,** had the idea for the jet pilots' ejection seat, first installed on the P-38; and Fred Null, a San Francisco Bay Area sculptor whose mother, **Rose Baren Null,** independently conceived reclosable tops for detergent boxes and children's books that play music, among other inventions.

On a scale of one to ten for both knowledge and helpfulness, David Doughan of the Fawcett Library in London must rank twelve or fifteen. Over a period of at least five years he cheerfully added a voluminous correspondence with me to his regular library duties (at one time his letter files contained one binder for me and one for all other correspondents), answering endless queries, unearthing such fascinating women as **Amelia Lewis,** 19th-century inventor of the People's Stove and cooking system, and providing precious encouragement from abroad. He also arranged for me to speak on my work to the Friends of the Fawcett in 1985. He read parts of a complete draft of my book, and filed that manuscript for posterity on the shelves of the Fawcett. His own work with Denise Sanchez, *Feminist Periodicals, 1855–1984: An Annotated Critical Bibliography of British, Irish, Commonwealth and International Titles,* appeared in 1987.

Librarians, in fact, will always hold a special place in my heart as a result of this project. The reference librarians at Menlo Park, CA, City Library were an ever-present resource, cheerful and efficient. Joanne Hoffman, a reference librarian at Stanford University's Green Library, deserves special mention. She was consistently helpful, patient, and inspiring in her quiet competence. Among other things, I will always remember her brilliant detective work on the elusive Charlotte Smith, the 19th-century reformer and advocate for women inventors to whom I dedicate this book. Judit Brody of the Science Museum Library was a great help with British patents, and shared with me her findings on British women inventors.

Public librarians in the smaller towns and cities willingly answered queries about inventors who had lived in their towns as much as a century earlier. Some became interested in my project, or in their particular local star, and did research for me far beyond the call of duty, or passed my inquiry on to a descendant or to a knowledgeable local historian. The town librarian of Jennings, LA, for example, put me in touch with descendants of **Emily Duncan;** Margaret Harding and Marjorie Kober of the Crete and Wilber, NE, libraries, respectively, searched newspapers and cemetery records for data on **Anna Dunder;** and Dennis Clark of the Meriden, CT, public library searched city directories for information on **Lura Brainerd.** Los Altos, CA, school librarian Nancy Tucker showed me the Katherine Lee thesis.

Deserving special mention among historical society librarians are Helen Ransom of the Connecticut Historical Society and Helena Wright of the Merrimack Valley Textile Museum, both of whom joined my search for the industrial loom improvement invented early in the 19th century by **Lydia Hotchkiss Crosby Atwood,** grandmother of the first notable U.S. opera singer, Clara Kellogg. We never found technical details or firm patent information, but the family and biographical data and clues about the invention that appear in my text are largely due to their efforts. Ruby Kelsey of the Troy, OH, Historical Society did extensive and fruitful research, well beyond the call of duty, on **Bertha Baumer.** Others of these valuable helpers will be footnoted at appropriate spots in the text.

Mary Elena Goodan of Menlo Park, CA, was one of my first research assistants, and one of my best, finding things I hadn't even asked her to find, persisting on what I did ask, and providing typewritten reports. She was also very encouraging about the project, which must have seemed amorphous at that point. Another early and very fine helper was Sharon Refvem, now an architect doing postgraduate work with a prominent architect in Switzerland. Stanford students Laurie Foster and Ellen Lehman were both very much interested in the project, and among the few who worked long enough to become really useful. Ellen now works in publishing; at last report, Laurie intended to become a midwife. Other research assistants over the years were Cheryl Baker from UC–Santa Cruz, a precise worker who typed my first bibliography; Connie Klopsch; Deborah Bruseth, who had not only a very sharp brain and excellent typing skills, but undeniably the most beautiful handwriting of the lot—far more important than one might imagine; Susie Barrett; Belinda Hopkinson; and Lois Blosk.

Edith Axelson, already a fine genealogical researcher, has learned the skills of archival patent research, to become, quite simply, indispensable since 1988.

Outstanding among my male assistants was Matt Cantor, an architecture graduate from UC–Berkeley, who helped me for several months at Wilbur Hot Springs in the early 1980s. At that point I was struggling to do scholarly research while living two hours from the nearest decent library, in a place that lacked electricity, plumbing, and central heating. Matt not only did good work, but cheered me through some very dark days. Eric Baldwin, Tim Coshow, and Corrie Robb made up in intelligence for what they lacked in handwriting, and were helpful for short periods of time in the mid-1980s and in 1991.

Perhaps the helpers who deserve the greatest thanks, however, are those who worked without pay, out of interest in and desire to help the project, and realizing that all research costs came out of my own pocket. Noteworthy among this number are Margaret Hunt, who told me early on about the inventor of the chenille-tufting process, and tried to track her down for me; and Ann Bassetti, who pored through issues of *Inventive Age* in the Library of Congress basement, unearthing several women inventors of the late 19th and early 20th centuries.

My mother provided information about early 20th-century sanitary pads, and did some very helpful census and local-history research on the fascinating Ohio inventor **Nancy Graham;** and my three younger children have worked at various times as research assistants. Holly typed my first master list of women inventors, and did some in-depth research on a Washington State woman. Iris found some books for me at the University of Oregon, and later helped scan Patent Office name lists for women's names. Kevin did some excellent city-directory and other research for me in Chicago, and surveyed ten years of Stacy Jones' columns for women patentees.

Among the many delights and blessings of this long-term learning experience have been the many people I have met because of it, most of whom have helped me in one way or another. To all of them, named and unnamed, my heartfelt thanks.

# Introduction:
## "Women Hold Up Two-Thirds of the Sky"

In 1981 I wrote a paper with this Mao-revising title. The two-thirds fraction was chosen advisedly, to reflect findings from United Nations studies of time budgets for the two sexes worldwide, from anthropology, and from personal experience and observation to the effect that women do considerably more than their share of the work of the world. Two-thirds seems a reasonable estimate.

Equally important is the subtitle of that paper (Stanley/Rothschild): "Notes for a Revised History of Technology." I have adopted it as the subtitle of my book, for my purpose in writing the first substantial, serious treatment of women inventors is nothing less than to make a beginning at revising the history of technology.

The nature of that beginning is twofold: one, to focus for the first time on women's contributions to human technology; and two, to revise the received definition of technology, particularly the way it is *applied* in histories of technology, in college physics, engineering, and history courses, and in everyday life. Indeed, I hope to show that including women's contributions will not merely revise but transform the history, and especially the prehistory, of technology. When technology is no longer just what men do, but what people do,[1] both the definition of *technology* and the definition of *significant technology* must inevitably change.

In short, *Mothers and Daughters of Invention* is a pioneering and, I hope, a pathbreaking work. In writing it, I walked a precarious line between Lynn White's dictum that a historian should be bold, and the ever-present warning that fools (read "novices") rush in where angels (read "experts") fear to tread.

It is also a work of compensatory history, and unashamedly so, though it is now much else besides. *Mothers and Daughters of Invention* began in the late 1970s, when no book had ever been done on women as inventors.[2] Not only were there no books, but even book chapters and articles on women as contributors to technology were vanishingly rare. The few that did exist tended to date from the 19th century, with that era's reticence about names and sources, or to be superficial treatments—bare lists of inventions with no inventors' names—"dancing dogs" articles, or brief profiles of an inventor or two of the author's acquaintance. Even such welcome exceptions as Mozans' chapter on women inventors was early (1913) and contained serious errors. The only known periodical devoted to women inventors (Charlotte Smith's *Woman Inventor* of Washington, DC;

---

1. I am indebted to the late Lynn White, Jr., for this liberating overstatement.
2. Otis T. Mason's *Woman's Share in Primitive Culture* is an isolated exception. However, as its title indicates, it is limited to primitive cultures, and invention is not its sole focus. It has not really been superseded, though it appeared in 1894. As for today, a few small books on women inventors have now appeared. One is in French (Moussa, [1986]), a chatty, popular volume, dealing exclusively with living women. As this book goes to press Moussa announces another book, in English, but I have not seen it. Janet Panabaker of the Women Inventors Project of Canada has published a small but creditable volume of biographical sketches titled *Inventing Women* (1991). Vare & Ptacek (1988) is also a popular treatment, also biographical. Liberally pockmarked with such serious errors as crediting **Marie Curie** with inventing the Geiger counter, it is further marred by deliberately conflating discovery and invention (leading to such bêtises as crediting women with "inventing" clay!), and padded out with material on discoveries as well as inventions. Suffice it to say that the need for a compensatory history in the field of women and technology is not appreciably diminished by the appearance of this latter book.

1891) lasted just two issues. My first computerized search of the subject turned up exactly two newspaper articles, one of which I was never able to find.

Librarians did not even use *Women inventors* as a category for filing information. A computerized search of the holdings of the University of California Library on the subject, done at Berkeley as late as 1986, also came up with nothing. And histories of invention or technology are still appearing that mention either no women or only a token few. Ernest Heyn's *Fire of Genius* (1976), for example, mentions only three, all patronizingly. Valérie-Anne Giscard d'Estaing's *World Almanac Book of Inventions* (1985), mentions more women than the standard history—over twenty, all told—but still a disappointing number, and often hides the inventor's gender by using last names or first initials only.[3] Most disturbing, she omits virtually all of women's highly significant medical inventions, such as the Nobel Prize–winning radioimmunoassay (**Rosalyn Yalow**).

*Notable American Women* (*NAW*) lists only three women as inventors in its first edition (1607–1950), and none at all in its second (through 1975). Two different kinds of omission are involved here. First, the *NAW* compilers omit several noteworthy women who were probably professional inventors—or the 19th-century female equivalent—such as **Eliza Alexander, Maria Beasley, Helen Blanchard, Mary Carpenter** (later **Hooper**), **Mary Jane Montgomery,** and **Emily Evans Tassey**—to name only several who specialized in machines. Particularly striking here is the omission of the woman who invented the automatic dishwasher, **Josephine Cochran** (or **Cochrane**) of Shelbyville and Chicago, IL. Second, the compilers often fail to mention the inventive achievements of women included for other accomplishments. **Dr. Mary Edwards Walker** is a glaring example here. Her biographer seems more concerned with casting doubt on her medical credentials and debunking her Civil War service than with researching her real achievements and the complex richness of her mind—which not only conceived the first Equal Rights Amendment, virtually as it now reads, and foresaw the limitations of the focus on suffrage, but invented several things, some with great commercial impact even today. Her three reform garments are now forgotten, but Mary Walker also invented the inside neckband for men's shirts, the return postcard for business letters, the return-receipt postcard for registered mail, and the idea of printed return addresses on envelopes. She is also credited with suggesting to the Post Office a method it duly adopted for handling third-class mail—and, in the field of medicine, with new preventive methods for tuberculosis. Not one of these inventions and innovations is mentioned in *NAW*.[4]

In other words, the field of women and technology is at least a decade behind such fields

---

3. She also misses women credited in other sources. She omits any mention of **Lady Lovelace** (whom she wrongly calls Lady Ada Byron) in connection with Babbage's proto-computer (293). Noting that the computer language Ada was named after her, and calling her the first computer programmer (299), she then omits her from the index. **Grace Hopper**'s early contributions to computer technology are virtually ignored; the discussion of COBOL does not mention her at all, the first assembly language is credited to a British team rather than to her, etc. (cf. my Chap. 5). Indeed, no women are mentioned in connection with any of the major computer languages or hardware developments discussed under data processing and information systems. **Catherine Greene** is not mentioned. Giscard d'Estaing dates the "mechanical" automobile windshield wiper to 1916, ignoring **Mary Anderson** (c. 1866–1953), who patented hers in 1903; and credits the hang-glider to Francis Rogallo without mentioning his wife and co-inventor, **Gertrude Rogallo** (W. Garrison). These are just a few glaring examples that leap to mind.

4. A welcome exception to this kind of omission is **Harriet Hosmer,** whose childhood mechanical and inventive bent and later experiments with perpetual-motion machines are both mentioned in her *NAW* biography. Her artificial marble compound, however—her only commercially successful invention—is omitted.

as women and art, literature, and music—or even women and science. There is still a need for what might be called collecting the tokens and naming the names.

In any such pioneering work it is doubly incumbent on the author to make explicit her underlying views and assumptions from the start. Since these views and assumptions will inevitably shape both the research and the arguments made from it, they cannot be ignored.

## Inventive Capacity

First and foremost, I assume that men can be every bit as inventive as women. In other words, *inventive ability is probably independent of biological gender.* If inventive ability correlates with any other human trait or capacity, I suggest it is artistic talent—an impression tentatively supported by my ongoing study of women inventors' profiles. Another obvious correlate would be so-called set-breaking thinking. Suppose that both artistic ability and the ability to break sets ultimately come down to an ability (or willingness) to receive intuitions, inspirations, clues, visions, pictures, or whatever one chooses to call messages from the right brain—and use them to see problems in a new way. Inventive ability might then consist, on the biological or neurophysiological level, of access to the right brain and, on the psychological level, of receptiveness to right-brain input, plus the ability and persistence to solve the newly re-visioned problems in a practical way.

Controversy still rages over possible sex differences in connections between the right and left human brains. Common sense suggests that if females do indeed, as some have claimed, have more of these connections than males, and if inventive talent consists in any significant part of openness and access to right-brain thinking, then, all else being equal, women should be *more* inventive than men. But more recent work casts doubt on the original findings of gender difference in this regard. Moreover, of course, all else has not been equal in Western and proto-Western culture since the Bronze Age, and it starts being unequal from the moment of birth; so this thesis is impossible to test at the moment.[5]

Among the many things not equal for the two sexes in modern Western culture are three pertinent here: freedom to explore the physical environment, freedom to engage in rough-and-tumble play, and free time to play in general. Females have less of all three than males. This is true for little girls as well as for adult women. I suspect that it has been true in most human cultures, at least from sedentism onward. Indeed, even among other primates the females probably have less time for play (Boulding, 62). The first two of these freedoms are sometimes thought to influence the development of spatial-visualization ability, and if so could affect some kinds of inventive capacity, as well as inventive *opportunity* (more about play itself below). I suggest that women inventors will often, like **Harriet Hosmer,** have been tomboys in their youth.

No one really doubts that women are ingenious and creative. Anyone harboring such doubts need only ponder the tale of the woman bootlegger in India who reportedly taught monkeys to guard her jungle camp, warn her of approaching intruders, and then, if the intruders proved to be revenue police, to rush over to them on her signal, and pull down their pants to give her time to escape! (It scarcely matters whether this delightful story from a tabloid weekly is apocryphal or not; attribution of the strategem to a woman is

---

5. Had the original "imperfectly lateralized" brains been male rather than female, would the researchers have called them "better integrated" instead? It will be interesting to see how this controversy is resolved, and what happens to the terminology as a result. On the question of how early, and how unconsciously, parents begin to instill gender differences in their babies, see Maccoby & Jacklin.

significant in itself.) Or consider the American woman, **Dr. Mary Willard,** who has taught monkeys to serve as aides to quadriplegics—or the French-Canadian nun **Jeanne Mance,** who devised the world's first prepaid health plan, in the 18th century—or the Chinese woman physicist who invented a suitcase-size xerography machine costing about $1,500 as compared to Xerox Corporation's machine the size of two tables and costing thousands of dollars just to rent. Or consider the genetic index of grandparenthood created by **Dr. Mary-Claire King** at UC–Berkeley and **Dr. Ana Maria Di Lonardo** in Argentina, giving grandparents—particularly the organized Grandmothers—the necessary evidence to sue in court to reclaim their "disappeared" grandchildren. The mother of *Spycatcher* author Peter Wright kept him alive as a premature infant in war-time Britain by improvising an incubator from glass chemical jars and hot water bottles. He was born during a zeppelin raid, and no medical services were available (12).

Female creativity, then, can scarcely be denied. It is merely that our sex-role stereotypes seek to confine that creativity to such "acceptable" areas as art, music, dance, writing, and cooking, whereas "real" invention and technology have to do with weapons and machines and chemical compounds created in laboratories. Thus most of us—including virtually all historians of technology to date—manage to hold in our minds the two contradictory ideas that women are highly creative beings and that women do not invent. Cognitive dissonance, I suggest, is held at bay by the traditional identification of *invention/men/ machines,* a linkage I hope to help break by devoting an entire chapter to women's achievement in mechanical invention.

## Inventive Motive and Opportunity

I assume that, all else being equal, for much of human prehistory the workers created their tools. That is, whoever did a given activity devised, made or assembled, and improved the necessary tools and procedures. Thus in pre-agricultural and nonhierarchical—or what I sometimes call pre-Takeover—societies, women's motive and opportunity for invention and innovation varies only with their economic tasks and the pressure on performance of those tasks.

Even later, the same basic variable of sexual division of labor is of primary importance in determining women's motive and opportunity to invent. Unfortunately, from about the Bronze Age to the mid-19th century in Western culture, women's areas of work, at least for the middle and upper classes, were ever more narrowly circumscribed. Perhaps even more important—for women continued to *labor* in such fields as agriculture, mining, construction, etc., long thereafter—they were no longer in any sense "in charge" of any area except what came to be called the domestic one.

## Inventive Achievement

This principle—that women invent wherever they work at a dignified level, and particularly where they are exclusively or primarily charged with that endeavor—means that in proto-human and gathering-hunting societies, and even as late as early horticultural times, if women were in charge of butchering the animals, or slicing and drying the meat, or of debittering the acorns or detoxifying the cassava and processing its starch, preparing the animal skins, or of whatever cooking was done, then they most likely invented and improved the processes and compounds and tools used in those activities: knives and scrapers, tanning methods and chemistry; food-preserving and -processing technology—

including refinements in the exploitation of fire, storage containers, preservative agents, steatite cooking surfaces or vessels, cooking stones, ovens, mortars, pestles, grain-grinders, etc.

Indeed, one anthropologist has suggested that the apparatus and techniques used for processing sago starch should be classed as engineering.

If women were in charge of medicine, especially herbal healing, which seems likely considering the number of primitive societies where women are the exclusive or primary gatherers of vegetal foods, then women probably first *discovered* the healing properties of the several herbs, and then *invented* the remedies based thereon. If women were in charge of weaving, I assume that they invented the first looms, with loom weights, combs, and all other attendant technology—which of course demanded an earlier invention of spinning and techniques of obtaining fibers from nature—and improved them thereafter. Consider, for example, the ingenious backstrap loom found anciently among the Moche women of Peru (Alva, 44–5). Wherever women were the first potters, they most likely developed kilns from ovens, or from cooking pits. They soon learned to mix the clay with sand or straw or bits of broken pots, and invented means of turning the pots as they worked (at first, turntables or simply a rounded stone imbedded in the bottom of the pot rather than true wheels).

In other words, through early horticultural times, or until sedentism allowed specialization to run amok—and perhaps even later for some persistently female activities—we can look at the sexual division of labor in a given society and get a fair idea what women invented. And since, in significant numbers of groups, the men reportedly hunted and/or fished and then sat down, while the women did everything else, women's contributions to technology up through the early Neolithic is most likely tremendous. Indeed, the three traditional hallmarks of the Neolithic Revolution—agriculture, pottery, and weaving—are all generally agreed to be women's work and women's inventions.[6]

Notice that two kinds of argument are possible here: from the division of labor in a given society at a given level of specialization; and from statistical considerations of numbers/percentages of primitive societies that assigned a given task to one sex or the other, whether exclusively, primarily, or "equally."[7]

I assume further that women's achievements in technology have virtually always been underreported. For example, one can read innumerable archaeological reports focused entirely on arrowheads,[8] as if the people could eat, sleep, wear, and worship arrowheads, rather than just kill with them. Women did, of course, hunt, as Agnes Estioko-Griffin and

---

6. Controversy has lately arisen over whether women can indeed be credited with inventing agriculture, doubtless part of a backlash in response to recent revisionist work by feminists in anthropology. In any case, my own assumption is that women are the likelier inventors of intentional cultivation, for reasons that will become clear in my Agriculture chapter.

7. As will be seen particularly in the case of agriculture, men sometimes receive credit for "equal" participation when all they do is clear the fields before cultivation begins. This may not even happen every year, and even where it does, the hours of labor involved are minuscule compared to the hours spent by women in planting, weeding, watering, and harvesting the crops throughout the rest of the year.

8. Part of this, of course, came about because stone arrowheads, stone axes, and the like survive far better than almost all the materials that Neolithic and later women were usually thought to work with, except of course for pottery: wood, cloth, bone, plants, leather, and fiber and textile materials. However, women undoubtedly also created stone tools, and this has been hidden by the stereotype. Otis Mason makes this clear as early as the 1890s; and Flannery's work (Ucko & Dimbleby) unwittingly reveals that ground (as opposed to chipped) stone tools are more closely associated with women's work than with men's at the time of the transition to grinding technology.

her husband learned from the Agta people of the Philippines (Dahlberg, ed.), and this, too, has been underreported. However, projectile hunting seems to have been more of a male than a female activity in most cultures to which Western scholars look for antecedents. Indeed, some have argued that it was projectile hunting by males—and the control of fire by females—that paved the way for the initial sexual division of labor in human technology. Note, however, that shot for slings could be baked clay as well as stones (did women use the one and men the other, or did women produce the missiles for everyone?), and that the North European periglacial culture may be as exceptional among the world's cultures several thousand years ago as the traditional Eskimo are today.[9] Sadly, our view of prehistory is largely determined by this apparently hunting-obsessed culture.

Even when the people are still around, and being studied by anthropologists rather than archaeologists, and even in the late 1980s, the accounts are often androcentric in their focus on the hunting bows, arrows, other weapons—and whatever it is that the males do—while the women's activities are taken for granted as part of the landscape. Indeed, in unconscious testimony to this attitude, anthropologists may speak of women's activities—if at all—in the passive voice, as if they had no agent. Or they trivialize women's work by unconsciously confusing primitive women with modern Western women who are not in the paid workforce. One would like to report that this is no longer happening, but unfortunately it is still with us. In John Pfeiffer's 1986 *Smithsonian* magazine article on the Cro-Magnons, the author focuses on the men going bravely off hunting—and making tools, doing the art, calendric, and other admired activities—while the women cooked and "tidied up the hearth"![10]

Given such persistent trivialization and underreporting, it seems arguable that what we know or can deduce from archaeology and anthropology about women's inventions is merely the tip of an iceberg.

Naturally, the same underreporting or total omission plagues history as well as archaeology and anthropology. Paradoxically, indeed, the danger may be greater here; for we tend to trust written sources and not use our eyes, whereas art work and artifacts have a stubborn reality of their own. For example, if the text accompanying an ancient seal carving or vase painting speaks of a god and *his* consort, we might just fit that neatly into our Western concept of religion without noticing that the female is larger than the male, clothed while he is naked, seated while he is standing, etc. In other words, what might really be represented above that androcentric caption is a goddess and her son/consort or even a suppliant.

The problem is so severe, in fact, that it is tempting to search for women's inventive achievement in the silences of history. In other words, whatever no male claims credit for—those inventions whose descriptions begin with the words "No one knows who invented the . . ."—may well be the work of a woman. A possible case in point here is the spinning wheel, which appeared as early as the 11th century AD in China, reaching Europe in the Middle Ages, and there being improved by various male inventors whose names *are* recorded (or their unsung wives or sisters) to create the form we know of it today.

9. Peoples living north of the Arctic Circle are the only known exception to the rule that among fully gathering-hunting societies women provide 60–80% of the group's food by weight (Lee & DeVore).

10. Pfeiffer does credit the women with repairing (did they also invent and originally make them?) reindeer-skin tents for the summer. In this work they use bone needles (76). However, when he talks about the invention of the needle, he speaks of "people" who "plied the world's first needles to fashion 'tailored' goods, warm form-fitting clothing" (77–8).

## The Matriarchy Perplex

It is tempting to sidestep this question, since it seems to evoke knee-jerk responses on both sides. However, with every passing year of research, the links between technology and power become clearer. Sophie Drinker, after studying women's achievements in music for two decades, was adamant that being in charge mattered tremendously to women's motivation for innovation.

Copper tools did not replace obsidian (where they did so) in the Chalcolithic Near East because they were uniformly better, for in many cases the obsidian tools they replaced were equal or superior to the copper ones (Wertime, in Schmandt-Besserat, ed., 20–1). And recent work has shown that obsidian blades are sometimes superior even to steel, for certain kinds of surgery (Press, "*Cutting*"). There had to be other reasons why copper was adopted, having to do with power, or at least prestige. Likewise, the modern steel axes being introduced into tropical sago-processing areas are not superior, at least in speed, to the traditional stone axes for felling sago palms. Yet steel is rapidly replacing stone (see, e.g., Townsend). The introduction of steel axes has demonstrably upset the power relations within Australian aboriginal society—in the case I've seen described, to the detriment of *male* power (L. Sharp).[11]

Again, modern Western medicine is not uniformly superior to local herbal remedies. Yet—perhaps partly because of the fearsome magic of the guns behind the doctors—Western medicine has not only displaced local medical systems, but paved the way for Colonial power and the Christian religion as well, most notably in Africa in this century.

Thus the question whether women were ever "in charge" at least of their own work, if not of society, becomes inescapable. As my thought has evolved in the course of writing this book, I now welcome the question, for it gives me the opportunity to do a little stereotype-bashing, and at the same time to address the question of re-periodizing history so that it makes sense for women.

My own assumptions are as follows: In proto-human and early gathering-hunting times, relations between the sexes would have been roughly egalitarian, with age, ability, skill, and experience more important than gender in conferring whatever power may have existed in our terms. Decision-making was consensual or shifting with expertise or special circumstances. If either sex had a power advantage at this time, it was probably the females, as they had the awesome power of bleeding without dying, and they alone could give birth to young. Even after birth, the young would die unless women fed them. Women also provided 60–80% of the food of the group except for peoples living north of the Arctic Circle or around the glaciers. Even among the Eskimo, where male hunters provided virtually all of the group's food, survival depended on the marvelous fitted fur clothing created by the women; and this would have been obvious every time the man emerged from the igloo to hunt.

Women may have gained a further edge where they were, as Elise Boulding speculates, the first to control fire. However, the males' generally greater physical strength could have been a counterbalancing factor here.

In short, at this stage of human development the sexes pursued complementary and often interdependent occupations and were recognized for these contributions to the group. As Eleanor Leacock so persuasively argues ("Women"), -*archy* as we know it

---

11. Interestingly enough, this was the *only* example of the introduction of a new technology affecting power relations in a society to be used in a Stanford University Values, Technology, Science and Society course I audited and guest-lectured for, in 1985–6, even though my research indicates that the opposite power shift, from female to male, is in fact overwhelmingly more frequent.

simply did not exist.[12] And as Peggy Sanday convincingly shows (*Female Power*), women's status and power relations between the sexes are best where the economic contribution of the two is approximately equal. To Sanday's formulation I would add only that this may also work if the contributions are *felt* to be equal, whether they actually are, when closely analyzed for time and effort expended, or not.

The first shift toward a male advantage in this early balance may have come with the invention in the Upper Paleolithic of projectile hunting—the slings and stones, the spears and spear-throwers, and eventually the bows and arrows—that allowed hunting to be done individually or by small parties, and at chosen times rather than in large cooperative group drives or irregularly when opportunity presented itself. These inventions paved the way for what has almost uniformly been reported as a male specialization. (We should keep always in mind, however, the corrective case of the *Agta women, who hunt as much and with the same equipment as men,* except during very late pregnancy and very early lactation. See also Ruth Landes on Ojibwa women, Jane Goodale on Tiwi women, and Leacock on women in egalitarian societies [Bridenthal, Koonz, & Stuard, e.g., 26].)[13]

Indeed, Lila Leibowitz has argued that projectile hunting was the *first* specialization, and the first real division of labor by sex, whereas females and males probably originally foraged equally for vegetal foods. Marek Zvelebil finds evidence in a Mesolithic cemetery of northern Europe for a male specialization in hunting as the only exclusively gendered role. Other specialized roles in the society, as indicated by grave goods, burial position, etc., were occupied equally by both sexes; that is, approximately equal numbers of females and males are found in such graves. All of the skeletons buried with indications of a specialized hunting role, however, were male.

The second shift toward male advantage in proto-Western cultures probably came—or at least the groundwork was laid for it—whenever men learned the male role in human procreation. It has been argued that this happened as early as the representations of bisons mounting in the famous cave art of Altamira in the Upper Paleolithic, about 13,000 years ago. Of course, we cannot be certain that the artists were male,[14] or that these depictions had the meaning we give them; perhaps they simply reflect observations of bison behavior, which would obviously include mounting. Moreover, it seems by no means inevitable, even if the Altamiran artists understood the bison pictures much as we do, that they generalized from bison males to human males. At least, it does these people no insult to hypothesize that, like certain Christian theorists today, they distinguished between themselves and (other) animals.

My own assumption is that at some point people—probably females first—began to suspect the male role in both animal and human reproduction; but for a long time they held simultaneously in their minds the contradictory idea that women conceived by the wind, the moon, the sea, or by some divine or supernatural agency. A transitional view may be,

12. My thinking here, and throughout my book, owes much to the work of Leacock, Ruby Rohrlich, Peggy Sanday, and the authors of the excellent articles collected by Frances Dahlberg in *Woman the Gatherer*.

13. If the idea that women might have invented spear-throwers and other weapons as well as or instead of men seems difficult to handle, see the well-researched fiction of Jean Auel, whose *Valley of the Horses* posits just such a scenario.

14. Indeed, recent work on the technology of the cave paintings opens the question. To whatever degree women were more expert, by virtue of their daily tasks, with the technology of exploiting fire, they may have been the ones who produced the necessary pigments, for several of these were heated, and reheated, to very high temperatures during production (Leroi-Gourhan, "The Archaeology of Lascaux Cave," *Scientific American,* June 1982: 104–12).

as reportedly held by some modern-day primitive people, that intercourse is necessary but not sufficient, that it *opens the way* for the fertilizing agent or principle, or builds up the fetus once it begins to develop. Perhaps for some groups it was not until animals were routinely domesticated, several thousand years later, that the relationship became unmistakable. Only then could gestation periods and breeding behavior be observed at close quarters and over several seasons; only then did the negative effects of getting rid of the troublesome males become apparent.

Once this male role was unmistakable to a whole group, in any case, it could become an enabling or psychological element in a male bid for power.

To recapitulate, then, humans began from a position of relative sexual equality with perhaps some female advantage, and moved through a shift toward male power where projectile hunting became important and a male specialization. The next sharp shift—the first really great imbalance—toward female power, came with the advent of horticulture. Whatever arguments may be advanced against the idea that women invented "agriculture," women seem everywhere to be the first gardeners. In most horticultural societies ever studied, women are exclusively or primarily in charge of cultivation. Even where researchers report "equal" participation by females and males in horticultural work, the men usually just clear the fields, whereas the women do all the rest of the work throughout the cultivative year—or years. Among the Tupi-Guarani of South America, for example, the men cleared gardens that were then used by women, who did virtually all the other cultivative work, for the next four to six years. In short, the men worked about two months every five years, devoting the rest of their time to hunting, fishing, feasting, drinking, and warfare (Murdock & Provost; Coontz & Henderson, eds, 79)! How this came to be defined as "equal" participation seems an important question.

At one of the few sites where we possess a completely excavated series of habitation levels from a mixed economy of early horticulture supplemented by gathering and hunting, to the end of hunting and full dependence on cultivation—Çatal Hüyük in the Near East (modern Turkey)—James Mellaart reports that representations of the male hunting god become less frequent and gradually disappear, while representations of a female deity become more frequent and finally dominate as cultivation triumphs, about 5800 BC. He finds clear evidence of women's high status in this society, including the larger, permanent sleeping platforms for women in the houses, with the children buried under them; the red ochre on female skeletons, often buried in shrines; and artistic representations.

If this means what it seems to mean, and if Çatal Hüyük is at all representative, women reached the zenith of their power in the early Neolithic, when their cultivative specialty provided the economic basis of the whole culture; when the major deity was conceived as female; and men, though beginning to help with cultivation and to develop their own specializations—including the eventually fateful ones of warfare and trade—were relatively underemployed.[15] Perhaps males felt undervalued and somewhat at a loss.

Then at some point in the later Neolithic, and particularly as the Bronze Age began, the pendulum began to swing the other way. The reasons are complex, and not yet all known. Among them were a population (or at least a birth) explosion, and a resulting greater dependence on agricultural products; technological changes—the introduction of the metal-tipped, animal-drawn plough and long-distance, large-scale irrigation—and the

---

15. Çatal Hüyük is not the only evidence, nor Mellaart the only proponent, for this view. See, for example, Marija Gimbutas' *Goddesses and Gods of Old Europe* and other writings, and Elise Boulding's chapter on the Neolithic agrovillages.

resulting professionalization of agriculture; a postulated drying trend in the climate (or changes in local river courses)[16] that made irrigation ever more necessary and conferred new power and prestige on the holders of its secrets; religious changes coming with the incursion of nomads worshipping a single male deity and bearing a patriarchal culture; and economic changes attendant upon a market economy and long-distance trade, with its need for surplus goods and agricultural produce. When the population—and water needs—in the Near East reached the point where nomads could no longer pasture their flocks among the farmers in winter, the nomads could either fight for their ancient rights or settle among the farmers. In some cases they probably did one and then the other. If they attacked, or threatened attack, and their leaders were bought off with high positions within the agricultural society, they would bring their patriarchal values and their male god into influence with them; and since the priests of at least one such male god, Jaweh, suppressed all other god(esse)s and their priest(esse)s—who, coincidentally, must have held a significant share of the economic power of the agricultural society, if Sumer and its temples are at all typical—the religious takeover had inevitable political and economic implications for women.

Even if the nomads had not been a factor, war might have broken out over rights to irrigated land, and warfare was bad for women. Not only did it give males a new power base within their own society, eventually giving rise to kingship, when war leaders refused to step down at war's end, but it led to the enslavement of women whose men were killed in battle. Women were probably kept alive, it should be noted, not only for sexual reasons or because they were more docile than men, as usually suggested, but because of their skills in weaving, pottery, dyeing, jewelry-making, brewing, cooking, farmwork, etc., which contributed to the wealth of their captors. Once women from another group were seen as slaves, it may have been easier for men to see women of their own group, and women in general, as somehow inferior. Perhaps even more important, once women were seen as "other," as these captured women inevitably were, abuse seems thinkable and control seems necessary.

This is not the place to analyze all the factors in women's loss of status during the Bronze Age in the Near Eastern cultures giving rise to our own. Scholars do not by any means agree, in any case, either about the precise nature of women's previous (late Neolithic) status, or about the causes of the loss in status that seems indisputable from the decline of female deities, and from the law codes of the third and second millennia BC.

To recapitulate my own view in brief: women started out fairly equal, maintained this general balance through men's specialization in projectile hunting and their own specialization in uses of fire; reached an all-time status high in the early agricultural societies of the late Neolithic, and then, perhaps partly because of that imbalance, suffered a sharp decline in status and power, which I call the Takeover, essentially complete by the end of the Bronze Age in the Near Eastern cultures giving rise to our own. The situation between the Bronze Age and the present has continued dismal, with new declines initiated by the Christian takeover of Europe, the Colonial takeover of Africa and America, and the 19th-century Cults of Domesticity and Invalidism in England and the United States. Only with the second wave of the women's movement beginning in the late 1960s has there been any significant upturn in women's political and economic status and achievement, and it remains to be seen how far that will go.

16. As Andrew Sherratt points out, the question of climatic changes in the prehistoric Near East is a minefield, where it is nearly impossible to be right. However, if a city that depends on irrigated fields loses its source of irrigation water (as in Sumer, when an all-important river suddenly changes course), a preoccupation with water becomes understandable.

However, isolated bright spots and exceptions must be noted. In the Near East, specifically, even as specialization grew, and agriculture became a profession rather than something the women did along with all the rest of their work, the early city-states were probably "ruled" by the high priestess in religious and some economic matters, and in other matters by a council of elders of both sexes who would, for example, choose the war leader.

Worldwide, there was at the very least a time when descent was matrilineal, residence was matrilocal, and women had far more status and power than they have now. This is *not* the same as matriarchy; but it differed from the present state of affairs in ways crucial for women's contribution to technology, as for all aspects of their lives. Even after the advent of kingship—which I prefer to call *monarchy,* since it allows for female rulers—some women could and did rule in their own names; and these events, like women's other accomplishments, have been glossed over, misinterpreted, and underreported. For example, the Ashanti head of state in pre-Colonial Africa was a woman. She was neither the king's wife nor his mother, but what we might call a High Queen: she appointed the king, and was above him in the hierarchy. Yet Western observers of Ashanti society mention this order of things only in passing and call this powerful woman by the misnomer of queen mother (Coontz & Henderson, eds, 23). Consider, too, the fierce Woman Chief of the Crow (1806–54) who led her adopted tribe not only in the hunt, but in war. Eventually, her village of 160 lodges named her their greatest warrior and third highest chief (Starr).

Again, one would like to be able to say that such underreporting or omission of women rulers is a thing of the past. Unfortunately, it is not. As a glaring recent example, see the May 1986 *Smithsonian* article on the Maya (W.W. Johnson). Scanning either text or illustrations, one would get the impression that Maya rulers were exclusively male. Females appear only in erotic or subordinate positions relative to males. Buried in a parenthetical phrase on page 45 of the article is the information that *two of the nineteen rulers at Palenque were women.*

Finally, there may even have been societies that should be called matriarchal. *At least,* and this is crucial, *they cannot be called patriarchal.* Since our stereotypes so blind us to what is before our eyes, and a patriarchal and misogynist stereotype has informed most of what we call history, I submit that we do not really know whether there ever were matriarchies or not, and now may never know. But one of the best candidates for such a classification, as Ruby Rohrlich has convincingly argued, is prehistoric Crete— where there were female deities *and no dominating male deities.* Says Sarah Pomeroy with characteristic super-caution, "There is, to be sure, a dearth of clear evidence establishing the matriarchy, but there is equally, as no reviewer pointed out, a lack of conclusive evidence proving the existence of patriarchy in Minoan Crete" (cited in Downing, 10n).

All of which brings me to the position I really wish to take on this vexed question, and the one that I would have espoused without preamble or defense were it not, as I said, for the inextricable connections between technology and power in all stratified societies:[17]

> Once we see that Crete might have been neither a matriarchy nor a patriarchy, we are free to ask what Anne Barstow calls "a more realistic question about prehistoric societies, namely, did men and women ever relate in ways other than dominance/ subjection?"

---

17. I am indebted for the wording here, and for much clear thinking on a tortured subject, to Marymay Downing, who challenges the received view of prehistoric religion by cogent reference to Crete.

## *Why, Then, So Few?*

If women have at least equal inventive capacity with men, and were in the early days of human economic development probably the premiere technologists by virtue of their larger areas of work, why, then, are there so few great women inventors? The answer is that they have not been few; it is only that history has not recorded them. In other words, the real question is not, Why so few? but, *Why do we know so few?* The easiest and most obvious answer for modern times is that women hold far fewer patents than men. Most people equate patents and inventions, and though I will always distinguish between them here, the connection is inevitable for purposes of this question. Estimates of the percentage of patents granted to women in the United States in the 200 years of our patent office to date average about 2%. Almost 95% of all U.S. patents are still granted to men; and in Britain and other developed nations, with the possible exception of the former Soviet Union, the situation is markedly worse.

Since patents are such an important factor in the stereotype, let us pause to consider briefly why women hold so few patents. The simplest reason is that patents cost money, and women have historically had, and still have, less money than men. Second, patents take time. The less legal and business experience an inventor has, the more time is required, especially if she tries to proceed without a patent attorney. And women have not only less disposable income but less disposable time than men.

Third, in the 19th century in both the United States and Britain, until the Married Women's Property Acts were passed, even if a married woman invented and patented something, it would be her husband's property. Thus she might simply abandon the idea, or allow the husband to patent it in his name to begin with.

Fourth, both inventing and patenting require some technical expertise. Since women have been systematically excluded from technical training for more than four thousand years in Western and proto-Western cultures, it should be no surprise that women inventors often lack the chemical, mechanical, metalworking, or electronic expertise to put their ideas into working, practical form. Susan McDaniel and her colleagues report from their Canadian sample (8) that "Lack of expertise . . . in certain domains . . . was a serious challenge to many of the women we interviewed."

Faced with this problem, the woman with an invention idea has four choices: 1) learn the skill herself (which may create time, money, and self-image problems); 2) hire an expert; 3) try to sell the idea to an invention-development group or to a company in the field; or 4) give up. A frustrated inventor from Burbank, CA, **Gina Piediscalzzi,** began studying electronics and plastics technology in the mid-1970s, in order to be able to give form to her invention ideas (Weisberg, L-8/16/76).

The best alternative is probably #2, but problems arise even here, as many women lack the ready cash to hire highly paid experts, and thus find themselves signing over a part interest in the invention for the help they need. Nevertheless, many successful women inventors have chosen this route, hiring expert consultants to help complete their inventions. Early in the 19th century, for example, **Martha Coston** enlisted a chemist in the last stages of completing her husband's night-signalling system, eventually used in most of the world's navies. **Dorothy Rodgers** abandoned the patent process for her individualized people-cooling bracelet after her chemist-collaborator died. **Bette Graham** experimented extensively in her kitchen laboratory until her correction fluid was working well, but hired a chemist at the end to standardize the formula that would become Liquid Paper®. **Maria Baker** of Los Angeles consulted an electronics engineer in completing her device for selecting children's TV shows (Weisberg, L-8/16/76).

Even **Beulah Henry,** the so-called Lady Edison, among the most prolific of all American women inventors, used professional model-makers to give form to her many inventions. Her original models were often carved in soap, or made from materials at hand, such as hairpins, adhesive tape, buttons, clips, and spools—or, as in one case, a rusty nail, a board, and a rock. "I know almost nothing about the laws of physics, mechanics, or chemistry," she declared, "and the only technical term I can remember is *periphery.* . . . Sometimes when I am trying to explain an idea to model-makers and engineers, they tell me it is against all the laws of physics, but as that means nothing to me I just have to carve my ideas from bits of soap, complete them with rubber tubing, tape, etc., and keep talking until I get it over" (Yarbrough).

Lack of technical/mechanical expertise can be a double handicap wherever not only an invention but the tools or machines to make it, or some part of it, must be invented. **Emma Mills** was noteworthy in the 19th century, and Beulah Henry in the 20th, for overcoming this double obstacle. Mills' typewriter improvement demanded new tools, which Charlotte Smith says Mills invented (*WI*/1). Beulah Henry's soap-holding bath sponge demanded a new machine to cut the sponge as needed; Henry devised the machine.

Sometimes the share taken by the expert is exorbitant, and sometimes the women lose control of or recognition for their invention. (This happens to male inventors as well, but is more likely to happen to women.) **Lillian Malt** conceived a radical new keyboard that would increase typing and data-entry speeds by 40%, with less fatigue for the operator, but had to collaborate with an electronics engineer in producing a working model of her Maltron keyboard. Unfortunately, in England, at least, the keyboard is now sometimes known as "Stephen Hobday's keyboard." Likewise, **Maria Luck-Szanto** conceived the idea of the tailor-knitting machine. She first worked with a mechanical engineer, but as the project revealed its true complexity, she was bought out and an electronics engineer brought in to computerize the machine's operation. Though she did receive a financial reward, the machine is now known by the names of these two engineers, her original collaborator and the computer expert.

A fifth reason why women hold so few patents is the self-image problem. This has several aspects. In the 19th century, and to some extent even today, women hesitate to claim credit for achievements, because to do so seems immodest or unladylike. Also, if a woman has internalized our cultural stereotype that women do not invent, it may not even occur to her that her original idea is (*a*) an *invention* and (*b*) *patentable*.

Worse than this, women have internalized the idea that it is not okay to be an inventor—that it is not fitting, unfeminine, somehow basically in conflict with their proper role. Even **Mary Spaeth,** who invented a reclosable cereal box at age eight, and later invented—and patented—the tunable dye laser for the Army, does not really think of herself as an inventor (K. Brown, 181–2). This is in spite of the fact that elsewhere in the same interview she says, "Among the three people working with dye lasers, I would say that I invented it and the other two discovered it." She also says in so many words, "I was never much interested in patents."

Finally, women are socialized to be generous and giving, and may freely pass their ideas on to relatives and friends, rather than seeking to patent and profit from them.

A less obvious but major reason why we know so few great women inventors or great women's inventions is that so few people know any anthropology. A significant share of human technological achievement belongs to prehistory; and a large part—arguably the larger part—of that share belongs to women. All the basic human food staples, including grains and root crops, for example, were domesticated and improved—and detoxified or otherwise processed for edibility—at a time when women were very likely still the exclusive or primary cultivators, and were certainly the food-processors. Yet because so few of us study anthropology, this is not generally known. Moreover, anthropology itself has not escaped the androcentric stereotype of the rest of the culture.

Primate ethology is also too little known; or rather, the species studied have tended to be those which reinforced androcentric and hierarchical stereotypes of human behavior. Much more research remains to be done, and findings must be applied with great caution; but the only two (nonhuman) primate inventions actually observed in process and documented in the scientific literature, to my knowledge, were the work of a female monkey (Lampe, 38).

Beyond this, one very basic answer should by now be apparent: women's areas of work and authority over work, and thus their motive and opportunity for invention, have been progressively narrowed in our culture from at least the Bronze Age onward.

An additional answer is that both the areas of endeavor open to women and women's achievement within those areas have been underreported. Nor need we wait for history to see distortions of women's role and contributions. Julie Matthaei points out that Colonial American women's role in agriculture was underreported in England right at the time, in order to encourage young women to indenture themselves and emigrate to the Colonies (339 n9). Insofar as these reports became part of history, in any case, they influenced it.

The fact of the androcentric distortion of our history, and the way in which stereotypes blind historians to what is before their eyes, would seem to need no repeating. Yet the history of technology itself is so new, and the compensatory history of women and technology so embryonic, that a few horrible examples may be in order:

In the 19th century a feminist economic reformer named Charlotte Smith—to whom I've dedicated this book—persuaded the Patent Office to issue a list of women who had received U.S. Patents from the opening of the office to that date (late 1880s), just before the Centennial of the Office in 1890. This list is one of the few useful sources on women inventors in the United States, and is therefore the one used by any scholar of the subject. Unfortunately, the list is marred by two kinds of errors that shed further light on why we know so few great women inventors.

First, significant numbers of women and their patents are omitted. For example, the compilers list 124 women's U.S. patents for 1876, but *omit at least 27 others*. The percentage of omission here is shocking enough in itself. Worse yet, the omissions do not seem to be random. Machines, particularly such "nontraditional" ones as dredging machinery and machinery for raising sunken vessels—are disproportionately represented among the omissions (Stanley, "Conjurer").[18] In other words, the list-compilers were more likely to miss women's patents for "unexpected" inventions.

Second, the compilers misclassified women's inventions in trivializing ways, even with the descriptions of the inventions before their eyes. Two glaring examples are as follows: **Harriet R. Strong**'s storage dam and reservoir system—first of its kind, exhibited at the Columbian Exposition, and nearly built on the Colorado river during World War I—is classified as a container for kitchen debris! Another woman's architectural building block is classed as a toy! With such friends, women inventors needed no enemies.

Finally, we know so few women inventors because technology has been defined to

18. Since writing this paper I have learned that the first section of the compilation, into which the listing for 1876 would fall, may have been the work (or under the charge) of one man, then-Commissioner R.G. Gill.

exclude women's work. And even where women's work has been discussed in the history of technology, it has been relegated to an inferior or insignificant position. This, too, is still happening. In the early 1980s I participated as one of the experts in a dissertation project to identify "foundational" technologies and inventions in human technological history. Wherever I tried to insert technologies having to do with women's work—even after proving them indeed foundational according to the criteria required—they were vetoed by the other (male) panelists (Bindocci).

Which brings me to two more aspects of my proposed revision in the history of technology: redefining technology, and redefining significant technology.

## Redefining Technology

Lynn White says that technology is what people do, and reminds us that there is even a technology of prayer, at least in the Orient, where prayer wheels carry the prayers of the faithful to heaven. Unlike White, however, most historians of technology and most anthropologists, particularly males, before the 1970s seemed to define technology as what men do. In fact, as late as the mid-1970s, the authors of a cross-cultural study of factors in sexual division of labor present an index of participation in which exclusively male activities rank 100 and exclusively female activities rank zero, and state in so many words that though certain activities are exclusively female, *none of these are technological.*

Even the dictionary definition of *technology,* from Webster's *Unabridged Dictionary,* second edition, senses 1 and 3, gives ample if subtle warning that technology has something to do with the male preserves of professional science and large-scale manufacturing (emphasis mine):

1. Industrial science; the *science or systematic knowledge of the industrial arts,* especially of *the more important manufactures,* as spinning, weaving, metallurgy, etc.
3. Any practical art utilizing *scientific knowledge,* as horticulture or medicine; applied science contrasted with pure science.

Note, however, that the first two examples under the first meaning are spinning and weaving.

This book, on the other hand, includes such things as sex aids, menstrual technologies, fertility and anti-fertility aids; childbirth technologies, including herbal preparations and massage; the technologies of and aids to nursing, weaning—such as weaning foods—and childcare and socialization (such as baby massage, toys and games and other educational devices) in general; new foods, food chemistries (such as fermentation and leavening), and new types of foods in the definition of technology, and thus within the concept of invention.

## Redefining Significance

Moreover, I intend to give these things their proper weight and importance in human technological history. This endeavor will take four forms:

1) showing the foundational nature of many of women's major early inventions and innovations;

2) "renaming" and/or reclassifying certain of women's inventions both to show their proper ontological place in technology, and to revalue them in the interim until the new view of significance can take hold;

3) documenting women's achievement in areas already defined as significant by males; and

4) proposing a new and arguably more valid definition of significance that will include rather than exclude women's work to date.

In the course of this book I try to show rather than tell what I mean by all this, but some preliminary examples may be in order.

Regarding the foundational nature of women's early inventions, consider that women were the first builders in many cultures, and that the first architecture in some areas was the tent (cf. Faegre) or the storehouse, and owed much to women's technologies of tanning, weaving, basketry, felting, and pottery. The first chemistry was that of cooking, fermentation, dyeing, tanning, and cosmetics, all traditionally associated with women. Glue probably originated in the stewpot—or in the dough receptacle (wheat paste). According to one authority on early technology in the Near East, the process of heating flint for easier knapping may have arisen from the heating of pigments for art, and the first smelting of metals probably occurred in the process of firing pots (Schmandt-Besserat, ed., *Early Technologies*). Recently, as noted above, it has been suggested that women's setups for processing sago starch could be considered the beginnings of engineering in that society. The possible examples are myriad.

My second proposal concerns women's more recent inventions, and my first two examples are taken from the above-mentioned 19th-century list of women patentees in the United States (hereinafter LWP). Many women became concerned late in the century about the harmful effects on women's health of tight lacing, and began to design—and patent—Dress Reform corsets and other items of clothing whose avowed purpose was better health. I consider these patents in my Health/Medicine chapter, not as Wearing Apparel, as the LWP classifiers did. The practice of the early Patent Office supports me in this, classifying many corsets and abdominal supporters under a medical/surgical heading.

Equally important, I classify women's washing-machine and sewing-machine patents under Machines, whereas the LWP classifiers placed them under such inevitably trivializing headings as "Washing and Cleaning" and "Sewing and Spinning."

A 20th-century example that delights me because it comes from the pen of a scientist is this: The famous couturière **Madeleine Vionnet** (1877–1975) is credited with inventing the bias cut in dressmaking. J. E. Gordon, in his fascinating book *Structures; or, Why Things Don't Fall Down*, points out that, *technically speaking,* she "exploits the low shear modulus and high Poisson's ratio of certain square-weave fabrics in the 45-degree direction" (cap. to pl. 17; pp. 251ff). Which description—which *naming*—is likely to gain Vionnet the more respected place in the history of technology?

Gordon also points out, incidentally, that cloth is properly classified as a *structure* rather than a material, and that the making of cloth is therefore properly classed as building or engineering.

My third proposal—the matter of much of my book—needs no discussion here. I will only point out that in addition to having to their credit all three foundational inventions usually associated with the Neolithic Revolution, women were also the likely inventors of herbal medicine and contraception, and the first to introduce rotary motion into human technology.

My fourth proposal goes to the heart of the matter, for significance, like beauty, is in the eye of the beholder. If that eye has a misogynist cast, then everything it beholds will be distorted in that way. If the beholder believes that women do not invent, there will be much that his (or even her) eye simply will not see, and therefore can never record. Thus the battle is a daunting one, but as time is a major enemy—the hidden iceberg of knowledge about women's contributions relentlessly melts with each passing year of distance from the events—I propose to make a start.

Even the most traditional historians of technology would probably not dispute two of my main criteria for the significance of an invention: 1) technological impact and 2) economic impact. I propose, however, to add a third criterion: 3) human impact—which would at one stroke include much of women's contribution to technology in the past 4,000 years. By "human impact" I mean the effect on people's comfort, convenience, and quality of daily life as well as on human welfare in a more general sense.

Which is more significant, for example: a sophisticated new weapon that may be understood by a dozen people in the world—and (we hope) never used at all—or an improved menstrual pad or a remedy for cramps or premenstrual tension that could contribute to the comfort of half the world's population for one week out of each month during their entire reproductive lives? Which is more significant: a new hormonal contraceptive that alters the entire hormonal balance of a woman's body, or an ancient system of contraception by acupressure? Which is more important: a small improvement on the doomed dinosaur we call the internal-combustion engine—allowing it, say, to propel a car much faster than the speed limit—or the invention (by **Bette Graham**) of Liquid Paper®, which transformed the working lives of literally millions of secretaries before the days of word processors?

Even with regard to economic impact alone, women's inventions, especially if dismissable as "domestic," do not receive the credit they deserve. Consider, for example, the invention of lace, which, aside from its beauty, probably saved the economy of entire regions of Europe in early modern times. Consider, too, the invention of the chocolate chip in the United States a few decades ago. In an experiment designed to create a new chocolate cookie, **Ruth Wakefield,** new co-owner of the Toll House Inn in Massachusetts, cut bits of solid chocolate into the dough. The chocolate did not melt and flavor the whole cookie as she had expected, but the effect was pleasing, and chocolate bits (chips/morsels) were born. Nestlé soon began manufacturing them and, considering the popularity of the original inventor's Toll House cookies, both their sales and hers must have been enormous.

Now, moreover, **Debbi Fields** of Palo Alto has built a multi-million-dollar corporation on her new chocolate-chip cookie and other innovative products, to the point where she now buys a significant share of Hawaii's entire macadamia nut crop. *One quarter of all cookies now reportedly contain chocolate chips.* (This figure is probably from the United States alone, though my source did not so specify; but even here the dollar figure for cookie sales gives this "domestic" invention considerable impact.)

The same could be said for commercially available diet foods, for such diet and exercise programs as Weight Watchers and Jazzercise, which were also women's inventions; and for a great many other things we take for granted and do not connect with technology.

Considering the human impact allows me to include in my definition of *invention,* if not of *technology,* and certainly to rank as significant (though for the most part they will not be discussed here), many social innovations that have been women's. Again, space allows for only a few examples. One of the most noteworthy, **Jeanne Mance**'s invention of the prepaid health-care plan, has already been mentioned. Social services in hospitals were **Dr. Marie Zakrzewska**'s idea, and a German management consultant named **Christel Kammerer** came up with the idea of flextime in work. Much of Franklin Roosevelt's famed welfare plan of the 1930s, including apparently the Social Security System as finally adopted, was the work of his Secretary of Labor, **Frances Perkins.** Adoption services with formalized procedures seem to have been women's ideas, with a significant contribution from **Dr. Gladys Dick,** otherwise famous for her work on scarlet fever in the 1920s. Noise-reduction stratagems for urban environments were the brainchild of a

19th-century American woman, **Julia Barnett Rice,** and many systems for aid to handicapped people were women's ideas, such as simian aides for quadriplegics, an odd-shoe exchange, canine aides for the deaf, etc.

**Jane Addams** instituted profit-sharing in her enterprises—which was at least an independent innovation or re-invention, if not a first. **Lore Harp** and **Carole Ely** in a computer business in Southern California in the late 1970s provided not only child care but *housecleaning* help as a worker benefit in their corporation; and **Linda Wachner,** President of Warnaco, a Fortune 500 company, counts child care as an expense-account item for her staff when they have to work late (*Ms.,* Jan. 1987: 80). The idea of doctors clustered together in office buildings was the economic brainchild of a U.S. woman physician, **Dr. Eliza Mosher.**

I suspect that most of the many hotlines that now save and comfort so many victims or potential victims and perpetrators of everything from suicide to drug and child and wife abuse, were women's ideas, but I have firm information on only one—a teen-loneliness hotline in San Diego. This is an area—like so many others in this field—that deserves more research.

## Note on Stereotype-Bashing

If I could give this section its own epigraph, I would invert Dante's famous sign over the gates of hell: Let all who enter here take hope—and abandon their deepest stereotypes. The psychological literature on conscious attempts to revise or destroy stereotypes is not sanguine. As with the revisions in the history of technology, however, the attempt must be made.

The central target is obviously the stereotype that women do not invent, but this is only one of the many heads of the Hydra, and all of its many interlocking and supporting stereotypes and myths must also be chopped off, or it will simply grow back again, strong as ever. One of the most important of its other heads is the myth of Man/male the Hunter/Provider. Not to spend an undue amount of time here—since in a sense the whole book is an unmasking of the falsity of the stereotypes separating women and technology— and not to reinforce the very stereotypes I seek to destroy by stating them yet again, I will simply state the opposite, positive form of the most important stereotypes that need to be abandoned for a fair assessment of (*a*) this book and (*b*) women's contributions to technology:

Women hunt.

Women provide.

Women invent.

Women rule.

Women heal.

Women minister to deities.

Further, early deities are more likely to be female—or to embody both female and male principles—than to be male.

Early human societies are gathering-hunting societies (if the name is to reflect the relative economic contribution of the two activities).

Early human societies are neither matriarchal nor patriarchal, but relatively egalitarian and consensual in their decision-making.

Power relations between the sexes are most likely to be egalitarian where the economic contribution of the two is—or is felt to be—equal.

Females are as mobile as males except during the brief periods of very late pregnancy and very early lactation.

Fully nomadic gathering-hunting females are not constantly pregnant, but have only about four children, of whom only two survive.

Child care is done right along with other work—not as a full-time occupation.

Females are very nearly as strong as males, *if trained from early childhood* to carry siblings, water, firewood, gathered vegetable food, garden produce, etc., while males are not burden-bearers.

## Note on Organization of the Book

*Mothers and Daughters of Invention* operates on four levels of organization. On the first level, it is divided into five important areas of human endeavor wherein I document women's contributions to technology: agriculture and food-related technology; health/ medicine and related technology; sex, fertility, and anti- fertility technology; machines; and computers. These are the five mega-chapters of the main text.

Within these five areas—except for that of computers, which is too recent—the organization is roughly chronological, beginning in prehistory, barely touching on the period from approximately the Bronze Age through the 18th century, and then dealing in some detail with 19th- and 20th-century achievements. Within each time zone the approach is first to divide the area of endeavor into meaningful subcategories (as contraception into physical/barrier methods, herbal/chemical methods, etc.). Within those categories, the organization is biographical/chronological, with brief biographies (in rough date order) of women inventors who contributed to this area of endeavor. For some women no biographical information was available except for what could be deduced from the patent records, but these are carefully named, and the available information for them is given.

For women with good biographical information available elsewhere—as with women profiled in *Notable American Women* or in full-length biographies—I refer readers to these sources, and focus my own treatment on factors potentially relevant to their becoming inventors. I am especially interested in such things as sibling order, presence or absence of brothers, encouragement from parents to achieve and to be creative, tutoring at home or schooling in an all-girls' environment, technical training or a father with a mechanical or technical career, other inventors in the family, a tomboy girlhood, presence of artistic talent, and left-handedness or ambidextrousness.

Women inventors named in the book appear in the index, incorporated within a single alphabet rather than in a separate name list. However, the inventors' names are in boldface type.

References appear in parentheses within the text, keyed to a single-alphabet bibliography at the back of the book. Footnotes appear on the pages within the chapters. Chapters 1–4 have appendices, containing lists of inventors and their inventions and other important supporting materials that might disrupt the flow of the main text. A list of abbreviations for sources and organizations appears at the head of the Bibliography.

*Mothers and Daughters of Invention*

# 1 Daughters of Ceres, Songi, and Corn Mother: Women Inventors and Innovators in Agriculture and Related Technology

Women and men had different kinds of teachers, and therefore were learning different kinds of things. The plants taught the women; the animals taught the men.
—Elise Boulding, 1976

It is generally agreed that gardening developed as a gradual elaboration of gathering activities by women.—Martin & Voorhies, 1975

Subsistence economy is in the hands of the women and gives them real power and status, which are not stressed in the ideological system.—Cora DuBois, 1944

With the invention of the plow, which, like the wheel required animal traction, men passed over into agriculture.—Ralph Linton, 1957

Sustenance is in a plow!—Sumerian "Farmer's Almanac," c. 2000 BC

But the little red hen said,

"All by myself / I planted the wheat,
I tended the wheat, / I cut the wheat,
I took the wheat to the mill / to be ground into flour.

All by myself / I gathered the sticks,
I built the fire, / I mixed the cake.
And
all by myself
I am going to eat it!"—"The Little Red Hen"

## INTRODUCTION

Anthropologists now generally agree that women invented agriculture. More precisely, in their role as main food-providers in gathering-hunting times, they became the plant experts of the species, and thus the likely inventors of gardening, which became agriculture. But since the myth of Man the Hunter/Provider dies so hard, I will review some of the support for my opening statement.

First comes evidence from myth and religion. The earth itself is often deified and worshipped as a goddess. Women's fertility has long been associated with earth's fertility, and such beliefs pervade European folklore. Significantly, in Old Europe the fertility aspect of the Great Goddess may have arisen with agriculture as something new in the 7th millennium BC (Gimbutas, *Goddesses,* 201).[1]

Many peoples' myths tell of a goddess or female spirit who gave them their staple crop plant and/or the knowledge, tools, and technologies needed for cultivation. Often this is the central figure of the Goddess trinity. Familiar examples are the Greek and Roman myths crediting Athena/Minerva with inventing agriculture, and the Demeter story, in

---

1. "Old Europe" encompasses southern Italy and Sardinia, Greece, western coastal Turkey, the Balkans, Czechoslovakia, southern Poland, and the western Ukraine, from about 7000 BC to 3500 BC. The civilization that flourished there climaxed in the 5th millennium BC.

which Demeter holds all growing things within her power. Throughout the British Isles and Europe the Corn Mother was celebrated at the end of every harvest until the 20th century; and many New World cultures had a Maize Mother, Quinoa Mother, or the like (Frazer, *Bough,* 399ff). Pachamama, Goddess of the Crops, was the main deity worshipped among rural Peruvians in Inca times; Chicomecoatl was the Great Corn Mother of the Aztecs; and Bachue was the divine ancestress and protector of agriculture among the Chibcha of Colombia (M. Stone, *Mirrors* I: 76, 85, 89). One interpretation of Demeter's name is "Barley Mother" (Graves, *Myths,* Index). Familiar, too, is the origin of our word *cereal* in the Latin *cerealis,* "pertaining to Ceres."

Isis of Egypt traditionally invented agriculture and gave grain to her people. Less familiar to Western readers are the analogous Sumerian, Indian, and African myths. The Sumerian goddess Ninlil gave her people their understanding of planting and harvesting methods. The Sumerians also revered a specific grain goddess, Ashnan, as well as Uttu, Goddess of Plants (M. Stone, *When God,* 3; Kramer, *Sumerians,* 173, 219–22, and *Mythology,* 42, 72). Inanna of Sumer is closely associated with the date from which her name derives (Rohrlich, "State," 86); and the date palm was sacred to Isis and Lat (Graves, *Myths* I: 86). Bengali women still dance together in worship of Gauri the Corn Goddess (Dudley, 113). Songi, Great Mother of the Bantu, notched the teeth of a young woman named Nsomeka, and from these notches sprang both livestock and fruit trees. The Bambara credit TijWara with introducing agriculture, and ensure good harvests by pleasing Her (Spretnak, *Politics,* 59). Like Isis and Demeter, Ala of Nigeria is protector of agriculture and teacher of peace and order (M. Stone, *Mirrors* I: 89).

The Andaman Islanders associate their ancestress Bilika with yams (Evelyn Reed, 146). Trobriand Island men usually plant yam gardens, but always in a woman's name; on Dobu seed yams cannot be bought, but stay within the matrilineal line; and yams symbolize women's sexuality (A. Weiner, 138, 139, 210). The Aztec earth goddess, Civacoatl, is "She who plants root crops" (Leacock, *Myths,* 258). Throughout Asia the spirit of rice is female, and the many legends of the "invention" of rice feature various princesses as the inventor (Bray, L-5/30/79).

G.E. Fussell, in his history of farming techniques, states the argument baldly (2–3):

> The agricultural, or rather food-providing, revolution that took place in those remote days . . . must have been the work of the women in the first place, because most ancient peoples worshipped a female goddess, the giver of good bread. Her names were many; Isis, Cybele, Agdistis, Dindymene. . . . An attenuated memory of her works was the effigy that used to be erected on the last wagon to carry the final sheaves home.

In addition to such origin myths, we find cultivation deeply entwined with religious practices where work has not been artificially separated from worship. Among the Kayans of central Borneo, in order to "secure and detain the volatile soul of the rice," a priestess brings several ritual instruments into the field: a miniature ladder, a spatula, and a basket holding hooks, thorns, and cords (Frazer, *Bough,* 414–5).

> With the spatula the priestess strokes the soul of the rice down the little ladder into the basket, where it is naturally held fast by the hooks, the thorn, and the cord; and having thus captured and imprisoned the soul she conveys it into the rice-granary.

This procedure is reminiscent of the seed-harvesting practices of Panamint Indian women in North America, as described by Otis T. Mason (*Share,* 16).

Robert Graves speaks of a time "when only women practiced the mysteries of agriculture," when ploughing was taboo for men, and Erysichthon could be punished with insatiable hunger for daring to plough without Demeter's consent (*Myths* I: 92, 93).

In Canaan, grain had to be desacralized for ordinary use (Gray, 77; cf. Leviticus 2:14).

Beans were sacred to the White Goddess, and in primitive times only her priestesses might plant or cook them. This taboo, reflecting the close association between beans and the birth and rebirth of souls, seems to have outlasted most cultivation taboos on men. The men of Pheneus in Arcadia, for example, traditionally could plant anything but beans (Graves, *Goddess,* 69–70; *Myths* I: 28, 95). And the Luttrell Psalter from 14th-century England shows men ploughing and planting grain, but women sowing beans by hand.

Grafting of date and fig trees, known before Hammurabi's time and treated in his code, was a ritual, possibly resembling the one for grafting lemon branches onto laurel or olive. This "unnatural" joining, described with repugnance by a Jewish writer, demanded correct timing with the solar and lunar cycles, and a simultaneous ceremonial act of intercourse between the grafter, who was a beautiful maiden, and a male assistant or co-celebrant (J. Lindsay, 281).[2]

These rites must have been prevalent in the Middle East, and not an isolated case, since in the struggle between the Goddess and Jehovah, Jehovah's priests forbade their followers to eat grafted lemons in order to keep them from being drawn into "orgiastic activities of the neighboring peoples" (*ibid.*).

Possible examples of agriculture as ritual or worship range from spring-planting, harvest, and first-fruits festivals worldwide to the ancient Greek Thesmophoria, a women's festival in which the rotting flesh of pigs (earlier thrown sacrificially into deep chasms) was mixed with the seed-corn to ensure fertility. Not only was this festival dedicated to Demeter, but the pig flesh may have been a substitute for women's menstrual blood, the original fertilizer in the ritual. At Syracuse women baked cakes for Thesmophoria in the shape of the female pudenda (Thomson, 127, 221–3, 231). According to Judy Grahn,

> As the mother makes the child with her own flesh so does the Mother-Earth make us with Her flesh—in the form of grain. To teach this principle to the people of their communities, the women [or menarchal girls] made bread in the shape of a female, or . . . the yoni, or the round moon. This special bread . . . was and is eaten in a public ceremony.

Citing Lawrence Durdin-Robertson, Grahn continues: "The Eucharist of Isis consisted of the bread which she had given humankind and milk from her bosom, drunk in a breast-shaped chalice. In some ceremonies blood—originally menstrual blood—was used" (Spretnak, *Politics,* 271).

Second, there is evidence from division of labor. This comes primarily from anthropological reports on division of labor in surviving preliterate groups, but also from ancient historians and travelers, graphic representations, and other sources. Anthropology shows that women were—and are in surviving foraging tribes—the primary gatherers of plants for food, supplying 60–80% of the diet by weight except in the Arctic (Dahlberg; Lee & DeVore; N. Tanner, *Becoming*). *Of 185 societies in Murdock and Provost's standard cross-cultural sample, 107 assigned such gathering predominantly or exclusively to females* (207, 216ff).

Through this work, women gained a wide and deep knowledge of the location, life

2. Just as with the Goddess's son/lover, the male presence here may mean that the rite was already shifting from a purely female mystery. Compare, in any case, Lindsay's male-oriented wording: "A very beautiful maiden must hold the bough while a man from behind has anal intercourse with her; and while he is thus enjoying her, the girl grafts the branch on the tree" (J. Lindsay, 281).

cycle, growth habits, and requirements of the edible plants of their territory. Though men might have known these things, their knowledge was passive. Moreover, in bringing roots, cuttings, and seeds of favorite food plants over the years to a series of camps, women had ample opportunity to observe and exploit accidental germinations, re-rootings, and sproutings, and to notice the best locations for these chance gardens—where water was repeatedly spilled, where excrement or refuse was deposited, where the ground had been disturbed by their digging sticks.

As an extension of division-of-labor evidence, what might be called *technical clusters* can also provide possible insights into the history of innovation. For example, where the same plant was used as food, medicine, and contraceptive—as with jack-in-the-pulpit in North America—women, the virtually universal cooks and food-processors, and the exclusive arbiters of contraception, could logically be connected with the medicine as well. Such a connection seems especially strong where the only difference in preparation among the three is the degree to which the plant is detoxified by cooking, leaching, etc. Another good example here is the ancient use of Icelandic moss (a lichen) for food, medicine, and dye.

Third, there is evidence from archaeology. This includes tomb carvings and paintings representing division of labor, and results of new studies on the archaeology of behavior, as well as the familiar evidence of grave goods. For example, if in a horticultural or early agricultural society, women are buried with hoes and men are not, as at the important Neolithic site of Çatal Hüyük in Anatolia (Mellaart, *Town*, 208), women were probably the primary cultivators in that culture.

Fourth is the evidence from language. To take a hypothetical example here, if the word *plough* in several ancient languages meant literally "woman tears earth," we would have a powerful argument for female origins for the plough. Likewise, if several modern languages took over this word at successive identifiable times, we would have some idea of how the invention spread. We do know, for instance, that although Sumerian scribes had written words for two kinds of plough by Uruk IV, the Sumerian word for the plough is not of Sumerian origin. Neither are their words for farmer or for furrow. Thus, Sumerologists conclude "that the basic agricultural techniques . . . were first introduced in Sumer, not by the Sumerians but by their nameless predecessors," sometimes called proto-Euphrateans or Ubaids (Sherratt, 116; Kramer, *Sumerians*, 41).

The folk names of many medicinal herbs suggest not only their function, but their connection with women: squawvine, motherwort, and cuckold herb[3] leap to mind.

Finally, the very word *dairy* means "place of the maids," just as *faery* is the land of the fays and *aerie* or *eyrie* the place of the eggs. Since cognates of the root *da'i* or *dey* for nurse, maidservant, or milkmaid appear in Indo-European languages from India to Iceland, what went on in the dairy was most likely women's work from very ancient times (*OED;* Kettner's, 152–6).

Fifth, there is evidence from primate ethology. Nonhuman primates have not developed horticulture, and primate-human comparisons are fraught with danger. However, two food-processing inventions have been documented among Japanese snow monkeys (*Macaca fuscata*). These are noteworthy not only in themselves, but because both of them originated with a single female, Imo, and because they reveal the transmission patterns of this new technology in the group.

The monkeys, closely observed by Japanese scientists for some time, had overpopu-

---

3. "Cuckold herb" is a composite known to science as *Bideus connata*, traditionally used as an emmenagogue (Hutchens, 272), or medicine to bring on the menses (presumably most often "blocked" by a pregnancy, in this case a pregnancy where the father was not the husband).

lated their island and were being provisioned. In 1953 the scientists began leaving sweet potatoes on the beach. The tubers were good food, but a problem for the monkeys because they became coated with sand. Imo began washing her potatoes in a nearby stream. When the scientists began leaving grain on the sand as well, grit became a worse problem, and Imo came up with her second and more imaginative innovation: she threw the grain into a pool of water. The grain floated, and the sand sank. The two food-washing methods spread through the group, starting with Imo's closest relatives and also spreading from playmate to playmate among the young monkeys (Lampe, 38–9).

To whatever degree the males were peripheral in this or other primate groups, transmission would be less efficient both to and from them.

Among chimpanzees, with whom we share more than 99% of our genes, *females spend significantly more time using tools than males do,* and most of the tools in question are food-gathering tools. Chimpanzees do use digging sticks, for instance, but a better example is the *termiting* tool. *Hunting,* by contrast, a largely male pursuit, is done without tools or weapons. Most interesting of all, in a study of three food-processing techniques for two different nuts eaten by chimpanzees, females chose the two more difficult and complex techniques significantly more often—and used them more efficiently—than the males (Goodall, "Behavior"; Boesch & Boesch, "Sex Bias" and "Sex Differences"; N. Tanner, *Becoming,* 73, 190).

Anthropologists do not generally agree exactly how the domestication of plants came about, and of course it may have come about in various ways in different areas. The scenario just sketched under division-of-labor evidence seems plausible, but there are several categories of possible explanation, among them the one reflected in this Malagasy myth (Madagascar): "Rice was given to woman as a consolation for a lost child—it grew from where the child was buried" (Parrinder, 24–5). If seed grains were buried with the dead, women would be likelier to observe the results, as they usually prepared bodies for burial (or other disposal). Or if a seed-gathering woman was buried with baskets of her seeds, as appropriate to two of her important occupations—or if select stalks or seeds were buried as offerings to the Goddess Earth[4] especially if mixed with menstrual blood—the resulting lush stands of the group's favorite plants would have taught a vivid lesson.

Whatever the details, by goddess-given role and lifelong training, women were fitted to be the first farmers, and such they became.

In the course of so doing, they inevitably invented many specific tools and technologies. Before looking at some of these, I pause to make explicit my basic assumptions here. In one study of 104 horticultural societies, 50% have exclusively female cultivators, another 33% have the sexes sharing cultivation, and only 17% have exclusively male cultivators (Martin & Voorhies, 283). From these unexamined figures alone, the probability of a woman's having invented a cultivation tool or process in these 104 societies would be high. Moreover, *among societies depending on cultivated foods for less than a quarter of their diet, nearly 70% have exclusively female cultivators;* and at a dependency rate *between 26 and 45%, nearly 60% still have exclusively female cultivators (ibid.,* 215). Thus in the early days of horticulture, when many of the basic tools and methods were being devised, women were evidently highly likely to be exclusive cultivators.

Second, there arises the question of who was in charge.[5] This, too, seems likely to have

4. Possibly in the form of a corn doll or puppet, a custom that survived in rural Greece till classical times (Graves, *Myths* I: 93).
5. For a persuasive comment on the role of this factor in invention and innovation in music, see Drinker (59).

been women for a very long time. As M.K. Martin and Barbara Voorhies note, "The monopolization of cultivation by women probably continued until either special environmental circumstances or the sheer scale of required labor favored the part-time recruitment of males" (*ibid.*, 216). Women's original dominance in the technology implies that men came in to help and did what women outlined for them to do. Moreover, men's first work was usually preliminary—ground-clearing—rather than central. Indeed, "Equal participation of the sexes" usually means that "males become responsible for the clearing of plots preparatory to planting" (*ibid.*). If measured in hours spent over the whole cultivation season, this was far from half the work.

Thus, *I assume that,* in the absence of clear evidence to the contrary, *if a tool or technique appears in the horticultural tool-kit or technology complex, especially at an early stage, women most likely invented it.*

As horticulture developed into agriculture, women's role declined. However, in a cross-cultural sample of 515 societies (including the 104 horticultural societies just mentioned, plus agricultural groups), women dominate cultivation in about 41% of the sample, and are "equal contributors" (note again the meaning of this phrase) in a further 37%, leaving only 22% of such activities dominated by men (*ibid.*, 214). Thus *even in the early agricultural phase* (again, in the absence of evidence to the contrary and especially at the earlier stages associated with lesser cultivation-dependency), *there should be a greater-than-even chance that women invented a given tool or technology.* By this stage, of course, the developments must be examined individually and culture by culture. At the very least, women's role in cultivation from the beginnings of agriculture to the present is far greater than previously imagined.

Underlying both these assumptions is a still more basic assumption made throughout this book for preliterate groups: *in the absence of evidence to the contrary, the workers invent and improve their tools.* In other words, wherever one sex is exclusively or primarily responsible for any activity within a society, and specialization has not created an exclusive group of toolmakers, we can assume that that sex has invented and steadily improved the tools and processes involved. As Frances Dahlberg puts it (13–4), "in most societies, the sex using the tools makes them; indeed, usually the individual makes his or her own tools."

With these assumptions in mind, let us look at women's cultivative inventions and innovations, ancient and modern.

## BEFORE HORTICULTURE

Women's most significant invention here, of course, is the idea of horticulture itself—bringing favored food plants to grow in appropriate areas near the camp instead of foraging where they grew by chance.

Two very early physical inventions underlying this major invention—often and rightly called the most significant single invention in human history—are the digging stick and the carrying basket (or other carrier or container). A digging stick may not sound like much of an invention to one who has not dug for mariposa lilies in the hard ground of a California summer, or excavated a big yam.[6] The digging stick was for early woman an all-purpose tool and weapon, useful for knocking fruits or nuts off trees and for defense as

---

6. To get a yam half an inch in circumference and a foot long out of the ground, a woman might have to dig a hole more than a foot square and two feet deep (Evelyn Reed, 107). And yams can grow several feet long (Chapple & Coon, 174–5).

well as for digging up food. Indeed, so ubiquitous did the digging stick seem to incoming whites that "Digger" became their generic term for certain California Indians (Kroeber, *Two Worlds,* 18–20). Sticks were also useful for transporting fire, and this may be how women discovered the benefits of a fire-hardened point on a stick (and perhaps of fire-hardening for wood in general); in any case Elise Boulding considers that women probably invented the fire-hardened point for the digging stick (78).

!Kung women of Southern Africa created another improvement: a rock weight "to supplement push." This weighted digging stick was part of their "characteristic equipment" (Charles Singer I: 59, 172). But who invented the digging stick in the first place? It must have been extremely early, as even chimpanzees use sticks for digging and for levers (N. Tanner, *Becoming,* 140). In fact, the argument for a female inventor of the true digging stick (modified for the purpose and kept for long periods of time) rests partly on observations of chimpanzee behavior: female chimps consistently use tools more often for feeding—as, for example, in collecting termites—than do males (*ibid.,* 73; McGrew/Dahlberg, 50); and the termite-collecting tools of chimps and presumably of hominids "may have been the basis for developing digging sticks" (Zihlman/Dahlberg, 94).

The carrying sling, basket, or other container enabled early women to gather enough food in one trip for themselves and their offspring while, if necessary, carrying an infant along. Nancy Tanner and Adrienne Zihlman argue convincingly in their pathbreaking *Signs* article (see also Tanner, *Becoming,* 144–5 and *passim*) that in the extremely early period when the carrying sling or other container was being invented, females were far likelier than males to need and thus to create such a device since they would have been gathering not just for themselves as males did, but for themselves and their offspring, probably at least one infant and one juvenile (see Dahlberg). Also, already by the Australopithecine stage, the evolving human foot was no longer suitable for grasping, so that foraging mothers could no longer expect infants to cling unsupported.

Animal skins and slings or "bags" of loosely woven vines were probably the first carriers (along with such found objects as giant leaves, ostrich egg shells, and wild gourds). Early or proto-human females may have used weaving skills first learned in making sleeping nests to fashion the first slings, gradually refining these skills over millennia to create both basketry and textiles. Native American women used baskets for gathering, but had other means of carrying infants. !Kung women, still foragers in the 1960s, used a combined child-and-food-carrier of antelope skin, called a *kaross* (Shostak, "Life," 51). These slightly built women, in fact, carried some 30 pounds of food, plus an infant or small child, in their *kaross* over several miles. As a result they needed to forage only about three times a week. Dani women of Highland New Guinea use long carrying nets for children and sweet potatoes (Lindholm & Lindholm, 84, 86), doubtless little changed from their foraging days.

Any people who become seed-gatherers, as did certain North American and Near Eastern cultures, especially need containers. Because wild grasses, including the ancestors of modern wheat and barley, shed their ripe seeds at a touch of the wind, the heads had to be caught just before peak ripeness and cut with a reaping knife or sickle or shaken or beaten into a carrier. Panamint women of Death Valley used a small funnel-shaped basket and a wicker-work paddle in harvesting the sand-grass seed (*Oryzopsis membranacea*). With the paddle they beat the grass panicles over the rim of the basket, causing the seeds to fall inside (Mason, *Share,* 16).

Another improvement made in gathering technology by Native American women was the tumpline, a carrying strap passing across the forehead. Otis T. Mason reports Ute women using both gathering wands and tumplines (*ibid.,* 146–7).[7] So thoroughly identified with women's work is this burden strap, indeed, that in the puberty dream of the young boy who will become *berdache* (man/woman), he must choose between the bow and the burden strap (Whitehead, 87–8).

These wild grasses need not be threshed, unlike their high-yielding, non-shedding descendants, but the seed had to be freed of its hull or glume either by roasting or by pounding in a mortar, and then, as the English word *winnowing* suggests, letting the wind blow away the resulting chaff. Ute women had wood and stone mortars and roasting and winnowing trays (O.T. Mason, *Share,* 146–7); the foraging !Kung women had wooden mortars and pestles (Draper, 80) among their very few household goods; and as early as the 14th millennium BC, forerunners of the Natufian[8] peoples in what is now Turkey had sickles and querns (grain grinders), mortars, pounders, and pestles. They even had bone spatulas possibly used for cleaning flour off the quern (Mellaart, *Neolithic,* 30).

These early sickles were probably bone, with inset flint blades, although according to Michael Partridge (134), the earliest Egyptian sickles were actually jawbones with the teeth sharpened to form a cutting edge. (Samson found one sharp enough to kill a thousand Philistines; Judges 15:14–7). The querns were of the saddle type still used in Mexico today, consisting of a flat stone base and a stone roller or muller for rubbing and crushing the grain on the base.

The size of some of these early Natufian mortars, the fact that they were built into house floors or pecked out of cave terraces, and the use of worn-through mortars in shrines and as grave markers testifies to the importance of this new agricultural tool kit to these people. Indeed, this may be its first appearance in the world (Mellaart, *Neolithic,* 30).

By the height of the Natufian culture, 10,000–8000 BC, possibly even before domesticating any plants, Natufian women had straight-sided bone reaping knives with elaborately carved handles and V-shaped grooves to take the flint blades; pestles shaped like animal hooves; and what Mellaart calls "sickle blades with lustre" (*ibid.,* 33, 34).

A major technical advance of the pre-horticultural period was the development of ground stone tools as opposed to chipped or flaked tools. In fact, Kent Flannery lists this as one of three "pre-adaptations" for early cultivation, along with an expanded subsistence base that he calls the "broad spectrum revolution"and the development of storage facilities (77–9). Women may well have been on what might be called the grinding edge of technology here. In any case, *the earliest ground stone tools seem more often connected with women's work and roles than with men's.* At some of the earliest Near Eastern sites where such tools appear, for example, they were evidently used mainly for milling ochre for cosmetic, ritual and mortuary uses. At Mureybat (Syria), the ground stone tools consisted "mostly of querns and mortars." Interesting finds here were flint balls that may have been rolled around inside the querns (Flannery, 78; Van Loon, 216).

As Flannery observes (78), "the ground stone technology was there," and when people began collecting cereal grasses, they "had only to adapt and expand a pre-existing technology in order to deal with grain processing."

Simultaneously with this early food-gathering and -preparation tool kit in the 14th millennium BC there also appeared clay-lined storage pits (Mellaart, *Neolithic,* 31, 36). By 9000 to 7000 BC many sites report subterranean pits, some of which, as at Ain Malla-ha, were even plastered (Flannery, 78). Native American women made storage baskets for

---

7. An American ethnologist who was unusual in focusing on women's contributions to technology.
8. For a quick reference on unfamiliar names of ancient cultures, peoples and archaeological sites, see Sherratt.

acorns while still gathering and hunting—for example, the Yahi pine-root storage baskets that stood as tall as a ten-year-old child (Kroeber, *Ishi*, 14, 21). Women had also devised many ingenious methods for preserving food, notably cheese (which probably originated in southeastern Asia about 8,000 years ago) and North American Indian women's pemmican (O.T. Mason, *Share*, 27–9). This area of women's invention will be discussed in more detail below.

## HORTICULTURE

### Cultivation: Tools and Processes

As horticulture gets under way, before 8000 BC in Palestine and by the 7th millennium BC in Old Europe (Hawkes, *Atlas*, 40; Gimbutas, *Goddesses*, 17, *passim*) women created new tools and modified old ones to meet new needs. Widened into a blade at one end (or with a blade of stone, bone, or shell) and with a projection part way up the handle to allow the foot to push it into the dirt, the digging stick becomes a spade; with two handles, and pulled through the light soil by two women, it even becomes a plough. The hoe may have been a modified digging stick or a separate invention, but it gives rise to a plough, called the hoe-plough.

### The hoe

No one knows how early this was invented, though virtually all horticultural societies seem to have one, and "hoe agriculture" is one term for horticulture itself. Hoes are so widely wielded by women that it seems safe to assume women invented them. Edward Hyams posits this scenario (9):

> The inventors of gardening were people—almost certainly women—of the New Stone Age. To these Neolithic "peasants" and villagers we owe the fundamental tools and techniques of the craft. The digging-stick, its point hardened by charring, foreruns the spade and the plough. A stick with a truncated crotch makes a hoe; if, after centuries of using it, you suddenly have the bright idea of lashing an oyster-shell or the shoulder-blade of a game animal to the shorter leg of the crotch, your hoe is much improved.

One Neolithic culture where, as mentioned, women are buried with hoes and men are not, is Çatal Hüyük in Anatolia.

A powerful visual image is "Woman Hoeing the World Snake," a bauxite sculpture created by the Cahokian Culture of North American Woodland Indians between AD 1000 and 1250. This so-called Birger figurine, found in Illinois, probably reflects division of labor as well as cosmology. And the woman's hoe at this stage would be her own artifact (see Brose *et al.*).

Hoeing not only loosens the soil, but breaks up large clods to create the proper *tilth* or fineness, and kills weeds. Mundurucu women (Brazil) use a hoe for weeding (Murphy & Murphy, 61).

The Inca hoe appearing in Pomo de Ayala's agricultural calendar consists of a straight stick fastened to a curved one. The drawing for January shows a woman and a man hoeing,[9] cultivating a field in the rain; June shows a woman and a man digging potatoes,

---

9. Francesca Bray suggests classifying the Inca as agricultural rather than horticultural (L-5/30/79). From the standpoint of their degree of dependence on cultivation, she is certainly correct. For this book, however, I follow Martin & Voorhies in using plough cultivation and large-scale irrigation to distinguish the agricultural stage.

the woman on her knees using a hoe. The December drawing for potato-planting shows a man making a hole in the earth, a woman planting the tubers and a second woman holding a short-handled hoe to smooth the soil over the planted seed (J. Leonard, *First,* 70ff). Iroquois women farmers used hoes with a deer-scapula blade (Bridenthal & Koonz, 25).

## The spade

Still the mainstay of cultivation in the New Guinea Highlands, where women are the primary cultivators, the spade has been an important cultivating tool for thousands of years. Mount Hagen New Guinea women cultivate their garden plots carefully by spade (men clear new land and are involved in the cultivation of bananas and sugarcane). Until recently, this spade would have been made of wood, and doubtless by the women themselves. Now it is metal (Brookfield, "Highland," 437–8; Martin & Voorhies, 250ff, 253).

The English captions of Ayala's calendar in Time/Life Books' *First Farmers* volume, incidentally, epitomize a major research problem in women's history: Western androcentric distortions in observing other cultures. Both sexes worked as cultivators among the Incas; and as Ayala's text elsewhere makes clear, women were very important both in the founding of Inca culture and in cultivation. Indeed, the Incas believed women must do the planting or the crops would not grow. As already mentioned, the main rural deity was Pachamai or Pachamama, Goddess of Crops, and the Empress supervised many women farmworkers on the royal estates. Ayala's drawings themselves probably depict women's role in Inca cultivation rather fairly, although Ayala was himself male, half-Spanish, and writing for the Spanish king. However, Time/Life's caption for September's corn-planting picture begins as follows (J. Leonard, 70–1): "With an ornate digging stick, *a farmer* punches holes into which *a woman* scatters corn seeds." A second woman holding another tool is not even mentioned. Likewise, for May: "Harvest time brings a division of labor. While a man cuts stalks, *his female helper* carries bundles of them on her back to the field where she lets them dry" (emphasis mine).

## The plough

The plough is one of the two marks conventionally distinguishing agriculture from horticulture. However, the earliest ploughs as developed by women from digging sticks or hoes probably belong as much to horticulture as to agriculture, or at least should be considered transitional between the two.

Women are virtually never credited with inventing the plough, primarily because modern writers, thinking of modern ploughs and modern women, assume women would not have been strong enough to use them. Since female strength is relevant to female capability in general, let us look at this idea more closely. When the earliest ploughs appeared, women were the workhorses of society—the main cultivators and preparers of food; the carriers of firewood and other fuels, water, clay for pottery, harvested grain, and other burdens; and often the house-builders. Sustenance, in short, was in a woman. Trained from childhood for this work, they would have been very strong. Says a 19th-century observer in British Guiana, "Moreover, the women, . . . accustomed to hard labor all their lives, are little weaker than the men. If a contest arose between an average man and an average woman, it is very doubtful with which the victory would be" (O.T. Mason, *Share,* 8). Such observations become almost a refrain in anthropological reports, presumably because the observers' cultural stereotypes are challenged.

Indeed, in some cases, women were reportedly the stronger sex. Otis Mason devotes an entire chapter of *Woman's Share in Primitive Culture* to woman as "The Beast of Burden," with examples worth citing here. One 19th-century expedition failed because the would-be leader refused to bring any women (127):

"For," said [an informant] . . ., "in case they meet with success in hunting, who is to carry the [game]?" "Women," added he, "were made for labor; *one of them can carry, or haul, as much as two men can do."* [emphasis mine]

An observer in the Malay Archipelago, reported (128–9):

A Dyak woman generally spends the whole day in the field, and carries home every night a heavy load of vegetables and firewood, often for several miles, over rough and hilly paths; and not unfrequently has to climb a rocky mountain by ladders, and over slippery stones, to an elevation of a thousand feet. Besides this she has an hour's work every evening to pound the rice with a heavy wooden stamper, which violently strains every part of the body. *She begins this kind of labor when nine or ten years old,* and it never ceases but with the extreme decrepitude of old age.

An unnamed writer in *Woman's Work for Women* (Nov. 1888) marveled at the strength, competence—and singing—of Kurdish women in their daily routine (130):

In the evening they spin and make sandals; [then they lie down briefly]. After midnight they go up [the mountain] to get loads [of grass and wood]. In the early morning I often saw the women, looking like loaded beasts, coming down the precipitous mountain path, one after another, spinning and singing as they came.

This same observer also reports seeing "women with great paniers on their backs and babies on top of these or in their arms, going four days over that fearful Ishtazin pass, carrying grapes for sale and bringing back grain" (130–1). But perhaps most dramatic was an episode from the road (130):

Soon we came to a place where the road was washed away, and we were obliged to go around. We saw a woman there with a loaded donkey which could not pass with its load. The woman took the load on her back and carried it over and led the donkey over. She also carried a load of her own weighing at least one hundred pounds, and she had a spindle in her hands. Thus she went spinning and singing over the rugged way which I had passed with tears and pain.

Lest observations antedating the modern discipline of anthropology be discounted, note that a 1970s study of sexual division of labor showed "Burden carrying" allotted either exclusively or predominantly to women in 70 out of 185 societies, and shared equally by the sexes in another 46. Only 18 societies allotted this work exclusively to men (Murdock & Provost, 207, #34). And a 1970s report from the New Guinea Highlands quoted "one of the more perceptive" male informants as declaring, "Women are our tractors" (Tiffany, 21).

Should any such woman turn her head to inventing a plough, there seems little doubt she could turn her hand to it, too. The point here, then, is not who was genetically stronger, but *who was trained for strength.* Hunting may require great endurance, and some strength, certainly. But a man does not hunt every day, nor does he kill on every hunt. Even when he kills, he may hold an instant feast for himself and his hunting companions, who will then help him carry home the remaining meat. Or, as in Mason's example just cited, he may leave the carrying entirely to the women.

Likewise, land-clearing may be laborious unless done by fire; but again, it is done only every few years. Women, on the other hand, carry heavy loads of water, firewood, and children every day, and often do all the other work of cultivation. Thus, when the earliest ploughs were invented, women were probably not often called the weaker sex.[10]

Moreover, as Jacquetta Hawkes (*Atlas,* 161) and others have pointed out, in the light soils of the Middle East, ploughing was not the arduous task it was in the heavy soils of northern Europe, and the earliest "scratch" ploughs did not turn the earth, but scratched a shallow furrow for the seed. Such ploughs were functional, too, in the arid climate, creating a "dust mulch" that prevented evaporation and allowed the seeds to sprout (Linton, *Tree,* 115).

Assuming, then, that women had the strength to invent—and use—a plough, it remains to ask whether they did. According to Otis Mason (*Share,* 24, 25), "Long before the days of discoverers and explorers who wrote about them, women in America, Africa, and the Indo-Pacific were farmers, and had learned to use the digging stick, the hoe, and even a rude plough." He cites Livingstone's report from Africa (Angola), of a double-handled hoe dragged through the ground by two women. Since the earliest known representations of Middle Eastern ploughs date from about 3000 BC (*EB/79,* S.V. "Agriculture"; Steensberg, 5), when the ard-plough was already a pictograph sign and the technology already considerably more complex than described here, I cite Livingstone's observation in full (408):

> We saw the female population occupied, as usual, in the spinning of cotton and cultivation of their lands. Their only instrument for culture [i.e., cultivation] is a double handled hoe which is worked with a sort of dragging motion.

These early wooden ploughs seldom survived to be found by archaeologists—or were not recognized even when found. When historians, even historians of technology, say *plough,* they generally mean the much later and heavier sod-turning ploughs—iron-tipped, male-guided, animal-drawn—documented in the Near East after 3000 BC. References to the earliest hoe-ploughs and ards are surprisingly scarce in nontechnical sources. The *Sunday Times of London* illustrated encyclopedia of invention (De Bono, ed., 20, 104–5) is a pleasant exception: Michael Greenhalgh, writing on the yoke, says "men and women probably pulled the first ploughs."

The word *ard* does not even appear either in the *OED* or in *Webster's Unabridged Dictionary* (2nd ed.). The word *plough* does not appear in the index of Jacquetta Hawkes' *Atlas of Early Man.* The representations of ploughs nevertheless appearing in Hawkes (92, 99, 125), though crude-looking affairs, are already male-guided, animal-drawn (ox, ass), and late—after 3000 BC. With this cultural background, it is scarcely surprising that women virtually never get credit for inventing the plough. The 1979 *Encyclopaedia Britannica* article on the history of agriculture illustrates the problem perfectly: "The first

---

10. Even modern European and American women are by no means so much weaker than men as generally supposed, if we look at all classes; nor is the basic genetic difference quite so great as once believed. I can personally attest to the strength of American farm women from my rural Ohio childhood and to the strength of Greek village women from observations made in the Peloponnesus in the 1970s. See also Susan Bell, 153, 207, 268; Branca, *Women,* 19, 20; Kleinberg/Trescott, ed., 189–91; and Selma Williams, 46. Recall, too, the American slave experience as crystallized by Sojourner Truth in "Ain't I a Woman?" (*NAW*). For a report of women's continuing role as water-carrier and burden-bearer in a less developed country see Forché, 93, 95. Finally, see the results of modern sports-medicine research and the ever-more-successful participation of modern women in sports (Crittenden/Twin).

cultivating tool was a hoe pushed or pulled by a man or (more commonly) a woman. Later, *two oxen were attached and the hoe became a plow*" (I: 972; emphasis mine). Between hoe and animal-drawn plough is a long silence.

Mythology, however, supports the existence of an earlier, woman-invented plough. Both Athena and Minerva are specifically credited with inventing the plough, and Ceres is supposed to have invented both plough and ploughshare. Both Ceres and Athena/Minerva are credited with being first to tame oxen and accustom them to the yoke (Avery, S.V. "Athena"; Boccaccio, 11ff; *EB*/58, S. V. "Athene"; M. Gage/1: 27; Michael Grant, 89; Graves, *Myths* I: 96, 213).

Robert Graves cites a less familiar myth in which a nymph named Myrmex claims to have invented the plough instead of Athena (I: 213), commenting (215–6):

> If Myrmex was, in fact, a title of the Mother-goddess of Northern Greece, she might well claim to have invented the plough, because agriculture had been established by immigrants from Asia Minor before the Hellenes reached Athens.

(Or is this actually a claim for indigenous European invention?)

Archaeology supports mythology here. From Satrup Moor in southern Denmark comes a Neolithic ard dated to c. 4200 BC. This find could easily have been mistaken for a spade. However, as Axel Steensberg points out, the blade bears perforations for traction ropes, and convenient handgrips have been burnt into the handle. In Steensberg's reconstructions, one person would have steered it (free end of shaft under armpit, right hand high in an overgrip, left hand low in an undergrip) while another pulled it through the soil by the ropes fastened through the blade (22–3).

Light soils suitable for scratch ploughs are rarer in northern Europe than in the Near East, and this 6,000-year-old Satrup ard seems as yet an isolated find. However, Andrew Sherratt finds "simple ploughs or ards," presumably of the more usually recognized beam-fitted and animal-drawn type, in northwest Europe by the 4th millennium BC; thus the Satrup ard represents a poorly documented earlier stage of development. Cultivation probably arose first where soils were light enough for early ploughs, and the women of Old Europe, no weaker in the 5th millennium BC than their Near Eastern sisters, could doubtless have wielded them. Moreover, at this time the earth itself would have been seen as a goddess, her fertility properly the province of women. Thus it seems highly likely that women invented these earliest ard-ploughs, and continued to wield them at least until the advent of animal traction.

Likewise in the Near East, though no Neolithic ard is documented there, "The antiquity of the plow shows that an elemental form was developed very early in conjunction with planting grains," probably first in Mesopotamia (*EB* 58 VI: 853). George Ewart Evans suggests a three-person system for the earliest hoe-ploughs of the Middle East, noting that two could haul a lengthened hoe-handle along the surface while a third guided it with a stick fixed to the foot of the hoe (DeBono, ed., 7).

Ralph Linton traces the ard plough to Southwest Asia between 5000 and 3500 BC (*Tree*, 226–7). If by this he means the beam-fitted and presumably animal-drawn "crook" ard he depicts on page 115, the origin of forms drawn by humans using ropes—or dragged or pushed through the ground by means of handles—is presumably pushed back to a still earlier time. Such forms are indicated by rope-traction ards, very similar to the Satrup ard, documented by Steensberg from Bronze Age Syria. Lest anyone underestimate this ancient tool, incidentally, Steensberg notes that it created a furrow 5–6 cm deep, and could completely cover the seed sown in the previous furrow (22–3).

*Irrigation*

Irrigation is often the second defining characteristic of full-scale agriculture; however, the reference there is to large-scale or "scientific" irrigation (cf. Hyams, 10). Small-scale irrigation, on the other hand, definitely arose with horticulture in some areas—in fact, wherever horticulture developed in an arid or semi-arid climate, or wherever early crops demanded a great deal of water. Thus women were most likely the earliest irrigators. It is probably no accident that springs and rivers so often have female deities, or that, as Graves points out, colleges of priestesses were charged with watering the land by rainmaking, irrigation, and well-digging (*Myths,* I: 204).

There are two basic types of irrigation: bringing plants to the water, and bringing water to the plants. Around Çatal Hüyük in the 7th millennium BC, women sowed crops in the flood pools of the river, just as Navajo women sowed their squash and beans in the fans of washes in the Arizona desert. Çatal Hüyük, with 6,000–10,000 people, practiced the "most efficient and advanced" cultivation of the Neolithic Near East, based on this "natural irrigation" (Mellaart's term) or floodwater farming. Women, depicted in the city's art with symbols of their farming role, and buried with their hoes and adzes, were the cultivators of Çatal Hüyük (Rohrlich, L-2/20/80, and "State," 78, 79; Mellaart, *City,* 202, 98).

Women have most often been associated only with the first type of irrigation. According to Jacquetta Hawkes, however, by 8000 BC the inhabitants of Jericho in Palestine were leading perennial spring water to their grain fields and vegetable plots (*Atlas,* 40). Although Hawkes does not say so here, these early cultivators would have been women.

These two early examples may be exceptional, but irrigation is now generally conceded to have begun in the Near East as early as the 6th millennium BC, possibly even earlier (Flannery 6, 89ff; Hawkes, *Atlas,* 58–9; Sherratt, 41, 109ff). In the carefully excavated and dated levels of Çatal Hüyük, this is the millennium (c. 5800 BC) when agriculture triumphs, representations of the male god decline and disappear while representations of the Goddess become more frequent, and the power of women increases (Rohrlich, "State," 79; Mellaart, *Town,* 176, 100). Whether or not anything precisely parallel was happening at Choga Mami northeast of modern Baghdad, women would surely have been in charge of cultivation there when "artificial water-spreading" began in the 6th millennium BC (Sherratt, 110):

> Its position on an alluvial fan, where water enters the plain, was exploited by the villagers, who cut trenches across the branching streams to spread the water to a wider area and thus increase its effective use for agriculture.

Women were still cultivators and society relatively egalitarian in the Ubaid period that gave rise to the Sumerian, and the Ubaid "certainly practiced irrigation" (Rohrlich, L-2/20/80, "State," 180). Moreover, modern studies of Iraqi irrigation archaeology indicate that essentially the same watercourses were in use from "the onset of agricultural life in the Ubaid period (about 4000 BC)" till just before 500 BC. This long-lived system relied heavily on natural watercourses, with only short branch canals that "could have been cleaned easily or even replaced without the necessary intervention of a powerful, centralized authority." Quite possibly the irrigators of this period simply breached the levees along local watercourses to water a given field. With this arrangement, silting would not have been the serious problem it later became. Salinization, too, was evidently avoided until about 2400 BC,[11] when it appeared and plagued southern Iraq for 700 years (Jacobsen & Adams, 1254, 1252).

The semi-mythical **Semiramis** (Sammuramat) is credited not only with building the famous Hanging Gardens of Babylon,[12] which as the first example of overhead irrigation must have required enormous amounts of water—but with inventing tunnels and causeways over morasses. The fascinating point in this far later example is not so much what Semiramis actually did or did not do, or how many women's accomplishments she may have swept into her legend, but the fact that in at least one successful long-distance irrigation system (Iran's) the canals or *qanats* are underground—*tunnels*, in effect— greatly reducing losses from evaporation. And tunnels are attributed to a woman (Boccaccio, 6–7; *EB*/1973, "Semiramis"; M. Gage/1:22–3; Mozans, 341; see also Gore, 616; Lewy).

In Malaysia, Dyak women created bamboo aqueducts to bring water to their houses (O.T. Mason, *Share,* 25). Nineteenth-century observers credit the Dyaks with "a system of irrigation . . . that would do credit to far more civilized people. The whole valley is dug out in squares, the earth heaped up, forming good paths, and the water is conducted by drains and bamboo pipes from one field to another" (H.L. Ross, 406). Could the women bamboo-pipe experts be the creators of this system? In dry areas of the New Guinea Highlands, where, as mentioned above, women are still the primary cultivators, fields are irrigated with bamboo piping (Brookfield, "Highland," 438). Marilyn Strathern finds men now doing the ditching and fencing, but could this system have been women's invention, or an elaboration by men upon a system devised by women?

Arguably, at any rate, women invented irrigation and brought it to varying levels of development in many cultures before this part of farming was taken over, as I will suggest it was in the Near East, by men armed with new technologies and the idea of long-distance irrigation.

### Fire as land-clearing tool

Gatherer-hunters early used fire to drive game, to improve grazing for their game animals, and to encourage certain plants at the expense of others (see, e.g., Jarman *et al.,* 136). Since both sexes participated in these earliest hunting and related activities (Dahlberg, 11), either women or men may have thought of intentional fire-clearing first and both may have participated in reducing it to a system in any given group. However, Elise Boulding has some provocative thoughts on the likelihood that women were first to tame fire, at least in interglacial Europe (79–80). Otis Mason explicitly claims for North America that "it was the genius of the women that invoked the aid of the fire fiend to devour the forests" (*Share,* 147) and clear land for their gardens. Among the Mundurucu of Brazil,

---

11. Interestingly enough, if Urukagina's Law Code is a fair indication, this is about the time that Sumerian women suffered a severe loss of status (Rohrlich, "State," 97), and may no longer have played important or decision-making roles in agriculture. Perhaps, too, population—or the grain trade—had increased to the point where alternate fallow years (during which deep-rooted weeds created a dry-zone barrier to the rising salts; Jacobsen & Adams, 1251–2) were no longer possible.

Sources disagree not only about the date of the code (Kramer; c. 2350 BC), but about its portent and value. Writers conscious of women's interests find in it an ominous loss of previous rights and freedoms, whereas Kramer calls it "one of the most precious and revealing documents in the history of man and his perennial and unrelenting struggle for freedom from tyranny and oppression" (*Sumerians,* 79).

12. One of the Seven Wonders of the Ancient World.

interestingly enough, although the men cut down the trees and clear away the heaviest brush for a garden plot, it is "the gardener" who fires the downed vegetation, reducing all to ashes and returning minerals to the soil (Murphy & Murphy, 60).

*Fertilizer*

Women seem to have invented fertilizer as well, at least in several cultures. The first fertilizer of human origin, of course, would have been the accidental enrichment afforded germinating seeds by midden heaps and feces; but women in their food-gathering and -processing and child-training roles would have had the greater opportunity to observe both of these phenomena.

Menstrual blood was another early fertilizer. Available only to women, this chemically and symbolically powerful life-giving fluid was mixed with the seed grain before planting, underscoring the religious nature of the cultivative act in early times. As late as Medieval times, European peasant women carried seed grain to the fields in rags stained with their menstrual blood, in a continuation of rites or practices dating back to Eleusinian fertility priestesses (*WEMS*, 644). Penelope Shuttle and Peter Redgrove speculate (182–3) that this menstrual-fertilizing rite may be the origin of the ancient Greek Thesmophoria. In this great autumn festival, from which men are excluded, women gathered the naturally decomposed remains of suckling pigs (thrown into caves or chasms early in the summer), mixed this material with the seed grain, and scattered the mixture on the fields (*EB*/11, S.V. "Thesmophoria"; Grahn/Spretnak, *Politics,* 269; Thomson, I: 221–2).

The Massachusetts Indian women who buried a herring in each hill of corn (Linton, *Tree,* 90–1) supplied the growing plant a richer combination of nitrogen and phosphorus than found in modern commercial fertilizers. The material tests so high, in fact—about 8% nitrogen and 7% phosphoric acid—that it acts like dried blood when composted (Rodale's, S.V. "Fish"). U.S. school children learn about the fish, but are seldom told that Native American women were the cultivators (cf. J. Jensen, "Native").

Burning kelp for fertilizer is done by both women and men in the Orkneys, and by women alone in the Shetlands (Fenton, 58–60). Considering the many food and dye and anti-fertility uses of seaweed in the world's coastal and island cultures, women were probably the original seaweed experts.

H.C. Brookfield ("Highland," 438) tells us that in the Mount Hagen and Wabag areas of the New Guinea Highlands, "the soil is heaped up into large mounds within which green manure is placed." The main purpose may be warming the soil against frost, but mulching is widespread. Women also plant casuarina trees in a deliberate program of land improvement. And in the traditional Hagen society, women are the primary producers. Each new bride receives specific garden plots, in which "she is expected to plant and cultivate a variety of crops . . . to support her husband and offspring" (Martin & Voorhies, 253, 255). Males, by contrast, have a "limited twofold involvement with cultivation"; they clear garden land, and are "involved in the cultivation" of such "male" crops as bananas and sugar cane (considered male because of the heavy labor involved in harvesting). Women may even plant some of these crops, but not harvest them. Women, in short, are the primary cultivators. As cultivators of sweet potato, maize, green vegetables, yams, and taro, they would seem the likelier originators of fertilizing and mulching technology.

Among the Bemba of Africa, both sexes clear land for planting. That is, the men indulge in a highly prestigious, dangerous, and ritualized form of tree-cutting unique to their tribe, and the women pile the heavy branches for burning. The women aim to collect the most brushwood to burn, because the ash itself is valuable (Tiffany, 59). Presumably the women know the fertilizing effect of the ash as well as the sterilizing and weed-killing effects of the flames.

### Domestication and selective breeding of plants

By far the most impressive achievement of horticultural woman was domesticating and improving by selection all the basic food plants, including those that still feed the world today. As Gordon Childe points out, "every single cultivated food plant of any importance has been discovered by some nameless barbarian society," among them wheat, barley, rice, millet, maize, yams, manioc, squashes, and taro. During the first Green Revolution, usually called the Neolithic Revolution, "mankind, or rather womankind, had not only to discover suitable plants and appropriate methods for their cultivation, but must also devise special implements for tilling the soil, reaping and storing the crop, and converting it into foods" (56–8).

By 6000 BC Çatal Hüyük had already developed bread wheat (*Triticum aestivum*) from emmer wheat and a naked six-row barley from the simpler, hulled, two-row barley (Mellaart, *Town*, 202). Both of these were hybrids. Women may not have deliberately hybridized these plants, but they recognized the hybrids at harvest time and saved the seeds for planting, gradually increasing them until they formed the whole seed stock. Obviously women in many cultures, in following this process for grains and other plants, developed the concept of seedling beds where seeds were sprouted and grown to a certain stage before transplantation to garden or field. The Irigwe of Nigeria, for example, did this (Leibowitz, 136).

Noel Vietmayer points out (26) that of the hundreds of thousands of known plant species, only about 150 have been commercially cultivated for food. Of these, 30 provide almost 85% of our food by weight, and 95% of our calories and protein. Some 75% of all human food energy, in fact, comes from just eight cereals—wheat, rice, maize, barley, oats, sorghum, millet, and rye—and the first three of those account for three-quarters of that 75%. *All were domesticated in prehistory.* Vietmayer also discusses several high-protein grains and other plants domesticated millennia ago by women in South America, New Guinea, and elsewhere, and only now being rediscovered by modern agriculturalists: quinoa, the winged bean, and tarwi, for example.[13]

Pre-Incan Peruvians must have developed a technique of propagation by cuttings, because they grew certain sterile squashes and melons impossible to propagate any other way (Hyams, 115). The Mundurucu women also know this method for growing manioc (Murphy & Murphy, 124).

Perhaps the earliest change women made in their grain-bearing grasses was to select for non-shattering heads. This selection could have been accidental at first: more of the non-shedding mutants would have been left for women to gather. But having noticed the difference, they could have selected for it in their seed grain (Hawkes, *Atlas,* 39, 49).

---

13. Note a few of the amazing features of the winged bean from New Guinea (Vietmayer, 26–7). It provides six different foods, all valuable: the pods, which may be as long as a man's forearm, can be eaten raw, steamed, boiled, or stir-fried as a crisp and chewy vegetable if picked young; the leaves, cooked and eaten like spinach, contain enough vitamin A to correct the deficiency that blinds thousands of children each year in the tropics; the tendrils, looking like lacy asparagus, are also edible; the flowers make a sweet-tasting garnish that looks like steamed mushrooms; the tuberous potato-sized root has firm, ivory-white, nutty-tasting flesh that can be eaten boiled, baked, fried, or roasted, but is four times richer in protein than potatoes and ten times richer than cassava; and the seeds are 40% protein and 17% edible oil, rich in iron and vitamin E after cooking—approximately equivalent in nutritional value to soybeans. Best of all, the winged bean is a legume and thus needs no artificial fertilizer.

Women's achievement in root-crop areas is no less striking. Wild yams, such as those gathered by the Australian aborigines, for example, are about the size of a cigar, whereas the cultivated yams grown by Melanesian women may be six feet long and a foot or more thick (Chapple & Coon, 174–5)!

## Food-Processing

Second only to woman's achievement in domesticating all the world's major food plants is her achievement in making plants edible. A major and relatively well-documented part of this work deals with the seeds of grasses like corn, wheat, rice, and barley. Here I will focus on a less familiar but equally major area of women's food-processing work:

## Detoxification/debittering

The staple or important foods of many cultures are poisonous in their natural state, and need extensive treatment to make them safe to eat. Others not actually poisonous taste bitter or can cause nausea or diarrhea: the acorns that supported several Near Eastern and Native American groups; olives, so important in Mediterranean cultures; the Pacific Island and Far East staples taro, sago, and breadfruit; certain yams important in Mexico, South America, Africa, New Guinea, and Asia; and the manioc that feeds hundreds of millions in tropical Africa and South America. Several plants less economically important but equally interesting in their detoxification technology could also be listed. Although several of these methods were developed before or in the absence of horticulture— notably acorn debittering—I group them here to show the scope and significance of these early women's technological achievement.

*Acorns.* "The acorn," says Theodora Kroeber, "meant to Indians what rice means to Cantonese Chinese or maize to Mexicans" (*Two Worlds,* 21). Indeed, the Ohlone of California reckoned their new year from the acorn harvest. Native Americans recognized at least half a dozen edible varieties, both sweet and bitter. Bitter acorns contain tannins. Also found in oak bark, tannins are used in dyeing, tanning, and healing—all women's work. The debittering technology is basically a leaching process—achievable by leaving the crushed or whole acorns in net bags in a stream for a season (cf. Coon, 59) or, more laboriously, by reducing the acorns to meal and washing the meal—followed by cooking.

Ohlone women followed the more laborious technique. After hulling several handfuls of acorns with her *hammer and anvil stones,* each woman

> put the kernels into a *stone mortar* or . . . *mortar basket* (a bottomless basket glued to a rock). Sitting with the other women of the village, she pounded the acorns with a *long pestle,* pausing now and then to scrape the acorn flour away from the sides of the mortar with a *soaproot fiber brush.* Then she pounded some more. The rhythmic thumping of the women's pestles filled the air.

When the kernels were all crushed, she put the flour through a *sifting basket,* returning any coarse particles to the mortar for further pounding.

The meal was fine now, but still very bitter. The Ohlone woman's leaching technology consisted of a *hole scooped in creekside sand* or a special *leaching basket,* both lined with fern fronds. She dumped the meal in and poured water through it repeatedly till the tannins were gone. Knowing that *hot water* speeded the process, she used it if firewood was plentiful.

The usual cooking method among both the Yahi and the Ohlone was to mix the meal with water in a *watertight cooking basket,* place heated *cooking stones* in the mixture, and stir constantly with a *looped stick* or *wooden paddle* until the mixture boiled, creating acorn mush. To make the "deliciously rich" acorn bread, the woman boiled the mush a little longer and spread the batter on a *hot slab of rock* or baked it in an *earthen oven.* Acorns are nutritionally equivalent to wheat and barley (Kroeber, *Ishi,* 21; *Two Worlds,* 181; Margolin, 41–5). Note the many separate inventions involved.

If the acorn harvest failed, the Ohlone gathered buckeyes. The women had to roast, peel, mash, and then leach the meats of these large brown nuts to make them edible. The leaching took eighteen hours, and must be faithfully done, for buckeyes were not merely bitter but poisonous, containing prussic acid. Indeed, like the soaproot, which had to be roasted before eating, unprocessed buckeye mash was used to stun fish (Margolin, 50, 40).

Acorns were also once an important food in Greece and, as already mentioned, in the Near East. As late as 6000 BC, in fact, the settled cultivators of Çatal Hüyük still gathered acorns from the hills (Mellaart, "City," 202).

Native North American women adapted their acorn-processing technology to use European materials in the 19th century. Here, we have at least one name: **Bertha Norton,** a woman of Maidu-Wintuk ancestry, debittered her acorn meal in a gunny sack by pouring water through it until the bitterness was gone (Westin, 19).

*Olives.* The olive drupe or fruit contains considerable amounts of vegetable acids as well as a bitter glucoside called olivil. Two 19th-century pickling/debittering methods will illustrate the difficulties involved (Hayne, 195ff):

> The olivil can be removed either by a very prolonged treatment with pure water or by a briefer treatment with a lye solution. The water treatment is probably older and sounds simpler, but actually demands great patience and care, since the fruits must be layered only about two feet thick, and the water must be pure and changed often enough to frustrate the otherwise rampant bacteria. The vats or containers need false bottoms or some other means of drawing impurities off from the bottom, and floating covers or some other means of keeping the olives submerged at all times. The olives must be tasted daily after the 20th day and at the proper tartness put into brines of ascending salinity . . . for a year or until used (if more than a year, the brine must be changed again). The salt acts to re-firm the softened flesh of the olive and also to slow or stop bacterial growth.

This method conserves both the flavor and the oil of the olive, but takes up to two months.

The faster, lye-soaking method also has its difficulties: it must be carefully watched, the purity of lye is hard to control, the concentration of the solution must be less than 6%, etc. Moreover, the lye neutralizes the bitter principle but does not remove it, so that several pure-water rinses and the same series of finishing brines are needed. Lye also bleaches the olives more than water does, may soften them so much that they become unpalatable or fall prey to fungi, and may saponify (turn soapy) the olive oil.

Concludes Arthur Hayne, writing from a California State Agricultural Experiment Station at the University of California in 1896 (196):

> The process of pickling the olive is one of the most delicate of the agricultural arts; . . . so many complicating circumstances enter into the problem, requiring the exercise of so much close observation and judgment, that all the Experiment Station can do is indicate the general line of procedure and warn the beginner against common errors.

Yet women of Northern Africa had devised this "most delicate" process before the advent of written records. And **Martha Laurens Ramsay** re-invented it in South Carolina in the

18th century (A. Earle, 85). Ramsay's *Notable American Women* biography omits this achievement, concentrating on her piety, submissiveness, and ideal motherhood.

Women are anciently associated with the olive from the mythical time when Athena took possession of Athens by planting the first olive tree beside the city's well (Graves, *Myths,* I: 60). Virgil calls Athena "inventress of the olive" (Mozans, 335).

*Taro.* The plant commonly called taro belongs to the arum family. A major source of food in the Pacific Islands and the Far East, *Colocasia esculenta* is now cultivated in some 1,000 varieties. The characteristic bitterness of arums, from calcium oxalate or salicylic acid, must be removed or de-activated before eating. Cooking is the usual technique (*EB*/58, S.V. "Arum"; Linton, *Tree,* 67, 96).

Americans may know taro as the basis of Hawaiian *poi.* Hawaiian women not only peel and pulverize the tubers, but slightly ferment the resulting paste in the course of their preparation.

As already noted, Native Americans ate a North American relative of taro called Jack-in-the-pulpit (*Arisaema triphyllum*), inspiring two of its other names, bog onion and Indian turnip. However, they did so only after pounding the root to a pulp with water and allowing the mass to dry for several weeks before using it as flour, a sequence reminiscent of manioc-processing (see below). The unprocessed root or corm contains calcium oxalate. Dried and powdered roots were used in medicine (the corm was official in the *U.S. Pharmacopoeia* from 1820 to 1893), and (unprocessed) in contraception—again, women's work (M. Weiner, 64).

Likewise, two plants the Australian aborigines use for food after a tedious cooking process with many changes of water—the Queensland matchbox bean and a tuber called *tjarri* or *ka'ata*—are used unprocessed for contraceptives (Himes, 28–9).

*Sago.* Starch prepared from sago palm pith is an important tropical foodstuff in both the New and the Old Worlds, from the Orinoco Basin to Southeast Asia, India, the South Pacific, northern Australia, and New Guinea. The pith is subjected to a complex process of preparation which sometimes "almost assumes the scale of light engineering, with the construction of an elaborate trough and conduit system and the fabrication of an assortment of filtration baskets and vats" (Ruddle *et al.,* 22). The operation consists of cutting the palm, removing and chopping or pounding the pith, then washing it repeatedly. Starch particles washed out with the running water are then filtered and allowed to settle. During washing the pith must be kneeded or trampled; some groups do this with the pith in a bag or in a trough with a filter at one end; some trample the pith in a wide-mesh basket resting on a stand built over a river or pool, etc. Melanau women of Sarawak had special trampling platforms for sago.

A great many individual inventions are involved in the harvesting and processing of sago—the *stone cutter* now being replaced by the steel axe (though timed studies show that the stone tool is very nearly as efficient; *ibid.,* 16); an *adze* for removing the pith; *pounders, strainers, filters, trough* and *water-conduit systems* of varying complexity, etc.

In some groups women do all the sago work, in others men do it all; in still others the work is divided, men cutting, pounding, or rasping the pith, women extracting the starch. Interestingly enough, women's degree of participation is directly related to the importance of sago in the group's diet, and *is higher where sago is more important* (Ruddle *et al.,* 937):

> In general, where sago is the primary carbohydrate staple (complemented by hunting, fishing, gathering, and only small amounts of gardening), men may fell the palm, but both pounding and washing are the exclusive responsibility of the women.

By contrast, where sago is a minor, famine, or feast food—or where women now cultivate rice or cassava, men may do all the sago work.[14] In the intermediate case, where sago is but one of several staples, or now considered a supplementary food, and women are ever busier with rice- or tuber-gardening, men come in to help the women in sago-processing, usually with the pounding (*ibid.*, 93). A fascinating group not specifically mentioned by Kenneth Ruddle, Patricia Townsend, and their co-authors is the Agta (Philippines), where both women and men hunt and fish, but where the processing of the caryota palm— including felling it—is definitely women's work (Estioko-Griffin & Griffin, 133).

Since women are in charge in two of the three possible cases—the two in which sago is more important in the group's economy—women probably created most of the inventions and invention systems used in this complex process.[15]

*Cassava/manioc.* If judged by numbers fed, the most significant of women's detoxification technologies is that for making edible starch out of manioc, or cassava. Species *utilissima*, indeed, these tubers yield the staple food of 200 million people. As with acorns, users recognize bitter and sweet forms, though the difference here may be environmental rather than genetic (Sokolov, 34, 38). In any case, bitter manioc contains prussic (hydrocyanic) acid, which must be removed before eating. Otis Mason describes in some detail the processing of bitter manioc by Guiana Indian women, using their combination press and sieve called a *matapie* (*Share,* 38–40):

> No scene is more characteristic of Indian life than that of women preparing cassava. One woman, squatting on her hams and armed with a big knife, peels off the skin of the root and washes it. Another woman, grasping one of the roots with both hands, scrapes it up and down an oblong board or grater studded with small fragments of stone like a nutmeg grater . . . The rhythmic "swish" caused by the scraping is the chief sound in the house, for the labor is too heavy to permit talking. The grated cassava is placed into a long [basketry] sieve or *matapie* so woven that a weight on the bottom will compress and open the sides, and we have press and strainer in one. The cassava, saturated with its poisonous juices, is forced into this matapie and suspended from one of the beams of the house. Through a loop in the bottom of the matapie a heavy pole is passed, one end of which rests on the ground. A woman now sits on the pole, and her weight stretches the strainer and forces the poisonous juice, which is caught in a vessel below. . . . The cassava is taken from the matapie, broken, sifted, and baked into griddle cakes, which are dried in the sun.

Mundurucu women make farinha (manioc flour) instead of cakes. They use two methods (Murphy & Murphy, 7–9, 123–7), one substantially as just described, and the second a little easier. Instead of peeling and grating the tubers, the women sometimes put them into a stream and leave them there to soften for three to eight days. Then they bring them to a trough in the farinha-making shed, where they skin and trample the soft tubers to break them up completely. Then they put the moist pulp into a *tipiti*, much like the invention Mason describes (Murphy & Murphy, 9):

---

14. Interestingly enough, this—essentially a phenomenon of gathering and processing—is precisely opposite to what happens to women's participation in cultivation: women's participation declines as the importance of cultivated food in the diet rises (see below).

15. Sago is not universally considered toxic. However, according to the Sanio-Hiowe of Papua, New Guinea, certain people always vomit when they eat sago pudding cooked by the stone-boiling method. This group also considers sago unsuitable for infants. Where pottery or metal cooking pots are used, allowing the starch to be cooked more fully, these claims do not seem to arise, indicating that remnants of toxin or indigestible material are destroyed by thorough cooking (Ruddle *et al.*, 61).

The tipiti was a long tube made of loosely woven palm leaves, wide with an open mouth at the top and closed at the bottom. The top end was suspended from a rafter, and a long pole was placed through a loop at the bottom. Two of the women sat on the end of the pole, the other end of which was secured near the ground, and the resultant lever pulled powerfully down on the tipiti. This caused it to elongate and constrict, squeezing out the water from the pulp and leaving the contents still moist, ready to be sieved. When only a dribble of water came from the tipiti, the women emptied the pulp into a large sieve placed over a shallow basin and gently worked it through the mesh with their fingers. It dropped into the receptacle as a coarse, damp cereal, and the pieces that did not go through were taken by another of the women and pounded with a wooden mortar and pestle.

The resulting very fine pulp still contains enough prussic acid to cause severe gastritis. It must be toasted on a griddle and constantly turned with a spatula until heat drives off the remaining volatile acid. The resulting farinha flour, taking most of a day to make even by this "easier" method, will last one small household only about a week, and is less nourishing than the so-called *farinha seca.* Farinha-making is the major chore of the Mundurucu women.

When they make farinha by the more laborious peeling-and-grating method, the Mundurucu women must wash the pulp in several changes of water right in the farinha shed. Pure tapioca settles out of the first wash water. The women carefully save it for two days, then sieve and toast it (Murphy & Murphy, 125).

Other byproducts of the manioc detoxification process are *cassareep,* a sauce and preservative for meat made in Guiana by boiling the milky toxic juice squeezed out of the pulp by the matapie/tipiti; and a half-gelatinous cake made by Carib women from the starchy residue in the matapie (O.T. Mason, *Share,* 38–9).

Notice the several inventions involved: *stream-leaching; the grater,* once a board studded with stones but among the Mundurucu now often the perforated half of a five-gallon metal can (Murphy & Murphy, 125); *the sieve,* of basketry; *the matapie/tipiti* itself, which Mozans calls "probably the most remarkable culinary invention of woman in the state of savagery" and "for the inhabitants of the vast basins of the Amazon and the Orinoco . . . the most important ever made" (340); *the toasting spatula* (the Mundurucu use a canoe paddle); *the griddle* (the Mundurucu use griddle and oven exclusively for cassava; otherwise, they boil, ember-roast, or cook food over the fire on the babricot, a tripod topped with a rack of green wood strips; Murphy & Murphy, 128, 16); the *detoxification of the extracted toxin* itself, and so on. Though Mason and other observers have usually focused on the matapie, note that the grater is also astonishingly ingenious. Consisting as it does of stone fragments fastened to a wooden board, it demands, among other things, the invention of a glue both waterproof and strong enough to withstand the heavy pressure of the grating motion for large tubers.

From a close look at women's techniques for debittering and detoxifying food plants, then, some patterns begin to emerge: water-soluble glucosides and alkaloids are leached out; other undesirable substances are drawn out by brines; volatile oils and other substances not soluble in water are driven out by heating and cooking; acids are neutralized by strongly basic solutions. The women even use fermentation where necessary. Tubers, roots, fruits, nuts are peeled or hulled, crushed, ground, or sliced in order to speed up the process. (Enzymes capable of destroying or deactivating the glucosides sometimes exist in the plants themselves, but in separate cells; thus the crushing or grinding operation in itself can be partially effective.)

Women also know other means of speeding up these processes, such as using hot water instead of cold or salt water instead of fresh. All this sounds suspiciously like chemistry.

Finally, women's detoxification technology involves many physical inventions, such as knives, graters, sieves and strainers, griddles, mallets, and of course the amazing *matapie/tipiti*.

This discussion leads naturally into a discussion of cooking. However, let us note in passing an instance in which an ancient (presumably women's) processing method proved strikingly superior to a modern "improved" method. In 1969, peasants of Hopei Province in China began to suffer mysterious symptoms of fever, general malaise, and burning sensations, eventually followed by heart damage and infertility. It turned out that the ills were caused by an impurity in the cottonseed oil they consumed daily. The traditional processing method of boiling the cottonseed before pressing it for oil had inactivated this impurity. In the 1960s, however, a new commune system streamlined the pressing operation, substituting cold-pressing for the accepted process. Thus the impurity, a phenolic compound called gossypol, remained unaltered and fully toxic in the resulting oil.[16] This story should warn us not to discard traditional methods without considering why they were adopted in the first place—and carefully recording them for the future.

The foregoing examples illustrate the multitude and quality of horticultural woman's achievements in food-processing and—since some of the detoxification processes continue into agriculture—provide a transition to woman's achievements in that later stage.

## Food Preparation: Cooking

*Pace* Charles Lamb, few would deny that women—probably Neanderthals—invented cooking.[17] Of the 185 societies in the standard cross-cultural sample, 117 assign cooking exclusively to women, another 63 assign it predominantly to women, and only four assign it partly or largely to men. It is the only activity in the sample not exclusively male in any culture (Murdock & Provost, 207). What is usually denied by omission, or at least not emphasized, is, first, that *cooking is technology,* and as such involves a great many physical and chemical inventions; second, that *cooking was once the cutting edge of technology*—so much so that baking at least was once a sacred activity; third, that *cooking is significant technology;* and finally, that *cooking gave rise or contributed to many other technologies*—smelting, pottery, herbal healing, dyeing, and—much later—alchemy, for example—and thus to metallurgy, ceramics, medicine, and chemistry.

If dictionary definitions of technology as "industrial science . . . " or "any practical art utilizing scientific knowledge" (*Webster's Unabridged,* 2nd ed.), reflect prevailing views, then it would scarcely be surprising if pre-1975 histories of technology devote little attention to cooking. What is surprising is that we seem to be going backward in this regard. Singer's classic history of technology, though devoting few pages to "culinary arts," and speaking of "control of fire" rather than cooking *per se,* does contain two striking statements: Gordon Childe points out (Charles Singer I: 41) that discovering how to make and control fire allowed humans to utilize as food a wider range of nutritive substances than any other living creature. In other words, cooking freed us from the strict

---

16. This blunder in the name of progress may have led inadvertently to the discovery of a new male contraceptive; and researchers may have learned how to prevent or minimize gossypol's toxic effects on humans (Djerassi, *Politics,* 202–5); but this is small comfort to the peasants who lost their health.

17. In the "Dissertation upon Roast Pig," the son Bobo first tastes burnt flesh when the family hut burns down with the family pig inside, and keeps on burning down the house to get more. I did once see cooking seriously attributed to Man the (Ice Age) Hunter trying to thaw frozen meat out of the ice (Halsbury, 12; Singer I: 31).

dependence on the natural food supply that limits other animals. And R.J. Forbes notes, "The kitchen is the birthplace of many technical operations and apparatus. To it we owe furnaces and ovens, apparatus for grinding and crushing, the use of alcoholic fermentation, methods of preservation, and the extraction of liquids by pressure from seeds and fruit" (*ibid.*, 271).

In a 1983 survey conducted for a doctoral dissertation at West Virginia University, by contrast, a panel of technology "experts"[18] omitted cooking from their list of "foundational technical developments" in human technological history (Bindocci, 143–8). Since cooking appeared as a choice in one round of the survey, this was an active rather than a passive omission.

To state what should be obvious, cooking is technology because it involves techniques and processes that women not merely discovered, but experimented with, elaborated upon, and reduced to effective, repeatable systems. Anyone could drop a piece of meat or a yam into a fire, retrieve it later from the ashes, taste it, and observe that it tasted good, was easy to chew, or failed to cause a stomach-ache, as the case might be. But only an inventor would later experiment with deliberately placing food in the fire, varying the duration, the stage or heat of the fire—perhaps even the fuel itself—until she arrived at a satisfactory and relatively uniform technique that we would call roasting.

Anyone could spill water on a hot rock and observe the steam and sizzle, or kick a hot rock from the fire circle into a shallow puddle and notice the water begin to boil. Anyone, too, could notice that certain kinds of rocks did not burst when heated. But only an inventor would later experiment with various kinds and shapes of rocks, and with various deductions from her original two observations, leading to inventions as diverse as soapstone griddles and cooking pots, or cooking stones stirred constantly to boil acorn mush. Otis Mason points out that women even added legs to their soapstone pots, which must have complicated the carving enormously, but stabilized the pots and allowed a draft to circulate beneath them (*Share,* 34).

As early as the Upper Paleolithic, East European sites featured substantial dwellings, including fireplaces with "intentionally corrugated floors" or small grooves or ditches leading outward from the ash area. "Modern experiments have shown that this modification increases oxygen flow and produces a hotter flame" (Sherratt, 90). The list of physical inventions already connected with cooking by horticultural times is, in fact, enormous, and for that reason beyond the scope of this chapter. Suffice it to say that before horticulture in many cases, and certainly by the end of the horticultural phase of human technological development, women had invented most of the basic forms of cooking: parching, roasting, griddle-cooking or toasting, baking and boiling, and early but not unsophisticated vessels, techniques, and devices for accomplishing them.

In an era when fire was newly tamed, cooking would have been an advanced technology. In addition to the developments already mentioned, meat could be smoked, dried, or partially cooked to preserve it even when there was no sun. Whole new classes of food, particularly the easily stored roots and tubers, as well as certain tough seafood, could now be exploited.

All kinds of cooking must have seemed mysterious transformations and may have been linked in people's minds with the many transformations they saw in the year's round and attributed to the Earth, earth/womb, or the Goddess Earth. Evidence of this very early belief is necessarily fragmentary, but baking seems once to have been a sacred activity.

---

18. Drawn from among 1) directors of science-and-technology centers, 2) members of the Society for the History of Technology, 3) scholars in the area of women and technology, and 4) business and industry.

According to Sjöö and Mor, only women could use the early beehive ovens, because they were seen symbolically as the "Belly of the Great Goddess" (10). And Marija Gimbutas reports from Old Europe that temples of the 6th and 5th millennia BC seem to have been conceived of "literally as the body or the house of the deity." Both actual temples from this era and clay models of temples found at altars contain figurines of women performing "various religious activities—ritual grinding, baking of sacred bread, attending offerings—or . . . seated on the altar. . . ." In a Ukrainian shrine dating from the mid-5th millennium BC, sixteen female figurines sit on chairs at the altar, while another sixteen are baking, grinding, or standing by the remains of a bull sacrifice. These are not isolated findings. By 1982 some fifty such models from various cultures and phases are known from Old Europe. Dr. Gimbutas considers them more informative than actual temple remains, because "they present details of architecture, decoration, furnishings, otherwise unavailable to prehistoric archaeology" (Spretnak, *Politics,* 25).

Certain ancient rituals incorporating baking have come down to us disguised as patriarchal church rites (*ibid.,* 271):

> As the mother makes the child with her own flesh so does the Mother-Earth make us with Her flesh—in the form of grain. To teach this principle to the people of their communities, the women, or sometimes the girls at menarche, made bread in the shape of a female, or specifically of the yoni, or the round moon. This special bread, whether made of rice, millet, corn, wheat, or rye, was and is eaten in a public *cere*mony. . . .

Observing that "the raising of a vessel is a common practice in ritual celebrations" (cf. elevating the Host in the Catholic Mass), Gimbutas reports that even today in Bulgaria, at New Year's, "the three eldest women of the family thrice raise the trough containing the sacred New Year's bread-dough" (*Goddesses,* 82–3).

Cooking is significant technology because it affects the daily life of everyone on earth, influencing growth, health, and even psychological well-being at the most fundamental level. In gathering-hunting times, cooking allowed us to exploit new environmental niches by enabling us to eat previously inedible foods. It helped toothless elders to survive longer by softening their food and making it more easily digestible, thus indirectly furthering the survival of the people, who benefited from their spirit, knowledge, and wisdom. As Ishi, the last Yahi, put it, "It is good to have Old Ones at the firepit; they laugh and tell stories, and then we forget the saldu [whites]" (Kroeber, *Ishi,* 8).

Abundant examples of the contributions of cooking to other technologies could be given. Suffice it here to note that many of the vessels and processes of alchemy—including stills and double-boilers—came straight from the kitchen and were invented by women (though mostly later than the era being discussed here; see e.g. Alic, "Women," 307); that cooking must accidentally have produced the first baked clay objects and that kilns were modified (or unmodified) ovens or covered fire-pits, etc.; and that cooking gave women the extensive experience with use of fire and heat that carried over into many other technological areas, including prehistoric art (cf. Leroi-Gourhan).

## AGRICULTURE

As a society depends more heavily on cultivation, and approaches the large-scale cultivation of true agriculture, women begin to lose out as exclusive or even primary cultivators, especially in the Near Eastern and European cultures giving rise to our own. As societies become more complex, of course, the question is no longer so much who does the work, but who is in charge—and who makes the tools. If women farm, but as slaves or hired workers rather than owners or trustees for the clan, for example, they no longer have the same incentive to innovate. Martin and Voorhies detail how labor is transformed between the two periods. As their Table 9-2 shows, women, exclusive or "equal" cultivators in 83% of the horticultural societies in the sample, are exclusive or equal cultivators in only 19% of the agricultural societies (283):

**A Comparison of Sexual Labor Division in Cultivation Between Sample Horticultural and Agricultural Societies**

| Cultivation Type | Technology | | | |
|---|---|---|---|---|
| | Horticulture | | Agriculture | |
| | # | % | # | % |
| Female | 52 | 50 | 15 | 16 |
| Male | 17 | 17 | 75 | 81 |
| Both sexes equally | 35 | 33 | 3 | 3 |
| Totals | 104 | 100 | 93 | 100 |

Before looking at women's inventive achievements in agriculture, let us examine this dramatic shift, which I call the Takeover.

## Takeover Theory

Why do men participate more and women less as cultivation becomes economically more important and larger in scale? Why in so many proto-Western cultures did men take over what women had invented and developed into an efficient, productive, and adaptive system? It is a crucial question. In the Near East, for example, barley was money (cf. Friberg; Rohrlich, "State," 90). The agricultural surplus was the basis for all other wealth. The very title *ensi* or *ensik,* for the head of the city-state in Early Dynastic Lagash (Sumer) meant "administrator in charge of plowed land"; and a stylized copper or gold sickle was a symbol of power in the Near East and the Caucasus as late as the 3rd millennium BC (Boulding, 174; Gimbutas, *Goddesses,* 83). The threshing flail was one of the two insignia of power in the kingdom of Egypt. Thus the agricultural takeover seems relevant to the political takeover in progress about the same time, c. 4000–2000 BC. Unfortunately, these processes coincide roughly with the invention of writing, so that we know much more about the *status quo post* than *ante.* Ignorance has deceived some into arguing that male dominance has always been the order of things—lowered from heaven, as it were, like the kingship at Kish in Sumerian mythology. Only recently have we begun to insist on knowing how and why the change occurred.

Various theories have been advanced. The Marxist explanation, based on Lewis Henry Morgan's insights into the economic bases of social development, has merit, but also some failings. That is, storable food surpluses and the new concept of land and domestic animals as private property could certainly lead to economic inequities. Irrigation, too, makes some parcels of land more valuable than others. Compounded by the development (or imposition) of a market economy, these inequities seem to have given rise over many centuries to the stratified societies whose successors we see in the developed world today. But perhaps this was not inevitable. Stored food surpluses could exist, for example, without creating markedly stratified societies—and without destroying the high status of women. As religion became institutionalized in the Near East, surplus grain and other farm produce were evidently offered to the Goddess and stored in Her temples for drought relief (cf. Charles Singer I: 48). When this voluntary offering became an involuntary tax, society may have reached a turning point in the development we are trying to explore.

The new concept of private property is much more dangerous, as Morgan correctly says (Leacock, *Myths,* 122ff). In combination with a market economy encouraging production for exchange, it could easily alter the economic relations of a society. What none of this really explains, though Engels tries,[19] is how men emerged as a superclass in each resulting socioeconomic class.

In other words, even if clan egalitarianism rested on the collective ownership of productive property, and even if the new private ownership of that property shattered this egalitarianism, why did *males* become the owners of the new property? Engels himself considered that women were administrators of the "old communistic household."

Men may have been entrusted with the *distant* herding of a group's large domesticated or semi-domesticated animals, especially if these were the group's former main prey animal. But women probably played an important role in the actual domestication process, as providers of milk and special "infant" food to the orphan baby animals that were likely the first domesticates. And as mothers of the clan, women could have been considered the "owners" of the animals, and the men merely assigned to do the herding (*the herder and her husband*), unless some other and very powerful factor intervened. The shift to male ownership of land is even more difficult to explain.

Marxists are right, too, in seeing that power usually flows to those who control the means of production. But as the Little Red Hen of the English folktale (Galdone) pointed out long ago, controlling the *distribution* of production is the ultimate economic power. And nothing has so far explained how and why men became primary producers instead of women and/or got control of distributing women's production.

M. Kay Martin and Barbara Voorhies attempt an ecological approach to the question, arguing that the new division of labor came about and persisted because it worked, was adaptive; that because men were, on the average, physically stronger and spent less time with child-rearing, they were more mobile and thus were chosen for the herding of large animals when these were domesticated. The use of animal power to draw ploughs then seemed an extension of these duties and ploughing could be considered a form of land-clearing, a part of cultivation often already assigned to males (286).

Recent research has undermined most of this argument. As already noted, women trained for strength from early childhood may be very nearly as strong as men, if the men are not so trained. Nor are hominid males significantly more mobile than females. Although primate, hominid, and early human males may range farther than their corresponding females, female foragers did and do range over considerable territories, with or without infants (Draper, 83; Shostak, "Life . . . ," 53), and even in an advanced state of pregnancy (Zihlman/Dahlberg, 92). Among the Agta people of the Philippines, where women regularly hunt, *women range as far as men* (Estioko-Griffin & Griffin, 11). Women's gardens may lie far from their dwellings—six or seven miles among the Irigwe of Africa, for example (Leibowitz, 136), and water or firewood may be even farther away. Women travel these distances several times each day with heavy loads of food. In fact, Adrienne Zihlman concludes flatly (/Dahlberg, 89):

---

19. It should be noted here that Engels was wrong about the respective contributions of the two sexes to food production, considering that in pre-agricultural times the males were the food providers (cited by Sacks in Reiter, ed., 204). I am indebted here not only to Karen Sacks, but to the brilliant critique and updating of both Morgan and Engels by Eleanor Leacock in her *Myths of Male Dominance.*

To postulate that early human females were sedentary denies their primate heritage as well as evolutionary continuity. . . . In fact, there were no such sedentary females in gathering and hunting societies.

Furthermore, ploughing is arguably less analogous with ground-clearing than with soil-preparation, once done (by women) with digging stick, spade, or hoe.

Elise Boulding argues that women in the Near Eastern agrovillages were simply too busy to take on such additional tasks as herding. Likewise, and even more important, they were too preoccupied with their heavy and varied daily workload to deal with—perhaps even clearly to see—the developing problems of scale in their societies' food-production systems. The men, on the other hand, gradually deprived of their hunting role in favor of the tamer work of herding, which employed only a few of them in any case, had less to occupy them *on a daily basis* than women did. Thus they would have had more mental and physical leisure to focus on and deal with new problems.

According to Boulding, male traders[20] would have been the first to see advantages in higher production—production, that is, of still larger surpluses for exchange. As cattle-herders, she says, they would also be the more likely to figure out that an ox could drag a hoe through the fields and thus speed up the soil-preparation process. Women may have lost out rather quickly once the men had two farming specializations (ploughing and cattle), retaining only the "subsidiary" tasks of weeding and carrying water to the fields (126, 162–3, chaps. 4,5).

Boulding's point about the time budgets of the two sexes is well taken. But it is not clear to me how, in the first place, such important tasks as weeding and watering (cf. irrigation) became classed as "subsidiary" to soil preparation. Nor does it seem self-evident that herding, often at considerable distance from the cultivated fields, would equip men to invent the animal-drawn plough—unless by this time some men had already been recruited to draw the earlier rope-traction ploughs found in Syria and elsewhere in the Middle East. Also, Ruby Rohrlich contends that Neolithic women in Çatal Hüyük and Sumer, at least, "participated in the long-distance trading exchange" ("State," 78). However, men did *eventually* become the *long-distance* trading specialists and the *large-scale* ploughing specialists, at least in cultures giving rise to ours.

We may never know the answer for sure. For one thing, answers differ for the various areas.[21] For another thing, most of the groups that could have enlightened us have vanished. But the available evidence seems to me to support a scenario something like the one outlined in Appendix A-1.

Suffice it to say here that we must look for developments that affected the two sexes

---

20. But cf. the Negrito Agta of the Philippines, where women from the mountain groups not only trade but are considered the more aggressive traders since men dislike confrontation (Estioko-Griffin & Griffin, 134).

21. I focus here on seed- and grain-based cultivation, but an entirely different system arose in the tropics, based on roots and tubers. Yam and taro cultivation have been dated as early as 13,000–9,000 BC in southern Asia (Ucko & Dimbleby, 423; Ayensu, 54). If these estimates prove correct, the real "cradle" of agriculture may have been Southeast Asia.

   This root-cultivating technology was also evidently women's. Edwin Loeb, for example, specifically credits women with (303–4) "most of the inventions of the period including the use of bamboo utensils for cooking and holding liquids, the steeping and beating of tapa to make cloth, the cultivation of root plants and above all, the domestication of small animals as household pets." There are definite parallels between the two systems—Loeb, for example, finds a matrilineal context for the development of cultivation (Boulding, 151–2).

differently after sedentism and particularly during the transition to agriculture. Primary among these, as I see it are 1) a birth explosion, 2) a change from female-centered to male-centered religion, and 3) the professionalization of the cultivative role. All three are discussed, with other factors, in Appendix A-1, but note that a *birth* explosion can take place without causing a *population* explosion, if child mortality rises at the same time—as it is likely to do when larger numbers of people settle permanently together. A birth explosion, however, demands a *pregnancy explosion,* creating more of those otherwise few periods during a woman's life when hard work is difficult or dangerous for her (very late pregnancy, early lactation)—and, at the same time, greatly increasing her workload in food-providing and processing, and in child care, especially in sick-child care. Moreover, caring for several infants rather than one is geometrically more stressful, not to mention the emotional trauma if several children die.

The precise how and why of the overthrow of Goddess-worship in the Middle East, and the precise degree to which cultivation was still entwined with that earlier religion, are controversial, but the gradual—or sometimes sudden—triumph of male deities in the area after 3000 BC is undeniable. The negative psychological effects on women—and the converse effects on men—though intangible, seem equally undeniable.

For a purely secular explanation, the factor of choice is professionalization. Here, the importance of women's already heavy workload becomes crucial. When gardens became fields that had to be ploughed and cultivated many times each year with animal-drawn ploughs, and when irrigation became a complex matter linked to a region-wide system of canals and ditches—in short, when farming became a full-time rather than a part-time job, involving specialized technical skills that a girl could not learn from her mother—women were arguably just too busy to take it on. This pattern of women's inventing a technology and developing it to a satisfactory level for the family- or even village-scale economy, only to lose out in that activity as it becomes commercialized and professionalized, is, I think, a widespread and important one. The same thing can be observed in women's work in medicine, contraception, and machines, and should be kept in mind when reading the following three chapters.

### Takeover Through 18th Century

Women continued to contribute to agriculture, of course, more in some cultures than in others. In Africa and Asia, for example, as Francesca Bray points out (L-2/20/80), women do two-thirds to three-quarters of the work. But either they are no longer in charge—as at Umma in Sumer, where female slaves did much of the heavy agricultural work (Rohrlich, "State," 93)—or they are *seen* as not being in charge, and thus modernized out of charge, as in Africa, where European colonial intruders often taught males "improved" agricultural technology.[22] And we do not learn of their contributions, for the writing of history became a male role whose holders defined history as what men did. To learn what women's contributions were, or even to make an educated guess supported by statistical probabilities, would involve a detailed correlation of the sexual division of agricultural labor with developments in agricultural technology for all groups for whom such data are available. I hope others will undertake this massive study; it is beyond my scope here.

---

22. The opposite can also happen, as where colonial powers draw the men off to labor in mines or in cities for wages; any new agricultural technology then introduced falls to the province of the women who stay at home in the villages.

To show just how complete the darkness can be, between the Takeover, or about 2000 BC in the Middle East, and the 18th century AD for cultures giving rise to our own, I have the names of only three women inventors or innovators in agriculture and related fields. Two of these are semi-legendary, and the achievements of all three are couched in the most general terms. The earliest of the three is **Semiramis/Sammuramat** of the late 9th–early 8th century BC, already discussed. With regard to the plausibility of attributing engineering and irrigation innovations to Semiramis, it is interesting to note that in the 1st century BC, a woman named Phyle held the highest office in her city and constructed a reservoir and aqueduct for it (Bridenthal & Koonz, 75–6). The Arabian **Queen Zubaidya** or **Zubayda** (d. 831), is credited with building an aqueduct in Mecca at her own expense. In fact, says Elise Boulding, this wife of Harun al-Rashid not only "rebuilt Alexandretta after it was destroyed by the Greeks," but "generally seems to have played the role of 'development expert' for Arabian cities" (389). Other sources credit her with an aqueduct leading to Aleppo or with general improvement of the wells and water supply on the 900-mile route from Iraq to Mecca (e.g., Uglow).

The next name after Sammuramat is 1500 years later: **Princess Eanswith** of Kent, in the 6th–7th century AD in England. Refusing her intended (pagan) groom, Eanswith founded a monastery and devoted her life to it and to "improving the agricultural practices of the farmers" on the monastery estates (Boulding, 397).

From South America comes the third named agricultural innovator of this period. According to Sarmiento, one of the Inca queens, **Chimpu Urina** (Chinbo Urma), was an outstanding zoologist, naturalist, and inventor. She not only introduced several new plants to cultivation, but studied plant toxins to devise arrow poisons for hunting and war (Hyams, 116–7).

Of course if the many significant inventions that support daily life, usually dismissed as "domestic," were properly studied, these centuries would teem with women inventors. But setting the record straight on daily-life inventions is another book or two in itself.

## 18th-century inventions

In 18th-century Pennsylvania we find **Sybilla Righton Masters** (d. 1720), a Quaker woman who invented a maize-stamping mill and also a new way of working and staining in straw and palmetto leaf to produce women's bonnets with New World materials. Interestingly enough, stamping is the modern process of choice here.

Although her British patents of 1715 and 1716 were granted in her husband's name, we know of Sybilla because her husband was prominent, and because she became an entrepreneur as well as an inventor, selling her "Tuscarora Rice" in Philadelphia, securing a monopoly on importing palmetto leaf to England, and marketing her patented hats and bonnets "against Catherine Street in the Strand," London (*NAW*).

In order to appreciate the economic significance of Masters' bonnet-making invention—and those of **Betsy Baker, Mary Kies, Sophia Woodhouse Welles,** and others after U.S. independence—recall that straw hats and bonnets were important items of dress for both sexes in the 18th and 19th centuries, both in England and here. Their importation became controversial during the War of 1812, and their manufacture, once these inventors had shown the way, became big business in New England. By 1869, for example, a single town, Franklin, MA, produced 1,500,000 hats and bonnets valued at more than $1 million (Tharp, 10). Multiply this by the several similar towns to gain some idea of the scale involved.

As early as 1722 an Englishwoman named **Frances Willoughby**[23] received British Patent #1855 for "Thrashing grain" (*WI*/1; Woodcroft). Unfortunately neither source gives any details.

A potential landmark innovation in veterinary medicine was the proposal of **Lady Mary Wortley Montagu** (1698–1762) that livestock as well as people be inoculated against infectious disease. Unfortunately, her ideas met violent opposition. This English traveler, writer, and feminist, niece of the famous novelist Henry Fielding, was perhaps best known for her daring lifestyle, but obviously had a fine and original mind (M. Lee, 22; Uglow).

**Eliza Lucas Pinckney** (1722–93) is often recognized for managing a plantation from the age of sixteen, and for her success after extensive study and experimentation—where others had failed—in introducing indigo cultivation to South Carolina. Though seldom called an inventor, she managed in the 1740s to re-invent the complex process for producing indigo dye from the indigo plant despite misleading information supplied her. She also devised a method of preserving eggs in salt. This colonial Renaissance woman also tutored slave girls, reflected on the future economy of South Carolina and planted oak trees for shipbuilding, studied the skies and sighted a comet predicted by Newton, undertook silk culture, and performed many kinds of agricultural experiments. In fact, according to one source, in the course of these experiments she became a natural scientist. When she died in 1793, George Washington asked to be one of her pallbearers (*NAW;* Ravenel; E. Pinckney, Letterbook; Hymowitz & Weissman; Selma Williams, 32–3, 181–2).

**Martha Laurens Ramsay** (1759–1811), another South Carolina woman, re-invented the process for debittering and preserving olives. She also experimented with the growing of olives. A brilliant woman who learned to read at three and eagerly absorbed all available education, she nevertheless subordinated herself to husband, children (11 in 16 years), and religion, and thus becomes a shadowy figure to us. Her *Notable American Women* biographer speculates that the "besetting sin" of Mrs. Ramsay's diary was struggle against this lifelong repression, and concludes, "In the next generation, she might have been a Sarah Grimke" (*NAW*; A. Earle, 85).

Though Masters, Pinckney, and Ramsay were certainly not unique in inventiveness, they were unusual and nearly unique among American colonial women in securing lasting recognition for their achievements; and Masters was evidently unique in securing patents. All three were upper-class women.

One **Mrs. Paulet** is rare among 18th-century Englishwomen in securing provisional credit for inventing Stilton cheese: According to René Lecler, travel editor of *Harper's Queen,* London, "The British . . . produce at least one cheese of world renown—the delicious Stilton . . ." (77). Stilton is a blue-veined cheese made golden with double cream; that is, the cream from one milking is added to the whole milk of the next. Although the origins of Stilton can be debated, Lord Ernle's article on Stilton for the *Britannica* (14th ed.:V:333) says:

> Mrs. Paulet of Wymondham in the Melton district of Leicestershire, is said to have been the first maker of Stilton cheese. She supplied them to Copper Thornhill, who kept the Bell Inn at Stilton (Hunts) on the great north road from London to Edinburgh, and they became

23. This name appears in *The Woman Inventor* as "Francis Willoughly." Typographical errors plague Smith's publications, which often seem done in a rush. Woodcroft's list is, however, correct (Brody, L-3/18/85).

famous among his customers, and throughout England. The manufacture of Stilton cheeses became an industry of the district. Mrs. Paulet was still living in 1780.

A Frenchwoman named **Marie Harel** is credited with inventing the gourmet cheese named for her native village of Camembert. In one version of the story, Marie was a 19-year-old cheesemaker. One day in 1810, her thoughts more on the boy next door than on her cheesemaking, she omitted the blue mold culture from one batch of cheese. As with all such happy accidents, however, what matters is what the observer does next. In Marie's case, the rest is gourmet history. Upon the tasting of that one batch with its soft creamy texture and delicious flavor, her fame was made.

In this version of the story, Marie was born in 1791 and died in 1845. Other versions claim that Camembert cheese was already known at the beginning of the 18th century; that Marie Harel invented it, but at least by 1791–3; that Marie did not live in Camembert— and even that her own village had 28 women named Marie Harel! One interesting source says that Marie Harel's 1791 contribution was to grow the white mold *P. candidum* on a well-known cheese in a controlled manner. So who was the real Marie Harel? Dr. Frank Kosikowski, Professor of Food Science at Cornell and long-time student of cheesemaking, simply says in a recent *Scientific American* article that Marie Harel developed Camembert cheese in 1791 ("Cheese," 99). Elsewhere he says that she grew up in another famous cheesemaking area, the Brie region near Meaux, but came to Normandy and married a farmer at Camembert. On her farm "she originated the miniaturized Camembert wheel" cheeses, sold in nearby Vimoutiers as Angelots, or Little Angels or Angelotts (referring to Pays d'Ange in Normandy) (*Cheese,* 8).

Whatever version we accept, there seems little doubt that Camembert cheese was invented or given its modern form and flavor by a Frenchwoman named Marie Harel. Her achievement is immortalized on a tombstone in the Camembert churchyard, and by a statue erected in her memory in the nearby town of Vimoutiers.[24] The epitaph on the tombstone reads, "Marie Harel, 1791–1845 / *Elle inventa le Camembert* / R.I.P." The statue, destroyed in the Normandy invasion of 1944 but later restored by Americans, reigns today over the local marketplace—as well it might, since as of 1972 the industry Harel made possible was producing half a million cheeses a day.

My own favorite version, because it accounts for the conflicting dates, is that Marie Harel the elder (born Marie Fontaine), invented Camembert in the late 18th century and taught the method to her daughter or granddaughter, also named Marie, who, with her husband (Paynel) improved or industrialized the process about 1810. The monument evidently commemorates the first Marie for it shows a date of 1761 (*Larousse Gastronomique,* 237); the tombstone is that of her daughter or granddaughter, a later innovator (Lecler, 77; Koskiowksi, "Cheese," 99, *Cheese,* 8, and L-5/21/85: Thom & Fisk, 117; *EB*/1967, S. V. "Cheese": 363; USDA *Agriculture Handbook* 54:22; J.G. Davis, 674; N. Eikhof-Stork, 83; *Fromages de France* [1953], 27; Plume, 88).

Before leaving the 18th century, we must discuss an agricultural machine that changed American history, and the disputed role played in its invention by a Rhode Island–born woman, **Catherine Littlefield Greene** (1755–1814). Greene would be one of the most significant agricultural inventors of the 18th century—and indeed of modern times—if we could confirm the persistent reports that she really invented the short-staple cotton gin, or if we could even be sure that she was co-inventor as well as muse and patron to Eli Whitney.

The truth may never be known, though it is hard to imagine such acres of smoke without

24. The gift of cheesemakers in Van Wert, Ohio (Plume, 88).

at least a small fire. What we can say for certain is that Whitney's alleged failure to credit Mrs. Greene no more proves she did *not* make a crucial suggestion for the original invention than his paying her royalties from his patent proves that she *did*. One source suggests some girlhood technical knowledge for Catherine; another attests to her quick and intuitive mind; and still another reports that Whitney actually did once publicly acknowledge her help (see Appendix A-2).

My own view is that, although Catherine Greene almost surely did not invent the cotton gin "instead" of Eli Whitney,[25] she probably did make that intuitive leap from familiar brushes—lint brushes, hairbrushes used to clean combs, one hairbrush used to clean another, etc.—to a bristled cleaner for the hopelessly clogged teeth of Eli's first model, and that she grabbed up a hearth brush or a hairbrush or whatever was handy to show him when the inspiration struck her. Moreover, *without this suggestion, Whitney might have failed to invent the gin*. Even M.B. Hammond, writing on the cotton industry in 1890, and firmly opposed to any idea of co-invention by Greene, accepts this scenario. He considers the suggestion insignificant relative to the invention as a whole. But then so is a horseshoe nail, relative to a war.

### 19th and 20th Centuries

Nineteenth- and 20th-century women have invented and innovated in all fields of agriculture: *plant culture,* including planting, cultivating, harvesting, irrigation, and fertilizer techniques, selective breeding and creation of new plants as well as treatments for plant diseases; *animal husbandry,* including innovations in livestock care and management, creation of new breeds, and inventions in veterinary medicine; *irrigation, drainage, and related; planting, cultivation, and harvesting; processing and storing of crops; waste-recycling and exploitation of agricultural products; nutrition and new foods; agricultural machines; fiber-processing;* and *genetic engineering*. Not all can be discussed here, and I will look more closely at some areas discussed—dairy and poultry inventions; food-processing machines and devices; and cultivative machines, for example—than others. Indeed, this whole massive chapter is but a brief and preliminary sampling of women's contributions to agricultural technology.

Women, after all, were farmers, not just farmers' wives. They were also homesteaders on their own or with other women and unrelated males. This item from a Milwaukee newspaper of 1882 is obviously not unique ("Young Ladies Locating Lands"):

Miss Eva Hanson . . . left yesterday in company with a young lady and gentleman friend to "take up" land in Dakota. . . . *Not a few* of the friends of the parties *of both sexes* are anxiously awaiting the result of the venture, and if it turns out well a large number of young folks will leave Milwaukee for the wilds of the North west. . . . *Word from the two lady clerks who left Chapman's store for Dakota to take up land a short time ago, is that they are doing well, and are contented.* [Emphasis mine]

### Plant culture

*Seed-treatment.* Coating seeds with fertilizer or pesticide rather than treating the whole field is like using an unerring rifle rather than a shotgun. If menstrual blood was the first

---

25. Claiming that she was the "real" inventor does almost as much harm to compensatory history as failing to raise the question at all. Unfortunately, this rash claim continues to recur, long after it should have been laid to rest. See, for example, Vare & Ptacek (p. 15; contradicted later in their own text), published in 1988.

seed-treatment, then women are the likely inventors of the process, and **Dr. M. Claire Shephard** of England is operating in an ancient tradition. Dr. Shephard holds an advanced degree in plant pathology and more than twenty patents for novel agricultural chemicals. She was involved in the invention and development of two systemic fungicides, dimethirimol and ethirimol, applied to the seed (and soil) to control foliage diseases (Shephard, L-12/10/81; see also "Plant Diseases" section below).

*Planting and cultivating techniques.* Once the seed has been selected and treated, and the ground prepared, planting can begin. But planting requires arable land, already scarce and becoming steadily scarcer. Among the more significant contributions to late-20th-century agriculture are techniques for producing food in areas once thought unsuitable. Several women should be mentioned here, working in cities, deserts, the frozen north, and other undeveloped areas. **Miranda Smith** (c. 1943– ) of Washington, DC, and New York City, was involved in Canada's Rooftop Gardens Project, researching and developing rooftop cultivation technologies. Returning to Washington, she worked with the Urban Agriculture Project of the Institute for Local Self-Reliance, investigating the food-production capacity of urban areas, and providing technical assistance to those who wish to experiment with urban agriculture. In the late 1970s Ms. Smith worked for the Bronx Frontier Development Corporation's project using the city's vegetable waste to create a high-grade growing medium. Miranda Smith grew up in a family that ran an organic nursery. She graduated from American University in 1964 (*WBWRA,* 36).

Significant examples from around the world include **Hsing Yen-tzu** (c. 1940– ) and her women's planting team from Hopei Province, who in the 1950s devised a method of planting spring wheat in hard-frozen ground, and **Yekaterina Novgorodova** (c. 1930– ) who devised hotbed techniques, combined with bonfires, sprinkling routines, and special ripening shelves, for growing vegetables in Siberia. **Mariya Gocholashvili,** head of the Plant Physiology Laboratory in the All-Union Research Institute of Subtropical Crops, worked on winter-hardiness in crops from the mid-1930s to the mid-1940s. **Yelizaveta Ushakova,** Director of the Gribovo Vegetable Selection Station beginning in 1948, wrote "many papers on advanced vegetable production." Austrian **Elizabeth Bokyo** and New Zealander **Wendy Campbell-Purdie** have both experimented with desert-cultivation techniques, Bokyo in Israel and Campbell-Purdie in Algeria (Sheridan, 75–80; Kupry, 88–95; Macksey & Macksey, 167).

*Fertilizer.* But perhaps the most exciting innovation in cultivative techniques of the 20th century, an idea that makes the Green Revolution look like a wrong turn, is the algal inoculation technique devised by **Elisabeth Bryenton** (1961– ) of Lakewood, OH, while still a teenager.

Elisabeth (Betsy), firstborn of two daughters, was always rewarded for creativity and achievement. As mentors she took her parents, "gifted intellectuals in complementary areas" and two high school teachers. (One, her biology teacher, Josephine Chrysler, noticed Betsy's research potential by her sophomore year.) As a role model she took Amelia Earhart. Bryenton has artistic talent in both painting and sculpture, which she pursued after college, at the Mainline School of Art in Haverford, PA.

During seventh grade she became fascinated by plant physiology, particularly the nutrient requirements of plants. After six years, working two hours a night and most weekends in a home lab, she created a mixture of up to 12 strains of green and blue-green algae that provides plants as much nitrogen as synthetic fertilizers at a fraction of the energy costs in production. *Only about a gallon per acre is needed.* Since the inoculant can be grown in water heated by the sun or by various industrial processes, virtually no fossil-fuel energy is needed. In 1980, Bryenton noted that "the energy saved by using seven gallons of my product instead of an equivalent amount of commercial fertilizer . . . could heat an average Midwestern home for an entire winter." She estimated the potential worldwide savings in natural gas alone at 1.5 trillion cubic feet a year and, assuming inoculant costs of about $0.66 per gallon, an overall savings of $18 billion a year. By 1982 the figures were 2 quadrillion cubic feet and $40 billion (R. Preston, 92).

Once applied to the soil around plants, the algae can fix nitrogen from the air, just as the *Rhizobium* bacteria associated with legumes do. But Bryenton's inoculant works for any kind of plant. Bryenton has successfully tested it with wheat and soybeans.

The inoculant will probably be marketed as a liquid concentrate to be diluted with water and applied to the soil. It can be stored easily in plastic jars. The fertilizing effect lasts a whole growing season, the algae reviving even after drought. A natural-selection process occurs after application, so that the best-adapted algae for a given area survive and reproduce. Thus after two or three years' applications healthy populations of the nitrogen-fixers should become established in the soil, producing *a renewable fertilizer* without further cost or effort.

An unexpected bonus of Bryenton's invention is that plants grown in algal-inoculated soil have a much lower rate of disease. Germination and growth rates seem to be faster.

In one sense algal inoculation is not really new. Researchers in Rothamstead, England, have been experimenting with blue-green algal crop dressings since the 1850s, for instance, with good results. For at least 300 years, Vietnamese and Chinese farmers have used a blue-green alga to fertilize their lowland rice paddies. Seaweed—a giant alga—has been used as fertilizer for millennia in many cultures. Elisabeth Bryenton's major contribution was combining several species in the most effective proportions for varying conditions. Further, she has determined the right mixture of water and nutrients to maintain the algae in a potential commercial preparation. Although some sources speak of a patent pending on the inoculant, Bryenton herself says that it is not and will not be patented, for the same reason Coca-Cola® is not patented.

At the International Science Fair in San Antonio (as one of two winners from the Northeastern Ohio Science Fair) in May of 1979, she won not only the U.S. Army Grand Prize, but six other awards for a total of seven, the most ever won by a single entrant. All this before graduating from high school. Immediately upon graduating, she was invited to Washington to meet with the Secretary of the Army and a group of top generals. She has been featured in magazines as diverse as *Seventeen* and *The Tennessee Farm Bureau News*. She has met the Japanese imperial family and has been chosen one of the 82 most interesting People in Town by *Cleveland Magazine* (1982).

Many companies sought her for their research staffs, but Bryenton opted for a liberal arts education at Princeton, where she graduated *cum laude* in 1983. It remains to be seen whether she will seek venture capital to found her own company, or sell or license the perfected secret formula to a company that will manufacture and market the inoculant.

Meanwhile, her interests have shifted from plant to human physiology. She now plans a PhD in neuroscience, and has an "interactional neurotransmitter theory of depression," which she considers her second most important achievement (Bryenton, Q:3/7/85; "First Class of '82"; Fisher & Kovach; Fitch; Kovach; Lempert; Shuck; R. Preston).

Elisabeth Bryenton thinks women are natural inventors, possessing "strong vision in scientific research if given the opportunity and acceptance." She considers herself an inventor by nature, but says it is difficult to be an inventor today because so few companies will hire anyone purely to search and invent. Some day she intends to create a place that will do just that, an Advanced Learning, Research and Invention Center (Q-3/7/85).

*Irrigation, drainage, and related inventions.* A scattered but distinguished 19th-century sample here. According to Caroline Dall, a French peasant woman invented a system of field drainage so successful that the French government sent her to French Guiana to introduce the system there (*College,* 157n).

According to Orison S. Marden's *Consolidated Encyclopedic Library* of 1903 (4087), "A California woman whose first invention, in early youth, was a corset, has lately patented several inventions relating to reservoirs and irrigation." Marden probably refers to the most noteworthy woman inventor in irrigation in the 19th century—**Harriet W.R. Strong** (1844–1929). Harriet showed inventive ability even as a child. During a summer's illness, for example, she invented a means of opening and closing her windows from her bed. No corset appears among her known patents, but since Harriet had back trouble, she may have created a corset for her own needs without patenting it.

Harriet Strong invented a system of dams and reservoirs for water storage and flood control. Catapulted into the economic arena by her husband's suicide in 1883, she moved with her four daughters to a debt-ridden ranch near Whittier, CA, and set out to make it pay. In 1884 and 1886 she patented window-related inventions: a Device for raising and lowering windows and a window-sash holder (presumably her childhood invention, perfected), and a Hook and eye (1884). She must have intended to exploit the hook and eye commercially, for she showed it at the World's Industrial and Cotton Centennial in New Orleans that same year ("Something for Everyone").

So successful was Harriet Strong with her business ventures and her ranch that she became the first woman member of the Los Angeles Chamber of Commerce ("Work of One Woman," 31), and is known not only for her civic and cultural work, but as an entrepreneur and agriculturalist. The secret of her success was doing her homework. For four years she studied farming methods and horticulture, irrigation and water control—and marketing. During those years, challenged by growing walnuts, pomegranates, and citrus fruit on semi-arid land, she devised her Strong System of water-storage in river canyons. She patented her design in 1887 and exhibited a working model of it at the Columbian Exposition in Chicago in 1893. Since this model won awards in both the Agricultural and the Mining Departments at the Fair, it may have incorporated the modifications reflected in her later patent of 1894, a Method or means for impounding debris and storing water. A Central American engineer took the model back to his own country for use there (*ibid.*).

This invention established Harriet Strong as an authority on water control and irrigation, and led her *DAB* biographer to call her, among other things, an engineer. According to this article, "Her name is intimately connected with the development of irrigation, and she was among the first, if not the first, to advocate the conservation of water by building storage dams near the source of mountain streams" (IX/1935–36: 147).

Strong had long advocated the building of her dams, a series of three curved and double-curved structures, on the Colorado River. She argued that they would provide both irrigation water and electricity, the succession of dams ensuring safety in case of a break, and the back pressure from the lower dams helping to hold the upper. During World War I she proposed the project again as a war measure, to increase the world's food supply; and the dams might indeed have been built had the war not ended when it did (*NAW;* Radford, L-2-3/80); "Work of One Woman," 32).[26]

---

26. The compilers of a 19th-century official list of American women patentees classified this important dam and reservoir system as a culinary device!

For the 20th century, the Women's Bureau study has several examples in irrigation and water control as part of Category E under Agricultural inventions (Farm buildings, fence materials, water and drainage equipment). Listed are a cistern, a flume gate, pumping apparatus, a pump drain, a tank, a water bed fence, and a windmill. Category D (Planting, tilling, and harvesting machinery and equipment) also contains some items that may belong here: an irrigating device, an irrigating system, a portable irrigating stand, a diffusion block for subsoil irrigation, a time-controlled dam gate, and a soil-reclaiming means for water-current machines. As with other inventions catalogued by this study, their numbers and significance are unspecified, and the inventors unnamed (WB/28).

*New plants.* This area of innovation becomes much easier to trace in the 20th century, when new varieties and species become patentable. Then we find several women following in the footsteps of Chimpu Urina and our unnamed foremothers who domesticated and improved all major grains and fibers in prehistory. One 20th-century American woman innovator whose work is known before plant patents is **Elizabeth White** of Lebanon State Forest, Whitesbog, NJ, developer of the commercial blueberry. Daughter of cranberry growers, she knew the business and was aware of the cranberry's neglected cousin the blueberry, growing on the edges of her family's bogs. After reading a bulletin on experiments with blueberry culture in 1911, she began working with the Department of Agriculture, using Whitesbog for field tests. From among the successful varieties developed in these trials, White hybridized and domesticated the blueberry we now know, always looking for bigger, bluer, and hardier fruit. In 1916 she marketed Tru-Blu-Berries. Another of her innovations was protecting her berry baskets with cellophane (Sherr & Kazickas, "Landmarks"; Wiser, "Women," 5).

Early among female plant patentees was **Josephine D. Brownell** of Little Compton, RI, who in 1937 patented two roses. In the early 1940s she patented two wilt-resistant hybrid tea roses (#458 and #459). Brownell's tea roses were not only wilt-resistant, but continuously blooming, winter-hardy, and resistant to black rust ("Nine Patents"). This Rhode Island woman was not only one of the earliest but also one of the more prolific American woman plant patentees. Between 1949 and 1955 she patented 31 roses, unnamed in the patent records, but quite possibly becoming named varieties in the gardening world.

Other U.S. women receiving early plant patents were **Esther C. Johnson** of La Mesa, CA, for a lobelia (1937); **Anna L. Miller** of Nashport, OH, and **Elizabeth A. Briggs** of Encinitas, CA, for gladiolus (both 1937); and **Marie Schaefer,** Yakima, WA, for a walnut tree.

In the 1950s **Esther G. Fisher** of Woburn, MA, patented at least eight new roses between 1950 and 1956, including a hybrid tea rose named Merry-Go-Round (1952), assigned to Arnold-Fisher Company; an unnamed lavender rose (1955), assigned to Conard-Pyle of West Grove, PA; and the Capri Rose (1956). Since Fisher appears in Patricia Ives' cryptic 1980 article on inventive achievements by American women and blacks (126), she may be an important plant developer and breeder. Indeed, she may have developed the first lavender rose, long sought by breeders, and causing a sensation when it finally appeared. Ives, however, gives no details.

The Meilland family of France and Antibes, creators of the famous "Peace" rose and leading rose-breeders for five generations, have always numbered women among their working ranks (M. Green, *passim*). **Claudia Meilland,** for example, built the original nursery business with her husband and ran it while he fought in World War I. **Marie Louise Meilland** co-patented, with the present company head, Alain A. Meilland, and **Michele Meilland Richardier,** a new and distinct variety of rose, a hybrid tea with double flowers, coral inside and yellow and orange and pink outside, that is supposed to last very well when cut (1976).

One of the Meillands' most noteworthy recent creations, reaching the United States market in 1980, is "Ambassador," an "orangy" hybrid tea that survived from a group of 60,000 crosses. **Louisette Meilland** (mother of Alain), created "Ambassador" in 1967.

Any rose introduced under the Meilland name will have survived a ten-year elimination process, starting as one of tens of thousands of new-season crosses and undergoing years of observation on its shape, color, perfume, novelty, hardiness, and suitability for the amateur garden. Only two or three of these myriad crosses will ever be named and sent out to compete for awards and buyers.

Developing a new rose like "Ambassador" can cost up to $45,000, but can generate half a million dollars or more in income, since Meilland roses dominate the Western European flower market—75% of all cut flowers cultivated in France, 90% in Spain, 70% in Italy, and 50% even in Holland. If, as Maureen Green says (46), the French invented the modern rose, several French women obviously share in the credit.

Aside from the world-famous Meilland women, **Marguerite Denoyel** of Venissieux near Lyons, patented three roses in 1947 and '48, assigning all three to Jackson & Perkins Company of Newark, NY.

More recently, **Kathleen K. Meserve** of Nissequogue, Long Island, has developed six popular hybrid Blue Hollies. Between 1963 and 1978, Ms. Meserve created and patented Blue Girl, Blue Boy, Blue Prince, Blue Princess, Blue Angel, and Blue Maid, by crossing English holly (*Ilex aquifolium*) with a low-lying Japanese holly (*Ilex rugosa*). These hardy and brightly colored varieties, now common in both American and European nurseries, are classed as a new race or species and named *Ilex* x *meserveae* after their creator. A university horticulture department once spent thousands of dollars trying to create a new strain of holly; Kathleen Meserve did it with about $15.

During the harsh winter of 1977, her Blue Hollies were sometimes the only ones to survive in gardens. The Meserve hollies are important enough botanically and horticulturally to appear—with their creator—on the front page of the *New York Times* "Leisure" section in 1978. Their value and interest seem enduring, for when in late 1985 I queried *Sunset* magazine about a rose patentee, a senior garden editor volunteered: "There is another woman who has made a considerable contribution to horticulture—Mrs. Kathleen Meserve, the breeder of the 'blue hollies.' . . ." The Conard-Pyle Company bought the rights to them and introduced each group with extensive press releases (Conard-Pyle press releases; DeLatiner; Dunmire, L-11/8/85).

**Dr. Catherine H. Bailey** of Rutgers University, Department of Horticulture, is part of a research team creating new hybrid cultivars of peaches, nectarines, apples, and pears. A joint 1969 winner of the National Peach Council's Annual Recognition Award for outstanding contributions to the peach industry, Dr. Bailey has introduced at least 39 new varieties of fruit since joining Rutgers in 1948. Many of these are patented, including several new apples (*WBWRA*).

**Alice Vonk** (1908– ) of Sully, IA, fulfilled one of my youthful dreams by winning the $10,000 prize offered by the Burpee Seed Company for a true-breeding snow-white marigold 2½ inches across. Fourth daughter of a farm family, Alice finished eighth grade at 11, and took no further schooling—except from plants. She reports that she just loved to grow things, began gardening at 9, and has not stopped since.

She read about the white-marigold prize in a 1954 Burpee catalogue and decided to try for it. Starting with a primrose marigold, she kept her project a secret from everyone, including her husband, who would, she said, have discouraged her. After twenty years of trials and repeated selection of seed from the sturdiest of her near-white plants, she sent in the winning entry in 1974. In so doing she succeeded where the giant Burpee Company itself and over 8,000 other amateur gardeners had failed. Newspapers and gardening magazines throughout the United States reported David Burpee's announcement from Fordhook Farm on August 28, 1975, that an Iowa widow and home gardener named Alice Vonk had won the $10,000 prize.[27]

This payment may make the white marigold the costliest flower in the world, including Burpee's own research and the judging of the thousands of hopeful entries over the years. The company stands to recoup those costs, however, since marigolds rank first in its sales, and the whites are highly popular. Alice Vonk's marigold, named "Snowbird," was introduced on a trial basis in 1981 and as a regular item in 1982.

Alice Vonk began working at the age of 8 or 9, helping with chores and fieldwork—shocking oats and picking corn—at home, and lighting the fire at the local school. By age 11 she was hiring herself out to neighbors. She had little time for play, either then, or while bringing up eight children and cleaning the local church to help make ends meet. But she has found time over the years to paint and write poetry.

As of 1983, at age 75, she still had a large garden and two lawns to mow. With the Burpee prize money, she bought a garden-tiller and a new lawn-mower—and went to Holland. No fancy cars, rings, or fur coats, she says laughingly. As the only concession to her age, she invested the rest of her money.

Currently doing exciting agricultural research at Stanford University is **Dr. Virginia Walbot.** Though mainly a basic scientist (see "Genetic Engineering" section below), she deserves brief mention here as a member of a team that developed a new sweet corn (Shurkin, "Walbot"; Lynch).

The United States has no monopoly on female plant wizards. In 1951 a plant geneticist named **Raina Georgieva** became Professor and Head of the Genetics and Selections Department of the G. Dimitrov Higher Agricultural Institute of Bulgaria. In that post she developed (or led a team that developed), among other things, tobaccos resistant to mosaic virus, new varieties of flax, and winter-foliage barley. She has also worked on techniques of plant hybridization (*WBWRA*).

While head of a laboratory at the Southeastern Agricultural Research Institute of the USSR in the early 1950s and early 1960s, **Valentina Mamontova** won a Lenin Prize with two other scientists for developing high-yield strains of winter and spring wheat and barley. Since the former Soviet Union is marginal for growing wheat and often has to import grain, this is important work. During the same period, **Maria Bordonos,** a senior scientific worker at the Ukrainian Academy of Agricultural Sciences, won a Lenin Prize with **Olga Kolomiets** and four others for producing a new single-seed form of sugar beet (Dodge, 222).

**Mariya Gocholashvili** and **Yelizveta Ushakova** have been mentioned as leaders in horticultural research for the underdeveloped world. Their positions suggest leadership in developing new varieties or cultivars as well as new cultivative techniques. This would be particularly likely for Gocholashvili in her work on winter-hardiness of subtropical crops.

But perhaps one of the most economically significant of women's 20th-century plant

27. Some sources reported that she also developed a purple bean and a lemon cucumber; she did not, but was one of the first to grow them in her area (Vonk, Q-7/19/83; Lowe, L-7/21/83; Burpee press releases, 8/28/75; Burpee catalogues, 1976, 1981, 1982; David Burpee, "In Search . . ."; "Mrs. Vonk's Victory"; "The $10,000 Marigold"; Neary, 20-8; L. Harris, L-8/18/82; San Francisco *Chronicle,* ad for *People* magazine, Sept. 1975).

innovations—at least prior to genetic engineering—is the large-boll cotton developed by an anonymous Chinese woman. Genetic engineering, of course, will produce plants we do not now dream of, and women as well as men are involved (see below).

*Plant diseases.* No one can say what women may have been doing in their own gardens or on their own farms, unsung or under the aegis of their husbands, to control plant pests and diseases. Their methods were usually handed on freely to friends and daughters, and have passed into gardening custom. At least two U.S. women, however, did patent such remedies in the 19th century: **Joanna B. Tribble** of Middleborough, MA, for an Improved composition for preventing disease in vegetables (1865); and **Rebecca J. Walker** of Goshen, OH, for an Improvement in the mode of protecting fruit trees from curculio, etc. (1870).

Even before the end of the century, work in this area of agriculture was moving into the laboratory. The Department of Agriculture's Bureau of Plant Industry, for example, opened in 1887. Although women were excluded from many Department labs, Erwin Frink Smith, the plant pathologist in charge of BPI, made it a point to hire women. By 1920s, he had hired more than twenty, among them such talented scientists as **Effie Southworth Spalding, Nellie Brown, Clara Hasse, Charlotte Elliott, Agnes Quirk, Della Watkins,** and **Mary Bryan.** Although kept at the assistant rank, these women earned what Margaret Rossiter calls "modest fame" for outstanding work on crown galls, citrus canker disease, and corn and chestnut blight (61). Since their lab was mandated to serve agriculture, their research almost certainly had practical applications.

In the later 20th century's professionalized attack on plant diseases, women have made several interesting contributions. **Helen K. Tobol** of Concord, CA, patented a Method of controlling rice blast in 1974. Considering that the world produced 417.9 million metric tons of rice in 1983 (USDA estimates), to feed more than half of the global population (Higham, 100), the potential importance of this invention is great indeed. Ms. Tobol and her co-inventor Ronald Sbragia assigned their patent to the Dow Chemical Company.

At least three British women scientists have done important work. A senior Government scientist, **Dr. Joan Moore,** has pioneered in plant-disease-forecasting systems. In the 1940s and '50s **Dr. Mary Glynne** pioneered in several areas of cereal pathology at Rothamsted Research Station. And **Dr. Margaret C. (Claire) Shephard** (1931– ) of Maidenshead, Berkshire, England, holds at least twenty patents with various co-inventors for 1972–81, covering fungicides, pesticides, and agents that regulate plant growth. Nurtured in a home where "defeat was not encouraged," where both parents and grandparents encouraged curiosity, hard work, and high standards, and where books, ideas, and materials for practical experiment were always available, she had a good basis for her future career. Her educational background included competitive grammar schools in Wales and London, a BS in botany (with honors) in 1952 from University College, London; studies in plant pathology at the Imperial College of Science and Technology (1952–4), followed by a PhD in plant pathology in 1957.

Attracted by the idea of finding chemical aids to food production in a hungry world, Dr. Shephard entered the research department of Imperial Chemical Industries, Ltd (ICI), in 1954. There she has stayed, as plant pathologist on a team engaged in inventing and evaluating new agricultural fungicides. For some idea of the significance of this work, note that fungal diseases destroy billions of dollars' worth of world crops yearly, and British studies show 10–16% yield increases for spring barley from seed treated with Milstem[28] to control mildew (1968–74 figs; Shephard *et al.*, "Sensitivity," 59). One of the

28. ICI's trade name for etherimol.

fascinating things about Claire Shephard's work is that these are usually new chemicals, with unknown properties. As she puts it,

> This has entailed working at the frontiers of knowledge, designing test systems capable of picking out the properties which will be useful in tomorrow's world. It involves painstaking comparisons of the properties of closely related chemicals to work out the relationships between molecular structure and biological performance in order to improve on these features.

Specifically, Dr. Shephard's inventions are as follows: First, she has shared in the design of new chemicals, some of them patented. Second, she helps transform untried compounds into practical, effective fungicides. This involves learning how the chemicals will behave in plants. In order to do this, she has pioneered in (a) techniques for handling many plants and plant diseases in the glass house, (b) miniature tests enabling compounds to be tested on tiny potted plants in the laboratory rather than on whole fields, and (c) design and construction of controlled-environment rooms that were, as she puts it, "novel in their methods of producing the required conditions." In order to survey the sensitivity of powdery mildew on British barley crops to the fungicide ethirimol, she had to create (d) new testing methods.

Dr. Shephard was an important part of the research team that created the systemic fungicides dimethirimol, ethirimol, and bupirimate. Though not the earliest systemic fungicides, dimethirimol and ethirimol were the first commercial products that could be applied to either soil or seed to control foliage diseases. She also worked on a product called "Vigil," used to control rusts and powdery mildew on cereals in France. She continues to innovate. As of 1985 she reports several new European patents granted or pending, and mentions "an interesting new fungicide," PP969, which she was involved in creating. This fungicide is the first to control fungal diseases in the plant crown after being applied to the soil or to the stem (L-3/16/85; Shephard *et al.,* "PP969").

Dr. Shephard herself considers her glass-house techniques and her miniature tests among her most important contributions to her field (L-12/10/81; 3/16/85). However, she was the first to recognize the effects on plant growth of some of ICI's triazoles, with the result that one of them, paclobutrazol, is now being developed worldwide as a growth-retardant (Shephard, L-3/16/85, "Factors"; Shephard *et al.,* "Sensitivity").

*Herbicides.* Claire Shephard's work on growth-regulators leads naturally to another important area of agriculture invention—herbicides or control of plant pests. This area, not discussed at length here, is exemplified by **Florence Gleason**'s 1986 patent for a Cyanobacterin herbicide, which works by inhibiting plant growth. As the patent is assigned to the University of Minnesota, Gleason is probably an academic researcher. She lives in Chaska, MN.

## Animal Husbandry

*New breeds, varieties.* To a remote quarter of the American Southwest known as the Land of Refuge came three women from an Eastern city. They bought an abandoned homestead claim. Later, each of the three sisters, Salina, Maria, and Elena—not surnamed by my source—took up a homestead and lived on it for the required period. By 1912, when Mary Gay Humphreys wrote about them for *Century* magazine, the women had enlarged their Beulah Ranch until it stretched for almost five miles along their canyon, with timber, water rights, and possible underlying coal.

The sisters raised cattle in a personal way, knowing each animal by name and personality and able to communicate with each. When any member of her "family" was missing, Salina would search until it was found, dead or alive. Unlike neighboring ranchers, the women fed their cattle in winter, and made smudge fires against flies in summer. Many of their cattle obeyed voice commands and, in blizzards, headed for the ranch house.

**Salina,** the eldest sister and cattle manager, was "upgrading on the old range cattle, which are bigboned, but hardy, with the less hardy white-faced Herefords." Salina told Humphreys (529), "I've been breeding a variety of my own that I call the 'diamond white tails.' " Jim Brinks of Colorado State University notes that a diamond-shaped white patch does indeed occur over the head of the tail in longhorn-Hereford crosses, and is a dominant trait (PC-3/28/85). Thus Salina was probably on a profitable track, and it would be interesting to learn the fate of her experiment.[29]

The first authentic new breed of cattle created in North America, the Santa Gertrudis, was developed on the King Ranch in Texas. Santa Gertrudis cattle resulted from crosses between the tall, large-framed, but ill-tempered Brahman cattle, with their resistance to heat, drought, insects, and disease and their ability to fatten well on grass, and the shorter, better-tempered, beefier, but far less hardy Shorthorns. Introduced about 1930 and featured in *Fortune* magazine by 1933, Santa Gertrudis cattle were officially recognized as a breed in 1940.

Although **Henrietta King** (1832–1925) ran the ranch from 1885, when she inherited a debt-ridden half-million acres, until she died at the age of 92, a power in Texas economic life, she is unaccountably omitted from *Notable American Women.* And although an innovator in several areas, she is not generally credited with any role in the Santa Gertrudis development (Lea II: 724, 725). Jean Blashfield, however, says, "The development of hardy but meaty Santa Gertrudis cattle, for which King Ranch is now renowned, started under her direction" (*Hellraisers,* 25). The breeding program did begin during Henrietta King's lifetime. And crossbreeding with Zebu cattle had long been done on the ranch, an idea that might or might not have been hers but which she would have had to approve. As Tom Lea himself says (II: 513), her manager "kept her informed and he made no decisions of consequence without her knowledge and approval. She was at all times the highest authority in the conduct of her properties' affairs, and she was always consulted."

According to Lea, the breeding program was not formalized until the year of Henrietta King's death. Of course something less "formalized" but obviously successful had been going on for years; and had King been a man, she would doubtless receive the kind of credit that laboratory heads usually receive for research done in their labs.

Whatever may have been her role in the Santa Gertrudis development, Henrietta King was an innovator. She donated land for a new town, then planned the town. She was among the first to use a chemical dip to protect cattle against tick fever, and she developed, possibly with the collaboration of her son-in-law, "some of the earliest scientific techniques for beef production" (Blashfield, *ibid.,* 25; McDowell & Umlauf, 181).

A contemporary woman whose contribution to an important new cattle breed is undisputed is **Sally Forbes** of Sheridan, WY. She and her husband built the Becton Stock Farm together. Just as they were beginning to form the Red Angus breed, her husband died, so that her joint role became the sole role. The Red Angus is now an established breed, with a breeding association in Denton, TX (Brinks, PC-3/28/85).

An Australian perfectly placed to contribute to new breeds or varieties, or to new

29. Sandra Myres suggests Arizona for the location of El Refugio (L-6/24/85). I would be grateful for further information on this interesting episode in ranching history.

breeding techniques, or all three, is **Helen Newton Turner.** She has written several textbooks and published more than 100 scientific papers on sheep-breeding and production, and has a worldwide reputation in her field. According to Lois O'Neill of *WBWRA* (42), Helen Turner has contributed to the famous and economically important Merino breed of sheep: "Of particular importance in Dr. Turner's work is selection in the breeding of Merino sheep." She has received the FAO Ceres Medal in recognition of her long-time work in sheep-breeding.

*Care of livestock—methods, apparatus, research.* 19th century: as Ada Bowles reminds us (Mozans, 346–7), many women patented inventions having to do with horses. However, since these seem to pertain more to transportation than to agriculture, they will not be discussed here.

As the story of Salina and her sisters suggests, women seem concerned with humane treatment of livestock. In the 1880s at least three women patented improvements in livestock or cattle cars. It would be interesting to know whether these women were responding to an animal-cruelty scandal, or just to a new market. In any case, in 1881 **Nancy P. Wilkerson** of Terre Haute, IN, patented a Cattle car; and in 1882 **Annie K. Pentz** of Clearfield, PA, patented a Stock-car. In 1885 **Geneva Armstrong** patented a mechanism for feeding cattle on trains. By 1891, according to Charlotte Smith, Armstrong was actively promoting her invention to railroad executives for their livestock cars (*WI/1*, 4). She also exhibited a model stock car incorporating her feeder at the Columbian Exposition of 1892–3, in the Inventions Room of the Women's Building (Handy, ed., Pt. I, Section on Women's Bldg: 9). Interestingly enough, LWP omits this potentially important invention. More surprising, Caroline Romney also omits Armstrong from her article on women's inventions at the Exposition. Miss Armstrong will be discussed further in Chapter 4.

In the late 1880s and early 1890s three Midwestern women patented livestock feeding troughs: **Elizabeth A. Craig** of Arlington, KS (1886); **Susan F. Smith** of Minneapolis, KS (pig trough, 1890); and **Cora B. (and J.) Shellenberger** of Paulding, OH (1892).

20th Century: WB/28 has a category of Stock-raising equipment, comprising 18 inventions of 14 types for the ten chosen years:

| | |
|---|---|
| Animal blanket | Automatic feeding hopper |
| Animal nose yoke | Feeding regulator |
| Animal poke | Feeding rack |
| Cattle guard | Feeding trough |
| Cattle stanchion | Stock-feeding device |
| Corn-shield for horses' heads | Hog waterer |
| Device for applying insecticide | Tag for ears of animals |

As with my nonsystematic 19th-century sample, there seems to be a focus on feeding the animals.

Though none of the inventors is identified, at least one feed rack was invented by a Nebraska woman "because conservation of foods as encouraged by the Government during the World War made it necessary to prevent all waste of grains and produce the greatest amount of food within the shortest time possible. This could be better accomplished by means of proper self-feeders or racks for livestock, which not only saved grain but also gave quicker returns in the marketing of live stock" (WB/28: 18). **Ingabrit T. Ingebritson** patented a Watering trough in 1914.

Early among my scattered but interesting 20th-century sample is the improved sheep shears invented by **Mary A. Lipscomb** of Eureka, CA, and patented in 1907 (*Inventive Age,* Apr. 1, 1909: 6).

**Miriam Louis A. Rothschild** (1908– ) is a brilliant entomologist and parasitologist. First to analyze the flea's remarkable jumping mechanism, she has published more than 200 papers and co-authored a book, *Fleas, Flukes, and Cuckoos* (1952). Bowing to family pressure, she took no formal degrees, but nevertheless won a professorship of biology and an honorary DSc degree from London University. In 1977 Dr. Rothschild hosted an international conference on the flea at her estate in Northamptonshire. Hers was one of the early voices warning against the use of toxic insecticides. She belongs in this chapter, however, for another distinction: she has innovated in humane methods of handling livestock (Uglow, 400; *WBWRA*). A television documentary featured her in 1985.

In 1959 **Helen R. Skeggs** of Montgomery County, PA, patented a new compound whose derivatives work as growth-promoters in farm animals. The compound would be added to the animals' diet in a concentration not less than 50 units per kilogram, but making up less than 1% of the diet by weight. Ms. Skeggs and her co-patentees assigned the patent to Merck & Company of Rahway, NJ.

Last but certainly not least, an example from the forefront of animal husbandry technology:

**Melinda L. Boice** was a member of the team, with Benjamin G. Brackett, that produced the first successful calf by in-vitro fertilization. The work, using an egg from a hormonally treated Holstein cow and sperm from a Holstein bull, took place at the New Bolton Center for Animal Research at the University of Pennsylvania. The calf was born in 1981. This method allows valuable cows to have far more offspring (by implantation of their fertilized embryos in less valuable animals), and valuable bulls can sire millions of calves each year instead of about 50,000 as before (" 'Test Tube' Calf," 7).

*Dairy and poultry inventions.* I looked initially at this constellation of inventions and innovations because my Ohio farm background made me aware how often women did the milking and milk-processing and cared for the chickens on a farm. I also learned about butter-and-egg money. A farm woman's money from selling her butter and eggs was sometimes the only cash income of a small farm in a given year. It was almost always the only income the women themselves controlled. Despite its obvious economic importance—impact on the local retail economy, for example—the phenomenon has been little studied.[30]

Dairy inventions: On the 40-acre farm that was my favorite spot in a southern Ohio childhood, my grandparents usually kept about seven milk cows. They ran the farm aided only by a niece. The twice-daily milking was a three-person job unless the orchards or the hayfield called my grandfather off, in which case the two women handled the milking alone. My grandmother made her own butter and cottage cheese, neither bearing any resemblance to the supermarket species. She bartered her fine brown eggs for the only things she bought from town—sugar, coffee, salt, a few spices, flour, matches, and the like. In so doing, she was following a time-honored tradition.

In thinking about the significance of the 19th-century dairying inventions appearing in Appendix A-3, it is well to remember that milk did not always come from the supermarket in cardboard cartons, already pasteurized, homogenized, and carefully graded by butterfat content. It came by twice-daily effort from cows that were none too clean, and it had to be warmed in winter and cooled in summer, usually without benefit of refrigeration. Cream had to be separated off; butter had to be churned from it by hand or with small mechanical

---

30. See, however, Joan Jensen's *With These Hands;* Zimmerman, ed., 1983: 136, 138; Sandra Myres, *Westering Women;* Charles Clarke, ed., 264; and especially the bibliography in Kleinegger. Jensen estimates butter-and-egg money at 40% of a Colonial farm's income ("Cloth, Butter," 17).

churns, then worked to remove remaining buttermilk, shaped, and salted; cottage cheese had to be made from the sour milk, and so on.

According to Joan Jensen ("Churns," 68), the first (U.S.) woman to patent a butter-related invention was a Quaker farm woman from Bucks County, PA, named **Lettie Smith** (b. c. 1814).[31] She applied in 1850 at the age of 36, and received U.S. Patent #9,959 on August 23, 1853. Her invention was important enough to be reported in *Scientific American* (Dec. 21, 1850: 106). Shortly thereafter, she went to medical school. Since she was single, and thus presumably self-supporting, her butter-worker may have been a financial success.

At least three Canadian women patented dairying inventions during the 19th century as well: **Sophia Wilson** of Windsor, Ontario, patented her "method of making one pound of butter from one pint of milk and other ingredients" in 1868. These "other ingredients" included salt, eggs, and egg whites. One year later, in 1869, **Margaret Duff** of Hamilton, Ontario, patented a synthetic butter called "Duff's Patent Butter." On June 29 of the same year, **Mary Taylor** of Huron County, patented a new method for the "manufacture of cheese from sour milk."[32]

During the 19th century (through February, 1895) at least 20 U.S. women patented 21 churns or related inventions, accounting for 42% of the 50 dairy-related inventions tabulated for 44 women in Appendix A-3. Worth mentioning here, perhaps, are **Mary J. Bridges** of Texas, whose 1883 invention was a Combined clothes-washer and churn; **Eliza Brough** of Greenville, MI, who received two churn patents in a single year (1877); **Effie L. Prall** of Wellington, KS, who patented a Motor for churns in 1887; and **Sarah E. Saul,** who had other noteworthy inventions to her credit. **Gabrielle De Nottbeck** of New York City and her Reciprocating churn of 1875 were omitted from LWP.

In addition to the 21 churn inventions, which can all be considered mechanical (even the four churn-dashers are part of the churn's power-transmission train), these women patented a milk-aerator, a separator (cream from milk), and an early milking machine (see below). Thus machines account for at least half of the 50 dairy inventions noted for the 19th century. The nonmechanical inventions in this group were milk-coolers (4), butter-workers (5, including **Lettie Smith**'s early one and **Hannah Warner**'s 1879 Improvement in coolers for butter-workers), milk pails, vessels, strainers, and skimmers (10), and three cheese- and butter-making processes.

Here, as usually elsewhere in my text, systematic information on 19th-century women's inventions comes from the source I call LWP: *Women Inventors to Whom Patents Have Been Granted by the United States Government, 1790–July 1, 1888* (with two updates carrying the list through February 28, 1895), originally published by the Government Printing Office. The list appeared as a result of persistent efforts by feminists, notably Charlotte Smith, to mark the Patent Office Centennial. Many women's inventions were undoubtedly left unpatented or patented by men. This would have been especially likely before passage of the Married Women's Property Acts, as a woman's property—including her inventions—all belonged to her husband.[33] Consider just one group,

---

31. Katherine Lee reports an earlier churn invented by Adna Norcross of Hallowell, ME, but Norcross is a man (Hallowell, ME, Vital Records, IV).

32. I am grateful to Canadian researcher Barbara-Anne Eddy for calling my attention to this information (Nostbakker & Humphrey, 25).

33. Beginning with Mississippi in 1839, states and territories began to pass reforms in the area of married women's property. As late as 1850, however, married women had legal control over their own property in only 17 states, several of them in the West (Myres, *Westering,* 258), where women's contribution to the economic life of the society was more evident.

American westering pioneer women, and just one item, the churn. Churns were fairly large and heavy, sometimes made of glass. They may have been left behind at the start, thrown out to lighten the load at a mountain pass, or broken along the trail. In any case, several women's journals mention "make-do" churns.

One woman told of a "cradle churn" that she could work with her foot while doing other chores. Most overland journals describe an "automatic" churn the women invented by hanging a milk container from the wagon tongue so that the jolting motion churned the butter during the day (Myres, *Westering,* 133, 147; L-3/15/80 and 4/80). Some of the pioneer women's "make-do" may not have been patentable, but I suspect that some of it was, and was patented by their sons and husbands in less hectic days. In any case, the subject deserves further research. Pioneer women's journals have largely been studied as sociocultural documents, not as sources on 19th-century technology.

To the best of my knowledge, none of these women won fame or fortune for their inventions, though **Anna Baldwin** may have won a bit of anonymous and posthumous notice in the form of a television quiz-show question in the late 1970s: "A woman invented the milking machine: true or false?" The answer, which the contestant missed, was "True," but the announcer gave no name. Baldwin's machine of 1879, which used a hand-powered pump to provide suction, was indeed an early one, though obviously not the first,[34] as indicated by its designation as an "Improved milking machine." Since she held three other dairy-related patents, Anna Baldwin was probably a dairywoman, possible operating on her own.

**Katharina Seifert** seems also to have been connected with a commercial dairy, and her two cheese-making-machine patents suggest an active role. The Seifert family had been in Cincinnati at least since the 1860s. In 1883, the year of Katharina's first patent, and in 1887 the directory shows "Andrew Seifert, dairy," on Knowlton Street. By 1900, though Andrew still appears (no occupation listed), a Frank Seifert (same address) operates the dairy. By 1907 no Seifert dairy is listed. Katharina Seifert is probably the sister or wife of Andrew Seifert; if the wife, they may have had three sons: Andrew Jr., a distiller; Edward, an engineer; and Frank, "dairy."

An unpatented dairy invention with possible economic promise is the ice-less milk cooler and refrigerator exhibited by **Caroline Westcott Romney** (b. c. 1850) at the Columbian Exposition. This was only one of at least fourteen inventions Mrs. Romney showed at the Fair: "more articles of domestic utility on exhibition . . . than any other man or woman at the Fair." Romney was not a farmer or even, to my knowledge, a rancher, though she had lived in Colorado. Rather, she was a journalist who had traveled in Mexico and founded a newspaper in a snow-covered tent in an early mining camp. She was also a magazine editor. In the late 1880s she edited—and published—the trade journal of the porous earthenware industry, *The International Fire-Proofer.* Her editorials show considerable technical understanding of the product and its properties, an understanding she exploited in her cooler and several other inventions.

At the time of the Fair, Romney lived in Chicago and edited *The Journal of Industrial Education,* for which she wrote a two-part article on women's inventions at the Exposition. Unfortunately, the magazine evidently died before Part 2 appeared (inventors surnamed after *H*), which might have listed her own inventions (Handy; Romney/Eagle,

---

34. A common mistake of novice patent researchers is to assume that someone who patents, say, *a* milking machine, has invented *the* milking machine (see Warrior and Vare & Ptacek).

579n and *passim;* Romney, "Women . . ."; Weimann, *Fair,* 430–1; *International Fire-Proofer* 1:1 [Sept. 1888]: 3ff).[35]

Marie Harel's and Mrs. Paulet's respective inventions of Camembert and Stilton cheese have already been discussed. Nineteenth- and 20th-century women's important new cheeses will be discussed under New Foods.

20th century: At the turn of the century, in 1900, **Bertha Northcutt** of Maryville, MO, patented a churn. This rather complicated-looking item, co-invented with Joseph Northcutt, featured a very large geared wheel and crank handle that might have functioned partly as a flywheel. Turning the crank apparently imparted motion to the churning blades inside at a 90-degree angle to the rotation of the large outer wheel.

According to WB/28, during ten selected years between 1905 and 1921 American women patented 35 items of "Dairy supplies and equipment" of the following 20 types:

| | |
|---|---|
| butter worker | vacuum milk-shipping can, maintaining |
| churn and churn dasher | temperature for 24 hours |
| motor to operate churns | milk-carrying tank |
| combined churn and butter worker | milk cooler |
| churn cover | milk pail |
| cow-milker | milk strainer |
| cow-tail holder | skimmer |
| cow-udder protector | apparatus for washing separators, |
| cream can | skimmers, etc. |
| cream dipper | cream separator |
| milk can | cream separating bottle |

The 35 inventions in this area constituted 15.8% of American women's 221 agricultural patents in the ten years covered.

Manufacturing Category III.8, Food products and beverage processes and apparatus, contains a cheese-making invention and a composition to facilitate the whipping of cream (two out of 26 subcategories totalling 40 inventions).

Beyond this, my information is fragmentary. One very notable 20th-century woman multiple inventor, who will be discussed at more length for her poultry and refrigeration inventions below, innovated significantly in the American dairy industry. As head of New York City's bacteriological lab early in the century, **Mary Engle Pennington** developed milk and dairy inspection standards and a checklist procedure for dairy herds that became the national standard (*NAW*/mod.).

To the best of my knowledge, Dr. Pennington had no physical inventions to her credit in the dairy field. One example of women's continued innovation in this area, however, is the Udder-protector patented in 1928 by **Carolyn T. Jenkins** of Dunbar, NE.

A 20th-century British woman has innovated significantly in the field of milk-coolers. In the 1950s, Sussex dairy farmer **I. Violet Dawson** invented a cooler that rapidly cools fresh milk to 40°F, enabling her in the late 1970s to sell the best milk in Sussex. Since not only the degree but the timing of cooling (and warming) has proven crucial for all kinds of organic compounds from wines and cheeses to frozen sperm, there seems no reason to doubt that immediate cooling makes a difference. Among other things, it should keep bacteria from getting a start.

35. The exact date of the magazine's demise is unclear, though I have found no copy later than Sept. 1894 (vol. 9, no. 1). I would be grateful to any scholar who could locate for me Romney's papers or the Oct. 1894 issue of *JIE*—or both.

Other milk producers were impressed, but, she says, "the Milk Marketing Board refused to consider the new idea because they were committed to using their current stock of bulk coolers" (G. Taylor, 35). New ideas routinely meet resistance, but women's ideas seem to meet more dogged resistance than usual. Says Violet Dawson, "The inventor's lot is not a pleasant one, especially if that inventor also happens to be a woman."

As the 20th century advanced, and the subsistence or family farm became rarer, rural youth no longer automatically learned to milk cows. In fact, milking could be considered a disappearing art. **Dianne Prestridge**'s Milking-training device would have been ludicrous in the 19th century, but today it meets a need. Mrs. Prestridge (c. 1936– ), who lives on a 20-acre farm in rural Kitsap County, WA, created the device for her 4-H youngsters in 1977. For fun, she made her cow—at first just the back half with a rubber udder full of Kool-Aid—a purple one. Later, she molded a whole cow of fibrous plaster and named it "Little Squirt." After "Little Squirt" had taught hundreds of children to milk, she hired a fiberglass manufacturer to mold six of them—of which Hickory Farms of Ohio bought three, displaying one in its Bremerton, WA, store. The purple-cow trainers have become popular at the Kitsap County Fair and other fairs in Washington State, and as of 1981 Mrs. Prestridge was negotiating with another manufacturer. Of her inventive skill, she says, "I'm a fiddler by nature . . ." (S. Blair).

Women agricultural inventors and innovators should be far more common in less thoroughly urbanized countries than in the United States, and **Leida Peips** of Viliandi, Estonia, can exemplify them. Graduate of an agricultural school and a specialized dairy course, as of 1980 she managed 110 cows at a collective farm. Consulting at first with an agricultural scientist, she devised techniques for super-efficient milking (using time-and-motion studies) and meticulous procedures for monitoring the cows' nutrition (including tests for blood levels of critical elements). Her key innovation, however, lies in actually *rearing and training* young heifers in ways that suit them to the new milking methods (Kupry, 43).

Using her techniques, Peips achieved milk yields of 5300 liters per cow per year, bringing her Union-wide publicity. Not satisfied with these celebrated yields, or with her Certificate of Honor (and Moskvich car) from the USSR Exhibition of Economic Achievement, she set a goal of 6000 liters. A workaholic and "superwoman," Leida Peips nevertheless sets aside time for her children in the idealized portrait of my source (Kupry).

Pursuing a high-tech approach, U.S. scientists have found a bovine growth hormone that increases milk production. Contributing to the technology of such hormone treatments is **Barbara J. Bruce** of Portage, MI, who with Thomas H. Fraser devised and patented a means of preparing the hormone. Both these scientist-inventors work at Upjohn in Kalamazoo, and their 1984 patent was featured in Stacy Jones' "Patents" column (*NYT,* Apr. 21, 1984: 28).

Poultry inventions: Women also invented for the chickens that laid their farms' golden egg: e.g., **Julia P. Clement**'s Improvement in Hen's nests (Williamston, SC; 1875); two Improvements in chicken coops, by **Nancy J.** (and Robert L.) **Todd** of Shamrock, MO (1877), and **Pamela Millar** of Kenton, OH (1879). **Dicie Croan** of River View, KY, patented a compound for treating chicken cholera (1881), and **Margaret N. George** of Evansville, IN, an Improvement in medicines for both chicken and hog cholera (1871; for further information on women's veterinary inventions, see Appendix A-4). **Elvira Hulett** of Pawlet, VT, patented an Egg tester (1894).

Apparatus and compounds for preserving eggs will be discussed under food-preservation and storage, below.

Both British and American women patented improvements in the all-important incubators that allow hundreds of baby chicks to be hatched artificially rather than by sitting hens, freeing the hens to lay more eggs. For example, **May Arnold** of Acton, Middlesex (now Greater London), England, patented or applied for patents on an Incubating and an Artificial hatching apparatus in 1880 and 1882, respectively. Arnold also patented a means for Preventing the escape of poultry (1880). In 1888 another Englishwoman, **Gertrude Hoddinott,** applied to patent a Self-regulating incubator (Appln. #15,891); no patent is on file (J. Brody L-11/19/85). In 1893 **Eliza J. Bray** of Fair Forest, SC, patented an incubator. Though some years later than the British women's inventions, Bray's was early for the U.S., only seven incubator patents preceding hers.

In the ten years of the WB/28 study, 1905–21, American women patented 34 inventions in the category of Poultry-raising supplies and equipment, 15.4% of their 221 agricultural inventions for those years. The 34 inventions fell into the following 20 categories:

| | |
|---|---|
| anti-vermin perch | incubator |
| brooder | incubator egg-turning |
| brooder coop | incubator heater attachment |
| chicken feeder | incubator thermometer |
| device to keep hens from setting | incubator temperature regulator |
| food compound to promote rapid growth of poultry | incubator trap |
| seed-sprouter for chicken feed | poultry dip holder |
| egg-case filler | poultry duster |
| hen's nest | poultry harness |
| housing device | poultry roost disinfector |

As with the dairying inventions—and indeed throughout—WB/28 provides no names.

Inventing during these years was Iowa's "Poultry Queen," **Rebecca Johnson** (1853–1921). Born Rebecca Salina Veneman in Illinois, she had the ordinary quota of rural schooling and then married. Not until she was widowed with three young children to support did she think of trying to make a living from poultry. At that time (1870s), she said, there were no poultry journals, and the "agricultural papers never said a word for old biddy." Not only were there no incubators, but "if anyone had hatched chickens by artificial means they would have been arrested for witchcraft." She had to learn everything from scratch, through experience. She made her hens a warm house where she kept them all day in cold weather—and fed them cabbage, beets, and other vegetables so that they would lay all winter. Soon she was selling enough eggs to pay all her living expenses and buy feed for the cows and pigs as well as for the chickens themselves.

Johnson's second marriage, to a frail man with no farming skills, did not improve her finances, especially as four more children arrived. When he fell ill after a time of poor crops, and they were about to lose their house, she "went into the poultry business in earnest." She invented her first incubator and hatched about 5000 chicks in a season. In 1906 she published a book to answer the hundreds of queries she received after her story "got into the papers." By 1909, when she wrote about her success for the *Iowa Year Book of Agriculture,* she was earning $300 a month from her poultry business in the busy season, and thought nothing of hatching 2500 chicks in a single day.

Rebecca Johnson created her original incubator by the rational expedient of watching a setting hen for hours on end. Noticing the hen get up periodically and shift the eggs around, she wondered whether the eggs were being turned. By making pencil marks on the eggs, she confirmed that they were, and made this part of her incubator protocol. But perhaps the most interesting feature of her incubators was a thermostat, a bimetallic strip that responded to a temperature change of plus or minus one degree from the ideal. In early models, the temperature change rang a warning bell; in later models, it actually raised or lowered the wick of the heat lamp. Her Incubator alarm was patented in 1908 from Maxwell, IA. The State Historical Society of Iowa has her latest model.

In addition to her incubator inventions, Johnson also created a "Magic Compound" that would prevent coccidiosis when added to chicken feed. She had discovered that chickens were tolerant of arsenic, and that arsenic kills the protozoan responsible for this costly poultry disease.

This remarkable woman was featured in 1990 in a series of televised vignettes on Iowa history (Shotwell). She is also very much alive for her grandson John Combellick, who still lives in Iowa. His mother, a child of Johnson's second marriage, vividly recalled for him the mother who knew healing roots and herbs for most ills, who repaired farm machinery, and who did surgery on her own daughter's misshapen finger (Combellick, PC-6-8/91; R. Johnson, "A Woman's" and *How to;* Shotwell).

In 1926 **Lizzie H. Dickelman** of Forest, OH, patented a brooder, a heated house (next stage after the incubator) for rearing chicks without the mother birds. Her patent provided a multi-purpose small farm building, specifying a "Frame with adjustable sash for convertible grain bins and brooder houses."

According to a *Christian Science Monitor* article of 1942 ("Ideas . . . ," 15), it was a woman concerned about breakage and loss of her precious source of cash who invented the cushioning pasteboard liners (divided into squares) for egg boxes. Minnie J. Reynolds noted in 1908 (7) that the pasteboard insert for shipping eggs, each in its separate compartment, was the idea of a farmer's daughter. These could be the same woman, but neither Reynolds nor the *Monitor* writer gives a name. Nor can one be sure that the invention(s) were patented. But **Ellie H. Florie** of Abbeyville, LA, had patented an Egg carrier as early as 1890.

In the 20th century, **Mary Engle Pennington,** a food scientist and egg specialist, invented a new kind of egg-packing case that reduced breakage (Yost, *AWS,* 94). WB/28 lists an "egg case filler."

To get some idea of the significance of these egg-cushioning inventions, note that 5–10% of all U.S. eggs break before reaching market, costing the poultry industry more than $200 million in 1981. A possible new approach is to try to develop eggs with harder shells ("Shell-Shocked," 34).

As Deborah Fink and Dorothy Schwieder note in their paper on Iowa farm women of the 1930s, "While poultry raising was expected of farm women and was not usually considered a technically sophisticated enterprise, it actually entailed a high degree of skill. *Women learned the trade first from their mothers* and second from extension programs and farm journals" (5; emphasis mine). Keeping chicks warm during spring storms and power failures, without benefit of thermostats—evidently Rebecca Johnson's invention was not universally available—required considerable care and ingenuity, as did culling the flocks to maximize the ratio of feed to egg production. One woman developed her own quantitative culling method: She "spent an entire day in the hen house counting each egg laid and marking each hen which laid an egg." Other women had developed, over the centuries, effective qualitative methods, evaluating the hens' pelvic-bone position, comb color, and foot color, among other features (*ibid.,* 6).

In 1945, **Luella P. Dreyling** of South River, NJ, patented an improved Poultry litter.

The "Ice Woman": One of the most outstanding woman agricultural inventors of the 20th century thus far, with innovations in both dairy and poultry fields, was **Mary Engle**

**Pennington** (1872–1952). Although profiled in *NAW,* and by the *New Yorker* in her lifetime, she is little known today, and her background is of some interest for the woman inventor's profile.

As early as age 12, Mary gave intimations of what she would become. She read chemistry for pleasure, and startled her private-school teachers by demanding to be taught chemistry. After three years of chemistry and biology at the Towne Scientific School, University of Pennsylvania, she was denied a BS because she was a woman. Undeterred, she continued her studies and later received a PhD from the school. Her parents, though surprised at her scientific bent, were supportive. Her father, lacking sons, even rather liked the idea that his elder daughter was preparing for a career.

After postgraduate fellowships in botany at the University of Pennsylvania and in physiological chemistry at Yale, Dr. Pennington returned to Philadelphia in 1898. Finding no academic or professional openings for women, she started her own analytical laboratory for local physicians. Her precise work won her such a reputation that she soon headed the City Health Department's bacteriological laboratory. She also lectured at the Women's Medical College of Pennsylvania.

Passing Civil Service examinations as M.E. Pennington, she won appointment as a bacteriological chemist with the Department of Agriculture in 1907 and in 1908 became Chief of the new Food Research Laboratory. The mission of this laboratory, located in Philadelphia at Pennington's request, was to enforce the 1906 Pure Food and Drug Act. As mentioned above, her checkup procedure for investigating the condition of dairy herds supplying milk to the city became the national standard. Noting that children were dying from contaminated hokey-pokey ice cream[36] sold by street vendors, Dr. Pennington showed the peddlers microscope slides swarming with bacteria from their wares. Sincerely horrified, they agreed to boil their pots and ladles. Throughout her public health career she showed this double ability to find scientific answers to a problem and then win cooperation from sellers and producers.

Dr. Pennington next worked with poultry, gathering firm data and establishing shelf-life guidelines for frozen chickens. In 1911 she tackled the egg industry. In one of her most spectacular achievements, she developed sanitary methods for dealing with the cracked, scarred, or outsize eggs destined to be broken, dried or frozen, and sold to bakers and confectioners. She persuaded three Midwestern plants to adopt her method (involving refrigeration and conditions of near-surgical asepsis). When these three were flooded with orders, the whole industry followed suit, and a threatened Agriculture Department crackdown was avoided. Eventually she co-authored a book called *Eggs* (1933).

A gardener by avocation, and a Quaker by religion, Dr. Pennington devoted most of her life's energy to her work. She never retired. At the time of her death she was vice-president of the American Institute of Refrigeration, and still maintained her New York consulting office.

Several other inventions and innovations place Dr. Mary Engle Pennington here. Before World War I, she invented a new method of slaughtering chickens. After thousands of experiments she devised a technique of piercing the fowl's brain with a small knife and cutting the jugular veins from the inside. For unknown reasons, this makes plucking easy and leaves the body in excellent shape for traveling and storage (Heggie, 26). She also invented the necessary special knife (*NAW*/mod.). To get her method accepted—as it still was in the 1940s—she had to persuade several states to repeal their

36. According to one source, "hokey-pokey" is Victorian London children's corruption of the Italian ice-cream vendors' cry of "*Ecco un poco,*" or "Here's a bit" (De Bono, 107).

laws against transporting undrawn chickens. She argued that the birds suffered less bacterial invasion if they remained "sealed packages."

Dr. Pennington also developed an inexpensive system for cooling poultry immediately after slaughter and before dry-packing for refrigerated shipment. Her method replaced the old system of packing poultry in ice for shipment. Since the ice usually melted before journey's end, most poultry arrived at best waterlogged and at worst, spoiled.

For the egg industry, aside from her procedures for frozen and powdered eggs, she developed cooling and storage procedures for whole eggs. She also, as mentioned above, designed a new kind of breakage-reducing packing case (Yost, *AWS,* 94).

All in all we can thank Dr. Pennington whenever we choose a frozen chicken at the supermarket or break an egg into the frying pan.

### Food-processing

By a curious schizophrenia, our culture trumpets that we are what we eat, yet devalues women's work and inventiveness in food preparation and its technology both ancient and modern. By rights, as here, food-processing is part of agricultural and related technology. By rights, too, its tools and machines are properly classed as mechanical rather than "domestic."

Since restoring full credit to women in this field would demand volumes, I limit my discussion to a few areas and to selected examples in each. During the 19th century women still processed foods extensively in their homes, prior to cooking. Their inventions in this area, patented and unpatented, are legion, but especially interesting in our machine-worshipping culture are the food-processing machines and devices.

*Machines, devices.* 19th century: LWP provides a fascinating glimpse of 19th-century women's U.S. patents for food-processing machines and mechanical devices. Women invented and patented at least two ice cream freezers, the first as early as 1843; seven egg beaters in the 1880s alone; and various other mixers and beaters, including a mayonnaise mixer, a cake mixer, six dough mixers and/or kneaders, a dual-purpose pastry fork, a salad-mixing device, and two unspecified mixing-beating machines. They also patented at least three improved apple-paring, coring, and quartering machines; two ice-shavers, three devices for processing meat (slicer, mangler, tenderer); two vegetable slicers, three grape or raisin stoners (seed-removers); and five fruit presses or juice extractors. Finally, they patented three flour-sifters, three corn-cutters, a corn-popper, and a pea-shelling machine.

My survey of LWP was not exhaustive, and LWP is not complete, of course, but women patented *at least* 46 food-processing machines and devices in just over fifty years, between 1843 and early 1895 (see Appendix A-5).

Noteworthy among these women are the earliest, **Nancy M. Johnson; Margaret Wilcox,** cooking-range patentee and multiple inventor; and **Annie Mangin,** the only black woman on the list, to my knowledge.

Before **Nancy M. Johnson** (c. 1795–1890) came along, says Charlotte Smith, ice cream had to be made by constant hand-stirring with a spoon (*WI/*1: 2). Whether or not Mrs. Johnson, a life-long resident of Washington, DC, actually invented the first crank-actuated freezer, as sometimes claimed, she certainly originated the standard American type: a wooden bucket for salted crushed ice, a covered metal can rotated inside it by means of an external crank, and an internal dasher to agitate the cream mixture. Though recent popular sources (Fabricant; Ungerer, C-6) claim the invention was never patented (and erroneously date it to 1846), it was patented in 1843—the first U.S. patent for this device. Unfortunately, however, Johnson sold her patent rights for only $1500. The buyers undoubtedly made millions (M. Reynolds, 3). The *Encyclopedia Americana* article on ice cream (1978 edition) mentions Johnson's invention but misdates it 1846, and fails to name her. (Some of these errors may go back to Leggett, who omits Johnson from his compilation of U.S. patents, 1790–1873.)

Mangin's pastry fork, useful both as a whip and as a flour mixer—"the only thing of its kind known at the patent office" (Weimann, *Fair,* 269)—appeared in New York State's Afro-American Exhibit at the Columbian Exposition in Chicago, 1892–3.

Two other noteworthy inventions shown at Chicago were **Priscilla M. Burns**'s Perfection Sifter and **Mrs. A.J. Hambel**'s Tivolia Beater. Mrs. Burns's sifter, patented in 1893, not only measured the flour, but encased the sifting process to keep flour from flying about (Romney, "Women" [see "Barnes"], 23, 25). Mary Logan includes Priscilla M. Burns of St. Louis in a 1912 list of "Prominent American Women Inventors" (887).

Mrs. Hambel must have gone home from the Fair a happy woman, as the exhibit brought her 60,000 orders for her beater (Weimann, *Fair,* 432). She was probably the Allie J. Hambel listed in the 1892 Chicago directory as a dressmaker, widow of Melvin Hambel, living at 709 N. Halsted (the only Hambel listed for that year).[37] Caroline Romney gives a different address (4727 Evans Ave.) in her article on women inventors at the Fair; but Hambel could have moved by 1894 when Romney's article appeared. The profits from her Fair sales alone should have enabled her to move up in the world if she wished to do so.

The Tivolia could be anchored air-tight in a quart glass jar with a convenient handle, for mixing fruit juices and the like, or used independently for mixing bread, mashing potatoes, whipping cream, beating eggs, etc., accomplishing all such tasks faster and more easily than hand-beating (Romney, "Women," 25). It would be interesting to know whether this distinctive beater design—frames enclosing horizontal prongs—became standard for the late 19th and early 20th centuries. I have found no patent for this invention, which accordingly does not appear on the appended list.

Too late for LWP were **Mary C. Aguero**'s patented fish-scaler of 1896 (New York City) and **Eulalia V. Brake's** 1898 Machinery for cutting noodles (Toledo, OH).

Most noteworthy here is what this 19th-century U.S.-patent sample—which could be expanded to include the dairy machines already listed—indicates about women's technical achievement in the mechanical end of food-processing.

British women were not idle in the food-processing field, either, though I have only scattered examples:

A British inventor and entrepreneur who patented an ice cream freezer and an "ice cave" or storage box for ice cream in 1885 was **Agnes B. Marshall** of London. Mrs. Marshall must have possessed ferocious energy. Proprietor of a cooking school, creative retailer of culinary utensils, gadgets, and equipment; importer of specialty foods, author, lecturer, publisher, and proprietor as well of an employment agency for "superior cooks," she also wrote several cookbooks, of which the first (1885) has recently been reprinted as *Victorian Ices and Ice Creams.* According to the London Directory for 1889, she was located at 30 and 32 Mortimer Street, Cavendish Square.

Her broad, shallow freezer design is in use again for ice cream freezers built into household refrigerators (Agnes Marshall, viii). She may have held several patents in addition to her freezer and her ice cave (Brit. Pats. #2911 and 13007, appl. 1884). Pertinent here is her Ice-breaking machine (#13928, appl. 1885).

Two of Marshall's contemporaries who received British patents for food-processing

---

37. I am indebted to Debbie Bruseth of Chicago for this information.

and related inventions are **Louise Steffen,** who patented a process for Making compounds of sugar and lime from molasses, etc. (1880), and **Denise C. Ralu,** who applied for three patents in 1881—#3995 for Transporting, preserving, etc., pulpy fruits; #4100 for Beermaking, and #4532 for Confiture preserves.

Much earlier was **Elizabeth Dakin,** who in 1848 patented Apparatus and machinery used for cleaning and roasting coffee (#12,198). In 1847 one William Dakin had received a related patent (Woodcroft).

20th century: The Women's Bureau Study—quite properly—shows women's food-processing inventions under manufacturing as well as household categories. Of the 40 invention types in Manufacturing Category B, Food product and beverage processes and apparatus, the following seem relevant here:

Apparatus for preparing food products
Mixing machine
Nut-shelling and separating machine
Tool-holding means for food machines

If all the invention groups in this category were equally numerous, these four groups would represent about six inventions.

Category A, Kitchen equipment, under Household inventions, contains the following relevant invention groups:

| | |
|---|---|
| Biscuit and doughnut cutter | Juice extractor |
| Egg washer and slicer | Meat tenderer |
| Fruit, meat, or vegetable grinder | Mixer, egg beater |
| Fruit seeder and corer | Sifter, sieve, strainer |
| Grater or masher | Vegetable tapper |
| Ice cream freezer | |

If the 80 groups in the category were all the same size, these eleven types would number some sixty inventions.

Belonging to the 20th century, but earlier than the first year covered by the Women's Bureau study, are two inventions by **Annie M. Furrow** of Washington, DC: a Green corn husker/silker and a Corn carver (both of 1903).

Versatile inventor **Beulah Louise Henry** patented an ice cream freezer in 1912 (not covered by WB/28). Henry's freezer, her first patent, was noteworthy in that it was a vacuum freezer.

In 1916, also not covered by WB/28, **Madeline M. Turner** of Oakland, CA, patented a fruit press. Patent Examiner Patricia Ives calls Turner's method of extracting juice from fruits "ingenious," and reproduces the patent drawing. Ives' description does not reveal how the device was powered. Clearly, however, the press was a complex machine with a geared mechanism for advancing the fruit from hopper to plungers. Turner is noteworthy as one of the few known black women patentees in the U.S. system ("Creativity," 47).[38]

**Lillian Gilbreth, Christine Frederick,** and **Mary Pattison**—particularly Gilbreth—may well have designed food-processing machines for their model kitchens or consulting jobs. Joan Marlow specifically credits Gilbreth with developing an electric food mixer

---

38. Indeed, Ives calls her only the fourth, after **Sarah Goode** (1885); **Miriam Benjamin** (1888), and **Lyda Newman** (1898). However, since race, like gender, is not recorded on patent applications, and some black women passed for white, this figure is undoubtedly much too low. I have encountered at least one woman not on Ives' pre-1916 list: **Annie Mangin,** mentioned just above for her pastry fork (1892).

(and other "electrical devices") for the efficiency kitchen she designed for the Brooklyn Gas Company in the 1920s (224). Since the first electric motor appropriate for such a device did not appear till 1910, the first electrically powered mixer in 1918 and the first successful stationary mixer, the Sunbeam Mixmaster, not till 1930 (Harpur, 56; *WABI*, 151), this was early days indeed.

A Maryland woman who innovated in commercial food-processing is **Anna E.E. Schneider** (b. 1889) of Baltimore. As a young woman she had intended to study medical illustration at Johns Hopkins, combining her own artistic talent and her father's urging toward a medical career. Influenced, however, by watching her mother regain sight and health after adding whole grains to her vegetarian diet, Anna determined to make whole wheat flour, like that ground in her home, available to the public. She is sometimes known—surely wrongly—as America's first woman miller. Although her father, an aircraft designer and inventor, created the original hand mill that inspired her, she seems to have developed both the commercial process and the electrically powered commercial-scale mills. Like the mills of the gods, Anna Schneider's mills grind slowly, one pound of flour each nine minutes. Unlike the divine models, however, her mills grind not so fine—producing granules rather than powdery flour. All this is calculated to preserve nutritional value: the slower milling process is cooler, less likely to affect the volatile essential oils of the grain; and the granules resist oxidation better than powder, keeping the grain's natural nutty flavor. She opened for business in Baltimore in 1926.

As demand for her "Eugenia Whole Wheat Flour" grew, without advertising, Miss Schneider opened more mills, but always refused to speed up production. Schrafft's and Macy's in New York, Marshall Field's in Chicago, the Waldorf-Astoria and Astor Hotels, West Point, the U.S. Naval Academy, and a South African steamship line were among her customers. Still active at the time of the U.S. Bicentennial, she was featured in *Notable Maryland Women,* one of the Bicentennial state volumes on women's achievement (Helmes, 330–5).

A post–World War II example is the doughnut machine invented by a Japanese woman. According to Reiko Matsutoya, vice-president of the Japanese Women Inventors Association, this was a fine machine, and the woman patented it, but someone nevertheless copied the idea. Lacking the necessary legal experience to fight the infringers, she made no money from her invention. Her experience, however, inspired the founding of the Association, which now pursues "vigorous and wide-ranging activities" on behalf of Japan's inventive women (Murray, 1). Unfortunately, Matsutoya fails to name the doughnut-machine inventor.

Scattered American examples for this century include **Joy C. Adams'** fruit-juice extractor of the 1930s and **Florence Anderson**'s semi-automatic ice-cube maker of 1960. As Anderson's obituary credits her with a "host of household gadgets and appliances," she may well have other food-processing machines to her credit, but I have found no such patents. Her employers may have kept them as trade secrets.

One of the very few Australian women in my sample is **Mrs. E. Adams** of Ingham, Queensland, who invented a fish-scaler in the 1970s (Australian Broadcasting Commission series, 1978–9).

A minor but intriguing example from the late 1970s is **Pauline Berke**'s hot-dog-bun shaper. Ms. Berke's device creates a bun with a long cavity for hot dog and condiments. No splitting is necessary, and since the bun is open only at one end, the hot dog can be eaten without dripping. Pauline Berke, from the Los Angeles area, is a member of Inventors' Workshop International (Weisberg, L-8/16/76).

The invention with the most name recognition in this field is an early one for the century, the Melitta Coffee Filter. This filter, and the popular drip system that goes with it, including a special pot, was invented by **Melitta Bentz** of Dresden, Germany, in 1908, because she wanted better coffee for her family. At the Leipzig trade fair of 1909, she sold over 1200 of her "coffee makers." Today they are used in a hundred countries (Vare & Ptacek, 42–3; *WABI*, 151).[39]

*Tools.* Too numerous to detail are the food-preparation tools invented by women. Noteworthy as the work of a famous chef are the Dione Lucas Gourmet Knives. **Dione Lucas** (1909–71) was born in Italy, daughter of a British artist. Trained as a jewelry-maker and as a cellist, she switched to cooking, which she calls a comparable art, completed two courses at the famous Ecole du Cordon Bleu in Paris and, in the 1930s, she ran a restaurant and cooking school in London. When her marriage failed, she came to the United States with two children, a trunkful of pots and pans and no money. Soon her Cordon Bleu restaurant and cooking school in New York City was attracting celebrities.[40] By the mid-1940s she launched one of the first television cooking schools. According to *Time* magazine, her show, *To the Queen's Taste,* stood out from the other recipe shows "like a glistening grape in a flavorless aspic" ("Airborne Recipes," 61). During the next two decades she continued to teach and write about cooking, creating many cookbooks and opening at least four restaurants. As a late venture, she opened the Dione Lucas Gourmet Centers in New York City and elsewhere.

Lucas invented the Gourmet Knives when the knife she was using proved an unsatisfactory de-fatting tool for her Suprême de Volaille. In frustration she created these knives with their excellent balance, molybdenum-alloy blades, and strong rosewood handles curved to fit the fingers. In 1974 four knives sold for $19.98, with a lifetime free-replacement guarantee ("Airborne Recipes"; Gibbs; Lindheim; "You have ruined . . .").

*Methods, techniques, processes.* Our foremothers' work in debittering and otherwise improving the edibility of staple foods is now largely professionalized and transferred to the laboratory throughout the West. Examples of women's patents from the first two decades of the century, before this development was complete, come from the Women's Bureau Study's Category III.B, Food products and beverage processes and apparatus for making same (WB/28, 22):

| | |
|---|---|
| Beverage-producing material | Preparing whole rice |
| Bread raiser | Purifying sugar juice |
| Preparing carrot flakes | Treatment of grain for beverages |
| Preparing citrus powder | Wine-clarifying |

Since these are only 8 of 26 types totalling 40 inventions in this class, probably about 12 inventions are involved (assuming equal numbers in each type).

A few later examples are scientists working in the USDA laboratories. **Linda C. Brewster** at the Pasadena laboratory has received four 1970s patents for reducing bitterness in orange juice and other citrus products (Jett, L-5/29/80). **Marie S. Gray,** a Research Chemist in the New Orleans laboratories, has patented a Treatment of

---

39. Unfortunately, Vare & Ptacek's work is riddled with errors and oversimplifications. Their introduction alone could set back the state of knowledge on women's contributions to technology at least a century. Perhaps the worst howler is the statement that Marie Curie discovered radioactivity (contradicted later in their own pages). Pertinent and painful here is the first reference to Melitta Bentz (on p. 16, not on p. 14 as indicated in their index), crediting her with inventing "the modern coffee pot." Anyone reading this could be pardoned for thinking she had invented the electric percolator. The actual nature of her invention becomes clearer on pp. 42–3.

40. Including fellow-inventor Dorothy Rodgers.

cottonseed meals, followed by extraction with solvents, to remove gossypol (1981). Recall that gossypol is an impurity in raw cottonseed oil that causes sterility and cardiac problems. Since cottonseed oil is so widely used for cooking and animal feeds, Dr. Gray's process could have enormous economic significance worldwide.[41]

**Maura M. Bean,** a Research Food Technologist at the Western Regional labs in Berkeley, CA, has patented Novel methods for improving the baking properties of cake flour without chlorine (1979).

## Cooking

Cooking was thousands of years old when the first known patents for cookery were granted on the island of Sybaris about 500 BC (Gomme, 4). Few new types or systems of cookery have therefore been invented in modern times. Women have been inventive in adapting cooking to gas and electric stoves, and thence to microwaves and toaster ovens, to say nothing of woks; they have learned to prepare low-calorie, low-salt, low-sugar, and low-fat diets, among others, and to steam vegetables to retain their nutrients. But the central body of cooking knowledge and methods has remained fairly stable.

Even seemingly new methods or systems are often modern re-inventions or refinements of ancient methods. Consider fireless cookery as described by Heidi Kirschner. The Kirschner fireless cooker is a plywood box, insulated with hay, straw, or blankets. The cook brings a pot of food to a boil, simmers it briefly, then covers and sets it inside the box, adds insulation, and tightly closes the lid. By evening, the dish is deliciously done. Fireless cookery uses less energy than a crockpot, and food cannot overcook. Important for the developed world where energy costs are high, and crucial for the Third World where firewood is disappearing, this energy-saving method deserves more attention. For Alice Walker's charming account of her first taste of fireless cookery, see "The Strangest Dinner Party. . . ."

*Cookware.* A prolific field of women's invention that must be mentioned in passing is cookware—pots and pans. Women's early inventive and manufacturing monopoly in this area was undermined wherever pottery became commercialized or iron vessels were introduced. However, as always, women's role continues to be greater than imagined (cf. E. Soule). Two noteworthy 20th-century examples of cookware inventions involving women's practical genius are Teflon and Pyrex. Teflon (Brit. Bluron) was accidentally discovered by Dr. Roy Plunkett of Du Pont in 1938, but no one thought to apply it to skillets for more than fourteen years. The idea of a nonstick cookware surface apparently occurred independently to several people in the 1950s. One such person was the wife of French engineer Marc Grégoire. When she saw him using Teflon on fishing lines, she suggested that her pans could profit from such a substance. Seeing the genius of her idea, he took it up, and founded the Tefal Company in 1955 (Onanian, L-9/8/76).

Pyrex ware was born when a Corning Glass Co. engineer took home a heavy glass container to use for battery acid. His wife borrowed it to cook her pot roast, and a whole new line of cookware was the result. She is still known in the business as the "**Pyrex Housewife**" (Peters & Waterman, 124).

In the 19th century, **Ellen Swallow** put a pouring lip on each side of the saucepans in her Center for Right Living, to accommodate left- as well as right-handed cooks (R. Clarke, 66–71).

---

41. As discussed below in the context of contraception (Chap. 3), gossypol has been tested as a male contraceptive in China. Ways may be found to mitigate or even prevent its cardiac and other bad effects.

*Stoves.* As already noted, Elise Boulding thinks women probably tamed fire. Myths on the origin of fire often deal with the first cooking of such things as manioc cakes (Frazer, *Myths . . .* ). Otis T. Mason (*Share,* 235) and Mozans (341) credit women with inventing both oven and chimney, fitting the division of labor in preliterate human groups. According to Sjöö and Mor (1981: 11), only women could use the early beehive- or dome-shaped ovens, as they symbolized the belly of the Great Goddess. Mason points out that women invented steatite griddles or "stove-tops" in very early days. Although the familiar names in Western stove technology are male—e.g., Rumford's scientific fireplaces, and Franklin stoves (Heschong, 35)—women have also made important improvements in stoves and related devices.

As always, patent records underrepresent this contribution. According to one stove manufacturer in the foundry city of Troy, NY, for example, the greatest Franklin-stove improvement he ever saw was a woman's invention, though patented and profited from by a man (M. Gage/1: 18; M. Logan, 883). Could this unsung woman be the enterprising Victorian doll designer and jill-of-all-trades **Izannah Walker** (1817–88) of Central Falls, RI? Creator of Walker dolls, skilled carpenter and inventor of presses and dies for shaping doll heads, tinkerer in household gadgets and dabbler in real estate, Izannah also, according to her niece, designed a parlor heater "that beat Ben Franklin's" (Formanek-Brunel, n13; Bessette, L-10/1/91).

Unpatented, too, are the many ingenious trailside improvisations of American pioneer women. In addition to adapting to an entirely new fuel (buffalo chips), westering women often needed to improvise some sort of stove. Sandra Myres reports that many women dug a trench about a foot deep and a yard long, put their crane across the trench, and suspended from it their coffee pot and camp kettle. One woman called these trench cookers "a very good substitute for a stove" (Myres, 106).

Baking was harder. One determined Oregon-bound pioneer held an umbrella over her skillet for nearly two hours to bake bread for the party's supper (Myres, 147; Wertheimer, 252).

Women's make-do continued at their new homes, which often bettered trail conditions only by standing still, and these ad hoc stoves and ovens are no less noteworthy for being unpatentable. A pioneer woman named Harriet Upton reported (cited in Baxandall, 73):

> . . . One sturdy mother became so enraged because her husband procrastinated in building an oven, saying that she could no longer bake bread and do all her cooking in one big iron kettle, that she fashioned some bricks of mud and built an oven which was such a success that people travelled out of their way to see it in action. . . .

19th century: Nevertheless, even the patent records tell an impressive story, especially for the 19th century. *LWP alone* (1809–March 1, 1895) *lists at least 116 women holding 139 patents for stoves or related inventions,* counting toasters but not counting several patents for ash-sifters. Omitting the china kilns, smelting furnaces, heating stoves, fireplace improvements, and other inventions not specifically connected with cooking, we still find 39 women holding 41 patents. Of course since fireplaces and heaters often doubled for cooking, and poor working girls sometimes used lamps for heating or cooking, or both (See Ruggles', Galer's, and Rathbun's patents), these figures are highly conservative. Detailed research would undoubtedly expand the list found in Appendix A-6. Omitted by LWP, for example, is **Marion J. Wellman**'s safety water-back for cookstoves (New York City; 1867).

**Harriet Farnam**'s 1880 Cooking and heating apparatus was calculated to save fuel when broiling or heating irons. Farnam herself is noteworthy as a multiple inventor and entrepreneur, and particularly as the inventor of the Hoosier Flycatcher, which seems to have been the 19th-century equivalent of the No-Pest Strip®. Though born in New York and living in Indiana when she patented her flytrap, her stove patent listed her residence as St. Paul, MN.

Possibly also noteworthy is **Margaret Owen**'s Cooking range or stove (1891). Since Owen lived in Lambourn, England, she must have felt that the invention had real commercial potential to justify obtaining a U.S. patent.

Among the inventions most interesting to the Board of Lady Managers' Patents Committee at the Columbian Exposition of 1892–3 was "The Wilcox & Laura Nevins range and furnace, which utilizes the wasted heat of the cooking stove" (Official Minutes, 122).[42] I have found no **Laura Nevins** in the patent records, but **Margaret A. Wilcox** of Chicago patented several stoves and related inventions, including the one probably referred to here: Combined cooking and hot-water-heating stove (1892). Since two of her nine inventions were assigned to the Wilcox Water Heater Company of Chicago, and since one **Olive Wilcox** from nearby St. Louis also patented an oven invention (see Appendix A-6), one might posit a family heating business, possibly run by Margaret's husband or father. However, Margaret Wilcox was already a widow by 1883 (city directories). Her first patent, for a Combined clothes and dish washer, is dated 1890, and the Wilcox Company first appears in the directory in 1892. Thus rather than inheriting a business, she may have founded one to exploit her invention. And although the business appears in the directories for only three years, she had not remarried by 1905—and she received a patent as late as 1907. Perhaps she made enough money from the company or from her inventions, or both, to support herself.

Also shown at the Chicago Exposition was **Louise D. Fein**'s Independent Steam Cooker (1892). This ingenious energy-saving device made one burner into four by fitting a column of water over it and flaring the container into a double-walled steam tray with room for four pots. Tubes at the sides of the custom-fitted cooking pots brought direct steam, as well as heat from below the tray, to the food.

According to Caroline Romney, who describes the cooker in detail ("Women," 24), Miss Fein, of Cincinnati, also invented a coffee pot using steam tubes, as well as one-pot steamers. The entire apparatus was light-weight, and allegedly eliminated cooking odor. Thus far I have located only the one cooker patent, although Fein had earlier (1887) patented a fire escape.

Fein's invention is reminiscent of the People's Stove and Cooking Utensils patented by **Amelia Lewis** of London, England, almost twenty years earlier (an ad in her journal, *Woman's Opinion,* for January 21, 1874, says that the stove was already patented), except that the latter included a specially designed stove in addition to the steamer and other custom utensils. A self-styled socialist and a member of the National Association for the Promotion of Social Sciences,[43] Mrs. Lewis intended her cooking system to save fuel and to burn cheap peat or straw. In fact, she claimed in the *National Food and Fuel Reformer* for December 5, 1874 (80), that the stove could use coal, coke, compressed peat, or any other fuel, consuming only a third or even a quarter as much as an ordinary small grate. Moreover, the stove would heat the room, and cook food according to scientific

---

42. I am grateful to Jeanne Weimann, author of *The Fair Women,* for this reference. Could Laura Nevins have been related to Blanche Nevins, a sculptor who exhibited at the Fair (Weimann, *Fair,* 295)?

43. I am grateful to David Doughan of the Fawcett Library, London, for my information on Amelia Lewis and her publications. See David Doughan and Denise Sanchez' annotated bibliography of feminist periodicals in British repositories, 1855–1984.

principles, retaining flavor, juices, and nutrition. In addition to fish, meat, and vegetable steamers and a meat roaster, Mrs. Lewis provided a "frizzler," doing "all the work a usual frying-pan is designed for," but acting "in such a manner that no delicate meat essences are lost, and that the fibre is not hardened, as in the ordinary frying-pan."

Predating all the rest is another British invention. As early as 1789 **Mary Howson** received British Patent #1685 for Stills and boilers.

Omitted from LWP, but patented, and exhibited at the Philadelphia Centennial in 1876, was **Mrs. Orrin Smith**'s Cooking range (1866). Mrs. Smith, wife of a Chicago commission merchant, had her stove featured in *New Century for Women* (114) under "Exhibits of Illinois." Fitted with fireboxes for coal and for wood, a central oven heated by either box and two small side ovens with gentle-heating compartments, nine covers or cooking plates (seven on top and two on the hearth of one firebox), and a grate for broiling, this stove is unusually versatile. "The range seems . . . particularly adapted to family use," observed *NCW,* "from the fact that, at will, it may serve to cook large, or small quantities of food with corresponding amounts of fuel and heat." A small model appears in the Smithsonian's "1876" Exhibit (Warner/Trescott, 117).

Also omitted from LWP, but patented, according to *New Century for Women,* was another Centennial invention, the Parlor cook-stove adapted for gas by **Mary A. Browne** of Chicago. Since no explanation appeared with the model, says *NCW,* "its special claims to superiority over other parlor cook stoves are not easily seen" (114; Warner/Trescott, 111). Patent records (1869–78) reveal no stove for Mrs. Browne, though in 1873 she patented an Attachment for sewing-machine tables.

Patented, but too late for LWP, is the oven thermometer invented by **Lizzie S. Barndollar** of Trinidad, CO (1896).

A great many 19th-century women's inventions of cook-stoves and related devices, of course, were never patented. Important among these were the many innovations for the experimental kitchen and consumer-testing laboratory **Ellen Swallow** established in her home in the late 1870s and 1880s. She called it the Center for Right Living. Pertinent here are her substitution of gas for coal and wood in cooking; determining optimum cooking times to save both money and nutritional value; and placing a hood with a ventilating fan over the kitchen stove. Whether or not Ellen Swallow was, as Robert Clarke implies, first to do all of these things (and whether alone, or with her husband), she certainly pioneered the scientific testing of air quality, nutritional quality, fuel economy, and time management to justify the improvements (R. Clarke, 66–71).

The Rumford Kitchen she masterminded at the Chicago Exposition may well have included improvements in stove and oven design (*NAW;* Weimann, *Fair,* 462).

Among **Caroline Romney**'s 14 inventions at the Columbian Exposition were "Oven fixtures for conservation of heat." These items appeared in the United States Exhibit under Manufactures, rather than in the Woman's Building (Handy, 41). No details appear in the official catalogue; but presumably, like most of her Fair inventions, they incorporated porous earthenware.

One of the most noteworthy cook-stove inventions of the 19th century occurred among the Shakers. In 1876 **Eldress Emeline Hart** of the Canterbury, NH, community invented a revolving oven. This commercial-scale brick oven featured pierced metal shelves to circulate heat evenly, an isinglass window for monitoring doneness, four separate compartments with a door for each, and a temperature gauge—a real innovation in those days. It would hold 60 pies or 70 loaves of bread at a time. The revolving shelves allowed the Sisters to remove finished items without burning their hands.

According to June Sprigg, this oven was one of the few inventions the Shakers patented.

If so, it must have been done in a man's name, as I find no patents for Emeline Hart in the records. Hart's oven, however, represented such an obvious advance that many "worldly" bakeries purchased it. The Shaker *Manifesto* of April 1899 calls her "our valued Eldress Emeline Hart." She was a top administrator, a religious leader, and a businesswoman (manufacturing the Shaker cloaks invented by another Sister, **Dorothy Durgin**). I know of only one other Hart invention—her improved windmill of 1858—but others may have gone unrecorded.[44]

20th century: The best available systematic source for the early 20th century, as usual, is the Women's Bureau Study of 1923. Since the study had no stove or cook-stove classification, no numbers are available. Several classes, however, contained cook-stoves or related items: Category IV.D, Heating equipment and appurtenances, contains a Combination furnace and cook-stove, and many of its 62 types of devices could apply to cooking as well as to heating stoves. Category XII.A, Kitchen equipment, for instance, contains the following 14 groups of cook-stoves or related items among its 440 inventions:

| | |
|---|---|
| Apparatus for boiling eggs | Lamp stove |
| Broiler | Oven and warming closet |
| Candy cooker, hot air | Protector for stoves |
| Cooking device | Sanitary cooking apparatus |
| Cook-stove and attachments | Shield from oven heat |
| Device for cooking asparagus | Toaster |
| Fireless cooker | Waffle iron |

If all 80 groups in the subcategory were of equal size, these 14 groups would contain 77 inventions for the ten years.

The Fireless cooker here may be one or many, but one **Ada T. Hill** of Salem, IL, patented a fireless cooker on October 10, 1910. **Jennie Jones** of Winchester, KY, patented a Toaster in 1919.

Category XII.E, Dining-room equipment, includes one Electric toaster group that could contain as many as four inventions.

Beyond these early years my sample scatters: **Helen L. Borza** of Yonkers, NY, patented an Electric stove (1948) with Raphael Borza. The Borzas patented a Windmill toy that same year. Also in 1948, **Gladys Gowen** of Ashstead, Surrey, England, patented Improvements relating to toasting devices. Her improvement allowed the toast to be turned while still in its holder.

America's "Lady Edison," **Beulah Louise Henry** of North Carolina and New York City, patented a Baster oven in 1962. Its technological significance is unclear, but most of Miss Henry's inventions were actually manufactured and sold, by her own or other companies (*WWAW*/1).

**Jackie Jackson** (c. 1957– )[45] of New York City holds two patents (1975, 1978) for

---

44. She also tried her hand at didactic poetry. The August 1899 *Manifesto* (182–3) published her "Colloquy Between Procrastination and Alacrity" (Deming & Andrews, 157; Sprigg, 174–5, 184, 187; Irvin, L-1/21/80 and "Machine," 315; White & Taylor, 313–14). I am indebted to Helen Irvin, author of *Women in Kentucky* and of excellent research on Shaker women, for information on Shaker women in general and Eldress Hart in particular.

45. Jackie's given name, Crowell (she is a direct descendant of Oliver Cromwell), appears on her patents. This illustrates one of the problems with patent research. If Dorothy Stephenson of the New England Inventors Association had not sent me her name, with nickname, as a woman inventor, I would have dismissed her as male.

all-purpose Household lighters. Replacing matches for lighting barbecues, fireplaces, candles, ovens, etc., the lighters are long wands using disposable butane cartridges.

One of a pair of identical twins, Jackie grew up in a small town with an artist father and an inventor uncle in the family. As a child she loved to draw and took art courses at Sarah Lawrence College (where she took her AB). Her primary school was much like the one-room schoolhouse of pioneer America—three grades in one room and a grand total of five in her eighth-grade graduating class! Jackie is a confirmed believer in the entrepreneurial spirit, considering intuition, determination, and persistence essential to success. New-product ideas come to her easily. As a result, she shares the frustration of many creative people—that marketing steals time from creation. The main culprit is the NIH (Not invented here) syndrome, as inventors call the discrimination by large corporations against new-product suggestions from outside.

Jackie intends to concentrate on marketing her lighters before pursuing her many other ideas. In the late 1970s she worked in New York City for a foreign policy research institution (Jackson, Q-1978; L-11/15/78, 4/25/79, 1/29/85).

It would be astonishing, however, if women's ideas and advice were not incorporated into the many cook-stoves and related devices created and marketed by United States corporations in this century. Any female design engineers at such corporations are still unsung, but at least two examples of women who *were* consulted leap to mind: Lillian Gilbreth and Christine Frederick. Though her focus was time-and-motion studies directed toward efficiency, **Lillian M. Gilbreth** (1878–1972) also considered the happiness, comfort, safety, and general well-being of the workers involved. She carried these concerns into her work on household appliances, and according to Edna Yost (*AWS*, 119–20), few such appliances were marketed during the 1930s and '40s without her input and imprimatur:

> Many home-makers today reap the benefit of her ideas here without being aware of it, for she has served as expert consultant for many years for organizations that have made an imprint on homes through more efficient distribution of space in home planning, better placing of sinks, stoves and other equipment . . . , and better design of equipment as well.

Alva Matthews tells how Gilbreth once stopped 4,000 women on the street and measured the distance from their elbows to the ground to determine the correct height for stoves (and sinks) in her most efficient kitchen (44). Whether this impromptu survey was for the Brooklyn Gas Company model kitchen or another project is not clear; but there and elsewhere she would obviously have considered stove improvements. Says Martha Trescott (/Rothschild, 31):

> Gilbreth, Inc. often consulted with firms that either employed mostly female workers (both clerical and factory operatives) or made products that would be used mostly by women. No doubt having a woman at the helm was instrumental for the Gilbreth firm's obtaining these contracts.

Lillian Gilbreth also became concerned with the needs of handicapped workers and homemakers. *Notable American Women* speaks of "her innovative use of the techniques of motion analysis to design special equipment and routines to make housework possible for handicapped persons." Specifics on this important work are missing from *Normal Lives for the Disabled* (1944), which she wrote with Edna Yost, among other sources. As a consultant to New York University's Institute of Rehabilitation Medicine, Gilbreth developed a model kitchen specially adapted to the requirements of the handicapped. This kitchen became an internationally known training center. She would necessarily have modified the stove for safe use by variously handicapped cooks (A. Matthews, 44; *NAW*/mod.; Trescott/Rothschild, 23–37; *WBWRA,* 102; Clymer & Erlich, 1–11; Yost, *AWS,* 99–121; J. Marlow, 218–25)

Less well known, and even more focused on efficiency, was the slightly younger **Christine Frederick** (1883–1970). Moving to Greenlawn, Long Island, with two young children, she created the Applecroft Home Experiment Station in her home. This was a model efficiency kitchen (and laundry) where she tested household equipment, utensils, and products. She helped standardize the height of kitchen sinks and other working surfaces at comfortable levels, something that Gilbreth had also worked on. Otherwise, my sources do not specify her innovations; but as with Gilbreth, in creating her model kitchen she could not have ignored the stove (*NAW*/mod.; Giedion, 526, 614, 620; Torre, ed., 20–2; Handlin, 418–20).

**Mary Pattison,** the New Jersey housewife who wrote *Principles of Domestic Engineering,* was another disciple, like Christine Frederick and the Gilbreths, of pioneer efficiency expert Frederick Winslow Taylor. In Colonia, NJ, she started an agricultural experiment station, where she applied time-and-motion studies to the home (Handlin, 416–8). Pattison's work probably led to suggestions for improved stove design.

One of the unsung design engineers actually employed by industry was **Florence Beebe Anderson** (1910–79) of Chicago and Libertyville, IL. Called "a professional engineer and inventor" in her Chicago *Sun-Times* obituary of April 1979, Mrs. Anderson had been Design and Feature Coordinator for Borg-Warner Corporation's Norge Division and Chief Engineer for Duncan Meter Corporation, among other jobs. Though most of her patents applied to refrigerators, Anderson's obituary credited her with parking meters and "a host of household gadgets and appliances," including a toaster. Some of her unspecified (unpatented?) household appliances may have included stove improvements (obit., Chicago *Sun-Times,* Fri,. Apr. 20, 1979; Tobin, L-4/26/79; Poyner, PC-5/15/83). An alumna of Northwestern University and the Illinois Institute of Technology, Mrs. Anderson was a member of the American Society of Mechanical Engineers.

**Jessie Cartwright,** employed by Norge, was involved in the development of that company's famous Radarange, part of the microwave breakthrough in cooking. Cartwright is profiled in Chapter 4.

Another employed engineer was **Joy C. Adams** of Bridgeport, CT. Her Electric toaster, her two Coffeemakers, and her Fruit juice extractor (1937, 1938) were all assigned to GE, as was her earlier Fuse for an electrically heated device (1935).

A contemporary inventor and engineer, self-employed rather than working for industry is **Frances GABe** of Self-Cleaning House fame. In addition to her self-cleaning fireplace hearth, she has created a cook-stove improvement. She has not yet decided whether to include it in her House—or, as she puts it, to "sponsor" it (L-6/13/85).

Arguably, the greatest need in 20th-century cook-stoves is fuel conservation and a shift to renewable fuels. As Third-World populations soar, even theoretically renewable resources such as wood prove non-renewable in practice. Forests and brush cover are being cut or grazed over faster than they can replace themselves. The wasteland advances.

An obvious solution is to use the sun for cooking as women did before they tamed fire—only better. The world's most famous woman inventor in solar energy is **Dr. Maria Telkes.** One writer called her the Sun Queen (Behrman, 98ff). In the 1950s and '60s Dr. Telkes invented simple solar cookers and ovens that will roast, broil, bake, or broil food without using wood, fossil fuels, or animal dung. (If animal dung could be spread on the fields rather than burnt, crop yields could double.) Her cookers can be adapted to any national or tribal cuisine, are safe enough for a child to handle, and will not scorch foods. Moreover, they free the cook from constant stirring.

Telkes' cookers and ovens are simple to use but sophisticated in design, with triangular and round models, and models with heat storage in addition to heat capture—in short, with all of modern physics behind them. In 1954 the Ford Foundation granted Dr. Telkes $45,000 to develop her solar cooker (Behrman, 98ff; Rau, 76ff; Telkes, "Solar . . . Ovens," 1–10).

Swedish-born **Countess Stella Andrassy,** who holds one of her several patents jointly with Dr. Telkes, is also prominent in the solar energy field. Concern for the energy needs of the world's poor has motivated her lifelong work. As of 1979, Andrassy held ten patents on solar water heaters, stills, stoves, and food driers. She has created easy-to-assemble kits of both her stove and her drier. After years of working alone, Countess Andrassy and her husband, a mechanical engineer, joined the Princeton Solar Engineering Group, to mass-market her fuel-saving devices (Andrassy; Moynehan, 23; *WBWRA*, 750).

*Food storage and preservation*

Women have been innovating in food storage and preservation since they invented pemmican and granaries and clay-lined storage pits thousands of years ago. Doing belated justice to this enormous achievement would fill volumes. Here I can give only highlights, scattered modern examples, and a few statistics. Important modern examples include vacuum and other canning processes and equipment, solar and other food driers, and modern work with refrigeration and frozen foods, vacuum canning, etc.

**Amanda Theodosia Jones** (1835–1914) was the fourth of twelve children in a book-loving family. She grew up to have several lives—as teacher, poet, spiritualist, inventor, and entrepreneur. In 1872, Amanda Jones conceived a vacuum-canning process, which she worked out with the help of Professor LeRoy Cooley of Albany, NY. Five 1873 patents covered this original process; two to Jones alone, one to Jones and Cooley jointly, two to Cooley alone with one of these assigned to Jones. This Jones process, also called the Pure Food Vacuum Preserving Process(es), became the standard canning method in the United States; and although Amanda Jones invented other things, and devoted years to writing and other pursuits, she was always interested in this most important invention of hers. In 1905 she patented two more related inventions and she twice founded companies to exploit her process and apparatus.

Of these, the Woman's Canning and Preserving Company of Chicago (1890) was the more interesting, for it took only women as workers—or stockholders. Although Jones left the company after three years because she lost the support of her fellow officers, the company lasted until 1921.

Other inventions pertinent here are Jones's vacuum process for drying food (1906), and her Ready-Opener Tin Can (Appleton's III: 463; Mainero; *DAB* X; *NAW;* Willard & Livermore, 424; *WWAW*/1914–15: 438)

As might be expected, since most food was still preserved at home during the 19th century, Amanda Jones was only the most noteworthy of many women inventors in canning, preserving, jelly-making, etc. According to Martha Rayne, **Loretta Brownlow** of Illinois patented "a simple and convenient invention for crushing and straining fruit required in making jellies" (115). Pioneer women, of course, had no such apparatus as Jones's or Brownlow's, but improvised by sealing used tin cans with a soldering iron, or invented their own containers. **Mary Hopping** recalled putting up jelly in tomato cans, but "we found a way to break the necks from beer bottles to make glasses for the jam . . ." (Myres, 158).

An isolated British example is **Katharine Jane Dance**'s 1882 application (#5429) for a patent for Storing and treating grain for bread-making.

In less than forty years, between 1857 and 1895, American women patented at least 40 compositions, processes, and items of canning equipment, including fruit jars, funnels, a boiler for heating fruit in jars, an elastic cap for sealing cans and bottles, a jar-filler and -holder, compositions and apparatus for preserving eggs, etc. With one exception, these women appear on LWP (see Appendix A-7).

**Helen Bierer**'s 1896 egg-preserving process, too late for LWP, was an interesting one. It consisted of coating the eggs with a mixture of lard, formic acid, and tincture of benzoin, and packing them in a body of "dry air-excluding substance." Bierer lived in San Francisco.

At the threshold of the new century, also too late for LWP, was **Anna Compton** of Ft. Worth, TX, with her Apparatus for canning fruits and vegetables (1900).

*Drying.* 19th century: At least six 19th-century American women patented food-driers, and **Lydia J. Cadwell** of Chicago seems to have specialized in driers. She received the following patents: Process & apparatus for desiccating substances (1881); Method & apparatus for desiccating eggs, etc. (1884); Drier (1886).

Cadwell also patented a fertilizer-drier in 1886, and held at least seven patents all told. Both she and **Minnie Lloyd** (see below) seem to have used brewer's or distiller's by-products to produce animal food. It would be interesting to know whether these women operated brewing or distilling companies, or came from families that did so.

In any case, Lydia Cadwell looks like a Renaissance woman. Appearing in Chicago directories, 1876–93, she is variously (or simultaneously) a photographer, proprietor of the Lydian Art Gallery and of the Gentile Photo Studies (same address), secretary and then manager of the Mexican Ricolite Company (or Mexican Ricolite & Green Onyx Quarries). These listings suggest that she began as a photographer, worked for or joined forces with another photographer, Charles Gentile (at 103 State Street in 1875), and prospered enough by 1878–9 to open her own gallery. The Lydian Gallery persisted, sometimes expanding along State Street, until 1883 or so. From 1884 to 1888 she appears in the directories with only a residence listing. Perhaps by then she had income from her patents of 1881–6. Her two 1889 patents for "Pavement" may reflect her affiliation with the ricolite/onyx operation, which began about that year according to the directory (cf. ricolettaite).[46]

For purposes of the woman inventor's profile, it is significant that Lydia Cadwell was a "femme sole," and thus probably inventing at least partly to make money—and that she had artistic talent.

To get some idea of the possible significance of Cadwell's and Lloyd's inventions, consider that one of the main by-products of brewing is the spent grains. Rich in protein and fat, these grains have long been used to feed livestock. They can be used wet, as at the end of brewing, but are much more valuable at 7% moisture or less—more stable, easier

---

46. Cf. Caroline W. Romney and her porous-earthenware inventions. The census and trade-journal research required to shed further light on this complex 19th-century entrepreneur is beyond the scope of this volume, but Cadwell will be included in my proposed biographical dictionary of 19th-century American women inventors. Meanwhile, I would be glad to hear from descendants or other relatives, or from other scholars who know about the market for art—and Mexican ricolite—in Chicago in the late 19th century. I am grateful to Kevin Simmons for Chicago directory research.

to store, and cheaper to ship—all especially useful where breweries are far from agricultural areas (*EB*/67, S.V. "Brewing").

The five other food-drier inventors were **Minnie B. Lloyd** of New York, NY, Improvement in malt and grain driers (1874); **Amelia Mey,** of Buffalo, NY, Improved fruit drier (1878); **Mary M. Burchfield** of Kingston, IL, Steam fruit drier (1885); **Mary J. Richmond** of Republic, MO, Fruit drier (1893); and **Jennie P. Duval** of Richmond, MO, Fruit drier (1894).

A related invention was **Anna Brockway**'s Fruit-bleacher of 1883. Brockway lived in Ingal's Crossing, NY.

20th century: Early-20th-century food-preservation inventions listed in the Women's Bureau study under Manufacturing (III.B, Food products and beverage processes and apparatus for making same) are Composition for use in preservation of meat, Compound for preserving eggs, Drying foods (presumably a food-drier or driers), Preserving compound for eggs, and Vegetable-curing plant. As usual, no numbers are available; probably about six inventions are involved at most.

Under Household inventions, kitchen equipment (XII.A), the Women's Bureau study noted Canning and preserving container; Canning and preserving method and equipment; Canning and preserving fruit jar holder; Food drier; and Sealing press. If all categories of this class were equally numerous, some 30 food-preserving inventions would be involved.

Women have sun-dried food for millennia, but the solar food driers of **Stella Andrassy** and **Maria Telkes,** which focus, amplify, and control the sunlight, are a quantum design leap forward from ancient sun-drying, and a quantum leap in energy-saving technology as well. Crop-drying is a further promising use for some of Dr. Telkes' solar-energy inventions. Since 1978 Countess Andrassy's food drier has been mass-marketed in the form of an easy-assembly kit (SWE 1952–76; Behrman, 99; Moynehan 23).

As with so much of women's innovative work with foods, recent systematic work in food-preservation and storage has moved from farm kitchens into the laboratories of government and industry. Two mid-century examples appropriate to close this section are **Ann S. Hunter** of the USDA and **Harriet Louise Burns,** whose connections seem to lie with agribusiness.

In 1955, while working in the Eastern Regional Research Center, Dr. Hunter of Philadelphia patented a process for Dehydration of potatoes by diffusion (Jett, L-5/29/80). Hunter and her co-patentee devised a process for dehydrating cooked mashed potatoes by adding ethanol (alcohol) to form a semi-liquid slurry, placing the slurry in a very thin-walled cellulose-film container, exposing the container to air currents until diffusion through the walls of the container has removed 60% of the water in the slurry, and then recovering the remaining solids.

In 1951 Harriet Burns of Houston, TX, co-patented a process for Treating rice bran and rice polish. This was a heat-treatment designed to stabilize the oils in the grain, retarding the development of fatty acids. Burns and her co-patentee Millard Cassidy assigned their patent to the American Rice-Growers' Cooperative Association of Lake Charles, LA.

*Refrigeration/freezing.* 19th century: Women also made advances in refrigeration and cooling. A noteworthy 19th-century example would be the Refrigerators and cooling room for slaughtering houses shown by **Caroline Westcott Romney** at the Columbian Exposition. This exhibit appeared not in the Woman's Building, but with the United States Exhibit, Dept. H (Manufactures), Group 116, Refrigerators, etc. In the Inventions Room of the Woman's Building Mrs. Romney exhibited several smaller devices exploiting the insulating properties of porous earthenware: the Milk cooler and refrigerator mentioned under dairying inventions, and a Water cooler and refrigerator (Handy, 9, 15, 21, 28, 41, 42).

The other milk-coolers mentioned as dairying inventions, as well as the ice-cream freezers classed above as food-processing machines, could also be listed here.

In the 1860s **Harriet Randal** of Powersville, MO, and Pottersburg, KS, invented a provision safe used as a refrigerator (*WI*/1), and in 1868 **Julia W.D. Patten** of New York City patented an improved ice-preserver. A few other patented 19th-Century women's inventions in cooling and refrigeration (from LWP) are[47] **Margaret L. Brisbane** of New York, NY, Improvement in dishes for refrigerating and preserving food (1879); **Nancy A. Lamon** of Bertram, TX, Refrigerator (1890); **Mary M. Harris** of Chicago, IL, Refrigerator (1893); **Julia L. Fitzgerald** of Terra Alta, WV, Refrigerator, (1893); and **Louisa B. Saddine** of Hot Springs, AR, Water-cooler (1893).

Among this latter group **Mary Harris** is noteworthy. Her refrigerator was quite popular at the Columbian Exposition in 1892–3, winning 600 orders from fairgoers (Weimann, *Fair,* 432).

20th century: Too early for the Women's Bureau Study is the Hand refrigerator and lunch box invented by a **Miss Phillips** of Dorchester, MA. This invention was considered important enough to be mentioned in the *Scientific American* in 1903 ("Women Inventors," 247).

An ill-starred figure[48] in the history of American invention was **Mabel Hunt Slater** (1864–1942), daughter of the famous artist William Morris Hunt and millionaire widow of Horatio Nelson Slater II. As early as 1904 she patented a Refrigerator (ice-cooled); and in 1914, only one year after the first functional domestic refrigerator, the Domelre, was manufactured in Chicago, she patented an automatic ice gauge for refrigerators. A leader in New York and Boston society, with a country place at Bar Harbor, Mrs. Slater got at least a dozen other patents between 1915 and 1926, the profits going to charity. She also had a significant (unpatented?) invention, mentioned in her *New York Times* obituary: a sleeping bag that doubled as a garment. This was used in the trenches in World War I.[49] She made her home for many years in Webster, MA "Presidents, Actors . . .," 57; *WABI,* 154).

WB/28 has no separate category for refrigeration and cooling. However, the study shows a few scattered examples under Dairy and Kitchen equipment, possibly as many as 20 inventions.

---

47. Not intended for food-preservation, but technologically related, is the photographer's refrigerator patented by **Mary A. Thornton** of Perrysburg, OH, in 1871.

48. Her wealth brought great troubles. In 1914 her brother sued her in a money dispute; in 1917 an anarchist tried to shoot her because of her wealth; and in 1930, after long wranglings over a fortune of $25,000,000, her children had her committed to an asylum ("Roosevelt's Aid . . ."). There, despite repeated attempts by her brother, her sister, and friends to free her—attempts that included pleas to then-Governor Roosevelt of New York and eventually reached the Supreme Court—she died in 1942. The wrangling did not even stop with her commitment; see the many *New York Times* stories on these battles (Mar. 6, 1914:14; Nov. 11, 1917:16; Nov. 12, 1917:9; May 22, 1923:21:6; Nov. 4, 1930:28:5; Jan 27, 1931:26:2; June 21, 1931:6:2; July 28, 1932:9:5; Sept, 27, 1932:2:1; Nov. 18, 1932:14:4; Nov. 22, 1932:14:6; Mar. 3, 1933:36:3; May 28, 1933:16:1), and Mrs. Slater's obituary (Nov. 28, 1942).

49. This may be the same invention unhelpfully described as a "military blanket" in a 1928 article on famous inventors ("Presidents, Actors . . ."). Note that Mrs. Slater's husband, Horatio Nelson Slater II (d. 1902), was the grandson of two inventors, Samuel and **Hannah Wilkinson Slater,** and the son of a third. He continued the profitable and highly reputed Slater textile industry, which made, among other things, Army blankets (Mabel Slater, *NYT* Obit., Nov. 28, 1942).

In 1936, **Martha Kohler** of Landsberg Warthe, Germany, patented a Beer-cooling apparatus. The fact that Kohler troubled to get a United States patent suggests some significant commercial potential for it.

**Dorothy Rodgers** (1909– ), well known as the wife of composer Richard Rodgers, and as an author and businesswoman in her own right, is also a multiple inventor. In the late 1930s she conceived an individual people-cooler, consisting of a bracelet filled with dry ice (for men, a watch bracelet). This invention was never manufactured because she abandoned it when her chemist collaborator died. A more significant invention, the one placing her in this chapter, was her device to rescue children trapped in refrigerators. Though this invention was produced and found to cost less than a dollar, no refrigerator manufacturer would touch it (Heyn, *Fire,* 47).

**Esther D. Bookman** of New York City patented a Refrigerating and air-conditioning apparatus in 1948. This is probably for space-cooling, but the technology may be cross-applied.

A Polish inventor, **Irena Malinowska** of Warsaw, in the late 1960s attacked the problem of bacterial infestation in refrigerator systems. Her U.S. Patent (1969) for a Method of disinfecting refrigerator rooms allows these spaces to be disinfected without warming, for the freezing point of her disinfectant is below −20°C.

British dairywoman **Violet Dawson**'s important new milk-cooler of the 1950s has already been mentioned. Two other noteworthy mid-century women are **Jessie Cartwright** and **Florence Beebe Anderson** (1910–79). Both worked in-house for large corporations. Cartwright's best-known achievements lay in the field of laundry machines, and will be discussed in Chapter 4. However, according to her favorite granddaughter, Whitney Cartwright Brown (PC-4/21/83), she had "something to do with the automatic ice-maker in Norge refrigerators."[50] This cryptic umbrella may cover several noteworthy inventions over a period of years.

Florence Anderson seems to have concentrated on refrigeration. Her semi-automatic ice-cube maker (1960) has already been mentioned. It would be interesting to know whether this was the first such device. Anderson was Design and Feature Coordinator for Norge, and Design Engineer for the Deepfreeze Appliance Division of Motor Products, and she held at least the following other refrigeration patents: Refrigerating system, incl. defroster drainage receptacle and support (1954); Evaporator door assembly (1955); Breaker strip and control assembly (1955); Refrigerator evaporator and shelf structure (1956); Liquid dispenser for refrigerator cabinets (1956); Evaporator door and hinge assembly (1956); Refrigerator cabinet load-supporting shelf and mounting therefor (1956).

Some of the more notable 20th-century achievements in refrigeration were the work of **Mary Engle Pennington** (1872–1952). Unlike Cartwright and Anderson, Pennington was concerned less with the "hardware" of refrigeration than with its theory and technique as applied to an entire national system of food distribution, and particularly to frozen foods. Nevertheless, she had specific technical inventions to her credit. Dr. Pennington improved the design and construction of both household and commercial refrigerators, and of cold-storage units. She also had a hand in the complex technology that brings a vast variety of frozen foods safely to the supermarket, including determining the best processes

---

50. Despite persistent attempts, neither Whitney Brown nor I have been able to get specifics on most of Jessie Cartwright's in-house inventions. I would be grateful for information from former colleagues, researchers into the history of household technology, business historians, or anyone knowledgeable on this subject.

and temperature progressions for each food, and which variety of pea or strawberry will taste best when frozen.

Although Clarence Birdseye usually receives credit for "industrial freezing" of food about 1924, he had several predecessors, of whom Mary Pennington was one. As early as 1905 unnamed persons "on the East Coast" of the United States were using a cold-pack process (freezing packaged foods in brine cooled to 2°F.) for strawberries, ice cream, etc. Pennington may or may not have been involved here, but as early as 1913 she was experimenting with freezing sweet-corn kernels in Minnesota (*WABI,* 78). Perhaps her most important achievement here was solving the technical problem of humidity control during freezing, so that foods do not dry out as they freeze,[51] yet do not stay so moist that they mold.

So strongly was Pennington identified with frozen foods that her 1941 *New Yorker* profile was titled "Ice Woman." From her early interest in wholesome dairy foods, throughout her public-health career, she studied refrigeration and freezing systems. For the egg industry, aside from her procedures for frozen and powdered eggs, she developed cooling and storage procedures for whole eggs (Yost, *AWS,* 94).

For the fish industry, Dr. Pennington developed a process for scaling, skinning, quick-freezing, and dry-packing fish fillets immediately after the catch arrives. Like most of the processes she developed, this one became standard in the industry.

During World War I, Dr. Pennington advised the Government on railroad refrigerator cars, riding thousands of miles in cabooses—and ice cars—testing them in operation. Only 3,000 of the 40,000 cars passed her inspection. Using her work, officials laid down principles of insulation and ventilation so sound that they changed little between that time and the advent of mechanical cooling about 1950.

After the Armistice, Dr. Pennington entered industrial research at twice her federal salary, then started a highly successful consulting firm. She was the first woman member of the American Society of Refrigerating Engineers and one of the first dozen women to join the American Chemical Society. In 1940 she received the Garvan Medal. At the time of her death she was vice-president of the American Institute of Refrigeration.

At her own table, Dr. Pennington served mostly frozen foods. Not even Lucullus with all his lavish banquets, she liked to point out, could set a better table than the average American worker, who could have strawberries from Ohio, celery from Florida, broilers from Kansas, eggs from the Mississippi basin, beef from Arizona, and fish from the Pacific Ocean off Vancouver—all thanks to refrigeration. Indeed, she might have been speaking for herself when she quoted from Alexander Pope to express the aim of refrigeration: "To take to the poles the products of the sun, / And knit the unsocial climates into one." In less flowery terms, it is this dedicated Quaker scientist and inventor, as much as any other one person, who brought us raspberries in December (Heggie; *NAW/* mod.; Yost, *AWS,* 81–98).

*New foods*

Women have been creating new foods since they invented cooking—and long before, as they brought new plants and other items into their edible repertoire, and learned new ways to detoxify and prepare them. In fact, from prehistory till modern times, in all but the most

---

51. Sometime, of course, drying is the main objective of freezing. Freeze-drying is an ancient method of food-preservation invented independently, and probably by women, both in the Arctic and in the Andean highlands. In the Andes it is used for potatoes. Women have also used freezing to dry laundry in the winter (Hopson, 91).

rarefied circles of all but the most urbanized and highly stratified cultures, the list of new foods invented by women and the list of a society's new foods would be the same list.

An example from prehistory is cassareep sauce made by Guiana Indian women from the poisonous juices of the cassava tuber, boiled down to a syrupy consistency. According to Otis Mason, this is the origin of the "far-famed pepper pot" of the West Indies (*Share*, 38, 39).

In addition to new individual foods or dishes, women have also invented whole new types or categories of foods for special needs and purposes. Infant formulas, trail or travel foods, warpath or survival foods, and weaning foods are only the most obvious examples from prehistory.

With the professionalization of cooking evident at least as early as the royal households of Greece and Rome, men began to enter the field. By the mid-17th century in France, for instance, the chefs of the royal household were male. From that time onward in French cuisine, in fact, most of the famous names have been male. But women continued to do or supervise most of the cooking at all but those topmost levels, and naturally continued to innovate. Béchamel sauce, for example, is usually attributed to one Louis de Béchamel, Marquis and steward in the household of Louis XIV—or to one of the chefs of Louis' kitchens. According to Martha Rayne, however (252), it was actually invented by the **Marquise de Béchamel.** If the lady did indeed conceive the exquisitely simple formula of equal parts cream and velvet-down sauce, table historians would no longer have to wonder, as E.S. Dallas does, how such an arrant fool as her husband could have devised what had eluded far more learned men and far more cunning cooks than he—an invention, that has, every day since, "contributed more to the peace and satisfaction of mankind than any other relish in the range of cookery."[52]

Important modern examples of women's new-food inventions, categorical and individual, are cakes as we now know them, English mustard, and several of the world's fine cheeses. The modern cake was invented by a Catholic nun named **Marie Mofort** (Herter & Herter, II: 245). No date is given, but the first recorded use of the word *cake* in this sense in English is about 1420 (*OED*).

Before 1729 mustard consisted of mustard seeds either whole or coarsely pounded and boiled in vinegar (or *must*). Then "an old lady," **Mrs. Clements** of Durham, invented the dry English mustard. Grinding the seeds in a mill, she sold the resulting fine powder. Says E.S. Dallas, "She kept her secret and made a little fortune out of it, trotting about from town to town on a packhorse for orders, and contriving to secure the patronage of George I" (Kettner's, 308). Strictly speaking, since it was no longer sold in vinegar, it was no longer mustard, but the name stayed.

**Mrs. Paulet** and **Marie Harel,** inventors respectively of Stilton and Camembert, already celebrated above, are far from isolated examples.

Pennsylvania Dutch Pot or Cup cheeses, for instance, were invented and are still made by Amish women of the Northeastern United States. **Mrs. Marius Boel,** working on her family's farm, deserves much of the credit for the high-quality Danish blue cheese called Flora Danica. And in England as of the 1970s, **Dr. Helen Chapman** of the National Institute for Research in Dairying at Reading was one of the outstanding researchers on new cheeses.

In France, more than anywhere else, women have received credit for their cheesemaking inventions and innovations. Normandy has been particularly rich in this area of female

---

52. After repeating an anecdote that shows Béchamel's vanity and foolishness, Dallas calls him "the most fortunate of ninnies and the most superfluous of puppies" (Kettner's, 60–1).

achievement. In the 1920s **Emma Hutin Roustang** went to Normandy where, with technical aid from Professor Massey of the Pasteur Institute, she created the first pasteurized Gervais cheese. At 24 she was the production director of the Gervais Company, supervising 600 workers. Her mother-in-law had earlier introduced Brie into the Muese Valley, the major production area for this cheese. **Mme Charles Pommel** originated Petit Suisse cheese; **Mme Veuve Le Petit** developed Le Petit cheese; and **Mme Veuve Buguet** of Chambois created a soft cheese. **Mme Depaudry** of Trouville made important contributions to Pont l'Evêque.

Cheese historian Dr. Frank Kosikowski of Cornell considers that women specialized—and excelled—in the softer paste cheeses made in the kitchen, as opposed to the harder and heavier cheeses made at the mountain pasture sites by the men. A Gruyère cheese, for example, traditionally weighs 125 pounds (Ehle, 145). In addition to Camembert and the cheeses just mentioned, then, these women's cheeses could thus include Brie, Coulommiers, Carré de l'Est, and chèvre or ripened goat's cheeses (Kosikowski, *Cheese*, 8), among others.

19th century: Nineteenth-century American pioneer women invented their own trail or travel foods in response to the necessities of the Western migration. One young woman cooked her leftover supper beans down into a "mush"—presumably resembling refried beans—and formed them into a roll. Then when her family stopped at noon the next day, she had only to cut slices from her bean roll and put them between leftover biscuits to create "a nutritious and tasty lunch" without having to build a fire (Myres, L-3/15 and 4/80, and *Westering*, 123).[53] By my definition this clever woman—**Mary Hopping** according to pioneer narratives—had invented a vegetarian luncheon loaf and an early convenience food, to say nothing of a new kind of sandwich. The sandwich itself (or at least the word; see *OED*) was only about a century old, and more British than American.

Faced with trail-like conditions even after reaching their new homes, the women became wizards at "make-do," their self-deprecatory term for *invention.* They learned to bake with "shorts" (a milling by-product containing wheat germ, bran, and a bit of flour) instead of flour, created "coffee" from browned grain, made "vinegar" pies, refined sugar out of watermelon syrup, and baked cakes without butter or eggs (Myres, *Westering*, 148). The woman who created vinegar pie was an inventor, indeed, but we do not know her name.

Also anonymous is the St. Louis World's Fairgoer who invented the ice cream cone in 1904, thus starting an enormous American industry. Annoyingly, we have her escort's name (Charles E. Menches),[54] but not hers. In one version of the story, Menches precipitated the invention by offering her a bouquet and an ice-cream sandwich at the same time. Equal to the challenge, this inventive young woman rolled one of the waffle-like sandwich layers into a container for the flowers and the other into a container for the ice cream. (De Bono, 107; McDowell & Umlauf, 372).

---

53. I am grateful to Sandra J. Myres, Professor of History at the University of Texas–Arlington, for alerting me to these inventions even before her fine book *Westering Women* appeared. On the bean loaf, see Hopping, 5.

54. In some versions of the story, Menches was an ice-cream vendor at the Fair. The other known claimant to this invention was Abe Doumar of Jersey City, NJ, a souvenir vendor at the same Fair. Hearing an ice-cream vendor complain about running out of dishes mid-day, Doumar allegedly suggested he make a cornucopia out of a waffle, wrap it around his ice cream and sell it for twice as much ($0.10!; Wallechinsky & Wallace, 912–3). I would be grateful to hear from descendants of Menches, his inventor-companion, or witnesses to this event, or from anyone else with information leading to the name of this young woman.

20th century: WB/28 has no New-foods category, but lists recently patented Food products and beverage processes (and apparatus) as Category B under Manufacturing and related inventions (22). Pertinent-looking types of inventions are as follows:

| | |
|---|---|
| Beverage-producing material | Health Food |
| Composition for use in making piecrust | Macaroni or like food product |
| Dry shortening flour | Nut butter |
| Food product, and method of making same | Pie-filling composition |

Assuming equal numbers for the 26 types of invention in the 40-item category, these 8 types could involve as many as 12 inventions.

An outstanding inventor of new foods from the period just after the Women's Bureau Study is **Anna Eugenia E. Schneider** of Baltimore and Catonsville, MD, mentioned above.

In addition to her mill and Eugenia Whole Wheat Flour, Anna Schneider created several important new foods during her long career. Her campaign to provide whole-grain nutrition to the American public coincided with an anti-white-bread campaign mounted by Dr. Harvey Wiley, former USDA Chief Chemist, sometimes called the father of our Pure Food and Drug Act. One of the doctors and scientists supporting her work urged her to contact the planners of Byrd's 1939 expedition to the South Pole. Dr. Paul Siple, the supply officer, asked her to develop a trail biscuit for the expedition. She originally suggested encapsulating a spoonful of honey in the middle of each biscuit; when this idea was rejected, she simply incorporated as much honey as possible into her final formula. The requirements were exacting: the biscuits must taste good, and must keep exceedingly well, withstanding tropical heat on the way to the Antarctic, and sub-zero temperatures after arrival. After four different labs tested and approved the Eugenia biscuits, she produced several hundred of them, which traveled to the Antarctic on the expedition's two ships. The men found the biscuits both tasty and nourishing. Their keeping qualities were certainly superb, as some tins were opened after 25 years and found still "delicious."

During World War II Anna Schneider created "Eugenia's Eskimo Biscuits" as a military survival food, modifying her original product to include still more honey. The American Friends Service Committee and other organizations shipped her biscuits to the hungry in Europe after the war. At home, Miss Schneider marketed her survival biscuits to the general public as "Eugenia Honey Biscuits."

As Marianne Alexander observed in her *Notable Maryland Women* biography, Miss Schneider's career represents "more than a business success. . . . She made a product of superior nutritional value available, not only to private citizens, but to the men serving the nation on the Byrd Expedition and during World War II. She must also be credited with increasing the public's awareness of the nutritional value of whole wheat in the diet" (Helmes, 334).

Other notable American food inventors or innovators are **Rose Markward Knox** (1857–1950), who transformed gelatin from an invalid food into a desirable dessert, **Ruth Wakefield,** who invented Toll House or chocolate-chip cookies, and **Debra J. Fields** (1957– ), who has created a top-selling line of chocolate chip and other cookies, brownies, and beverages.

Born in Mansfield, OH, third daughter of a druggist, Rose Markward worked before her marriage in the glove industry. In 1890 Rose and her husband Charles Knox invested their savings of $5,000 in a business to introduce gelatin desserts. She not only prepared and tested all the original recipes, producing a booklet titled *Dainty Desserts* (1896), but accompanied her husband on selling trips. When Charles died in 1908, Rose Knox sold off the more diversified branches of the business, creating an operation with 60% of its sales in food (surpassing all competitors in unflavored gelatin) and 40% in industry, photography, and medicine. By 1915 the company was incorporated for $300,000; by 1925 it was listed at $1,000,000. From the first she determined to run her business in "a woman's way." She directed her advertising toward women, founded an experimental kitchen, and established fellowships at the Mellon Institute to research new uses for gelatin. As early as 1913 she gave her workers a five-day week (although expecting as much work as for five and a half); closed the back door and announced that everyone would enter by the front; gave two weeks' vacation and time off for illness. Though employees allegedly feared to cross her, as of 1935, 85% of them had stayed with the firm for twenty-years.

Long a member of the American Grocery Manufacturers' Association, she became in 1929 the first woman elected to its board of directors. This canny entrepreneur and inventor of gelatin desserts continued to go to her plant every morning at 9:30 until, at 88, arthritis forced her to work at home. Even then, she continued as president until age 90, when her son succeeded her, and she moved to chair of the board. She died just short of her ninety-third birthday (*NAW; WBWRA*).

**Ruth Wakefield** created the most commercially successful cookie of its day, and one of the most successful of all time, in the 1930s. Soon after she and her husband converted the old Whitman, MA, Toll House into a restaurant and inn, Mrs. Wakefield was trying to invent a new chocolate cookie. When she cut a bar of Nestlé semi-sweet chocolate into the batter of an old American butter drop cookie, she succeeded beyond her finest fantasies— inventing not a chocolate cookie, for the chocolate did not diffuse throughout the batter, but the original Toll House cookie, and, thus, chocolate-chip cookies.

In fact, she can be said to have invented the chocolate chip, whose name reflects the manner of its invention rather than its present commercial form as a chocolate *drop*. Nestlé printed Wakefield's recipe on the package when they marketed semi-sweet chocolate chips or "bits." *Mrs. Wakefield's Original Toll House Cookbook* went through 26 printings. Ruth Wakefield is now dead, and the Toll House Inn destroyed by fire, but Toll House cookies may live forever (Amariani; "Fire Destroys . . ."; Strouse).

The latest contender for Ruth Wakefield's title is **Debra ("Debbi") Fields** of Palo Alto, CA, who invented Mrs. Fields' chocolate-chip cookies. Her experiments with chocolate-chip cookies began at age 13 with Ruth Wakefield's original Toll House recipe and continued till so many people raved about her unique soft and chewy, super-chocolatey cookies that she decided to go into business.

Married to a chocoholic, and mother of two children, Debbi Fields is known for treating employees "with great tenderness and respect." Free-lance writer Alan Furst, after experiencing her kindness, gave her an unusual accolade for the proprietor of a $30-million-a-year business: She is "a lady who would, if she could, tiptoe into the world's room at night and throw a giant blanket over it" (328). Mrs. Fields and her cookies have been featured in *People* magazine as well as *Fortune,* and she appears on the cover of the charter issue of *Utah* magazine ("Cashing in"; "Debbi Fields"; Furst; Richman, "A Tale"; "Sweet Success"; "A Visit to the Kitchen").

In less than a decade she expanded from her original tiny Palo Alto shop to a multi-million-dollar, international food empire with corporate headquarters in Park City, UT. Mrs. Fields has created several other new cookies in addition to her original semi-sweet-chocolate-chip cookie—including macadamia-nut-chocolate-chip, white-chocolate-chunk, and milk-chocolate-chip cookies—as well as double-fudge brownies with chocolate chips, a new chocolate-macadamia-nut soda, and a line of giant muffins. Fanatical about quality, she uses real chocolate, pure vanilla, butter, and no preservatives and has been known to walk into a store and throw 200 less-than-perfect cookies into the garbage. She refuses to franchise, and recently withdrew her cookies from a profitable outlet in the Stanford University Bookstore because she could not guarantee *warm* cookies there.

Of unknown significance is the Beverage composition and process for producing same patented in 1955 by **Rose A. Weiss** of Miami Beach, FL. This new cola contains such exotic ingredients as chestnutleaf extract, oil cassia, and oil myristica, in addition to the kola-nut extract, phosphoric acid, and sucrose.

Potentially quite important in a protein-starved world are such new protein foods as those made from soybeans. During the 1950s and '60s a talented Minneapolis chemist named **Twila Paulsen** (1914– ) worked for the Archer Daniels Midland Co. Described by Roy Erickson of Archer Daniels as "a gifted research chemist," Paulsen received three patents for a soy flour and other new items made from soybeans, one in 1963 and two in 1968. Her most important research was on soy proteins, but she published nearly a dozen papers on various topics in agri-science. Dr. Paulsen's first job—in Depression days—was testing milk for a dairy; she lived mainly on the milk samples. Now retired from Archer Daniels, Twila Paulsen Okken lives in San Bruno, CA (Elizabeth Reed, PC-6/25/88, L-8/8/88; Berg, PC-6/25/88; Erickson to Reed, L-8/88; Paulsen, L&Q-8/30/1988).

Similarly important are the "Protein isolates with improved properties prepared from sunflower seeds" patented by **Dr. Antoinette A. Betschart** in 1978. As of 1980 Dr. Betschart was a Research Leader at the USDA's Western Regional Research Center (Jett, L-5/29/80). Several other recent women inventors of new foods and related technology or components have worked for the USDA. **Faye C. Barnes** of the Northeast Regional Research Center is the co-inventor of a Preparation of baked goods having sourdough flavor without a sourdough starter (1974). **Susan B. James,** also at the Northeast Center, is co-patentee on a 1974 patent for Nutritional iron protein complexes from waste effluents.

At the Western Regional Research Center in Berkeley, CA, **Maura M. Bean** has patented Novel methods for improving the baking properties of cake flour without using chlorine (1979). As of 1980 Mrs. Bean was a Research Food Technologist. Three other Berkeley scientists have patented new foods. **Mary L. Belote** received a 1968 patent for a dough composed mainly of mashed potatoes that can be formed into a sheet, cut into desired shapes, and baked. **Catherine J. Dunlap** is the co-patentee of Novel food products prepared from legume seeds (1978). Patent Advisor **Margaret Connor** is co-inventor of a technique for extracting protein concentrates from grain millfeeds (1975).

*Diet foods.* An important new category of foods for the late 20th century with its mania for fitness and trim figures, at least in the industrialized West, is diet foods. Women, of course, have been in charge of invalid and convalescent diets for millennia, not only as healers, but as wives and mothers, and have innovated in the area in myriad unsung ways. In this century, when diet (low-calorie) and dietetic foods have both gone commercial, at least two important innovators have emerged: **Myrtle ("Tillie") Ehrlich Lewis** and **Jean Nidetch.**

Tillie Lewis (1901–77) is a Horatia Alger from Brooklyn. At 12 she was folding kimonos in a garment factory; at 50 she owned the fifth largest canning factory in the United States. By 1973 she was one of the ten top businesswomen in the country. She was known in the business world as the Tomato Queen—for introducing pomodoro tomatoes to this country when botanists and growers both told her it couldn't be done, and making a fortune from canning and retailing them.

Her importance for this book is twofold: When she became sole owner of Flotill Industries in 1937, she expanded, building new canning plants for asparagus, spinach, and other vegetables; as well as meat products; establishing more warehouses, and setting up an experimental laboratory staffed by chemists and agricultural experts. In all these plants she gave her workers high wages and unheard-of benefits: day-care centers, incentive plans, transportation for workers who lived far from the factory, and rest periods for elderly workers. In 1940, after inspecting conditions at Flotill, the American Federation of Labor made an exception for Tillie Lewis's plants, allowing workers to return to work there in the midst of a California-wide cannery strike.

An innovator in products as well as in employee benefits, Tillie Lewis introduced her Tasti-Diet line of foods in 1952. This "enormously successful" new category of food made her, as Joan Marlow puts it, a "pioneer in the production of artificially sweetened foods" (32; 316–21). Although she did not invent such foods, she probably deserves credit for standardizing them and making them commercially available for the first time.

**Jean Slutsky Nidetch** (1923– ) of Weight Watchers deserves credit for devising, in the 1960s, the first truly successful technique or program for losing weight and keeping it off in modern sedentary society.

Now Weight Watchers is Weight Watchers International, Inc., a multi-million-dollar business that has expanded to include a magazine, summer camps, and, most relevant here, diet foods. These tasty, nourishing, and calorie-counted foods include dozens of different frozen low-calorie meals or main dishes, reduced-calorie breads, mayonnaise, cottage cheese, desserts, and many other products, and are generally calculated to make dieting and continued weight control easy and pleasant. Weight Watchers has both a Director of Nutrition and a Manager of Nutrition Services (both women in 1985), and Hayden Davis of WWI says Nidetch's "background" is such that "she was not in a position to play a major role in the formulation of commercial food products" (L-6/21/85; translated, she didn't know enough chemistry). However, Jean Nidetch must have created many of the early recipes. She is, in any case, the author of several successful cookbooks.

Uncounted thousands of other women have probably innovated in diet foods, responsible as they have usually been for the nutrition of most families. Nathan Pritikin, for instance, credits at least three women for "the menu plans and many of the recipes" for the famed Pritikin Diet. These three are **Esther Taylor,** the "talented head cook" at the California Pritikin Center in its early days, and "two able supervisors": **Jana Trent** and **Ilene Pritikin.** The unnamed dietician on the team may also have been female (Pritikin, 75). This story could probably be repeated many times over, no matter whose name appears on the book covers.

One final commercial example is the diet pizza dough created by Cornell University scientist **Sally O'Brien** and colleague in 1982. The new product tackles two American problems: surplus weight and surplus egg whites, for the new dough is 85% egg whites. The conventional pizza dough, by contrast, consists mostly of flour, and the average 12-inch pizza contains about 900 calories. The egg-white crust has only about 200 calories. O'Brien, a research specialist with the Poultry and Avian Science Department at Cornell, worked with a professor of food science to create the diet crust. Frozen versions were scheduled for market by 1983 ("Diet Pizzas . . . ," 16).

Another important subcategory of new foods for the 20th century is *convenience foods.* Women have been creating such foods for a very long time—**Mary Hopping**'s bean loaf is a relatively recent example—but convenience foods have become a major industry today. Of literally thousands of possible examples, I will cite only scattered instances, among them the work of two USDA scientists from the Western Regional Research Center in

Berkeley, CA. **Lorraine R. Fletcher** has patented pre-cooked poultry products pre-coated with a tightly adherent batter (Pat. #3,169,069); and Research Chemist **Dulcie M. Hahn** is co-inventor of a process for making beans quick-cooking (#3,869,556; Jett, L-5/29/80).

A natural concern about convenience foods is their nutritional value. In quick-cooking rice, for instance, processes used to make the rice absorb water more quickly also affect its nutritional quality. **Jeanne Cox** of Coquitlam, British Columbia, and her husband have devised a Quick-cooking rice process that both minimizes the loss of nutrients from the rice and adds calcium, phosphorus, and citrate to it (U.S. Pat., 1975). This is a boiling or steaming process in which the water contains the nutritional additives desired. According to David Sires of the Cox Group of Lynden, WA (a professional society for the inventive arts), Jeanne Cox has invented "several other processes." Regrettably, I have no details on these.

## Nutrition

A major concern of mothers and other women inventors for millennia has been nutrition. Two noteworthy modern inventors in this area are **Julia Halberg** and **Dr. Kamala Sohonie** (c. 1912– ) of Bombay. Halberg, recognized at 16 in 1971 as one of "Tomorrow's Scientists and Engineers," went on to study the circadian rhythms in plants and animals in the Department of Pathology and Laboratory Medicine at the University of Minnesota. Awakened to world hunger by travels in India, she seeks to correlate daily rhythms and food intake in such a way that people get maximum benefit from what they do eat. Doctors—and women healers before them—have long known that the human body absorbs different amounts of the same drug, depending on the dosage time. Halberg has applied this idea to the ingestion of food—what might be called a Rhythm Method of nutrition (*WBWRA*, 34).

Dr. Sohonie graduated from Bombay University in 1933 with a BS in chemistry and physics. Encouraged toward research by her father, she became, after much pleading, the first woman admitted to the Indian Institute of Science at Bangalore. There, and at Cambridge, she did important basic research in food and plant chemistry. She isolated and purified for the first time an elusive potato enzyme, and worked out the oxidation-reduction processes in plant tissues, isolating the enzyme systems associated with plant cytochrome *c,* and demonstrating that the processes are the same in plants as in animals.

After 1942 Dr. Sohonie first studied nutrition at the Nutrition Research Laboratory in Coonoor, where she worked on developing enzymes to release vitamins in foodstuffs, studied the destruction of B vitamins in foods, and also studied anti-vitamin factors. Her later research in the Biochemistry Department of the Institute of Science in Bombay, on the trypsin inhibitors and hemagglutinins of popular Indian legumes, had important nutritional implications. However, the major work that places her in this chapter is her experiments with enriching the Indian diet with finely ground *dhanata* (paddy flour usually lost in the milling, but containing valuable vitamins, minerals, proteins, and fat) and with *neera,* or palm juice. This national drink of India had never been studied before Dr. Sohonie took it up in 1947. Her demonstration that *neera* contains sizable amounts of A and C vitamins and iron, and that the vitamin C can survive the concentration of *neera* into palm jaggery and molasses, paved the way for using the latter as cheap dietary supplements.

In 1949 Dr. Sohonie became Professor and Head of the Biochemistry Department at the Bombay Institute, staying there until her retirement in 1969 and serving for the last five years also as Director of the Institute (Richter, 18–23).

*Agricultural machines*

The stereotype enshrined in the phrase "the farmer and his wife" is so tenacious that women's 19th-century achievements in agricultural machinery may be surprising. Even the Patent Office clerks who complied an official list of women's patents evidently found agricultural machines an unexpected category, for they classed Nell Westburg's compli-cated-looking Band-cutter and feeder for a threshing machine (1893) not under Agricul-tural Implements, but under Sewing and Spinning!

Three of the earliest known women patentees in this area were English. As already mentioned, **Frances Willoughby** patented a grain-thrashing invention in 1722. In 1811 **Ann Hazeldine** received British Patent #3422 for a Plough for cultivation of land. On July 7, 1842, **Lady Ann Vavasour** received #9405 for Machinery for tilling land (Woodcroft). She was a member of the Anglo-Irish gentry who wrote a travel book to raise money for an agricultural school for the Irish poor. Here Lady Vavasour describes her tiller and how she came to invent it, saying she made the first model, blistering her hands in the process (Brody, L-3/28/85, and "Patterns," 6–7; Vavasour, 452–5). Since the finished machine would be large and expensive, she intended for each parish to buy just one and rent it out to individual farmers.

She had often observed, she said, how effectively Italian vineyard workers broke up the soil around the vines with their long, bent forks, and wished that a cultivating implement could be made to perform that same action—and to reduce the hard labor of ploughing. "After much thinking," she made her first model in the woodworking town of Rippoldsau, and took it to a carpenter for completion. The carpenter was eager to know what it would be used for. When further changes were needed, she enlisted "a clever engineer," saying, "I hope I now have an efficient instrument, that will dig and pulverize 6 acres at least of land a day," to a depth of ten inches with two or at most three horses, and "in a much better manner than any plough could perform it." Noting that the plough leaves hardpan only a few inches below the surface, she hopes "My implement will leave [the soil] open and broken up, so that the rain can sink in," and on the other hand the soil will stay moist longer in dry weather, "for early in the season, the roots will not have gone so deep as to get into the wet soil" (453–4).

Unfortunately, the prototype manufactured by Cottam & Hallen failed in its trial at the 1842 Agricultural show at Bristol. According to a report in the *Journal of the Royal Agri-cultural Society,* whether the "enormous machine," was used before or after the plough, and even in favorable soil, "the earth adhered to the spoon-shaped teeth of the barrel, closed the spaces between them, and accumulated until the machine became an im-mensely heavy roller, consolidating the ground, and completely reversing the intentions of her ladyship" (vol. 3, 1842; 345).[55] Whether this failure was due to Lady Vavasour's design, to Cottam & Hallen's interpretation, or to saturated soil, we cannot be sure; but her initiative in seeing a problem, designing a machine to solve it, creating a model with her own hands, and convincing a machinery firm to manufacture it is remarkable. Lady Vavasour herself was rather remarkable, not least for her lively thought and writing style.

According to Charlotte Smith, a **Mrs. L.E. Whitten** of Mountain Hill, GA, invented a

---

55. I am grateful to Judit Brody of the Science Museum Library, who called my attention to Vavasour's book, and to Nigel Harvey, librarian of the Royal Agricultural Society, who supplied information on the fate of her invention. The tiller was not an isolated instance of Lady Vavasour's interest in technology. On the same journey, she records in detail the construction and insulation of an icehouse, and a German method for hardening and fireproofing the wood of a tree before it is cut down (450-2).

machine for threshing and cleaning peas and garden seed (*WW,* Jan. 5, 1891: 4). A search of the U.S. Patent Office name lists from 1885 through 1894 reveals no such name. However, since Charlotte Smith had inside information on inventions (possibly through her son, a Patent Office clerk in 1887, 1891, and 1892), Mrs. Whitten's invention clearly existed, at least as a patent application, by 1891. Not clear is whether the application was denied, dropped for lack of funds, or completed by someone else. Omissions in the name lists are rare.

According to Mozans (352), "The earliest development of the mower and reaper, as well as the clover cleaner, belongs to Mrs. A.H. Manning of Plainfield, New Jersey." Mozans refers to **Ann B. Harned Manning** (c. 1790–1870s). Her husband patented the clover cleaner, Mozans continues, but "as he failed to apply for a patent for the mower and reaper, his wife was, after his death, robbed of the fruit of her brain by a neighbor. . . ." Mozans was wrong on two counts: this was *not* the earliest reaper,[56] and Mrs. Manning's husband William patented both machines, the clover cleaner on November 24, 1830, and the reaper on May 3, 1831. Ann Manning *was* "robbed of the fruit of her brain," or at least of official credit for these two early and important machines, but evidently by her husband. She may also have been cheated out of any benefit from her later improvements on the reaper, to the best of my knowledge not patented (see below).

Mozans was quite right to emphasize the Manning reaper as he does. Though not the first, it was an early functional machine, possibly America's earliest entry in an international race to mechanize the harvesting of hay and grain. It was important enough to be mentioned in several agricultural histories (e.g., Steward, 49, 50, 59, 92), and evidently the forerunner of most successful later machines (Blandford, 134–5). Hubert Schmidt, for example, thinks the Manning Reaper influenced both Hussey and McCormick (140). L.H. Bailey in his *Cyclopedia of American Agriculture* points out that among the factors transforming American agriculture in the 19th century, the most important was the new farm machinery. The Manning mowing-machine is his first supporting example (62). Continues Bailey, "The invention of the reaper was one of the most important events in the history of modern industry."

Note, too, that the Manning machine predates the McCormick Reaper, having been invented in 1817–8 (M. Gage/1: 9). Even its patent predated McCormick's patent of June 1831 by a month.

But what about Mrs. Manning's role in this invention? We have the report of prominent suffragist Matilda J. Gage, who knew Ann personally and was interested in women as inventors. Gage's two accounts of the achievement, separated by some thirteen years, differ significantly. In a series of centennial letters written for the Fayetteville, NY, *Recorder* in 1870, she says (M. Gage/1, cited in Hanaford, 623):

> One of the earliest mowing-machines was perfected by a lady of my acquaintance, now over eighty years of age, who aided her husband in bringing that and a clover-cleaner to perfection.

In her revision of this material for the *North American Review* (May 1883), Gage gives Ann Manning sole credit for both reaper and cleaner, and mentions other mechanical improvements of hers as well (M. Gage/2: 345):

---

56. There are conflicting claims. Suffice it here to note that Pliny the Elder described a reaping machine in the 1st century AD. The *Encyclopaedia Britannica* credits the Englishman John Common with inventing the (modern) reaping-machine in 1812, and the Scotsman Rev. Patrick Bell of Carmylie, Fifeshire, with the first *successful* reaping machine in 1826 (11th ed., s.v. "Agriculture," 396; "Reaping," 945). The history of the clover-cleaner needs further research.

A third great American invention, the mower and reaper, owes its early perfection to Mrs. Ann Harned Manning, of Plainfield, New Jersey, who, in 1817–18, perfected a system for the combined action of teeth and cutters, patented by her husband, William Henry Manning, as "a device for the combined action of teeth and cutters, whether in a transverse or revolving direction." Mrs. Manning also made other improvements, of which, not having been patented, she was robbed after her husband's death by a neighbor whose name appears in the list of patentees upon this machine. Mrs. Manning also invented a clover cleaner, which proved very lucrative to her husband, who took out the patent.

If Ann Manning was over 80 in 1870, then this 1883 version almost certainly postdates her death. Could Gage have visited Manning during her last illness, attended her funeral and talked with relatives, read an obituary, or all three, and gained a clearer idea of this extraordinary woman's real contribution? Ann's role in the Manning machines and her sole invention of the "other improvements" mark her as one of the most significant American agricultural inventors of the 19th century, male or female.

Personal details on Mrs. Manning are scanty. Thanks to Gage, and to Martha Rayne (116–7), we know she was born Ann Harned, about 1790. Thanks to information from the Westfield Memorial Library, Westfield, NJ, and from a collateral descendant,[57] we know that she was the sister of a well-known 19th-century physician and reformer, Clemence Sophia Harned Lozier (c. 1812–88). Dr. Lozier joined Gage and Lillie Devereux Blake in petitioning the New York legislature in 1875 for women's right to vote in Presidential elections, confirming the family connection with Gage. From Lozier's *NAW* biography and other sources, we can infer the following:

Ann or Anna B. Harned was probably born in Plainfield, NJ, one of the older, possibly the oldest, of the 13 children of David Harned, a Methodist farmer of Huguenot descent, and Hannah Walker Harned, of English Quaker stock. Ann may have preceded her sister Clemence at the Plainfield Academy. As the family produced a noted woman physician and suffragist who divorced her husband and resumed her own name, a woman inventor, and a male physician as well (Dr. William Harned), it was obviously no common family. The Quaker background of Hannah Walker Harned, herself a "doctoress" in New York City in the 1820s, undoubtedly shaped the upbringing of her daughters.

At an undetermined date Ann married William Henry Manning (b. 1783), son of Joseph and Providence Cooper Manning. The Mannings were a prominent pioneer family in the Plainfield region, Baptists, and linked with another prominent early family, the Laings. Ann and William had three children: James, Ann Eliza, and Hannah (O. Leonard, chap. 28). Orphaned at 11 (1824), Clemence may have gone to live with Ann.

In any case, the sisters probably maintained contact, since Matilda Gage knew them both. Ann Manning may even have moved to New York State to be near Clemence, since Gage seems to have lived all her married life in or near Fayetteville, NY (*NAW*), and had strong contacts in Troy. The 1871 Troy directory shows an Anna Manning, widow, living at 156 Fifth Avenue (Andersen, L-4/17/87).[58]

Oliver Leonard's "Pioneers of Plainfield" series appeared in the Plainfield *Courier-*

---

57. I am grateful for superb detective work by Pamelyn Ferguson, Reference Librarian at the Westfield Memorial Library, and for family information sent me by the great-great-granddaughter of Dr. Clemence Lozier, Ann Manning's younger sister. Thanks are also due to Thomas E. Ricketts, a Manning descendant and President of the Plainfield Historical Society, for putting me in touch with this relative.

58. The Mrs. Wm. H. Manning who shows up in 1873 at 3 Washington Place, is probably *not* the same, since this was an elegant address on the south side of the private Washington Park, to which only the surrounding residents have keys. Also, this woman apparently moved to Saratoga Springs, NY, in 1891; Ann Manning would then have been at least a hundred years old.

*News,* 1916–8. Curiously enough, Leonard's several Manning chapters never mention the Manning Reaper, and tell very little altogether about William Henry Manning's branch of the family. Further research will be necessary to illuminate Ann Manning and her background.

The famed McCormick Reaper itself might never have seen the light of day except for a West Virginia woman with a problem-solving genius (Russell Conwell, *Acres of Diamonds,* 46–7, cited in Boulding, 686):

> According to Mr. McCormick's confidential communication, so recently published, it was a West Virginia woman, who, after his father and he had failed altogether in making a reaper and gave it up, took a lot of shears and nailed them together on the edge of a board, with one shaft of each pair loose, and then wired them so that when she pulled the other way, it opened them, and there she had the principle of the mowing-machine. If you look at a mowing-machine, you will see that it is nothing but a lot of shears.[59]

Less than two decades after Ann Manning's inventive work with agricultural machines, **Ethel H. Porter** of Lincolnton, NC, received a U.S. Patent for "Straw-cutting and fodder" (1834). The wording of LWP's brief description is ambiguous; but since patenting a *technique* for cutting straw and fodder seems unlikely, Porter's invention was probably a straw- and fodder-chopping machine or improvement.

In 1860 **Elizabeth M. Smith** of Burlington, NJ, patented an improvement to the Manning Reaper. This was a device for changing the knives without stopping the wheels (*WI/1*; LWP). According to Matilda Gage (Hanaford, 623), yet another New Jersey woman still alive in 1870 invented a safety device for a mowing machine that throws the knives out of gear whenever the driver leaves the seat. I have not found this patent. Perhaps the device was never patented. Or perhaps this was one of Ann Manning's unpatented improvements claimed and patented by the unscrupulous neighbor after William Manning's death.

**Catherine Greene**'s possible role in the creation of the short-staple cotton gin is discussed in Appendix A-2, though the patent came so late in the 18th century (1794) that the invention's major impact belongs to the 19th. Whatever may be history's final verdict on Catherine Greene, late in the 19th century two women patented cotton-ginning improvements. In 1889, **Jane C. Osborn** of St. Louis received a patent for "Ginning cotton"; and in 1892, **Susie B. Hall** of Little Rock, AR, patented a "Brush for cotton gins." Osborn's improvement involved applying oil to the cotton during ginning, which sounds useful. Hall's invention was not just a brush, but a revolving cylinder armed with brushes pointing in two different directions. This sounds like an improvement on the brush-cylinder allegedly suggested to Whitney by Catherine Greene.

Also, in 1890, **Lucy B. Easton** of Rochester, MN, patented a Flaxseed-separating machine.

Women's numerous inventions of dairy machines have been discussed above. Suffice it here to recall that 25 of the 50 dairy inventions tabulated for the 19th century were machines or parts thereof.

Many other 19th-century American women now lost to history must have invented or suggested improvements to agricultural machines. About forty such women, whose contributions ranged from plough and harrow improvements to hop-strippers and wheat-cleaning machines, managed to patent their ideas (see Appendix A-8). Noteworthy

---

59. This was also the principle used by the Rev. Bell in 1826: a series of scissors fastened to a "knifeboard" (*EB*/11, s.v. "Reaping," 945).

among the planting, cultivating, and harvesting machines and agricultural vehicles here are the twelve patents for ploughs. Five of these belonged to one Iowa woman, **Elma M. Mitchell.**

To the best of my knowledge, none of these women became famous for her inventions. **Emma Mills** may have received some notice (Chap. 4), but not for her hop-harvesting machines listed here.

Swelling the total for 19th-century agricultural machines are the straw-working and other fiber-processing machines, the dairy and other food-processing machines, the windmills, etc., discussed elsewhere.

*20th century.* For the ten early years of the 20th century covered by the Women's Bureau Study (WB/28) alone, women invented and patented at least 71 pieces of Planting, tilling, and harvesting machinery and equipment (Category D), falling into the following sub-categories:[60]

| | |
|---|---|
| Beet harvester | Fruit picker |
| Cane stripper | Grain cleaner for threshing |
| Corn holder for grain bundles | Grain-lifter attachment for harvesting |
| Corn planter | machines |
| Corn thinner | Grain-picking machine |
| Corn-harvesting machine | Grain header and conveyer |
| Seed corn stringer | Gravity grain separator |
| Sod and corn-stalk cutter | Harvester |
| Cotton planter | Hay baler |
| Cotton scraper | Hay spear |
| Cotton-picking machine | Machine for cutting chaff |
| Cotton chopper | Marker for planters |
| Cotton ginner | Peanut plow |
| Gin saw filer | Plow |
| Portable comb. cotton-picking, ginning | Disk plow and jointer |
| condensing, and compressing machine | Reversible disk plow |
| Cultivator | Plow attachment and jointer |
| Double-row cultivator | Motor-driven plow traction wheel |
| Cultivator tongue | Rotary plow share |
| Cultivator attachment | Rice-hulling machine |
| Decorticating machine for fibrous plants | Seed planter |
| Disk sharpener | Tobacco-handling implement |
| Draft yoke for oxen and cattle | Tractor |
| Fertilizer, insecticide (spreader?) | Tractor-hitch allowing farm machinery to |
| Fertilizer distributor | be pushed by tractor |

As usual, WB/28 gives neither numbers nor names, but **Augusta P. McKay** of Sheffield, AL, with her Reversible disk plow of 1911 must be one of them. Unfortunately, because the categorized listing of inventions in the Name List volumes uses only initials, the task of trying to match these inventions with women's names in the given years is a monumental one. However, *these 71 inventions form an impressive 32.1% of the 221 agricultural inventions located by the study.*

Too early for WB/28 was **Sallie S. Pharr** of Marshallville, GA, who patented a (cotton) Planter in 1903. Sallie was born Sallie Slappey, in Marshallville, where her relatives still live today (A.D. Smith, L-4/1/85). She appears on Mary Logan's 1912 list of Prominent

60. Excluding the six having to do with irrigation and water control.

Women Inventors (888). The Women's Bureau seems to have omitted **Mary Jane Francis** of Galeton, PA, and her Huckleberry picker of 1911, a mechanical aid if not a machine, and certainly belonging here in Category D. **Anna J. Meeks** of Jennings, PA, patented a Rice-header in 1915 (a year not covered by WB/28).

Beyond the scope of WB/28, **Tyra F. Rich** of El Campo, TX, patented a Tractor implement attachment in 1932. Tyra (or Thyra) is a Scandinavian female name; the presence of a co-inventor named Calvin E. Rich suggests a wife-husband team. Another such team may be **Andrean G. Ronning** and Adolph Ronning of Minneapolis, who patented a harvesting machine in 1928. Andrean is a female name related to Andrew (Dunkling & Gosling).

In 1953 **Valerie Cary** of Opelousas, LA, with two men, patented a Sugar-cane harvester and loader. Since the machine was assigned to International Harvester, Cary may have been an in-house designer for the company, and the machine may have been manufactured and sold.

Presumably, as the small family farm gave way to agribusiness, and farm machinery tended increasingly to be the costly products of design engineers at Allis Chalmers, International Harvester, and the like, women's formal and documented role in this area of agricultural technology hit a new low. Exemplifying the contributions they nevertheless make are the unpatented "adaptations to farm equipment" created by **Annilea Weeks** (1953– ) of Winnipeg, Canada (Women Inventors Project of Waterloo, Ontario, 1989). London born, Weeks works for Manitoba Province. She is a first child, right-handed but partly ambidextrous, and has no formal technical training. She says she invented what she did out of necessity and frustration, and because of "living with a man who takes everything apart if it doesn't work!"

*Hand tools.* Women in 19th-century America also improved hand-cultivating tools. The most noteworthy of several such patents was probably the one granted to **Elizabeth H. Buckley** of Colchester, CT, on February 28, 1828. This was not just a better shovel, but a revolution in the mode of making scythes, shovels, etc. The patent seems in fact to have covered *the first sheet-iron shovel and the process for making shovels and other tools out of sheet iron rather than forging each one individually.* Eli Whitney and **Sister Tabitha Babbitt** have both been highly praised for an analogous development with iron nails. Paving the way as it did for quantity production of such tools, Buckley's patent appears on a list of important women's inventions reprinted in a Shaker *Manifesto* article of July 1890 (20: 7: 164).

Much later in the century, **Luna J. Aderhold** of Wedowee, AL, patented an Improvement in hoes (1870). In 1886, **Eliza B.** (with Albert J.) **Wilcox** of Maple Rapids, MI, patented a Combined pitchfork and rake. Eliza Wilcox also held an earlier patent, again with Albert, for a Wallpaper-hanging device (1885). **Alice J. Chamberlain** of Bridgeport, CT, patented a Rake (1893).

For the early 20th century, the Women's Bureau researchers found 27 patents issued to women for Garden tools and equipment in the ten years covered. The items fell into 18 types of which the following seem relevant here (WB/28: 20):

| | |
|---|---|
| Adjustable rake | Shears |
| Combined, rake, and hoe | Strawberry runner cutter |
| Hoe blade | Weeding tool |
| Rake attachment | Weed-cutter |
| Lawn mower | |

If all 18 types contained equal numbers of inventions, the 9 types listed here would represent 13 or 14 inventions. One of them may be the Weeder patented by **Sarah E. Ball** of Ritchey, IL, in 1910. The category accounted for 12.2% of the 221 agricultural and related inventions.

*Genetic engineering*

The most exciting frontier in agricultural technology, as in so many other technical areas, is genetic engineering. Women as well as men are contributing here. Remember, for example, as the Lasker and Nobel committees did not, that **Annie C-Y. Chang** worked with Stanley Cohen at Stanford on the gene-splicing techniques that make possible all of genetic engineering. Her name appears second (after Cohen but before Boyer) on the 1973 paper (Q.V.). And without Esther Lederberg's landmark discovery of what happens in viral induction (when a virus directs a bacterial cell to copy the viral DNA instead of its own), the very concept of genetic engineering would not exist. Following are select examples of more recent inventions/innovations:

**Mary-Dell Chilton** (1939– ) of Washington University in St. Louis is a member of a team of university and industry scientists that developed the first method for introducing foreign genes into plant cells and reliably producing normal fertile plants. Chilton and her colleagues moved a gene from a yeast into tobacco plant cells, using the bacterium that causes crown gall disease as the carrier of the alien genetic material. Their innovation, announced in 1983, was to alter the bacterial DNA to keep it from causing tumors in the tobacco plant cells. Thus the cells could be grown into intact plants.

The next step is to get the transferred genes to operate in their new setting (the yeast gene originally failed to work because its control region did not "make sense" to the tobacco plant). Three groups, including Chilton's, have now solved this problem by attaching plant control regions to foreign genes to produce "chimaeric genes" that now operate correctly in the new plant host. Chilton nevertheless foresees "plenty of significant problems" to be solved before genetic engineering becomes a routine part of the plant-breeding process (see Chilton refs.; "Plants Bearing," 184). This genetic-engineering technology, which Dr. Chilton considers her most important work, has not been patented.

Mary-Dell Chilton, brought up in a family with two brothers, attended a private school for her early years of schooling. The family had a strong work ethic and success orientation, and both parents encouraged her toward achievement. Dr. Chilton received her PhD from the University of Illinois in chemistry, in 1967, did postdoctoral research at the University of Washington in Seattle, and has held both research and teaching posts, most recently that of Associate Professor of Biology at Washington University, St. Louis. As of the mid-1980s she was Executive Director of Agricultural Biotechnology at Ciba-Geigy Corporation, Research Triangle Park, NC. She has published numerous scientific papers, and has been honored with the Bronze Medal of the American Institute of Chemists and an honorary doctorate from the University of Louvain, Belgium (1983) (Chilton, Q- & c.v./8/1/85).

**Dr. Virginia Walbot,** Associate Professor of Biological Sciences at Stanford University, is a corn geneticist whose work has practical applications in agriculture. As of 1981 corn accounted for $25 billion of the total American farm product of $87 billion, or nearly 30% of all farm income. It was, moreover, the largest single source of sugar consumed in the United States, corn sugar being cheaper than either beet or cane sugar, and also a source for cooking oil, fabric sizing, and coatings for photographic film. Thus the potential importance of such research is great.

Dr. Walbot's major work is basic research. She is investigating, for example, the very nature of plant growth, and particularly the two cells in one stage of the corn embryo that form the apical meristem. This structure organizes the shoot and directs all further growth. The knowledge gained here may be pure knowledge. However, she is involved in several other research projects with clear practical applications: research on what enables an oxygen-starved plant to produce food without oxygen, by fermentation, may lead to knowledge of what turns genes on and off. Work on plant immune systems may lead to new kinds of disease-resistance in plants. Research on the roles of specific genes in corn's growth and development, using recombinant DNA techniques to mark the genes, may be of use to commercial corn breeders. Work on the genetic causes of male sterility in corn plants will be of enormous value in the production of hybrid corn varieties.

In a minor but intriguing facet of her work, Dr. Walbot was part of a team of University of Missouri researchers and commercial corn breeders that developed a corn genetically incapable of losing its sweetness and turning starchy. (Unfortunately for lovers of corn on the cob, this is a field rather than a garden corn.)

A Stanford graduate who returned to her alma mater in the early 1980s, Dr. Walbot continues to work on improving corn. As of 1981 she was engaged, among other things, in work on garden (sweet) corn (Shurkin, "Walbot"; Lynch).

The frost that paints windowpanes is a killer in the fields. Frost damage costs U.S. farmers more than $1 billion each year, and estimates of potential damage run as high as $5 billion. Genetic engineering may offer a way to prevent some of this enormous loss. Plants can actually withstand temperatures of −6 to −8°C for several hours—unless ice crystals form. Two common types of plant bacteria, *Pseudomonas syringae* and *Erwinia herbicola,* often serve as nuclei for the formation of such crystals, thus causing most of the frost damage. To get rid of these two bacteria, scientists have considered spraying fields with streptomycin, (which works, but is not environmentally sound), or releasing a series of viruses that attack only these two bacteria (criticized because viruses sprayed on plants often prove unstable).

**Cindy Orser** at the University of Wisconsin leads a team taking a genetic-engineering approach to the problem. In August of 1983 Orser and colleagues announced success in altering one strain of *P. syringae* so that it no longer triggers ice-crystal formation. Plans to test the "microbial anti-freeze" by spraying altered bacteria on crops early in the growing season so that they can become the established strain—which worked in the laboratory with pinto beans—have been snarled in an environmental controversy.[61] Not till 1985 was field testing finally approved for a remote Northern California area; not till 1987 did the tests begin.

In the aftermath of the Green Revolution,[62] with the world hungrier than before and

---

61. The environmentalists' objections are not trivial. These same ice-nucleating bacteria that are such a nuisance on crops also occur in the upper atmosphere, and may be important in precipitating both rain and snow. If the genetically engineered "ice-minus" bacteria should reach these upper levels, and there become dominant, the world's climate might be altered (Julie Miller, "Microbial").

62. As G. Tyler Miller, among others, has pointed out, the Green Revolution not only demanded enormous amounts of water and nitrogen fertilizer (which was costly in energy to produce and too expensive for poor farmers to buy), but tended to reduce genetic diversity in basic crops. It also had social consequences, since in many areas only well-to-do landowners could take advantage of it (174–80). As a

nitrogen ever more of a limiting factor in food production, agricultural scientists have turned to natural organisms that can fix nitrogen from the air—the *Rhizobium* bacteria and certain blue-green algae. Some of these blue-green algae have long been known to serve a valuable purpose in the rice paddies where they flourish. Since one such alga, *Nostroc commune,* is used as a food in China (Chapman & Chapman, 23), it is interesting to reflect how much women may have intuited and put to use about the fertilizing role of the organism.

In any case, the scientists have taken two basic approaches: physical and genetic. The physical approach, usually called inoculation, consists of placing the algae or bacteria in the soil around the plants, where they do their work, and the plants pick up the nitrogen with their roots, just as with a soil application of fertilizer. This is **Elisabeth Bryenton**'s approach. The genetic approach, which concerns us here, involves transferring the nitrogen-fixing genes from bacteria or algae directly into the genetic material of the plants so that the plants can in effect create their own fertilizer. Needless to say, this would be a green—or a blue-green—revolution indeed.

**Mary-Dell Chilton**'s group, with its new technique for rendering a harmful carrier harmless, has taken one step in the long and complex process of conferring nitrogen-fixing capability on higher plants. Two other women contributing significantly to this field are **Ada Zamir** of the Weizmann Institute and **Sharon Long** (1951– ).

Dr. Zamir is part of a Cornell team that in 1981 succeeded in transferring all 17 nitrogen-fixing genes from a bacterium into a yeast ("Yeasts Receive"). Unfortunately, the transferred *Klebsiella pneumoniae* genes did not fix nitrogen in their new environment. Since the transfer was successful and complete (genes transferred back to the bacterium after 40 generations *do* function there), it seems likely that the group had encountered the problem recently solved by Dr. Chilton's group. But an important step had been taken.

Although Sharon Long's work is basic research—she is a molecular biologist at Stanford University—her studies in plant genetics could have important practical applications. Recognized in 1984 as one of America's Top 100 Young Scientists (*Science Digest* eds, Dec. 1984: 50), Dr. Long is known for identifying and cloning the genes allowing bacteria to locate and enter certain plants. She has worked with the *Rhizobium* bacterium that, as its name implies, invades the roots of such legumes as alfalfa, soybeans, and peas, where it lives symbiotically with the plant, receiving moisture and protection, and producing nitrogen for the plant's growth. Long's work opens the possibility of genetically altering the bacterium to make better invaders.

If either of the two approaches described here should prove successful, the worldwide savings in mineral and energy resources and in the cost of producing food would be counted in the billions of dollars.

## SUMMARY

Clearly, agriculture, food, and related technologies are important areas of invention for women. This continues to be true in the late 20th century, even in the United States, where mechanization, professionalization, and agribusiness have diminished farm women's role.

The most recent U.S. Patent Office study of women's patenting behavior (OTAF) finds

---

high-tech development introduced into Third World countries from outside, I suspect it may have worsened the position of women agriculturalists in particular.

two agricultural/food-related classes among the active classes[63] where women's participation was highest for the study period (1977–88). Class 426, Food or edible materials: processes, compounds, and products, ranked eighth among the 15 active classes with the highest *numbers* of U.S.-origin women-inventor patents (372), and fourth on a parallel list of classes with the highest *percentages* of such patents (10.9%). Class 119, Animal husbandry, ranked fourth on the percentage list (7.25%).

63. An active class is one gaining at least 1200 new patents as "original" classifications for the period 1977–88 (OTAF, 22).

# 2  Daughters of Isis, Gula, Hygieia, and Brigit: Women Inventors in Health and Medicine

The earliest doctors among the common people of Christian Europe were women who had learned the virtues and use of herbs. The famous works of Paracelsus [1493–1541] were but compilations of these "wise women" as he himself stated . . .
—Matilda J. Gage, 1893

The reason for [Mother Seigel's] extensive knowledge and information concerning a great variety of diseases was plain to me. She had remained with the patients day and night, watching every symptom and the effect of every remedy used, while the doctor stays with the sick but a few moments every day.—Anna J. White, 1886

Not until the second decade of this century did a random patient taking a random ailment to a random doctor have even a 50-50 chance of "benefiting from the encounter."—Martin Gross, 1966

Even as we humble and destroy and control disease after disease, *illness* paradoxically multiplies.—Thomas E. Gaddis, 1925

I think men are just as capable as women in medicine.—Estelle Ramey, 1970

## INTRODUCTION: PREHISTORY

Women are the once and future healers of the world. According to a student of Native American medicine, the medicine man was "the ancestor of the priest, the antagonist of the physician for centuries. If there is any ancestor or colleague of the physician in primitive society, it is the lay healer, usually a woman, the midwife" (Vogel, 22). Women's early medical expertise grew, like horticulture, out of their deep knowledge of plants. Since women gathered for children as well as themselves, they were the more likely to learn about plants' properties—to notice, for example, which plants other animals never ate. Where women owned or cared for the small domesticates, as often they did, they might also notice what these animals ate when they were sick.

Women also, of course, tried unfamiliar plants themselves, especially during droughts or migrations, perhaps tasting progressively larger portions of a plant until they declared it food, famine food, or not-food.[1] Suppose that in the progression, a woman noticed multiple effects: intestinal cramping followed by a worming effect; mild dizziness, but less joint pain; or a terrible taste followed by a respite from coughing. As food-processors, women had the better opportunity to note the effects of eating certain parts of the plant as opposed to other parts; and, after they invented cooking and other processes, to notice which plants caused vomiting or drowsiness or even death if eaten raw but were harmless if processed.

In fact, some of the strongest evidence for women's early pre-eminence in medicine is the number of plants that are both foods and medicines (and toxic in some forms or

---

1. Recent research on rats—second only to humans as omnivores—seems pertinent here. A rat given an unfamiliar food will taste a bit; if it feels any ill effects in the next 24 hours, it will never touch that food again. Rats also choose foods to correct experimentally induced dietary deficiencies, and communicate information about approved new foods to other rats in their group (Roper, 30, 32). With regard to animals' abilities to link cause and delayed effect, note that Japanese snow monkeys transplanted to Texas learned to avoid coyotillo berry in their new environment, even though its neurotoxic effects do not appear for 50-150 days (Lampe, 40).

dosages). Just a few examples: rhubarb is a food but also a laxative, and its leaves are poisonous; clover is a heart stimulant, but also a food. Melanu people of Sarawak add pearl sago to tea or boiling water to cure diarrhea (Ruddle *et al.,* 67). The Tiwi of Melville Island near Australia regard the toxic nub of certain cycads as famine food or medicine (Goodale, 180). The Latin species name of asparagus *(officinalis)* reveals that apothecaries once stocked it, and even lettuce had its medicinal uses, e.g., for slowing the pulse (Dallas, 47, 275). Fungi and lichens can be either food or medicine, as truffles, manna (Sharnoff, 135), and penicillin remind us.

Lichens, in fact, are striking examples of the connections being suggested here. A dark brown lichen, *Bryoria,* or "tree-hair," becomes a food among Native Americans of the Pacific Northwest, after the women bake it in pits and shape the resulting black paste into loaves. Though white observers found it distasteful, many tribes used it as a famine food. The Nez Percé Indians use this lichen to treat digestive ailments, and the Okanagan make it into an antiseptic salve for the navels of newborns (Sharnoff, 142).

Medicinal plants are also linked with other areas of women's work, with lichens again good examples. Consider the stringy, gray-green *Usnea* lichens. In cultures as widely separated as ancient Greece and China, as well as in South America, East Africa, Europe, and Malaysia, the lichen has been used medicinally: for gynecological disorders (always a female preserve), for stomachaches, colds, and antibiotic salve. The Dakota and Wylackie of North America used it to dye porcupine quills and tan leather, respectively. Porcupine quill art is a woman's invention and in some groups done only by members of a female secret society (Berry, 113); leather-working is an ancient specialization of women (Mason, *Share,* chap. 4). *Usnea* is used in bedding, and the Simla Hills people of India stuff cushions with it (Sharnoff, 141–2).

*Bryoria,* both medicine and famine food, was also used by the Thompson Indians to make both clothing and shoes. Oakmoss lichen *(Evernia prunastri)* has long been harvested by the ton in Morocco and Yugoslavia for perfume-making, another ancient female specialization (Samuel warned the Israelites that a king would take their daughters away to be perfume-makers; I Samuel 8: 10ff).

Many lichens were anciently used as dyes, continuing into the early modern period in northern Europe (Anderson & Zinsser, 100–1), and even into the 20th century.

Despite this obvious connection between medicinal plants, food, and women's other work, little has been written on women's role in medicine before the Middle Ages. Remedying that omission is far beyond the scope of this chapter. Here, however, I shall present evidence from mythology, anthropology, archaeology, literature, pharmacology, and early history for the importance of women's early role, since that evidence underlies my two basic assumptions here: 1) that women were the first healers, serving until displaced by men in Western and certain other cultures (often after Western contact), and 2) that any remedy, treatment, surgical procedure, or instrument invented before this male takeover in any culture—or where women are still the healers—can reasonably be attributed to a woman or women.

*Mythology*

Let us look first at a few of the world's many ancient goddesses of healing.[2] The four who preside over this chapter—Hygieia, Isis, Gula, and Brigit—represent Greece, Egypt,

---

2. A count to determine what percentage of all known deities of healing are female might be interesting, but not definitive, since so many goddesses were eliminated or treated to involuntary sex-change operations in ancient times.

Mesopotamia, and Ireland, respectively. Hygieia, whose very name means "health" and gives us our word *hygiene,* is variously known as the daughter of Aesclepius and as one aspect of Athena. A goddess of healing seen as daughter of a god of healing has surely been demoted, but a close look shows that both versions support women's early pre-eminence in medicine.

Robert Graves (*Myths* I: 174ff) refers to a pre-Hellenic medical cult presided over by Moon-priestesses at local shrines. Athena, patroness of this cult, received the epithet "Hygieia" because of her cures. Her all-heal or universal remedy[3] was the oak mistletoe, and Graves suggests that Aesculapius (Lat. form of Aesclepius) means "that which hangs from the esculent oak"; in other words, the mistletoe. Thus, says Graves, Aesclepius is merely a personification of the mistletoe's healing power,[4] originally dispensed by the priestesses in curative rites. Note that mistletoe was an ancient abortifacient, strictly woman's medicine.

Understanding Hygieia as one aspect of Athena sounds a theme common to early medicine in many Near Eastern and European cultures: the Great Goddess worshipped there under various names until the end of the Bronze Age usually had healing attributes, and might be invoked by a special name in that role. Athena Hygieia is one example; Minerva Medica another (Hurd-Mead, 32). In Mesopotamia, surgery and medicine were both divine functions exercised by the Goddess Bau (Rohrlich, "Transition," 53–4).

Isis, Great Goddess of Egypt, was not only the giver of agriculture, but also the Egyptians' greatest deity of healing (Hurd-Mead; Diodorus, i.25; W. Jayne, 64–8), invoked as follows (Ebers Papyrus, W. Jayne: 66–7):

> May Isis heal me as she healed her son Horus of all the pains which his brother Seth brought on him when he slew his father Osiris. . . .

Aside from her general healing attributes, Isis is specifically credited with originating the science of medicine and with discovering or inventing such remedies as her favorite verbena and one similar to the Japanese *dosia,* as well as with inventing the art of embalming. She mummified the dismembered Osiris in order to save the pieces intact (M. Gage/1:21–31; Hurd-Mead, 15, 37n; Mozans, 335), and then breathed life into his mouth (W. Jayne, 65).[5]

Even older than Isis, though sometimes amalgamated with her, and also a goddess of healing, was Neith. Adopted by the Egyptians in the First Dynasty, probably from Libya, she was sole parent of the sun god Re. Her temple at Sais was a famous healing sanctuary, *with a medical school attached* (W. Jayne, 71–2).

Gula, an early Assyrian goddess, was honored also in Babylon and Sumer (W. Jayne, 121; Rohrlich, "State," 88). Walter Jayne identifies her with Ma-Ma, or Mummu. Thus

---

3. Note that Panacea (another Greek goddess of healing and sometimes daughter of Aesclepius; Hurd-Mead, *HWM,* 33) means "all-heal" or "cure-all." Unless otherwise noted, all citations of Hurd-Mead refer to her *History of Women in Medicine (HWM).*

4. The story in Graves is more complex than this. According to Druid tradition, and perhaps in other ancient goddess-worship, mistletoe was the oak tree's genitals, lopped off with a golden moon-shaped sickle in a castration/sacrifice ceremony having to do with the death of the year-king. In any case, the mistletoe gained its power through a sacrifice ordered by the queen/goddess.

5. Breath and life are so universally linked that a purely symbolic interpretation is possible, but this myth could indicate that early healers used mouth-to-mouth resuscitation. Midwives faced with a baby who did not breathe might be first to try the technique.

The Finns also have a myth of a woman reviving a dismembered male. In the Kalevala, after the hero Lemminkainen is torn to shreds and scattered in the waters of the black river, his mother gives him life for a second time (Aldington & Ames, 313; Lönnrot, 91).

Gula, too, may follow the pattern of the Great Goddess with healing aspect. Ishtar possessed healing attributes, but healing was not her central role. Gula, on the other hand, was primarily a healing deity, often called the Great Physician. Gula appears in bas-reliefs enthroned with a wand in her hand, receiving processions of priests and priestesses bearing petitions from their patients. As late as the 7th century BC, King Assurbanipal petitioned Gula to cure his little son from some insect-borne disease (Hurd-Mead, 11).

Two other Sumerian goddesses worth noting here are Ninisinna, patron deity of the art of medicine and healing (Kramer, *Sumerians,* 206), and Ninhursag. In one myth Ninhursag instructs the Plant Goddess in creating eight (healing) plants, and gives birth to eight healing deities, one for each of her patient's pains (Kramer, *Mythology,* 57–9; Rohrlich, "State," 86).

One of the three persons of the Celtic triple Goddess Brigit, whose name means the "High One," presided over physicians and healing. Brigit's priests and worshippers were female. She is not only goddess of healing, but giver of all crafts and arts, as Minerva was among the Romans (Graves, *Myths* I: 80–1; W. Jayne, 513). In Irish folklore, healing is one of the gifts from the Land of Faery to mortals; Faery is peopled almost entirely by women.

Many other examples could be given: Tsenabonpil, ancestral heroine of the Melanesian New Irelanders, who gave her people all knowledge, including medicine; Ixchel, the Maya Moon Goddess—and first woman—who gave her people their knowledge of healing; Po Yan Dari of ancient Cambodia; Eir, the Scandinavian goddess of mercy and "best physician," who sat on her hill of healing granting health to all women who could climb to her; Habetrot, the English healing deity and spinner; Kamrusepa, the Hittite goddess of magic and healing; and the White Goddess Leprea or Alphito, whose prerogative it was to cure (or cause) leprosy (Powdermaker, 34–5; Stone, *Mirrors* I: 92; Monaghan, 94, 112, 126, 244; Hicks, 136–7; Graves, *Myths* II: 63). However, the foregoing should suffice. Appearing worldwide, often identified with the most ancient and central forms of the Great Goddess, these deities point to an early pre-eminence for women in healing.

Intermediate between deities and earthly women are such fabulous females as the Buschfrauen of Germany and the Dziwozony of Poland (Monaghan, 52–3, 90), both of whom knew herbal healing secrets.

Earthly women as well as female deities may become legendary for their healing skill. In an interesting reversal of the explanatory tale of healing as a gift from Faery land, **Biddy Mannion,** a midwife of Inishshark off the Irish coast, was stolen by the fairies to heal their rulers' sickly child. So great was Biddy's reputation for healing that it had spread even to that realm. She restored the child to life and returned safely to her home (Monaghan, 45–6).

Revered among the Chams of Cambodia was the "divine priestess" Pajau Tan, a healer who was too successful. Finally, she was sent to live in the moon "because she kept raising all the dead" (Monaghan, 244).

Sometimes these women—and goddesses—are linked with specific medicinal plants or instruments. The Andean Goddess of Healing and Happiness, Cocomama, for example, is the resident deity of the coca plant (Monaghan, 69), which South American native peoples had used ritually and to fight fatigue long before cocaine was refined from it. To Hecate, the Moon Goddess, were attributed aconite teas for teething and for children's fevers. In fact, Hecate had a medicinal garden, visited by Medea the sorceress, where grew "many a magic herb, black poppy, similax, mandragora, aconite." Artemis was linked with mugwort *(Artemisia),* another ancient all-heal[6] and birth drug. Rhea invented liniments for relieving the pains of children (J. Harrison, 37–9).

An Austrialian aboriginal myth credits women with inventing the stone circumcision knife and showing men how to use it (cited in Evelyn Reed, 295).

From Africa comes an Ekoi explanatory tale in which the daughter of Sheep gets "the 'medicine' " from "the Nimm woman" (S. Feldman, ed., 173).

## Art and crafts

The graphic evidence has too seldom been noted. An example from South America occurs on ancient pots of the Moche people of Peru (whose potters were women), which "generally . . . portray the curer as a woman wearing a shawl." In this area today, the curers are all or mostly male (Alva et al., 28, 46–7).

## Literature and history

From scattered literary and early historical sources, we can get some idea of women's specific contributions to the technology of healing. Clues to female origin are either that the remedy or procedure in question is so ancient that women were likely the sole or principal healers when it was invented; that the plant involved is a food[7] or otherwise connected with women's work; that the remedy is attributed to a goddess or woman; or that it pertains to what we now call gynecology, obstetrics, or pediatrics, all women's work until modern times.

The earliest known medical document is a Mesopotamian clay tablet dating from the late third millennium BC, when Gula was still goddess of healing in Sumer and most physicians were probably still women. Unlike later medical documents, it contains not a single magic spell or incantation, demon or deity, but records herbal and other natural remedies. A document from Hammurabi's time (c. 1750 BC), when most physicians were male, by contrast, blames diseases on demons and recommends incantations as cures. The same contrast appears between the work of the wise women (witches) of Medieval Europe and the usurping male doctors (i.e., holders of university degrees), learned in theology but ignorant of plant wisdom (Rohrlich, "State," 88–9). The men, as someone aptly put it, were portent-wise and plant-foolish.

Among the general historical references are Hittite texts describing "Elderly Women" who held "prophetic and advisory positions," and were linked with both physical and mental healing (Stone, *When God,* 131). In pre-colonial times Goba women of South-Central Africa were polyandrous healers and oracles (also descent-group leaders, land-shrine priestesses, and queens) and so remembered as late as the 1970s (C.S. Lancaster, 541).

*Remedies and treatments.* This category is so massive that to treat it fairly would demand several books, and others have written on it far more learnedly than I could (e.g.,

---

6. Modern medicine has isolated artemisin not only from *Artemisia* species, but from the ancient Chinese fever herb *Quing hao,* and used it successfully in several thousand cases of malaria, including some that were resistant to chloroquine and other synthetic quinines ("New Old Remedy").

7. Curiously enough, the connection between food and medicine seems to have survived the male takeover of medicine and come straight down into early modern Europe. In Norwich, England, for example, in the 16th century the barber-surgeons often served food and particularly drink in or just outside their shops; and the London apothecaries' companies did not break off from the grocers until 1617 (Pelling, "Occupational Diversity," 489–90).

Hurd-Mead; Chaff *et al.*). Furthermore, women are rather generally acknowledged to be the first experts in herbal medicine. What is not always acknowledged is the degree to which modern medicine rests on ancient women's medicine; virtually never mentioned is the possibility that in focusing on isolating a single active principle from some of these complex herbal remedies, medical scientists may have discarded elements that reduced or compensated for side effects, controlled absorption rates, etc. Here I will cite just a few particularly interesting examples.

According to Dr. Dan McKenzie, women discovered the properties of pine tar and turpentine, and of chaulmoogra oil, used in leprosy. Leprosy still afflicts some 15 million people, mostly in developing countries (*Science News* 119 [Mar. 21, 1981]: 184). Women also invented remedies from acacia, almond, asafoetida, balsam, betel, caffeine, camphor, caraway, foxglove, linseed, parsley, peppers, pomegranate, poppy, senega, sugar, and "hundreds more" (cited in Forfreedom, 9).

The Hebrew Bible speaks of female ointment-makers. By Nehemiah's day (5th century BC), they had formed an apothecaries' guild with their male colleagues, and lived in a special alley in Jerusalem (Nehemiah 3: 8; I Samuel 8: 13; Chas. Singer I: 29).

According to the *Iliad,* the Egyptian woman Polydamna gave her pupil Helen the secret of the famous nepenthe, inducing a pleasant forgetfulness (Thomson cites *Odyssey,* 4.125–32, 220–30; see Hurd-Mead, 37n, for possible ingredients). The contributions of the early botanist **Artemisia of Caria** (4th century BC) have entered history partly because her name survives in some of the herbs she discovered or studied (and partly because of the splendid tomb she built for her husband Mausolus). Several *artemisia* species, as already noted, have medical uses: in the bath to prevent an abortion, as a fumigation in retarded menses or retained placenta; or (another species, used with elaterium) to cause abortions. According to Pliny it was Artemisia who discovered the value of wormwood— and the delights of absinthe. Praised for her medical skill by many ancient historians, she was said to know every herb used in medicine (Hurd-Mead, 32, 40; W. Jayne, 345).

For further examples of this sort, see Hurd-Mead's chapter, "Medical Women in Ancient Times," in her *History of Women in Medicine;* see also Sandra Chaff's bibliography.

*Surgery.*  According to Hurd-Mead, Hebrew women had considerable surgical skill, notably in obstetrics. Midwives used a vaginal speculum or dilator to make structures visible, knew occipital version (turning the baby face-down in the birth canal to ease birth), used probes, scalpels, cauteries, and tenacula (delicate hooks for lifting arteries and other structures). The Talmud and the Niddah reveal that women did embryotomies and Caesarean sections (before Caesar). Zipporah, wife of Moses, circumcised her son—and other babies—with a flint knife. Egyptian women also circumcised, performed Caesarean sections (before Caesar), and even removed cancerous breasts. Herodotus found these women especially skillful with their stone knives and also expert at splinting broken bones (Hurd-Mead, 25, 19).

Thorkild Jacobsen discusses specific midwifery instruments and equipment in his notes on the Mesopotamian Great Goddess Nintur (Aruru, Ninhursaga) and her power to give form to the embryo in the womb (290–1): a head-covering (hairnet?), the birth chair, a slender rod for piercing the membranes, a copper water pail, a flint knife or special reed knife for cutting the umbilical cord, and a "birth brick" presumably used as a cutting "board" or surface.

No one can be certain that women invented these flint or obsidian knives, but division-of-labor evidence seems to favor that view, and flint is specifically mentioned in pre-Hellenic myth as the gift of the Goddess (Spretnak, *Lost,* 42). Obsidian, interestingly enough, is sharper and holds a better edge than steel for certain uses, and is returning today to some areas of surgery (Press, "Cutting"). As for the birth chair and obstetrical instruments, the likelihood that they were women's inventions approaches certainty. These and other aids to birth will be discussed more thoroughly in the chapter on fertility and anti-fertility technology.

Aside from midwifery, and game-butchering, women probably gained useful knowledge by preparing bodies for burial. The most elaborate such preparation in the Near East was the Egyptian practice of mummifying bodies. As noted, Isis is specifically credited with inventing the process for Osiris; and in Old Kingdom Egypt, priestesses were in charge of mummifying bodies (Hurd-Mead, 4; Boulding, *Underside,* 185).

In another part of the world, Scandinavian women's graves contain surgical instruments not found in men's graves (Hurd-Mead, 16).

## Anthropology, Archaeology

Based on her anthropological and herbal literature search for *Clan of the Cave Bear,* Jean Auel hypothesizes that the use of willowbark and of datura for pain, clover-leaf tea as a heart stimulant, and pokeweed roots and pokeweed berry juice for rheumatism pain and lumps and growths—among many others—probably date to Neanderthal women (Auel, PC-1981). Willowbark, of course, contains the analgesic of modern aspirin, and in 1981 scientists found a possible anti-viral agent in pokeweed; a protein that inhibits the multiplication of herpes and some other viruses, but is not toxic to cells at the necessary concentrations.[8]

But we need not depend on fiction, no matter how thoroughly grounded in anthropological research. As just noted, Hurd-Mead reports, in an article on Scandinavian medical women, that ancient Scandinavian women's graves contain surgical instruments not found in men's graves. Among the Maidu Indians of North America as late as the 1930s, women were the only shamans. With their drums, they officiated as both doctors and magicians. The Chinese Wu shamans were originally female and, in pre-Zhou times, revered both as healers and as priestesses (Colegrave, 42). Nor was ancient women's medical knowledge exclusively herbal. California Indian medicine women used a technique that modern medicine has just rediscovered for fighting cancer and other diseases: visualization. The medicine woman took her patient to a lonely place in the forest, sweated and pommelled her, then sucked the disease from her mouth and put it in a basket to be taken home. Pain was "visualized by color and length." The Seneca (Iroquois) had teams of shaman-healers who came to the patient, always led by a woman (Hurd-Mead, 4).

Early circum-Mediterranean women healers showed knowledge of other facets of the mind-body connection as well, using a technique called incubation, or temple sleep, in the temples of female healing deities such as Isis (W. Jayne, 67). Patients came to the temples to be healed, underwent purification rituals that were also relaxing, and were then told to sleep in an outer part of the temple for the night. Attendants moved among the sleepers, touching them and making quiet suggestions for their cure (W. Jayne, vi, 30–31). Interestingly enough, dream-incubation is being revived by at least one clinical psychologist, Karen Hagerman, for treating rape victims (Spretnak, *Politics,* 465–6).

---

8. *Clan of the Cave Bear* is obviously not a scholarly work, but the research underlying it was extensive, and the work has some interest as a compendium of otherwise scattered references, a hypothetical Neanderthal pharmacopoeia in a single woman's medicine bag. The pokeweed discovery appears in *Science News* 119 (Mar. 21, 1981): 184.

Healing was also a function of pre-Hellenic oracular temples, where the one consulting the oracle "slept in a holy shrine with her/his ear upon the ground." Gaia was the first patron of the Delphic oracle, and according to Charlene Spretnak, oracles themselves were a female invention (*Lost,* 30).

Another ancient treatment was heat therapy (hyperthermia). Considering the ancient uses of hot springs—over which nymphs and goddesses like Sula of Bath often presided—and the great comfort from heat in general, there seems no reason to doubt that early healers would use heat, and that they might have tried it for many different conditions. Heat therapy for cancer, for example—though only recently rediscovered by modern medicine—is evidently thousands of years old (Grady, "Heat"; "Hyperthermia's Hot Spot," 141). The extreme age of heat therapy certainly puts it within the time when women were still in charge of healing. The first forms of treatment may have been wet or dry hot packs and hot poultices of various sorts.

Interestingly enough, the Roman deity of healing by (external) fire was a goddess, Feronia. Once a nymph of springs in Central Italy, she was originally a deity of the Sabines and other central Italian tribes, but eventually became a Roman State deity with a temple in the Campus Martius. The ancients also knew the value of inducing fever (internal heat) to fight various kinds of illness and malaise (W. Jayne, 425).

## Pharmacology

Until recently, modern medicine usually ignored anthropological reports of primitive herbal medicines. With the advent of medical anthropology, however, pharmacologists have begun to analyze the plant preparations represented in these reports. In 1962, for example, the Bombay laboratory of Ciba pharmaceutical company reported a study of 175 plants anciently used as drugs. The researchers found 64 of them "physiologically active, 35 apparently in significant ways, worthy at any rate of a good deal of further study" (W. Jayne, v–vi). Since laboratory conditions can never perfectly reproduce either the full regimen specified by the ancient healers or conditions inside the human body—especially the effects of long-term use—and since certain trace elements may have gone undetected, this is a good score indeed.

More recently, a worldwide research team coordinated by the United Nations and reporting from the University of Illinois College of Pharmacology has been analyzing all reported herbal contraceptives. The results of this study—again finding many active ingredients worthy of further investigation—will be discussed in more detail in Chapter 3. Suffice it to note here that Norman Farnsworth, head of the Pharmacology Department at the University of Illinois Medical Center, considers about half the traditional remedies used by Chinese rural doctors as having "a rational basis" acceptable to Western minds. One of the top drug-evaluators in the United States, Dr. Farnsworth has developed a theory of convergent evolution for folk remedies, which says that if widely separated peoples use a given plant for the same purpose, this constitutes "prima facie evidence that the putative use is valid" (cited in Seaman & Seaman, 435).

## Health Maintenance

One of the most neglected areas of medical history—as of medical education—has been the maintenance of general good health, including, among other things, nutrition. Two suggestive examples in passing:

Yahi Indian women of Northern California saved their salmon bones, ground them with mortar and pestle, and baked the bonemeal into their acorn cakes or other food. These semi-nomadic gatherer-hunters had no milk at all and no greens for much of the year. Even their meat and fish consumption was seasonal, although they did dry some deer meat. Thus without this practice, calcium might have been deficient in their diet (Kroeber, *Ishi,* 56).

Judging by the center of gravity best suited to many of the postures, physical yoga may have been a women's invention. For correct execution of the postures, the body's center of gravity needs to be in the hips, as it is in women (Spretnak, *Politics,* 522).

## TAKEOVER THROUGH THE 18TH CENTURY

In a great many groups there was a male takeover of women's healing work, partly analogous to the agricultural takeover posited in Chapter 1. Using that earlier analysis as a paradigm, consider the following sketch of events and scattered sample of myths and tales from various cultures:

In the Near East, male physicians became more prevalent during the Bronze Age. These men now treated upper-class patients, while women healers treated mainly the lower classes. Not that the male brand of medicine was superior. In fact, as mentioned above, male healers seemed to depend largely on magic spells and incantations. Says Ruby Rohrlich, "As medicine became a male profession, serving mainly the elites, these 'old husbands' tales' seem to have become incorporated into the medical lore, although the women healers ministering to the lower classes probably continued to use the herbal medicines" ("State," 89).

Elise Boulding speculates that, "as we know was the case later in Europe," women healers may have prepared lists of herbal remedies for the new male doctors (*Underside,* 180). Since the male intrusion represented a direct conflict with a time-honored role for women, it is hard to imagine why the women would have acted as they did unless to save patients from suffering at the hands of ignorant practitioners. The early medical tablet mentioned above may have been such a practical pharmacopeia.

A complex Sumerian myth written down about 2000 BC details the origins of the Plant Goddess Uttu. Though a female deity gives birth to the healing deities, the degree of male participation—and instigation—is striking. Ninhursag, one of the most ancient Sumerian deities, perhaps to be identified with Ki (Earth) herself, or with the cosmic mountain representing the united heaven and earth before their separation, does not give birth unaided, but only after receiving the "water of the heart" which Enki, the water god, causes to flow upon her. From this union springs Ninsar who, impregnated in turn by Enki, gives birth to Ninkur who, again with Enki's help, gives birth to Uttu. Uttu gives birth to the eight important plants after impregnation by Enki—who then proceeds to eat the plants. Then Ninhursag curses him, and disappears.

Persuaded by the fox to return when Enki is near death, Ninhursag gives birth to eight healing deities, one for each of Enki's ailing parts. But even this she achieves only after putting Enki into her vagina (Rohrlich, "State," 86).

Note that Ninhursag provides healing *deities* instead of healing *plants.* Kramer comments on a "superficiality and barren artificiality of the concepts implied in this closing passage of our myth." It is clear, he says, from the Sumerian original that "the actual relationship between each of the 'healing' deities and the sickness which it is supposed to cure, is verbal and nominal only. . . ." In other words, "it is only because the name of the deity *sounded* like the sick body-member that the makers of this myth were induced to associate the two . . ." (emphasis Kramer's; *Mythology,* 59). Old husbands' tales were indeed creeping in.

The transitional nature of this myth is emphasized by Ruby Rohrlich, who considers it a takeover myth of another and more fundamental kind: "This seems to be the first mythic attempt in human history to wrest from the female deity the principal role in procreation. Although Enki fails here, Ninhursag continues to lose rank throughout the second millennium, 'until she seems completely supplanted by Enki' " ("State," 86). At the very least, the mythic importance of the male role in procreation is unmistakable.

The process is better documented for Greece. According to Robert Graves, the Aesclepius myth refers to the suppression, in Apollo's name, of a pre-Hellenic medical cult, presided over by Moon-priestesses at various oracular shrines (*Myths*, I: 176–7). Graves suggests that the curative rites were secret, and it was death even to investigate them. Such a harsh penalty could mean that the priestesses foresaw, and tried to prevent, the coming takeover.

Elise Boulding mentions colleges of priestesses connected with various female deities or legendary figures—Carmenta, Nicostrata, Blodeuwedd, and Samothea—as well as a Minoan-Cretan college, and a Vedic college where Kali invented the Sanskrit alphabet. Boulding connects these colleges with women's legendary invention of the alphabet; but colleges of priestesses can also be associated with medical schools, as noted above regarding Isis. When Athena's gift of wisdom was displaced by the new order, at least one group of fifty maidens drowned themselves rather than be ruled by Apollo. Comments Boulding, "The stories of mass drownings of colleges of priestesses hint at dramatic takeovers by the men" (*Underside*, 193–4).

In addition to the suppression in Apollo's name of the Moon-priestesses' medical cult, Graves mentions "brutal" repression of a college of priestesses by Hittite invaders, and seizure of sanctuaries and forced marriages of Moon-priestesses by Hellenes (*Myths*, I: 320, 129, 204). Graves connects a college of priestesses with the medical cult suppressed in the Aesclepius myth (*ibid.*, 177): Apollo's cursing the crow, burning Coronis to death for her love affair, claiming Aesclepius as his son, and then, with Cheiron, teaching him the healing art means that "Apollo's Hellenic priests were helped by their Magnesian allies . . . to take over a Thessalian crow-oracle, hero and all, expelling the college of Moon-priestesses and suppressing the worship of the goddess." Graves makes clear elsewhere that fifty Moon-priestesses were the normal complement of these colleges, which seemed to have a range of roles in addition to healing, from watering the land to prophecy to ensuring good fishing (*Myths* 1: 81, 129, 204).

An interesting bit of graphic evidence suggests a possible Greek medical takeover or transition myth: an 11th-century manuscript shows Artemis giving the powerful healing herb mugwort to the centaur Cheiron, the ancient physician who lived on Mount Pelion in Thessaly (J. Harrison, 37). As we have just seen, Cheiron was one of Aesclepius' teachers.

By the time of Pericles, medicine was among the professions specifically forbidden to women in Athens. Records survive of women who practiced illegally, and were punished (Boulding, *Underside*, 260, table 6-1).

In Welsh legend, a "maiden from under the water" marries a man, bears him three sons, and then leaves because he breaks the conditions of her staying. One day she appears to her eldest son and charges him with a great work—healing the sick and caring for the helpless. When he asks how, she hands him a satchel full of recipes and prescriptions, saying "Here is knowledge for thirty, much less three." The brothers became the most skillful healers in Wales and were known into modern times as the Meddygon Myddfai, their family continuing the tradition for eight hundred years (G. Jones, 188–94). This tale seems to me analogous to the myths in which Athena instructs Triptolemus in agriculture, or gives the fig tree to Phytalus and his family.

In most of Europe, the takeover may begin with the introduction of Christianity. One of the Church's darkest chapters is the witch-burning holocaust of late Medieval and early modern times.The overwhelming majority of those killed were women who still knew the ancient herbal medicine—or who did not wish to marry and live ordinary lives in settlements—or both. Fragments like the Polish tradition of Dziwozony may show how these wise and non-conforming women were transformed into "other," which alone can account for what happened (Monaghan, 90):

> Dziwozony . . . was the Polish name for the wild women of the woods, whom the Bulgarians called Divi-Te Zeni and the Bohemians, Divozensky. They lived in the forest in underground burrows, seeking to understand the secrets of nature, especially those of herbal medicine. The Dziwozony were said to have large square heads, long fingers, and very ruddy bodies.

In Africa, the takeover may have been delayed till the 19th or even the 20th century, with the arrival of white invaders bearing a male-monotheistic religion and Western medicine. Or it may have happened earlier, arising independently or with Arabic or other patriarchal influences. An anthropology student in the late 1970s reported to me a medical takeover myth from East Africa. This myth stated explicitly that the women of the group had once been in charge of healing and medicine, but the men had stolen the secrets and the women were no longer allowed to be healers.

The male takeover of healing in proto-Western societies may have been less complete than surviving records would indicate, since those records were usually written or compiled by men. Hurd-Mead suggests, for instance, that Soranus of Ephesus (AD 98–138), whom Norman Himes calls the greatest gynecologist of antiquity, may have relied on the work—described in a treatise now lost—of a woman gynecologist called Cleopatra (Hurd-Mead, 59). Nevertheless, the loss to us is scarcely less stunning to contemplate.

From centuries just before and just after the beginning of the Christian era, we have only a few names of women healers or students of medicine, but some of them are inventors. **Mary the Jewess,** for example, an important early alchemist who probably lived and worked in Alexandria in the first century AD, invented a prototype of the autoclave (F. Taylor, 39).[9]

In Rome, **Octavia,** sister of the Emperor Augustus (1st century BC), studied medicine and practiced in her own home, inventing "many useful remedies." Scribonius Largus cites her toothache remedy: barley flour, honey, and vinegar mixed with salt and baked, then pulverized with charcoal and scented with spikenard flowers. **Messalina** (AD 21–47), wife of Claudius, invented a toothpaste and devised several remedies using burnt staghorn. Hurd-Mead suggests that she studied medicine with Vettius Valens, founder of a new medical sect (59–61).

Galen (c. AD 130–c. AD 200), greatest medical writer after Hippocrates, revered for over 1400 years, praises several medical women for specific remedies: **Origenia** for remedies for diarrhea and spitting blood; **Eugereasia** for her remedy for nephritis or inflammation of the kidneys (containing squills, bryonia, white pepper, cedar berries, iris root, myrrh and wine); **Xanita** for something like a Dover's powder. A colleague of Galen was a woman doctor named **Antiochis,** specializing in diseases of the spleen, in arthritis, and in sciatica. Antiochis was also interested in techniques for preserving female beauty. Galen

---

9. The airtight autoclave sterilizes surgical instruments using steam brought under pressure, thus providing moist heat at temperatures above boiling. (For Mary's other inventions of early chemical and protochemical apparatus, see Alic, "Women.")

copied her prescriptions for plasters to use in dropsy and sciatica, chest pain, and gout. This is probably the Antiochis to whom the town of Tlos in Asia Minor erected a monument in appreciation of her medical ability (*ibid.*, 62, 63).

Hurd-Mead mentions several medical women working during the next 800 years (*ibid.*, chap. 2), but as they are often known only from a tombstone inscription, she cannot detail their possible inventions or innovations. She does establish, however, that women worked as physicians and not just as midwives during this period. She also mentions that because of Galen's preference for medicine, surgery was left largely to women (and barbers) for the next millennium. Thus if surgery advanced at all during this period, women have a good chance of being responsible for the advances.

Then in the 11th century came Trotula of Salerno,[10] writer of several books on medicine, and a member of the new faculty when the medical school at Salerno was reorganized in mid-century. Trotula wrote a book on general medicine, as well as her complete compendium on obstetrics and gynecology, and is also credited with some of the earliest writing on children's diseases. She used an infusion of aloes in rosewater for skin diseases and blemishes, and advocated a highly nourishing diet for persistent cases of female sterility. Her criteria for wet nurses show that she knew the dangers of syphilis and tuberculosis. Trotula is credited, in fact, with being first in Western medicine to recognize the skin eruptions and other outward manifestations of syphilis. Trotula's operation for a completely torn perineum seems to have been original with her. Her care to avoid postoperative contamination with feces in this procedure indicates some understanding of the origins of infection (Hurd-Mead, *HWM,* 144ff; 150; "Trotula").

## Middle Ages

Kate Campbell Hurd-Mead, Muriel Joy Hughes, Mary Daly, and Barbara Ehrenreich and Deirdre English have given us considerable insight into Medieval women's role in healing and medicine. As Christianity spread, Europe began to experience a male takeover of an older, female-dominated, herbally based practice of medicine arguably analogous to that the Near East had experienced centuries before. Again, the intruding males knew more about religion than about remedies, more about heresy than about herbs. As Hurd-Mead drily puts it, "we cannot believe that the mere study of scholastic philosophy and of astrology, or even of the Seven Liberal Arts . . . could have made a better doctor of a monk than deep herbal knowledge and practical bedside study of a patient made of women" (4).

Again, too, a religious change is involved. We need not rehearse the story at length, but a brief reminder may be in order. The Catholic and Lutheran Churches waged war on women healers or wise women—branding them as sorceresses or witches and burning literally millions of them alive between AD 1400 and 1700,[11] for two reasons: to advance the economic interest of their university-trained male doctors, and to stamp out the old (Goddess-worshipping) religion, which many of the wise women still practiced—or were accused of practicing.

A further motivation for the murders may have been that the women healers were too successful. Life on earth was supposed to be painful and full of suffering, so that people would focus on the afterlife. These women often succeeded in curing the diseases that

---

10. After extensive research, Hurd-Mead became convinced not only of Trotula's existence, but of her medical eminence and its persistence for 500 years (*HWM,* 123–42, and "Trotula"). For modern discussions see Mason-Hohl; Stuard, "Trot."

11. Mary Daly suggests 9 million, but even some conservative estimates range as high as 3 million; 183n.

otherwise scourged humankind. Worse yet, Mary Daly suggests, the women's success might have shown up the incompetence of the male professionals (*Gyn/Ecology*, 193–4; cf. Sjöö & Mor, 44).

Our loss in this horror is not limited to the sufferings of the women burned alive and their families—particularly daughters forced to watch—or even to the curtailment of women's power and prestige. The witch-burnings have left a legacy of fear and superstition that plagues us still. Most pertinent here, they struck a blow at medical knowledge from which Western medicine has still not recovered. A few women healers survived, and some knowledge had been communicated to male relatives—like Paracelsus, for example. But literally hundreds of the more intricate remedies—to say nothing of the complex instructions for gathering times, drying methods, best solvents or other extraction methods, etc., and the accompanying dietary and other regimens for certain treatments—would by their very nature be likely to be lost. They could not be conveyed in the quick-and-dirty herbals that survive.

Especially prone to loss, I speculate, would be the emollient or other modifying herbs that women had devised over the millennia to soften the harsh effects of certain plant drugs. Recall, for example, that the original prescription for dropsy, whose "active" ingredient modern science determined to be digitalis, contained *some twenty herbs.* George Devereux notes that Amazonian natives of Brazil add to all their drugs a "corrective" second drug. When Amazonas women have had the desired one or two children, they take the *imdi rau* or "remedy of the blood" to become permanently sterile. This is a decoction of the leaf of a small forest shrub, mixed with the juice of another plant as a corrective or auxiliary. Moreover, while taking the remedy, the woman "observes a very rigorous diet, barely nourishing herself with fruits and tubers." According to Devereux's firsthand observers, the 114 women of the group had only a total of 110 children among them (38, 197). Note, too, that the toxic effects of the male contraceptive gossypol, from cottonseed oil, can be avoided or reduced by taking it with iron salts, which bind to and inactivate gossypol (Djerassi, *Politics,* 203–4).

The cryptic and possibly confused Celtic myth of Miach/Midach may reflect the loss of traditional medicine: Miach was a son of Diancecht (a prominent healer of the Tuatha De Danann), and one of the four who sang charms over the healing well that revived the fallen after the second battle of Moytura. Miach restored the hand of Nuada, which had been cut off nine days before, and for which Diancecht had substituted a silver hand. Diancecht became angry and struck his son on the head with a sword. Miach healed himself after the first three blows, but the fourth penetrated his brain. From his grave sprang 365 herbs "according to the number of his joints and sinews." His sister Airmed gathered these herbs, but Diancecht confused them so hopelessly that "no one knows their proper cures unless the [Holy] Spirit should teach them afterwards." The myth may be late enough to reflect a takeover, for Miach's name is borrowed from Latin *medicus,* "physician" (W. Jayne, 518).

Sigrid Undset, a trained anthropologist as well as a Nobel laureate in literature, paints a vivid portrait of a Medieval woman healer—and of the conflict between wise women and the Church in Norway—in *Kristin Lavransdatter* (I: 34ff). When Kristin's younger sister Ulvhild is crushed by a falling beam, the family sends for both a priest and a "leechwife." Priests were also doctors, or "leeches"—a point not often brought out—and liked to be called in case of illness. Their priest is fairly competent, but Undset makes clear that this is not always the case. The name "leech" reflects the doctors' favorite and sometimes only treatment method—bleeding. Both priest and leechwife being temporarily unavailable, Kristin's mother sends for Lady Aashild, the local "witch-wife" (I: 37–8):

It seemed to [Kristin] now as if stony hands were pressing on her heart. . . . Her mother would not . . . send for Lady Aashild even when she herself was near death's door at Ulvhild's birth. . . . [Lady A.] was a witch-wife, folk said—the Bishop of Oslo and the chapter had held session on her; and she must have been put to death . . . , had [she] not been . . . of such high birth and . . . like a sister to Queen Ingebjorg. . . .

She was said to have bewitched her young second husband, and

None of the great folk in the Dale would have to do with them; but, privily, folk sought her counsel—nay, poor folk went openly to her with their troubles and hurts; they said she was kind, but they feared her, too.

At this transition point in Scandinavia, as in the Near East long before, male doctors treat the wealthy, and women healers treat the poor, often without fee.

Hearing of the accident, the priest arrives before Aashild. He gives Ulvhild an anaesthetic, examines her carefully, and concludes that nothing is to be done but pray.

Undeterred, Lady Aashild stays with the family for the summer, taking full charge of Ulvhild's care. She treats the child with herbs (including preparations for restful sleep)—and obviously with love, reading to her daily, and creating for her a light back-brace or corset of leather and twigs. Eventually, contrary to the priest's prognosis, Ulvhild becomes strong enough to walk about with a crutch.

In the absence of women healers' own records, we cannot be sure what they inherited from their foremothers and what they invented for themselves, but we have some interesting indications. Peasant women of Eastern Europe, for example, knew that binding a piece of moldy bread over a wound would keep it from festering. In short, they had devised primitive antibiotic therapy. Women healers warned against refined white sugar, whose dangers are only now being pharmacologically confirmed (Raper, "Decade," 1; Dufty, 28–40, esp. 32–6).

Siberian peasants had a palliative herb for leprosy, which the heroic Englishwoman Kate Marsden journeyed fourteen thousand miles by sledge and horseback across Europe and Russia to find (Marsden, "Leper," *Sledge*). Though Marsden brought knowledge of it to the West only in the last century, this was very likely an ancient remedy.

Medieval literature contains many episodes in which women act as healers (see Hurd-Mead, 227–9; Hughes, *passim*). In the tales of the Round Table, "fayre Elayne" gathered herbs to make for Sir Launcelot "a bayne," and the squire's beautiful daughter treated the injured Gawain. The damsels in the Tristan stories used plasters of egg white, fennel, salt, and plantain leaves on infected wounds, gave mandrake for pain, and, with mandrake as a narcotic, even removed cataracts. Isolde herself is credited sixteen times in the English versions of the stories with both medical and surgical skill (Hurd-Mead, 264), and Isolde's mother is even more skilled and powerful.

Three 13th-century Spanish medical writers admit their indebtedness to women: Arnald of Villanova (1234–1311); Petrus Hispanus (later Pope John XXI; d. 1277); and Raymond Lull (1235–1315). Arnald taught medicine to pupils of both sexes at Montpellier, and said that he learned many of his remedies from medical women. He cites specifically a Roman woman with a remarkable plaster to cure sore throat and a Montpellier woman with a secret remedy for hemorrhage (Hurd-Mead, 236).

In the 14th century a medical professor at Oxford rode forty miles to get a woman doctor's prescription for jaundice (*ibid.*, 262).

**Catherine Sforza,** Countess of Forli, was an Italian Renaissance woman known, among other things, for doctoring her soldiers and her subjects in time of plague (Stanford).

Paracelsus (1493–1541), first to disagree openly with Galen's humoral theory, was the child of two physicians. His mother, superintendent of the hospital at Einsiedeln, Sweden, taught him until he entered the University of Basel. Later in life, he admitted that he learned some of his best prescriptions from women, especially one "wonder-drink" from an old Swedish woman. Did he learn his successful treatment of syphilis with mercury from a woman? His statement in 1527 that he burned his pharmaceutical texts because their knowledge came from the Sorceress is famous (Hurd-Mead, 375–6, 341; Ehrenreich & English, 17).

While the unequal battle for control of medicine blazed across Europe in the death fires of wise women, women continued to work as healers during the 16th through 18th centuries. As before, literature gives us some idea of their work. For example, Erminia, heroine of Tasso's 16th-century *Gerusalemme Liberata,* cures Tancred with remedies from her own garden or bought for the purpose. Her mother had taught her how to use all kinds of herbs for healing wounds and quieting pain (Hurd-Mead, 351).

History holds examples of real women healers. In 16th-century Italy **Loredana Marcello-Mocenigo** (d. 1572) was known for the formulas and recipes she devised for plague victims. These were presumably palliatives rather than cures, but nevertheless noteworthy (Forfreedom, 119).

**Louyse Bourgeois** (1563–1636) was the most famous French obstetrician of the 16th century, friend and pupil of the great Ambroise Paré, and midwife to the queen. Discussed in more detail in Chapter 3, she is noteworthy here as the first to treat the form of anemia called chlorosis with iron (Hurd-Mead, 356; Chaff *et al.,* 12).

Margaret Pelling of the Wellcome Institute for the History of Medicine, Oxford, has studied medical practitioners in Norwich, England. As for male practitioners learning remedies from women, she says, "Remedies exchanged between [laypersons] and medical men are regularly attributed to women." A specific example is the treatment for scaldhead adopted (and taught to men) in St. Bartholomew's Hospital in the 17th century.[12] In a study covering 1550–1640 (roughly Bourgeois' lifetime), Dr. Pelling finds at least 30 recorded women medical practitioners[13] connected with the barber-surgeons' company (Norwich had 292 practitioners, all told; Pelling, "Tradition," 163).

Some of these women had formal qualifications, such as licenses in physic and surgery, and "were also involved in apprenticeship." Although some 16th-century industrial legislation had excluded women, an early constitution of the Norwich Barbers' Company referred to "brothers and sisters," implying some equality of women in the Company (Pelling & Webster, 222). By the 16th century, however, Company ordinances implied, if they did not actually state, that women could run businesses and indenture apprentices only in the right of their husbands. And none of Pelling's 30 women practitioners figured in her Group A, those "who bore some part in the formalities and responsibilities of the city and the barber-surgeons' company." Thus, although Pelling contends that medicine was by no means fully professionalized, except possibly in London, at this period, one of the more important elements of the takeover is clearly well advanced even in the provinces ("Occupational Diversity," 507–9).

At the same time, more than a third of the medical practitioners hired by Norwich to

---

12. Scaldhead is a scalp disease characterized by pustules, scaling, and falling hair (L-12/23/86).
13. This figure, which includes some midwives, is probably too low. As Pelling notes, if as many midwives were active in each twenty-year period of the study as have been documented for one decade, it could rise as high as 70.

make the poor fit for work were women, and the City hired practitioners of either sex even if they were neither "free of the city," native-born, nor licensed (by the Church) to practice (Pelling, "Healing," 121–2).

From another part of England, in the same century, comes a rare instance of a woman medical practitioner known by name. This was **Elizabeth Talbot** (d. c. 1651), Countess of Kent, granddaughter of the famous Bess of Hardwicke. In all likelihood we know her work because she wrote *A Choice Manuall of Rare and Select Secrets in Physick and Chirurgery.* She may have devised or improved many remedies, but was apparently best known for the "Countess of Kent's Special Powder," or "Gaseous Powder." Grains of the powder dissolved in various liquids were used for complaints from phlegm to flatulence and "fits of the Mother," a form of hysterical convulsion afflicting women. The remedy was also considered good for smallpox, measles, spotted or purple fever, "swoonings and passions of the heart," and "Plague or Pestilent Fevers" (Fraser, 47).

**Hannah Woolley** (b. 1623) is another Englishwoman roughly contemporary with Pelling's Norwich period. Best known for supporting herself as a writer at this early date, she also learned "physick and chirurgery" from her mother and older sister and devised and sold her salves and ointments, "sovereign pills" and "cordials" (Fraser, 318–20; Wallas, 30). Indeed, she listed the making of salves and ointments among her accomplishments as early as age 14. So far as I know, Woolley did not patent these remedies, though by the 18th century some few Englishwomen were doing so. On March 26, 1760, for instance, **Ann Pike** received British Patent #748 for an Ointment for itch and all scorbutic humours.

Englishwomen of this period sometimes recorded medical inventions in their diaries. **Elizabeth Freke** (1641–1714), for instance, listed 446 of what Harriet Blodgett calls "home remedies" in her diary from 1671 to 1714. **Lady Eleanor Butler** (1739?–1829) had created an "ingenious remedy" for severe physical trauma, a "poultice of Bread and Milk and Sweet Oil as Hot as he could bear it," which she dispatched in 1789 to a man run over by a wagon. And **Lady Fortescue** had "a receipt to preserve the teeth," consisting of a tobacco-water wash, recorded in 1788 (Blodgett, 67).

In 1775 a French widow, **La Dame des Granges,** invented a barley flour (poultice?) that cured chest ailments. The recipe was advertised for sale in an 18th-century French periodical by one Demoiselle Durand, a seller of fashionable clothing (Hafter, L-5/12/1980).

The great **Marie Colinet,** whose work spanned the late 16th and early 17th centuries, originally worked as a midwife in Geneva, but became more noteworthy in other fields. She married the renowned surgeon Fabricius of Hilden, who taught her surgery. By his own admission, she excelled him. In one especially difficult case, of a man with two shattered ribs, she had to open his chest and wire together the fragments of bone. On reclosing the wound, she covered it with a dressing of oil of roses and a plaster of barley flour, powdered roses, and wild pomegranate flowers, mixed with cypress nuts and raw eggs; then bandaged it with padded splints. Afterwards, she regulated his diet, staying with him for ten days. He was well in four weeks (Hurd-Mead, 361, 433).

Marie Colinet was the first to use a magnet to remove steel fragments from the eye. Medical historians usually credit Fabricius with this invention, but he states explicitly that it was hers (*ibid.,* 361).

France's Canadian colonies saw one of the most significant medical innovations of our period. In 1655 a medical missionary and devout laywoman named **Jeanne Mance** (c. 1607–73) devised the first prepaid medical plan in North America—perhaps in the world.

This was in what was to become Montreal, in a hospital appropriately named Hôtel Dieu (Bannerman, 29–30).

In England's American colonies, a **Mrs. Edwards** in effect founded the New York cosmetics industry when she advertised her miraculous beautifying wash for hands, face, and neck in 1736. It was not merely beautifying, however, but claimed to take away "skin-burnings or pimples" and "cure pustules, itchings, ring worms . . . and other like deformities of the face and skin," without any "corroding quality." Mrs. Edwards also sold a lip salve and a tooth powder (Selma Williams, 196).

Julie Matthaei reminds us that women then were credited with "magical" natural powers. These powers undoubtedly had more to do with the "folk knowledge of curative herbs and other medicines" inherited from their mothers than with anything supernatural, but the attribution of magic is itself significant, for it suggests both efficacy and specialization. Among the many women who had specific remedies to their credit was **Mary Bannister,** whose Drops of Spirit of Venice was "one of the most advertised remedies in colonial papers." A **Nurse Tucker** advertised an ointment to cure "the Piles, Rheumatism strains, all kinds of Pains, Ring-worm, Moths, Carbuncles, Sun-burning, Freckles, and chopping [chapping?] of the Skin; and . . . sore Breasts . . ." (Matthaei, 60).

It was in England itself in the late 18th century that one of the most important remedies ever credited to a woman came to light: a cure for dropsy used and kept a secret by "an old woman in Shropshire, who had sometimes made cures after the more regular practitioners had failed." This same cure was also known to countrywomen of Yorkshire; and Margaret Pelling several times mentions that women had cured some poor patient of dropsy. Though the medicine contained twenty-odd herbs, Dr. William Withering, who saw a patient it had miraculously helped in 1775, established that the active ingredient came from foxglove.

This showy flower, *Digitalis purpurea,* was familiar to every English gardener, and often mentioned in the botanical and medical literature from the 16th century onward, but not for this usage—and in fact, not for any correct usage. It took Dr. Withering ten years to discover (or rediscover) the proper gathering time, which part of the plant to use, the best drying method, and the proper dosage to avoid such unpleasant side effects as vomiting and purging. Mollifying these effects, of course, may have been the purpose of some of the other ingredients in the original remedy. In any case, digitalis is still saving lives, and the preferred form is still the dried and powdered leaves. The even more powerful digitoxin comes also from the foxglove leaf. Dr. Withering became famous for his *Account of the Foxglove . . .* (1785); the Shropshire woman's name is not known (N. Taylor, 121ff; Ehrenreich & English, 12).

Some of the most outstanding of women's medical achievements in these centuries— the work of the great French and German obstetricians and midwives—will be discussed elsewhere.

## Anatomy

Several women contributed significantly to medicine's supporting branch of anatomy during this period. Whether professors, demonstrators, or model-making technicians, they were all inventors. The earliest known is **Alessandra Giliani** of Bologna (1305–24), assistant to Mondino or Mondinus de Luzzi, often called the father of anatomy. She was Mondino's demonstrator, or *prosector,* which means that while he lectured to medical students from an elevated platform behind her, she did the actual dissections. Exquisitely skilled, she was also a pioneer in anatomical injection. She devised a way of drawing the blood from even the smallest veins and branches without tearing them, and then of injecting various colored liquids that hardened to make the vessels easily visible. Presumably the different colors distinguished between arteries and veins or emphasized particular pathways and networks. By this means she could show the circulatory system in all its complexity and detail from a model that was absolutely true to nature. Regrettably, none of my sources gives any idea what her injectable liquid may have been (Hughes, 86–7; Hurd-Mead, 224–5; Mozans, 237–8n).

Naturally she was quite useful to Mondino; in fact, according to a history of Bologna's medical school, she brought him great fame and credit. When she died at nineteen, Mondino's other assistant, probably her fiancé, erected a tablet to her memory in a local church (Hughes, 87):

> In this urn enclosed, the ashes of the body of Alexandra Giliani, a maiden of Periceto, skilful with the brush in anatomical demonstrations and a disciple, equaled by few, of the most noted physician Mundinus of Luzzi, await the resurrection. She . . . died consumed by her labors March 26, in the year of grace 1326. . . .

Lydia Farmer mentions the work of a "young female student" from the 15th century, wax models of human anatomy, now in the Bologna Museum, but presumably shown at the Columbian Exposition. Regrettably, Farmer does not name her.

The 18th century boasts three distinguished women anatomists who were also inventors—**Angelique du Coudray** of France (1712–89), **Anna Morandi Manzolini** of Bologna (1716–74), and **Mary Catherine Biheron** of Paris (b. 1730). These three are credited with inventing, respectively, the female torso mannikin for demonstrating obstetrical anatomy; anatomical preparations in wax of separate parts of the body (clearly invented earlier, but possibly lost); and a new modeling material for anatomical facsimiles.

Born into a prominent medical family in Clermont Ferrand, du Coudray became a licensed midwife in 1740 and head *accoucheuse* at the local hospital the same year. Her lecture methods were unique, for it was she and not the Englishman William Smellie who first used a female torso—which she called "une femme artificielle" in obstetrical lectures.[14] She also used an actual fetus to give her pupils practice in delivery (Hurd-Mead, *History,* 499). One reason her achievements have come down to us is that she wrote a textbook on obstetrics (1759) that went through five editions, and published her collected works in 1773 (Chicago, 176; Dall, *Historical,* 153–4; M. Gage/1: 19; Hurd-Mead, 497, 499–500; Marks & Beatty, 66–7).

Anna Morandi found herself, after only five years of marriage, with a sick husband and six children to support. She turned to the making of anatomical models in wax, an art she probably learned from the husband, Giovanni Manzolini, Professor of Anatomy at the University of Bologna. In the course of her studies she discovered the true termination of the oblique muscle of the eye (Hurd-Mead, 509n). In fact, Mozans credits her with "a number of important discoveries made in the course of her dissections of cadavers" (236). Upon her husband's death, Anna Manzolini became lecturer in her own right. In 1760 she became Professor of Anatomy, with the added title *modellatrice*.

Many other honors came to her during the last years of her life. The City of Milan tried to woo her away from Bologna by offering to let her name her own conditions (Mozans, 237). Emperor Joseph II of Austria bought one of her models. Catherine II of Russia

14. Matilda J. Gage erroneously calls her the first (anatomical) lecturer to use a mannikin, which she invented and then perfected. Gage may be correct in saying that her (female torso) mannikin was approved by the French Academy of Surgeons on December 1, 1758 (Gage/1: 19; Chaff *et al.,* 11).

invited her to Moscow to lecture and made her a member of the Russian Royal Scientific Association. The British Royal Society elected her a member and invited her to lecture in London. Several biographies have been written of her, and we even know something of her physical appearance and personality: tall and blonde and handsome, she was described as a tireless worker.

One of her models is in London's Wellcome Museum of the History of Medicine. A beautiful figure of a young woman's body, it reveals the viscera, made of "some preparation of clay and wax colored to imitate nature" (Hurd-Mead, 509n). This unidentified preparation or the coloring method—or both—may have been her own invention, improving on what she had learned from her husband. According to Caroline Dall, Anna Morandi is honored in Italy as the inventor and perfecter of anatomical preparations in wax. Considering the rapid decay of anatomical specimens from cadavers in the days before refrigeration, we can easily imagine the importance of this invention to the study of medicine. At about the same time, to show the state of the art in anatomical models before her work, Dr. William Smellie was lecturing in London, using a wooden woman "with an abdomen of leather, in which a vessel of beer, a cork stopper, and a bit of pack-thread imitated the impulses of nature" (Dall, *Historical,* 113ff, 155ff; *College,* 155n).[15]

Whether because she had invented a new material, or because of her sculptural skill, Anna Morandi Manzolini's anatomical models were "most highly valued and eagerly sought for on all sides." They could scarcely be distinguished from the parts of the body they imitated. Her own collection went to the Medical Institute of Bologna after her death, and became one of its most precious possessions (Mozans, 236).

Mlle Biheron's road to success was much harder. Born the daughter of an apothecary, she loved anatomy but was too poor to attend dissections. She nevertheless began her studies at the age of sixteen. She got specimens from body-snatchers and kept the corpses concealed in her room. Frustrated with their rapid decay, she learned to imitate the parts in wax. Eventually her work came to the notice of the Academician Jussieu and the celebrated Paris physician Villoisin, both of whom aided her. Over some thirty years, she perfected the anatomical mannikin. She also invented a special modeling material that was more permanent than wax. She kept its formula secret, preferring to sell her models instead. Her facsimiles of body parts were so lifelike that, as one wit exclaimed, they lacked only the odor of the originals.

Gradually her work brought honors. When Prince (later King) Gustavus of Sweden visited Paris in 1771, Mary Catherine Biheron was invited, along with Lavoisier, to lecture before him at the Academy of Sciences. The Prince was so impressed that he offered her the position of demonstrator of anatomy in the Swedish Royal University. She taught John Hunter and many other famous anatomists, and according to Dall much of Hunter's later work was based on Biheron's observations.

Her collection of waxworks was open to the public during her lifetime. Eventually, Catherine of Russia bought the entire set (Dall, *Historical,* 155–6, and *College,* 155n; Mozans, 238; Hurd-Mead, 492).

The American reformer and women's-rights activist Paulina Kellogg Wright Davis (1813–76) was apparently the first to import a female anatomical mannikin into the United States. She had studied anatomy during her first marriage and lectured to women on anatomy and health reform after her husband's death. When she unveiled this

15. Dall notes but rejects the conflicting claims of Gigoli Tumnio and Lelli to having invented wax models.

anatomically accurate *femme modèle,* some of her hearers fled the room or fainted (Hanaford, 677–82; *NAW*).

One American example of an inventor of anatomical models is **Dr. Eliza M. Mosher** (1846–1928), mentioned here because I do not discuss anatomy for the 19th century. Best known, perhaps, for her work on human posture, Dr. Mosher created a "Posture Model" of the human spine, with the vertebrae so articulated as to allow various natural—and presumably unnatural—spinal positions for demonstration purposes.

## 19TH AND 20TH CENTURIES

### *Apparatus, Instruments*

#### *19th century*

A Frenchwoman, **Mme Rondet** (b. 1800), invented a tube for resuscitating newborn infants deprived of oxygen during labor. Chaussier's earlier tube was awkward to use, but the Royal Academy continued to credit him alone for the invention (Dall, *Historical,* 161–1; *College,* 156n).

At least one Englishwoman, late in the century, received an American patent for medical apparatus: **Elizabeth Rowland** of London, for her inhaler (1899).

Among American women who patented medical or related apparatus in the 19th century should be listed **Clarissa Britain** of St. Joseph, MI, and Chicago, who patented an Improvement in ambulances in 1863; and three women who patented improved hospital beds: **Hannah Conway** of Dayton, OH (1868); **Isabella Waller** of Cleveland, OH (1872); and **Elizabeth Staples** of West New Brighton, NY (1888). Staples' invention was called a Hospital transfer-bed. Britain's ambulance patent is only nine years later than the first recorded use in English of the word *ambulance* to refer to an "ambulance wagon" (*OED*).

**Clarissa Britain,** though totally unknown today, may be of some interest. Her residence is listed on her patents as St. Joseph, MI. She held seven patents by 1864, making her one of the most prolific American women inventors at that time, certainly the most prolific *patentee.* Ida Tarbell mentioned her in an 1887 article on women inventors for the *Chautauquan* magazine (357). A schoolmistress (head of the Niles Female Seminary in 1843), Clarissa Britain was the sister, and housekeeper and hostess, of Calvin Britain, founder of St. Joseph and later Lieutenant Governor of Michigan. In 1872, though living in Chicago, she was still a landowner and mortgage-holder in Berrien County, MI. She was listed in her own right in the Chicago city directory of 1871, though her household contained two males (Burnham, L-3/2/82; Pugh, L-5/4 and 6/4/82; *Niles Democrat,* Sat., Apr. 20, 1872: 2; 1850 Michigan census).

Some thirty years later, another woman patented an interesting-sounding ambulance. **Cornelia H.W. Larrabee** of New York City patented a Combined ambulance and tourist-wagon on December 6, 1898.

The several women inventors of electro-medical apparatus are discussed below. Noteworthy among them is **Dr. Elizabeth French,** who exhibited her electrotherapeutic apparatus at the Philadelphia Centennial, and wrote books on electrotherapy.

**Mary F. Henderson**'s Movement-cure machine is of interest primarily in contrast with the rest-cures so often advocated for 19th-century women, as in the *Yellow Wallpaper* (Gilman). Henderson, who lived in St. Louis, received her patent in 1889.

**Mary Florence Potts,** entrepreneur and inventor of safer and more comfortable sadirons, also invented a "remedial medical appliance," presumably using heat therapeutically. By then living in Austin, IL, she received this 1892 patent jointly with her husband Joseph.

**Dr. Charlotte Amanda Blake Brown** (1846–1904) was physician, surgeon, hospital founder, inventor, and medical innovator all in one. Born in Philadelphia, a first child, she graduated from Elmira College (in New York) in 1866, married Henry Adams Brown in 1867, and had three children in rapid succession. During her pregnancies, following a family tradition and with her husband's encouragement, Brown began reading anatomy with the family doctor. In 1872 after the birth of her third child, she left her family in California to enter the Woman's Medical College of Pennsylvania, receiving her degree in two years.

In San Francisco in 1875, with the support of several other women, she founded a dispensary for women and children that was the forerunner of San Francisco Children's Hospital. Its purpose was twofold: to provide female doctors for women patients and to give those doctors professional experience. Although not known as a feminist, Dr. Brown was focused on the welfare of women and children.

Her medical career was marked by innovation. She designed a milk-sterilizing apparatus for her own hospital. In 1887 she suggested the establishment of a tumor registry, the bank of data on cancer patients that underlies so much of our progress against cancer. In 1893 she told her colleagues about incubators that were saving premature babies in France.

Her spare time during her last years was devoted to the California branch of the National Conference of Charities and Correction, which she had helped organize. She died in 1904, at fifty-seven. The San Francisco Children's Hospital now carries on her work (*NAW*).

Among the medical-apparatus inventions of less famous women may be listed **Mary J. Smith**'s Press for preparing fomentations. This New Yorker received her patent in 1899. Charlotte Smith gives us some other examples in *The Woman Inventor* (#1), some of which do not appear on LWP:

In the late 1880s **Dr. Josephine G. Davis** of New York City invented a portable means of administering a hot-air or vapor bath. Using a "suitable lamp" as a source of heat, the device suspended a canvas or rubber covering from the shoulders to contain the heat, while keeping the heat source from touching the back.

Nurse **Annie Caller** of Albany and Troy, NY, saw the need for a better surgical splint, and proceeded to invent one, patenting it in 1888. Likewise, the meat broilers of her day did not preserve the flavor and juices that would make food tempting to invalids, so she invented and patented an improved meat broiler in 1889. Smith called her surgical splint "very useful."

Caller's meat broiler was listed in *Inventive Age;* and the Shaker *Manifesto* for July 1890 (20: 7: 164), citing *Inventive Age,* listed the broiler as an example of several "really valuable devices" recently patented by women in their "steady march . . . to the front, as inventors."

Both *Inventive Age* and the *Manifesto* also mention **Emeline C. Post** of Yorktown, VA, and her Therapeutic bottle (Design Patent) of 1889, and **Caroline Drake**'s Hospital bed attachment. Not having found Drake's patent, I do not know her residence or the date of her invention, except that it was obviously before July 1890. Charlotte Smith calls Post's heat-administering bottle "handsome in design and of unquestioned utility." The bottle was used in hospitals to replace hot cloths and mustard plasters (*WI/2*: 2: 3).

**Hannah Milson** of Buffalo, NY, patented an Improvement in ozone machines in 1876. Five years later the patent was re-issued, which indicates continued interest on Milson's part in improving and possibly commercially exploiting her invention. It is not clear to me how this device worked, but it is worth noting that as late as 1956 two Los Angeles men patented an Ozone therapeutic device.

LWP lists 15 other women receiving 16 patents for medical and therapeutic apparatus: an aspirator/cauterizer (uterine), disinfecting and purifying apparatus (2 patents), a facial symmetry device and a finger compress (same patent), an inhaler, a medical mask, a poultice protector, a respirator, a druggist's sieve, a speculum, four syringes, and an improved vapor bath.

Three years too late for LWP, in 1898, **Jennie C. Harrington** of Ossian, IA, patented a Clinical thermometer shield.

## 20th century

Among 20th-century inventions of medical apparatus must be mentioned **Marie Curie**'s mobile X-ray units, the "Little Curies" she created during World War I by putting X-ray machines into Renault cars commandeered from her friends, to be driven onto the battlefield. Indeed, she drove one of them herself. Considering her dedication to physics—and her pre-war inability to drive—her achievement is doubly amazing (Eve Curie, 290ff). A generation that was weaned on M*A*S*H takes timely care of wounded soldiers for granted. In 1914 battlefield care itself was an innovation.

**Dr. Sara Josephine Baker** (1873–1945) is noteworthy as a child-health pioneer, for catching "Typhoid Mary" Mallon, and as a medical inventor. Taking her MD in 1898 at the Women's Medical College of the New York Infirmary for Woman and Children, she opened a private practice, supplementing her income by working as a city medical inspector. Appalled by the high infant mortality—1500 deaths a week in summer—in 1908 Dr. Baker took a team of thirty trained nurses into one East Side district. She taught bathing, proper ventilation, and proper clothing for infants, and urged breast-feeding to avoid unpasteurized milk. By summer's end her district showed 1200 fewer infant deaths than for 1907. Other districts showed no significant change. Her work led to the establishment of a division of child hygiene in the City Health Department, with her as director.

As head of the world's first tax-supported agency entirely dedicated to child health, she had no precedents; thus her innovations were many. She emphasized education, distributing hygiene pamphlets and establishing a free training school and licensing procedures for midwives. She also set up Baby Health Stations, dispensing pure milk and health advice.

She was the first woman to receive a PhD in Public Health (1917). By the time she retired from the Child Hygiene Bureau, New York City had the lowest infant mortality rate of any major city in the United States or Europe.

Although Dr. Baker qualifies as a medical inventor and innovator on several grounds, the specific invention that places her here is the foolproof container she helped design for applying the medicine used to prevent blindness in newborns (*NAW*/mod.).

A whole class of apparatus that probably should be considered here is massage devices. A single example from early in the century must suffice: **Anna Dunder**'s Massage implement, patented in 1910. Worth at least a passing mention, in view of the vogue enjoyed by colonic therapy as late as the 1950s, is **Emma Adamson**'s Colonic therapy apparatus. Adamson, a resident of Denver, received her patent in 1930.

An important medical innovator who also invented a piece of medical apparatus is **Sister Elizabeth Kenny** of Australia (1886–1952). In addition to devising the only effective treatment for polio before vaccines, this Australian healer invented and patented the so-called Sylvia stretcher. According to some sources, Kenny invented an improved field stretcher during World War I, a wheeled and canopied conveyance that greatly eased the transport of emergency patients (Uglow; Cohn; Marlow, 263). Kenny's autobiography, however, mentions no invention until 1926, when she was called to a grisly accident in which a little girl named Sylvia had fallen under an eight-horse plough. The plough had broken both her legs and torn off two toes. Kenny used her own stretcher, a rigid cot equipped for treating shock. At the hospital, with shock controlled, surgery could begin at once, and Sylvia recovered. The surgeon was impressed: he had just observed a parallel case in which a little boy, not pre-treated for shock, had died after surgery. He suggested that Kenny patent her stretcher and make it available in all Australia's remote areas.

Characteristically, Sister Kenny presented the stretcher to the Queensland Women's Association, offering to let them patent it and profit from it. Luckily for the world, however, the Association's legal advisor insisted that she retain the royalties. It was these royalties that supported her long fight to get her polio treatment accepted. (To avoid giving her medical enemies ammunition, she never took money for any of her treatments.)

The War Offices of Australia, Britain, the Crown Colonies, France, and the United States were all interested in Kenny's stretcher (Kenny, 75–7; Marlow, 263, 264). My sources, however, leave unclear whether there were actually two different stretchers, or whether the Sylvia (shock-treatment) stretcher was an improvement on her original battlefield invention.

A Norwegian ex-nurse has invented a piece of emergency medical equipment that belongs alongside Dr. Kenny's invention, though it comes sixty years later. **Halldis Aalvik Thune** (1936– ) of Hønefoss, Norway, has patented a stiff support for the back and neck that allows accident and injury victims to be moved safely. Her Neck-Aid has wide, strong cloth wrappings that reach around to fasten in front of the torso and across the forehead with Velcro® (or similar) fastenings. Immobilizing the head and upper body, Neck-Aid protects the injured neck and spine from further damage until the victim can be evaluated and properly immobilized in a hospital. She got the idea after witnessing a car accident in which a driver with an injured back was trapped behind the wheel. She thought of the grooved sink drainboard in her kitchen, and visualized that kind of stiff support lengthened and placed under victims of spinal injury before moving them. Made of plastic, for lightness, Neck-Aid entered the market in 1986.

Halldis Thune is the oldest of seven children. Her father is an inventor. She has some artistic talent, and there are many artists in her family, one of them a nationally known sculptor. She is partly ambidextrous. Ms. Thune has an unusual background for a nurse, having minored in history and ethnology before entering nurse's training. When she got her idea for Neck-Aid, she left nursing, quit school, and formed Norhall Company to market her invention. At Klekk '86, the first Norwegian Inventors' Fair, she won 15,000 Norwegian kroner (about $2500) as the best female inventor at the Fair. In 1987 she won the World Intellectual Property Prize at the International Inventors' Exposition (Salon Internationale des Inventions) in Geneva (WIPO, S116,2).

Ms. Thune holds U.S. and Canadian patents, among others. The Norwegian National Health Service considered adopting her invention. At a medical conference held in Oslo in June 1986, doctors from Poland, Sweden, and other countries took a great interest in Neck-Aid, and one Swedish doctor decided to test it.

Halldis Thune was the second woman admitted to the Norwegian Inventors' Association, and the first to serve on its council. Concerned about getting girls and women into invention, she has proposed a school or workshop for women inventors, a joint effort with the Swedish Inventors' Association. She has been interviewed on radio and written a paper on women inventors. As of 1986 she had several other inventions in progress, two of them also medical. At age fifty she found her niche (Thune, L-4/24 and 7/18/86, and "Women As Inventors"; Kolbjørnsen & Apneseth; Warn & Nilsson).

Three examples of medical apparatus for hospital use are **Isabelle D. De Bella**'s apparatus for attaching a manometer (device to measure blood pressure) to a hospital bed, **Mary E. Nelson**'s surgical strap, and **Janet Brousseau**'s Orthopedic traction harness. In 1962 De Bella, of Washington, DC, patented a device with highly adjustable positioning, designed for hospital beds with hollow end-posts open at the top. Nelson, from Portland, OR, patented a strap to keep the patient's leg flexed during knee surgery (1978). The strap is continuously adjustable, and designed for quick release. Brousseau, from Los Altos, CA, co-patented the apparatus with Jon A. Brousseau.

In the early 1970s, **Reatha Wiggins** (c. 1935–   ), a black technician at New York's Metropolitan Hospital, patented a multiple aspirator (device for removing mucous or foreign material from the throat, especially needed on children's wards).

Born in Dublin, GA, Wiggins grew up in Youngstown, OH. Beginning as a hospital volunteer, she became a hematology technician. Often called to the children's wards, she saw that if the technicians and nurses could always carry several aspirators, they could save many trips and precious time. She created her invention in spare hours, and patented it with the help of an experienced inventor named Solomon Harper.

To this point Reatha Wiggins' story is a success story. But here, as so often happens, things began to go sour. Although at least one New York physician, Dr. Alex McCown, called her multiple aspirator "a useful piece of equipment," she could not interest a manufacturer, and watched helplessly as other technicians copied and used her idea. Interviewed in 1972, she feared that companies would simply wait until her patent expired and then start manufacturing the aspirator. Her comfort must be that her invention might some day save a child's life (H. Bernard, B-10).

One of England's early women engineers, a founder of the Women's Engineering Society and first woman member of the Institution of Mechanical Engineers, **Verena W. Holmes** (c. 1900–1964) had a number of medical and surgical inventions to her credit. Her 1920s invention of an artificial pneumothorax apparatus for the treatment of tuberculosis may be mentioned here. A joint invention with Dr. Wingfield, a family friend, the device was one of her most successful inventions. The "Holmes & Wingfield" was used, among other places, in the British prison system, and Holmes once received a letter from a grateful prisoner (Parsons, 4).

A minor but useful piece of pediatric apparatus is the child's urine-specimen collector devised by **Gertrude Müller** (1887–1954) of Fort Wayne, IN. Best known as co-founder of the Toidey Company and inventor of the Toidey Seat (1924), she also invented many other child-health and -safety items from the 1920s to the '50s. Her specimen-collector was acclaimed by pediatricians (Stewart, n.p.; *NAW*/mod.).

One of the more important pieces of 20th-century medical apparatus is **Dr. Jessie Wright**'s Respir-Aid rocking bed for polio victims who need help in breathing, but are not completely confined to an iron lung.

English-born Jessie Wright (1900–70), daughter of a headmaster, migrated with her family to the Pittsburgh area at age six. Becoming a U.S. citizen, she took both BS and MD degrees at the University of Pittsburgh (1932, 1934). After her internship she became medical director of the University's D.T. Watson School of Physiatrics, later serving as Instructor in Orthopedics (1936–51), Associate Professor of Physical Medicine and Rehabilitation (1952–67), and eventually trustee. Specializing in the study and treatment of crippling diseases, she received many honors and awards, including the presidency of the American Academy for Cerebral Palsy (1962), membership in the American Academy of Physical Medicine and Rehabilitation, honorary membership in the American Physical Therapy Association, Pittsburgh Woman of the Year (1950), and Distinguished Daughter of Pennsylvania (1952). She contributed articles to professional journals, served as a consultant to Argentina in setting up its rehabilitation program, and attended many international conferences.

For all her professionalism, however, Dr. Wright never forgot the importance of psychological and social factors in illness, and the vital role of the patient's family in rehabilitation. She was an elder in the Presbyterian Church, and took time for civic projects.

She got the idea for the fast-rocking bed as a breathing aid in 1944 when she and her staff were trying to free a patient from the respirator. The Sanders oscillating bed used in circulatory disease alternated too slowly and had other deficiencies for Dr. Wright's purpose. She communicated her idea and requirements to Charlotte Baron, a public health nurse from Toledo, OH, who was studying with her at the time, and Baron agreed to get a bed built to those specifications.

The basic unit is a sturdy, two-crank adjustable hospital bed mounted on special rockers connected to a remote electrical unit. The unit requires alternating current, but can use either 110 or 220 volts, and can be operated by hand if power fails. It can be adjusted to rock 12–26 times per minute, and to incline at any angle up to 45 degrees. An adjustable knee rest, a back support, a divided footboard, and padded shoulder guards keep the patient from sliding as the bed rocks. As the head of the bed rises, the abdominal organs shift downward, pulling the diaphragm down and drawing air into the lungs. Conversely, when the head falls, the organs press upward against the diaphragm, pushing air out.

First to try the bed was an eighteen-year-old girl stricken with polio just as she entered college. She was a quadriplegic, paralyzed in her trunk, and lacking power in the diaphragm. Now exhausted, undernourished, and suffering from pneumonia, she could not leave her respirator for more than three minutes. On Dr. Wright's rocking bed, however, she stayed out for three hours the first time. Penicillin cured the pneumonia, and nourishment was supplied intravenously. Within eight months, the girl was free of her respirator all day and till past midnight; twice she had stayed out for twenty-four hours.

As a contribution to life quality for polio's worst victims, the importance of The Respir-Aid can scarcely be overestimated. When the bed was invented, these victims were much more numerous in the United States than today. The incidence of polio had risen from 7.4 cases per 100,000 in 1937 to 14.3 per 100,000 in 1944, the year of Dr. Wright's idea—and major epidemics had hit Los Angeles in 1934 and Detroit in 1940.[16] The rocking bed not only aids breathing, but improves circulation and eases nursing care (as compared to an iron lung). The rocking motion and periodic pressure of the abdominal contents may even help reactivate the breathing response. The Respir-Aid is useful not only in polio, but in vascular and cardiac disease, for sedated psychiatric patients, and for others with temporary breathing problems. Thus the bed would still have a place in physical medicine and rehabilitation even if polio were conquered worldwide as in the U.S. Unfortunately, this is by no means so. As of 1981 fewer than 10% of the 80 million children born each year are immunized, and world health authorities note an "ominous increase" in paralytic polio in particular ("Five-Year Checkup," 82).

Dr. Wright's original article describing the bed, its operation, and its uses for polio victims appeared in the *American Journal of Nursing* (Jessie Wright, 454ff).

16. See Hand; WHO statistics, *Science News,* Apr. 5, 1986.

When technology made possible a portable breathing machine, **Helen Johnson Lubensky** was asked to help design it. She had worked for Schiaparelli in Paris, and had designed the WAVE uniform in World War II (D. Wright, PC-3/15/88).

**Anne R. Chamney** (1931– ) of London, England, is a chartered engineer, with a MS in biomechanics, and a PhD in physiology. Between science degrees she took a BA at the Open University. She is a mechanical engineer specializing in medical apparatus. After an apprenticeship with de Havilland Aircraft (1953–8) and classes at Hatfield Technical College, she became a Technical Assistant at the British Oxygen Company, makers of anaesthesiology equipment (1959–61). There, in the Medical Development section, she evaluated competitors' equipment, designed new machines and improved existing ones, but took out no patents.

Wishing to work more directly with physicians, Anne Chamney then became a Senior Technician at the University College Hospital Medical School. In 1966, on a Clayton Fellowship in Medical Engineering awarded her by the Institution of Mechanical Engineers, she designed a new oxygen tent. This is her best-known invention, winning her a Supplementary Award from the Engineering Materials and Design Association. Its advantages over conventional oxygen tents, says Chamney, are adequate oxygen concentrations in intensive-care situations; easier nursing access; much lighter weight (15 *vs.* 150 pounds) and thus greater ease of transport and set-up than for older tents; less than one-tenth the cost of the standard model. Chamney's device was patented by the National Research Development Corporation.

At the hospital she has been involved in many projects, including "several new designs for equipment," but the oxygen tent remains her only patent. As she explains, "In some cases, items I considered suitable [for patenting] were too close to existing, but loosely worded, patents." Like many other inventors, she feels patents are rarely worth the time and trouble, in view of the almost inevitable litigation. "Much of my design work," she says further, "has been on items I needed for other projects, and thus is neither published nor easy to assess."

Dr. Chamney feels that her most important work is her PhD research on the effects of carbon-monoxide exposure during pregnancy in rats. Since carbon monoxide occurs in cigarette smoke, this work is relevant to recent findings on smoking and human pregnancy. In any case, here is a multiple inventor honored by her peers for a major invention of medical apparatus. Several elements in her background correspond to my proposed inventor's profile: her father was an inventor; she went to an all-girls' school from nine to 16; and she was ambidextrous as an infant (Chamney, Q-1/18/84, L-5/9/84, and "Bioengineering"; Laverick, L-4/7/83; "Miss Anne Chamney").

Two recent apparatus inventions are noteworthy in that their inventors have no medical or technical training. **Helen Smit** and her daughter **Julie Smit** (1955– ) of Evanston, IL, have come up with at least three medical inventions. Patented are an anti-absorption lining for part of the intestine, and a skin vacuum cleaner. With their intestinal-lining invention, and their unpatented idea for a disposable (i.e., absorbable) balloon to be inserted in the stomach, these two women have independently arrived at two of the three or four main approaches to control of extreme obesity today. The skin cleaner could be useful for astronauts, for workers in "clean rooms" or hazardous occupations who need to remove all particles of a dangerous substance from their skins before leaving a work area, for health-care workers whose patients must avoid allergens, etc. The Smits have nonmedical inventions to their credit as well—for a total of at least five by 1985 (Crossen).

**Dr. Oon Tian Tan** of Boston Medical Center deserves a brief mention here. In devising the organic-dye-laser treatment for hemangioma birthmarks (see below), she almost certainly modified the existing apparatus. RN **June Abbey** may deserve more than a brief mention, but her work comes to my attention just as the book goes to press. In 1981 she developed an electronic monitor for bladder fullness (Sherrod, 11).

I will close this apparatus section with two highly significant inventions, one from the 1930s and one from the 1980s.

As early as 1935, Dr. John Gibbon and his wife **Mary Hopkinson Gibbon** (c. 1911–86) foresaw the possibility of open-heart surgery. Working at Massachusetts General Hospital in Boston, they devised a pump-oxygenator that could temporarily replace the human heart and lungs. This was the prototype of the heart-lung machine first used successfully almost twenty years later. At Jefferson Medical College in Philadelphia in 1953, the machine took over heart and lung function in a teen-aged girl for 26 minutes, while John Gibbon repaired a hole in her heart ("Heart-Lung . . ."; Harpur, 100–42).[17] Obviously, this is one of the most important apparatus inventions in recent medical history, making possible all open-heart surgery, and lengthier surgical procedures in general, to say nothing of heart and heart-lung transplants.

Mary came to John Gibbon's lab from Beacon Hill, Boston, via Bryn Mawr. She had previously worked as surgical research technician to Dr. Churchill, one of the joint chiefs of the West Surgical Service at Massachusetts General Hospital, and was present at the all-night vigil with a pulmonary embolism patient that spawned the idea for the heart-lung machine in February of 1931. Mary participated fully in other physiological experiments during that year, with published results, and when Churchill renewed John's research fellowship, he agreed to employ her part time as John's assistant. "We had two children," says John, "but in those days we were able to get good and inexpensive household help to care for the children while we were at work" (Gibbon, "Development," 233). The main project now was to be a mechanical replacement for the heart and lungs during surgery. John uses the pronoun "we" throughout his story of the development of this pathbreaking machine, often clearly referring to Mary and himself, and mentions specific things that she did (*ibid.,* 242).

Mary and John Gibbon, married in 1931, had four children, all told. After John's death in 1973 she married Lowell Thompson of Ipswich, MA, and lived there until her death in 1986 (Gibbon, "Application," "Development"; Gibbon, John, Obit.; "Mary . . ."; Thompson Obit.).

One of the most exciting inventions in 20th-century medical apparatus by either sex is the xeroradiography machine developed by **Dr. Wu Huai-Yin** and her colleagues in China and reported in the early 1980s. The invention, requiring 13 years of work, is significant in itself, but particularly in contrast with the comparable equipment developed by the Xerox Corporation. Dr. Wu's machine is the size of a suitcase, creates first-rate breast "xerographs," and costs about $1000. The Xerox technology involves two machines, the image-processor and the conditioner—each the size of a table—and leases for $10,000 per month. The story goes that Dr. Wu solved some problems that had vexed the Xerox engineers with a small fan, some dust, and cat's fur. However she did it, this xeroradiography machine makes the technology much more widely available.

As of 1982 Dr. Wu was Director of the Department of Physical Chemistry in the

---

17. Interestingly enough, though the 1930s entry in Harpur's *Timetable of Technology* attributes the invention to both spouses, the 1952 entry credits John Gibbon alone (Harpur, 200, 142). Marks & Beatty also give Mary joint credit, but mistakenly date the work to 1937. The most authoritative secondary source on Mary's contribution is Bordley & Harvey, 507–8.

Pharmaceutical Sciences Faculty at First Medical College, Shanghai. When contacted, she refused to take individual credit for the invention, for, as she put it, "honor belongs to the researching group" (Wu Huai-Yin, L-4/5/82; Bolemon, L-1/30/82; Cameron, L-2/16/82).

## Treatments/Remedies

As late as the 19th century, Western medicine was imperfectly professionalized. Though dominating both what they were redefining as medical education and the legislatures that would put teeth in the new definition, male professionals had not yet won full public confidence. Women particularly distrusted organized medicine, and continued to have babies at home, use their foremothers' remedies, and consult healers not trained in medical schools (see, e.g., Stage, "Poisoning Century" chap.). Many included healing in their work—as Rosalyn Baxandall and her co-editors put it, "making cheese and curing ringworm was all part of the general category of a woman's household manufacture" (16)—while others became known outside their households. In Harriet Beecher Stowe's vivid prose (*Pearl of Orr's Island,* 17–8):

> Miss Roxy and Miss Ruey Toothacre were two brisk old bodies of the feminine gender and singular number, well known in all the region of Harpswell Neck and Middle Bay, and such was their fame that it had even reached the town of Brunswick, eighteen miles away.
>
> They were . . . , in the Old Testament language, "cunning women"—that is, gifted with an infinite diversity of practical "faculty," which made them an essential requisite in every family for miles . . . around. It was impossible to say what they could not do: they could . . . braid straw, and . . . upholster and quilt, could nurse all kinds of sicknesses, and in default of a doctor, who was often miles away, were supposed to be infallible medical oracles.

That people had good reason for distrusting 19th-century medicine is clear from hindsight. Many prescribed remedies were harsh, and doctors knew little about the real causes of disease. Bleeding to syncope (fainting) was a common prescription, and the poisonous calomel (a chloride of mercury) was another. An eight-year-old girl applied in her own handwriting to several medical schools—noting that she had none of the requirements for admission—and over half the schools admitted her (Dufty, 261).

That women had special cause for distrust is clear from the statistics on childbed fever alone, but can be emphasized by two brief examples: the first human artificial insemination with a donor other than the husband was done on a chloroformed woman without her knowledge or consent in 1884, in order to save her husband from confessing his sterility. And as late as World War I, child-health pioneer **Dr. Sara Josephine Baker** pointed out that it was "six times safer to be a soldier in the trenches of France than to be a baby born in the United States" (Robertson, ed., 19; *NAW*).

In the 20th century, professional medicine improved, and the age of the specialist arrived with a vengeance. In this atmosphere, though women's home remedies still survive in odd corners, more women began to join the medical establishment, and to achieve on its terms. Several of them, however, have also worked to humanize the new medicine and created social innovations within it.

With all this in mind, let us take a look at women's achievement in medical invention and innovation since 1800.

## 19th century

Women, like men, bought patent medicines at a great rate in the 19th century. They also invented or improved remedies of their own for everything from dyspepsia and coughs to ague and piles.

*Patent statistics.* As usual, the patent records reveal only a fraction of women's achievement. The anonymous black woman (and probably mothers of many colors through the ages) who improvised a decongestant by letting a child smell urine-soaked day-old diapers from a laundry pile or hamper (Tul, PC-1984) would scarcely consider this an invention, for example, even if it had been patentable and she had had the necessary money. For remedies and treatments, the patented fraction may be smaller than usual. As already noted, most women "doctored" as part of caring for their families, not for money. They handed remedies down from mother to daughter, passed them on to neighbors and friends, and wrote about them, if at all, in cookery books, diaries, letters, books of household management, or women's magazines. Thus they continued to be the healers and major dispensers of remedies for much of the population for most of the century, but their work was hidden.

Even those few women who saw healing as a business may have hesitated to get patents, for patenting a cherished remedy meant revealing its formula. The patent records nevertheless tell a fairly impressive tale. According to LWP, between 1863 and 1895 alone, women received U.S. patents for 27 ointments, liniments, salves, and lotions; 19 unspecified "Medical compounds"; two cures for lung, chest, and throat diseases, two cough remedies and one remedy for catarrh; a tonic, an "alterative remedy," a cure for bilious complaints; and three bitters or remedies for dyspepsia; one "topical remedy," one remedy for rheumatism;[18] a cure for skin diseases, two remedial cosmetics, a hair restorative, an eyewater, and two remedies for corns and bunions; a remedy for worms, a remedy for piles,[19] and cures for cholera, diphtheria, diarrhea, scarlatina, and sore throat.

**Sara W. Brown** of Willoughby, OH, patented one of the ointments with Jerome B. Brown in 1875.

The patent for a skin disease remedy was granted in 1871 to **Elizabeth Angeline Shewell** of 159 Warren Avenue, Boston. A physician, she appears in Samson's Boston Directory for 1876 both under Shewell and under Secor, probably her maiden name. One of the medical compounds was invented by **Jennett Cooper** of San Francisco, who patented an extract of yerba santa and *Grindelia robusta* in 1874.

Nineteenth-century British women also undoubtedly continued as healers. Woodcroft's index (to 1852) shows **Rebecca Ching**'s worm-destroying lozenges (Pat. #3129 of May 7, 1808), and **Mary de Lima Barreto**'s method of treating and curing ruptures (Pat. #2644 of Aug. 30, 1802). Patented with Joseph de Oliveira Barreto, the method comprised an ointment followed by a truss, and rest (D.E. Walker, L-10/29/81). Some British women who received U.S. patents are included in the figures given above for 1863–95.

*Unpatented but famous.* Some of women's best-known 19th-century remedies and treatment regimens were never patented: Mother Seigel's Curative Syrup, the Healolene ointment invented by a Shaker sister (White & Taylor, 317), and Lydia E. Pinkham's Vegetable Compound.

The Shakers sold seeds, herbs, and herbal medicines to the outside world, and have sometimes even been credited with inventing or independently reinventing the pill form of medicine. The first recorded use of the word *pill* in this sense in English is late 15th century. However, pills are probably as old as the first woman who wrapped her child's bitter dose of medicine in a bit of fruit or dough or bread. They may have been independently invented many times before the knowledge became general. One **Mrs. Benedict,** for instance, was known in the 19th century for devising a method of wrapping

18. In addition, one of the liniments just listed was for rheumatism.
19. One of the ointments just listed was specifically for piles.

a thin wafer around bitter medicine. One prepared the wafers in advance, using a thin flour paste and two hot flatirons; then at need softened them with water and closed them around the powder "like a turn-over pie" (Juster, 116).

Women definitely worked in this Shaker enterprise and may have run it in some communities, although a skilled botanist and chemist, Brother Alonzo Hollister, directed the Mt. Lebanon operation at one point (Pearson & Neal, 105). One possible name here is **Faith Whitcomb.** In the 1880s she was the proprietor of the Faith Whitcomb Medicine Company in Boston. Since advertising for Faith Whitcomb's Shaker Liniment and her Balsam Bitters appears in *The History of the Shakers* (1881), in a brief text with illustrations showing Whitcomb in Shaker costume and the Harvard, MA, herb garden, she probably either was or had been a Shaker (Richmond, #237).

Healolene was a Shaker product, mentioned by Eldress Anna J. White and Leila Taylor in their book *Shakerism: Its Meaning and Message* (317). According to an undated advertisement sent me by the Shaker Museum at Old Chatham, NY, it was "Excellent for chapped hands and to use after shaving," and guaranteed not to soil the finest fabrics. The Shaker Sister who invented the ointment, whom White and Taylor regrettably fail to name, was living in 1904 in the North Family of the Mt. Lebanon (NY) Shaker community, where Healolene was manufactured.[20]

It was also the Mt. Lebanon Shakers who produced and sold a version of Mother Seigel's Curative Syrup in the United States, where it was sometimes called Seigel's Syrup or the Shaker Extract of Roots (Richmond, 47ff). Eldress White's article, "The Mystery Explained" from the *Shaker Almanac* for 1886, tells how this came about, and also gives some insight into the life and personality of **Mother Seigel** (b. c. 1793). In July 1868, White sailed for Germany "to learn from Old Mother Seigel herself the composition of her marvelous remedy" (6). She found her, a fine, benevolent-looking old Quaker lady of about seventy-five, in her home near Berlin, where she had a wide reputation for curing those given up by doctors as incurable.

Mother Seigel willingly told how she came to devise her remedy. In 1852, after nursing a very sick patient, she broke down. Her nerves and digestion were shattered, and she suffered with shifting pains all over her body. Despite consulting physicians, she grew steadily worse for two years, until at last she longed to die.

Finally, one day while "tottering" near a spring on her land, she picked a sprig of a plant and chewed it on the way home. Her children and grandchildren wanted to vomit her because the plant was considered poisonous, but she insisted on sleeping, and had her first restful sleep in two years. On waking, she felt better. Secretly gathering more of the plant, she chewed it during the night and felt stronger in the morning. Now she made a tea of the roots, leaves, and twigs, and drank it daily, gradually improving until she felt entirely well. And "from the year 1855 neither my children nor myself have ever suffered one day's illness."

Mother Seigel's first taste of the plant was, of course, a chance *discovery* (possibly guided by an unconscious childhood memory of something her mother or a neighborhood healer used). Her subsequent *invention* lay in devising the most effective method for gathering, drying, and processing the herb and in formulating the syrup. Considerable experimentation and careful recording of results must have been involved in determining

---

20. If she did not die between 1904 and 1906, she was probably one of the following nine women living at North Lebanon in 1906: Eldress Julia A. Scott, Amelia Calver, Amy Reed, Anna Maria Graves, Eunice Cantrell, Adaline Cantrell, Corinne Bishop, Sarah Cutler, or Nellie Watts (information from Old Chatham Museum, 4/16/81).

the best month for taking the plant, which part(s) contained the desired ingredients, and the form the remedy should take. In short, identifying a useful plant as Seigel initially did is a discovery; transforming that useful herb into an effective, reproducible drug or treatment is an invention.

Mother Seigel used the syrup in her nursing practice, where she found it helpful with all diseases, and the best blood-purifier ever. It became so popular that she built a separate building for sorting and drying the plants, compounding and bottling the syrup. She became wealthy from its sales, but always gave the remedy freely to the poor.

Likewise, she freely shared it with White, who stayed in Germany for six weeks to learn Seigel's methods. Said Seigel, "I am rich; my aim is to do good; everyone in Germany knows me and my Syrup; but if I can give the sufferers of the New World something for which to be thankful to Mother Seigel I shall die feeling that I have left nothing undone which I could do." White, in turn, resolved to publicize the syrup throughout the world. Back home, she entrusted it to the Mt. Lebanon Shakers, "who have had experience in growing herbs and extracting from them their best medicinal qualities. The medicine is now known as the Shaker Extract of Roots, or Seigel's Syrup."

Since White evidently became a retailer of the remedy in New York and London, selling some 15 million bottles in four years through the London office alone (*Shaker Almanac,* 1883), she is not an entirely unbiased source. On the other hand, as a Shaker Eldress, she is neither the typical gullible consumer of patent remedies nor an unscrupulous profiteer (presumably all her profits went to the Church in any case). Upon analysis, the plant used in the syrup was found to contain five alkaloids with the following effects on the human body: a soporific, soothing the nerves and allaying pain; a "laxative," acting upon the liver to aid in the elimination of poisons; a sudorific or sweat-producer, opening the sweat glands and pores to eliminate poisons by another route; a diuretic, acting on the kidneys to eliminate their share of poisons; and an "alterative," producing a proper acid-alkaline balance in all bodily fluids, strengthening the liver, improving nutrition, and building up the flesh. Nature, White concluded, had so harmoniously blended all these five qualities in the structure of this one valuable plant that it should indeed aid in the cure of any disease, and she echoed Mother Seigel in calling it "the best purifier of blood that ever came under the notice of the medical profession."

In later advertising material, White claimed further that Seigel's Syrup could slow the aging process. It would be interesting to read a modern pharmacological analysis of the Syrup—to learn at last the name of the plant—and to see how long it remained in the pharmacopeia. Certainly, as mentioned above, its sales were phenomenal.

Also phenomenal were the sales of **Lydia Pinkham**'s Vegetable Compound—over $3 million a year at its peak popularity in the 1920s. Although Lydia Pinkham died before becoming wealthy, she was already becoming nationally known by the time of her death. Immortalized in the 1880s by a ballad of the Dartmouth College Glee Club,[21] **Lydia Estes Pinkham** (1819–83) of Lynn, MA, is in the tradition of the 19th-century women who knew home remedies for all but the most serious maladies, and who shared the recipes for these remedies just as for cakes, cookies, or casseroles. If her husband Isaac had not gone bankrupt in 1873, she might never have sold her Vegetable Compound.

21. Sung to the tune of "Our Redeemer," it ended with the refrain, "Oh, we'll sing of Lydia Pinkham / And her love for the human race, / How she sells her Vegetable Compound / And the papers, the papers, they publish her face" (Stage, 9, 41-2). One verse went: "Tell me, Lydia, of your secrets / And the wonders you perform. / How you take the sick and ailing / And restore them to the norm . . ." (cited in D. Jackson, 107). In addition to this song, she has inspired at least three biographies (Stage; Burton; Washburn), many articles, and mentions in untold numbers of books.

Yet once her business was launched she acted with spirit, and seems to have had a genius for advertising. "Only a woman can understand a woman's ills," she declared in hundreds of newspaper advertisements, and her advice, both printed in the ads and written in answer to the letters that poured in, boosted sales tremendously. Sources differ as to who first thought of using Lydia's picture on the bottle (Caroline Bird says no one knows, 122; *NAW* and Sarah Stage [40] say it was her son Dan), but Lydia had a special photograph taken for the purpose, and the idea made advertising history.

Sources also differ on the development of the famous Compound itself. According to Sarah Stage (27–8), Lydia Pinkham received from Lynn machinist George Todd the basic recipe for a medicine purported to cure "female complaints" from painful menses to prolapsed uterus, and then modified it, perhaps by referring to John King's *American Dispensatory* or to her own notebook of herbal remedies, entitled "Medical Directions for Ailments." The *American Dispensatory,* of which she owned a well-thumbed copy, was the most complete listing of pharmaceutical botanicals of its day. Other sources imply that she took the entire basic recipe directly from King, modifying a formula for a "uterine tonic and sedative" from his 1870 edition, though Donald Jackson notes that Lydia, "like many housewives[,] concocted her own medicines from herbs and other plants" (*NAW;* Bird, 123–4; D. Jackson, 110).

However arrived at, the formula contained the following ingredients, recorded in Lydia's journal: unicorn root, life root, black cohosh, pleurisy root, and fenugreek seed, macerated and suspended in about 19% alcohol as a preservative, then filtered through cloth and bottled (Stage, 32; Bird, 124).[22] The preparation of the formula also evidently involved some cooking or heating. When the alcohol content of the remedy became a temperance issue, Pinkham made pills or lozenges of the herbal ingredients alone. However, it should be noted that boiling water enhances the properties of black cohosh root, but dissolves it only partially; alcohol dissolves it entirely (Hutchens, 56; D. Jackson, 110).

The medical profession dismissed her compound along with the rest of "patent medicine" (e.g., Cramp). However, traces of estrogenic materials occur in several of the Compound's original ingredients (Stage, 243), and the true verdict on many folk remedies is only now beginning to come in, as more sophisticated analyses reveal unsuspected trace elements in them. Indeed, a traditional and still popular remedy for menstrual problems, the Kiehl Pharmacy Herb Tea No. 42, contains two of the same ingredients as Pinkham's original formula—pleurisy root and black cohosh (Stage, 32; Seaman & Seaman, 179). The ailing women friends to whom Lydia Pinkham first gave her compound free of charge all praised it, as did a great many users over the years. Whether these satisfied users were helped by expecting help, by a mood lift or relaxation from the alcohol base, by the Compound's botanical ingredients—or by some combination of the above—is still open to debate.

In her pioneering Department of Advice, however, Lydia Pinkham made an indisputable contribution. She herself considered it no less important than the business side of the Lydia Pinkham Medicine Company. Personally answering each letter received, she urged good diet, exercise, and cleanliness. As *NAW* puts it, "the confident tone of her advertisements and the genuine concern evident in her letters were in themselves therapy for some." Her advertising copy opposed the evils of fashion, which in those days

---

22. A later label listed crystalline Vitamin B-1, gentian, both true and false unicorn root, dandelion, and chamomile in addition to the life and pleurisy root and black cohosh, but no fenugreek seed, and only 15% alcohol "added solely as a solvent and preservative" (Stage, picture section).

included the harmful practice of lacing women's corsets so tight as to restrict breathing and deform ribs and inner organs. She also urged mothers to inform their daughters about sex and reproduction, and invited schoolgirls too shy to talk with their mothers to write directly to her. Many did. She wrote a widely circulated sex-education booklet that was frank but in good taste.

And when, as late as 1881, a suffering woman wrote to her describing horrifying symptoms,

> It is an affection of the gums and the mucous membrane of the mouth—the gums turn white and a layer easily rubs off leaving them very red and angry—the inside of my cheeks and corners of my jaw are white and look and feel hot and parboiled and contracted—it has extended to my tonsils and seems to have started down my throat . . . ,

Lydia Pinkham diagnosed calomel (mercury) poisoning (Stage, 45–6). The woman also reported "womb trouble," constant back pain, and frequent agony from an enlarged urethra. Though she slept well and had a good appetite, she felt weak and lacked endurance. Many doctors continued to prescribe calomel over long periods of time, even though warned of its cumulative effects, which were much as this poor woman detailed. Her physician told her he had never seen the like of her malady before.

The face on the Vegetable Compound label is unquestionably sincere. Whether Lydia Pinkham most deserved fame as entrepreneur, advertising and promotion innovator, herbalist and medical inventor, or counselor of women and young girls, it is gratifying to recall that the most famous female face of the 19th century belonged to a feminist.

*Patented but not on LWP.* Certain lesser-known but possibly useful remedies are omitted from LWP, even though patented. According to Katharine Lee (60), **Charity Long** patented a cure for consumption in 1812, whereas LWP records no medical remedies until 1863. In 1881, **Dorothy L. Martin** of Warren Co., NJ, patented a Medical compound. Also, according to Charlotte Smith (*WI*/1: 3), a **Mrs. E.J. Austin** of Carthage, NY, marketed several well-known medicines. She finally patented two of them, the Magic Fluid, and the Gold and Silver Surprise Paste. Neither of these remedies appears in LWP. Moreover, considering the many tonics, cough remedies, worm medicines, and cures for corns and bunions still circulating in rural Ohio during my childhood, I suggest that the LWP listings represent a tiny fraction of women's real achievement in this area.

*Unpatentable.* Other new or improved treatment regimens were probably not patentable (though today they might be). Victor Priessnitz's water cure was in the United States most often conducted (or at least staffed) and largely patronized by women. It was far more humane than, for example, the rest and sensory-deprivation "cure" made infamous by Charlotte Perkins Gilman in *Yellow Wallpaper;* and much more sympathetic and realistic in its treatment of women's reproductive ills than the male-dominated medical profession of its time (K. Sklar, xviii). Some of the women who ran water-cure establishments or were water-cure practitioners undoubtedly adapted the original regimen.

**Mary Gove Nichols** (1810–84), for example, was a well-known health reformer and a leader of the American water-cure movement. At 22, ten years before learning of Priessnitz's system, she devised cold-water treatments and began administering them to neighborhood women. Moreover, by 1845 she had her own water-cure house, where treatment with a variety of baths and cold wet packs was combined with a vegetarian diet. In 1851 she opened, with her second husband, a short-lived coeducational water-cure medical school called the American Hydropathic Institute. Nichols spent the last years of her life in England, where she and her husband operated a water-cure institute at Malvern from 1867 to 1875 (*NAW*).

*Electrotherapy.* Among the most interesting new treatments in 19th-century medicine was electrotherapy. The therapeutic use of electricity took several forms, including simply stimulating various body areas with small shocks. Such stimuli could be used both in diagnosis (as of nerve damage or death), and in treatment. The best-known 19th-century American woman in this field was **Elizabeth J. French** (1821–1900) of Philadelphia, New York, and Boston, who exhibited her electrotherapeutic appliances at the Centennial in 1876 and wrote several books on electrotherapy, including *A New Path in Electrical Therapeutics* (Philadelphia, 1873). In 1875, Mrs. French had patented an Improvement in electro-therapeutic appliances. At the time of the Centennial she was listed as the widow of Joseph French, a civil engineer.

Born Elizabeth Poorman in Mechanicsburg, PA, she studied and practiced medicine with her father, but upon marrying French she moved her practice to New York. By the time of her death, according to her *New York Times* obituary, she had achieved "almost world-wide" fame in electrotherapy "receiving commendation from the most distinguished members of the medical profession in her age and time."

Her marriage was evidently unhappy, as Dr. French had left her husband long before his death in 1871. Her temperance work might indicate that he drank. In any case, by 1860 she was living in a boardinghouse in New York City. Interestingly enough, many years later she was buried, not with her husband, but in the same Philadelphia cemetery plot as Thomas Culbertson, one of her fellow boarders in that house.

Two of her three children also had distinguished careers. May French Sheldon, noted African explorer, first woman fellow of the Royal Geographical Society, owned publishing houses in Boston and London. Belle French Patterson became a physician. Dr. French spent her last years in Boston living and practicing with this younger daughter. Elizabeth's obituary appeared in four Boston papers, a Philadelphia paper, and the *New York Times* (Jan. 12, 1900), among others (Gribble; Eliz. French titles, NUC/pre-56; Sheldon, 131n; Warner/Post, 165, 169; Warner/Trescott, 114).

Between 1870 and 1895 the following other women patented electrotherapeutic devices (LWP):

**Mary A. Hayward,** Brooklyn, NY—Improvement in electric and vapor chair (1871)
**Emily A. Tefft,** East Otto, NY—Electrical therapeutic appliance (1880)
**Louise Epple,** Chicago, IL—Galvanic abdominal truss (1888)
**Mary E. Thomas,** Cardington, OH—Electromagnetic abdominal support (1890)
**Mary E. Thomas,** Cardington, OH—Voltaic insole (1891)
**Marion A. MacMaster,** Utica, NY—Electro-medical chair (1890)
**Sophia Hetherington-Carruthers,** Sydney, NSW, Australia—Electro-medical appliance
    (1893)
**Lizzie Lane,** Dunellen, NJ—Electrical head clamp for relieving pain (1893)
**Flora A. Brewster,** Baltimore, MD—Electric belt (1894)
**Friederike Fritsche,** Berlin, Germany—Galvanic chain (1894)

The therapeutic uses of electricity so popular in the late 19th century, later dismissed as quackery, take on new respectability in light of recent discoveries about the role of electrical stimulation in promoting healing, relieving severe and chronic pain (cf. **Dr. Judith Walker**'s work [Raloff, "Zapping"]), stimulating bone growth, and speeding the movement of teeth in orthodontia. But it was not in patent remedies or in electrotherapy that women made their most important contributions to healing and medicine as the century turned. Aside from continuing—and revitalizing—their time-honored role as herbal healers, they contributed significantly to establishment medical treatments as well, inventing, co-inventing, or contributing to the development of many kinds of regimens and remedies, including sulfa drugs and antibiotics.

## 20th century

Despite the medical profession's own successful propaganda, as Martin Gross wrote in 1966, "Not until the second decade of this century did a random patient taking a random ailment to a random doctor have even a 50-50 chance of 'benefiting from the encounter.' "

Women continued to invent remedies after the turn of the century and, especially in rural areas and in hard times, to use their mothers' home remedies. Jeane Westin's book on women surviving the Great Depression of the 1930s credits an interesting remedy for poison oak to one **Laura Buhekuhl:** equal parts whiskey and sour cream, applied three times a day (17). As always, some women patented their remedies. **Jeannette W. Nesmith** of Gelpin, CO, for example, patented an Ointment in 1907. Not even the rich and famous were exempt from the urge to do something useful in this field, as witness **Papal Countess Hedwig Von Redwig** (d. late 1930s–early 1940s), who was credited with inventing a famous cough remedy. According to her "nephew," she was a chemist who compounded the syrup using "various antianalgesics and the root of a little-known South American plant" (Rand, 81ff).

However, the "invention" or development of most modern drugs is a complex process involving many steps and several people, often beginning with the laboratory synthesis of new compounds with unknown properties. Such was the history of tryparsamide, the drug that controls African sleeping sickness. However, one name has become connected with this strikingly successful remedy: that of **Dr. Louise Pearce** (1885–1959).

Born in Massachusetts, the elder of two children, Louise Pearce took her AB at Stanford University in physiology and histology (1907) and her MD at Johns Hopkins (1912). In 1913 she joined the Rockefeller Institute for Medical Research, as a fellow in the laboratory where Simon Flexner was seeking new arsenical drugs to treat sleeping sickness. Dr. Pearce and a male colleague were assigned to test newly synthesized compounds found capable (in the laboratory) of destroying the microscopic parasite, *Trypanosoma gambiense,* that causes the disease and gives it its medical name of trypanosomiasis. In 1920 Pearce went out alone to the Belgian Congo to do clinical testing.

A meticulous worker, she treated patients at all stages of the disease, beginning with graded single doses of tryparsamide, following the results with microscopic tests, and keeping careful records of dosage and effects. Her success was spectacular. Patients' blood was free of parasites within days, their bodies totally free of parasites within weeks. Mental and general physical health usually returned even in the severest cases. The grateful Belgian Government awarded her the Order of the Crown of Belgium and, in 1953, the $10,000 King Leopold II Prize with a second decoration, the Royal Order of the Lion.

Appointed an associate member of the Rockefeller Institute in 1923, Dr. Pearce continued to do valuable scientific work, making important contributions to the understanding and treatment of syphilis, cancer, and smallpox. She also did much to advance the cause of women in medicine and science, serving, for example, as board member and then president (1946–51) of the Woman's Medical College of Philadelphia. Friends interviewed for the *NAW* biography remembered her as cordial and hospitable, fond of fine clothes, jewelry, books, and art; holding her own in conversation with male colleagues, and generally enlivening the staid atmosphere of the Rockefeller Institute. Dr.

Pearce spent her last years at Trevenna Farm, Skillman, NJ, the home she shared with novelist Ida A.R. Wylie (*NAW*/mod.).

*Antibiotics.* Antibiotics are a prime example of the complexities involved in inventing a modern drug—beginning, however, not with the synthesis of a new compound, but with ancient women's knowledge.

Antibiotic therapy is one of modern medicine's most powerful tools, exploiting weapons evolved during eons of competition among earth's teeming micro-organisms. The textbook story of antibiotics usually opens in 1928 when blue bread mold (*Penicillium notatum*) attacked Alexander Fleming's staphylococcal bacteria cultures. Antibiotic knowledge, however, is actually very ancient. Hyssop (*Hyssopus officinalis*) has been used for thousands of years—among other things, as a tea for respiratory infections. The leaves—where "penicillin mold grows and thrives"—are laid on wounds to cure infection and promote healing (J. Rose, 69). The likelihood that women knew antibiotics first seems high since the daily uses of both fungi and lichens are far more frequently connected with women's work than with men's. The fungi are foods or contaminants or catalysts of foods, including the vital bread and beer—manna itself may have been a lichen. Lichens are dyes with built-in mordants; they yield potent fertility and anti-fertility substances and childbirth aids; they occur in perfume and in insecticides (Sharnoff, 135, 142).

Sir James George Frazer reported (Pauwels & Bergier, 106) that "savages suffering from infectious illnesses eat *Penicillium notatum.*" Australian aborigines knew the value of "molds" taken from the shady side of eucalyptus trees in treating wounds, but were ignored when they brought the specimens to a research institute (Bickel, 61). At least one eucalyptus shelf fungus is also superbly edible if properly prepared (Robert Baldwin, PC). More than 2500 years ago, the Chinese knew the curative properties of moldy soybean curd (a common food), and used it to treat boils, carbuncles, and similar infections (*EB*/58, S.V. "Antibiotics"). Both the Chinese and the ancient Greeks used *Usnea* lichens in medicine, particularly for gynecological disorders (Sharnoff, 140). Native American women used mold from acorn meal to heal boils, sores, and inflammations (J. Rose, 88).

University of Massachusetts researchers report that a Nubian people living in the Sudan fourteen centuries ago were getting large doses of tetracyclines, probably from the mold-like Streptomycetes growing in their mud-lined grain-storage pits.[23] The researchers had already deduced a remarkably low rate of infectious disease among these people when graduate student Debra Martin discovered tetracycline in their skeletons. Another of the researchers, Margaret Keith, said that although there is no direct evidence the tetracycline was taken therapeutically, the population may have made the connection between the mold-like substance and their health (Ancient People's . . . , 11).

Eastern European peasant women had for centuries bound moldy bread over wounds (Raper, 1). In fact, fungi, often in the form of slices of moldy bread used dry or in poultices to prevent wounds from festering, are prevalent in the primitive pharmacopeia from the Ukraine to Mexico (Halsbury, 20). These women could probably identify most of the common molds of their region by sight, smell, and/or taste. Women's knowledge of the ergot fungus growing on rye grass, with its abortifacient and other uses, is ancient. The

---

23. Newspaper reports of this discovery described these as "mud grain bins," but clay is likely as the kind of "mud" for several obvious reasons. Of interest in this context, however, is the rapid inactivation of antibiotics in soil by bacterial activity and by adsorption to clay. Clay adsorbs basic antibiotics but not acidic or neutral ones (Doetsch & Cook, 178).

potent hallucinogenic ritual drink of the Eleusinian Mysteries contained carefully measured amounts of ergot (Wasson; Kerenyi, App. I).[24]

Thus if fewer wise women healers had been burned as witches in early modern Europe, we might have had antibiotics sooner—and in more variety—than we did. After all, thousands of different molds can grow on bread alone, and more than half of all lichens are thought to have antibiotic properties (*EB*/58, S.V. "Penicillin"; Sharnoff, 140).

Working in a long tradition of wise women was 19th-century healer **Elizabeth Stone** of Peshtigo, WI, the only early antibiotic therapist I know by name. Seventy years before Fleming's "discovery," Stone worked among the North Woods lumberjacks, treating their many injuries. Her pharmacy was a round yellow tin box full of moldy bread, regularly replenished with scraps from her table. Whenever she opened the box, a cloud of green puffed out. For dressing wounds she made poultices of the bread with warm milk or water, which she applied directly to the skin. Healing was rapid and clean. Elizabeth Stone never lost a patient from a timber-related injury (Stellman, 87).

Readers may object that these early women used antibiotics in their crude form. As the following story of the development of penicillin shows, however, modern researchers, with all their scientific techniques and equipment, did much the same for decades after Fleming's report.

Penicillin: Penicillin is the first[25] and in some ways the greatest antibiotic of them all—attacking the previously untouchable staphylococci and virtually nontoxic to humans—and women's role in its development is unusually well documented (e.g. Bickel). Although three men—Fleming, H.W. Florey, and Ernest Chain—received the Nobel Prize for this drug in 1945,[26] Fleming merely saw a new side to the old observation that bacteria would generally not grow on moldy agar. He did determine that a substance from the mold diffused into the supporting broth, and that it killed a broad spectrum of bacteria, including staphylococci. But after a few tries at isolating the pure antibiotic, which he named penicillin, he returned to his other work.

Howard Florey, whose Oxford research team took up penicillin where Fleming dropped it, always said that the development of therapeutic-quality penicillin was a group effort. Florey's people appeared in alphabetical order on the original paper (*Lancet*, Aug. 24, 1940: 116), including the team's two women, **Jean Orr-Ewing** and pathologist **Margaret Jennings.** Chemist Ernest Chain was listed first. Florey also always credited his first wife, **Ethel Florey,** MD, without whose painstaking clinical trials penicillin would not have become available when it did (Royal Society of Arts Lecture, 1944; Bickel, 118–22).

**Dr. Gladys L. Hobby** (1910– ) was on the first U.S. team to test penicillin clinically. The first clinical paper, bearing her name and two others, appeared in 1941 (Bickel, 118). Born in New York City, **Gladys Lounsbury Hobby** studied at Vassar (AB, 1931), then at Columbia (MA, 1932; PhD, 1935). After a research assistantship at Columbia's Presbyterian Hospital, she worked as a microbiologist for Pfizer from 1944 to 1959,

24. I am grateful to Christy Baldwin, filmmaker of *Presence of the Goddess* (1986), for this information.
25. Actually, several antibiotic substances predated penicillin (see pyocyanase below), but penicillin was the first of a new class of wonder drugs. Because it acted against three of humankind's worst enemies—wound-infecting staphylococci, the pneumonia bacterium, and the syphilis spirochaete—without harming human cells, it transformed medical care throughout the world.
26. Penicillin is often thought of as having been discovered, but as the briefest glance at the story will show, its development to therapeutic levels was a process consciously undertaken, and demanding several inventions both of organisms and procedures (which today would be patentable), and years of hard, goal-oriented work along the way.

becoming Scientific Director of the company's Infectious Diseases Research Institute in 1959 (*AMWS*/12; see "Other antibiotics" below).

But this is getting ahead of the story. Florey and his team knew that *Penicillium notatum* released into its culture broth a powerful anti-bacterial substance. They had concentrated and purified it, and had developed small-scale production methods. But they did not know exactly what the substance was, and they could not produce it in quantity. Florey's first penicillin patient died, after a striking initial improvement, because the drug supply ran out. The output of the laboratory flasks was pitifully small. Britain was now at war with Germany, her scientific resources strained to the breaking point. The situation became desperate when Germany cut off the Allies' supply of sulfa drugs.

Florey did the only sensible thing under the circumstances—he brought the problem to the United States, particularly to the Department of Agriculture's Northern Regional Laboratory in Peoria, IL. The work continued on both sides of the Atlantic, following three lines of attack, all involving women: finding or creating higher-yielding strains of *Penicillium;* inventing mass-production methods; and isolating and determining the chemical composition and structure of the active molecule. The ultimate goal was to synthesize penicillin for unlimited production.

European researchers also contributed to the effort: **Sara Dath,** at the Pasteur Institute in Brussels, gathered and examined monumental numbers of molds and bacteria. A young woman researcher, **Dr. Kaplan,** at the Russian Institute for the Biochemistry of Microbes and Immunology, working under **Professor Yermolieva,** collected and cultured a mold from the wall of an air-raid shelter. It turned out to be a new *Penicillium.* Dr. Kaplan won the Stalin Prize for her work (Bickel, 60, 213). Most of the work, however, took place in the United States.

In Peoria, an unnamed woman[27] brought in the moldy cantaloupe carrying the first really promising strain. This mold yielded fifty times as much penicillin as Fleming's original strain. **Dr. Dorothy I. Fennell** (1916–77) was the Peoria microbiologist who in 1942 isolated the strain and discovered this promising yield. She continued her work on penicillin through the mid-1940s, and in 1946 co-authored a paper on increasing yields of penicillin X in submerged culture (Raper & Fennell). In fact, she played an important role in developing mass-production methods for penicillin. Fennell received the Lasker Prize in 1946 for her work on penicillin, and went on to become a world-renowned expert on *Aspergillus* and *Penicillium,* winning the Federal Women's Award in 1976 and giving her name to several fungi (Ribando, L-3/26/80; USDA, press rel. [5] & Nomination, 1976 Fed. Women's Award, 1; USDA int. circ. TC 76-28 of 8/17/76: 1; *WBWRA,* 32).

Not content with seeking naturally occurring higher-yielding strains of *Penicillium,* researchers began creating mutant strains, irradiating cultures with X-ray and ultraviolet radiation. It was **Dr. Elizabeth McCoy** (1903–78), a bacteriologist at the University of Wisconsin, who produced the ultraviolet mutant X1612, eventually used for all further drug production. Without Dr. McCoy's strain, *yielding 900 times as much penicillin as Fleming's,* penicillin could never have been released to civilians during World War II. Dr. McCoy's research interests ran from fermentation bacteria to staphylococci to the effects of pollution on spawning fish. She devised screening methods for new antibiotics that

27. Mary Hunt, a Peoria lab technician assigned to collect moldy vegetables from local markets, is usually credited with finding this cantaloupe, and has entered history as "Moldy Mary." Kotulak, however, says this specimen was actually brought in by someone else, name unknown (1,4). The "pretty golden look" of this melon was caused by a healthy growth of *Penicillium chrysogenum Thom,* one variety of which doubled the previous drug yields (Bickel, 184).

were later used in industry, and she patented oligomycin, an important enzyme inhibitor (O'Hern, 200). After her penicillin work she was promoted to full professor at Wisconsin, the second woman outside home economics and nursing to attain that rank. Continuing her research even after retirement, Dr. McCoy had three books and 103 articles to her credit when she died. Recognized worldwide as a truly great microbiologist, she was granted the ScD posthumously by her university (Bickel, 185; O'Hern, 195ff; *WBWRA*, 219).

In the second line of attack, most of the work was done at Peoria, where new growth media of corn-steep liquors and lactose enriched with trace metals increased the penicillin yields tenfold (Bickel, 146–7, 232). However, at least one Australian woman also contributed to this research approach. **Dr. Nancy Atkinson** read the Oxford reports in 1943, and experimented with growing the molds from cultures shipped from England. She supplied F.H. Faulding & Co., Ltd, with samples, growth data, corn-steep-liquor formulas, and her extraction and purification methods.

The first pure sodium salt of penicillin, when isolated in the United States, turned out to be a four-member ring never before seen in a naturally occurring compound. It was a woman, X-ray crystallographer **Dr. Dorothy Crowfoot Hodgkin** (1910–94), who determined the precise structure of the penicillin molecule. Teams of chemists and crystallographers worldwide attacked this problem as soon as pure crystalline salts became available, but Hodgkin's team first determined the orientation of the sulfur atoms in the crystal lattice (Opfell, 219):

> For this complex project Dorothy Hodgkin used all her imagination and insight, which her colleagues said amounted to genius. As an associate remarked, she thought about the complex molecule in mental stereo. . . .

Halfway through World War II, Dr. Hodgkin applied the computer to the exceedingly complex calculations needed for the structural determinations. In 1964 she won the Nobel Prize in Chemistry for determining the structure of both penicillin and Vitamin B-12. The Nobel citation described her success with penicillin's structure as "a magnificent start to a new era of crystallography" (Bickel, 216; Opfell, 211, 219).

Now synthetic penicillin was within reach. But because of penicillin's unique structure, the old production methods using fermentation broths were more economical.

Modern researchers continue to refine the production and therapeutic characteristics of penicillin, and women continue to be involved in this innovative work, though these later developments are less well documented. In 1979, for example, **Magda Huhn** and **Eva Somafi** of Budapest, Hungary, received a U.S. patent for a Process for preparation of reactive penicillinic acid and cephalosporanic acid derivatives. They and their two male co-patentees assigned their patent to Chinoin Gyogyszer es Vegyeszeti Termekek Gyara RT.

Other antibiotics: Women have invented or been involved in the development of other antibiotics as well as penicillin. Furthermore, women were involved in the invention of the sulfa drugs that preceded penicillin. Unlike penicillin, the sulfas are antibacterial in the human body, but not in the test tube, a crucial difference for research. In 1936 Prof. and **Mme Tréfouël** and their colleagues at the Pasteur Institute in Paris split the azo dye compound into two molecular rings, one the clinically useless dye, and the other a colorless bacteria-killer known chemically as para-aminobenzenesulfonamide, or sulfanilamide (Bickel, 50; cf. *EB*/68, S.V. "Sulfonamides," which mentions only J. Tréfouël). At least two British women, **Maeve Kenny** and **Dora Stephenson,** participated in separate clinical trials of an early sulfa drug called Prontosil in the 1930s (*EB*/68, S.V.

"Sulfonamides"). In the United States **Dr. Eleanor A. Bliss** (1899–1987) became famous by 1937 as "one of the two Hopkins doctors to perfect the marvelous new medicine that cures streptococcus infection." In extensive studies with white mice, Dr. Bliss and Dr. Perrin H. Long demonstrated that the antibacterial agent was the chemical base, sulfanilamide, and not the dye. Bliss and Long proceeded to clinical trials in humans, achieving the first known cure of a case of streptococcal meningitis in 1936. They also worked with sulfapyridine and other sulfas, and in 1939 presented their work in a book, in which they suggested that these drugs would cure 33 different diseases, including childbed fever, gas gangrene, gonorrhea, pneumonia, streptococcal meningitis, and certain rheumatic and kidney infections (Stevenson, 35–6). **Helen Dyer**'s investigations of amino acid analogues played a role in the development of sulfa drugs (*WBWRA*, 220).

A 1930s antibiotic was pyocyanase, derived from *Bacillus pyocyanase.* **Miss Schoental,** working under Ernest Chain at Oxford, extracted three different antibacterial substances from this bacillus, also called *Pseudomonas aeruginosa.* Pyocyanase had a brief vogue for treating infections (Bickel, 62; *EB*/68, S.V. "Penicillin").

Colorado-born **Dr. Odette Shotwell,** in 1948 a research chemist specializing in molds for the USDA at Peoria, created two new antibiotics—duramycin and azacolutin—during her first assignment there. By inventing new methods for separating antibiotics from fermentation by-products, she played an important role in the development of two other antibiotics, cinnamycin and hydroxy-streptomycin. Author of 118 scientific papers, and holder of three patents (Benedict *et al.;* Ribando, L-3/26/80 & PC-4 1980; Stodola *et al.*), Shotwell has made important contributions to knowledge of aflatoxin, a cancer-causing toxin produced by molds.

A candidate for the Outstanding Handicapped Federal Employees Award in 1969, Dr. Shotwell was no ivory-tower scientist, but entered into many community activities, serving as co-chair of the Peoria NAACP education committee, president of the Peoria League of Women Voters, consultant on an inner-city education program, board member of a local arts-and-science center, and subcommittee chair of the Mayor's Commission on Human Relations (*WBWRA*, 32).

During World War II, Russian protein chemist **Marina Glinkina** organized a laboratory to produce a new anti-gangrene antibiotic that was used in the surgical divisions of Leningrad hospitals. Trained in the physics and chemistry of polymers, proteins, and antibiotics, she returned after the war to the Physico-Technical Institute at the USSR Academy of Sciences, doing highly successful research with Professor S.E. Bresler in protein chemistry. As of the 1960s a senior scientific worker, Glinkina was trying to create a general theory of the interaction of organic ions with polyelectrolytes. Her work has already been applied to industrial production (Dodge, 226).

In 1950 **Dr. Gladys Hobby** co-authored the original paper announcing "Terramycin, a New Antibiotic" (Finlay *et al.*), though her name does not appear on the patent (#2,516,080; Raper, 58). Micki Siegel credits her as "co-discoverer" of terramycin in a 1984 article (78), but *inventor* is more accurate for a drug of which Kenneth Raper remarked in his 1952 survey of the past decade of U.S. antibiotic research, "I think it can be safely said that in no other instance has the discovery of an antibiotic resulted from a planned campaign more extensive in scope or singular in its ultimate objective." A broad-spectrum antibiotic, terramycin was considered one of the five major antibiotics in a field of some 300 (Raper, 27, 35).

As this "wonder drug" proved effective against more than fifty diseases, it is fitting that *Mademoiselle* magazine gave Dr. Gladys Hobby one of its Merit Awards in science for 1950, stating that she had "coordinated the research that proved terramycin . . . safe and effective for treating human beings." The laboratory and clinical trials were obviously extensive: "Dr. Hobby guided all the laboratory experiments, in test tubes and on animals, then organized a world-wide program of clinical work which involved over two thousand patients and more than 150 clinics" (*Mademoiselle*, Jan. 1951: 71).

Dr. Hobby also worked on the development of streptomycin and other antibiotics at Pfizer (Siegel, 78).

During years of research, Gladys Hobby held several concurrent medical positions, including one as clinical instructor and assistant professor at Cornell's Medical College (1959–74). She also became editor (1967–70) and then editor-in-chief of the journal *Antimicrobial Agents and Chemotherapy*, a diplomate of the American Board of Microbiology, and a member of several professional organizations in her field, including the American Academy of Microbiology, the American Society of Microbiology, and the American Thoracic Society. Dr. Hobby has done research in the areas of bacteriophage, bacterial variation and enzymes, streptococci, pneumococci, tubercle bacteria and experimental tuberculosis, sulfonamides, immunizing agents and germ-free life, among others. As of 1972 she was affiliated with the Infectious Disease Research Institute in East Orange, NJ. By 1978 she had more than 125 articles and six book chapters to her credit (*AMWS*/1941/1972; Bickel, xiii, 128; O'Hern, 81ff; *Mademoiselle*, Jan. 1951: 71; Siegel, 78; Stuart, 85).

In 1952 **Annie M. Brown** of London, England, received a U.S. Patent for Manufacture of polymyxin, an antibiotic derived from the soil bacterium *Bacillus polymyxa*. She assigned her patent to Burroughs Wellcome (USA) of Tuckahoe, NY.

In 1976 **Alla F. Teteryatnik,** a Senior Scientific Assistant at the All-Union Antibiotic Research Institute, won one of the 50 national prizes given each year by the All-Union Society of Inventors and Rationalizers (VOIR). Although my source (*Vsesoyuznoye*, 76–8) gives no details, since Dr. Teteryatnik worked at an antibiotic research institute, her prize-winning invention or "rationalization" likely dealt with this area of medicine. Teteryatnik and three other women winners of 1976 donated their prizes to a Soviet Peace Fund.

Antibiotics, though wonder drugs indeed, are not without their problems. In addition to their sometimes serious side effects, antibiotics generally do not kill dormant bacteria. When these later become active, the symptoms return. In 1986, however, **Elaine Tuomanen** and colleagues at Rockefeller University found that a new group of antibiotics called penems were effective against dormant bacteria, at least in the laboratory ("Killing . . .").

Bacterial resistance: As antibiotic therapy became routine, doctors noticed that some infections and diseases no longer responded to antibiotic treatment. Bacteria were developing resistance to the drugs. **Dr. Hattie Alexander** (see below) was one of the first to understand this. Sometimes the immediate problem could be solved by changing antibiotics, but in other cases the patient was allergic to the next-best drug, or the course of the disease was too rapid to wait and see whether resistance developed. Thus **Esther** and Joshua **Lederberg**'s invention of a method of detecting and isolating drug-resistant strains of bacteria proved invaluable (Keenan & Laine, 5).

Fungicidal antibiotics: Doctors also began to see more fungal infections. Antibiotics, in fact, were too good at their job, wiping out not only their targets, but the bacteria that usually control the body's resident fungi. For example, women quite often get "yeast" infections after taking antibiotics for a bladder infection. By the late 1940s the search for better fungicides was on. This search gained new urgency with the advent of open-heart surgery, and especially heart transplants, since fungal invasion of the chest cavity often followed such a procedure. In 1950, however, when **Drs. Rachel Fuller Brown** (1898–  ) and **Elizabeth Hazen** (1885–1975) announced their joint invention of the first really safe and effective fungicidal antibiotic, heart-transplant surgery was scarcely dreamed of.

Nystatin would be an invention, even unpatented, because Brown and Hazen *applied* an earlier discovery—that microscopic soil organisms produce chemical weapons against their competitors, both bacterial and fungal. Beginning with a definite end-product in view, they moved toward it through several technical steps. Hazen's name appeared first on the 1950 paper, and *NAW* gives her top billing in this work; but, as Edna Yost makes clear, Brown's collaboration was important (*WMS*, 64ff).

These two women came to scientific eminence from unlikely backgrounds. Elizabeth Hazen was the daughter of Mississippi cotton farmers who died when she was four. She attended public schools and an Industrial Institute and College. Unable to afford graduate school, she taught high school science and took graduate work in the summer. She got her MS in bacteriology from Columbia at 31 and her PhD in microbiology at 42 (1927). Rachel Brown was the daughter of a widow living in Springfield, MA. She went to Mount Holyoke with the aid of a scholarship and money from a wealthy friend of her mother. Originally a history major, she was inspired by the great Emma Perry Carr to double-major in chemistry. Her fairy godmother also supported her through her MS in chemistry from the University of Chicago. Rejecting further aid at this point, she became a junior-college science teacher and saved money for a PhD program in organic chemistry. In two years Rachel passed all necessary courses and presented her thesis. But then, as Yost puts it, "for some reason still unknown to her, acceptance of the thesis was delayed and the orals could not, of course, be given until it was accepted" (*WMS*, 69).

While waiting for a decision she ran out of money and took a job with the New York State Health Department at Albany. Not until seven years later, when her work had brought national recognition, did her professor contact her to suggest that she extend her upcoming stay at a professional meeting in Chicago and take her orals in her current research area. Her earlier thesis was then accepted, and the PhD granted. Dr. Brown was 35 years old.

Elizabeth Hazen had worked with vaccines and serums and the microbiology of infectious diseases at the New York State Health Department. During World War II she studied human fungal infections, building a collection of cultures and slides for the State. Inspired by penicillin, she began to look for a naturally occurring fungicide.

Dr. Hazen had been focusing on actinomycetes—mold-like soil organisms that yield antibiotics—gathering soil samples, isolating organisms, and testing to see whether they killed human-disease fungi. Her most promising candidate, from a Virginia dairy farm, was not described anywhere in the literature. Her new biochemist, Dr. Rachel Brown, was to isolate its antibiotic. The process involved growing broth cultures of the organism, straining out the surface growth or pellicle after five or six days, and then by a complex process—involving solvents calculated to dissolve the antibiotic but not other constituents of the broth—separating out and crystallizing the active antifungal agent.

In fall 1950, after extensive tests in mice, Drs. Hazen and Brown announced successful production of two anti-fungal agents from a soil actinomycete. One of the two, called Nystatin in honor of the New York State Labs, was unlike any known antibiotic. Preliminary tests showed that it attacked fungi—including close relatives of those causing human disease—but spared some common bacteria attacked by many antibiotics. In other words, its therapeutic potential looked good. Pharmaceutical houses clamored for the pending patent, but Brown and Hazen assigned it and all royalties (over $13 million) to the

Research Corporation, to further research in the natural sciences. E.R. Squibb & Sons licensed the patent, and eventually marketed Nystatin, which has proved both remarkably nontoxic and effective against vaginal yeast (*Candida albicans*) infections, as well as fungal infections of the intestine, skin, and mucous membranes. It also combats molds in food and livestock feed, and was even used in restoring murals and manuscripts after the 1966 flood in Florence (*NAW*/mod.).

Work had always been Elizabeth Hazen's passion, and so it continued after 1950. In 1958 she became an associate professor at Albany Medical College. After retiring from teaching in 1960, she became a full-time guest investigator at Columbia's Mycology Laboratory.

Honors came to Drs. Hazen and Brown, though perhaps not commensurate with their achievement. They shared the Squibb Award in Chemotherapy in 1955. In 1968 Hazen received a Distinguished Service Award from the New York State Health Department. And just before she died, she and Dr. Brown became the first women to win the Chemical Pioneer Award of the American Institute of Chemists. Rachel Brown has been honored by Mount Holyoke.

Two more recent examples in passing: In 1979 **Marcella Masoero** of Milan, Italy, became a co-patentee of new chemical salts with fungicidal action, receiving a U.S. patent. Much more recently (late 1986), **Elaine Tuomanen** and her colleagues at Rockefeller University have made important advances in meningitis. Sometimes even when antibiotic therapy is "successful" (kills all nondormant bacteria) the patient still dies. On the theory that the killed bacteria could worsen the inflammation of already inflamed nerve tissue, they gave anti-inflammatory agents along with the antibiotics. In rabbits with induced meningitis, the double treatment improved survival rates ("Killing . . .").

*Buffered aspirin.* Women invented aspirin, in the form of a willow-bark extract, in prehistory. This ancient connection is preserved in the names of salicin and salicylic acid, aspirin's active ingredient—both coming from the Latin name for the willow, *Salix.* Though eventually refined and standardized by modern chemistry (first synthesized, 1853; first used by doctors for arthritis, 1890s [*EA; EB*/58]), this old-wives' remedy was not greatly improved until the invention of buffered aspirin in this century. Her daughter Jane R. Shipley reports that **Mrs. Shipley** of Annapolis, MD, invented the buffering process for aspirin while working for Bristol-Meyers (V. Beauchamp, PC-3/14/88). Given the almost universal use of aspirin for pain, fever, and arthritis—and the new findings on heart-attack prevention—Shipley's invention, designed to prevent stomach irritation, becomes the more significant.

*Infectious diseases.* **Dr. Elizabeth Hazen**'s early work on antisera at the New York State Health Department (state of the art before antibiotics) and **Dr. Gladys Dick**'s 1920s work on scarlet fever toxin and antitoxin are pertinent here (*NAW*/mod.).

In the 1930s distinguished pediatrician **Dr. Hattie Alexander,** discussed in some detail below, conquered the once uniformly fatal meningitis caused by *Hemophilus influenzae* bacteria (*NAW*).

**Dr. Louise Pearce,** though most acclaimed for her work on sleeping sickness, also deserves credit, with Dr. Wade Hampton Brown and others, for isolating the virus of rabbit pox. Since this disease resembles human smallpox, this work is pertinent to human disease.

*Syphilis.* Dr. Pearce and Dr. Brown also thoroughly investigated experimental syphilis in rabbits. Since rabbit syphilis resembles human syphilis, their observations were useful in research and treatment of human patients (*NAW*).

**Justine J. Wanger** and colleagues took three years to develop the modern intravenous-drip form of drug delivery now used in all hospitals (see "Drug-delivery systems" below). In the process, among other things, they made possible the greatest advance against syphilis since Ehrlich's discovery of salvarsan—their so-called "Five-day Cure for Syphilis Found Effective in 85 per cent of Patients," announced by the New York Academy of Medicine in 1940 (Hobson).

**Dr. Carolyn S. Pincock** (1910–   ) of Washington, DC, was the first to use penicillin to treat childhood syphilis, early in World War II. For this work she was named Medical Woman of the Year (Chaff, 306).

Chicken pox: With many bacterial diseases under control and viruses increasingly implicated in devastating human ills, the hunt is on for effective anti-viral agents. Results announced late in 1991 of a treatment effective against varicella-zoster (chicken pox virus) are thus doubly significant. Some 3.5 million children in the United States alone develop chicken pox every year. The economic cost has been calculated at $400 million, most of it attributable to parents' lost wages while staying home with sick children. Researchers led by **Lisa M. Dunkle** of St. Louis University tested the drug acyclovir (best known for its use with genital herpes) in a double-blind study of 815 otherwise healthy children infected with chicken pox. They found that immediate treatment results in briefer rashes, fewer blisters or lesions, less fever and malaise, fewer respiratory symptoms, and less itching than in the untreated control group. All told, the children are ready to go back to school (or normal activity) about three days sooner than previously advised. Co-author of the study, which appeared in the *New England Journal of Medicine* on November 28, 1991, is **Ann M. Arvin** of Stanford University (Raloff, "Antiviral").

*Emergency treatments and blood-related inventions.* In 1932 an unpaid research assistant at the University of California devised the antidote for two deadly poisons, cyanide and carbon monoxide. **Matilda Moldenhauer Brooks** took a test-tube antagonism between methylene blue dye and the two poisons, applied it to the human system, and invented a practical, life-saving remedy. Dr. Brooks studied at the University of Pittsburgh and at Harvard, where she took her PhD in 1920. She had worked as a biologist for the Public Health Service before coming to UC. Her work in Berkeley was fully professional, but because of anti-nepotism rules (her husband was on the faculty) she was not paid (*WBWRA*).

Also noteworthy here is **Justine J. Wanger**'s work in resuscitation. During the early 1930s Wanger worked with S.N. Blackberg of Columbia on resuscitating animals killed by anaesthesia or electric shock. The New York Telephone Company saw the possible implications for high-voltage line workers, and lent special electrical apparatus for her experiments. The result was a series of precisely calibrated resuscitation procedures, using various combinations of artificial respiration, direct heart massage, and injections of epinephrine into the femoral vein, to be pumped toward the inactive heart by a glucose solution. So successful were Wanger's techniques that some of her animals were "killed" and revived a dozen times. When "Studies in Revivification" was published in the *American Journal of the Medical Sciences* in 1932, it bore her name as first author.[28]

**Dr. Allene Jeanes** (1906–   ) has been noted for her agricultural achievements. Her most important medical invention was her development of a blood-volume expander using the carbohydrates she had worked with for so long, dextrans. The Korean War created a demand for such a substance, for emergency transfusions when whole blood was

28. My latest source on Wanger is 1941 (Hobson). Her later life deserves to be investigated as well; indeed Wanger would seem to deserve a biography.

not available. Apart from this, there is a normal emergency demand for transfusion material free of viral contamination.

"Invention" is the right word here, not just because Dr. Jeanes holds ten patents in carbohydrate chemistry, but because the dextrans' potential as a blood-volume expander was discovered elsewhere, but reduced to practice and adapted for mass production by Dr. Jeanes' team. It was for this work that she became the first woman to win the Department of Agriculture's Distinguished Service Award (1953). She also won the Garvan Medal in 1956 and the Federal Women's Award in 1962. At her retirement in 1976 she was honored by her colleagues for her career-long achievements (*WBWRA*).

**Judith Graham Pool** (1919–75) died of a brain tumor just after her fifty-sixth birthday, depriving medical science of a brilliant worker. She had made major contributions to knowledge in two different fields of physiology, one of which had highly significant clinical applications. In 1942, in a landmark paper written with Dr. Ralph W. Gerard of the University of Chicago, she reported the first measurements of membrane potential within a single muscle fiber.

Moving into human blood research, Dr. Pool focused on the clotting factor, often called AHG or Factor VIII, lacking in hemophiliacs. In 1959 she discovered that Factor VIII can be precipitated by cold from normal plasma. She then used this discovery to devise a simple, cheap method of preparing the factor and administering it to hemophiliacs, or teaching them to inject it themselves. Her definitive paper appeared in 1965, with Angela E. Shannon, BS, as co-author. Dr. Pool's invention allowed hemophiliacs for the first time in history to live nearly normal lives. Her preparation can stop a hemorrhage, prepare a hemophiliac for surgery—unthinkable before—and prevent bleeding into the joints, a crippling complication.

Let us examine Dr. Pool's invention in more detail. In order to stop hemorrhaging in a hemophiliac patient, doctors must establish Factor VIII levels in the bloodstream at least 10–35% of normal, and maintain those levels for up to twelve days. Since frozen plasma has only about 65% as much Factor VIII as fresh plasma, and the half-life of the transfused factor in the bloodstream is only eight to eleven hours, the volumes of plasma needed to maintain the necessary clotting-factor levels often overloaded the patient's circulatory system.

English researchers tried concentrates of clotting factor from animal blood, but immune reactions to these preparations made them useless for more than one emergency dosage. Swedish researchers had devised a fractionation process for concentrating human plasma, and American workers had adapted this small-scale process to their large-batch programs. The injectable material resulting from this approach, however, was only about seven times as potent as whole plasma, and very costly.

Dr. Pool took an entirely new approach. Separating the plasma from the whole blood, she then quick-froze the plasma with dry ice and ethanol, and placed it in a refrigerator for cold-thawing over about 24 hours. Next she cooled the thawed plasma to 4 degrees centigrade, and centrifuged it at maximum speed for about 15 minutes. At this point, plasma proteins (including Factor VIII) came out of solution and clung to the bottom of the bag, so that the plasma could simply be poured off the top. The resulting "cryoprecipitate," was many times as concentrated as any previous preparation, and lasted longer in the blood-bank freezers. Best of all, the cost is extremely low. The blood bank where Dr. Pool did most of her work, in fact, accepted replacement blood as payment at a rate of one unit of blood per four cryoprecipitate units, with no service charge. This is because the blood bank still had a unit of whole blood essentially intact, except for Factor VIII and fibrinogen, after the Pool concentrate was removed.

Dr. Pool also invented the method of determining the concentration of Factor VIII in human plasma. Her co-inventor was J. Robinson.

Dr. Pool was born Judith Graham in New York City. She did both her graduate and her undergraduate work at the University of Chicago, receiving her PhD in physiology in 1946. Married and the mother of three children, she lived a full and varied professional life, teaching physics and English and working at Stanford Research Institute in Menlo Park, CA (1950–3), before coming to Stanford University's School of Medicine. There she remained from 1953 until her death, rising from research associate to full professor. She wrote some fifty papers in at least two subfields of physiology. In addition to her research, she served in several advisory and editorial positions, including membership in the Medical and Scientific Advisory Committee of the World Federation of Hemophilia and a position on the editorial board of *Transfusion.*

Honors came to her, though fewer, perhaps, than the importance of her work would suggest. She won a Fulbright Research Fellowship to Oslo in 1958–9, the Murray Thelin Award of the National Hemophilia Foundation in 1968, the Elizabeth Blackwell Award of Hobart and William Smith Colleges in 1973, and the Paul M. Aggeler Memorial Lectureship in 1974. Her professional memberships included the American Physiological Society, the American Society of Hematology, the American Association for the Advancement of Science, the International Society of Blood Transfusion, the Council on Thrombosis of the American Heart Association, and the Society for Experimental Biology and Medicine.

Colleagues remember Judith Pool as an incisive and stimulating peer; students remember her as an inspiring teacher. Just before her death she became increasingly involved in improving the role of women in science, working, as one admirer put it, "with the same quiet, tough effectiveness which characterized her scientific work" (Creger *et al.*). She founded and chaired the Professional Women of Stanford Medical School in the early 1970s, founded and served on the steering committee of the Association of Women in Science (AWIS) and served as trustee and then co-president of the AWIS Educational Foundation (1972–3). She also chaired the Faculty Affirmative Action Committee at Stanford the year she died (*NAW*/mod.; Pool & Bunker; *NYT* Obit., July 15, 1975; Pool & Robinson; Pool & Shannon; Dallman & Pool; *WWA*/1974–5).

The Memorial Resolution published by her Stanford colleagues contains a fitting tribute (Creger *et al.*):

> Judith Pool not only had the clearest of minds, but she was a willing and stimulating scientific conversationalist. She inspired many of her colleagues and students of medicine with her ability to pose a question, design the right experiment to answer the question, criticize the results, draw only the most appropriate conclusions, and do all of this in a gentle, unassertive, but at the same time leader-like way which will be truly missed. In addition to her superb qualities in informal scientific conversations, she was also the most lucid of writers and lecturers.

**Martha Hainski**'s 1973 method of preparing gamma globulin (see below) is also relevant here.

In 1986 **Dr. Ruth Benesch** and her husband reported developing a method of inducing blood to give up its hemoglobin more easily. The two Columbia chemists have incorporated an analog of Vitamin B-6 into hemoglobin stripped from red blood cells. The compound works as does the body's natural compound (DPG) for this purpose, but roughly ten times as well. Moreover, it keeps the free hemoglobin from splitting in half and being excreted. The Benesches' modified free hemoglobin, already tested on rats and rabbits, could serve as a carrier of extra oxygen for trauma patients or those whose tissue is blood-starved (Raloff, "Making Blood Share . . .").

*Pediatrics.* Some might have predicted that women's medical inventions would cluster in pediatrics and related areas. My research has not borne this out. Of course, many inventions discussed under other headings do relate to child health—several of the vaccines and techniques to be discussed under Prevention, for instance, deal with childhood diseases; **Dr. Helen Taussig**'s blue-baby operation is a surgical procedure for infants; **Gertrude Müller**'s many inventions for child health and safety included a child's urine-specimen collector, etc. But if the category is smaller than expected, the women and their inventions are outstanding, as the following examples reveal.

**Hattie Alexander** (1901–68) of Baltimore worked for three years as a bacteriologist in order to save enough money for Johns Hopkins Medical School. There she made a brilliant record. Receiving her MD in 1930, Dr. Alexander took a year's internship in pediatrics at the Harriet Lane Home in Baltimore, where she developed her lifelong interest in bacterial meningitis.

After a second internship at Babies' Hospital of the Columbia-Presbyterian Medical Center in New York City, Dr. Alexander joined the clinical faculty as an instructor in pediatrics. Her first research was on the fatal meningitis caused by the *Hemophilus influenzae* bacteria. Many other bacterial diseases responded to antiserums developed by inoculating horses with the bacteria in question and then withdrawing serum containing the horse's immune bodies; but horse serum was useless in this meningitis. Learning that rabbit antiserum had proved effective in pneumonia, she set about developing a rabbit antiserum for *H. influenzae.* In 1939 she announced that her new serum had cured infants critically ill with this type of meningitis, and within two years, fatalities had dropped from 100% to about 20%. Dr. Alexander became world-famous. Later she experimented with sulfa drugs for the disease and eventually perfected an antibiotic treatment. Fatalities dropped to 10%.

Over the years at Columbia, Alexander did both basic and clinical research, making several important original contributions. As noted above, she was *one of the first to realize that bacteria develop antibiotic resistance by genetic changes*—in other words, one of the first to understand antibiotic resistance in bacteria. Likewise, she immediately realized the significance of the 1944 Rockefeller Institute report that scientists had succeeded in changing the genetic composition of pneumonia-causing bacteria by using DNA. In 1950, with her collaborator **Grace Leidy,** Dr. Alexander became the first to alter the *H. influenzae* bacterium with its own DNA, devising the necessary techniques in the course of the research. Later, she extended her work to cover other bacterial species and viruses.

At the same time, however, she kept up a pediatric practice and also did clinical studies on tuberculosis, reporting her research in some 150 articles. Her prolific writing was precise and organized, winning her the 1954 Stevens Triennial Prize for best medical essay. Many other honors came to her, including a full professorship at Columbia in 1958. She also won the E. Mead Johnson Award for Pediatric Research (1942), became the first woman to win the Oscar B. Hunter Memorial Award of the American Therapeutic Society (1961), and the first woman president of the American Pediatric Society (1964).

In private life Dr. Alexander was a fun-loving woman who liked to travel, cultivated exotic flowers, and kept her own speedboat. For many years she lived in Port Washington, NY, with her long-time companion Dr. Elizabeth Ufford. She died of cancer in 1968 (*NAW*/mod.).

**Dr. Carolyn Pincock**'s 1940s work with childhood syphilis comes into perspective when we recall that the death rate from syphilis in the United States was then still 12.7 per

100,000, and thousands of cases of congenital syphilis were reported each year (Chaff, 306).

**Dr. Roberta Ballard,** Chair of the Pediatrics Department at Mt. Zion Hospital in San Francisco, developed the drug used to get premature infants' lungs to develop. She also created the first hospital alternative birth center in the United States.

Premature infants now survive in far greater numbers, partly because of work like Dr. Ballard's. However, the price is high. Every year, according to a study reported in the *New England Journal of Medicine* in December 1981, the very oxygen that allows them to breathe causes moderate vision loss in 1300 infants and severe loss in 250 others in the United States alone. **Dr. Helen Mintz Hittner** of Baylor College of Medicine in Houston and her colleagues studied the use of Vitamin E for preventing such visual loss. In a controlled study with 101 premature infants, they found that a daily "pharmacological" dose of the vitamin (100 mg/kg of body weight) significantly reduced the severe visual loss. The new treatment saves over a thousand infants yearly from a life of blindness or near-blindness ("Vitamin E . . .").

**Dr. Janet Sterling Baldwin** has pioneered in complete-care programs for children with cardiac disease. Dr. Baldwin recognized that, perhaps particularly for children, it was not sufficient to give medical or surgical treatment to the physical heart problem; psychological and emotional care must be added, or the result would be poor (Chaff, 304).

**Dr. Josephine Hilgard** (1906–   ) of Palo Alto, CA, uses hypnosis to relieve the pain of children undergoing repeated procedures such as bone-marrow aspiration. Dr. Hilgard, a psychiatrist, now Clinical Professor Emerita of Psychiatry at Stanford, has worked for more than a decade with these children at Stanford Children's Hospital and Presbyterian Hospital in San Francisco. Her method is to spend enough time with a child to ease anxiety, then induce hypnosis and evoke pleasant fantasies. If she stays during the operative procedure, giving further suggestion as necessary, the child can stay hypnotized and focus on the pleasant fantasies rather than on the pain.

Dr. Hilgard's studies, reported in her book *Hypnotherapy of Pain in Children with Cancer,* indicate that her technique can significantly reduce the pain experienced by more than half the children undergoing bone-marrow aspirations. It can also help with persistent pain.

Josephine Hilgard had a long career in psychology and psychiatry before turning to hypnosis. After graduating with honors from Smith in 1928, she took a PhD (1933) in child psychology at Yale, studying with Dr. Arnold Gesell. She met her husband Ernest at Yale, and moved to Stanford with him in 1933. Jobless because of Stanford's nepotism rule, she decided to take an MD (1940). After a Stanford internship and training at the Institute of Juvenile Research in Chicago, she worked with Frieda Fromm-Reichmann and Harry S. Sullivan at Chestnut Lodge near Washington, DC. Her research dealt with intergenerational effects in mental illness and with affiliative therapy, in which she pioneered.

Dr. Hilgard received several awards for her publications on hypnosis, including one from the National Society for Clinical and Experimental Hypnosis for the best book published in 1970 (*Personality and Hypnosis*).

*Cancer.* Having "conquered" smallpox, polio, and several once-deadly childhood killers, medical researchers declared war—rather optimistically, as it turns out—on cancer. Many women have been involved in this ongoing double search for basic understanding and effective treatments. Only a few of these women can be discussed here, some of the earliest of whom, of course, were working before the "war" was overtly declared.

**Justine Wanger**'s basic research on hibernation contributed to the frozen-sleep approach to cancer therapy. Both the hibernation research, reported at an American Medical Association meeting in May of 1939, and the effects of long-continued freezing on cancer tissue are featured in the *Britannica* yearbook for 1940 (439), though Wanger herself is not mentioned. As of 1941 her ongoing research included work on cancer as well as in endocrinology (see below; Hobson).

Another striking early example here is **Dr. Virginia K. Frantz** (1896–1967). Beginning as a surgeon, she moved into pathological research, and did important work on cancer. In the late 1920s at Presbyterian Hospital (Columbia), she did important studies on pancreatic, thyroid, and breast cancers. In 1935, she received national recognition for the first description (with Dr. Allen Whipple) of insulin-secreting tumors of the pancreas. In the 1940s she did the work that brings her to this section: she became one of the first to demonstrate that radioactive iodine was effective in identifying and treating metastatic thyroid carcinoma.

Dr. Frantz received many honors during her career, some more relevant to her surgical years and thus discussed below (Surgery). Noteworthy here are her term as first woman president of the American Thyroid Association and her winning of the Elizabeth Blackwell Award for distinguished service in medicine. She died in 1967—of cancer (*NAW*/mod.; Frantz).

Also engaged in cancer research in the 1940s was **Dr. Alice Leigh-Smith.** A pupil of **Marie Curie** and wife of a British diplomat, Dr. Leigh-Smith was the first British woman to receive a doctorate in nuclear physics. Though best known for her co-discovery of a new radioactive element, anglohelvetium, in 1943, Dr. Leigh-Smith also did cancer research, first in the British Empire cancer campaign (from 1936), and then in Switzerland during the War. The 1943 news stories mention that Dr. Leigh-Smith is "noted also for cancer research" and that she was investigating a new treatment (London *Times,* Jan. 6 and 14, Jun. 18, 1943). Since she worked at the Radium Institute and was using a radium compound when her observations led to the isolation of anglohelvetium, the new treatment presumably involved radiation therapy.

In 1968 **Gertrude B. Elion** (1918–   ) received the Garvan Medal for her work in chemotherapy. At Burroughs Wellcome, where she had worked since 1941, Elion became a leader in the field of anti-cancer agents, including drugs for leukemia. She was spurred on, as she revealed in a 1991 interview, by watching someone close to her die of cancer. In addition she synthesized and studied drugs to treat gout and organ-transplant recipients. Her work led to the development of the major AIDS drug AZT.

Born in New York City, she attended Hunter College. Despite graduating *summa cum laude* (1937), Gertrude Elion could not get a graduate assistantship and was forced to take low-level laboratory jobs, high-school teaching jobs, etc. Receiving her master's degree in 1941, however, she was hired by Burroughs, where by 1967 she was Director of Experimental Therapy. She holds adjunct professorships of pharmacology at Duke University and the University of North Carolina, and several honorary degrees. Her professional memberships include the American Association for the Advancement of Science, the American Association for Cancer Research, the American Society of Hematology, the American Chemical Society, the American Society of Hematology, the American Chemical Society, the American Society of Biological Chemistry, the New York Academy of Sciences, and the British Chemical Society (*AMWS*/12; *WBWRA,* 222). In 1983 she retired to become a scientist emerita at Burroughs. In 1988 she won the Nobel Prize for medicine. Her 45 patents are seldom mentioned in the Nobel publicity, but in 1991 Gertrude Elion became the first woman to be inducted into the National Inventors

Hall of Fame (Allen; *AMWS/* 12; Bouton; "First Woman"; "Inventors Hall"; " 'Maverick' Woman"; "Nobel Pair"; NPR; "Three Researchers"; "3 Share Nobel"; *WBWRA*).

Also at Burroughs during the 1950s was **Elvira A. Falco.** She co-patented chemother-apeutic agents with Elion and with Elion's frequent co-patentee George Hitchings, as for example her 1956 patent for 3-Amino-4-cyanopyrazole and a method of preparing it; and her 1957 patent for Oxazolopyrimidines and method of preparing same.

In 1981 researchers at Baylor College of Medicine announced an experimental treatment for advanced breast cancer: diverting some of the patient's own blood (plasma) through "a column containing a protein with the unique ability to change the character of the body's immune defenses," and then returning the blood, now containing "several different potent immune system components" newly activated, to the patient. One of these researchers was **Lisa Miller, RN.** The treatment is not a cure in itself, but stimulates the immune system and renders surviving cancer cells more vulnerable to conventional chemo- and radiation therapy. In the small series reported, it produced remissions in three out of five women, two of whom were apparently free of cancer at the time of the report. The third had a small recurrence treatable by chemotherapy.

Commenting on the treatment, which was published in the *New England Journal of Medicine* (Nov. 12, 1981), Drs. Karl and **Ingegerd Hellstrom** of the Fred Hutchinson Cancer Research Center in Seattle note that the Texas work builds on basic research they did more than a decade earlier. They consider it an exciting development, but call for further testing to see how lasting the remission is ("New Weapon"; Terman *et al.*)

In July 1985 a new cancer chemotherapy was reported by a team of scientists from three research centers—Stanford Medical School, M.D. Anderson Hospital in Houston, and the University of Vermont. The team, led by Dr. Branimir Sikic of Stanford, included at least one woman, **Dr. Nina Friend,** also of Stanford. Adriamycin (doxorubicin), the leading drug used against a broad range of cancers, damages the heart in long-term use. Consequently, scientists have long sought to engineer a form of adriamycin that would kill cancer cells but not damage heart muscle. MRA-CN, synthesized in 1982 at SRI International in Menlo Park, CA, is the first promising drug of this type.

The drug's power is clear. Tested against cells from nine different human tumors, it proved 100 to 15,000 times as potent as the standard drug. And in early results it spared heart-muscle cells, at usual therapeutic dosages. Further, MRA-CN seemed to work against cancer cell lines resistant to adriamycin. However, this 1985 report estimated that the drug was at least two years away from clinical testing.

In addition to modifying the drugs themselves, scientists are pursuing another approach to making cancer chemotherapy both more effective and less toxic to healthy tissue—targeted drug-delivery systems. Very promising is liposome packaging, in which drugs encased in tiny membrane spheres are aimed at specific tumor sites in the body by attaching the spheres to monoclonal antibodies (proteins designed to recognize specific cellular "addresses"). Monoclonal-antibody research, often cancer-directed, goes on at several centers, among them a laboratory at the University of California, San Francisco, where Drs. Demitros Papahadjopoulos and **Kate Mathay** collaborate with Dr. Sikic's Stanford team (Hofstadter, "Experimental Drug").

Although the most famous names in monoclonal-antibody research are thus far male, at least one other woman, **Dr. Ellen Vitetta** of the University of Texas at Austin, has done important innovative work there (Fjermedal, *Bullets*).

An appropriate closing example here came to my attention as the book went to press. Three Stanford women, **Dr. Elaine Zeman, Marilyn Lemmon,** and **Kate Hirst,** worked with Dr. J. Martin Brown of the Therapeutic Radiology Division to produce another type of targeted drug. The new chemical, called SR 4233, selectively attacks tumors because it is metabolized to its toxic form upon reaching the oxygen-deficient areas that are characteristic of tumors. In mice SR 4233 destroyed 99.9% of the cancerous cells. Another innovative feature of the work was the use of a blood-pressure drug to divert oxygen-rich blood away from the tumor to be attacked, making it more severely oxygen-poor and thus more vulnerable to the drug (Hofstadter, "Chemical").

*AIDS/immune system.* The original animal studies with interleukin-2 were done by two Scripps Institute researchers, Dr. William Weigle and **Marilyn Thoman.** One of the biggest medical news stories of 1984 involved interleukin-2 as a possible AIDS treatment, exploiting the hormone's stimulating effect on the immune system. Interleukin-2, as produced in the healthy body by helper T-cells, is a signal molecule, stimulating T-cells to grow. Though these were laboratory tests only—using lymphocytes from 16 AIDS patients and 10 controls—the hormone dramatically improved the activity of "natural killer cells" and also stimulated growth and division of T-cells from some AIDS patients. Just as in the animal studies, the interleukin restored sick T-cells' ability to attack abnormal and foreign cells. Clinical trials have not borne out all of its original promise, but the approach should be on the right track in an otherwise very bleak treatment picture.

At least one woman was involved in this exciting research. **Claudia Benike,** with Dr. Jeffrey Lifson at Stanford, tested the hormone's effects in the laboratory (Hofstadter, "Protein").

A preventive for AIDS is every researcher's dream. Late in 1986 **Dr. Candace Pert, Joanna Hill,** and colleagues at NIH reported that by adding a short peptide found on the HIV virus envelope to brain-cell cultures, they had blocked the virus's entry into the cells. Pert's group planned next to try a long-lasting analog of the peptide in human AIDS patients (Silberner, "AIDS").

AIDS, of course, is only one of many possible problems with the human immune system. Sometimes newborns lack a crucial enzyme, adenosine deaminase (ADA), without which their immune systems simply shut down. Such infants' only hope has been a bone-marrow transplant (about 50% success), or living in a sterile bubble, as did one Texas boy until he died after surgery at 12. Advances in genetic engineering suggest that the missing gene could be supplied to the infant's own bone marrow, a technique successful in monkeys by 1985.

Meanwhile, direct replacement of the enzyme itself seems to offer promise. At first trial, the enzyme broke down too soon in the infants' bodies. Now, researchers hook polyethylene glycol (PEG) onto the ADA molecule in such a way that it shields the enzyme from short-term destruction without blocking its active site. The new replacement enzyme was developed at Enzon, Inc., of South Plainfield, NJ. **Rebecca Buckley** of Duke University directed the successful clinical trials in 1987 (Silberner, "Sneaking . . .," "Enzyme Ruse . . .").

Auto-immune disorders are being studied worldwide. **Michelle Boulanger** of Marly-le-Roi, France, takes the chemical, new-compounds approach. In 1981, with two male colleagues, she patented Piperidylbenzimidazolinone compounds and a method of treating auto-immune, allergic or anti-inflammatory diseases, as well as asthmatic dyspnea (difficult or pathological breathing caused by asthma). They assigned their U.S. patent to Science Union et cie, Société Française de Recherche Medicale, Suresnes, France.

*Multiple Sclerosis (MS).* This baffling disease affects over 250,000 people in the U.S. alone. A promising treatment that entered clinical testing in 1985 also involves monoclonal antibodies. Injections of a monoclonal antibody called GK1.5 prevented mice from developing an experimental allergic encephalomyelitis (EAE) with MS-like effects to which they were then exposed; and reversed the effects of the disease in already-paralyzed animals. Two women were involved in this experiment: **Dr. Leonore Herzenberg** of the Stanford Genetics Department and **Mae Lim,** a neurology research associate.

To avoid rejection problems in human MS patients, researchers will probably use a new type of genetically engineered monoclonal antibody that is part-mouse and part-human. This "chimaeric" antibody was developed by researchers from Columbia, Becton Dickinson Monoclonal Center in Mountain View, CA, and Stanford. The team included two women: **Dr. Sherie L. Morrison,** Professor of Microbiology at Columbia, and **Dr. M. Jacqueline Johnson,** a former graduate genetics student at Stanford (Hofstadter, "Combination Antibody . . ."; Waldor *et al.*).

The history of this approach to multiple sclerosis dates to the 1950s, and prominently involves at least one other woman, **Dr. Marian Kies** (c. 1915–   ) of the National Institutes of Health. Dr. Kies' previous biochemical research had been in plant enzymes and growth factors at the Department of Agriculture. Coming to NIH in 1953, she got a new laboratory and a new problem. "It was thought," she notes, "that [EAE] might be the experimental prototype of multiple sclerosis (MS) and that an understanding of the mechanism of EAE would help to explain the cause of MS, or at least suggest some means of therapy." She and her co-workers posed four questions: 1) Which brain component induced the disease (whole-brain or spinal-cord tissue had been used to date)? 2) How did it do so? 3) How can we prevent induction? And 4) What will all this tell us about MS?

First, Dr. Kies and her team needed to be able to measure the encephalitis-causing activity both in whole tissue and in various fractions isolated from it. In devising the necessary bioassay they worked with an NIH neuropathologist, Dr. E.C. Alvord, Jr., and neurochemist **Dr. Elizabeth Roboz-Einstein** of Georgetown Medical School. The first papers appeared in 1957. After quantifying the relationship between whole tissue and the appearance of disease in guinea pigs, they used the assay to monitor the purification of the encephalitogen. This led them to identify the disease-causing substance as a protein located in the myelin. This discovery, incidentally, changed the whole scientific view of myelin, once considered relatively inert. They named the substance myelin basic protein (BP). BP is unique among central-nervous-system proteins in causing a destructive auto-immune reaction in sensitized cells rather than in antibodies.

Thus armed, Dr. Kies' team tackled the basic mechanisms involved in inducing and preventing EAE. By 1974 they had successfully treated EAE in guinea pigs, using BP itself in a vehicle called incomplete Freund's adjuvant (IFA). Of course, treating an experimental disease in a highly susceptible inbred strain of guinea pigs is very different from treating it in human beings. Many questions—such as the exact mechanism of disease-induction—still remained unanswered. However, it was a beginning. Dr. Kies found that BP therapy also worked on primates, but only with antibiotics or steroids.

Graduating from the University of Illinois in 1936, Marian married a research metallurgist, and bore three children without interrupting her scientific career. She had worked at George Washington Medical School, Oak Ridge, the Agriculture Department, and the National Science Foundation before coming to NIH. Some of her moves accommodated her husband's career (though apparently both careers were usually considered), and at NSF she had "an uneasy feeling that I might have had a better offer had I been male." Thus she was already aware of discrimination against women in chemistry when asked to write one of the chapters of Derek Richter's book, *Women Scientists: The Road to Liberation*. Her account is matter-of-fact, supported by pertinent statistics, and totally without bitterness. As of 1982, when it appeared, Dr. Kies was at the Laboratory of Cerebral Metabolism, National Institute of Mental Health, Bethesda.

Dr. Roboz-Einstein later went to the University of California, where as Professor of Neurochemistry she continued her EAE research.

*Arthritis.* **Dr. Katherine Tepperman** of the University of Cincinnati announced an improved treatment for rheumatoid arthritis in 1984. The injections of gold salts long used to treat intractable arthritis can cause kidney damage. Her therapy involves Auranofin, a gold-based drug that can be taken by mouth. It has two major advantages over the previous regimen: dosages are smaller, with less risk, and patients can take it themselves. At first report, side effects were limited to diarrhea. With more than 36 million rheumatoid arthritis victims in the United States alone, the significance of a successful treatment would be enormous ("Gold Drink").

Much earlier (1950s), in Europe, **Dr. Ana Aslan** (c. 1900– ) developed a treatment for rheumatic diseases. Consisting mainly of novocaine, it can be taken orally or by injection. In the United States it is approved for experimental use only—as an anti-depressant; but in Roumania, Dr. Aslan's Gerovital H-3 treatment is well accepted and extremely popular. Its use as an anti-aging treatment has apparently overshadowed its anti-rheumatic uses, however, as its name indicates. According to Dr. Aslan, Gerovital has proved effective against depression, high blood pressure, general weakness, and aging, defined as "the deterioration of the capacity to feed the cells." In 1974, 20,000 patients came to her Institute of Geriatrics in Bucharest to take the treatment (W. Ellis, 703–4ff).

*Heart and circulatory problems.* Since it was an English witch—a woman healer—who discovered the effects of the foxglove herb and developed from it a potent digitalis-containing remedy for dropsy and other symptoms of heart disease, it seems appropriate that a modern woman is developing an important adjunct drug for digitalis. Digitalis has been the treatment of choice in congestive heart failure for centuries, but patients with kidney problems cannot use it. **Dr. Georgia Lesh-Laurie,** in research supported by the American Heart Association, has discovered a new heart stimulant lacking these side effects. Dr. Lesh-Laurie's work builds on results of a jellyfish sting: victims noticed a sudden neurological and cardiovascular response. Her new stimulant is a protein found in the toxin of the hydra, small freshwater cousin of the jellyfish. The sting victims made an observation which Dr. Laurie turned into a discovery; her next step, partly completed, is to turn it into an invention—a pure and standardized drug with predictable effects (Amer. Heart Assoc., Research Report 5, *Scientific American,* May 1985: 42).

Further research would show many other examples, but suffice it here to mention **Dr. Elizabeth M. Smith** of Schering Corporation in Bloomfield, NJ. Dr. Smith is co-inventor of an angiotensin-converting enzyme inhibitor that was in clinical trials in 1986. The only known source of this inhibitor was the venom of a South American ant. As its name suggests, the compound will be used to treat cardiovascular disorders, and to reduce or control high blood pressure (Magatti, L-9/5 and 10/17/86).

*Skin diseases/conditions.* Psoriasis: The "heartbreak of psoriasis" is much talked about but little eased by medicine thus far. Recently, however, the Psoriasis Research Foundation was founded in Palo Alto, CA, and several promising treatments have been announced. Noteworthy among them is a treatment devised by a biologist: **Elaine K. Orenberg,** PhD, formerly a Senior Research Associate in Dermatology at the Stanford

University Medical Center, announced some success with localized heat, used in conjunction with other treatments. The heating, induced by ultrasound, and used for 30 minutes three times per week on chronic psoriatic plaques, produced remissions lasting three months or longer, and the treatment caught the imagination of the popular as well as the scientific press (Orenberg, Deneau, & Farber; Malkinson; Napoli).

Dr. Orenberg, who took her PhD in reproductive physiology at the Pennsylvania State University in 1965, and who was Director of Laboratories and Vice-President of the Psoriasis Research Institute (1982–7), is now a consultant to private industry. She has more than fifty scientific publications to her credit.

Acne: For years one of the few useful treatments for acne was the so-called peeling treatment developed by **Dr. Emmie List** (or **Litz**) of Vienna, friend and teacher of Helena Rubinstein. But this was a drastic and costly procedure that did not work for everyone. Meanwhile, the suffering and shame of millions of teenagers continued. One source, estimating that 70–80% of all adolescents are afflicted, notes that in severe acne, "In addition to the progressive scarring which such lesions may produce, rather marked psycho-neurotic changes may occur because of the young patient's concern about his or her pimply appearance . . ." (*EB*/58, s.v. "Skin Diseases"; *NAW*/mod., s.v. "Rubinstein"; Rubinstein, 32, 38, 43).

**Margaret H. "Peg" Sherlock**'s anti-acne compound was featured by Stacy Jones in his regular Saturday "Patents" column in the *New York Times* (Feb. 24, 1979). Ms. Sherlock, a resident of Bloomfield, NJ, assigned the patent rights to Schering Corporation of Kenilworth, NJ, for whom she has created many other novel compounds, including at least two other anti-acne agents. Indeed, by the time she retired in 1986, after forty-odd years with the company, she had 53 U.S. patents to her credit.

However, according to a Schering patent counsel, Anita W. Magatti, Margaret Sherlock is probably best known at Schering for her work on a product whose patents do not even carry her name. She played an important role in developing chlorpheniramine, the antihistamine in Schering's widely used CORICIDIN® and in certain other companies' cold and allergy remedies as well. In five or six years of antihistamine work, says Sherlock, "we obtained four or five good compounds. . . ." Sherlock's name does appear on the patents involved in at least two other Schering products: a veterinary analgesic, and the tranquilizer TIN-DAL® (acetophenazine maleate, Pat. #2,985,654). She also worked on "Prantal," intended as an ulcer treatment, but eventually used mostly for poison-ivy creams and anti-perspirants; on two other potent tranquilizers, "Trilafon" and "permitil"; the diuretic drug "Naqua," and the anti-inflammatory analgesic "Clonixin" (Magatti, L-9/5/86; "Peg Sherlock").

A co-patentee with Margaret Sherlock on one of the anti-acne agents is **Dr. Heide Roebke** of Belleville, NJ, a former Schering employee who died about 1978 at a young age. Another team member in the early 1970s was **Marjorie Behrens,** a chemist from California ("Peg Sherlock," 3–4).

Even more disfiguring than acne, though rarer, are the so-called port wine birthmarks. Until recently these marks could be removed only by surgery, which sometimes left ugly scars. Since the 1960s, argon lasers have been used to remove them by closing the abnormal blood vessels. However, this technique may cause burns. Early in 1986, **Dr. Oon Tian Tan** of the Boston University Medical Center reported treating a series of 30 patients with pulsed, organic-dye lasers. These lasers can be tuned to hemoglobin's peak absorption frequency to reduce scarring. Dr. Tan reports no scarring in her series of 30 (J. Hecht).

*New compounds.* Research continues on new chemical compounds with treatment applications. Pertinent here are several women already discussed, such as **Gertrude Elion, Elvira Falco,** and **Margaret Sherlock.** Not previously mentioned is **Karlene W. Salamon,** patentee of the unpronounceable new compounds called 7-[tetrahydro-4-(3-Hydroxy-1-octenyl)-2,6-Dioxo-4H-Cyclopenta-1,3-Dioxol-5-YL]-5-Heptanoates. Salamon, who lives in Chicago, assigned her 1979 patent to G.D. Searle & Company of Skokie, IL.

During the 1970s and '80s, several European women received American patents for chemical compounds with therapeutic applications. A few examples: In 1970 Swiss inventor **Hedwig Besendorf,** with others, patented new phenethyl compounds useful as muscle relaxants and anti-convulsants. In 1971, **Berit Margareta Olsson** of Transgund, Sweden, patented an antifibrinolytically active isomer. French inventor **Arlette Chosson Adam** co-patented water-soluble compounds having immunodepressive properties in 1978. And in 1978, four Russian women—**Eva Evseevna Mikhlina, Valentina Yakovlevna, Anna Dimitrievna,** and **Nadezhda Andreevna Komarova**—were part of a team that patented a Medicated compound for treating allergic diseases.

In 1981 **Irmgard Bottcher** of Berlin, Germany, with a male co-inventor, patented anti-inflammatory compounds. The two chemists assigned their patent to Schering Corporation in Berlin.

*Drug-delivery systems.* The best drug is useless unless delivered in effective form and strength to the correct part of the body. Women have been involved with this area of healing for millennia, with their ointments, plasters, and poultices. Since women were also involved in two of the most effective drug-delivery developments in the 20th century, this seems an especially effective closing for the Remedies/Treatments section.

Slow intravenous drip: A less likely training ground for medical research and innovation than Broadway and the Ziegfeld Follies can scarcely be imagined; yet such was the background of **Justine Johnstone Wanger** (1895—    ). Born in Hoboken, NJ, Justine Johnstone seemed headed for show business at an early age, appearing as a mannequin for Mme Sherry in New York City at fifteen, and in the Folies-Bergères at sixteen. Then came two years of vaudeville, followed by a Broadway role in *Watch Your Step.* She played Columbia in the Ziegfeld Follies of 1915, acted in two other Broadway musicals—*Stop, Look, and Listen* and *Oh Boy!*—and finally starred in *Over the Top* in 1917–8.

Like others before her, however, "the girl who owns Broadway" longed for more meaning in her life. Her eventual solution was unusual. After trying legitimate theatre and marriage (to Walter Wanger of Paramount Studios) she turned to the study of science. At the suggestion of her husband's doctor, Samuel Hirshfeld, she enrolled at Columbia University. Her professor was so impressed with her plant research that he suggested her as a research assistant to Dr. Hirshfeld and his colleague Harold T. Hyman, who were studying the effects of the speed of injection or transfusion on patients. Hypothesizing that too-rapid injection caused some post-transfusion deaths, the two doctors coined the term "speed shock" for the phenomenon. Since her colleagues were practicing physicians, much of the ongoing research was left to Justine Wanger. When the final report came out in 1931, she appeared as third author.

Hirshfeld and Hyman, with Wanger's aid, had confirmed their original hunch that injection rate was a crucial factor in the success of intravenous treatments and transfusions. A dog would die if its own healthy blood was injected back into it too fast; harmless salt or sugar solutions could kill if injected at high speeds, while toxic drugs could be given safely if dripped slowly into the veins over several hours, etc.

Wanger and colleagues had in effect invented the modern I.V.-drip method of drug delivery now universal in hospitals. One of the many improved treatments made possible by their work, as noted, was a great advance in the treatment of syphilis.

After the intravenous-drip publication, Justine Wanger continued her research, studying, among other things, resuscitation, melanin (dark human pigment), and hibernation, leading to advances in cancer research. When the Wangers moved to Hollywood in 1933, Justine again became associated with Samuel Hirshfeld. Her pleasant Hollywood house, with swimming pool and garden, had a research lab in the cellar. With her husband and two adopted sons, her music and other interests, she had by 1941, at age 46, created a full and meaningful life for herself, including ongoing research in the field of endocrinology and cancer.

My last source on Justine Wanger is dated 1941 (Hobson). But even if she did no more than reported there, she was being ridiculously modest when she said, "There is as much to tell about me as there is about a filterable virus. I just got by." Closer to the mark, surely, is Dr. Hyman's verdict: he called her both great and good.

Transdermal drug delivery: The slow-intravenous-drip method researched by Justine Wanger and her colleagues revolutionized drug therapy in the 1930s. The comparable revolution in the 1980s may come with large-scale adoption of transdermal (through-the-skin) delivery of drugs, conceived in its modern form by Alexander Zaffaroni of ALZA Corporation, Palo Alto, CA, and developed or reduced to practice by **Dr. Jane Shaw** (1939–   ) of nearby Atherton CA, then on the ALZA research staff. Not every drug is suitable for this form of delivery, of course. Drugs that irritate the skin, or do not act at plasma concentrations achievable with this method, cannot be used.

This modern work refines and quantifies an ancient woman's method—gathering data on skin permeability, testing for the most permeable skin site, finding a workable vehicle for the drug, and devising a mechanism for controlled release, allowing delivery of the drug at a known constant rate over long periods, etc.

In 1978 Dr. Shaw and her collaborator announced an in-vitro test for skin permeability to various drugs, and a mathematical model of the permeation process. In the same paper they reported determining that the skin behind the ear was the most permeable to the test drug (scopolamine). These findings enabled Shaw and her colleagues to identify drugs suitable for transdermal delivery.

Scopolamine was the first drug for which a transdermal therapeutic system (TTS) was invented. In addition to the requirements already mentioned, Dr. Shaw and her colleagues had to find or create a nonirritating adhesive, to hold the drug in place but easily release it, a TTS construction allowing the drug to travel one way only, and a means of suppressing bacterial growth at the site. But perhaps most ingenious of all was the ALZA researchers' idea for a drug-delivery rate lower than the absorption rate of the least permeable skin observed from any individual, so that the system, not the (variable) permeability of the patient's skin, determines the actual rate of drug delivery to the blood stream. Scopolamine, useful in motion-sickness, may also help prevent the nausea and vomiting associated with cancer chemotherapy, and alleviate vertigo.

TTS drug delivery has definite advantages in chronic conditions such as hypertension, where patients may forget to take medicine when they have no symptoms; where drugs would otherwise have to be given very frequently, etc. By 1986 Dr. Shaw had five patents, four dealing with the scopolamine TTS, and her professional publications already numbered 108.

Born in England, the younger of two children, and privately educated through secondary school, Jane Shaw got a BSc with honors and a PhD from Birmingham University. Her title at ALZA as of 1981 was Vice-President, Product Research and

Development and Program Director, Transdermal Therapeutic Systems. By 1986 she was Executive Vice-President, ALZA Corporation; President, ALZA Research; and Chairman of the Board, ALZA Ltd (UK). She continues to work in research and development of transdermal permeation of drugs (J. Shaw, L-12/29/81 and 7/30/86, "Drug Delivery . . ."; Shaw and Chandrasekaran). With regard to her place in medical history, sad to say, recent articles on ALZA and its achievements fail to mention her. See, for example, the 1986 *Fortune* article (Bylinsky, "Alza").

Vaginal drug delivery: In the late 1930s, a ballet dancer named **Leona Chalmers** invented and patented a rubber menstrual cup called the Tassette, which she registered as a trademark in 1960. In the late 1960s, a nurse named **Barbara Waldron** and her husband John collaborated on a plastic disposable model of the cup, which they called Tassaway®, patented by Barbara in 1968. This cup, it now turns out, is effective for retaining medications within the vagina to treat conditions ranging from venereal disease to vaginitis (Oram; "The Cup"; Waldron, L-1/10/83, encl. p. 11).

## Surgery

For centuries after Galen, who devalued surgery, only women and barbers were surgeons. Since the Church frowned on autopsies and dissection, university training in anatomy was pitiful. Until the rise of the modern surgical specialty, many surgical advances were probably made by women. Unfortunately, such advances were often not recorded until a male doctor rediscovered or reinvented them—or learned them from a woman—at which time he often got the credit. **Marie Colinet**'s magnetic removal of metal slivers from the eye is a striking example (Hurd-Mead, 361, 433).

Margaret Pelling's work on barber-surgeons and other medical practitioners of Norwich, England, 1550–1640, enriches our sparse knowledge of early modern medicine and surgery. It is a welcome corrective to the usual assumptions that barbers were male and that medicine was fully professionalized by this time. If it was not so in Norwich, the second city of England until 1650, it was probably not so anywhere in English medicine except perhaps in London. As noted above, Pelling has also located at least 30 women, including midwives, among Norwich medical practitioners between 1550 and 1640, who were connected with or members of the barber-surgeons' company. The barber-surgeons were generally "more tolerant" of women than some of the other guilds, and therefore contained a greater proportion of women. The barber-surgeons, a guild or "company" in Norwich as elsewhere, were only one of several kinds of medical practitioners. They had lower status, less wealth, and fewer civic responsibilities than the apothecaries, as a group. Yet—and this may be Pelling's most surprising finding—they bore the "main burden of general practice in the towns" (Pelling, "Occupational," 489, 490).[29]

Despite the fact that medical licenses were already issued by Church authorities, some of the women had formal qualifications and licenses to practice. At least one substantial woman surgeon is known by name: Alice Glavin (d. 1597). That she practiced in her own right is indicated by the fact that she was hired by the City to heal several poor patients (so they could work and would no longer need public charity) before her surgeon husband's death. Wife of one surgeon and mother of another, she shows no connection with the barber-surgeons' company (Pelling, "Tradition").

---

29. Note that the burden consisted of what women—i.e., ordinary wives and mothers—could not handle at home (cf. Pelling's citation of Katherine Paston and a patient who came to a neighboring practitioner only after her arts and remedies had failed to help him; "Tradition," 166).

Once men entered surgery in large numbers and began to control it—once the specialty became fully professionalized—male professionals excluded women from it even more rigidly than from medicine. This situation continued into the 20th century. **Dr. Virginia Apgar,** after two years of a prestigious surgical internship, abandoned surgery in the belief that as a woman she could never make a living at it. Even today, women are scarcer in the surgical specialties than elsewhere in medicine. The following 19th- and 20th-century examples suggest what women could have done had they been admitted freely into the field:

**Dr. Eliza M. Mosher** (1846–1928), is most noteworthy in preventive medicine, as the inventor of numerous posture-improving devices. Her incentive for creating a new knee operation was personal: after a severe knee injury in 1883 she was in constant pain, sometimes bedridden, walking only with crutches. This situation became intolerable for such a dynamic woman, and about 1890 she devised an operation to be performed on herself. The operation was successful (*NAW*). Dr. Mosher wrote an article titled "A Critical Study of the Biceps Cruris Muscle and Its Relation to Diseases in and Around the Knee Joint" (1891) that undoubtedly describes the operation and the theory behind it (*NCAB*/1916).

**Mathilde Schott** of New Haven, CT, patented a Surgical knife in 1890. The distinguishing feature of this knife seems to be its detachable blade (Knight, 126, 129).

**Helen Augusta Blanchard** invented a surgical needle, as did **Dr. Mary Harris Thompson** of Chicago. During Blanchard's long career (her patents span 1873–1915), she was best known for sewing machine inventions—and as an entrepreneur. However, she was a versatile inventor who also patented a corset-cord fastener, a pencil sharpener, and an improved elastic goring for shoes, among other things. Mary Logan, writing in 1912, includes Helen Blanchard and her surgical needle of 1894 in a list of "Prominent American Women Inventors," with a brief biography (887, 889).

Logan also mentions, but does not name, a Maryland woman who has "distinguished herself by many inventions," among which was "the eyeless needle now used so largely by surgeons" (885).[30] This was **Ella Gaillard,** whose delicate mechanism for screwing catgut or cord into the blunt end of a needle (patented 1874) avoids the bulky doubled cord demanded by eyed needles. Gaillard is a versatile multiple inventor discussed at more length in the "Machines" chapter.

**Mary H. Thompson** (1829–95) graduated from the New England Female Medical College, beginning her practice in 1863. She chose Chicago in order not to compete with the women doctors already practicing in New York. She founded the Chicago Hospital for Women and Children, renamed the Mary Thompson Hospital of Chicago after her death. She also played an important role in founding the Women's Medical College in Chicago, where she became Professor of Hygiene (1870–7) and later Clinical Professor of Obstetrics and Gynecology.

Dr. Thompson was the best-known woman surgeon in the Midwest, and one of the best-known in the country. Specializing in abdominal and pelvic surgery, she was the first—and long the only—woman doing major surgery in Chicago. The achievement that brings her to this chapter, however, is her invention of a widely used abdominal needle, probably in the late 1870s or early 1880s. Noted surgeon Albert J. Ochsner said Thompson "convinced many of us that it was possible for a woman to be a real physician and surgeon." And indeed she eased the way for later women physicians. By 1881–2 she

---

30. Logan gives no dates, as well as no names; but although her book was copyrighted 1912, her wording follows Lydia Farmer (420), who wrote in 1893.

was vice-president of the Chicago Medical Society. To her undeniable skills she added charm, kindness, and generosity, often giving poor patients food and shelter. Patients of all classes became her friends for life.

She believed firmly in woman suffrage, but devoted little time either to it or to the city's social life, contenting herself with work and a circle of friends (*NAW;* Chaff, #1253, #268).

**Fannie M. Garies** of New York City invented an instrument for chiropodists (specialists in conditions of the hands and feet), assigning half her 1894 patent rights to Mary E. Bustard, also of New York.

Trained in the 19th century, but creating or at least publishing her inventions in the early 20th century, was **Dr. Alice G. Bryant.** First woman to specialize in otolaryngology (ENT), she invented medical apparatus in the specialty. But ENT is also a surgical speciality, and some of her several inventions are surgical instruments that became standard in the field (Bever, 62–3, index). Her new tonsil tenaculum and her nasal polypus hook, instruments used in tonsillectomies and in the removal of nasal polyps (both reported in *Laryngoscope,* Dec./Jan., 1905–6), deserve mention here, as do her tonsil separators, her tonsil snare cannula, and her bone-gripping forceps. Articles announcing these inventions appeared over the decade 1905–15 (Bryant).

Alice Bryant attended M.I.T. in 1882–3, but received her undergraduate degree from Vassar in 1885, and her MD at the Women's Medical College of New York in 1890. She practiced at the New England Hospital for Women and Children and at the New England Deaconess Hospital. Dr. Bryant was active in science as well, publishing more than 75 articles in her field. She was also a pioneer in establishing evening clinics for working people at Boston's Trinity Dispensary. Dr. Bryant was one of the first two women admitted to the American College of Surgeons, in June 1914 (*WBWRA,* 213).[31]

During the 1920s, **Verena Holmes** worked for Allen & Hanbury's, Ltd, a London firm of surgical instrument-makers. While there she patented at least four surgical instruments or related inventions. Holmes received British Patents #261,856 for Improved scissors, forceps, etc., and #276,564 for an Aspirator. She also patented Headrests (#261,511) and a Personal headlamp for surgeons (#278,827) (Brody, L-8/22/84). Verena Holmes was a talented engineer and a versatile inventor whose patented inventions range from steam and internal combustion engines to medicine and surgery, and even to office machines. She is, in fact, probably best known for her safety paper cutter (see below).

An essential aspect of modern surgery is anaesthesia. In 1923 **Dr. Isabella Herb** (c. 1868–1943) and colleagues introduced the use of ethylene as an inhalation anaesthetic. Though just as explosive as ether, it works more quickly (Harpur, 75). Herb is mentioned first among three introducers in Patrick Harpur's *Timetable of Technology,* with Dr. Arno B. Luckhardt and one other man. Luckhardt wrote the *Britannica* article on ethylene (1960s editions) without mentioning Herb, or himself, but citing a paper he wrote with the third colleague. The report of Dr. Herb's retirement from Rush-Presbyterian Hospital in Chicago in 1941 solves the puzzle: "Dr. Herb is known to the medical profession throughout the world as the anaesthetist who first administered ethylene gas at actual operations, and did much to develop methods for the safe and effective use of this now generally favored anaesthetic. The anaesthetic property of ethylene gas was discovered by Dr. Arno B. Luckhardt, Professor of Physiology at the University of Chicago, and Dr. J. Bailey Carter, then a medical student . . ." ("Dr. Isabella Herb," 1). Dr. Luckhardt, in other

---

31. The other was another Boston physician, Dr. Florence West Duckering, a Tufts Medical School graduate (1901).

words, was the discoverer, and Dr. Herb the *inventor* or developer of this new anaesthetic. She was first to use it not only in general surgery (Mar. 14, 1923), but in obstetrics (May 1, 1923).

As early as the 1890s, while associated with Dr. Albert Ochsner at Augustana Hospital, she created another innovation in anaesthesia: with Dr. Lawrence H. Prince, an obstetrician, Dr. Herb developed the "open drop" method of administering ether, which eased the induction process for patients and also greatly reduced the toxic after-effects. This technique became the universally accepted method of inducing ether anaesthesia.

Dr. Herb, who lived in Winnetka, IL, and practiced in Chicago, received her MD from the Women's Medical College of Northwestern University in 1892. She wanted to be a surgeon, but "events" led her toward other fields, such as pathology, bacteriology, and anaesthesia. After postgraduate work at Johns Hopkins and study abroad, she worked with the Mayo brothers in Rochester, MN. From 1909 until her retirement in 1941, she was Chief Anaesthetist at Chicago's Rush-Presbyterian Hospital and Professor of Surgery at Rush Medical College. She was the first woman appointed to the medical staff at Presbyterian ("Dr. Isabella Herb"; *NYT* obit., May 30, 1943: 27).

Even a well-anaesthetized patient may suffer muscle spasms that could disrupt surgery. **Enid Johnson**'s pioneering use of curare as a relaxant to prevent such spasms was considered worth mentioning in Harpur's *Timetable of Technology* (125).

As mentioned, **Dr. Emmie List** (or **Litz**) of Vienna, invented the peeling treatment for acne and other deep-pitting skin blemishes. Depending on the skin-removal technique used, this treatment may be surgical. Weekly abrasions over a six-month period expose new skin underneath and obliviate the scars. After teaching Helena Rubinstein in 1908–9, Dr. List worked in the Rubinstein clinic in London (Rubinstein, 32, 38, 43 ; *NAW*/mod., "Rubinstein").

**Dr. Else K. La Roe** (1900– ) has two quite different innovations to her credit. Born in Germany, niece of a doctor and granddaughter of a scientist, she determined at an early age to do something unheard of: to become a surgeon. She was the only girl in her gymnasium—and valedictorian of her class. Educated in medicine and surgery at Heidelberg (MD, 1923), Vienna, and Paris (with her grandmother's aid), she worked first in a clinic specializing in the reconstruction of tubercular scar tissue. After working at a police station, she became deeply interested in social problems, becoming director of a leading birth control clinic, lecturing on women's health, and consulting to the Russian Government to improve its existing birth control clinics and to open such clinics in Central Asia. She also wrote an international birth control film and several books on women's hygiene. To avoid discrimination, she opened her own plastic surgery clinic.

A leader of the German women's movement (also a peace movement), even while with increasing revulsion she entertained Hitler and Goebbels for her banker husband, Dr. La Roe edited a women's magazine in Germany in the early 1930s. Shortly before Hitler came to power, he visited her editorial office. When he told her women were good for two things only—cooking and bearing children—she ordered him out. Not surprisingly, she soon left Germany.

In the United States, Dr. La Roe once again opened her own clinic, specializing in reconstructive surgery for women. Her goal was not so much beauty *per se,* but avoiding the psychological and professional consequences of deformity or old age. She developed her own breast-reconstruction techniques while still in training, and wrote *The Breast Beautiful.*

Chemical research connected with her surgical experiments led her to develop anti-fatigue compounds—including a strong stimulant used by European military forces, by the Swiss Himalayan Expedition, in the pre-Olympic contests in France and Switzerland, and in the Tour de France cycling competition, among other places. Developed in collaboration with Ernest Fuhrman, this was trebiotine, an extract of tryptophan (La Roe, 212–3).

Dr. La Roe lectured widely and wrote several other books, but always found time to indulge her interests in tennis, golf, squash, swimming, shooting, riding, and travel (Chaff *et al.,* 301; La Roe; *WA,*36; *WWAW*/3).

Older of two children and only daughter of a prominent New York family, privately educated through secondary school, **Dr. Virginia K. Frantz** (1896–1967) majored in chemistry at Bryn Mawr, graduating first in her class in 1918. Allowed to enter Columbia's College of Physicians and Surgeons because of a shortage of male applicants, she ranked second in her class (MD, 1922). Upon graduation she became the first woman ever to win a surgical internship at the Presbyterian Hospital. From 1924 to 1927 she worked as a surgeon. Housekeepers enabled her to meet the double obligations of work and family (married as a medical student, two children by 1927). Although she switched to research in surgical pathology partly in order not to compete professionally with her husband, and had a third child in 1930, the marriage ended in divorce in 1935.

Happily, Dr. Frantz did important research and received wide recognition in her new field. Bringing her to this chapter is her World War II work with Raffaele Lattes on the control of bleeding in surgery. This work led directly to the innovation of using gauze-like oxidized cellulose to stop bleeding. Absorbed by the body during healing, this substance could be packed directly into wounds. The innovation gained wide acceptance, and in 1948 Virginia Frantz received the Army-Navy Certificate of Appreciation for Civilian Service. She was president of the New York Pathological Society in 1949 and 1950, and in 1957 won the Elizabeth Blackwell Award for distinguished service to medicine. From 1924 till 1962 she taught at Columbia's College of Physicians and Surgeons, rising to full professor in 1951. When she died, she rated an obituary notice in the *New York Times,* and she is one of very few scientists included in *Notable American Women.*

Arriving at the mid-20th century, we find another surgical needle and a new type of surgical knife in which women have played significant innovative or developmental roles. As a research metallurgist with the Bundy Corporation of Norwalk, CT, **Dr. Laurence Delisle Pellier** of Westport, CT, worked on such innovations as developing techniques for applying electron microscopy to metallurgical problems, and tailoring metal alloys for use in chemical plants and fine instruments. A charter member of the Society of Women Engineers (SWE), Pellier received the Society's 1962 Achievement Award for significant contributions to metallurgy. She has written many technical papers and has worked and published with the American Society for Testing and Materials.

After a BS in Chemical Engineering from the College of the City of New York, she took a master's degree in metallurgy from Stevens Institute of Technology. She became a member of the American Institute of Mining and Metallurgical Engineers, the American Society for Metals, the Société Française de Metallurgie, and the Electron Microscope Society of America, among others. Although her research seems patent-prone, she evidently patented only her surgical needle; specifically, her process for gold-plating such needles. This achievement is mentioned in the SWE citation (SWE Achievement Award Booklet, 1952–76; 1962).

**Janet Marie Dearholt Esty** (c. 1940–　) is President and co-founder of Neomed, manufacturer of the electronic surgical instrument called Neoknife. Although my source is ambiguous on Esty's possible role in the technical development of Neoknife—which improved on an existing category of electronic-scalpel-with-cautery by adding a coagulator feature—clearly the new company would not have emerged without her. Also clear is Esty's penchant for innovation. At Univac, she says casually, she "built a new thing called data management," which was taking care of all the paperwork and reporting requirements for government contracts (Rich-McCoy, 127ff, 136).

In the late 1950s or early 1960s, **Valentina Nokitina,** a "junior scientific worker," won a Lenin Prize for perfecting an operation to restore hearing to victims of otosclerosis (a stiffening of the tiny middle-ear structure that must move in order to transmit sound). My source (Dodge, 222) gives no exact date and no details except that she had four collaborators.

In the 1970s two other Russian women co-patented a Magnesium-base alloy for use in bone surgery. **Nina Mikhailovna Tikhova** and **Vera Fedorovna Terekhova** received U.S. Patent #3,687,135 with a male colleague on August 29, 1972.

Yet another Russian woman deserves mention here. On September 30, 1975, **Lidia Alexandrovna Potekhina** of Moscow, with two male co-inventors, received a U.S. patent for a Device for the eversion of hollow organs and a vascular stapling instrument incorporating same. This instrument, in other words, enables surgeons to turn the cut end of a blood vessel inside out and form a cuff for stapling during surgery or repair.

An earlier Russian invention that seems conceptually related is the work of **Tatiana L. Ivanova** and four male colleagues, all of Moscow. Receiving a U.S. patent in 1971, this is an Instrument for circular suturing of hollow body organs. Although the hollow body organs in question here are intestines rather than blood vessels, the "suturing" in both instances is actually stapling.

**Henrietta Krzewinski**'s invention would never make the news, but might contribute significantly to the comfort of patients and/or the convenience of doctors. On March 4, 1975, listing her residence as Old Bridge, NJ, she patented a Disposable surgical gown with a water-proof front panel.

A minor but interesting surgical invention is the Suture and needle holder patented in 1981 by **Annette M. Chapel** of Chassell, MI, and a male co-patentee. This invention is a disposable storage block for surgical needles and sutures that holds them selectively available, first for use and then for disposal.

A small but significant innovation was the idea of **Dr. May Owen** of Fort Worth, TX. One of the most honored physicians in Texas, she gained recognition fairly early in her career. She noticed that surgeons' glove powder fell into open wounds on the operating table, causing scar tissue to form around it during healing. As a result of her intervention, glove powder was thereafter made in a form absorbable by the body. As of 1975, Dr. Owen, who received her MD in 1921 from Louisville Medical School, had been in practice for over fifty years. Among her many honors was having Fort Worth's medical museum named the May Owen Hall of Medical Science (Chaff, #4026; Owen, 92).

In 1982 a woman gynecologist, **Vicki Georges Hufnagel** (c. 1950–  ) reported a breakthrough in gynecological surgery: using ultrasound for "visual navigation" during the procedure. Dr. Hufnagel, whose research interests lie in reproductive genetics, predicted that the new technique will greatly improve the safety of such routine procedures as dilatation, dilatation and curettage, and locating or removing IUDs (B. Sloane, Jan./Feb. 1982). Dr. Hufnagel has also pioneered a laser technique for correcting pre-cancerous conditions of the uterine lining, thus avoiding hysterectomy. Considering that 670,000 hysterectomies were performed in the U.S. alone in 1985, perhaps 90% of them unnecessary, and that many had tragic consequences for the women's health and sexuality, this becomes a highly significant invention indeed (Schumaker, 57, 54, *passim;* Hufnagel).

Nurses as well as doctors have created surgical inventions, though their work is underrecognized. One nurse who managed to gain at least monetary recognition is **Karen Tate,** a surgical nurse at the University of Minnesota Medical Center in Minneapolis. Nurse Tate invented prefab, pre-packaged "suture boots," rubberized coverings for the clamps that hold the prolene sutures used in open-heart surgery. Without such protection—which nurses used to improvise, by cutting up rubber catheters to fit over the clamps—the sutures were often broken by the serrated edges of the clamps.

In 1979 Tate improved on the improvisation, creating a device that precisely fits the clamps, and took her idea to a St. Paul surgical supply firm. The firm liked the idea and promised her a contract. However, although she not only invented the protector, but aided the company in researching and testing it, and although it has been on the market since 1981, by 1987 the company had paid her nothing. Finally exasperated, Nurse Tate took them to court, where a jury awarded her 30% of the sale price or half the net profit from sales of her suture boots, for a total of over $245,000 in past profits and over $275,000 in projected profits. In April of 1987 an appeals panel upheld the jury award. As of this writing, the case was before the Minnesota Supreme Court ("RN Awarded").

A nurse who fought for direct technical recognition—and won—is **Mary Louise O'Grady** (1954– ) of Chicago. Head nurse in cardiac surgery at the Michael Reese Hospital and Medical Center, she often found herself taking the brunt of surgeons' frustrations with imperfect surgical instruments. In 1984 she decided to do something about it. With the encouragement and consultation of Dr. Robert Replogle, Director of Cardiac Surgery at Reese, she drew up improved designs and models for three crucial instruments—a forceps, a suction instrument, and a dilator. Her search for the perfect design took her as far as Germany, where she observed the work of master surgical instrument-makers.

She created three improved instruments, that were tested in several American hospitals and eventually marketed by Walter Lorenz Surgical Instruments Company of Jacksonville, FL. All three bear Nurse O'Grady's name, engraved directly on the stainless steel of the instrument: the O'Grady/De Bakey forceps, the O'Grady coronary artery suction, and the O'Grady coronary/vascular dilators.[32] Her name also appears in the company catalogue. In the operating room the surgeons call the instruments "Mary Lou's." She rejected one company's offer to make prototypes and pay her for her inventions because they refused to put her name on the instruments or list it in their sales catalogues. "It's probably hard for some people to understand," she says, "but I wanted to see my name and have that feeling of satisfaction. I wanted to be right up there with the rest of them. I didn't want to be bought out."

Nurse O'Grady's forceps is actually a group of instruments, made in one pediatric and two adult sizes. Her improvement on De Bakey's design lay primarily in making the instrument easier to grasp and handle. "Before," she points out, "the shaft was smooth, and when you handed the forceps to the surgeon he would bounce the instrument on the operating table and pick it up wherever his fingers felt comfortable." O'Grady's design contains "indentations and holes on the shaft of the forceps to give surgeons the perfect fit during surgical procedures." Hospitals across the country use the O'Grady instruments.

Mary Louise O'Grady graduated from Cook County Hospital's School of Nursing, and went immediately into surgical nursing. By 1987 she had been in that field for more than

32. The first bears Dr. Michael De Bakey's name as well because she improved instruments on which he holds a patent.

ten years, a head nurse for two years. She saw her inventions as a new beginning, enabling her "eventually" to hang up her nurse's cap for good. "Eventually" came fast. By 1991 O'Grady had her own consulting business—and the gratification of seeing her instruments used in her father's quintuple bypass, and in her baby niece's heart surgery as well (Sherrod; J. Blackburn).

Two other nurses' inventions of the 1980s will be discussed elsewhere.[33]

As in other technical areas, women without medical training occasionally have ideas for valid surgical inventions, including operations. A striking example is a mother-daughter inventor team, **Helen** and **Julie Smit** of Evanston, IL. "Some years ago," while out for a drive, the two women were discussing how wonderful it would be if obese people could eat all they wanted and still lose weight. Their first idea was a disposable bag to be inserted into the stomach. Essentially, that idea has since been put into effect by surgeons—presumably as an independent invention—and some doctors now consider it the treatment of choice for severe obesity, far safer than stomach-stapling or removing part of the intestine. A second idea, which the two women have patented, is an intestinal coating or lining that keeps some food from being absorbed. The Smits have several non-medical inventions to their credit as well (Crossen).

*Transplants*

**Ruth** & Reinhold **Benesch**'s modified hemoglobin could have important surgical applications. Since it can unload oxygen at low temperatures, it could be useful in oxygenating organ transplants during shipment or supplying oxygen to hearts during surgery (Raloff, "Making Blood").

The bone-marrow transplants used to combat leukemia can, if rejected, be as fatal as the disease itself. In 1987 **Dr. Georgia B. Vogelsang** of Johns Hopkins reported devising a thalidomide treatment, probably most useful in conjunction with cyclosporine, to combat GVHD (graft-versus-host disease). Charles Coltman of the Cancer Therapy and Research Center in San Antonio calls Vogelsang's treatment "one of the most important advances in bone marrow transplantation" (D. Edwards, "Thalidomide"; "Thalidomide . . . Comeback").

Another promising anti-rejection treatment was announced in 1987. **Dr. Judith Shizuru, Anita Gregory,** and **C. Tien-Bao Chao** and colleagues reported results of experiments with mice carried out in the labs of Dr. C.G. Fathman at Stanford University School of Medicine. They created permanent acceptance in the mice of transplanted pancreatic tissue in a single course of treatment with monoclonal antibodies, without any other drugs. Not only is this "a pretty exciting model," as Fathman put it in an interview, for creating transplant tolerance in general, but it is a further step toward the possible goal of using tissue transplants to cure Type 1 diabetes in humans, a condition affecting about a million Americans (Hofstadter, "Monoclonal"; Shizuru *et al.*).

The current frontier of transplant surgery is nerve and nervous system tissue transplants. In pioneering work published in 1987, **Victoria Luine** of Rockefeller University reported that fetal brain cells transplanted into rats whose brains have been damaged in areas controlling sexual behavior can restore normal sexual activity, at least for a time. (D. Edwards, "Nerve Transplant").

One of the most significant modern surgical operations devised by a woman, taken out

---

33. For further information on inventions by nurses, contact Olga Church, a nurse-historian at the University of Illinois in Chicago, and the American Nursing Association (ANA). I am grateful to Church and to Linda Bernhard for material on O'Grady.

of chronological order to close and climax this section, is the blue-baby operation[34] invented in the 1940s by pediatric cardiologist **Dr. Helen Brooke Taussig** (1898–1986) and first performed by Dr. Alfred Blalock. (Taussig had done a surgical internship, but was not a practicing surgeon.) This operation, often called the Blalock-Taussig operation, has not only saved the lives of thousands of children since 1945, but opened the way for many other forms of heart surgery. If an oxygen-deprived child could tolerate heart surgery, so could most other patients, and surgeons could venture where none had dared before. The result, as one prominent cardiac surgeon noted, was much of today's cardiac surgery.

Watching Blalock tie off a vessel leading to the lungs, Taussig decided such a vessel could also be built, repaired, or enlarged. Blue babies' narrowed artery between heart and lungs limits their oxygen so that their color is a sickly blue rather than a healthy pink. When Dr. Taussig's operation is successful, the baby changes color before the surgeon's eyes on the operating table. The operation was among the five or six significant new developments in surgery mentioned in the *Britannica* yearbook for 1946 ("Medicine," 466).

In direct recognition of this milestone operation, Blalock was elected to the National Academy of Sciences in 1946. Taussig had to wait until 1973. Honors did accrue, however delayed, and she came a long way from the days when Harvard allowed her to take medical courses but not a degree—and forbade her to speak to any of the male students in her histology class. She became a full professor at Johns Hopkins in 1959, first woman president of the American Heart Association in 1964, first woman master in the American College of Physicians in 1972. A Chevalier of the French Legion of Honor and recipient of the United States Medal of Freedom, Dr. Taussig also received 20 honorary degrees. Though officially retired from Johns Hopkins in the late 1970s, she continued to work an eight-hour day (Diamonstein, 408–12; Gilbert & Moore; J.H. Stevenson, "Blue Baby" in Helmes; Ireland; *WBWRA*).

If Helen Taussig had done nothing more than create the blue-baby operation, her place in medical history would be assured. However, she and Dr. Frances O. Kelsey of the Food and Drug Administration together kept thalidomide off the U.S. market, thus sparing American parents the heartbreak suffered by thousands of European parents in the 1950s and '60s (an estimated 3,500 deformed babies born in Germany alone by August 1962).[35]

## Prevention

Since the male takeover of medicine, female practitioners have often become identified with preventive medicine, whereas male practitioners have focused on drastic and costly measures applied after illness strikes. Noteworthy in the late 19th century for her focus on

---

34. Lest anyone question the inclusion here of a new surgical operation, note that such operations are patentable (cf. Stacy Jones' "Patents" column, *NYT*, May 23, 1987: 18).

35. Alerted to the developing tragedy by German students, Dr. Taussig went abroad in January 1962 to assess the situation for herself. Satisfied that thalidomide was indeed causing the deformities, she wrote a lead editorial for the May 25 issue of *Science*, warning doctors and the public about the danger. The editorial also appeared in the *New England Medical Journal*. As each passing year shows us more commonly used drugs and substances that can damage the fetus, Dr. Taussig's 1962 words sound prophetic: "In many instances the damage is done before the mother knows she is pregnant. Therefore, young women must learn to be cautious about new drugs . . . [and] physicians must bear in mind that sleeping tablets, tranquilizers, and other apparently innocent drugs may do terrible harm to the rapidly growing embryo and the unborn child" (*Science* 136: 3517; *WBWRA*).

preventing illness was **Dr. Eliza M. Mosher** (1846–1928). She patented what appears to be a Reform underwaist in 1893 (see below), but most of her physical inventions dealt with preventing back problems by ensuring good posture. She invented a desk with a built-in book rest for reading without neck-strain (1899), an infant's chair (1915), a "hygiene" school chair, and the Mosher kindergarten chair, used throughout the United States. She also designed and supervised the construction of back-supporting subway seats for the Brooklyn subway line (*CAB* XI; 140; *NCAB* XV: 305; "Too Busy," 322). But her most significant innovations in the area of preventive health were perhaps the unpatentable ones: as the first woman faculty member at the University of Michigan, she created the first university health service, and pioneered an improved and more rigorous physical education program for women, systematic record-keeping, and preventive health programs in general (*NAW*).

**Dr. Mary Edwards Walker** is not usually celebrated for any medical achievements, being better known as America's only woman winner of the Congressional Medal of Honor—and as an eccentric. But she evidently devised techniques for preventing tuberculosis and at one time ran a sanitarium (Levin, 71).

**Dr. Lilian Welsh** (1858–1938) was born in Columbia, PA, the daughter of a merchant. After a brief teaching career, and an MD at the Women's Medical College of Pennsylvania (1889), she studied physiological chemistry and bacteriology at the University of Zurich, the latter in a pioneering course taught by a student of Robert Koch. After a residency at a state mental hospital in Norristown, PA, Dr. Welsh joined a friend and classmate, Dr. Mary Sherwood, in practice in Baltimore. Partly as a result of volunteering at the Evening Dispensary for Working Women and Girls, they both decided their true interest lay in preventive medicine. As it happened, the first modern public health crusade in the United States—against tuberculosis—was just getting under way. Appointed to the Maryland State Tuberculosis Commission, Dr. Welsh was one of the originators of the so-called Maryland Plan, a blueprint for public health programs based on Maryland's initial attack on tuberculosis and adopted across the country.

As Professor of Physiology at the Women's College of Baltimore (later Goucher College), Dr. Welsh pioneered in giving every entering student a physical examination and keeping a permanent record of the results, used for assigning levels of activity in the gymnasium. She was also innovative in adding solid science content to the college's hygiene course.

Dr. Welsh was part of the group that got women admitted to Johns Hopkins Medical School by making an endowment conditional on that promise, and later worked to get women into all areas of graduate study at Hopkins. Active in the suffrage movement, she joined the famous march staged in Washington, DC, just before President Wilson's first inauguration (Helmes).

Another important figure in preventive health/medicine in the first half of the 20th century is **Martha Eliot, MD.** Dr. Eliot served in the Federal Children's Bureau during the 1920s and '30s, becoming its assistant chief in 1934. She did important pioneering work in preventing rickets in children (Rossiter, 233).

**Dr. Florence Sabin** is the American woman doctor perhaps most strongly identified with preventive medicine, having undertaken to reform the health program of her native Colorado as a third career at an age when most would have been thinking of retirement. This was in the mid-1940s, when she was in her seventies. Dr. Sabin raised the consciousness of the nation so far that with only a little poetic license one could credit her with re-inventing preventive medicine.

Another prominent American physician who pursued a second career in prevention is **Dr. Virginia Apgar.** Most famous for her infant screening procedure, the Apgar Score (see below), Dr. Apgar devoted the second half of her long medical career to the prevention of birth defects. During her tenure with the March of Dimes, the foundation's income increased from $19 million to $46 million (Diamonstein, 15).

*Preventive medicine*

Other medical women may not define their work primarily in terms of prevention, but have to their credit noteworthy inventions or innovations designed to prevent specific diseases or conditions. Some may not be doctors at all, yet do work with important applications in health and preventive medicine. For example, **Diane Oliver**, President of Idetek, San Bruno, CA, has invented a test for trichinal infestation in pigs. This ELISA (enzyme-linked immunosorbent assay) test is pertinent here since trichinosis is still considered a public health hazard in the United States, with 152 cases reported between 1983 and 1985. Oliver's test, which may someday screen all hogs intended for commercial slaughter, can easily be adapted to detect the more serious parasite *Toxoplasma gondii* as well (Thomsen, "Do Vets").

An important category among the preventive inventions of medical women is infant and childhood infectious diseases. **Dr. Anne Yeager** (see below), working with other researchers at Stanford University Medical Center in Palo Alto, CA, improved screening techniques for cytomegalovirus in donor blood. This work eliminated the danger of a virus that formerly killed one out of twelve premature infants. Dr. Yeager also established new childbirth techniques to prevent mothers with herpes from transmitting the disease to their infants at birth (Yeager, Obit.).

Another important category of preventive medicine is birth defects. Striking some 4,000 infants each year in the United States is hydrocephalus, or "water on the brain." In this condition, fluids around the fetal brain and spinal cord become blocked and instead of circulating smoothly, build up in brain cavities and put pressure on the developing brain. Doctors can now detect the condition as early as the 20th week of pregnancy, and surgeons can implant a pressure-relieving shunt after birth, but brain damage may occur long before, causing mental retardation, paralysis, convulsions, and even death.

In 1981 **Maria Michejda** and Gary Hodgen at the National Institutes of Health reported successfully implanting a valve in the skulls of hydrocephalic monkey fetuses. Whenever pressure built up to a certain level, the valve automatically shunted excess fluid into the surrounding amniotic sac. Of these fetuses, 80% survived birth, and nearly all seemed healthy and normal nine months later. Of the unoperated controls, by contrast, only 10% survived birth, and all of these died within two weeks. At the time of this report, one such operation had already been done on a 24-week human fetus. The surviving monkeys must be followed carefully, but the initial results here are promising ("Preventing Hydrocephalus . . .").

Yet a further category of desired prevention is allergic responses, particularly the potentially life-threatening ones such as bee-sting shock and asthma. As early as 1958, **Dr. Mary Loveless** (c. 1900–  ) perfected the single-shot immunization against allergic reactions to bees and wasps. A Stanford MD, she did the research at Cornell. With her treatment, approved by the FDA in 1979, only annual boosters are needed (C. Baldwin). In work reported in 1987, Dr. Richard Moss showed that allergy shots could actually prevent asthma. Among his colleagues in this research were **Dr. Margaret Sullivan** of Stanford, and **Yao-Pi Hsu,** a research assistant in Moss's lab at Travis Air Force Base, CA (Hofstadter, "Allergy").

*Diet*

**Sylvia Wassertheil-Smoller**'s research indicates that Vitamin C may protect against cervical cancer. She and her colleagues at the Albert Einstein College of Medicine in New York City studied a sample of 98 women, 49 "cases" (one positive Pap smear or two consecutive suspicious smears) and 49 "controls" (two negative smears), comparing the two groups' diets on a three-day food record showing intake of 71 different nutrients. The cases and controls were matched for age, race, and childbearing. Results reported at the Second Annual Bristol-Myers Symposium on Nutrition Research in Washington, DC, in the early 1980s, found significant differences for two of 19 possibly cancer-protective nutrients: beta-carotene (a precursor of Vitamin A) and Vitamin C. Women with positive or suspicious Pap smears ate significantly less of both than did controls. Even when corrected for such other cervical-cancer risk factors as low income, early sexual intercourse, frequent intercourse, and numerous sexual partners, the differences were still significant.

During the 1970s, researchers reported protective effects of Vitamin C against lung, colon, skin, and stomach cancers. In 1983, **Elizabeth Bright-See,** a Canadian nutrition scientist at the Ludwig Institute for Cancer Research in Toronto, was studying Vitamin C as a protective against colon cancer. Dr. Bright-See finds Wassertheil-Smoller's results on the vitamin's cervical-cancer protection "interesting." Obviously, however, as she points out, the study now needs to be extended to measuring the Vitamin C intake physiologically for both cases and controls, rather than relying on food diaries (Arehart-Treichel, "Vitamin C . . .").

**Barbara Reed** has noted an important possible connection between diet and behavior, and instituted an innovation that could have significant social as well as medical results. Ms. Reed, an experienced probation officer in Ohio, noticed a dramatic improvement in her own health and well-being when she cut sugar-filled junk food, white flour, and canned goods out of her diet. Since much of the improvement was psychological and behavioral—no more recurring nightmares, mental lapses, fatigue, or violent mood swings—she naturally wondered whether some of her probationers would experience the same benefits. Beginning in 1971, she advocated this dietary change to all her clients, and those who followed her system reported feeling better, more energetic, and more emotionally stable. Most important—and this was quantifiable—*the rate of repeat offenses among her group dropped so dramatically that a judge in her area* (Cuyahoga Falls, OH) *noticed it.* Said he, "I was amazed at the dramatic results in persons who were placed on probation to her through my courts."

Even though Reed had a sample of 1,000, scientists considered her work anecdotal and have done their own studies. A small 1980 study, done with juvenile offenders at Virginia's Tidewater Detention Center, was designed specifically to test the relationship between sugar consumption and "antisocial behavior." Without telling either the staff, the cooks, or the youths the real reason for the change, the investigator drastically restricted white sugar in the detainees' diets, and then measured behavior changes according to counselors' daily records of disciplinary incidents. Comparing the four months before and three months after the diet change, the researcher found *a 45% drop in antisocial behavior in the institution.* Further extension and refinement of the data indicated that violent behavior was the most likely to show improvement. So impressive were these findings that as of 1983 the State of Virginia was contemplating instituting the Tidewater diet throughout its correctional system.

Also as of 1983 Diana Fishbein, a criminologist at the University of Baltimore, focused on sugar in her study of diet and maladaptive behavior. With colleagues at the University of Maryland, she was investigating how diet affects brain function and sugar metabolism. Preliminary results from these and other studies indicate that Barbara Reed's insight will be confirmed and her innovation in the handling of criminal offenders and institutionalized patients validated ("Diet-Behavior").

### Vaccination/vaccines

Although modern medical women may have focused on prevention partly by default, taking the less lucrative public-health positions shunned by their male colleagues, the focus is also an ancient one. Some evidence indicates, for example, that women may have invented vaccination.

*Smallpox.* Since the word *vaccination* comes from inoculation with *Vaccinia* (cowpox) virus to prevent smallpox (caused by the closely related *Variola* virus), it seems appropriate to start with smallpox. According to a Chinese researcher at the East Asian History of Science Library in Cambridge, England, Sung Dynasty documents (AD 960–1279) show a Taoist nun using variolation (inoculation with a weakened form of the smallpox virus itself). Since the patient was the Prime Minister's child, and women practitioners often treated children, the female forms found in the passage are probably correct, and we can infer, though we cannot be certain, that women were in charge of such work (interview, May 30, 1979).[36]

Likewise, when Lady Mary Wortley Montagu (1689–1762) observed variolation in Turkey early in the 18th century, she reported that women were the variolators. As she wrote to a friend in 1717 (italics mine):

> Apropos of distempers I am going to tell you a thing that . . . will make you wish yourself here. The smallpox, so fatal and so general among us, is here entirely harmless by the invention of ingrafting . . . the term they give it. There is *a set of old women who make it their business to perform the operation* every . . . September when the great heat is abated. . . . They make parties for the purpose. . . . The old woman comes with a nut-shell full of the matter of the best sort of smallpox, asks what veins you please to have opened, etc.

The woman put a little of this "matter" or pus into the open vein and bound a hollow shell over the wound. The patient had a little fever for two days, but no scarring. The variolator always took pus from a healthy child on the thirteenth day of the disease, carrying it from patient to patient in a clean glass vial kept at body temperature between her breasts. Mothers sometimes inoculated their own children simply by putting pus on the skin and pricking it in with a pin (Hurd-Mead, 469).

Although Lady Mary met opposition and slander from male doctors when she introduced variolation to England in 1722 (*EB*/58, s.v. "Vaccines"), she managed to convince many people of its value, including members of the royal family and the nobility. Smallpox house parties became common both in England and in America, a welcome interlude of rest and merriment, as the inoculated infection was generally mild.

---

36. Both Cotton Mather and a correspondent, Dr. Woodward of the British Museum, had black servants who had undergone successful "Inoculation" for smallpox in Africa before being enslaved. No indication of the sex of the inoculators appears in my source (Beall & Shyrock, 98), but various African tribes reportedly have a goddess of smallpox whose temple orientations are determined by her function in removing the disease (Evans-Pritchard, xx). Judith Goldsmith reports that the Djen of Nigeria practiced smallpox vaccination "with great success" and weathered epidemics without any deaths, centuries before Western medicine was introduced (51).

This was more than fifty years before Edward Jenner's work. "Lady Montagu was not a medical woman," says Kate Campbell Hurd-Mead, "but she probably did more for mankind, medically, than most of the physicians of her time" (Hurd-Mead, 468–70; 469). Since the smallpox mortality rate was then about 50% (60,000 deaths a year in England alone), and the fear and loathing of the resulting pock-marks nearly as great as the fear of death, Hurd-Mead's statement seems reasonable. Indeed, smallpox has long been one of our great scourges. According to other 18th-century figures, it struck 95% of the adults and half the surviving children of Europe, and killed, overall, 10% of the population. About half the survivors were badly pockmarked, and many were in poor health (Burnet, 3–4).

No one denies the value of Jenner's work. However, his "discovery" of the antagonism between cowpox and smallpox was based on something local wisewomen had long known: that children who had had cowpox almost never got smallpox.[37] Even if women had not still been healers in rural areas, milkmaids obviously had the best chance to observe the protective results of "milkmaid's disease" (cowpox); and according to one version of Jenner's story, a dairymaid named Sarah told him the connection.

At least one other woman was involved in the continued development of smallpox vaccine. In the 1920s **Mary C. Maitland** worked with her husband Hugh at Manchester University on cultivating cowpox virus. The Maitlands showed in 1928 that this virus would grow in minced-up chicken kidney suspended in a fluid medium (chicken serum, mineral salt solution), without elaborate culture methods (Andrewes, 24; Maitland & Maitland). Indeed, the Maitlands invented the most valuable virus-culture method before John Enders' Nobel Prize–winning work of 1948. They even solved the problem of bacterial contamination in their Maitland flasks by subculturing the virus from flask to flask every three to four days (G. Williams, 143, 171).

*Vaccination* has now expanded to include any kind of inoculation that triggers the body's immune response to a given disease, whether the inoculant be bacterial or viral, live, killed, or weakened; whole or part; or even isolated antigens that stimulate the production of specific antibodies. Work is now well along toward cancer vaccines, triggering the body's immune system to recognize cancer cells as foreign and destroy them. The history of this expanded concept of vaccination begins with work on toxins and antitoxins in treating bacterial disease.

Before the antibiotic era, the only treatment for many dangerous bacterial diseases, such as pneumonia, was to identify the causative bacterium, inject it into a rabbit or a horse, wait for the animal to develop antitoxins, isolate the antitoxins from that animal's blood and—if the patient was still alive—inject the antitoxin. **Dr. Rachel Fuller Brown** did vital work in this field in the 1920s and '30s, identifying the specific carbohydrate that makes each pneumococcus (pneumonia-causing bacterium) unique, so that standard antitoxin serums for each could be prepared in advance. Then, as soon as a physician's specimen could be matched to a serum, treatment could begin (Yost, *WMS*, 64ff).

*Childhood diseases.* **Gladys Henry Dick** (1881–1963), working with her husband George, isolated the streptococcus that causes scarlet fever in 1923, after ten years of research. Other scientists had suspected a streptococcus, among them the English researcher Josephine Pratt and her colleagues in 1914 (*EB*/68). The Dicks then prepared

---

37. Note the slight difference between the way the *Encyclopaedia Britannica* ('58) reports this foreknowledge in its Jenner article ("When an apprentice, he had noted a popular belief in Gloucestershire as to the antagonism between . . .") and the way Kate Campbell Hurd-Mead reports it in her *History of Women in Medicine* ("Old women in the country had often found that children who had had cowpox never had smallpox . . ."; 468).

the toxin and antitoxin that prevent and cure the disease, respectively. Before their work, scarlet fever was a killer with a mortality of up to 25% (*NAW*/mod.). It caused an estimated 11.8 deaths per 100,000 in 1900–4 in the United States, and was still causing about 4 deaths per 100,000 in 1920–4 (DHEW/CHS). Even where it did not kill, it could maim, leaving children deaf from abscesses of the ear or mastoid, making them carriers with chronic sinus infections, or causing such complications as nephritis, arthritis, or rheumatic fever.

As a result of the Dicks' work, the death rate from scarlet fever dropped to .4 (1940–4) and then to .01 (1945–9) per 100,000 (DHEW/CHS). This work was done at the John R. McCormick Memorial Institute for Infectious Diseases (Chicago), founded by Edith Rockefeller McCormick and her husband after losing their son to scarlet fever in 1901. There Dr. Dick stayed until her retirement in 1953. Gladys and George Dick took the then-unusual step of securing U.S. and British patents for their methods of producing scarlet-fever toxin and antitoxin (U.S., 1925). In response to the resulting criticism, the Dicks said their aim was not financial gain but quality control. Gladys Dick later sued Lederle Laboratories for patent infringement and improper manufacture of toxin.

In 1925 the Dicks were nominated for the Nobel Prize, but no prize was given in medicine that year. However, they received the University of Toronto's Mickle Prize (1926), the University of Edinburgh's Cameron Prize (1933), and several honorary degrees.

Continuing to innovate, Gladys Dick became one of the founders of the Cradle Society of Evanston, IL, reportedly the first American professional adoption agency, and worked in it from 1918 to 1953. When she was 49 she and her husband adopted two children. Meanwhile, she continued her research, now largely on polio. She died in California in 1963 (*EB*/68, s.v. "Scarlet Fever"; *NAW*/mod.; Rossiter, *Struggles,* 143; 292; *WBWRA*).

Other notable cripplers and killers of children are diphtheria, whooping cough (pertussis), and tetanus. A World Health Organization report (WHO, 6th Report on the World Health Situation) for 1973–7 found that these three diseases, plus tuberculosis, measles, and polio, still caused some five million deaths a year among the world's children under five, blinding, crippling, or otherwise disabling five million more. Even today, these six diseases kill 3.5 million children each year ("Five-Year Checkup"; *SN* 129; 14 [Apr. 5, 1986]: 212).

In the United States, thanks to **Dr. Pearl Luella Kendrick** (1890–1980) and her colleague **Dr. Grace Eldering,** among others, the story is quite different. In 1900–4 the U.S. death rate from diphtheria was 32.7 per 100,000; from whooping cough, 10.7 per 100,000. By 1920 the figures had dropped to 13.7 and 8.9, respectively (DHEW/CHS), but at that rate the two diseases still killed some 25,000 people a year, mostly children. Tetanus was much less of a problem, with only 1472 cases reported in 1920, but killed nearly half its victims when it did strike (*EB*/58). In Tehchow, China, in the 1920s, for example, half the newborns died of tetanus in their first week of life (Chaff, 370).

In 1935 Dr. Kendrick received a National Research Council grant to study whooping cough. By 1939, when the disease still killed some 6,000 children yearly in the United States (National NOW *Times,* Dec./Jan. 1980–1: 12), she and Dr. Eldering had devised a whooping-cough vaccine suitable for mass production. As their vaccine became prevalent (1940–4), the death rate from whooping cough dropped to 2.2 per 100,000; by 1945–9 it was 1.0.

Dr. Kendrick then went on to invent the standard DPT vaccine that now routinely protects American infants against diphtheria, whooping cough, and tetanus. Thanks to this unassuming researcher—who refused to have the vaccine named for her, and omitted her name from her *Encyclopaedia Britannica* article on the vaccine—both diphtheria and whooping cough have virtually been eradicated in the United States.[38]

Pearl Kendrick was born in Wheaton, IL, on August 24, 1890. After taking a BS from Syracuse University (1914), she worked as a high school teacher and principal, and then as a microbiologist in the Michigan State Health Department labs, to finance her graduate education. She finally got her ScD from Johns Hopkins in 1932, at age 42. Although she held a Rockefeller Foundation fellowship (1929–32), at no time during her years of graduate study was she totally free of work responsibilities.

She continued to work in the Michigan State laboratories, serving as an associate director in charge of the Grand Rapids branch from 1926 to 1951. After World War II, when nations turned to civilian health both at home and internationally, Dr. Kendrick began to receive worldwide recognition for her work. Between 1949 and 1964 she served as a consultant for WHO, UNICEF, and the Pan American Health Organization, working with Great Britain's Medical Research Council (1949), participating in the WHO conference on diphtheria and pertussis vaccination at Dubrovnik in 1952, heading a UNICEF immunization project for Colombia, Chile, and Brazil and a program of pertussis vaccination in Mexico in 1961; and working with the national laboratories of Chile, Uruguay, Brazil, Venezuela, Panama, and Guatemala in 1964. She was the immunology delegate to the US-USSR Exchange Agreement Conference in 1962–3.

At home she became a diplomate of the American Board of Microbiologists, a member of the Microbiology Society and of the American Academy of Microbiologists, and a fellow of the Public Health Association, serving as vice-president in 1944. From 1951 to 1960 she was a lecturer in public health practices at the Department of Epidemiology, University of Michigan; after 1960 she was Lecturer Emerita. In 1960 the Michigan legislature passed a resolution acknowledging her contributions to the health of the state.

In addition to her landmark work in immunization against childhood diseases, she also did research on the serology of syphilis, on antigenicity in bacteriophage, and on multiple antigens.

Regrettably, I have few personal details on Dr. Kendrick—and none at all on Grace Eldering—though one can imagine the perseverance and dedication that brought Kendrick, without family support, from village school-teaching to the ScD from Johns Hopkins. She never married. She was a churchgoer; her favorite recreations were books and her dogs; her politics were Independent. Her work was done under fairly primitive conditions by today's standards. After a long and exceedingly useful life, she died in Grand Rapids at the age of 90 (See Kendrick obits., *Time*, Oct. 20, 1980: 105; Nat'l. NOW *Times*, Dec./Jan. 1981; Kendrick; also *AJPH* Curr. Index, 1911–35, Eldering, Kendrick; *AMS*/11; *AW;* "Calling the Shots"; Coulter & Fisher, 24; Hofstadter, "New Techniques," 15; Keenan & Laine, 5; Robin; *WWAW*/10).

Another woman doctor who worked on pertussis vaccine was **Dr. Margaret Pittman** (1901–    ). She came to the National Institutes of Health in 1936 after working at the New

38. This triple vaccine has recently become controversial, with reports of central-nervous-system damage and other serious side effects in some children. In order to put this controversy into proper perspective, we must keep in mind not only the advantages (in terms of numbers of children immunized) of a single injection as opposed to three or more, and the deadly nature of the diseases it protects against, but the actually very high safety factor of the vaccine. Estimates on the odds of a bad reaction vary from one in 300,000 to one in a million. Even after almost half a century, the Kendrick-Eldering achievement stands high indeed. Unfortunately, that achievement is still little known, and the present controversy has not corrected the omission. Reports I have seen fail to mention either Kendrick or Eldering by name (e.g., Robin; Stanford University *Campus Report,* October 23, 1985).

York State Health Department, and became famous in her lifetime for her work in standardizing this vaccine (O'Hern, 171ff; Rossiter, *Struggles,* 230).

Among the women involved in today's ongoing effort to improve the DPT vaccine and/or its pertussis component are **Dr. Alison Weiss** at the University of Virginia, and **Nancy Adelman** and **Mae Lim** working at Stanford with Drs. Stanley Falkow and Lawrence Steinman. These researchers are using genetic engineering to create strains of *Bordella pertussis* that produce defective toxin. Tests in mice indicate that the defective strains still cause an immune response, but are 100 times less likely to cause neurological damage. By late 1985 the Stanford group was already "half a step away" from having a vaccine ready for human testing. These modern researchers face a daunting task in demonstrating superiority to the Kendrick-Eldering vaccine, however, for proving that the new vaccine has less than a one-in-300,000 chance of causing harm would demand massive clinical trials (Hofstadter, "New Techniques," 15).

*Polio.* One of the big five child-killers, polio once terrorized the United States with summer epidemics. It is still a menace in the Third World, where fewer than 10% of the 80 million children born each year are immunized. A 1970s WHO report found that paralytic polio, in particular, showed an ominous increase ("Five-Year Checkup," 82; see also Pfaff). The names most often connected with polio vaccines are Salk and Sabin. However, their work rested on earlier work, and several women are noteworthy here. As noted, for example, Dr. Gladys Dick's later research focused on polio.

**Lilian Vaughan Morgan** (1870–1952) invented the first polio vaccine for primates, a crucial step toward a human vaccine. Born Lilian V. Sampson, she grew up in Philadelphia, and did both graduate and undergraduate work at Bryn Mawr, advised by M. Carey Thomas and supervised in her graduate work by T.H. Morgan. She studied zoology with E.B. Wilson. Her career falls into two parts, as she focused on zoology in the 1890s, and on experimental genetics in the early to mid-20th century, after her four children were grown. Lilian Morgan was an important geneticist, having discovered the attached-X chromosome in *Drosophila* in 1921. This discovery not only further confirmed the chromosome theory, but became a major tool for analyzing mutations in the X chromosome and for demonstrating crossing-over at one stage of meiosis. Though married to the more famous geneticist T.H. Morgan, she usually worked totally independently. Even in his labs at Columbia and at the California Institute of Technology, she always maintained a distinct laboratory area and researched and published separately from him. She published 16 papers on her own.

Nevertheless, she is obscured by him. She appears only in the 1949 edition of *American Men of Science,* and not in *Notable American Women.* Margaret Rossiter mentions her, but gives no idea of her work. Happily, a biographical article on Lilian Morgan has appeared, written by Katherine Keenan, herself a biology PhD (Rossiter, *Struggles,* 16, 22, 143, 212; Keenan abstr. in *Biological Bulletin* 161 [Oct. 1981]: 320; Keenan, "Lilian"; Keenan, L-July 1982; Keenan & Laine, 5).

Interestingly enough, Lilian's daughter **Isabel Morgan** (Mountain) also worked on polio, joining the poliovirus group at Johns Hopkins in 1944. According to a 1948 article in *The Independent Woman,* she was the only woman in the group. She had recently reported successfully immunizing monkeys against one strain of active virus (Rossiter, *Struggles,* 212; T. Whitman).

As early as 1931, **Dame Jean Macnamara** and Dr. Frank M. Burnet had found differences between two strains of polio virus. These differences were such that a successful vaccine would have to immunize against both of them (Hand, 55).

**Dr. Dorothy Horstmann** is another woman who should be mentioned in any history of polio vaccines. In 1952 she demonstrated the presence of the polio virus in victims' blood at very early stages, a milestone in the search for a vaccine. She also established many of polio's important characteristics, including its highly communicable nature, and proved live vaccine effective in various populations. Appointed an instructor at Yale Medical School in 1943, Dr. Horstmann became in 1963 the first woman to hold two full professorships at the university: Epidemiology and Pediatrics (*WBWRA,* 220).

Recent work on polio has focused on the genetics of the virus, partly with an eye toward better vaccines and new anti-viral drugs. In 1981 **Naomi Kitamura** and colleagues at the State University of New York at Stony Brook reported the exact sequence of the 7,433 nucleotide subunits making up the polio virus. As the first exhaustive analysis of this unusual type of RNA virus, Dr. Kitamura's research not only contributes significantly to basic knowledge, but has implications for improved vaccines and treatment. Using this work, researchers may be able to create viral strains incapable of causing polio, yet provoking an immune response. And knowing the exact cleavage sites may enable scientists to create specific inhibitors to serve as new anti-viral agents ("Poliovirus Read").

*Chicken pox.* This is usually considered a minor childhood disease, but it is highly contagious (more than 90% of all adults have had it) and can have such serious complications as encephalitis. The chicken pox virus has also been linked to later attacks of the painful nervous-system affliction called herpes zoster or "shingles." Chicken pox can be life-threatening for children with leukemia and other forms of cancer. According to 1984 figures, it strikes 3 million Americans yearly, causing an estimated 4500 hospitalizations and 100 deaths. In 1984 researchers at the Children's Hospital of Philadelphia reported that they had developed and tested an experimental chicken pox vaccine in a group of 914 healthy children ages 1–14. None had previously had chicken pox. Half of the children received the vaccine; the others a placebo. In the next nine months, not a single case of chicken pox occurred in the vaccinated group, whereas 39 cases appeared among the controls. Side effects of the vaccine were mild to nonexistent. Several women—**Beverly Neff, Barbara Kuter, Carol Rothenberger, Alison Fitzgerald, Karen Connor,** and **Arlene McLean** were involved in the research effort led by Dr. Robert E. Weibel (Weibel *et al.*). Unlike the San Francisco *Chronicle* article ("Chicken Pox. . ."), which named only Weibel, the KGO radio news story (San Francisco, May 31, 1984) featured McLean.

*Measles.* Even before the recent discoveries of slow viruses and their delayed havoc, measles was a serious disease. With these new findings, it becomes a major health hazard. Luckily, a measles vaccine does exist. But medical scientists do not rest easy just because a vaccine has been invented. A vital, ongoing part of the development of any vaccine is the clinical trials that determine, among other things, the best time or age for administering it. **Dr. Anne Yeager** (1939–84), whose untimely death at age 45 brought her to my attention, worked in this area in the 1970s. Her research, showing that measles vaccine was more effective if given at 12–15 months of age rather than earlier, changed medical practice. Daughter of a doctor, Yeager was a versatile scientist. Taking her MD at Cornell in 1965, she worked there till 1969, and then went to Stanford, where she was an Associate Professor of Pediatrics at the time of her death in 1984 (Yeager, Obit.).

*Yellow fever.* Among the deadliest enemies of humankind, plaguing the tropics all year and devastating temperate regions periodically, was yellow fever. Yellow fever wiped out Napoleon's whole army in Haiti in 1801, and killed far more American soldiers than did Spanish bullets during the Spanish-American War. Epidemics struck Philadelphia in 1793, killing 4,000; New Orleans in 1853, killing 11,000. In Memphis in 1878, although

25,000 of its 45,000 inhabitants fled the city, 18,000 caught the disease, and 5,000 died. Bodies lay in the streets as cemeteries overflowed. The city went bankrupt, lost its charter, and for fourteen years was managed by the State (J. Leonard, "Ghost," 20, 21). Yellow fever was a major obstacle to a Panama Canal. The disease still kills many people each year in the underdeveloped nations, and indeed seems to be increasing in prevalence in South America.[39]

Early attempts to control yellow fever focused on eradicating the mosquito that transmits it.[40] In the 1930s, however, Max and **Lillian Theiler** and colleagues, including **Nelda Ricci,** developed a yellow fever vaccine (Geof. Lewis, Oxford/Berkeley Program, 1979; G. Williams, 136ff). In 1930 Max Theiler had shown that passing the virus through mouse tissue cultures could greatly reduce its potency for monkeys (Theiler; G. Williams, 166–7). The ultimate vaccine, however, a strain known as 17-D, was prepared for use by multiplication in chick embryos, though Maitland flasks were used in the early stages (G. Williams, 144). For safety and effectiveness, this vaccine has only one rival—smallpox vaccine. Lillian Graham Theiler, originally Max's lab technician, became his wife and collaborator (G. Williams, 167), but is not usually credited in the vaccine work (see, e.g., *EB*/58, s.v. "Yellow Fever"; *CE*/3, s.v. "Theiler").[41] Nor did she share his 1951 Nobel Prize in medicine.

*Malaria.* A far greater world health problem is malaria. This disease, neither viral nor bacterial but parasitic in origin, is such an ancient and serious enemy that it has influenced human evolution.[42]

In 1955 malaria struck 250 million people, of whom 2.5 million died and many more were left weakened and debilitated. In that year the World Health Organization launched a program to wipe out malaria—primarily by broadcast spraying with DDT to kill the malaria mosquitoes, but also by the use of dieldrin and other antimalarial drugs. At one time WHO thought malaria was under control. However, the mosquitoes developed resistance to DDT, and came back stronger than ever, while the internal parasitic forms of several strains have developed resistance to the protective pills. Today, malaria once again threatens half of humankind. It strikes 110–150 million people a year, killing 1–2 million, and is the single most serious health problem in the tropics and subtropics. In tropical Africa alone, it kills a million children under 14 each year (G.T. Miller, E-83; Howard *et al.,* 29; "Five-Year Checkup," 82; Rennie, 24).

Now researchers are seeking a vaccine against the parasite. Prominent in this line of research is **Dr. Ruth Nussenzweig,** working with her husband and others at the New York University School of Medicine. She reported in 1980 that parasites inactivated by

39. Mainly because of failure to eradicate the *Aedes aegypti* mosquito; see Jonathan Leonard's update.

40. A Trinidadian woman alerted Dr. Andrew Balfour to the fact that monkeys get yellow fever and thus can serve as a reservoir of infection. Her servants could always predict a yellow fever outbreak when they found dead and dying monkeys in the nearby woods (Andrewes, 100).

41. Another heroine of the fight to control yellow fever was Clara Maass (1876–1901). She was among seven volunteers who agreed to be bitten by mosquitoes from an infected area in order to settle the question whether the disease was transmitted by filth or mosquito bites. She survived the first trial, but volunteered again a year later, and died of yellow fever at 25. The hospital where she trained as a nurse was renamed after her, and the U.S. Postal Service has issued a stamp in her honor. In spite of this earlier recognition, she was not mentioned in a 1981 *Harvard Magazine* article on yellow fever (J. Leonard, "Ghost").

42. In Africa a serious blood abnormality, the sickle-cell trait, survives in the gene pool because in heterozygous form it provides some protection against malaria. The same may be true of a sex-linked gene producing another blood abnormality (glucose-6-phosphate dehydrogenase deficiency disease), commonly known by one of its symptoms, favism, intolerance for eating fava beans (Ucko & Dimbleby, 537).

radiation while infective can arouse an immune response that will protect mice, monkeys, and humans from infection.

Work continues elsewhere, using various life stages of the parasite and pure surface proteins from one stage. Ruth Nussenzweig's breakthrough gave promise as early as 1981 of a malaria vaccine in the foreseeable future ("Antibodies"; Nussenzweig, "Progress"). Dr. Nussenzweig, chair of the Parasitology Department at NYU, announced further key results and obtained a patent—in 1984. These developments were considered important and generally interesting enough to be mentioned in Stacy Jones' "Patents" column (Aug. 25, 1984: 32), and in *Ms.* magazine (Jan. 1985: 23). In 1985 the Nussenzweigs succeeded in creating a synthetic vaccine, using one of the sporozoite coat proteins as an antigen. Also involved in this research are at least two other women, **Joan Ellis** and **Pamela Svec** (Godson; Godson *et al.*). The vaccine consists of a string of 12 amino acids mounted on a protein carrier. Human testing was awaiting FDA approval. Scientists polled by the editors of *Science Digest* ranked the Nussenweigs' new vaccine among the 100 top innovations of 1985 (Howard *et al.*, 29–30).

Final success was still elusive as late as 1991, when researchers reported a successful mouse vaccine (Rennie).

*Hepatitis.* Some of the most scientifically advanced vaccine research being done today uses genetic engineering. **Angelica Medina,** a biochemist at the University of California at San Francisco, was part of a team led by Dr. William Rutter that in 1981 created a new yeast designed to produce a complex particle that may immunize people against the chronic form of hepatitis (B). This severe and liver-damaging disease affects more than half a billion people throughout the world. The new technique will be discussed under Genetic Engineering, below.

*Typhoid.* In the mid-1980s, Dr. Bruce Stocker and colleagues at Stanford University reported a new typhoid vaccine developed by genetic-engineering techniques. The existing injected vaccine is only partly effective, produces some undesirable side effects, and requires yearly boosters. The Stanford vaccine is given by mouth and should have no unpleasant side effects. It should be more effective, since it is a live vaccine that survives in the body long enough for both immune systems to react to it. However, it should also be safe, since the typhoid bacteria—*Salmonella typhi*—used have been genetically disabled so that they cannot cause disease (needing two growth factors not found in human tissue, they cannot multiply). Further, the Stanford researchers and others are considering inserting other genes into the vaccine bacteria to produce "piggyback" vaccines, immunizing against several diseases at once.

Multiple vaccines already exist, of course, such as Kendrick and Eldering's DPT vaccine, but that was a *mixture* of the three antigens, not a genetic unit.

Piggybacking experiments are already under way, in Stocker's lab and elsewhere, with genes that could prevent traveler's diarrhea, schistosomiasis, and malaria. Perhaps the cleverest and most innovative aspect of this work is that the *Salmonella* bacteria in the vaccine would themselves manufacture the other immunity-rousing proteins, thus solving several production and packaging problems simultaneously. Women researchers contributing to the development of the Stanford vaccine include **Susan Hoiseth** and **Mary Frances Edwards,** former graduate students (Hofstadter, " 'Piggyback,' " 11).

*Cancer vaccines.* One of the dreams of modern medicine is a cure for cancer. Even more exciting would be a means of preventing it. Women have been involved in this search from the early 1900s when **Maud Slye** studied the heritability of cancer (see App. HM-2), through the 1960s when **Dr. Helen Dyer** of the National Cancer Institute won a Garvan Medal for her work on cancer-causing mechanisms (*WBWRA*, 164, 220), till today when **Betty J. Dabney** and colleagues recruit sprouts in the battle against cancer-causing chemicals ("Sprouts"). Dabney and her co-workers Chiu-Nan Lai and Charles R. Shaw at the Texas System Cancer Center reported to a regional meeting of the American Chemical Society in Corpus Christi, TX, in 1978 that extracts from sprouts of wheat, mung bean, and lentils may give some protection against chemically induced cancer. In the Ames bacterial test, the sprout extracts lowered the potency of three mutation-causing chemicals: 2-acetylaminofluorene, benzo(a)pyrene, and aflatoxin B-1. The sprout extracts are effective only against mutagens that need to be activated by body enzymes. The extracts were next to be tested on mice exposed to a chemical that produces colon cancer.

The viral theory of cancer's origin gives the most hope of prevention. For if a virus could be shown to cause any type of cancer, a vaccine might be developed against it, just as for polio and smallpox. **Dr. Charlotte Friend** (1921–87) pioneered in this approach. Indeed, she is recognized as one of the founders of viral oncology. Receiving her BA from Hunter College in 1944, she joined the WAVES and was put in charge of the chemical pathology laboratory at the U.S. Naval Hospital in Shoemaker, CA. At war's end she took a PhD in bacteriology at Yale (1950), and became a virologist at the Sloan-Kettering Institute for Cancer Research in New York City (1949–66). There she discovered the first virus that consistently causes leukemia in mice.[43] In 1957 she went on to develop the first successful vaccine against cancer in mammals, for which she won a *Mademoiselle* Magazine Merit Award for 1957. As the *Mademoiselle* citation noted, by inventing this mammalian vaccine, Dr. Friend took a "long, solitary step toward an eventual cure for cancer" (Jan. 1958: 68).

Starting in 1966, Dr. Friend was Professor and Director of the Center for Experimental Biology at Mt. Sinai School of Medicine. Among her other awards and honors are the Alfred P. Sloan Awards for 1954, 1957, and 1962, the American Cancer Society Award for 1962, Hunter College's Presidential Centennial Award for 1970, and NIH's Virus-Cancer Progress Award for 1974. She was elected to memberships in the National Academy of Sciences, the American Association of Cancer Research, the American Association of Immunology, and the American Association of Hematology, among others (*AMWS*/11, 12, 14; Friend, obit.; Moller & Friend; *WBWRA*).

Another *Mademoiselle* award winner in cancer research was **Patricia Bath.** Her award came in 1960 (Jan. issue).

In work contemporary with Dr. Friend's landmark work on the mouse-leukemia virus, **Dr. Sarah Stewart** (1906–76) and **Dr. Bernice Eddy** (1903–  ) first characterized the polyoma virus. Drs. Stewart and Eddy showed that this virus—as its name indicates— causes many types of tumors in several different animals. Perhaps most important, they showed that tumors could be spread from one animal to another.

Mexican-born Sarah Stewart came to this country at age five. She took her PhD from the University of Chicago in 1939, then worked as a microbiologist for NIH. Barred from cancer research for lack of medical expertise, and from Georgetown University Medical School for her sex, she became an instructor at Georgetown so she could take medical courses on the side. When women were at last admitted in 1947, Dr. Stewart became Georgetown's first female MD. She was also the first woman ever to be featured in the "Medical Men of Georgetown" column of the University's *Medical Bulletin*. She went on to do work that has been described as the cornerstone of viral oncology. Dr. Sarah Stewart

---

43. So-called Friend erythroleukemic cells were still being used in cancer research in the 1980s (Podo *et al.*).

died in 1976 of the disease she had battled to defeat (O'Hern, 161ff; Rossiter, *Struggles,* 230–3; *WBWRA* 88, 220–1).

Dr. Eddy took her PhD at the University of Cincinnati (1927), and joined the Public Health Service in 1931. She and Dr. Stewart were good friends before they became collaborators. It was Dr. Eddy who developed the early tissue-culture techniques for tumor viruses that made their work—and that of many others—possible. She retired in 1973 and as of 1979 was living in Maryland (*WBWRA*).

Working with Dr. Leon Dmochowski of the University of Texas, **Dr. Elizabeth Priori** of Baylor's Texas Medical Complex isolated a human cancer virus in the early 1970s. The researchers extracted the virus from the cells of a five-year-old boy who died of Burkitt's lymphoma, and succeeded in growing it in human lymph cells in the laboratory. The pinpointing and isolation of this virus not only gives cancer virologists a great tool, but should be an important step toward an anti-cancer vaccine (Levin, 152–3).

Also at Baylor is **Dr. Janet Butel,** of the Virology Department. By 1975 she had been working for a dozen years on the same hamster-tumor virus. She keeps a framed electromicrograph of "dear old SV40," magnified 50,000 times, hanging on her office wall. Dr. Butel is married to another Baylor physician, and has two children (Levin, 151).

**Dr. Ariel Hollinshead** (1929–   ) of George Washington University in Washington, DC, has taken a more direct and important step by developing vaccines for all four major types of human lung cancer—squamous, large cell, adenocarcinoma, and small cell. By 1980 lung cancer was the largest cause of cancer deaths in men and the second-largest for women (*SN* 118 [July 12, 1980]: 26). Eighty percent of its victims die within two years of diagnosis. Hollinshead's vaccines work on the principle that if the human body recognizes cancer cells as foreign, the resulting immune reaction can destroy them. The trick is to mobilize that immune reaction, since a tumor develops in the first place because no immune reaction destroyed the original cancer cells. Earlier researchers had tried injecting cancer patients with BCG (Bacillus Calmette Guérin, attenuated tuberculosis bacteria) as a scattershot approach to activating their immune systems. In the early 1970s Hollinshead succeeded in isolating surface antigens from the four major types of lung cancer cells, making possible a vaccine that would provoke antibodies to those specific cells.

As of the late 1970s clinical tests looked extremely promising. In a study begun in 1973 at the University of Ottawa, researchers used the vaccines in 52 patients with all four types of lung cancer. Fifty patients had had squamous, large-cell, or adenocarcinoma tumors removed by surgery, and two patients had early-stage small-cell cancer (deadliest of the four, usually inoperable). Twenty-eight of the 52 received injections of Hollinshead's antigens monthly for three months, and 24 patients served as controls, receiving the standard treatment. Among the group receiving antigens 80% were still alive five years later, as compared with only 49% of the controls.

The Ottawa results have been partially replicated at Roswell Park Memorial Institute in Buffalo, and a much larger trial with 300 patients began in June 1978 at more than a dozen medical centers in the United States, Canada, England, and France.

Dr. Hollinshead has overcome many difficulties, both financial and scientific, during her career. She began as a young pharmacologist at George Washington University, separating cancer viruses into various pieces and injecting them into hamsters to see whether they would stimulate the animals' immune systems against implanted tumors. To provide a control substance, she injected membranes from virus-induced tumor cells into part of the population. Unexpectedly, it was one of the membrane fractions that stimulated the immune reaction.

Realizing what she might have discovered, she set to work to prove it. This meant, first, repeating the experiment on inbred hamsters to make sure that genetic differences between donor and recipient animals had not caused the immune response. Her original funder lost interest when his expectations were not borne out, and no one was interested in her cell-membrane approach. Dr. Hollinshead went without research money or salary for eight months, until she found an interested scientist at the Ohio State University School of Veterinary Medicine. The experiments with inbred hamsters confirmed her original results: all the hamsters injected with cell-membrane extracts fought off implanted tumors, whereas 90% of the untreated hamsters failed to do so.

One of the largest remaining obstacles was isolating the antigens intact from the membranes. Here Dr. Hollinshead proved inventive—thus doubly assuring her place in this book. After exploring several possible isolation methods, she devised a new technique using low-frequency sound, which gently separated out the antigens without damaging them.

When she moved into the human tumor field, receiving cancerous lung tissue from her new collaborator, Tom Stewart, at Ottawa, the problems multiplied: separating lung-cancer cell antigens from cancer cell-membrane blocking factors; identifying the membrane antigens as hormones, enzymes, or other kinds of proteins; testing the antigens to make sure they reacted only with lung-cancer cells, etc. Not until 1973 was she ready for the clinical trials mentioned above; but by then she was exquisitely ready, and her work has received high praise. As Ben Papermaster of the University of Missouri Medical Center wrote to her in 1980, "Your brilliant work in the field of human cancer antigens is among the most advanced and exciting in the world" (Arehart-Treichel, "Tumor-Associated Antigens," 27). Late in 1987 came a further report of success in the clinical trials: combined results of two series of trials showed that 69% of the vaccinated patients were alive and free of cancer, as opposed to 49% of those treated by surgery alone (Jerry Bishop).

Hollinshead has now moved into the field of ovarian cancer. As of 1983 her ovarian-cancer vaccine had been cleared by the FDA for use in human subjects with advanced ovarian tumors already treated by surgery. The vaccine seems free of major side effects; therapeutic effects were still being assessed (Budoff, *Flashes,* 144–5). Hollinshead will be a name to watch when the history of cancer vaccines is written (*AMWS/*12, 14; Arehart-Treichel, *ibid.;* Budoff, *ibid.;* "Cancer vaccines").

*Common cold.* A cold vaccine has long eluded researchers, partly because the virus is such a chameleon. Early in 1983, however, researchers at Purdue and the University of Wisconsin–Madison announced a giant step toward the elusive vaccine: the creation of prismatic crystals of one cold virus—rhinovirus 14. This was the second such successful crystallization, but the first to create crystals large enough to examine by X-ray diffraction. Each crystal contains more than ten thousand million viruses. This innovation is clearly a research technique, but equally clearly an advance toward prevention. **Elizabeth Frankenberger,** though specifically disclaiming any innovative role, was nevertheless a member of the Purdue team. In 1983 she was working on a PhD in soybean genetics at Purdue (Julie Miller, "Crystals").

A practical approach, in the absence of a vaccine, is to prevent the spread of colds. In 1967 **Erna Memmel** of Munich, Germany, patented a Protective hygienic cover for sound-transmission and listening apparatus (e.g. telephone mouth- and ear-pieces).

*AIDS.* Perhaps only an AIDS vaccine could excite the world more than a cancer or cold vaccine. In 1988, Stanford scientists and visiting Japanese scientists announced development of a mouse with human blood-forming and immune cells. The mice accept transplants of human tissue because they lack an immune system. Discovered in 1983, these so-called SCID (Severe Combined Immune Deficiency) mice—analogous to the

Texas "Bubble Boy"—are easy prey for any infection. Without special care and regular antibiotics, they have a very short life. Implanted with portions of human fetal liver, thymus gland and lymph nodes, however, the mice become healthy and live five times as long as untreated littermates. Thus they have presumably created some sort of immune system basically human in nature. The fetal liver contains the stem-cell precursors of all our blood and immune cells. The thymus gland somehow "educates" stem cells about their defensive function; and the lymph nodes are the site of the complex reactions making up an immune response.

These chimeric creatures are expected to make possible new studies not only in AIDS, but in leukemia and many other diseases. A better understanding of the human immune system will almost certainly result. Advances towards an AIDS vaccine and AIDS treatment drugs are among the most immediate benefits expected. Until now, researchers had no feasible animal model for studying AIDS, as only chimpanzees could be infected. The research, initiated by Joseph McCune, took place in the laboratory of Dr. Irving Weissman. Women on the team were **Dr. Reiko Namikawa** of Aichi Medical University and **Dr. Miriam Lieberman** of Stanford. The University has successfully patented the process for producing the SCID-hu mouse (Hofstadter, "Mice"; McCune *et al.*).

*Sleeping sickness.* Like the work of Frankenberger's colleagues with rhinovirus 14, the 1960s invention by **Nina Kljueva** and other Russian scientists of a Process for cultivating trypanosomes is both a research technique and a vital step in inventing a vaccine against or a cure for an important human disease: trypanosomiasis, or sleeping sickness. Kljueva and her colleagues evidently worked at a combined medical- and basic-biological-research institute, for they assigned their 1968 American patent to such an institute. The team included at least three other women: **Rina V. Fradkina, Rita V. Krzhevova,** and **Natalia N. Sukhareva,** all co-patentees.

*Evaluation of vaccines.* **Claire Broome** (1949–   ) is an epidemiologist at the Center for Infectious Diseases in Atlanta. Chosen as one of America's Top 100 Young Scientists for 1984, Ms. Broome has studied the public health aspects of toxic shock syndrome and Legionnaire's disease, as well as less newsworthy diseases such as pneumonia and meningitis. The work that brings her to this chapter—and her own proudest achievement as of 1984—is a novel approach she devised to estimate the effectiveness of a pneumococcal vaccine. Her technique consists of comparing the distribution of serotypes (organisms distinguished by different surface antigens) in vaccinated and unvaccinated persons showing pneumococcal disease. Broome's method has proven crucial in defining the appropriate use of the vaccine in the United States (*Sci.Dig.* eds, 47).

*Gamma globulin/generalized immunity.* Gamma globulin is the blood fraction containing most of the antibodies. It is used wherever immunity or resistance needs to be bolstered, especially where no vaccine exists. It can also prevent infectious diseases in those exposed to them, as for example, infectious hepatitis (*EB* yearbook, 1946: 466 col. 2). **Martha Hainski** of Glendale, CA, is co-patentee of a method of preparing gamma globulin using polyethylene glycol. Hainski and her co-inventors assigned their 1973 patent to Baxter Laboratories, Morton Grove, IL.

## Health maintenance

Prevention has an aspect even more central than warding off specific diseases: maintaining body and mind in a healthful state. It is here particularly that the male-dominated medical establishment is found wanting and women have often excelled, partly because of their responsibility for family diet and general well-being, and partly because of the different focus this role-training gives them as healers. The group of nurses who formed

Alternative Healthcare Associates are a prime 1980s example here. Led by **Alison Williams** and **Sara Levine,** they have left more conventional careers to focus on health and preventive care.

Smoking is such a major health problem, and recent findings on the danger of secondhand smoke so disturbing, that one preventive invention deserves a quick mention here: **Jeannette V. Orel** of Los Angeles has invented a Smoketrapper® to help people who must live or work with smokers. A graduate of USC's Pollution Control Institute, and a member of the State Attorney General's Los Angeles Environmental Task Force, she has even developed a "conference" model and a model to deal with cigar smoke (P.L. Brown; Orel, L-8/6 & 8/19/87; Orel, Q-8/5/87; Orel, "Smoke Sorbing," and "Marlboro"; Seidenbaum; Sheraton; Skelton, 34; Zyda).

This area of prevention is so enormous and so diffuse, potentially encompassing all aspects of life, that I will conclude with just two extended examples here—dress reform for the 19th century and weight control for the 20th. Both are particularly pertinent to women, and both lend themselves to inventiveness. The first was surpassed only by suffrage as an issue in 19th-century feminism, and was a hot topic in 19th-century medicine as well; the second is both a 20th-century obsession and a feminist issue.

*Dress reform.* Anyone who has seen *Gone with the Wind* knows that 19th-century fashion dictated a tiny waist—preferably 17–19 inches. With some 22 pounds of skirt and petticoats suspended from that same tortured waistline, the fashionably dressed 19th-century woman must have been in agony. It is no wonder she often fainted or had the "vapors," and a small miracle that she was capable of any exertion at all. Noted lecturer and Civil War Sanitary Commission officer Mary Livermore reported that at autopsy many habitual tight-lacers showed grooves in their livers deep enough to lay an arm in. "Corset liver" was a common medical diagnosis. The lower rib cage and upper abdomen were also deformed, with predictable effects on breathing, digestion, and menstruation (Beecher & Stowe, 160–4). One of the associates of popular health leader Sylvester Graham (inventor of the graham cracker) estimated in the 1830s that "folly in dress" killed 80,000 Americans every year (Stage, 61).[44] Catharine Beecher declared that the "protracted agonies" endured by tightly corseted women were worse than "the horrible torments inflicted by savage Indians or cruel inquisitors on their victims" (K. Sklar, 213).

There was more rhetoric than good research on the effects of tight lacing, but apparently one woman doctor in London did experiment with corseting monkeys into wasp waists. Those that did not die at once grew weak and sickly (O. Jensen, 137). In the United States at the turn of the century, Dr. Clelia Mosher of Stanford University studied the correlation between dress and menstrual difficulties. Comparing tight corseting to foot-binding, Dr. Mosher quantified the question by weighing the skirts, petticoats, etc., worn by a sample of college women, and correlating this weight—often more than 15 pounds in her California sample—with waist measurements and reports of menstrual pain. She noticed a definite correlation (C. Mosher, "Normal . . . ," 178–9).[45]

Even if hourglass corsets and heavy skirts did no more than immobilize women,

---

44. These were almost entirely women. However, men took up the corseting fad to a limited degree and suffered accordingly.

45. For further reading on contemporary views of these evils, see Catharine Beecher and Harriet Beecher Stowe, *The American Woman's Home* (*AWH*), 1869, 160–5, and the publications listed in Deborah J. Warner, "Fashion Emancipation, Reform, and the Rational Undergarment," 24–9, esp. 25. See also Mosher's book, *Women's Physical Freedom* (1923).

deterring them from vital exercise and undermining their sense of competence, this combination would have been a menace to health both mental and physical. In such a context, dress reforms intended to reduce that menace emerge as health-related inventions. Women doctors spoke out on the harm done by tight corsets and clothing, among the first of them Dr. Bethenia Owens-Adair (Chaff, #694). At least three women doctors were actual dress-reform inventors: Mary Edwards Walker, Sara Josephine Baker, and Mary Beard Harmon.

Dress reform was more than just a cause with **Dr. Mary Walker** (1832–1919); it was a personal crusade. "The greatest sorrows from which women suffer today," she said, "are . . . caused by their unhygienic manner of dressing. The want of the ballot is but a toy in comparison!" (Woodward, 281). This sounds like an extreme position, but Walker believed women already had a de facto right to vote and need only exercise it. Tight lacing and heavy skirts may not have made women frigid, as Walker believed, but such clothing surely inhibited their movement and behavior every waking moment, not just at election time. To the degree that little girls' clothing was more restrictive than little boys', recent research on the relationship between childhood freedom to explore and later intellectual capabilities may bear her out.

Always a feminist, Dr. Walker volunteered for the Union Army early in the Civil War, receiving permission to wear "bifurcated garments" (trousers) during her service. She had adopted the bloomer costume in the 1850s, and found male garb so convenient that she wore it the rest of her life. She was not deterred even when her fellow women clerks at the Treasury Department demanded that she use the men's lounge, or when rotten eggs were thrown at her or dogs set upon her in the streets. For other women, whom she did not expect to make the same choice, she invented the American Reform Dress, the Mary Walker undergarment, and the Mary Walker Reform Pants. The undergarment—a one-piece high-necked suit with loose waist, whole drawers, and long sleeves—sounds cumbersome to us, but was a miracle of freedom compared to 15–20 pounds of petticoats. Stockings were drawn up over the legs of the drawers to keep the ankles warm and eliminate garters. The Reform Pants or trousers were buttoned to the undersuit to eliminate suspenders. This combination was to be worn under the Reform Dress, a frock hanging free from shoulders to knees, with no tight waist and no heavy, floor-length skirts to step on (*NAW*; Manning, 239, Woodward, 282).

Dr. Walker's reform outfit—like many others—struck American women as dowdy. Her real tragedy, though, lay in alienating people so thoroughly that her farsighted reforms could gain no hearing. Almost a hundred years ago, she wrote and lobbied for a bill virtually identical to today's Equal Rights Amendment. She was among the first women to try (successfully) to vote; she lobbied both Congress and the Albany legislature, and spoke in Washington, DC, for woman suffrage; she went to Europe and influenced 10,000 women to vote in the fall of 1869 (Woodward, 281, 294). In 1866 she was elected president of the National Dress Reform Association. Yet later in life she was often dismissed as eccentric at best and freakish at worst, and is still mainly remembered for wearing men's clothing. Said Mary Logan (579), "Doctor Mary Walker has been made the subject of abuse and ridicule by people of narrow minds. The fact that she persists in wearing the attire in which she did a man's service in the Army blinds the thoughtless to her great achievement and to her rights to justice from our government."

In the 19th century, not even babies escaped overdressing. They wore long dresses with petticoats that restricted their movements and made mountains of laundry for their mothers. They even wore corsets![46] Child-health pioneer **Dr. Sara Josephine Baker** created patterns for sensible and healthful baby clothing that were adapted for commercial use by McCall's Pattern Company (*NAW*).

**Dr. Mary Beard Harmon** won a $100 premium offered by the World Health Association in 1865 for the best reform dress (Warner, "Fashion," 25). Regrettably, I have no details on this garment, or on Dr. Harmon.

The most famous reform garment in the United States, of course, was the bloomer costume, invented by Elizabeth Smith Miller[47] and popularized by Amelia Jenks Bloomer in the early 1850s. **Elizabeth Miller** (1822–1911), oldest child of philanthropist-reformer Gerrit Smith, and a cousin of Elizabeth Cady Stanton, was a high-spirited and intelligent woman. Herself a reformer, she married a man who joined her in signing the call for the first Women's Rights Convention in 1843. One day in 1851, disgusted by the tight, heavy clothing that hampered her gardening, she went to her sewing room and designed a costume consisting of a short dress falling about four inches below the knees, and turkish trousers fitted with loose elastic at waist and ankle. She was inspired partly by the garments she had seen in a rest home where women were recuperating from the effects—skin welts and cracked ribs—of tight lacing. She called her creation simply the "short dress."

Luckily for Elizabeth, both her husband and her father supported her decision for comfort. Her cousin Elizabeth Cady Stanton, Amelia Bloomer (editor of the *Lily*), and many other feminists and suffragists took up the design as well. Focusing on Mrs. Bloomer's defense of the garment, people began calling it the bloomer costume, or simply bloomers. Wherever women wore the costume, it caused a furor, and eventually those suffragists who often spoke in public felt forced to abandon it. Elizabeth Miller wore it somewhat longer, but she, too, gradually lengthened her skirts and finally stopped wearing the trousers.

The precise impact of the bloomer costume is hard to assess. Aside from its famous advocates, hundreds of Lowell mill girls wore bloomers to work, and Antioch College women in Ohio wore them to class. Women did not always take off their bloomer-like gymnastic costumes when exercise classes were over. The bloomers' existence may have eased the way for the eventual abandonment of long, cumbersome skirts and heavy petticoats (*NAW;* Warner, "Fashion"; L. Tucker).

Temporarily frustrated in trying to reform outer garments, reformers turned to undergarments, insisting that weight be reduced to a minimum, skirts and stockings be supported from the shoulders rather than the waist, and vital organs be unconstricted. With this less visible target, the movement began to catch on. In 1873 the New England Women's Club of Boston established a Dress Reform Committee chaired by Abba L. Woolson, which chose to act rather than merely talk. Sponsoring a series of five lectures, four of them by eminent women doctors, they followed up by publishing the lectures in book form and by holding a women-only exhibition. More important, they began to inspect garment-makers with definite criteria in mind, giving some manufacturers'

---

46. Three sample patents: #76,354—**Martha Solomon,** Charlestown, MA (1868)—and #124,891—**Harriet Emery** and **Margaret C. Fuller,** Boston, MA (1872)—both for child's corsets; and #149,692—**Catherine Tardy,** Paterson, NJ, Improved baby-exercising corsets (1874).

47. Fanny Kemble allegedly wore a similar costume in 1849, and the bloomer also bears several similarities to Dr. Mary Walker's American Reform Dress and Pants, as well as to the Oneida women's costume.

garments their seal of approval. Further, they went into the business of distributing such garments nationwide.

**Mrs. H.S. Hutchinson** (with Mrs. E.R. Horton) originally presided over the Boston store established by the Dress Reform Committee to distribute approved garments. By 1876, however, Mrs. Hutchinson was evidently in business for herself, winning a Centennial Award at Philadelphia for her skirt-supporting waist and undergarments (Warner/Post, 165; "Fashion," 29).

As early as 1869,[48] **Catharine Beecher,** long a vocal opponent of "murderous fashions," published drawings of a "jacket that will preserve the advantages of the corset without its evils." Beecher advocated fitting this underwaist to the body while wearing corsets, stiffening it only with delicate whalebone to give it the proper shape, and then doffing the corsets—an ingenious technique for achieving a gentler support. The waist had two rows of buttons at the bottom to support skirts and petticoats. "By this method," says Beecher, "the bust is supported as well as by corsets, while the shoulders support from above, as they should do, the weight of the dress below. No stiff bone should be allowed to press in front, and the jacket should be so loose that a full breath can be inspired with ease while in a sitting position" (Beecher & Stowe, 164).[49]

**Susan Taylor Converse** of Woburn, MA, patented her Emancipation Suit of underwear in 1875 (two patents) as an improvement on the 1868 Emancipation Union underflannel, the first "hygienic" undergarment approved by the Dress Reform Committee. The original garment covered the body from neck to wrists and ankles, and was made of cashmere and merino wool. Converse's improvement could have short or long sleeves and was made of cotton or linen. Among its special features were a full, gathered bodice to free the breasts from constriction or irritation, and sets of buttons at various levels, for attaching several skirts or petticoats in layers. The top half was sold as the Emancipation Waist. The Converse garment, manufactured by George Frost & Co., was shown at the Centennial in 1876, but among the regular commercial exhibits and not in the Women's Pavilion (LWP; Warner/Trescott, 108; Warner, "Fashion," 26–8).

Starting in 1876, **Miss Emmeline W. Philbrook** of Boston, MA, patented several reform undergarments. The most famous of these was her Equipoise Waist, patented in 1876,[50] shown at the Centennial, sold by dress-reform advocates across the country, and advertised in women's magazines as late as 1900. Philbrook patented other undergarments in 1888 and 1893 (LWP; Warner/ Trescott, 115–6, /Post, 165, and "Fashion," 29). She appears in the 1890 Boston City Directory under "Corsets & Bustles," located at 175 Tremont Street, indicating a stable business career of nearly fifteen years. A versatile inventor as well as an entrepreneur, she also had non-apparel inventions to her credit.

Corset-makers, too, embraced the Dress Reform movement, creating boneless corsets with elasticized straps for beauty with comfort. In fact, according to one source, women patented 179 corsets between 1815 and 1895, "in the sad attempt, perhaps, to get a good one" (M. Reynolds, 5–6). Prominent among these inventors was **Mme Catherine Allsop Griswold** of Connecticut and New York. As early as 1866, while living in Willimantic, CT, Mme Griswold patented an Improvement in skirt-supporting corsets.

Mme Griswold may or may not have participated directly in the Dress Reform

---

48. Possibly earlier, since *The American Woman's Home* is in part a revision of Beecher's earlier *Treatise on Domestic Economy* (1841).

49. Beecher does not claim this underwaist as her own design. However, her usual practice in this book seems to be to credit the designs of others when she presents them and to say nothing when the idea is her own or her co-author's.

50. All three of her 1876 patents, two for underwaists and one for a spring clasp, are omitted by LWP.

movement, but she appears in a list of American designers and manufacturers who began, once the movement's hygienic principles were clearly defined, to market "a great variety of healthful garments" (Warner, "Fashion," 29). And skirt-supporting corsets, an early specialty of hers, generally have the virtue of supporting skirts and petticoats from the shoulders rather than from the waist. At the Centennial, the Worcester Corset Company won an award for Griswold's abdominal skirt-supporting corsets. At least one of these had a very modern-looking comfort feature: an elasticized insert in the shoulder strap just where it joined the bodice. Mme Griswold patented 19 corsets and related improvements (plus 12 design patents) between 1866 and 1889 (Warner/Trescott, 114, and "Fashion," 28, 29). She appears under "Corsets & Bustles," 495 Washington, in the Boston City Directory's business directory for 1890, documenting nearly 25 years of entrepreneurship.

Prevention of disease is specifically mentioned in connection with the work of another reform corset-maker, **Olivia P. Flynt** of Boston. Said the *New Century for Women* (No. 3: 1),

> Prevention being better than cure, the improved underclothing for women and children, invented by Mrs. O.P. Flint [*sic*] of Boston, is worthy of attentive consideration. Doctors may think they know, but only those of our sex really do know, the amount of harm done by a few mistakes in the underwear of women, especially in the weight of dragging skirts. . . .

Continually confronted, as a dressmaker, with "the evils resulting from tight lacing," Mrs. Flynt became so distressed that she invented the "True Corset" or "Flynt Waist" of 1876 as a preventive. In fact, she designed reform undergarments well before the Boston Dress Reform Committee existed, receiving her first two patents in 1873. She described the Flynt Waist as (*NCW, ibid.*)

> A scientific garment which proves more than a substitute for other corsets, for it not only obviates the necessity for several other under garments . . . but it comfortably and perfectly sustains the heaviest bust from the shoulders . . . , permits natural circulation, perfect respiration, and freedom for every muscle. . . . From its construction every Waist is a Perfect Shoulder Brace, a fact of great importance to growing girls and children.

Mrs. Flynt showed her undergarments, eventually covered by a total of four patents, not only at the Centennial, where they won a bronze medal, but at the Massachusetts Charitable Mechanics Association in 1878, at the Cotton Centennial Exposition in Atlanta in 1884–5, and at the Columbian Exposition in Chicago in 1893. *NCW* specifically commended Mrs. Flynt because "there is no sacrifice of fashionable appearance involved in the modern improvements." Olivia Flynt not only invented reformed underclothing, but wrote about it, publishing her *Manual of Hygienic Modes of Underdressing for Women and Children* in Boston in 1882 (*ibid.;* Weimann, *Fair,* 431; Warner/Trescott, 113, /Post, 165, "Fashion," 24, 29).

Like Philbrook and Griswold, Mrs. Flynt appears as a maker of corsets and bustles in the Business Directory of the 1890 Boston City Directory. She also appears in the regular listings under "Flynt Waist or True Corset," at 319 Columbus Avenue.

In October 1874, **Miss Clara P. Clark** of Wakefield, MA, patented her Combination Shoulder Brace Corset, constructed without any bones or wires, and the Dress Reform Committee approved it (Warner, "Fashion," 29). Later she patented an Improvement in skirt-supporters (1875).

Best known as the arbiter of American fashion in the mid-19th century, through her fashion house and *Demorest's Monthly Magazine*—and as the inventor of tissue-paper patterns—**Ellen Curtis Demorest** (1824–98) of New York also created "Health Corsets." Always insisting that the prime requisite of beauty is health, Mme Demorest maintained that although corsets were essential for support at certain periods of a woman's life, they should accomplish their purpose without harm to the body. Her corsets were advertised as "constructed on thoroughly scientific principles" (Warner, "Fashion," 29). Demorest evidently did not patent her elegant yet comfortable corsets, stocking suspenders, and skirt-supporters, but showed them at the Centennial, where the judges commended them (Warner/Post, 165).

Another enterprising and reform-minded corset-maker was Vienna-born **Madame Binner.** She had started her craft at age eleven, and lived to corset the early Hollywood stars. Coming to the United States in the 1880s, she was appalled at the craze for the hourglass figure. When X-rays became available, she began to send her clients to doctors to have films made of the effect on their ribs. In return, doctors sent women to her for light, sensible corsets. One of Mme Binner's famous customers was Lillian Russell,[51] first to wear her gartered corset. Miss Russell's $3900 model had diamond buckles. Mme Binner's fame eventually reached Hollywood, where she designed the special corsets worn by Alice Faye when she played Lillian Russell in the movie biography. Working for 20th Century–Fox, Binner made corsets for many of the younger stars, including Judy Garland, tailoring her work to the demands of each picture. Known in the 1940s as the "Dean of American Corsetières," she corseted Mrs. John D. Rockefeller, Mrs. George Gould, and Mae West, among others (*WA,* 62).

Twentieth-century women's dress still needs reform. Waists are less often corseted, but most fashions still have tight, inelastic waistbands. Hobble skirts deserve their name, and any skirt can restrict activity for modesty reasons. High heels deform the pelvis and contribute to shin splints in athletes and back problems in all who wear them. They make women more vulnerable to street violence because no one in high heels can outrun a mugger. However, 20th-century feminists have been less organized than their 19th-century counterparts. I have only scattered examples:

**Martha E. Larrabee** of Anita, IA, patented an interesting Skirt- and hose-supporter in 1911. It looked like a light shoulder harness with extra-long garters, so that all weight was supported from the shoulders with no constriction at the waist.

According to one source, Germany's dress-reform movement did not begin till this century. Designer **Gunda Beeg** was its founder, creating a loose-fitting dress to replace restrictive Victorian clothing (Chicago, 187).

Danish inventor **Anne Kalsø**'s Earth Shoes® with their recessed or negative heel—the opposite of high heels—have a health rationale behind them. Ms. Kalsø got the inspiration for her design while studying yoga in South America in 1957. She observed that by flexing her foot so as to lower the heel, she achieved a feeling similar to that attained in the lotus position. Admiring the graceful walk of the local Indians, she examined their footprints in sand and confirmed that their body weight fell most heavily on the heel, causing it to sink more deeply into the sand than the rest of the foot. Returning to Denmark, she designed a shoe to create that kind of walk and to stretch the Achilles tendon. She tested her models on 500-mile walking trips. Now Kalsø holds the first patent for a negative-heel shoe (#3,305,947).[52]

Proponents of the shoe maintain that by exercising and strengthening the Achilles tendon, and returning it to its intended length and flexibility, the shoe will help the entire

51. Russell was herself an inventor, patenting a traveling Dresser trunk in 1912.
52. There have been accusations that Anne Kalsø stole a design from marine engineer and ice skater Alan Murray. Both designers hold patents.

body fall into line, improving posture, and helping the wearer walk gracefully and comfortably. Critics talk about pulled muscles and overall agony. Whatever the physiological facts, Anne Kalsø is undeniably a successful entrepreneur as well as an inventor. She formed her own company in Denmark, Kalsø Systemet, Inc. During the 1970s she retailed her shoes only from Earth Shoe® Stores at an average price of $40 (1976), and she had many imitators. As she points out, however, lowering the heel of a shoe is not enough. During years spent perfecting the shoe, she also modified the arch, widened the toes for comfort and freedom, and shaped the front of the sole to allow a "natural" rolling motion in walking (Earth Shoe® ad, *Ms.* magazine, 1976; Zehring, 31). This tapered-front sole has also been widely imitated, and is indeed more comfortable, not to say necessary, for thick-soled shoes.

*Weight control.* In the tortured field of weight control, several women are noteworthy innovators. **Tillie Lewis** has already been mentioned as a canny entrepreneur who rose from penniless runaway at 12 to owner of the fifth largest cannery in the United States—a millionaire—at 50—and for introducing her Tasti-Diet foods in 1952. Containing artificial sweeteners instead of sugar, they were far lower in calories than most canned products. That year, the women's-page editors of the Associated Press voted Tillie Ehrlich Lewis Woman of the Year (Marlow, 316–21; *WBWRA*).

In the battle of the bulge, people need every possible choice of weapon. Physiological psychologist and taste researcher **Dr. Susan Stolte Schiffman** has provided two more choices. Chicago-born Dr. Schiffman graduated from Syracuse University, and took a PhD in psychology at Duke University (1970). After postdoctoral study at the Duke Medical School Center for the Study of Aging, she joined the Duke Psychiatry Department in 1972. Her research in the physiology and theory of taste has several practical applications: dealing with taste and smell abnormalities in old age, the cereal and perfume industries, the search for new non-fattening sweeteners, and fighting obesity. The first of the two highly practical inventions bringing Dr. Schiffman to this chapter is her weight-loss kit, Better Than a Diet®, calculated to help people eat less by eating more slowly. The kit appeared in the late 1970s. By the mid-1980s Dr. Schiffman was testing a line of flavor-sprays designed to satisfy cravings for such high-calorie foods as pretzels, chocolate, and pizza by simply spraying a synthesized form of the taste into their mouths. The sprays, if successful, will be marketed by Nutri/System, Inc., a national chain of weight-loss centers with more than 700 outlets (Kalat/1980, 27–9; Kalat/1988, 172–3; "Diet Chain . . .").

**Sally O'Brien**'s low-calorie pizza dough ("Diet Pizzas . . .") and **Jean Nidetch**'s Weight Watchers of America have also already been mentioned.

Although Jean Nidetch certainly did not invent the confessional/mutual-support approach to changing behavior, she seems to have been the first to apply it systematically to weight control. Making a profitable business of the approach was apparently the joint brainchild of Nidetch and two of her friends, **Felice Lippert** and her husband. As of 1985, Felice Lippert was still a consultant and advisor to Weight Watchers International.

Other successful innovators of diet regimens are **Sybil Ferguson** with her Diet Center Diet, and the creator of the **Jennie Craig** System.

**Julia Halberg**'s rhythm method of nutrition could be used to help people get the *least* out of their food rather than the most. Used in this way, her innovation could become a method of weight control.

Unusual in the field of diet-related innovations is a patentable invention—what the U.S. Patent Gazette calls a Diet Control Apparatus. This is the Diet Conscience, or Scolding fridge, patented by **Carol L. Kiebala** (1941–   ) of Palatine, IL, in 1978. This diabolical invention, mounted inside a refrigerator, is activated by opening the door, and can only be stopped by closing the door again. "Are you eating again?" it shrieks. "Shame on you! No wonder you look the way you do! Ha! Ha! Ha! You'll be sorry, Fatty! Do yourself a favor—SHUT THE DOOR!"

Mrs. Kiebala, tall and beautiful, and always struggling with her weight, got the idea during one such struggle in 1976. Using her daughter's talking toy telephone, she rigged the refrigerator so that when her husband opened the door to get a beer after work, a little-girl voice piped out, "Hello, what's your name?" His shock convinced her that she had a good idea, and she spent the next two years perfecting her device, getting financing, finding a manufacturer and a distributor. In its first year on the market, Diet Conscience racked up 200,000 sales, and Carol Kiebala and her invention were featured in *People* magazine (Connie Singer).

Two other patentable inventions in the weight-control field are the work, already discussed, of two Illinois women, **Helen Smit** and her daughter **Julie Smit** (c. 1936–   ): a coating or lining for the intestine to keep some food from being absorbed (Crossen); and the Gastric pouch patented in 1983 by Lawrence H. and **Lanette H. Wilkinson** of Albuquerque to reduce stomach capacity.

Exercise: Vital to weight control, but conferring other health benefits as well, is exercise. The more sedentary one's life, the more necessary this preventive becomes. Although many examples of women's innovations in this area could doubtless be given, two must suffice here:

**Judi Sheppard Missett** (1944–   ) invented Jazzercise in 1969. Missett was a professional jazz dance teacher from Southern California. Finding that many of her students valued dancing as exercise but were intimidated by formal training, she blended dance technique with exercise physiology to create a new entity that is now an internationally syndicated business. It is also highly profitable ($40 million revenues in 1986) and often imitated. Jazzercise has some 400,000 students in 15 countries (Aldige; Ellis-Simons).

In the early 1970s, **Callan Pinckney** (c. 1951–   ) of Savannah, GA, and New York City, created Callanetics by combining exercises she had learned in Europe with her own innovations. Her concern with exercise is deeply personal: born with scoliosis, one hip higher than the other, and severely in-turned feet, she wore full leg braces for seven years as a child. Nevertheless, doubtless partly because of a devoted mother who massaged her legs three hours a day for five years, she managed to recover sufficiently to study ballet with a pupil of Fokine. Later, traveling around the world for more than ten years, carrying a backpack almost as heavy as herself, she badly injured her back and knees.

Pinckney's system is, above all, gentle, insisting on specified thorough warm-ups. It demands nothing that would hurt a back. Callan Pinckney claims that it will resculpt a flabby body in a very short time, maintaining that one hour of Callanetics is worth at least seven hours of conventional exercise in terms of muscle tone, stamina, and visible results. Her writer on the *Callanetics* book, Sallie Bateson, calls Callanetics one of the most innovative exercise systems around today. Photographs of students doing the exercises suggest that they combine the most useful ballet exercises with certain yoga positions, modified by the needs of a delicate back.

Callan Pinckney comes from the prominent Colonial Pinckney family of South Carolina, plantation-owners, colonial administrators, and diplomats. Indeed her book seems to imply that she is descended from **Eliza Lucas Pinckney**, who married into the family in the 18th century (C. Pinckney, 13).

*Miscellaneous*

Vital to preventive health are those measures taken in infancy or early childhood to prevent serious disease or conditions later in life. An example would be applying silver nitrate to newborn infants' eyes to prevent gonorrheal blindness. **Dr. Sara Josephine Baker**'s improved container for administering this medication is worth recalling here.

Another example is detecting and correcting hearing problems in babies and very young children. If a child hears no normal speech to imitate by age three, s/he will never speak quite normally, even if the defect is discovered, say, at school entrance. Yet, for obvious reasons, infants' hearing can be hard to test.

On February 13, 1979, **Barbara Franklin** of Palo Alto, CA, patented her method of testing children's hearing. Stacy Jones mentioned her patent in his February 24 "Patents" column. Her method is especially useful for testing infants and other hard-to-test subjects (see "Diagnostic Tests" below). As of 1981, Dr. Franklin was Associate Professor of Special Education at San Francisco State University.

Detecting vision defects early is also extremely important, not only because some can be corrected only then, but because poor vision may make children clumsy or accident-prone, interfere with their reading, or cause other problems. But children's vision, like children's hearing, can be very difficult to test. Coming to the rescue here is **Kiyo Sato-Viacrucis** of Sacramento, CA, a public health nurse specializing in pre-school children, with her Blackbird "E" screening test (discussed in detail below).

A minor but interesting invention that should help prevent the athletic injury or condition called tennis elbow is **Madeline Mishel Hauptman**'s new tennis racquet, the Mad-Raq. The patent (1980) was issued under Hauptman's maiden name, Mishel, because she filed the application before her 1978 marriage and while still a graduate student at the University of Pennsylvania. The Mad-Raq, inspired by a pair of snowshoes hanging on the wall, has not two but three sets of strings, one parallel to the handle and the other two laced diagonally to the first and at right angles to each other. The design is calculated to give a better grip on the ball while reducing vibration of the player's arm and cutting strain by 15%. Ms. Hauptman has protected the Mad-Raq by a British patent (1979) as well as her American one. As of early 1980 she had other foreign patents pending as well, and manufacturers were testing prototypes. She hoped to sell her patent rights and see the racquet produced (Stacy Jones, *NYT*, Jan. 26, 1980).

*Dental health*

Since women dentists are so much rarer than women doctors, I made no systematic search for possible dental inventions; but one 19th-century achievement is too significant to be omitted:

According to one source, **Elizabeth Morey** of New York City invented the dental cap or, as it was called in the 19th century, the "skeleton tooth." Mrs. Morey learned dentistry from her husband, with whom she practiced for several years. She created the first dental cap for a woman patient who had a pin tooth. Leaving the tooth intact, as it was sound, she modeled and fitted over it a hollow artificial tooth, making it uniform with the others (Rayne, 107–8). Regrettably, Rayne gives no details, but the invention obviously predates her book, which appeared in 1893.

Although the original impetus may have been cosmetic, the health benefits of preserving rather than pulling a tooth are now clear, and dentists go to great lengths—gold inlays, capping, and root canals—to avoid extractions. Mrs. Morey was ahead of her time in believing that the first principle of dentistry is to save, not destroy, teeth. When asked whether she thought women fitted for dentistry, she said she thought them better fitted than men in at least one way: men "use too much force, and often crush a tooth or injure the jaw, in taking one out. When I am obliged to pull a tooth I take it out whole." She added that although her practice among "ladies" was large, her husband's was larger because, sadly enough, "many women object to being treated by one of their own sex ...; but I think their prejudice would be easily overcome, as it has been in the case of female physicians, if ladies knew that [female] practitioners ... had graduated at a regular dental college" (Rayne, 108, 109).

## Aftercare

### Patient or invalid care

As women have usually been the home nurses and caretakers of aged parents, as well as the hospital nurses, it is not surprising to find numerous nursing and invalid-care inventions by women. Most of these were never patented; some, like the Shaker cradles for the dying, might have been unpatentable. Although Shaker women were freed by celibacy from child-rearing cares, the sect preserved a fairly traditional 19th-century division of labor. Thus, it was probably a woman who thought of this tender way of comforting an elder's last days (Rudofsky, 92, 157–8).

A group of 19th-century British women's inventions were shown at the Chicago Exposition in 1892–3. These "invalid appliances and furniture designed by nurses," according to the official catalogue, were exhibited as Entry #337 in Great Britain's exhibit, but in the Nursing Section rather than as medical inventions *per se*. Their patent status is unknown.

Some British and American women, of course, did receive aftercare patents. Just at the end of the 18th century (1794), **Henrietta Caroline Bentley** received a British Patent for a "Bed for invalids, which may be made and the linen changed without inconveniencing the patient" (*WI/*1; Woodcroft). Two other British patents were **Alicia M.T. Amherst**'s Chair for invalids (1881), and **Eliza M.** and **Emma S. Howell**'s Portable urinal (1882). For U.S. patents, LWP records 9 invalid or hospital beds, mattresses, and attachments; 3 adjustable pillows, body supports, or bed-rests; 1 patient elevator and perambulator, 1 invalid lift, 6 invalid chairs (one folding) and 1 invalid table, 15 bedpans and related inventions, and 4 invalid-feeding inventions, among others (see App. HM-3).

Known outside the patent records are **Hannah C. Bailey, Annie Caller,** and **Caroline Drake.** Mrs. Bailey was the wife of the editor of the Utica *Observer* in the 1880s and early 1890s. Her invalid bed, featuring an attachment allowing patients to raise and lower themselves unaided, was mentioned by Charlotte Smith in the *Woman Inventor* (#1). Drake's hospital bed appeared in a list abstracted from *Inventive Age* for the columns of the Shaker *Manifesto* of July 1890 (XX: 7: 164). The *Inventive Age,* said the Shaker writer, has contained "illustrations of a number of really valuable devices," of which Drake's bed was one example. Annie Caller's broiler was mentioned both by Charlotte Smith (*WI/*1) and by the *Manifesto*.

A few of the listed inventions seem technically interesting, judging by the brief LWP descriptions. One of the invalid chairs was a Combined cabinet, commode, wardrobe, and invalid chair. Patented by **Agnes Hardie** of Leetonia, OH, in 1889, this sounds like a space-saving unit for a small bedroom. The flexible urinal patented in 1883 by **Carrie S. Murphy** of Dayton, OH, sounds useful in especially difficult situations, as well as more comfortable for many invalids.

Women's patient-and invalid-care inventions were shown at the Columbian Exposition in Chicago in 1892–3. In addition to the already-mentioned "invalid appliances and furniture" designed by anonymous nurses, the British Exhibit contained models of the Pridgin-Teale [system?] for excluding dust from rooms, cupboards, etc. (Entry #337, Nursing Section; Exhibition Catalogue).

In 1896, just outside the scope of LWP, **Florence E. Asper** of Chillicothe, MO, patented an Invalid's table.

*20th century.* The U.S. Labor Department Women's Bureau 1920s study (WB/28), sampling women's inventions for 1905–6, 1910–11, 1913–4, and 1918–21, had a Sick-room Equipment subcategory, roughly comparable to this group of invalid-care inventions. Of the 227 patents issued to women for Medical, Surgical, and Dental Equipment (Category XVII) during these years, 73 or 32.2% were for sick-room equipment.

Included were invalid beds, sanitary mattresses, mattress holders, bed tubs, bed attachments (including those for bedpans), bedpans (including elastic and cushioned types), pillows and cushions, supports and bed-rests, including limb-rests; douche tips, heating pads, hot-water bottles and heaters and tube attachments for same, ice bags, invalid chairs, medicine bottles and dispensers, mercury-vapor lamps, sputum cups, and sterilizers (WB/28, 46).

As invalid- and eldercare moved from the home to institutions, women may have invented fewer such aids. Their recent invalid-care inventions are low-and high-tech extremes: **Blanche V. Wood-Minor**'s Bed-protector apron, **Lucille Hathaway**'s diaper, **Julia Butcher**'s invalid-exerciser, and **Bessie J. Blount**'s automatic invalid feeder. Brooklynite Wood-Minor patented a "Sanitary device for adult bed-patients" (1956; a wide band encircling the waist and abdomen, and a waterproof "apron-like member" that folds around the buttocks, through the crotch, and up to the front band). Properly fastened, the apron leaves an opening for a colostomy bag. Hathaway, of Toledo, OH, received a 1975 patent for an adult diaper, adapted to fit comfortably without bulk or pressure on bony points (WB/1975: 2). Butcher's experience caring for a bedridden son led to her Body motion system (1975), providing support for much-needed muscle stimulus and therapy for severely retarded and other immobilized patients. A Swedish firm was interested but failed to get government approval (Butcher, L-11/5/81).

One of the most interesting of all the women's inventions in this category is the automatic invalid-feeder invented by **Bessie J. Blount** in the 1940s. This electrically powered device allows patients or invalids to eat or drink any solid or liquid food while sitting or lying in any position, without the aid of attendants (beyond the initial set-up). The feeder dispenses bite-sized portions of solid food down a tube to a spoon-shaped mouthpiece, shutting off automatically after each portion to avoid choking or overfeeding. The user signals for another portion by biting on a switch. Thus even quadriplegics and quadruple amputees can feed themselves, if they can bite and swallow.

Blount, who is black, was born in Hickory, VA, about 1914. In 1951 she was living in Newark, NJ, working as a physiotherapist and teaching physiotherapy at the Bronx Hospital in New York City. She had attended Union Junior College in Cranford, NJ, and Panzer College of Physical Education in East Orange. She had also studied physical therapy in Chicago. Her invention grew out of her rehabilitation work with World War II amputees. In order to evaluate her work properly, we should recall how many thousands of these battle victims crowded the VA hospitals in those early post-war years.

Bessie Blount was so good at rehabilitation that people called her a Wonder Woman. She had even taught some of her armless veterans to type with their feet. Such a success suggests great sensitivity to the problems of the disabled and to their deep need for self-reliance. Thus it was that when a doctor said to her, "If you really want to do something for these boys, why don't you make something by which they can feed themselves?" she decided to do just that. In the kitchen of her second-floor apartment, she worked in spare hours from her demanding job, with, as she put it, "a dream and the help of God." Five years and $3,000 of her own money later, on April 24, 1951, she got her patent. This would appear to be Patent #2,550,554, issued in the name of Bessie V. Griffin, and listed by Ives ("Patent . . .," 120) among her important patents issued to blacks and women. If so, however, Ives' brief description, "Portable Receptacle Support," is unrevealing, and neither Ives nor my others sources deal with the discrepancy in names.

While still working on her invention, she had tried to discuss it with the Veterans Administration. She finally contacted her Congressional representative, who got her an appointment with Paul B. Maguson, then Chief Medical Director. Maguson called her invention "impractical," and told her that the VA had enough nurses and aides to give personal care to veterans.

Ms. Blount was shocked, not only because the doctor who originally inspired her had felt otherwise, but because Director-Emeritus Malcom E. MacEachern of the American College of Surgeons and other hospital authorities had praised her invention. After trying briefly to market the feeder herself, she finally offered it to the French Government, proudly signing the agreement at the French Embassy in 1952. Her press statement said she felt she had contributed to the progress of her people by "proving that a black woman can invent something for the benefit of humankind" (A. Carter, 6).

Blount also invented a special wash basin to aid patients recovering from surgery. A commercial firm was evidently interested in this one (Adamczyk, L-5/21/81; A. Carter).

A more recent example to close this section comes from Southern California. Though perhaps the most frequent use of the Head Hammock® invented by **Agnita Britzman** (1921– ) of Long Beach, CA, will be recreational—for comfort while sunning, getting a massage, traveling in a car, or floating in a pool—this U-shaped inflatable pillow does have its medical uses; among them, supporting the neck of patients in wheelchairs or being moved on a gurney, and supporting an ankle, back, or other part of the body during recovery from surgery. It could also help to immobilize the head. Ms. Britzman has specifically designed a quick-change, disposable cover for the Head Hammock® that should make it more acceptable to hospitals (Des. Pat., 1980).

Britzman has also invented a simple wound-cover that protects recent and healing wounds, or any sore or tender spot, both from clothing and from further injury. Her motivation for this invention was personal: She taught herself to ski at 49 and one day gouged her elbow in a fall. Any least bump to the sore elbow was agonizing, so she improvised a prototype wound-cover from a styrofoam cup cut in half and taped to her arm. The curve of the cup left an air space, and room to spray medication on without removing it. This was in 1973. She had interested one company in licensing and manufacturing her invention, until a change in personnel broke off the negotiations.

When Agnita Britzman was a child in Dunkirk, NY, she wanted to be a doctor (L-7/12/83), but was given the choice of being a teacher or a nurse. She chose teacher. She is also a talented artist.

### Aids to the handicapped and elderly

Obviously, aftercare inventions overlap extensively with inventions helpful in long-term care. The invalid beds, sick-room furniture, and other items listed above could just as easily serve in prolonged invalidism. In this section, however, I will focus on inventions particularly pertinent to the specialized and long-term care demanded by the aged, disabled, and chronically ill.

Nineteenth- and 20th-century American and British women have made important contributions to the health, comfort, self-reliance, and care of the handicapped and elderly. Earliest of the 19th-century American inventors known to me in this category are **Sister Tabitha Babbitt** (c. 1784–1854) and Mary Ann Boughton. Sister Tabitha, sometimes known as Sarah, was a Shaker from the Harvard, MA, community. She invented, among other things, a new method of making false teeth. **Mary Ann Boughton** of Norwalk and Bridgeport, CT, devised an Improvement in the modes of forming the air chambers in dental plates (Knight/ Seymour *et al.,* 127). Boughton patented her denture improvement in 1871, along with an Improved cookstove. The Shakers, however, did not at first believe in patents, and I have found no record of any patent issued to Sister Tabitha (Deming & Andrews).

At the Philadelphia Centennial Exposition in 1876, **Sarah Hare Bancroft** and **Sarah W. Tucker,** both of Media, PA, showed a Bathing chair for ladies or invalids. They patented their invention and showed it at the Franklin Institute that same year. Sarah Bancroft was the wife of cotton-mill owner Samuel Bancroft (Warner/Trescott, 111).

A few years later, several women patented devices to assist the handicapped: **Mary E. Moore** of New York City patented an Ear battery for the deaf (hearing aid) in 1888. In 1889, **Ida V. Ready** of Brooklyn patented a Device for assisting in putting on garments. In 1890 **Hedwig M. Kolbe** of Philadelphia patented a complex and interesting-looking artificial leg, with a skeleton construction for light weight, a length-adjusting screw for the foot and a spring between foot and toes. This invention is omitted from LWP.

At least three 19th-century women patented eyeglasses: **Amelia Kahn** of St. Joseph, MO (1888); **Augusta Schultze** of New York City (1888); and **Mary D. Hanway** of Baltimore, MD (1892).

In the early 1890s, **Elizabeth (Lizzie) Sthreshley** of Austin, TX, patented both a typewriter and a slate for the blind (1890, 1892). About the same time (c. 1890), an anonymous Swiss woman invented a watch for the blind. Instead of hands, the watch had twelve tiny projecting pegs to represent the hours. One peg would sink for every passing hour. My contemporary source, Charlotte Smith's *Woman Inventor* (#1), gives neither the woman's name nor her patent date or number, but says that the watch had just been invented.

At the turn of the century, too late for LWP and too early for the Women's Bureau Study (WB/28), **Frances MacDaniel** of New York City patented her Electrical ear-trumpet (1900). This device, still trumpet-like but reduced to a cup no bigger than the ear itself, contained certain modern features: means not only for collecting external sound, but for producing a sound within itself and "for transmitting both of said sounds to the ear of the person using the instrument."

Bridging the 19th and 20th centuries with her long and creative life was **Lillian Moller Gilbreth** (1878–1972). Engineer, designer, efficiency expert, and industrial psychologist, she has been profiled so many times that the details of her life need little space here. Too often, however, biographers have treated her as an adjunct to her husband, or as a SuperMom combining a career and twelve children, instead of dealing with the unique and original force that she was. For a fairer evaluation of her work as an engineer, see Martha Trescott's article on Gilbreth in Rothschild, ed. *(Dea)*, and her forthcoming study of women engineers.

The aspect of Gilbreth's work that concerns us here is her inventions and innovations for the handicapped. Her interest in this area goes back at least to 1920, when she wrote

*Motion Study for the Handicapped* with her husband Frank. In the course of her later, more intensive study of applying industrial management methods to household efficiency, she did not neglect people with special problems. Using motion analysis in innovative ways, she designed special equipment—and special routines—to allow the disabled to do their own housework and thus to be independent. She also collaborated with her friend Edna Yost on a book titled *Normal Lives for the Disabled* (1944), which is regrettably more inspirational than factual, and gives no information on her own work. Later, as a consultant to the Institute of Rehabilitation Medicine at New York University's Medical Center, she developed an entire model kitchen for the handicapped, which became a world-famous training center. The step-on lid-opener for trash cans, an invention of hers now used everywhere, was probably among its features. She was still developing "new concepts in devices for the disabled" in her 80s (*WBWRA; NAW*/mod.; Marlow, 218–25; Matthews, 44; Yost, *AWS,* 99–121).

Like Gilbreth, **Sarah Pillsbury Harkness** is best known in another field, but has also done important work for the handicapped. Like Gilbreth, too, she is the mother of a large family (seven children). Sarah Harkness is an architect. With her partner-husband and several colleagues, including Walter Gropius, she founded the Architects' Collaborative in Cambridge, MA. Modeled on Germany's famed Bauhaus, the Collaborative deeply influenced American architecture. According to my source, Harkness has been involved in "several design-for-the-handicapped projects" (*WBWRA*).

**Ray Kaiser Eames** (fl. 1940s) of Santa Monica, CA, is a woman of many talents. A dancer who studied with Martha Graham, a painter and sculptor who studied at Cranbrook, she is also an innovative moviemaker and a prize-winning designer of furniture. With her husband Charles and architect Eero Saarinen, she designed one of the world's most famous modern chairs: the revolutionary molded plywood chair that won the organic design competition at the Museum of Modern Art in 1941. What brings Ray Eames to this chapter is her work, again with Charles Eames, on devices for wounded Navy veterans of World War II. This work took the Eameses from Michigan to California in 1941 (*WBWRA*).

In 1928 **Harriette E. Hodgson** of New York City patented a Plastic artificial hand. The same year she patented a Foot cap, and four nonmedical inventions.

In 1936 British researcher **Mary Jameson** received the MBE for her role with Dr. Fournier Albe in developing the Otophone, a device enabling the blind to read print by ear (*Electrical Age,* Apr. 1936). The Otophone works by directing a ray of light onto the lines of a book and, using a selenium receptor whose conductivity varies according to the intensity of the light coming back, turns the print into a series of sounds that a trained person can read.

Another Englishwoman, **Helen Craig-Davis** (c. 1894–1959) of London, invented— among other things—"several devices to help handicapped people." Unfortunately, the brief death notice in the London *Times* ("Woman Inventor Found . . .," Nov. 5, 1959) focuses on her demise in an apartment fire, listing only a few of her inventions in the most general terms. One is a device—for offices, but also quite useful to a one-armed person—for holding a telephone receiver to the ear, leaving the hand(s) free. Mrs. Craig-Davis had also patented a book-holding device (1952; Rimmer, L-9/15/81).

**Dr. Jessie Wright** (1900–70) was also born in England, but came to the United States as a child. She worked as a physical therapist before studying medicine, and had a lifelong interest in the problems of the disabled. Her Respir-Aid rocking bed has already been discussed as Medical apparatus, but deserves mention here as one of women's most striking contributions to long-term care of paralytics. When Dr. Jonas Salk was ready to test his polio vaccine, she provided the first group of patients. Moreover—and this further secures her place in this chapter—she devised other special apparatus and orthopedic appliances for handicapped patients (*Amer. Acad. for Cerebr. Palsy Bull.* 14: 1(Feb. 1, 1962): 1–3). Much honored in life, she was a candidate for inclusion in *Notable American Women.*

Devised for her own lengthy hospital stays, but useful for anyone who might have trouble dressing, was the hospital gown with Velcro® closure invented by **Rose Baren Null** (1913–80) of Chicago. Part of the design was to make it attractive, unlike the standard gown. Rose Null was ill for some eighteen years of her life, but despite this—or perhaps partly because of it—came up with many ingenious ideas for inventions. Two 1950s examples are her musical books for children, incorporating tiny Swiss music boxes, and a reclosable metal top for detergent boxes tight enough to allow an opened box to be dropped without spilling. None of her inventions ever reached the market, although a model still exists of the reclosable boxtop, which, according to her son Fred Null, was offered to Lever Brothers about 1957 (Null, I-10/28/89; Mayer, PC-11/15/89).

Perhaps the most famous British inventor of aids to the handicapped is **Josephine Banner** of Ambleside, Cumbria. She is famous, however, not for her inventions, but for her prize-winning sculptures, done under her professional name of Josephina de Vasconcellos, Fellow of the Royal Society of British Sculptors, Hon. Litt D. Since she also invented a new sculpting method, a word about her artistic achievements is in order. She was the first woman to rank second in the Prix de Rome, and her sculptures have graced many churches and other important sites, including Bradford University and the National Memorial to the Battle of Britain in Aldershot (Sandilands, 9–10).

As of 1980, she had to her credit three inventions to aid the handicapped or elderly: a textured Contact Carpet to be used as an educational device for blind or near-blind children; the "Track-a-Gear" to allow the severely handicapped to ride horseback, and to help the less disabled to mount for riding; and an adjustable stool or "standing seat" with a support rail to allow elderly or frail people to sit and stand unaided. Ms. Banner conceives of this last invention as especially beneficial to the lower-income elderly, as it might keep them self-sufficient longer (Banner, L-12/12/79). Mr. Beverley Silvanov, a London inventor and consultant, writing in 1976 for *The Inventor,* organ of the British Institute of Patentees and Inventors (IPI), credited Banner (anonymously) with "one of the greatest advances in medical invention." Elsewhere, he calls both the carpet and the Track-a-Gear "superb medical inventions." This judgment seems to have been confirmed by other experts, as in 1974 the Cognitive Research Trust gave her an award for her inventions. However, commercial success has not followed. According to Silvanov, the inventions do not appeal to the mass production industry; and the entrenchment of traditional rehabilitation and recreation methods for the blind makes it unlikely that this new and radical approach will be adopted—"a little like trying to convince an anaesthetist to try hypnosis or acupuncture in a surgical operation" (Silvanov, "Medical Invention," L-11/14/79, 7/10/86).

As of 1980 Josephine Banner held only one patent, at least in her own name: a 1969 British patent for her Contact Carpet—"Improvements in or relating to floor coverings [for assisting the blind]." The Carpet, a rubber mat to be placed in blind or near-blind babies' cribs, features varied textures applied in geometric patterns (corresponding to different colors for partially sighted infants). The babies learn the textured patterns by touch, and this learning provides a basis for further learning, especially in language (Rimmer, L-9/15/81; Banner, L-12/12/79). The underlying principle here seems to accord with recent findings in brain physiology, the physical basis of learning, and rehabilitation theory. Josephine Banner's arriving at the idea intuitively, quite without training, is remarkable indeed.

Apparently the carpet has uses for blind adults as well, for Silvanov notes that the Banner carpet is "without doubt the safest and most exciting method known for blind people to indulge in uninhibited body movement without fear of banging anything or anyone," and to work up a good sweat. He also assures me that Ms. Banner has continued to invent in the years since I last contacted her (Silvanov, L-7/10/86).

Still another British inventor who has tackled the problems of the handicapped is **Doreen Kennedy-Way** (c. 1935– ) of the Midlands area near Birmingham. A graphic designer by training, Ms. Kennedy-Way was the only woman inventor mentioned in a six-month sample of newspaper clippings solicited by IPI in the 1970s. The sample contained 1100 clippings. She has been mentioned in an article on women inventors in a popular British women's magazine, and on a BBC television show on inventors and invention. She is probably one of the three or four best-known women inventors in the country.

Altogether, Doreen Kennedy-Way has more than thirty inventions to her credit, and runs her own company in Birmingham. The invention that brings her to this chapter is her battery-operated three-wheeled vehicle to give mobility to spina bifida children (G. Taylor, 35). This may be her MOGO 1, described in her Kennedy Kinetics Co. brochure as a "Small Battery Electric 3-wheeler, designed at the request of the British Association for the Handicapped, for use by physically handicapped children on pavements [or] sidewalks; Speed 4–6 m.p.h." (5).

A Swedish counterpart of Banner and Kennedy-Way is **Maj-Britt Renndal Bergsvik** (1928– ) of Linköping. Artist, designer, inventor, and entrepreneur, she is the owner of MB Design-Idé HB and MIBO Konsult in Linköping. Only child of "very talented, artistic and inventive" parents, she began woodcarving at age 3, and says that "drawing, painting and sculpture are a way of life." Combining artistic with technical and practical concerns, she studied at the Helsingborg Art School and the Academy of Fine Arts in Copenhagen, but also at the Fredriksberg Technical School in Copenhagen and the Oslo Municipal Technical School in Norway. After working for some years in architecture, building, and town planning in Helsingborg, Ms. Bergsvik now looks to the future. No one, she says, is planning adequately for the year 2000, when there will be 600 million people over sixty. She wants them to have housing, furniture, and other items that not only meet their physical needs, but look attractive and make them feel good. She also has a personal motivation: rheumatic pain, and a serious back injury suffered in the 1970s.

Her first invention in her new program, a specially designed furniture-and-activity-trolley combination with various attachable equipment, won second prize among 350 entries in the Lions International 1976 competition for better technical aids and medical-care equipment. She has sketches and design drawings for a stable Mini-MIBO trolley for children of pre-school age and the MIBO Transport. The latter is a practical aid for everyday use both indoors and outdoors. On this invention, her father—who at 80 suffers from heart trouble and aching joints—built the prototype and supervised the testing "in collaboration with experts within the family, including doctors." Her father, incidentally, manages life despite his medical problems, largely because of "our inventions."

A fourth invention is the MIBO Shower Cabin, making it easier to take a shower. Ms. Bergsvik claims for it "a new approach with a great improvement for every user," noting that she produced the design with the collaboration of an engineer colleague who was once Development Manager at General Electric (USA).

A fifth invention is the MIBO postbox. It is larger, more convenient, and safer, by an ingenious design, than the ordinary box. Ms. Bergsvik also mentions "a few everyday innovations dealing with modern domestic developments, greater mobility, and increasingly long journeys."

Here is a richly creative mind coupled with a caring spirit. Note, too, how well she fits the hypothetical inventor's profile—only child, encouraged in creativity and independence by parents who were themselves artistic and inventive; possessed of noteworthy artistic talent, educated at least partly in an all-girls' school, etc. Maj-Britt Renndal's inventions have already enriched the lives of the handicapped and elderly of Scandinavia, and we can expect further inventions and innovations from her in the future (Bergsvik, L-9/5/86).

At the high-tech end of things, in the field of bionics, is **Dr. Mildred Mitchell.** Working in aerospace research, since 1958 specializing in adapting electronic devices to perform human functions, she has invented, among other things, an electro-mechanical "muscle" that can substitute for human muscle in space emergencies, and a "nail-bender" that uses air to bend iron. She and her team have created substitutes for animals' internal clocks, and are working on touch-sensing systems. Although the original uses are in aerospace, Dr. Mitchell's inventions will later prove useful to the handicapped and/or elderly as well (Vare & Ptacek, 169–70).

A California woman, **Carole Carrothers,** has made a minor but ingenious contribution to the field: a water-jogger. Astride its floating saddle, a person can "jog," exercising leg muscles while the body's weight is supported by the water. The water-jogger can be useful not only to the handicapped, but to the aged or convalescent—and for children and healthy adults it can be used in water play. In the 1970s, Ms. Carrothers was a member of Inventors Workshop International (IWI) in the Los Angeles area (Weisberg, PC-8/16/76).

**Agnita Britzman**'s Head Hammock® can also be used by elderly or handicapped people for their comfort or proper support while sleeping or sitting in a wheelchair.

Britzman, a versatile multiple inventor and artist with many inventions to her credit, has also created the "Spillway" Bib, a slip-over (no ties to tie), disposable bib for the convenience of handicapped mothers of infants (Design patent, 1981), and "Shoe Buck," a shoe or sandal buckle that can be buckled with one hand (Weisberg, L-10/22/80; Britzman, Q-3/7/83).

Two appropriate inventions mentioned briefly to climax and close this section are the talking wheelchair developed by **Dr. Carol Simpson** and colleagues, and the listening wheelchair made possible by **Martine Kempf**'s Katalavox computer. Simpson's wheelchair, bearing a device called a "versatile, portable speech prosthesis," speaks for the 1.5 million people silenced by cerebral palsy, strokes, and other brain damage in the United States alone.

Dr. Simpson was a winner in the Johns Hopkins First National Search for Applications of Personal Computing to Aid the Handicapped, and will be discussed, along with several other female winners, in the Computers chapter. Martine Kempf's invention, as one of the most exciting computer inventions of recent years, will also be discussed there. Suffice it to say here that her invention has made possible the first-ever voice-controlled wheelchair.

*Mastectomy aids.* Not a handicap in the usual sense, unless significant muscle is removed, but extremely distressing to its victims, is the loss of a breast in cancer surgery. Total mastectomy has now become thoroughly controversial, yet in the 1970s doctors were still doing 90,000 a year. By one 1977 estimate, one out of 13 American women had had a mastectomy (Marmon, 80), and were left to cope as best they might with their feeling of mutilation and unloveliness in a culture suffering from a breast fetish.

In this situation a really good breast prosthesis is a significant invention. As early as the

1950s **Mildred A. Wright** and **Dora E. Gates** of Bevier, MO, patented a method of Forming artificial breasts (1951).

In the 1970s **Dolores Guggenmos,** a designer and sewing teacher from Downey, CA, took a self-help approach. Mastectomy victims in her sewing classes asked for her help in reshaping their bras and swimsuits. Grateful women recommended her to their fellow patients, and she found much of her time consumed in this endeavor. Finally, she designed a sample bra and kit so women could create their own new shape. Delta Medical Industries Corporation signed a licensing agreement for her kit and as of the mid-1970s planned to market it nationwide. Ms. Guggenmos, a wife and mother, is a member of IWI, and at last report had other inventions in progress (Weisberg, L-8/16/76).

Entering the breast-prosthesis field on a much larger scale about the same time was entrepreneur **Ruth Handler** (c. 1917–   ), co-founder of Mattel Toy Corporation, and developer of the Barbie Doll.[53] Spurred by necessity—she lost her left breast to cancer in 1970—and drawing on her successful business experience (Mattel became the largest U.S. toy manufacturer), she led a team of designers, chemists, and engineers to create the Nearly Me® prosthesis.

To a 1977 interviewer, Ruth Handler recalled the depressing experience of shopping for her first prosthesis. "I tried every breast on the market," she said. "They were globs. There were no lefts or rights. You wouldn't think of putting your right shoe on your left foot, would you?" (Marmon, 80). Finally, she paid $400 for a custom-designed breast from a leading designer of artificial limbs.

At loose ends after resigning from Mattel in 1975, she taught a few management courses at UCLA and USC, but then decided to pursue an idea simmering since that depressing shopping trip: a realistic prosthesis for "mastectomees." With the designer who had created her own custom prosthesis five years earlier, Peyton Massey, she formed the Ruthton Corporation. Handler and Massey hired a team of chemists and engineers, but also worked with doctors, and examined actual breasts of many different sizes and shapes. One doctor became so excited by Ms. Handler's idea that he opened his files so that she could call all his patients. She promised each one a free prosthesis for coming in.

The final product, the Nearly Me® artificial breast, is so life-like that Ruth Handler challenges people to squeeze both her breasts and see if they can tell which one is real. Nearly Me® consists of a sculptured foam core surrounded by sealed compartments of liquid silicone for natural softness and resiliency, and encased in a polyurethane outer skin. Very light weight, it comes in 30 sizes, left and right, from 30A to 42DD. In 1977 it sold for $98 to $130.

In a brilliant stroke, Ruth Handler chose a staff composed entirely of mastectomy survivors. Women shopping for breast prostheses are terrified, she says, of everything from death to rejection by husbands and lovers. Her instinct to hire saleswomen sensitive to these fears had already paid off handsomely by her 1977 interview. By the next year, she predicted, she would be selling Nearly Me® in 15 cities and would soon expand into specially designed bras and swimsuits. She herself felt free to wear a swimsuit on Malibu Beach in California—something she thought she would never do again.

In 1987, her invention still on the market, Handler herself came to I. Magnin in San Francisco and Oakland to meet customers during a sale. The sale ad, offering a $25.00 discount, called Nearly Me III® "the lightest and most comfortable prosthesis," and called Handler's nationwide success "lengendary" (S.F. *Chronicle,* Sept. 12, 1987: 3).

---

53. Ranked by one 1985 source as one of the ten best toys of all time (L. Goldberg, S-1).

*Diagnostic tests*

Since proper treatment demands accurate diagnosis, the importance of effective diagnostic and evaluative tests is paramount. Women have been devising such tests since very ancient times. A striking survival of this ancient wisdom can be seen among the granny-women of American Appalachia, who taste the feces of infants in order to find out what is wrong with them.

*Screening tests*

RIA. One of the most powerful research techniques, and at the same time one of the most powerful diagnostic tools, of the 20th century is radioimmunoassay (RIA). The Nobel Prize Committee called RIA "the most valuable advance in basic research directly applicable to clinical medicine made in the past two decades" (Levin, 135). As its name suggests, RIA uses radio-isotopes and the body's immune response (forming antibodies specific to a given antigen) for measurement purposes. It is, in fact, one of the most sensitive techniques ever developed for measuring substances such as hormones in human blood and tissues. RIA has allowed doctors for the first time to measure the amount of insulin circulating in a diabetic's blood, a feat comparable to detecting a teaspoonful of sugar in a lake 62 miles long, 62 miles wide, and 30 feet deep (E. Stone, 31).

Let us look more closely at how RIA works. To detect a hormone, for example, researchers first label a known sample of the hormone with a radio-isotope, then prepare a solution containing a known amount of the labeled hormone, along with its antibody, which join to form a hormone-antibody complex. When a blood sample containing an unknown amount of the natural form of the same hormone is added to the solution, part of the labeled hormone is displaced from the complex because the antibody molecules "prefer" the natural hormone. The amount of radioactive hormone displaced is proportional to the amount of the natural form added and thus, upon separating the hormone-antibody complex from the solution after a time, the displacement can be measured to reveal how much natural hormone was present in the sample.

RIA was the joint invention[54] of **Dr. Rosalyn Sussman Yalow** (1921–  ) and Solomon A. Berson, MD. Five years after Dr. Berson's death, Dr. Yalow received the 1977 Nobel Prize in Medicine for her role in devising the technique (the Nobel Prize is confined to the living). She was the second woman ever to win the prize in that field; the first was Gerty Cori in 1947.

Announced in 1959, RIA did not catch on at first. Yalow and Berson almost had to promote it in the early days. One of its first contributions was a better understanding of diabetes, revealing that adult diabetics do not lack insulin, but their insulin is somehow inactivated, and that diabetics eventually become immune to the bovine insulin they take.

Now doctors and researchers continually find new uses for RIA. Pediatricians can prevent one kind of mental retardation by detecting and treating an infant thyroid deficiency. The test uses but a single drop of the baby's blood, and costs only about $1. Blood banks now screen their blood with RIA to prevent transfusion hepatitis. RIA can also detect deficiencies or surpluses in human growth hormone early enough to prevent certain kinds of dwarfism and gigantism; can help explain high blood pressure and infertility; can detect hormone-secreting cancers and other endocrine-related disorders. It can detect heroin, methadone, and LSD in the bloodstream; it can gauge circulating

---

54. RIA is *not*, as Vare & Ptacek call it (130), a *discovery*, but an *invention*, consciously created over a period of years, based on a discovery of quantifiability of certain immune-system behavior. RIA, though not patented, would have been patentable.

vitamins and enzymes to shed light on human nutrition. It can make antibiotic treatment more precise—and even help catch criminals.[55] In the early 1980s RIA was used to diagnose Legionnaires' disease at an early stage, by detecting *Legionella* antigen in the urine ("Rapid . . . ," 358).

Had Yalow and Berson patented RIA, they could have been millionaires. Laboratories selling RIA kits do some $30 million in business each year. Thinking like scientists, however, instead of like entrepreneurs, the two freely published their work.

Rosalyn Yalow is a study in contradictions. She is a Nobel laureate who spent up to 100 hours a week in the laboratory; yet she is a wife and mother of two who does all the shopping and cooking, and keeps a kosher kitchen for her physicist husband. In her office at the Veterans Administration Hospital in the Bronx, she displays a sign reading, "Whatever women do, they must do twice as well as men to be thought half as good. Luckily this is not hard." Yet in her professional partnership with Solomon Berson she often deferred to him, making lunch in the laboratory, making the plane reservations, letting him write most of their joint papers and many of her speeches. As a child she played with dolls and always intended to marry; yet by age eight she knew she wanted to be a scientist. After reading Marie Curie's biography at age 17, she knew just what kind of scientist she wanted to be.

And nothing has stopped her since—not even being refused a graduate assistantship at Purdue because she was a New Yorker, a Jew, and a woman; or being told at Columbia that she must start as secretary to a medical school professor. In February 1941 she entered the University of Illinois on an assistantship, and took her PhD there, studying under Prof. Maurice Goldhaber among others. After working briefly as an electrical engineer at Federal Telecommunications Laboratory in New York, she returned to Hunter College to teach physics. It was then, partly through her husband Aaron, who worked in medical physics, that she learned about the medical uses of radio-isotopes.

**Irène** and Frédéric **Joliot-Curie** had laid the foundation for nuclear medicine with their Nobel Prize–winning work in artificial radioactivity. The field took on a new dimension when Georg von Hevesy showed that radioactive phosphorus, fed to rats, could be used to trace their metabolic processes; and it became economically practical when Enrico Fermi built a nuclear reactor to generate isotopes inexpensively. Then, with World War II over, radio-isotopes became commercially available, and nuclear medicine was on its way. Dr. Rosalyn Yalow decided to go with it. "Nuclear medicine," she says, "was another good choice. In physics, I'd be one of a large group, but here my talents were very much required." In 1947 she joined the staff of the Bronx Veterans Administration Hospital, where she still worked at the time of her Nobel award, and soon began collaborating with Solomon Berson. Their early work applied radio-isotopes to disease diagnosis and to iodine metabolism. In that early work *she and Berson sometimes had to design, build, and calibrate instruments, because what they needed simply did not exist.* Some of these inventions might also have been patentable.

Later, they became interested in the smaller peptides, especially hormones. Insulin, as the most easily available, became their focus. Their work on resistance to insulin led directly to their development of radioimmunoassay. From then until Berson's death, both scientists were intensely occupied with applications of RIA clinical medicine and medical research.

Dr. Yalow's public deference to Berson deceived some colleagues into considering her

55. Part of the evidence against Dr. X, Mario E. Jascalevich, was curare detected by RIA in the victims' exhumed bodies (E. Stone, 100).

a less-than-equal partner in their 22-year collaboration. However, since he left the VA Hospital in 1971, Rosalyn Yalow has been more productive than before. During the next five years she produced 60 publications, and between 1972 and 1976 won a dozen medical awards in her own right. In 1976 she became the first woman ever to win the Albert Lasker Award for basic medical research (E. Stone, 103–4).[56]

Today she continues her work in nuclear medicine. She also shows a strong social consciousness—in 1978 she spent a month helping Indian researchers develop radioimmunoassays for infectious diseases—and a growing feminist consciousness as well. In her Nobel acceptance speech she said,

> We still live in a world in which a significant fraction of people, including women, believe that a woman belongs and wants to belong exclusively in the home; that a woman should not aspire to achieve more than her male counterparts, and particularly not more than her husband. . . . But if women are to start moving toward their goal, we must believe in ourselves, or no one else will believe in us; we must match our aspirations with the competence, courage and determination to succeed, and we must feel a personal responsibility to ease the path for those who come afterward.

Dr. Yalow now holds the chair of Distinguished Professor of Medicine at Mt. Sinai School of Medicine in New York. She has come a long way from the South Bronx where she was born, and has years of creative work ahead of her (Levin, 133–7; Opfell, 224–33; E. Stone, 29–34ff; Yalow, APS/APT speech, 1979).

RIA is one of those foundational inventions that inspire further inventions for decades to come. An example of the many spinoffs is **Mary M. Garrison**'s 1974 patent for Preparation of dry antiserum-coated solid-phase for radioimmunoassay of antigens. Garrison, a resident of Silver Spring, MD, and her male co-patentee assigned their patent to the Department of Health, Education and Welfare.

Useful, like RIA, for screening donor blood are procedures for detecting cytomegalovirus based on the work of **Dr. Anne Yeager** and colleagues at Stanford Medical School. This virus killed approximately one premature infant in 12 before the test was available.

*Fluorescent antibody-staining technique.* Bacteriologist **Dr. Cora Downs** of Kansas is perhaps best known for her work on tularemia, an infectious disease transmitted by insect bites or by handling infected animals, such as rabbits. At the University of Kansas, Dr. Downs began her study of tularemia with the case of a rabbit hunter who had contracted the disease. After her 1930 report she went on to study tularemia in depth, contributing greatly to scientific understanding of the nature, immunology, epidemiology, and causative organism of the disease.

During the 1950s and '60s she studied rickets. Then she turned, with others, to the work that brings her to this chapter: perfecting the fluorescent antibody-staining technique that became a useful tool in both diagnosis and research. Made a full professor of bacteriology at the University of Kansas as early as 1936, Dr. Downs was named Summerfield Distinguished Professor of Bacteriology in 1972. This was her favorite honor, because the chair's founder had said no woman would ever receive it (*WBWRA*).

### Infectious disease

Noteworthy here is the Dick test for susceptibility to scarlet fever, developed by **Dr. Gladys Dick** and her husband in the 1920s. Scarlet fever, virtually conquered in the

---

56. Other women had won Lasker awards, but for clinical work.

United States, thanks to the Dick anti-toxin, still rages in much of the world. Since the disease can leave children with significant damage, diagnosing it for treatment is still important.

Because of the devastating effects of untreated syphilis, and the numbers of its victims, a diagnostic test for the disease was urgently needed. Nor did the search stop with the announcement of Wassermann's test in 1906. By the 1920s the more reliable complement-fixation test became available. However, testing procedures were nonstandard and still fairly crude. At the New York State Public Health Department Laboratories in Albany, **Elizabeth J. Maltaner** worked with her husband Frank to standardize and improve the reliability of this test. **Rachel F. Brown,** who had joined the Department about 1925, worked briefly with the Maltaners on this project, but soon turned the work over to **Mary C. Pangborn,** a PhD chemist from Yale. One problem was that the researchers had not identified, and thus could not fully control, the active principle in the antigens used to stimulate the antibodies read in the test. Dr. Pangborn eventually isolated this active substance, which she called cardiolipin, making possible a standardized antigen and thus a vastly more reliable test (Richard Baldwin, 54–7).

The complement-fixation test, even with the Maltaners' improvements, was still complex, lengthy, and expensive. Rachel Brown later used cardiolipin in her invention of the far simpler precipitation test for syphilis. This test was not a total substitute for complement-fixation, but its conclusive negative results left only about 10% of the samples to receive the more complex and costly test to eliminate false positives. A graphic illustration of the superior speed of Brown's new test came at a serology conference in Washington, DC, when both procedures were demonstrated. Brown, working alone, finished all her tests by early afternoon and went out sightseeing. The three-member complement-fixation team, with identical samples to check, worked late into the night (Richard Baldwin, 57).

Coming to my attention just as the book goes to press is the work of **Anna Wessels Williams** (1863–1954) and of **Rhoda Benham** (1894–1957) in the diagnosis of rabies and fungal diseases respectively. Williams shortened the diagnostic time of rabies from weeks to hours. Benham developed cornmeal agar for rapid growth of chlamydospores, which became standard diagnostic procedure in yeast infections. She also pioneered in the use of taxonomic, metabolic, and serological techniques for diagnosis. Appropriately enough, she developed the first course in medical mycology in the United States (O'Hern, 36–7, 87ff).

*Hereditary diseases/conditions*

A vital area of medical diagnosis is detecting hereditary or congenital diseases, conditions, or deficiencies in infants. Some of these are treatable if caught at birth, but will otherwise lead to mental retardation, handicap, or even death. **Rosalyn Yalow** and RIA have already been mentioned in this regard. Other women scientists have achieved in this field as well.

A simple but effective test now applied to virtually all newborn infants in the developed world is the Apgar Score System, a series of reflex and other checks to establish the basic normality of an infant. Named for its formulator, **Dr. Virginia Apgar** (1909–74), the test measures *A*ppearance, *P*ulse, *G*rimace, *A*ctivity, and *R*espiration—or overall color, heart rate, reflexes, muscle tone, and breathing—to yield an APGAR score indicating which babies will need special medical attention in the crucial first minutes of life. The test results can also reveal certain long-term problems.

Dr. Apgar had originally intended to be a surgeon. After a stellar career at Mount

Holyoke—where she majored in zoology and minored in chemistry, worked to support herself, played on seven varsity sports teams, wrote for the newspaper, acted in plays, and played violin in the orchestra—she entered Columbia's College of Physicians and Surgeons in 1929. Receiving her MD in 1933, she won a prestigious surgical internship at the Columbia Presbyterian Medical Center. After two years, however, convinced that she could never make a living as a surgeon, she moved to the new field of anesthesiology. In 1938 she became head of the Anesthesiology Division at Columbia Presbyterian, the first woman department head there. Ten years later, she was appointed Professor of Anesthesiology, making her also the first woman to hold a full professorship at the medical school.

Dr. Apgar's best-known work occurred, however, after 1949, when she left administrative responsibilities and began to study childbirth anesthesia. Her years of clinical observations of newborns gave her the basis for her now-famous Apgar Score System, which she announced in 1952. Although unmarried ("I've never found a man who can cook") and childless, Dr. Apgar was sensitive to and involved with children. She once carried a child patient up nine flights of stairs because he was afraid of elevators.

Dr. Apgar won many honors late in life, including several honorary degrees. She won the Ralph M. Waters Award of the American Society of Anesthesiology, became the first woman to receive the Gold Medal for Distinguished Achievement in Medicine from the alumni of the Columbia College of Physicians and Surgeons, and was named Woman of the Year in Science and Research by the *Ladies' Home Journal.* She herself considered her newborn-scoring system her most worthwhile accomplishment (*NAW*/mod.; *WBWRA,* 209; Diamonstein, 15).

As of 1980 **Dr. Alaine Berman** was working at Hadassah–Hebrew University Medical Center on a simple diagnostic test to screen for hereditary childhood diseases. Dr. Berman has a background of years of research in these diseases (Levin, 38–41).

Biochemist **Dr. Sara Ratner**'s classic studies of the biochemistry of amino acids and protein metabolism led to tests for metabolic defects in human infants. As mentioned just above, these defects if uncorrected could lead to severe mental retardation and other handicaps or even death. As of the mid-1970s a member of the Public Health Research Institute of New York, she had received such honors as the Garvan (1961) and Neuberg medals, and in 1974 was elected to the National Academy of Sciences (*WBWRA,* 145, 164, 224).

**Dr. Elizabeth F. Neufeld** received a 1982 Lasker Award for clinical research. Specifically, she and Roscoe O. Brady, both at NIH, were cited for their contributions to the diagnosis, potential treatment, and screening of lipid-storage diseases and mucopolysaccharide-storage disorders which cause severe mental retardation and other handicaps. The Lasker Awards are the highest United States honor for medical research, and many Lasker winners go on to Nobel Prizes ("Lasker Awards").

In April 1981, **Lois Bleicher Wilson** (1947–  ) and John Wilson of the Medical College of Georgia in Augusta announced a new sickle-cell[57] diagnostic test, using a DNA probe and an enzyme that will cut the normal hemoglobin gene but not the mutant one. Finding the test 100% accurate in adults, the Wilsons then adapted it to create a successful prenatal test. This involved reducing the amount of DNA needed for the test so

---

57. Sickle-cell anemia is a severe, hereditary blood disease caused by the tendency of a variant form of red blood cells to assume a "sickle" shape and thus clog tiny blood vessels. Researchers have long sought to improve diagnostic tests for this painful and sometimes lethal condition. Its severe, homozygous form (inherited from both parents) strikes 2 out of every 1,000 blacks born in the United States (NSF, PR-4/27/82). It has been reported from widely separated parts of Africa, from India, the Mediterranean islands, Greece, and Turkey.

that it could be done on fetal cells harvested from the amniotic fluid, rather than on a blood sample. Lois Wilson, a microbiologist, and her biochemist husband joined forces in their post-doctoral work at Yale. Soon thereafter, they cloned the first human gene—for hemoglobin. Developing the prenatal test for sickle-cell anemia seemed a logical extension of that work, but required two and a half more years of research.

Lois Wilson is a firstborn child (one brother) in a home where college and a professional career for her were givens. Realizing that jobs were scarce in history and logic, her first loves, she entered science instead. She took her PhD in molecular biology at the University of Arizona in 1974. As of 1981 she was Assistant Professor of Cell and Molecular Biology at the Medical College of Georgia, Augusta, and had already published fifteen scientific papers. The grant supporting the Wilsons' research was a grant to Lois Wilson from the National Science Foundation ("Diagnosis," 12; Lois Wilson, c.v.; L-9/8/81; Geeyer *et al.*).

Dr. Yuet Wai Kan of the University of California medical center has also created a successful prenatal test that needs no fetal blood sample. At first, this test recognized only the West African sickle-cell variant. In January 1982, however, Kan and research associate **Judy Chang** created a universal test. They announced their finding at the annual meeting of the American Association for the Advancement of Science (D. Perlman, "S.F. Researcher").

## Cancer

One of the most welcome diagnostic tests in all of medicine would be a perfectly reliable early cancer test. At least four women working in this field have recently announced progress, and at least one of them has received a patent.

**Sally D. Bolmer,** working as a graduate student with Eugene A. Davidson at Pennsylvania State University in the 1970s, developed a blood test for cancer. In patients with known malignancies it proved 96% accurate, with only 2% false positives. The test works as follows: the researchers isolated a glycoprotein given off only by cancer cells. If an antibody to this protein is added to a blood sample, the antibody attaches to the protein and reveals its presence. Since the amount of glycoprotein found in the blood correlates with the amount of cancerous tissue present, Bolmer and Davidson, who patented their test in 1979, believe it will reveal cancer before any malignancy is suspected.

But perhaps the most exciting thing about their test—which was mentioned by Stacy Jones in the *New York Times*—is that it appears to detect most cancers, including carcinomas, melanomas, sarcomas, and Hodgkin's disease. By contrast, *the sixteen previous diagnostic tests for cancer were all specific to a particular kind of cancer.* Bolmer and Davidson, like Elizabeth Hazen and Rachel Brown before them, assigned their patent to the New York Research Corporation, which in turn has licensed it to Warner-Lambert for development and marketing (S. Jones, Mar. 31, 1979: 30; "New Cancer . . . Test," 37).

As early as 1971 Israeli immunologist **Chloe Tal** reported detecting a protein she called "T-globulin" in the blood of cancer victims—and also in the blood of pregnant women—but not in the blood of patients with various other diseases. Dr. Tal examined the blood of 520 hospital patients whose diagnosis was unknown to her. She found T-globulin in 356 of the samples. Of those patients, 350 turned out to have proven cancers, 3 were suspected of having cancer, and 3 were pregnant (Levin, 38–41).

**Dr. Elizabeth Stern** (1916–80), a pathologist at the University of California's School of Public Health, did a monumental study of 11,000 women in the Los Angeles area. On the basis of this study, she devised a series of 250 distinct degrees of cell progression in the cervix, from normal cells through advanced cancer. This series provided the basis for a screening test for cervical cancer, one of the most frequent types of cancer among women. Dr. Stern was also the first to discover that long-term use of birth-control pills could cause cervical cancer. She died at 64 of the very disease she had battled so effectively in her research ("Lost Women," *Ms.,* Dec. 1980: 44 ; Jane Brody).

**Dr. Lea Cercek** of Manchester, England, with her husband Boris developed a simple and inexpensive blood test for cancer in the 1970s. The Cercek test not only signals the presence of very early cancer—when it is only about a millimeter in diameter and undetectable by X-ray scans—but reveals where the cancer is located.

Dr. Lea Cercek, an eminent radiobiologist, reports accuracies of 97% in experiments with nearly 700 patients. The Cerceks' results have been tested and confirmed in medical and university laboratories at Velindre Hospital in Cardiff, Wales; in Germany, and in Japan. Success rates in some of these series were 100%. The Cercek test, developed at the Patterson Research Institute in Manchester under a British Government grant, works by mixing lymphocytes isolated from a subject's blood with protein tissue from known cancer victims. As the mixture is examined under special instruments, using polarized light, the tester sees the lymphocytes stimulated if and only if the sample subject has cancer. Moreover, the lymphocytes react most strongly to protein from the type of cancer the subject has. That is, if the subject's lymphocytes react most strongly to protein from a lung-cancer victim, that subject probably has lung cancer.

Another use of the Cercek test is to tell doctors whether they have succeeded in eradicating cancers by surgery. As early as 14 days after surgery the test can reliably be given; if it shows negative, the cancer is gone. If it shows positive, however, both doctor and patient have an early warning that more treatment will be needed.

The appeal of blood tests for cancer is that they avoid the hazards of X-rays and exploratory surgery, as well as the pain of biopsies and other older diagnostic procedures. They are also usually much cheaper. The Cercek test, for example, cost only about $5–6 in 1976, and was projected to cost even less when perfected (Cercek & Cercek articles; Lewin; Parmiter).

In 1976 the Cerceks were working to automate the test, making possible a mass screening program. With the necessary funding, said Dr. Lea Cercek in an interview at that time, the test could be ready for widespread use in six months to a year.[58]

*Miscellaneous tests*

Children receiving radiotherapy for Hodgkin's disease or other cancers often become hypothyroid. In the early 1980s, using RIA—which makes such sensitive measurements possible—Dr. Joseph Kriss and **Monika Bayer,** PhD, of Stanford University's Division of Nuclear Medicine developed tests for this condition. With prompt diagnosis the children can receive artificial thyroid hormone, which quickly restores normal functioning. Sometimes the condition is reversible.

In another project, Drs. Kriss and Bayer found that routine thyroid testing failed to measure thyroid levels accurately enough in the very ill, but instead gave uniformly subnormal readings. They devised a new test that measures only the active fraction (3 parts in 10,000) of the hormone. This is a far more accurate measure of function, and thus clinically extremely valuable. Monika Bayer is a biochemist ("Two Studies").

58. Lack of further reports on the Cercek test could mean that this funding was not forthcoming, that it is used in Britain but not approved in the U.S., or that the test proved less successful in practice than in the early tests.

**Ethel Florey, MD,** appears elsewhere in this chapter because of her prominent role in the development of penicillin. She also collaborated with H.W. Florey and Albert Szent-Gyorgyi on methods of testing for the presence of adrenal cortex hormone (Bickel, 39).

## Ultrasound

The use of ultrasound to visualize organs and tissues for diagnostic and research purposes is becoming commonplace. Recent improvements use computers to record the ultrasound waves digitally. At least one researcher reports significant enhancement in his analyses of diffuse liver and spleen disease. This medical application of the computer, discussed more fully in Chapter 5, began at Stanford University with the work of electrical engineer **Lynda Joynt,** PhD, and a colleague ("Two Studies," 6).

## Vision problems

In 1981 two Harvard neurobiologists proved that in order to function normally, the brain's visual cortex must be stimulated in early childhood (*SN* 120: 16 [Oct. 17, 1981]: 244). Although this finding probably surprised no one, it emphasizes the importance of screening tests such as the Blackbird Vision Screening System devised and copyrighted by **Kiyo Sato-Viacrucis** (1923–   ) of Sacramento, CA. A survivor of the tragic imprisonment of Japanese-Americans during World War II, Ms. Viacrucis tries to live without bitterness, but neither does she feel that what happened should simply be buried. The wounds must be examined, cleansed, so that healing can take place. Reparations, which she thought might confer awareness and a sense of historical responsibility on all Americans, finally came in 1988.

Kiyo Viacrucis is a highly popular nurse in the Sacramento City Unified School District. She is also a crusader for small family farms, a mother of four, a world traveler, and a former teacher of English. During years of administering the traditional vision-screening test, the "Snellen E", to very young children, she had ample opportunity to observe its defects. The most obvious defect was that since it required understanding of the alphabet, it worked for only about half the three- to four-year olds tested. Most children were not screened until age five or six, too late for the correction of amblyopia, or "lazy eye." This easily correctable condition causes an estimated 100,000 children to lose sight in one eye every year. Some 5,000,000 Americans are probably affected.

Frustrated, like many public health nurses before her, with trying to describe the "E" of the traditional tests to nonreading children, Kiyo Viacrucis was using an old trick of representing it as a three-legged table. Suddenly she got the idea of clipping the outer two legs into "wings." A four-year-old boy decided it was a bird, she asked the children to name it, and the Blackbird E was born. This was in the 1970s.

With Ms. Viacrucis' invention, children need not read or even recognize individual letters; they merely tell which way the blackbird is flying. She has created a simple storybook eye test that parents can administer at home, with step-by-step directions and a pair of "screening specs" (cardboard glasses with flip-out lenses) to prevent peeking during the test (copyright 1975). With this ingenious invention, the tester no longer has to hold a card over one of the child's eyes while trying to record responses from the other eye. Her Blackbird Vision Screening System provides charts and kits for use in clinics, schools, doctors' offices, and health departments. The system is particularly effective for very young, autistic, retarded, or non-English-speaking children and adults.

By 1981 it had already been used to screen more than a thousand preschoolers in

Sacramento, and had detected myopia, astigmatism, color-blindness, eye injuries, and hemophilia, in addition to lazy eye. After adopting the copyrighted kit, Sacramento schools reported the nation's highest response rate in early vision-testing. The kit is also being used in Samoa, Canada, and Texas (Department of Mental Retardation), and Ms. Sato-Viacrucis reports inquiries beginning to arrive from Indian reservations. Dr. Gerald Portney, late Chief of Ophthalmology at the University of California Medical Center in Sacramento, called the Blackbird System "a great service to mankind" (Jerry Cox; Rosner, 480–1; Viacrucis, I-8/25/81; L-9/8/81).

Estonian-born **Dr. Ingeborg Schmidt** (1899–   ) wrote her MD thesis on binocular color vision at the University of Tübingen, Germany, and stayed on to work there after her degree in 1927. Forced out by the Nazis, she worked as a color-vision specialist for the Institute of Aviation Medicine at the Air Ministry in Berlin. After the war, she was hired in the same specialty by the School of Aviation Medicine at Randolph Field in Texas. In the 1950s she did research in ophthalmology at Columbia University and at Indiana University, where she also taught. Retired in the late 1970s, she continued to research and write. She had published over 80 papers on color vision, night vision, and astronautics, and contributed to four books on optics. Her most famous paper discussed a diagnostic sign for a color-vision defect—later named Schmidt's sign (Levin, 43–4).

## Psychological tests

**Grace Helen Kent** (d. 1974) deserves at least a passing mention here. Neither her excellent early research nor her outstanding clinical work from the 1920s onward brought her any professional recognition, as her biographer sadly notes. However, she has to her credit, in addition to years of inspiring work at two Massachusetts State Hospitals (Worcester and Danbury), the Kent-Rosanoff Association Test (1907–10; Rossiter, 246–7, 382).

## Humanizing medicine (social-structural inventions/ innovations)

It has become almost a truism that as male doctors consolidated their hold on the medical profession in Western society during the 20th century, they became more concerned with medicine than with health, an unfortunate dichotomy reflected in the title of this chapter. Doctors have also been accused of caring more about technology and technique—not to say money—than about their patients' life-quality and feelings. The accusations have enough merit to bring reformers, often women, forward with social or structural inno- vations and inventions designed to redress the balance. In 1975 for example, **Mary C. Howell** outlined her vision of a women's health school dedicated to training women physicians to give health care in an atmosphere of mutual support, and reaffirming the values of service and respect of patients and co-workers alike (Chaff, 425).

These women's innovations and inventions, of which only a few examples can be given here, have repeatedly humanized or rehumanized medicine. Though not often ranked as such, they are among the most important advances in modern medicine.

Birth and death are the two great and unavoidable passages of life. In primitive cultures, women were usually in charge of both. Attended with suitable rituals and taboos, both were made as comfortable and as supported as possible. So much has been written about the mechanized and dehumanized horrors of high-tech hospital births and deaths that little more need be said here except to note the specific achievements of women in rehumanizing both events.

*Birth*

**Dr. Roberta Ballard** of San Francisco is credited with creating the first in-hospital birth center. In such centers the birthing woman can have her baby in more natural and comfortable surroundings, and she can direct the process; yet emergency equipment is near at hand.

Women began in the 1960s and '70s to rebel against the appalling coldness and indignity of hospital high-tech births. They sought obstetricians who would attend them in natural childbirth, or turned to midwives who would come to their homes. They also began once more to train as midwives, and these modern midwives found some obstetricians willing to co-operate with them in emergencies.

Growing out of this rebellion are the free-standing birth centers, so called because they are not physically attached to a hospital. Women come to these centers for sexual and reproductive counseling, artificial insemination, and other health services, as well as for prenatal care and childbirth. The births are attended by licensed midwives backed up by obstetrical nurses and physicians with local hospital privileges, both on call.

Many women's names doubtless belong here, since among the creators of the nonhospital or woman-directed birth movement, at least in the United States, were the Women's Self-Help Movement and the consumer health movement—the one entirely and the other largely pioneered by women.[59] (In Europe, where midwifery survived more strongly, free-standing birth centers had existed for decades. In England and Wales, for example, in 1969, such centers accounted for 12% of all recorded births; Eakins & Richwald, 16). As early as 1972 a Texas center offered nurse-midwifery to the rural poor. A California center, now closed, was operating by 1962, making it one of the first in the United States (Eakins, 50; Eakins & Richwald, 22).

At least one woman birth-center pioneer has come to national prominence; **Suzanne Arms** (1944– ), author of *Immaculate Deception* (1975). In her book, Ms. Arms weighed the American way of birth in the balance, found it wanting, and presented her own vision for a free-standing birth center: a vocal parents' group gets a bond issue passed, raises money to buy a solid older home downtown, forms a nonprofit corporation, refurbishes the home and staffs and equips it for births (294–5):

> An obstetrical nurse [is] hired to supervise the home. . . . The nurse handles all the business of clean and sterile linens and instruments, all on rotating loan from the city hospital twenty minutes away. . . . The home includes a small emergency room with a resuscitator, an incubator, oxygen, and a small supply of hormones and IVs. The nurse-administrator could help in a pinch, or until one of the on-call obstetricians or the ambulance arrived. . . .

Significantly, the emergency equipment stays out of sight, and the three large birthing rooms remind the women of home.

Moving beyond vision to action, Suzanne Arms began to organize in her own community of Menlo Park, CA. When she and six other women of the Birth Action Group called a public meeting in 1977, sixty people came. Frustrated in their plan to establish an alternative birth center in a local hospital, they raised money, formed the nonprofit Birth Place, Inc., and opened as a Resource Center in 1978. By 1980 the Center had 500 paying members, and handled more than 5,000 informational visits and phone calls on "alternative" childbirth and early parenting each year. By 1979 the Birth Place became a state-licensed and fully equipped facility for unmedicated childbirth. Arms' vision, minus

---

59. Some sources credit Grantly Dick Read's *Childbirth Without Fear* (1959) as well as Ferdinand Lamaze and Robert Bradley (1950s), with inspiring or giving focus to the movement.

a few features, had become reality. By 1984 over a hundred babies were born there each year, and the numbers were increasing.

In 1984 Pamela Eakins of the Institute for Research on Women and Gender at Stanford University assessed the "medical outcomes" of Birth Center births for a three-year period (57ff). Of the 251 women in labor admitted to the Center, 200, or 80%, gave birth there; only 51 were transferred to the hospital because of complications. Obstetricians attended 94 of the deliveries, nurse-midwives 85, and family physicians 72. Physicians assisted midwives on six occasions.

More than half the women who completed their labor at the Center gave birth without any medication at all. The most common form of medication used was xylocaine applied to the perineum for suturing after birth (just over one-third of cases). Most births, however (182, or 91%) took place without episiotomies, and an intact perineum was preserved in 26% of the cases, so that no stitches were needed. Pitocin was used, not to induce or speed labor, but only to ease delivery of the placenta or to control postpartum bleeding. Says Eakins, "There was no major maternal or neonatal morbidity or mortality at The Birth Place during its first three years of operation" (57).

The overwhelming majority of the 200 babies born at the Center (186 babies, or 93%) had high Apgar scores (7–10 on a scale of 10) by one minute after birth; by five minutes, 198 of the 200, or 99%, scored in that range. They also had good birth weights, ranging from 5.5 to 10.7 pounds. Only nine babies had significant complications; these included tight nuchal cords, improper heart beat, and respiratory distress. Two babies transferred to the hospital for respiratory distress stabilized on the way there; the total elapsed time from transfer decision till arrival at the hospital was 11 minutes and 13 minutes, respectively.

In all these categories, the Menlo Park Birth Place compared quite favorably with local and national hospitals. Its rate of Caesarean sections, for example, was only 2%. Although accounted for in part by pre-selection for low-risk births, this is still an excellent record. And between 1979 and 1982 its fees were about half the average local hospital fee for a birth. By 1983 its fees were more comparable with hospital fees, but clients could barter work at the center for part of their fees.

By 1984 there were more than a hundred free-standing birth centers in the United States. In a nationwide study of these centers, the newborn death rate was only 4–6 per 1000, compared with the national average of 13 per 1000 (Eakins, 62).

In striking conformance to the health-medicine dichotomy, by 1983 the American Public Health Association had adopted guidelines for birth-center deliveries, whereas the American Academy of Pediatrics and the American College of Obstetricians and Gynecologists said, "Until scientific studies are available to evaluate safety and outcome in free-standing centers, the use of such centers cannot be encouraged"—except, they added, in isolated geographical areas (Eakins, 51).

One of the more exciting results of women's stand against techno-birth is that many hospitals have transformed their obstetric services. Husbands, mothers, or other chosen partners now stay in the labor and delivery rooms, mothers keep their babies beside them, and birth suites look more like homes. Unmedicated childbirth is allowed, often encouraged. Indeed, by 1986 half of all hospitals had alternative birth centers. Obstetrics still has a long way to go, of course. Caesarean sections and episiotomies are still far too frequent; fetal monitors, drugs, and forceps are overused, and the anti-gravity birth position is still prevalent. The revolution will not be complete until birth is no longer a medical event, but once again a woman-directed, joyous life event. One step in that direction would be putting all routine births back into the hands of midwives knowledge-able in their foremothers' recently rediscovered herbal, psychological, musical, and physical methods for easing birth. Ironically, a recent impetus in this direction has come neither from the midwives nor from the women's movement, but from the insurance companies, whose sky-high malpractice insurance rates have forced many physicians to abandon the obstetrical part of their practice. Unfortunately, this same crisis has now begun to affect the free-standing birth centers.

## Death

Death, like birth, has become a technician's dream and a human nightmare. Reactions to the horror of dying in a glaring white room surrounded by strangers—or totally alone—and hooked to tubes and monitors, have at last begun to have their effect, ranging from the innovation of the Living Will to the development of hospices. One of the most prominent names in the hospice movement is that of **Dr. Cicely Saunders,** who founded St. Christopher's Hospice in London in 1967. A hospice is a free-standing facility or a hospital section with arrangements for home care for the terminally ill. There, no longer lied to by professionals, the dying can at last face the truth about their condition and begin to prepare for death. They receive pain medication and other help and nursing care as needed, but are otherwise left in peace, to pursue whatever thoughts and activities they wish, and receive care and comfort from their families.

England already had 25 hospices by the mid-1970s, but the situation in the United States, where more than 250,000 people need such care, has proven more difficult. Blue Cross and Medicaid originally balked at paying for it. Hospices were not accredited by the Hospital Accreditation Board and thus were not eligible for public funds. Licensing regulations have also proved a problem. If these problems can be overcome, no one need suffer techno-death unwillingly. As Dr. Josephine Magno, a cancer specialist at Georgetown Medical Center, puts it, the hospice alternative allows the physician to say, "Okay, we cannot cure you any more, but we can start caring for you" (*WBWRA,* 229; "Hospice . . .").

Appropriate optional furniture for hospice rooms might be the above-mentioned Shaker cradles for the dying. Since they make explicit the childlike dependency of illness and old age, it would be interesting to see how many of the residents requested one.

## Medical care

Fears of giving birth and dying are as old as the human race. But the nightmare of being denied medical treatment for lack of money is only as old as the professionalization of medicine; and the atrocity of being turned away from an emergency room for lack of medical insurance is newer still. The socialist answer to this problem is still resisted in the United States, and is simply not possible in many other countries. However, even in the United States, the elderly and the very poor can receive free or subsidized health care; and most corporations of any size have some health-insurance for their workers.

All this is well known. What is not generally known is that the first prepaid health plan in North America—perhaps in the modern world—was devised by a Frenchwoman in Canada. **Jeanne Mance** (c. 1607–73) was born in Langres, France. She took vows of poverty and chastity at seven, and devoted her life to a healing ministry. When she was thirty, invaders from Lorraine ravaged her town in Alsace, killing more than 5,000 people. Her own home was demolished. Jeanne Mance got her first taste of nursing at that time, living in the streets and tearing up her clothing for bandages. Inspired by her priest to go to Canada, she at first hesitated because of frail health, but eventually went to found a hospital in the wilderness. Queen Anne became aware of the project, and persuaded a wealthy widow to fund the venture, which was to be part of a new colony called Ville Marie.

Despite attempts by the Quebec colony to stop it, the founders departed, arriving in May 1642. The first hospital, Hôtel Dieu, was a log building; Mance's first patient, a colonist scalped by the Iroquois. This patient, like all her others, was treated "with the motherly tenderness for which she became renowned."

As late as 1653 the Ville Marie colonists were still in constant danger from hostile Indians as well as from famine, since men feared to till the soil far from their homes. Such conditions make it all the more amazing that in 1655 Jeanne Mance's Hôtel Dieu started its unique medical prepayment plan. Twenty-six families enrolled, gaining the right to have their wounds dressed and drugs prescribed for them for 30 sous a year.

Jeanne Mance died not long after laying one of the foundation stones for the Church of Notre Dame. One of the church's stained glass windows portrays her and her beloved hospital. Her hospital continued to provide free medical aid for many until 1903, and Mance herself was commemorated as late as 1909 by a bronze statue erected on the hospital grounds. As Jean Bannerman rightly says in *Leading Ladies of Canada,* "No priest, soldier, statesman, or explorer had left a more heroic record than 'The Mother of Lay Nursing in Canada,' who many feel was the true founder of Montreal" (26, 29–30).

Interestingly enough, self-employed women of modern India have designed a health system that makes the best medical care available even to the poorest and most isolated persons (Sheth, PC, 7/85).

Like obstetrics, gynecology has come in for much-needed criticism and reform as a result of the American women's movement reborn during the 1960s and '70s. **Carol Downer** of Los Angeles founded the first women's self-help clinic in 1971, the Feminist Women's Health Center, now located in Santa Ana. Though Downer and **Lorraine Rothman,** a member of the original group and inventor of self-help apparatus, focused more or less exclusively on reproductive health, they should probably be considered the "inventors" of the Women's Self-Help Movement as well as part of a reforming and innovative force in American health care.

The Center was raided in 1972 through the efforts of opponents, including a female undercover police officer and the city attorney. Both Downer and Wilson, a group member, were indicted for practicing medicine without a license. Wilson pleaded guilty to one of eleven counts against her, and the other ten were dropped. Downer, represented by attorneys Jeannette Christy and Diane Wayne, chose to fight, contending that in diagnosing a yeast infection and recommending a home remedy like yogurt, she was in effect doing no more than any mother might do for her daughter at home in her kitchen, and that restricting this time-honored right would be a dangerous intrusion into family rights. She won. But of course this was not the last attack. Early reactions from physicians ranged from backlash—exaggerated reports of the dangers of such procedures as menstrual extraction, for example—to co-optation. In both cases, says Ellen Frankfort in *Vaginal Politics,* doctors "were more upset at the independence that [menstrual] extraction in particular and self-help in general gave women . . . than at the dangers of either . . ." (Rothman & Punnett, 52).

Other important pioneers in the self-help movement for women are a mother-daughter team from Connecticut, **Lolly** and **Jeanne Hirsch.**

The Women's Health Care Movement is alive and well in cities as large as Los Angeles and Philadelphia and as small as Santa Cruz and Chico, California. The Boston Women's Health Collective continues to update *Our Bodies, Ourselves.* And the effects of women's greater consciousness of their health-care rights continue to be felt in the American medical establishment (Fishel; Rothman & Punnett; Boston Women's Health Collective, 600).

An important branch of the consumer health movement is the so-called Wellness movement, emphasizing staying well as opposed to treating illness once it occurs, and acknowledging people's right to direct their own health care. This movement has involved both men and women, and **Sara Levine** (1947–  ) of San Francisco has not excluded men from her Alternative Healthcare Associates, founded in 1982. (However, no men were enrolled as of 1983.) AHA is a group of sixteen nurses dedicated to treating those willing to assume responsibility for their own health. Specializing in stress-management, cancer, and family-community counseling, they make house calls and construct personalized health plans after a long personal interview, or "lifestyle assessment," with each client, usually in the client's own home or office.

*Personalized* may be the key word here. Levine stresses that AHA strives to complement, not compete with, existing health-care programs. By 1983 the group was receiving referrals from doctors, therapists, and visiting nurses. After the detailed health-needs assessment, the nurse whose skills best meet those needs is assigned to that client. Bodywork is often recommended, and the AHA nurses are trained in acupressure, jin shin, polarity, reiki, shen, shiatsu, therapeutic massage, and therapeutic touch. Seldom is establishment medicine able to use much touching. AHA also recognizes the basic importance of nutrition, and part of the personalized health plan is a personalized nutrition program. All sixteen nurses meet regularly to discuss their work.

Sara Levine was educated in English and nursing (RN from Youngstown State University in Ohio). She came to San Francisco in 1976, and began working in a hospital rehabilitation department where she did some home care. As of 1983 she was AHA's coordinator, main publicist and spokesperson, and the one who covered the follow-ups. She hopes eventually to make AHA a profit-making venture (Mildred Hamilton).

## Hospital reform

Women's hospital reforms and innovations could make a book in themselves. Indeed, they have been written about before, but usually with an emphasis on caring rather than on creativity. Some of these reforms and innovations—the work of the great French midwives, for example—will be discussed in Chapter 3. Among the more striking examples that belong here is **Dr. Marie Zakrzewska**'s institution in the 19th century of the first hospital social services. As already noted, Dr. Zak—so called because Americans could not pronounce her name—also introduced the idea of keeping records on patients (Levin, 66, 72). Much as we may criticize hospitals today, we would be shocked at the primitiveness of hospital care in the 19th century. Before the work of such innovative women as Marie Zakrzewska and others already mentioned as innovators in record-keeping, either nothing at all, or only name and address, might be recorded for hospital patients.

German-born Dr. Zakrzewska (1829–1902) was a pioneer on several counts: first professor of obstetrics at the New York Female Medical College; organizer of the New York Infirmary for Women and Children (first hospital in the nation to be staffed entirely by women); founder of the New England Hospital for Women and Children (first in the United States to have a nursing school), etc. "Intelligent, persuasive, and persevering, if somewhat opinionated and quick-tempered," as *Notable American Women* described her, Marie Zakrzewska insisted equally on access to medical training for women, and on high standards in that training. She may deserve more credit than any other single person for the success of American women in medicine (Levin, 64–6, 72, 86, 97; *NAW*).

Another criticism leveled at 20th-century medicine besides its obsession with technology and its disregard of patients' dignity, life quality, and feelings is that it has become too specialized and compartmentalized. Techno-medicine, in short, treats the separate body parts while ignoring the whole human being. **Dr. Connie Guion** (1883–1971) of New York Hospital founded the Comprehensive Care Program, coordinating all aspects of a patient's medical care, and bringing the personal attention of the general practitioner into big-city clinics. Born on a North Carolina plantation and educated at Wellesley and Cornell Medical College—where she graduated first in her class of 1917—she worked at New York Hospital for more than fifty years. In addition to founding Comprehensive Care, Dr. Guion was a pioneer in several ways: first woman to become a professor of clinical medicine (1946), first woman to win New York University's alumni Award of Distinction, first woman to be an honorary member of the New York Hospital Board of Governors, and the first to serve on the hospital's medical board. She practiced until age 87, just a year before her death (*WBWRA,* 209).

As already mentioned, **Dr. Janet Sterling Baldwin** is credited with the analogous innovation for children with cardiac diseases: she pioneered in complete care programs for them rather than treating the heart in isolation (Chaff, 304).

Among the most exciting of women's recent hospital reforms is the Planetree Unit of Pacific Presbyterian Medical Center in San Francisco, a whole new approach to hospital stays and to patient care. This first "consumer-oriented hospital unit," named for the plane tree under which Hippocrates taught, was the 1986 innovation of **Angelica Thieriot** after a disillusioning experience with modern medicine. Planetree considers that patients—and their families—will do better if they have more choice about and participate in their own care; that illness need not mean losing value and respect as a human being.

The room lights have dimmer switches, and patients can establish their own sleep schedules. The whole environment is designed to be positive and soothing—after study of ancient traditions as well as the latest neurological research on art and healing—with soothing colors, intimate textures, and soft, non-fluorescent light. Spouses or other family members can be trained—and wear badges—as Care Partners, giving as much of the patient's care as desired by both, including even such things as cleaning the plastic suction tube connecting the patient to a respirator. Patients can keep medications at bedside, take them as scheduled or as needed, and note time and dosage on a record later entered onto the chart. Planetree is collecting paintings, which patients can choose for their walls; music tapes are available for portable players issued to each patient on admission. Touching and hugging by the professional caregivers are not taboo.

Although not all doctors at Presbyterian accept Planetree, the number of doctors admitting patients to it has grown from fifteen to 85 in just one year. A University of Washington study will evaluate the safety, therapeutic value, and cost-effectiveness of the model project, with results available by 1988. Meanwhile, the first year appears to be a great success, and according to the Unit's administrator, "organizations like the American Hospital Association are looking to Planetree as a benchmark for the future of medical care" (Denney).

### Health-care delivery

A further criticism of modern medicine is that its practitioners cluster in wealthy cities and modern countries, leaving millions of people without access to it. Men as well as women, and physicians as well as other health-care workers, have devised programs to bring advanced medicine to developing countries—or critically ill patients to advanced medical centers. One such plan, called Heal the Children, is the brainchild of **Chris Embleton** of

Spokane, WA. Mrs. Embleton got the idea of helping children as a way of dealing with the death of her adopted Korean daughter in 1979. The specific idea came when she learned about a three-year-old Guatemalan girl who had only three weeks to live unless she had heart surgery. It took three days of persistent telephoning to arrange the operation, which took place at Children's Orthopedic in Seattle. Chris Embleton now has an organization that can call upon 40 doctors and 27 hospitals in 15 states. Members donate money to fly the ailing children from Guatemala (about half the referrals come from Canadian Naomi Bronstein, director of a home for abandoned children in Guatemala City), and elsewhere. Foster parents care for the children before and after surgery ("Tiny Pilgrims").

**Dr. Vivian Garrison** and two colleagues reported in 1981 on a five-year study of ways to develop cooperation between folk healers and more orthodox medicine in the area of mental health and psychiatric care. Although humanizing psychiatry was not the avowed purpose of this study, based in predominantly black Newark, NJ, its findings could have that effect. Dr. Garrison and colleagues located 86 folk practitioners involved in "spiritual" or "occult" healing and counseling in Newark, and made videotapes of sessions with them. These tapes vividly illustrate how such cultural factors as the person's self-concept, view, and beliefs on spirit control of mind and body can play an important part in mental health treatment.

One of Dr. Garrison's main findings was the diversity in "brands" of folk healers found among various ethnic groups, and the tendency of each group to use its own favorite type of healer. Therefore, the study concluded, no blanket policies for achieving cooperation between folk and establishment practitioners were possible at this time. However, cooperation could be approached case by case, with benefits for both doctor and patient. Doctors might get a more accurate diagnosis, and feel less frustrated in the treatment process, while patients would not only run less risk of being misdiagnosed and thus improperly treated, but should feel less alienated and fearful and more affirmed when their culture is thus respected. As Dr. Garrison noted, "For many patients, folk healers fill gaps where community mental health centers' clinical staff cannot distinguish psychotic delusionary and psychosomatic states exhibited by patients because of the psychological and social dynamics in their lives." In other words, the end result could be a more humane brand of mental-health care for those caught between two worlds.

As of 1981, Dr. Garrison was at the New Jersey Medical School. The research was supported by the National Institute of Mental Health, and Dr. Percy Thomas of the NIMH Staff College called it "one of the most important thrusts in mental health today" (V. Garrison; "Learning from Local Folk Healers").

*Self-help and the resurgence of herbal healing*

The return to herbal remedies and a heightened awareness of one's own body and its needs is not limited to menstrual or other female complaints, but women have led the way, partly because they have felt most poorly treated by modern medicine and most keenly aware of its dehumanization. Perhaps, too, the movement fits an ancient role of woman as healer that has never been totally lost.

An instructive example of self-help recommendations is the following for trichomonas vaginal infections from the Chico Feminist Women's Health Center in California (Nelson, 7–8): As a preventive measure, keep the body's pH balance slightly acidic by drinking cranberry juice or taking Vitamin C. Spermicidal jellies are also acidic and inserting one applicator-full every two to three weeks can be a preventive. Once an infection takes hold, use garlic suppositories alone or with a vinegar-and-water douche in three- to five-day cycles repeated until symptoms disappear; or a two-week regimen of twice-daily douches with solutions or "teas" made with either of the following herbal mixtures: witch hazel, comfrey root, and goldenseal; or myrrh, goldenseal, and bayberry bark. The garlic suppository itself is an ingenious invention: a peeled clove of garlic wrapped in gauze, which is then twisted and left as a tail for removal.

This same publication also contains herbal remedies for vaginal "yeast" infections, bladder infections, and other bacterial infections as well as palliatives for herpes flare-ups. It lists 17 herbal diuretics, including cornsilk, juniper berries, and wintergreen.

Even better than rediscovering herbal remedies is not losing them in the first place. Thanks in part to a dynamic Eskimo woman named **Della Keats** and her Della Keats Institute in Kotzebue, native healers are working with modern hospitals in Kotzebue and Bethel, Alaska, in an innovative program to preserve this knowledge and incorporate it into modern treatment regimens.

## Battlefield care

Women have also been concerned with humanizing battlefield medical care, and improving delivery of such care to soldiers and disaster victims. Following in the footsteps of **Marie Curie** and her mobile X-ray units is another Frenchwoman. From the time she got her pilot's license in 1910, **Marie Marvingt** crusaded for her aerial ambulance service. Nurse as well as pilot, she saw the value of prompt aid to wounded soldiers and to disaster victims—and of flying certain cases to distant hospitals. She urged the French Army to create a medical aviation branch and to outfit special rescue planes with nurses, stretchers, and surgical supplies. *As early as 1912 she designed and ordered a demonstration ambulance plane* from the Deperdussin factory, but the company failed before completing her order. During World War I the Army used only a few planes for occasional emergency medical service, but Marvingt did not give up. In the early 1930s she trained nurses and pilots for aerial medical work.

Finally, in 1934 the Army invited her to organize a civilian air-ambulance service in Morocco. Thus, after more than twenty years' campaigning, her dream of "the Wings that bring Mercy" came true, and flying nurses began bringing speedy medical aid to French soldiers in North Africa (Marvingt; Moolman, 30).

## Support groups

Last but certainly not least among our instances of humanizing ways with medicine are the many support groups for victims of various diseases and conditions, for parents of children suffering from life-threatening diseases, etc., that have sprung up over the past ten years. Further research would be necessary to show how many of these were founded by (or the ideas of) women. One noteworthy example that can be mentioned here is the Psoriasis Research Association of San Carlos, CA, founded by **Diane Mullens.** Herself a psoriatic, she founded the PRA in 1952 with a twofold purpose: first, to give patients a way to meet each other, learn ways of coping and receive support; and second, to raise money for research on the disease. For more than thirty years, Ms. Mullens ran the organization, personally answering letters and phone calls from her fellow-sufferers. In 1986 she donated her membership list to the more purely research-oriented Psoriasis Research Institute of nearby Palo Alto (Orenberg, L-3/23/88; PRI Newsletter 7(1988): 3: 2).

Although obviously not patentable, these support groups are innovations—social inventions, if you will—that have contributed as much to survival rates in certain life-threatening diseases, particularly cancer, as many patentable treatments.

*Genetic Engineering*

As already noted—and as the press reminds us daily—*genetic engineering* is transforming medicine and will soon make it practically unrecognizable. Two male researchers patented the basic technique, and one of them (Stanley Cohen) received a Nobel Prize for it in the 1980s, with considerable publicity. Not usually mentioned, however, is that as early as the 1940s **Dr. Hattie Alexander** and her colleague **Grace Leidy** were doing an early form of genetic engineering. Using their own techniques, by 1950 they were producing hereditary changes in the bacterium *Hemophilus influenzae,* with DNA obtained from this organism. Dr. Alexander continued to work in genetics till the end of her life, all the while maintaining her hospital practice and clinical research (*NAW/* mod.).

Also routinely omitted from glowing review articles on genetic engineering are two women without whom the technique might still be just a Frankensteinian fantasy: **Dr. Esther Lederberg** and **Dr. Annie C-Y Chang.** **Esther Lederberg** is the wife of Joshua Lederberg, and her reputation is overshadowed by his; but it was she who discovered the phenomenon of viral induction, in which a virus enters a bacterium, lies dormant for a time, but then suddenly becomes active, directing the bacterial genetic machinery to make copies of its invader—in such numbers that the bacterial cell bursts, releasing many new viruses to begin the cycle again. Before Esther Lederberg's crucial discovery, there was no reliable way to introduce foreign genetic material into a cell.

**Dr. Annie Chang** co-authored with Stanley Cohen and two others the landmark 1973 and '74 papers reporting the use of an enzyme to cut DNA at precise locations for inserting a desired section of genetic material (Cohen *et al.;* Morrow *et al.*). This technique, now known as gene-splicing, is the basis for all of genetic engineering. It—along with the resulting organism—is covered by two precedent-setting and lucrative patents assigned to Stanford University in 1980 and '84 ("Genetic Licensing . . ."; "Patent . . . Awarded Stanford," 1)—on which Dr. Chang's name does not appear.

Why she was omitted from these vital patents, from the 1980 Lasker Award, and from the 1986 Nobel Prize for this work, when Cohen's *Scientific American* article of 1975 makes her full participation so clear, is yet to be explained. The fourth author on the paper, Robert Helling, did challenge the granting of the 1984 patent to Cohen and Boyer alone, but was not upheld. Chang's name does not appear in the Stanford *Campus Report* article on the patent (Hofstadter, "Patent . . ."). A 1981 Stanford *Observer* article does mention Chang's co-authorship of the first report, but suggests that the original idea for the work arose in informal conversations between Cohen and Boyer in a Waikiki delicatessen in 1972 ("Licensing Plan," 4). Even if this is true, as repeatedly shown in this book, an idea for an invention is only the beginning.

In any case, Drs. Chang and Lederberg undeniably deserve a prominent place when the history of genetic engineering is written (cf. Wade)—and thus, indirectly, a prominent place among the pioneers and innovators of medicine.

Genetic engineering is involved in several inventions and innovations already treated in this chapter. However, since medicine is one of the main targets of this technology, it may be worthwhile to gather a few striking examples here.

*Prenatal diagnosis*

This controversial field uses new apparatus and techniques for detecting abnormalities in the fetus.. Often a problem can be diagnosed early enough to allow safe abortion of a hopelessly deformed or defective fetus, and herein lies the controversy.

*Sickle-cell trait.* Victims of the severest form of this hereditary blood condition (inherited from both parents) can suffer pain, paralysis, and early death. The work of **Dr. Lois Wilson** and her husband John, and Dr. Y.W. Kan and **Judy Chang,** enabling prenatal diagnosis of this condition, has already been discussed. Dr. Lois Wilson's long-term research plan calls for using this same methodology to detect other defects or inborn errors of metabolism as early as the 15th week of gestation ("Diagnosis by DNA"; L. Wilson, L-9/8/81; Geeyer *et al.*).

## Vaccines

Since live vaccines always walk a thin line between preventing and causing the disease in question, researchers have experimented with disabling the organism in various ways—killing or chemically treating it. As biochemistry, genetics, and immunology became more sophisticated, however, scientists began to isolate various parts of disease-causing organisms and their products to see which provoked an immune response, and which caused undesirable side effects. They also began to experiment with altering the disease-causing organism genetically as in the following example:

*Whooping Cough.* In recent years the pertussis (whooping cough) part of the DPT vaccine that protects some 99% of all U.S. infants from diphtheria, whooping cough, and tetanus has been found to cause brain damage in a few children. Even though the risk is only about one in 300,000, Stanford researchers, among others, have turned to genetic engineering to improve the pertussis vaccine. They have modified the *Bordetella pertussis* bacterium genetically so that it produces a defective toxin—100 times weaker in its ability to cause neurological damage. The plan is to create a *Bordetella* strain that still provokes an immune response, but avoids the present danger. As of 1985, at least three women were or had been involved in this important research: **Dr. Alison Weiss, Nancy Adelman** and **Mae Lim** (see Prevention section above).

*Hepatitis-B.* Now, in a further level of sophistication, scientists can isolate the immunity-provoking substance in (or produced by) a virus or bacterium, then, by genetic engineering, have that complex produced by a totally separate organism, thus removing all chance that the vaccine will cause the disease it should prevent. The earliest attempts to do this with Hepatitis-B involved transferring the genes for virus surface proteins into bacteria. The genes did act, and the bacteria produced the proteins, but then proceeded to degrade them with enzymes.

Scientists have now succeeded in transferring the same genes to a yeast. This breakthrough in vaccine technology was reported at the International Virology Conference in Strasbourg in August 1981. Dr. William Rutter's group at the University of California–San Francisco, including at least one woman, **Angelica Medina,** reports that the yeast host is highly productive, producing more than 10,000 molecules of the desired product per cell. Moreover, the yeast produces not just the naked surface protein, but an "immunizing particle" that is identical in size, shape, and sedimentation properties to the immunizing particle isolated from the blood of hepatitis patients. At the time of the 1981 report, the researchers were not certain that the particle contained the appropriate sugar. However, the new product looked so hopeful that Dr. Rutter predicted clinical trials before 1982. Assuming successful trials and FDA approval, the vaccine can be produced in unlimited quantities, at a cost of $30 or less for the necessary three shots.

Hepatitis is one of the world's major health problems, affecting half a billion people. The estimated 200 million carriers could each infect four or five others. In the United States alone, some 150,000 people a year suffer from it (1980 figures), and several thousand die. In Asia the disease is endemic in many populations. Many researchers blame it for the high rates of liver cancer in Asia and Africa ("Genetic Research"; "Hepatitis Vaccine"; Petit; "Yeast Engineered").

*Piggyback typhoid vaccine.* Much less devastating than hepatitis on a world scale but still a nasty problem with some 12 million cases yearly, is typhoid fever. Where medical care is primitive, the mortality is 10%. The existing vaccine, as already discussed, has several defects. A new oral vaccine, using live but genetically altered *Salmonella typhi* bacteria (unable to grow or reproduce without two substances lacking in human tissues), promises to confer lasting immunity. As of 1985, clinical trials in Chile were "soon" to begin.

More exciting than the new vaccine *per se* is the so-called piggyback concept whereby the vaccine bacteria would carry added genes so as to confer immunity to several diseases at once. Currently under consideration for riders are malaria, hepatitis, schistosomiasis, and traveler's diarrhea. The piggyback idea is the brainchild of Dr. Bruce Stocker of Stanford. Among the contributors to the vaccine's development, as noted above, are at least two women, former graduate students of Stocker: **Susan Hoiseth** and **Mary Frances Edwards** (Hofstadter, " 'Piggyback' ").

*Cell cultures.* Two scientists from the University of California Medical Center (Los Angeles), David Golde and **Shirley Quan,** created and patented (1984) techniques for producing a cell culture using cells from the spleen of a victim of a rare form of leukemia. Such cultures have been used for some forty years to study disease processes and to test new drugs. This particular culture could have enormous medical and commercial value, because it produces "significant amounts" of interferon and other scarce substances, including a substance that boosts blood-cell growth and, possibly, immune function. It might, incidentally, also gain a certain notoriety, as the leukemia victim in question, one John Moore, sued the scientists for using his cells without his explicit permission. He has asked for a share of the revenues from licensing the patent rights. As of early 1985, the cell line had been sold only to scientists, and no company had come forward to license it for its commercial potential (J. Stone; "Whose Cells . . .?").

*Monoclonal antibodies.* As described above, researchers at Stanford Medical School, Columbia University and Becton Dickinson Monoclonal Center, Inc., in Mt. View, CA, have created and tested a mouse-human antibody that may prove effective against multiple sclerosis (MS). Trials in mice show that the antibody can prevent the experimentally induced allergic encephalitis that has been considered the animal model of MS. Although clinical tests have yet to be completed, the possible new treatment was considered highly promising. At least four women were involved in this research: **Dr. Leonore Herzenberg,** a senior research associate in genetics, and **Mae Lim,** a neurology research associate, both at Stanford; **Dr. Sherie L. Morrison,** Professor of Microbiology at Columbia, and **Dr. M. Jacqueline Johnson,** a former Stanford graduate student in genetics.

*Late developments*

At the February 1988 meeting of the American Association for the Advancement of Science, **Susan Marqusee,** an MD/PhD candidate at the Stanford University School of Medicine, reported that she had designed and synthesized from scratch a protein fragment that coils itself into a helix—what might be called a synthetic helix. The accomplishment is of basic interest for elucidating the folding of proteins and thus for understanding biological processes, but has important potential clinical applications as well. Understanding protein folding, for example, could lead to custom-designed proteins, making possible "designer drugs" against many diseases. Marqusee's research, supported by National

Science Foundation and NIH grants, took place in the laboratory of Dr. Robert Baldwin of the Biochemistry Department (Hofstadter, "Medical").

## SUMMARY

In summary, then, what can we say about women's contribution to invention and innovation in the field of health and medicine? The foregoing pages amply support my early subjective impression that the contribution was—and continues to be—a great one. But what about objective measures? What do the numbers say? Not a great deal yet, but we can make a beginning. The most obvious place to begin is the patent records, primarily the United States patents.

First, how many of the approximately 4,000 American patents listed by LWP for the 19th century (May 5, 1809–February 28, 1895)[60] were granted for medical or health-related inventions or improvements? Not counting either contraceptives, menstrual aids, or the reform undergarments included in my suggested new definition of medical inventions, I find 200 health-medical patents listed by LWP. Although LWP covers essentially a century, these patents cluster together in just over three decades, between February of 1863 and March 1, 1895. This total of 200 "straight medical" inventions and improvements amounts to a little over 5% of all the listed women's patents, close to the 4.5% found by the Women's Bureau study for ten select years from 1905 to 1921.

If corsets are added to LWP's medical total (as the Patent Office's own classification might justify doing), and using Minnie J. Reynolds' count of 179 corsets patented between 1815 and 1895, that total rises to 379 or almost 9.5% of all LWP inventions. Even if we include only the 164 corsets patented after the beginning of the Dress Reform movement (i.e., 1866 onward) or, more strictly, the 37 corsets specifically labeled as abdominal supporters, we increase the total of patented health-medical inventions to 364 (9.1%) or 237 (5.9%), respectively, of the 4,000 listed inventions.

If contraceptive devices, menstrual aids, and reform undergarments other than corsets are considered, the figures are as follows:

| | |
|---|---|
| Contraceptives/feminine hygiene | —13 patents |
| Menstrual aids | —51 patents |
| Reform undergarments | —12 patents |

If all these categories of invention are accepted as medical and related, then (restricting the corset category to abdominal supporters) we have

200 straight medical
37 abdominal supporters
13 contraceptives/feminine hygiene
51 menstrual aids
12 non-corset reform undergarments
_____
313 health/medical total = 7.8%

Interestingly enough, the LWP compilers themselves, in their classified listing for the third installment of the list, included all of these categories except the non-corset reform garments in their "Medical Appliances" category.

_____

60. Not counting design patents or reissues.

In short, to the degree that LWP is a complete list,[61] women's patented inventions and improvements in the area of health and medicine for the 19th century United States number almost 8% of all their patented inventions for that period.

## 20th Century

In 1922, under the directorship of Mary Anderson, the Women's Bureau of the Department of Labor published a bulletin (No. 28) entitled *Women's Contributions in the Field of Invention,* cited herein as WB/28. The authors studied Patent Office records for ten select years between 1905 and 1921, categorized the women's inventions they found, and analyzed and presented their work statistically. Although the Bureau's categories differ from those of the LWP compilers, and from my own, we can gain a good idea of women's technical contributions in health-medicine from the Bulletin's figures:

In Category XVIII: Medical, Surgical, and Dental Equipment, the Bureau found 227 inventions for the ten years studied, or about 4.5% of the 5,016 patents granted to women during those years. These 227 inventions were subdivided as follows:

| | |
|---|---|
| Instruments and apparatus | —48 or 21.1% (of 227) |
| Sick-room equipment | —73 or 32.2% |
| Bandages, dressings, belts, supports | —59 or 26.0% |
| Stretchers, invalid carriers | —10 or 4.4% |
| Defective foot & limb corrections & aids [prostheses] | —21 or 9.3% |
| Manipulating & flesh-reducing mechanisms & equipment | —9 or 3.9% |
| Dental equipment | —7 or 3.1% |

The WB/28 compilers did a rather thorough job of gathering health-medical inventions into Category XVIII. However, a certain number of items were otherwise classified:

Category XVI, Personal wear and use, is another possible source. The Undergarments subcategory here contains 154 patents of 10 types, among which the following seem pertinent according to my own classification: bracing garment, maternity corset, and invalid bed gown and commode. If each type in the subcategory contained the same number of inventions, these three types would add 46 patents to the total.[62]

The Garment appurtenances subcategory contains 378 patents of 28 types, two of which—catamenial (menstrual) appliance and sanitary handkerchief container—seem

---

61. Among the nearly 30 women's patents omitted by LWP for 1876 alone, for example, were two abdominal supporters (Stanley, "Conjurer").

62. As noted above, since WB/28 gives no numbers for the types in its subcategories, and no inventors' names at all, and since the Patent Office's yearly lists of inventions give inventors' names by initials only and some categories include literally hundreds of inventions, every name of which would have to be cross-checked for gender in the alphabetical name list, this equal-numbers assumption seems the only expeditious way of proceeding. (Indeed, in some cases, the WB/28 terms for invention types do not seem to agree with the Patent Office terms, making the search not merely excruciating but actually impossible.) The only alternative would be to search the entire alphabetical name list for the ten years in question, involving more than 350,000 patents. I hope that for my purposes here, the very frequent and infrequent types will balance each other, so that my proposed additions to the central health-medical category will not be exaggerated. In any case, readers now have the necessary information to make their own determinations.

health-related. Under the equal-numbers assumption, these two types would number 27 patents. Doubting, however, that there would be more than one or two sanitary handkerchief containers, I estimate these two groups would add only about 15 patents to the total.

Under toilet articles we find 113 inventions of 33 types, which means that the two toothbrush types, the fountain tonic applicator, and the massage device types could add 12 patents to the total. It seems unlikely that there would be more than one Fountain tonic applicator, but there might be many toothbrushes.

Category XVII, Beauty parlor and barber supplies, etc., contains 46 inventions of 18 types, only three of which seem relevant here: facial massage implement, appliance for removing facial defects, and hair treatment. If all types were equally numerous, this category would add 7 or 8 patents to the total.

The Sanitary equipment subcategory (58 patents) under Category XIX, Safety and sanitation, includes 8 possibly relevant types of patent among its 19—antiseptic drinking cup, apparatus for purifying and filtering water, apparatus for sterilizing liquids, disinfecting device, filtering drinking cup, sanitary container cap, sanitary cuspidor, and sanitary toilet seat or cover—possibly numbering as many as 24 patents.

If these adjunct and health-related patents are added to the "straight medical"–dental total, we have

227
46
15
12
7
24
———
331

or about 6.6% of WB/28's 5,016 patents.

As the OTAF study shows, health and medical inventions are proving even more attractive to women in the last quarter of the 20th century than in the first. Nearly half of all women's U.S. patents granted 1977–88, for example, are chemical, and a significant percentage of those are medical or potentially medical in nature. Of the 15,136 U.S.-origin utility patents granted to women during these twelve years, for example, 1,181 are Class 424 (incl. 514): Drug, bio-affecting, and body-treating compositions. This is 10.08% of all patents granted in that class during the period. This group of patents ranks first among the 15 active classes with the highest *numbers* of patents granted to women, and sixth among a parallel list of classes with the highest *percentages* of patents to women.

Class 128, Surgery, with 694 patents, is fourth in the numerical ranking of the 15 active classes, and thirteenth in the percentage rankings, with 5.81% of the patents in the class granted to women. Other potentially pertinent groups of patents are Class 430, Radiation imagery chemistry, eleventh both numerically (229 patents) and percentage-wise (7.08%); Classes 534–570, Organic compounds, second numerically (852 patents); Class 435, Chemistry: molecular biology and microbiology, fifth numerically (479 patents) and second percentage-wise (15.89%); and Class 436, Chemistry: analytical and immunological testing, fifth percentage-wise (10.86%).

It is clear at a glance that these numbers and percentages add up to a participation significantly higher than that for agriculture and related inventions.

# 3   Daughters of Hera, Eileithyia, Prokris, and Teteu Innan: Women Inventors and Innovators in Sex, Fertility, and Anti-Fertility Technology

Holy Mother of Birth and Healing,
Matron spirit of all Midwives,
Goddess of the sacred totoixitl herb. . . .
It is Teteu Innan who fills the wombs
of those who wish to bear new children.
It is Teteu Innan who empties wombs
of the women who do not,
remembering always to ease the pain of childbirth
for Her daughters upon the earth.
—Pre-Aztec Mexico (from Merlin Stone, *Ancient Mirrors of Womanhood*)

If men menstruated, there would be a National Institute of Menstruation within a year.—Dr. Penny Wise Budoff, 1980

If men could get pregnant, abortion would be a sacrament.
—NYC woman cab driver; attrib. to Florynce Kennedy, 1970s

## INTRODUCTION

Women's inventions and innovations in what has been their domain for all but the last few eye blinks of history would fill several books. Among the many topics to cover would be aphrodisiacs, attractants, self-pleasuring techniques and devices, techniques and devices for increasing sexual pleasure, remedies for infertility, pregnancy and childbirth aids, menstrual technology (including remedies for menstrual problems), and anti-fertility technology, including both abortion (see App. SFA-5) and contraception. Here, in the interest of reasonable length, I will focus on menstrual and childbirth technology and contraception, with only the briefest of nods toward sex technology.

## SEX TECHNOLOGY

One of the saddest losses in a sex-for-reproduction ethos is women's sexual pleasure. Men, as often observed, can find pleasure or at least release in the briefest encounter. Women, on the other hand, take longer, needing more subtle stimulation from a talented and caring lover. The Gypsy coaxer, using exquisitely skilled foreplay to pleasure the Queen or tribal mother and bring her ready to the embrace of her consort,[1] and the

---

1. The matriarch's husband was a "drone," charged only with siring perfect children; if he did not, he would be "accidentally" killed, and replaced by another "stud-chieftain." Much closer to the tribal mother emotionally was her "coaxer" (*WEMS*, 360ff),

a man trained from an early age to control his own sexual responses and concentrate completely on his partner's pleasure. He was taught to know all the sensitive and erotic zones of the female body. In this

Polynesian creation myth in which the male is created for the sexual pleasure of the Goddess Hina, seem equally alien to our culture.[2]

Over the millennia, of course, even in cultures where men are trained to please women in physical love, there were less talented or less willing or absent males, postpartum intercourse taboos, etc., leaving women celibate for various periods of time. Consequently, women have invented various aids to sexual pleasure for themselves.

## Self-Pleasure Technology

The example I wish to discuss briefly here, the dildo, is interesting for its possible ritual origins and later commercialization. Dildos of varying levels of sophistication are known from the ancient Near East to Polynesia to modern Appalachia. Havelock Ellis discusses auto-erotism at some length in the first volume of his monumental *Studies in the Psychology of Sex*, which was banned on its appearance in 1897. After noting that self-stimulation "is found among the people of nearly every race of which we have any knowledge," he concludes that for non-European races "it is among women, and especially those . . . subjected to the excitement of a life professionally devoted to some form of pleasure, that the use of the artificial instruments of auto-erotism is chiefly practiced" (Ellis, I: 166, 169; Ezekiel 16: 17; Randolph, 97; Suggs, 45).

The apparent great age and common phallic shape of dildos suggest that they may have originated in defloration ritual and technology. In cultures where the Goddess was powerful and the Mothers her earthly models and representatives in all matters of birth and fertility, and where a woman's blood was a powerful and sacred substance, the taking

---

curious three-sided relationship, the coaxer gave the mother her physical fulfillment without ever penetrating her. Instead, by a combination of caresses, words, and breathing, he made her . . . ready to have an orgasm as soon as her husband took over.

These ideas so inimical to the priestly Christian attitude toward sex, and the Gypsies' position as worshippers of a triple Goddess (late and active preservers of Goddess-worship in Europe) make it clearer why Gypsies became targets of both Holocausts: the one mounted by the Christian Church against "witches" in the early modern period, and the one mounted by the Nazis in the 20th century.

2. Though it is hard to imagine anyone seriously doubting that the matters covered in this chapter were originally the province of women, it may be worth mentioning some of the goddesses connected with sex and reproduction, especially since some of them are unfamiliar to Western culture. Ermutu is an Egyptian birth goddess. Eileithyia is an Aegean birth goddess, later linked as an attribute to Artemis, and perhaps to other Greek goddesses as well. Hina is a great Polynesian goddess—for whose sexual pleasure, as noted above, the first male was created. Hera was particularly the deity of women and their sexuality. Juno, an ancient Italian goddess whose early myths are mostly lost, was also definitely a woman's deity. Under her different names or attributes, she ruled woman's entire reproductive life, including marriage: as Pronuba she arranged appropriate matches; as Cinxia, she oversaw the first undressing of the bride by the bridegroom; as Populonia, she was goddess of conception; as Ossipago, she strengthened the fetal bones; as Sospita she oversaw labor; as Lucina, she led the child to light (i.e., was the birth goddess). Juno was also connected with the menstrual cycle as an indicator of time (Monaghan, s.v. names).

Among the Ona of Tierra del Fuego, it was Kra, the Moon Woman, who came to earth, lived with women, and taught them women's ways (Drinker, 64). The Aztec Mayuel seems to have been a peasant woman, deified as the inventor of pulque and tequila, intoxicating liquids made from the maguey (*metl*) plant, and originally used to ease childbirth pains. Among the Maya, the Moon Goddess Ixchel was revered as the first woman of the world, and the one who gave the people of the Yucatan, Campeche, and Guatemala the gift of means to ease childbirth (Stone, *Mirrors* I: 91, 92).

of a virgin's hymen would have been far too sacred a matter to entrust to chance, and far too dangerous for a man to undertake. In such cultures defloration would be one of the blood rituals, along with first menstruation and birth.[3] Women would have been in charge—the girl's mother, a midwife, or a priestess, depending upon the culture—and women would have invented the technology.[4] In a Salish Indian myth from British Columbia in North America, an old woman "has intercourse" with young women by means of a horn she wears as a penis.[5]

However, by the 3rd century BC, Greek (Coan?) women were evidently buying dildos from cobblers, and preferred the wares of some above those of others (Hero[n]das, Mime VI):

> "We're alone, so let me tell you about these dildos:
> Harder than men, softer than sleep and laced with wool.
> Now there's a cobbler who likes women!"

These artificial phalluses were apparently rose-colored, euphemistically described in at least one translation as a "red belt," and in another (Herodas/Knox) called "the scarlet baubon." Women of Lesbos had ivory and gold dildos (Bloch, 260).

However, in this intimate matter, the takeover was incomplete. Women also continued to make their own dildos in Western culture. Aristophanes says that Milesian women had leather ones (H. Ellis, *ibid.,* 167–9). This Aristophanes passage (*Lysistrata,* v. 109) is particularly pertinent, as it specifically shows that women manufactured the leather "olisbos" themselves. Interestingly, Marylin Arthur says that free lower-class Greek women were sometimes cobblers (70).

Throughout the Middle Ages in Europe, women continued to make and use artificial phalluses, called *godemiches* in France, *passatempos* or *dilettos* in Italy and *dildo(e)s* in England, as we know partly from clerical protests, Burchard's Penitential, for example, assigns penalties to the woman who makes a phallus for herself or other women (H. Ellis I: 169n). Note again that the women are not merely using the aids, but making and perhaps selling them.

European nuns of the 16th century used glass dildos filled with warm water "to calm the sting of the flesh and to satisfy themselves as well as they can." Widows and other women anxious to avoid pregnancy also used these devices (H. Ellis, I: 169). In Elizabethan England, Lucia, a character in one of John Marston's satires, prefers "a glassy instrument" to her husband's "luke-warm" efforts. In 17th-century France *godemiches* were made of velvet or glass (Bloch, 261), the former probably made by the women themselves; the glass ones probably not.

---

3. Indeed, in several cultures the menarche may itself have been considered a defloration, or at least the two were inextricably linked and the girls' puberty rite included a ritual deflowering (Delaney *et al.,* 24ff).

4. Even in the later patriarchal cultures, where, as among the Brahmins of Kerala in India, males had taken over the rite, the power/danger of defloration was such that it still was not the province of any ordinary man, but of a priest, ceremonial husband, king, or medieval seigneur (cf., e.g, Delaney, 25; Gilgamesh [Sanders, 62, 68]).

5. Another hint of possible ritual origins of dildos lies in the ancient and still imperfectly known Eleusinian Mysteries, where ecstatic initiates symbolically consummated Demeter's love affair with Iasius/Triptolemus/Zeus "in an inner recess of the shrine, by working a phallic object up and down a woman's top-boot." Other participants, dressed as shepherds, then rushed in, displaying the child of the symbolic union (Graves, *Myths* I: 94).

One branch of women's self-pleasure technology, then, is devoted to imitating the penis, and thus to reproducing heterosexual intercourse.[6]

### Fertility-Promoters

Human fertility most often takes care of itself, but the exceptions can be heartbreaking. Women through the ages have devised fertility preparations for such cases. Jivaro women of the Amazon, for example, allegedly know a fertility herb so potent that it not only assures women of conceiving, but actually reverses the effect of their long-term herbal contraceptive or sterility plant (Maxwell, 344).

Three traditional fertility plants are the pomegranate *(Punica granatum),* the date palm *(Phoenix dactylifera* var. *samani),* and the moghat *(Clossostemon brugieri).* Egyptian women used the pollen grains of the date palm to induce fertility, and moghat was a postpartum hot beverage. Modern pharmacology has shown that all three do indeed contain estrogenic hormones, specifically estrone (Farnsworth, 717–8). Ginseng, a traditional menopause herb and general tonic for older people, can produce estrus in castrated animals. It evidently contains estradiol and estriol (Seaman & Seaman, 435).

Estrogens occur, in fact, in many human food plants, such as summer wheat *(Triticum aestivum;* estrone) and beans *(Phaseolus vulgaris;* estrone, estriol, and estradiol). Primitive women could well have observed the effects on their own bodies of eating the fresh spring plants, perhaps especially on their milk supply if they were lactating. Later, they could certainly mark the "spring flush" of their dairy animals.

The tricky thing about estrogens, of course, is that they may show pro-fertility effects in one dosage or set of circumstances, and anti-fertility effects in another. The same plant may also affect different animal species differently. A pomegranate extract, for example, has reportedly caused infertile matings in both rats and guinea pigs (Farnsworth, 539, 553). This duality underscores once again the importance of knowing as fully as possible what the women were recommending—not merely the dosages, but the season for gathering, the plant part to be used, the means of preparation and the form in which the herbs should be given, and any modifying herbs or accompanying dietary regimen.

Women have also long known the vital role of nutrition in fertility, both primitive healers and early women physicians often recommending a richer diet in cases of infertility. Pliny mentions a Theban Olympas who wrote on sterility. Her dates are uncertain, possibly 4th century BC (Hurd-Mead, *HWM,* 41, 56–7).

Ancient women's remedies often followed principles only now being "discovered" by modern pharmacology. For example, their star grass *(Aletris farinosa)* prescription for preventing miscarriage contains Vitamin E, and the Arsesmart or water pepper recommended in amenorrhea (see below) contains Vitamins C and K.

## MENSTRUAL TECHNOLOGY

In connection with a unique ability to bring forth life, women menstruate monthly during their reproductive years. Unless ill, pregnant, lactating (with simultaneous hard work or strenuous physical activity and nutritional stress), doing heavy athletic training, geneti-

---

6. It should be emphasized that this is only one branch. Lesbian women's sex-pleasure techniques and technology may include this branch, or may take a different, woman-identified approach. Information on this area of women's inventiveness, however, has until very recently been even more thoroughly hidden than that on sex technology in general and women-as-inventors in general, and thus cannot be satisfactorily dealt with here.

cally abnormal, or surgically altered, every woman bleeds, but does not die, roughly once a month for thirty years or more between puberty and menopause. Inventions and innovations related to this reproductive phenomenon therefore affect more than half the world's population for roughly half their lives.

Although women of at least one group, the Siriono of Eastern Bolivia, allegedly let menstrual blood flow unstaunched (Holmberg, 171), most women have preferred to prevent this.[7] Women's techniques for dealing with the menstrual flow and with menstrual disorders are vital aids to the comfort and well-being of all women of reproductive age. As such, they constitute an ancient and highly significant, but much neglected, part of human technology.

Menstrual blood, once the proud mark of a woman's power, may originally have been caught for religious purposes. The words *rite, ritual,* and *sabbath* all derive from words for the menstrual period, and communicants probably drank menstrual blood in some early ceremonies of communion with the Goddess. The Church accused witches of doing this as late as the Middle Ages in Europe, and it may have been true (*WEMS,* 637).

Certainly women used menstrual blood in hunting and trapping, to attract or repel various animals at will. (This puts a new light on the many taboos against women and the hunt, and on hunters' associating with women, particularly menstruating women, before a hunt.) Later, in early horticulture, women used menstrual blood as a potent fertilizer. Judy Grahn suggests, in fact, that the first clothing may have been a menstrual belt and fiber pad used not for cleanliness or comfort, but to catch a valuable substance (Spretnak, *Politics,* 269–71).

However this may be, ancient women's inventions to catch the menstrual flow are many and ingenious, adapting natural materials or, after sedentism, using specially altered materials. Native American women created tampons out of soft moss; Mediterranean women used small sponges; women of Alor used dry porous banana bark as menstrual pads. Ancient Japanese women used 8–12 paper tampons a day, held in place by a bandage called *kama;* and Indonesian women made tampons from vegetable fiber. Roman matrons used cloth bandages or tampons of soft wool; African and Australian women used bandages of grass or vegetable fiber, while other African women used tampons made of rolls of grass and roots; Egyptian women used rolls of soft papyrus (Delaney, 115–6; DuBois, 102).

In at least one North American Indian group, the Arikara, women made sanitary pads from buffalo skin tanned by smoking (M. Weiner, 88; Vogel, 238) and presumably softened, as these women know well how to do. Presumably, too, the pads would be attached to or form part of a belt, perhaps much like the one of softened antelope hide used in Africa by the women of the !Kung San. Robert Graves mentions a goatskin "chastity tunic" worn by Libyan girls (*Myths* I: 47). Was this the original "chastity belt," a sign devised by women themselves that they were in a powerful and possibly dangerous state? It would be important to know what percentage of known peoples made sex taboo (in the original sense) during menstruation, and if so at what stage of their development. We would also need to know whether the taboo originated with the women and was later taken over and perverted by the men (or misinterpreted by early Western observers). Graves does not say what the Libyan women wore under the chastity tunic; only that it was death

---

7. Female chimpanzees sometimes wipe periodic blood away with leaves, according to a report cited by Jane Goodall. Goodall herself did not see this, but did see leaves used to wipe away feces, urine, ordinary blood from wounds, sticky fruit juice, etc. ("Tool-Using," 234).

to a man to remove it without the wearer's consent. But a chastity belt could have had a simultaneous practical function as a menstrual absorber.

In order to learn more about primitive menstrual technology, we can also look at what women used to absorb the birth blood and fluids. Women of Guam, for example, used coconut husks wrapped with cloth (Goldsmith, 79).

Much of this early technology may have been lost, or hidden, after the advent of patriarchy. In fact, a proper analysis of the history and prehistory of menstruation would show that a takeover has occurred here as surely as in agriculture or medicine, though I will merely sketch it here:

The earliest attempt at male appropriation of woman's power to bleed but not die, the blood that makes babies, was probably simple imitation. The subincision rites and practices of primitive Australia, New Guinea, the Philippines, and Africa are the clearest examples. In this drastic procedure, males slit the underside of the penis from root to tip, so that when the bleeding organ is pulled upward against the lower abdomen, it resembles a menstruating vagina. In some parts of New Guinea, the very word for a subincised penis means "the one with the vulva"; elsewhere, the blood that the men periodically cause to flow anew from the wound is called "man's menstruation." Many male initiation rites involve the shedding of blood, often, though not always, from the penis. Our own practice of male circumcision, though done in infancy rather than at puberty, may have the same origin.

Eventually, if men gain more economic control and/or begin to dominate through their warfare roles, and especially with the onset of patriarchy, they seek not merely to imitate menstruation, but to take control of it, to redefine it as shameful and "unclean." Though because of their patriarchal assumptions and their reliance on male informants, most Western anthropologists working before the 1970s undoubtedly exaggerated and perhaps even misinterpreted the concept of menstrual fluids and menstruating women as "polluting" in primitive societies (Goldsmith), it seems clear that this takeover or attempted takeover was at the very least a psychological reality, and did actually occur in many places, including cultures giving rise to our own. As Judy Grahn reminds us, "One of Herakles' (Hercules') tasks in acquiring control of the Mothers' powers was to steal the female belt—that is, the menstrual offices—of the Amazon queen Hippolyta" (Grahn/ Spretnak, 272ff). If the earliest explanatory myths of fetal development maintain that the baby grows from accumulation of the menstrual blood, retained during pregnancy, it would be interesting to see at what point in a given culture—or many cultures if indeed this happened—the mythology changed to say that the baby is made up of semen, accumulated through repeated intercourse during pregnancy.

Although in Western culture men did not take over and professionalize or commercialize menstrual technology until the 20th century, they had much earlier transformed menstruation from a source of woman's pride and strength to something shameful and hidden, into one of the many ways in which women had to feel inferior to men.

## Women's Contributions, 19th-20th Centuries

As late as the early 20th century in Western Europe, England, and the United States, women still created their own pads or tampons from sponges or soft cloths or rags. My own mother (1908–   ) and grandmother (1886–1980) in rural Ohio folded and shaped old cotton knit undergarments into pads which they pinned to their underwear or to home-made muslin belts with tabs (D. Stanley, L-11/10/1979). The British menstrual napkin of the 1890s was made of the same bird's-eye-weave cloth as a baby's diaper, but slightly smaller; it was rinsed out in the water closet and put to soak in a copper washtub (Vicinus, *Suffer,* 41).

Even while many women were improvising in this individual way, other women were patenting inventions in "catamenial" technology (from the Greek word for the menses; first recorded use of the adjective in English, 1851), beginning at least by 1883. In fact, LWP shows 16 women who patented catamenial sacks, monthly protectors, suspensory bandages, and the like between 1883 and 1894 (App. SFA-4).

A seventeenth woman, **Emma H. Carpenter** of Springfield, VT, patented a Catamenial sack or bandage in 1892, unaccountably omitted by LWP. Some of the several other "Dress-protectors" listed by LWP (e.g., **Elizabeth Higgins,** Boston, MA, 1862), as well as one "Ladies' safety belt" (**Mary G. Porter** of Charlestown, MA, 1870), may be menstrual in nature. The "Bandages" listed by LWP as patented in 1842 and 1847, respectively, by **Maria P. Dibble** and **Frances Carter,** are not for menstrual use, though "bandage" is a frequent term for such a device. They are an Umbilical supporter combined with a corset (Dibble) and a Uterine supporter (Carter), respectively, and classified in Class 20, Surgical instruments (Ciarlante, L-11/24/86).

Did 19th-century women patent menstrual aids in disproportionate numbers; that is, more frequently than their percentage of all patents granted? Vern Bullough found "at least 20" such patents between 1854 and 1914, of which he lists 18.[8] The 19th-century part of his listed sample contains 14 patents, of which two were granted to women (Bullough/Trescott, 245, 250–1, n32). This gives a figure of 1/7, or more than 14%. Compare this with the share of U.S. patents granted to women in the 19th century, nearing 1% at the end (soaring from about .4% in 1873)—though it has been estimated by Patent Commissioners themselves as low as 1 in a thousand (Commissioner's Report, 1900).

Bullough's sample may or may not be representative, but it is certainly not complete for women's patents. As just noted, my preliminary search for 19th-century women's patents on menstrual technology turned up 17 between 1883 and 1894 alone. If we could assume that Bullough located all the male patentees (12), and I located all the female ones (17, including his two), we would have a total of 29, of whom 17, or more than 58%, were female.

In the 20th-century part of Bullough's sample, which contains only four patents, two, or 50%, are women's.

What we will never know, of course, is what women invented, improved, adapted, and passed on to friends, daughters, and neighbors without benefit of either patent or trademark. The menstrual sponges and cups of the 20th century, for example, are re-inventions and adaptations to new materials of very ancient inventions (see, e.g., **Doris Moehrle** and Contraception section below). Even women who patented one invention may have failed to patent others. **Louise Epple** of Chicago, for instance, patented a Galvanic abdominal truss in 1888, but she exhibited supporters, trusses, and bandages (some possibly menstrual) at the Chicago Exposition, apparently without taking out any further patents (Weimann, L-12/82).

---

8. Bullough's list gives patent numbers only, no names (Bullough/Trescott, 245, 250 n32). My total of 17 includes only indisputably menstrual devices. If all of the dress-protectors and -shields had turned out to be menstrual-related, the total could have reached 51. Though some are protectors for the hems of dresses in rainy weather, and **Mary Dewey**'s Dress shield (Chicago) was for the under-arm, etc., others will almost certainly turn out to be menstrual technology. Readers who wish to follow up on Bullough's list should note two minor errors: the patent listed as #57,664 is actually #57,665; and the last patent on the list dates from 1915 rather than 1914.

## 20th Century

Moving into the 20th century, we find the same mixed pattern of home improvisation and patented, hopefully commercial menstrual devices, especially early in the century. The spread of information was frustrated by the prudishness of the advertising industry—and, indeed, of the whole culture—forbidding "sanitary products," as they were now beginning to be called, to be advertised. For example, Johnson & Johnson developed a disposable sanitary napkin called "Lister's Towel(s)" as early as 1896, but dropped it when it could not be advertised and failed to sell (Delaney, 116; Bullough/Trescott, 245).

The same three basic technologies prevailed, for the most part, in the new century: absorbent pads, absorbent tampons, and retentive cups.

### Menstrual receivers and related technology

*Pads and belts.* Menstrual pads require some means of suspension or anchoring, usually an elastic belt with fasteners ranging from pins to ingenious means of locking the long tabs at both ends of the pads into a special hook or buckle to prevent accidental release. The first widely accepted breakthrough in pad technology was the use of cellulose fiber for disposable pads, pioneered by French Army nurses in World War I.[9] During the war, the Kimberly-Clark Corporation produced surgical dressings from cellulose wadding. Noticing that this wood-fiber wadding was far more absorbent than cotton, the nurses adapted it to menstrual absorbers, thus inventing the modern form of disposable sanitary napkin.

Left with a surplus of the bandage material, Kimberly-Clark exploited the nurses' invention, and Kotex® pads were born. The pads as finally marketed in early 1920[10] consisted of several rectangular layers of wadding sandwiched together and wrapped with gauze. The result was five times more absorbent than cotton. Even though advertising was still taboo—it took the company four years to persuade the *Ladies' Home Journal* to accept its ads—and the pads cost ten cents each, this development caught on, perhaps partly because of its superior absorbency.

MODESS® sanitary napkins—designed in response to what the (obviously female) consumers said they wanted—appeared from Personal Products Corporation (Johnson & Johnson) in 1927.

By 1976 the "feminine products" industry was a half-billion-dollar industry. By 1978 sales of feminine napkins had grown to $348 million in the United States alone. This represented six billion napkins. If the 5.4 billion tampons sold in the same year, again in the United States alone, are included, the sales rise well over half a billion dollars. (Bullough/Trescott, 246; Delaney, 115, 116; Kimberly-Clark Corp., Kotex Feminine Napkins Product Fact Sheet (PFS) and 1978 Suppl. Data, 6, 8; Personal Products).

In the years just before the new Kotex pad, we find the two 20th-century patents mentioned by Vern Bullough (Trescott, 250)—**Jennie L. Bornstein,** Dayton, OH, Sanitary support, (1912), and **Elfreda J. Corbin,** Minneapolis, MN, Napkin, (1915)— plus the very few listed anonymously by the Women's Bureau study for 1905–21. There is no separate category for menstrual-related devices, and I find only one certain listing (covering an unknown number of inventions): in Category XVI (Articles of personal wear

---

9. These nurses are unnamed in all my sources. I would be grateful if any of these women—or their daughters or granddaughters—would communicate with me.

10. Delaney says 1921; information sent me in 1979 by Kimberly-Clark Corporation (PFS: 1) says, "Early in 1920, Kimberly-Clark marketed the first disposable sanitary napkin. . . . Absorbency was the great advantage and early advertising carried this message. The initial product was very crude compared to present-day standards, but it performed admirably, nonetheless."

and use), a listing for "Catamenial appliance." This would include the Catemenial bandage patented by **Ida M. Argo** of Dallas, TX, in 1910.

Possibly menstrual-related devices are the "Dress shield" of Category XVI; and the "Bandage," "Supports," "Supporting belt" and "Suspensory" listings under Category XVIII (Medical/surgical/ dental), subcategory C. Nothing relevant appears under Category XIX (Safety/sanitation). Thus WB/28 is virtually useless here.

According to a 1913 *New York Times* article on women inventors, **June Haughton** was not only an expert rifle shot and a traveler, but the inventor of a practical motoring veil and "several small sanitary devices" ("Women Who," Oct. 19, 1913). The author's reticence suggests that Haughton's inventions were menstrual in nature. If we can infer a c. 1913 date for Haughton's patents, they should be counted in WB/28.

Three years too late for WB/28 is the Catamenial sack or napkin-holder patented by **Anna Brand** of Chicago in 1924.

Even after Tampax appeared in 1933 (Delaney, 117), many women continued to prefer—and to invent—pad-and-belt technology. In 1940 **Aniela Majewski** of Chicago patented a Sanitary pad holder. This was not a belt or fastening system, but a casing (analogous to the gauze casing of commercial pads) with an opening allowing the pad of one's choice to be inserted, and a means for keeping the pad from slipping.

In 1956 two patents were issued to women, one for an improved elastic belt with adhesive pad-attachments, and one for an improved "receptacle" or pad: **Beatrice Kenner** of Washington, DC, and **Anne M. Landy** and **Neola Seidler** of Philadelphia, PA. Ms. Kenner's invention is of particular interest, not only because she is one of the few black women inventors I have identified, but because she appears in a 1980 article on black and women inventors that tends to focus on relatively important inventions (Ives, 122). The author gives no details; but if Beatrice Kenner's belt was the first to provide adhesive tabs for attaching the pad, her invention could be a significant improvement indeed.

Landy and Seidler's invention is not just a pad, but a pad combined with a "fluid-impervious" layer and rim; designed wider in front and narrower in back; and contoured to fit the female anatomy so as "to lie in substantially flat contact" with the wearer's body. The Gazette description sounds, in fact, like some 1980s ads for the newest in absorbent menstrual pads.

The technology continued to evolve through the 1960s and '70s, with adhesive pads that cling to underclothing without belts, "light-days" pads, "maxi-pads" that are both thinner and more absorbent than earlier models, pads with waterproof outer layers, etc. According to a 1979 Kimberly-Clark ad, the first beltless pads came out in 1972. This evidently refers to the first Kimberly-Clark beltless pads; for Personal Products introduced STAY-FREE® beltless mini-pads in 1969, and STAYFREE® beltless maxi-pads in 1971. Developmental work on beltless products had begun at Personal Products (a Johnson & Johnson affiliate) in the early 1960s. The idea for the beltless or self-attaching pads—and for the smaller, more streamlined pad for light-flow days—came from women, during the course of "consumer panel tests"; and women were involved in their development at Personal Products; but Personal Products would not divulge those names (Keithler, L-1/5/83; Personal Products, n.d. [1983]).

Luckily, Kimberly-Clark has no such policy.[11] Nine female employees received patents for "feminine care products" between 1970 and 1986. Of these, four inventors, accounting

---

11. I am grateful to this company, and particularly to Tina Barry, its Director of Corporate Communications, for the names of women inventors at Kimberly-Clark and, rarest of all, for names of specific products incorporating the women's inventions.

for seven patents, worked in sanitary-pad technology, which received a new impetus after the toxic-shock scare of the early 1980s:

In 1971 **Carolyn R. Mobley** of Appleton, WI, patented a Tab construction for a sanitary napkin. Her improvement was to add a pressure-sensitive adhesive strip near the end of the tab, covered with a fairly stiff cover strip. This stiffness makes the tab easier to thread through the clasp or grip of a sanitary belt. The woman removes the cover strip and uses the adhesive patch to roll up the napkin for disposal.

In 1983 **Eleanor J. Fendler** and Leo J. Bernardin, also of Appleton, patented an Absorbent pad including a microfibrous web. In a cogent background discussion, the inventors point out that conventional absorbent materials are bulky and uncomfortable, and that "substantial portions of the absorbent layer are not used at all." One solution was to insert a layer of glass microfibers as a wicking agent between two layers of conventional absorbent material. This worked, but had its own problems, as the glass fiber was expensive, caused machining problems, and tended to come apart during use. Fendler and Bernardin's invention was to treat so-called superfine thermoplastic fibers, previously considered "inimical" to absorbing water solutions, in such a way that they can replace the glass wicking layer. This invention has been used both in the New Freedom® and the Thin Maxi Pads®.

Later that same year, **Pamela F. Baum** of Neenah, WI, patented a Sanitary napkin with disposal means. The improved "baffle" for enclosing the used napkin is attached so as not to interfere with adherence to the wearer's undergarment, and so that the user need not touch the napkin surface in order to wrap it for disposal.

The most prolific of the Kimberly-Clark inventors is **Billie J. Matthews** of Menasha and Winnebago, WI. Between 1980 and 1986, Ms. Matthews received five patents, four of them for sanitary napkins. First of these four was a Sanitary napkin with improved panty adhesive (1982). This "panty-liner" napkin conforms to body shape, with sides curving inward at the center. Her improvement lies in the placement of the adhesive, particularly in ending it before the napkin begins to curve outward again near its two ends, since, as she puts it, "adhesive at the ends of the napkin . . . may end up attached to the wearer rather than the undergarment."

Her second (1983), with four male co-inventors, is a napkin using thermoplastic fibers to form a distortion-resistant comfort layer. Also provided are special "transfer zones" to aid in absorbing "viscous" menses. This invention is incorporated into the Kotex Overnite pad®.

Billie Matthews' third and fourth napkin patents deal with increased resilience and shape-retention, and thus also with comfort: a Sanitary napkin with following baffle (1985); and a Resilient shape-retaining sanitary napkin (1986) (Barry, L-12/3/86).

Kimberly-Clark women continue to invent. In 1987 **Barbara Oakley** co-patented a Three-dimensional sanitary napkin having absorbent material contoured on the baffle side. Her invention will be used in Kotex's new Profile® feminine pad (Barry, L-1/27/87).

Procter & Gamble's 1986 entry in this field is "Always Plus"® ("If Your Maxi . . .").[12]

---

12. Interestingly enough, Procter & Gamble's name does not appear anywhere in the ad here cited except for the inconspicuous initials "P&G" with the copyright date. This uncharacteristic reticence may be connected with the still controversial implication of Rely® tampons in the toxic shock syndrome scare of 1979–80. Although all superabsorbent tampons were found more likely to cause TSS than their regular counterparts, Rely® tampons came in for the lioness's share of the blame—partly, as one perceptive newspaper columnist pointed out in late 1980, because they were so good at their job (Friedman, 60; Moskowitz).

This maxi-pad features a stay-dry layer next to the skin, a contoured shape wider at the ends and narrower in the middle, with adhesive flaps extending from that narrowed middle to fold over and attach to the outside of the panty, thus protecting the edges of the panty legs from stains, and attaching the pad more securely. This pad also comes in three lengths and at least two thicknesses for versatile protection.

With any kind of menstrual receiver, but absolutely with the beltless pads, undergarments become an essential feature of the technological system. Even with the pad suspended from a belt, special protective undergarments may be needed for days of heavy flow. **Sharon Washington** (1950–   ) of Pacoima, CA, has invented such a menstrual undergarment. A black born in St. Louis, Ms. Washington is also a versatile multiple inventor, with a woman's protective headgear and a teaching aid also to her credit. As of 1981 she had not patented any of her inventions, though she had tried with the headgear (Washington, I-1981).

*Tampons, sponges, etc.* Active women have long preferred internal menstrual absorbers that allow free physical movement, whether it be dancing, running, tennis, skating, hiking, or any outdoor sport or work. Among the earliest and most popular of such absorbers must have been natural sponges shaped into tampons. American women of the 1960s and '70s seeking to live more naturally and reduce their dependence on Establishment medicine, revived the use of natural menstrual sponges. This development was promoted by women's health centers and by small entrepreneurs like Sally Stevens of Sally Simmons Sponges, Comptche, CA (Lewallen, 16).[13]

Modern women have also adapted cellulose sponges to absorb menstrual flow. **Alicia Bay Laurel** is one who reduced the idea to specifics and thus to an invention: she suggests cutting cellulose kitchen sponge measuring 3-$\frac{1}{2}$″ × 4-$\frac{3}{4}$″ "width-wise" into four strips. In comparative testing with a popular brand of tampon (Pursettes Plus®), Janice Delaney and her co-authors found the sponge tampons lasted for 6-$\frac{1}{4}$ hours of heavy flow, whereas the commercial tampon lasted only 4 hours. The sponge tampons can also be reused after rinsing and can be boiled to sterilize them after several uses (Delaney, 118–9).

In the mid-1970s a women's self-help publication described the use of an artist's sponge, with a string attached to make removal easier (Nelson, 5). **Doris Moehrle** tells of a specially shaped menstrual sponge handed down from mother to daughter in her family, a kind of cup/tampon; this "Nature's Answer®" tampon, which she patented in 1978, also had a ribbon attached to facilitate removal. Called a "Reusable tampon" in its patent description, the modified commercial product is made of synthetic foam, covered with an open-mesh fabric, capable of being closed or removed by a drawstring. Like cellulose sponges, it can be rinsed and re-inserted, then sterilized by boiling (Moehrle, L-10/19/79). At last report, Ms. Moehrle was still struggling to market her product, partly because she wanted it recognized also as a contraceptive device.

Women have also been involved in inventing the better-known commercially available tampons, first marketed by Tampax in 1936 (Langone, "Riddle," 26). The Meds® ad in *Good Housekeeping* for June 1943 shows a white-coated woman wearing a stethoscope,

---

13. Unfortunately, the world's oceans have become so polluted that some naturally harvested sponges may be unsafe for vaginal use. In October of 1980, after one case of toxic shock syndrome was traced to such a sponge, the University of Iowa's Hygienic Laboratory found bits of sand and other debris, as well as traces of hydrocarbons (possibly from oil spills) in twelve Mediterranean silk sponges sold by the Emma Goldman Clinic for Women in Iowa City. The Clinic stopped selling sponges and recalled those already sold. Other women's health centers either followed suit, or prominently posted the results of the Iowa study. The UC–Davis Health Collective then launched a study of sponge data from around the world (Friedman, 60, 62).

and is captioned "Perfected by a woman's doctor," implying that a woman deserves credit for the Meds'® "insorber" feature, whereby the tampon absorbs 300% of its own weight in moisture. In the process it expands to a leakage-reducing cone shape with a "safety well" to hold unabsorbed fluid ("Perfected . . .," 100).

**Dr. Judith Esser-Mittog** of Düsseldorf, Germany, designed the o.b.® tampon for Johnson & Johnson. As late as 1981 the company's magazine ads credited an unnamed "woman gynecologist" (e.g., *Ms.*, March 1981). This frustrating anonymity was evidently Dr. Esser-Mittog's own choice (Oram, L-12/13/82). In more recent advertising, however, including some television ads, Dr. Esser-Mittog does appear, not only by name, but on camera (e.g., KGO-TV, San Francisco, June 26, 1984, *Hart to Hart* show prime-time commercial).

Consisting of rolled fiber-pad layers, the o.b.® supposedly expands uniformly from all sides, filling the vagina more completely than less flexible tampons, and guarding against leaks. The cotton-and-rayon layers achieve the necessary absorbency without any of the "super-absorbent" fibers implicated in toxic shock syndrome (Podesta).[14] The o.b.® tampon comes in three sizes and three absorbencies. Since the 1981 ads mention thirty years of use-testing, the invention must date to the early 1950s.

The o.b.® tampon is inserted "naturally," as the 1981 ads put it, i.e., without the insertion tube favored by some manufacturers—and many women. These tubes themselves, made of plastic or slick-coated cardboard, were an invention—or rather, several inventions, as there are various designs. Cleverly engineered with parts sliding closely inside each other, the tube adapted the syringe principle to "inject" the roll of absorbent material smoothly into position within the vagina.

As of 1976, at least two of the o.b.® tampon research and development staff—as well as two of Johnson & Johnson's marketing staff assigned to it—were women (Mahaffy, L-12/2/76).

At Kimberly-Clark, five women (positions unspecified) received eight patents for tampons or related inventions between 1970 and 1986 (Barry, L-12/3/86). **Virginia A. Olson** of Appleton and Oshkosh, WI, received three patents: Tampon applicator (1970; covered applicator tip to prevent absorption or displacement of body fluids during insertion, thus easing insertion); a Catamenial tampon and wrapper (1972; to keep the tampon clean before use and to wrap the used tampon for disposal; wrapper is attached to tampon in easily removable way that yet helps keep user from dropping it during unwrapping); an absorbent tampon (1972; also concerned with preventing drying of vaginal fluids during insertion). This last invention is used in the Kotex® stick tampon (Barry, L-1/29/87).

Five other Kimberly-Clark women also received tampon patents during this period (Barry, L-12/3/86, 1/29/87, 3/13/87): **Virginia A. Corrigan,** of Appleton, WI—Coating to aid tampon insertion and tampons coated therewith (1971; technology involved here applies a lubricant to a compressed tampon without either decreasing its absorbency or causing it to expand); **Patricia J. McKelvey,** of Appleton, WI—Wrapper structure for

---

14. For a reasoned review of the TSS scare, see Radetsky. Although a CDC study connected TSS preferentially with one brand of tampon, Proctor & Gamble's highly efficient Rely®, two studies done after this brand was withdrawn from the market showed no decrease in incidence of the disease. James Todd points out that premenstrual girls, postmenopausal women—and men—also get the disease. Ten of his original 35 cases were men, in fact. Rely® may have been a scapegoat. As of Radetsky's 1985 article *the mechanism of the disease was unknown,* though tampon use may be associated with increased *risk* among menstruating women, and though the bacterium involved, *Staphylococcus aureus,* was pinpointed early on.

tampons containing super-absorbent material (1977; this wrapper retains enough fluid at the tampon's surface to ease withdrawal); **Lin-Sun Woon,** of Appleton, WI—Folded Tampon pledget (1982; the folding method used here exposes more absorbent surface area, and creates a containment pocket for absorbent material, a tapered leading edge for ease of withdrawal, and a wrapped surface to prevent sloughing of fibers); **Billie J. Matthews,** of Menasha, WI—Tampon containing fusible portions (1981; designed for ease of manufacture and assembly; provides for variable absorbency); **Dawn M. Huffman,** of Winnebago Co., WI—Outer tampon tube with recessed finger grip (1986; provides grip for fingers on outer of two telescoping tubes of tampon-inserter without creating an uneven inner wall).

*Menstrual cups.* A 1970s women's self-help pamphlet recommends using an old diaphragm to catch menstrual flow (Nelson, 5), thus epitomizing both the ancient connection between menstrual and contraceptive technology and women's perennial inventiveness. In this case, since commercial preparations are available, the motive is likely to be either economic (pads and tampons are expensive) or health-related (tampons have been implicated in toxic shock syndrome, and also contain substances potentially harmful to delicate vaginal tissues).

However, as with menstrual pads and tampons, women have also been involved in inventing and developing commercial cup-type menstrual devices.

As a ballet dancer, **Leona Chalmers** of New York City found conventional menstrual absorbers inadequate. In the late 1950s, with the aid of her gynecologist brother, she developed a rubber menstrual cup, the Tassette® (Fr.: "little cup"). Tassette operated on the principle of catching and holding the menstrual flow rather than absorbing it. This is not really a new idea, for women have long adapted their cervical caps and diaphragms to catch menstrual flow. But Chalmers' cup was new in shape and placement, in forming a partial vacuum with the sides of the vagina to prevent leakage, and in being intended for mass production.

The 1960 trademark rested upon two 1937 patents, for a "Catamenial Appliance" and a "Vaginal Seal."

Chalmers leased sales and production rights to Tassette, Inc., of Beverly Hills, CA, which marketed the cup starting in 1959 and sold 100,000 units in just a few years. Unfortunately, however, two of Tassette's® greatest virtues—its reusability and its ten-year product life—made it unprofitable for a company manufacturing nothing else, and by 1964 the search was on for a disposable version. At some point, presumably by the mid-1960s, Leona Chalmers died.

**Barbara Waldron,** RN, Professional Services Director at Tassette, collaborated with John W. Waldron of Plastic Applications, Inc., to create a disposable version of the cup, which they named Tassaway.® Tassaway® is made of a very resilient rubber-like plastic that is self-lubricating and does not cause chafing. Barbara Waldron holds both the first Tassaway® patent (1968) and an improvement patent (1971).

Janice Delaney calls Tassaway® the first truly new menstrual product since the fig leaf:[15] it holds more than the average tampon, doesn't leak, and as an inert plastic, supports no bacterial growth. Moreover, although the manufacturers discourage this, it can actually be re-used if desired. Tassette, Inc., changed its name to Tassaway, and began marketing the new product, but went out of business in 1973. Barbara Waldron became vice-president of Plastic Applications, which acquired all rights to the cup, and still hoped

---

15. Tassaway® even found its way into a poem, written by Ellen Bass in a *Chrysalis* special issue on menstruation (1979; Bass, 48).

as of 1983 to market Tassaway®. The company holds enthusiastic letters from thousands of women asking for the product, which the Waldrons consider far less likely to cause toxic shock syndrome than other forms of internal protection, and far more truly sanitary than so-called sanitary pads.

Interestingly enough, presumably because of Barbara Waldron's role, Delaney found Tassaway's® ads factual and respectful of women's intelligence, using terms like *menses* and *uterus* and speaking in longer and more complex sentences than usual. Moreover, the Tassaway Company was an important part of the effort that finally broke the U.S. ban on radio and television advertising of sanitary-protection products—in November 1972. They ran ads on stations not subscribing to the National Association of Broadcasters' code; when little or no adverse reaction followed, they convinced the NAB to lift its ban (Delaney, 113, 117; Oram, L-12/13/82; Plastic Applications, 1,5, *passim;* J. Waldron, L-1/10/83).

## *Menstrual extraction*

The most revolutionary development in the field of menstrual technology is menstrual extraction, developed by the women of the Los Angeles Self-Help Clinic in 1971. Using a flexible plastic cannula and a plastic syringe, the women found that they could suction out the contents of the uterus and cut the menstrual period for many women from five days to five minutes. Interestingly enough, primitive women were before us here as well, though their methods were herbal rather than mechanical: Chippewa women reportedly used blue cohosh, often called squawroot *(Caulophyllum thalictroides),* to ease *and speed* menstruation (De Laszlo & Henshaw, 628). As of 1977, menstrual extraction, though potentially a great boon to women athletes, especially swimmers, and to women with long and heavy or painful periods, remained experimental, partly for legal reasons, partly because the procedure itself can be painful, and partly because the long-term effects of regular use were not yet known (Seaman & Seaman, 295–6).

The names most often associated with this innovation are **Carol Downer** and **Lorraine Rothman.** Downer is one of the founders of the women's self-help health movement, and a leader of the Los Angeles group. Rothman invented and patented the safer instrument now used by the clinics, the Del-Em (1974). To the original improvised equipment, Rothman added a by-pass bottle and a valve-and-tubing system to prevent air embolism (air injected into the uterus, potentially fatal) and to keep too much extracted material from collecting at once. Using this device, women have done much-needed research on menstruation. Unlike many research subjects, the women who participate here are not only fully informed and genuine volunteers, but actually in full or partial control of any procedure used (Fishel, 56–7, 58; Gage, 17ff; Rothman & Punnett, 44–8).

## *Remedies for menstrual pain and difficulties*

Women's herbals are full of preparations to deal with menstrual problems. Every culture had its herbs and teas, its massage techniques and pressure points, to help any woman who suffered in her menses, or failed to menstruate. Hurd-Mead, citing Pliny, mentions several ancient Greek medical women whose names, if not their works or their exact dates, have survived: Lais, Elephantis, Olympas, and Sotira, all of whom wrote on menstruation, among other things. Olympas used hyssop and nitre with bull's gall on a pessary to promote menstruation. They may have lived as early as the 4th century BC (*HWM*, 41, 56–7). Only the sketchiest idea of this rich lore can be given here—in Western culture much was lost with the witch-burnings of the early modern period—but these examples should suffice to indicate the quality and scope of the achievement:

*Amenorrhea (absence of menstruation).* Women all over the world have developed menses-promoters or emmenagogues to be used in amenorrhea, whether to bring on the first flow or to restore lost flow. According to Beatrice Medicine, for example, the Dakota Sioux know herbs to bring on menstruation in cases of delayed puberty (PC-5/1980). When the reason for lost menses was early pregnancy, of course, the emmenagogues functioned as secret abortifacients. Indeed, a list of traditional herbal emmenagogues and a list of traditional abortifacients would overlap considerably, though dosages and accompanying regimens would differ. Nineteenth-century medical entrepreneurs took advantage of this overlap to advertise openly their potions "to begin the monthly cycle" (Sklar, *Beecher*, 208, 318–29 n11).

Among women's many emmenagogues are aloes *(Aloe socotrina)*, angelica *(Angelica atropurpurea)*, arsesmart or water pepper *(Polygonum hydropiper)*, black cohosh *(Cimifuga racemosa)*, blue cohosh *(Caulophyllum thalictroides)*, castor bean leaves *(Oleum ricini)*, celery *(Apium graveoleus)*, centaury *(Sabbatia angularis)*, cotton root *(Gossypium herbaceum)*, feverfew *(Chrysanthemum* or *Pyrethrum parthenium)*, horehound *(Marrubium vulgare)*, and motherwort *(Leonurus cardiaca)* (Hutchens, s.v. common names). A modern self-help publication further lists red sage, basil, lemon balm, parsley, camomile, American saffron and elecampane (mild); yarrow, ginger, rosemary, mugwort, thyme, southernwood, false valerian or squaw weed, wintergreen, gentian, and ephedra or Mormon/squaw tea (moderate); and pennyroyal and myrrh (strong). This publication also mentions high doses of Vitamin C for five days before the period is due (Nelson, 4). Pennyroyal, according to Michael Weiner, was "extensively used in domestic American medicine" to promote menstruation; dried pennyroyal leaves remained in the *U.S. Pharmacopoeia* from 1831 to 1916. Notice how many of these herbs, like bay laurel (J. Rose, 41–2), are also condiments or seasoning herbs, and thus could be grown openly in women's gardens.

*Pre-menstrual tension.* Among women's remedies for pre-menstrual tension are camomile, spearmint, and lavender (mild); catnip, mugwort, alfalfa, kola nuts, and celery seeds (moderate); valerian, vervain, skullcap, passion flower, and hops (strong). Women have also used ladyslipper, but it can be poisonous. *All the herbs listed as "strong" must be used with extreme caution.* Valerian, for example, is a muscle relaxant that could affect the heart; and passionflower contains a morphine-like narcotic.

Contemporary women's herbals show awareness of the relationship between calcium intake and menstrual difficulties, recommending an increased calcium intake before the period is due, whether in the form of alfalfa (high in calcium) or a dolomite (calcium-plus-magnesium) supplement (Nelson, 5).

*Dysmenorrhea (painful, excessive menses).* Women also developed many remedies for painful, excessive, or otherwise abnormal menstruation. For painful menstruation, or cramps: black cohosh, blue flag *(Iris versicolor)*, camomile *(Matricaria chamomilla)*, cotton root, elecampane *(Inula helenium)*, and wild ginger *(Asarum canadense)* (see Hutchens). Chris Nelson (6) mentions dolomite and the following herbs used for cramps in addition to those already listed: sweetbalm, mugwort, motherwort, squawvine, spicebush, parsley, ephedra, and sassafras. Stronger remedies, to be used with caution, are black haw or cramp bark (a specific uterine sedative), skullcap, ginger root, blue cohosh, vervain, valerian, lobelia *(ibid.)*. Arikara women used *Artemisia* species during their periods: an infusion of the big wild sage *(A. gnaphalodes)* or of the roots of the little wild sage *(A. frigida)*. Rappahannock women relieved menstrual pain with a tea of fresh or dried pennyroyal *(Hedeoma pulegioides)* or a tea made from split twigs of spicebush *(Benzoin aestivale)* (Vogel, 238).

For excessive flow, known to modern medicine as menorrhagia, the remedies were often astringents. Cora Du Bois reports that the women of Alor had "medicines" to reduce the menstrual flow (108); and Minoan Cretan women used "the lily" to "check menstruation" (Bridenthal & Koonz, 47). Also reportedly effective for decreasing flow are: yarrow and horehound (mild); *Uva ursi* and red sage (moderate); and red raspberry, shepherd's purse, black haw, witch hazel, white oak bark, bayberry bark (strong). A combination tea for decreasing flow uses prickly ash, red raspberries, blue cohosh, wild yam, and cinnamon (Nelson, 4). One or both of the two sages used by Arikara women may have been intended to control excessive menstruation as well as to reduce pain (Vogel, 238).

Jeanne Rose's herbal lists further: amaranth *(Amaranthus hypochondriacus)* and lady's mantle *(Alchemilla vulgaris)* (38, 74). Michael Weiner notes that "several tribes" of North American Indians used blue-cohosh-root tea to reduce profuse menstruation (90). Alma Hutchens, with her unique Native American/Russian perspective, confirms amaranth, shepherd's purse *(Capsella bursa pastoris),* and witch hazel *(Hamamelis virginica),* and adds, among others, birth root (trillium or wake robin; *Trillium pendulum; T. erectum*); goldenseal *(Hydrastis canadensis),* life root *(Senecio aureus),* nettle *(Urtica dioca),* Solomon's seal *(Polygonatum commutatum),* and star grass *(Aletris farinosa)* (Hutchens, 23, 248, 306; 34, 144, 180, 205–6, 260).

The barberry *(Berberis vulgaris)* is reportedly useful for both menstrual pain and excessive flow (Hutchens, 25).

While some women study and attempt to recapture ancient herbal lore, others contribute within the medical establishment. **Dr. Penny Wise Budoff,** a family practitioner and medical school professor (SUNY), pioneered the use of prostaglandins for menstrual disorders, and developed a dietary regimen to combat premenstrual tension. Reading about Dr. Budoff's work, I was struck by something that should have been self-evident: she experimented first on herself. This time-honored strategy of medical investigators would not be available to a male physician.

Since 35 million women in the United States alone suffer some form of dysmenorrhea, Budoff's remedies are significant indeed (Budoff, *Cramps;* Thorpe, 36).

**Tamara Dejneka**'s three patents in prostaglandin research also deserve a brief mention here. Dejneka and two male co-patentees received two patents in 1979 and one in 1978, all assigned to E.R. Squibb & Sons of Princeton, NJ. Both of the 1979 patents involve a Method for inhibiting prostaglandin dehydrogenase.

Perhaps the biggest name in the field of pre-menstrual tension is British physician **Dr. Katharina Dalton,** who recognized and treated pre-menstrual syndrome (PMS) as early as 1948. When her own pre-menstrual headaches were relieved by monthly doses of progesterone, she began prescribing it to her patients. Although she agrees that eating starches and avoiding birth-control pills does alleviate PMS, she sees little value in some alternative approaches, considering vitamins overemphasized, and finding exercise irrelevant except in helping the patient's outlook (Eagan, "Selling," 30).

Other women physicians, however, are wary of progesterone and advocate more natural methods. Animal studies have shown increased rates of breast and cervical cancer; some women develop chest pains after a time, and vaginal and rectal swelling are common. Some women bleed every day, and some fail to menstruate at all. Moreover, the effect of a given dosage diminishes after a time, and the dose must be increased to have its effect. Roger Eastep of the FDA pointed out in 1983 that so-called Phase I studies—determining how much of a particular substance is absorbed by the body and how it works—have not been done for progesterone *(ibid.).*

Like Dr. Budoff, **Dr. Michelle Harrison,** a gynecologist practicing in Cambridge, MA, advocates dietary treatment for PMS. Most of her patients respond to a hypoglycemia diet and regimen: whole grains, no caffeine, lots of water, no sugar, frequent small meals, with 800 mg. of Vitamin B-6 during the pre-menstrual phase itself. Her self-help booklet contains good advice about diagnosis and treatment, charts for monitoring symptoms, and a look at the social and political questions PMS raises. She prescribes progesterone only for women "whose lives are shattered by PMS, who've made repeated suicide attempts or are unable to keep a job." Dr. Harrison is a spokesperson for the National Women's Health Network.

Likewise, **Dr. Marcia Storch** and her associate Dr. Shelley Kolton find that reducing salt intake helps limit water retention and prevent the resulting headaches. This simple precaution, plus 300 to 500 mg. of Vitamin B-6 daily, works in about 80% of all cases, though it may require a few cycles to take effect. Dr. Storch is the author of *How to Relieve Cramps and Other Menstrual Problems* (Eagan, "Selling," 30).

On the West coast, **Dr. Susan Lark** of Los Altos, CA, has developed a diet-and-exercise treatment for PMS symptoms. Dr. Lark, who is on the clinical faculty of Stanford University Medical School, published her *PMS Self-Help Book* outlining the regimen in 1984.

At the same time, women are looking to their roots as healers, combing herbals, consulting their grandmothers, and studying anthropological reports, in an attempt to reclaim ancient wisdom about the menstrual period. Herbs used for PMS in various cultures include—in addition to a dietary increase in calcium—camomile, spearmint, mugwort and vervain, among several others (Nelson 4).

## PREGNANCY AND CHILDBIRTH TECHNOLOGY

### Prehistory: Primitive groups

Anything that eases pregnancy and childbirth probably affects a group's fertility, and definitely affects the happiness and well-being of its women. I will divide my discussion here into sections on pregnancy care, birth care, and aftercare.

### Pregnancy care

Among a multitude of possible examples, women created ointments to help stretch the skin (Goldsmith, 12) and herbal preparations to ease or prevent morning sickness, for various complications of pregnancy, to ease and speed childbirth, and to prevent miscarriage (Nelson, 8–10):

*Anti-morning sickness.*

| | |
|---|---|
| alfalfa | anise |
| blue flag *(Iris versicolor)* | cloves |
| colombo root | knot grass |
| lavender | peach leaves |
| peppermint | sassafras |
| spearmint | star grass *(Aletris farinosa)* |
| sweet balm | |

*Preventing miscarriage.*
blue cohosh *(Caulophyllum thalictroides)*
star grass

*Strengthening pregnancy, preventing complications.*
motherwort *(Leonurus cardiaca)* (kidney complications)

## Birth care

*Easing, speeding labor; killing pain.* One of the several perversions of Western culture is its underlying attitude that birth *should* be painful. Noteworthy exceptions in 19th-century American medicine were the hydropaths, or water-cure specialists. The hydropathic view can be seen in this 1846 article cited from the *Water Cure Journal* by Kathryn Sklar (318 n8): "It is very certain that woman's suffering in labor can be in a great degree prevented, and that she need not endure that weakness after childbirth which is so common. Under the water treatment the constitution is not injured from bearing children. . . ." Among the water-cure innovations were the cold hip bath for preventing hemorrhage, and postpartum sitz baths. Although I cannot attribute specific therapies to women, it is worth noting that hydropathic establishments did have women doctors on their staffs—including, as noted elsewhere, **Dr. Mary Edwards Walker** (who had several nonmedical inventions to her credit as well as innovations in the treatment of tuberculosis) and **Mary Gove Nichols** (*NAW;* Sklar, 318 n9).

Although primitive women certainly considered childbirth painful, observers of both sexes consistently report that they give birth with relative ease (e.g., Goldsmith, 21ff; Murphy & Murphy, 163; Holmberg, 178; Du Bois, 31; Bates, 235). Tlingit women of Alaska have reportedly given birth while sleeping (Goldsmith, 23)! Even the Greek and Hebrew myths underlying our own culture contain hints of a time when birth was less painful: Artemis's mother Leto bore her without pains, leading the Fates to make Artemis patroness of childbirth (Graves, *Myths* 1: 83, 84); and the Expulsion from Eden condemns Eve to painful childbirth, implying that in the Garden things had been otherwise (Genesis 3: 16: ". . . in sorrow thou shalt bring forth children . . .").

Not the least of the reasons for these persistent reports of easy childbirth could be the mother's high level of physical activity right up to the beginning of labor. Such reports are too universal to need repeating here, but it is interesting to note that in most cultures where birth takes place in a special structure, the woman builds her own birth hut,[16] a spurt of extra activity near her time.

Another likely reason for easier childbirth is the squatting or semi-upright position assumed for birth in these earlier cultures,[17] supported either by one of women's ancient inventions, the birth stool or chair, or by her mother or other women, adding psychological to physical support and soothing massage—or both. From Southeast Asia come reports of a knees-to-chin position that should also be quite effective (Goldsmith, 21).

*Physical aids.* Far too little attention has been given to women's important massage technology. Its role in pregnancy and childbirth has been almost entirely ignored by

---

16. Judith Goldsmith (26–7) points out that presence of these birth huts does *not* mean that the society secludes its birthing women as "unclean" or "polluted." Rather, women seclude themselves. It could be argued that women consider birth a woman's mystery, that they feel a need to either exclude or protect men from it, or men feel a need to protect themselves from the power of this blood ritual.

17. Modern doctors belatedly analyzing the physics of childbirth and birth positions have now announced that *when a woman is standing, her contractions are 100 percent more effective than when she is lying down,* and moreover, that in the position most often adopted by primitive women (curved back, knees bent), the pelvis opens as much as 30 percent farther than in the supine, anti-gravity position favored by doctors for their own convenience (Kitzinger, 21, 93)! Modern Western industrial culture is the only one ever known to have adopted this posture.

establishment medicine, though a late-19th-century student of birth customs concluded that "there is hardly a people, ancient or modern, that do not in some way resort to massage and expression in labor." Modern midwives still use it. Primitive women commonly massaged pregnant women throughout pregnancy, as for example, among the Nama Hottentots, where a mother-to-be received a massage "several times a week." Ugandan women practiced a pre-confinement massage to make the pregnant woman more supple and give her an easier delivery. Expert masseuses in the Gilbert Islands gave passive exercise and massage during the eighth month to train the muscles for the strain of delivery. (Goldsmith, 41). Javanese women of Surinam visited the midwife regularly for massage after their seventh month (Goldsmith, 39, 41).

Judith Goldsmith points out that the main duty of the birth attendant in traditional births, as the Navajo term meaning "one who holds" partly conveys, was *to provide physical support for the mother and to massage her"* (38; emphasis in orig.). Among traditional peoples in the Philippines, birth attendants "continuously" massage the abdomen of the laboring woman; and among the Semai of Malaya, both relatives and neighbors stop by during a birth and "help to massage the mother's belly" (Goldsmith, 31, 28).

A specialized and ingenious birth-massage technique, exploiting women's knowledge that an infant's nursing causes the uterus to contract, is breast massage or manipulation of the nipples during labor. This is reported from the Lapcha women of Sikkim and the Siriono women of Bolivia (Goldsmith, 54).

Massage has uses besides the mother's comfort, vital as that is. To prevent both perineal tearing and the episiotomy that causes contemporary women so much postpartum pain, ancient women (and modern midwives) recommend massaging the perineum throughout pregnancy with olive oil, wheat germ oil, or sesame oil to help tissues stretch. During the labor process itself, primitive women apply heat and massage to the perineum to ease pain, promote relaxation, and to help the tissues stretch as needed without tearing (Hurd-Mead, *HWM,* 20; Kitzinger, 90, 105). See, for example, Jane Goodale's description of Glenda's labor on Melville Island, Australia (146–7). I have personally observed a contemporary midwife continually massage the gradually stretching and thinning perineum of a laboring woman with warm oil during the last stage of a birth. Though the woman was 36 years old, and this was her first child, no tearing at all occurred.

If tearing does occur despite these measures, or episiotomy becomes necessary, women have used ground-ginger-and-comfrey-root poultices for help with pain and for tissue-building; also comfrey-root or comfrey-leaf teas (Nelson, 14).

A strong abdominal kneading or manipulation can directly help to expel the infant or to turn it if it threatens to present otherwise than head down. Birthing women of one Brazilian tribe reportedly perform this labor-speeding massage themselves. According to Goldsmith, the prevalence of massage accounts for the rarity of breech and transverse births in what she calls the "tribal world" (39–42). Massage is also used to expel the placenta (Goldsmith, 158).

Zulu women had developed breathing exercises to be done throughout pregnancy as well as during labor, long before the world heard of Dr. Lamaze (Kitzinger, 108).

*Heat.* Birth attendants in so-called primitive or tribal cultures also exploited the relaxing effects of heat. The building of a special birth fire is a prevalent custom, and most peoples considered it important to keep the birth room warm. Indeed, concern to keep the birthing woman from taking a chill may have been one rationale for birth huts (Goldsmith, 80, 31).

Modoc Indian women gave the birthing woman large quantities of warm water to drink if her muscles failed to relax. They also placed heated stones on her abdomen and beside her bed. Both Mexican women of Tepoztlan and Filipino women might administer a "smoke bath" (in the Mexican case, by directing smoke and heat under a blanket). Interestingly enough, one of the smoke sources was laurel leaves burned in a clay pot.[18] Among the Kwakiutl of British Columbia the women applied hot seaweed to the mother's stomach and lower back, whereas the Tiwi of Australia pressed hot leaves against her back, groin, and legs as each contraction began, to relieve cramps. Many traditional cultures relaxed the perineum for birth by steaming (Goldsmith, 31, 45–6, 151). However, as will be seen below, perhaps the major use of heat techniques comes immediately after the birth.

Another ancient aid to birth, rather more "high-tech" than most, was prenatal dilation of the cervix. On Yap (Caroline Islands) women inserted tightly rolled leaves into the mouth of the cervix a month before birth was expected. Like the dried laminaria (seaweed) plugs used in abortion, these slowly dilated the cervix. Dilation by hand was also occasionally practiced (Goldsmith, 225 n71).

*Music.* Midwives from very ancient times used music to aid in childbirth. Judith Goldsmith, for instance, reports the practice from four African cultures and one South American native culture (75, 96, 100, 149, 440). Evidently traditional birth attendants used music not just haphazardly, as today a midwife might put on a woman's favorite record, but systematically and knowingly. Sophie Drinker points out, in fact, that the Roman Carmentes personifying the mother's fortune in childbirth were also "projections" of the midwives' incantations (chanting songs, not magical formulas). Music, says she, was the midwives' "chief means of assistance at childbirth" (Drinker, 119; Spretnak, *Politics,* 46–7).

*Mechanical aids.* Here must be listed the birth ropes, foot-rests, abdominal bands, and birth stools and chairs developed by birthing women and/or their midwives over the millennia. Goldsmith sketches a Southwestern North American birth with the woman kneeling on a sandpile and grabbing a rope strung from the rafters during contractions, but walking around between pains; and a Southeast Asian birth in which a woman clings to a rope hanging from the ceiling and holds her knees tight to her chin. Other reports of this method come from Brazil (where Kalapalo women use their old hammock as a rope), from Bolivia, and from other parts of South America. Anthropologist Ellen Basso noted that the combination of the hammock-rope and the woman's female supporters allowed her pushing efforts to be concentrated in her abdomen, rather than partly wasted in raising her legs and lower body. Presumably the wooden foot-rests for the woman to push against (Philippines) served an analogous function (Goldsmith, 19–21, 33, 34).

Abdominal binders, worn from about the fifth month onward, support and protect the pregnant woman's belly during active work or other exertion. These or other binders then come into play during labor and delivery, being bound high around the woman in the early stages, and gradually tightened and lowered as the baby moves down, to prevent its rising again, and thus to speed labor by making contractions more effective. This device was also called a birth sash, belt, or bandage. The birth attendant herself played this role in some cultures, grasping the woman around the abdomen and squeezing (Goldsmith, 13, 39).

The birth chair or stool is one of women's more familiar inventions, and thus will be covered only briefly here. It was used in much of ancient Europe and the Near

---

18. Vapor from steamed or burning—or chewed—laurel leaves was the intoxicant used by the oracle at Delphi and elsewhere (cf. Graves, *Myths* 1: 81).

East—among the Hittites, the Egyptians, the Greeks, and the Romans—and until recently in Turkey, Greece, and Sicily. A medieval European midwife's main tool was her birth chair, which she brought to the house for a birth. Birthing in a chair is also reported from West Africa. There were many variations, but the basic design was a low, short-legged chair or stool with a horseshoe-shaped seat or a seat with a hole in the middle. The height of the seat was low enough to permit the woman to bear down effectively, but high enough to let her rest between pains.

The hole in the chair allowed the baby to drop through and fall for a short distance (if only into the midwife's waiting hands), just as in the most primitive known cultures. Analogous to this is the Brazilian woman's hammock with its lengthwise split to allow the baby to drop through (Goldsmith, 36, 33, 34). The short drop, however achieved, virtually always to a soft (and sometimes antiseptic) substance such as ashes, served the same function as the harsher modern-Western spank to the baby's bottom.

*Herbal aids; difficult births.* In addition to these physical aids, primitive women had devised many herbal preparations to ease and speed childbirth, as well as specifically to combat the pain (Hutchens; Maxwell, 225; Nelson, 12–5; M. Weiner, 32–4):

balsam fir bark
birth root *(Trillium pendulum; T. erectum)*
black cohosh *(Cimifuga racemosa)*
blue cohosh
cotton root *(Gossypium herbaceum)*
partridgeberry or squaw vine *(Mitchella repens)*[19]

Women in several North American Indian groups drank a tea made from red raspberry leaves throughout pregnancy to ease their eventual labor, and drank preparations of partridgeberry and of blue cohosh for some weeks before their births were due. Bantu midwives administer the root of the *Um-kanzi (Typha latifolia)* to strengthen contractions and facilitate labor. Mundurucu women used at least two infusions during pregnancy to ease the coming birth: from bark of the *biribo* tree and leaves of the *puwo* tree. Used at the onset of labor in North America were red raspberry, partridgeberry, cedar bark (inhaling the steam), pennyroyal, bayberry bark, and licorice. Women also used lady fern, wild cherry inner bark (Cherokee), wild yam, black haw *(Viburnum prunifolium),* and corn smut to ease labor pains (Hewat, 98, Murphy & Murphy, 163; Nelson, 13; Goldsmith, 52).

Women also had herbal preparations for the rare cases of difficult birth. Western observers have not always been able to learn what they are. W. Ten Rhyne wrote of the Hottentots in the 17th century that (Goldsmith, 44)

when a difficult labor sets in, they invoke the aid of a plant of wonderful efficiency, which happily expels the fetus. I could not succeed in purchasing this plant at any price either directly or through others. Their reply was always that they are forbidden by law to share it.

Later observers had somewhat better luck, and the wonderful plant may have been Hottentot fig *(Mesembrianthemum edule)* or *Euphorbia restituta* (Goldsmith, 52). Analogous reports came from Surinam—"the sap of a tree, which insures . . . an easy and prompt delivery"—and from the Delaware Indians of North America. Corn smut

---

19. (Nelson; Hutchens). Jamaican midwives give thyme tea to speed early labor, and have a spice tea for ineffective contractions. Thyme contains a cardiac glycoside that increases both the efficiency of contractions and uterine muscle tone; the spice tea has a proven oxytocic effect, according to a Jamaican doctor. These midwives, or *nanas,* also use castor oil to speed labor, but this is the wild variety, milder than commercial preparations (Kitzinger, 104–5).

*(Ustilago zeae),* listed above, is one of the better-known birth aids, perhaps because of its similarity to ergot and perhaps also partly because of the excellent ethnopharmacological work of Matilda Coxe Stevenson among the Zuñi. Corn smut was adopted by Euro-American doctors in the late 19th century, and was official in the *U.S. Pharmacopoeia* from 1882 to 1894 (M. Weiner, 65).

Judith Goldsmith lists a fair sampling of these preparations from various parts of the world, some of which, such as black cohosh, are already listed above. Two of the most effective from North America have ergot-like effects: decoction of pulverized rattlesnake rattles, as used by Sacajawea in 1804; and cotton-root bark; and Goldsmith cautions that many of the preparations work so effectively because they contain abortifacients (44, 51ff).

A whole category of childbirth aids were the alcoholic beverages women brewed or distilled. Specifically noted for their uses in easing labor pains are the pulque and tequila made from the maguey cactus in the American Southwest and Mexico (Stone, *Mirrors* I: 87).

Some of these aids may be very ancient indeed. Paleolithic cave paintings portray game animals with fat, pregnant bellies, and in birth-giving positions. Sometimes the painting includes an internal diagram showing the animal's womb. Branches lying across the belly have usually been interpreted as spears, but Judy Grahn argues that they more likely represent plants with birth-aiding properties (/Spretnak, 267).

### Aftercare: postpartum remedies and technology

Ancient women developed analogous technologies for the birth of the placenta—which in some groups receives great attention—to stop postpartum bleeding, heal the perineum in the rare cases of tearing, and otherwise restore the new mother. Among these many (Goldsmith, 59–62, 73–4; Kitzinger, 106, 108; Maxwell, 225; Nelson, 14) I will mention only ergot. This ancient rye grass fungus *(Claviceps purpurea),* which could have been known from women's earliest gathering of the rye heads for food, yields powerful alkaloids still used in establishment obstetrics today. They specifically affect uterine muscle, inducing strong contractions and increasing muscle tone. Ergot is especially useful for expelling the placenta and preventing loss of blood (*EB*/58: s.v. "Ergot").

*Care of the newborn.* Mundurucu women cut the umbilical cord with a sliver of arrow cane (Murphy & Murphy, 164). Since this method prevails even though the group now knows metal, perhaps it has ritual significance, or perhaps the cane has antiseptic properties. Also, cutting the cord with the blunter cutting edge provided by a plant stem cutter, as used in this and many other primitive groups in preference to metal, crushes the arteries and prevents bleeding from the cord. Ancient preferences for leaving the cord uncut until the placenta is born, and has given up all possible blood, are also evidently better for the child (Goldsmith, 63–4). Jamaican *nanas* treat the stump of the cord with finely powdered nutmeg, which is slightly antiseptic (Kitzinger, 106).

Instead of slapping a newborn infant to elicit the first breath or cry, as already noted, primitive women often arrange for the infant to drop a short distance below the birth stool, chair, or hammock, or simply from the squatting or partly supported birth position to the ground. Among the Siriono of eastern Bolivia, for example, the women give birth in their hammocks with no other aid than a birth rope to hang onto. A woman who feels labor beginning softens the ground under her hammock with her digging stick, possibly adding a cushioning layer of ashes. The newborn slips through the hammock strings and falls a few inches to the soft earth (Holmberg, 178; Goldsmith, 63). Ashes also serve an antiseptic function.

Mundurucu women squat to give birth, supported by an older woman. Their infants fall to the ground, "a short distance, but usually enough to help breathing begin" (Murphy & Murphy, 164). Jane Goodale reports the same practice among the Tiwi (147).

Some primitive groups leave the vernix caseosa on to protect the newborn until it wears off. According to Sheila Kitzinger, this would be a good idea (94). But most groups have some sort of cleansing or washing procedure, which often has demonstrable physical benefits as well as ritual significance. Many groups use a cold bath, as for example in Jamaica and among the Mundurucu (Murphy & Murphy, 164; Kitzinger, 106), which could be regarded as an ordeal survived only by the stronger infants, or as a shock to establish deeper breathing and fully inflate the lungs, or both. Often, however, the baby is well greased after the bath (Goldsmith, 67).

Alorese midwives clean newborns with the thick juice squeezed from banana bark (Du Bois, 30). Australian Aboriginal women massaged their newborns tenderly with soft ashes and charcoal, and as Daisy Bates noted, "Soft white ash was an excellent substitute for boracic powder" (235, 234). Hopi mothers sprinkled the still-damp baby after its bath with "baby ashes," volcanic ash brought from a nearby mountain (Goldsmith, 66). As already noted, Siriono babies are often born on a soft bed of ashes. Tiwi mothers rub their newborns with milk and then with charcoal, or use a mixture of charcoal, dust, and milk (Goodale, 148, 29).

Primitive peoples also developed techniques for caring for premature infants, mainly involving constant warmth (Goldsmith, 67–8).

*Recovery care of the mother.* Primitive women usually return to normal activity soon after birth—sometimes immediately. Their effective techniques for getting back into shape include massage, exercise and bathing, abdominal binding, herbal treatments, special diet, and heat treatments, including steaming. Particularly noteworthy for today's abnormally isolated new mothers is the use of massage. These reports come from Java, Thailand, Mexico, and the Philippines, where the midwife visited every morning for as much as thirty days after birth, to massage the mother's arms and legs along the major veins, as well as her back and chest, and especially her abdomen. The massage, using coconut oil, lasted about an hour.

Women also had postpartum tonics—black haw *(Viburnum prunifolium),* partridgeberry or squaw vine, pennyroyal, juniper berries—and special diet items. Among the Zulu, if a woman bleeds too much in giving birth, she gets calves' blood to drink, and a special dish of spinach to eat (Kitzinger, 107). Women also devised poultices to shrink and prevent soreness in the perineal area, as one observer reported from Cambodia in the 13th century AD. Some indication of the effectiveness of this treatment, a poultice of hot rice and salt applied for 24 hours, may be gathered from the fact that these mothers were reportedly ready for intercourse a day or two after birth—and unresponsive husbands might be discarded.

Also worthy of note are the afterbirth heat treatments, one of the most elaborate of which was the ritualized fire-rest of Thailand. Here the mother stayed for up to two weeks in the birth room, lying on a tilted bed placed beside a large fire, her infant in its cradle on the cool side of the bed. The mother controlled the heat by extinguishing parts of the fire, when necessary, but generally stayed quite warm. She nursed her baby on demand. She repeatedly rubbed her abdomen and the small of her back with a special salve, and drank as much warm water as possible. She took little food, but every two or three hours she drank a few swallows of a hot alcoholic medicine intended to stop the afterbirth bleeding. Every day for the first three to seven days, the midwife returned, massaged the mother's abdomen and thighs, and gave both mother and child sponge baths with a solution supposed to aid the circulation. The woman also took steam baths periodically. This kind of treatment could be quite effective if not carried to extremes.

In many cultures, heat was focused more directly on the perineal area, with the mother sitting on a chair with a slatted bottom over a bowl or stove of flowing coals, or on a hot rock or on a heated sand bed (Goldsmith, 72–82).

## Nursing technology

Once the child is born and seems likely to survive, women have numerous preparations to stimulate or increase their flow of milk, among them castor bean leaves (Canary Islands; *Oleum ricini*), wild ginger *(Asarum canadense),* fennel, barley water, saw palmetto berries (taken last month of pregnancy), maple bark, cottonseed tea, alfalfa, wintergreen, baneberry leaves, white poplar bark, tall blue lettuce, comfrey, skeleton weed, and atole (blue corn).

Women also devised preparations to decrease the milk supply if an infant died or was being weaned (red sage or wild alum root rubbed on the breasts, and teas of cinnamon, parsley leaf, walnut leaf, and huckleberry leaf); poultices for caked or painful breasts (elderberry flowers, grated raw potato, star root, poke root, alder leaves); and treatments for sore nipples (balsam fir resin; common spurge or milk purslane in fat; partridge berry [squaw vine] salve made with cream); and even a remedy—camomile tea—for uterine pain caused by nursing (Hutchens, 81; Nelson, 15).

Since so many weaning foods, especially in starch-based agricultural societies, are deficient in protein, anything that makes nursing more pleasant for women, and their milk supply more copious, becomes significant for the healthy development of the culture's children.

Women's inventions and innovations for the care and socialization of their children beyond the earliest or neonatal stages are worthy of a separate book. Weaning foods have already been discussed.

## Antiquity—18th Century

An isolated example from ancient Rome is the work of **Octavia,** sister of Augustus and wife of Mark Antony. She appears to have been a serious physician who not only studied medicine formally and practiced it in her home, but encouraged good physicians in their work and, as already noted, invented many useful remedies. Her book of prescriptions, cited by Scribonius Largus, contained, among other things, "a wonderful salve to soothe the pains of childbirth," which he called "quite analgesic." It contained lard, rose leaves, cypress, and wintergreen (Hurd-Mead, *HWM,* 60–1).

An Egyptian woman physician named **Cleopatra,** apparently a contemporary of Soranus (or Soranos, AD 98–138) wrote a book mentioned by Kate Campbell Hurd-Mead in her history of women in medicine. Soranus has been called the most brilliant gynecologist of antiquity (Himes, dedication). It is doubly interesting therefore that Hurd-Mead considers it "at least as probable" that Cleopatra was the source of Soranus' materials as that he was the source of hers (*HWM,* 58). Although Hurd-Mead mentions no specific innovations, rare is the noted physician who does not make some improvement, refinement, or true innovation in medications or treatment regimens. Theodorus Priscianus, court physician to Emperor Theodosius (AD 400) says that Cleopatra wrote her book for her daughter, Theodata, and drew partly on her own experience in writing it.

## Contributions from Western Culture

Medieval and early modern European midwives and women obstetricians made important contributions to pregnancy and childbirth technology. Usually, as the following instances will show, we know of these women today only if they attended royal births or wrote noteworthy texts, or both. Sometimes even then information is sketchy, or the woman's existence is still denied. **Trotula** of Salerno (d. 1097), for example, apparently wrote several books, the most important of which was *Passionibus mulierum curandorum* ("Diseases of women"), but the originals were lost. Coming from the noble di Ruggiero family, Trotula had a large medical practice. Upon the reorganization of Salerno's University, she joined the faculty with her husband (Johannes Platearius) and two sons who were medical writers. The four of them worked together on a medical encyclopedia. The classic medical text from Salerno *De aegritudinum curatione* included the teachings of all seven chief masters of the school, among them both Platearius and Trotula (Alic, *Hypatia's,* 51).

Trotula devised a much-admired operation for complete rupture of the perineum (Hurd-Mead, *HWM,* 150), a surgical repair for disastrous births. Although Singer and some other male historians of technology have considered her legendary rather than real, Trotula probably was indeed, as Kate Campbell Hurd-Mead and other researchers concluded after lengthy study of available documents, the head of an all-female department of women's diseases at the University of Salerno Medical School, and the author of a gynecology text that was cited for centuries, as well as a book on care and diseases of the skin (Alic, *Hypatia's,* 50ff; Hurd-Mead, *HWM* and "Trotula"; Stuard, 537–42).

**Louyse Bourgeois** (1563–1638) was a prominent French obstetrician[20] and surgeon. She was also a poet. Wife of an Army surgeon named Martin Boursier, she had three children by age 24. Widowed, she studied midwifery, and soon became known for her skill. Friend and pupil of Ambroise Paré, she was called in for the Queen's confinements beginning with the birth of the future Louis XIII in 1601. Her books on childbirth, published between 1608 and 1613, were so revered that the noted scientist Laplace attached an essay to one of them.

The most famous of Bourgeois' books is her *Observations . . .* of 1610. Based on more than 2,000 births, it covered the stages of pregnancy, handling of labor abnormalities, pelvic anatomy, theories of sterility, abortion, etc., followed by careful directions for the aftercare of both mother and child, including diet and criteria for choosing a wet nurse. She was the first to number and systematize the various birth positions or presentations, with methods of version (turning the infant) for easier delivery. According to Matilda Gage (/1: 19) this book "totally changed the method of treating certain diseases from which her own sex are the sole sufferers."

Bourgeois introduced new methods of delivery. She was the first to induce labor in cases of *placenta praevia* (placenta blocking or partially blocking the mouth of the uterus) with hemorrhaging. *She also invented delicate obstetrical instruments,* unfortunately not named or described in my sources.

Her method of resuscitating apparently lifeless newborns—injecting warm wine from

---

20. She is often described as a midwife, but her skills and publications suggest otherwise. Bourgeois' names are variously spelled (Louise; Borgeois), and she sometimes appears under her married name of Boursier. Her birth and death dates are also in question, sometimes listed as 1564 and 1636, respectively (Chaff, 23; Chicago, 176; Marks & Beatty, 58). The *Encyclopaedia Britannica* omits her (1911 and 1958 eds).

her own mouth into the infant's mouth to "cut its phlegm," rubbing its body vigorously and bathing it in warm wine and water—saved several of the royal babies. The grateful king (Henry IV) promised her a life pension (not paid after his death).

In 1635 the French midwives petitioned to have Bourgeois teach a public course in obstetrics, but the Paris Faculty of Medicine was too busy with internal conflicts to comply. Had this happened, says Hurd-Mead (422), "the whole course of higher education for women in medicine might have been changed for the better" (Chaff, 12, 23; Chicago, 176; Damour; Gage /1:19; Hurd-Mead, *HWM,* 356–7, 358, 373, 386, 388, 394, 419–23; Marks & Beatty, 58–61).

By the 17th century male midwives and doctors were invading the childbirth field. Though supported by most of the medical establishment, and glamorized by their new forceps technology, they inevitably had less practical experience. And women knew the difference, as several English diaries reveal. Elizabeth Freke's diary, for instance, tells of being rescued after four or five days' fruitless labor—and her baby saved—by a skilled woman when the male midwives attending her had given the baby up for dead (Blodgett, 174).

These early males were the advance guard of a takeover that was all but complete in the United States by the early 20th century (except for remote rural areas and urban ghettos)—with disastrous results for women. The tale of the countless thousands who died needlessly of childbed or puerperal fever, because male physicians were too arrogant to wash their hands, has often been told, yet so effective has been the propaganda of the medical establishment that most Americans think maternal mortality dropped with the trend toward male-managed hospital births. Thus we need to be reminded that, as Adrienne Rich points out, "In the seventeenth century began a two centuries' plague of puerperal fever which was directly related to the increase in obstetric practice by men," and that "in the French province of Lombardy in one year *no single woman survived childbirth*" (. . . *Born,* 143; emphasis mine).[21]

However, certain outstanding European women continued to work and receive recognition as professionals in this field. The 17th-century German obstetrician **Justina Dittrichin Siegmundin** (1648–1705)[22] deserves mention here as the possible re-inventor or improver of nooses and blunt hooks for turning or extracting the infant in difficult

21. Judy Grahn's angry poem is to the point (*Wands,* 68–9):

> I, when the wind spirit
> swole me up
> sisters surrounded
> to hold me up.
> Ours was the whole birth,
> and the power of blood
> and the bread and roses.
>
> . . .
>
> And then one day the Foe came.
> He with his forceps.
>
> he with his forceps
> to replace my muscles, his
> pincers, metal instruments
> instead of my grandmother's long fingers
> and my midwife, whom he killed. . . .

22. As with most of these early modern women, her name and dates are variously given: Justine, Dietrich, Siegmund(in), Siegemund; 1645, 1650.

births. She certainly pioneered the technique of puncturing the amniotic sac to stop hemorrhaging. Nevertheless, her writings warned against the overuse and abuse of instruments, particularly the new forceps (Rich, . . . *Born,* 137). She studied obstetrics from books and from practice, working among peasant women for twelve years, gradually gaining some upper-class clients, and finally becoming royal midwife of Prussia. In 1689 she published her first textbook for midwives, drawing on the notes and illustrations she had kept on each delivery. Her book was reprinted six times (Chaff, 25; Hurd-Mead, *HWM,* 427–8; Marks & Beatty, 66–7).

Though **Angelique du Coudray** (Mme Angelique Marguerite Le Boursier du Cou-dray, 1712–89)[23] belonged to a prominent French medical family, and became head *accoucheuse* at the Hôtel Dieu in Paris, her main claim to fame is the invention of the anatomical mannikin used in teaching gynecology and obstetrics. This invention is usually credited to the Englishman William Smellie. However, du Coudray invented and perfected her "femme artificielle," a lifelike female torso of a woman in labor, which she used with an actual fetus to give her students practice in delivery, before Smellie's model, probably at least by the mid-1750s. The French Academy of Surgeons approved it in 1758. She also wrote a treatise on obstetrics, "from pity of the miserable victims of malpractice."

Commissioned by the French Ministry of Health, Mme du Coudray visited other hospitals and lectured in the provinces. Altogether, she trained 4,000 pupils. In her old age she received a government pension of 3,000 livres (Chaff, 11; Chicago, 176; Dall, *Pictures,* 153–4; Gage/1: 19; Hurd-Mead, *HWM,* 423n, 497, 499–500; Marks & Beatty, 66–7).

**Marie Louise Duges Lachapelle** (1769–1821), was the daughter of the famous French midwife Marie Duges (1730–97) and granddaughter of another midwife. She attended at a difficult labor when only 15, saving both mother and child by turning the baby. Her husband was a surgeon who died after only three years of marriage. After her mother's death she became head of the Maternity Department at the Hôtel Dieu, later studying in Heidelberg and organizing a maternity and children's department at the Port Royal Hospital. Her eminent contemporary Baudelocque called her the most remarkable *accoucheuse* of the time.

Lachapelle is credited with several worthwhile innovations in patient care and midwives' training, regrettably not specified in my sources. However, it is known that she tried to exclude the hordes of observers from the delivery room, that she advised immediate repair of a torn perineum, that in cases of *placenta praevia* she quickly dilated the mouth of the uterus with tampons and extracted the infant by turning it, thus saving two lives. She also invented a method of deftly turning a face or oblique presentation so that the infant could be born without forceps, and of replacing a protruding arm or shoulder before it was too late.

Perhaps her greatest innovation, however, lay in realizing the value of collecting statistics on great numbers of cases.

Her great work was a three-volume treatise containing statistics on 40,000 cases. In this book Lachapelle introduced an improved classification of fetal positions, reducing Baudelocque's 94 positions to 22. She always insisted on *minimum* use of instruments.

23. Her name is variously spelled: Ducoudrai, Du Coudray; and she is sometimes listed as Mme DuCoudray-Leboursier. Her birthdate is sometimes given as 1713—or 1736 (in which case her daughter may have also have been a notable midwife)—and her birthplace listed as Paris rather than Clermont-Ferrand (cf. Chaff, #1087).

She herself had interfered with nature in only 1.73% of all these cases, using forceps only 93 times, and resorting to Caesarean section only once. Her tables settled many questions still debated in obstetrics up to her time: average length of pregnancy, average duration of labor; frequency of certain pelvic abnormalities, etc. (Hurd-Mead, *HWM,* 461, 490, 497–9, 500; Marks & Beatty, 67). She was the teacher of Marie Boivin.

## 19th Century

**Marie-Anne Victorine Boivin** (1773–1847) of Paris and Versailles, has been called the first really great woman doctor of modern times, an insightful investigator and a wise diagnostician. Dupuytren said she had "an eye at the tip of her finger." Left a widow with a daughter, she resumed the medical studies she had begun as a child, working as an assistant to Mme Lachapelle at the Maternité. Mme Boivin invented a new pelvimeter and a vaginal speculum. She was also one of the first to use a stethoscope to listen to the fetal heart.

A woman of strong character, noble intellect, and lovable nature, Boivin received an honorary MD degree from the University of Marburg in Germany in 1827. Earlier in life she had been denied admission to medical school. In 1837, with her professor son-in-law, she published a textbook on uterine diseases (correcting and superseding the one in use for 150 years), and won a Prussian gold medal of civil merit. She also had several other publications, both original works and translations. She was disappointed, however, not to be admitted to the French Academy of Medicine and not to have an honorary degree from a French university (Chaff, 54; Chicago, 190; Dall, *College,* 155n; Mozans, 293, 294–5; Hurd-Mead, *HWM,* 393, 460, 490, 497, 500–1).

Most of the other women who contributed significantly to childbirth technology and/or its underlying research in the early modern and modern periods have already appeared above (Health/ Medicine). Worth mentioning again here, in the European context, is **Mme Rondet** (b. 1800), who invented and perfected a resuscitation tube for infants born "in a state of asphyxia." Although the earlier tube invented by Chaussier was too inconvenient to use, the Royal Academy still credited him with the invention (Dall, *Pictures,* 161–2, and *College,* 156n).

In the United States, the situation was somewhat different, partly because the country was so vast and underdeveloped, and medical education had less of a tradition. Much of women's knowledge about pregnancy and childbirth was still passed on from mother to daughter, midwife to midwife, or midwife to client. If written at all, it appeared in recipe or household management books rather than in medical books. *The United States Practical Receipt Book,* published in 1844, is a good example. Collected helter-skelter here are "old, traditional" recipes for foods, and formulas for ink, dyes, hair-curling compound, priming powder for percussion caps, and medicines. Also included are at least one pregnancy test, using the woman's urine, and a remedy for vomiting during pregnancy, containing, among other things, "calcined magnesia" (Baxandall, 17–8).

A whole section could be written on the diagnosis of pregnancy, but space does not permit. One interesting note: according to Patricia Branca, in the 1880s women learned to watch for a bluish discoloration of the vaginal passage as a sign of pregnancy (*Silent Sisterhood,* 82–3; Blodgett, n47). This diagnostic technique is undoubtedly millennia older, once taught to all women by their mothers, but as patriarchy advanced in proto-Western cultures, limited mainly to midwives, abortionists, and prostitutes—or lost altogether in some areas. What is interesting here is that it is still attributed to women rather than to a male doctor.

An isolated 19th-century example of a woman's contribution to childbirth technology that found its way into the American patent records is the Accouchement (lying-in) couch patented by **Eliza L. Moore** of San Francisco in 1873.

## 20th Century

Also worth mentioning again here, in the context of achievement within the system, is the distinguished contemporary American pediatrician **Dr. Roberta A. Ballard** (1940–  ). Trained at Chicago (MD in 1965), and at Stanford, Dr. Ballard was Director of the Newborn Nurseries at Mt. Zion Hospital and Medical Center in San Francisco from 1972 to 1976. She has been Chief of Pediatrics since 1975. The achievements that brought her to the Health/ Medicine chapter, and to further mention here, are her contributions to the treatment of respiratory distress caused by immature lung development in premature babies, and her creation of the first in-hospital alternative birth center. She also created Mt. Zion's intensive-care nursery in the early 1970s, and has served on an NIH panel to develop a national consensus for the management of infantile apnea (episodes where the baby stops breathing, usually during sleep).

Although the effect of betamethasone in preventing respiratory distress syndrome (RDS) in premature infants was discovered in New Zealand, any new therapeutic drug must undergo extensive clinical testing and modification before becoming a standard treatment. In fact, as already emphasized, this latter process is part of the drug's *invention*. Dr. Ballard directed the longest-running clinical research program on betamethasone and RDS, at least in the United States, collaborating with her husband in the laboratory aspects of work on the drug. Her current research is an extension of this work, in collaboration with researchers at Yale, University of Texas at Houston, University of California at San Francisco, and UCLA–Harbor Medical Center. The betamethasone treatment, though often effective, does not always prevent RDS, still a major cause of "preemie" deaths. Dr. Ballard and her colleagues are testing thyroid hormone in conjunction with betamethasone, to stimulate still faster maturing of the lungs. This two-pronged attack shows promise.

The Mt. Zion Alternative Birth Center, opening in 1976, became a model for other such units. In a 1980 article, Dr. Ballard presented the outcomes for 1000 women giving birth in the Alternative Birth Center (ABC) with 1000 other low-risk women delivered in the regular labor and delivery suite at Mt. Zion. ABC mothers required fewer Caesarean sections (7.7% as opposed to 13.5%), and far less anaesthesia (4.2% as opposed to 68.0%). Forceps were used about a tenth as often (3.2% as opposed to 33%). ABC infants had fewer low one-minute Apgar scores and required fewer resuscitation tubes. Far fewer ABC infants (6.8% as compared to 18.7%) were admitted to the intensive-care nursery. Clearly, as Ballard and her co-authors concluded, "for women who *choose* to . . . give birth in an ABC, who meet appropriate medical criteria and prepare themselves for childbirth, and who receive careful, continuous evaluation during labor, an ABC is a safe alternative." Moreover, she continues, in the most significant statement in the article (Ballard *et al.,* 63–4):

> it is the qualities of *control* and *responsibility, nonseparation from loved ones,* and *support and individualization of care* rather than just the setting that are most important in humanizing the birth experience in the hospital. As health care professionals we must work to bring these qualities into the place of birth for all women and to provide care that is flexible and caring in any setting, guaranteeing a safe, sensitive birth.

Herself mother of two children, Dr. Ballard has focused her professional life on developing child and maternal health programs that are both academically excellent and family-centered, a rare and much-needed balance (Ballard, L-12/9/86, c.v.; Ballard & Ballard; Ballard *et al.;* Belanger, L-12/9/86).

Some twenty years earlier, **Dr. Hilda H. Kroeger** of Pittsburgh helped design "an efficient labor and delivery-room suite" in the Elizabeth Steel Magee Hospital in her home city. For this achievement, she was named one of the "Medical Women of the Year" in 1955 (Chaff, 305).

An isolated example of a woman's birth-related invention from the 20th-century patent records is the Umbilical cord tie recorded by the Women's Bureau compilers in their 1920s study of women's patents (WB/28: 46).

## ANTI-FERTILITY TECHNOLOGY

Contraception ranks with the taming of fire and the computer as a landmark in human technological history. Since it has seldom received this ranking, however,[24] its early history is poorly understood. In fact, this early history, which is usually prehistory, is often simply denied, as Richard Vann reveals (Bridenthal & Koonz, 197):[25]

> But in some places, particularly in the late seventeenth century, after the birth of the third or fourth child, births became less frequent. This appears to have been a matter of deliberate choice, and not only the result of women's aging. . . . *Since no contraceptive devices existed before the mid-nineteenth century,* the most likely explanation is that contraception was effected through the male's withdrawal before orgasm (*coitus interruptus*) or that couples engaged in sexual intercourse less frequently after several years of marriage. [Italics mine]

Perhaps one of the reasons a historian could make such a statement in all seriousness in the late 1970s is that the early history of contraception is woman's history and therefore unknown. Even the recent history is partly underground insofar as women's contributions are concerned.

A major and familiar reason for this ignorance is that so many anthropologists have been male. In the North African Riffian culture studied by Carleton Coon, for example, men were forbidden to enter the regular women's markets, where women sold not only their fruits and vegetables, eggs and pottery, but "magico-medical materials which are supposed to act as contraceptives and to produce abortions." Coon learned of this section of the market only through his wife, for "the sale and possession of these is kept secret from the men because the use of them is considered ample reason for divorce" (Coon, 110).

In addition to the usual problems of what males talking only to male informants never learn, we have the special problems caused by anthropologists from a male-dominated culture hearing what they want to hear (or what informants think they want to hear) about a subject that may be threatening to them both. It is almost as if a Martian sociologist who

---

24. Exceptional here—perhaps unaware he was talking about a technology—is the noted population scientist R.V. Short, who says outright in a 1984 article (23), "One of the keys to the success of *Homo sapiens* is the retardation of the species' rate of reproduction," which happens in two ways: our long delayed puberty and "the prolongation of the interval between successive births through the contraceptive effects of breast feeding." Prof. Short refers here to evolutionary success. He continues, "These reproductive restraints have opened up new horizons for human evolution."

25. Vann's article is omitted from the second (1987) edition of this work (Bridenthal, Koonz, & Stuard).

abhorred secret societies came to Earth to study the Masonic Order and interviewed only Catholic women.

We also have the problem of what male anthropologists may never even ask. For example, if they have not won the confidence of any females, or if abortion is openly performed in the society, if they see or suspect infanticide, if the postpartum sex taboo is long and firmly in force, or the infant and child mortality rates high, they may see little reason to ask about contraception.

In some cases primitive women consider their anti-fertility methods more secret than other parts of women's business and will not share them even with women anthropologists or indeed with any outsider. Yolanda Murphy encountered this reticence among the Mundurucu; and I have encountered it with a Dakota Sioux medicine woman who, though herself a trained anthropologist, declined to reveal contraceptive secrets to anyone outside her tribe (Medicine). Many women today, even in so-called developed countries, will not talk about contraception with interviewers of either sex.

I suspect further that *the more effective a contraceptive device or agent might be, the more likely the women would be to conceal it,* and that even if they told the secret, they might not tell it all, especially to an outsider. That is, they might reveal the name of a contraceptive plant, but not which part to use, how to prepare it, the best gathering time, the corrective adjunct herbs or dietary cautions. This possibility has implications for the modern pharmacological and clinical evaluation of traditional herbal contraceptives, to be discussed below.

The true story of "Fanisi's Choice" should serve as a cautionary tale to anyone who believes we know the full story of primitive or prehistoric contraception: Fanisi, a woman of the world's fastest-growing country, Kenya, was quoted in a 1976 documentary as saying she wanted 20 children. A decade later she had joined a local women's cooperative, and had an IUD inserted without her husband's knowledge. The husband told the male anthropologist that the reason his wife no longer conceived was that his mother had come and delivered the required ritual curse (Tierney, 26, 33).

A subtle but very important further reason for our ignorance of early contraceptive technology is a special case of the culture-bound problem: European and American anthropologists long discounted reports of herbal contraceptives because our own medical science had discovered none; thus the very idea was unthinkable. Even Norman Himes, whose 1930s *Medical History of Contraception* has not yet been superseded and is a major source of this chapter, dismissed the many reports of "using the local flora with contraceptive intent," saying, "These [preparations] are undoubtedly ineffective since no drug taken by mouth is known to Western science that will prevent contraception or abort" (25).

The situation in anthropology has now begun to change, post-Pill, but much is irrevocably lost.

My underlying assumption in this chapter is that *women invented all the basic types of contraceptive known today,* often before recorded history, and certainly before the Christian era. Modern medicine has merely refined, or re-invented, devices and methods that women knew centuries or millennia before. Since this may sound revolutionary to some, it may be well to explain:

First of all, as Adrienne Zihlman points out, a hominid infant's survival would depend on its mother's ability to limit births (Dahlberg, ed., 94). The same would be true for gatherer-hunters over many millennia, and women of these groups have repeatedly said as much to anthropologists.

Second, men are often hostile to contraception. Men of the Cape York (Australia) group spoke of childless women who had "shut themselves up" by eating medicine, and told an observer that they would be angry if they found their women using that medicine (Himes, 28–9). From the above-mentioned Rif tribes (where one informant told Coon that if he caught his wife using contraceptives he would kill her) to ancient Rome (where orators predicted race suicide and demanded soldiers and colonists to take over the world)[26] to the United States Senate (where a four-celled fetus may receive more consideration than living women and children), males have attacked the female reproductive monopoly with religious, legal, economic, and emotional weapons. Even among horticultural groups where women wield considerable economic power and firmly control their own fertility, males disapprove of contraception (and at the same time are partly ignorant of what is going on; see, e.g., Murphy & Murphy, 161). Among Highland New Guinea cultures, men see women's methods of preventing or terminating pregnancy as "withholding from males their right, as men, to beget children," and believe that "even in the matter of bearing children there is a fundamental opposition between the aims and interests of the sexes" (Read, 25, 29).

Anthropologists repeatedly find that men want more children than women do (cf. Hrdy, "Heat," 75). Women matter-of-factly point out that childbirth is painful, ruins their youthful beauty, and gives them children who are much trouble to raise. Nomadic and gathering-hunting women point out that they cannot carry more than one child (Devereux, 18–9, 178, 197, 259; Himes, 28–9). Of the many specific examples of this difference that could be cited here, an important one (since the book focuses on women and a female anthropologist is co-author) is the report by Yolanda and Robert Murphy that Mundurucu women are "resentful of the continued cycle of pregnancy and birth," realistically seeing it as the source of their workload. And although most of them are "quite affectionate" mothers, after three or four live births their desire for more children declines markedly. "Men want children more than do women, they say . . ." (161).[27] Given such a pervasive attitude, it defies logic to suggest that men created any significant number of early contraceptive methods and devices. As for oral contraceptives and abortifacients, only a woman could try them on herself, the original and more honorable form of "clinical testing."

Throughout myth and fairy tale, the sexual counselors sought out by heirless kings and barren queens are predominantly women—witches or healers or wise women. In one version of the Norwegian tale of Tatterhood, for example, it is an egg woman (!) who tells

26. The Emperor Augustus, in AD 9, defended his pronatalist laws to the Knights of Rome. First praising the fathers among them "because you have shown yourselves obedient and are helping to replenish the fatherland," he then speaks of the "private advantages that accrue to those who marry and beget children," and finally comes to his real point (O'Faolain & Martines, 46–8):

[B]ut for the State, . . . how excellent and necessary it is, if . . . you are to rule others and all the world is to obey you, that there should be a multitude of men, to till the earth in time of peace, to make voyages, practice arts and follow handicrafts, and, in time of war, to protect what we already have with all the greater zeal because of family ties and to replace those that fall by others.

Cf. also Tertullian (2nd–3rd century AD), Christian Church "Father" in Rome. Knowing that the population of Rome was dropping, that abortions were commonplace, and that some women made their living as abortionists, he "hurled invectives against the whole class of midwives and women doctors, saying, "It is your tubes, your speculums, your dilators and hooks that are to blame for causing this destruction of the fruit of the womb" (Hurd-Mead, *HWM*, 77).
27. Interestingly enough, a 1981 poll of entering students at the University of California (Berkeley) showed that "more men than women found it 'essential' to raise a family" (*Cal. Monthly*, Mar.–Apr. 1981: 5).

the queen what to do to conceive. And though the remedy sounds magical (whether because the tale is corrupt or because the true remedy was secret), we should note that two plants are involved (E. Phelps, 1–2). According to Hellenic tradition the Theban seer Tiresias had once lived as a woman, and this double knowledge was the source of his wisdom (Graves, *Myths* I: 11; Lefkowitz, "Princess," 1401). In one version of the Greek myths surrounding Minos and Pasiphaë, the childless Minos consults Prokris, the woman refugee at his court who co-presides over this chapter (Himes, 187).

Furthermore, many of the herbal contraceptives reported by primitive groups are foods (or condiments) if prepared one way and contraceptives or abortifacients if prepared another way or if unripe or left raw. Grated raw unripe pineapple (or its juice), for example, has been reported as an abortifacient, or as a substance that "interferes with implantation or gestation, but which might well prove to be [an oral contraceptive] if used in smaller concentrations" (De Laszlo & Henshaw, 626; Suggs, 22), whereas the ripe fruit is a delicacy and the basis for many modern desserts. Tansy, a well-known abortifacient, was once much used in cookery, and as a bitter tonic. Tansy was even the name of a custard-like dish. Alum and other astringents occur in various foods, especially unripe fruits and nuts and their peelings or husks. Their mouth-puckering qualities could scarcely be overlooked; women clearly knew of their vaginal-tightening qualities for simulating virginity, among other purposes (see App. SFA-3). It would be interesting to know in how many tribal and pre-tribal languages, as in Latin and in English, the opening of the womb is its "mouth" (*os*). Even such staples of life as maize and wheat and such common fruits and vegetables as peaches and peas may have anti-fertility effects (cf. Devereux, 277; "E-vole-ution"; Farnsworth *et al.*; Levin, 154).

Another cluster-of-technologies argument here is the correlation between vermifuges (wormers) and fertility-related preparations. As late as the 19th century in rural America, women were in charge of dosing their children for worms. A list of common wormers shows a significant overlap with traditional fertility and anti-fertility preparations. Savin, tansy, southernwood, and rue are examples that leap to mind (J. Rose, 103, 105, 107, 109), but several less familiar ones could also be cited, such as bay laurel, feverfew, plantain, male fern, horehound, thuja or arbor vitae, St. John's wort, and Kamala *(Mallotus philippinensis)* (cf. J. Rose, 42, 60, 69, 73, 80, 98, 110).

The multitudes of female fertility deities and women's general close connection with fertility in myth and religion need not be recounted here. More to the point, perhaps, are the female demons—Lilith among them—blamed for infertility and infant deaths in some cultures (Kilmer, 171; Lichty)—and the many anthropological reports of women's control of fertility. Here are examples from both horticultural and prehorticultural human economies. Among the Iroquois (Seneca), not only was birth-spacing in women's hands, but women took greatest joy in the birth of a daughter (Tiffany, ed., 54). Ruby Rohrlich and colleagues report that "Australian (aboriginal) women . . . have complete control over the reproductive function . . ." (Rohrlich-Leavitt *et al.*, 128). Says Brian Hayden in his 1972 survey of population control among gatherer-hunters in general, "In relating these factors [dominance behavior, per-capita workload, and frequency of camp movement] to hunter-gatherers, it is vital to realize that all the important cultural methods used to limit populations—infanticide, prolonged lactation, abortion, abstinence—are exclusively controlled by females" (212).

Both result and contributing cause of our ignorance about women's role in contraceptive technology are two mistaken stereotypes about primitive women: that they limited their families, if at all, only by abortion and infanticide; and that their lives were one repetitive cycle of childbearing. This chapter is an extended disproof of the first; the evidence to disprove the second has long been at hand. Many early ethnologists and anthropologists were amazed at the small family sizes they saw, and reported their amazement in no uncertain terms.[28] Elsie Parsons, writing of the Isleta Indians in New Mexico, reported two women childless after three or four years of marriage, and said that her informant Lucinda thought two children enough for her daughter to have (Himes, 15). Morton C. Kahn, studying the Djukas of Dutch Guiana, South America, reported that these people were just maintaining themselves, with an average of two to three children per mother (Himes, 18–9). Hortense Powdermaker reported that the New Irelanders had had "practically a stable birth rate for three generations: an average of 2.6 children per woman, or 2.9 per fertile woman. The modal number was three," not even replacement rate for these people (Himes, 16n).

The subjects of investigation, the so-called primitive tribes, were amazed in turn at the fecundity of their European investigators, though this reaction is less often recorded (Evelyn Reed, 134):

> I heard of a white man, who being asked how many brothers and sisters he had, frankly said "ten." "But that could not be," was the rejoinder of the natives, "one mother could scarcely have so many children." When told that these children were born at annual intervals and that such occurrences were common in Europe, they were very much shocked, and thought it explained sufficiently why so many white people were "mere shrimps."

The truth is—and Lee and DeVore's worldwide study of gathering-hunting peoples has recently confirmed it—that most prehorticultural groups grew very slowly indeed, in some cases simply maintaining their numbers. Otherwise, they would have come swiftly into conflict with the carrying capacity of their environment, or with neighboring peoples. Zero population growth, in other words, or something very like it, may be the "natural" state of affairs (see Deevey). Norman Himes concluded, after studying more than 3,000 references including most of the reliable anthropological literature up to his day, that *every tribe ever studied has wished to limit its fertility, and has used all the means at its disposal to do so.* In the 1970s, Angela Molnos found the following groups practicing birth-spacing in East Africa alone (Bleek, 188 and n85): the Chaga, the Meru, the Mbeere, the Maragoli, the Kajirondo, the Kgata, the Kaguru, the Giriama, the Lugbara, the Lango, the Tukeu, and the Somali.

Where anthropologists see women oppressed by childbearing, it is usually after sedentism and cultivation become ways of life, and/or after their traditions are already debased by contact with whites or patriarchal traditions, particularly with Christian missionaries who condemn their traditional methods of population control.

The question at once arises whether this wish to limit fertility has been with us as long as our humanness, or whether it arose at some later point; and if later, what were its origins. Let us distinguish from the start between the desire of individual women to limit their families, and the desire of a whole group to prevent (or correct) overpopulation as defined by that group. Among fully nomadic gathering-hunting peoples, individual women seem to have tailored family size to fit their subsistence methods. That is, since a woman can carry only one child while foraging or during seasonal moves from one area to another, she would not bear (or allow to survive) another child until the first could walk

---

28. Of course we must differentiate here between truly primitive, prehorticultural peoples and those who have settled down and are or are becoming agricultural; and particularly between those evolving their own forms and those extensively influenced by Western or North African patriarchal conquerors and/or missionaries.

several miles unaided.[29] In achieving this limitation she may have been aided by her high physical activity level,[30] relatively low calorie and fat intake, and long nursing period, as well as by some poorly understood periods of nonovulation related to drought (cf. !Kung data) and by herbal or other contraceptive technology. (Some women undoubtedly feared childbirth as painful or life-threatening and/or did not want the responsibility of children at all, but this minority has largely been ignored.)[31] At this stage, women were firmly in charge of all matters having to do with birth and fertility—including anti-fertility—and no one would have questioned an individual woman's right to limit the number of mouths she had to feed.

With the advent of agriculture, children become producers as well as consumers of food, a needed labor force. At that point, population begins to expand significantly. Like many truisms, the idea of the agriculture-fueled population explosion contains considerable truth. As already noted, however, this common-sense view is now being challenged, some researchers maintaining that the human population had already begun to "explode" in the intensive-foraging stage that preceded horticulture (e.g., M. Cohen; Flannery; Zvelebil). Elise Boulding, even while describing the older view as part of an ecological approach to human development, dates "significant population expansion" and a "change in attitudes regarding childbearing" to about 10,000 BC, earlier than all agriculture as usually defined, and earlier than most settled horticulture as well (*Underside,* 11).

Whatever the outcome of this debate, however, no one doubts that population did rise dramatically with sedentism and reliance on cultivated food. "Population control mechanisms . . . are apparently suspended, or at least practiced more selectively," says Boulding, "once nomadic existence has been replaced by a settled agricultural way of life" (*ibid.*). As already discussed, however, several factors besides women's conscious choice may have begun to operate at that time. Seasonal variations in the estrogen content of the group's food plants, which may, as in several other species, have governed cycles of fertility and infertility and influenced mating behavior, could be evened out by a stored

29. Dr. Marjorie Shostak points out that the !Kung women she lived among in Botswana did occasionally carry both an infant and an older child out foraging. However, since they themselves weigh only about 90 pounds, and normally bring back 30 pounds of food or more, this is not customary. Someone in the encampment will usually care for a woman's older child while she forages (Presentation, Institute for Research on Women & Gender, Stanford University, Dec. 3, 1986).

30. Research indicates that progesterone levels drop with even moderate aerobic exercise. Puberty itself may be delayed by "intense physical activity": of 38 Harvard University female athletes, the 18 who began training before menarche first menstruated at an average age of 15, whereas the 20 who began training after menarche had menstruated at 13. "Each year of training before menarche appears to delay menstruation by about five months," reported these researchers. Further, "Intense physical activity before menarche was also found to contribute to irregular or nonexistent menstrual cycles," with about 83% of pre-menarchally trained athletes reporting this "problem," as opposed to only 40% of post-menarchally trained athletes. However, the effects seemed reversible. Studies done on ballerinas have shown that when the proportion of body fat drops below 22% (for 25-year-olds) menstruation may stop, and that for adolescents levels below 17% will delay puberty. Among the explanations offered are (*a*) since peripheral body fat converts androgens to estrogens, a certain level of such fat may be necessary for maintaining the usual hormone balance; and/or (*b*) the beta-endorphin that pours into the bloodstream during stressful workouts has been shown to suppress female hormones that regulate the menstrual cycle ("Exercise and Late Menstruation"; "Running Woe"; cf. Djerassi, *Politics,* 10–1).

31. Interestingly enough, in Lillian Rubin's late-1970s interviews with midlife women, some 25% of her sample "admitted guiltily, shamefacedly, that they don't consider themselves natural mothers, don't feel they were 'made for motherhood' " (26).

food supply increasingly dominated by grain. Sarah Hrdy notes that orangutans in captivity may copulate on any day of the female's cycle, whereas wild orangutans are somewhat more cyclical in their mating behavior ("Heat," 74–5).

Lower levels of exercise combined with a higher-fat diet could have brought earlier puberty and higher female fertility even as the trend toward patrilocal residence began to interrupt the mother-daughter transmission of anti-fertility information and instruction. Greater exposure to artificial light may also have acted to promote fertility (cf. Hartung; Lacey, 108). Storage of grain may increase (relative) estrogen levels in the kernels—which may already contain more estrogen than those of the grain's wild ancestors (Hrdy, "Heat," 73, 74, 76; Tangley; "E-vole-ution"; Farnsworth *et al.* 2: 717–8, refs 930–58; Edell, Feb. 1984; Sjöö & Mor, 62).

After several millennia of such rapid growth, however, societies seem to come full circle to the original view that population must be limited for economic if not for personal reasons. The pertinent difference for our purposes here is that somewhere along the way, women lose control of their own fertility, and population becomes something that, for example, the early states of the Near East feel empowered to regulate. This attitude, though supported partly by patriarchy and patriarchal religion, would have had little success without a culture-wide consciousness of overpopulation. This consciousness can be seen not only in the Atrahasis myth cited by Anne Kilmer[32] but perhaps even in a seldom-quoted line from the Expulsion from Eden. The wrathful male deity says *unto the woman,* "I will greatly multiply thy sorrow *and thy conception . . .*" (Genesis 3: 16). In other words, Hebrew mythmakers, too, thought that the end of the golden/ garden age, when food was free for the gathering, and the beginning of cultivation, when food must be obtained by the sweat of the brow, brought a population explosion. The Atrahasis myth, though undoubtedly much older, was inscribed on clay tablets about 1600 BC.[33]

The usual history of these rapidly sketched developments deals very little, if at all, with women's role in early anti-fertility technology, or the changes in attitude and practice here postulated for the early Neolithic. Such history focuses on events after the invention of writing; indeed, usually as late as the 2nd millennium BC, when the takeover was virtually complete in proto-Western cultures. Actually, however, for this inquiry, as for so much else in woman's history, the crucial time may be the transition from nomadic or semi-nomadic life to settled horticulture. For Western and proto-Western cultures, then, we must study what Elise Boulding calls the agrovillages of the Near East.

Among the crucial questions to keep in mind are the following:

---

32. Dr. Erle Lichty, a noted Assyriologist at the University of Pennsylvania in the 1970s, agrees with Kilmer's interpretation (see below) of this myth (Lichty, 26).

33. Examples of such consciousness appear in cultures as distant as Polynesia. In the mid-19th century, Rotumans reportedly kept several large double canoes on hand at all times because they feared population pressure and wanted the canoes in readiness for expeditions to search for new lands. A late-1950s report quotes their legendary chief Ru: "I see that the valleys are thick with people and even the uplands are becoming crowded. I have selected a star, and beneath that star there is a land that will provide us with a peaceful home." The solution for the Polynesians, then, if infanticide fails, is migration. However, Firth reported in the 1930s for Tikopia (New Zealand), that the chief's annual "State of the Island" speech exhorted heads of families to practice *coitus interruptus* to keep families small and prevent strife (J. Tanner, 513, 515).

It is also suggestive, though by no means statistically significant, that an Isleta Indian woman (New Mexico) could receive a medicine to preclude childbearing *at her first menstruation* (Himes, 13–7). Was she destined for some special role in the group, or was the group concerned to limit population?

1. Who was in charge of anti-fertility measures in general and contraception in particular?
2. At what point, or under what circumstances, did women's and men's fertility interests diverge?
3. What is the impact of unmistakable knowledge of the male role in conception?
4. What is the impact of patriarchy?
5. When did fertility cease to be an individual or family matter and become a matter for group, eventually state, regulation?

### Antiquity Through 18th Century

In order to look systematically (but at reasonable length) at women's early contributions to anti-fertility technology, I will relegate abortion technology to an appendix[34] and divide contraceptive methods into four categories: cultural, mechanical (including barriers and other physical methods), herbal or chemical, and other. Regarding abortion, suffice it here to note that post-conception agents are probably the older in most cases, and often the forerunners of pre-conception agents. Suppose a pregnant woman eats an unfamiliar plant in time of drought or scarcity, or upon moving into an unfamiliar area, and miscarries (or becomes ill and then miscarries) soon enough to connect the two events. Indeed, if Japanese snow monkeys can learn to avoid coyotilla berries, whose neurotoxic effects are months delayed (Lampe, 40), arguably proto- and early human women could do as much.

Over centuries or millennia, women of many groups could develop a mental store of such plants and their locations or favored growth conditions. At some point it would occur to one or several women that if a food-sized portion of a certain plant could interrupt a pregnancy, perhaps a smaller portion taken periodically might prevent the pregnancy in the first place, or somehow make it unlikelier.

The proposed categorization of contraceptive techniques is less simple than it sounds, since, for example, physical or barrier methods like douches and sponges can become chemical methods as well if the douches or sponges contain spermicides. However, it brings some order into a very complex subject.

### Cultural methods

*Postpartum intercourse taboos.* Among several early peoples an important method of birth control was to make sexual intercourse taboo for the mother (or both parents) for a certain period after a birth, often at least until the child was weaned. This period could range from several weeks[35] or months to four or five years as until recently among the Hunza of the Karakorams (Michaud & Michaud, 664), six years as among the Dugum Dani or Jale (Leibowitz, 136), or even longer. William Graham Sumner says of the Plains Indians (*Folkways,* 315), "It has long been the custom that a woman should not have a second child until her first is ten years old." Beatrice Medicine of the Dakota Sioux confirms that among her people the taboo period could be as long as ten years, during which time the husband might take the wife's sister as a second wife.

Jewish couples in the first century AD evidently practiced not avoidance, but with-

---

34. No one disputes women's invention of abortion-causing drugs and techniques, and Philip Devereux has provided a useful compendium of abortion methods in primitive society up to the time of his writing.
35. The forty days customary in Jewish, Christian, and Islamic societies (Coon, 124; Lichty, 23) and among the North American Huron and Iroquois corresponds with the six-week period still recommended by Western medical practitioners today.

drawal—"threshing inside and winnowing outside"—during the first two years of lactation (Noonan, 10).

Although such practices are not usually discussed as contraceptive in intent or invented by women, it is hard to imagine their being a man's idea.[36] Evelyn Reed sees the postpartum avoidance taboos as merely the most stringent of a series of taboos instituted by females to protect themselves and their offspring from basically carnivorous males.[37]

This is an intriguing if unprovable idea. Among some primates an incoming male kills his predecessor's offspring who are still with the mother. Male chimpanzees do sometimes kill and eat infant chimps (Tanner & Zihlman, 595); and a frequent prey of chimpanzee hunters (almost always male) is infant baboons (Teleki, "Omnivorous," 34, 35).[38] Among some primitive groups, initiation rites include knocking out certain teeth symbolically or actually related to the threat of cannibalizing infants.[39] Consider, too the Greek myth in which the Curetes dance around the infant Zeus, clashing their spears against their shields to drown out his cries lest his father Cronos should hear them and come and devour the infant, as he was in the habit of doing with his children at birth; and the Egyptian myth in which Isis teaches women to tame men enough to live with them (Monaghan, 155). Neither of these myths is unique.

Whatever the origins and early intent of the long postpartum period of sexual abstinence, it obviously had a contraceptive *effect* not lost on women. Since the taboo also kept the group in balance with its food supply and ensured that infants received, in the prolonged nursing usually involved, the protein necessary for their early health and brain development, we might even consider it a survival factor.[40]

Moreover, such taboos would seem to me in themselves strong motivations for women to develop effective contraceptives and abortifacients (cf. Devereux, 18) to defer the birth of the first child, to conceal lapses from the taboo, and to space other children further apart

---

36. Indeed, observers often explain sororal or plural marriage or "perversions" as necessitated or justified by the taboos.

37. Reed's biology and her anthropology, as often in her book, have not kept pace with her leaping ideas. For despite the fact of the gathering-hunting division of labor and despite reports from societies in which women did not eat—or cook—meat, no one has proved—or even suggested—that women in general did not eat meat. Reed's further inference, that *because* women did not eat meat they had less sex drive than men, is even less supportable.

38. Compare, however, Brian Hayden, who specifically distinguishes between primates and gathering-hunting humans on this point, saying that among the former it is the males who limit population (by infanticide), whereas among the latter, the females are in control of all the means of limiting population.

When a baboon was the prey animal, one apparent stimulus to predation was the "vocalization of infant or juvenile baboons.... In more than half of the 28 episodes involving baboons that I witnessed, the crying of young baboons preceded any evidence of predatory interest on the part of the champanzees" (Teleki, *ibid.*, 35).

39. The question has been raised why human males lack the over-sized canine teeth of some primates and why they differ so much less in size from human females than certain male primates do from their females. Tanner and Zihlman (606) think it likely that at an early stage in human or proto-human evolution, when females chose their males absolutely, they selected for those less threatening to their offspring.

40. As already noted, the lowest infant mortality prevailing in Peru 5,000 years ago was among coastal tribes who got plenty of protein from shellfish. And the most successful early weaners of infants I have encountered—the Marquesans, who according to one source delivered infants at two to four months into the hands and care of their fathers—had invented a high-protein infant formula consisting of pre-chewed shrimp meat and coconut milk (Leibowitz, 135).

so as to continue enjoying their sexual pleasure.[41] The anthropological literature provides ample evidence of desires to prolong youthful periods of sexual freedom and pleasure— from the aboriginal Formosans who aborted all pregnancies occurring before their mid-thirties, to the Yap Islanders who aborted all pregnancies until they could no longer put off assuming full adult responsibilities, to the young Tiwi wives who want to postpone motherhood so as not to interfere with their love life (Devereux, 9, 25–6; Schneider, "Abortion"; Goodale, 145). However, as already mentioned, the very existence of a postpartum sex taboo, especially a long one, might keep anthropologists from searching for other contraceptive methods in that society.

*Breast-feeding.* As mentioned above, the postpartum intercourse taboos often seem to bear some relation to the need to breast-feed the child. Nursing itself has a contraceptive effect, and can serve to limit population even in the absence of a postpartum intercourse taboo. Says Carl Djerassi in *Politics of Contraception* (1979, 10), "It is likely that even now, when considered on a global scale, more births are prevented by breast feeding than by any other method of contraception." However, as well-fed modern women find to their sorrow, the protection is incomplete. In fact, it probably operates well only with a calorie intake so low that the nursing mother uses it all in lactating or for her own bodily processes, and thus has too little body fat to ovulate regularly.[42]

Obviously, women did not invent nursing as a contraceptive technique. But at some point in many cultures they began consciously to prolong it, wholly or partly in the hope of preventing further pregnancies. In fact, both Baffin Land Eskimo and Native American mothers reportedly prolonged the suckling of their children "with the *intention* of keeping the family small." Egyptian women, too, used prolonged lactation to space their children at least three years apart, as did American Colonial mothers (Himes, 16, 66–7; Norton, 232–3).

*Other intercourse taboos/regulations.* Another cultural technique is to regulate certain categories of human sexual intercourse, such as pre- or extra-marital love-making. Among the Mbuti of Africa, for example, young partners in premarital liaisons must not embrace fully, but must hold each other by the shoulders. A Western reader might assume this prohibition could work by awkwardness alone, though Colin Turnbull's informants were

---

41. Such taboos may also have inspired women to invent many "sexual aids" and to develop masturbation to a high art, but these subjects have been even less studied than contraception.

42. Sheila Kippley *(Breast-Feeding and Natural Child-Spacing)* maintains that frequent stimulation of the breast by suckling releases a hormone, oxytocin, that suppresses ovulation, and that if all the child's sucking needs are met by the breast, without supplementary bottles or pacifiers, nursing can protect the mother against pregnancy for about two years. In fact, for mothers who met her "natural mothering" criteria—no pacifiers, no bottles, no solids or other liquids for five months, demand feedings, including night feedings, and some lying-down nursing (at naps or night feedings), she found a mean of 14.6 months of amenorrhea (and the study was not yet complete, as some of the mothers were still nursing and still not menstruating) (App. II).

The hormonal mechanism is not completely understood, but two factors appear to be important: neural stimulation from the breast, and the mother's nutrition. That is, the mother who carries her baby around with her all day and feeds it on demand will be better protected than the mother who puts her child to sleep in a bed and feeds it on a rigid schedule. Also, in a Rwandan study of women depending entirely on nursing for contraception, the malnourished rural women (who also carried their babies continuously and fed them whenever they cried) received much better protection than their better-fed urban sisters (Djerassi, *Politics,* 10–1).

This finding fits with observations of the !Kung before and after settlement (Agric.), except that the !Kung were not *mal*nourished, but merely lived on a low-calorie, low-fat diet. Perhaps the crucial factor, assuming a woman is well enough nourished to produce milk at all, is how much body fat her diet allows in spite of nursing. A woman needs 15–17% of her weight in body fat to support ovulation.

not deterred. Interestingly enough, neither Turnbull nor Patrick Putnam, with fifty years' experience in the area between them, could record a single case of premarital pregnancy (Turnbull, "Mbuti," 209).

More interesting for our focus here are the techniques unmistakably evolved and applied by women. Among these are probably the original form of the chastity belt—a visible sign of unavailability. The goatskin chastity tunic worn by Libyan girls has already been mentioned. It was death for a man to remove this *aegis* without consent (Graves, *Myths* I: 47).

*Ridicule.* A time-tested cultural technique for influencing behavior is ridicule. In rural Ghana in the mid-1970s, for example, "a woman who brings forth quickly in succession is ridiculed as being 'primitive' and having no self-control" (Bleek, 149). Himes reports several analogous instances from such areas as Fiji, whose women "have a decided aversion to large families, and have a feeling of shame if they become pregnant too often, believing that those women who bear a large number of children are laughing-stocks . . ." (Himes, 24).

*Imposed barrenness/celibacy.* Yet another way of preventing excess births is to impose temporary or permanent celibacy or barrenness on certain groups in the society. Nuns, monks, and priests in Roman Catholic countries, young men not yet warriors among the Masai, and unmarried girls in most Western societies are examples that leap to mind. *Imposed* celibacy or barrenness demands a stratified society and some general consciousness of overpopulation, rather than individual women's desire to limit their own families.

As suggested above, both the dawning of such a consciousness and the inception of these cultural techniques may be much older than our first records of them in proto-Western cultures, and women may have had some choice at first. An intriguing example from the Near East appears in the Atrahasis Epic, partly paralleling the Noah story. In the Atrahasis version of the Flood myth, however, the reason the gods decide to destroy humankind is not human wickedness, but *overpopulation:* "Twelve hundred years had not yet passed [since humans were created] when . . . the people multiplied. The land was bellowing like a bull." Enlil, who seems to be the gods' earthly overseer, complains to the great gods, saying, "Let there be a pestilence [upon mankind]."[43]

The plagues bring terrible suffering and social disruption. By the sixth year, parents have sunk to eating their children, but still "the people are not diminished, / They have become more numerous than before." This time, the gods demand a "final solution": a Great Flood.

Only Atrahasis, warned in a dream, escapes the waters with his family. Furious at this survival, the gods are about to finish the job when a compromise allows humans to live, *but limits their fertility.* Pertinent here are lines establishing an unexplained "third group" among the people; decreeing that there shall be "bearing women and barren women"; and setting up five classes of priestesses, only one of which may have children.

Whether the male god Enki alone or Enki and Nintu, the Lady of Birth, together negotiate this compromise is unclear from my sources (cf. Kilmer, 171 and 172 n48), but Nintu must carry out the new system. Also, priestesses in this area (Sumer) before the written Atrahasis version just cited seem to have regulated their own fertility, as other

43. Although the immediate substance of Enlil's complaint is that humans are so noisy he can't sleep, as Kilmer notes, "in view of the whole poem now before us, I believe that our understanding of man's offense must be based primarily on his numerical increase, and only secondarily on his noisiness, which may be regarded as the natural consequence of the many lively beings" (167).

sources make clear. The *Sal-Me* priestesses, for example, did marry. They also bore children conceived in their temple rites. *But they bore no children to their human husbands* (Rohrlich, "State," 87–8).[44] In carrying out this cultural decision, of course, they may have used some of the methods soon to be discussed.

Among the Jews, also, a division of women into bearing and barren is linked with the time of the Flood. According to Genesis Rabbah (xxiii, 2):

> The men of the generation of the Flood . . . each took two wives, one for procreation and one for sexual gratification. The former would stay like a widow throughout her life, while the latter was given to drink a potion of roots, so that she should not bear, and then she sat before him like a harlot. . . .[45] The proof of this is that the best of them, who was Lamech, took two wives, Adah (so called) because he kept her away (ya'ade) from himself; and Zillah, to sit in his shadow (Zillo), i.e., for sexual purposes only.

Unfortunately, we have this fascinating tale from the male viewpoint alone, and indeed in the Bible, both Adah and Zillah bore children. In fact, Zillah was the mother of the mysterious Tubal-cain, "instructor of every artificer in brass and iron" (Genesis 4:22). Perhaps the two names were originally intended to be generic and descriptive or even metaphorical, as indicated by their literal meanings, rather than the names of particular women. The cup of roots itself, of course, is an herbal contraceptive (see below).

Rivkah Harris illuminates this ancient population-control stratagem, if such it was, in her study of *naditu* women at Sippar in the Old Babylonian period (beginning about 2000 BC). These women were evidently (*a*) chaste, (*b*) cloistered, (*c*) mainly upper class and affluent, and (*d*) not priestesses *per se* but (*e*) belonging to a special *class*. First, and most pertinent here, not only were they not "sacred prostitutes," as temple sexual celebrants are often mistakenly called, but they were forbidden to have sexual relations at all, either with worshippers in the temple or with their husbands in the exceptional cases where they did marry, as with the *naditu* of Marduk. Indeed their very name probably comes from the word for a fallow or uncultivated field. Thus, a *naditu*'s husband might take a *second wife for bearing children* (106, 108, 108n).

Secondly, the *naditu* lived in a special residence all their lives, though they evidently could go out or at least receive visitors upon occasion. Some sort of cloister is connected with the temple as early as Ur III, but its purpose is uncertain. The whole cluster of customs is reminiscent of the Roman Catholic nunnery, with the girl seen as bride, as the daughter-in-law of deity and consort; the dowry brought to the cloister, etc. Here, however, the dowry reverted to the family on the *naditu*'s death (R. Harris, 108, 109, 113).

That the *naditu* seem to have been elite women, including even princesses (*ibid.,* 109, 123ff) is important to note, for had the custom prevailed throughout the population, the effect on fertility would have been even more significant.

The *naditu*'s role is still debatable. Harris suggests that they held positions in the temple administration. They may have had religious reasons for entering the cloister, and their

44. Sarah, the wife of Abraham, was barren for many years. The Bible as now edited represents this as an unwilling and undesirable state. Suppose, however, that she was a high-ranking priestess in the Goddess-worshipping religion of her home. In this case, her leaving with Abraham and her eventual motherhood appear in a very different light (cf. Teubal).

45. It would be good to know the original Hebrew word here, as "mistress" or "lover," or some other less judgmental word might well be more accurate. Cf. the usual Judaeo-Christian slander of Goddess-temple sexual celebrants as "temple harlots" or "sacred prostitutes."

prayers and offerings may have been considered more efficacious than those of the ordinary person. In the end she calls them, as noted, a special *class* of women (106, 109, 116, 121, 108).

These are vital clues in one of the central mysteries of women's changing status in the ancient Near East.

*"Perversions."* Brian Hayden's list (212) of "all the important cultural methods of population control"[46] omits one: sexual practices stigmatized by Judaeo-Christian scholars as "perversions." These practices are fiendishly difficult to research. In the general absence of good data we cannot connect them certainly with birth-control motives or attribute their origins confidently to women. However, since in the Near East, at least, some of these practices are rather firmly linked with Goddess-worship, and since women clearly have a greater interest in avoiding births than men, this cluster of cultural methods deserves at least brief mention here. To whatever degree the sexual urges of any human group are satisfied by sexual practices other than heterosexual vaginal intercourse with intravaginal ejaculation, the fertility of that group may be reduced. Such practices as anal intercourse (whose presence in Near Eastern agricultural-religious ritual suggests that it was used in the temples by priestesses enjoined to remain barren), oral intercourse, masturbation, withdrawal—the "sin of Onan"—and homosexuality would all affect the birth rate according to their prevalence. Herodotus mentions that a 6th-century BC ruler of Athens wanted no children by his new wife and therefore "had intercourse with her not according to custom," which Noonan interprets to mean either withdrawal or anal intercourse (Noonan, 16). This gives us no indication of the prevalence of the custom, however.

It would also be useful to know the mythic context of these practices, i.e., which deity, if any, was said to have taught them.

Our knowledge of these practices is limited by the campaign of Jehovah's priests to extirpate them from Canaan—and from history—as part of the Goddess religion. So severe was this aversion that the modern Bible usually does not even name the practices, but lumps them together as "whoredoms." Christian missionaries, anthropologists, and librarians formed by this tradition have seen to it that we also know little of similar practices among surviving primitive peoples.

A rare example of a people who have had some success in resisting outside devaluation and destruction of their sexual practices is the Hopi (see, e.g., Duberman, Eggan & Clemmer). Note that the Hopi are matrilineal, and "Hopi women played a central role, owning the houses and the clan lands and caring for the sacred masks and paraphernalia of the religious societies." Although the Hopi had lived for at least a thousand years on or near their mesas, by 1900 they numbered no more than 2,000 (*ibid.*, 101). "Perversions" were by no means their only method of population control (according to Michael Weiner [40–1], they used Jack-in-the-pulpit, or Indian turnip, to achieve either temporary or permanent sterility), nor was this the avowed purpose of the rites described by Duberman's informants; indeed, quite the opposite. But these practices were known in the culture.

## *Mental/psychological methods*

Unfortunately, this area of contraceptive history has been almost totally neglected. Only with biofeedback's findings has modern science begun to re-examine allegations that

---

46. Infanticide, prolonged lactation, abortion, abstinence.

ancient women had mind- or dream-control over their own fertility. Such control would go far to explain Turnbull's findings in Africa and Malinowski's in the Trobriand Islands,[47] among others, of adolescent promiscuity without premarital pregnancy. However, any such discussion is beyond my scope here. Suffice it to note, according to Shuttle and Redgrove, "There is evidence that colleges of women in the past, by techniques of deep introspection and dream control, were able to influence the events of conception and birth . . ." (25; chaps. 4, 5).[48]

## Physical/mechanical methods

Here I will discuss the many and varied barriers to the passage of sperm; expulsion of sperm from the body after intercourse; douches; and other physical or mechanical anti-fertility methods. Barrier methods, of course, make little sense until women are aware of the role of seminal fluid in conception.

*Barriers.* Women worldwide have been ingenious in using grass, moss, sponges, and plant-fiber waddings or tampons to staunch their menstrual flow. Anything that blocks menstrual flow outward could also block sperm traveling inward, and it insults the intelligence of primitive women to suppose they missed this. The connection would have been particularly obvious whenever a barrier was left in place during intercourse. Thus menstrual technology gives rise directly to contraceptive technology. As time goes on, women seem to have invented almost every imaginable kind of barrier from simple packings and gums to cervical caps and internal female sheaths. Often the barriers were soaked or smeared with spermicidal, astringent, or irritating substances. Some examples:

Dahomey women of West Africa crushed a certain "tubercled root" and inserted the pulp as a vaginal plug. Bapinda and Bambunda women of Central Africa plugged their vaginas with rags or finely chopped grass, as did the Chaga and Lugbara of East Africa. The grass plugs can be dangerous if left in place too long,[49] yet women continued to use them, so strong is the desire to limit fertility. Slovak peasant women in recent times have used rags to plug their vaginas against conception, apparently without complications (Bleek, 188; Himes, 9–11, 182).

47. Malinowski himself said that although he was satisfied the unmarried Trobriand girls did not use birth control, i.e., contraception (chap. 16), he "cannot speak with the same conviction about abortion. . . ." However, he thought it was not practiced to any great extent, and he was sure that the abortifacient herbs mentioned to him "did not possess any physiological properties" (Seaman & Seaman, 292).

48. Premarital barrenness is also reported by Pitt-Rivers from the Eddystone Islands (Himes, 27). Paradoxically, as Pitt-Rivers suggested, modern research indicates that promiscuity itself may reduce fertility. Just as couples with incompatible blood types are more likely to be childless, prostitutes, who encounter dozens of different blood types in their careers, secrete sperm-inhibiting antibodies for incompatible blood types into their cervical fluid. In cultures where the blood type is not uniform and premarital sexual freedom is complete, this phenomenon could also help explain the long-term puzzle of premarital barrenness (*Ms.,* Apr. 1981: 15).

49. Two women near death were brought to a colonial medical officer. Both were suffering from retention of feces and urine, serious infection, and high fevers—due, as eventually discovered, to impacted grass plugs in their vaginas. Both husbands and several women companions of one of the patients denied any knowledge of the plugs (Himes, 10–1). This case shows just how little we probably know of primitive women's contraceptive technology. The women concealed their techniques, perhaps because they knew their husbands would be hostile, or because missionaries had told them such practices were wrong. But for these two instances of the method gone wrong, the group—or the custom—might have disappeared before any European learned of it.

Ancient Hebrew women knew a soft, spongy material called *mokh,* an absorbent substance like cotton or hackled wool, which they used as a vaginal plug or cervical cap (Talmud, cited in Himes, 72, 73n). They may have learned the method from Egyptian women during their captivity (Egyptian women used lint tampons). *Mokh,* by then called "*much,* a kind of wadding," was still being recommended in the 16th century for Jewish women too ill to tolerate a pregnancy (Himes, 73n).[50]

In fact, Jewish religious writings contain numerous references to "absorbents" and the circumstances under which they might be used for contraception: harlots use them; a slave woman soon to be freed uses them in order not to conceive before freedom so that her child will also be free; a captive uses them until she knows what her status will be; and a convert to Israel uses them until she is fully accepted so as to be sure her child will be accepted as well. Religious writers are also concerned with knowing the exact time when menstruation begins so that a woman's husband can obey the ritual taboos on intercourse during menstruation: if she is using a contraceptive absorbent, this may be difficult (Noonan, 10–1).

The sponge: Himes assumes that when rabbis specify three classes of women who "must cohabit with the sponge," they are referring to *mokh* rather than to actual sponges. Shirley Green assumes otherwise (101). Why suppose Hebrew women ignorant of the most effective barrier contraceptive available before modern times—the sea sponge, used alone or soaked in a spermicide or gummy substance? After all, sponges appear all around the Mediterranean from ancient times in functions exploiting their superb absorbency. Sponges have also been used as menstrual absorbers for untold millennia. Thus their contraceptive uses would probably have been discovered early. Says Hugh Davis, "Since ancient times, sea sponges have been used as a cervical barrier for contraceptive purposes." Cleopatra herself is supposed to have relied on such a sponge (4).

A 19th-century source mentions that Constantinople women insert a sponge before intercourse. American slave women used sponges to prevent pregnancy. In the 1930s, Polish and Ukrainian peasant women on their annual trips to market routinely bought a small sponge to prevent conception. Again, the reports are modern, but the practice, as part of folk medicine and peasant women's culture, is probably ancient (Himes, 182, 178; Sterling, 12).

According to a 1912 report, the women of Easter Island, before intercourse with foreigners, put a piece of alga or seaweed just in front of the mouth of the uterus (Himes, 22). This report is interesting in view of the modern use of a seaweed plug *(Laminaria)* to dilate the cervix for abortion and the likelihood that ancient Japanese women may have used the plug alone to cause abortion.

Oils; gummy substances: Various thick, oily, or gummy substances used alone, with spermicidal, astringent, or irritating substances, or with another barrier, were among the most effective parts of woman's early contraceptive technology. The Karo-Bataks of Sumatra introduced a small ball of opium into the vagina as a contraceptive (Himes, 38). Depending upon its degree of refinement, the opium might be a rather gummy mass that could indeed effectively immobilize sperm. Honey is recommended in Egypt, Greece, India, and Islam, and by a Roman army physician.

---

50. If the Jews seem enlightened or permissive towards women's use of contraception, it is because the command to be fruitful and multiply was given only to males. Although of course no male could obey the command without female cooperation, the omission of women from the wording of the command gave rabbis a legal way out.

The Kahun or Petri Papyrus of about 1850 BC mentions both honey and "auytgum." The Ebers Papyrus (1500 BC) recommends honey on a lint tampon. Dioscorides recommends honey (usually mixed with other substances). Soranus of Ephesus also recommends smearing the mouth of the womb with old olive oil or honey or cedar gum or juice of the balsam tree.[51] Indeed, cedar gum is at least as old as Aristotle (Aristotle's wife?), who recommends it. Old Sanskrit sources recommend smearing the vagina with honey and ghee (clarified yak butter; Himes, 61, 63, 64, 66, 80, 87, 90, 121, 151).

Greek women used olive oil as a barrier, and Aristotle mentions "anoint[ing] that part of the womb on which the seed falls" with olive oil (Himes, 66, 80). As late as the early 1930s **Dr. Marie Stopes** was teaching the use of olive oil, sometimes as sole preventive, and in 1931 cited a series of 2,000 cases with no failures (Himes, 80).[52] Dr. Stopes, discussed in more detail below, also suggested butter as a cheap and easily available barrier contraceptive for the poor (Himes, 152n). Other greasy or gummy substances used were oil of cedar, tar, pitch, gum arabic, and frankincense and myrrh, both gum resins. In this connection it is interesting to speculate on the meanings of the gifts of frankincense and myrrh brought to Mary after the birth of Jesus.[53]

Cervical cap; internal sheath: Two more sophisticated types of barrier contraceptive are cervical caps and female internal sheaths. As noted, Hebrew women sometimes molded *mokh* into a cervical cap. Likewise, Japanese prostitutes applied disks of oiled paper to the cervix. German-Hungarian women in Banat applied disks made of melted beeswax to the cervical opening (Himes, 126, 182). Casanova recommended using half a lemon, with most of its juice and pulp extracted, as a cervical cap (Himes, 17–8), a practice he surely learned from a woman. Dorothy Sterling reports the same practice ("hollowed-out halves of citrus fruits") among slave women in the United States (12), and Sheila Kitzinger found contemporary Colombian women using an orange half for this purpose (58).

On female sheaths, Himes has one mythical and one historical report. The mythical one is the more interesting: it shows a man going to a woman for sexual counsel; it may be an isolated report of a prevalent practice; and it shows how the condom may have been invented. Himes (187) cites Antoninus Liberalis:

> Now inasmuch as the semen of Minos contained serpents and scorpions, all the women who had cohabited with him had been injured. For this reason he married Pasiphae, who was immune against infection because she was the daughter of the King of the Sun. Inasmuch as this union remained sterile, Prokris [in refuge at Minos' court from her husband] sought a remedy in the following: She *slipped the bladder of a goat into the vagina of a woman.* Into this bladder Minos cast off his serpent-bearing semen. Then he went to find Pasiphae, and

---

51. Although the compilers of these papyri may have been men, and Soranus (or Soranos) of Ephesus was certainly male, I assume that the compilations represent much more ancient material gathered largely from women healers, patients, or family members. This seems a particularly reasonable assumption for obstetrical and gynecological material. As already noted, Hurd-Mead suggests that Soranus, whom Himes called the most brilliant gynecologist of antiquity, borrowed much of his material from a woman gynecologist named Cleopatra.

52. Considering male hostility to contraception, I wonder whether olive oil may not have succeeded as well as it did because women could tell their husbands it was a lubricant—and because they could buy and store it for another purpose—whereas with some other methods the purpose could not be concealed.

53. Myrrh, a traditional aphrodisiac, was also taken internally as an abortifacient.

cohabited with her. In this manner Pasiphae conceived not only Ariadne and Phaedra, but also two other daughters and four sons. [Emphasis mine][54]

**Prokris**'s invention may have occurred to other women as well. If many women used such sheaths, to prevent disease or conception, or both, they must occasionally have seen them withdrawn with the penis of their sexual partner. Seeing the penis thus accidentally sheathed could have given either partner the idea of the male sheath. These were first made, like Prokris's device, from animal bladders.

The historical report of a female sheath contraceptive comes from Dutch Guiana in South America. The Djuka women living there snipped off one end of an okra-like seed pod about five inches long, and inserted it into the vagina. The intact end pointed inward, the open end receiving the penis and becoming a sort of "vegetable condom." The Cornell bacteriologist who reported this invention did not know how widespread it was; but since the Djukas were sexually promiscuous both before and after marriage and yet were merely maintaining their numbers, some effective form of population control seems likely. Their child mortality rate seemed high to Kahn, but their postpartum intercourse taboo was only two years (Himes, 18–9).

Doubly interesting in view of these fascinating but fragmentary reports is the woman's condom announced in 1988. As for Fallopius, the main impetus may have been disease-protection (here, HIV infection), but it also has a definite contraceptive purpose, and it offers several advantages over traditional male condoms. The new sheath is under women's own control, and its polyurethane composition is much stronger than rubber.

The device, manufactured by Wisconsin Pharmacal of Jackson, WI, and called the WPC-333, consists of a soft, loose-fitting polyurethane sheath and two flexible rings like the ring that stiffens a diaphragm. Inserted like a tampon, it protects the entire vagina, the inner ring covering the cervix and the outer remaining outside to hold it open. Thus far, 500 couples in Denmark, Britain, and the United States have tested WPC-333 for comfort, safety, and contraceptive effectiveness. Given FDA approval, it could sell over the counter. Chemist **Mary Ann Leeper** headed the team developing the female condom for the American market ("New Alternative").

Chastity belts: A special class of barrier to conception is the so-called chastity belt, mentioned above as a cultural method. The usual image of this device in Western minds is the Medieval attempt by departing Crusaders to keep their wives chaste—and miserable—until their return. However, myth and anthropology record at least two other types devised by women for their own purposes. The legendary Near Eastern Queen **Semiramis** invented a chastity belt to keep female courtiers from gaining influence over her son, through whom she intended to rule. Her model bore a lock opened by a key (Boccaccio, 6–7). Cheyenne Indian women used a chastity rope to signify to males that they were not open to sexual advances (Berry, 112–3). Since certain primitive women, among them the !Kung, used soft leather straps as menstrual absorbers, any of them could easily have invented a chastity belt for personal or ritual purposes.

*Douches.* Douching is another very ancient and widespread woman's contraceptive technique. Where water alone was used, it would be a physical method; where a

---

54. In another version of the myth (Graves, *Myths* I: 298ff), Minos's noxious spermatic fluid is Pasiphaës's way of punishing him for his many infidelities. Here, Minos seduces Prokris as well, but she demands that he first take "a prophylactic draught—a decoction of magical roots invented by the witch Circe." Robert Graves suggests that Prokris may actually have been a priestess of Athena, and the seduction the Cretan king's demand for a legitimizing ritual marriage (*ibid.*, 302). Could the "prophylactic draught" reflect an early VD preventive or remedy invented by moon-priestesses?

spermicide or other additive was used, it would be an herbal or chemical method. Its effectiveness varied with timing—before or after intercourse; if after, how soon—and ingredients—spermicidal or not.

One ingenious variation on the douche, reported by Himes (82), is women's directing their own urine to flush out the vagina. Such a method could be used in deserts or wherever water was not handy.

*Expulsion of sperm.* One of the simplest contraceptive methods, when women know the role of the seminal fluid in conception, would be to expel the fluid from their bodies immediately after intercourse. Indeed, most of the known techniques are so simple that they need not be dwelt upon here. However, A.C. Rentoul, Malinowski's antagonist in the Trobriand Islands controversy[55] believed he had solved the mystery of premarital barrenness in the face of total promiscuity when he learned (Himes, 31)

> . . . that the female of the species is specially endowed or gifted with ejaculatory powers, which may be called upon after an act of coitus *to expel the male seed.* It is understandable that such powers might be increased by use and practice, and I am satisfied that such a method does exist.

His wording suggests consciously controlled spasmic or rhythmic muscular movements of expulsion. Similar reports from Port Darwin, Australia, indicate that the native women there used violent abdominal movements to expel sperm (Himes, 40).

The other noteworthy expulsion technique is that of the Kgat(t)a women of Africa, who reportedly use a drug to expel the semen (Bleek, 188; Devereux, 26). It would be most interesting to know the composition of this drug and whether it is also used in childbirth, to hasten labor or expel the placenta.

Expulsion techniques are ridiculed by most birth-control experts today. However, we might note in passing that in the 19th century the contraceptive sponge was seen not so much (or not only) as a barrier, but as a way of wiping the sperm out of the vagina when the sponge was withdrawn after intercourse, as Knowlton makes clear in his famous *Fruits of Philosophy* (33; cited in Hellerstein, Hume, & Offen, 190).

*Deep massage: tipped uterus.* Another physical method of contraception is tipping the uterus backward by strong abdominal massage or manipulation. Conception becomes more difficult if the uterus is tipped far enough, because the cervix can rise in the vagina until its mouth is covered by the opposite vaginal wall. Preliterate women in various cultures seem to have known this. Since this Eastern method is unfamiliar in the West, I quote at some length from an 1897 report (Himes, 36–7):

> First the abdomen of the woman, closely above the symphysis, was pushed downwards by the Dukun [midwife] with stiff, spread fingers. Then, with both hands close above the *Ligamentum Pouparti,* the midwife rubs upward very forcefully, causing the abdominal muscles to relax. Then both hands grasp the uterus from the side, pulling this forward; while the thumbs, with a forceful pressure, push the uterus downward. This latter movement, as well as the first pressure above the symphysis, causes extraordinarily severe pain.

This Javanese procedure, known as *ankatproet* in the Dutch rendering of the native word, was usually done immediately after an abortion or a delivery.

Ten years earlier, Cornelius L. van der Burg reported the same or a similar procedure

---

55. The debate raged over whether these people did or did not know the role of sexual intercourse—i.e., the male role—in procreation. Malinowski maintained that they did not.

from the Dutch East Indies. Here, the procedure was apparently done on young girls (Himes, 36n):

> In the girls the sexual impulse develops very early, and is gratified [regularly] without fear of consequences, when the services of certain skilled elderly women *(doe-koen)* have been requisitioned. . . . These women appear, in fact, to understand, by means of pressure, rubbing, and kneading, through the abdominal walls . . . how to induce anteversion or retroversion of the uterus, to such an extent as to prevent . . . conception. It is said that the only inconvenient consequences of this procedure are trifling pains in the lumbosacral and inguinal regions, and some trouble in passing water during the first few days after the manipulations. . . . Later, when a girl . . . wishes to marry and become a mother, by a reversal of the manipulations, the uterus is restored to its natural position.

Medical reports on the frequency of retroflexed uterus among Javanese women varied from 50% to 95.8% (Himes, 37n). If the sample populations were representative, this contraceptive technique must have been widely used.

Japanese women also deliberately tipped the uterus for contraception (Himes, 127). A modern massage practitioner (United States) confirms that such people claim to be able to correct a retroflexed uterus. Thus perhaps they could also induce one if so desired.

The Sumatran Malays also practiced a "massage of the abdomen" to prevent conception, as did the Karo-Batak women of the same region (Himes, 38). Women in many parts of the world also knew how to use massage to cause abortion. Interestingly enough, as late as the mid-19th century in the United States, the noted water-cure practitioner Russell T. Trall was recommending massage to prevent implantation of a fertilized egg by causing the uterus to contract quickly (Himes, 268).

*Acupuncture.* Perhaps related to this ancient knowledge of massage and manipulation is the acupuncture point for contraception, known to the ancient Chinese (PC, Science and Civilisation of China Project, Cambridge, England, 5/30/79). This point is located about two inches below the navel.

*Intrauterine devices.* Most vexed among mechanical contraceptive methods is the intrauterine device, or IUD. According to historians, the IUD was invented either in 1909 by a German doctor named Richter (a thread pessary) or in the 1920s by another German, Graefenburg (a hard intrauterine ring). Then, beginning in the late 1950s, these devices were rediscovered and developed in several modern forms.

However, in Hippocrates' time lead tubes filled with mutton fat were being inserted into the cervix and left there, allegedly to open the passage for sperm, but obviously plugging it and probably irritating the uterus as well. The same technique is recommended in 13th-century Italy and elsewhere in Europe during the 15th and 16th centuries (S. Green, 110; Wood & Sutters, 35). Midwives of the 14th century were advised to use pessaries and "other instruments" for removing uterine moles (Hurd-Mead, *HWM,* 283), which suggests a meaning analogous to modern instruments for dilatation and curettage. Note, in any case, that these "instruments" were definitely in the hands of midwives. By the 16th century stem pessaries already existed and were known to French midwives (Hurd-Mead, *HWM,* 359). Shirley Green reports 17th-century stem pessaries functioning like a combined cervical cap and long-stemmed IUD, supposed to promote conception by propping up a sagging uterus (110ff).

Then there is the famous story of the Tuareg camel drivers who inserted a pebble into the uterus of female camels to keep them from getting pregnant on long journeys, and the tradition that this practice of "stoning"—and European knowledge of it—are at least as old as Aristotle (Southam, 3). Clive Wood (2–3) thinks Aristotle was referring to some other practice in the passage in question. The real question here, it seems to me, is not whether "stoning" is as old as Aristotle, but whether it is not in fact far older, and invented by women. The debate as usually framed, for example, ignores a strong support for women's origination of the pebble IUD: aside from the obvious fact that women had far more reason to know of such a technique from their own experience and needs, among the Azgeur Tuareg women traditionally owned the camels (Campbell, 218; Murphy & Murphy, 228). At birth a Tuareg male receives a sword; a female, a camel.

Wood argues that the tradition has no relevance to human contraception: since "stoned" female camels refused to mate, the practice could not have become popular with women. Wood's argument rests on two assumptions: that female humans would react precisely as female camels did to a foreign body in the uterus, and that women learned the uses of the pebble from men. The first is clearly not true if modern IUD experience is any guide; and the rest of early contraceptive history argues against the second.

Another difficulty is the question of infection, especially since modern, "scientifically designed" IUDs have recently come under fire for causing pelvic inflammatory disease and other problems. However, Schneider points out, after observing the prevalence of instrumental abortion on the island of Yap ("Abortion," 214–5):

> Yap women live in their microbiological environment all their lives without the protection of antiseptics or the habits of cleanliness and hygiene common among us. Consequently, the resistance to infection on the part of those who survive is generally quite high. This does not mean that they never become infected. It only means that were an American woman to try to abort by the plug-and-injury method she would very likely incur a serious if not fatal infection, where the typical Yap woman probably incurs only a mild, localized infection.

Another tradition of foreign bodies in the uterus of domesticated animals comes from Japan, where balls of metal or ivory were evidently used to sterilize cows (Wood, 3). If women used these first on themselves, a possible origin of the famous *ben-wa* suggests itself: balls intended for uterine insertion may have been expelled into the vagina and the pleasurable effect noted. Two Indian sources declare that the *dhustura (Datura metel)* root inserted into the "yoni" (presumably the uterus) will prevent conception. These sources are 15th- and 17th-century works; but as their authors point out, "Techniques are older than formulated, recorded evidence" (Dash & Basu, 15, 11).

According to the second of the Hippocratic books on women's diseases, the treatment for prolapsed (sagging) uterus consisted of fumigation (introducing medicated vapors into the vagina) followed by opening of the cervix with wooden dilators of increasing thickness. Inserted into the uterine cavity to a depth of four fingers, these wooden dilators were to stay in place during the day and be replaced by a lead tube at night. If this treatment failed, the practitioner must dilate the cervix with a lead tube and insert through it into the uterus one or two small pessaries the size of an olive and smeared with rose or iris oil to ensure passage. The pessaries were made of resin (Wood, 4–5).

Of course, as Wood points out (5), victims of a prolapsed uterus are not likely to get pregnant in any case, so this treatment might not reveal the contraceptive value of a foreign body in the uterus. However, this seems to me scarcely the point. The important datum here is that three centuries or more before Christ, even male medical practitioners (who wrote most of the books) routinely used devices resting partly in the cervix and partly in the uterus, as well as some placed entirely within the uterus. Moreover, they knew a method of inserting things into the uterus. Thus, almost certainly Greek women had long before discovered such techniques—and their contraceptive uses.

One strong piece of evidence for the ancientness of the IUD—and thus for its invention by women—is that the word *pessary* itself comes from a Greek word meaning a small stone of the kind used in a checkers-like game (Wood, 5). *Pessary* has had varied and confusing meanings over the ages—British usage today calls a cervical cap a pessary, for instance—but if pessaries were originally stones, and were effectively used at all, they must have been used inside the uterus. Wood, blind to any possible role of women in their own contraceptive technology, says only, "This Greek origin [of the word] has again inclined some authors to the view that an intrauterine contraceptive pessary was well known to classical physicians."

Dr. Anna Southam, though agreeing that the first mention of intrauterine pessaries in written medicine occurs in Hippocrates, is much more positive and specific about the ancientness of the IUD than most writers. After retelling the camel story with some illuminating details (the practice was common in both Arabia and Turkey, the camels were saddle camels, and a hollow tube was used to insert the stone), she continues (3):

> For over 2,000 years intra-uterine devices have been used for every conceivable reason, including population control. Many different materials have been utilized. These include wool, ivory, wood, glass, silver and gold, ebony and pewter; they have been made of platinum and studded with diamonds.

As elsewhere in this book, I assume that the more ancient a contraceptive technique, the more likely it is to be a woman's invention. For this chapter I have assumed, and perhaps should emphasize here by amplifying upon it, that until males took over gynecology and obstetrics in a culture—if they did—women invented their own contraceptives and passed them on from mother to daughter, midwife to client, wise woman to suppliant or friend. In the West, even as males took over formal, institutionalized medicine, women's diseases and matters having to do with conception, abortion, and birth were the last to go. And even after male practitioners had begun to treat women's diseases and attend at some births, they would for centuries merely be collecting, using, and reporting remedies and devices created much earlier by women and their midwives and healers.[56] In other words, just as European peasant women knew the virtues of moldy bread for preventing wound infections long before penicillin was "discovered," the Hippocratic use of cervico-uterine and intrauterine devices probably rested on a long history of women's contraceptive technology.

Moreover, some of the Greek and Roman medical writers were women. One Aspasia, whose works are now lost, was evidently a much-admired physician. Her very effective-sounding contraceptive prescription was a wool tampon soaked with herbs, pine bark, nut galls, myrrh, and wine (Wood & Sutters, 37).

A modern woman contraceptive inventor (or inheritor of a long underground tradition) in 1950 "put her gold wedding ring to good use" by "somehow inserting it into her uterus as a contraceptive device." The golden IUD stayed in place for nearly three years, causing no problems, and allowing no pregnancies (Hugh Davis, 7). This woman may not be so rare as the medical literature would indicate.

Many primitive abortion methods could have led women naturally to discover the amazing benefits of a foreign body in the uterus. One of the commonest abortion techniques is to insert a probe or stick into the uterus, breaking the membranes and

---

56. Paracelsus was probably not the only one whose pharmacopeia came from "the Sorceress," i.e., a wise woman (or women) of his acquaintance (Ehrenreich & English, 17). Note, too, that from the time of the male takeover of midwifery between the Middle Ages and the 17th century, contraception as seen in the works of male doctors made no real advances over pre-existing techniques until the rubber caps and diaphragms and the vulcanized rubber condoms of the 19th century.

dislodging the fetus or so irritating the uterine lining that it will shed and expel the fetus. This must have been done in very early times with an ordinary stick or bone.

According to anthropological reports, for example, the Masai used a sharp stick, the Chagga the midrib of a leaf, the Hawaiians a kind of wooden stiletto, the Eskimo a walrus bone, the Japanese a pointed root anointed with musk, inserted between the fetal membrane and the uterine wall. The Batak used a kind of pea-shooter to propel the midrib of a leaf or a bamboo sliver into the uterus (Devereux, 36). Yap women introduced a thin rolled plug of dry hibiscus leaves (which expand when moistened) into the mouth of the cervix, then scratched that area with a bit of stick, stone, or iron, a fingernail, or other sharp object until blood appeared (the plug-and-injury method described by Schneider, 213). Post-industrial women of Europe and the United States use metal coat-hangers and knitting needles, among other things.

I speculate that in ancient Japan, women inserted a dry plug of seaweed *(Laminaria)* into the mouth of the cervix. Expanding as it absorbed moisture, it gradually dilated the cervix and eventually brought on an abortion. This ancient method—if ancient it is—has been taken over by modern medicine as a gentle day-long cervical dilator used before an early vacuum abortion. Dr. Susan Harlap and others report fewer miscarriages in subsequent pregnancies when this kind of dilation is used in women who have never borne a child.[57]

The midrib of a leaf would seem fairly fragile, as would bamboo slivers, and dried hibiscus leaves might be inclined to crumble. In the Batak method, there would be no way to pull the leaf midrib out again. Most of the methods mentioned could leave debris behind in the uterus, which might or might not be expelled with the resulting abortion. If any bit of debris large enough to cause an irritation became permanently lodged in the uterus, sterility would very likely follow. We need not assume that every woman to whom this happened made the connection; in each small, close-knit group, one—especially one midwife—would be enough.

In fact, we need not speculate at all in several cultures, for their women produced abortions by inserting foreign bodies into the uterus to stay. The Romans used papyrus or dry sponges; women of Mainate, India, a small stick; Turkish women, tobacco or an olive stem; Japanese women, a silk thread covered by musk. Arab medicine—in techniques I assume were learned from women—recommended a ball of wool containing irritants or a ball of tobacco or cotton soaked in the juice of *Colotropis procera* (Devereux, 36).

And at least two cultures report the use of intrauterine objects specifically for contraception. Zulu women used IUDs to avoid becoming pregnant during their long migrations, which continued into recent times. Interestingly enough, the decision to do this rested with the tribal elders—a political rather than a personal decision—who must sacrifice and ask the ancestors' forgiveness. The insertion itself involved a ceremony, using pebbles collected from a special sacred river (Kitzinger, 61). And Polynesian mothers sometimes inserted a hibiscus root into the uterus to avoid pregnancy (H. Fisher, "Prehistoric," 34).

Thus, however we interpret the Tuareg camel-driver story, women in cultures widely separated around the globe clearly invented intrauterine contraceptive devices long ago.

---

57. I have queried several of the modern clinical researchers testing *Laminaria* as a cervical dilator, and have either received no reply or been told the person knew nothing of ancient usage. However, dried seaweed (including *Laminaria* species) is an ancient food in Japan, and the method is so similar to the one just described from Yap, that I suggest women deserve credit for it.

As Wood and Sutters succinctly put it, "the story of intrauterine devices serves to show how well buried the history of a method can become" (107).

## Herbal or chemical methods

*Physico-chemical methods.* Douches become chemical methods when they contain spermicidal ingredients. Himes considers douching rare among primitive women, "completely unknown" to the ancient Greeks and Romans (96), and not very effective in any case, particularly if used after intercourse rather than before. However, douches containing spermicide, astringents strong enough to tighten the mouth of the cervix, or compounds causing uterine contractions can be at least partly effective. In the late 19th century black women of Guiana or Martinique reportedly used a douche of lemon juice "mixed with a decoction of the husks of mahogany nut." Since one or two tablespoons of lemon juice in a quart of water is an effective spermicide, and the mahogany husks are astringent, this should be quite an effective douche (Himes, 17).

Oribasios, a 4th-century Greek medical encyclopedist, in his chapter on contraception, mentions a "decoction of coronilla seed injected into the vagina" before coitus (Himes, 93). The plant is not identified here, but *Coronilla emerus,* or scorpion senna, is a cathartic; and *Coronilla scorpioides* yields a poisonous glucoside called coronillin. Compare also Dioscorides' recommendation of honey mixed with *Coronilla securidaca,* a sword-shaped sicklewort (Himes, 87).

Shirley Green speculates (65) that douching may have been rare because women found it hard to inject solutions into the vagina far enough and under enough pressure to accomplish their purpose. Here may lie the link with contraceptive tampons and pessaries, at least in some cultures. That is, if a woman hit upon the idea of flooding her vagina with an astringent or spermicidal fluid, or a combination of the two, she might also devise a tampon or other device to keep it from dribbling away again before having its effect. (Such an idea might logically arise from her means of dealing with menstrual flow.) Later, she might simply soak the tampon in the solution to start with, or include the desired ingredient in the pessary.

Douching may have been less rare than Himes and Green think. Several male compilers' recipes for precoital washes and anointments of the penis, ridiculed by these writers (e.g., S. Green, 57, 59), may represent male appropriations, or compilers' corruptions, of much older contraceptive douches invented by women and intended for female use. (On the other hand, as Noonan suggests, these may represent quite rational means of introducing astringent or gummy substances deep into the vagina at just the moment of need [15].) Two "preventive lotions" collected in a compilation of "old, traditional" recipes in the *United States Practical Receipt Book* (1844) probably represent a purer tradition handed down in the female line. One uses pearlash and the other bichloride of mercury as main ingredient (Baxandall, 17).

Most interesting of all, primitive women may actually have had the necessary technology for injecting fluids into the vagina under pressure, for Devereux reports that *many primitive groups had bulbed syringes* (reportedly for enemas; 37). Also, women as brewers and bakers were the custodians of fermentation, and could have put natural carbonation to some innovative uses. Should these women be thought any less inventive than the modern Texas prostitutes who use Coca-Cola® as a spermicidal douche, shaking the bottle to provide the necessary propulsion (J. Rose, 191)? In this connection the Kahun medical papyrus contains some interesting wording, at least in one translation: honey is to be "sprinkle[d] on her uterus" (Himes, 61), which would seem difficult to do without some means of injection under pressure.

The first modern documented reference I have seen to a vaginal douching syringe is from an inventory of a French prostitute's tools of the trade in 1600 (S. Green, 98), but the history of prostitution is still in the dark.

By the 19th century, spermicidal douches had become one of the three commonest contraception methods, at least among the women diarists and letter-writers cited by Sandra J. Myres in *Westering Women* (154ff). One Lizzie Neblett declares that no doom would horrify her more, including even prison or solitary confinement, than the knowledge that the next dozen years would bring her five more children. She urges her absent husband to get her some "preventitives," admonishing him, "[D]on't start home without a good quantity of pulverized Ergot and as good a syringe as you can find, Richardson's No. 1 that I described to you long ago is the best I know of" (155).

Whatever the actual frequency of douching among primitive and ancient women, and whatever the relationship of that method to the development of astringent- or spermicide-soaked tampons and pessaries, these physico-chemical contraceptives are among the most effective developed before modern times. Indeed, many women today, wary of hormones, are returning to cervical caps and diaphragms, the latter usually combined with spermicidal creams and jellies (Seaman & Seaman, 230).

One of the most effective varieties of the barrier-plus-spermicide was the sea sponge soaked in various sperm-killing or -slowing ingredients. Cleopatra's sponge was moistened with vinegar (Hugh Davis, 4), which Shirley Green calls one of the most effective vaginal spermicides of all time (59). The 19th-century women of Constantinople moistened their sponges with diluted lemon juice just before intercourse (Himes, 182). Citric acid is a highly effective spermicide; the practice, since it is cited as folk medicine, is probably ancient.

One improvement in the contraceptive sponge would be attaching a string or ribbon to it for easy removal. This simple but significant improvement was almost certainly invented by women many times over, centuries if not millennia before the following 20th-century description by Francis Place (Himes, 216–7):

> What is done by other people is this. A piece of soft sponge is tied by a bobbin or penny ribbon, and inserted just before . . . intercourse . . . , and is withdrawn again as soon as it has taken place. Many tie a piece of sponge to each end of the ribbon, and they take care not to use the same sponge again until it has been washed.

Another highly effective class of barrier-chemical contraceptives consists of tampons soaked in astringents, spermicides, sperm-slowing compounds, or some combination of the three. The tampons could be of plant or animal fibers—cotton, lint (scraped linen), or wool—and the ingredients they contained were many and varied, though a few recur persistently: honey and various gums and oils to slow the sperm and clog the mouth of the cervix; alum, natron (sodium citrate) and other astringents from unripe fruits or nut skins to tighten the mouth of the cervix; various acids—acetic, citric, tannic, lactic—and alkaline preparations and salt to kill sperm. Juniper oil on a tampon, using yet another chemical group, the aromatics, was known at least by the time of ancient Greece (Hurd-Mead, *HWM,* 47n).

The Ebers Papyrus recommends honey on a lint tampon with acacia tips, remarkable in that the tips of the shrub acacia contain gum arabic, easily transformed chemically into lactic acid, the effective ingredient of modern contraceptive jellies as late as the 1930s (Himes, 63–4). Soranus of Ephesus recommends several interesting and quite possibly effective substances and combinations of substances to be introduced to the mouth of the womb by soft wool: equal parts of pine bark and *Rhus coriara* (sumac)[58] pulverized with wine; equal parts chalky soil and panaseroot alone or mixed with water; fresh (pome)granate pulverized with water; dissolved alum and (pome)granate pulverized with water, for example (Himes, 90). Whether or not Soranus was indebted to the Egyptian gynecologist Cleopatra, the contraceptives listed by the Egyptian compilers and by Soranus would have been much more ancient than their own writings, and almost without doubt originally devised by women.

*Chemical pessaries.* Closely related, and also likely to be effective, are the suppositories or pessaries in which the vehicle is not a fiber tampon but one of the gummy substances themselves. Achehnese women of Sumatra used "a black mass in the form of a pill" inserted into the vagina before coitus. When analyzed, this mass proved to contain a large quantity of tannic acid, a stronger spermicide than lactic acid. Tannic acid in a solution of one part in a thousand immediately immobilizes sperm, whereas lactic acid at that strength takes nearly two hours (Himes, 38).

The Kahun medical papyrus recommends putting honey and natron (native sodium carbonate, an astringent) into the vagina. Dioscorides recommends honey mixed with juice of peppermint or sicklewort *(Coronilla securidaca)* or "Hedysarum," and used as a suppository or pessary before coitus. Soranus also recommends smearing the mouth of the womb with old oil or honey or cedar gum or opobalsam, mixed with white lead, or with alum or with galbanum in wine. Aristotle favors anointing "that part of the womb on which the seed falls" with oil of cedar, or with ointment of lead or with frankincense, commingled with olive oil (Himes, 61, 87, 80, 151). Several of the gums and gum resins recommended as barriers here obviously also contain aromatic ingredients that might have spermicidal properties.

The lemon-half cervical cap used also by American slave women—among others— should really be classed as a physico-chemical method. Any juice or pulp remaining in the cap would contribute a powerful spermicide, citric acid (Himes, 17–8, 182; Sterling, 12).

*Oral contraceptives.* Among the most interesting of woman's many anti-fertility achievements is her work with oral contraceptives, usually herbs or herbal preparations. Himes cites—and dismisses—many reports of "using the local flora with contraceptive intent." Perhaps Himes and others would have seen these women's pharmaceutical inventions in a new light had they reflected that, as John Noonan points out, "Potions are the first form of contraceptive mentioned by any of the classical writers, and the type most often mentioned. In the Hippocratic writings a potion is the only contraceptive described." This is a drink called *misy,* clearly distinguished from abortifacients, which produced a year's sterility (Noonan, 13).

With this in mind, let us take a quick look at some of the many anthropological reports dismissed by Himes: Yao women used plant sap. Natives of the Oasis of Sima in the Libyan desert drank infusions of gunpowder and also used the foam from a camel's mouth. Some of the things sold in the Riffian women's secret contraceptive market were herbs, but Carleton Coon was unable to identify them. In the Moroccan town of Fez, women ate castor beans, one for each year that they desired to avoid pregnancy. The Bapindas and Bambundas of Central Africa, in addition to the above-mentioned vaginal plugs of grass or rags, took "native remedies" by mouth in order not to have children oftener than every two or three years (Coon, 110; Himes, 7–10).

---

58. Sumac bark is both astringent and antiseptic and is also quite high in tannin (Hutchens, 266).

Shawnee Indian girls reportedly "drink the juice of a certain herb which prevents conception, and often renders them barren through life" (Himes, 12). Aleš Hrdliĉka reports that "The San Carlos Apache believe that artificial sterility can be induced, but the means is not generally known. It is supposed to be some variety of root." White Mountain Apache women who wanted no more children swallowed a little burnt earth from underneath the fire; and some Huichol women drank a decoction of "a certain plant" to prevent childbearing. Cora women swallowed scrapings of male deer horn (Himes, 13).

Cherokee women chewed and swallowed the roots of spotted cowbane (*Cicuta maculata,* also called musquash root and beaver's poison) for four days in a row, to become sterile for life. Plants of this genus do contain powerful cicutoxins. Michael Weiner cites a similar-sounding report for a plant he calls "water hemlock," also containing cicutoxin (40). Women of Neah Bay in Washington State drank a decoction of an unnamed herb to prevent conception. Among the Isleta Indians of New Mexico, a girl might receive *at her first menstruation* a medicine to prevent childbearing. Canelos Indian women of Ecuador habitually took a contraceptive medicine prepared from the small *piripiri* plant. The root knots of the plant were crushed and soaked in water, which the woman drank; afterwards she could eat only roasted unsalted plantain and small wild birds (Hern; Himes, 13–7: *Webster's Unabr. /3*)

Other Native (North) American women's herbal contraceptives are as follows: ragleaf bahia or "twisted medicine" (a tea of the roots boiled in water for thirty minutes), used by the Navajo; Indian paintbrush (a tea of the whole plant), used by Hopi women "to dry up the menstrual flow"; a strong decoction of blue cohosh to promote menstruation, used by the Chippewa; dogbane-root tea, tribe not specified; decoction of the root and rhizome of wild ginger, tribe not specified; a tea of whole thistle plants, used by the Quinault Indians of what is now Washington State; and a tea of American mistletoe leaves used by the Indians of Mendocino County, California. Navajo women had another preventive as well: one cup of a decoction of boiled antelope sage root, drunk during menstruation. Indians living in what is now Nevada drank a half-cup of a tea prepared from the root of false Solomon's seal daily for a week.

Some of these Native American women could produce permanent as well as temporary sterility, by using different plants or varying the dosage or strength of given plant preparations. Indians of what is now Quebec Province, Canada, drank an infusion of the pounded roots of a milkweed species *(A. syriaca)* for temporary sterility. Shoshone women (Nevada) drank a cold-water infusion of stoneseed roots daily for six months to produce permanent sterility. As will be seen, the anti-fertility effect of this plant *(Lithospermum* sp.) has been confirmed by modern science. The Shoshone also some- times used an infusion of deer's tongue for contraceptive purposes. And, as already noted, Hopi women used the Jack-in-the-pulpit, or "Indian turnip," to produce temporary or permanent sterility, depending on the dosage (M. Weiner, 40–1). This plant could be a food with proper detoxification.

Edward Hyams notes that Inca women grew *Stenomessum varietum* in their gardens as a contraceptive. Contraception could be a matter of life and death in this culture where intercourse between an upper-class Inca woman and a commoner was a capital offense (118).

William Pitt-Rivers collected several contraceptive recipes from Melanesian and Polynesian sorcerers and old women, containing native herbs and roots mixed together with magical substances, but he saw no reason to suppose them effective. Old women in the Torres Straits area south of New Guinea gave younger women the leaves of two trees and a large shrub: the *argerarager (Callicarpa* sp.), *sobe (Eugenia,* near *E. chisiacfolia),* and *bok.* They chewed the young leaves well and swallowed the juice until they felt their bodies wholly saturated with it. Once this happened, which took some time, they were supposed to be sterile indefinitely. Native women of New Britain (New Guinea) ate the leaves of an unknown plant to prevent conception (Himes, 19–21).

Linton reported that Marquesan women also used herbal contraceptives, but that he did not learn their nature (Himes, 23). Note how often in these reports, in fact, the plant used is "unknown" or "unnamed." Except for those areas where culture-destruction was far advanced, I suspect that the women knew very well what plants to use, but would not tell a male anthropologist such secrets.

Fijian women, too, use herbal contraceptives, preparing an infusion of scrapings from the peeled roots and bruised leaves of the roqa tree plus the leaves and root of the samalo. Interestingly enough, this seems to be a "morning-after" remedy (Himes, 24).

Dr. Hortense Powdermaker reported that the New Irelanders believed certain leaves could make a woman sterile if she chewed them and swallowed the juice, spitting out the pulp. (Here, men usually owned the secret, but traditionally received it, not from their fathers, but from a mother's uncle.)[59] Powdermaker, like most Westerners of her day, was skeptical about these preparations. However, perhaps because of the islanders' low birth rate, she brought back seven varieties of leaves for analysis. One of them, *Rubus moluccanus,* was also used as a contraceptive in India, and two others, *Acanthaceae* and *Curcuma,* contain ingredients that might make them useful as emmenagogues (Himes, 25). One of Powdermaker's informants, Tsangamor, chewed the sterility leaves (given her by an older woman) before her first menses and, though taking two husbands and several lovers, never had any children (Powdermaker, 279–80). As Tsangamor put it, after chewing the leaves her stomach "dried up."

Professor Radcliffe-Brown reported oral sterility preparations from many New Guinea and neighboring groups. Sending some of the plants in for chemical analysis, he reported that "at least one of these medicines does produce in a few days a shrinking of the female breast" (Himes, 25). Von Oefele suggests that if women take certain abortion-causing drugs, namely ergot and preparations from *Aristolochia* species, regularly, "a certain under-development of the uterus takes place; and that this in itself either causes sterility or acts as a hindrance to conception." Many other dangerous drugs, he points out, "also cause inflammation and catarrh of the cervix, which interfere with conception" (Himes, 25–6).

Women of Tumelo, an island in Dutch New Guinea, knew at least four contraceptive plants. Of one, *kakau,* the women ate the leaves with sago bread; of *natunmum* they ate both leaves and fruit with sago bread; another plant, unnamed, they dried and made into a cigarette and smoked, swallowing the smoke. The fourth, called *lapalet,* is the most potent. Of this, women peeled the root, cut it into little disks, and then ate it alternately with coconut kernels. They could not chew it, for the taste is too bitter, and the poison reportedly so destructive that it not only sterilizes but can also kill a three- or four-month fetus (Himes, 26–7).

In the *egoro* (barrenness) rite of the Eddystone Islands, women ate bark scraped from two nut-trees called *ngari* and *vino,* and from the *petepete* tree, mixed with scrapings of a special reddish stone from another island, all put inside a betel leaf. With this they ate an *anggavapiru* nut and lime, recited a certain formula, and wore a girdle of *molu* "taken

59. By the time of Powdermaker's visit (1930s), some of the men were receiving it from their fathers, but the tradition of matrilineal inheritance was still strong (Powdermaker, 205, 243). Note that Lesu's "doctors" did medical services for money.

where it crosses the path." Pitt-Rivers, who reported this rite, did not evaluate the possible chemical effectiveness of this mixture, but he did note that despite free premarital sex, not a single premarital birth was reported during his visit to the islands, and only one was mentioned in the interviews, occurring "many generations ago" (Himes, 27).

Australian aboriginal women had two poisonous plants that they leached and cooked in a tedious process when they wished to use them for food, but left untreated when they wished to use them as contraceptives: the Queensland matchbox bean *(Entada scandens)* and the wild yam *Dioscorea sativa,* var. *rotunda* (cf. De Laszlo & Henshaw, 628). Men said that the women have "shut themselves up" by eating medicine (Himes, 28).

More recent reports continue the theme:

Agta women (Philippines) reported to Agnes Estioko-Griffin that "they space their children," using "various herbal concoctions that supposedly prevent conception." They also have abortifacient and menstrual applications. Although the practice varies, all the Agta know these "medicines," and they are "frequently used." Significantly, the Griffins report, "Our census data indicate that some women seem to be successful in spacing births" (Estioko-Griffin & Griffin, 138–9).

At least three East African groups use herbal preparations for contraceptive purposes: the Maragoli, the Kaguru, and the Tukeu (Bleek, 188).

Many more such reports could be cited, but I want to mention only a few especially interesting ones before moving on to what modern chemistry is beginning to make of it all.

Legend has it that Attila's daughters drank decoctions of arum leaves and saxifrage roots to remain sterile (Himes, 175). A 3rd-century rabbi gives this formula for the cup of roots mentioned in the Genesis Rabbah (xxiii, 2): Alexandrian gum (of the *Spina aegyptia),* liquid alum, and garden crocus, a denar's weight each, mixed together. "Two cups of beer with this medicine," he said, "cure jaundice and sterilize." Interestingly enough, the corms of the autumn crocus *(Colchis autumnale)* contain two substances, colchicine and demecolcine, found to have anti-fertility effects in rats (Farnsworth *et al.,* 551).

Another cup of roots is the strong solution Mundurucu women (Brazil) make by grating ginger root or another, unidentified, root into water. They drink this for three days after a birth to prevent pregnancy for a year; for an entire month, to achieve permanent sterility. Yolanda and Robert Murphy could not verify the effectiveness of the preparation because so much secrecy surrounded it (161).

Both Soranus of Ephesus and 19th-century European peasant women—as well as many medical compilers in between—advocated continuous drinking, after each menstrual period, of a blacksmith's metal-quenching water. Thus recommendations of this method span at least 1800 years (Himes, 63, 92, 113, 162, 184). Such persistent recommendations seem to demand some investigation, and Himes wrote in the 1930s that research was in progress, but I have not located the results. One theory was that the water might contain lead or iron sulfate, one possible basis of the contraceptive prescription called *misy* (Himes, 82).

Several patterns become clear at once from reading Himes' collection of reports: First, many of the contraceptive agents were also poisons. Nor is this merely an ancient phenomenon. As late as World War I, East Austrian and British women were willing— even eager—to work in factories where lead was used because they thought it would keep them sterile (Himes, 173). Often it did. Unfortunately, of course, as Shaw's play *Mrs. Warren's Profession* made public, it also killed some of them. In Eastern Austria women used arsenic for abortion and contraception, creating a high rate of arsenic poisoning in that area (Himes, 178). As Margaret Sanger observed, women are willing to walk to the gates of death to control their fertility.

However, in addition to these generalized poisons, some of the preparations directly affect the reproductive system. When the work of modern pharmacology is done, those few of women's ancient anti-fertility preparations that have reached us may stand divided into these two large categories: general toxins and specific reproductive inhibitors.

Third, there is a regular and significant progression in the strength and effect of several of the women's preparations: in the milder form, they cause temporary sterility; in a stronger form they cause permanent sterility. In the mildest form of all, usually after considerable preparation, the plant in question is a food. As in the Australian example of the two plants that could be either foods or contraceptives, this progression gives a clue how the properties of the plants were probably discovered.

Another pattern is now emerging for women's ancient herbal contraceptives: more and more of them are being validated by sophisticated modern chemical analysis. Motherwort, for example, chosen as one of a core group of 370 especially promising plant sources of anti-fertility substances in a worldwide scientific study sponsored by the United Nations World Health Organization (WHO), has been known to Chinese women as a contraceptive for over 2,000 years (Vietmayer, 32). Its English common name marks its traditional use in fertility matters in our culture as well.

According to Michael Weiner, writing in the 1970s, "several tribes" of North American Indians used wild carrot species to "encourage delayed menstruation." Clinical experiment has shown that extracts of wild carrot can cause the uterus to contract (88).

Russell Marker isolated from Mexican wild yams the sapogenin called diosgenin, transformable by a two-step chemical process into progesterone, making possible the birth-control pill later perfected by Carl Djerassi and colleagues. Natives of Mexico and Central America had long used the toxic water-soluble forms of sapogenins, the saponins, from these yams as fish poisons, under the generic name *barbasco;* and Mexican women traditionally used "barbasco root" as a contraceptive. Likewise, as already mentioned, women of Cape York, Australia, used the wild yam *Dioscorea sativa* var. *rotunda* to produce sterility.[60] Dr. Djerassi questions whether the yams could actually have a contraceptive effect, since they contain no progesterone or other human sex hormones, but only the sapogenin raw materials for laboratory synthesis of the hormones (Djerassi, *Politics*, 234–5 and L-10/27/81; Himes, 28; De Laszlo & Henshaw, 628; *Webster's Unabr./2*, "barbasco"; *EB/58*, s.v. "Contraception"). Indeed, the yams may act by a general toxic effect rather than by hormonal mechanisms. Alternatively, the human body might perform its own synthesis. In any case, a traditional contraceptive receives a left-handed sort of confirmation.

Indian women from Paraguay's Matto Grosso region traditionally dried the leaves and stems of a plant known to Western science as *Stevia rebaudiana Bertoni,* and drank a water extract daily to prevent pregnancy after their families were complete. Tried on female rats for 18 days, the preparation reduced fertility 57–79%. The effect persisted nearly full strength for seven or eight weeks, while the rats seemed perfectly healthy (S. Green, 130: *EB/58*, s.v. "Contraception").

This may be one of the plants Carmem de Novais had in mind when she wrote that whereas Western science has not yet found a chemical contraceptive safe for women, "the Amazon people have been using medicinal plants as a successful contraceptive method for

60. For permanent sterility, they eat the tuber raw early in the morning on an empty stomach, lie down, and eat nothing else all day (De Laszlo & Henshaw, 628).

many thousands of years." Another might have been the Paraguayan form of motherwort (Vietmayer, 32). Writing in the Winter 1975 edition of *Indigena,* de Novais, herself Brazilian, condemns drug companies for taking the Indians' medical knowledge (coca, ipecac, quinine, curare, etc.), buying Indian land to produce drugs from the plants, then selling these drugs to the people who told them about the plants in the first place (Mander, 72–3).

As early as 1954, Henry De Laszlo and Paul Henshaw, respectively of the Fertility Research Center, Colnbrook, England, and Planned Parenthood in New York City, compiled a list of 100 "Plant Materials Used by Primitive Peoples to Affect Fertility." At that point, most of the reported substances were "almost completely unevaluated" (by modern Western scientific procedures), and available only in fragmentary reports from primitive cultures, from Western popular medicine, from 19th century books on medical botany, and from old *materia medica.* De Laszlo and Henshaw, who seem well in advance of their time in some ways, comment (626):

> Information from such sources is to some extent held in question by modern investigators who often dismiss the accounts as superstitions or "old wives' tales." Confidence in such sources of information may be strengthened, however, by recognizing examples of valuable drugs recently isolated from plants used by primitive peoples for generations. . . . Even greater confidence may be gained by referring to the recent scientific literature on the desert herb *Lithospermum ruderale,* used by Indians in the southwestern part of the United States for fertility control purposes.

Many of the plants listed by De Laszlo and Henshaw duplicate reports collected by Himes.

One late-1960s study (Hern, pub. 1976) sounded a negative note, concluding that (9) "[C]omparison of fertility in women who had and had not used herbal contraceptives revealed no support for the hypothesis that the herbal preparations are effective." However, in the body of the paper, this author gives details that call his unqualified conclusion into question. For one thing, he provides numerous striking instances of long-term contraceptive success among the Shipibo-Conibo women of this Peruvian village, especially with the permanent-sterility drug. For another, he excludes from his sample all those using modern contraceptives. At first glance this would seem justified. However, these might be the very women most determined to succeed with contraception and most attracted to a rational, results-oriented approach to the problem.

Hern also notes that the villagers have had outside contact for more than 100 years (with Catholic Spanish culture), that the culture is changing rapidly—from sororal polygyny to monogamous marriages, for example—and that he will examine fertility rates in light of other researchers' conclusions that "the high fertility seen in preindustrial societies today is not necessarily a matter of a sudden desire for more children but the consequence of Western colonial expansion and the adoption of Western economic systems" (10).

Hern says nothing about the adoption of Western diet and religion, but both could have had important effects. For example, one of the most knowledgeable women he talked to—who, incidentally, had recently been fitted with an IUD at the Amazon Hospital—emphasized that during the period when these traditional contraceptives are taking effect, in the case of *totimawaste,* one day during each of two successive menstrual periods, the woman must also observe a strict dietary regimen, *avoiding salt, sugar, and fat,* and must also observe sexual abstinence (15). Since these people had both a three-year and a permanent sterility drug *(navawaste* and *piripiri* or *totimawaste),* as well as other preparations, further research would seem to be in order.

As world hunger increased, and found a voice through the United Nations, modern science began to focus on remedies for overpopulation. Facing its own failure to find a truly effective and *harmless* contraceptive, Western science began belatedly to turn to ancient experience for possible new directions. Norman Farnsworth, Audrey S. Bingel, and others, in a two-part 1975 review of more than a thousand botanical and pharmacological references, list 225 plant species as "folkloric" contraceptives and/or interceptives.[61] Among these, some of the most promising (those from which purified active anti-fertility principles have been isolated), are the garden pea *(Pisum sativum)*, the stoneseed *(Lithosperum ruderale)*, the crape jasmine *(Taberenaemontana heyneana* or *coronaria?)*, the kamala tree *(Mallotus philippinensis)*, and the rue family (e.g., *Citrus* species).

As one 1960 study noted, rats fed on peas (30–100% of the diet, both sexes) failed to produce any litters at all; mice fed only 20% peas in their diet produced fewer offspring. Moreover, as the study's author noted, the population of Tibet, where the staple diet was barley and peas, had stayed the same for the 200 years previous to the study. An active principle isolated from pea oil and reported in that study was labeled "I." Pregnant rats given "I" during the first 8–10 days of pregnancy absorbed their fetuses. In a clinical trial among Indian women, those receiving "I" had only half as many pregnancies as those receiving a placebo, and the substance appeared nontoxic in the doses given. One possibility is that "I" acts as an antagonist to Vitamin E (Farnsworth *et al.*, 541).[62]

Medical circles first learned in 1941 of Navajo women's stoneseed contraceptive. Whole plants or extracts of *Lithospermum ruderale*, eaten or injected, turned out to have anti-fertility effects in mice, rats, and hens (long gaps between estrus periods, failure to ovulate, lowered pituitary activity, etc.). Research done in the 1950s showed pituitary effects (antigonadotrophic) not only in *L. ruderale*, but in several other *Lithospermum* species, as well as in several other members of the borage family (Boraginaceae). On this "active" list were several traditional plant contraceptives or abortifacients or both: the blessed thistle *(Cnicus benedictus)*, the blazing star (true unicorn root?; *Chamaelirium luteum)*, bugloss or ox tongue *(Anchusa officinalis)*, comfrey *(Symphytum officinale)*, *Chenopodium album*, and *Rubus idaeus*.

The active principle contained in all these plants, and concentrated in the roots, was first called lithospermic acid. Later work indicates that this "acid" is probably a mixture of at least three substances.

Extracts of crape jasmine *(T. heyneana* or *coronaria)* prevented pregnancy in rates when given by mouth. Researchers isolated several indole alkaloids, among which coronaridine turned out to be the active principle. At a dosage level of 5 mg/kg/day, it prevented pregnancies and was also estrogenic (Farnsworth *et al.*, 546–7).

The glycoside rutin, found in the leaves of the common rue *(Ruta graveolens)*, in other members of the rue family such as *Citrus* species, and also in tobacco leaves and buckwheat, is an active anti-fertility agent. According to one study a diet containing only 0.1% rutin lowered the fertility of female mice. Another study done in rats and at higher dosages did not support these results; however, the species and dosage differences make the two studies difficult to compare (Farnsworth *et al.*, 547).

Rue has other medical uses—such as strengthening blood vessels in high blood

61. The Farnsworth study defines conception as "the successful implantation of a blastocyst in the uterine lining" (541), and therefore classifies substances that interfere with implantation as interceptives, a subclass of contraceptives.

62. Chemically, "I" is m-Xylohydroquinone. One later study failed to confirm these results, but see discussion below of possible reasons for false negatives.

pressure—and rutin occurs in a dietary supplement marketed under the trade name of Biocrest. But rue has long been known as an anti-fertility agent, especially as an abortifacient, and witches are supposed to have used it extensively (*EB*/58; *PDR*).

The kamala tree of the tropical East Indies *(M. philippinensis)* yields a compound called *rottlerin* that produced 100% infertility at 20 mg/kg in an unnamed species (Farnsworth, 547). The orange-red powder from the seed capsules yields a dye and a worming agent.

The overall findings of Farnsworth *et al.* on primitive women's contraceptives are as follows: of 225 plant species occurring in the literature they reviewed (their Table II; by no means a complete list, and heavily weighted toward India), 63 had been evaluated either in animals or in humans for anti-fertility effects (their Table III). Of the 63, 49 were confirmed in at least one study as having some anti-fertility effect(s) on at least one animal species.[63] I counted a confirmation even though the plant part tested might be different from the plant part traditionally reported, or where the part used was not specified in one table or another, since plants may contain the same substance in more than one part.

Only five of the substances reported in their Table II as traditional contraceptives were tested on humans in the literature summarized in Table III. Of those, four showed anti-fertility effects (three of them in combination with each other); and the fifth, though listed as still experimental in humans, had produced sterility of both sexes in mice. *The combined preparation,* taken for 22 days while abstaining from intercourse, *prevented conception for an entire year.*

Regarding the 14 species for which fertility effects were not confirmed, it is important to note that there are many possible reasons for false negatives. Among these reasons, some of which Farnsworth's group themselves point out, are the following:

1. *Species differences.* As the thalidomide tragedy showed, the same drug can act differently in different mammalian species. Billed as a tranquilizer and sleeping aid for pregnant women, the drug failed to make rats sleepy.[64] Human females are unique among mammals in several aspects of their reproductive cycle. Rabbits, used in some tests, are also unique in at least one way, ovulating whenever they mate.

2. *Experimental design.* Several of the experiments reported in Table III evidently tested only one anti-fertility effect: that on implantation. Preparations could also work by preventing ovulation, preventing fertilization, or interfering with the transport of a fertilized egg.

3. *Wrong dosage.* Some substances are estrogenic at one dosage level and anti-estrogenic at another (Farnsworth *et al.,* 539).

4. *Timing of dose.* One compound studied caused pregnant rats to resorb their fetuses if given in the first 8–10 days, but had no effect if given later.

5. *Omission of auxiliary drugs or dietary prescriptions.* This could happen because of faulty anthropological data, corrupt traditions, or modern chemists' preoccupation with

---

63. This number includes 13 cases in which the species tested in Table III may or may not have been identical to the species cited in Table II as a traditional contraceptive, but was in the same genus (e.g., *Paeonia officinalis* on Table II and "*Paeonia* sp." on Table III). By the same token, I counted only one confirmation even though in some cases two or three other species in the same genus had confirmed anti-fertility effects.

64. In fact, it was this difference that alerted Dr. Frances Kelsey, sparking her resistance to a quick license for the drug in the United States and leading her to insist on further data, especially from pregnant rats. As already noted, it was this courageous woman who saved the U.S. from the epidemic of deformed babies that struck Germany. (The relatively few American thalidomide babies were born to mothers whose doctors gave them the drug on an experimental basis.) Dr. Kelsey deserves to be remembered as one of the major heroines of American history.

isolating the single active principle. Yet some auxiliary drugs might be quite important; for example, the action of lithospermic acid is strengthened by the addition of rutin or chlorogenic acid. (Compare Suzann Gage's report [41] of women's using rue in combination with other abortifacients.)

6. *Improper gathering, storage, or preparation of the plant materials.* One 1955 study of stoneseed showed the anti-fertility activity concentrated in the roots, and greatest between June and September. Root *extracts* of this plant could stay highly active after four years' storage—but if stored at the wrong temperature or acidity, *lost half their potency in one week.*

Different solvents produce extracts of varying strengths. For example, Hutchens notes (46), "Boiling water enhances the properties of the [black cohosh] root, but dissolves only partially; alcohol dissolves wholly."[65]

7. *Wrong part(s) of plant tested.* In some of the studies Farnsworth *et al.* reviewed, the researchers had tested a different part of the plant from that reported as a traditional contraceptive. This difference could be crucial. For example, both comfrey and water horehound *(Lycopus europaeus)* contain "lithospermic acid." However, in these two plants, the "acid" remains inactive until activated by a preparation of the leaves. Thus if women had traditionally used whole-plant preparations of these two species as contraceptives, but chemical researchers tested only root extracts (because the acid is concentrated in the roots of stoneseed and other borages), the negative results would not invalidate the traditional report.[66]

I have gone into this matter at some length because the principles are important to an informed reading of pharmacological evaluations of traditional medicines or anti-fertility agents.

In Part II of their review of herbal anti-fertility agents, Farnsworth *et al.* list several plants that contain estrogenic compounds, both steroid (Table VI) and nonsteroid (Tables VII and VIII). In Table IX they list plants with demonstrated estrogenic activity. Thirteen plants from the Farnsworth list of traditionally reported contraceptives (their Table II) appear in one or more of these tables, meaning that their anti-fertility activity or *potential* is confirmed. Eight of these are already confirmed in Table III; but five were either not listed or listed but not confirmed (see accompanying table). Thus five more plants from Table II can be regarded as evaluated for and confirmed in anti-fertility activity, for a total of 68 out of 225 evaluated, and 54 out of 68 confirmed.

The Farnsworth review is a valuable one, but could have been far more valuable had the authors thoroughly surveyed the literature on traditional contraceptives and abortifacients. They apparently did not, for example, consult either the Human Relations Area Files or Himes' excellent 1936 review of more than 3,000 sources.

Two monumental ongoing studies, one of folk medicine in general and one focused on fertility control, should give us a much clearer picture of women's contributions to anti-fertility technology. In response to the late Chairman Mao's directive to merge traditional and modern medicine, Chinese researchers at the Shanghai Institute of Materia Medica began systematically searching traditional medicine for birth-control drugs and treatments for heart disease, hepatitis, and tropical diseases. By 1979 they had isolated about 200 main ingredients—*30 of them with structures new to science*—from more than

65. Black cohosh is one of the ingredients of Lydia Pinkham's Vegetable Compound.
66. I know of no such reports. However, Suzann Gage (41) lists both comfrey and "horehound" as women's traditional abortifacients. She is probably referring to the common horehound, *Marrubium vulgare,* instead of this one, but both are in the mint family (Labiatae). *M. vulgare* evidently does cause uterine tissue to contract (Farnsworth *et al.,* 546).

80 herbs. They had synthesized 12 of these drugs and put more than 30 compounds into production and use (*Science News*, Oct. 27: 293). Unfortunately, my source did not specify how many of the validated drugs were contraceptives.

The Farnsworth team, located at the University of Illinois College of Pharmacy, was part of a worldwide study directed by the United Nations. In the study just discussed, they examined claims of fertility control for over 4,000 plant species. Using computer analysis of their 1975 results, they have now identified a core group of 370 herbal preparations as especially promising. Under WHO leadership, a worldwide network of laboratories is joining the research with 36 plants as their initial targets. Two impressive examples are motherwort *(Leonurus cardiaca)*, an ancient herbal contraceptive of Chinese women, and zoapatle *(Montanoa tomentosa)* a small Mexican shrub belonging to the sunflower family. Preliminary tests on animals and humans confirm folklore's claims for the latter plant, and as of 1979 a Swedish research institute was poised to start a formal clinical trial. Prostaglandin-like compounds from this plant may finally give us a safe and successful "morning after" pill, one of the most desperately needed forms of contraception (Vietmeyer, "Greening," 32; L-12/27/81).

*Cyclical or rhythm methods*

Methods relying on a safe period for intercourse, with abstinence the rest of the month, are notoriously unsuccessful for modern women; and primitive women do not seem to have used them extensively. However, Nanda women in East Africa and the Isleta Indians of New Mexico had at least hit upon the right time of month for a safe period (Himes, 6, 15).

A subcategory of cyclical methods is the mucus method. One of these, the Billings Ovulation Method, was originated and popularized by physician, author, and lecturer **Dr. Evelyn (Lyn) Billings** of Melbourne, Australia, and her neurologist husband, Dr. John Billings. The Billings Method, which dates to 1964, works on the premise that a woman is fertile each month only during the 100 hours immediately surrounding ovulation, and that her cervical mucus changes dramatically as this fertile period approaches. By examining this mucus each day, she can pinpoint—and receive forewarning of—her fertile period and take steps to avoid impregnation during it.

According to Mercedes A. Wilson, unpaid president of WOOMB, U.S.A. (World Organization of the Ovulation Method Billings) and one of the method's strongest advocates, the Billings method *when used correctly* is 98.5% effective. Endocrinologist James B. Brown has confirmed the proposed correlation between ovulation and mucus changes; and other scientists, such as Dr. Anna Flynn of England and Prof. Thomas Hilgers of Omaha, NE, have confirmed Brown's results. Moreover, the method was apparently endorsed by such eminent Catholic figures as Mother Teresa and Pope John Paul II. Ms. Wilson has written a textbook on the method for doctors and other professionals.

Planned Parenthood, however, has refused to endorse the Billings Method, as has the U.S. Agency for Development.[67] Dr. Louise B. Tyrer, Planned Parenthood's vice-president for medical affairs, pointed out in 1980 that although highly motivated and knowledgeable users of the method might obtain better results, her organization had to

---

67. Instead, Ms. Wilson angrily maintains, USAID is dumping outdated birth control pills on Third World nations (McCormack, 8). Along these same lines, Frances Moore Lappé, author of *Food First* and *Diet for a Small Planet,* and founder of the Institute for Food and Development Policy, pointed out in a 1981 interview that USAID has injected Depo-Provera into millions of women in over seventy countries. Depo-Provera was banned in the United States because it proved carcinogenic in laboratory animals (Spitzer, 10).

consider the average user; and the average failure rate for all such "natural" or "fertility effectiveness" methods is 19% (McCormack, 9).

One problem, of course, is that sperm live much longer than previously thought, sometimes up to five days. Another whole category of problems arises because so many things can disrupt the normal cycle—infections, fevers, other illnesses, and stress. Moreover, some women evidently ovulate more than once a month, and may even be stimulated to do so by intercourse itself.

Moreover, according to at least one study, 30% of women lack the distinct pattern of mucus changes on which the method is based. A period of abstinence is involved, which demands a cooperative partner. And the method is not for squeamish women who do not like to touch their genitals (McCormack, 8–9; Seaman & Seaman, 259–61).

Interestingly enough, the mucus method, like so many contraceptive techniques, is probably ancient. According to Connie Sutherland Hernandez, whose thesis deals in part with natural methods of birth control, Hawaiian, Cherokee, and Bantu women all knew about the "glistening" or fertile vagina. In Hawaii, the original name of the sacred and fertile island of Kahoolawe was "Kohe Malamalama," which means "red, bright, or glistening vagina." An old Hawaiian woman told Hernandez that she knew the glistening vagina was the fertile vagina "before she had the body of a woman." Among the Bantu of East Africa, women pass birth-control information to their daughters at puberty; there, each woman wiped the outer lips of the vagina with a smooth stone to collect the mucus. Cherokee women used the same technique (Hernandez, L-3/26/81; Sutherland, 25–6).

Gerald and **Selmaree Oster** have developed the Body Aware system of natural birth control, with simple saliva and urine tests to detect ovulation; specifically the surge of LH (luteinizing hormone) that signals the second half of the ovulatory cycle. Gerald does basic research in reproductive science; Selmaree has worked as a health professional in Third World countries, where she saw women beginning to reject Western contraception. In any case, any birth-control method that might contribute to vitamin deficiencies in an already undernourished population (as the Pill can do), should definitely be avoided.

To use the Osters' test—a sheet of paper dipped into urine or saliva changes color in response to the high estrogen levels that support ovulation—a woman must menstruate but need not be regular, a distinct advantage over many methods.

Selmaree Oster is also teaching women to recognize their own bodily signs of approaching and occurring ovulation. A heavy feeling, abdominal discomfort, or discomfort throughout the midsection and back are general reports, but each woman has her own signs. In an initial group of 15 women, all were able to pick up nuances they had never focused on before. The Body Aware system was scheduled for marketing about 1980 (Seaman & Seaman, 271).

*Surgical methods*

In addition to the absorbent *mokh,* the cup of roots, and probably the sea sponge, ancient Hebrew women also evidently knew surgical methods of producing permanent sterility. The Talmud mentions women who have been "split," without describing the operation, but making clear the sterilizing intent (Noonan, 11).

According to a 1935 report, the Malays practiced ovariotomy for contraception (Himes, 40n). Queensland aborigines of Northeast Australia allegedly performed ovariotomies on a few young girls "in order to produce for the young men of the tribe a special class of prostitutes, who would never become pregnant" (Himes, 40–1). Likewise, reports from Southwestern Australia indicate that "females . . . are operated on in some manner to prevent conception. These women have a mark or scar on the side above the hip," and

Western observers assumed that the ovary had been removed (Ronhaar, 380). Not clear in either of these two latter reports is who did the operations. However, women's business was—and still is—very strictly women's business among these aboriginal peoples. Also, wherever women butchered the meat from the hunt, women are the likelier sex to have the anatomical knowledge necessary to devise such operations. Wherever women were also the healers or charged with funeral preparations, or both, they would be even more likely to have such knowledge.

According to Strabo, the ancient Egyptians and Lydians knew how to remove human ovaries (Himes, 67).[68] The Egyptian report is especially interesting in that the necessary anatomical knowledge could have come from embalming practices (the abdominal organs were routinely removed in the creation of mummies). Since Isis is credited with teaching the Egyptians embalming, and Old Kingdom priestesses were in charge of embalming (Boulding, 185), the Egyptian embalmer's high art probably originated with women. Another indication that contraceptive ovariotomy was probably a woman's invention in most cultures where it occurred—aside from women's usual early monopoly in all matters of fertility—is that the earliest surgery in many cultures, apparently including the Hebrew culture, is obstetrical or gynecological surgery.

## Modern Period

In spite of the male domination of medical research, the dangerous hormonal direction of much of that research, and the underground nature of contraception even today in many subcultures of Western society, I have located several modern women inventors of contraceptive (or potentially contraceptive) devices, substances, or methods for the 19th and 20th centuries.

### 19th century

*Pessaries.* In the 19th century the U.S. Patent Office refused to patent anything overtly described as a contraceptive. Inventors' favorite euphemism for patent applications was "pessary" (Bullough/Trescott, 240). As early as 1867 **Emeline T. Bringham** of Philadelphia patented an Improved pessary. Two decades later, **Alice O. McCord** of Colfax, Washington Territory, patented an Automatic stem pessary (1887). Five years later, in 1892, **Eliza Kirwin** of Indianapolis, IN, patented something identified in LWP only as a Pessary, but whose illustration in the Patent Gazette reveals its contraceptive potential.

A group of physical/mechanical inventions that sheds further light on the word *pessary* is the patent class of "Abdominal and uterine supporters." As noted, the Patent Office classifies them as medical or surgical inventions; and as the 1897 Sears catalogue model makes perfectly clear, in this device a heavy belt supported the abdomen externally and held in place a hard rubber pedestal—called a pessary—that supported the uterus internally. Carl Degler points out that *pessary* was almost universally used in the 19th-century United States to describe "various medical devices for holding the womb in place . . ." (218). The Sears model could scarcely be worn during intercourse—indeed, it resembles nothing so much as a chastity belt. But a pessary, though no longer a stone, is still an intrauterine or at least intravaginal device. And devices classed as uterine supporters alone (rather than as combined abdominal and uterine) could have served a covert contraceptive function.

---

68. By the time the practice appears in written history, a king is ordering it used in his harem. This, however, says nothing about its origins.

*Vaginal syringes: contraceptive douching.* The earliest possibly contraceptive syringe patent I have located for a woman in the United States is the Improvement in combined aspirators, concealed uterine cauterizers and vaginal syringes patented by **Anna L. Palmer** of Chariton, IA, in 1879. This sounds like a medical or surgical instrument, possibly for stopping uterine bleeding, and/or performing abortions, among other things. However, as a vaginal syringe was one of its parts, it deserves mention here.

Syringes were almost certainly in use among knowledgeable women long before this, of course. According to Mary Beth Norton, the "certain small contrivances, probably syringes," sold in Philadelphia in the 1790s probably combined with prolonged nursing to produce the reduction in American family size observed in the late 18th century (233–4). And in the 1880s a male doctor opposed to contraception listed "the abuses of carbolic acid," an ingredient of a contraceptive douche, first in his roll-call of evils (Kitzinger, 60).

**Laura M. Adams** of New York City patented a vaginal Syringe early in 1881. **Margaret Hart,** also of New York City, patented a Syringe in 1885. **Annie H.** and M.G. **Collins** of Camden, NJ, patented a Syringe in 1890; and in 1892 **Lucy R. Meyer** of Chicago, IL, patented a Syringe and syringe box.

Interestingly enough, as late as the 1950s, a survey of several thousand Indiana women found that 20% of those using any contraception were still douching, and apparently reducing their conception rate by 80% compared with the unprotected (Degler, 216).

**Elizabeth Holcombe** of Syracuse, NY, patented a Vaginal irrigator and urinal in 1881. Presumably this was a bidet-like device for postcoital douches or baths.

*Cervical cap.* An unnamed 19th-century German midwife seems to have independently invented a cervical cap (Himes, 319–20):

> A certain section of the rural area showed a striking decline in the number of children. Most of the farming families had only two or three children, and then no more. More thorough inquiry disclosed that a midwife possessed this secret. Unknown to the woman, she placed, at the end of delivery, a foreign body in front of the os which occluded the entrance.

There are some difficulties with this 1823 report as quoted—such a "foreign body" would presumably interfere with postpartum and menstrual bleeding. But the context of the report makes it clear that the author of the 1838 gynecological treatise who cited the report was comparing this woman's device with the individually fitted rubber cervical cap.

As already noted above, the cervical cap is actually an ancient method, invented at least a thousand years ago (and probably far longer) by women in many cultures using everything from molded opium to silky paper to orange and lemon halves. For a useful review of its past and present, naming no women inventors but clearly showing women's role in testing its effectiveness as of the late 1970s, see Mary-Sherman Willis's article in *Science News* (116:25 & 26: 431ff).

*Diaphragm.* According to at least two sources (J. Dash, 39; Levin, 189), **Dr. Aletta Jacobs** (1854–1929) helped develop the vulcanized rubber diaphragm usually credited solely to (and often named after) the German physician W.P.J. Mensinga.[69] This device has been called the greatest advance in birth control since the condom (Forster, 259), and justly so, since it has the double advantage of being approved by doctors and, once prescribed, of being under woman's control. Even before receiving her formal MD, Jacobs had opened a clinic for poor women in Amsterdam. Appalled at her patients' suffering from repeated pregnancies, she determined to find a practical contraceptive, and

---

69. Sjöö & Mor give her sole credit (42), but since they cite no sources, I cannot evaluate this striking assertion.

joined forces with Mensinga to perfect the diaphragm pessary. Her role is not specified, but it seems logical that she would have done large-scale clinical testing and suggested design improvements. The date was 1883 (J. Dash, 39).

Norman Himes does not give Jacobs this credit, noting that the diaphragm was sometimes known as the Dutch pessary because "Dr. Rutgers and his followers in Holland did much to make it well known" (321).[70] But when Margaret Sanger visited Holland, she found diaphragms available in 14 sizes, and women being fitted for them *in shops* throughout the country (Forster, 259–60).

An innovation as important as her role in the development of the diaphragm is the world's first birth-control clinic, opened in Amsterdam in 1878.[71] Largely thanks to Dr. Jacobs' work in this clinic, and its offspring throughout the country, Holland had the lowest maternal death rate in Europe, *one-third that of the United States,* and, in its major cities, the world's lowest infant death rate as well. It was this record that brought Margaret Sanger to Holland—and brought the Dutch Neo-Malthusian League a medal of honor and a royal charter from Queen Wilhelmina in 1910 (J. Dash, 39).

An innovator in many areas, Aletta Jacobs was the first to investigate working conditions for Dutch women and girls, and the first to tackle the question of prostitution in the Netherlands. She also led Dutch women in their suffrage movement and, late in life, worked for peace (Chaff, 116).

*The perennial sponge.* In the 1870s **Annie Besant** (1847–1933), English free-thinker and feminist, suggested a sponge soaked in quinine solution and inserted into the vagina before intercourse as perhaps the best contraceptive device then available. Contraceptive douches using quinine had long been known, but at least one source implies that Besant originated the quinine-soaked sponge, pointing out that "Besant's homemade method" would have been more effective than William Rendell's better-known commercial pessary of cocoa butter with quinine sulfate, because cocoa butter inhibits the spermicidal properties of quinine (Seaman & Seaman, 239–40). Interestingly enough, the *Britannica* (*EB*/58) discusses Besant as Theosophist and Socialist, ignoring her birth-control work, though her 1877 trial established Englishwomen's right to birth-control information![72] Shirley Green, by contrast, calls her a birth-control militant (69).

## 20th century

*Pioneers.* **Margaret Sanger**'s life (1879–1966) and career as a birth-control pioneer have been well documented. Most people at least know her name. Several biographies are available, she appeared on the first-day covers of a family-planning commemorative stamp (1972), and a television special dramatized her life in the early 1980s. However, it may not

---

70. Himes, in fact, scarcely mentions Jacobs, defending his near-omission of her, Margaret Sanger, and one other pioneer on the grounds that they are already so well known, are either still living or so recent that they can scarcely be called historical, and are more important in the social and economic history of contraception than in its medical history (xviii). These rationalizations, however, do not keep Himes from mentioning Mensinga at least eight times.

71. Sources vary on this date from 1878 (J. Dash, 39; Levin, 189) to 1881 (Himes, 309). Himes wants to quibble about this credit as well, noting that if *clinic* is defined as a medical center staffed with physicians where medical students are instructed, then birth-control clinics did not start in Holland. "To be sure," he says, "Dr. Aletta Jacobs opened the first systematic office work in contraception in Holland about 1881. But until recently there was no dispensary or hospital work in contraception. Two out of fifty of the 'clinics' were served by midwives, the rest by workingmen's wives especially instructed, mainly by Dr. Rutgers" (309). This negative treatment, incidentally, is Himes' sole mention of Aletta Jacobs.

72. For more information on Besant, see Uglow, ed., and her *Autobiography,* cited there.

be generally known that she coined the term *birth control* (Marlow, 242), or that she and her second husband Noah Slee founded the first doctor-staffed birth-control clinic in the United States, the Birth Control Clinical Research Bureau in New York City. Staffed largely by women doctors and intended among other things to test and produce new contraceptive devices, the Bureau developed an *inexpensive version of the new German contraceptive jelly*. The Bureau was also innovative in another very important way: it began to keep *careful records* of the safety and effectiveness of the various contraceptive methods. By 1938 Sanger had established a nationwide network of 300 such clinics.

Sanger and Slee also persuaded (and subsidized) an American company to mass-produce the diaphragm. Perhaps most interesting of all, it was Margaret Sanger who in the early 1950s pushed Gregory Pincus into creating the first birth-control pill; she, too, who found support for the research—from another woman, heiress Katharine McCormick (Marlow, 244; *NAW*/mod.). As Dr. John Rock, Pincus's collaborator, recalled, "There was no government support, no foundation support. . . . There was nothing but Margaret Sanger's influence on a sympathetic lady with charity and plenty of money." Added Rock, "Mrs. McCormick . . . was giving me the world" (J. Dash, 111).

This is the tale as usually told, but **Katharine Dexter McCormick** (1875–1967) was not just a sympathetic rich lady. She was educated as a biologist in her own right, having graduated from M.I.T. with a BS in 1904, and she was committed to finding an effective contraceptive for women. Carrying her commitment into action, she smuggled diaphragms into the United States for Sanger's Clinical Research Bureau. Mrs. McCormick had sponsored endocrine research for years, partly in hope of finding a cure for her schizophrenic husband. She had also supported contraceptive research projects over two decades, and she recognized—as neither the Planned Parenthood Federation nor the scientists at Searle Pharmaceutical Company had done—the contraceptive potential of Pincus's earlier experiments with progesterone as an ovulation suppressant. There seems room for some question whether Pincus himself quite recognized this potential; at least, he seems not to have pursued it until urged to do so by the knowledgeable Mrs. McCormick (and by Sanger). *To whatever degree this is the case, Katharine Dexter McCormick deserves a share of the credit for the invention of the first birth-control pill* (Blashfield, *Hellraisers*, 15–6; *NAW*/mod.).

Before leaving Margaret Sanger, it is important to note that she always emphasized women's *freedom* as the motivation for contraception, and was not content to cite economic or health motives. Here is the crucial split between those who want women to control their own reproduction and those who abdicate control to doctors. As Sanger said in *Woman and the New Race*, "Women in all lands and all ages have instinctively desired family limitation. . . . Frequently [economic] pressure has existed, but the driving force behind women's aspiration *toward freedom* has lain deeper" (italics in original). Consistent with that philosophy, she proposed to make contraceptive devices and information available to all women, not just those whose health might be threatened by another pregnancy. Unfortunately, some of her less radical and less farsighted successors compromised on this vital point. She herself was forced to compromise to the degree of giving doctors control of the newly legalized contraceptives; and American women have reaped the bitter fruit of both compromises ever since.[73]

Among Sanger's fellow pioneers and co-combatants in the battle for legalized birth

---

73. Mary Ware Dennett, founder of the National Birth Control League, and a colleague of Sanger's in the legalization battle, refused to compromise with doctors, and thus has been forgotten. As Barbara and Gideon Seaman put it (240–1),

control in the United States was **Blanche Ames Ames** (1878–1969) of Massachusetts. Suffragist, artist, feminist, and inventor, Ames co-founded the Birth Control League of Massachusetts, an affiliate of Sanger's national League, in 1916. Completely committed, she used her family connections to gain the support of prominent New Englanders for the League. She researched maternal health statistics, initiated a parents' petition to the Massachusetts Medical Society, and publicly debated Catholic leaders. When in the 1930s, despite the League's best efforts, the law still forbade the dissemination of birth-control information, Blanche Ames reacted with spirit—and in an ancient tradition: "Women must resort," she wrote, "to their old expedient of self-help. They must tell each other how to regulate conception. Mothers must teach their daughters, since their doctors may not supply the means." She herself suggested—and illustrated with her artist's skill—methods for making a diaphragm using a baby's teething ring or a ring from a jelly jar, and provided formulas for spermicidal jellies (*NAW*/mod.).

The anonymous woman who in 1950 improvised an IUD from her gold wedding ring (Hugh Davis, 7) was acting in the tradition Blanche Ames sought to revive.

Roughly contemporary with Sanger and Ames, a distinguished scientist (botany) in her own right and equally a pioneer, was **Marie Carmichael Stopes** (1880–1958) of England. Inspired and enlightened by one of Sanger's lectures in 1915, she divorced her impotent husband and began her crusade to educate women about sex, conception, and contraception. She founded England's first birth-control clinic—still open today—and campaigned to reach poor as well as middle-class women. She wrote several books, among the first and most famous of which was *Married Love* (1918), which sold 10,000 copies in six months. Most pertinent here, she also invented and patented several contraceptive devices and preparations. Among them were her "Racial" brand contraceptive sponge (the reference was to the human race), her "Pro-Race" or "Racial" cervical cap, her diaphragm design, and her own spermicides (Neushul).

Stopes recommended combining the cervical cap with an effective spermicide—such as fat impregnated with quinine—and for all her devices maintained strict manufacturing standards. Of the London Rubber Company, which manufactured her high-dome cervical cap, she demanded an all-rubber construction with a soft rim. Later she modified her original design. Originally she opposed the diaphragm or Dutch cap because of its large diameter, the metal in its rim, and its generally poor safety record. Later she designed her own diaphragm, for which she demanded that the spring must absolutely be prevented from working through the rubber, and that it must be flexible enough always to spring back after being pinched together for insertion.

Marie Stopes' determined campaign breached the political and educational barriers to contraception for Englishwomen.

*Rational methods: barrier, chemical, combined, hormonal.* Not long after the turn of the century, **Louise Nelson** of Oakland, CA, patented a Pessary (1907). This is a year not

---

Dennett made a serious mistake. She failed to reckon with the economic interests of physicians. In the 1920s, obstetricians had pretty well won a long state-by-state battle to ban midwives, a victory that had terrible consequences for poor immigrant and country women, for there were not enough doctors to serve them. . . . Having staked out control over parturition, doctors were not about to support anything that might cut back on business, unless there was something in it for them.

While Sanger saw the dollar signs on the wall and called for birth control by precription only, Dennett demanded total repeal of the laws banning contraceptives, including those which people can use on their own.

covered by WB/28. Not surprisingly, contraceptive inventions are rare in this compilation. I found only 4 types of invention (syringe, syringe bidet, douche tip, and pessary), numbering 8 or 9 inventions under the equal-numbers assumption, and all classed as Medical or Surgical (Category XVIII.A, B, or C).

In the 1930s **Hannah Stone** of the Sanger Bureau pioneered in the U.S. in combining the diaphragm with spermicidal jellies (Seaman & Seaman, 277). This innovation is particularly significant because for many women the diaphragm comes loose during intercourse; thus for these women the main value of the diaphragm is containing the spermicide until needed.

Women have also contributed to contraceptive technology from within establishment medicine, after Sanger and her colleagues won their long battle. At least one woman, for example, took part in the research leading to the modern birth-control pill: **Dr. Connie Chambers Harris** of Modesto, CA. After graduating from Carlton College in Minnesota, she took a PhD in chemistry at the University of Nebraska. Interestingly enough, however, she became disgusted with the direction of this research after seeing the effects of the hormones on rats. Convinced that the dosage was far too strong, she left establishment chemistry, and has since worked in a home laboratory (PC, Stanford University VTSS class, 1985). Later versions of the Pill have contained lower dosages of hormone.

One of the more important women pioneers working within establishment medicine is **Dr. Edris Rice-Wray** (1921–    ). Educated at Vassar and Northwestern, Dr. Wray pursued a private practice for some years before becoming interested in public health and working in that field in Puerto Rico in the late 1940s. After taking a master's degree in Public Health at the University of Michigan in 1950, she worked for the World Health Organization, co-founded and directed the Mexican Association for Family Welfare, and finally became Assistant Professor of Population Studies at the University of the Americas (Mexico) in 1976. Dr. Rice-Wray directed the clinical testing that proved the worth of the birth-control pill; she has also tested a controversial long-lasting injection contraceptive that may produce sterility for six months (Depo-Provera). She was chosen as one of the 75 most notable women in the United States in 1970, and has won the Margaret Sanger Award for distinguished work in the population field (1978), as well as several other awards (*AMWS*/15; Gruson; Robinson, 71, 73).

Women outside the medical establishment, but enterprising enough to patent their inventions, can also be cited here.

The history of **Doris Moehrle**'s Nature's Answer® reusable sponge-tampon supports the theory that barrier contraceptives arose from menstrual absorbers. Doris Brooks Moehrle was born in Holyoke, MA, in 1931. For at least three generations the women in her family had used and handed down to their daughters a flexible sponge-like menstrual absorber. In her own words,

> The menstrual sponge was handed down. It was a blessing as I was a very physical, active person. When I married at age 22, I would use the sponge as a menses absorber during intercourse. Well, it worked so well as a sperm-blocker that somehow I felt this was a deep dark secret that unmarried young ladies were not to know about. . . . [T]his was 1952. My husband was in show business, and in our travels I made the absorbers and gave them to numerous friends, and explained the usefulness. . . . I derived so much satisfaction gazing at the amazement in their faces as I would unfold the simplicity and common sense of it all!

Convinced by delighted friends that she should market the device, she made some important modifications, not only to improve it, but to make it unique and thus patentable:

What I did . . . was to combine the two [menstrual sponge and contraceptive sponge as described in a 1954 sex manual] in the shape of the diaphragm, and because of its flexibility and conform[abil]ity it is unique.

Then, aided by a patent attorney, she applied for a patent. Thus began a long battle. Mrs. Moehrle got her patent in August 1978. Moreover, her Nature's Answer® reusable tampons—which are made of nontoxic synthetic material, last for months, and can be sterilized by boiling—are approved by the FDA. However, they are approved not as a contraceptive, but only as a menstrual absorber.

Late in 1979 Doris Moehrle was fighting an infringement by a California biochemist (who had been testing a product effectively identical to hers but with added spermicide), and still fighting, too, to get grant funds for the necessary testing to see her tampon recognized as a contraceptive. Rejected in turn by media representatives, doctors, and granting agencies, she refuses to give up, even though, as she puts it, "Bureaucracy reigns supreme." Meanwhile, the (male) biochemist went on to receive grant support, and eventually to market what became known as the Today® sponge with considerable fanfare in 1983. Today® has been featured, for example, in *People* magazine (Fischer), in two articles in *Ms.* magazine (McAdoo; Eagan), and in a *Changing Times* review article on the state of the contraceptive art in November 1983 ("Update on Contraceptives"). Despite a failure rate variously reported as 15% and 16.8%, and despite the denial in 1987 of a patent-renewal application, Today® continues to be promoted as this book goes to press.

Nature's Answer® would be a barrier contraceptive. Of $70 million recently spent worldwide on contraceptive research, Moehrle points out, only $50,000 went to barrier-methods research. She may have put her finger on another cause of her funding difficulties when she says Nature's Answer® "would be devastating to the medical [products] field, but joyous for females from [first] menstruation to menopause." What she did not mention, but undoubtedly a major handicap in the eyes of funders, is her lack of scientific or medical training.

Doris Moehrle formed her own company, Nature's Answer Tampons, Inc., of Las Vegas, NV. Keenly aware of the ecological benefits of reusable tampons, she mentions them among her product's advantages, but she is equally determined that the contraceptive advantages shall become known so that women will have an effective alternative to hormones and IUDs. About 2% of women who tested her competitor's Today® sponge experienced some irritation over the test period, and another 2% were allergic to the spermicide ("Update . . . ," 74). Moehrle calls her product an improvement on the diaphragm because its extreme flexibility makes it conform perfectly to the vaginal walls, and because it requires no creams, jellies, or drugs. Nature's Answer® comes in three sizes: small, medium, and large. "All of my strength and drive in pursuing this project," she says, comes "from the hope that young females such as my daughter Grace will be free from chemical-ridden tampons and [carcinogenic] fibers harmful to the cervix" (Masterson; Moehrle, L-10/19/79).

**Marjorie R. Murray** (with Daniel A. Murray) of Honolulu invented and patented a pillbox with timer and buzzer, called a Personal Pill Reminder in 1974. Not specifically a contraceptive pillbox, but useful for that purpose, this box can be set for any desired dosage schedule. It has a timer in the lid and a battery-powered buzzer that sounds off at the proper hour. Opening the box shuts off the buzzer, but it will buzz again when the lid is shut unless reset for the next interval. This device was featured in *Popular Science* in 1975 ("Pill Box").

*Male contraception.* **Dr. Martha Voegli** is a Swiss physician working in India. She has devised, and is working to get accepted, a male contraceptive technique: three weeks of daily 45-minute baths in water of 116°F. This method, which can make a man sterile for six months, costs nothing but the fuel for heating the water and, according to Dr. Voegli, is totally reversible (Djerassi, *Politics,* 140–1). Such inexpensive, self-administered, yet effective methods are desperately needed throughout the Third World.

Testifying before the House Select Committee on Population on March 19, 1978, **Barbara Seaman** suggested that condoms should be made in small, medium, and large sizes—to be labeled Jumbo, Colossal, and Supercolossal to assuage male egos. If this innovation is Ms. Seaman's own, even though it is actually a reinvention of an old idea,[74] she deserves a brief mention here. For, as she points out, contraceptive "failures tend to occur at the extreme ends of the scale. In men who are petite, [condoms] fall off, and in men who are extra well endowed, they burst" (Djerassi, *Politics,* 16–7). Better-fitting condoms take on new urgency and importance in the AIDS epidemic.

In addition to these technical innovations, Barbara Seaman proposes some radical social innovations as well: only women to be admitted to obstetrics and gynecology residencies; no more foundation money to be awarded to men for research on the female reproductive system; all lawmaking and administration of laws pertaining to female reproduction, abortion, and sterilization to be removed from the court and legislative systems and transferred to an agency modeled after the National Labor Relations Board; no United States or UN sponsorship of or participation in any national or worldwide conferences or activity on population unless women are represented in proportion to their numbers in every participating nation (Chaff, 572). Under these conditions, the atmosphere for women's reproductive freedom would be transformed indeed.

According to Dr. Carl Djerassi, the representation of women in establishment contraceptive research is improving. For example, he points out, nearly 20% of the scientific participants at a 1978 international symposium on male contraception were women, whereas a decade earlier, only 1–2% would probably have been women (*Politics,* 121). Thus perhaps a book on women inventors written ten years from now might be able to list many more women's names for the recent period.

*Natural or cyclical methods.* The Billings mucus-examination method has already been discussed above. With each new scare about harmful effects of the Pill, so-called natural methods of contraception enjoy a new vogue. For thoughtful people of both sexes, doctors and others, the hormonal methods are always a source of uneasiness. For women who cannot take the Pill, or do not wish to, the more refined these less intrusive methods can become, the better.

**Louise Lacey** is a California woman who took the Pill for several years, but was forced to stop after developing a fibroid mass in her breast. She suffered such severe hormone withdrawal effects that she determined to find a better method of contraception. Vitamin therapy seemed to help, but as soon as she stopped the vitamins, her symptoms reappeared. She had evidently suffered a permanent metabolic disturbance.

Lacey began her studies, including an investigation of reports of the moon's effects on human sexual and menstrual cycles. Prepared by these studies, she recognized the importance of Edmund Dewan's work on artificial light for regularizing menstrual

---

74. Condoms were being sold in several sizes in 18th-century London—by women entrepreneurs (S. Green, 84–5; 91ff). They were also available in various sizes in brothels, according to Casanova (cited by Green, 86).

periods, and thus as a way to "phase lock" and control ovulation. His goal was fertility; Lacey's was anti-fertility, or what she now calls discretionary fertility. Her innovation was to adapt Dewan's work to a contraceptive purpose.

Beginning in October 1971, experimenting on herself, she devised a system of using artificial light to entrain (regularize) ovulation and thus avoid it for intercourse. Her "Lunaception" system works as follows: First the woman determines what her cycle is, and what precise pattern of temperature changes signals her ovulation, by taking her temperature religiously every day at the same time for several months. Then she sleeps in absolute darkness for the first thirteen days of her cycle, counting the first day of her period as Day 1. On Nights 14 through 16, however, she sleeps with a source of white light (over 25 watts), burning in the room. On the remaining nights of the cycle, she returns to absolute darkness.

She can continue to use temperature readings and/or check her vaginal mucus to make sure she is on track; but barring severe stress, illness, or transcontinental travel, her cycle should remain regular. She can, from the start—and for Lacey this was part of the point—synchronize her cycle with the moon. (If the cycle is disrupted, the woman can entrain herself again by the same method.)

By the time her book appeared in 1974, Lacey herself had successfully used Lunaception as her sole form of birth control since March of 1972. Several of her friends experimented with the method as well. As of 1974, a sample of 29 women using Lunaception through 344 menstrual months experienced only 2 pregnancies, both easily explainable: one woman had intercourse during the lights-on time; another had just had a baby and her cycle was not yet readjusted. All but one of the 29 women were able to synchronize their cycles as desired. As Lacey herself notes, the method requires a fair amount of will power and commitment. But for those who wish to avoid hormones and chemicals in the years-long battle for contraception, its appeal is considerable. Moreover, she and the women in her sample reported many other benefits as well, among which are less pre-menstrual tension, shorter menstrual periods, and more awareness of all kinds of bodily rhythms, an awareness that can be used to schedule job interviews and other important events (Lacey; Seaman & Seaman, 265–9).

*Mental/psychological methods.* Shuttle and Redgrove suggest that modern women, like the ancient colleges of priestesses, could potentially wield mind- or dream-control over both conception and birth, positing hypnosis, biofeedback or yoga-like mechanisms for their doing so (25, chaps. 4, 5). These authors are neither the first nor the most scientifically rigorous to make this suggestion (although Peter Redgrove has worked as a research scientist), but their analysis is the most original, and the most woman-centered.

If, as **Barbara Brown**'s research indicates, the human brain can exert control over the body and its functions even down to the level of a single cell, a woman should theoretically be able to close her cervix against sperm until a spermicidal douche could be administered, or keep a fertilized ovum from implanting—or indeed to prevent ovulation itself. Brown herself, having observed a rhythm of change in the skin's electrical potential that seems linked to the menstrual cycle, sees the potential for a biofeedback-based form of natural birth control. If, she suggests, women could learn to monitor this "skin talk," then "learning to control these slow monthly rhythms might lead to a natural and simplified method for birth control. This possibility may not be such a long shot as it seems at first glance" (*ibid.,* 168–9).

Some mind-body or brain–reproductive system link, in fact, would seem one logical explanation for the already-mentioned findings of adolescent promiscuity without accompanying mass pregnancy among many primitive peoples, not to mention the frequently observed modern phenomenon whereby a supposedly infertile woman adopts a child and immediately gets pregnant.

*Research.* Basic research often produces findings with contraceptive applications. **Carol Grace Smith** and colleagues at the Uniformed Services University of the Health Sciences in Bethesda, MD, reported in 1983 that monkeys given THC (tetrahydrocannabinol), the active principle of marijuana, in doses equivalent to human "moderate use" failed to ovulate for periods ranging from 103 to 135 days. Less rigorous studies done on humans have shown similar menstrual disruption in chronic marijuana users. These findings could mean that an effective female contraceptive could be developed from THC. Strong cautions are in order here, however, since heavy marijuana use may cause "serious, possibly irreversible, menstrual disruptions in adolescents," and THC may be directly toxic to developing egg cells (Chen).

One of the more pressing research goals in the anti-fertility field in recent years is a good male contraceptive. The Chinese gossypol fiasco may have a silver lining, for gossypol is the active ingredient in one of the most promising male contraceptives yet developed. Chinese peasants inadvertently experienced this contraceptive effect—and other effects—in 1969. Scientists called in to explain found an impurity called gossypol in the cottonseed oil used for cooking in that region. Further experiments showed in 1971 that gossypol (not removed by the new cold-pressing method of extracting the oil from the seeds) was indeed the culprit and that men were more susceptible to its effects than women.

Clinical tests on 4,000 healthy men suggest that gossypol may be more than 99% effective in controlling male fertility, taking about two months to reach full effect, and about three months for the sperm count and viability to return to normal after treatment stops. Some side effects have been reported—temporary "weakness," appetite changes, nausea, and some decrease in sex drive (which usually returned after reassurance or treatment of symptoms). Gossypol is not a steroid,[75] and as soon as the infertility effect is established, the dosage can be cut to one-fourth the starting dose. According to Huang Liang, a scientist at the Beijing Institute of Materia Medica, the side effects seem to be less serious than those seen with the women's Pill.[76] "Why not let the men take it?" she asks (*Science News,* Mar. 3, 1979: 136 and Oct. 27, 1979: 295; Djerassi, *Politics,* 202–5). One arrangement suggested in China is that in stable relationships the woman could take her Pill for a year, then the man take his for a year; thus no one would suffer the effects of long-term use.

**Ann de Peyster,** at the University of California at Berkeley, showed in 1982 that gossypol lowers testosterone levels in male rats even at dosage levels low enough to keep the rats fertile (Langone, "Quest," 30). Work such as Dr. de Peyster's in elucidating gossypol's mechanism of action will be necessary before the substance can become a full-fledged invention of a male contraceptive.

Researchers are also pursuing an endocrine, or hormonal, approach to male contraception. I have no reports of specific successes, but it is worth noting that the list of participants in a symposium or workshop on the "Endocrine Approach to Male

---

75. As Carl Djerassi points out (*Politics,* 205), gossypol is a polyhydroxylated binaphthalene derivative, totally unlike any substance hiterto studied that interferes with sperm production.

76. These toxic effects can be avoided for humans by administering moderate doses in the presence of mineral salts such as iron salts, which bind gossypol and thus inactivate it (Djerassi, *Politics,* 203–4). This finding underscores once again the importance of the secondary, mollifying, and moderating drugs, dietetic regimens, etc. of the old herbal medicine.

Contraception" held in Norway in April of 1978 includes at least 17 women (possibly more—some names ambiguous; some initials only) in a total of 128. It is from these labs—in Germany, Canada, the United States, and Scandinavia—that an effective male contraceptive will most likely come.

## SUMMARY

Since most women spend 25 to 30 years of their lives trying to avoid pregnancy, a look at women's own contributions to anti-fertility technology is long overdue. My study underscores and amplifies what Norman Himes reported half a century ago, that women of every known group have wished to limit their fertility, and have invented devices and techniques for doing so. Where we seem to find no such technology, it is most likely because a male anthropologist could not learn of it, because the knowledge was already lost through previous contact with or conquest by patriarchal and/or "more advanced" Western medicine or Christian missionaries, or both, or because the group's earlier methods were being swamped by lowered death rates and greater female fertility induced by fattier diets and sedentism, among other factors.

Women in earliest times may have limited their fertility mainly by abortion (App. SFA-6), discovering abortifacient plants in connection with their food-gathering and -preparation specialties, and physical abortion techniques from injuries they sustained during pregnancy. However, to a greater degree than previously suspected, these women also had contraceptives. They probably developed herbal oral contraceptives by extension from their abortifacients, on the principle that a regular small dose of something that aborts a fetus may make the womb hostile to a pregnancy in the first place. It seems likely that abortifacients gave rise, by experimentation with amount and duration of dosage and accompanying dietary and herbal regimens, to preparations for both temporary and permanent sterility. Moreover, the cluster-of-technology analysis of plant preparations used for menstrual difficulties, pregnancy, childbirth, fertility, aphrodisiacs, contraception, and abortion would show that many of the same plants are used in various dosages and regimens in several of these areas.

Physical abortions could give rise to IUDs, whether the method was slow dilation of the cervix with laminaria or similarly acting "barks," or quick dilation with tubes or probes and scraping or other means of detaching the fetus. The persistence after abortion of intrauterine debris could at first be accidental but would become deliberate once the anti-fertility effects were observed.

Barrier methods could have been among the earliest of women's anti-fertility inventions from the standpoint of opportunity for making the necessary observations, since as soon as women took measures to staunch their menstrual flow, they had at hand the means for barrier contraception. However, these methods—like such cultural methods as prolonged nursing and postpartum intercourse taboos—would make no sense until women knew for sure the male role in procreation. The date of certain knowledge has been much debated, and perhaps will never be known. Women may have had the knowledge before men did.

In the ancient Near East women's control of their own fertility was disrupted by some of the same factors that had earlier undermined their control of agriculture—population pressures, warfare, patriarchy and patrilocal marriage, and the desire of a skygod's priests to destroy the ancient religion of the Goddess who was both Earth Mother and Queen of Heaven. Some women had begun to specialize as midwives before the reproductive takeover, however, so the battle for control of reproduction was in part a struggle between two sets of specialists rather than between emerging specialists and existing generalists as had been the case with agriculture.

For modern Western culture, the decisive battle was that waged in Europe during the 14th to 17th centuries AD for control of medicine and midwifery. For once the wise women were cast as witches, and either burned to death or made dangerous for virtuous women to consult; once midwives were slandered as untrained and unsafe, not to say unfashionable, and male midwives took over with their new forceps technology, there remained no widespread source of the ancient anti-fertility knowledge except the prostitute, and she was beyond the pale for other reasons. It is not only because certain ancient methods—e.g., the sponge soaked in spermicide—worked so well that contraceptive technology advanced so little from the end of the Bronze Age until modern times. Even Fallopius's early condoms—which in any case were not ostensibly for a contraceptive purpose—were a re-invention of an ancient use of sheep and goat bladders.

In the 19th century, even as male doctors consolidated their control of medicine and medicine's control of birth, women began to demand reproductive freedom once again. As the century turned, the cry became stronger, both from individual women and from organizations led by such pioneers as **Annie Besant, Marie Stopes,** and **Margaret Sanger.** Women also participated directly or indirectly in the greatest advances in contraception in many centuries, the diaphragm and the birth-control pill, as well as receiving patents for several minor improvements in pessaries and syringes. **Doris Moehrle** and **Blanche Ames Ames**—made prophets by current attacks on *Roe v. Wade* and on government funding for poor women's abortions—remind us that women can never count on men to safeguard their interests when it comes to contraception. Self-help must always be kept in reserve.

As always—but here perhaps to an exceptional degree—the patent records give little indication of women's inventive contribution. If a 1984 survey of nearly 7,000 American women is any indication, however, women tend to prefer the most effective methods of birth control. Voluntary sterilization was the most widely used method (almost one-third of the women in the sample), followed by the Pill with its failure rate of only 2%, the condom (10% failure rate), and the IUD (4% failure rate). Foams/spermicides and the diaphragm with failure rates of 15% and 13%, respectively, fell well behind the other methods in usage.

Spontaneity was the second concern after effectiveness, with age, income, marital status, and availability of health care as other factors in women's choice. Safety is also an important concern, as IUD use was declining at the time of the study, and usage of the Pill has risen and fallen with medical reports of harmful side effects, improvements in the Pill, etc. (A comparable study done today, therefore, would undoubtedly show a greater preference for condoms because of their protection against HIV infection.) Religion, concludes this study, "makes little difference in patterns of contraceptive use" (Kash).

Since we have no reason to think ancient and early modern women would be any less concerned with effectiveness, once their families were complete, we can suppose with some confidence that their achievements in anti-fertility technology were as impressive as they are unlikely to be fully known. Fanisi's husband, regrettably, is not the only one in the dark.

# 4 Daughters of Athena, Semiramis, Margaret Knight, and Wei-Feng-Ying: Women Inventors of Tools and Machines

Tools preceded machinery. The cylindrical log or rock with which a savage woman triturates foodstuffs is a tool; so is the muller, with which grain is reduced to meal on a metate. But it is next door to a machine—that is, a contrivance in which a vertical shaft is used, fixed in the upper stone and loosely piercing the nether. . . . [T]he earliest form of this meal-producing machine is found in the hands of women . . .—Otis T. Mason, 1894

As a child I never cared for the things that girls usually do. . . . I was always making things for my brother . . . I was famous for my kites, and my sleds were the envy and admiration of all the boys in town. I'm not surprised at what I've done; I'm only sorry I couldn't have had as good a chance as a boy and have been put to my trade regularly.—Margaret Knight

. . . I often heard [Elias Howe] say that he worked fourteen years to get up that sewing machine. But his wife made up her mind one day that they would starve to death if there wasn't something or other invented pretty soon, and so in two hours she invented the sewing machine. Of course he took out the patent in his name. Men always do that.—Russell Conwell, 1877

## INTRODUCTION

At least as old in the West as the Clockmaker God is a deep reverence for the machine and for mechanical invention. Women's mechanical inventiveness, however, is seldom discussed. Obeying a stereotype that separates women from machines as smoothly as oil from water, most historians of technology have ignored women's contributions to "important" machines and dismissed women's acknowledged mechanical inventions by classing them as "domestic."

## PREHISTORY

### The Evidence from Myth

When Prometheus was on Olympus stealing fire from Hephaestus, he stole mechanical skill from the senior partner in this workshop—Athena (Frazer, *Fire*, 194). Hephaestus and Athena shared not only a workshop on Olympus but temples in Athens. According to Robert Graves, in fact, the name Hephaestus may be a masculine form of *he apaista*, "the goddess who removes from sight"; in other words, Athena, the original inventor of all mechanical arts (Graves, *Myths* I: 87). If so, the emergence of the male Hephaestus as a separate deity could be a compressed takeover myth. The theft itself can be seen in the same way, of course, though the Prometheus myth has been described as late and literary.

Recall, too, that Hephaestus set up his first smithy under the protection and patronage of two females, Thetis and Eurynome, and that when he was restored to Olympus, Hera set him up in a finer workshop (*ibid.*). These tales may be analogous to the agricultural takeover myths in which the great goddesses who once gave grain directly to humanity later instruct a king or hero in grain-culture, who in turn instructs the people. When Hephaestus creates robots to help him in his smithy, he makes them female (*ibid.*).

Graves equates Daedalus, master artificer of the Mediterranean world, with Hephaestus, and says that Daedalus learned his craft—specifically the lost-wax method of bronze-casting—from Athena (I: 317).

The Roman **Minerva,** once an Etruscan deity (Menrva) of handicrafts, may always have been patroness of artisans. According to Boccaccio's sources she was the first to make iron weapons. Minerva is specifically credited with inventing the distaff and the needle, weaving (thus presumably the loom), olive "mills" and the technique of crushing olives to extract the oil, among other things (14ff; M.J. Gage/1: 27–8; Monaghan, 203–4). As already mentioned, some sources consider women's food-processing apparatus and systems the world's first venture into engineering.

Before **Brigit** was an Irish saint, she was a great Celtic Triple Goddess. In one of her three persons she was patron goddess of smiths. In the tangle of Christian and earlier mythic material surrounding Brigit, what interests us is not just that she had authority to appoint the bishops in her region—though this in itself hints at her former rank—but that she required all her bishops to be practicing goldsmiths (Monaghan, 49; cf. M. Stone, *Mirrors* I: 64)!

The first or divine ancestress of several primitive groups is credited with giving her people "all knowledge," presumably including early arts and crafts, tools, and machines. **Tsenabonpil** of Melanesia (whose two sons had no human father), **Miti-Miti** of Siberia, and **Sarasvati** of India are good examples.

## *Evidence from Anthropology: Division of Labor*

In addition to evidence from myth, we can look at the prehistory of mechanics. Historians of technology usually list five primary machines—lever, wedge, screw, pulley, and inclined plane—plus the wheel and axle, as the basis of further mechanical invention (e.g., O.T. Mason, *Origins,* 59–60). *Women used and probably invented all or most of them in prehistoric or early historic times.*

For example, though not often classed as such, the digging stick is a lever. Whether used primarily for gathering—as to pry large yams out of the ground—or for cultivation, the digging stick is overwhelmingly a woman's tool. (Murdock & Provost, 204, 207, 209, 216ff; Tanner & Zihlman, 599).

Early peoples used the wedge not only for felling and splitting trees, but for breaking masses of ore, lifting weights, and hafting stone tools (fastening stone blades to wood or bone handles, for example, with rawhide thongs; O.T. Mason, *Origins,* 65). Women certainly felled trees from earliest times (*ibid.,* 204), and were the usual burden-bearers. They also hafted axes and other tools (as Jane Goodale shows for the Tiwi, 155–6). Indeed, since women used tools for more of their earliest work than did men (earliest hunters probably used no tools or weapons), and since women were probably the first miners (in search of clay and red ochre), they are at least as likely as men to have devised wedges. Moreover, wedges are also used in setting up looms (O.T. Mason, *Share,* 279), invented and nearly exclusively used by women for millennia.

Women's burden-bearing role might also make them the likelier sex to have invented the inclined plane.

The *tipiti* or *matapie,* women's invention for detoxifying manioc (cassava), combines the principles of press and sieve with a lever of the third kind,[1] and comes, in the words of Otis Mason, "as close to the screw as we shall be able to find in savagery" (*Origins,* 60–1, and *Share,* 38–40; Sokolov, 34, 38).

---

1. These kinds refer to different relative placement of the fulcrum, the weight, and the power. In the third kind, the power is applied between the weight and the fulcrum.

Moving to a stage beyond "savagery," we note the Mayan screw-top ceramic jar recently found at Rio Azul, Guatemala. Pottery-making is predominantly or exclusively women's work in 80 out of 105 societies in a standard cross-cultural sample (Murdock & Provost, 205, 207). If, as in so many cultures, women were the potters here, then Maya women look like independent New World inventors of the screw at least by the 5th century AD (Begley, "Lost," 73).

According to Otis Mason, all early peoples knew the simple "dead-eye" pulley—"Any line drawn around a fixed object, as a tree, and pulled in one direction for the purpose of moving an object in another direction"—and used it not only for hoisting loads but for moving them horizontally. The nearest approach to the pulley among American Indians was "the woman's device for drawing the skin covering to the top of the tent poles" (*Origins,* 61–2). Sioux women used this device (*Share,* 279).

Women also evidently invented the parbuckle in their work as burden-bearers. Says Otis Mason, if dictionaries in defining "parbuckle" were to speak in the language of his own volume, they would define it as "a kind of purchase for hoisting upon a woman's back a bundle of fagots or other cylindrical load" (*Share,* 123).

According to Arthur Pound's brief history of transport, the American "aborigines" had used all five of the primary machines (not including the wheel; 3). Unfortunately, Pound does not elaborate on his statement. However, since among these people women were both builders and burden-bearers, it seems likely that he unknowingly refers here to women inventors in many groups.

Women's greatest contribution to the history of mechanics, however, is probably continuous rotary motion. Historians of technology generally agree that mechanics took a quantum leap forward when rotary motion replaced reciprocal or back-and-forth motion in many technical areas. Says E.C. Curwen (138), "Such an advance could only have been the product of a brilliant engineer or mathematician—some forgotten forerunner of Archimedes, who failed to achieve the immortality of the classics."

Consider the difference, for example, between oar and propeller, reciprocal saw and circular saw. This latter example is chosen advisedly, for, as discussed below, an American Shaker woman, **Sister Tabitha Babbitt,** independently invented a circular saw about 1810 in Harvard, MA.

Historians of technology have not usually credited women with inventing the wheel, either as a basic machine for transmitting or changing the direction of a force, or as applied (with axle) to vehicles. Nevertheless, by far the earliest example of continuous rotary motion—the spindle—belongs indisputably to women's work. Though some see the first such motion in grain-grinding apparatus, Otis Mason maintains, "The first continuous [rotary] motion . . . was by the spindle in the hands of women" (*Origins,* 77; *Share,* 156).

Moreover, says Mason, in devising the twine-twister and the spindle whorl, ancient women had not only introduced rotary motion to mechanics, but invented the flywheel for storing momentum. Says he, "The spindle with its whorl is a free wheel and axle, with the principle of the flywheel fully developed and the [twine- or cordage-]twister, well known to savages, is a still simpler fly-wheel" (*Origins,* 76).

Much discussion has been devoted to possible evolutionary relationships among several early examples of rotary motion, particularly axial rotation, and the wheel. British technology expert Lord Halsbury[2] mentions the spindle whorl, the quern, and the potter's

2. Though little known in the United States, Lord Halsbury has impressive credentials. Among other things, he was the first Managing Director of the National Research Development Corporation, and a Fellow of the Royal Society (Halsbury, 10).

wheel as "other primitive objects involving axial rotation," and says, "Whichever preceded the others almost certainly influenced its successors" (13). Spindle whorls are the earliest by thousands of years. H.S. Harrison finds it "not improbable" that rotary motion was discovered independently in several technical areas, and that there was no evolutionary relationship among spindle, fire-drill, and rotary quern. However, he thinks a relationship among the vehicular wheel, the potter's wheel, and the spinning wheel is "probable" (Chas. Singer I: 71–2).

No one seems to have written about the rotary churn used by Sindhi women of northern Pakistan. The rotation in this device, though incomplete, was not only axial, but belt-driven, actuated by a woman sitting on the floor and pulling with one hand and then the other on a leather thong twisted around the axle of the spindle which in turn rotated the beaters inside the containers of milk (Smithsonian Museum, 1987 exhibit).

The mystery of which sex invented the vehicular wheel may never be solved, but from present evidence we cannot rule women out. In Meso-America, wheeled vehicles appeared only on miniatures that may be either children's toys or religious objects (Doster *et al.,* 55). If the former, they may well have been made by women; as to the latter, the gender of the makers would depend on the division of labor in making such votive objects.

Rollers are often considered to precede and possibly give rise to the vehicular wheel. We find obvious examples of rollers in the woman's metate and muller, and the less familiar technique of Eskimo women for landing their boats on inflated sealskins (O.T. Mason, *Share,* 279). Of course the Eskimo culture neither gave rise to nor markedly influenced our own. But that culture, like ours, has been described as technology-ridden, "gadget burdened" (Bruemmer, 44); and if women could invent this roller technology in one culture, they could invent it, or something analogous to it, in another. The culture that built Malta's great temples had rollers for moving and maneuvering large building stones (Hawkes, *Atlas,* 101). Some have considered this culture gynocentric if not matriarchal, and suggest that the temples were built without coercion from a strong central government. In any case, the island was undeniably a center of Goddess worship, and myth credits "a giant Titan-woman with a baby at her breast" with building the great Cgantija temple singlehandedly in one day and night (Sjöö & Mor/2: 110–1).

Last, but scarcely least, myth credits Athena with inventing the chariot and Minerva with inventing the cart.

### The Evidence from Primate Ethology

A study on chimpanzees' use of natural hammers conducted by Swiss scientists Christophe and Hedwige Boesch "undermines," as one report put it, "the cherished belief that the first human tool users were men. It now looks as if the credit should go to our female ancestors" ("Sex Bias"). Observing nut-cracking behavior among wild chimpanzees in Ivory Coast, Africa, the Boesches found the animals exploiting both thin-shelled Coula nuts and hard, thick-shelled Panda nuts. The Coula nuts could be opened fairly easily on the ground, with a natural wooden club and a tree root or flat stone as an anvil, but were much harder to crack in the trees, where the process required "skilful coordination and the ability to anticipate the need for a hammer before climbing the food tree." The Panda nuts demanded a stone hammer, and "precise control . . . in order to extract the kernels without smashing them to bits." Although roughly equal numbers of males and females opened Coula nuts on the ground, "the other two techniques were performed almost exclusively by females." Moreover, the Boesches found the females more efficient (fewer hits per cracking operation) than the males (Boesch & Boesch).

Jane Goodall had earlier reported that female chimps spend more time using tools than males do. These were primarily the termiting tools, allowing the animals to exploit an important source of dietary protein.

## TAKEOVER THROUGH THE 18TH CENTURY

I will not attempt to trace the mechanical takeover in detail. Books could, and I hope one day will, be devoted just to the economic impact on women of the coming of iron to proto-Western cultures. But one need only compare the attributes and epithets of Athena to those of a 19th-century lady—or woman's share in the Neolithic division of labor to today's stereotype about women and machines—to see that a very thorough takeover occurred.

An intriguing figure from this long period, transitional between the mythical and the historical, is the Middle Eastern queen known in the West by her Greek name, **Semiramis.** Before attempting to distinguish between the two different Assyrian queens whose qualities and achievements survive together in Semiramis, let us look for a moment at what she is credited with doing. The most familiar tradition, of course, is that she built the famous Hanging Gardens of Babylon. Second of the Seven Wonders of the Ancient World, these gardens are often called simply the Gardens of Semiramis (e.g., *EB*/58). According to Matilda J. Gage, Semiramis invented military "engines" that ended the long siege of Bactria. Gage also credits this queen with devising means of reinforcing the tunnel she built underneath the Euphrates, and declares that "all modern tunneling can trace its origin back to Semiramis" (/1: 22–3). According to Mozans (341), Semiramis invented canals as well as bridges over rivers and causeways over morasses. Boccaccio reports (6–7) that Semiramis invented the first chastity belts and forced all her court ladies to wear them so that they could not gain influence over her son, the royal heir. Allegedly fastened with lock and key, this device could arguably be classed as a mechanical invention.

In one sense it scarcely matters whether one or several actual women are involved; the fact that such important engineering and mechanical innovations and inventions are attributed to a female is significant enough. However, with the help of Hildegard Lewy, we can distinguish between the two queens, learn their probable reign dates and the names of their husbands and sons, and even tentatively credit the various technological contributions (once clarified as to what was really involved) to one or the other with some degree of confidence.

The two queens in question were the Assyrian **Sammuramat,** Queen of Samsi-Adad V and mother of his heir, Adad-nirari III (ruled 810–738 BC). Samsi-Adad's reign is usually listed as 823–11 BC, but during the last four years of that period (from 814 onward) Sammuramat ruled. The other queen, five generations later, was **Naqi'a,** "the Pure One" (Zakutum in Assyrian). She was probably an Aramaean princess from the city of Lahiru in Babylonia. Daughter of one of the princes subjugated when Sargon conquered Lahiru about 711 BC, she may have been given to Sargon's son Sennacherib[3] as a bride or concubine.

---

3. See Lewy's solidly documented and fascinating article for her argument identifying Ninos, the name given in Greek sources for Semiramis's husband, as Sennacherib. Briefly, the name means "inhabitant of Ninevah," and it was Sennacherib, not Samsi-Adad V, who lived and built in Ninevah. Diodorus's informant was Babylonian; since the Babylonians hated Sennacherib for destroying their city, they refused to say his name, calling him just "the Ninevite," which Diodorus rendered as "Ninos."

Lewy suggests that Sammuramat's accomplishments were primarily war-like and Naqi'a's generally peaceful. The military campaigns attributed to Sammuramat's son Adad-nirari—against Media, Egypt, and Ethiopia—actually occurred before he ascended the throne. Thus Lewy concludes that Sammuramat probably ordered these campaigns and may have commanded them herself (278). It is Sammuramat, too, though Lewy does not make this connection, who is linked with the Bactrian siege described by Diodorus and mentioned by Gage. Thus if either queen did indeed invent or improve her armies' siege engines, it was probably Sammuramat.

As to the chastity belt, from current information I can only say that this gruesome instrument is generally thought to have been invented in the Near East, and brought to Europe by returning Crusaders; and that such a war-like queen might well have taken such a heavy-handed—and ingenious—approach to a political problem. Nor did Sammuramat neglect the problem of male influence over or seduction of royal women, for Roman sources call her the first to castrate men (Boulding, 364); i.e., presumably as palace eunuchs. The attribution, true or not, contributes another piece to the puzzle.

Naqi'a, on the other hand, is best known as a builder. Her husband Sennacherib had destroyed Babylon by the brutal expedient of diverting the Euphrates through it. She ruled much of the Assyrian empire even before Sennacherib died, almost certainly the entire eastern and southwestern portions of Babylonia. This included her home, Lahiru, the temporary capital pending the rebuilding of Babylon. That rebuilding, as Lewy shows, took place under Naqi'a's administration, and in just about eleven years. Therefore, Naqi'a may well deserve credit for numerous engineering achievements, even if not all the things often ascribed to her.

Diodorus says that the Hanging Gardens were not Semiramis's work (Lewy, 271n). Edward Hyams notes (12) that the ziggurat as artificial hill with garden was already very ancient by the 9th century BC. If such gardens were already ancient in Sammuramat's day, still less they can be attributed to Naqi'a, in the 7th century. However, to whatever degree such gardens were an essential part of the religious and administrative complex of Babylon, Naqi'a would have been responsible for recreating them, along with everything else; and it is not impossible that in the process she improved—or innovated in—the design of their watering and drainage works.

Bridges, canals, and causeways over morasses were surely older than Naqi'a's reign; a region-wide system of irrigation was in place in Mesopotamia at least by 3000 BC, for example, involving numerous canals. However, a necessary step in the rebuilding of Babylon was to create a large flood-control reservoir outside the city and rebuild and strengthen its floodwalls.

Note, too, that Sennacherib found the Ninevites ignorant of irrigation. As part of transforming Ninevah from a provincial town to a great city, therefore, he created an irrigation system relying partly on natural stream channels and partly on new canals. Moreover, on the newly irrigated land, he planted marvelous gardens with both native and exotic plants. For the grateful people, this would have been a major innovation. Since Sennacherib usually receives credit for the rebuilding of Babylon, whereas Naqi'a was evidently actually in charge, perhaps Naqi'a played an unsung role in these Ninevite projects as well?

Diodorus specifically credits Naqi'a at Babylon (ii.7.2–9.8) with building "a mighty fortification wall, a bridge across the Euphrates, two palaces, a reservoir outside the city proper, and the temple of Bêl in the very center of the metropolis" (Lewy, 271). If indeed she built two palaces, one on either side of the Euphrates, she might well have designed a passageway under the river to connect them, and might have suggested new ways to strengthen the construction. Though tunnels were scarcely new, tunnels under rivers are another matter, and may have been rare, if known at all, in that era.

A further evidence of Naqi'a's power and outstanding character is Sennacherib's choice of her son, Esarhaddon (Assurah-iddina), to succeed him, after his eldest son died (Boccaccio, 6–7; *EB*/73: s.v. "Semiramis"; M. Gage/1: 22–3; Lewy; Mozans, 341).

Naqi'a is called **Nitokris** in Herodotus. The name is sometimes spelled "Nitocris." Without really resolving the confusion for the nonspecialist, Lewy says that Nitokris was actually an Egyptian princess. Matilda Gage lists Queen Nitokris as the builder, in whole or in part, of the third pyramid, "considered by Pliny to be much more beautiful and wonderful than those [of Cheops and Chephrenes]," and as the inventor of a certain kind of arch, used for the first time in the building of this pyramid. According to Gage, both Eusebius and Africanus credit Nitokris with this achievement. Gage calls her the "last sovereign of Manetho's VIth dynasty," and notes that Manetho described her as flaxen-haired and rosy-cheeked" (/1: 21–2). In an intriguing link with Naqi'a's alleged tunnel-engineering innovations, Nitokris reportedly revenged herself on the murderers of her husband Merenra II by causing "a large subterranean hall" to be built, ostensibly for celebrating festivals. She invited the guilty noblemen to a feast in the hall, and as they sat at table, "the waters of the Nile flooded the artificial cave through a secretly constructed canal, and the guests were all drowned." The story has been thought more mythical than historical, but the combination of tunnel-engineering and waterworks attributed to a woman is striking. Nitokris also resembles Semiramis and Naqi'a in that she ruled in her own right, at least after Merenra's death (181).

Equally intriguing is **Hypatia** of Alexandria (c. AD 370–415). Though her independent writings are lost, and most sources devote more space to her beauty and her grisly death than to her intellect and her inventions, the intervening sixteen centuries have not obscured her brilliance.

Daughter of Theon, a famous professor of mathematics at the University of Alexandria, who set out consciously to produce a perfect human being, Hypatia spent her childhood surrounded by scholars. As a young woman she traveled to Athens where she may have studied under Plutarch the Younger and his daughter Asclepigenia. Reportedly she preferred Plotinus's more scientific Neo-Platonism. Returning to Alexandria, she was invited to teach at the University. She lectured on Platonic philosophy, mathematics, astronomy, and mechanics, and her home became an intellectual center. Her lectures— and her intellect—were famed. Students traveled from afar to hear her. Indeed, she became so famous that a letter addressed simply "The Muse, Alexandria," would reach her. (Alexandria had nearly a million people at this time.)

Margaret Alic calls Hypatia "primarily an algebraist" ("Women," 309), and notes that she wrote a commentary on the *Arithmetica* of Diophantus (later incorporated into the Diophantine manuscripts), did a popularization of the *Conics of Apollonius,* and probably collaborated with her father in revising Euclid's *Elements*—an edition still in use today. Hypatia and Theon also collaborated on at least one book about Euclid and on a commentary on Ptolemy's astronomical work, the *Almagest.* Suidas' list of her works includes no philosophy, only mathematical and astronomical writings (*ibid.*).

What brings Hypatia to this chapter, however, is neither her brilliance nor her lost writings, but her interest in mechanics and practical technology. Surviving letters from her pupil Bishop Synesius of Cyrene contain Hypatia's designs for several scientific "machines" or instruments, two of which Alic describes in some detail: an astrolabe (used to simulate the apparent daily rotation of the celestial sphere, and thus to solve astronomical problems); an apparatus for distilling water; an instrument for measuring water levels; and a hydroscope or hydrometer, for determining the specific gravity of a liquid. Mozans mentions also a planisphere.

The story of her murder—by Christian fanatics—is almost unbelievably brutal. In March of AD 415, "certain heady and rash cockbrains" pulled Hypatia from her chariot, dragged her into a church, stripped her, and killed her by scraping the flesh from her bones with sharp shells. When she was dead, they quartered her body and burned it to ashes. "For historians," writes Margaret Alic, her death "often symbolizes the end of ancient science. . . . [A]fter Hypatia, there were no significant advances in mathematics, astronomy or physics anywhere in the West for another 1000 years" (Abbott & Winston; Alic, "Women," 309; *DSB;* Goodwater, 15, 64; Mozans, 135ff).

The silence on women's contributions to medieval technology is oppressive. Though female blacksmiths, and thus presumably makers of tools and machines, did exist, as English place names like Ladysmith reveal, little or nothing has been recorded of their work.

Russell Conwell contends that farm women invented the printing press (45–6; Boulding, 686). In the cryptic style typical of his popular 19th-century lectures, he does not elaborate. What he may mean is that in inventing food-processing presses, such as olive, cheese, apple, and grape presses, and certain kinds of dough rollers, women had discovered basic principles and created apparatus that need only be adapted to forcing inked blocks against paper, and eventually to printing from cylinders. Recall, for example, that both Athena and Minerva are credited with inventing the extraction of olive oil. Whether Conwell knew of the Sumerian cylinder seals from the 4th millennium BC is debatable; they impressed a continuous combined image of pictures and ideographs on moist clay when rolled over its surface, and might be considered an early form of printing. Still earlier, of course, is women's incised or impressed decoration of pottery—which according to Marija Gimbutas takes the form of script as early as the Neolithic in Europe (*Goddesses,* 85). Textile printing, too, is an ancient women's art and industry that has applications to the imprinting of text on paper.

Whatever may have been women's role in the printing *press,* women are credited with at least three important inventions in printing. One falls in Gutenberg's own century; the other two, though much later, I will also mention here. In the 15th century, **Isabella Cunio** and her twin brother invented wood engraving (M. Gage/1: 16–17). In the 19th century, according to Charlotte Smith, a woman invented syllable type (*WI*/1: 4: 2). **Helen Bruneau Van Vechten** solved the tricky problem of perfect registry of recto and verso in sheet-fed printing with hand-made papers (V. Brown, 32–3; Lamperti, L-1/28/84).

One very important machine that may well have been a woman's invention—if only because no male seems to take credit for it anywhere—is the spinning wheel, which appeared in the East and traveled West. According to Lynn White, it was in use in China at least by AD 1035, perhaps earlier, and reached the West in the 13th century ("Eureka!" 4). Once brought to Europe, it became the subject of numerous improvements, and the "modern" form of the wheel is often credited to a German named Johann Jürgen, about 1553 (Grigson & Gibbs-Smith, 353).

In more recent times we find further evidence refuting stereotypes about women and machines. Transitional between the 18th and 19th centuries is the famous short-staple cotton gin patented by Eli Whitney. The impact of this landmark 18th-century machine fell mostly in the 19th century, where, as already discussed, it transformed the economy of the American South and probably contributed to the outbreak of the Civil War. Catherine Greene's much debated role in the creation of this machine (see App. A-2) may never be proven. Suffice it here to note that Whitney did arrange to pay Greene royalties and, according to a Shaker writer, once publicly admitted her help (*Shaker Manifesto,* July 1890, n.p.).

## 19TH CENTURY

Victorian ladyhood notwithstanding, the 19th century provides many female machine inventors. These are individual women who invented or contributed to the invention of highly significant single machines, as well as groups of women who patented important *types* of machines. Let us look first at a few landmark machines and the women, some anonymous and virtually all unsung, who were involved in their creation. Then, for an idea of women's overall achievement in the century, we will look at some *classes* of important machines and the women, mostly patentees, who invented and/or improved them, followed by a survey of women's mechanical inventions in several areas from agriculture to transportation. Finally, we will look at a prolific and arguably professional group of women inventors who specialized in machines. The section closes with two such machine specialists whose careers continue into the 20th century, **Helen Blanchard,** and **Margaret Knight.**

The first woman to be discussed here is anonymous. And the second most prolific woman inventor of the century, with over fifty inventions to her credit, is known only by her first name and last initial, **Mary S.**[4]

### Landmark Single Machines

These four highly significant machines might almost have been chosen purposely to show women's versatility as mechanical inventors. All four are usually attributed solely to men, yet woman evidently invented or contributed a crucial idea to each one.

### McCormick reaper

American schoolchildren probably still learn about Cyrus McCormick and his reaper, which played such a significant role in the mechanization of American agriculture. What children don't learn is the story told by Russell Conwell. Conwell, who is full of information of this kind, cites "a recently published interview with Mr. McCormick," in which the inventor admitted that after he and his father had tried to create the reaper and failed, "a West Virginia woman . . . took a lot of shears and nailed them together on the edge of a board [with one blade of each pair loose]. Then she wired them so that when she pulled the wire one way it closed them, and . . . the other way it opened them. And there she had the principle of the mowing machine" (R. Conwell/1968, 45–6; Boulding, 686).

The schools do not teach, either, about **Ann Manning** and her role in the earlier Manning Reaper. Invented by Ann and William Manning in 1817-18, and patented in May of 1831, it probably influenced both Obed Hussey (inventor of another early reaper) and McCormick (Rayne, 116; H. Schmidt, 140).

---

4. By some criteria, this woman could be considered the most prolific 19th-century American woman inventor; Charlotte Smith says (*WI/*1) that all of her fifty-odd inventions were patented—though not in her name. Unfortunately, this will be virtually impossible to document, for the inventions were sold to various unnamed patent agents in three cities (St. Louis, Washington, and New York) at unspecified dates probably mostly in the 1870s. Mary S., moreover, died young; presumably if she had lived, she would have created even more inventions. I would be grateful for information leading to the location of any of Charlotte Smith's papers; to issues of *The Working Woman;* and/or to the identity of Mary S. of St. Louis (see below).

*Small electric motor*

"Tearfully but bravely the young wife handed to her boyish inventive husband, 'Tom,' the silk dress in which she had been married only eight years before." Thus Waldemar Kaempffert trivializes **Emily Goss Davenport**'s contribution to Thomas Davenport's electric motor—the sacrifice of her wedding dress to wind the coils of an electromagnet (106). Actually, her contribution was by no means limited to tearful sacrifice. "Emily Goss," says Thomas's biographer Walter Davenport, was by all accounts "an extraordinary woman . . . well fitted by nature and training for the position of help-meet for a struggling inventor" (47). Thomas's brother Oliver's account of those crucial days in the 1830s make it clear that Emily participated in the work. Thomas brought home an electromagnet, and took it apart, over Oliver's protests. Emily, who "had a fine education and was as enthusiastic as he was, . . . wrote down exactly the way the wire was wound on, and all about it, from beginning to end."

Then Thomas, a blacksmith, made a larger "horse shoe" (soft iron core), and "he and Emily wound it."[5] After coating the iron with glue, they had to wind it with silk. "Emily tore up her silk wedding gown into strips, and used that." Note the subtle but important difference from Kaempffert's account: the change from passive to active confirmed by other investigators and witnesses (Davenport, 79–80; Gies & Gies, 213–4; Doster *et al.*, 194). Then wire had to be wound around the core. It was delicate work, for the wire coils could not touch each other, but must be separated by silk. Oliver Davenport's account suggests that Emily did most of it. In any case, he concludes, "the magnet was a grand one, and it would 'lift a ton a minute.' " Says Walter Davenport, "This picture of Thomas and Emily Davenport, together with 'brother Oliver,' sitting up practically all night long, learning the secret of how the original magnet was made and then constructing a larger and better one before they slept . . . is worthy of the best saga the greatest poet of the times could have written" (55).

Then began the months of work and study to get a wheel to revolve by magnetism. The wheel would start to turn easily enough, as the underlying magnets drew the magnets on the wheel; but as soon as these were opposite the underlying ones, the wheel would stop, and Davenport couldn't break the current fast enough to keep the wheel moving. He told Emily, "It's no use; there is no power short of the Almighty quick enough to do that." But Emily, undaunted, and obviously knowledgeable, said, "I wonder if quicksilver isn't a conductor?" They tried it, and "after working . . . till three o'clock in the morning, they made the wheel turn" (62). This was in July 1834.

The original machine as Walter Davenport describes it (63) "proved to be the key to . . . the problem of the electric motor." Later, Thomas Davenport devised an automatic current-breaker (64).

Thomas knew that his motor was just a beginning. However, he did make a seven-inch wheel revolve at 30 revolutions per minute (later, three heavier wheels at 50 revolutions per minute); and before he died, his motor powered not only his own shop machinery, but the world's first electric printing press (Doster *et al.*, 194).

The original idea was undoubtedly his alone. When he first saw the electromagnet in operation (Davenport, 57–8), "Like a flash of lightning the thought occurred to me that here was an available power . . . within the reach of man. . . . 'In a few years,' I said to some gentlemen present, 'Steamboats will be propelled by this power.' " But without Emily's informed suggestion at a crucial moment, he might have given up. Instead, in

5. Many years later, Franklin L. Pope remarked that this and later new magnets were copied "with Chinese fidelity" from the original (Davenport, 69).

1891 the noted patent expert Franklin Pope concluded, after exhaustive investigations, that Thomas Davenport deserved "the chief credit for the invention of the electric motor now in use in the electric car lines of the world, as well as the motors commonly used to drive all kinds of machines in our shops and factories" (Davenport, 68–9).

Pope ignores what may yet prove the most revolutionary application of the small electric motor: labor-saving household appliances. The automatic washing machine, the food-mixer, -blender, and -chopper; the vacuum cleaner, and others of their class could help liberate full-time homemakers from their 60- to 96-hour work weeks.[6] Without Davenport's motor, none of these things was practicable. As Oliver Jensen wrote in the 1950s, washing machines were patented in England as early as 1691, but the average household had to wait well into the 20th century for power-driven washers—all for lack of "the light power unit" (7). **Lillian M. Gilbreth** saw the liberating potential, and was involved in designing several such machines, partly to enable handicapped women to manage households or live on their own.[7]

Walter Davenport's portrait of Emily Davenport outside the workshop is sketchy. Daughter of a prominent farmer named Rufus Goss, and descendant of the celebrated New England traveler Thomas Carver, she and her husband lived at Brandon, VT, with their boys (number unspecified). Thomas Davenport became obsessed with his motor, neglecting his smithy and impoverishing the family. Yet Emily always encouraged him to go ahead. Though she remarried after his death, her tombstone proclaims her the Wife of Thomas Davenport, the Brandon Blacksmith (47–8, 49 [likeness], 79, 92, 95).

*Jacquard loom*

Both Russell Conwell and his wife, Jennie Hayden Conwell, maintain that Mme Jacquard actually invented the famous loom attributed to her husband in the 1840s (Boulding, 686; J. Conwell, 143). Regrettably, neither author gives any details.

*Burden horseshoe machine*

This quintessentially American machine caused a sensation in its day, and could be seen as an important forerunner of mass production. Usually credited solely to the prominent inventor Henry Burden (d. 1871) of Troy, NY, who patented it in 1857, it is a major source of his fame.[8] Sources vary on the speed at which this machine turned out horseshoes, from one every three seconds (Hanaford, 623; M. Logan, 883; Mozans, 351; Rayne, 116) to one per second (*EA*, s.v. "Henry Burden"). The precise speed, however, is less important than the economic impact: the Burden horseshoe machine mechanized an important process in horse-drawn transport, drastically cut the cost of horseshoes, and doomed the blacksmith

---

6. As Joann Vanek and others have shown, however, the technology is not sufficient by itself. American women who do not work outside the home spend as much time on housework and family-related duties as their grandmothers did. Only when women decide they deserve free time and adjust their standards and negotiate participation in the work of the home by men and children, can they be free, even with the aid of technology.

7. I would be grateful for specifics on these innovations of Gilbreth's. The ordinary biographical details are available from many sources, including *NAW*/mod.; see also Martha Trescott's forthcoming book on women engineers.

8. The Report of the Patent Commissioner for 1891 (v) lists Burden, with specific reference to "horseshoe machinery" among "American inventors whose reputation has become national and whose improvements have formed the foundation of manufacturing industries of great magnitude." The passage places Burden in the company of Whitney, Howe, and McCormick.

long before the advent of the horseless carriage. It provided horseshoes for the North in the Civil War. According to the successful patent-renewal application (1871), the machine had already saved the American public $32 million.

Nineteenth- and early-20th-century feminists, however, claimed confidently that this machine was actually a woman's invention. Said Matilda Gage in 1870 (Hanaford, 623):

> Many of woman's inventions have been patented under men's names. The largest foundry in the city of Troy is run to manufacture horseshoes, one . . . every three seconds. The machine which does this work was invented by a woman; but the manufacture is carried on under a man's name, and will be exhibited [at the Philadelphia Centennial] as man's work.

Ada Bowles, Mozans, Martha Rayne, and Phebe Hanaford (who cites Gage) all repeat the story; but this anonymous woman inventor has thus far proved elusive. Paul Uselding, who studied Henry Burden, knows nothing of such a tradition, nor does the Rensselaer County Historical Society. Burden's daughter and biographer, Margaret Burden Proudfit, says nothing of it either (R. Andersen, L-10/26/81; Uselding, PC-1982). This silence, of course, does not necessarily invalidate the tradition. Indeed, the mysterious woman inventor or contributor might be Margaret herself. She had a close relationship with her father, and continued living in Troy after her marriage.

Matilda Gage, the apparent source of the story, lived in New York State. Only a year after Burden's patent, she entered public life by speaking at the National Woman's Rights Convention held in Syracuse. By the time of the patent renewal, Gage was working actively for suffrage at both state and national levels. In 1869, as vice-president and secretary of the New York State Woman Suffrage Association, she had correspondents in 47 counties. Presumably her work required some state-wide travel as well. Perhaps she had an informant in Troy? The various versions of her 1870 pamphlet on women inventors contain what sounds like inside information about Troy's manufactures and the inventions underlying them (Gage/1,/2; Logan, 883; Mozans, 351; Rayne, 116; Report of the Patent Commissioner for 1891, v).

It is intriguing to speculate that this informant may have been **Ann Manning,** whose sister, Dr. Clemence Sophia Harned Lozier, has a documented suffragist connection with Gage (*NAW:* "Gage," "Lozier"). In one of her Centennial letters to the Fayetteville, NY, *Recorder,* Gage calls Manning (without naming her) "a lady of my acquaintance, now over eighty years of age," mentioning her hard on the heels of her reported conversation with a Troy foundry-owner and stove manufacturer.

This man, too, is a candidate for Gage's informant on the underside of invention in the foundry city: he was the one who told her that the "greatest improvement he ever met in stoves" was a woman's invention—though patented, and profited from, by a man (Gage/1: 18; cf. Hanaford, 623).

### Landmark Classes of Machines

#### Sewing machines

The sewing machine provides a good transition here, involving as it does one woman's unpatented and unsung contribution to Howe's machine, as well as the work of about a hundred women who received sewing machine patents during the 19th century.

Said the U.S. Patent Commissioner in his 1891 report, "No greater labor-saving device than the sewing-machine was ever invented, or is ever likely to be . . ." (vi). The sewing machine enabled garments to be sewn ten to thirty times faster than by hand, and became *the world's most widespread machine.* If the "iron seamstress" initially put thousands of flesh-and-blood women out of work,[9] it also started whole new industries, and transformed the relationship between women and clothing (G. Cooper, 57–9; K.R. Gilbert, [9]; cf. Offen, 93, 95).

"It has sometimes been tauntingly said to woman," notes Matilda Gage (/1: 18), "that she has no mechanical genius, as, if she had it, the sewing-machine would have been her invention, instead of man's. . . ." It now appears that this taunt is not merely unkind but untrue. In his most famous lecture, "Acres of Diamonds," Russell Conwell tells us that Mrs. Elias Howe completed *in two hours* the sewing machine her husband had struggled with for years. Conwell's source was impressive—Elias Howe himself, in campfire conversations as they served together in the Civil War (46; Boulding, 686):

> Who was it that invented the sewing-machine? If I would go to school tomorrow and ask your children, they would say, "Elias Howe." He was in the Civil War with me, and often in my tent, and I often heard him say that he worked fourteen years to get up that sewing machine. But his wife made up her mind one day that they would starve to death if there wasn't something or other invented pretty soon, and so in two hours she invented the sewing machine. Of course he took out the patent in his name. Men always do that.

**Elizabeth J. Ames Howe** of Boston came from an old New England family. She married Elias on March 3, 1841, when he had a machinist's salary of $9 a week. Conwell does not mention her precise contribution to the machine that made her husband famous, but one source—a woman who had heard Conwell lecture— suggests it may have been Elizabeth's idea to put the eye of the needle in its point, and this was the breakthrough that allowed him to finish the machine (Kitchen, PC-8/24/83).

Lord Halsbury maintains that the invention of the sewing machine required much more mechanical ingenuity than the invention of the knitting machine, the vacuum cleaner, the domestic refrigerator, the household air conditioner, the deep freezer, the washing machine, or the dishwasher (15). Since he had no idea he was speaking of a woman's co-invention, we have no reason to doubt his statement.

Conwell's account fits the traditional story of Howe's invention (cf. Grace Cooper, 183ff) in at least two ways. First, by 1843 Elias's health was so poor that he was often out of work, and Elizabeth had to take in sewing to support the family. Months of hand-sewing under pressure of poverty could create a strong interest in easing such labor. Just watching Elizabeth work spurred Elias. Second, Howe does seem to have completed the invention during one very short period, creating a model in a few months, and sewing his first seam in 1845. However, the usual explanation given for this burst of speed after years of labor is the timely intervention of George Fisher, whose investment of $500 in return for a half interest in the patent enabled Howe for the first time to concentrate on his invention. Fisher also agreed to board the Howe family during this period.

Conwell's account conflicts with the traditional story in ways that may or may not be critical; e.g., the 14 years Conwell alleged for work on the machine. According to Cooper, Howe probably got the idea of creating a sewing machine about 1837, which would mean more like seven or eight years for the invention process, at least to his first working model, or nine years to his patent, and some sources suggest as little as five years (*EB*/11; /58).

---

9. American Quaker inventor Walter Hunt is credited with inventing the first lock-stitch sewing machine in the early 1830s. According to one story, he suggested to his daughter that she manufacture the machine, but she objected that it would put seamstresses out of work (Wallechinsky & Wallace, *People's,* 914). For whatever reason, he dropped the idea after proving that his machine worked, and sold his invention for a small sum without even patenting it (G. Cooper, 11).

We may never know, at this remove, the full story, but Conwell's story follows a familiar pattern and is buttressed by firsthand information.

Literally dozens of women invented and patented sewing machines, improvements, and accessories during the 19th century. Aside from one example of a British patent from the 1880s—granted on June 11, 1881, to **Marie LeCoeur** for a Sewing-machine gearing—my focus here is American women. None of them achieved lasting fame; none even appears in *NAW*. But several were at least important enough to be mentioned in contemporary sources or in histories of the sewing machine. Notable among these is **Helen Augusta Blanchard** of Maine, whom I discuss at the close of the 19th-century section. Suffice it to say here that from the standpoint of *regular patents hold in her own name*, she was one of the most prolific American woman inventors of the century, that 22 of her 28 known patents were for sewing machines or related inventions,[10] and that she invented zigzag sewing.

Blanchard seems not to have exhibited at the Centennial, at least not in the Woman's Pavilion (Warner/Trescott). Other American women inventors, however, were quick to take advantage of the Exposition's commercial opportunities. Notable among these was **Mary P. Carpenter** of New York City, whose exhibited machine presumably incorporated her three previous patents (1870, '71, '72) for a sewing machine needle and arm, a feeding mechanism, and a third, unnamed, improvement. Her self-setting and self-threading needle-and-arm mechanism was lauded by the *Scientific American* for September 10, 1870. The needle was so easy to thread that it could be threaded in the dark, and was attached to the arm in such a way that it was less likely to break. Carpenter secured European as well as American patents, and founded the Carpenter Sewing Machine Needle Company to manufacture her invention. Interestingly enough, this invention is omitted by LWP.

Several years later, remarried, as Mrs. Mary P.C. Hooper, she invented a machine for sewing straw braid. A model of one of her sewing machines is in the Smithsonian's collections (G. Cooper, 210, 232).

**Hannah G. Suplee** of Philadelphia exhibited an open-eyed easy-threading sewing machine needle. Unable to find instrument-makers skilled enough to make her needle, she journeyed to San Francisco to find a Chinese artisan. Evidently she succeeded, because the *New Century for Women* called her needle "the joint result of Yankee training and Chinese skill" (91; Warner/Post, 167). Interestingly enough, the San Francisco Directory for 1869–70 shows both A.H. Suplee, agent for the Elliptic and American Button-hole Sewing Machine, at 203 Kearny, and a Suplee Needle Company, with an office at 716 Montgomery. (All but one of the listings under Sewing Machines in the 1869–70 business directory were located either on Kearny Street or on Montgomery.) Albert Suplee was Hannah's husband; his dwelling, and thus presumably hers, was listed at 217 Bush (Langley's Dir.). By the time of the Philadelphia Centennial (1876), Mr. Suplee was reportedly "not much interested in business" (Warner/Trescott, 106). Perhaps his wife made up for what he lacked.

Hannah Suplee held four patents altogether, two pertaining to sewing machines. One of these covered her needle (1869), and one an unspecified machine improvement (1871)

---

10. Some of her patents were granted in the early 20th century, but Blanchard was primarily a 19th-century figure. If design patents are counted **Catherine Griswold**, with 19 regular and 12 design patents for corsets and related items, surpasses her. **Margaret Knight** had a great many more *inventions*, but only about two dozen *patents*.

co-patented with John H. Mooney of San Francisco. Mooney was a machinist with premises near Mr. Suplee's on Kearny.

As the wife of a sewing-machine agent, Hannah Suplee probably knew what the market wanted. And indeed, at the 1875 Fair of the American Institute in New York City, she received a bronze medal and sold 15,000 of her needles (Warner/Post, 167). Hannah Suplee is mentioned not only in *New Century for Women* but in the *Woman's Journal* (6 [1875]: 299), and in Ingram's description of the Centennial. Her patent model is in the Smithsonian's "1876" Exhibit today.

Several sewing-machine histories, such as those by K.R. Gilbert and the Singer Company, mention no women inventors at all. Indeed, the Singer history claims zigzag sewing for its own engineers, and fails to mention Helen Blanchard. However, it does refer to a patent that had to expire before they could do anything in the zigzag area. Grace Rogers Cooper's history of the sewing machine, written in 1968, mentions the following women in a list of patent models held by the Smithsonian: **Elisa Alexander, Kate C. Barton, Helen Blanchard, Mary P. Carpenter, Clara P. Hoffman.** Jean E. Richard, also mentioned, turns out to be male.

**Elisa** (or **Eliza**) **H. Alexander** of New York City held a patent (1857) granted only eleven years after Howe's. This is the machine whose model is in the Smithsonian's collection (G. Cooper, 207, 232). All but one of Alexander's seven patents, spanning nearly thirty years, pertained to sewing machines. Three were braiding attachments. One (1884) was an Apparatus for and method of embroidering. It featured a modified presser-foot with an opening for the passage of a thread or core, and a carrier adapted to carry embroidering material or thread. Ida Tarbell also mentions Alexander (357).

**Kate Barton** of Philadelphia held two patents for sewing machines in the 1870s, dated 1872, and 1876. The earlier one, mentioned in the *Woman's Journal* for December 14, 1872, and in Phebe Hanaford's biographical work on 19th-century women (632), adapted sewing machines to heavy goods such as sails. The Smithsonian holds a model of her 1876 patent (G. Cooper, 212, 232).

Blanchard and Carpenter have already been discussed. **Clara P. Hoffman** of Buffalo, NY, patented two sewing-machine improvements with Nicholas Meyers (same city). Her first patent was an improvement in blind-stitch machines (1878); her second, an improvement in hemmers (1879). Grace Cooper (212, 233) is the only outside source I have found who mentions her, and then only in her list of Smithsonian-held models (the 1878 patent only).

**Kate Beaird** of Tyler, TX, patented a hand-propelling device for sewing machines in 1888. According to Charlotte Smith, whose son was a Patent Office clerk in 1891, the device was a popular seller in Washington, DC, and indeed throughout the country (*WI*/1).

Beaird evidently did not show her machine at the Columbian Exposition in 1892–3. In fact, the only woman inventor of a sewing machine who seems to have exhibited at Chicago was **Harriet Ruth Tracy** of New York and New Brighton, NY. Tracy received 15 patents, 11 of which pertained to sewing machines, and most of which were mechanical (loop-takers, feeding and actuating mechanisms). Her first sewing machine patent (1889) covered a rotary-shuttle machine capable of a new kind of lock- and chain-stitch. Its bobbin held more than a thousand yards of thread (Weimann, *Fair*, 431). The Board of Lady Managers' Patents Committee at the Exposition was particularly interested in this invention (*Official Minutes*, 121–2).

Charlotte Smith mentions several sewing-machine improvements in her 1891 lists of noteworthy women's inventions. She calls particular attention to a sail-sewing machine— possibly **Kate Barton**'s—a Magic Ruffler, and a leather-sewing machine. The Magic

Ruffler—which threaded the needle while the machine was running—was the invention of **Abby H. Price** of New York City, as her 1863 patent for an Improvement in machine-made ruffles was assigned to the Magic Ruffle Company of that city. The leather-sewing machine may be **Eliza Alexander**'s. Smith calls the inventor "a practical woman machinist who for many years carried on a large harness manufactory in New York City" (*WI/*1: 4:2).

**Nina H. Piffard**—later **Piffard-Frances**—of Piffard, NY, showed a self-threading sewing-machine needle (1893) at Chicago, possibly in the Woman's Building, as did **Eva J. Hall** of Stillwater, MN. Hall held three patents for sewing-machine needles, all dated 1892. An unnamed woman inventor—probably one of these two—was denied space in Machinery Hall because she could not afford a large display (Weimann, *Fair,* 232–3; L-10/12/82 and 1/83). Piffard-Frances was mentioned in a 1913 *New York Times* article as the inventor of several other useful and commercially successful items (Janeway, 55f).

In 30-odd years, between 1862 and 1894, at least 84 women *besides those already mentioned* got 92 patents pertaining to sewing machines. The overwhelming majority are mechanical in nature—attachments for marking, quilting, tucking, felling, spinning, hemming, ruffling, and all the many functions of a sewing machine; improvements in tension devices, in driving, starting, and feeding mechanisms, in bobbins, presser feet, treadles, looping attachments, bobbin-winders, and hand-operating attachments. There is even an Improved buttonhole-sewing machine (1877). **Emma F. Briggs** of San Marcos, TX, and **Anna H. Clayton** of Louisville, KY, both patented sewing-machine motors (1888, 1892). **Elizabeth Cheshire** of Cincinnati, with Edward Cheshire, patented a book-sewing machine and needle (both 1885). None of these women appears in any outside source, to my knowledge; but the sheer numbers and variety of their mechanical inventions are impressive (App. M-1):

Though none of these women is known today, one of the earliest, **Caroline Garcin** of Colmar, Alsace-Lorraine, was part of a *cause célèbre* in her own day. When sewing factories appeared in France during the 1860s, some of the seamstresses complained of "extreme genital excitement" after working their treadle machines all day. An 1866 report by a Paris physician, Eugene Guibout, warned of "the potentially nefarious masturbatory effects" of two-pedal industrial sewing machines, and their dangers to women's health. The report, a sensation throughout Europe, was extensively quoted by opponents of women's working. Although a later report disagreed with Guibout, the question remained.

In an effort to resolve it by eliminating any harm to women workers, Mlle Caroline Garcin enlisted the aid of a skilled clockmaker in Colmar to develop "an automatic sewing machine" without foot pedals. The American patent, on which her name appears first, calls the invention a driving mechanism or motor for sewing machines. Its exact mode of operation is difficult to decipher from the patent drawings, though spring-loading seems an important feature. Suffice it to say that Mlle Garcin and her co-inventor were honored by the French Academy of Sciences in 1872 (Offen).

All in all, these 97 women, ten with some degree of recognition and 87 totally unrecognized, to my knowledge, received at least 154 patents for sewing machines and related inventions by mid-1899.

## Dishwasher

The first commercially available power-moded dishwasher was invented by **Josephine G. Cochran** (1839–1913) of Shelbyville, IL. According to Patrick Robertson's *Shell Book of Firsts* (50), a "Mrs. W.A. Cockran" developed this machine over ten years, completing it in 1889 only after the death of her husband, who had kept her too short of money to perfect her invention.[11] The first of her six U.S. patents is dated 1886. She kept improving her machine until her last patent in 1911.

Cochran exhibited her dishwasher at the Columbian Exposition in 1892–3. Both family and restaurant models were available, the larger models steam-driven. The Cochran dishwashers were not only shown at Chicago, but used in most of the major restaurants there, and for a very good reason: they could wash, scald, rinse, and dry five to 20 dozen dishes of all shapes and sizes in just two minutes (Robertson, 50; D.E. Walker, L-5/24/82; Weimann, *Fair*, 431).

Josephine was born in Ohio, daughter of an engineer and great-granddaughter of a steamboat inventor. In 1870 her husband was a circuit clerk. The Cochran Dishwasher was manufactured by a Chicago machine firm (Robertson, 50). Mrs. Cochran was still active in 1908, for she showed her dishwashers at the Martha Washington Hotel Suffrage Bazar in New York City in November of 1908 (9th Census, Shelby Co., IL; M. Reynolds, 2–3).

Cochran evidently received some recognition in her own time. She appears in Mary Logan's list of prominent American women inventors (886ff), and her invention came to the attention of the Russian revolutionary Kropotkin, perhaps through the Columbian exhibit. Wrote Kropotkin, who had great faith in the possibilities of machines, "If there is still work . . . disagreeable in itself, it is only because our scientific men have never cared to consider the means of rendering it less so," and he was excited because a "Mrs. Cochrane" in Illinois had invented a [dish]washing machine (Warrior, 87).

Considering the prevalence of dishwashers today, Josephine Cochran's name should be a household word. It should also be a schoolhouse word, taught right along with Eli Whitney (and Catherine Greene), Elias (and Elizabeth) Howe and Cyrus McCormick (and Ann Manning—and the anonymous West Virginia woman). Moreover, if the claims for the Cochran Dishwasher are valid, we may need to look backward instead of forward in dishwasher technology, as studies indicate that many modern dishwashers save only about 12–37 minutes a day (Bose & Bereano, 87–8). However, the 19th-century volume of Singer's classic history of technology omits the dishwasher entirely, as do Grigson and Gibbs-Smith. Harpur (92, 172) and Giscard d'Estaing (151f) discuss it, but do not credit Cochran. Harpur, in fact, dates the first practical model as late as 1932.

At least 30 other women patented 31 dishwashers and improvements between 1870 and 1895, including the intriguing Combined dish- and clothes-washer patented in 1890 by **Margaret A. Wilcox** of Chicago (App. M-2). To the best of my knowledge, all but one of the earliest (1870), **Glory Ann Wells** of Luzerne, NY, are lost to history. Even Wells is known only as an exhibitor at the Philadelphia Centennial of 1876, briefly mentioned in a 1970s article by Deborah Warner of the Smithsonian (Warner/Trescott, 119). LWP omits her.

*Washing machine*

Early washing machines were little better than washboard and tub. Indeed, women were still patenting washboards in 1892 (e.g. **Mary C. Burke** of Montpelier, ID) and later.[12] Women still had to heat, carry, and empty all the necessary water for early machines, for

---

11. Robertson's date is puzzling. It may be a British marketing date; I find no British patent 1888–1900. Since his account contains at least one other error besides the "Cockran" spelling (locating Shelbyville in Indiana instead of Illinois), this date may simply be wrong.

12. I have found a washboard patent as late as 1925, which makes sense if we recall that many rural areas of the United States did not receive electricity until the late 1930s, or even until after World War II (e.g., southern Ohio).

example, both for washing and for rinsing. At first the machines did not agitate the clothes; women plunged a dolly up and down in the tubs by armpower. Even with electric machines, popular after about 1918, women sometimes had to carry water.

In fact, commercial soap powder may actually have saved women more time than these early electric machines, since no one now need make soap at home, or tediously shave or boil bars of store-bought laundry soap. As late as 1899, **Juliet L. Ryan** patented a Soap-grater, and she lived not in a rural area, but in New York City. Washing fluids seem to have preceded powders, at least among women's patents. **Martha H. Sanderson** of Fremont, MI, patented a Washing fluid as early as 1872, followed by **Annie Rhoads** of Baraboo, WI ("Washing compound") in 1884; **Maggie E. Minnick** of Galveston, TX, in 1887; and **Mary Schneider** of St. Louis, in 1888. **Elizabeth F. O'Neal** of Philadelphia patented a Washing powder in 1892.

When wringers were attached to the machines, women still had to hand-feed the heavy wet clothes through the rollers between washes and rinses and again at the end of the process—very carefully, to avoid broken buttons, mashed fingers, or worse.

Not until 1927 did the agitator machine appear with its motor under the tub, and not till the mid-1940s did the fully automatic machine appear (*EB*/58 XI: 813, 815; Harpur, 49; Doster *et al.*, 194). Even then, as Ruth Cowan points out, the automatic machines were developed for commercial rather than home use (Trescott, *Dynamos*, 40).

No woman's name is popularly connected with the first invention of the washing machine, patented in England by 1691 and several more times in the 18th century (O. Jensen, 71). But it would be astonishing if women's ideas were not incorporated into some of these earliest patents. And if **Zina Phinney** of Cairo, NY, is female (Phinney does not appear on LWP, but at least one other Zina does), an American woman received a patent as sole inventor of a Washing and churning machine as early as 1812.

According to Orison S. Marden, writing in 1903, many of the hundreds of 19th-century washing-machine patents were actually useless; and one of the most successful machines was invented by a woman (4089). Marden may refer to **Margaret Plunkett Colvin** (1828–94) of Battle Creek, MI. Mrs. Colvin showed her Triumph Rotary Washer at the Philadelphia Centennial, where it was called "the successful result of years of experiment by a practical woman" (Warner/Post, 167). This "practical woman" did not stop experimenting after her 1871 patent and Centennial success, either, for she later patented three clothes-pounders, two in 1878 and one in 1881, and also showed her Triumph machine at the Columbian Exposition in 1893. Her husband Ashley Colvin manufactured and marketed her machines (Warner/Trescott, 111).

During the thirty-odd years from 1862 through 1894, at least 48 other American women patented washing machines or related inventions, including a clothes wringer, a clothes strainer, and a clothes drainer (LWP) (see App. M-3).

These, of course, are just the women who had the time, money, and self-confidence to patent their inventions, and to do so in their own names. More than a few others may have been patented in the names of male relatives or agents. Still others may have been fully developed and even marketed, but not patented. Some machines may have been patented, but omitted from LWP, which is the source of this list. **Ella Goodwin** of Chicago, IL, for example, does not appear on LWP. Yet we know that she invented a washing machine, for she showed it both at the 1884 World's Industrial and Cotton Centennial Exposition in New Orleans, and at the Columbian Exposition of Chicago in 1892–3 (Romney, "Women," 25; "Something for Everyone"). And indeed she patented it in 1885.

Like most inventions, the Goodwin Washer stands on the shoulders of previous technology rather than leaping into the totally new. Early washing machines tended to mimic one of the major pre-existing ways of getting dirt out of cloth—pounding followed by water-rinsing; rubbing followed by water-rinsing, etc. The so-called clothes-pounders obviously used the former technique, whereas certain other machines, Goodwin's among them, attempted to reproduce the washboard action of 19th-century hand-laundering (cf. Giedion, 150). In Caroline Romney's contemporary description ("Women," 25):

> [T]he Goodwin Washer . . . consists of a tub with a wash-board in sections, rising from the bottom in the form of a low cone to a common apex, and a similar combination of washboards to fit over this cone, the upper one being supported from a frame-work across the top and operated by means of a crank, thus rubbing the clothes between the upper and lower boards.

The so-called washing "mills" of the 18th century, as the name indicates, also adapted existing (but nonlaundry) technology to laundry work. English diarist Mary Hardy, a prosperous farmer's wife, rejoices in 1791 at borrowing a neighbor's washing mill and doing "3 weeks linnen without a work woman" (Blodgett, 33). **Margaret P. Colvin**'s washer, however, is described as a *rotary* washer, which may represent something new. This is another reason for thinking Colvin's "Triumph" machine is the successful model Marden had in mind.[13]

Only one of these 48 names is familiar to me—that of **Semele Short** of Cincinnati, OH—and hers only because she, like Colvin, appears in Deborah Warner's pioneering article on women inventors at the Centennial. According to *New Century for Women,* Short's blanket-washer "appears to be the thing" (No. 2, May 20, 1876: 2). Mrs. Short also exhibited two other inventions—a mangle-and-ironer simple enough for unskilled workers and a drying frame for lace curtains—and received a Centennial Award. All three of her patents were granted in 1873. Short also showed these inventions at Cincinnati Industrial Expositions of 1874 and 1879—where they won a silver medal, a bronze medal, and a premium, respectively—and again in 1883). A model of the curtain-drying frame is in the Smithsonian's "1876" Collection (*NCW, ibid.;* Warner/Trescott, 117; Weimann, L-1/28/83).

**Julia C. Smith** and **Rachel P. Smith** may also have exhibited at the Centennial, as models of their inventions appear in the "1876" Collection (Weimann, *ibid.*) However, as they are not among the 80-odd exhibitors listed by Deborah Warner, these women would have exhibited, if they did, outside the Women's Pavilion. As already noted, the "1876" Exhibit contained some inventions not present at the Centennial, and their patents are later—1879 and 1882, respectively. Jeanne Weimann locates them both in Chicago, though Julia Smith's patent lists her address as Ashton, IL. A Julia C. Smith does appear in the Chicago directory by 1890, however.

Some of these women's washing machines sound at the very least ingenious, notably **Sarah Sewell**'s Combined washing machine and teeter- or seesaw of 1885, **Mary J. Bridges**'s Combined clothes-washer and churn of 1883, and **Margaret Wilcox**'s Combined dish- and clothes-washer, already mentioned.

Too late for LWP is **Cecilia A. Brewer,** mentioned in the *Patent Record* as having patented "an improved and quite intricate washing machine, with the wash-board at the bottom," in 1899 (G. Walsh).

Clothes-wringers may also be considered here, though they were long a separate item

---

13. Colvin's machine was not, however, as Vare & Ptacek claim, automatic. Such machines were not available till well into the 20th century. Nor do I find the six patents for washing machines that they attribute to her (37).

attached to the washtubs. As early as 1871 **Jessie H. Murray** of Kirkwood, NY, patented an Improvement in washers and wringers combined. Black inventor **Ellen Eglin** of Washington, DC, invented but did not patent an improved wringer that must have represented a significant technological advance, for it made a fortune for the unscrupulous agent who bought it from her for only $18 in 1888. When asked why she sold it so cheaply after months of hard work, she said, "You know I am black and if it was known that a negro woman patented the invention, white ladies would not buy the wringer. . . ." Eglin was a member of Charlotte Smith's all-female labor union, the Woman's National Industrial League. Her economic fortunes seem to have fluctuated, for the DC directories list her variously during the 1880s and '90s as federal clerk (Treasury; Census Bureau) and charwoman. Charlotte Smith wrote in 1891 that Eglin was working on another invention which she intended to patent *and market,* and was looking forward to exhibiting it at the Woman's International Industrial Inventors Congress, where women could participate regardless of color.[14] (*WI*/1: 3: 2).

   **Mildred Lord**'s is the only washing machine mentioned in a list of "Prominent American Women Inventors" published by Mary Logan in 1912. **Margaret Wilcox** appears on Logan's list, but for another invention. Wilcox, indeed, has several inventions to her credit, and deserves notice as a multiple inventor and possible entrepreneur.

*Typewriter*

Another transformative machine in modern economic life was the typewriter. Although no woman's name is connected with the original invention of this machine—except for the blind Italian countess who inspired the creation of one for her use—between 1888 and 1894 at least 13 American women invented typewriting machines or related inventions, 12 of whom got patents (see App. M-4):

   The thirteenth woman, **Emma D. Mills,** created a typewriter improvement for which she had to devise her own special tools. This must predate 1891, when Charlotte Smith wrote about it in the first issue of her *Woman Inventor*. Smith says that Mills patented the invention, and in her own name; but if so, LWP omits her. Nor do I find the patent in the annual Patent Office name lists, though one Anson Mills, U.S. Army, patented a typewriting machine in 1887. If she is, as seems almost certain, the Emma De Losse Mills of Clinton, NY, who patented a hop-stripping machine in 1885 and a hop-picking machine in 1894—also omitted by LWP, she becomes an interesting inventor indeed. Smith calls "Miss Mills . . . one of the strongest advocates of women's entering formerly male occupations." Taking her own advice, Mills was, at the time Smith wrote, preparing to manufacture and sell her own invention (cf. M.J. Reynolds, 5; "Women Inventors" [1903]).

   Some of these 13 inventions sound possibly significant, such as **Lizzie Sthreshley**'s Typewriting machine for the blind (1890), **Anna M. Rothert**'s Upper-case treadle attachment (1891), and **Adelaide H. Woodall**'s Reverse-movement attachment (1889). Emma Mills' invention was considered significant enough to be mentioned by *Scientific American* (Apr. 4, 1903: 247), though without any useful description. Charlotte Smith calls it an attachment. None of the 13 names is familiar today except possibly that of **Lizzie J. Magie,** and that not for her typewriter, but for her Landlord's Game, prototype of Monopoly® (see Suid & Harris).

---

14. I would be grateful for information about this Congress—whether it was ever held and if so, when and where; names and inventions of exhibitors, etc. It was probably one of Charlotte Smith's projects.

*Machines by Category of Endeavor*

Nineteenth-century women invented a great many other machines and mechanical improvements in addition to the landmark individual machines and classes of machines just discussed. Just how many—or indeed, the extent of woman's contribution to mechanics in general in this century—has never been accurately determined, even where records exist. LWP, for example, covers over 85 years of women's patents, but classifies them for less than three years (Oct. 1892–Feb. 26, 1895), and this classification is seriously flawed. Some inventions are ludicrously misclassified, as when **Harriet Russell Strong**'s dam and reservoir system appears under Culinary Utensils, and **Nell Westburg**'s band-cutter and feeder for a threshing machine appears under Sewing and Spinning (Stanley, "Rose," 12, 14).

Errors aside, the categories chosen often differ from mine. Washing machines, for example, appear not under machines (LWP has no such category), but under a Washing and Cleaning heading that also includes a horse brush and a soap receptacle.

Katherine Lee, who alone has attempted a statistical survey of United States patents granted to women since 1790, uses the Patent Office's broadest division of all patents into General Mechanical (actually General *and* Mechanical), Chemical, Electrical, and (later) Designs and Plants. In the even-decade years she sampled (1790–1970), Lee counted over 58% of women's patents as "General Mechanical"; but this simply means the inventions were neither Chemical nor Electrical, not designs, and not new plants. Aside from the other difficulties with Lee's totals and her misreading of the category name, such a catch-all category as "General Mechanical" is useless here, as it contains so many nonmechanical inventions. In the earliest years, for example, it includes patents for "Balsam lavender," Cream of tartar, and a Shirt (LWP; K. Lee, 17). On the other hand, the Patent Office classes some apparatus and machines as Chemical or Electrical.

Everybody—including the LWP compilers—has found it too daunting a task to go back to the patents themselves, relying instead on the brief descriptions found in the name lists or the Gazettes. The name-list descriptions do not always reveal whether an invention is a machine or not. Take for example the much cited 1809 patent granted to **Mary Dixon Kies** of Killingly, CT, for, as LWP describes it, "Straw-weaving with silk or thread."[15] Mozans calls her invention a process (344). However, Lynn Sherr and Jurate Kazickas, who visited Killingly, among many other hometowns of noteworthy American women, while compiling their Bicentennial *American Woman's Gazetteer,* call it a straw- and silk-weaving *machine,* as does Louise Tharp in her "Bonnet Girls" article (6). A *Scientific American* article of 1903 calls it a *device* ("Women Inventors," 247). A Schenectady, NY, patent attorney named Arthur Kies (no known relationship to Mary's husband) who has researched Mary Kies, speaks of her patent as granted in 1809 "for a process or machine . . ." (H. Warner).

One problem in Kies' case, as in several others, is that her original patent was destroyed in the Patent Office fire of 1836 (*ibid.*). Some patents were restored after the fire, as inventors or their descendants sent duplicates to Washington; but others were never found. The lost patents make up a very small percentage of the total; but in some of these early years American women received only one patent.

With these caveats in mind, let us sample women's contributions to the mechanical technology of various broad areas of human life and work:

---

15. This is allegedly the first United States patent granted to a woman.

*Agriculture*

Women's 19th-century agricultural machines have been discussed in some detail in the Agriculture chapter. Worth a brief mention here, since she is omitted from LWP, is **Iris Hobson** of Fort Scott, KS, who patented a potato-digger in 1876.

*Architecture*

The myths of Minerva and Hestia—and many anthropological reports—indicate that women probably invented architecture. At least, women were the first builders in many cultures, as among the !Kung gatherer-hunters of Africa, primitive people of Southern Australia, and the Agta of the Philippines (Shostak/Stanford; B. Hayden, 213; Dahlberg, ed., 133, 135; Boulding, *Underside*, 10, 74, 75, 79–80, 105, 118–9). At Çatal Hükük the mud bricks for houses were made in wooden molds squared with adzes; women there were buried with hoes and adzes (Mellaart, *Town*, 207, 209). Women were the weavers (or felters or tanners) and erectors of tents among the nomads; and tents can be seen as exquisitely adapted architecture (Faegre). In the intervening millennia, women have been responsible for many technological advances in building in general and housing in particular. Noteworthy among them for the 19th century in the United States are **Catharine Beecher**'s efficiency kitchen (O. Jensen, 74–5) and **Harriet Irwin**'s patented design for an octagonal house. **Mary Nolan** of St. Louis[16] exhibited blocks made from her new building material, Nolanum, at the Philadelphia Centennial in 1876; and **Julia Meyenberg** of Chicago patented a structural support (1895) and at least two building blocks (1894, 1895).[17] Nolan intended her new kaolin composition and new shape for building blocks to revolutionize the construction industry, and the Centennial judges also expected "important results" from her invention. Meyenberg's 1899 patent was mentioned in the *Patent Record* as an example of the variety of women's recent inventive achievement. This building block had "beveled ends and alternate rows of tongues and grooves for locking them together." Nolan's blocks were also interlocking (Warner/Trescott, 106–7, 115; G. Walsh).

Though it is not traditional to speak of architectural machines, some components of houses and other buildings are certainly mechanical, and some 19th-century women's inventions related to housing or construction are strikingly mechanical in nature. A few examples can be given here: **Elizabeth Bell**'s machine for making the pottery pieces used in chimney construction (Brit. Pat. #3019, 1807); **Harriet Tracy**'s safety elevators, **Emma Walsh**'s elevator improvement (Elizabeth, NJ, 1884); **Sarah Winchester**'s crank for opening screens or shutters while the windows are closed, and her burglar-proof catch for windows, patterned on the Winchester rifle catch. The reclusive Mrs. Winchester had several inventions to her credit: a porcelain washtub or sink with molded-in washboard and soap dish; a kitchen drainboard with molded-in grooves for draining water into the sink, a hinged, fold-down kitchen counter, metal dust-excluders for stair corners, tilting shower heads, an underground drainage system, fireplaces with drops for ashes, a drainage system that collected and re-used water from upstairs plants, a watering system for her greenhouses and window boxes that included an installed faucet above each box, and stairs with two-inch risers to accommodate her arthritis. All of these were installed in her home, once the largest strictly private residence in the United States (Rambo, 3), and

16. Nolan was a versatile inventor and entrepreneur who probably deserves a brief biography. Among other things, she joined with Charlotte Smith in the early 1870s in founding the *Inland Monthly*.
17. The LWP compilers classed the 1894 block as a toy!

can still be seen today on the Mystery House tour in San Jose, CA. Though Sarah Winchester's name does not appear in the patent lists, some of these inventions were evidently patented and/or sold in her interest by her lawyer, because the revenues from them added $250,000 to her already enormous fortune (Rambo, 9–10, 12–3; Stanley, "Mothers"/1979, 65, n44; J. Stark, 4; tour pamphlet and spiel, Winchester Mystery House, San Jose, CA, 1970s).

*Windows and ventilation-related.* Nor was Mrs. Winchester the only woman to invent things pertaining to windows and ventilation. Another famous 19th- and early-20th-century inventor, **Harriet R. Strong** (1844–1929), while still a young girl, invented a device allowing her to open windows from her sickbed (not patented till 1884). She also patented a Window-sash holder (1886; Sherr & Kazickas, *AWG;* Amram, "Invention As Problem . . . ," 7–8). At least 23 other 19th-century women received 22 U.S. patents for window constructions or parts and ventilation-related inventions between 1867 and September 30, 1892, most of them obviously or arguably mechanical.

Seven of the 22 inventions classified as "Building appurtenances" by the compilers of the third installment of LWP (October 1, 1892, through February of 1895) pertain to windows, and at least four of these are mechanical. One of the four, incidentally, is patented by the most famous 19th-century American woman inventor, **Margaret Knight** (1894).

The only other woman in this group of inventors, aside from **Harriet Strong,** who received any recognition for her work was **Kate Duval Hughes** (b. 1837). Born in Philadelphia to a wealthy family of French descent, but thrown on her own resources after an unfortunate marriage, Hughes was recognized in her day as a woman of many parts. In addition to her two window-fastening inventions, she evidently re-invented the process of extracting the essential oil of frankincense (c. 1890), and incorporated it into an ointment for skin diseases that was used in many hospitals. She also wrote and published religious books for children, and held a government job (Willard & Livermore, 400; Sherr & Kazickas, Women's Calendar, June 15, 1979; Amram, "Innovative," 31).

This nonexhaustive list alerted me to window and ventilation-related devices as a favored area for women's 19th-century inventions in architecture and related technology. Indeed, a whole section could be written on women's inventions for both cooling and heating buildings, such as **Charlotte A. Van Cort**'s 1883 Apparatus for cooling . . . buildings (New York City). Moreover, as will be seen below, this predilection seems to continue into the first quarter of the 20th century.

*Fire escapes.* Fire escapes seem to be another favored area for women inventors. Eight of the 22 Building appurtenances mentioned just above as patented by women between October 1, 1892, and March 1, 1895, were fire escapes. Two of these were patented by one woman, **Barbara Fox** of Napoleon, IN.

Earlier in the century, women received at least 24 U.S. patents for this safety device. Such disasters as the Great Chicago Fire of 1871, which killed 250 people, destroyed 17,450 buildings, and did nearly $200 million damage, and the Brooklyn Theatre Fire of 1876, which alone killed 300 people, may have been fresh in these women's minds, along with the many smaller factory and residential fires that plagued the newly crowded cities. Omitted by the third-installment classifiers of LWP is the Balcony hoist and mechanical fire escape patented in 1893 (#501,455), by **Julia B. Chetwynd** of London, England.

Thus a total of at least 33 fire escapes were patented by women in less than twenty years, between 1877 and early 1895. This is 3.49% of the 945 fire escapes patented during this period, significantly more than women's share of all patents at that time (less than 1%).

Too late for LWP is the Fire escape patented by **Harriet D.T. Wilson** of New York City in 1899.

Noteworthy among these women are **Harriet Tracy,** whose sewing machine and elevator inventions are discussed elsewhere in this chapter, and the several women who created fire escapes in combination with more ordinary devices—multipurpose inventions more likely to be given house room, perhaps. In this latter group are **Mary Jones** of St. Paul, MN, whose bedspring fire escape was essentially a long rope stored in the room where it was most likely to be needed, as the lacing to support a mattress; **Anna Dormitzer** of New York City, among whose dozen or more patents was a Combined window-cleaning chair and fire escape; and **Amalie Schaefer** of Chicago, who seems to have invented a similar combination. **Eleonore Eder** of St. Louis, as will be seen below, also patented a baby-carriage improvement.

These, of course, are only the patented examples. Charlotte Smith mentions in the *Woman Inventor* (/1: 4) a **Mrs. D.M. Barer** of New Decatur, AL, who invented a fire escape. In a "recent" trial, says Smith, "the persons descending let themselves down . . . as easily as if they had wings." This invention is probably c. 1890, as this issue of *WI* appeared in April 1891. The *Englishwoman's Review* for January 15, 1893 (65), mentions **Matilda Stewart Barron**'s Child's fire escape (or means of lowering a child safely from a window) doubling as a dress stand. No information is given as to patent status.

*Space-saving construction/furniture.* I also class as architectural the built-in, folding, extension, and multipurpose furniture designed to save space, of which folding and extension items, as well as some multipurpose ones, are relevant here. Women contributed significantly to this type of furniture, perhaps responding to the loss of living space as more people moved into small urban apartments and the size of the average American house decreased. Examples are **Augusta M. Rodgers**' Folding chair of 1870; **Josephine P. Barr**'s Roller-top extension table of 1900; and **Sarah Goode**'s Folding cabinet bed of 1885. **Carrie E. Brower**'s Reclining high chair of 1898 was probably not a space-saving device; but many folding high chairs doubtless originated so.

Augusta Rodgers is a prolific and versatile inventor, holding seven patents granted between 1870 and 1874, and known from contemporary sources as an inventor of safety and heating devices for railroad cars. Charlotte Smith mentions that Rodgers has protected her folding chair and her earlier mosquito canopy for beds with British as well as American patents, which would indicate some expectation of profit from the inventions (Hanaford, 632; Smith, *WI*/1). Rodgers is mistakenly credited with inventing the automobile heater (see below).

**Netta G. Rood** of Chicago and Evanston, IL, patented both a Wardrobe bed (1880) and a Portable summer house (1882). The hinged and adjustable wardrobe bed fit into a wall recess, drawn in and out with pulleys. The summer house received some publicity in its day because its folding wire-screen sections could be taken apart and shipped to the beach for a family's summer holiday. Cloth shades rolled up from the bottom of each section gave privacy. It also featured folding furniture. A portable, wire-screen, insect-proof summer house, which may well have been this one, was exhibited at the 1884 World's Industrial and Cotton Centennial Exposition (Farmer, 419; "Something for Everyone").

## Art

Even art has its machines. The most obvious example is photography, but art also spawns mechanical devices when it enters the service of science, engineering (drafting), or commerce (illustration or advertising art), or needs to be taught to the untalented. A few examples may be given here: **Hertha Ayrton**'s drafting line-divider of 1885; **Emma**

**Allen** and **Laura O. Girvin**'s Photographer's retouching machines of 1883 and 1889, respectively; **Gertrude King**'s Ellipsograph of 1894; and **Sarah Harrington**'s much earlier Machine for taking and reducing likenesses. Both Ayrton and Harrington were British. Another example from photography is the Focusing attachment for cameras patented in 1896 by **Margaret C. Booth** of Haverford, PA.

*Commercial/retail*

Coin-operated machines apparently go back at least to Roman times, but came into their own in the 20th century with its Automats and video arcades. In the 1880s and 1890s, however, at least four women patented coin-operated machines. **Clara E. Patterson** of New York City patented a Coin-controlled opera glass (1889) and a Coin-operated stereoscope (1889). Coin-freed opera glasses are still to be found in London theatres today. Stereoscopes, however, are just collector's items. We can hardly imagine the economic and recreational importance of the stereoscope craze, but stereoscope-viewing was the television of that era. Patterson also patented an Automatic picture-exhibitor in March of 1889.

**Lizzie J. Darr,**[18] also of New York City, invented and patented a Postage-stamp-vending apparatus (1891). **Bertie Hallett** of London, England, took out U.S. Patent protection for her Coin-operated locking mechanism (1891), as did **Helen A. Burt,** also of London, for her Coin-freed or actuated machine (1892).

Other machines involved in retail sales and general commerce were **Sarah A.W. Reinheimer**'s Barrel-tapping and -emptying device (Winchester, IN; 1893); **Katharine J.C. Carville**'s Parcel-delivery system (NYC; 1894); **Edith E.L. Boyer**'s Wrapping machine of 1896 (with Israel D. Boyer, Dayton, OH); and **Madame M. Yale**'s facial steaming machine. **Ethel Ray** had earlier patented an animal trap (1890). Katharine Carville may be the New York store employee praised anonymously for inventing a profitable parcel-delivery system (Marden, 4086; *Sci.Amer.Suppl.,* May 25, 1901, 21241–2). Edith Boyer is an entrepreneur as well as an inventor, whose commercial-machine patents continue through 1924.

I have found no patent or invention date for Mme Yale's machine, but it existed at the time of the Columbian Exposition in Chicago in 1892–3. Its inventor, described by the Pittsburgh *Leader* as young and lovely, with masses of blonde hair, a "priestess of the cosmetic art," was obviously considerably more than this. She was the sole proprietor of a $500,000 company with main offices in New York and Chicago, and several large-city branches. In March of 1892 she had testified in favor of cosmetics—opposed by Charlotte Smith—before the Agriculture Committee of the House. Back from Washington, lecturing on cosmetics at the Chicago Opera House on March 17, she revealed a clash between herself and the formidable Bertha Palmer, head of the Board of Lady Managers at the Fair and ultimate arbiter of exhibits in the Women's Building. Mme Yale wished to exhibit her steaming machine and cosmetics; Mrs. Palmer refused. Then she apparently denied refusing.

Now Yale was laying her case before the public, even bringing Charlotte Smith—whom she spotted in the audience—up to the podium to speak. Though divided on cosmetics, they were united on Mrs. Palmer.[19] In this episode Mme Yale showed herself

---

18. LWP erroneously says "Dar*e*"; cf. her patent application (July 10, 1890).
19. Smith rather accurately saw Mrs. Palmer as the epitome of the elite women who ran the Women's Building with little knowledge of or true regard for the needs of working women. She would probably have

an inspired publicist. Her pro-cosmetics argument to feminists was equally inspired: training and skills being equal, the woman who looks better will get the job, so why not make the most of your appearance? (Weimann, *Fair,* 507–11; "Mrs. Potter Palmer"). Obviously, Mme M. Yale could also be formidable, and it would be instructive to learn more of her career.[20]

A bizarre entry into this field, an early mechanical advertising sign, was invented by **Black Susan Winslow** of Chicago for the prostitutes living in her house in the 1880s and '90s. To relieve them from scratching or tapping on the window or ringing a cowbell to attract the attention of passing males, she devised a mechanical woman, whose hand automatically struck the window. The hand was made of metal, its tapping action powered by a battery. Winslow enters history because of a crusading anti-vice officer named Clifton Wooldridge. Black Susan weighed 449 pounds. Too fat to get out any door of her house, she repeatedly defied the police until Wooldridge took her back door off its hinges, sawed out a two-foot section of wall, and dragged her up a splintery plank to the waiting wagon (Asbury, 106–8; Mark, 14–5). Wooldridge's version is slightly different, but the essentials are the same. He credits Black Susan alone with the invention (289–92).

Most of the complaints against Winslow and the other madams of this brawling city in the late 19th century were for robbery, as many of them ran "panel houses" equipped with another madam's "paracommercial" mechanical invention—the sliding panel built into the wall of each room, through which an accomplice in the next room could steal the customer's money from his clothes, hung on the chair set in front of the panel. Credited with this diabolical invention was **Moll Hodges** of New York City, who also had houses in Philadelphia and Chicago (Asbury, 124).

## Education

Education in general, like art education, sometimes has mechanical aids, of which a few may be noted here: **Elizabeth Oram**'s Globe for teaching geography (1831); **Agnes Engdahl**'s Penmanship-teaching machine shown at the Centennial of 1876, and **Mary Huntman**'s Tellurian of 1893. (A tellurian apparatus illustrates the causation of day and night by earth's rotation on its axis in relation to the sun and the moon.) Engdahl, "head instructress" at the Stockholm Polytechnic, had earlier exhibited her device at the London Exposition of 1871. *New Century for Women* described her model (May 27, 1876: 34). Since she is not mentioned by Deborah Warner, she probably did not exhibit in the Women's Pavilion, but found a place among the general exhibits at the Fair.

## Electrical and related

Since so many electrical devices can be classed as machines, this is probably a logical place to mention a few examples of women's 19th-century electrical inventions. Katherine Lee found the following numbers of electrical patents issued to women in the 19th century: none before 1880, 1 in 1880, and 5 in 1890 (19). The 1880 patent could be **Emily Tefft**'s

---

agreed with Palmer's public position that no cosmetics should be shown in the Building, but according to her information Palmer had actually given exclusive space to a French cosmetics firm, "for a consideration" ("Mrs. Potter . . ."). If *any* cosmetics were to be shown in the Women's Building, she felt, they should be American ones. Moreover, Mme Yale should be entitled to space in the Women's Building if only as an inventor.

20. At least five women patented seven items of vapor-bath or facial steaming apparatus in the 19th century (1862–92), including **Mary Hayward**'s Electric and vapor chair. One woman, **Carrie Aurelia Munro(e)** of Salt Lake City, received three such patents.

Electrical therapeutical appliance. I cannot identify the five 1890 patents because Lee gives no names, and because after 1880 she sampled alternate weeks. However, **Mary E. Thomas**'s Electro-magnetic abdominal support and **Agnes Sommerville**'s Male genital electrode are likely candidates, as are **Mary Riggin**'s Railway crossing gate, **Cecile Puech**'s defurring machine, and **Elizabeth E. Bell**'s telephone mouthpiece.

At least eight other women received 10 United States patents for electrical and related inventions during the late 19th century (App. M-4). Possibly noteworthy among these is the one British inventor, **Mary McMullin** of London who took the trouble to obtain an American patent for her invention. McMullin's invention seems to be a battery-powered, scalp-stimulating hairbrush with metallic bristles (U.S. Pat. Gazette, Sept. 16, 1884).

A New York City music teacher named **Clara M. Brinkerhoff** patented a telegraph key in 1882. With her co-inventor George Cumming, she took out both British and U.S. patents for a Telegraphic electrode. She and Cumming, a former telegraph operator, redesigned the contact points on an earlier Cumming key. Their "periphery contact points" design—which was also applicable to other telegraphic devices—received several awards. The two inventors founded the firm of Cumming & Brinkerhoff to sell both the key and the design (P. Israel, 144–5).

Unusual in holding three electrical patents at this early date is **Ida Himmer** of New York City. The battery is her sole invention, but two electric-clock patents are issued to her and a male co-patentee. Since her winding device was assigned to the Standard Electric Clock Company of New York City, presumably it was to be manufactured and sold. A fourth invention, a circuit-closer for electric clocks, patented by Vitalis C. Himmer, also of New York, was assigned to her. As Vitalis holds other electric-clock patents, some assigned to Standard Electric, he and Ida may have been a married pair of inventors, sometimes collaborating. Perhaps he assigned this patent to her because she played an active role in its invention. They may have worked for Standard Electric Clock.

LWP omits Ida's Secondary electric clock, even though this patent was issued the same year as her battery patent, and misspells her name as "Hemmer" in its listing of her 1886 patent.

The versatile **Ella Gaillard** of San Francisco and New York City is discussed below as a professional inventor. Her 1886 Electric lighting apparatus, fairly early for such devices, is the fourth of eight inventions she patented between 1874 and 1892.

**Nina F. Whitney**'s electric alarm clock, though by no means the first (1840s and 50s; *EB*/58), is rather early. **Mary Shaffer**'s noninductive cable, though totally unheralded, was at least in an important technical area, as her better-known contemporary Edwin G. Acheson became known partly for his anti-induction telephone or telegraph wire (*DAB;* Trescott, *Rise*).

*Health/medicine*

Women's inventions to promote health and healing have also already been treated at length, but some are so strikingly mechanical as to deserve at least a brief further mention here. Obvious examples are **Hannah Milson**'s Improved ozone machine of 1876, **Mary Henderson**'s Movement-cure machine of 1889, and the whimsical device reportedly patented in 1854 by **Elizabeth Flautt** for propelling invalids out of bed and onto the floor (Weimann, *Fair,* 421).

Two sickbeds patented by British women have prominent mechanical aspects. The first, patented by **Henrietta Caroline Bentley** in 1749, allows the bed linen to be changed "without incommoding the patient." The design shows a bed in three layers, two of them movable by pulleys. One of the two movable layers consists of a webbing on which the invalid lies. Raised to allow sheets to be changed underneath, it can then be lowered again onto the fixed mattress.

The second was one of four patents issued to **Sarah Guppy** of Bristol between 1811 and 1831. Her bed (1831) was a four-poster with metal posts disguised as wood. Her design included under-bed storage drawers, doubling as steps to help the patient into bed, and an exercise machine suspended from the top. Sarah Guppy was the wife of a prosperous merchant—which is just as well, for prior to 1852 a British patent cost almost £95, more than the annual income of most of the population. Equally important, perhaps, as Judit Brody notes, Mrs. Guppy came from a family of industrial entrepreneurs, "where a tradition of invention had already been established"; and one of her other three patents (1811) pertained to the construction of railroad bridges (Woodcroft; Brody, "Patterns," 5, 7–8).

## Labor-saving (domestic)

Here are considered women's many mechanical inventions to ease their daily round of work at home. By my definition of significance, one could argue that this category of machines is the most important of all, since it has the greatest impact on the most people for the greatest part of their lives.

Women have often been taunted with not having invented even the tools and machines of domestic life or housewifery. Ludicrous for prehistory, this taunt is far less valid even for recent times than existing histories of technology would suggest. In addition to their work with washing machines, dishwashers, and the like, women have created or improved upon a great many labor-saving devices used every day. Although women have too seldom patented their ideas, even the patent records show a substantial achievement here. My examples, patented and unpatented, but drawing mainly from LWP, will appear under five headings: food-processing, laundry, cleaning, sewing, and dressing.

*Food-processing.* A large category of such machines is *food-processing devices,* already discussed under Agriculture. As a refresher example, consider the Tivolia beater invented by **Allie Hambel** of Chicago. As will be recalled, she took 60,000 orders at the Chicago Exposition of 1892–3 (Romney, "Women," 25; Weimann, *Fair,* 432).

*Laundry machines and related.* A second very large category of women's labor-saving inventions would be laundry machines. Women's contributions to washing machines, wringers, etc., have already been discussed. But washing was only part of the laundry process. If washing occupied all of Monday for many 19th-century housewives, ironing occupied all of Tuesday. Women patented several irons—most notably **Mary F. Potts**—but most of these, though made of metal and falling into the General and Mechanical category, are obviously not machines. Certain special kinds might be worth an example here: as early as 1894, **Nannie G. Davis** of Washington, DC, patented a steam-pressing attachment for sadirons. Davis, incidentally, also held two design patents for spoons. However, ironing technology did have its mechanical aspects. At least three women patented mangles, for example, between 1873 (**Semele Short** of Cincinnati, OH) and 1893 (**Susan Wiggins** of Stewartsville, IN, and **Rosa Koenig** of Milwaukee, WI).

Ironing and fluting machines: Women patented significant numbers of ironing and fluting machines to help them restore after laundering the intricate tucks, pleats, and ruffles dictated by 19th-century fashion. Between 1859 and 1883 alone, at least 12 women received U.S. patents for such devices.[21]

---

21. Some of these machines, of course, may have been intended for commercial laundries. But improvements often appeared first on commercial machines and were later translated to domestic use.

Noteworthy among these women are **Henrietta Cole, Mary Carpenter,** and **Susan Knox.** Cole held at least four patents altogether, only one of which—the least significant—appears on LWP. According to Mary Logan (885n), the last patent granted by Congress on March 2, 1861, was to Henrietta H. Cole for a fluting machine. Cole showed her improved "Pony Fluting Machine" at the American Institute Fairs in New York in 1873 and 1874, winning a diploma and a bronze medal, respectively; then took it to the Centennial in Philadelphia in 1876, where she won a Centennial Award (Warner/Trescott, 112). Her "Pony" machine is mentioned by Esther Berney in her *Collector's Guide to Pressing Irons and Trivets* (56; illus.). Presumably both Fluting-machine patents apply to this well-known model. One of Cole's other two inventions was machine-related, a Piston-packing. LWP shows the reissue of this patent (#4,349 of Apr. 25, 1871), but not the original.

**Mary P. Carpenter** has appeared before in this book, and will appear again. What brings her to this section is her Ironing and fluting machine of 1862. This was her first patent, issued while she still lived in Buffalo, NY.

**Susan Knox** also appears in Esther Berney's book. Berney notes that she invented the "popular fluting machine manufactured by Henry Sauerbier of Newark, NJ," and that this was an improvement on an earlier patent issued jointly to her and W.D. Corrister, also of New York City. Knox also received a reissue patent, on April 26, 1870 (44, 53, 55, 131). Susan Knox exhibited her fluting machine at the Centennial, in 1876, the model bearing her picture (Warner/Trescott, 105). That model, now in the Petersen collection in Los Angeles, was recently on view in Fred Amram's Minneapolis exhibit on women inventors, "Her Works Praise Her" (1988).

**Abby Price** and **Abby Urner** patented the first lamp-heated fluting device (Berney, 45, 174).

**Hannah Luchs** is not otherwise noteworthy, so far as I know, but she did receive a reissue patent, #8,338, for her fluting machine on July 1878, which she assigned to Daniel E. Paris of W. Troy, NY. Since Troy was such a great foundry city, this may mean that her machine was to be manufactured there by Paris.

Plaiting machines: Closely related are the pleating or "plaiting" machines, used to iron in the many pleats demanded in a day of elaborate dressing. Sometimes, as in **Martha E. Wilson**'s Improvement in fluting and plaiting machines, the fluting and plaiting functions are combined (Galesburg, IL; 1873). Starting with Wilson, at least 10 women patented 14 plaiting machines between 1873 and 1890. At least two of these women would seem to repay further research, as they each received three machine patents: **Mary Sallade** of Philadelphia, and **M. Addie Maybee** of Syracuse.

**Nancy Graham** (c. 1837–89) of Jackson, OH, is also interesting because she received two stove patents (one omitted from LWP) in addition to her plaiting machine, and because she seems to have been an entrepreneur and manufacturer as well as an inventor. Born Nancy Jane Dobbs in Ohio, she married Christopher Graham before 1859. Her first two children were born in Indiana in 1859 and 1862. By 1865 she was back in Ohio, where her other six children were born, the last three between 1870 and 1880. As of 1870 Christopher Graham was a tanner. The family seems to have belonged to the Methodist church, as the Jackson-born children were baptized there.

Nancy Graham died of cancer in her early 50s, just a few months after her second stove patent was granted. She was in Lamar, MO, in the last days of her illness; she died and was buried there, late in 1889. At the time of her death, she owned real property in Jackson and made a will to dispose of it and her other assets. Among her bequests is $250 to her eldest child and executor, Pelow Graham, intended, she says "as compensation for his service connected with my business at Jackson, Ohio." She leaves $50 each to her three middle sons, and "all my real estate, . . . particularly my real estate in Jackson County, Ohio" equally to her three daughters, her youngest son, and her husband, Christopher's share to go to these four children at his death, "share and share alike."

It is interesting to note that Graham makes her eldest son her executor, and leaves her husband only a life interest in a fifth of her real property. This, combined with her patented inventions and her entry into the business world, suggests that her husband may have been a poor provider or, at the very least, less ambitious and successful than his wife. An estrangement between Nancy and Christopher could also explain her move to Missouri just before her death. The other obvious explanation would be that her eldest daughter Martha had moved there, and Nancy went to her because she needed care. Interestingly enough, Martha does not seem to have been married by 1889, although she would have been 27 years old.

The nature of Nancy Graham's business is not yet certain, but it may have been a stove-manufacturing and/or -sales enterprise.

Nancy J. Graham was recently honored in her home town of Jackson, OH. The Herizon Women's Collective included her in an exhibit, "Women to Be Proud Of—Historical Portraits of Notable Jackson County Women," mounted at the City Library (D. Stanley, L-3/21,23,24, 6/11 & 27, 11/29/82; Jackson County, OH, Birth & Death Records, 1867–1908; Will of Nancy J. Graham, Nov. 21, 1889; 1870 & 1880 censuses, Jackson County, OH; Herizon, "Women to Be Proud Of . . . ," November 1982; T. Tucker, L-8/20, 8/25, 9/1, and 9/4/84).

Other laundry aids: Not machines *per se,* but definitely mechanical and thus pertinent here, are the folding ironing boards, the adjustable pant(aloon) stretchers or shapers, the curtain stretchers, and the reels and pulleys allowing clothing to be hung on a line inside and then conveyed out through an upper-floor window to dry, among other laundry aids that women invented during the 19th century. Following are single examples (all patented, all from LWP) in each group mentioned:

**Belle West,** Winfield, KS, Combined step-ladder and ironing board (1888)
**Ellen B. Viets,** Boston, Improvement in steaming and drying apparatus for shaping pantaloons (1873)
**Semele Short,** Cincinnati, Improvement in lace-curtain-stretchers (1873)
**Caroline Rosenthal,** Philadelphia, Improvement in reels for clothes lines (1873)

Possibly noteworthy here is **Ellen Viets,** who received seven patents for pantaloon and coat-shapers in just five years (1873–77, inclusive). She may be an early clothes-cleaning entrepreneur. In any case, her seven patents make her one of the more prolific women patentees of her day. As noted elsewhere, **Semele Short** showed three laundry inventions, including her curtain-stretcher, at the Philadelphia Exposition and won a Centennial Award (Warner/ Trescott, 117).

Since so few black women inventors are known to history, the special sleeve-ironing board patented in 1892 by **Sarah Boone** of New Haven, CT, is worth mentioning here as well. Padded and curved for its purpose, it is also hinged to its support (U.S. Pat. Gazette; Stanley/Zimmerman, 60).

*Cleaning machines/devices.* Carpet sweepers: **Anna S. Bissell** (1837–1934) did not, as Vare and Ptacek claim, invent the famous Bissell carpet sweeper of 1876. But she did help her husband build both the original model and the many sweepers sold before large-scale manufacturing began. She also ran the company for many years after her husband's death, and she did invent the "Little Helper," a toy sweeper that rescued the company from a slump in 1893 (Cleary, 13f; Hambleton, 54–6; Huhn, L-4/25/88; Kane, 163; *NCAB,* "Melville Bissell"; Vare & Ptacek, 37–8). The Bissell sweeper has survived the onslaught of the vacuum cleaner, and is now being touted as an energy-saver, sold in the Pacific Gas and Electric Company's One-Stop Energy Shop (PG&E *Progress,* May 1988).

**Sarah B. Stearns** of Duluth, MN, did, however, patent an Improved carpet cleaner that centennial year. In 1883, Stearns patented a Ventilating screen for windows.

At least three other women patented carpet-sweepers or attachments in the 1890s. **Martha B. Holden** of Chicago, for example, showed a carpet-dusting attachment for the standard carpet-sweeper at the Columbian Exposition in Chicago in 1892–3. This was a cloth-covered roller to be dampened with ammonia and attached to the front of a sweeper (or to a separate handle). As it rolled across the carpet, it picked up dust (Romney, "Women," 26). Patented in 1892, this device was called simply a Carpet-sweeper by LWP. According to Mary Logan (888), **Alice A. Pyle** of Richmond, VA, patented another carpet-cleaning attachment that same year; and two years later **Emma H. Raymond** of Grand Rapids, MI, patented a Carpet-sweeper.

Dustpans: Dustpans may seem insignificant inventions until one recalls how useful they are, even in today's mechanized households—and how much money is to be made from such small ideas as bobby pins, Liquid Paper®, and Post-It Notes®.[22] Dustpans are not machines, of course, but those with hinged or pivoting or adjustable parts, or wheels, must be counted mechanical. A woman recognized in her day for a lucrative dustpan invention was **Hannah V. Shaw** (b. 1851) of Lawrenceburg, IN, who patented her Dust-pan with pivoting bail in 1886. Ida Tarbell predicted in 1887 that the socially prominent Mrs. Shaw would probably make a fortune on her invention (355–6).

Born Hannah Virginia Fitch, she studied at Indiana Asbury University (Class of 1873), and was one of the founders of Kappa Alpha Theta. She married Archibald Shaw in 1873, and had nine children. She was still alive in 1920 (Cahill, L-8/16/83; DePauw University *Alumnal Record,* 1920:52; *History of Dearborn,* 905; Tarbell, 355–6).

About the same time, **Emma G. Nunn** of Vassar, MI, patented an Adjustable handle for a duster. Ten years earlier, **Elizabeth M. Page** of Philadelphia showed her large Dirt- or dust-catching pan on wheels at the Centennial (Warner/Trescott, 115). Unlike Mrs. Shaw's invention, this was apparently not patented. At least ten other women received nine dustpan (or duster) patents in the 19th century, but I have not determined whether these meet my criteria for being classed as mechanical. Suffice it to say here that the earliest of them was **Amelia B. Hoffman** of Roxbury, MA (1866), and the most noteworthy were **Emma L.** and **Mary A. Dietz** (1821–1918) of Oakland, CA, known for their "Cyclone" snowplow (see below).

Mop-wringer pails: Another mechanical cleaning aid is the so-called housemaid's pail or mop-wringer pail, a sturdy, often galvanized bucket fitted with a device for wringing out string mops. These floor-mop wringers sometimes resembled miniature laundry wringers, with a hand crank turning double rubber-covered rollers. I find five such patents granted to four women in the century, the earliest to the versatile multiple inventor **Mary**

---

22. Says Mozans, "Often what are apparently the most trivial inventions prove the most lucrative. Thus a Chicago woman receives a handsome income for her invention of a paper pail. . . . The gimlet-pointed screw, which was the idea of a little girl, has realized to its patentee an independent fortune" (351). Indeed, according to the *Scientific American,* by 1890 that gimlet-pointed screw has produced "more wealth than most silver mines" (Oct. 11: 225).

**P. Carpenter** in 1866.[23] This was the second of her 13 original patents, granted while she still lived in Buffalo, NY.

Window-washing aids: **Miss Marietta Tremper** of NYC showed her Window-washing machine at the Philadelphia Centennial in 1876 (Warner/Trescott, 118). LWP omits her, and I find no such patent in the surrounding years; her device may have been commercially exploited without patent protection.

At least 9 other women, including one from London, England, patented window-cleaners, and window-washing chairs, scaffolds, and ladders which, though not machines, are hinged and folding and thus mechanical. A tenth woman, **Anna Dormitzer** of NYC, received 12 patents between 1878 and 1891, at least 8 of which were for window-cleaning chairs or step-chairs. One of these, like **Amalie Schaefer's,** was combined with a fire escape. Two of Dormitzer's other patents, listed only as step-ladders, may also have been intended as window-cleaning aids. In sum, at least 10 women received 17 patents in this specialized area.

*Sewing aids.* The fourth large category of domestic labor-saving machines and devices lies in the area of sewing or garment-manufacture and repair. In addition to needle-threaders for sewing machines, women invented needle-threaders for hand sewing. Two fairly early ones were patented by **Sophia L. Mercer** of Washington, DC, and **Julia Foster** of Philadelphia in 1872 and 1876, respectively. Mrs. Foster had at least three 1876 patents altogether, including a Bill file and the Bed-dusting rack she showed at the Centennial (Warner/Trescott, 114).

A whole section could be written on pattern-drafting instruments and apparatus and garment-cutting devices. A few examples are **Mary Blauvelt,** Ithaca, NY, Dressmaker's marking and cutting gauge (1873); **Mary E. Sinnott,** Bakersfield, VT, Improvement in cloth-cutting machines (1877); **Eunice L. Crosby,** Plymouth, WI, Cloth-measuring and cutting machine (1884); and **Sarah A. Cooke,** Chicago, Machine for cutting and making garments (1892). Mrs. Blauvelt (b. 1832), wife of a blacksmith, showed her gauge at the Philadelphia Centennial (Warner/Trescott, 111).

Tracing wheels are related to this stage of garment manufacture, transferring the pattern to the cloth. **Martha E. Kellogg** of Flint, MI, and **Elisabeth C. Gefoert** of Brooklyn patented such devices in 1882 and 1886, respectively.

Also obviously mechanical are the following examples of buttonhole-cutters and gauges, the scissors, combinations of scissors with sewing-machine repair tools and the like invented by 19th-century women:

**Sarah H. Brisbane,** Fordham, NY, Improvement in scissors (1868)
**Margaret J. Stubbings,** Youngstown, OH, Improvement in combined scissors and tapelines (1873)
**Eliza E. Emery,** Boston, Improvement in gauges for marking cloth for cutting (1873)
**Sarah L. Faucett,** NYC, Improvement in scissors for use with sewing machines (1874)
**Phebe L. Carter** and James W. Rhoades, Clyde, OH, Combined shears, button-hole cutter, etc. (1875)
**Elizabeth Wiggins,** Brooklyn, NY, Improvement in scissors-gauges (1876)
**Clara Rogers,** New Orleans, Improvement in scissors and shears (1876)
**Mary T. Lotz,** Loveland, OH, Button-hole cutter (1881)
**Alice L. Campbell,** Chicago, Buttonhole-scissors gauge (1891)

---

23. The other three women were **Emma C. Wooster** of NYC (1875); **Lavina Grube** of Liberty Mills, IN; and **Eliza A. Wood** of Philadelphia (1889, 1891). **Jennie E. Zeiser** of Schenectady patented a Scrubbing pail in 1892, which may or may not have had a wringer attached.

**Louise Austin,** Whatcom, WA, Pinking shears (1893)
**Abby S. Vose,** Providence, RI, Buttonhole cutter (1895)

**Carter** and **Rhoades'** device was manufactured and sold, and is mentioned—and illustrated—in a 1980s article on women inventors (Amram, "A Woman's Work," 78). It is, however, omitted from LWP. **Stubbings** had at least one other invention to her credit in 1873, an Improved coffeepot. **Clara Rogers** showed her scissors—incorporating five repair tools for sewing machines—at the Philadelphia Centennial. Praised in *New Century for Women* as solving the problem of disappearing repair tools, these Machine Scissors combined screw-driver, glassfoot remover, three-inch measure, wrench, and needle-straightener (*NCW* 135, 136; Warner/Trescott, 116).

**Louise Austin**'s Pinking shears of 1893, although not the first pinking device, may have been the first pinking *shears,* as earlier patents seem to refer to a pinking *machine* or *device* or *tool.* Austin, the sole patentee, lived in Whatcom, WA.

Perhaps most noteworthy of this group is **Abby Vose,** whose buttonhole cutter (see below) may well have been used in industry as well as in the home (*Patent Record,* Dec. 1899: 9; "Women As . . . ," *Literary Digest,* Jan. 6, 1900: 14, 15; *Scientific American Suppl.* #1325, May 25, 1901: 21241; G. Walsh). Each of the claims for this invention describes it as a "*machine* for cutting button-holes" (U.S. Pat. Gazette, July 30, 1895; emphasis mine).

*Dressing aids/devices.* Fifth and last are mechanical improvements in shoe and garment-fastenings, and aids to donning complicated garments. Nineteenth-century female clothing was so elaborate that dressing and undressing involved some effort on the part of the woman or her servants—or some technology. Gloves, shoes, and dresses (both in the body of the dress and on the sleeves) bore literally dozens of buttons, snaps, or hooks, all of which had to be fitted into tiny stiff loops, or accurately snapped—often, in the case of dresses, down one's back). Women had to dress the children, whose clothing, at least for some occasions, was equally elaborate. As might be imagined, therefore, garment-fasteners and aids to fastening in the form of glove-buttoners, shoe-buttoning hooks, etc., were patented in some numbers. Although I will give only a few examples, it should be noted that hooks and eyes, safety pins, snaps, and spring clasps are all mechanical, and that at least one woman's glove-buttoner is supposed to have made her a fortune ($5,000 a year in 1899; G. Walsh). In typical Victorian style, this reference is left anonymous, but at least two women patented glove-fasteners in the 1870s—**Letitia Ferris** of San Francisco in 1873, and **Louisa Bonesteel** of Rochester in 1877—and **Ida E. Mushette** of Oakland, CA, patented a Combined shoe- and glove-buttoner in 1894.

According to Phebe Hanaford (623), the "Self-fastening button" was a woman's idea, though patented and shown at the Centennial under a man's name.

**Ann** or **Anna McKevit** of Chicago received four patents in all, three in 1883 and one in 1890. Two of her 1883 patents were for Button-fasteners. One of these included a button improvement, and her third 1883 patent was for a "Separable" button. McKevit also patented a Combined protection case and drip ferrule for umbrellas (1890). Women patented 28 other dressing aids between 1868 and 1897, including **Harriet W.R. Strong,** Oakland, CA, Hook and eye (1884).[24]

*Dress-* or *skirt-elevators:* Skirt elevators may seem a surprising inclusion here. Before dismissing these devices as either nonmechanical or insignificant, however, we should note that they functioned to transmit force over some distance, sometimes with a change

---

24. This is the same Harriet Strong—inventor, entrepreneur, and civic leader—who invented an important dam and reservoir system, among other things.

of direction, using long bands or strings and pulley-like parts. We should also try to imagine walking through dusty, muddy, or slushy, unpaved streets in floor-length skirts that weighed 15–20 pounds dry. A woman needed both hands to hold up such skirts. Unfortunately, she might need to carry an armload of packages or hold a child's hand at the same time. In such a predicament, a skirt elevator was a godsend. Several different types were invented, and some were patented, in the late 19th century.

I have chanced upon only one British patent—**Jeanette Fleurman**'s Patent #1508 for 1881, for [a means of] Raising and supporting ladies' dresses—but this does not mean British women were inactive in this area. At least one Englishwoman was among the 10 women who received 11 American patents for skirt elevators, 1864–92:

**Zera Waters,** Bloomington, IL, Improvement in ladies' skirt-lifters (1864)
**S. A. Moody,** NYC, Improvement in ladies' dress-elevators (1866)
**Margaret H. Bergen,** Brooklyn, two patents: Improvement in skirt-elevators (1872) and dress-elevators (1874)
**Mary Dewey,** New Albany, IN, two patents: Improvements in dress-elevators (1872, 1874)
**Elizabeth E. Norton,** Bridgeport, CT, Improvements in skirt-elevators (1874)
**Emily Stewart,** NYC, Improvement in dress-elevators (1874)
**Phoebe G. Baker,** S. Yarmouth, MA, Improvement in skirt elevators and adjusters (1876)
**Nannie C. Green,** Brooklyn, NY, Skirt adjuster (1881)
**Annie Lange,** Leicester, England, Skirt-elevator (1891)
**Josephine Asten,** Washington, DC, Skirt-elevator (1892)

None of these women seems to have won any lasting recognition for her invention. It may be that the best-selling dress elevator of the century was unpatented: **Ellen Demorest** was not only an inventor and a canny entrepreneur, but the principal arbiter of 19th-century American fashion, with her adaptations of French styles. She is one of the inventors of standardized tissue-paper patterns—the main inventor, according to Ishbel Ross's biography.[25] She is also credited with inventing (but apparently not patenting) the Imperial Dress Elevator, with weighted strings to raise or lower a skirt and keep it clear of street or sidewalk (Bird, 56), which she probably sold in her well-stocked emporiums and by mail.

Many further examples could be given here. The study of women's contributions to their own domestic labor-saving technology is still in its infancy, but some idea of that contribution should be deducible from the foregoing. I will close this section with one striking invention: **Elizabeth Howell**'s Self-waiting table of 1891. Howell's table is not called a Lazy Susan in the Patent Gazette, but Caroline Romney's article on women inventors at the Columbian Exposition of 1893 makes clear that such it was: "The Self-Waiting Table . . . has a revolving disc, located in the center, ten inches back from the edge, upon which all of the edibles are placed, each person serving himself, by simply

25. **Ellen Butterick** and Ebenezer Butterick are often credited with inventing standardized paper patterns, originally for men's clothing, though the Butterick patterns—made of stiff paper—were not marketed commercially until June 16, 1863, whereas the Demorests had introduced theirs in the 1850s. Within the Demorest establishment, Ellen Demorest's sister **Kate Curtis** and her husband William J. Demorest deserve part of the credit for this invention, which included not only standardized sizes and a mathematical system for cutting the patterns out, but the use of tissue paper (which allowed mass-production, cutting many copies of the same pattern at once; reduced bulk for mailing and storage, etc.), and the flat-folding in envelopes decorated with colored plates of the fashions like those in *Godey's*. Furthermore, the pattern pieces themselves were color-coded—violet for a bodice, gray for a sleeve, etc.—for quick recognition of the various parts even by those who could not read (Bird, 53–5; *DAB,* "Ebenezer Butterick"; I. Ross, 20–3).

turning the disc until the article he wishes is in front of his plate" (26). *Lazy Susan* does not appear in the *Oxford English Dictionary,* which suggests that the term if not the device was an American invention. According to Dolores Hayden, who discusses inventiveness in the American communal societies of the 19th century (*Grand,* 48), "Members of the Oneida Community produced a lazy-susan dining-table center," among other things, but she omits the date and the inventor's gender.

This useful device has been adapted not only to home dining tables but to family-style restaurant tables, and to storage cupboards for easy access to small items. Howell lived in Maryville, MO, at the time of her self-waiting table patent as well as her other two patents, one of which—her Accordion-plaiting (-pleating) machine of 1890—was also mechanical. Romney calls her Mrs. Howell, and she was probably the Mrs. E.E. Howell of the 1906 Maryville directory, living at 301 E. Third, and proprietor of the Howell Millinery Company on S. Main Street.[26] D.J. Howell, listed at her address and identified from other sources as a 1905 Maryville High School graduate, may well have been her son. A history of Nodaway County, where Maryville is located, mentions Mrs. Howell's millinery shop, located on S. Main, as employing 3–4 women (Airy, 125–8).[27]

## Manufacturing and industrial machines

A seldom-acknowledged subcategory of women's mechanical achievement is that of machines used in manufacturing and other industries.

*Early examples.* Women received patents in this area before the 19th century. **Sarah Jerome,** for example, received two British patents for an Engine for cutting timber into pieces for making band boxes, and an improvement on the same. These 1630s patents make her the earliest recorded British woman patantee (Woodcroft). Unfortunately, there is some question about Jerome's actual role in these production saws. According to the British patent records the inventor of the earlier of the two was one John Lane who died in debt to Sarah Jerome, and she hoped to recover her money by patenting his invention (Brody, L-3/18/85). On the second patent, however, issued to her and a man named Webb, there is no mention of any debt. Though Webb's name comes first on the patent, and Judit Brody suggests that Webb invented the improvement (L-4/16/85), Jerome should not be discounted as co-inventor without some further research.

Jerome's first patent raises interesting questions. How was the debt incurred, for example? Women often have an invention idea, and develop it to a point, but then need the help of a chemist, blacksmith, machinist, engineer, draftsman, or lawyer—usually male—to bring it to patentable form. Relations between inventor and expert can become strained, and the men's names may end up on the patents. **Julia Butcher**'s draftsman, for example, became her co-patentee just for doing the drawings of her invalid-exerciser (Chap. 2)—for which she also paid him. This was her lawyer's advice (Butcher, L-11/5/81).

Suppose for a moment that Sarah Jerome has the original idea for the saw. She pays or contracts to pay for all necessary tools and materials, as well as the patent costs and fees,

---

26. LWP gives her middle initial as *B* on her 1890 patent, but *E* on her other two (1891 and '94), whereas the Gazettes give *H* on the 1890 patent and *E* on the other two. Patentees might use different middle initials on separate patent applications if they had two middle names (e.g., married women with given middle and maiden names). Since the variation here refers to the same patent, however, either the LWP compilers or the Gazette compilers, or both, must have erred.

27. Margaret Kelley, Coordinator of the Heritage Collection, Maryville Public Library; and Martha Cooper, a local historian and researcher from Maryville, kindly provided this information (Kelley, L-3/24 and 4/2/87).

but agrees to let Lane appear as inventor if he will create the working model and secure the patent. Lane contributes no money. He succeeds in building the "engine," but then dies before perfecting it or before the patent is granted. Jerome might well consider that he died in debt to her, and she would have had every right to apply for a patent or complete an application process Lane might have started.

In the 18th century, one **Elizabeth Taylor** received British Patent #782 for a "Set of engines, tools, and instruments for making blocks, sheaves, and pins" (Woodcroft; C. Smith, *WI*/1).

One of the first ten women patentees listed by LWP is **Phoebe Collier** of Catskill, NY, who patented an Improvement on the machine for sawing wheel-fellies (wooden wheel rims) in 1826. Collier's listing is another example of the difficulty in locating women's mechanical inventions from a quick reading of LWP. From the LWP entry—"Wheel-fellies, sawing"—even informed readers might consider that Phoebe Collier improved an earlier *process* or *technique* for sawing such rims, rather than a *machine* for doing so.

Before mid-century, a French woman inventor is connected with an important type of industrial saw; the bandsaw. A bandsaw is a narrow bit of metal made into an endless strip, with teeth on both edges. Like the circular saw, it employs rotary motion. However, the bandsaw's narrower cut saves wood. Bandsaws also saw faster and use less power. Waldemar Kaempffert notes that the French have always claimed they invented the bandsaw, although William Newberry of England has a strong claim with his 1808 patent solving the problem of welding the ends of the band together. Among the French inventors who worked on the bandsaw Kaempffert lists **Mademoiselle Crépin** of Paris. Though of high social standing, she "did not care for a life of fashion, and devoted herself to mechanical pursuits." In 1846 she patented a bandsaw. A French saw-manufacturer named Perin, himself an inventor of saws, realized its merits and purchased her patent rights. He exhibited a bandsaw, which Kaempffert calls "a composite invention," presumably meaning that it incorporated features of Crépin's invention, at the French International Exhibition of 1855 in Paris. Kaempffert gives no technical details on Crépin's bandsaw. However, he does say in passing that the saw was "among her inventions," a casual reference to multiple creativity that in itself speaks volumes.

Kaempffert's Mlle Crépin is **Anne Pauline Crépin,** a *découpeuse* (woodcutter or -carver) in the Faubourg Saint-Antoine outside Paris, who took out a French patent for a circular saw on August 29, 1846. This is the present-day bandsaw with a thin soldered (or welded?) blade. It revolutionized the technique for thin-cutting wood in general, and particularly for cutting veneers.[28]

Considerably later in the century **Henriette Tuyssuzian** received British Patent #2835 of 1881 for Endless or band saws.

Many 19th-century manufacturing machines were belt-driven. That is, the power to run them was transmitted from its source to the moving parts by a belt. Such belts, then, become an important part of manufacturing, and since they need to be endless when in operation, a strong, safe, and long-lasting means of joining their two ends also becomes quite important. Such an invention was that of **Maria Schultze** or **Shulze** of Berlin, who received British Patent #5578 of 1880 for "Uniting ends of machine-driving belts." She

28. Alphandery's French, from the *Dictionnaire des inventeurs français,* is:

[E]lle prit le 29 août 1846 un brevet pour une scie circulaire qui révolutionna la technique et qui était l'actuelle scie à ruban à lame mince soudée, utilisable grâce à des guides en bois. Cette scie facilitait considérablement le débitage du bois de placage.

I am grateful to Belinda Hopkinson of UC Berkeley for this reference.

then took out a U.S. patent (1881), for a Belt connection. Her method combined sewing and gluing.

*Straw-weaving, hat-manufacturing.* As already mentioned, the first patent granted to a woman by the United States Patent Office went in 1809 to **Mary Dixon Kies** of S. Killingly, CT, for what may well have been a manufacturing machine: a straw-weaving device. If Kies' invention was indeed a machine rather than or as well as a process, then the invention patented twelve years later, in 1821, by **Sophia Woodhouse Welles** (or Wells) and her husband Gurdon of Wethersfield, CT, must also have been mechanical, for her improvement is said to have superseded Kies's (LSS: Sherr & Kazickas, *AWG*). Welles sent one of her redtop-hay-and-spear-grass bonnets to the Society of Arts in London in 1821, and won "the large silver medal" and 20 guineas (Tharp, 6: another source [Klein & Klein, 304] speaks of a gold medal). Rather surprisingly, LWP omits Welles's patent (cf. LSS, 1821).

**Lucy Burnap** of Merrimack, NH, created a further improvement in 1823. This invention, adopted by hat manufacturers, made possible a big business in women's straw hats. By 1828 **Henrietta Cooper** had patented a process for whitening the leghorn straw (Klein & Klein, 304–5; LWP).

However, this is getting ahead of the story.

As early as 1798, **Betsey** or **Betsy Metcalf** (1786–1867) of Providence, RI, started the American straw-weaving industry by reinventing the braid and weave of a Dunstable bonnet she saw in a Providence store window. This she did at age twelve, and without unraveling a sample of the braid. She and her aunt also devised a way to bleach the bonnet-straw by holding it over hot coals sprinkled with sulfur (Tharp, 4). Often told that she should have taken out a patent, Betsy replied that she did not want her name sent to Congress, and that she had no regrets about freely teaching others her braiding technique. She wrote her own account of the incident at age 72, and made a replica of the original bonnet for posterity.

Betsy's portrait, several of her bonnets, and other artifacts are on exhibit at the Rhode Island Historical Society and at the Westwood, MA, Library and Historical Society. (Betsy married Obed Baker and moved to West Dedham, MA, now Westwood). Until the 1960s her grandson Ernest Baker still lived in the Baker house in Westwood, which is now open to the public (E. Baker, 59–61; Bowen, 71–3ff; Dall, *College;* Gage/1: 12–21; A.N. Meyer, 278–9; H. Miller, "Betsy . . ." and L-1/16/80; Westwood, MA, Historical Soc.; Tharp, 2–5, 6).

It is difficult for us today to imagine the scale and commercial importance of the 19th-century American straw-braiding and -weaving industry. From Betsy Metcalf's circle of friends and neighbors just before 1800, the manufacture mushroomed. Children took braiding to school; women carried bundles of straw on journeys and even to social occasions. Eli Whitney's mother was a bonnet-making entrepreneur (Tharp, 3). In 1810, only twelve years after Betsy's first bonnet, the estimated value of the straw bonnets manufactured in Massachusetts alone was over half a million dollars. As of 1870, 10,000 people worked in the Massachusetts bonnet industry, producing 6 million hats and bonnets a year, or four hats for every resident of the state. And Massachusetts was not the only hat-manufacturing state. Pennsylvania produced another $600,000 worth of hats and bonnets yearly at the same time (Gage/1: 11), for example, and the entire population of the United States in 1870 was under 40,000,000 (Stat.Abstr.U.S.: *EB*/11).

According to Betsy Baker, the straw-splitter used in this industry was a woman's invention. For her first bonnet, Betsy used her thumbnail to split the straw (later she owned two splitters, inherited by her grandson). Unfortunately, Baker did not name this woman inventor (Tharp, 7, 5). Sewing the braided straw together was part of the hat-making operation; and some of the sewing machines mentioned above were probably used in the hat-making industry. One of **Mary P. Carpenter**'s inventions, for example, was a machine for sewing straw braid (#171,774; 1876).

These early straw-weaving inventors received national recognition and esteem, shown directly by two First Ladies. Dolly Madison sent a letter of congratulation to Mary Kies (Larned, 432; Sherr & Kazickas, *AWG*, "S. Killingly"; H. Warner), and wore one of Sophia Welles's hay bonnets in support of American industries during the embargo against British goods (1812; Welles evidently created her invention some years before patenting it). Louisa Adams, wife of John Quincy Adams, ordered several of Welles's Wethersfield hats while she was First Lady in the late 1820s, and John Quincy praised them as "an extraordinary specimen of American manufacturing." Welles's house still stands on Main Street in Wethersfield today, and several examples of her bonnets are on display at the Old Academy Museum there (Sherr & Kazickas, *AWG*, "Wethersfield"). Such recognition makes LWP's omission of Welles's patent all the more surprising.

*Textiles.* For millennia women have been acknowledged experts in cordage and textile production, inventing cord-twisters, spindles and spindle whorls, looms and loom weights. Their early achievements in this area are fairly well acknowledged and will not be reviewed here, except for a few recently discovered or little-known items: the date of the first weaving has now been pushed back to c. 30,000 BC (evident on the famous Venus of Le Spugue); knitting frames were used by North African nomads ("Eureka," 22); and the spinning jenny may have been invented by a woman.

Modern mechanized weaving, however, is usually attributed to men. As with agriculture, women invented weaving but men took over and professionalized it, and women's contributions were buried. **Mme Jacquard** is a latter-day case in point. An early American example is **Lydia Hotchkiss Crosby** (1795–1878). This remarkable woman was descended through her parents Ambrose and Sarah Hotchkiss of Guilford, CT, from old "Connecticut Yankee" families. She evidently married young, for when her husband Orrin Crosby died in 1818, leaving her with three young children to support, she was only 23. As her granddaughter, Clara Louise Kellogg, tells the story (Kellogg, 2),

> As for my maternal grandmother—she was a woman with a man's brain. A widow . . . with no money and three children, she chose, of all ways to support them, the business of cotton weaving; going about Connecticut and Massachusetts, setting up looms—cotton gins they were called—and being very successful.

This granddaughter, the first really famous U.S. opera star, wishes the world to know that her grandmother was a fine musician (other sources call her a talented violinist) and a sought-after teacher of music theory. What Clara does not say is that this loom or (en)gin(e) Lydia Crosby was installing in cotton mills in two important textile states *was her own invention.* Unfortunately, neither of the two 19th-century sources of this information (Stowe, ed., 360; Hanaford, 572) tells exactly what the invention was; and even *Notable American Women* merely calls it "a mechanical device used in many New England cotton mills" (*NAW*, "Clara Louise Kellogg"). However, had the invention not been indeed, as Phebe Hanaford calls it, "a very valuable invention," she could scarcely have succeeded in getting it used in so many mills. Harriet Beecher Stowe says, in fact, that she "superintended the erection" of her invention in "most of the large cotton mills. . . ." I have located no patent for Lydia Crosby. This could mean that she allowed a male relative or a patent attorney to take out the patent. In this case, a possible if very early candidate would be the loom patent issued to Richard Crosby of Newark, NJ, in 1812

(Ciarlante, L-11/24/86). It could also mean that her patent was not restored after the Patent Office fire of 1836, or that she never patented her invention, preferring to keep it a secret and sell the machines instead.

Lydia was surrounded by inventors. Several of her Hotchkiss ancestors, her first husband Orrin Crosby, her second husband Charles Atwood, and her son-in-law George Kellogg were all inventors. No further inventions are mentioned for Lydia after her marriage to Atwood, which made her financially comfortable again. However, she may well have played a role in some of the many inventions usually credited solely to Atwood after that date—a new way of making steel pens, the technique of making German silverware, a hook-and-eye machine, and a simple machine for making jack or scale chains, among others. Orcutt and Beardsley, in their history of the Town of Derby, CT, 1642–1880, call Lydia "by nature an artist in mechanics, whose judgment about [Atwood's] inventions had great weight" (356–7). Charles Atwood died of typhoid fever in 1856, widowing her for the second time.

Lydia Hotchkiss Crosby Atwood died in Perth Amboy, NJ, at the age of 82 (Family Bible records of Ambrose Hotchkiss, CT State Lib., Hartford; Hanaford, 572; T. Hopkins, 1091–2; Stowe, ed., 360; *NAW*, "Clara L. Kellogg"; Ransom, L-7/23, 9/15, 11/30, and 12/1981, and 4/6 and 7/8/1982; Derby, CT, Hist. Soc., L-11/14/ 1981; Orcutt & Beardsley, 356–7).

An important 19th-century textile-manufacturing invention was the loom for weaving seamless bags, used to manufacture pillow cases. It was **Lucy** (or **Hannah**?) **Johnson,** a skilled hand-weaver of Rhode Island, who in 1824 figured out, after long experimentation, how to weave seamless pillow cases, and wove seven pairs. She exhibited these cases and won premiums at several fairs. One sample was finally hung in the Rhode Island Society for the Encouragement of Domestic Industry. Unfortunately, Johnson never patented her invention, and died impoverished near Providence at 78. A man adapted her work to power looms and "reaped a pecuniary harvest," as the *New Century for Women* put it in 1876 (June 24: 51; Hanaford, 633–4; M. Logan, 883–4). Interestingly enough, the patent records show two later woman inventors in this specialized area: **Addie M. Pont** of Warwick, NY, patented a loom for weaving tubular fabrics (1880); and **Harriet E. Hinchcliffe** (née Emerson) of Rockford, IL, received two 1895 patents for Tubular knit fabric.

In 1887 **Mary C. Crompton** (with Wyman H. Crompton) received three British patents for looms, on one of which, #7313 of June 1, she appears as Administratix for G(eorge) Crompton. Crompton is a famous name in British and American textiles, Samuel Crompton (1753–1827) having invented the spinning mule that combined the operations of Hargreaves' jenny and Arkwright's frame. George Crompton (d. 1886) came to the United States in 1839. Son of another inventor, William Crompton, and perhaps related to Samuel as well, George is credited with improving "practically every part of the loom," adding 60% to its productive capacity, and also inventing "many new textile fabrics." He held 200 patents. Mary Christina Pratt Crompton was George's wife, as her appointment as administratrix suggests (*DAB*/1930). More research is needed here to clarify Mary Crompton's contribution to the family reputation.

Charlotte Smith credits women with three important 19th-century textile-producing and related machines: a spinning machine with 12–40 threads, a rotary loom doing three times the work of an ordinary loom, and a wool-feeder and weigher described as "one of the most delicate machines ever invented, of incalculable benefit to the wool manufacturer." (*WI*/1). Unfortunately, Smith does not name these three inventors. On the spinning machine, she may have been referring to **Louise D. Croes** and **Margaret F. Rodenbach** of Roulers, Belgium, who patented a Spinning machine (with J. Lapin) in 1892 (U.S). Smith could have had advance knowledge of this patent-in-process through her son who worked at the Patent Office. Croes also patented an attachment for this machine in 1893 with Lapin and another male inventor. Alternatively, though Smith usually focused on contemporary patentees, she could have had in mind **Margaret Hulings** of Indianapolis, who patented an Improvement in spinning machines in 1865.

Other 19th-century spinning inventions by women are **Sister Tabitha Babbitt**'s improved spinning wheel head and the (means for?) Accelerating spinning wheel heads patented by **Diana H. Tuttle** of Williamson, NY (1824). One of the sewing-machine improvements mentioned above was a spinning attachment.

**Catherine Greene**'s possible role in the invention of the short-staple cotton gin has already been discussed, as have the gin improvements patented by **Jane C. Osborn** of St. Louis (1889) and **Susie B. Hall** of Little Rock (1892). **Eliza Jane Harding** of Denver, CO, patented a Depurator, or machine to cleanse cotton before spinning, in 1887.

An early patent (1834) granted to **Margaret Gerrish** of Salem, MA, for Manufacturing cloth from milkweed fibers (*Asclepias syriaca*) appears in the patent records under Factory Machinery (LSS).

LWP lists Florence Veerkamp of Philadelphia, who patented an Improved machine for weaving cords and cord-coverings (1864), as well as other machines, and Philancy D. Tifft of Eagleville, OH, who patented an Improvement in warp stands for looms (1870); but these inventors are actually male.[29] **Mary Gordan** invented rods for drawing in warp threads (Centennial Exhibit, Smithsonian); and **Cornelia E. Hill** of Floyd, IA, patented a Machine for winding warp (1891). **Caroline Wimpfheimer** of Philadelphia showed her Loom for making hair ribbon at the Centennial. I have not located a patent. She was a widow who employed 60 people in her ribbon-making enterprise (Warner/Trescott, 119).

**Sally M. Levy** of Milwaukee and **Alice M. Perkins** of Lacrosse, WI, patented knitting machines in 1885 and 1892, respectively; Perkins' machine was for circular knitting.

*Garment industry.* Women also contributed mechanical inventions to the garment industry in the 19th century. Indeed, so firmly was garment-making linked with women in the stereotypical view that, as already noted, the LWP compilers of the 1880s and '90s classed **Nell O. Westburg**'s Band-cutter and feeder under "Sewing & Spinning" whereas it actually belonged to a rather complex threshing machine! **Irene Hoyez** of Stephen, MN, also patented a Band cutter (1894), which may actually belong to a sewing machine.

As mentioned, **Helen Blanchard**'s sewing machines were mostly industrial. Research in the business and patent-assignment records might show that several others of the listed machines were used in factories as well.

Between 1877 and 1892 at least three women patented cloth-cutting machines, one of which measured the cloth at the same time: **Mary E. Sinnott** of Bakersfield, VT (1877); **Eunice Lodeski Crosby** of Plymouth, WI (1884); **Sarah A. Cooke** of Chicago (1892). Other garment-related inventions (a small sample among many), probably the work of enterprising dressmakers or seamstresses, are **Annie A. Waldie**'s Combined cutting gauge and marker (Philadelphia; 1893), **Nevada Steen**'s Gauge for bias cutters (Corpus Christi, TX; 1894), and **Marie Horn**'s Garment-measuring and drafting apparatus (both 1894). Since Horn was from Germany, her seeking an American patent indicates

---

29. Before Florence Nightingale's fame, the given name Florence is often male, especially with Dutch surnames. This is one of relatively few instances where LWP mistakenly includes a name. Errors of omission are far more frequent.

considerable commercial potential for her machine. **Louise Schaefer** of Oneida, NY, patented a Method and apparatus for making patterns in 1894.

Most garment production in the United States in the 19th century was still done at home; thus there is inevitably some overlap with the sewing aids listed above. **Abby Vose,** for example—whose buttonhole cutter was featured in the *Patent Record* for March 1899 (G. Walsh) and December 1899; in a *Literary Digest* article of January 6, 1900 ("Women As . . .": 14, 15), and in the *Scientific American* a year later (Suppl. #1325, May 25, 1901: 21241)—has already been mentioned. The *Literary Digest* includes Vose anonymously as the creator of one of a group of profitable inventions patented by women, 1895–9; "A Rhode Island woman invented an improved buttonhole-cutting machine that measures the distance between the button holes automatically, with much profit and convenience to garment-makers" (14). Thus her device was evidently not only manufactured, but adopted in the garment industry. Each of the claims for this invention describes it as a "*machine* for cutting button-holes" (U.S. Pat. Gazette, July 30, 1895; emphasis mine). Vose may actually have worked in a garment factory, as she is the first example appearing after the following statement ("Women As . . ."):

> . . . a large proportion of the more successful women inventors are those who have had only medium or limited educational advantages, but have been daily toilers in the various lines of industry.

*Candy-making.* Food is usually processed rather than manufactured, at least before the 20th century, and food-processing falls under the heading of Agriculture in my scheme. However, candy is a special case, arguably not even food—though in the ancient Near East it was food for the gods—and usually thought of as manufactured. A noteworthy invention by a woman in this field is the eccentric gauge for cutting candy lozenges devised by **Mrs. Robert Shields** and shown at the Centennial in 1876. Mrs. Shields evidently did not patent her invention, but sold the machine or rights to it, and saw it installed in several large candy factories (Warner/Post, 165; Warner/ Trescott, 117). **Emily Rochow**'s centrifugal sugar machine was probably for producing spun sugar or cotton candy (Brooklyn, NY; 1876). If cotton candy, as some sources maintain, was first introduced at the St. Louis World's Fair of 1904—where is was called "Fairy Floss Candy"—and if the necessary machine was not patented until 1899 (by two men), then Rochow's machine could have been for producing icings or a local, unnamed, confection. On the other hand, especially since the 1904 confection is not actually called cotton candy, perhaps we need not rule out a Brooklyn precursor of the St. Louis delight, and a female-invented precursor of the gasoline- and electric-powered machines.[30]

In 1897 **Edith E.L. Boyer** of Dayton, OH, patented a Machine for making caramels with her husband, Israel D. Boyer. Edith Boyer, a versatile inventor of both machines and mechanical toys, is more fully discussed below.

*Paper and paper products.* Nineteenth-century women contributed at least three interesting inventions to the paper-products industry: **Mary A. Martin**'s scam for

---

30. Indeed, spun-sugar confections may date back to Roman times, the art lost during the Middle Ages and later rediscovered. Henry Weatherly's *Treatise on the Art of Boiling Sugar* (1884) describes sugar-based confections from around the world. He mentions "fine silken threads" made by spinning sugar from forks, but these were evidently a garnish for a caramelized-sugar delicacy (Leitner, 35). However, spinning sugar by hand must have been tedious work. Candy-making in general did not become a major industry until machines were invented for making the various types of candy ("Hey . . ."; Leitner, 35). Another possibly pertinent term in the patent records is "silk candy."

wire-cloth for paper manufacture (Springfield, MA; 1882); **Rebecca Sherwood**'s Improvement in reducing straw and other fibrous substances for the manufacture of paper pulp (Fort Edward, NY; 1864); and **Anna M. Zimmer**'s Improvement in wood-grinders for paper pulp (Elkhart, IN; 1875). **Margaret Knight**'s best-known invention was the one that made possible the manufacture of square-bottom paper bags. She held three patents on this machine, all told.

*Containers/bottling/packaging.* As early as 1819, **Sarah Thomson** received British Patent #4370 for a Machine for cutting corks (May 15). **Amalia Donally**'s bottling and labeling machines are good American examples here. In the years 1891–3, Donally received five patents for bottling apparatus—a Bottle-filling apparatus, a Bottling apparatus, a Bung for bottlers, and a Labeling machine, plus a reissue of her first patent, #445,882 (1891), for bottle-filling apparatus. Both this first patent—omitted by LWP— and its reissue were assigned to the Donally Manufacturing Company of New York City.

**Maria E.B. Miller** of Omaha, NE, patented an Improvement in machines for filling and corking bottles (1879). One of **Mary P. Carpenter**'s many inventions was a barrel-painting machine, and **Ina C. Davis** of Newark, NJ, patented a Machine for printing the peripheries of spools and other circular bodies (1891). Carpenter's machine, incidentally (#174,477 of 1876), was omitted from LWP. **Maria E. Beasley** of Philadelphia, discussed below with the machine specialists, received at least seven patents for Barrel-making and related machines between 1881 and 1891. In 1886 **Emma Billette De Villeroche** of Concarneau, France, received a U.S. Patent for the manufacture of metal cans and boxes. And in 1890 **Nettie Henry** of St. Louis patented her Excelsior-cutting machine. **Edith Boyer** of Dayton, OH, patented a Wrapping machine in 1890.

Late in the century (1898) **Eureka C. Bowne** of Trenton, NJ, already known for her street-sweeper (see below), patented a thumbscrew device to be incorporated into the neck of bottles, for breaking them and thus preventing their being refilled. If, said the *Crockery & Glass Journal,* Mrs. Bowne "has really struck the key of this bottle business, it will be more than a remarkable invention . . ." (Sept. 1, 1898 : 21).

*Leather-processing and shoe-making.* In 1884, **Marie Garnier** of Paris, France, received a U.S. Patent for a Leather-dressing machine. In 1890 and 1894 **Cecile Puech** of Mazamet, France, received U.S. patents for defurring or dehairing animal skins. LWP specifically describes the 1894 patent as a Process of and apparatus for unhairing skins. Puech's inventions probably had some commercial importance since she took the trouble to get American patent protection. She may or may not be the inventor referred to in the *Literary Digest*'s turn-of-the-century article "Women As Inventors" (14–15): "A woman has just perfected a valuable apparatus for removing wool from skins by electricity, showing that women are quick to adapt the modern facilities of the age to practical purpose. . . ."

The shoe-making industry was enormously important in New England's economy during the late 19th century. Though women are seldom connected with this industry, several examples of women who invented shoe-manufacturing machines or parts therefor can be given. **Margaret Knight** invented at least four sole-cutting machines, and **Maria Beasley** patented a Machine for pasting shoe uppers. **Mary A.C. Holmes** of Newport, RI, patented three improvements in Machines for channeling and pricking the soles of boots and shoes (1879–81). **Ellene Alice Bailey** of St. Charles, MO, and New York City had some 14 inventions to her credit, the best known of which was her "Dart" needle for sewing on shoe buttons, first patented in 1884. The Dart needle worked for machine-sewing of other buttons as well, but was primarily a Shoe-button needle (Malotke *et al.*,

3). Bailey patented improvements on this needle in 1886 and 1888. **Julia R. Wood** of Foxboro, MA, patented a Lasting and sole-laying machine (1893).

*Printing and publishing.* One of the more noteworthy mechanical inventors in this field was the unnamed woman inventor of Syllable type with adjustable cases and apparatus mentioned by Charlotte Smith in the *Woman Inventor* in 1891. **Jane Austen** and **A.C. Semple** both invented machines for cutting books and trimming pamphlets, presumably in the 1870s. Possibly shown at the Centennial, their machines now appear in the Smithsonian's "1876" Exhibit (Weimann, L-1/28/83). **Elizabeth** and Edward **Cheshire** of Cincinnati, OH, patented a Book-sewing machine and a book-sewing needle (both 1885). **Fannie K. Crane** and Harry Bradshaw of Topeka, KS, patented "an apparatus for feeding paper in single sheets to any machine in which paper is to be operated upon by printing, ruling, folding, or otherwise . . ." (1891). **Mary C. Walling** of Washington, DC, patented a Stamping machine in 1885. This is a machine for printing or stamping patterns on fabric rather than on paper, but the technologies overlap.

*Wig-making.* An early American patent in this field was **Lydia A. Seago**'s Improvement in machines for weaving hair for wigs (1869). **Emma Hough** of Brooklyn, NY, patented two Improvements in hair-heading machines (1875). **Ella J. Crosby** of Sabula, IA, patented another hair-heading machine in 1876.

*Metal-working/metal articles.* One of the few American women known during her lifetime as a professional inventor was **Mary Jane Montgomery** of New York City. Among four known patented inventions the one that concerns us here is an Improved apparatus for punching corrugated metals (1866).

Noteworthy, especially if she indeed held only four patents, is the recognition Montgomery received. She is mentioned both by Ida Tarbell in the 1880s (356) and by the Shaker *Manifesto* of July 1890 (164). The *Scientific American* says of her in 1903, "There seems to be one professional woman inventor in America and her name is Miss Montgomery." She may well have several *inventions* to her credit, patented or not. The Shaker writer mentions her locomotive-wheel improvement and says she "has invented other devices"; the *Scientific American* writer mentions both the war vessel and the locomotive wheels,[31] and then says, "A number of other patents have been issued to her." LWP omits one of Montgomery's patents (an 1868 patent for a Bridge, granted jointly to Richard and Mary J. Montgomery), and may omit others.[32]

Most recently, Montgomery has been mentioned by Anne Macdonald, who notes in her *Feminine Ingenuity,* that Montgomery's war vessel was one of 125 women's patent

---

31. The *Scientific American* article exemplifies the problems with sources in this research. The writer's name is not given, nor are sources specified, and the date order of the two noted inventions is reversed. Worse, the first paragraph of the article contains a glaring error. It states, "Up to the present time almost 3,958 patents have been granted to women," whereas actually, as LWP shows, that figure was reached by early 1895—not counting design patents, reissues, or the significant numbers of patents omitted by LWP (Stanley, "Conjurer"). Moreover, by the date of this article, women were inventing at such a rate that in ten years between 1905 and 1921, they received more patents than they had in the entire 19th century (WB/28)!

The most distressing thing about this particular error is that if anyone should have known something about women inventors at this time, it should have been the people at this journal, which focused on invention.

32. As I point out elsewhere (Stanley, "Conjurer"), if 1876 is typical, the omissions of the LWP compilers from their supposedly complete list of women's inventions 1809 to March 1, 1895, are extensive and not entirely random. Women's machines, particularly "nontraditional" machines (i.e., having to do with heavy industry, etc.), rather than with women's traditional work) are disproportionately represented among the omissions.

models exhibited at the Cotton States and International Exposition in Atlanta in 1895 (15, 188). Biographical information on this inventor has been elusive; but if co-patentee Richard Montgomery was her husband, we may yet find her in a census. The *Scientific American* article speaks as if she were still alive in 1903.

**Mary Shay** of Fordham, NY, patented a Machine for rolling axles, spindles, and other metal articles (1885), and assigned her right to another woman, **Mary A. Seitz** of New York City. Unlike Montgomery, however, Shay and Seitz (who may have been manufacturer, investor, or co-inventor) are totally unknown.

*Miscellaneous manufacturing machinery.* In 1870 **Mary J. McColl** of Chicago patented a Machine for cutting wax for artificial flowers. McColl also held several other patents related to the wax-flower industry, including an improved wax compound, two improved processes, and an improved tool for veining wax leaves (1870–7). Mary McColl could have been an entrepreneur, making and selling the popular wax flowers of her day, though she does not appear in the special Chicago directory that reported 1870 census information, and none of the male McColls listed is in that or a related business, either.

LWP lists Augustine D. Waymoth of Fitchburg, MA, and Florence Marmet of Cincinnati, OH, as inventors respectively of a (wood) Lathe (1889) and an Improvement in elevators and conveyors for coke crushers (1877). Both, however, have proved to be male.

**Frida Kunz** of Donaueschingen, Germany, secured American patent protection for two Processes and apparatus for preparing fiber for the manufacture of brooms and brushes (both 1884). These machines presumably had a certain commercial potential.

**Jenetta V. Bohannan** of Washington, DC, and Baltimore, MD, invented and patented two Cigarette machines. (LWP misspells her name on both of these entries.) Though LWP's brief description could cover either vending or manufacturing machines, these are both manufacturing machines (1890, 1891). That is, they represent a new and/or improved technology for rolling tobacco in cigarette papers. Bohannan's third known patent was for a meat-slicing machine.

**Sarah C. Krause** of Bethlehem, PA, patented a Milliner's steamer in 1891.

**Eliza J. Smith** of San Francisco patented a Marker, cutter, and polisher for plastic stone-work (1890). The Gazette reveals that this invention, which appears to be a tool (rather than a machine) for use with artificial stone, was patented in Mexico on August 13, 1888. This unusual entry suggests that biographical research on Smith would be rewarding.

**Minnie E. Wilson** of Indianapolis patented a Gas-making apparatus (1893).

*Office machines*

In addition to their typewriter improvements, 19th-century women invented and patented several useful office machines. For example, women patented three paper-punching or perforating machines: **Rosalie W. Mackall** of Washington, DC, in 1878; **Winifred C. Utley** of Milwaukee, WI, in 1882; and **Anna M. Parks** of Albany, NY, in 1893. Two women patented stamp-cancelers: **Ellen D.W. Hatch** of Rolla, MO (1878), and **Ann M. Fortier** of San Francisco, with Shadrach Fortier (1884). **Victoria I.H. Bundsen** of London, England, thought highly enough of her stamp-affixer to seek American patent protection for it in 1885. In 1884, another Englishwoman, **Frances Hogg,** received a British patent for a Combined knife and measuring paper cutter. In 1882 **Mary E. Winter** of Galesburg, IL, patented an Adding machine. Her male co-patentee, Cushman W. Crary of Chicago, assigned his share of the rights to her.

Probably the most noteworthy of this nonsystematic sample, however, is the Check punch patented by **Ellie N. Sperry** of Bridgeport, CT, in 1890. Shown at the Columbian Exposition in Chicago in 1892–3, this is the punch used in banks the world over to protect checks against alteration, by actually punching the numbers of the check amount into the paper (K. Knight, 126, 129).

*Power-generation, engines, and related*

The isolated writers who discussed women inventors in the 19th and early 20th centuries were fond of listing their mechanical inventions, especially such things as motors and generators, because they were somehow the most surprising. The writers seldom name names, although **Maria Beasley**'s steam generator (1886) was important enough to get mentioned by Ida Tarbell (357), and **Margaret Knight** gained considerable recognition for her rotary factory-driving engines in her own day. The following are other women who invented steam engines or generators and related devices:

**Lydia F. Renshaw,** Cohasset, MA, Sectional steam generator (1876)
**Virginia M. Ring,** New Orleans, Improvement in automatic water cutoffs (1878)
**Lilly B. Tubbs,** Philadelphia, Cutoff for hydraulic and other engines (1885)
**Marie L. Russell,** Chicago, Overflow indicator (1888)
**Marie M. Forestier,** Brussels, Belgium, Stopcock (1889)
**Elizabeth E. Jepson,** Flensburg, Germany, Speed governor (1891)
**Virginia S. Miracle,** Janesville, WI, Fuel-feeding device (1893)

Some of the inventions listed above under Electrical and related inventions could appear here as well. **Amelia H. Lindsay**'s Rotary engine of the 1880s (Pittsburgh, PA) definitely belongs, as do the three motors listed by the LWP compilers for the early 1890s:

**Nancy L. Turner,** Washington, DC, Motor (1893)
**Sarah J. Rollason,** London, England, Windmill or wind motor (1894)
**Amy F. Robinson,** Colorado Springs, CO, Variable driving gear (1895)

Also belonging here, of course, would be the sewing-machine, washing-machine, churn, and other motors patented by women but listed with those machines (Q.V.) rather than as motors.

A noteworthy feature of **Josephine Cochran**'s dishwasher was that it was power-moded (steam). **Nelly P. Levalley** of Milwaukee, WI, patented a Drive chain in 1887. As its Gazette description mentions no application, the invention is probably intended for general use with many kinds of motors and engines. Levalley was an early woman student at M.I.T. (Bever; Trescott, *New . . . Paths*).

Two British women received patents for "obtaining motive power," one as early as 1838: **Ann Bird Byerly** received British Patent #7871 of that year. In 1881 **Anna DiViani** received Patent #2178 for Obtaining or applying motive power (May 18).

Power generation for propelling vehicles, ships, etc., will be discussed under Transportation.

*Perpetual motion machines.* A special category of power generation technology is the tantalizing quest for endless, self-regulating motion. As early as 1825, Carnot saw that there was a fundamental obstacle to perpetual motion, which was that whenever energy is converted from one form to another, something is always lost, observations that were confirmed and codified in the more recent laws of thermodynamics. Yet long after Carnot, inventors seek to evade these laws. Most of these seekers seem to have been men—for reasons that may bear on the question whether women invent for different reasons than men (Stanley, "Once and Future . . .").

One American woman who entered the lists for perpetual motion, after a successful art career, was the first great U.S. woman sculptor, **Harriet Hosmer** (1830–1908). Hosmer is discussed as a professional inventor, below. Suffice it to say here that while pursuing perpetual motion, she invented a magnetic motor that was evidently taken seriously by at least some scientists in its day ("Miss Hosmer's . . . ," n.p.).[33]

Hosmer may not have been alone. George Walsh, writing in the *Patent Record* in 1899, says both sexes send in "more silly and irrational inventions . . . than can be imagined," adding that "one or two women have even sent in models for machines claiming to solve the mystery of perpetual motion," but all such patents are automatically disallowed.

## Transportation and related

Women's inventions in this line are numerous and varied, ranging from items dealing with horseback riding, horse-drawn vehicles, railroads, and streetcars to baby carriages and bicycles, ship-propulsion, snow-removal, and street-cleaning—in short, all the major forms of 19th-century transportation and related technologies. The following examples should give a good sense of women's achievement in this area.

*Horseback riding, horse-drawn transport and related.* Rev. Ada Bowles, who studied women's inventions for twelve years, was aware of women's many devices relating to horse transport. Once while out driving with an old Vermont farmer who doubted women had ever invented anything, she snapped, "Your horse's feed bag and the shade over his head were both of them invented by women." The old man was so surprised that he could barely gasp, "Do tell!" Mozans, who reports this exchange, continues (347),

> Had he investigated further, he would have found that the flynet on his horse's back, the tugs and other harness trimmings, the shoes on his horse's feet and the buggy seat he then occupied were all the inventions of women. He would, doubtless, also have discovered that the curry-comb he had used before starting out on his drive, as well as the snap hook of the halter and the checkrein and the stall-unhitching device were likewise the inventions of members of that sex . . . for women have been awarded patents—in some instances several of them—for all the articles . . . mentioned.

The patents referred to do exist—Mozans has obviously seen LWP—and though most of them are not the first patents for the items mentioned, they may have been in use at the time. Not all are notably mechanical (though virtually all are arguably *parts of a mechanical system);* but because transportation is an important field of women's invention,[34] I include a fairly extensive listing of women's 19th-century U.S. patents for inventions relating to horse transport as Appendix M.6.

Strikingly mechanical among these inventions are **Ivy Hart**'s Wagon jack of 1878, **Hannah Bowker**'s Unhitching device for stable-stalls (1880), **Leila Harrison**'s Hitching device of 1884, **Martha Hancock**'s Wheel attachment for sleighs, **Carrie Grant**'s Vehicle axle, **Clara Turney**'s Horse-controlling device, and **Matilda Busby**'s Brake for carriages, all of 1890. **Harriet Carmont**'s Starter or brake for vehicles (1894) is noteworthy not only for its mechanical nature, but in that she as an Englishwoman took the trouble to get a U.S. patent. Busby's patent is omitted from LWP.

I did no systematic survey of British women's patents relating to horse transport, but

33. I am grateful to the Schlesinger Library for calling this material to my attention and allowing me to use it.

34. As will be seen below, women's transportation and related inventions constitute over 6% of their recorded patents for ten sample years in the first quarter of the 20th century (WB/28).

note that three women received British patents in this area in a single year (1888): **Sarah B. Beswick** for a Varnish for harness, **Adele Engstrom** for Curry combs, and **Elizabeth Lange** for a Hansom cab. Only the latter, of course, is mechanical.

**Annie H. Chilton** of Baltimore has two patents for a Combined horse detacher and brake (1891, 1894), representing the safety orientation of several of the devices in this group of inventions. (Runaway horses were an ever-present danger in horse transport, and a rapid and reliable means for loosing the carriage from the horses and bringing it to a stop were much in demand.) Chilton, who had earlier patented a safety pin (1887), exhibited her horse-detacher in the Woman's Building at the Columbian Exposition in Chicago in 1893 (Weimann, *Fair,* 428, 431). The runaway horse also endangered a rider, and as early as 1769, **Anne Still** received a British patent for a stirrup-release.

All told, considering only American patents, 52 women received 59 patents in this important transport technology between 1864 and 1894.

Charlotte Smith tells us in the *Woman Inventor* (#2: 2) that a woman has discovered how to make horsehoes out of rubber. Without naming the woman, or revealing whether or not the invention was patented, she notes that these shoes are light and durable, and improve the hooves. Next, quips Charlotte, we shall have rubber horses run by electricity!

Perhaps the most startling example, however, of women's contributions in this area was patented in a man's name, and has never been attributed to a woman except in 19th-century feminist writings. This is the famous Burden horseshoe machine, already discussed above.

The other major 19th-century means of private or personal transportation in addition to horseback riding or horse-drawn conveyances were baby carriages and bicycles.

*Baby carriages.* Ruth Cowan, among others, points out that the history of technology has completely neglected not only the necessarily female technologies (except for contraception), but the child-rearing technologies, which need not be but have been female-dominated in recent history.[35] We know far more, for example, about bicycles than we do about baby carriages—even though wheeling the baby about in this contraption instead of carrying it next to one's body on walking journeys inevitably has significant physical and psychological effects—on both mother and child.

Thus it is doubly interesting to find the following contribution by 19th-century women to this technology, considering only U.S. patents (LWP):

**Harriet D. Ambler**, Brooklyn, NY, Improvement in children's carriages (1874)
**Anarella A. Winsor**, Boston, Improvement in children's carriages (1876)
**Alice A. Hodges** (w/ George), Baltimore, Brake for children's carriages (1885)

---

35. Cowan's eloquent statement is worth quoting at length ("From Virginia Dare . . . ," 30–1):

Women menstruate, parturate and lactate; men do not. Therefore, any technology which impinges on these processes will affect women more than it will affect men. There are many such technologies and some of them have had very long histories: pessaries, sanitary napkins, tampons, various intrauterine devices, childbirth anesthesia, artificial nipples, bottle sterilizers, pasteurized and condensed milks, etc. Psychologists suggest that those three processes are fundamentally important experiences in the psycho-social development of individuals. Thus a reasonable student of the history of technology might be led to suppose that the history of technological intervention with these processes would be known in some detail.

That reasonable student would be wrong, of course. The indices to the standard histories of technology—Singer's, Kranzberg and Pursell's, Daumas's, Giedion's, even Ferguson's bibliography— do not contain a single reference, for example, to such a significant cultural artifact as the baby bottle.

**Pauline Austin**, Brooklyn, NY, Child's carriage attachment (1889)
**Salome P. Davis**, San Jose, CA, Baby carriage (1891)
**Eleonore Eder**, St. Louis, MO, Baby carriage body (1891)
**Eleanor Lotinga**, Ely, England, Convertible child's carriage (1892)
**Eveline L. Campbell**, San Francisco, Child's carriage (1893)
**Jennie R. Lyle**, Dowagiac, MI, Nursery case for children's carriages (1893)
**Mary L. Barr**, Indianapolis, Baby carriage (1894)
**Selma Kehlenback**, Pleasant, IN, Baby carrier (1894)
**Ora Orr**, Westport, CA, Combined child's carriage and cradle (1894)

At least one of these women got rich from her invention. Mozans (351) and Rayne (117) report that "a San Francisco woman"—presumably **Eveline Campbell**—received $14,000 for her patent. Lydia Farmer (419), Logan (884–5), and Charlotte Smith in the *Woman Inventor* (/1), credit her with $50,000. Either figure would have been a fortune in the 1890s.

*Bicycles.* Though bicycles were invented early in the century, they were little more than a curiosity until driving mechanisms were added (1839) and frames made lighter and cheaper. Early machines weighed 50-60 pounds. The velocipede of 1850, first to have a drive mechanism (front wheel), would have cost skilled British workers a quarter of their annual wage, and would actually have exceeded the annual wage of a female domestic servant (Grigson & Gibbs-Smith, 42). Not until the rear-chain drive was marketed (1870s), tubular frames and the drop frame for women were introduced (1884), and Dunlop's pneumatic tire appeared (1888), did bicycles become really popular. (This British chronology is roughly applicable to the United States, where, according to *Inventive Age* [July 6, 1892: 4], the first velocipede was built in Connecticut in 1865 and bicycles came into "general use" about 1878.) Soon after that, they became a craze.

As women began to ride in numbers in the 1880s, a controversy raged over whether such brazen and unladylike behavior should be allowed. The bitterness of the battle astounds us until we recall the great freedom that bicycling gave to women. Not only a sudden mobility and escape from chaperonage, but a certain freedom of dress was involved, for obviously no one could ride a bicycle in floor-length skirts with 15-20 pounds of petticoats in the way. Indeed, more than a third of the 19 women's bicycle-related patents found in a nonsystematic survey of LWP and other records were for cycling shoes and garments, including a British patent as early as 1881.[36]

36. Seven out of 19: **Louisa Morton**'s Boots and shoes for bicycle-riders . . . (1881); **Margaret H. Lawson**'s Bicycle garment (Boston, 1893); **Emma Dryfoos**'s Bicycle habit (1895); and three cycling skirts (**Florence Ahern**, NYC, and **Catherine V. Berry**, Washington, DC, both 1896; and **Carrie G. Childe**, NYC, 1897). **Lena Sittig**'s Skirted trousers (1895) are evidently for cycling: Best & Co. exhibited Mrs. Sittig's "bicycle goods" at the Atlanta Exposition of 1895. Sittig, who lived in Brooklyn, held three other patents (1892–4), and was credited by the *Scientific American* (Apr. 4, 1903: 247) with inventing a waterproof garment called the "duckback." Listed as Mrs. Frank Sittig, she showed this waterproof garment at the Suffrage Bazar held at the Martha Washington Hotel in New York City in November 1908 (M. Reynolds, 3).

Note the very practical garb of the gymnast who rode the bicycle for Alice Austen's photographs in Maria Ward's 1896 book *Bicycling for Young Ladies* (see below)—full trousers tucked neatly into slim knee-high boots. What we need to remember in evaluating the significance of these patents is that, just as human beings could not have survived in the Arctic without the fitted fur clothing made by the women, women could not take advantage of this excellent form of private transportation in the 19th century without clothing that preserved the female image to some degree while conferring freedom of movement.

When a woman could be considered a hussy[37] for even riding a bicycle, it is doubly interesting to find the following patents credited to women:

**Catherine A. Williamson**, St. Louis, MO—Seat for bicycles—two patents (both 1887)
**Kate Park**, Chicago—Bicycle lock (1890)
**Martha E. Slocum**, Meadville, PA—Velocipede (1892)
**Sadie M. Bates**, Montclair, NJ—Luggage-carrier for velocipedes (1893)
**Sarah L. Naly** (w/ M.S. Jones)—Philadelphia, Bicycle saddle (1893)
**Dell M. Hawes**, Ortonville, MN—Pneumatic tire (1894)
**Effie M. Battenberg**, Decatur, IN—Umbrella attachment to bicycles (1896)
**Beatrice L. Bagot** (w/ Charles)—Devices for locking velocipedes (1896)
**Alice A. Bennett**, Elgin, IL—Bicycle canopy (1896)
**Eliza A. Crossley**, Auckland, NZ—Vehicle wheel (metallic wheel with twin spokes for bicycles, etc.) (1898)
**Emma J. Dunn**, Humeston, IA—Velocipede (1899)

Although several of these inventions can be classed as appurtenances, and the bicycle seats or saddles (unless adjustable or fitted with springs), the luggage carrier, and the tire are not mechanical, they are listed here as essential parts of a transportation machine. The canopy and the umbrella attachment are folding or collapsible; the rest are fairly obviously mechanical. Notice that Dell Hawes' pneumatic tire is only six years later than Dunlop's (Grigson & Gibbs-Smith, 42).

It may be that, as often happens, women's most important bicycle inventions were left unpatented, or patented by men. An invention described as a "practical alteration" to the bicycle's mechanism, "an improvement which was almost universally adopted," was the work of **Maria Emily ("Violet") Ward** of Grymes Hill, Staten Island, NY. Violet Ward was an avid amateur inventor, and daughter of an inventor (her father, General William Greene Ward, had co-invented the Ward-Burden breech-loading rifle). She wrote a popular book titled *Bicycling for Young Ladies* (1896). My source gives neither the date nor any details on the nature of her invention, and I have located only one patent for Violet (listed as Maria E.G.M. Ward of NYC), for a Bodkin (1893). It is tempting to hypothesize that Violet Ward—or some other woman—may have suggested the lightweight tubular construction or the drop frame for women introduced in 1884, especially since Grigson and Gibbs-Smith in their chronology of bicycle invention name the inventors of most of the improvements mentioned, but not these two (41–2).

Ward might have slipped out of history altogether, except that as a friend of photographer Alice Austen, she figures in Austen's recent biography (Novotny; see also O. Jensen, 18, 19, 121, 126, 127).

*Railways.* In the pre-automobile era of the 19th century, the major means of public land transportation in addition to horse-drawn vehicles was the railways, cross-country in rural areas and between distant cities, and street railways in the cities. Women's contributions to the two kinds will appear in Appendix M.7, where it will be seen that at least 62 women received 96 patents in this area in the three decades 1864–94.

Noteworthy among these many women is the earliest, **Mary Jane Montgomery**. Once

---

37. Or worse. The vitriol of the anti-bicycle crusade has to be seen to be believed. Obviously part of the threat was the sexual freedom—and stimulation—that cycling allegedly gave to women. Sad to say, some of the opponents of women's cycling were other women. The Boston Women's Rescue League, for example, warned grimly that 30% of all fallen women had at some time been bicycle riders (O. Jensen, 126).

identified by *Scientific American* as the only professional woman inventor in the country, she is mentioned above and will be discussed below as a specialist in machinery.

Several of the women would seem, by the numbers and nature of their inventions, to merit further research: **Eliza Murfey** (1837–79) of NYC and **Harriet(t) Devlan** of Jersey City with their many improvements in packings and bearings for railway journals;[38] **Catharine Gibbon** of Albany, NY, with her four rail-construction patents; and **Augusta Rodgers**.

Connecticut-born Eliza Murfey, in fact, if *Inventive Age* is correct, may turn out to be one of the most prolific American women inventors of the century, and by far the most prolific of her day.[39] Between 1869, when she "commenced inventing devices for steam packing," and her death in 1872 (*sic*), she "took out about thirty different patents, beating the record made anywhere on the globe." I have thus far found only 23 patents 1870–5, plus one reissue, but even this number places her near the top. In addition to several materials and processes for packing journals and bearings, she patented an Improved packing for pistons (1872), a Bearing surface for spindles, and a preparation for same (both 1875). It would be fascinating to know how she arrived at these inventions from her occupation as a physician, which she seems to have entered after the death of her husband Henry, about 1869.

Rodgers received seven patents in the years 1870–74, including a bedstead canopy and a folding chair, her four railway patents, and yet another heater improvement that may also apply to her railway-car heater, but which I have not listed here. She was mentioned by the *Woman's Journal* in 1872, when she took out British patents for her mosquito canopy and her folding chair. Sad to say, her only modern recognition has been a mistaken credit in Wallechinsky and Wallace's *People's Almanac* for inventing the *automobile* heater (911).[40]

The popular 19th-century lecturer Russell Conwell (47) mentions that a woman invented a trolley switch, his context implying a complex technological feat. This is probably **Alice Middleton**'s Transfer apparatus for traction cable cars (Philadelphia; 1890).

A strong theme of safety runs through this group of women's inventions. **Eleanor A. McMann** of Cleveland, for example, received three patents for (Safe)guards for sleeping-car berths; three of **Augusta Rodgers'** patents were improved spark arresters and cinder and smoke-conveyors; and the whole rationale for her car heater was fire-prevention. **Ellen E.L. Woodward** of Chicago patented a Safety-car in 1888.

A striking cluster of women's inventions here is car-couplings and coupling pins: between 1872 and 1894, women received at least 14 patents for car couplings. One of these, **Laura J. Gott** of La Grange, OH, also patented two fire escapes. A major motivation here was safety. Nor was it only a female search, for by 1884 the patent records showed some 4000 attempts to invent an automatic coupler, coupling cars on contact so

38. These puzzling-sounding inventions refer to a technology for lubricating railroad-car axles, consisting of solid bronze bushings lubricated by oil-soaked flax in journal boxes. This arrangement held the oil in place, but did not always deliver it where needed, in which case the axles seized up, and the cars went off the tracks. Thus the various improvements proposed by Murfey and Devlan, among many others. This technology hung on even after roller bearings were invented, because it was cheaper (Waldron, L-1/10/83, encl.).

39. Catherine Griswold, Helen Blanchard, Margaret Knight, and Mary S. surpassed her either in total patents or in total inventions (see below), but are all later. Moreover, Mary S. held no patents, and some of Blanchard's and Knight's patents fall into the 20th century (*IA*, May 5, 1891: 5).

40. The principle of Augusta Rodgers' fireless heater may indeed have been adapted for automobiles, but Wallechinsky & Wallace are merely citing her 1870s patents, and thus have wrongly given the word *car* one of its 20th-century meanings.

that workers need not risk getting crushed as they went between cars to couple them (*EB*/11, vol. 22: 855–6). However, a technical struggle to dominate the large market for a uniform coupling was also in progress. According to *Inventive Age* ("Car Coupler"), the battle between advocates of the "vertical plane" and "link and pin" technologies was still raging in 1891.

**Jane Grey Swisshelm** (1815–84), the well-known 19th-century journalist and reformer, is credited with the idea that railroad cars should bear red tail-lights as a safety measure.[41] Too late for this group and for LWP, but also safety-oriented, is **Maria Beasley**'s 1898 patent for a Means for preventing the derailment of railway cars.

Outstanding in the group of women railroad patentees, receiving national and even international recognition in her lifetime, is **Mary Walton** of New York City, who invented a means of deadening the clangor of elevated railways. Some of the greatest machinists and inventors of the country—including Edison himself—had evidently tried without success to solve this major urban nuisance. Walton had an extra incentive: she lived near the El tracks (*WI*/1), and daily suffered its assault on the ears. She devised a system involving, among other things, bedding the rails in sand stabilized with asphalt. Her invention, patented in 1881, was immediately adopted by the New York Metropolitan and other elevated railway companies. Says Matilda J. Gage, "The benefit to human health and life likely to accrue from this invention can scarcely be realized" (/2 : 346).

The benefit to Walton was considerable as well, for she received $10,000 and "royalty forever" for her silencing invention.[42]

Mary Walton had a second railway-related patent to her credit, which also looks like a boon to the environment. In 1879 she had patented an Improvement in locomotive and other chimneys, consisting of a smoke-burner that consumes all the smoke from a fire, furnace, or locomotive, as well as certain kinds of dust. Her burner also destroyed the offensive odors emitted by factories and gasworks. When she traveled to England, British officials congratulated her, calling the burner "one of the greatest inventions of the age."

It would be interesting to know how Mary Walton gained her technical knowledge. Was her father perhaps a railroad engineer? The only clue is her own statement from 1884: "My father had no sons, and believed in educating his daughters. He spared no pains or expense to this end. . . ." (*WI*/1; Lexington, KY, *Weekly Transcript*, Aug. 28, 1884; Tarbell, 356; Gage/2 : 346).

Also noteworthy is **Geneva Armstrong** (c. 1858–1930) of Elmira, NY, and Chicago, who according to Charlotte Smith invented "a machine for feeding cattle on trains" (*WI*/1). This 1885 invention is omitted from LWP; the Gazette describes it briefly as a Stock-car. Armstrong, the owner of a farm in western New York, is more versatile and interesting than appears at first glance. She not only patented this invention, but marketed it to the railroads, personally urging its adoption at a meeting of railroad men. She also exhibited it at the Columbian Exposition in Chicago in 1892–3. In the 1890s she moved to Chicago, where she lived until her death. No occupation is listed for her in the directories,

---

41. Walking at night along the railroad tracks not long after a serious train wreck had occurred on the Pennsylvania Central line in the late 1840s, Swisshelm was struck by the contrast between the brilliant white headlight announcing the train's approach, and its sudden disappearance into the dark on passing. The next day she wrote a letter to a Pittsburgh newspaper suggesting that a red light should be mounted on the rear platform of all trains. The idea was taken up in an editorial, and within a week all Penn-Central trains had such tail-lights (Convis). Swisshelm had a patented invention to her credit as well (a hot-air chamber for a stove; 1876), which is not mentioned in her *Notable American Women* biography.

42. I am grateful to Helen Irvin, formerly of Transylvania College, for calling my attention to this item in the Lexington, KY, *Weekly Transcript,* Aug. 28, 1884, datelined New York City.

nor did she ever marry—which could indicate that she succeeded in selling her invention to one or more railroads. However, in the 1900 and 1910 censuses she listed herself as an art teacher or critic. These censuses show Armstrong as a boarder and "partner," respectively, of Alice J. Leland.

Another Columbian Exposition exhibit, evidently unpatented, was the Footstove for railroad cars invented by **Caroline Romney**, a 19th-century newspaper and magazine editor, writer, and inventor. She showed more than a dozen inventions at Chicago, and wrote an article on the other women inventors who exhibited there.

British women also contributed to railroad technology, beginning as early as 1811 with **Sarah Guppy**'s patent for "Erecting and constructing bridges and railroads. . . ." Guppy's intent in her construction "without arches or sterlings" is to keep the bridges from being washed away by floods. Another early British patent also pertains to the railroads themselves rather than to the cars: **Annet Gervoy**'s Patent of 1852, for a Means to prolong durability of rails on railways. As already noted, Sarah Guppy, wife of a Bristol merchant, came from an entrepreneurial family with an established tradition of invention. She received at least four patents between 1811 and 1831, for a diverse group of inventions; and had many ideas she never developed, some of which appear in her letters (Brody, "Patterns," 5, 7–8 and PC-8/85).

*Snowplows.* At least two of the three snowplow patents I have discovered for 19th-century women are for railroad use, and thus could be added to the previous category:

**Jane A. Bryant**, Pottsville, PA, Improvement in devices for removing snow from railroads (1873)
[Andrew E. Dietz for] **Emma L.** and **Mary A. Dietz**, Oakland, CA, Railroad snowplow (1887)
**Harriet L. Fraser**, NYC, Snow plow (1888)

Noteworthy here are the **Dietz** sisters, **Emma Louise** (1836–1909) and **Mary Amanda** (1821–1918), who were recognized for their "Cyclone" snowplow, even though it was patented for them by their older brother. Perhaps the story was first told in the pages of *Inventive Age,* an inventors' periodical published in Washington, DC, from 1889 into the early 20th century. The sisters' achievement was picked up from there by the Shaker *Manifesto* for July 1890, which noted that the "Cyclone . . . sends the snow in great white waves to each side of the tracks." It was also mentioned more than 20 years later by Mary Logan (885), who says the plow is to be attached to the engine's cowcatcher, and further notes that "the proceeds from this have well repaid the time and ingenuity given to perfecting their patent." Despite letting their brother patent their snowplow, the Dietz sisters were by no means fading Victorian flowers. Born in New York State into a family of 8 children, they were entrepreneurs, running their own bathing establishment in Alameda, CA, with another brother. Emma Dietz appears as Head of household in the 1900 census of Alameda County, even though her sister Mary is older, and one of her brothers lives in the house.[43]

*Street-sweepers.* Charlotte Smith tells us in 1891, "A street sweeper of great merit was devised and patented by a New York lady" (*WI*/1: 4). I originally thought she must be referring to **Mary S. Kjellstrom** of that city. Kjellstrom's street-sweeper patent, though

43. I am grateful to Wilber Leeds, G.R.S., of Oakland, who provided me with much useful genealogical and professional information on the Dietz sisters and their family (Fred Dietz, comp.; Federal census, Alameda County, CA, 1880, 1900; Quinn II: 573; Husted's Oakland, Alameda, and Berkeley, CA, city directories; Bus. Dir., Alameda Co., 1889: 172; Geo. Dietz, Obit., Oakland *Tribune,* Mar. 19, 1900: 2: 6; Superior Court of California Probate Records (Alameda County).

not granted till 1892, would have been applied for by 1891, and could have been known to Smith through her son (then a Patent Office clerk).

However, Kjellstrom seems to be totally unknown, whereas another woman inventor of a street-sweeper—**Eureka Frazer Bowne**—(b. 1858) seems to have won considerable recognition. Indeed, according to the *Crockery & Glass Journal* for September 1, 1898, "it is said of her that she is the most gifted mechanically of any woman in the United States." She would have to dispute that title with **Helen Blanchard, Margaret Knight,** and **Harriet Hosmer,** among others, but the repute is interesting in itself. And Bowne was living in New York City at the time of the sweeper invention. Thus she seems more likely to be the inventor Smith had in mind.

Bowne's street-sweeper was patented in 1879, some years before the *C&GJ* article— which reports a new Bowne patent, for a Non-refillable bottle (1898). (This bottle also involved a mechanical invention.) Born in Trenton, NJ, Eureka is described as "a tall, willowy blonde," who lived on West State Street until she married Mr. Bowne, "a swarthy hatter of Trenton." They evidently moved to New York when he became manager of Macy's book and stationery department. "Mrs. Bowne made a hit," continues the *C&GJ* writer, "by inventing the first successful street sweeping machine used in New York City" (21). This sweeper, incidentally, is one of LWP's more striking omissions. By 1898 she is back in Trenton. By 1900 she is a widow, lodging in a home where there are two servants, and the only other lodger is a lady physician (1900 census). Nothing in this brief sketch suggests how Eureka Bowne came to invent these two unrelated and remarkable items; no occupation is mentioned for her, and her husband's work is not pertinent. Perhaps, as her given name suggests, she was truly an inventor, and not just a person with another occupation who happens to invent something.[44]

Nor are these the only street-sweepers invented by women around the turn of the century. According to Martha Rayne (118), writing in 1893, "A Hoboken lady, having had her dress spattered with mud by a clumsy street sweeping machine, invented the Eureka street sweeper." The "Eureka" brand name suggests that Rayne may actually refer to Eureka Bowne, though Bowne is so far located only in New York and Trenton.

Rayne may mean **Florence W. Parpart,** an inventor-entrepreneur from Hoboken (also Pittsburgh and New York). The *Patent Record* (Sept. 1900:1) says she has already patented "several things . . . very profitable to her, as well as very creditable." But Rayne wrote in 1893, and I find no Parpart patents before her "sanitary street sweeper" of 1900 (and 1904). In any case, this machine is "far ahead of anything in its line," and heartily approved by sanitary experts and the public. Miss Parpart (*PR* omits her co-inventor/husband-to-be) has arranged to manufacture and sell these machines, and moreover will herself engage in the street-contracting business in New York and other cities. As of fall 1900 she was negotiating to introduce her sweeper in Boston, Chicago, Omaha, Milwaukee, San Francisco, Cambridge, MA, Camden, NJ, Greensburg and Altoona, PA, and Portsmouth, OH.

Not a street-sweeper *per se* but a widely adopted adjunct to street-cleaning was the invention of **Cynthia May Westover** (1858–1931) of New York City. Her 1892 patent, for a Dumping cart with movable body and self-emptying derrick, was granted while she worked for the New York City Street Commissioner. Sources are unclear on this point, but she may also have invented a smaller cart also widely used by city street-cleaners (*NCAB* 22: 52). In any case, she won a gold medal and diploma from the Parisian Academy of Inventors for her dumping cart, and was made an honorary member.

---

44. Bowne is the first inventor I have encountered with this given name. I am grateful to Beatrice Smith of St. Paul, MN, for calling her to my attention.

Westover had a bewildering array of talents—and occupations. Horsewoman, naturalist, and mathematician, she also taught geology and vocal music. She became a factory superintendent before age 20. Moving from Colorado to New York City in 1882, she refused several offers of operatic roles. She studied languages, passed a Civil Service Exam and worked as a customs inspector, and finally entered the Street Commissioner's office, the first woman so appointed. Later she embarked on a journalism career, married John Alden of the Brooklyn *Eagle*, worked for the *Ladies' Home Journal*, and wrote several books, including one called *Woman's Ways of Earning Money* (1904). Late in life she turned to philanthropy, founding and running the International Sunshine Society for the benefit of blind babies and children. She herself had no children (Appleton's; *CAB* 8: 142: *DAB* Suppl. 1; Ireland; M. Logan, 846; *NCAB* 22: 52; *WWWA*/I (1897–1942); *WWW* NA Authors 1921–39, I; Willard & Livermore, 761–2).

*Ship transport.* Women's inventions in water or ship transport are not numerous, but several are substantial, including propulsion devices and even heavy dredging machinery:

One of the earliest and most often mentioned in the rare writings on women's inventions is **Sarah P. Mather**'s Submarine telescope and lamp (1845). Intended to allow ship's hulls to be examined and defects diagnosed without the labor and expense of dry-docking, this is the first U.S. patent for such a device. Shipwrecks can also be examined, obstructions to navigation located and removed, and lost torpedoes sought for, all amounting to considerable savings for the navy and merchant marine. Says Matilda Gage,

> We deem the telescope which examines the heavens a wonderful invention . . . ; but how much greater scientific genius must these women possess who have invented an instrument for bringing to view the depths of the ocean, and making its wonders as familiar to us as those of the dry land.

Later improved by Mather's daughter, the telescope is still being mentioned well into the first quarter of the 20th century (Gage/1: 20, 21; Gage/2: 346; Mozans, 345; Shaker *Manifesto,* July 1890: 164; *WI*/1; Tarbell, 356).

When Mather, who was from Massachusetts, patented her invention, only about 30 women had received U.S. patents. Another relatively early invention is **Mary Jane Montgomery**'s Improved war vessel of 1864.

At least five other American women received patents related to ship transport during the 19th century:

**Temperance P. Edson**, Dedham, MA, Inflator for raising sunken vessels (1865)
**Cornelia E. Beaumont**, Cleveland, OH, Improvement in lifeboats (1878)
**Maria E. Beasley**, Philadelphia, Life-rafts (1880, 1882)
**Katharine C. Munson**, Winthrop, MA, Berth for vessels (1890)
**Charlotte V. Morozowiez**, Pittsburgh, Steering vessels by steam (1891)

A sixth, an Englishwoman, received an American patent: **Elizabeth Myers** (w/ C. Myers, J. Davis)—screw propeller, (1892); and a seventh, **Emily E. Tassey** of McKeesport and Pittsburgh, PA, received five patents in just four years: an Improvement in apparatus for raising sunken vessels (1876), two Improvements in siphon propeller-pumps (1876, 1880), Propulsion of vessels (1876), and a Dredging machine (1876). Her patents for ship-propulsion and for dredging machinery were omitted from LWP (Stanley, "Conjurer"). She will be discussed below as a machine specialist.

Edson exhibited her invention at the Philadelphia Centennial in 1876 (Warner/Trescott, 113).

**Maria Beasley**, whose other patents mostly pertain to manufacturing machines, will also be discussed below as a machine specialist. Her life-raft (two U.S. patents) was patented in England as well as in the United States.

**Charlotte Morozowiez**—whose invention, like Tassey's, is omitted from LWP—appears twice in the *Patent Record,* a periodical for inventors, in 1900: she has sold 1% of her 1891 patent for steam-powered steering to Charles F. Skibbe of Pittsburgh for $500 (Jan. 1900: 12; Feb. 1900: 11). If this is indicative of the true worth of her invention, it must have had enormous commercial potential indeed.

**Florence Parpart,** of street-sweeper fame, is also credited with a collapsible boat, which she successfully marketed to prospectors headed for the Alaskan gold rush ("Woman's Remarkable . . ."). This may be unpatented.

At least three unsung women received British patents related to ship transport, beginning as early as 1818 with **Félicité Raison Selligue**'s patent for Propelling machinery. **Sarah Coote** patented a technology for Caulking ships, boats, and other vessels in 1844, and **Mary Honibull** patented Anchors in 1838. Honibull's patent, however, was an extension of W.H. Porter's earlier patent, which leaves open the question of her role in the invention process.

But perhaps the most important nautical invention of a woman in the 19th century was the Vansittart-Lowe screw propeller invented and patented by **Henrietta Vansittart** (1833–83) of Richmond, England. Acting from a thoroughly Victorian sentiment, she did a thoroughly un-Victorian thing (for a lady): she created and patented a ship-propulsion invention. Her father, James Lowe, had invented a stern-mounted propeller to replace the partly submerged, side-mounted paddle wheels of his day.[45] Although the Lowe propeller was used on various British warships, infringement battles robbed him of financial reward, and he died impoverished in 1866. Determined to win for her father the recognition that had eluded him in life, Henrietta began to experiment and only two years later (September 18, 1868) received British Patent #2877. She claimed that her propeller enabled ships to go faster with less fuel, less vibration, and better reverse steering.

How did a woman with no known technical training achieve this coup, becoming perhaps England's first woman engineer? Obviously she was a highly intelligent and talented inventor. Also, she had no children. Third, since she created her improvement so quickly, she must have been familiar with her father's work. Indeed, at age 17 she was aboard HMS *Bullfinch* during its propeller trials. Her ability to oversee her propeller's detailed design specifications, its manufacture, and its installation on such British ships as the gunboats *Decoy* and *Coquette* argues a considerable expertise. Finally, Frederick Vansittart must have been a supportive husband, for Henrietta took out American (1869) and European patents for her invention and showed it at expositions as far away as Australia, winning medals and diplomas.

She attended the meetings of the Naval Architects' Society and of the London Association of Foremen, Engineers, and Draughtsmen, before whom she read a paper in

---

45. James Lowe claimed to be the originator of the fully *submerged* stern-mounted screw propeller, and won a court decision upholding this point, although John Ericsson and Francis Pettit Smith are more familiar names in the field. I am grateful for this and other information underlying this discussion to B.M.E. O'Mahoney of Wetherden, Suffolk, England, and Dr. Elizabeth Laverick, editor of the *Woman Engineer* (London), for allowing me pre-publication access to Mr. O'Mahoney's article cited here. The article contains a likeness of Mrs. Vansittart, illustrations of some of the ships fitted with her propeller, and information on her family background, recognition, and later career, with references to published and unpublished sources. A brief monograph on Vansittart by O'Mahoney is now on deposit at the Fawcett Library in London.

1876. When her brilliant career was cut short by mania and death (1883), the LAFED Journal printed her obituary (O'Mahoney, *passim*, and L-1/30 and 3/20/1983).

*Other forms of transport.* Women also invented in various minor forms of public and private transport and related fields. At least three women received patents for paving,[46] but these are not mechanical. I will close this section with three intriguing patents and one literally garden-variety: two for balloon transport, one for a "Hand-carriage," and one for a Market truck (wheelbarrow). Of **Teresa Martinez y Perez**, who received a British patent for Navigable balloons in 1888, I know nothing. **Mary H. ("Carlotta") Myers** of Mohawk, NY, however, was one of the first American women balloonists to make a solo flight. Billed as "Carlotta, the Lady Aeronaut," she won fame as an exhibition flier. In 1885, with her husband Carl, she patented a Guiding apparatus for balloons. One year later, she set a world altitude record that is still astounding: without benefit of oxygen, she soared four miles above Franklin, PA. In addition to exhibition flying, Carlotta and Carl ran a balloon factory and flying school in Frankfort, NY (McCullough; Moolman, 10–1, 14).

**Maria Webster**'s Hand-carriage (1882) looks like an early hand-propelled wheelchair, though the propulsion is by means of cranks rather than by turning the wheels. The rider sits very low, and the wheels are relatively tall. The Gazette's extract from the patent claims does not mention invalids or paralytics, however, so perhaps this is just a personal means of transport. Webster lived in Farmington, MI.

The humble but useful wheelbarrow is a form of local and personal transport that should not be omitted. **Amanda Morrison**'s 1883 patent for a rather elaborate-looking compartmentalized model called a "Market truck" can serve as an example of women's contribution here (Wilmington, NC).

From the foregoing approximately 210 patented inventions, it can be seen that transportation is a significant area of invention for women. Just the patents listed here total more than 5% of the nearly 4,000 inventions patented by women in the United States between 1809 and early 1895.[47]

Moreover—and most pertinent for the present chapter—150, or over 70%, of these inventions are either machines or mechanical in nature.

## Miscellaneous machines

In this group will be considered women's inventions of machines too striking to be omitted, but from areas not covered in depth here (or of a kind where women seem to have made only occasional or isolated contributions).

A fascinating device, all the more so because it was invented by a black woman, is **Miriam E. Benjamin**'s Gong and signal chair. Miriam Benjamin appears in the Washington, DC, directories between 1888 and 1895, after which she drops from sight.

---

46. **Mary H. Alexander**'s Wood pavement (1873), and **Sarah M. Hoyt**'s two patents for Paving blocks for walkways (1887, 1888). **Lydia J. Cadwell**'s two Pavement patents (1889) apparently pertain to building-floors rather than roads. Cadwell, a versatile multiple inventor, also held several patents for grain driers.

47. The total of 210+ transportation patents assumes that the anonymous Hoboken lady patented her street-sweepers—and is not Eureka Bowne. It also includes a few patents omitted by LWP. The total of women's patents from LWP, by my reckoning, for U.S. patents granted to women between 1809 and February 28, 1895, is 3980, not counting design patents or reissues. As discussed elsewhere (Stanley, "Conjurer"), this figure may be significantly too low. However, since correction is still in progress, I will use the LWP figure for now. The result, nearly 5.3% of the LWP total, agrees rather well with the WB/28 figures for the first quarter of the 20th century (Transportation category: 6.9%).

Whether she moved, died, or married at that point I am not certain, but she was one of four women included on a "list of Colored Inventors in the United States as furnished for the Paris Exposition, 1900" (Diggs, 24). Miriam is listed as a teacher in 1888 (the year of her patent) through 1890, then as a Census Office clerk[48] in 1891, and as a teacher again in 1892, 1894, and 1895. She does not appear in 1893.

The brief title of Miriam Benjamin's invention in the Patent Gazette for July 17, 1888, describes it as a "Gong and signal chair for hotels, etc.," presumably allowing those seated in the lobby to signal porters, bellhops, desk clerks, etc. Whether any hotels adopted her chair is unknown.[49] Benjamin is remembered because, according to Irene Diggs, the United States House of Representatives adopted her chair for use by members wishing to signal pages (24):

> Formerly pages were signalled by members clapping their hands and "the noise incident to this method was frequently a great disturbance to the House proceedings. The gong system involved merely pressing a button on the member's chair which rang a small gong while displaying a signal on the back of the chair."

Patricia Ives of the Patent Office mentions Benjamin in a 1980 article on black and women inventors, but gives no details (Ives, "Patent," 114, 124). It would be fascinating to know how this black schoolteacher came up with such an idea—and what happened to her after 1895. Perhaps the royalties from her chair enabled her to retire from teaching and/or move out of the Capital.

On the other hand, according to the Capitol Architect's Office, the House had already installed desks with electric buzzers for summoning pages by 1873, and the files do not indicate that Benjamin's chair was considered (Thayn, L-10/31/86). This is not surprising, since her patent was not issued till 1888, but it destroys Diggs' proposed motivation for the adoption of Benjamin's chair. The *Afro-American* newspaper for September 7, 1895, says Miss Benjamin *hopes* to get it adopted by the House of Representatives,[50] so any adoption that did take place would be later in the century.

Two intriguing British patents in this category are **Marie R. Marelle**'s Automatic registering apparatus of 1888 and **Louisa Hunt**'s keyless locks of 1881.

In a pass through LWP, we find:

**Maria Procope**, Stockholm, Sweden, Improvement in adjustable tongues for organ reeds (1873)
**Mary A.E. Whitner**, Philadelphia, Improvement in stereoscopes (1875)
**Louise A. Vaidis**, Chicago, IL, Revolving trapeze (1889)
**Ada Van Pelt**, Oakland, CA, Permutation lock (1890)
**Ethel A. Ray**, Florence, SC, Animal trap (1890)
**Amanda J. White**, Wilkesbarre, PA, Mining machine (1892)
**Ethel A. Ray**, Florence, SC, Crane for suspending mail bags (1892)

To the best of my knowledge, none of these women won lasting fame for her invention, but **Ada Van Pelt**'s Permutation lock (1890) is the lock with 3000 combinations

---

48. Interestingly enough, one Samuel Benjamin, living at another address, is listed in 1891, 1892, 1894, and 1895 as a Pension Office clerk.

49. I would welcome information on this point from other scholars. An interesting typo in one of my sources makes it a "Gong and signal chair for *motels* and the like" (Ives).

50. The *Afro-American* mistakenly calls Benjamin the only colored woman to have obtained a patent thus far. On the contrary, at the very least, **Sarah E. Goode** of Chicago had patented her folding cabinet bed by 1885 (Stanley/Zimmerman, 60), and I suspect there were others.

mentioned anonymously by the writer of "Clever Women Inventors" in the *Patent Record* for December 1899, the *Literary Digest* ("Women As Inventors") in 1900, and Marden's *Consolidated Encyclopedic Library* in 1903 (4086). In the *Scientific American* ("Women Inventors," August 19, 1899), Van Pelt, wife of the Oakland, CA, Postmaster, is mentioned by name. As both *SA* and *PR* point out, Van Pelt also patented a widely used House-door letter box (1892) that signals the postal carrier when there is a letter to collect.

Too late for LWP, and issued in her co-inventor's name but assigned to her, are **Dr. Mary Josephine Alsbau**'s Torpedo and Gun patents, both issued on the same day in December 1899. The following "Personals" item in the *Patent Record* (May 1900: 4) reveals that Alsbau was the co-inventor:

> Dr. Mary Josephine Alsbau, principal stockholder in the Just-Alsbau Torpedo Co., and joint inventor of the projectile, denies ever having had the intention of marrying Captain Just, as reported a few months ago. It develops that the captain has a wife and children in Australia, and his present whereabouts are unknown.

Alsbau, however, is also featured in at least one *New York Times* article ("Woman Perfects"), which gives a long technical description of the workings of the torpedo, and credits Dr. Alsbau with its steering mechanism. The projectile is designed to carry a charge of 2000 pounds of dynamite, traveling in two linked but separate parts so as to cause havoc both above and below the water line. The guidance method is said to use Marconi's principles (radio control?). A million-dollar company was formed to manufacture the torpedo.

## Professional women inventors/machine specialists

For a third perspective on women's contribution to mechanics in the century, it seems appropriate to look at some 19th-century Athenas. These notable "contrivers" meet the obvious and objective criterion of having several inventions to their credit, patented or unpatented, all or most of which are machines or mechanical inventions. In addition, several of the women were certainly entrepreneurs, and others may prove to be so on further research. Some of them deserve to be classed as professional inventors, though they might not have given that response to a census-taker.

Most of my examples are American, but one of the earliest is French: **Mlle Crépin**, who, as noted earlier in the chapter, patented a bandsaw in 1846. Further research will be needed, but Kaempffert's statement that she "devoted herself to mechanical pursuits," and his reference to other inventions, is significant, especially since he mentions only two women in his entire book. (The other is **Emily Davenport**, to whom he fails to give appropriate credit.)

An early American woman inventor of decidedly mechanical bent, who might have become a professional inventor and patentee had she not been a Shaker, is **Sister Tabitha Babbitt** (1784–1858). In addition to her circular saw of 1810 (Sprigg, 174) and her still-earlier improvement in spinning-wheel heads, Sister Babbitt independently invented cut (as opposed to individually forged) nails—a contribution to mass production—and a new method of making false teeth (Deming & Andrews, 153, 156, 157; Horgan, 72; White & Taylor, 312; Irvin, "Machine . . . ," 314–5). Both of the latter two inventions may have had a tool or machine component; but because of the Shakers' early aversion to patents, documentation is lacking. Likewise, she may have created many other inventions which she, or chroniclers of the community, failed to mention.

One of the earliest of my documented American Athenas (see accompanying table on page 343) is **Eliza** (or **Elisa**) **Alexander** of New York City, who received at least seven patents between 1857 and 1884. Six of the seven are sewing-machine improvements or attachments. Her earliest invention, patented in 1857 when only forty-odd women had ever received U.S. patents,[51] was a mechanism for guiding the finished sewing work away from the operator. Any machine seamstress would appreciate this device; in commercial sewing it would be indispensable. The speed at which the material was pulled away was regulated by the speed of the sewing mechanism. Her other sewing-machine patents were three braiding attachments (apparently facilitating the attachment of decorative braid to cloth or garments), an embroidery attachment, and a remarkable-sounding device "adapted to guide the material and lay a cord in a fold therein so that when [the cloth is] stitched and straightened, the stitching cannot be seen."

The fact that Alexander's inventive career spanned nearly thirty years, as evidenced by granted patents alone, argues that she took herself seriously as an inventor. Moreover, she may well have had still other inventions to her credit. This is a common pattern for multiple inventors of both sexes.

The fact that all but one of her patented inventions pertained to sewing machines suggests that she had a sewing-related business. Perhaps she owned a small garment factory. At the very least, she probably sold her inventions for profit. Ida Tarbell uses Alexander as an example of women inventor-entrepreneurs (357):

> It must not be inferred . . . that one invention is the limit of a woman's power. Many women have *made a business of inventing and putting their goods on the market.* Eliza Alexander of New York has taken out patents for a sewing-machine, for several braiding attachments for machines, and for lawn tennis apparatus.[52] (Emphasis mine)

**Mary Jane Montgomery**, also of New York, is another early specialist in machines. Her patents—at least those known to me—were few in number (four) and all granted in a short span of time (1864–68). Nevertheless, they cover inventions that are not only quite striking for a woman of her era, but of a sort—an Improved war vessel, an Improvement in locomotive wheels, an Improved apparatus for punching corrugated metals, and a Bridge—that is hard to imagine her whipping up in the kitchen, or creating in moments stolen from housewifery. And, indeed, as already noted, the *Scientific American* in 1903 called her the "one professional woman inventor in America . . ." ("Women Inventors").

Both the *Scientific American* article and the Shaker *Manifesto* for July 1890 (164) noted that Montgomery made money from her inventions—"considerable money" according to the Shaker writer—certainly one criterion for a professional inventor. These sources mention "other devices" or "a number of other patents," suggesting that Montgomery may have had other inventions to her credit besides the four I have found. She may have collaborated with Richard Montgomery (relationship unknown) as on their 1868 joint patent for a Bridge, and on the 1866 patent Richard assigned to her. The *Scientific*

---

51. LWP lists only 39 previous patents; however, LWP omits several women, even at this early date. Noteworthy among them are **Charity Long** and **Elizabeth Wadsworth**. Neither my Storrs paper (Stanley, "Rose") nor its published version ("Conjurer") deals fully with these earliest omissions. I intend eventually to publish a corrected and annotated LWP as a data base, and would be grateful to scholars for information on women who should be included.

52. The lawn-tennis apparatus was not a machine, but a device attached to the bottom of the racquet handle, for retrieving tennis balls from difficult spots. Patented in 1882, it was omitted by LWP.

**Thirteen\* 19th-Century U.S. Professional Women Inventors (Specializing in Machines)**

U = Unknown   P = Patents   I = Unpatented Inventions   T = Total

| Name | Patent or invention dates | Residence | P | I | T+ |
|---|---|---|---|---|---|
| **Eliza Alexander** | (1857–82) | NYC | 7 | U | 7 |
| **Maria Beasley** (c. 1847–?1904) | (1878–98) | Philadelphia, Chicago | 15 | U | 15 |
| **Helen Blanchard** (1840–1922) | (1873–1914) | Portland, ME; Boston, Phila., NYC | 28 | U | 28 |
| **Mary Carpenter** | (1862–94) | Buffalo, SF, NYC | 13\*\* | U | 13 |
| **Ella Gaillard** | (1874–92) | SF, NYC, MD | 8 | 7+ | 15 |
| **Harriet Hosmer** (1830–1908) | (1830–1908) | Watertown, MA; Italy, London | 5++ | 2(U) | 7 |
| **Amanda Jones** (1835–1914) | (1873–1914) | Clinton, WI; Chicago; Junct. City, KS; Brooklyn | 12 | 2(U) | 14 |
| **Margaret Knight** (1838–1914) | (1871–1912) | Framingham, Boston, MA | 22¢ | c. 67 | c. 89 |
| **Bertha Lammé** (1869–1943) | (1893–1905) | Pittsburgh, PA | 0 | U# | U |
| **Mary J. Montgomery** | (1860s–'70s) | NYC | 4 | U | 4 |
| **Mary S.** (c. 1851–80) | (c. 1870–80) | St. Louis, MO | 53#+ | U | 53 |
| **Emily E. Tassey** (1823–99) | (1876–80) | McKeesport, Pittsburgh, PA | 5 | U | 5 |
| **Harriet Tracy** (1834–1918) | (1868–93) | NYC, New Brighton, NY## | 16 | 1 | 17 |

\*Mildred Blakey of PA, who appeared in earlier versions of this table, proves to be male.

+This fig. is a minimum, for usually I know only number of patents held; even where I know of unpatented inventions, I do not know how many. Knight's total comes from an interview in old age, and may be a low estimate.

\*\*Not counting her several reissue patents.

++Hosmer was a childhood inventor, and spent her last 30 yrs. working on magnetic motors and perpetual motion. She patented her synthetic marble (US), and applied for 4 Brit. patents for magnetic motors (one finalized). She may also have patented her mechanical page-turning device for musicians. Her new modeling technique in sculpture was unpatented. Thus her total is highly uncertain, but should be *at least* 7.

¢Knight is usually credited with 2 other patents assigned to her.

#Lammé held no patents in her own name, but worked on a corporate innovation team (Westinghouse) for 12 yrs.; the numbers of important motors and generators she helped design could arguably equal the total of patents received (or applied for) by her brother Benjamin Lammé and other teammates during those years.

#+No patents in own name; sold 53 inventions to agents who patented them.

##Tracy was born in Charleston, SC, and died in London, but patented from NY State.

American writer speaks of "Miss Montgomery" as if she were not only still alive, but still active in 1903. If she had done nothing in technology since 1868, she would likely have been forgotten.

Beginning her career about the same time, and far more prolific in patents held, was **Mary P. Carpenter** (later **Hooper**) of Buffalo, San Francisco, and New York City. Over a period of more than thirty years, from 1862 to 1894, Carpenter received at least 13 patents, not counting her several reissues. At least six of these were for machines or mechanical inventions, several of them pertaining to sewing machines.

Mary Carpenter's first patent, granted in 1862 while she lived in Buffalo, NY, was for an Ironing and fluting machine, for creating folds or soft pleats in ladies' garments. Esther Berney mentions this machine as one of three particularly interesting devices from the 1860s. Carpenter's machine had "a portable (handled) cast iron stove that burned coke or charcoal. To this stove could be connected twelve tubes, more or less, of different forms and sizes. Flutes could be made as quickly as the operator could pass the fabric between and alternately over and under the tubes" (Berney, 45).

Just after the Civil War, Carpenter patented an Improved mop-wringer, and in 1870–1 the invention that brought her to the pages of the *Scientific American* and eventually to the Philadelphia Centennial: her Sewing machine feeding mechanism, needle and arm. In the late 1860s she moved West, for two of the three patents that presumably covered this important invention list her residence as San Francisco. Did she come, as **Hannah Suplee** came, in search of Chinese artisans to create her sewing-machine needle? Or in search of greater opportunity in general? Is she the Miss Mary P. Carpenter listed in the Langley Directory of San Francisco in 1869–70 as an assistant teacher at the Greenwich Street Cosmopolitan School?

In any case, by 1872 Carpenter was in New York City, where she seems to have stayed for the rest of her patenting career (1894). In addition to her three original patents for the Carpenter needle-and-arm mechanism, Mary Carpenter also patented a Machine for sewing straw braid (1876) and a Work-holder (1894) classed under "Sewing and Spinning" by the LWP compilers.

Carpenter also patented a Barrel-painting machine, in 1876. In that same year she exhibited a sewing machine at the Philadelphia Centennial (Warner/Trescott, 111–2). At some point between these 1876 patents and 1885, when she patented her Grated shovel, Mary Carpenter married, becoming "Mrs. Mary P.C. Hooper," under which name she received her last five patents.

Mary P. Carpenter was also an entrepreneur, founding at least two companies: the Carpenter Sewing Machine Needle Co. and the Carpenter Straw Sewing Machine Co. The former was in business in 1870 when the *Scientific American* article went to press; the latter existed in 1885, for Carpenter assigned to it her Reissue Patent #10,600 for her 1876 Machine for sewing straw braid.

Regrettably, Deborah Warner says nothing about Carpenter except to note her marriage and list several of her patents. Three of them, incidentally—including what is probably the most important, her first sewing-machine patent—are omitted by LWP. One of the salient features of her sewing arm and needle was its extreme ease of threading. The operator guided the thread through a very large hole in the needle near the top—so large that she could use touch rather than vision if necessary—and then drew the thread down and into the eye through a split in the needle. This needle was not only "self-threading," but self-acting (or self-setting) thus avoiding the complex adjustments required on many machines before sewing could begin ("Carpenter . . .").

**Harriet Ruth Tracy** (1834–1918) of New York and New Brighton (Brooklyn), NY, received at least 16 patents between 1868 and 1893. Virtually all of them are unmistakably machines, or mechanical in nature. In addition, her Sewing-machine table has mechanical parts. Her crib attachment for bedsteads and her fire escape are probably either mechanical or partly mechanical in nature; only her Design Patent, #22,544 of 1893, for a Sewing-machine frame, seems likely to be nonmechanical. Even her Stove (1882) fits in front of a fireplace grate, and looks collapsible.

Her other ten patents pertain mostly to sewing machines, including feeding mechanisms, loop-takers, and actuating mechanisms, as well as two Shuttle-actuating mechanisms. Three, however, deal with elevators: an Automatic elevator hatch, an Elevator and a Safety device for elevators (1890, 1892). Harriet Tracy evidently did not exhibit at the Centennial in Philadelphia in 1876, though by then she already had two inventions to her credit. It was at the Columbian Exposition in Chicago in 1892–3 that she shone. The Board of Lady Managers praised her for having invented a "rotary shuttle, lock, and chain-stitch sewing machine—whose lower bobbin carried more than a thousand yards of thread, 'an achievement . . . that has baffled a generation of masculine inventors.' " Moreover, Tracy's safety elevator was installed as the public elevator in the Woman's Building at the Fair. This so-called gravity elevator featured "automatic platforms that keep the shaft constantly closed, and prevent any person falling through it, or flames ascending it" (Weimann, *Fair,* 431).

Her safety elevator, in fact, may be her most significant invention. It was evidently invented some years before it was patented, for in December of 1881 it rated an entire column in *Demorest's Monthly.* At that time, an elevator incorporating "Tracy's safety automatic accident- and fire-proof floors for elevator shafts" was already installed in "the large new building" at No. 80 E. 14th Street [New York City], "where all interested can witness their operation." According to the *Monthly* writer, the Patent Office had pronounced Tracy's invention "entirely original." The *Monthly* writer gives welcome details on the nature and operation of Tracy's solution to the dangers posed by elevator shafts:

> The invention consists of a double set of light, yet substantial, automatic floors, one series being arranged above, and the other below the elevator. The floors are strung on small wire ropes at each corner. These wires being fastened at the ends, above and below the elevator, are run over wheels at the top and bottom of the shaft, so that when the elevator ascends, the floors below it are caught by balls arranged at proper intervals on the wires; these balls vary in size, the upper ones being small enough to pass through the floors that are below, until they reach the one to be carried up, and being too large to pass through this floor, it is carried to its proper place, and the same with all the floors below, each in its turn. The floors above the elevator are all carried with it to the top of the shaft. As the elevator descends, they rest respectively on projections, placed in suitable positions in the shaft, being so arranged that they pass in succession until the projections intended for each particular floor are reached, where they remain on a level with the main floor until the ascent of the elevator, when they are carried to the top as before.

The arrangement was so simple and inexpensive that it could be applied either to new elevators or to elevators already in use. Even if it had been expensive, it should have been worth installing for its safety benefits alone. Particulars on the Automatic Floors were available either from G.C. Tracy, 82 E. 14th Street (husband?) or from W. Jennings Demorest, 17 E. 14th Street, New York.

Tracy lived in New York City for more than two decades, 1868–89, patenting at least six inventions during that time. One of the six was an Automatic elevator hatch of 1882, but not the automatic floors as described. This invention is patented instead in the name of **Ellen Demorest** herself (#259,105 of June 6, 1882: Automatic floor for elevator shafts). Thus the Demorest article raises an intriguing tangle of questions: Why did Harriet Tracy allow Demorest to patent her invention? Were their husbands business associates, with G.C. Tracy in debt to W.J. Demorest? Did Ellen Demorest lend Harriet Tracy money to help her develop her invention? Was W. Jennings Demorest not only an inventor, but an inventor's agent or consultant, helping others to market their inventions? Did the Demorests use their influence or financial backing to get the elevator manufactured and tried in a large New York building?

Perhaps we shall never know. At any rate, Tracy seems to have found her independent feet as a commercial-scale inventor at some point in the 1880s, for during an amazing burst of creativity between 1890 and 1893, she received ten more patents. And as already noted, she received further technical and commercial recognition during the Columbian Exposition. During this latter period she lived in New Brighton, NY.

Doubly noteworthy, in that her five known patents are mechanical and all pertain to ships or waterways, was **Emily E. Tassey** (1823–99) of McKeesport and Pittsburgh, PA. In July 1876 she patented an Improved apparatus for raising sunken vessels. Later that year, on one December day—she received three further patents: an Improvement in siphon propeller-pumps; one, mentioned above, for Propulsion of vessels; and a Dredging machine. LWP omits the last two of these three same-day patents (#184,997 and -998) which is puzzling, since it includes the first, and all three would have appeared immediately together whether the clerks were using an alphabetical name list or a numerical-order Gazette.

Tassey's 1876 pump is designed "to provide means of discharging large portions of water—as from a coffer-dam—into the river." It appears to be a combination of the siphon principle and the water-raising screw, the screw consisting in this case of a steam-driven screw propeller (Patent #184,996 of December 5).

In 1880 she received a second patent for a Siphon-propeller pump, which the LWP compilers did include.

Her first and last patents list her residence as McKeesport, whereas the other three (granted only months after the first) place her in Pittsburgh. The distinction may be of minor significance, since McKeesport is only 14 miles from Pittsburgh and may have been seen even in the 19th century as a suburb or satellite of the larger city. Both cities are on the Monongahela River, McKeesport at the mouth of the Youghiogheny. Beginning possibly as early as the 1830s, McKeesport was an important boat-building center, and Pittsburgh had a dry dock for lifting steamboats out of the water for repairs as early as 1836 (*EB*/58; Abbot & Harrison, 31).

Emily's middle initial, *E*, stands for Evans—a family name—and it is tempting to think that she might be related to the famed American inventor Oliver Evans (1755–1819), who also had ideas for steam propulsion of ships (*DAB*). At any rate, the Evanses were prominent in early McKeesport history, and thus some information is available about Tassey.[53]

Emily Tassey was born Emily Evans Rowland, daughter of Hugh Rowland and Hannah Evans (and granddaughter of James and Emily Alexander Evans), December 15, 1823, at the home of her grandfather Evans just outside McKeesport. Educated at Steubenville Seminary, she taught in the McKeesport School before her marriage. In 1845 she married

---

53. I am grateful to Eunice Emery, Reference Librarian at McKeesport's Carnegie Free Library, for the following information.

a lawyer, William D. Tassey, who died in 1857. The three children she had then to support may partly explain her later enterprise and achievement. She taught in the public schools of both Pittsburgh and McKeesport, and in the Pittsburgh High School. My sources record no shipbuilding business she bought or founded, but she did conduct a "young ladies' school" in McKeesport, and was "well known for her business ability." She lived in McKeesport until her death just before the turn of the century (Abbot & Harrison, 31–2; Emery, L-3/9/87; *Old Home Week,* 326).

Virtually unsung in her own century, as now, is **Ella N. Gaillard.** Between 1874 and 1892 she patented eight inventions, at least four of them mechanical. By 1893, according to Lydia Farmer—who discusses her anonymously (420)—she had "distinguished herself by many inventions." Farmer lists a dozen, seven of them distinct from the eight patents. Thus by 1893, Gaillard had at least 15 inventions to her credit. Nine of these are rather obviously mechanical in nature (dates are patent dates):

| | |
|---|---|
| Improvement in musical tops (1876) | automatic toy |
| Fountain (1885) | folding basin |
| Musical watch (1886) | folding flatiron |
| Comb. paperweight, calendar, and | lock revealing whether it is locked or not |
|     musical watch (1889) | carriage telephone |

One other, the "novel bird-cage chain," may be mechanical, depending upon the mode of attachment to the cage (e.g., with swivel), but I have not counted it here.

The spring-powered Fountain "renders music while throwing a stream of water . . . with such precision that not a drop escapes to spoil the carpet" (Farmer, 420). The carriage telephone is nearly a hundred years ahead of its time.

Less obviously mechanical is Gaillard's eyeless needle. Much used in surgery, this 1874 invention used a delicate internal screw mechanism to attach catgut or other cord to a needle without doubling. If this use of the screw is counted, ten of 15, or two-thirds, of Ella Gaillard's inventions are mechanical. Since her known patents span nearly twenty years, and her inventive career doubtless several more, she was at the very least a serious inventor.

Gaillard lived in San Francisco at the time of her first patent, the eyeless needle of 1874. By 1876 she was in New York City, where she remained at least till 1892, patenting inventions dealing with clocks, music boxes, and the like. Her other inventions include, in addition to the ones mentioned above, a dress shield, an Electric illuminating apparatus (1886), and a sweatband for hats. By 1893 she was apparently in Maryland, where Farmer learned of her. She may have belonged to a Swiss watch-, toy-, or music-box-making family, where she became interested in the family enterprise and entered the work herself.

Between 1878 and 1891, **Maria E. Beasley** (c. 1847–1904?) of Philadelphia received 14 patents, eight of which had to do with the barrel-making industry. Six of these eight were machines—two designated simply as Barrel-making machines, two Barrel-hoop-driving machines, a Barrel-stave-shaping machine, a Barrel-setting-up machine. A seventh patent cryptically described as "Notching and cutting hoops for barrels," turns out to be a process for cutting several parts at once. Two of the six patents not related to barrel-manufacturing were also machines, or mechanical in nature: a Machine for pasting shoe uppers, patented in 1882 and a Steam generator (1886). Later, in Chicago, she patented an anti-derailment device for railway cars (1898). Thus, nine of Mrs. Beasley's 15 patents, spanning 21 years from the late 1870s to the late 1890s, were for machines.

Deborah Warner does not mention Maria Beasley as a Centennial exhibitor, at least not in the Women's Pavilion—and of course her first patent was not granted until 1878. Since she lived in Philadelphia, however, the Centennial with its impressive exhibit of women's inventions may have inspired her to go ahead with an inventive career she had been considering. She apparently exhibited her barrel-hooping machine at the World's Industrial and Cotton Centennial Exposition in New Orleans in 1884 ("Something for Everyone").

Maria Kenny married Samuel Beasley in 1865. Though the census-takers have her "keeping house" in 1880, and list no occupation for her in 1900, city directories tell a different story. In the Philadelphia directories from 1880 (her first appearance) through 1890, with one break in 1887, she is a dressmaker. And in the Chicago directories 1891–6, she lists her occupation as Inventor! Moreover, Beasley was mentioned in several contemporary and nearly contemporary sources. Charlotte Smith says that Maria Beasley turns out hundreds of ready-made barrels in a day—implying that she is a manufacturer as well as an inventor. Both oil and sugar refineries, according to Smith, are obliged to this inventor, and pay her "profitable royalties" from her invention (*WI*/1: 4). Maria Beasley is also mentioned by Ida Tarbell (357), Minnie J. Reynolds (7), and Mary Logan, who notes that her barrel-hooping machine alone brings her $20,000 per year (885). Logan wrote in 1912, when Beasley's known barrel-making patents would already have expired. However, Beasley's royalty agreements with the oil and sugar industries may have outlasted the patents; she may have kept other inventions or improvements as trade secrets; or Logan may be citing 19th-century figures. Clearly, however, in those pre-income-tax days an annual income of $20,000 was an enormous fortune.

Beasley's inventions were not merely profitable, but diverse—she patented a foot-warmer, a baking or roasting pan, and two improved life rafts in addition to the machines already mentioned! It should be noted, too, that she took the trouble to obtain British patents for at least two of her inventions—her Life-saving raft (1880) and her invention for Hooping casks, etc. (1882). Thus by achievement, by repute, and by her own evaluation, Maria Beasley could certainly be considered a professional inventor.

It is long past time that we heard these stories of women who achieved in the most quintessentially "male" part of a man's world—the machines arena—in the most unlikely of centuries. Also long overdue is the realization that such achievement is by no means so rare or outlandish as we have been led to believe. At the same time it would not be fair to tell these stories of competence rewarded without balancing them with two lives that represent darker but equally possible outcomes for a 19th-century Athena. These are the lives of **Mary S.** of St. Louis and **Bertha Lammé** of Westinghouse.

Mary S. (d. 1880) was one of the most prolific woman inventors of the 19th century, and we do not even know her last name. She followed what has usually been seen as a male pattern, starving in a garret room in order to invent; yet, to the best of my knowledge, not a single patent bore her name when she died. Something about that stark story speaks for all the unsung women inventors of her day, just as Sadie Sachs later spoke to Margaret Sanger of all the women desperate to control their fertility. Mary S. is the reason Charlotte Smith became interested in women inventors. Her story and others like it may even have been what politicized Smith and made the rest of her life a battle for women's economic rights; for, as Smith puts it, Mary S.'s case "I regret to state is only *one of my similar hundred ones,"* and "best illustrates the position of industrial women, whom I have the honor to represent" (emphasis mine). As Charlotte Smith tells the story in her short-lived periodical *The Woman Inventor* (/1: 2), Mary S. was the daughter of an inventor. This man, like his daughter after him, had a passion for inventing. Unlike her, he also liked to drink. He sold his inventions for five or ten dollars, with the result that the family was soon destitute. When his wife died, the younger of the two daughters was placed in a "charitable institution," and Mary went to work in a drygoods store for three dollars a week.

According to Smith, Mary had all of her father's inventive genius. Unfortunately, she seems also to have inherited his lack of business sense, perhaps exacerbated by a female lack of confidence. Says Smith,

> She patented in a lawyer's name a valuable invention, which has since proved a grand financial success, and yet she received in full payment the sum of five dollars for her invention. Her reason for not patenting the same in her own name, was that if it had been known the invention of a woman, it would have been regarded as a failure.

Nevertheless, she kept on inventing. Charlotte Smith gives us some idea of Mary's achievement, though regrettably without describing any of the inventions: Mary S. "perfected and executed thirty-seven different inventions that her father had commenced," and she created and sold to agents in New York, Washington, and St. Louis, "sixteen different inventions of her own." This gives her the rather astonishing total of 53 patented inventions in a brief life. (Since she was 24 when Smith met her, about 1875, as a result of an *Inland Monthly* article on women's occupations, including that of inventor, she can scarcely have been over thirty when she died.) In including her here I am assuming that, since male patent attorneys and agents were willing to buy her inventions and patent them in their own names, at least a significant share were machines or mechanical in nature.

The "grand financial success" all went to the above-mentioned lawyer and his ilk, not to Mary S., who as Smith notes "furnished the brains for man to create wealth, when he himself had not the genius to invent, but only the craftiness and cunning to take away the products of that woman's inventive labor." Meanwhile, Mary herself, in her attic room, "suffered often for the necessaries of life," lacking even tools, as Smith reports giving her the only tools she had at Christmastime in 1878. By 1880 Mary S. was dead. "Her secrets, her genius, her skill," says Smith, "are buried with her."

But, like Sadie Sachs (whose death of a botched abortion in the arms of Margaret Sanger inspired Sanger's birth-control crusade), perhaps Mary S. did not die in vain. Before she died, she begged Smith "to see that in the future justice was done to woman inventors"; and Smith seems to have accepted the charge: "as I resided in Washington, and could do so much for my sex, I have been creating public opinion, as well as demanding justice for women as breadwinners, women as creators of wealth; women as industrial factors, for the past twelve years." It was because of Mary S., then, and her unsung sisters, that Charlotte Smith lobbied so tirelessly for the creation of a list of American women patentees, which finally appeared in three installments, 1888–95 (LWP). Flawed as this listing is, it is still a priceless source on the inventive achievement of 19th-century women.

**Bertha Lammé** (1869–1943) seems at first glance to be an entirely different case. And indeed her family background was much more comfortable both in childhood and in adulthood, her education far more advanced, and her professional life far more rewarding, at least financially, than Mary S.'s. We know Bertha Lammé's full name and a certain amount about her personal life; her daughter was still alive as of 1983. However, despite the fact that Bertha worked on a famous team in the early days at Westinghouse, with a famous brother who wrote his autobiography, we know very little about her precise contributions to the work of that team, except that she was in charge of the mathematics in a program of design by mathematics. Guenter Holzer's research at Westinghouse provides the only clues except for what can be deduced from the patents granted to Benjamin during the years of Bertha's employment. Holzer points out that Bertha worked in the direct-current field, and that in 1894, one year after she joined the Company, Benjamin completed work on the induction motor for commercial use. He was also closely involved at about this time in designing the rotary converter. "Based on their previous close relationship," says Holzer, "one can assume that Bertha Lammé's first engineering projects involved the induction motor and the rotary converter" ("Bertha," 6–7).

Presumably she also worked closely with Russell Feicht, another Westinghouse engineer, who became her husband. He designed a three-phase, 80-horsepower AC motor to propel a boat-pulling machine for use on the Miami Canal. Feicht also developed the first large two-pole induction generator, used with rotary converters in the copper industry. Just after the turn of the century, he designed a 2000-horsepower induction motor, the largest then in existence, to run the water cascade at the 1904 World's Fair in St. Louis (*ibid.,* 7).[54]

Bertha Lammé was well enough known in her own day to be mentioned as "the particular star among American women electricians" (i.e., women who work in the field of electricity and electrical engineering or technology) in an article titled "Women in Electricity" that appeared in the *Woman's Journal* in December 1899 (based on a story in the New York *Sun,* November 26, 1899). Calling her a "full-fledged electrical engineer, with five years' practical work to her credit," the writer notes that Miss Lammé "designs machinery, makes calculations, and does exactly the work of a male electrical engineer."

The giant motors and generators created during this fertile early period formed the basis for the Westinghouse empire. Bertha Lammé was a part of the excitement for twelve years, from 1893 to 1905. According to Alva Matthews, herself an engineer who has studied the history of women engineers in this country, Bertha "collaborated closely with her brother on the development and invention of many of the company's basic patents." Matthews also says that Bertha herself held several patents.[55] Yet her name does not appear on any patent I have found, and her daughter maintains that Bertha never spoke to her of her work (Boyer, L-8/27/83). Her brother and co-worker says only that "she took up the work of calculation of machines" (Lammé, 91).

Bertha Lammé's story, in a nutshell,[56] is this: She was born Bertha Avanella Lammé on December 16, 1869. She grew up in rural Ohio, near Springfield, in a family of Dutch ancestry. In her early school years she had a reputation as the brightest and most studious in her class. Much influenced by her older brother Benjamin, who directed her reading, she followed him into engineering. In 1893 she graduated from Ohio State University in mechanical engineering—the first woman to do so—with a specialty in electricity (Matthews, 44). Benjamin was already chief engineer at Westinghouse. Before the year was out, Bertha came to the Company at the request of one Mr. Schmid, joined her brother's team and, as already noted, stayed for twelve years. In 1905, at 36, she married

---

54. In fact, to get the possible universe of patented machines that Bertha Lammé may have worked on, we could list the patents obtained—or filed—during her years of employment by the entire team.

55. "Some Pioneers," an important unpublished speech on early women engineers in the United States (44), kindly sent me by Martha Trescott. As of 1980, Westinghouse had no information beyond Holzer's. I would be most grateful for any further concrete information on this point.

56. I am grateful to Guenter Holzer of St. Cloud, MN, an independent researcher who generously shared with me his information on Bertha Lammé's life and professional contribution (L-12/24/82; 1/12, 1/20, 1/22/83; PC-11/15/82), and to Martha Trescott for directing me to him. For more light on Bertha Lammé's life and the perplex of her possible inventive achievement, see Martha Trescott's forthcoming book on the history of women engineers.

another team member, Russell Feicht, whereupon she left engineering, had one daughter, Florence Lammé Feicht, in 1910 and seems never to have looked back.

A clue to the mystery may lie in one of the two sentences Benjamin devotes to Bertha in the main text of his autobiography: "She had taken an Engineering degree in the Ohio State University more for the pleasure of it than anything else . . ." (Lammé, 91). Perhaps she was actually the more brilliant of the two, and unconscious envy accounts in part for his near-total silence about her contributions at Westinghouse.

No one would deny Benjamin Lammé his own brilliance and his multitude of patents, of course. Nor would Schmid have entrusted Benjamin with his (Schmid's) pathbreaking idea of design-by-mathematics—forerunner of computer-aided design—had he not had impressive gifts. Moreover, Benjamin appears to have been a loving and helpful, if somewhat domineering, brother during Bertha's childhood and up through her undergraduate days. For example, when she lost most of a year at Ohio State because of illness, he tutored her so that she could still pass the examinations and graduate with her class, and Bertha reports that he had big plans for the two of them to go into the business of designing mechanical toys after she graduated. Holzer finds mutual respect for each other's intelligence between the brother and sister (Lammé, appendix; Holzer, "Bertha," 6, 5).

Nevertheless, a real question remains, pending further evidence, whether Bertha Lammé did not actually contribute more to the early achievements at Westinghouse than revealed by "taking up the calculation of machines"; and whether Benjamin Lammé as chief engineer could conceivably have given her such an important charge if she did not possess something more than a good head for numbers. An unpublished typescript from Westinghouse files calls her a "designing engineer," and "one of the few women who have made a notable success of this work" ("Benjamin," 6).

It is also interesting to note that Ohio State University named an important laboratory after *both* Lammés, not just Benjamin.

What we can say for certain is that she held a professional position which by its very nature would seem to dictate a basic contribution to many of the most famous of Westinghouse's early motors and generators, machines that ushered in the modern industrial era in the United States. And, as her work for twelve years, and the work of her team, was to invent these machines, she can be considered a professional inventor. Holzer considers her the first representative of a new breed of engineers who could succeed by mathematical ability and book learning alone, without extensive practical experience (8).

Finally, let us look at three women whose professional lives, like Bertha Lammé's, span the turn of the century from 19th to 20th, but on whom we have patent information: **Helen Blanchard, Harriet Hosmer,** and **Margaret Knight.** Blanchard and Knight are two of the most prolific American women inventors of the 19th century, whether considered from the standpoint of patents held or of inventions credited to them.[57] One of them, **Margaret Knight,** when her unpatented inventions are considered as well as her patented ones, is one of the most prolific American women inventors in U.S. history. Except for our so-called Lady Edison, **Beulah Henry** (who holds more patents, but probably has about

---

57. **Mary S.** of St. Louis, as just discussed, created more *patented inventions*—53, all told—making her the most often patented woman inventor of the 19th century, though none of the patents was in her name. **Catherine Griswold,** a highly prolific corset inventor and patentee, held 31 patents, thus surpassing both Margaret Knight and Helen Blanchard in *numbers of patents.* Indeed, *considering only number of patents held in one's own name,* Mme Griswold is the most prolific American woman inventor of the 19th century, though several of hers are Design Patents. Another very inventive and enterprising corset-maker and

the same number of inventions to her credit), Knight scarcely had a rival until the last quarter of the 20th century.

Mildred Blakey, a prolific Pennsylvania inventor originally included here because listed by LWP (1890), proves to be male.[58]

**Harriet Goodhue Hosmer** (1830–1908) of Watertown, MA, was the only child to survive infancy. When her mother also died, her father focused all hopes on her, teaching her to ride, swim, and shoot in order to strengthen her health. An independent-minded tomboy, Harriet was often in trouble in school. Luckily, her father sent her to Mrs. Charles Sedgwick's school, where her individuality and energy were not crushed.

In childhood she had showed a mechanical bent, and by her teens, encouraged by Fanny Kemble, set up a studio at home where she practiced drawing and sculpting, and experimented with "little machines and household devices," as her *NAW* biographer put it. I have no details on these youthful inventions.

Art—specifically sculpture—was Harriet's first love, however, and her immediate problem was to find some way to study human anatomy. A schoolmate's father persuaded a physician friend in St. Louis to give her daily instruction equivalent to his medical school lectures. Eventually she studied in Rome, becoming the first world-famous U.S. woman sculptor. Such was her success that replicas of her statue of "Puck" sold for $1000 each in the late 1850s. The State of Missouri commissioned a colossal statue of Thomas Hart Benton, the City of Dublin purchased a work, and she was widely exhibited in London.

In the late 1870s Hosmer's interest shifted back to mechanics and invention, and after 1880 her sculpture output slacked off. She got her first patent for "motive power" in 1878 (communicated from Rome to Henry E. Newton, whose name appeared on the patent), and continued with three others on the same subject (two provisionally protected only) dating to 1880 and 1881. In addition to these British patents, she got a U.S. patent for Processes of making artificial marble in 1879. The marble, created by subjecting limestone to moist heat and pressure, was not sculpture-quality, but very good for interior architectural uses, and apparently a commercial success. The *New York Evening Post* correspondent who interviewed her in London in 1878 had high praise for the product— "as perfect marble to all appearances as ever was quarried"—and for its varied and beautiful colors: pure white, deep black, a delicate antique cream, and "rarest green and warmest rose-colored marbles."

Harriet also invented a page-turning device for musicians, and a new sculpting method involving a wax-coated plaster-of-paris model for the statue. The inch-thick wax allowed fine-finishing work to proceed without the constant care necessary for a clay model. I have found no patents for these two inventions.

The U.S. Patent Office was not receptive to perpetual-motion apparatus. However, her

---

designer, **Mme Lavinia Foy**, received 19 patents, thus ranking below **Helen Blanchard** but above **Harriet Tracy** and **Maria Beasley**. If I excluded Blanchard and Knight's early-20th-century patents, they would be surpassed by a few more 19th-century women—such as **Eliza D. Murfey** with 23 patents before 1880—but these two women seem essentially 19th-century figures, and thus entitled to have their whole inventive achievement discussed in terms of that century.

58. Since writing my papers "A Rose by Any Other Name . . ." and "The Patent Clerk as Conjurer," I have researched literally hundreds of 19th-century patentees with ambiguous first names, some listed on LWP and some not. This research has revealed many more errors of omission and a few, like this, of inclusion. Having considered Mildred unambiguous, however, I am indebted for this correction to Anne Macdonald (412n141).

work evidently impressed some British scientists, and the *Post* reporter noted that "if the testimony of some of the best informed scientists of England is worthy of trust, she has placed her name on a level with those of Fulton and Morse (*NAW;* Brit. pat. records; "Miss Hosmer's . . .").

**Helen Augusta Blanchard** (1840–1922) lived a long and productive life. She held at least 28 patents dating from 1873 to 1915, founded at least one company to manufacture and/or exploit one of her most important inventions, and was well known enough in her day to be mentioned by several contemporary writers and publications—Ida Tarbell,[59] Willard & Livermore's biographical dictionary, the *Scientific American,* Minnie J. Reynolds, and Mary Logan, for example—though curiously enough, not by Matilda J. Gage or Mozans or the *New York Times* obituary columns.

Helen Blanchard is totally unknown today. Grace Rogers Cooper's sewing-machine history mentions her only as having one model in the Smithsonian collections, for example, and she fails to appear in a list of "all" United States sewing-machine manufacturers. What she should be famous for—*her invention of zigzag sewing and the machine to do it*—is usually credited instead to the Singer Company in its industrial applications and to Pfaff and Necchi for home machines. Presumably, one or more employees at these companies independently, later, invented the necessary machine modification to produce this kind of stitch. During the customary patent search, the companies (who may have formed a royalty-sharing agreement) became aware of a pre-existing patent (probably Blanchard's; possibly she threatened to sue), and rather than fight an expensive legal battle or license her machine and pay royalties, they simply waited until her patent expired and then marketed their own machines. "Numberless annoyances and obstacles" beset Blanchard as an inventor, say Frances Willard and Mary Livermore, "in the shape of . . . mercenary infringement of her rights and unscrupulous assaults upon the products of her brain" (97; Garbutt, L-2/5/83). Perhaps this was one such case.

Probably few achievements of her life gave her more satisfaction than buying back the family homestead—which still stands in Portland, ME—and the business block lost some years earlier by her father. This looks small to us beside her inventive achievement, but may have been one of her major incentives for entering the business world. The patents tell their own story:

| | |
|---|---|
| Boston: Sewing machine | (1873) |
| Boston: Sewing machine | (1874) |
| Boston: Improved sewing machine | (1875) |
| Boston: Improved sewing machine | (1875) |
| Boston: Elastic seams for garments | (1875) |
| Boston: Improvement in methods of uniting knit goods, etc. | (1875) |
| Boston: Improvement in elastic gorings for shoes | (1875) |
| Boston: Welted and covered seam | (1876) |
| Boston: Crochet attachment for sewing machines | (1877) |
| Phila.: Sewing machine needles | (1882) |
| Phila.: Spool-case | (1883) |
| Boston: Pencil-sharpener | (1884) |

59. Tarbell says (357) that Helen Blanchard has received 11 patents by the time of her writing (1887), but actually she had 12 patents by then, her twelfth having been granted in 1884. Missing one patent in a hand-eye count of LWP, however, is easy to do, particularly if Tarbell was looking, as she indicates, at a MS (complete for patents through Dec. 14, 1886) rather than the printed result, which was not published until 1888.

| | |
|---|---|
| Phila.: Corset-cord fastener | (1888) |
| NYC: Securing reeds and cords to edges of material | (1893) |
| NYC: Sewing needle | (1893) |
| NYC: Surgical needle | (1894) |
| NYC: Making split needles | (1896) |
| NYC: Split needle | (1898) |
| NYC: Sewing machine | (1898) |
| Phila.: Sewing machine needle | (1900) |
| Portland: Hat-sewing machine | (1901) |
| Portland: Over-seaming machine | (1901) |
| Portland: Seam for sewed articles | (1901) |
| Portland: Hat and applying sweatbands thereto | (1901) |
| Phila.: Hat-sewing machine | (1901) |
| Phila.: Hat-sewing machine | (1907) |
| Phila.: Selvage sewing | (1914) |
| Phila.: Hat-sewing machine | (1915) |

At first glance, Helen Blanchard seems most noteworthy for the sheer numbers of her patented inventions—more, if we count her 20th-century patents (and exclude Griswold's design patents) than were held by any other 19th-century woman, including even the much better known Margaret Knight! However, although numbers are significant, Blanchard's achievement has more striking claims to our notice: one, that her inventions are mostly mechanical; two, that her machines are mostly commercial—that is, intended for the factory rather than the home; and three, that many if not most of them were actually used in the industries for which they were intended.

Helen Blanchard is also noteworthy for finding time in the midst of such a busy life to secure a "grateful welcome into the social life of the metropolis" (New York), and for "aiding with open-hearted generosity the meritorious efforts of struggling women wherever she has found them." In fact, according to Willard and Livermore, "she has distinguished herself as a benefactor of her sex." It would be gratifying to know some specifics here—as indeed it would to know much more about her in general—particularly about her childhood and schooling, and the influences that fostered her mechanical and inventive ingenuity.

What we do know is that a model of her first patented invention, the so-called over-seaming machine (1873) is in the Smithsonian's collections. We also have a certain amount of personal and family information.

Helen Augusta Blanchard was born in Portland, ME, in 1840,[60] daughter of a wealthy shipowner and respected businessman named Nathaniel Blanchard and his wife, Phoebe Buxton Blanchard. There were five other children: two daughters, Louise Phoebe and Persis E., and three sons, David H., Augustus, and Albus R(ea?). According to Willard and Livermore (97), Helen was "a lineal descendant" of the Huguenot exile Thomas Blanchard, who came to the United States in 1639. She was therefore very distantly

---

60. Sources conflict on the date of her birth. Her obituary in the Portland *Evening Express & Advertiser,* January 13, 1922, says she was born on November 25, 1839. However, Pat (Mrs. John) Curry of Richmond, California, who has been studying the Blanchard family genealogy in some detail, cites Portland Birth Records, Bk. 6, p. 143, stating that Helen Blanchard was born on October 25, 1840 (PC-1/30/86; L-3/30/87). The 1870 Portland census lists Helen as 26 years old, which would have her born still later, in 1844! In such conflicts, it would seem that the city birth records should prevail. I am grateful for this information to Arthur Gerrier of the Maine Historical Society and to Joan S. Hayden and Pat Curry, genealogists of Portland, ME, and Richmond, CA, respectively.

related to another descendant of his, the prominent early American inventor Thomas Blanchard.[61] This male inventor was descended from the exile's son also named Thomas; presumably Helen was descended from a different son (Nathaniel?).

Already in girlhood Helen had begun to show what Willard and Livermore call "that inventive power which has made her name famous." However, there is nothing to indicate that she received any kind of mechanical or technical training,[62] and she patented nothing until she was over thirty years old—after her father had suffered business reverses in the panic of 1866 and died leaving the family in financial trouble. Perhaps she needed the stimulus of economic necessity to overcome her 19th-century inhibitions against taking credit for her own achievements and entering the business world as a woman. Be that as it may, she certainly succeeded admirably once she began.

Patent records suggest that Helen moved initially to Boston, where she created her first sewing-machine inventions, including her overseaming machines, an improved method and/or machine for sewing knit goods, which may be what Willard and Livermore call "the machine for simultaneous sewing and trimming on knitted fabrics," and a crochet attachment for sewing machines, plus at least one invention for the shoe industry—an Improvement in elastic gorings—among other things. The Boston City Directory for the years of these patents list many Blanchards, but no Helen, and no Louise—the unmarried sister with whom she lived after leaving Portland.

Helen may have stayed with her brother David at 529 Columbus Avenue while in Boston rather than taking a house of her own. She may have gone to Boston to find financing and a manufacturer for her machine. Charlotte Smith says that Helen had to borrow money to pay her first Patent Office fees (*WI/1*). I note a firm of mechanical engineers called Holmes & Blanchard on Haymarket Square, E.O. Holmes and John W. Blanchard, proprietors; and a John Bigelow, President of the Collateral Loan Company at 117 Summer Street (1876 directory). Helen Blanchard assigned her 1875 Patent for an "Improvement in methods of uniting knit goods, etc.," as well as an 1877 patent (for her crochet attachment), to J(ohn) Bigelow of Boston.

Two sources agree that in 1876 Helen Blanchard founded the Blanchard Overseaming Company of Philadelphia to exploit some of her sewing-machine inventions. According to Willard and Livermore (97), who wrote less than twenty years later,

> She applied her powers to the intricacies of machinery, and in 1876, by the results of her inventions, she established the Blanchard Over-seam Company, of Philadelphia, which was the originality of what is now called zigzag sewing, both inside and outside of material sewed, and which achieved a signal success.

Mary Logan says essentially the same thing. However, the Philadelphia directories show no such company until 1881, which accords better with the patent and other records.[63] Philadelphia appears as Helen's residence on patents granted in 1882 and '83 for improvements in sewing-machine needles. Property deeds in their names suggest that

---

61. According to Carolyn Cooper, a Yale University researcher in the history of science who researched Thomas Blanchard for a dissertation on his wood-working inventions, the great inventor had no direct descendants, but many nieces and nephews (L-8/30/85).

62. Unless a clue lies in an interesting datum from the 1870 Portland census, which shows two civil engineers living with the Nathaniel Blanchard family at that time. Was the once-wealthy merchant reduced to taking in boarders, or were these young men related to the family? Helen is listed without occupation (Curry, L-3/30/87).

63. I am grateful to Joan Garbutt of Philadelphia for directory and local business-history research on Helen Blanchard in that city.

Helen and Louise Blanchard lived, or at least maintained a residence, in Philadelphia during the 1880s (J. Hayden, L-5/10/82).

The company appears at 916 Chestnut Street in the directories for 1881–5, with slightly varying listings: as the Blanchard Overseam Machine Company for 1881 and 1883–5, and as the Blanchard Hosiery Machine Company for 1882. This was an area of Federal-style brick buildings about five stories tall, well within the central city and near other "shops" having to do with textiles, textile manufacturing, sewing machines, etc. It was just a few blocks from the Reading Railroad, and from some of the best in-town housing in the city. The building itself was destroyed by fire in the 1970s, but others of the type still stand. By 1886 and 1887 the directory listing reads simply the Blanchard Machine Company, with James Greenwood and John Bigelow, machinists, listed at the address. (Interestingly enough, there is a Blanchard Machine Company in Cambridge, MA, as late as 1914, the year before Helen Blanchard's last known patent.)

New York City appears as Blanchard's residence on patents for 1893–8. Portland, ME, appears on patents granted in 1901, and Philadelphia reappears intermittently after 1883 and on her latest patents (she evidently often wintered in Philadelphia and summered in Maine). Helen Blanchard died in 1922 in Providence, RI.[64] She was buried in the family plot in Portland's Evergreen Cemetery.

Much more remains to be discovered about the Blanchard Overseam Machine Company, for, according to Willard and Livermore, "A number of great industries have sprung from that company, and the benefits of that invention have spread through the country." Further research in the business history of Philadelphia, the Dun & Bradstreet records, and the detailed history of commercial sewing machines in the late 19th and early 20th centuries should be illuminating.

Much remains to be discovered about this 19th-century Athena as well. Among the more obvious questions is what happened to her fortune. Charlotte Smith says in 1891 that she "now owns great estates, a manufactory, and many patent rights that yield her a large income in royalties" (*WI/1*). All indications are that Helen Blanchard was tremendously wealthy. Yet no will was probated for her estate, either in Providence or in Portland. Perhaps after her stroke of 1916 she was declared incompetent, and her estate apportioned at that time among surviving nieces and nephews, including Louise Merrill of Providence. That is just one of the many mysteries awaiting further research (Boston Dir. 1876; C. Cooper, L-8/30/85; G.R. Cooper, 211, 232; Curry, PC-1/30/86) and L-3/30/87; Garbutt, L-2/5/83; Gerrier, L-3/1 and 3/26/82; J. Hayden, L-3/8, 4/2, 4/14, 5/10/82; M. Logan, 889; Blanchard Obit., 1/13/22; NYC dir. 1890–1901; Phila. dir. 1876–87; Probate Ct., Providence, RI, L-1/24/83; M. Reynolds, 5; *WI/1*; Tarbell, 357; Tulk, L-9/3/82; Willard & Livermore, 97; "Women Inventors" [*Sci. Amer.* Apr. 4, 1903]).

**Margaret Knight** (1838–1914) is one of only three women listed as inventors by the editors of *Notable American Women* (early period) in their classified list of occupations.[65] She has often been profiled—though not always perceptively, and never in a full-length biography. On her childhood we get the amazing-tomboy perspective—the story of a girl

---

64. This datum puzzled me because of Blanchard's great love for her home in Portland. However, it appears that after losing her lifelong companion, her sister Louise, in 1914, followed in a short space of time by the deaths of her married sister Persis and all three brothers, she suffered an incapacitating cerebral hemorrhage in 1916. Her niece, Louise Merrill, who lived in Providence, may have been a favorite relative—or the only one willing or in a position to care for Helen.

65. The other two are **Amanda Theodosia Jones** and **Sybilla Masters**, both treated in an earlier chapter (Agriculture). Masters, of course, lived and worked before the Revolution (d. 1720). It should be noted that

who thought it was silly to "coddle bits of porcelain with senseless smiles," and preferred instead to whittle and build things—kites and sleds that made her the envy of all the boys in town.

At twelve she got the idea for her first invention, a mechanism to prevent accidents to workers in the cotton mills. Her invention occurred to her when, on a visit to the mill where her brothers worked, she saw a girl wounded by a spindle flying loose. This invention has been described in conflicting terms—as a stop-motion mechanism, as a means of preventing spindles from flying loose from their moorings, and as "a covered shuttle which is in general use today in cotton mills, where it is a great protection to the operators" (Janeway, ed., 55). These three descriptions, though at first glance sounding like three different inventions, are not actually conflicting. However, Ernest Heyn says Knight's invention was "a contrivance to shut down a textile loom automatically when a steel-tipped shuttle accidentally fell out" (*Fire,* 46). This makes no sense. It is the flying off of the shuttles that creates the danger.

Most illuminating is Naomi Smith's version: the accident that inspired Margaret happened when a broken thread snagged a moving shuttle and pulled it off the machine. One of the girls working below was hurt by the falling shuttle's sharp tip. Margaret saw that such injuries could be prevented if the machine stopped instantly when a thread broke. After some experimentation, she made a workable device to do just this.

No one seems to have considered that Knight might not have stopped with a single safety invention. One of the primary tasks of her biographer will be to reconstruct this first invention. What we know is that (a) it was a safety invention, (b) it was mechanical, and (c) she did not patent it. Nevertheless, it appears to have been adopted.

Many such tasks of clarification remain, for except for her invention making possible the square- or satchel-bottom paper bag (and later improvements), amazingly little seems to be known about Margaret Knight's technical achievement. That invention alone made her famous, partly because others had been trying to do it for years, and partly because of its sheer usefulness. Anyone who ever grocery-shopped in England or France before the advent of plastic bags can appreciate Margaret Knight in a very immediate way. Getting home before the flimsy, flat envelope bags tore was a real challenge.

The usefulness and significance of this invention, then, is not in debate. To what degree Margaret Knight profited from it, however, is not at all clear. Mozans says (351) that she received $50,000 for it shortly after patenting it. Charlotte Smith says that she *refused* $50,000 for it at that time. In any case, she received a decoration from Queen Victoria in 1871.

The number of inventions created by Margaret Knight is also unclear. Some have concluded, upon hearing her called a female Edison and credited with more than eighty inventions, that her accomplishments were greatly exaggerated. Much more likely, I suspect, is that writers have simply failed to distinguish between patented and unpatented inventions. For instance, the writer of a *New York Times* story of October 19, 1913 (Janeway, ed., 55), credits her with 89 inventions; the writer of her obituary in S. Framingham, MA—where she had lived for at least twenty years—with 87. Both of these writers are speaking of total inventions, patented and unpatented. Indeed, they may be taking her own count, which I suggest would be low. Her *NAW* biographer, on the other

the classified list of women in various occupations is a *selected* group (no criteria given) of the women profiled in these three volumes. However, I do not recall any other women characterized as inventors in the brief description of life work that follows each name.

hand, evidently unable to believe that a woman inventor could be so prolific, and seeking to refute what he saw as an exaggeration in the obituary estimate of 87, speaks only of patents, and suggests "a figure more like 27."

I find 22 patents issued to Margaret Knight, solely in her name or with male co-patentees, dating from her first patent for the paper-bag machine in 1870 to a Resilient wheel patented in 1912. This figure could increase if we count reissues, if the LWP compilers omitted some of her patents (cf. Stanley, "Conjurer"), or if—as I suspect—some patents issued to others but assigned to her are actually co-inventions (cf. #748,689 and #777,832). The *New York Times* story, based on a 1913 interview with Knight, says (Janeway, 55):

> The time has come now, however, when men must look to their laurels, for the modern field is full of women inventors. The oldest of them and the one having the most to her credit, is Miss Margaret E. Knight, who at the age of seventy[66] is working twenty hours a day on her eighty-ninth invention. . . . Among labor-saving devices in cotton and woolen mills, rubber and shoe factories, her latest is a motor for driving them.

Sometimes, too, an inventor's most important and potentially profitable inventions are kept as trade secrets rather than patented.

We do not know the technological or commercial significance of many of her inventions. As mentioned, Pound's *Chronology of Significant Dates and Achievements* in automotive (and General Motors) history mentions that "Knight invented sleeve-valve engine" in 1905. As also already noted, this may be Charles rather than Margaret Knight. If Margaret, the date should be earlier, as her engine of 1902 would have predated his by three years, and she received seven more patents for rotary engines between that date and 1904. Ernest Heyn says that she "designed, and patented, a sleeve-valve automobile engine as an alternative to the poppet valve" (*Fire*, 46). Her *NAW* biographer mentions that her particular contribution to rotary engines was her work on a sleeve-valve engine, and that she assigned some of her inventions to the Knight-Davidson Motor Company of New York; but he then goes on to say "there is no indication that [the engines] were widely used." One wonders whether there is hard evidence that they were *not* widely used.

Except for the tomboy stories, we know little about Margaret's childhood. She was born in Maine, moved to Manchester, NH, and received little formal education—"at least not beyond secondary school," according to *NAW*. Of course, in the 1850s very few Americans of either sex went beyond secondary school; indeed most left school after the eighth or even the sixth grade. Important areas for further research, in addition to a meaningful evaluation of her technological contributions to American industry are details about her education, any technical training she may have received, either formal or informal; whether she had any mentors or anyone who encouraged her inventive bent; and, perhaps most important of all, what she was doing between the time she invented her safety device for textile mills at age 12, and her first patent at age 32.

What we can say for certain is that Margaret Knight held at least 22 patents in her own name, eight of which dealt with rotary engines. The following others were also machines or mechanical in nature:

---

66. Either Margaret Knight's age or the date of this story as cited by Janeway is wrong; for Knight would have been seventy-five in 1913. Perhaps we have either some coyness on Knight's part or a bit of harmless gallantry on the writer's part.

| Impr. in paper-feeding machines | (1870) |
| Impr. in paper bag machines | (1871) |
| Impr. in paper bag machines | (1879) |
| Spit | (1885) |
| Sole-cutting machine | (1890) |
| Sole-cutting machine | (1890) |
| Sole-cutting machine | (1891) |
| Sole-cutting machine | (1893) |
| Window frame and sash | (1894) |
| Reel (spinning or sewing machine) | (1894) |
| Feeding Apparatus for sole-cutting machine | (1894) |
| Numbering mechanism | (1894) |
| Automatic tool for boring or planing concave or cylindrical surfaces | (1903) |

Her last patent, for a Resilient wheel (1912), is also mechanical.

Her inventive career of more than thirty years can be divided into three major periods, in which she produced inventions dealing with 1) the paper bag industry, 2) the shoe-manufacturing industry, and 3) rotary engines. She seems to have been employed by various companies during the three periods, and apparently sold her machines to those companies. At least, she assigned her third paper-bag-machine patent to the Eastern Paper Bag Company, and several of her rotary engines to the Knight-Davidson Motor Company of Saratoga, NY. Whatever she may have received for her paper-bag machine, she certainly did not die a wealthy woman. Her total estate at the time of her death was under $300. Although her body of granted patents is not the largest for any woman of the century, Margaret Knight has one of the best claims of any 19th-century American woman to be called a professional inventor. As the *New York Times* writer noted, at seventy years of age she was working twenty hours a day on a new invention.

Said Knight herself, "I'm not surprised at what I've done. I'm only sorry I couldn't have had as good a chance as a boy, and have been put to my trade regularly." And yet, continues Phebe Hanaford (632–3), "she knows as much about machinery as though she had made it a study all her life. It is a genuine gift; and she can no more help making machinery than Anna Dickinson[67] can help making speeches" (Hanaford, 632–3; Heyn, "Ten . . . ," 20, and *Fire,* 46–7; Janeway, ed., 55; R. Kennedy; J. Marlow, 126–8; Mozans, 351; *NAW;* Pound, 50; Sherr & Kazickas, "Landmarks," 100; N. Smith, 192–3; Tarbell, 356; Weimann, L-1/28/83; *WI*/1:3:2).

## 20TH CENTURY

In the 20th century the universe of women who patented machines or mechanical inventions is at once larger and more elusive than in the 19th. There is no LWP, and until 1988 there was no useful systematic study for this century beyond 1921. Even the latest studies give no names. Moreover, the most prolific women patentees—and thus the more likely to be noticed and profiled—tend now to be chemists rather than mechanics. **Ruth Benerito** and **Giuliana Tesoro,** both textile chemists, leap to mind. Dr. Benerito, who works for the Department of Agriculture, has some 55 patents to her credit, mainly dealing with wash-and-wear fabric treatments; Dr. Tesoro, who worked for private industry, well over a hundred.

67. Civil War orator and lyceum lecturer (1842–1932).

Invention came increasingly under the aegis of corporations, as Research and Development (R&D) or New Product Development. The work was also professionalized in the sense that training in math, chemistry, physics, engineering, metallurgy, metalworking—and now in computers for computer-aided design—was required for the jobs in question. Women were sometimes hired or consulted on new household machines, but their work tended to be invisible unless they were already famous, like **Lillian Gilbreth.** Even when women won these design-engineering or product-innovation jobs, they were often submerged in teams whose leaders patented the team's work—as with **Bertha Lammé.**

For comparative purposes, however, I will look at the same four landmark machines (the sewing machine, the dishwasher, the washing machine, and the typewriter); then treat inventors of machines by category of endeavor; and close with a few Lady Edisons—one of whom, **Beulah Henry,** was actually known by that title.

### Landmark Classes of Machines

#### Sewing machine

Twentieth-century women continued to improve sewing machines, but received, to my knowledge, even less notice for their work than did their 19th-century sisters. According to Minnie J. Reynolds (5), by 1908 women's patents for sewing-machine improvements since 1841 numbered "some 125," but she names no names. Nor does she give any breakdown between the two centuries. Since the occasion of Reynolds' article was an exposition—the Suffrage Bazar held at the Martha Washington Hotel, New York City, on November 6–7, 1908—some of the names may be retrievable.

As to the numbers themselves, by my count women already had at least 150 sewing-machine patents by 1894. If they continued to invent (and patent) in 1894–1908 at the same rate as during 1885–94—by no means certain, but a place to start—women's sewing-machine patents by 1908 should have numbered nearly 240.[68]

The next indication we have of women's 20th-century sewing-machine inventions is the 1920s Women's Bureau Study (WB/28), covering, as will be recalled, patents granted to women in 1905–6, 1910–1, 1913–4, and 1918–21. The Bureau compilers had a category for Sewing and embroidery machines and parts, which contained 61 inventions of the following 39 types:

| | |
|---|---|
| Connecting and disconnecting mechanism, embroidery machine | folding attachment |
| | footrest |
| Expanding frame for embroidery machine | corded tuck attachment |
| Plaiting machine | gauge |
| Portable hand sewing machine | guard |
| Sewing machine | head |
| Sewing-machine attachments: | leaf |
| belt | positive timer for lockstitch machine |
| braid-applying attachment | presser foot |
| bobbin-unwinding attachment | lifter for presser foot |
| bobbin-stripper | ribbon guide for tucking blade |
| cloth guide | ruffler |
| edge guide | spacing attachment |

68. Calculated at the rate of 58 per decade, based on my actual count for 1885–94. Since LWP probably omits a few, and since some of the plaiting and braiding machines I treat separately may actually belong here (cf. WB/28: 40), my figure is a conservative one, even though it includes a sewing-machine seat and a sewing-machine cover or two.

spool-holder and thread guide
stocking-holding band for darning
stopping device
side guider for sewing on skirt braid
tension mechanism
threading attachment
thread cutter
thread cutter and holder

thread cutter and hem creaser
thread guide and holder
tucking guide
trimmer
other attachment
folding receptacle for sewing machine
implement holder for sewing machine

Curiously, the Women's Bureau compilers placed these inventions neither under Manufacturing nor under Household inventions, but in their Category XIII, "Supplies for use in industry, agriculture, commerce, and the home" (WB/28: 40, 15).

As always, WB/28 gives no names. *Inventive Age* for June 1, 1911, however, mentions one **Louisa Boyd** of Stuart, IA, who invented a Sewing-machine attachment. This was a device to guide the work and aid in gauging tucks and seams on either side of the presser foot. Since Boyd's patent was granted in 1910, hers should be one of those listed. From the brief *IA* piece we learn that she got her patent with the aid of a patent attorney ("Mechanical Inventions . . ."). **Annie Livingston** of Winnipeg, Manitoba, Canada, patented a "threading attachment" in 1911.

Beyond the Women's Bureau study I have scattered but interesting examples of women's sewing-machine inventions. Of course, the sewing-machine was by now a "mature" machine, and except for replacing foot-treadle power by electricity, saw no fundamental changes until the recent computerized machines.

**Beulah Louise Henry** (1887—  ) received several sewing-machine and related patents: for a Double chain-stitch machine (1936); for a Seam and method of forming seams (1941); for a Sewing apparatus (1941); and for a Method of forming a lock stitch (1959). The writer of a 1937 *Literary Digest* article on women inventors praised her lock-stitch bobbinless sewing machine, as did the writers of an earlier *Coronet* magazine profile (1939), and of a 1940 *New York Times* article ("Lady Edison," 22; "Beulah Henry," 159; "Women Gaining," D-3). And indeed the bobbin, with its frequent need for rewinding, its awkward winding process, etc., has been a major bottleneck in sewing-machine technology. One source maintains that "Miss Henry's greatest boon to modern manufacturing is perhaps her bobbinless sewing machine," explaining the invention as a new type of thread-feeder (Yarbrough).

In 1940 she received a reissue of her 1936 Double chain-stitch machine. This machine may be the bobbinless model acclaimed just above, the underslung bobbin being replaced by a duplicate needle-and-arm mechanism. Miss Henry got the idea for this machine while watching "a ribbon of electric light" playing back and forth across a theatre marquee. Working all night with her usual model-making tools—hairpins, spools, and other odds and ends—by morning she had solved the basic problem and had her first sketches of the machine (Yarbrough, col. 2).

According to the *New York Times* article, Henry already had 60 patents to her credit by that time. She undoubtedly had *at least* 60 *inventions* to her credit—her title of Lady Edison is sexist but not inapt—but only 24 *patents,* at least in her own name.

**Dorothy Young Kirby** (c. 1909–  ) retired to Clermont, FL, in her sixties, and started making her own clothes to keep busy. Unhappy with the home-made look of the garments, however, she invented the Dot Young Sewing Guide to enable her to produce perfectly straight seams and professional-looking top-stitching with her machine. Marketing the Guide through six different mail-order catalogues, she had sold more than 250,000 of them by 1984.

Now a widow, she has joined the Central Florida Inventor's Club, and patented a magnetic version of the Guide that need not be screwed to the sewing machine. As of 1984 her company reported nearly $50,000 in profits. Says Kirby of her inventive work, "it's been a great outlet for me when I needed something. . . . I've had a chance to travel, and I have no money worries. Besides, inventing keeps me young" (Siegel; Tilden, L-10/14/80).

**Rosa Csüdor-de Valerio** (1894–   ) is remarkable not only for her invention itself, a new embroidery technique and the sewing-machine attachment for doing it, but for the age at which she conceived it—81. Nor was this to be her last invention. In 1986 she was contemplating a second embroidery invention, a new technique for "painting pictures with wool," as she calls it.[69]

Rosa Csüdor was born in Transylvania (modern Roumania). A popular and lively girl, she became an entertainer and then a skating champion. She married and reared a family—her daughter, granddaughter, and great-granddaughter all celebrated her 90th birthday with her. She now lives in Switzerland.

When asked about her invention, she describes herself as clever with her hands and having a pretty good brain. About 1975 she conceived a new embroidery technique and worked out the necessary mechanism to allow it to be done by machine. (Or perhaps she long ago conceived the technique, but now adapted it to the sewing machine, and devised the necessary attachment.) Friends advised her to seek a patent and, determined not to burden her family in her old age, she did so.

Thus was born the Ro Ri Ro, a device manufactured by Langenthal of Switzerland, that produces a "decorative textile without fabric base support." Its operation is difficult to visualize, but it seems to create coils or curls of wool, then sew them together into a web of joined loops, which are cut to form a nap. Rosa Csüdor holds Swiss, Canadian, and U.S. (1979) patents. The Ro Ri Ro has won medals at expositions in Geneva, Basel, and Nuremberg, and has also been shown in the United States (at St. Louis).

This great-grandmother, far from becoming a burden to her family, has received international attention. The late Princess Grace of Monaco was one of her admirers. She appears in an article by a United States Patent Examiner on inventions by women and blacks (Ives, "Patent," 821), and in a book by an officer of the World Intellectual Property Organization (Moussa, 146–7).

## The dishwasher

Minnie Reynolds' account of the Martha Washington Hotel Suffrage Bazar of 1908 mentions that **Josephine Cochran** exhibited her dishwashers there. This suggests that Cochran continued to improve her machines after the turn of the century; otherwise, in the decades since her first patent, her invention would surely have been superseded. What is certain is that Cochran's dishwashers were used in hotels all over the country at the time of the Bazar, and that Mrs. Cochran herself managed the company that manufactured and marketed them. Reynolds calls her a Chicago woman (2–3).

Another woman inventing in the dishwasher field early in the century is **Anna Dunder** of Crete, NE, who patented her dishwasher in 1903, too early to be represented in the Women's Bureau study. Dunder received a second patent in 1910—for a Massage implement. This item, patented in one of the study years, is probably the "Massage

69. Farag Moussa's French is ". . . une nouvelle technique pour 'peindre des tableaux avec de la laine . . .' " (146).

device" of Category XVI, Articles for personal wear and use, Subcategory I, Toilet articles.

By 1910 Dunder has moved to Wilber, NE—like Crete, settled mainly by people from Bohemia (modern Czechoslovakia). Neither town's newspaper seems to have recorded the granting of Dunder's patents. However, the Dunder family lived in Crete at least by 1887, for one Joseph Dunder is listed as a founding member of the Bohemian-American Gymnastic Association in that year. Two Anna Dunders lie in the Wilber cemetery, and she may be either the one who died in 1926 at age 84, or the one who died in 1915 at either 38 or 48 (Gregory, 107, 187; Kobes, L-5/11/87).

The Women's Bureau compilers have no Dishwasher category. Category VII, Hotel and restaurant equipment, lists a "Cleaning-off machine," which may refer to dishes, and a Machine for drying and polishing glasses. Listed under Kitchen equipment (Category XII.A) is "Dishwashing machine," but as always no count of patents per type. If the 440 Kitchen-equipment inventions were distributed equally among the 80 types so listed, then we should find 5-6 dishwashers for the ten years. I have located two women patentees of dishwashers from the WB/28 years. The first is **Alma T. McHargue** of Columbus, MT, who patented a Dishwasher and drainer (with Oscar F. McHargue) in 1911. The second is **Maria Doujak** of New York City, who patented a Dish-washing machine in 1920. Doujak is interesting in that earlier the same year she patented a Life-saver automobile attachment. My search has not been exhaustive, and there may be others.

One of the most prolific women inventors of recent years, another Lady Edison, is **Frances GABe** of Oregon. Noteworthy among the dozens of inventions involved in her Self-Cleaning House are her dish-washing cupboards. Frances GABe has made one of the homemaker's dearest dreams come true: putting dirty dishes away in the cupboard, and finding them sparkling clean again as if by magic when they're needed for the next meal (GABe/ Zimmerman, ed., 77–8).

*Washing machine*

The Women's Bureau Study lists eight relevant types of invention under Category XII (Housekeeping inventions), Subcategory C, Laundry equipment:

| | |
|---|---|
| Apparatus for washing | Clothes wringer and attachments |
| Bucket clothes washer | Heater for washing machines |
| Clothes catcher and drainer | Washing machine |
| Clothes pounder | Washing machine attachment |

If all the 31 types of invention in this subcategory contributed equally to the category total of 165, women's washing-machine-and-related inventions in these eight types would number some 42 over the ten select years. All we can say for sure is that there were *at least* 8 washing-machine-and-related patents granted to women in the years studied.

As usual, of course, WB/28 mentions no names. Katherine Lee, however, notes a washing-machine patent granted to a woman in 1920, one of the study years (43). This turns out to be **Agnes M. Hills** of Stanfield, OR.

A quick survey of the 1925 patent records shows nearly a hundred (98) washing machine patents. This includes three clothes-washers, all the machines listed as washing machines *per se*, plus a few related items such as a gearing, a motor, a cylinder, and an "attachment" for washing machines. Not included are washboards or related items—of which, interestingly enough, 11 were still patented in this year[70]—or wringers. At least

---

70. **Marie Olpe** of Jamaica, NY, for example, patented a Washboard rest in 1925.

one of these 98 inventions is a woman's: **Ethel R. Blake** of Toledo, OH, patented a Washing machine with William F. Blake on November 10. Also, Erick Moe of Astoria, SD, assigned half of his washing-machine patent to **Bertha Moe,** which could mean that she had some role in the invention. Thus just over 1% of washing-machine patents were granted to women in 1925.

**Agatha K. Bremer** of New Washington, OH, patented a washing machine in 1940.

As with dishwashers, women's 20th-century contributions to washing machines were increasingly limited to a consulting or testing role, or to participation in R&D teams for the large washer manufacturers. If such teams were led by men, the women's contributions might never be known.

One woman on a corporate team who did win some recognition is **Jessie Whitney Cartwright** of Oak Park, IL (1897–1977). In 1972, for example, she was elected to the City of Chicago's Hall of Fame. Cartwright, who delayed her career until her three daughters were grown, had graduated in home economics at Michigan State University and done graduate work at Columbia and at the University of Minnesota. After a stint in market research, she served for many years as National Home Services Director for the Norge Division of Borg-Warner, starting in 1948. She had also worked briefly for Bendix.

Nationally known as a home economist and as an authority on appliances, she ran the women's division of the United States pavilion at the 1966 International Trade Fair in Bari, Italy. According to Andrea Hinding (#4,064), Cartwright "developed a number of patentable improvements on home appliances." Two of Cartwright's three Chicago newspaper obituaries say that she held patents on numerous home appliance inventions, and the *Sun-Times* obituary calls her "Inventor Jessie Cartwright." In spite of Cartwright's status as a public figure and her 25 years with Norge, however, shockingly little technical information is available. I find only one patent, and her papers, donated to the University of Illinois, though not fully catalogued at the time of my inquiry, evidently contain no technical items (Ritzenthaler, L-1/3/79).

From information supplied by her granddaughter, Whitney Brown of Alaska, and by a former colleague, Edna Poyner, comes a sketch of her professional achievement. Whitney grew up with her grandmother, and sometimes went to work with her at Norge. Her grandmother told her she either patented or worked on at least 6 major innovations, mainly at Norge: Norge's microwave oven, the Radarange;[71] the stop cycle on Norge dryers; the icemaker on Norge refrigerators; the back panels on washers and dryers; the delicate cycle in washing machines; and the Ripplette, a small agitator that could replace the large agitator of a washing machine for woolens or delicate items. Because Norge required employees to sell all patents to the company for $1.00, she never got any of the profits from her inventions. According to Brown, Jessie made a lamp out of her old Ripplette.

The Ripplette is the one patent (1954) I have located for Jessie Cartwright. As for the other inventions, the Radarange was a major seller among microwave ovens. The stop cycle in the dryer was useful for indoor drying of sweaters and other knitwear that would be ruined by tumble-drying. The dryer models in question had an insert rack to hold the garment for drying by heat alone. We now take for granted the so-called delicate washing cycle; but once, all lingerie, hosiery, and such had to be washed by hand because the vigorous agitation of the regular cycle would damage them. We also expect metal back panels on our laundry machines, but they were a striking innovation—and a marked safety improvement, specifically shock-prevention—when Cartwright thought of them. They

---

71. The first microwave cooker, a spinoff of radar, went on sale in the United States in 1947. This was a commercial model; the home model appeared in 1965 (1972 in Britain; Harpur, 134).

undoubtedly also made the machines quieter, and protected their motors from laundry room lint.

She also invented an industrial dyeing machine for the Putnam Dye Corporation of Effingham, IL. This machine featured a special dye-release compartment that dispensed the dye compound gradually for even coloration.

As Director of Norge's Home Institute, Jessie Cartwright was responsible for testing Norge's home appliances to make sure they met specific household requirements. Her specialty was laundry equipment. In this role, Cartwright was in a position to influence the design of a great many new and proposed Norge machines. Edna Poyner, once Cartwright's assistant at Norge, thinks that the discrepancy between her reported and actual patent output may be accounted for by, among other things, a number of products she invented but Norge never marketed. There may have been patent disclosures or applications that were dropped.

Ms. Poyner notes that Jessie invented a washing machine at Bendix (presumably before 1948). This would mean a total of at least eight important machine inventions to her credit in mid-century, four pertaining to washing machines.

Looking at Jessie Cartwright's life through the eyes of Whitney Brown and those who met her in her 70s and 80s, one can almost forgive the focus on her nontechnical accomplishments. They are impressive. At Michigan State University in 1912, she was a suffragist, and one of the first women in Michigan to drive a car. A concert-quality pianist and a trained and gifted actress, she was also a published writer and a teacher of creative writing. When her children attended a private school, she taught there, at both elementary and secondary levels. Cartwright had a special relationship with this granddaughter, who was evidently much like her, giving the girl "a personal integrity and confidence in my accomplishments" (Whitney Brown in Cartwright papers).

Cartwright seldom joined any organization unless she could be president of it. After serving three terms as president of the Chicago Electrical Women's Round Table, she was finally made honorary (permanent) president. As founder of the WAC Mothers organization during World War II, she naturally became its president. She was one of the first sponsors of the Cordi Marian Society settlement houses in underprivileged neighborhoods of Chicago.

Like Eleanor Roosevelt, she championed victims of injustice. After a years-long battle she won written apologies from Senator Joseph McCarthy and J. Edgar Hoover for a friend accused of being a Communist. She also persuaded the William Penn Hotel (Pittsburgh) to give Lena Horne a suite in the days before blacks were welcome at such establishments.

Red-haired and lively, she had a keen sense of humor that led her to call herself an "accomplished washerwoman," rather than a laundry expert or an engineer. Her first attempt at retirement lasted exactly one day; then she went back to Borg-Warner for ten more years. She tried retiring again in 1966, and the State Department sent her to the Bari Trade Fair. In her 80s she took up yoga and became a hospital volunteer, visiting patients who had no visitors, and circulating foreign-language magazines to patients hungry for reading material in their own languages. She continued this work until a mugger struck her down at a bus stop one day. She died soon thereafter. At her death she willed her body to the Northwestern University School of Medicine (W. Brown, PC-4/29/ 83; Cartwright Papers; "Jessie . . ."; Obits., Chicago *Daily News, Sun-Times,* and *Tribune,* June 17, 1977; Poyner, PC-5/15 and 5/18/83; Tobin L-4/26 and 5/7/79).

The noncorporate inventor in the latter years of the century is exemplified by another Chicago woman, **Belle Tucker,** who patented a lint trap for a washing machine in 1970.

An unnamed Peruvian woman invented a small portable washing machine that can be fitted to the edge of a washbasin. Exhibited at the WIPO Exhibition in Geneva in 1985, this important invention adapts modern technology to Third World conditions (SWIA, 3).

Unusual for inventors of either sex is a real breakthrough in machine technology. Such a breakthrough—and perhaps the most original contribution to washing-machine technology in the 20th century—is the combination closet and clothes-washer invented by **Frances GABe** of Newberg, OR, for her Self-Cleaning House. Soiled wash-and-wear clothing can simply be hung in this self-cleaning closet and the control dial set. An hour later, the clothing will be fresh and clean and dry, hanging where it belongs (M. Fuller, 11; GABe/Zimmerman, 79).

## *Typewriter*

In the earliest years of the century, **Emma D. Mills** may still have been active in the typewriter field, as the *Scientific American* article mentioning her appears in 1903 ("Women Inventors"). As noted, she not only devised a typewriter invention, but created her own special tools for the task (*WI*/1).

The Woman's Bureau study has a category (X) for Office supplies and equipment comprising 71 inventions, of which 22 (Sub-category A) are for Office machines and attachments. These 22 patents are of 12 types, of which 8 pertain to typewriters:

| | |
|---|---|
| typewriter machine | typewriter keyboard |
| typewriter paper guide | typewriter line indicator |
| typewriter ribbon spool | typewriter underscorer |
| typewriter attachment | typewriter spacing and securing anchor |

If all the inventions in the subcategory are equally distributed among the 12 types, which seems reasonable here, this listing would represent about 14 inventions. Since the electric typewriter was invented about 1901 (though not fully accepted until IBM's 1935 models; Harpur, 13, 100), it would be interesting to know whether any women patented electric machines this early.

In addition, Subcategories B and C under Office supplies . . . , covering 41 patents for Stationery and miscellaneous equipment, and eight patents for (office) Furniture, contain the following three types of typewriter-related inventions: an Eraser shield for typing, a Paper-clamp for typing, and a Noise-reducing typewriter stand, possibly another six inventions.

In the WB/28 era, but in a year not covered by the study (1907), **Sarah M. Hawley** patented a paper-feeding mechanism for typewriting (typesetting?) machines. The mechanism involved three rolls of paper.

WB/28 gives us no idea what percentage of all typewriter patents were taken out by women. My systematic survey for a later year, 1940, shows that of 93 patents for Typewriters, Typewriting machines and related inventions, one was patented by a woman. (Not all of the men's patents pertained directly to the typing mechanisms—included were a carrier and an ink fountain, for example—but most were mechanical). **Martha F. Cooke** of Chicago patented her Typewriter key locking attachment on February 27. Obviously intended for an electric machine, it involved "a power-driven actuating mechanism including an electric motor" and used an "energizable electromagnet" in the actual locking operation.

**Beulah Henry** also turned her attention to typewriters, receiving a total of 11 obviously mechanical patents in this field over three decades (1932–64). She patented a supplemental ribbon attachment, several duplicating attachments, two feeding and aligning devices, and two other unspecified improvements. One of her duplicating attachments produced several originals, a significant feature in the era before photocopiers.

In her aligning devices, to keep multiple copies from slipping, she used a principle adapted from the Swiss music box—a row of tiny pins around the edge of the platen, making possible "a billing process the business world had long sought." Her Continuously attached envelopes of 1952 and her Direct and return mail envelope of 1962 are probably geared to typewriter or other machine billing or mass mailing for businesses.

According to a 1937 article, Beulah Henry had solved for National Cash Register Company the problem of making their registers write like typewriters,[72] and was working on a *silent typewriter*. By about 1940, she claimed to have perfected the silent typewriter, using her own small machine to work out the details ("Beulah Henry," 159; "Lady Edison," 22; "Women Gaining," D-3; Yarbrough, cols. 2-3). If we add this silent typewriter (unpatented?), her 1949 patent for a Key for key-operated machines, and her Sweetheart Key Tops (fingernail-saving cushions for typewriter keys, possibly trade-marked rather than patented),[73] her typewriter and related mechanical inventions number at least 14.

Britain's **Lily Pavey** (c. 1918–   ) patented her Musigraph or Musikriter about 1961. The London *Observer* reporter who interviewed Pavey in 1963 noted that she had devoted all her spare time and energy to the invention for the past thirteen years. Born in a circus and sketchily educated, Lily Pavey has an intuitive mathematical ability—and a dream Voice that guides her toward achievement. It was the Voice that told her to invent a music typewriter, but she got the crucial technical idea for the machine while riding in an elevator. If an elevator could stop at different levels, she thought, a typewriter could be made to write vertically. Then came years of working at night to convert a regular typewriter, figuring out how to give vertical elevation without moving the paper, and how to create 8000 combinations with 46 keys.

Although music typing and typesetting are now done by computer, the music typewriter was an important and long-sought invention. Pavey, an office worker, won such distinguished backers as the Gulbenkian Foundation for her work; and her initial success inspired a rash of newspaper stories with headlines like "Little Miss Pavey has done something which has always baffled the world's engineers and musicians" (Gill, n.p.). (An American machine did exist, but hers was the first British music typewriter, and invented independently.) Lily Pavey also won the Gold Medal at the Brussels Inventors' Exhibition for 1963. Just a month after her Brussels triumph, Pavey started her own company at Dorking to manufacture the machines, which were to sell for £100.

Later, the machines were manufactured by Imperial Typewriter Company (though she had offers for it from France and Japan), at a cost of £260; but by 1967 very few had been produced, and Pavey was embroiled in royalty disputes. The machine had cost her £8,000 to develop, and she was receiving very little return on that, for her, enormous investment.

By 1961 she already had a prototype of her electronome, a device enabling a composer to hum into a microphone and have the typewriter print that note. It also sounds out each note as it types, making the Musigraph ideal for blind musicians.

By 1967 she had improved the Musigraph with a device called the spherigraph. With clip-on type faces, and a new kind of shift key, it could be used for complicated math and

72. Today the company has no records of this work (see below)—which of course does not mean that it did not happen.

73. A 1930 newspaper article referred to "the typewriter that saves all the well-manicured lady stenographers' polished and immaculate fingernails" ("Miss Henry").

chemistry symbols, ideographic languages, shorthand, and even ballet choreography notation. With the spherigraph, words can be added to the music.

These typewriting inventions are just two of Pavey's many brainchildren. This remarkable woman, once-widowed and once-deserted mother of two, has also taught herself algebra, chemistry, and electricity, written a textbook of permutations and combinations of notes to make it easier for children to learn musical scales (unpublished as of 1963), and mastered the violin, cello, clarinet, trombone, piccolo, saxophone, sousaphone, musical saw, and bagpipes. She has invented means of recording herself playing sixteen different instruments as if at the same time. She wants a fortune not for herself, but to repay the National Assistance for its aid. When people doubted her ideas, she took it in stride. "Like that Kipling chap said," she told an interviewer, " 'If you can meet with triumph and disaster and treat those two impostors just the same . . . ,' well, you're all right" (Dal Molin, L-3/14/80; Doughan, L-1/15/79, 1/11/80; Gill, cf. Harpur, 74, 82; "Her Master's . . ."; Knowles; "Musigraph Brings . . ."; Orton; Pavey, flyer, 1987; Rosser, L-4/25/88; SWIA, 3).

Seemingly undaunted by computerization in the music-typing field, Lily Pavey was still promoting and improving her inventions in the 1980s, appearing at the 1985 WIPO Exhibition in Geneva (SWIA). She maintains that her system can be used by the blind, and that it demands less compromise in the musical notation, allowing composers or musicians, for a small example, to vary the direction of flags on the notes to increase clarity (Rosser, L-4/25/88).

A major advance in typewriter and electronic keyboard technology is the Maltron electronic keyboard, named for **Lillian Malt** (1925– ) of Lillian Malt Training Services, Godalming, Surrey, England. In arranging his 19th-century typewriter keyboards, printer Christopher Sholes imitated typesetters' cases, and separated letters most often occurring together in English words as widely as possible, to keep the keys for these letters from clashing in rapid typing. As a result, the shorter, weaker fourth and fifth fingers have a great deal to do, especially on the left hand, where they handle both *a* and *s*; whereas the left thumb is idle, and the right thumb merely operates the space bar. With electronic keyboards, however, no keys strike paper; thus no collisions. Only letter frequency, strength and nimbleness of fingers, and the most efficient and comfortable hand position need be considered.

Lillian Malt has done just that in creating her Maltron keyboard. The Maltron features concave left- and righthand key wells, with key heights fitting the varied lengths of human fingers. Between and below these two wells lie keys operated by the two thumbs. The left thumb, for example, takes *e*, the most frequent letter in English. Typists using this keyboard and the Maltron Mark II letter layout should increase both speed and accuracy by 40%, and Malt predicts many will be able to type 160-200 words a minute, or about twice as fast as most good typists do today. Even if they keep the QWERTY arrangement and merely change to the new shape, they can expect a 25% increase in speed after only five to ten hours' practice.

Lillian Malt was born in South Africa, where she took a Commercial Teacher's Degree at Witwatersrand Teacher Training College. Though one of five children, Lillian was separated by six years from her nearest siblings. Her mother, an omnivorous reader, encouraged Lillian's creativity and independence by allowing her to help but requiring her to decide what needed doing and then propose to do it. Likewise her father, who had a brilliant mathematical mind, would tell Lillian the answer to difficult math problems, but make her work out how to arrive at it. Also, since the family lived 15 miles from the nearest village, the children often had to create their own toys and games.

Immigrating to England in 1955, Malt opened her secretarial training business. The idea for the Maltron keyboard grew out of training typists and other keyboard operators, and then training tutors to use her training methods. From analyzing common error patterns, she saw defects in the QWERTY arrangement. Conceiving a keyboard with keys designed to fit different finger lengths, she found no manufacturer was willing to tackle the problem. Then, about 1974–5, Stephen Hobday brought her a one-hand keyboard he had designed for handicapped people. When she pointed out its limitations, he asked her to give some thought to an ideal keyboard. She told him she already had, but had been told it couldn't be made. Said Hobday, "You tell me what you want and I will tell you whether I can build it or not." Thus out of Lillian Malt's expertise in keyboard use and training, her design for the shape and letter positions, and Hobday's electronics knowledge, the Maltron keyboard was born.

In addition to its more comfortable, finger-fitting shape, which helps to prevent wrist, finger, neck, and back problems in its operators, the Maltron takes account of neurological research pinpointing vowels as the largest single source of spelling confusions in English. In the Maltron layout, Malt points out, the vowels are separated and placed so as to reduce such errors. Unfortunately, QWERTY is so well entrenched that the Maltron has not been widely adopted, at least not in the United States. However it is still very much alive, and was still being mentioned prominently in analyses of keyboards in 1985 (Johnston). A 1978 analysis (Culkin) considered the Maltron superior to the other major contender in reform keyboards, the Dvorak (Gies & Gies, 307; Hobday, PC-6/79; Malt, L-8/6, 27/80 and "Keyboard"; "QWERTYUIOP?").

Lillian Malt and the Maltron provide a striking example not only of women's technological achievement, but of how that achievement gets dropped out of history. *The Inventor,* journal of the British Institute of Patentees and Inventors (London) featured the Maltron in at least two articles during the late 1970s. The first (July 1978: 13) calls the Maltron "a major British advance in typewriter technology," without naming its inventors. The second (April 1979: 5) calls it "S.V. Hobday's completely new keyboard"! With only such sources to work from, tomorrow's historian of technology might well list Hobday as the Maltron's sole inventor.

*Word-processors.* One of the ultimate business machines of the 20th century is the word-processor. Beginning with memory typewriters in the 1960s, and progressing through dedicated systems, it finally arrives late in the century as one feature of personal computers. This technology will be discussed primarily in the computer chapter, but at least one woman, **Evelyn Berezin** (1925–   ) of E. Setauket, NY, should be mentioned as a word-processing pioneer.

A business major who switched to physics, Ms. Berezin worked for several corporations designing early computers and pioneering on-line systems before deciding to form her own firm, Redactron, with two male associates. Their product idea was a competitor for IBM's editing typewriter, whose first models had appeared in 1964. Although Ed Wolf was the chief design engineer for the Redactron machine—the Data Secretary—and Berezin as company President was already shifting from design into management, she "had a lot to say about the design of the word processing system." The Data Secretary had new capabilities (for example, allowing more changes in recorded text than IBM's editors), and it undersold the IBM model by about $2000. Thus Ms. Berezin and her colleagues rightly felt it would be a successful challenger. She sold Redactron after only six years for a spectacular profit (Berezin, c.v., L-7/6, 8/1/84; Ives/1980: 124; Powell; "She's Invading . . ."; L. Sloane, 257).

## Machines by Category of Human Endeavor

In this section we will consider women's mechanical inventions by field. As in the 19th-century section, I treat the fields in alphabetical order.

### Agriculture

Agricultural machines were discussed in Chapter 1. However, since the contribution of America's agriwomen[74] to farm economy and technology is so often ignored, it may be worth reemphasizing here that women's contributions to farm machinery did not stop at 1900.

The Women's Bureau compilers found that 221, or 4.4% of the 5,016 patents in their sample pertained to Agriculture, which they made Category I. And of those 221, at least the following types were machines, tools, or mechanical in nature (WB/28: 14, 18–20):

*A. Poultry-raising supplies and equipment*
Incubator egg-turning device
Incubator temperature-regulator
Incubator trap

*B. Dairy supplies and equipment*
Apparatus for washing separators, skimmers, etc.
Butter-worker
Churn and churn dasher
Comb. churn and butter-worker
Cow-milker
Cow-tail holder
Cream separator
Motor to operate churns
Vacuum milk shipping can

*C. Stock-raising equipment*
Device for applying insecticides
Automatic feeding hopper
Feeding regulator
Stock-feeding device
Hog waterer

*D. Planting, tilling, and harvesting machinery, equipment*
71 inventions of 54 types, listed in Chapter 1

*E. Farm buildings, fence material, water and drainage equipment*
Flume gate
Pumping apparatus
Windmill

*F. Plant enemy exterminators*
Animal trap
Boll-weevil exterminator
Tree-protecting apparatus
Weed exterminator

*G. Garden tools and equipment*
Adjustable rake

---

74. Their own term for themselves, at least in one organized faction.

Lawn mower
Shears
Gardener's wheel chair
Half-spray sprinkler

Thus, among United States agricultural patents granted to women during the ten years sampled from 1905 through 1921, at least 100, or 45.2% were machines or mechanical inventions.[75] Most if not all of them, too, were invented by farm women themselves, not by urban inventors with good ideas, or by women working for corporations (WB/28: 17).

One pertinent invention classed by the WB/28 compilers under Manufacturing is a Seed-grading machine. Since only a plant can make a seed, this must have been used either by farmers wanting to improve their crops or by wholesale or retail seed-sellers before packaging their seeds.

As farm machinery became more sophisticated and costly, and the work of designing it increasingly professionalized and "corporatized," women's contribution either dropped drastically or became buried under the names of others. Examples beyond the WB/28 compilation are scattered (see Chapter 1). Probably as the century advanced, and farming became increasingly a corporate rather than a family enterprise, something to learn in college rather than from parents, women's contribution to the technology of farming began to be squeezed out. Virginia Fink's 1991 study of Southeastern Ohio farms would seem to bear this out.

Two interesting examples of women's farm inventions from 1940 are **Marie Huschka**'s Fowl-dressing appliance (NYC) and **Dora E. Warfel**'s Poultry-feeding appliance (Conestoga, PA).

*Architecture*

As houses got smaller, women probably continued to patent (or suggest to others who patented) folding, extension, multipurpose, or otherwise space-saving furniture. Scattered examples here, before and after the WB/28 years, are the Roller-top extension table patented in 1900 by **Josephine P. Barr** and Charles D. Barr of Nebraska City, NE, and **Ethel A. Cooper**'s Folding ironing board of 1938. Cooper, who lived in Denver, also invented a knee shield for men's trousers that was mentioned in a *Christian Science Monitor* article that same year ("Tinkerers End . . .").

Extensive examples of mechanical inventions pertaining to architecture or construction come from the Women's Bureau Study. These inventions pertain both to the construction of buildings, and to mechanical devices used inside them, particularly in houses. Patents found in at least four of the Bureau's categories—Agriculture, Manufacturing, Structural equipment and materials, and Household inventions—pertain to building, and at least the following are machines or mechanical and related: Under Agriculture, subcategory of Farm buildings . . . , we find a Portable smokehouse invented by a Minnesota woman. To the Bureau compilers she wrote, "As a farmer's wife my duty was to cure meats for

---

75. Because of the way WB/28 presents data, as previously noted, it is impossible to tell exactly how many machines are involved. In each subcategory the total number of inventions listed exceeds the number of types of invention, so that each type could conceivably contain more than one. In no case here, how-ever, would the average exceed two. Thus one invention per type, though a conservative figure, is not badly off.

The discrepancy between this figure and that for 20th-century agricultural machines in Chapter 1 arises because there I discuss only the 71 tilling and harvesting machines of Subcategory D.

summer use[,] and smoked meat is very much favored in my family. I tried to make mine without expense and after I completed this device I used it successfully two years before I obtained a patent on it" (WB/28: 17).

Under Manufacturing, subcategory G, Miscellaneous, we find a concrete-block-molding machine.

The most directly applicable categories of the Women's Bureau study, however, are IV, Structural equipment and materials, and XII, Household, with 208 and 1,385 inventions, respectively, of which at least the following types seem clearly or probably mechanical:

*IV. Structural equipment and materials*

*B. House-building parts, etc.*
Drain valve
Knockdown or portable frame and hoist
Fountain brush
Paint spray
Wallpaper-hanger appliance

*C. Door and window fixture*
Awning sash construction
Bell
Door
Door check
Door lock and key
Door lock, circuit-closer for
Door lock-hook
Door-operating mechanism
Hinge
Sash
Sash, comb. holder and locking device
Sash fastener
Sash lift and lock
Sash tightener
Screen door
Screen door brace
Screen, rolling, window
Shutter and fastener
Ventilator
Window and screen attachment
Window construction
Window fastener, for ventilation
Window fixture
Window frame, revolving
Window grating, adjustable
Window lock

*D. Heating equipment and appurtenances*
Coal chute
Draft equalizer
Fireplace flue
Flue cleaner
Flue stopper
Fuel door, furnaces

Gas attachment for heating furnaces
Gas igniter
Gate for draft chamber, furnaces
Heat-distributing attachment for stoves
Hot-air register
Radiator attachment
Radiator repair device
Thermostatic pressure valve

*XII. Household*

*B. Ash, garbage, and trash receptacles*
Ash distributor
Chute for garbage

*H. Bathroom equipment and conveniences*
Bathtub and fixtures
Bathing apparatus
Bath spray
Controlling mechanism

*K. Hangers, brackets, household hardware*
Curtain bracket, fixtures
Shade fixture
Shade guide

As always, the numbers of patents in selected types of invention can only be crudely estimated. Assuming that the concrete-block machine and the portable smokehouse are single inventions, we can estimate the total number of architectural-mechanical and related inventions recorded by WB/28 as follows:

| | |
|---|---|
| Portable smokehouse | 1 |
| Concrete-block-molding machine | 1 |
| House-building items | 7 |
| Door and window fixtures | 63 |
| Heating equipment, etc. | 23 |
| Ash, garbage apparatus | 10 |
| Bathroom equipment | 9 |
| Hangers, brackets, etc. | 10 |
| TOTAL (est.): | 124 |

Note the continuing popularity of window- and ventilation-related inventions. An early-century example from a year not covered by WB/28 is **Bessie F. Jackson**'s Shade-roller of 1904 (NYC).

In 1940 **Olive M. Doud** and **Ruth V. Otto** of Newell, SD, patented their machine for circulating clean, deodorized, and sterilized air.

Heating equipment is another fertile field for women's invention. I have done no systematic survey, but one noteworthy example is the Heating furnace patented by black inventor **Alice Parker** of Morristown, NJ, in 1919 (Ives, "Creativity," 19).This should be one of the WB/28 inventions.

Both **Lillian Gilbreth** and **Sarah Pillsbury Harkness** designed furniture and appliances for the handicapped, as did **Ray Kaiser Eames** and her husband after World War II (*WBWRA*). Gilbreth designed appliances and indeed whole kitchens, and Harkness created housing designs suitable for the handicapped, which must have involved many mechanical aids.

However, the ultimate example of the house as machine is the Self-Cleaning House (SCH) being designed and built in Newberg, OR, by **Frances GABe** (1915–   ). The GABe dishwasher cupboards and clothes-washer closets have already been mentioned. On a larger scale, but similar in principle, is the house's ability to spray-wash, rinse, and dry itself with air currents in an hour's time, all at the touch of a button. The water comes from floor and ceiling in a fine, warm mist. It does not harm floors, walls, beds, or other furniture because these are made of waterproof material, as is the upholstery (another of GABe's inventions), and the beds are covered by plastic shielding sheets that roll out like window-shades from the head of the bed, to fasten below the foot. In line with Frances GABe's determination to make the SCH economical in cost as well as time, the General Room-Washing Apparatus and other self-cleaning features are powered not by electricity, but by the cleaning water itself. Her transparent protector boxes to be set on table- and desktops to hold papers, etc., are hinged in such a way that the No-Bother lid folds out of the way when the box is open.

Even more impressive is GABe's means for making the SCH floor joists perpetually adjustable, maintaining the proper slight slope of the floor for drainage of the cleaning water even after the house settles. She recounts with some delight the time she showed one of the local builders her SCH floor joists:

> He looked and gulped as though he'd swallowed a bug. He stared at me for some unbelieving seconds, then shook his head to clear it. He cracked up, guffawing loudly, and said at last, "Well . . . why in the hell not?!!!" What could he say? It *is* good construction.

Most impressive of all is that the SCH meets all building-code restrictions, and will be available for a low cost, estimated at about $50,000 in 1984. Doing all the work and financing herself, Frances GABe has built her own model for about $15,000.

By 1983 she had already been working on the SCH for 27 years. By 1985 Frances GABe held (or had applied for) 68 patents connected with the House. She has invented all her life, she says, but never patented any other inventions, primarily because of the cost.

GABe's background is interesting from the standpoint of the woman inventor's profile. Born in the desert near Boise, ID, Frances was the third of six children (four sisters and one brother). She describes herself as an artist-inventor, and says she has several kinds of artistic talent. She is ambidextrous. Since her mother actively discouraged her, for fear she'd "get the big-head," Frances spent much time alone or with her father, a building contractor and designer, who sometimes took her with him on jobs. As for schooling, she was partly tutored at home and partly enrolled in public schools, but left home at age 13 and enrolled herself in a Girl's Polytechnic School. Always a loner, she considers that she has no mentor except possibly her father, and is convinced that inventors are born, not made.

In her marriage she always resented housework as a thief of time with the children. Though her divorce was amicable and her husband lives nearby, he refuses to discuss the SCH with her. Some neighbors, and several engineers, are unsympathetic and she uses the pseudonym GABe to protect her relatives' privacy. Ten years ago she was told she was twenty years ahead of her time, but she remains unperturbed, convinced that hers is an idea whose time has come—not only because so many women now work outside the home, and because older people can remain independent longer in a Self-Cleaning House, but because she feels no one should have to spend life, as she puts it, throwing dirt out the door faster than the family can bring it in. And indeed her view would seem to be confirmed by a barrage of publicity in the mid-1980s.

Though 70 years old in 1985, GABe reports working at least half of each day on the SCH, and spending other hours on her autobiography and on a textbook on the House. In the evenings she may relax with one of her arts, jewelry-making (M. Fuller, 13; GABe, Q-5/19/85, PC-6/13/85; GABe/Zimmerman, ed., *Technological*, 77ff; McMurran; Siegel, 80).

**Alice C. Austin**, a self-educated architect from Santa Barbara, CA, had a related vision earlier in the century, as her kitchenless houses designed about 1914 for Llano del Rio, CA, reveal. Austin provided for hot meals that would arrive in special containers from central kitchens, to which the dirty dishes would then be returned to be washed by machine. She designed built-in furniture and roll-away beds, heated tile floors to replace dusty carpets, and decorative window-frames to replace curtains. An underground supply system using small electric cars connected each house to the central kitchen, and in turn to the city center for deliveries of laundry, shopping parcels, and other items. Austin also advocated the undergrounding of all electric, water, gas, and telephone lines. Had Alice Austin's vision prevailed, Los Angeles might be a livable place today (D. Hayden, *Grand*, 242ff).

An interesting example of a built-in mechanical device is the Fireplace damper actuating tool patented by **Virgie M. Ammons** of Eglon, WV, in 1975. Patricia Ives included her in a 1980 article on black and women inventors; Mrs. Ammons describes herself as Afro-American (PC-1983). The purpose of her invention is twofold: to prevent a fireplace damper from fluttering noisily in the wind, and to prevent heat loss from an imperfectly closing damper (caused by pressure differences between room and chimney).

## *Art*

We might assume that mechanics comes to the aid of art mostly in photography and drafting, or applied art. However, a little thought reminds us that sculptors, and etchers and other print-makers, regularly use machines, tools, and mechanical techniques and devices in their work; painters may use spray techniques, etc.

The Women's Bureau's Category XXI.B, Artist's and sculptor's devices and equipment, is most pertinent here. Of its 11 types of invention, four seem likely to be mechanical in nature:

    Adjustable artist's desk and easel
    Drawing table
    Etching apparatus
    Reproducing outlines of a form

Photography: The Bureau classes photographic inventions under Category XIV, Scientific instruments. Pertinent here are at least the following eight types of invention:

    Attachment for motion picture machines
    Automatic controller for camera shutters
    Box with removable slides for packing pictures
    Camera
    Camera attachment
    Machine for treating moving-picture films
    Motion picture machine
    Shifting camera back

Since the 76 inventions in this category fall into 54 types, these eight types could represent some 11 inventions, with even distribution among types.

Drafting: The Bureau classes drafting apparatus under Category XIV as well:

Adjustable measuring and ruling device
Combined rule, trisquare, and calipers
Combined rule and adjustable compass
Drafting tool
Measuring instrument

If distribution is even for these five types, they would represent seven inventions.

## Commercial/retail-trade inventions

The Women's Bureau compilers recognized Trade as a separate category (VI), containing 71 patents. In various subcategories, this category contains several machines and mechanical devices: For example, the 36 patents pertaining to Store equipment and furnishings fall into 26 types, of which 11 look mechanical:

Automatic delivery system
Automatic stock counter
Box- and package-elevating apparatus
Bundling machine
Coin-receiving and delivering apparatus
Collapsible foot-measuring device
Collapsible rack
Combined ribbon-holder and measuring mechanism
Registering machine for coin-controlled apparatus
Tag-attaching device
Value-printing and indicating device for scales

If the 36 patents were evenly divided among the 26 types, this would amount to little more than one patent per type, or some 13-14 inventions.

Subcategory B, Advertising devices and equipment, contains 14 patents of 12 types, of which only two look mechanical (not counting the Electric sign and the Illuminated changeable sign): Advertising car construction and Advertising machine. This probably means no more than two inventions here.

Subcategory C, Measuring and dispensing devices, contains 21 patents of 14 types, of which half are machines or mechanical in nature:

Liquid dispensing device
Liquid vending machine
Machine for applying labels
Shoe-shining machine
Stamp and ticket vending machine
Vending machine
Vending machine, coin-operated device for

At an average of 1.5 patents per type, this list could refer to ten or eleven inventions.

This survey shows 25-27 patents granted to women for trade-related machines or mechanical devices.

The Automatic cut-off for water glasses and the Tray mechanism (Hotel and restaurant equipment), the Dyeing machine, the Folding machine for flat work, and the Machine for folding fabric articles (Steam laundry and dyeing-and-cleaning . . .), and the seed-grading machine mentioned above, would add a further six inventions or so to our business/commercial machines total.

Within the WB/28 span, but in a year not studied, is **Edith E.L. Boyer**'s Labeling machine of 1907 (Dayton, OH). Discussed below as a manufacturing-related device, it could also have been used in the retail trade, and WB/28 so classifies one label-affixing machine. Definitely belonging here are her 1920 Value-printing and indicating device for scales, which appears in WB/28, and her Mechanical vernier of 1924 (having to do with the same process in scales), which does not, as it is too late.

Aside from WB/28 the examples as usual are scattered but interesting. Very early is the Coin-freed mechanism for goods-delivery machines patented by **Alice M. Argles** of Wimbledon, England, in 1901. As always, her troubling to protect her device in the United States argues for some commercial potential.

Only quasi-commercial—if, as one source hints, it was used mainly on parking meters (Obit., Chicago *Sun-Times,* 4/20/79)—is **Florence Anderson**'s much later Coinbox for coin-operated machines (1952). The Patent Gazette description, however, sounds more generally applicable. Anderson, a corporate inventor and engineer from Chicago, is discussed above and in Chap. 1 in connection with her refrigerator inventions.

**Barbara Brown** of Springboro, OH, assigned her 1979 patent for a Constant velocity driving means to National Cash Register of Dayton. The Patent Gazette drawing suggests that it is part of the printing mechanism of a cash register. Her co-inventor is William Dugan of Memphis.

*Education*

The Women's Bureau Study's Category XX.A, Mechanical aids to teaching, comprises 40 inventions of 19 types, at least the following majority of which sound mechanical:

Appliance or device for teaching reading
Collapsible globe for geography teaching
Inflatable globe for geography teaching
Counting apparatus
Device for teaching fundamental operations with numbers
Device for teaching number work
Drill device for visual instruction, very young children
Means of teaching the alphabet
Means of aiding teaching of spelling
Means for producing a correct writing habit
Educational device
Educational toy
Kindergarten educational device
Kindergarten loom
Method and apparatus for teaching manuscript forms

*Electrical and related*

Some early-20th-century women inventors in electricity and related technology began work in the 19th century. **Harriet Hosmer**'s experiments with electromagnetic motors continued until her death in 1908, but she has only 19th-century patents, and is thus discussed above. **Bertha Lammé**'s work at Westinghouse continued until her marriage in 1907. Thus, whatever her role in the great electrical motors and generators being produced there, it would have continued. Her brother Benjamin's design team continued to patent at a fierce rate.

Some have suggested that **Nora Stanton Blatch DeForest,** who studied engineering specifically in order to understand (and collaborate in?) her husband Lee DeForest's early radio work, may have contributed to that work. I cannot corroborate this suggestion, but do not dismiss it out of hand (cf. Rossiter, 91). DeForest began selling radio apparatus to the public in 1910 (Harpur, 35).

Too early for WB/28 is **Charlotte R. Manning** of Meriden, CT. Her Electroplating patent (1903) was significant enough to be mentioned in the *Electrochemicals Industry* journal (Trescott, *Rise,* 399). Manning also held two other patents, one relevant here: Enamelling of metal hollowware (1907).

Other early-20th-century inventions not covered by WB/28 are **Sabella G. Doherty**'s Mirror-decorating and illuminating device of 1903 (NYC); and **Mary Aitken**'s Semaphore signal of 1912 (Chicago). **Dr. Alice G. Bryant**'s medical instruments and devices have received some notice, though probably not as deserved, particularly in her own medical specialty. Pertinent here is her handy electric foot switch for the diagnostic or therapeutic devices in an otolaryngologist's office (1916).

The Women's Bureau compilers found electrical inventions in many categories in 1905–21. Under Category XIV, Scientific instruments, etc., for example, appear a Sound-recording and -producing instrument and an Electric light attachment for levels, certainly electrical, and an Exposure meter and a Register meter which may well have been so. Also listed here are two alarm-clock inventions and a Printing-ribbon-shifter for time clocks.

Category V, Transportation, also contains some electrical types of invention, as for instance an Electric engine starter, a Combination tail and auto-license light, a Dimmer for headlights, an Electric lighting system, and an Electric tire-attaching means. The Aircraft Subcategory contains a Dirigible headlight, and the Boats Subcategory, an Electrically propelled boat (31, 30).

Category IV, Structural equipment and materials, contains a Subcategory E, Lighting equipment and appurtenances, covering 26 inventions of 23 types, several of which are electrical or related:

Arc-light carbon
Electrolier
Electric lamp bracket
Electric lamp, portable
Incandescent electric lamp fixture
Supporter for electric globe shades

Category IV.D contains an Electric heater.

Category VI.B, Advertising devices and equipment, contains an Electric sign and an Illuminated changeable sign. Category II, Mining, smelting, etc., contains a patent(s) for Electrolytic separation of metals (20). Under toilet articles (Category XVI, Personal wear and use) appear an Electric hairbrush and a Rotary toothbrush; under Beauty parlor and barber supplies appears an Electric needle, as well as various hair-drying and -waving apparatus (44, 45). Under Medical inventions appear at least three types of electrotherapeutic devices, one unspecified, one Electrothermal blanket, and one Medical battery electrode, to say nothing of a Foreign body localizer for X-ray work (46). Under Safety devices appear several safety devices for elevators as well as an Automatic elevator-door closer and a Safety control for power presses (47).

Category XIII, Supplies for use in industry, agriculture, commerce, and the home, however, is probably the main repository for electrical inventions. It contains an entire

Subcategory B, Electrical equipment, apparatus, and supplies, totaling 45 inventions of the following types:

| | |
|---|---|
| Air-cooling fan | Float-operated circuit closer |
| Attachment plug | Housing for electrical apparatus |
| Automatic circuit-breaker | Impulsion motor |
| Automatic circuit-controlling motor | Incandescent lamp socket |
| Battery plate and terminal connection and | Insulator |
| fastening device | Magneto controlling device |
| Burner | Motor control |
| Call service apparatus | Portable motor |
| Casing for electric cells | Protective container for dry batteries |
| Circuit lock | Rheostat |
| Circuit-closer | Safety transport carrier for electric |
| Conducting cord holder | incandescent lamps |
| Contact device | Socket-key extension |
| Control of electric motors and apparatus | Spark plug |
| therefor | Switch |
| Cross-arms for electrical construction | Switch box |
| Dry battery cell | Testing device for ignition systems |
| Electrical resistance | Vapor electric apparatus |

Thus, during ten select years in the first two decades of the 20th century, women received a considerable number of useful and interesting electrical patents.

The Switch mentioned in the foregoing list may be the Electric switch patented in 1910 by **Delma E. Lee** of Pontiac, WA. This invention was mentioned in *Inventive Age* in 1911 (25: 7[July 1]: 6).

Likewise, this periodical credits **Emadia J. Grenier** with two inventions in telephone switchboards, giving her residence as Chicago (Nov. 1, 1911: 6). One of these Grenier patents, a Combined jack and restoring trip for telephone switchboards (1910), lists her residence as Menominee, MI. This invention should appear in WB/28, but is not under XIII.F, Telephone equipment . . . , or under XIII.B, Electrical equipment. Nor is this second patent on the Patent Office Name Lists just before 1911, though *Inventive Age* is usually reliable.

**Angelica E. Post** is almost certainly the inventor of the Dry battery cell listed (assuming it was only one).[76] Her invention, patented in 1906, was considered important enough to be mentioned in the *Electrochemical Industry* journal (Trescott, *Rise*, 339). Post, who lived in Boston, also held two other patents, one of them relevant here: an Electric pocket lamp (1904). **Amélie D'Espéon Coste** of Chicago had patented a pocket lamp as early as 1891, but this was not electric. Post's is surely one of the earlier electric pocket lamps, forerunner of the modern flashlight.

The Carbon arc patent(s) of Category IV may have belonged to **Hertha Ayrton** (1854–1923). Though born in the mid-19th century, Phoebe (Hertha) Marks Ayrton was in many ways a modern figure. After graduating in mathematics from Girton College, Cambridge, she studied electricity and magnetism at Finsbury Technical College in the 1880s, attending lectures given by her future husband, William Ayrton. She devoted her life to research in electricity, and in 1899 became the first woman member of the Institution of Electrical Engineers. Among her known inventions were a Pulsemeter and a

---

76. As the 45 inventions fall into 33 types, each type would, if the distribution of inventions is equal per type, contain approximately 1.36 inventions.

new kind of Carbon arc light for movie-projection (1913). This grew out of a 1912 request from a "kinematograph company" for help in improving their projectors. She improved not only the carbons, but the design of the lamp housings. However, Mrs. Ayrton had been working on carbon arcs since 1895, or possibly as early as the 1880s, and her book, *The Electric Arc,* was published in 1902. Thus she may have had other inventions or innovations in carbon-arc lighting to her credit well before 1913.

During World War I, Hertha Ayrton contributed searchlight improvements, the famous Ayrton anti-gas fan (designed to repel poison gas from the trenches), and studies on the ventilation of gun emplacements. In 1903 the War Office and the Admiralty consulted the Ayrtons on improving and standardizing their electric searchlights. At first with her husband and then alone, she worked on the project until 1910, suggesting, among other things, an improvement in the shape of the arc that ensured a quick return to steady burning after re-carboning. Says Minnie Reynolds in 1908, Mrs. Ayrton "is now receiving the recognition of the scientific bodies of England for her new searchlight" (7). The anti-gas fan, based on her well-known work with water currents, was first tried at the front in 1915 or 1916.

Hertha Marks was fortunate in many ways, not the least of which was finding a supportive husband. Her university education was subsidized by George Eliot and Barbara Leigh-Smith Bodichon. The latter became a lifelong patron, leaving her protégée an independent income in her will. Hertha named her only daughter after Bodichon. She worked on significant projects, got her work published, and even read a paper before the Royal Society. But she was cheated of her cherished dream of membership in the Society. Hertha Marks Ayrton received the Hughes Medal of the Royal Society and was nominated for membership in 1902, but, as Mozans put it, "she failed of election because the council of the society discovered that 'it had no legal power to elect a married woman to this distinction' " (212).

Though an accomplished physicist, she was always more interested in application— i.e., in invention and technology—than in theory. While still a student she invented a drafting line-divider, and showed it at the Columbian Exposition (Ayrton, *Electric,* "Origin"; Blackburn; Davidson, 39–40); *EB*/58; Mozans, 212, 230; M. Reynolds, 7; E. Sharp; Trotter, 48; *WWWS*).

The great American actress **Maude Adams** (1872–1953) is sometimes said to have started a second career at age 49, as a lighting designer at General Electric. Working there beginning in 1921, according to *Notable American Women,* she developed by about 1923 "an incandescent bulb widely used in color film [projectors], but failed to secure the patent." She was advised to sue to protect her rights, but did not wish to do so. According to one source she worked with Charles Steinmetz himself at GE. Evidently Adams became wiser later, for she did patent three Illuminating devices (1932, 1934, 1935) with male co-inventors. Adams's address on her patents was New York City.

Apparently her original improvement in projector lamps was an improved capability for showing color film in full daylight. This must have been invented before 1928, for in that year *Literary Digest* noted ("Personal Glimpses"):

Among the more or less famous persons in other fields of activity who figure also as inventors, we find Maude Adams, actress, working in the laboratories of the General Electric Company at Schenectady to perfect a device for showing motion-pictures in full color by daylight.

Born Maude Kiskadden in Salt Lake City, she took her actress mother's maiden name of Adams for the stage. She began with juvenile parts in San Francisco—as La Petite

Maude, she received a salary from age five—joined her mother's traveling stock company at eleven, and reached the New York stage at sixteen. Her greatest successes were *The Little Minister* and *Peter Pan.* James M. Barrie, in fact, was inspired to complete *The Little Minister* when he saw Adams on stage and instantly visualized her as Lady Babbie. She played Peter in *Peter Pan* more than 1500 times. The Peter Pan collar she designed for her costume became a favorite in the wider world of women's fashion. Adams had a special affinity for Barrie's work, but her success was not limited to Barrie: at the opening of Rostand's *Chantecler,* for example, she received 22 curtain calls.

Even after leaving the stage she never lost interest in theatre. After her stint as a lighting designer, she became Professor of Drama at Stephens College in Columbia, MO, where she worked until retirement.

Maude Adams, great actress and ingenious inventor, must also have been a canny businesswoman. By 1897, at 25, she headed her own theatre company. By 1900 she owned large estates on Long Island and in the Catskills, both still hers at her death more than fifty years later. After leaving the stage in 1918, she lived for nearly twenty years essentially without working. *Notable American Women* and other accounts of her life suggest that her financial success was entirely her own (*EB*/58; Ireland; M. Logan, 772–3; McHenry, ed.; *NAW*/mod.; "Personal Glimpses").

It is also worth noting that as recently as 1984 a history thesis from the University of Wisconsin dealt with Adams, calling her "Maude Adams, an American Idol . . ." (Kuehnl).

Roughly contemporary with Maude Adams's early GE years is the work of **Lura Brainerd** of Meriden, CT. In 1919 Brainerd received three patents, for an Apparatus for making motion pictures, Techniques for "Making motion pictures," and "Producing moving pictures." Lura Brainerd appears as a dressmaker in the Meriden City Directory for 1911–3. She has rooms in a well-to-do neighborhood near the center of town (D. Clarke, L-3/18, 5/4/87).

Also roughly contemporary with Adams's early GE work is an important series of contributions by **Edith Clarke** (1883–1959): her 1920s–'40s innovations in analyzing large electrical systems. Clarke graduated from Vassar in mathematics and astronomy, then went on to take the first master's degree in electrical engineering ever granted to a woman by M.I.T. (1919). She later became the first woman fellow of the American Institute of Electrical Engineers. She pursued a varied career, including several years as a professor in the Engineering College of the University of Texas at Austin.

For more than twenty years, however, Edith Clarke worked in the Central Station Engineering Department of General Electric in Schenectady, NY. She may even have overlapped with Adams at GE, though in a different department. There, she developed calculators used to monitor and predict the performance of electrical transmission lines (1925) and the 60-cycle performance charts so widely used by engineers monitoring such systems. She also developed a system of asymmetrical components for analyzing unbalanced power systems. She received at least three patents: for a Calculator (1925); for Electric power transmission (1927); and for an Electric circuit (1944).

Just then in the United States, small power systems were being linked together into ever larger ones, and her work was highly significant. In 1954 the Society of Women Engineers honored her with its Achievement Award for "her many original contributions to stability theory and circuit analysis" (Award Booklet, 1952–76; *NAW*/mod.; Peden, 271). Several of these "many" contributions may have been unpatented inventions or innovations.

As mentioned in Chapter 1, another General Electric employee, **Joy C. Adams** of Bridgeport, CT, patented several small electric appliances during the 1930s—a toaster, two coffee-makers, and a juice extractor in 1937 and 1938. She also patented a Fuse for an electrically heated device (1935).

The writer of a 1940 *New York Times* article on women's gains in the patent rolls found an example in **Margaret Conant** of Kansas City, KS, and her Rectifier for changing alternating into direct current without leakage of electricity ("Women Gaining"). Her 1940 patent applies, in part, to the process of applying carbon to one face of the unit.

**Jessie T. Pope** of Detroit is noteworthy as a black woman inventor of this period. Eleanor Roosevelt considered Pope's thermostatically controlled electric curling iron important enough to help her secure both U.S. and Canadian patents for it (1946). This iron—perhaps the first or one of the first to incorporate a thermostat—produced a steady, even heat, curling hair of any texture in only 15 minutes. A beautician by trade, Mrs. Pope at first manufactured the irons herself, by hand, but could fill only 10% of her orders. *Hue* magazine took note of her when, twelve years after obtaining her patents, she finally found financial backing for mass production at a factory in River Rouge, MI ("Curling").

The early 1940s were war years, and at least three women made important contributions to the war effort in electrical or related fields: **Dr. Florence Van Straten, Dr. Chien-Shiung Wu,** and the actress **Hedy Lamarr.** Florence Van Straten (1913– ) generated several new ideas for the Navy. She invented a sonic device to prevent icing of airplanes (patented, but not developed by the Navy) and a radar-facsimile system that automatically plots out the scope information at the radar set, or displays it simultaneously at any number of air stations (developed and used). In fact, she became known as the one to call when Navy scientists had a particularly tough technical problem. Says Edna Yost (*WMS*, 135–6), "She might suggest having a new piece of equipment designed, or work out a new technique using existing means, or show that if new data were gathered the problem could be solved." Many of these were doubtless inventions and might have been patented under different conditions. Van Straten also devised improved techniques for weather balloons.

Florence Wilhelmina Van Straten was born in Darien, CT, the only child of Dutch immigrants. She attended public elementary schools, but went to Girls' High School in Brooklyn, and also studied in a girls' lycée in Nice, France. She originally wanted to be a writer, but realizing that writers seldom make a living, she took a PhD in physical chemistry. She had before her a good model of female achievement: her mother was a brilliant linguist fluent in six languages, the highest-salaried woman in Holland before emigrating.

Joining the Waves when World War II broke out, Florence entered the Naval Weather Service, staying on as a civilian employee at war's end. In 1956 she won the Navy's Meritorious Civilian Service Award and became a Commander in the Naval Reserve. Much of her later work dealt with the meteorological aspects of missile development, and parts of it were still secret when Yost wrote about her in 1959 (Phelps, ed., 251; Yost, *WMS*, 124–39).

**Chien-Shiung Wu** (1912– ) has been called the world's foremost female experimental physicist, and is perhaps best known for designing the landmark experiments that demonstrated nonparity in the universe. But she is also an inventor. A member of the Manhattan Project in the 1940s, Dr. Wu worked on improved instruments to detect radiation. She has also devised a method for separating two types of rays during the decay of radioactive elements—a total of at least five inventions. Dr. Wu, who came to the United States in 1936 and completed her education here, has been profiled often enough in connection with her achievements and awards in physics to need no further biography here.

It is worth noting, however, since several of her more scientific profiles do not, that her father believed very strongly in women's emancipation. He founded a girls' school that later became very well known. He taught all his children to ask questions and to solve problems. When Chien-Shiung was chosen from her graduating class to enter the National Central University, with highest honors, but was ill-prepared for her real love, physics, her father brought her textbooks in higher algebra, physics, and chemistry, and encouraged her to study during the summer.

Dr. Wu in her turn has become concerned with getting women into physics, and with their experience after they enter. In a 1981 interview she blamed the scarcity of women in physics both on high school guidance counselors who don't encourage girls to take the crucial math and science courses, and on social scientists who, she said, "really don't do justice to women." She cited an especially subversive series of statements by Bruno Bettelheim in a 1965 M.I.T. symposium on women in science and engineering. Of a Russian woman engineer he said that she loved her work with "a womanly embracing" of her tasks rather than a "masculine conquering" of them. After quoting Balzac about a woman given the advantages of a "man's" education, he opined that a "woman's point of view" may be advantageous in education and social sciences but not in physics and mathematics "where we strive always for objectivity" (Gilbert & Moore, 70).

In 1963 *Newsweek* called Dr. Wu the "Queen of Physics" (92f). In 1974 she was honored as Scientist of the Year by *Industrial Research* ("The 1974 I-R Award Winners," [3], 29–30; *AMWS; WBWRA;* Yost, *WMS*).

So far ahead of its time that it was never used during the war was the anti-jamming system developed for radio-controlled torpedoes by **Hedy Lamarr** (1913–    ). The Austrian-born actress, whose real name was Hedy Kiesler Markey, had discovered a flair for invention—and gained considerable technical knowledge—while married to munitions manufacturer Fritz Mandl. Coming to the United States early in the war, she wished to contribute to her adopted country's war effort, and considered leaving M-G-M to offer her services to the new Inventors' Council. Composer George Antheil, a friend of Hedy's, speaks of "various secret weapons, some of which she had invented herself," but she obtained only one patent, for a Secret communication system, jointly with Antheil in 1942. One of its features—Hedy's idea—was frequency hopping, whereby a broadcast signal carrying, for example, torpedo commands, bounced from frequency to frequency at split-second intervals. Only a synchronized receiver could receive the entire signal; thus the broadcast could not be "overheard" or jammed.

Poorly understood by popular authors—partly because it was at initially classified as secret—the invention is significant, for even though not used during World War II, frequency hopping is the main anti-jamming technology used in today's billion-dollar defense systems. Lamarr, now living in New York City, has never received either recompense (Sylvania engineers developed a similar system after her patent expired) or due recognition for her invention (Antheil, 330–1; Blair, L-1980–83; Blashfield, "First Nude . . . ,"; *CB* 1954, "George Antheil"; McDowell & Umlauf, 372; Panabaker, 14–5).

Among minor but interesting examples as the century progresses may be listed **Helen L. Borza**'s Electric stove of 1948. Borza, who lived in Yonkers, NY, patented both her stove and a Windmill toy jointly with Raphael Borza.

**June Sylvia Thimblethorpe** of London, England, attacked a small but annoying problem whose solution could have had commercial value. With Wilfred S. Thimblethorpe, she received a British patent for Improvements in electric heating units to make the elements less breakable, and thus of course more portable (1948).

Two Southern California women were inventing in the sound-recording and -

reproduction field at mid-century: **Dorothy Blaney** and **Golda Duncan. Dorothy L. Blaney** of Los Angeles received ten patents for sound-recording and -reproduction apparatus between 1950 and 1960. Since all her patents were assigned to Radio Corporation of America, she probably worked for RCA, and some of her devices were probably incorporated into the company's products. In 1950 she received three patents: a Magnetic head testing system, a Monitoring system for magnetic recording, and a third titled Editing and inspection of invisible sound tracks. In 1952 she received her only joint patent (with Arthur C. Blaney), for Magnetic recording and reproduction. In 1953 she patented a Combination sound recording and record and a Combination photographic and magnetic sound record and producing same. In 1954 she received only one patent, for Magnetic record copying, described as "a system for producing duplicates of a magnetic sound track." In 1957 she again received three patents: a Magnetic volume-control system, a Magnetic compression system, and a Magnetic compression method. If Arthur C. Blaney was her husband, it is interesting to note that he held related patents, also assigned to RCA, as early as the 1940s.

In the late 1960s **Golda A. Duncan** of North Hollywood patented two sound-reproducing devices, one called simply a Phonograph device (1968), and one called a Sequential-play phonograph (1969). Interestingly enough, in 1960 Theodore R. Duncan, also of North Hollywood, patented two related items—a toy Phonograph and tone arm and an unspecified Phonographic device seemingly pertaining to a phonograph arm, assigning 26% of both these patents to Golda A. Duncan. Indeed, Theodore received a dozen other patents between 1950 and 1961 (including other toys), at least two of which he also assigned partly to Golda. If Golda and Theodore were spouses (or former spouses), it is further interesting to note that Golda's two patents are granted jointly not with Theodore, but with Robert Aleson of Sylmar, CA, and that at least the Sequential-play phonograph is assigned to Lorraine Industries, Inc., of Bridgeport, CT, suggesting that it was to be manufactured and sold.

A minor but possibly commercially interesting invention of the 1960s was the Portable hair dryer patented on March 6, 1962, by **Harriet J. Stern** of Rockville Center, NY.

*Lasers.* Perhaps the ultimate electrical "machines" of the 20th century are lasers. Two women's laser inventions have been discussed with regard to medical treatment. **Mary Spaeth** (1940– ), a research director at Lawrence Livermore Laboratories in California, invented the tunable dye laser—producing light at different parts of the color spectrum as desired—while at Hughes Aircraft (1966). For the Army, she invented the resonant reflector that made ruby laser range-finders—and modern supermarket check-out technology—workable. Spaeth has also done pioneering research on Q switches for lasers, and helped develop new isotope-separation technology, using dye lasers (K. Brown, 168, *passim*). According to Kenneth Brown, who interviewed her for his *Inventors at Work,* Spaeth received only two patents (assigned to Hughes and the Army) for these several significant inventions. By her own report, however, she is not much interested in patents, but in "getting things to work together."

Born Mary Dietrich, she was a tomboy abetted by her father. She loved to build things, and was given a jig-saw and other carpentry tools at age three (*ibid.,* 169, 181). She also began inventing very young. By age eight she had devised a way to reclose cereal boxes—now used on most cereal boxes—and she sketched an atomic bomb rather accurately before anyone knew what one looked like (181–2). She always thought of herself more as a person than as a girl, and thus never dreamed that there were things she couldn't do. (Nevertheless, she felt discrimination at Hughes, and left partly because of an unwritten policy against line-manager jobs for women [177, 179].) Spaeth studied nuclear physics, and holds a master's degree in physics from Wayne State University in Detroit.

*Professionals.* Turning now to a group of professional women scientists and engineers—of whom Mary Spaeth, of course is one—we find several examples of talented inventors and significant electromechanical work.

Among the most interesting is **Mabel MacFerran Rockwell** (1902–81). She helped design the power system for the Colorado River Aqueduct project, and was the only woman actively involved in designing and installing the power-generating machinery for Hoover Dam.

Born in Philadelphia to Quaker parents, she attended the Friends' School at Germantown, and Bryn Mawr, but transferred to M.I.T., where she took her BS in 1925. In 1926 she was awarded a degree in Electrical Engineering from Stanford University. She ranked first in her class at both M.I.T. and Stanford, becoming a member of Sigma Xi and receiving Stanford's Elwell Fellowship.

As a technical assistant for Southern California Edison, she worked in electrical transmission. Her innovation here was "one of the earliest applications of the method of symmetrical components to practical power system relay problems." She devised a method of locating breakdowns or faults in transmission when more than one occurred simultaneously in different parts of the system. Miss MacFerran's method had great practical value in a system whose tower lines carried as many as twelve circuits.

Her second position was the Metropolitan Water District of Southern California, where she did her work with the Colorado River and Hoover Dam projects. She later worked for the United States Bureau of Reclamation, designing the transmission and distribution systems to bring water from the Central Valley Project dams to irrigation districts and other public agencies in California. She also worked for several years with the Air Force, designing electric substations.

By 1946, when Alice Goff wrote about her, Mabel Rockwell had four technical papers to her credit, all published in the *Transactions* of the American Institute of Electrical Engineers, and three of the four winning awards.

Later in life she became a consulting engineer. In the many years unaccounted for in my sources, she undoubtedly had further inventions and innovations to her credit. Just one example here is the so-called "Serjdetour" telephone protector, mentioned in a 1980 publication of the Society of Women Engineers.

Mabel Rockwell's life seems outwardly smooth. Certainly it was full of professional achievement. She may even have experienced relatively little discrimination. According to one source, for instance, she was promoted faster as an engineer than her husband. In the eyes of some, at least, however, her personal life was less secure—though probably no such observation would have been made had she been a man. A long-time friend of Rockwell's called her "more man than woman," and noted that, though tender-hearted with children and animals—once she almost drowned in an attempt to save a cat from drowning—she found it difficult to be a mother to her own daughter. Once when Mabel received an orchid corsage at a speaking engagement she seemed not to know what it was, and gave it to another woman in the party. She was a night person who worked all night, and ate in all-night restaurants rather than cook. She divorced her husband when she was sixty (Goff, 94–112; Morgan & Roden, PC-5/16/86; SWE, *Technical Contributions 1980,* 28; *WBWRA,* 188).

During the 1950s and '60s, **Elizabeth Zimmerman,** a Research Specialist at Philco's Advanced Technology Center in Blue Bell, PA, was making contributions to high-frequency germanium devices, negative-resistance diodes, parametric amplifiers, and the active elements used in microelectric circuits (Peden, 273).

**Margaret H. Fisher** of Mytholmroyd, England, assigned her 1967 U.S. patent for a Multi-channel time and sequence electrical control apparatus (1967) to Process Units (Halifax), Ltd, Halifax, England. As briefly described in the Patent Gazette, the invention consists of two banks of switches controlled by a program card, allowing both the timing and the sequence of a series of operations to be controlled. The unit sounds as if it might have applications in scientific experiments, in manufacturing, and in security. The fact that she and her co-inventor (whose name, Leslie Newsome, could be either male or female) protected their invention with an American patent probably indicates some expected commercial potential.

New Jersey-born **Dr. Beatrice Hicks** (1919–79) was one of the founders of the Society of Women Engineers, and its first president. She decided to become an engineer at age 13, when her engineer father took her to see the Empire State Building and the George Washington Bridge, and she learned that it was engineers who built such structures. After taking a BS in chemical engineering at Newark College of Engineering, she studied electrical engineering and physics and got her MS at Stevens Institute of Technology.

Dr. Hicks' career included both technical work and business administration. She became the only woman technician and then the only woman engineer at Western Electric, where she worked on the design and development of telephone equipment and where, as an early award citation put it, "the quality of her work became legend." In 1945 she became Chief Engineer at Newark Controls, Bloomfield, NJ, a firm specializing in environmental sensing devices. Here she worked on the design, development, and manufacture of pressure- and gas-density controls for aircraft and missiles. Her best-known invention is the gas-density switch, a key component in systems using artificial atmospheres.

As early as 1952, *Mademoiselle* magazine chose her as Woman of the Year in Business, recognizing her work as Vice-President and Chief Engineer at Newark Controls. By 1955 she was President of the firm. In 1959 Dr. Hicks and her husband represented the National Society of Professional Engineers on Project Ambassador, a fact-finding and goodwill tour of South America. By 1967, at her husband's death, she became the owner of Rodney D. Chipps & Associates consulting firm.

Other honors besides the *Mademoiselle* award and the goodwill ambassadorship were to be hers. In 1963 the Society of Women Engineers gave her its National Achievement Award for "her significant contributions to the theoretical study and analysis of sensing devices under extreme environmental conditions, and her substantial achievements in international technical understanding, professional guidance, and engineering education." She became the first woman to receive an honorary doctorate in engineering from Rensselaer Polytechnic. She also received an honorary ScD from Hobart and William Smith Colleges. She died of a heart attack in 1979 (*Mademoiselle,* Jan. 1953: 77; Peden, 272; SWE Award Booklet; *SWE Newsletter,* Nov./Dec. 1979: 5–6).

Seattle ceramics engineer and entrepreneur **Patricia Corwin** is an outstanding example of combining family life with a successful engineering career. Irene Peden's "vivacious" is obviously an understatement; Corwin's energy, in fact, is reminiscent of Lillian Gilbreth's. She and her husband have seven children and as of the 1960s when Peden wrote about her, were joint proprietors of a firm making specialized scientific and research devices from ceramic materials. At that time Patricia Corwin worked in her home laboratory, designing ceramic transducers. Her husband, also a ceramics engineer, worked with her at home in the evenings. When a contract deadline neared, the whole family got involved, the older children helping with simple testing or babysitting the younger ones. The children kept their own accounts and presented them when payment arrived for the contract.

Oceanographic researchers and fisheries were the Corwins' main customers. On at least one occasion the Corwins' work led to a commercial product: a fathometer transducer used on boats to detect rocks, schools of fish, etc. It works as an electroacoustic device, converting electrical energy to mechanical energy and transmitting pressure waves at audible frequencies to the water around the ceramic element (Peden, 282–3).

While working for Itek, **Ilsa Abrahamson** (later **Koch**) patented a Photo-sensitive cell (1968). This invention was considered worth mentioning in communications on women inventors between Homer Blair, Vice-President for Patents and Licensing at Itek in Lexington, MA, and Dorothy Stephenson of the New England Inventors Association in the late 1970s.

Miniaturization became the goal after mid-century not only for computers, but for space vehicles and missiles; and microelectric or printed circuits were the technological breakthrough that made that goal reachable. **Elise F. Harmon,** a physicist and chemist of New Bedford, MA, devised an entirely new method of creating printed circuitry: a hot die stamp method of infusing silver conductors on polymerized thermo-plastic and thermo-setting materials. For this she received her single known patent (1953). Other inventions may have been secret for security reasons (see below). In the 1970s she headed the printed-circuit activities at Aerovox Corporation in New Bedford.

Harmon also worked on printing resistors as part of the printed circuit, on electrical equipment for aircraft, and on proximity fuses. During World War II she worked at the Naval Research Laboratory in the Aircraft and Electrical Division. In fact, it was largely due to her work on carbon brushes that American airplanes were able to fly as high as German planes and successfully combat them during that war. In 1956 she received the Achievement Award of the Society of Women Engineers. A graduate of North Texas State College, she took her MS at the University of Texas, and has done further work at George Washington University and the University of Maryland. She is a member of the American Chemical Society, the Institute of Radio Engineers, and the Texas Academy of Sciences (Malotke *et al.,* 28; SWE Award Booklet, 1952–76; *WBWRA*).

**Edith Olson,** an inorganic chemist working with the Army, devised photographic and lithographic techniques for printed circuits that are now used worldwide. Olson was assigned the task, with three other scientists, of drastically shrinking the electronic parts required for missiles—a little, she observed, like printing all the books in the Library of Congress on a single grain of rice. Succeeding in this impossible-sounding task, and thus saving the Department of Defense an estimated $200 million a year, she became, in 1959, the first woman to receive the Department's $25,000 prize (Blashfield, *Hellraisers,* 172).

**Irene Carswell Peden** was born in Topeka, KS, in 1925. An electrical engineer, she has succeeded both in industry and in academia. After getting her BS from the University of Colorado in 1947, she worked for two years as a junior engineer for the Delaware Power and Light Company. In 1953–4 she was a research engineer at the Midwest Research Institute in Missouri. Moving to California, she worked on antenna problems at the Stanford Research Institute. She took her MS in electrical engineering at Stanford University in 1958, then worked as an acting instructor (1959–61) and as a research assistant (1958–61) at the Hansen Microwave Laboratories, studying measurement techniques for microwave circuits. She received her PhD in electrical engineering in 1962.

Dr. Peden has a number of firsts to her credit. She was the first woman to get a PhD in electrical engineering from Stanford, the first woman engineer/scientist to do research fieldwork in the Antarctic interior, and the first woman member of the Board of Directors of the Institute of Electrical and Electronic Engineers. Moving from Stanford to Seattle and the University of Washington, she became in the late 1960s the only woman faculty member in that university's Engineering Department, and vice-chair of the Seattle section of IEEE. By 1971 she was a full professor at Washington, and later served as an associate dean. In 1973 she received the Society of Women Engineers Achievement Award, partly for her work in radio waves.

Dr. Peden's other strong professional interest is the role of women in her field. She has written on women in engineering careers, does career counseling in high schools, and received special recognition for her work in engineering education from the Society of Women Engineers.

Married to a lawyer (1962), Dr. Peden enjoys water sports in the lake-studded Seattle area (*AMWS*/15; Peden, 269; *WBWRA*, 189, 191–2).

**Martha J. Thomas** (1926–   ) is a chemical engineer who specializes in phosphor research. Born in Boston, the oldest child of two teachers, she attended public schools. Both parents encouraged her to achieve. An honors graduate of Radcliffe College, she took her PhD in chemistry at Boston University in 1952. From then until 1970 Dr. Thomas taught in the evening division at Boston while working in increasingly responsible positions at Sylvania, moving to head the Phosphor Research and Development Section, and then to manage the Technical Assistance Labs (1972). At the same time she was carrying family responsibilities as a wife and mother of four daughters.

One of the few women then working in phosphor chemistry, Dr. Thomas has received at least seven patents in the field. She developed, for example, the phosphors that made possible Sylvania's natural-daylight fluorescent lamps, and made mercury lamps 10% brighter. She is a member of the American Chemical Society and the Electrochemical Society, and has published numerous articles in professional journals in her field. In 1965 the Society of Women Engineers honored her with its National Achievement Award for her contributions to chemical engineering (*WWAW*/77–8; *AMWS*/14; M. King; SWE Award Booklet, 1952–76, Thomas, Q-1983).

In 1961 **Phyllis Hersh** of Teaneck, NJ, patented a Bi-directional magnetic tape recording apparatus. Applications for magnetic tape recording have proliferated in recent years, most obviously in the fields of music and computers. Hersh's invention, classified as Electrical by the Patent Gazette, appears to pertain to information storage, and thus perhaps to computers. The patent is assigned to Curtiss-Wright Corporation of Carlstadt, NJ, though Hersh does not seem to have been a Curtiss-Wright employee.[77] Such an assignment history would seem to mean that the patent was of some technological and/or commercial interest, however. Ms. Hersh's other patent, held jointly with **Evelyn Berezin,** will be discussed in Chapter 5.

**Mary Beth Stearns** was born Mary Beth Gorman in Minneapolis around 1925. Receiving her BS at the University of Minnesota in 1946, she took a PhD in nuclear physics at Cornell in 1952. She has worked as a research physicist at the Carnegie Institute of Technology (1952–6), at the University of Pittsburgh (1957), at General Dynamics Corporation (1958–60), and at Ford Motor Company's Scientific Laboratory (1960–81), where she became a principal scientist, setting the direction of research at Ford. Dr.

---

77. According to Arthur L. Frederick, General Patent Counsel of Curtiss-Wright (L-8/1/84), the wording "by mesne assignments" on the patent means that Ms. Hersh had already assigned the invention to someone else before Curtiss-Wright—perhaps to her actual employer—who would then assign it to Curtiss-Wright.

Stearns was the company's first top female executive. She devised the High Resolution Electron Spectrometer and used it to study atomic structure, properties, and behavior, concentrating on metals and their alloys. Her work has applications to developing new alloys, electronic devices, and magnetic research techniques. In 1981 she became Professor of Physics at Arizona State University at Tempe (*AMWS*/15; *WBWRA,* 184, 520).

A late-1970s example of the professional (i.e., corporate) inventor, is **Muriel Abrash** of Paterson, NJ, who patented an Electrical circuit element in 1978. She and her male co-patentee assigned the patent rights to Engelhard Minerals & Chemicals Corporation of Iselin, NJ. The element consists of a thick film resistor of preformed glass frit with lead ruthenate or iridate dispersed within it, bonded to a conductor also made of glass frit, but containing enough of an electrically conductive component to provide the desired electrical properties.

An isolated British example of the professional woman engineer and inventor is **Dr. Elizabeth Laverick** (1925– ). As of 1983 Dr. Laverick was Editor of the *Women's Engineering Society Journal.* She was also an electronics engineer with at least five inventions or innovations to her credit: an absolute method of measuring alternation at microwave frequencies, a rotary standing-wave indicator, a construction for a Cassegrain Aerial, an application of microstrip to an aerial problem, and a form of microwave circular polarizer. Since the average professional inventor in Britain has only four patents to his[78] credit, Dr. Laverick need not hesitate to call herself an inventor (Q-12/14/83).

She is the second daughter (no brothers) of a manufacturing chemist who was inventive though without patents, and a wife who could assist him. Elizabeth attended state rather than private schools. Encouraged toward achievement by both parents, she took an undergraduate BSc and a PhD in physics and radio, and then went into the electronics industry.

A Russian example is **Katerina E. Sirotkina.** As apparently was usual with Soviet invention teams, the whole team appears on the 1978 patent, in this case 17 co-patentees, 11 of whom are women. Besides Sirotkina herself, they are **Rita M. Kogan, Sofya I. Kudinova, Ljubov S. Sizova, Svetlana S. Reznikova, Nina A. Tsekhanovskaya, Larisa V. Randina, Svetlana L. Bocharova, Galina P. Gulyaeva, Raisa I. Bondarenko, Galina I. Rybalko,** and **Yanina A. Adomanite.** The invention is a Photosenstive polymeric material and an electrophotometric material prepared by chemical means.

A fitting final example of the professional woman inventor of electrical and related devices is **Dr. Betsy Ancker-Johnson,** holder of perhaps the highest posts in both government and private industry ever achieved by any American woman inventor. Dr. Johnson has been Associate Director for Physical Research at Argonne National Laboratory, and served in the late 1970s as Assistant Secretary for Science and Technology in the United States Department of Commerce. She has also worked as an engineer for Boeing in Seattle, and as of 1980 had become Vice-President for Environmental Affairs for General Motors.

While at Boeing, from the mid-1960s to the mid-1970s, she received at least four electrical or related patents: for Signal generators using semiconductor material in magnetic and electric fields (1966); for a Fast sequential switch with adjustable delay (1974); for an Infra-red optical detector-amplifier (1974); and for a Detection-Modulator for an Optical Communication system (1975). On all but one (the Fast sequential switch), she was sole inventor.

---

78. In the study this "average" inventor was a white male, age 44.

Dr. Ancker-Johnson also had at least three other inventions at Boeing, which were protected by Invention Disclosures: Non-equilibrium plasmas in solid density probe (1967); Multi-mode solid signal generator (1969), and Solid-state amplifier and phase-detector (1969). On all of these she had at least one co-inventor.

Although this brief synopsis makes her career sound like a smooth success story, Dr. Ancker-Johnson reveals in a published interview (Kundsin, 44–9) that she experienced both overt and covert discrimination, especially early in her career, leaving her convinced that "a woman in physics must be at least twice as determined as a man with the same competence, in order to achieve as much as he does." After four idyllic years at Wellesley she found graduate school at Tübingen, Germany, a shock. Her professors and fellow students repeatedly told her that women cannot think analytically and thus she must be there to get a husband. Like Rosalind Franklin in England, she was cut off from all the informal sessions where so much intellectual stimulation and exchange occurs.

Another shock came during her first research job. She was laid off during her first pregnancy, and for the last three months could not even enter the laboratory to hear a lecture or get a book without special permission from the lab director. During her second pregnancy, her paycheck stopped eight weeks before her due date and resumed six weeks after delivery, even though she kept on working throughout. Just a few weeks later, when the wife of one of her male assistants had a baby, he received a week's leave *with pay*.

Since Dr. Ancker-Johnson has four children, all told, two by birth and two by adoption, she has more on her mind than most of her male competitors in management and research. Although she enjoys her work, she feels she must, and she does, compete "vigorously," because that's the way the research and management games are played. Here again, she points out, a woman is "damned if she does and damned if she doesn't. If she does not compete, it proves women can't do physics; if she does compete, she isn't feminine and hence, presumably, is some sort of freak." Dr. Ancker-Johnson has managed to commit herself simultaneously and successfully to both career and mothering goals. As of 1979 she was one of five women members of the National Academy of Engineers (Ives, *Creativity,* 124; Kundsin, 44–9; F. Reynolds, L-9/10,23/80; *WBWRA,* 141, 142, 190).

Although my major focus here is on Dr. Ancker-Johnson's own inventions in electrical and related fields, her impact on another field of invention is worth noting. In 1977 as Assistant Secretary of Commerce for Science and Technology, she took innovative steps to bring researchers into compliance with the NIH guidelines for recombinant DNA research, offering to speed the patent process for cooperative corporations ("Update: DNA," 7).

*Independents.* Independent women inventors of the 1960s and '70s also had some interesting electrical inventions to their credit: **Marie Van Brittan Brown** is a black woman inventor from Jamaica, NY, who patented a Home security system in 1969. She received the patent jointly with Albert L. Brown, but her name comes first. This ingenious invention has both video and audio components. It allows the occupant of a house to scan a would-be visitor outside the front door without getting out of bed, to talk with the caller, and even to release the door lock if desired (Ives, *Creativity,* 119; Stanley/Zimmerman, 63).

As yet unheralded inventions being developed in the early 1970s by women members of the Inventors Licensing and Marketing Association (ILMA) of Southern California are **Mari Harper**'s U-turn signal device and **Maria Baker**'s device for selecting the television shows children can watch. Harper, a resident of La Mirada, CA, is a versatile inventor with several other inventions to her credit; and Baker's device is relevant both to the environment and to childcare/socialization technology. Less significant but possibly commercially interesting are a self-cleaning fish tank and an illuminated pet collar adaptable to hold a flea collar, being developed during the same period by another ILMA member, **Gina Piediscalzzi** of Burbank, CA. At least two of these women have felt the lack of technical, specifically electrical or electronics, expertise in developing their inventions. As of 1976, Piediscalzzi was studying both electronics and plastics technology. Baker was working with an electronics engineer on her device (Weisberg, L-8/16/76).

A controversial independent inventor of the early 1980s is **Barbara J. Hickox** of Sacramento, CA. Hickox, who claimed that her invention (an Electrical dynamo, 1981), could harness the electricity surrounding the earth, broke into the news in the spring of 1980 when she filed a $2.5 billion lawsuit against the United States Department of Energy, claiming that she was refused an application for federal funding to research her device. Ms. Hickox was acting as her own attorney in the U.S. District Court. Unfortunately, the news media dropped the story, so the outcome is still a mystery ("Woman Fights"; KGO [San Francisco] Radio News, Mar. 22, 1980).

*Medical applications.* Two 1970s electrical inventions with medical applications are the X-ray luminescent screen patented by **Agnes D.M. Schrama–de Pauw** of Eindhoven, Netherlands, and Albert L.N. Stevels (his name first), and the Electromagnetic device and method for cleansing blood vessels patented by **Shirley Goldman** of Monsey, NY, with others. Both patents were granted in 1978. Ms. Schrama–de Pauw may have been working for U.S. Philips Corporation of New York City, for she and her co-inventor assigned the rights to the invention to that company.

*Satellites and satellite-tracking.* Since the USSR launched the first satellite, the famous *Sputnik* of 1957, and since **Dr. Alla G. Masevich** (1919– )[79] devised the original tracking system for that project, this Soviet woman must be credited with the highly significant invention of the first satellite-tracking technology. She must further be credited for the speed with which she did it; she was given six months from her notification to the scheduled launch date!

Born in Tiflis of Georgian and Polish ancestry, she was the eldest of three children. Inspired by reading popular books on science by Prof. Perelman at Leningrad Polytechnical Institute, she corresponded with Perelman for six years, and upon graduating from secondary school wished to go to Leningrad to study with him. Bowing to her parents' objections, she went instead to the University of Moscow, majoring in physics and graduating in 1941. After a detour into metallurgy dictated by World War II, she returned to her first love, astrophysics. At 33 she was Deputy Chair of the Astronomical Council of the Academy of Sciences of the USSR; at 37 she became Professor of Astrophysics at Moscow University. When the Academy of Sciences charged the Council with tracking *Sputnik,* Dr. Masevich as Deputy Chair took it on.

Although her first tracking system was primarily optical in nature (calling for "entirely new techniques of precision timing and velocity measurements" as well as new telescopes), radio may have been used even in that first system. Certainly, subsequent systems have used radio telescopes and other more sophisticated electrical and electronic devices. Dr. Masevich continued for many years to run the USSR's satellite-tracking operation, along with her own research and teaching, and activity in the society for popularizing science in her country. In the 1960s she was vice-president of the Institute for Soviet-American relations, and a member of the Presidium for Soviet-Austrian Relations.

---

79. Sir Bernard Lovell anglicizes her name as Massevich in his 1963 *Saturday Review* portrait (Q.V.). However, Dodge and Blashfield agree on the single-*s* spelling.

Foreign scientific bodies have honored her as well: she is a Foreign Member of the British Royal Astronomic Society, and Brazil's Scientific Council for Astronautics has made her a member. As of 1963 she chaired the Tracking Group of COSPAR, an international Committee on Space Research that emerged from the International Geophysical Year.

In private life the wife of a metallurgical engineer, Joseph Fridlander, and the mother of a daughter, Natasha, Dr. Masevich is portrayed as charming, witty, boundlessly energetic, fluent in four languages including English—and altogether a star. As Sir Bernard Lovell of Jodrell Bank put it in 1963, "Any country would be fortunate to have Alla Masevich as a citizen. The Soviet Union is especially fortunate to have such a dynamic and charming person as its ambassador in the spheres of astronomy and space research where international collaboration is so vital" (Blashfield, *Hellraisers,* 171–2; Dodge, 225; Lovell).

An American scientist involved in satellite projects is **Marjorie Rhodes Townsend** (c. 1930–   ). Younger than Masevich, and taking a different route to an interest in satellites, Townsend entered the field at about the same time, going to work for NASA in 1959. At that time she directed the design of the ground data-processing system for the *Tyros II* weather satellite.

Born in Washington, DC, she entered college at 15, and became the first woman to receive an engineering degree from George Washington University. An electrical engineer, she has pioneered in the use of satellites to study X-rays in space. In 1948 at only 18 she joined the National Bureau of Standards as a physical science aide. She has also worked with the Naval Research Laboratory and more recently with the Goddard Space Flight Center, where she is director of the Small Astronomy Satellite Program. In that position she is responsible for the origination, design, construction, and testing of the satellites as well as for the actual launches of the instruments.

Marjorie Townsend has at least one specific invention to her credit: she is co-inventor of a digital telemetry system.

In private life she is married to an obstetrician and the mother of four sons. Married at 18 (already a junior at George Washington), she worked to put her husband through medical school while attending night school herself. During her eight years at the Naval Research Laboratory she worked her way up from junior engineer to one of the ten youngest GS-12s there: a senior engineer in the Airborne Sonar Branch of the Sound Division. During the same eight years she bore her four sons. Like Masevich, she has enormous energy. Colleagues describe her as methodical, well-organized, highly intelligent, a top-notch engineer—and well liked. Marjorie Townsend credits much of her success to the attitude of her father—also an engineer and inventor—who expected her to do great things, and specifically encouraged her to study engineering. Interestingly enough, she also credits her accelerated grade level in school, noting that in high school she was too young to be concerned with attracting boys (or for them to consider her really a girl), so she could devote herself to her studies.

Thanks in significant part to Marjorie Townsend's excellent work, the NASA Small Astronomy Satellite (SAS) program has been highly successful. *Uhuru,* the world's first X-ray astronomy satellite, was launched in 1970, opening up a whole new field of high-energy astronomy. In 1972 the second satellite in the program, a detector of gamma rays—was launched from Kenya; and the third scheduled launch was 1975 (continuing X-ray surveys). Honors for this work include NASA's Exceptional Service medal (1972), the Federal Women's Award (1973), and the Knight of the Italian Republic Order from Italy (the SAS program is a joint Italian-American venture) (N. Martin, "Beginning," 47; Holden; *WBWRA,* 90, 190).

Not in the same league with the two previous bodies of achievement but interesting in their own way are two 1970s patents: **Madalin Clark**'s Vibrating back scratcher shaped like a hand (1978) and **Rosa E. Martinez**'s 1979 patent for Audio-visual headphones. Martinez, who lives in San Antonio, TX, has created a combination earphone and goggle apparatus "for providing a user with [stereo] sound from a sound-input source and simultaneously providing the user with visible light in a pattern varying in accordance with variations in the sound . . ."

In 1980 **Cheryl A. Deckert** of Lawrenceville, NJ, patented a Method for inspecting electrical devices. Though not itself mechanical, it is a crucial part of the manufacturing process for such devices, and is therefore included here. The method, specifically for testing the quality of a protective "passivating layer" covering the conductors of a device, consists of coating the device with fluorescent dye, applying electrical energy to two conductors, and exposing the fluorescent dye to light and/or heat.

Highly significant again are **Dr. Esther Conwell**'s contributions to lasers and **Dr. Elizabeth Gross**'s work with living batteries. **Dr. Huai-yin Wu**'s amazing small xeroradiograph and **Dr. Judith Walker**'s cold laser treatment for pain have already been discussed in the Health/Medicine chapter.

**Dr. Esther Conwell** (1922– ) thinks of herself primarily as a physicist rather than an inventor, and her work as Scientist-in-Charge of the Electronic Materials program at the General Telephone and Electronics Laboratories in the 1960s was theoretical. The specific field of her research was semi-conductors, one of the hottest new fields of physics. However, by the 1970s she was a Principal Scientist in the Xerox Research Laboratories, Rochester, NY, where she received at least three patents: for Externally controllable miniature lasers (1977); for Elastomer Wave-Guide Optical Modulators (1978); and for a Waveguide imaging system (1979). She also invented a Microwave power detector (E. Conwell, c.v., 2).

Esther Conwell was born Esther Marly, one of three daughters of a photographer and his wife. Though reporting no mentors, she notes that her father did encourage her toward achievement. She took her PhD in physics at the University of Chicago and taught at Brooklyn College before joining the technical staff at Bell labs (1951–2). She is married and has one child.

Among her awards and honors are the Society of Women Engineers Achievement Award for 1960, appointment as Abby Rockefeller Mauze Professor at M.I.T. in 1972, and membership in the American Association for the Advancement of Science, the American Physical Society, the National Academy of Engineering, and the Institute of Electrical and Electronics Engineers. Currently (1981– ), she is a Research Fellow at Xerox. In addition to her patents she has to her credit over 120 scientific articles and one book (*AMWS*/15; E. Conwell, Q-1983, L-6/2/83, and c.v./1983; Peden, 272; Roth & O'Fallon, 24–5; Thomsen, 90–1; *WBWRA*).

A fitting inventor to climax and close this section is **Dr. Elizabeth L. Gross** (c. 1940– ) of Ohio State University. Her invention, an improvement in photovoltaic cells that use living organisms—"living batteries," as they are popularly called—looks forward to the 21st century and a world without fossil fuels. Dr. Gross is a biochemist, whose research specialties are photosynthesis and membrane biochemistry. She took her PhD at UC–Berkeley in biophysics in 1967, and after a stint as a research associate at the C.F. Kettering Research Laboratories, she was hired by Ohio State, where she became Professor of Biochemistry in 1979. One specific invention, for which she was cited in a popular magazine in 1982, is devising a way to use a cheap carbon electrode instead of platinum in the chloroplast solar battery. By 1982 she had achieved 4–5% efficiency, and her cells could produce 2,000 microvolts of electricity (*AMWS*/15; Maurer, 44–5; cf. Munson & Stambler, 12–3).

What does all this add up to? A systematic search for women's electrical patents, among the more than 4 million patents granted since 1900, was beyond my resources. At the very least, however, we can say that women have invented, sometimes significantly, in the field of electricity and electrical machines and devices.

### Health/medical

These devices have been discussed in the relevant chapter. Strikingly mechanical examples for a brief reminder here are **Dr. Jessie Wright**'s rocking bed for patients needing help with their breathing, **Dr. Anne Chamney**'s oxygen tent, **Doreen Kennedy-Way**'s battery-operated go-cart for spina bifida children, **Julia Butcher**'s invalid-exercising bed, and **Mary Gibbon**'s contribution to the prototype heart-lung machine. A noteworthy Soviet example is the Instrument for circular suturing of hollow body organs (an intestinal stapling machine) patented in 1971 by **Tatiana L. Ivanova** of Moscow and three male colleagues.

### Labor-saving, domestic

One of the largest categories of this kind of mechanical invention is discussed in the Agriculture chapter, under Food-processing. Examples that leap to mind as essentially labor-saving and only tangentially food-related are ice-makers for refrigerators, in which both **Florence Anderson** and **Jessie Whitney Cartwright** seem to have played a role, and Anderson's liquid dispensers for refrigerators. Corporate inventor **Joy C. Adams** of Bridgeport, CT, is noteworthy for her several 1930s labor-saving kitchen inventions, such as her Fruit-juice extractor of 1938. **Dame Caroline Haslett** in England, like **Cartwright, Lillian Gilbreth,** and perhaps **Christine Frederick** in the United States, applied time-and-motion studies to the home, likely resulting in the improvement of various appliances. Dame Caroline may also have been consulted specifically for this purpose by the appliance industry as her American counterparts were.

Wherever running water, cookstoves, and central heating eliminate women's carrying of water and firewood or coal, the remaining large categories of domestic labor aside from childcare are cleaning, laundry, and sewing, or clothing manufacture (and repair). Dishwashing and clothes-washing machines and certain other laundry devices have already been discussed, as have sewing machines, though clothing was increasingly manufactured outside the home as the century progressed.

There remains, nevertheless, a large field of mechanical inventions calculated to save labor in the home, such as cleaning devices, laundry apparatus other than washing machines, and miscellaneous smaller labor-saving devices and gadgets, such as can-openers, mop-wringers and sponge-mops with squeezing devices attached, and garbage-disposal apparatus. Like their 19th-century sisters, 20th-century women invented a share of these. Precisely what share is difficult to determine, even for patented inventions, without an army of researchers, but the Women's Bureau study gives us some overall numbers and the percentage of women's own inventions devoted to such machines and devices for the first two decades of the century.

Category XII, Household inventions, is the most pertinent here. Its Subcategory C, Laundry equipment, contains 165 inventions of 31 types. Excluding the washing machines, wringers, and other related inventions already listed above, we find the following eleven types of labor-saving laundry devices:

Clothes sprinkler
Clothes stretcher
Curtain stretcher
Drying frame and device
Iron and attachments
Iron attachment for stove

Ironing machine
Laundry strainer
Sleeve ironing board
Sleeve, waist, cap ironer
Spring hinge attachment for board and tubs

If the 165 inventions were distributed evenly among the 31 types, these 12 types would represent about fifty-eight inventions.

A woman's ironing machine from these years is the one patented by **Clara G.J. Simon** of Portland, OR, in 1913.

Category XII.D is House-cleaning devices. Its 81 inventions fall into 17 types, of which the following seem classifiable as mechanical:

Brush cleaner
Carpet and portiere beater
Carpet sweeper
Cleaning device
Cutlery cleaner
Device for holding material for cleaning
    wallpaper

Floor-scrubbing equipment
Oil-absorbing device
Step-ladder
Stovepipe cleaner
Wall cleaner
Window cleaner

By the equal-numbers assumption, the 12 types just listed would contain about 57 inventions.

Subcategory F, Bedroom equipment, also contains at least two examples of what might properly be classed as labor-saving cleaning equipment: a Bed-airing device and a Mattress-turner. Since this category totals 127 inventions of only 11 types, these two types could conceivably represent as many as 23 inventions. Some of the subcategory's Bedstead, Bedstead-attachment and other inventions not counted here might turn out to be both mechanical and labor-saving in intent. Beds that fold into the wall, for example, are not only space-saving (i.e., architectural as previously classed) but labor-saving, since they need never be made.

Subcategory H, Bathroom equipment . . . , contains one pertinent type of invention: a Toilet-cleaning device. This single type probably represents two inventions at most.

Subcategory N, Sewing and knitting . . . conveniences, contains the non-sewing-machine inventions for saving labor in the home manufacture of clothing, linens, carpets, etc. Definitely pertinent here are at least the following types: Needle-holder and threader, Quilting frame, Scissors and cutting guide, Skein-holder and -winder. Since this subcategory contains 119 inventions of only 22 types, the four types just given could represent more than 20 inventions.

Category XIII, Inventions covering supplies for use in industry, agriculture, commerce, and the home, is also pertinent here, as it contains not only "supplies" but machines and mechanical devices, some of which are obviously labor-saving and either definitely or potentially domestic.

Under Subcategory A, Cutlery, tools, and hardware, we find:

Barrel-tapping device
Can-receiving and -piercing device

Nail extractor
Saw sharpener

As these 37 inventions are highly varied, comprising 25 types, the four listed types probably represent only 4 or 5 inventions.

Beyond the Women's Bureau study I have only scattered examples, some of which, however, are interesting:

*Laundry.* In the laundry category I find as three early examples **Hannah Anderson** of Elgin, IL, who patented a Garment-stretcher in 1902, **Mournin N. Hall** of McPherson, KS, who patented a Clothes-drainer in 1904, and **Mary McLaughlin** of Philadelphia, who patented a Sadiron, also in 1904. McLaughlin assigned half her patent rights to **Anna H. Cassel,** also of Philadelphia, who may therefore have played some role in the iron's development. Hall's clothes drainer had several interesting features, including a design that directed water drained from the clothes back into the wash boiler. The attachment to the boiler was mechanical, with a latching device and a lever provided with an eccentric cam.

When electric irons appeared (c. 1920 in London, England [Harpur, 64]), and women no longer had to heat heavy sadirons on the fireplace fender or on a cooking or heating stove, Tuesday must have been a brighter day all across the nation. But this invention brought one new annoyance: the cord connecting the iron to its power source was always in the way. In the 1930s **Eleanor Uljee** of St. Louis, MO, invented and patented a Wire coil to support the cord of electric irons—one of those little inventions that makes a big difference in the quality of life. Unfortunately, even today irons and ironing boards are not routinely equipped with an Uljee coil. Gimmick catalogues are full of substitute solutions to the problem—devices that clamp to the ironing board and hold the cord high, etc. However, most of these shorten the cord too much to be really useful. Uljee's invention was featured in a *Christian Science Monitor* article of 1938 ("Tinkerers End").

**Ethel A. Cooper**'s Folding ironing board of that same year has already been mentioned.

A British example is the Machine for polishing or glossing collars, cuffs, and the like, patented by (Frank and) **Sophia J. Greenwood** of Bradford, England, in 1931.

*Cleaning.* An example from the early years of the century is **Corinne Dufour** of Savannah, GA, who according to one recent source invented a portable vacuum cleaner between 1900 and 1910 (Eliz. Reed, L-2/6/87; *Invention & Technology,* Spr. 1987: 58). The patent I find is in fact dated 1900, an Electric sweeper and dust gatherer. If this were truly a *vacuum* (i.e., suction) sweeper, Corinne Dufour would be noteworthy indeed, for she predates both Hubert Booth of England, who patented a commercial/industrial suction-cleaner in 1901, and Murray Spangler, also of England, who in 1907 sold the rights to his lightweight vacuum cleaner to W. Hoover. The Hoover Company in turn introduced the first relatively lightweight commercial model in 1908. The *OED* first records the use of the term *vacuum cleaner* in 1903 (Grigson & Gibbs-Smith, 421–2; Harpur, 13, 26; *OED,* s.v. "vacuum").

As it is, using a fan and rotating brushes to lift the dust and dirt, Dufour deserves recognition for inventing a labor-saver for women's home cleaning work rather than for industrial use, and for harnessing electricity to run it. As Ruth Cowan points out, vacuum cleaners were usually developed first for commercial use, and only later translated to the home (Cowan/Trescott, 40).[80] Cowan dates reasonably priced (electric) models on the retail market to 1910 ("Less Work," 58).

---

80. Grigson and Gibbs-Smith point out in *Things* that Booth was not the first to think of suction-cleaning, but he was "early in the field," and also "ignorant of unsuccessful devices patented before his own." His patent of 1901 was "a cumbersome, piston-operated vacuum pump." So large were Booth's gasoline-powered machines that they had to be "mounted in horse-drawn vans [and] . . . drawn up at the kerbside outside the premises to be cleaned. . . ." Household vacuum cleaners, according to these authors, developed in the United States (421–2).

Dufour, who lived in Savannah, GA, has at least four patents to her credit altogether. In addition to her sweeper, she patented a Fire escape (also 1900) and two garment inventions (1901, 1902).

At Sears, women classed as home economists worked anonymously on improving vacuum cleaners. Two such are recent retirees **Alice Stoltzner** and **Lou McClelland** of Chicago.

A noteworthy mid-century example in the field of cleaning conveniences is **Doreen Kennedy-Way** of Dudley, West Midlands, one of Britain's most prolific and best-known women inventors, with her Tidy-Flor, a magnetic device to be attached to broom handles for picking up pins, paper clips, tacks, or any small metal objects encountered in the course of cleaning. Her device is intended for the home, but also for hair salons, carpenters' workshops, and schools. After a varied background in engineering and design, including both corporate and independent work, she founded her own company, Kennedy Kinetics, where during the 1970s she concentrated mainly on medical-related devices (see Chapter 2) and household conveniences.

Another noteworthy example is a woman who might have been known simply as the wife of a famous husband. **Dorothy Rodgers** (1909–  ) is the widow of composer Richard Rodgers. Not content with reflected glory, however, she forged a life of her own, with a house restoration and decorating business called Repairs, Inc., a newspaper column and other writing—and several inventions. Barbaralee Diamonstein's *Open Secrets* calls her "Columnist, decorator, household-gadget inventor, needlepointer" (336). In addition to several "small gadgets for friends" not identified in Rodgers' memoir, *A Personal Book* (165), and presumably unpatented, Dorothy Rodgers has received at least two patents for toilet-bowl-cleaning devices. The more famous of these is the Jonny Mop (1946).

This invention, inspired partly by her World War II experience with tasks usually done by servants, must have been in the forefront of disposable cleaning devices. A later patent (1953), described as a Mop but consisting of a flush-away toilet-bowl-cleaning pad, may be an improvement in the Jonny Mop pad. Rodgers consulted a friend who was a patent expert and member of the National Inventors Council about her idea. He considered it patentable, and wrote the claims. Rodgers' ingenious design uses a material that will disintegrate in water, but holds together long enough to do the cleaning chore. Where the swab attaches to the handle, the disintegration is retarded, maintaining the connection until cleaning is complete, and then making it easier to disconnect the pad and flush it away without touching it. A further innovation was impregnating the swab with a cleansing agent. Rodgers sold this invention to Johnson & Johnson—though not without a battle,[81] and not until 1950 when almost five years of the patent's life had passed. Nevertheless, according to Ernest Heyn, the Jonny Mop made Rodgers "a small fortune" (*Fire,* 17). She held a registered trademark on the name (1951) as well as her patent (Ives, "Patent," 125).

As early as 1911, incidentally, another woman had invented and patented a so-called Sanitary toilet-bowl cleaner. Miss **Jane Anderson** of New York City patented this device, which, like Rodgers', consisted essentially of a handle or holder and a pad or swab

---

81. Personal Products Division of Johnson & Johnson claimed that their modified handle created a new product (i.e., not protected by Rodgers' patent). Rodgers took her case to the American Arbitration Association, where she won a unanimous decision. Thereafter, she collected royalties from Johnson & Johnson until her patent expired in the 1960s. She assigned half the royalties to her daughters Linda and Mary, and the Jonny Mop became quite an important item in the family's life. Her daughters called her the "Toilet Queen," and Mary made her mother a little gold-charm replica of the device, which Dorothy carries on her key chain.

material, in Anderson's case a metal tong-like handle and special paper. Anderson had other inventions to her credit, at least two (a fire extinguisher and a slipper rack for beds) on the market by 1913, when she was mentioned in the *New York Times* (Janeway, 55). This 1911 patent should appear on WB/28, and may be the Toilet cleaning device (or one of the type) classed under Bathroom equipment . . . rather than under House-cleaning devices.

Dorothy Rodgers has at least seven inventions to her credit. In the 1930s she invented a "people-cooling device," a bracelet or watchband holding dry ice or other refrigerant close to the wrist veins. She had hoped to exhibit this invention at the New York World's Fair of 1939, but the demands of Repairs, Inc., and the outbreak of war distracted her from completing it. Eighteen years later, another inventor patented essentially the same idea. Rodgers also created a potentially very important safety invention: a (mechanical) device to rescue children trapped in refrigerators or freezers. In her design, the doors would open in response to pressure from inside. Unfortunately, though the improvement would have cost only about a dollar, manufacturers refused to adopt it, and the patent expired. Her idea of combining soap and detergent to eliminate the ring around the bathtub, a patent disclosure, was appropriated by a soap-and-detergent manufacturer.

She had more success with two other inventions—her Try-on clothing patterns, which she sold to McCall's, and an educational toy, which she sold to Ideal Toy Company. The Try-on patterns allowed clothing patterns to be altered on a sewing machine and re-used. Part of the invention here is the material from which the patterns are made. Try-on's were adopted mainly by dressmakers and dressmaking schools. Her book-toy was designed to teach people to read, learn a foreign language, or do arithmetic (Rodgers, 162–5).

Despite the financial success, technological ingenuity, and human significance of her inventions, Dorothy Rodgers did not think of herself as a full-fledged inventor, or at least would not say so for publication. In 1977 she says, "It was during the years of Repairs, Inc.'s existence that early flashes of my career as a sub-standard inventor appeared" (*Personal,* 162).[82] Ernest Heyn, however, thought well enough of Rodgers' inventive talent to make her one of only three women he mentions in his *Fire of Genius* (1976).

With regard to the woman inventor's profile, note that Dorothy Rodgers was educated partly at home by governesses, that she wanted to be a sculptor and formally studied sculpture both here and abroad, and that she eventually became, among other things, an interior designer (Diamonstein, 336–9; Heyn, *Fire* and "Mousetraps"; Ives, "Patent," 114, 125; Rodgers, *Personal,* 162ff).

Also worth mentioning here is a cleaning machine that is the brainchild of a Cuban-born black woman, **Leonora Hafen** (1941–   ). Hafen, living in Jackson Hole, WY, in the early 1980s, has been inventing since childhood, when she conceived, among other things, a water-sweeper for leaves. As of 1981 she held only one patent, a design patent for a Cleaning brush that would reach into small cracks and crevices (1975), but had also invented a rug shampooer that would eliminate splattering and clean within an inch of the wall. A widow with four children to support, she feels keenly her lack of money, technical and other education, and market expertise. At last report she sought a backer for her ideas, while working at two jobs in order to save money to market her cleaning brush and other ideas in case no backer appears (Hafen, I-5/1981; Stanley/Zimmerman, 63).

---

82. Such exaggerated modesty or low self-estimation afflicts all too many women inventors, and plagues the researcher of women's contributions to technology. Two British inventors and one American inventor leap to mind who refused to give me any information about their inventions at all, not because of privacy but because of hypermodesty. So effectively have the male gate-keepers of technology done their work that women now keep themselves out.

Uncompromisingly focused on saving labor in housecleaning is contemporary multiple inventor **Frances GABe** with her Self-Cleaning House. Since her cleaning devices are all built in, they have been discussed under Architectural machines, but they are worth mentioning again here, especially since they are powered by the force of the water spray, and not by electricity.

In 1981, **Alma A. Hutchins** of Pasadena, CA, patented her Vacuum cleaner. The salient feature of this invention is a motorized bag-cleaning apparatus.

*Sewing aids.* A striking example here is **Elena Popescu** (1887–1944) of Roumania. As a girl of 14, home on a school holiday, she was touched by an old servant's efforts to thread a needle, and set her mind to devising something to make the task easier. She tried finer needles, animal and human hair, and spines of plants, finally succeeding with a kind of "crochet" needle made from very fine steel. Even this gave trouble until, with use, its hook became more elongated. Thus the "Roumanian needle-threader" was born. The year was 1901.

With the help of her younger brother, already called "the engineer," she made two or three more needle-threaders. People who saw them realized their usefulness immediately, and soon housewives and needlewomen alike were asking her for them. With her brother's help she made twenty or thirty more, but then went back to school and forgot all about her invention. Twenty years later, in a haberdashery shop in Bucharest, she saw essentially her needle-threader, advertised as "new from the U.S.A."

Her only son, Constantin Copaciu, now of Monaco, has written a vivid biographical sketch of his highly original mother (see App. M-8), who paraphrased Valéry on her deathbed. She was always out of step with her wealthy and conservative family. Next-youngest of eighteen children, she often slipped off to visit the Gypsies, whom she found fascinating. Her mother despaired of "this child which is no girl and not even a boy."

Elena Popescu is fascinating not only for what she was, as vividly described by her son, but for what she might have been. The industrial city of PloesËti, where she lived at the turn of the century, received information on the newest discoveries and inventions. She and her younger brother competed to see who knew more about the new science and the new technology—about Edison, the ideas of Jules Verne, and the successes of Howe and others. Only he became an engineer. In another era, the two of them might have gone off to engineering school together (Copaciu, L-8/19/81 and Memoir).

An isolated mid-century American example is the Button-attaching means patented in 1950 by **Anna M. Yohn** of Washington, DC.

*Miscellaneous labor-saving gadgets.* It would be surprising if our Lady Edison, **Beulah Louise Henry,** had never turned her attention to domestic labor-saving devices. Though she apparently lived in hotels most of her adult life, and most of her patents seem to cover commercial or business machines and toys, she patented an Ice-cream freezer (1912), at least one Can opener (1956), and a Baster oven (1962). The can opener features an attachment for opening cans with tear strips. Moreover, one source specifically mentions "a number of [devices to make] work easier in the home."[83]

Prominent here again is **Doreen Kennedy-Way,** who has patented a battery-operated cellophane-bag cutter and also a cling-film cutter, calculated to save frustration in dealing with today's armor-packaging of many foods and other household items.

---

83. This source (Yarbrough) comes from a Charlotte, NC, newspaper of about 1940, sent to me by the Alumni Director of Queens College, Charlotte (A. Davis, L-6/19/87). It is part of the Charlotte History Collection of the Charlotte Public Library, and bears no publication data or page numbers.

Contemporary with Kennedy-Way is **Ethel V. Hill** of New York City, who patented a Can opener with a male co-inventor in 1979. This opener is designed to produce a (re-)sealable edge in a can that has been opened.

As noted elsewhere, **Sarah Winchester** incorporated a great many labor-saving devices into her never-finished Mystery House in San Jose, California. Some of them were her own inventions; she probably had to her credit about a dozen inventions, all told. I have not located Sarah Winchester's patents (presumably obtained for her by agents or lawyers), but since she was building steadily from 1884 until her death in 1922, some of the inventions may fall into the 20th century.[84]

A young Frenchwoman has invented an apparatus for instant counting of rows of knitting. She exhibited it at the WIPO exposition in Geneva in 1985 and was featured (anonymously) in a flyer distributed at the show by the Swedish Women Inventor's Association (SWIA, 3).

*Manufacturing/industrial*

The 20th-century completion of a shift in manufacturing from the home to the factory is by now a commonplace, but it has never been traced in women's inventions. A thorough examination of the phenomenon is beyond my scope here, too, but the 19th-century manufacturing machines listed earlier in this chapter—and their inventors—are available for comparison.

About 1895, a Georgia girl, 15-year-old **Catherine Evans** (later **Whitener**) (1880–1964) re-invented the Colonial New England art of tufting, invented by women to beautify their darns. Discussed here because full development and commercialization came after 1900, the process involved stitching two layers of cloth together in patterns and then cutting the threads. Inspired by a beautiful old counterpane, Cathy decided to make one herself. This meant spinning her own cotton 12-ply yarn and *modifying a 4-inch ribbon-inserting bodkin by grinding the blunt end to a fine point.* She chose an Irish Chain pattern of simple squares, which offered a short length of yarn for cutting, and locked it into the material by washing and shrinking to produce a 24-ply tuft. She thought the finished surface looked like grass turf, and called the process turfing, but *tufting* was the name that stuck. When the process was mechanized in the 1930s, the cloth was called chenille.[85]

The new art caught on rapidly. Soon Catherine had more work than she could do, and was teaching others her technique. From the proceeds of her bedspreads she helped her family buy a larger farm near Dalton, Georgia, where they moved in 1909. Tufting became "a full-fledged commercial trade which raised the economic level of her home—Whitfield County, Georgia" (Lurd & Patton-Crowder, 100). Indeed, the craze spread until bedspreads for sale flapped on clotheslines along a fifty-mile "bedspread boulevard" north and south of Dalton. Tufted spreads were sold in Marshall Field of Chicago, and in Wanamaker's of Philadelphia, and Miss Evans' entire production was contracted for years ahead. The production staff of *Gone with the Wind* ordered a special bedspread for Scarlett O'Hara's bed. A movie, *The Magnificent Needle,* told Evans' story.

---

84. One of the main sources on Mrs. Winchester's inventions is Ralph Rambo, whose uncle, Edward B. Rambo, was her local financial advisor. It may even have been he who obtained her patents, though his nephew does not say so. Her attorney and the executor of her estate was Roy F. Leib (Rambo, 14, 15).

85. Presumably from the resemblance of the rows of tufts to French chenille yarn or cord (a cotton core with silk or wool fibers sticking out from it like the hairs of a furry caterpillar), used to decorate clothing. The word appears in English at least by the 1730s (*OED*). I am grateful to Tom Tucker, author of *Brainstorm! . . . Twenty American Kid Inventors* (Farrar, Straus & Giroux, 1995), for Evans' correct birthdate.

My sources give no hint that Catherine Evans, who had married Will Whitener in 1922, participated in the 1930s invention of mass-production machines for tufted or chenille bedspreads. As noted, in the beginning she modified a metal bodkin for her purposes. By 1905 she had cut production time in half by *devising a means to transfer the designs to the backing sheets, a kind of stamping iron.* Beyond this, she enjoyed the hand work. Lee Kary suggests that the husbands of some of the tufters entered the business and mechanized it. But "Miss Catherine" may have been consulted during the mechanization process, as for example where gooseneck sewing machines were being modified to do the new work. In any case, she continued to be honored as the founder of the industry. When the Tufted Textile Manufacturers Association opened its Dalton headquarters on March 14, 1958, the guest of honor was Catherine Evans Whitener (Bindocci, PC-3/2/84; Hunt, L-3/78; Kary; Lurd & Patton-Crowder, 100).

The Women's Bureau study has a category, III, for Manufacture, divided according to the following areas: chemical, food and beverage, foundry and metalworking, leather and shoemaking, power machinery (nonelectric), textiles, and miscellaneous. The 223 patents here comprised of 4.4% of the 5016 patents identified, almost exactly the same as for Agriculture (221 patents). In each area of Manufacturing, the compilers considered both processes and *apparatus*. Although the brief descriptions are sometimes ambiguous, at least the following may be classed as machines (apparatus) or mechanical in nature:

*A. Chemical products and processes and apparatus for making such products.*
Apparatus for generating gas
Burner for petroleum, hydrocarbons
Kiln and method of operating same
Lubricator for elevator guides
Pneumatic distributing system for fluids
Propulsion of air and other gases and fluids
Twyer-gate for gas-producing kilns

*B. Food products and beverage processes and apparatus for making same.*
Apparatus for preparing food products
Mixing machine
Nut-shelling and separating machine
Tool-holding means for food machines

*C. Foundry materials and apparatus, machine shop and other metal-working tools and devices.*
Cutting torch tip
Device for shaping ends of hollow cores
Die
Duplicating and swaging device
Dust-collector for small metal-grinding machines
Gauge attachment for shears
Lubricating bearing
Mold for casting metals
Molding machine
Multiple wire drawing machine
Oiler for air compressor
Reamer
Sand-molding machine
Skelp for making tubes

Soldering iron
Spindle bushing
Standing valve construction for well pumps
Temporary binder
Tubular armor-making machine
Valve construction
Wire twister
Wire-splicing tool

*D. Leather and shoe-making processes, machines, and tools.*
Appliance for inserting beading in boots & shoes
Combined punch and bottom set tool
Edge-trimming machine
Leather-working machine
Machine for affixing heels to shoes
Machine for boring and drilling wooden French heels

*E. Power machinery and apparatus other than electric.*
Air compressor
Apparatus for utilizing momentum
Belt drive
Flexible shaft coupling
Hoisting apparatus for logging
Internal combustion engine
Reversible turbine
Rotary engine
Single-acting steam engine
Solar heating plant
Steam boiler
Water motor
Water wheel
Wave motor

*F. Textile products, processes, and apparatus for making products.*
Fabric-making machine
Humidifier
Machine for making reinforced sheeting
Rug machine
Rug-making implement
Thread-measuring machine
Yarn reel, knockdown
Yarn-spinning apparatus
Looms:
    Automatic reshuttling apparatus
    Braking device
    Harness action
    Filling-detecting mechanism
    Protector rod holder
    Picker action
    Shuttle
    Thread-tensioning device for replenishing
    Other attachment

*G. Miscellaneous products, processes, machines, and tools used in manufacture.*
Apparatus for making leather-covered cushioned chair seats
Brush blank boring machine
Cigarette tip and forming same
Circular glass cutter
Clay-turning machine
Clothing ticket printing machine
Concrete-block-molding machine
Continuous kiln
Cord- and rope-making device
Cushion-stuffing machine
Device for arresting plaiting bobbins of plaiting machine
Finger ring expander
Glass-grinding machine
Glove-finger-turning device
Guided glass cutter
Jeweler's pliers
Jeweler's tool
Joint assembling machine
Machine for drilling and shaping
Machine for tonguing and grooving box parts
Paper vessel-making machinery
Paper-coating machine
Printing apparatus
Process and apparatus for saturating paper rolls
Pronged fastener-setting machine
Rope-measuring machine
Sack-lining machine
Strip-notching and slitting machine
Tobacco-ordering machine
Tobacco-stripping machine
Type clamp
Wrapper-sealing machine

Since Chemical products . . . numbered 35 inventions of 28 types, or an average of 1.2 inventions per type, the 7 types I have classed as mechanical could represent 8–9 inventions. The Food and beverage category showed 40 inventions of 26 types, or approximately 1.5 per type on the average; the four mechanical types just listed could therefore represent some six inventions. The Foundry and metal-working category numbered 30 inventions of 23 types, or about 1.3 per type if distribution were equal, the 22 mechanical types just listed, comprising the overwhelming majority of the category's types, could represent about 28 inventions. Leather and shoe-making accounted for 13 inventions of 12 types, so that the 6 types I classed as mechanical probably represented no more than 6 inventions at most. Power machinery numbered 18 inventions, of which all were by definition mechanical. The Textile category showed 34 inventions, in 28 types, or little more than 1.2 per type, giving an estimated average total of 20 inventions for the 17 types I classed as mechanical. Miscellaneous manufacturing machines and tools numbered 53 inventions of 49 types, or very little more than one per type. Thirty-two of these seemed likely to be mechanical.

In sum, if the women's patents were equally distributed among the types of invention listed, the estimated number of machines or mechanical inventions represented here would be about 119. The minimum, assuming each listed type to contain only one patent, would be 106. In either case, *the mechanical or apparatus patents comprise about half the patents listed for manufacturing.*

Since some if not all of **Beulah Henry**'s sewing-machine improvements were for commercial use, they should be noted again, briefly, here. Miss Henry's bobbinless machine has been called a boon to manufacturing, and "the job that will doubtless revolutionize the garment industry" (Yarbrough, col. 2). Miss Henry also invented the special machine to cut the sponge for her floating soap holder, and an automatic machine for attaching snap fasteners to garments (Yarbrough; "Study").

Loosely following the WB/28 classification, but continuing beyond 1921, let us look first at *Food and beverages.* Most of these items have already been discussed under Agriculture, but a few strikingly mechanical and mass-production-oriented items may be noted again here. **Zoya V. Zhavoronkova**'s 1970 Device for putting food-stuffs into cans seems worth citing again, since this Russian woman troubled to get a U.S. patent. The food-stuff in question is fish. Zhavoronkova and her male co-patentees (her name first) reside in Kaliningrad, Russia.

Two other examples, one major and one minor, leap to mind: the doughnut machine invented by a Japanese woman, and the hot-dog-bun shaper invented by a Los Angeles woman. According to Reiko Matsutoya of the Japanese Women Inventors Association, a Japanese woman invented—and patented—a really effective doughnut-making machine. When someone copied her idea despite the patent, she lacked the legal expertise to fight back, and never made any money from her invention. She became depressed and even suicidal. However, she decided to live, and her example inspired others not only to protect themselves more effectively, but to form the Association. Regrettably, Ms. Matsutoya— or her interviewer—failed to give this inventor's name or the date of her invention (G. Murray, 1).

In the 1970s **Pauline Berke** of Los Angeles, a member of Inventors Workshop International, developed a shaping machine for hot dog buns. Her device creates a bun with a long cavity for the hot dog. No splitting is necessary, and the popular sandwich can be eaten without mess, as the catsup cannot ooze out the side.

**Anna Eugenia Schneider**'s flour mill is a noteworthy earlier example (1920s). Her flour was first marketed in 1926.

*Leatherworking and shoemaking.* In 1902, too early for WB/28, **Margaret E. Morris** of St. Louis, patented a Machine for packing shoe uppers. She assigned 6/7 of her patent rights to six other individuals, which may mean that the invention was to be produced and put into use.

*Textiles.* **Alida Marcoux** of Milford and Hopedale, MA, is almost certainly one of the WB/28 inventors. Between 1902 and 1906, as an employee of the Draper Company of Portland, ME, and Hopedale, she patented eight looms or loom mechanisms: a Filling-supplying loom (1902), three Filling-replenishing looms (1903, 1904, 1905), a Take-up stop-motion for looms (1904), Crack-preventing means for looms (1905), a Filling-detecting mechanism for looms (1906), and a Thread-tensioning device for replenishing looms (1906). All four of the 1905–6 inventions should be on the WB/28 list, but I find only the Filling-detecting mechanism and the Thread-tensioning device for certain.

Hopedale was a Christian Socialist community founded in the 1840s by Rev. Adin Ballou of Milford, MA. All members had an equal voice in the use of property, though property was held individually. The community owned the means of production, except for the valuable patent rights of Ebenezer D. Draper. Gradually, the Draper family, with its textile mills, came to control the community, buying out its joint stock and paying off its debts. The community then ceased to exist except as a religious society. Hopedale became a separate town in 1886. I cannot say what remnants of Christian Socialism survived in the Draper Company as it entered the 20th century, but the 1911 *Britannica* notes ("Hopedale") that the village is "remarkable for the comfortable cottages of the workers."

Nor can I say what connection, if any, Alida Marcoux had with the religious community. This would be particularly interesting to know, since Ballou is described as "an ardent exponent" of women's rights. But in 1903, with one patent to her credit, she was listed in a book published by the Draper Company as one of the "Hopedale Inventors." This was a list of 72 inventors "directly employed by our various organizations and their predecessors up to Jan. 1, 1903," and having patents "turned over to Hopedale interests." In 1904 Marcoux appears in a similar list of "Principal Inventors" holding Northrop loom patents assigned to the Draper Company between January 1, 1886, and July 1, 1904. In this second list she is credited with three patents (her fourth was granted just after July 1). Since all her patents are assigned to the Draper Company, she presumably continued in their employ through 1906. It would be interesting to know why, after inventing so intensely for four years, she apparently patented nothing else after 1906 (Ballou; Draper Co., *Labor-saving*, 210, and *Textile Texts*, 309; *EB*/11, s.v. "Hopedale").

**Rosalia P. Hill** of Charles City, IA, was another early-20th-century loom inventor. In 1910 she patented a Tension for warp-reels and hand-looms. This patent appears in the Women's Bureau lists, but under Arts and Crafts equipment rather than Manufacturing.

Examples beyond the WB/28 years are scattered. In the 1940s one **Tempest Slinger,** who I assume was female, received a British patent for an Improvement in loom weft forks.

From 1940 come two inventions by three women: **Jessie A. Woods** of Ashland, KY, patented a Loom in May, and **Bessie S. Owen** and **Beulah Fay Meek** of Cleveland patented their Weaving machine in December. Meek had patented another invention—a hair-curling device—earlier that same year, listing her residence as Chicago.

Two examples from the 1960s: **Edith A. Honold** of New Orleans, with James N. Grant and others, patented another device for use in the textile industry, a Card assembly sampler (1967). This device enables quality-control inspectors to collect samples of staple fibers from rapidly moving machinery. In 1963 **Julia M. Sloan** of Metairie, LA, with others, patented a Wrinkle recovery testing apparatus. Both Honold and Sloan evidently worked for the U.S. Department of Agriculture, assigning their patents to the Secretary.

A recent loom improvement—that is also much else—comes from Canada, where **Gillian Loban** of Calgary, Alberta, was inspired by the needs of blind and otherwise handicapped weavers. In 1979 Ms. Loban was teaching weaving at the Canadian National Institute for the Blind. Her FREEHAND® Craft System, as its name suggests, leaves both hands free; it also allows the user to sit rather than stand while weaving. It consists of a hardwood clamp base, supporting frames appropriate to various crafts at any desired angle. The FREEHAND® system can be used not only for weaving, but for quilting, embroidery, drawing or painting—or even for reading, and has therefore been much appreciated in hospitals and nursing homes. Loban holds Canadian, British, and American patents, the American patent demanding three separate submissions between 1981 and

1985, and costing more than $3000. Her system is now widely used in schools, training workshops, hospitals, and homes, all across Canada, England, and the U.S.A.

Gillian Loban was born in England, the older of two daughters. Her early schooling took place there—sometimes in a bomb shelter; later she attended a Catholic girls' school. Her father, a sign painter, "was always drawing and making things." She is lefthanded and "somewhat ambidextrous"; at age five she was forced to write right-handed for a time.

Loban, who manufactures FREEHAND® with the help of a supportive husband, is delighted with her invention. No longer is the weaver the servant of the loom, says she; now it's the other way around. She is President of FREEHAND® Fibercraft Company, Ltd, and a founding member of the Women Inventors Network Group Society (WINGS), of Alberta (Lentin; Loban, L-11/29/88, FREEHAND® brochure).

A Russian example is the Apparatus for controlling the actuating mechanism of a loom dobby, patented by **Alexandra M. Mikhalina** (and others) in 1979. (A dobby is a small Jacquard-like apparatus for weaving small figures.)

Strictly speaking, perhaps, knitted goods may not be textiles. However, the knitting process can be used to produce varieties of cloth or fabric, and thus are discussed here. **Maria Szanto Luck** (1899– ) of London, England—**Mme Maria Luck-Szanto,** as she was known professionally—is an isolated but important example of women's contributions to knitwear technology. Born Maria Szanto in Budapest, she was the second of four daughters. Her father was a boot-and-shoe manufacturer, and her mother an embroiderer, so that she was exposed early in life to the business of manufacturing apparel. Her aunts also embroidered professionally. When she stayed with them she spent much time in the workroom where they created large ecclesiastical hangings. Aside from this artistic atmosphere, she was encouraged toward creativity and achievement by parents and teachers, and always enjoyed making things—toys, doll clothes, figures, book illustrations. She is ambidextrous, and feels that she has artistic talent in drawing. Her pre-secondary education took place in private girls' schools.

After studying art, dress design, and tailoring in Vienna and Paris, she became by 1932 the manager of a tailoring business in Vienna. Not satisfied with the knitwear fabric available for suits, she decided to experiment with creating a fabric that would look, cut, and wear more like other suiting materials. After considerable experiment she achieved the desired appearance and texture by using "different threads of wool and yarn of varying thicknesses" (Pinnell). For example, she twisted together wool and rayon yarns, to be knitted together with all-wool yarns. One such wool-and-rayon combination was called "Spintex."

In 1935 Mme Szanto went to England, where she continued her research using "the wide range and superior quality of British yarn," and by 1945 had created a dress fabric combining the strength and durability of tweed with the softness and elasticity of hand knitting. Satisfied customers boasted that they could spend the night in an air raid shelter and wear the same dress to the office next morning. They were also classic and timeless designs, shown in two shows a year and worn by royalty, society women, and such famous actresses as Irene Dunne, Vivien Leigh, Ava Gardner, and Lilli Palmer.

Some of her original knitting design patterns and a few of her classic garments from the 1940s are now in the Victoria & Albert Museum in London. A few appeared in the exhibit "Knit One, Purl One: Historic and Contemporary Knitting from the V&A's Collection" in 1985 (Hinchcliffe, 2, 22–3).

Mme Szanto's business became so successful that her German-born husband, Fred H.

Luck, left his own job after serving in the British Army during World War II, and joined his wife's enterprise. Both became British subjects in the 1940s.

Up to this point, Mme Szanto was using hand-knitted fabric in her dresses—fabrics designed by her and created by highly skilled women (only 10% of applicants accepted) working at home. These workers created the various pieces of the garments, which were then assembled under the direction of Mme Szanto or her partner. However, in the late 1950s, as an unexpected result of a beret-designing project she did for the firm of Kangol, where the so-called Basque berets were knitted in one piece, she got the idea for a knitting machine that would create garments by horizontal shaping—in one piece and completely shaped, or tailored. A machine that knitted a complete beret, for example, could produce a circular skirt, and so on. But she must adapt both the garment designs and the existing flat-bed knitting machine.

After trying her idea out in hand-knitting, she explained it to an engineer, Kenneth Macqueen, who undertook her proposed machine modification. Their names appear together on the original patent application of 1949, and she participated in the early experiments, which created a version of the machine.

However, Szanto and Macqueen soon realized that a more sophisticated version would be necessary if they were to sell it to the knitting industry. Because Mme Szanto lacked mechanical-engineering expertise and was busy with her dress-designing business—and because Macqueen needed further funding from another engineering firm—she allowed him to appear alone on the revised patent application. Legal agreements, however, protected her financial interest in the invention. Final patent specifications were filed in 1957 and published in May of 1960.

The developer-investors were finally ready to publicize the invention in the early 1960s. The machine was now a veritable monster, partly computer-operated, and, like the original computers **Grace Hopper** confronted, room-size. Macqueen died in 1965, but another engineer, Frank Robinson, became fascinated with the Tailor Knitting Machine, and eventually developed a knitting machine capable of the three-dimensional knitting that had been Mme Szanto's original idea. Just as with **Lillian Malt** and her Maltron keyboard, however, the Tailor Knitting Machine is often attributed solely to Kenneth Macqueen—or to Frank Robinson—without any mention of the original inventor (Luck-Szanto, I-4/25/ 84; Q-5/15/84; L-5/3 and 15, 6/25, 12/17/1984, 1/10, 6/10, 7/20, 10/10/85; 9/18, 10/8/86; "Electronic Translation . . ."; Hinchcliffe; "How to Knit a Yacht"; Macqueen; "Miss Szanto's New Fabric"; Pinnell; "Runs a Business").

*Metalworking and -forming machinery.* Taking another of the WB/28 categories, we find that one of Mildred Blakey's early-20th-century patents appears on the list, a Skelp for making tubes (1905). As mentioned above, this prolific inventor has, most surprisingly, proven to be male. Apparently the Women's Bureau compilers followed LWP, or made the same very understandable mistake as the LWP compilers, and included at least one of "her" three known 20th-century patents on the WB/28 listing.

Just too early for WB/28 is the impressive-looking Amalgamating apparatus for mercury patented by **Marguerite Bloume** of Paris, France, in 1902. The apparatus worked by electricity.

In 1944 **Dorothy G. Atkins** of Groton, CT, patented a drill jig. Between 1957 and 1964 **Inna Kazanskaia** received the Lenin Prize for developing a complex of mills for rolling graduated cylindrical sections. A laboratory head at the All-Union Research and Design Institute of Metallurgical Machine Building in Moscow, she was honored along with six team members (Dodge, 222). In the late 1960s **Tatyana B. Golovkina** received an

American patent for a Machine for tying coils and packs of iron, as for instance of rolled iron, with wire (1969). She and her male co-inventors assigned their patent to the same Metallurgical Machine Building Institute in Moscow, which was also Golovkina's residence at that date.

*Manufacturing machines for plastics and synthetic fibers, etc.* In 1981 **Sheila E. Widnall** of Lexington, MA, with a male co-inventor patented a Fiber-velocity imparter device for dry-forming systems. Since the device forms a continuous fibrous web of dry-laid fibers in a gaseous medium, it might seem to pertain to the synthetic-fiber industry, except that the patent is assigned to the American Can Company. Thus it may have to do with forming plastic or styrofoam containers.

Luckily, this patent is not our only datum on Dr. Widnall, for it gives no hint of her doctorate in aeronautics and astronautics from M.I.T., her previous position as Director of the Office of University Research for the United States Department of Transportation, or her position as of the mid-1970s on the faculty of M.I.T. Nor, of course, does it reveal any of her honors, such as a Ford postdoctoral fellowship at M.I.T. (first woman to win it), the Lawrence Sperry Award of the American Institute of Aeronautics and Astronautics, for "notable contributions to the advancement of aeronautics," and the 1975 Achievement Award of the Society of Women Engineers, for "significant contributions to the fluid mechanics of low-speed aircraft and hydrofoils."

I have found no other patents, but Dr. Widnall almost certainly has important innovations and inventions to her credit, patented or not. Some of them may have been secret when devised (she also worked for Army Research), and others are perhaps more properly discussed under Transportation. Suffice it here to note that she conceived and managed the development of M.I.T.'s new (1970s) wind tunnel, that wind tunnels are used not only in research, but in aircraft manufacture—and that the air-flow principles involved may not be irrelevant to her dry-forming device of 1981 noted just above.

In 1985 Dr. Widnall was one of the judges for the second Great International Paper Airplane Contest (*Science/85,* Feb.: 88–9; Malotke, 23; *WBWRA*).

*Miscellaneous manufacturing machines.* As technology advanced with the century, the list of types of manufacturing also changed, to the point where some types lumped under "Miscellaneous" by the WB/28 compilers deserve separate discussion. To preserve comparability with the WB/28 study, I will include them here, but will subdivide the category.

Paper goods, containers, etc.: Between 1916 and 1941, **Lydia B. Koch** of New York City held at least 19 patents of her own, and 16 more assigned to her by others, relating to the manufacture of paper bottles. Her own patents, ranging from 1924 through 1941, are as follows: Manufacturing milk bottles, etc. (1924); Bottle (1927); Constructing bottle receptacle elements (1928); Paper bottle, Manufacture of paper bottle containers (both 1931); two Paper containers (1932); a Container, a Bottle and its manufacture, and a Paper blank for paper bottles (all 1935); Manufacturing paper bottles, a Paraffining machine, and Apparatus for capping bottles (all 1936); a Spinner-head, an Apparatus for conditioning paper, and another Container (all 1937); Expansible spinner machine (1941).

On ten of these inventions, Koch is sole inventor; on nine others she has a male co-inventor; her 1941 co-inventor Irving F. Mandell assigned his right to Koch. Many of her patents are assigned to the Reinforced Paper Bottle Corporation of New York City or Chicago, for which Koch may have worked either in house or as a consulting designer-inventor.

The "Expansible spinner mechanism" of 1941 is a "mechanism for incurling the wall edges of conical paper containers," an obviously useful apparatus in the manufacture of paper cups for water-coolers, soft-drink and snow-cones vendors, etc. Likewise, Koch's paraffining machine, her apparatus for capping bottles, her spinner-head, and her apparatus for conditioning paper are unmistakably manufacturing machines, and some of her patents covering the manufacture of containers may also deal with machines.

Equally interesting are the several other patents assigned to Koch, though patented by others. As early as 1925 William E. Oliver of Detroit, MI, assigned to Koch half the rights to his patent for a Hardening and preserving composition for paper and other fabrics, *and manufacturing same.* In 1931 Ludy Vogelsang of E. Orange, NJ, patented Bottle-cap-forming dies, and assigned the patent to Koch. In 1936 Andrew Bodor of Newark, NJ, assigned two patents to Koch—for a Paper cap for bottles and a Nonrefillable bottle. In the same year Albert Karlsson-Ygger of New York City assigned to her his Combined roasting and grinding machine—which if she did help invent it, properly belongs above, among the Food and beverage manufacturing machines. Only one of her co-patentees and assignors was a woman: **Ruth Oliver,** whose name appears on at least two patents—a 1923 Paper milk bottle and a 1933 Closure cap for paper milk bottles.

Still earlier, judging by the patents assigned to her between 1916 and 1925, Lydia Koch was involved in the manufacture of cinematographic targets. Koch is just one of many fascinating women who were probably entrepreneurs as well as inventors, and whose careers would richly repay further research.

*Rationalizations in manufacturing.* The former USSR distinguished a category of innovation not covered by patents but nevertheless vital to efficient production, and thus deserving of recognition. This is the *rationalization:* a suggestion for an improvement in a manufacturing process or the organization of production, often including an idea for modifying a machine. People who make such suggestions are called *rationalizers,* and the Soviet inventors' organization was called the All-Union Society of Inventors and Rationalizers (VOIR). Naturally, some rationalizers are also inventors, and vice versa. In the United States, such ideas often arise more informally, via the company suggestion box or even just in conversations between management and employees; and the line may be drawn somewhat differently between patentable and unpatentable innovations. Certainly the view of individual as opposed to team achievement is different.

Several noteworthy examples are available from the former Soviet Union. In 1975 **A.A. Valusaitane** won one of fifty USSR-wide prizes given by VOIR to women inventors and innovators. A senior technologist at a silk mill in Kannas, Lithuanian S.S.R., she had to her credit 76 proposals for improving production in the mill. The annual savings to the P. Zibertas Order of the Red Banner of Labor Silk Combine as a result of those suggestions was 1.2 million rubles, or about $2 million (Mandel, L-7/8/80; VOIR, 76–8; 1970s figures).[86]

Another 1975 winner of the VOIR diploma and decoration was **Tatiana G. Podkorytova,** an electrician at the 50th Anniversary of the USSR Superphosphate Plant in Dzhambul. Her "dozens of suggestions for rationalization" resulted in a saving of over 40,000 rubles.

A third 1975 winner of this VOIR prize, always awarded on the eve of International Women's Day, March 8, was **Ye. A. Ozhina,** already a Hero of Socialist Labor and brigade leader of cake-makers at the Moscow Bolshevik Pastry Factory. Her 28 rationalizations saved the factory about 158,00 rubles (VOIR).

---

86. I am grateful to David Doughan, outstanding research librarian at the Fawcett Library in London, for a translation of the section titled "The Participation of Women in Technological Creative Work" from the VOIR interim report of 1978.

Winners of the same prize in 1976 were **K.S. Chernova,** a department head at the Moscow State Civil Aviation Exploitation and Repair Technology Research Institute, **F. Ys. Slutskaya,** an engineer-technologist at the Kalinin Coachbuilding Plant, **Alla F. Teteryatnik,** a senior scientific assistant at the All-Union Antibiotic Research Institute, and **V.I. Vatazhina,** a laboratory chief at the Moscow Polymer Research Institute. The VOIR publicity sheet gives no technological details, but notes that all four women donated their prizes to the Soviet Fund for the Defense of Peace (VOIR).

Since fifty women were honored each year with such a prize on the national (All-Union) level alone, not to mention smaller regional awards, Soviet records would reveal literally hundreds of women belonging in this category.

Though no such formal structure exists for honoring American workers for money-saving suggestions, here are at least three women who, if they had lived in Moscow, would undoubtedly be recognized as rationalizers:

**Hazel Hook Wolz** (1891–1974), a brilliant woman forced to leave school and work in a beauty shop to support her family, invented the bobby pin about 1916, when the flappers began to bob their hair. According to her granddaughter, Mary M. Whitman (d. 1989), she patented her invention, but it may well have been registered in the name of a lawyer or other agent.[87] She spent most of her savings—and years of hard work—developing and marketing it—making the pins by hand, testing them on her customers, improving the design, doing her own advertising, etc. Then, just as she began to make money, a big company introduced a slight change in her design, a crimp in one arm of the pin. Granted a patent on the basis of the change, and backing their design with a lavish advertising budget, they took away her market. She managed to sell her patent for what her granddaughter describes as "a bundle"—though of course far less than she would have received in sales royalties on bobby pins—to a man who wanted to set his son up in business, but later lost much of her money in the stock market crash of 1929.

Embittered by what she saw as the theft of her idea and the loss of her just reward, she went back to the beauty shop and refused even to try to develop later ideas for real estate promotion, Maine-chance-type salons, thin-hair restorer, and—pertinent here—mechanical and other improvements in factory production processes.

Born in Texas, Hazel Hook was the second of seven children, of whom five survived infancy. She was the granddaughter of an English-born shipwright named George P. Hook who came to Oneida, NY, in the 19th century. Her parents were William Henry Hook of Oneida and Johnnie Maude Hicks of Sercy, AR. Left as the sole support of the family when both parents and her older sister died about 1909, and handicapped by a speech block and third-grade education, she went to work as a beautician.

Moving to California with Mary's grandfather, an artist, she bore a daughter and later, with Fred Wolz, a son. Living first in Carmel and then in the Oakland-Alameda area, she continued to work as a beautician until late in her final illness. She died impoverished by medical bills. Mary Whitman sees her grandmother as a tragic figure, brilliant but hampered in reaching her potential by a deep sense of inferiority. She was "convinced that men were the rightful masters" and spoke of her accomplishment with the bobby pin as "mainly dumb luck." She was, continues Whitman, "the most self-effacing and manipulative (a slave's prerogative) person I had ever met" and at the same time "a flamboyant

---

87. As Mary Whitman put it, " 'Gra' . . . was very quirky about using her name for 'legal matters.' She would often use her maiden name—not from woman's lib leanings but to 'protect herself legally.' " Ms. Whitman says that she never understood what all this was about but has a sense that her grandmother felt unknowledgeable and therefore vulnerable in business matters (L-May–June, 1975).

character with mythic qualities" who devoutly believed in the myths of her Southern aristocratic origins (Whitman, L-8/16/74; 2/23, 4/20, and May–June/1975; I-10/23/84).

In the United States in the 1930s and '40s, under the influence of the Gilbreths and their pupils, and under the strong spur of wartime production needs, manufacturers were vitally interested in making production ever more efficient. As a 1937 winner of the Gilbreth Medal pointed out to a group of foremen about 1940 (Hard, 36):

> A few years ago the idea was to have all the thinking in a plant done by a little group of thinkers in a room by themselves. That's out. In a really modern plant everybody must be a thinker.

An appealing example comes from the Bristol-Myers plant at Hillside, NJ, where employees dumped used chemicals, then walked down two flights of stairs with an order for more, and back up two flights to their stations. But a sub-foreman came up with a better idea: the tray-dumper could let the order down through the floor on a fishing line with an ordinary rod and reel. This idea was put into effect, with significant savings of time (Hard, 38).

And then there was the **Work-Simplifying Girl.** A young employee of Larus & Brother tobacco manufacturers in Richmond, VA, she stood beside a system of conveyor belts carrying cartons of cigarettes. It was her job to turn the cartons at a certain point so that they traveled head-on. At first, she actively turned each one, and her boss pondered whether a machine could be built to do the turning instead. He decided it could, for about $300. By the next day, however, he observed that the young woman seemed not to be working (Hard, 38):

> There wasn't a motion in her. The first finger of her right hand, though, was projecting about an inch over the belt conveyor. Mr. Glazebrook marveled. The girl had found the strategic spot. Each carton, as it struck her finger, swerved exactly right and went its way head-on.

She had "automated" the process. The company installed a nail just where the young woman had held her finger. The nail then turned the cartons, and she moved to another job in the plant, with a raise. Unfortunately, however, her name does not appear in the article, and Larus & Brother cannot supply it (Robelen, L-11/7/1983).

Later, when the United States entered the war, awards were given to workers who improved production methods and efficiency as a contribution to the war effort. **Mary Fretch** of the Radio Corporation of America (RCA) in New Jersey was among eight workers honored by the War Production Board in 1944 with what the *New York Times* called "idea awards." At a Board luncheon in Washington, Fretch and her counterparts received certificates honoring their work and attended a special preview of the National Labor-Management Production Exposition. The exposition, demonstrating various "win the war" production ideas contributed by workers, was intended to focus national attention on the need to increase production by 20% in 1944.

The *Times* article gave no details on any of the workers' ideas, though the sub-headline mentions "time-saving *devices*." Donald Nelson, Chairman of the WPB, noted that thousands of such ideas, put into practice, "have saved untold millions of man-hours, have conserved vital materials and have sped the weapons of war ever faster to our fighting men." But in honoring these eight people, Nelson continued, "we are honoring the patriotic American workers whose contributions are above and beyond the call of duty." He called Mary Fretch the "nation's best woman suggester" ("Idea Awards").

Fretch also received one of *Mademoiselle*'s 1944 Merit Awards, and the magazine's awards issue (Jan. 1945: 128) gives a few further details: Fretch worked in the Parts

Preparation Department at RCA. She suggested time-saving changes in the manufacture of radio tubes used in military devices, changes that saved some 35,000 worker-hours each year! *Mademoiselle*'s term for Fretch and the other rationalizers is "suggestion-eers."

At a professional level, some of **Mabel Rockwell**'s work at Lockheed during the war years can be seen as a series of rationalizing innovations, many of which were probably not patentable. Joining Lockheed Aircraft at Burbank, CA, in 1938, by 1940 she was Production Research Engineer, directing 20 engineers doing research in airplane manufacturing processes. Her group's research made possible the use of spotwelding rather than riveting. In fact, as Rockwell herself noted in 1941, "the strength and consistency of spotwelded joints were so improved by Lockheed's research program that the Civil Aeronautics Authority gave its approval for extension of the use of spotwelding to some parts of the primary structure of airplanes . . ." (Goff, 97). Spotwelding is so much faster than riveting—a mass-production technique, in short—that production was greatly speeded and made less expensive at the same time.

Rockwell's group may have been involved in modifying the French stored-energy spotwelders (to avoid sudden enormous drains on the plant's power system) for use at Lockheed. They certainly demonstrated that roll spotwelders were faster and more consistent than fixed-electrode machines; that problems caused by erratic air pressures could be overcome by installing a large "air receiver" near the welders, and that surfaces to be welded had to be absolutely clean, including even removing part of the oxide film covering the surface of aluminum alloys and usually thought to aid the welding bond. They also proved that spotwelded joints compare favorably with riveted ones in fatigue strength under the small torsional vibrations sustained in airplanes, and developed methods for accurately measuring current and pressure during the weld. These accurate measurements led in turn to improved welding techniques. Rockwell also determined the most efficient assembly method for use with spotwelding (Goff, 96ff).

Later in life Mabel Rockwell became a consulting engineer. In the many years unaccounted for in my sources, she undoubtedly had several inventions and innovations to her credit, whether patented or not, substantial numbers of which, as appropriate to her consultant capacity, might qualify as rationalizations.

Another anonymous example, cited in the Peters and Waterman *MBWA (Management by Walking Around) Diary* for 1985 (110), is the woman assembly-line worker who suggested a new path for the wires inside a blender. Her suggestion saved *three feet of wire per blender.* Her manager admitted that he didn't know there was that much wire inside the machines.

But perhaps the ultimate example of a rationalizer comes from China. Born about 1933 to a poor family in Northeast China, **Wei Feng-ying** trained as a machinist and at 20 began work in a factory. By the late 1950s she had already come to the notice of her superiors for enabling her group to complete three years' production goals in one. Among her specific achievements was automating a machine that had cut off a worker's fingers. By 1958 her production group had 106 proposed technical innovations to its credit, 99 of which were realized. Two of them alone had saved the State 170,000 yuan. By 1965 over 80% of the technical processes in her workshop were automated, and she had *more than four hundred technical innovations to her credit.*

Married and the mother of two children by age 32, she is a member of the Central Committee of the Communist Party, and participated in the tenth Party Congress in 1973. In 1976 a Western writer considered her one of China's young women leaders (Sheridan, 59, 72ff).

*20th-century miscellaneous*

Early in the century, but falling between the cracks of the WB/28 study, is **Edith E.L. Boyer**'s Labeling machine of 1907. Edith Boyer has appeared in the 19th-century discussion as co-inventor with I[srael] Donald Boyer, of a Wrapping machine (1896), a Candy-making machine, and two mechanical toys, a Spinning top and a Locomotive toy (all three 1897). I. Donald Boyer is listed in the 1893–4 directory of Dayton, OH, as a Machine designer. By 1902 Edith Boyer is widowed and listed as a tag manufacturer. Already in 1900, however, she had patented a Triple-action jack. She is not listed as her husband's executor on the patent, nor is there any other suggestion that she was not the sole and true inventor of this jack. It is interesting to note that a Boyer & Radford Company, Manufacturers of Lever and Screw Jacks, etc., existed in Dayton at that time (1902 directory); the Boyer involved was neither Donald nor Edith, however, but E.C.

Edith Boyer appears in the Dayton directories, manufacturing tags, until at least 1920. She also appears in the business directories in 1904–5 and 1919–20 as one of Dayton's three tag manufacturers. In 1924 she got another patent (see above). By 1925, however, she appears simply as Edith E.L. Boyer, wid. Donald, so presumably she has now retired. In 1944 she fails to appear in the name listings, though the street directory still shows her at 102 S. Findlay, where she had lived for more than forty years. By 1946, however, the S. Findlay house is occupied by someone else, which probably means that Edith died in 1945 (Dayton, OH, city directories, 1893/4–1946).

Obviously Edith Boyer was an enterprising woman with access to some technical and mechanical expertise. To what degree she herself possessed this expertise is unknown, but her six patents are certainly suggestive, as is the fact that she continued to invent after her husband's death. Her tag-manufacturing business is a natural background for her invention of a labeling machine, but it would be interesting to know, among other things, how she translated her idea into metal, whether the machine became widely accepted in the industry, etc.

**Galina N. Artamonova** of Yaroslavl, then USSR, was part of a team that invented a Tire-building drum. Since the inventors took the trouble to get a U.S. patent in 1978, they presumably considered the machine to have some technological value, or commercial potential, or both.

In the early 1980s **Virginia L. Hammersmith** of Monroeville, PA, patented a Method *and apparatus* for controlled removal of excess slurry from organic foam (1981). She and her male co-patentee (her name first) assigned their patent rights to the Aluminum Company of America.

*Office machines*

In addition to typewriters (discussed above) the Women's Bureau study finds patents granted to women for other office machines and mechanical aids.

Under Category X.A, office machines and attachments, appear the following non-typewriter mechanical examples, potentially totalling some seven patents (WB/28: 33):

Calculator
Duplicating machine
Signature roll for duplicating machine
Alignment device for duplicating machine

Under B, Stationery and miscellaneous equipment, appear a Pencil-sharpener and attaching clip, a Pencil stamp, and a Stamp- and label-applying machine, possibly numbering six patents.

Under Subcategory C, (Office) Furniture, appears a Revolving chair, probably a single invention.

I have not located the inventor(s) of the Calculator(s) listed by the Bureau, but I have located a very interesting early-20th-century woman inventor of calculators. Just too early for WB/28, she is mentioned in a 1905 issue of *Inventive Age* (Feb.: 6): **Emily C. Duncan** (1849–1934) of Centralia (now Wisconsin Rapids), WI, and Jennings, LA, who patented two banking-related calculators (1903, 1904). Born Emily Cornelius Forbes in Coral, IL, she is interesting in several ways, not the least of which is that her husband, James E. Duncan, was also an inventor. In fact, the family thought James was the only inventor in their line. However, they knew that Emily always helped James with his inventions and other work in his home workshop, sometimes till the wee hours of the morning, and one granddaughter wrote to another in 1984: "Of course I do think that Grandma was the back bone of everything Grandpa did" (St. Clair, L-5/1/84). Duncan will be discussed more fully in the "Computers" chapter.

In 1908, not covered by WB/28, **Marie Heilbron** of New York City patented (with Jerome Walker) a Machine for opening letters. Miss Heilbron, who worked in the mail-order department of one of the city's largest stores, had good reason to desire such an aid. According to Minnie Reynolds, a fast worker opened about 30 letters a minute, but *Heilbron's machine could open 400.* Exhibited at the Martha Washington Hotel Suffrage Bazar of 1908, it was already on the market and being rapidly adopted by companies with a heavy mail volume. Interestingly enough, a woman was secretary and treasurer of the company that marketed the machine (Reynolds, 1, 3). Having operated the machine used in many offices today, I doubt it is as fast as Heilbron's.

*Paper cutter.* A British example of an office machine radically improved by a woman is the safety paper cutter patented by **Verena Holmes.** Holmes was first and foremost an engineer, who will be discussed under Transportation machines below, but she was also a versatile inventor with two medical inventions and a sound-amplifier, among other things, to her credit as well.

The British call the paper-cutter a guillotine, and Verena Holmes' model was manufactured and sold as the Safeguard Guillotine for paper and card. According to Claudia Parsons, writing at Holmes' death, this was "the first really satisfactory safety guillotine ever to be designed" (4). Many schools adopted it for student use, and Parsons says, "It was gratifying to Verena to think of the number of fingers she had saved from amputation." The device was noteworthy in another way as well, for it was produced at the all-women firm of Holmes & Leather, Ltd, founded by Verena in 1946 to give employment to women engineers.

Verena Holmes' patented Stencilling apparatus of 1958 is probably also intended for commercial use.

### Power-generation and transmission

At least three prominent women in the field of solar power are pertinent here—**Countess Stella Andrassy, Maria Telkes,** and **Dr. Elizabeth Gross**—along with **Barbara Hickox** and her controversial idea for tapping into electrical fields surrounding the earth. However, Telkes and Andrassy are discussed at some length in connection with the application of their work to cooking, and Gross and Hickox have both proved difficult to research. Barbara J. Hickox of Sacramento, CA, hit the news in 1980 when she filed a $2.5 billion lawsuit against the U.S. Department of Energy for failing to aid her in developing her energy-related invention. She claimed to have devised a means of tapping into the "electrical band" around the earth. No further stories appeared in any newspaper or in the broadcast media, to my knowledge. A call to the Sacramento *Bee,* which originated the story, proved fruitless, and Hickox did not appear in the 1980 Sacramento telephone directory. The lawsuit must have been dropped or thrown out of court, for no further word was heard of it. A patent was granted (1980) but, justly or not, authorities were apparently convinced Hickox's invention could never work.

Unquestioned is the achievement of **Dr. Elizabeth Gross,** Professor of Biochemistry at Ohio State University, mentioned above as producing 2000 microvolts of power at 4–5% efficiencies of as early as 1982. However, Dr. Gross seems to be omitted from survey articles on photovoltaics (e.g., Jeffries), and my communications to her have gone unanswered.

Perhaps the most noteworthy single example here is **Mabel MacFerran Rockwell** (1902–81). She helped design the power system for the Colorado River Aqueduct project, and was the only woman actively involved in designing and installing the power-generating machinery for Hoover Dam.

As noted above, she ranked top in her class at both M.I.T. and Stanford, and upon graduation found professional work as an electrical engineer. As a technical assistant for Southern California Edison, she worked in electrical transmission. Her innovation here was "one of the earliest applications of the method of symmetrical components to practical power system relay problems." In short, she devised a method of locating multiple transmission failures occurring simultaneously in different parts of the system. Miss MacFerran's method had great practical value for a company whose tower lines carried as many as twelve circuits.

Her second position was with the Metropolitan Water District of Southern California, where she did her work with the Colorado River and Hoover Dam projects. Later, for the United States Bureau of Reclamation, she designed transmission and distribution systems to bring power from the Central Valley Project dams to irrigation districts and other public agencies in California. She also spent several years designing electric substations for the Air Force (*AW/*III [1939–40]; Goff, 94–112; Morgan & Roden, PC-5/16/86; SWE, . . . *Technical,* 28; *WBWRA,* 188; "Woman Engineer Honored").

## Scientific/research machines and devices

Another area that would repay further research is women's contributions to scientific apparatus. The WB/28 compilers considered it worth a separate category, XIV, where they included nonsurgical scientific instruments, laboratory equipment, meters, scales, watches, optical and photographic goods, apparatus and supplies. The group numbered 76 inventions all told, or about 1.5% of the 5016 patents in the sample. This categorization could be disputed, as it includes 15 camera-and-related types of inventions which are probably not all scientific equipment; 6 types of clocks and related mechanisms that do not necessarily sound like scientific timepieces, and 5 types of optical inventions, some of which seem to pertain to spectacles for improving vision (and thus might belong under Health/medical inventions). Of the remaining 28 types, the following seem obviously or probably mechanical (as well as, in some cases, electrical):

Adjustable measuring and ruling device
Apparatus for estimating vapor pressure
Calipers
Circuit controller for liquid gauges
Comb. rule, trisquare, and calipers
Drafting tool

Exposure meter
Instrument for indicating ship stability
Marine compass
Recording apparatus
Register meter
Relay sounder attachment
Scales
Sound-recording and -producing instrument
Sun dial and compass
Temple measuring device
Weighing device, automatic

If the 76 inventions were distributed equally among the 54 types, the 17 arguably mechanical types would account for 23–24 inventions.

Noteworthy post-WB/28 inventions in this category are **Mlle Yvette Cauchois'** X-ray spectrometer and other scientific instruments, the apparatus **Rosalyn Yalow** had to devise in order to do the research leading to her invention of radioimmunoassay, and astronaut **Sally K. Ride**'s contribution to the giant mechanical arm used to launch and retrieve objects in space on the space shuttle flights (see below). According to an American physicist who knew Yvette Cauchois personally—Dr. Robert Smither of Argonne National Laboratories—she had many scientific instruments to her credit; the Cauchois spectrometer is only the most famous. Smither, himself an inventor,[88] used it in his own work in the 1970s, for measuring the energies and intensities of gamma rays. Moreover it is, he says, a common type of X-ray spectrometer used in many laboratories around the world (Smither, L-7/3/78). Since Cauchois was reportedly "near retirement" in 1969, she was probably born just after the turn of the century.

**Julie K. Randolph** of Antigua Island was a secretary at RCA when she invented a mechanism for automatically removing warning lights from the cable tethers of research balloons. These balloons are used for low-altitude observation and surveillance. Ms. Randolph was on medical leave from her job when the idea came to her, so she experimented with a piece of rope and a clothes pin. She assigned her 1978 patent to RCA of New York City. By the time the patent was actually granted (she got the idea in 1976, and filed for the patent in July 1977), Ms. Randolph had become a technical clerk working at Cape Canaveral (now Kennedy Space Flight Center). Stacy Jones featured her invention in his "Patents" column for the *New York Times* (Nov. 25, 1978).

## *Theatrical apparatus*

Since the acting profession has long been open to women, it should not be surprising that women have invented—and patented—numerous devices for stage effects and other theatrical apparatus, in the 20th as in the 19th century. The WB/28 compilers created a subcategory D, Theater apparatus, under their Category XXI, Arts and Crafts. Though containing only ten inventions, it is an interesting group, and nine of the ten are probably mechanical, or partly mechanical, in nature:

Apparatus for producing illuminated motion effects as of rain, snow, fire, or smoke
Cannon for acrobatic performance

88. Smither's patent (Crystal diffraction lenses for imaging gamma-ray telescopes, 1984) covers technology that may become crucial to the SDI ("Star Wars") program. I am grateful to him for calling my attention to Cauchois (see Smither, "New Method," n1; "Crystal," n5).

Device for producing stage rain
Fire illusion apparatus
Producing color effects and apparatus therefor
Production of lighting effect on stage by means of high-tension currents of high
    frequency
Scenic mechanism
Theater stage
Theater-seat hat-holder

A post-WB/28 example is **Raye Coplan**'s Marionette invention. Coplan received her 1950 patent for a means of simplifying the operation of puppets. Coplan, who lived in Jamaica, NY, assigned the patent to Peter Puppet Playthings, Inc., which paid her $25,000 immediately and agreed to pay her a 5% royalty on net sales. Royalty payments, becoming a matter of public record because of a dispute with the Internal Revenue Service, amounted to $29,668 in 1951, $37,926.17 in 1952, and $28,081.74 in 1953. This is apart from Coplan's salary as Treasurer of the corporation, and any value in the shares of the corporation that she owned (Tuska, 14).

## Toys

Toys in general are beyond my scope here, but a quick glance at mechanical toys seems appropriate. The WB/28 compilers created a subcategory A, Children's playthings, under their Category XXII, Amusement. Of the 170 inventions classed as Children's playthings, at least the following types seem likely to be mechanical:

| | |
|---|---|
| Aerial toy | Game, musical |
| Building toy | Magnetic basketball game |
| Buzz ball | Musical blocks |
| Camera | Railway |
| Coaster | Table football |

Because of a catchall category labeled "Other toys," we have even less idea than usual how many inventions might fall into each type, but it is interesting to note that the 170 children's playthings comprise over 3.3% of the 5016 patents included in the study. Thus if the ten mechanical types just listed should constitute nearly a third of the 170 patents (i.e., 10 of 33 types) located in the study, they would number just over 50, or about 1% of the 5016 patents in the sample.

As a post-WB/28 example in passing I will cite **Helen L. Borza**'s Windmill toy of 1948. Borza, of Yonkers, NY, patented this toy jointly with Raphael Borza, also of Yonkers. A much more famous example is **Beulah Louise Henry,** who patented several toys with moveable parts or mechanical parts of toys among her 49 known patents. Miss Henry will close this chapter as a 20th-century specialist in mechanical inventions. Suffice it here to say that she patented, among other things, an Eating toy animal (1953), a Movable lip for toy figures (and actuating means for same; 1943); and a Simulated dispensing device—a toy cow that gives "milk" (1951).

## Transportation

Women's contributions to transportation continue to be important. If in the 19th century women invented mainly for horse-drawn, railroad, and ship transport, with important work in baby carriages and at least one major bicycle invention, in the 20th century they seem to concentrate on automobiles, airplanes, and space travel—again, with at least one major bicycle contribution.

Of the 345 inventions placed by the WB/28 compilers in their Transportation Category, 152 deal with automobiles (bodies and parts, tires and tire attachments, and accessories), and 25 more deal with traffic signals and indicators—for a total of 177—all or virtually all of which are mechanical in nature. This is a little over half of the Transportation inventions, and thus about 3.5% of all inventions in the sample. This is an impressive record. The inventions themselves are impressive as well, including such things as a flexible axle, a carburetor, a clutch mechanism, a driving mechanism, an electric engine starter, a piston rod, a priming mechanism for explosive engines, and a starting mechanism (Stanley, "Inventing").

I have located what should be at least two of these WB/28 inventions: **Anna A. Brashear** of Santa Barbara, CA, patented a Tire in 1919; and **Maria Doujak** of New York City patented a Life-saver automobile attachment in 1920. ("Life-saver automobile attachment" does not appear on the WB/28 list in so many words, but several "attachments" are listed.)

Note also that among the flood of ideas submitted to Henry Ford for improving the Model T were letters from women. "Housewives," says Reynold Wik, "working over the kitchen stove dashed off suggestions on paper splattered with hot grease from the frying pan" (61). One 73-year-old woman from Ohio wrote touchingly to Ford (Wik, 62):

> I know you believe in helping people help themselves. I want to pay my way in the world. I have some debts I want to pay. Inventing is all I can do. I have invented a child's seat for the Model T. Are you interested?

Ford may or may not have used this woman's idea, but **Jean Aimes** of Cleaver Hulme International seems to have invented the first commerical child's car-seat in Britain in the 1960s (2 patents; Burleigh). Another woman suggested to Ford a reclining seat like the Pullman railway seats, another suggested openable windows, still a third suggested safety helmets in the form of inflatable rubber turbans, and a fourth suggested an anti-robbery mechanism (Wik, 62, 65, 67, 70). If these women's suggestions are representative,[89] it would seem that women were concerned with safety and comfort.

Rather surprising for these early years of the century is the WB/28 total of 19 women's patents pertaining to aircraft (15 types):

Aircraft
Aircraft, lighter than air
Airplane
Direction-indicator for air and marine craft
Dirigible airship
Dirigible headlight
Flying machine
Landing brake
Lift indicator for flying machine
Liquid container for airplane

---

89. Since, as I show elsewhere, compilers of women's inventions are influenced by a stereotype of the kinds of inventions they *expect* from women, and since this stereotype might be especially strong where the automobile is concerned, it may well be that they are not entirely representative. However, no one denies that men are also concerned with safety and comfort; nor do I wish to denigrate the importance of those areas of invention.

Mechanism for steering plane vertically and horizontally without human contact
Parachute
Propeller
Safety landing attachment
Wire reinforced fabric for wings

**Lillian Todd** may or may not be one of the patentees included on the WB/28 list (no patents found), but according to Valerie Moolman, she unveiled her first airplane in 1906. This glider-type aircraft never actually flew, but had some new features, such as pneumatic tires, and a unique system of fans and propellers that attracted serious attention from designers and engineers at the 1906 exhibition of the Aero Club of America. By 1908 Todd was showing an airplane with collapsible wings, enabling it to be reduced to one-third its flying size for easy transport and storage. The *New York Times* called her "the first American woman airship inventor" (Corn, "Making," L-8/3/85; McCullough, 39; Moolman, 12; "Woman Has New Aeroplane").

A possibly noteworthy post-WB/28 example is **Vera C. Hodges** of Los Angeles, who patented a Starter for an engine in 1935, and assigned it to Eclipse Aviation Corporation of East Orange. Such an assignment is likely to mean that the device would be manufactured and used.

A contemporary example too important to be omitted is **Yvonne Brill** (1924–  ) of Montgomery Township, NJ, and London, England, who began working on rocket propellants and propulsion systems in the late 1940s before leaving to rear three children. Returning to RCA at 41, she feels she has done "better and more intricate work since turning 40." There is concrete evidence to support her feeling: Ms. Brill has invented and patented a new satellite-propulsion system (the hydrazine/hydrazine resistojet system, 1974) that augments the performance of the propellant with electrically supplied heat, so that a smaller mass of propellant is required. Thus lightened, a satellite can either carry a larger payload into space or stay in orbit longer. Brill's system also enables satellites to change orbits in space. By 1984 it was used on RCA's communications satellites.

Yvonne Brill also managed the development, production, and testing of the Teflon solid propellant pulsed plasma propulsion systems aboard the RCA-built *NOVA I* spacecraft launched in May 1981. *NOVA I,* as Brill's 1951 nomination for membership in the National Academy of Engineering put it, "brought electrical propulsion [for satellites] to operational status in the U.S."

In 1986, Yvonne Brill, whose original training was in chemistry (MS, University of Southern California), received the Society of Women Engineers Achievement Award for outstanding contributions in the field of space propulsion. Specifically commended—in addition to her deriving the first industry standard for assessing the performance of rocket propellants, and enabling "new propulsive capabilities through innovative integration of R&D concepts"—is her implementing the "first operational electric propulsion system to keep satellites on station."

A member of the Scientific Advisory Board to the U.S. Air Force, a Fellow of AIAA and SWE, and winner of the Diamond Superwoman Award given by *Harper's Bazaar* and De Beers Corporation, she has nevertheless found time to become a "recognized leader in establishing women as contributors to space engineering." In the late 1980s Brill was in London, working on the INMARSAT project (Brill, c.v./1986, nomination for the National Academy of Engineering, L-2/21/89; W. Garrison; H. Morris, L-11/9/88, PC-11/15/88; N. Martin, ". . . Beginning," 48; Siegel, 71: SWE Achievement Award, 1986/1988: 5).

*Miscellaneous*

The WB/28 compilers found categories for virtually all the women's patents they discovered. Their Miscellaneous Category, XXIII, contains only 14 patents altogether, in three very small groups: Election and registration conveniences, Church equipment, and Burial equipment. The election conveniences number only 6 inventions of four types, three of which sound mechanical: a Folding booth, a Voting booth, and a Voting machine.

One of the more noteworthy of American women's recent patents was granted for a voting machine, to **Susan Huhn** (now **Eustis**) of Connecticut. Her machine was featured in Boston newspapers and in *People* magazine in 1976, while the patent was still pending (granted 1977). Although for some years after her Barnard graduation she worked in the sales and service of mechanical voting machines, her machine is electronic, and as such, will be discussed in the final chapter, on computers. Likewise, her current direction in both research and invention is entirely toward brain physiology and computers.

Truly miscellaneous is **Annette Fridolph**'s Clamping device of 1958, described by the Gazette as "A clamp for releasably securing two telescopically adjustable members one to the other." This sounds as if it could have applications in many areas from beach umbrellas to camera tripods to card-table legs. Fridolph is a resident of New York City.

Random examples of interesting miscellaneous machines are the Luggage carrier patented in 1978 by **Deborah Ratchford,** a flight attendant from Arlington, VA; the Fog-dispersing machine invented by **Mrs. J.E. Garner** of Blackpool, England, about 1953; and the Line-casting machine invented by an unnamed woman in the WB/28 study some time between 1905 and 1921 (WB/28: 34). Ratchford's invention, a hook-and-wheel setup capable of carrying several suitcases at once, was featured in Stacy Jones' *New York Times* "Patents" column (June 17, 1978: 30).

Mrs. Garner claimed that her invention, though only about 20 inches by 7 inches by 7 inches, would clear fog in a quarter-mile radius within only 20 minutes. Her invention rated a brief notice in the London *Daily Herald* for May 10, 1953, and the Ministry of Supply had agreed to test the machine when the weather was suitable.

The WB/28 compilers considered the Line-casting machine to be for commercial fishing boats. As they noted, "The small list in this classification of inventions patented by women requires no other comment than to call attention to the fact that the devices are not confined to fishing for sport. Some of the inventions are concerned with commercial fishing" (WB/28: 34). However, the only line-casting machines I found were typesetting machines. Either one is interesting in the context of women's inventions, of course.

Many inventions that would perforce be classified here because their precise nature is seldom recorded are the childhood inventions of women who might or might not later become well-known inventors, such as **Helen Blanchard** and **Harriet Hosmer** in the 19th century; **Amelia Earhart, Beulah Louise Henry,** and **Leonora Hafen,** among others, in the 20th.

### Machine Specialists: The Lady Edisons

One of the most noteworthy of this rarefied group—the only one, to my knowledge, to be called a Lady Edison in the Patent Office as well as in the popular press, is **Beulah Louise Henry** (b. 1887) of Raleigh and Charlotte, NC, and New York City. She began inventing as a small child, sketching mechanical gadgets before she could even draw a straight line, and long before anyone could understand what she had in mind. She did not receive her first patent, however, until age 25. This patent, for an ice cream freezer, was granted in 1912. A distinguishing feature was its vacuum seal.

Despite this official recognition, a "Charlotte-wide reputation" by 1912, and some national publicity in the 1920s–'40s, little is known about her after 1950. Born in Raleigh, NC, she was a direct descendant of Patrick Henry, granddaughter of Gov. W.W. Holden of North Carolina, and daughter of Col. Walter R. and Beulah (Williamson) Henry. Her father was an art connoisseur and lecturer, her mother was an artist, and her only brother, Peyton, became a songwriter.

Between 1909 and 1912, she studied at two colleges in Charlotte, NC:[90] Presbyterian, which became Queens, and Elizabeth. Queens College, which was reportedly "quite academically stimulating to young women of the time," still exists in Charlotte, the address given on her 1912 patent. In 1913 she patented a Handbag and a Parasol, still listing her address as Charlotte. She assigned half the rights to her parasol patent to one W.T. Woodley of that city (A. Davis, L-6/19/87; "Miss Henry").

In physical appearance she had "superb auburn hair" and a "commanding presence." All sources comment on her femininity, one (Yarbrough) mentioning "pale gold hair a mass of curls" and another comparing her looks to Mae West's ("Beulah").

Between 1913 and 1924 there is a decade's blank in her patent activity, but she can scarcely have been idle, for by 1922 she had filed for a patent for her spring-limb doll (granted 1925), and by 1924 she turns up in New York City with a second Parasol patent assigned to the Henry Umbrella & Parasol Company of that city. Perhaps she made enough money from her first three patents to set herself up in business. One source, James Nevin Miller in the *Washington Star* ("All Inventors . . .") notes, in fact, that she found a manufacturer for her original parasol—an umbrella with detachable, snap-on covers of various colors to fit a single frame.[91] This lucky manufacturer sold 40,00 umbrellas in sixty days (Yarbrough, col. 2). By 1924 the *Scientific American* reported that she had her inventions patented in four different countries, and was president of two newly incorporated companies. A fair success for a woman of 37 ("A Study").

Miss Henry does not name her two companies, but according to *Who's Who of American Women* (1958–9) by 1929 she was President of B.L. Henry Co. of New York City. And the 1924 patent assignment reveals the existence of the Henry Umbrella & Parasol Co., Inc., of New York. Perhaps the B.L. Henry Co. was her general inventions corporation, and Henry Umbrella & Parasol was, just as it sounds, to manufacture and sell her umbrella inventions.

While continuing to invent and patent throughout the 1920s and succeeding decades, she worked as an inventor for the Nicholas Machine Works of New York City from 1939 to 1955, and served as a consultant to various companies manufacturing products invented by her, such as Mergenthaler Linotype Co., Klinert Rubber Co., Averell Doll Co.,[92] International Doll Co., Henry Umbrella Co., Display Mannequin Co., and Ideal Toy

90. Among the many mysteries to be solved is why such a brilliant young woman should wait so late for advanced schooling. I am grateful to Adelaide Davis, Alumni Director, Queens College, for information on the name changes and other background.

91. Miller says this was her "very first invention." Other sources—and the patent dates—disagree, however, placing her vacuum ice cream freezer first (1912 as opposed to 1913 for the first parasol patent—which was, in fact, her third invention in point of patent date, following both her freezer and her handbag with snap-on covers).

92. Though spelled "Aver*i*ll" in *WWAW*, this was presumably the company of **Georgene Hopf Averell** (1876–1963), a prominent doll designer of the 19-teens through at least the 1940s, known, independently of any association with Beulah Henry, for introducing the "Raggedy Ann" and "Raggedy Andy" dolls and the "American Baby Doll" with its unbreakable head (Averell, *NYT* Obit., Aug. 28, 1963).

Corporation, among others. As early as 1930 she had already sold 40 of her "hunches" to large corporations ("Miss Henry").

She evidently never married, and lived in hotel apartments for convenience, a pattern that began when her parents moved with her to New York City in the 1920s. Three known addresses are the Imperial Hotel (1920s), the Hotel Victoria (1930s) and the Hotel Seville at Madison Avenue and 29th Street (1950s), all in New York City. Other interests besides inventing become evident in her organizational memberships: the New York Microscope Society, the Museum of Natural History, the National Audubon Society, and the New York Women's League for Animals. A 1930 interviewer reported that she originally wanted to be a writer ("Miss Henry"). She definitely had artistic talent. As she told another interviewer (Yarbrough), "I think literature and art are far above things mechanical. I have . . . painted many water colors, but the world calls me an inventor." Her living room walls were hung with her paintings.

Inescapably the most fascinating thing about Beulah Louise Henry, however, is her fertile inventive mind. She astonished scientists and patent officials alike, not only with the number and variety of her inventions, but with their mechanical and technical nature, particularly since she was completely innocent of technical training. It was this innocence, perhaps, along with her financial success, which freed her to be so candid about the intuitive, unconscious sources of her best ideas. In a 1937 interview, for example, Henry revealed that she had "color-hearing" or synesthesia; i.e., she was one of that 5% or so of people who envision a specific color for each tone of the scale. Colors take on outlines, which the mind fills in and transforms into objects. This she considered quite ordinary— as ordinary as inventing. All one needs for inventing, she said ("Lady Edison," 22),

> is time, space, and freedom. She just looks at the ceiling, at a bill-board across the street, and, lo and behold! there is her invention, three-dimensional, wheels whirring and everything.

She solved the problem of the snaps for her snap-on umbrella covers—which the "biggest umbrella men in the country" said could never be done—while getting ready to go out to a matinee ("A Study . . ."):

> That snapper had worried me for a long time. . . . [But] when I start out to do a thing I usually manage to get it done some time. I was putting my gloves on when that snapper in all its details visualized itself against a green drapery. That ended the theater party. Mother wanted to know if I was ill, but I told her it was only the snapper and I . . . just had to sketch it out on paper before I should forget.

And she got the idea for a new lock stitch in sewing machines from an electric movie sign ("Beulah Henry").

There seems no reason to doubt these statements, especially since analogous stories can be gleaned from the biographies of a great many inventors, major and minor. However, as Joseph Rossman, an assistant patent examiner, notes about the umbrella-snapper story (20), "She admits that the problem was before her a long time. . . . That thinking was going on during the interval can hardly be denied." In other words, in invention just as in science, solutions to problems may seem to come out of the blue, in a dream or vision, or from an inner voice; but they are far more likely to come to the prepared than to the unprepared mind.

Moreover, like Edison himself, Beulah Henry did not leave the realization of these ideas to chance. She had a "large laboratory," with "a staff of mechanics who are thoroughly familiar with her lay descriptions of scientific terms," and paid pattern-makers thousands of dollars to make her models ("Lady Edison," 22; Yarbrough, col. 3).

Her headquarters in New York's Hotel Victoria held a model of each of her patented inventions. In order to make room for them, she streamlined her living quarters, with a collapsible table that fit under the bed, and a small electric grill that popped out from "a little pigeon-hole in the wall."

As usual with prolific inventors, sources conflict on the number of her inventions. Indeed, they even disagree on the number of *patents* she held, which is a matter of public record. Once again we are undoubtedly dealing with distinctions the writers and interviewers failed to make, not only between patented and unpatented inventions in general, but between patents and design patents, trademarks, and patent disclosures; and between patented inventions and unpatented but successful inventions that are fully realized but kept as trade secrets, either by the inventor herself (manufactured and sold by her), or by a company to whom she sells that right, for a large initial sum plus royalties. Such inventions may be the most important and most profitable of all. Some confusion for the layperson in all of this is natural.

Thus we read in Rossman's 1927 article that Miss Beulah Louise Henry has 47 inventions to her credit. Rossman and the *Scientific American* writer do not make the mistake of confusing inventions with patents (by 1927 she held only 16 regular *patents*). A *Scientific American* writer of 1924 mentions several of these 47 inventions, all called "successful." Of those mentioned, the following do not seem to match any of Miss Henry's patents, and thus provide insight into her *unpatented but successful* inventions ("A Study . . ."):

a telephone index attached to the phone, opening fanwise to show the needed names
a pencil with "a heavy fitting eraser"
an electric-fan shield (packable in a trunk for traveling, presumably for use with hotel-room fans)
a rubber reducing garment
a machine for automatically fastening snap fasteners on clothing
a "Kiddie Klock" for teaching time
glove snaps
"Across Country," an educational game about the U.S. railroads

A 1930 newspaper article from Charlotte, NC, also avoids this particular confusion, mentioning only two *inventions:* a duplicating typewriter and a kicking, winking doll ("Miss Henry").

However, much of the popular press is confused. According to the 1937 *Literary Digest* article "Lady Edison," Beulah Henry had 52 patents to her credit by that date (May 3: 20). I find only 25 patents by then (plus a few reissues), and therefore conclude that the reference was to 52 *successful inventions,* 25 patented and the rest sold or being manufactured by the inventor herself, without benefit of patent protection. A case in point here may be an invention she had just recently completed at the time of the interview:

The National Cash Register Company disagrees about what Miss Henry [claims *not* to know, about science]. A fortnight ago, they called her in, showed her an adding-machine, begged her to make it write like a typewriter. She did. And when they asked her how she had managed it, she replied that it must have been an "inner vision."

No such patent appears in her name in 1937 or in the years immediately following, and a NCR patent counsel and historian I contacted in 1987 could not confirm this story. But if she actually created or contributed to the first printing adding-machine, this would be a significant invention indeed.

The next mention of Beulah Henry as "Lady Edison" is a 1938 article in the *Washington Star* (J.N. Miller). This article credits her with only 42 inventions, but goes on to say that she has been turning out inventions at the rate of about a dozen a year since the date of her "first" invention. As already noted, the writer of this article, James Nevin Miller, is confused about Henry's first invention, but if indeed she had turned out a dozen inventions a year since either 1912 or 1913, she would by 1938 have had over 300 inventions to her credit.

In 1939 *Coronet* magazine profiled Beulah Henry, "whose uncanny inventive skill puzzles orthodox scientists." This article mentions "52 workable inventions," ranging from a football-inflating device to a bobbinless sewing machine. Since at least one further patent had been granted to Henry by this time (her Feeding and aligning device for typewriters, 1938), we have further indication that these writers are not referring to patented inventions.

In 1940 the *New York Times* published an article titled "Women Gaining in the Patent Rolls." A section subtitled "Woman Holds 60 Patents" was devoted to Beulah Henry. Since only one further patent, an improvement on her Feeding and aligning device, plus one reissue, had actually been granted her by 1940, Henry's unpatented inventions were probably still surging ahead of her patented total, and confusion was surging right along behind. In fact, if the same ratio of total inventions to patented inventions prevailed throughout her inventive career as this 1940 article seems to reflect (60 to 27, not counting reissues), Beulah Henry had probably created over a hundred inventions by the time of her 49th and last known patent in 1970 (108–9 extrapolated unpatented total).

Note that the confusion continues right up to the present. A 1980 article written by a patent examiner credits Miss Henry with 58 patents (Ives, "Patent," 114). This must count, at the very least, all the reissues, and possibly some of the trademarks as well.

Note, too, that Beulah Henry's inventive career was quite long. From her first patent in 1912 to her last in 1970, is nearly sixty years. And since we know she was a childhood inventor—her "hunches" began about age 9 ("Miss Henry")—that is a conservative figure.

I have called Beulah Henry a machine specialist. One might look at her handbag patent, her parasol bag, her shoe, her dolls, her soap-holding sponge, her hair-curler, etc., and demur. However, a closer look shows that she deserves this classification. For one thing, as noted above, she worked as an inventor for a machine works from 1939 to 1955, some sixteen years. She also consulted to many firms that were producing her inventions—such as Mergenthaler, a manufacturer of typesetting equipment. And in order to create her clever Latho soap-holder—doubling as soap-holder and washcloth or bathing sponge, and snapping open to receive a cake of soap—as mentioned above, *she had to invent the machine to cut the sponges* in such a way that the soap could be placed inside and snapped in. (To understand the appeal of this invention, incidentally, we should recall the spectacular success of Ivory soap for the same reason: it floated.)

Many of her patented inventions have to do with typewriters, including the Protograph, an attachment mentioned as early as 1930 that makes an original and four copies of a letter or record without carbon paper. As of the 1937 *Literary Digest* interview, Miss Henry was working on a silent typewriter, as well as on her later-patented Feeding and aligning device for typing multiple carbon copies without slippage. She got the idea for this one—which uses a row of tiny pins around the platen—from her Swiss music box. Others have to do with sewing machines, including what is one of her most-often-mentioned inventions, a bobbinless sewing machine.

Even her dolls and other toys are often mechanical. To be more precise, her particular improvements are mechanical in nature. Her inventions in the field of rubber balls and toys often apply to improvements in the valve closures of inflatable bodies. Her "roulette" or Indicator top was "an improvement to the ordinary 'put and take.' " Her tops are the usual six-sided spinners, but have little windows cut in and six different-colored glass balls rolling around on the inside as they spin. The faces are numbered and any number of wicked combinations can be made with the turn of the numbers and the colored balls ("A Study").

Beulah Henry patented a doll with a radio inside it in 1925. A 1930 article described a doll she invented as a godsend to all Santas, a doll that could "kick, wink, and make eyes of regular 'It' appeal" ("Miss Henry").[93] This doll may have incorporated early forms of the eye structures for figure toys and for dolls that she would patent in 1935 and 1942. In 1951 she patented a toy cow that dispensed milk and in 1953 an Eating toy animal. In 1941 she patented a Device for producing articulate sounds. In fact, since she served as a consultant for the Averell Doll Co., she may be the "inventor of a small mechanical voice box" with whom **Georgene H. Averell** collaborated in inventing a talking and crying doll which she introduced to the market after World War I. (If so, this could account for part of the blank in Miss Henry's patents between 1913 and 1924.) If this Averell-Henry collaboration produced the first cry-baby doll and/or the first talking doll, it might have been ignored by histories of technology, but its commercial potential was enormous. Miss Henry patented an Eye structure for dolls, a Plural-syllable sound-originator, a Movable lip for toy figures *and actuating means for same,* and a Duplex sound producer.

A striking example of a doll with mechanical features is Beulah Henry's "Miss Illusion" doll. At the push of a button, the doll's eyes switched from brown to blue. With her dark wig snapped off, a blond wig snapped on its place, and her dress turned inside out, she looked like an entirely different doll ("Beulah Henry"; "Lady Edison"; J.N. Miller; Rossman, 18–9; "Women Gaining . . ."; WB/75: 3; WWAW/58–9).

All told, at least 32 of her 49 patents pertained to machines or were mechanical in nature, if we count—as we should—her snap-on umbrella and purse covers as mechanical. The press's treatment of these parasol and purse covers is revealing. Often the writers emphasize the idea of matching one's umbrella to one's gown, whereas what is really remarkable—and the basis for the patent—is the snapping *mechanism.* Also patented, according to Henry herself, as reported in the *Scientific American* interview are "the little steel pincers that fasten the [cover] seams to the frame" ("A Study").

As for her unpatented inventions, her long association with the Nicholas Machine Works, her consultancy with the Mergenthaler Co., and other evidence seems to indicate that many of them were also machines. Of the ones mentioned by the *Scientific American,* the pencil (if mechanical), the electric-fan shield (adjustable to any fan), the "Kiddie Klock" for teaching time, and the glove snaps should probably be counted as mechanical in nature. The "Kiddie Klock" works as follows: "By pressing a button the hands are rotated and in a little window in the clock will be seen a picture showing fairy characters, so that the child knows that when seven o'clock is turned to, he or she sees 'Little Red Riding Hood' and in a short time learns to tell time as well as a grown-up."

Beulah Henry is no longer our most prolific woman inventor, but only recently has she been surpassed. Even today only **Giuliana Tesoro** and possibly **Ruth Benerito** have more *patents in their own names* (about 125 and 50, respectively, as of 1983), and both these women are PhD chemists receiving patents for novel compounds. Moreover, all of

---

93. "It" was Elinor Glyn's euphemism for sex appeal in her 1927 novelette, *It.* Clara Bow (1905–65), the ultimate flapper, starred in the movie version of *It,* which gave birth to the "*It* girl" (*NAW*/mod., "Bow").

these two chemists' inventions are chemicals and all related to textiles. When we count both patented and unpatented inventions, assuming that Drs. Tesoro and Benerito have patented all of theirs, and assuming that Beulah Henry had, as hypothesized above, patented fewer than half of hers, the Lady Edison passes Dr. Benerito, and comes very close to Dr. Tesoro (about 108 to 125), and the untutored Miss Henry's versatile achievement becomes all the more striking.

Our second example of a machine specialist comes from England, where both opportunities and recognition for women inventors are less than here. **Doreen Kennedy-Way** (c. 1925–  ) of Dudley, West Midlands, already mentioned in scattered places as an inventor of medical-related, labor-saving and automotive machines, deserves an integrated discussion here so that the full scope and versatility of her inventiveness can be seen.

In stark contrast to Beulah Henry, Doreen Kennedy-Way received a rather thorough technical training. Her guardians owned an automotive-development plant, where she was initiated, as she puts it, into the "world of design and development." She started in the Drawing Office, and "after three very informative years, I was introduced to Machine Shop Practice, Sheet Metal Work, Welding, and thence to the Auto Development 'Test-Beds.' " While with her family's company, she gained not only design training and experience, but practical road experience, working on the company's pit crew at race meetings throughout the U.K., Europe, and the United States.

Transferring to another organization, E.R.A., at Dunstable, she became involved in the "378" project—designing and developing a revolutionary new car with stressed-skin body construction, fold-down rear seat, alloy wheels, rear engine, etc. The power unit was cast in magnesium and designed to lift out for repairs. Other unique features at the time were a magnetic clutch and push-button gear selection. The "378" car was later acquired by Austin, and "sported the Flying 'A,' " but never went into production as a whole.

Kennedy-Way's next project was the development of new two-stroke engine units, with flat-top pistons and cylinder heads without the normal hemispherical expansion area, but bearing a slot that ended in two smaller channels (at each end), leading to the spark plugs set in a horizontal plane. The engine used for these experiments was a Villiers 6E-197cc two-stroke, converted to water-cooling, and running at a compression ratio of 22 to 1. This model was successful enough that similar modifications were applied to a four-cylinder Hillman "Audax" engine. Before the development was complete, however, E.R.A. was acquired by the Solex Group, and she, with several others, was laid off.

Her background and experience enabled her to get another job right away, designing special taps and dies for Rolls-Royce and the Indian State Railways, among others. The name of her new employer was C.E. Johannsen, also in Dunstable.

Working in her free time, she had begun to design and develop small two-stroke engine units of 25 and 50cc, appropriate to motorcycles. When this work came to the attention of the motorcycle industry, she was commissioned by Stevenage Company to design and develop a 4-cylinder, 50cc racing motorcycle. When her design was successfully tested, she was invited to head Stevenage's Research and Development section, which she did. While there, she was once again involved in both design and testing—i.e., riding the new cycles—as for example during the testing of a Pirelli tire on a Rizzato-50 prototype racing motorcycle created for Cesare Rizzato of Modena, Italy. When Stevenage was merged with a Midlands company, she was asked to stay, but soon left in dissatisfaction.

Since the British motorcycle industry was declining, she changed direction, and joined a North Midlands company designing and developing food-preparation appliances. Once again a victim of a company reorganization, she was laid off.

Changing direction yet again, Ms. Kennedy-Way now joined a research-and-development company in Dudley. There she was "directly responsible for the design of the TRI-MO 'fun' machine for children," a three-wheeler with a two-stroke engine. She was also involved in the development of other wheeled devices, such as Small-Wheel Tandems, etc.

Finally, in 1973, she left this company to form her own company, Kennedy Kinetics. Since then, she has created at least the following inventions, some of which as of 1979 had already been taken up by manufacturers, and others were being evaluated for licensing:

1. the Tidy-joint—a domestic or catering device for netting small joints of meat
2. the Tidy-kut—a portable, battery-electric cling-film (plastic wrap) dispenser operating on the hot-wire principle
3. the Tidy-flor—a magnetic device to be attached to broomheads, for use in hairdressing, tailoring, and dressmaking establishments, as well as by carpenters, woodworkers, etc.
4. the Qwik-cut—a battery-electric (hot wire) poly-bag (plastic-bag) cutter
5. the Two-up—a twin-head attachment for office-type staplers, inserting two staples with each stroke, for brochures, booklets, etc.
6. the mini-KOMET—a 35cc two-stroke 3-wheel dragster for 5- to 8-year-olds
7. the KOMET—a 50cc two-stroke 3-wheel dragster for 9- to 15-year-olds[94]
8. the 315LT—a coin-operated, battery-electric trolley for use at air, sea, and rail terminals
9. the 2W20—a power-assisted, battery-electric 2-wheel trolley for Post-Light collections or deliveries; usable on pavements or sidewalks; 4–6 mph
10. the 4W20—a power-assisted, battery-electric 4-wheel trolley for interdepartmental use, in mailrooms, offices, etc.; pallet box can be removed and trolley used to move office equipment
11. the MOGO 1—a small battery-electric 3-wheeler, designed at the request of ASBAH, for use by physically handicapped children, on pavements or sidewalks; 4–6 mph
12. the MOGO 2—a spinoff from MOGO 1: a "fun" machine for 9- to 15-year-olds for off-road use; 10 mph
13. the RUNABOUT—a one-person commuter vehicle for adult use; a 3-wheeler for short journeys such as shopping, etc.; designed primarily to meet the needs of urban areas where public transport is virtually nonexistent; but with the "run-down" of fossil fuels, could fill a much wider need in all developed/developing countries; built to MOT Road-Highway specification

To have completed 13 such complicated mechanical inventions in less than six years, while also starting a company, is impressive indeed. Note, too, that Ms. Kennedy-Way's areas of invention reflect all her past areas of experience.

In addition to the foregoing completed inventions, at the time she published her company brochure in February 1979, she had other ideas in process of development, or, as she put it, "in embryo":

1. the UNI-KADDY—a battery-electric golf-club carrier
2. the SKU-BYKE—a two-stroke commuter 2-wheeler for town use
3. the SPARKY—a battery-electric 2-wheeler for town use; similar frame and forks to those of the SKU-BYKE
4. the PETITE—a battery-electric 3-wheeler car for two persons

94. Both the KOMET and the mini-KOMET, Kennedy-Way notes, feature pneumatic wide-wide tires, twin chrome meggas, chopper-type seats, gull-wing mudguards, tubular forks, cradle-type tubular frame, motorcycle controls, expanding-hub brakes, centrifugal clutch, etc. They evolved as a result of the upsurge in sophisticated toys.

Like many inventors—though this has too seldom been documented for women inventors—Doreen Kennedy-Way also designs special tools as needed to create her inventions. Examples she gives are Roll-pin insertion guides, Long drill guides, Deburrers, Tubular die-sticks, Step-jaw pliers for chain-clip removal, etc.

Even in the short time between publication of her brochure in February of 1979 and November of that same year, she had "innovated further items" in which an American entrepreneur was showing interest, and which were being manufactured under license in the U.K. She had taken a booth at the 1979 Energy Show, and had been invited to participate in the 1980 show.

Little wonder, then, that when the British Institute of Patentees and Inventors (IPI) surveyed their members in 1978–9 to see which ones were being mentioned in the media, Doreen Kennedy-Way's name showed up. Unfortunately, hers was *the only woman's name* located by the survey, in the six months covered (Gear, I-5/79). In December of 1978, within the survey period, Gabrielle Taylor published an article on women inventors in *Women's Realm,* a popular British women's magazine, mentioning not only Kennedy-Way, but at least one other woman already discussed in this book ("Agriculture"), **Violet Dawson.** However, apparently none of the IPI respondents (overwhelmingly male) saw this article.

Even more interesting than—and indeed related to—this gendering of audience is Taylor's selection among Kennedy-Way's inventions to discuss for her readers: the safety potato-masher with a D-shaped handle, the Tidy-Flor, the battery-operated plastic-bag cutter, the three-wheeled battery-operated vehicle for severely handicapped children, and the coin-operated luggage trolley for railway stations and airports. These are undeniably important and interesting inventions, but they give a far different picture of Kennedy-Way as an inventor than would a list containing—or focused on—her automotive, motorcycle-engine, and tool inventions.

Presumably, Kennedy-Way's total body of inventions and innovations is still increasing. The 18 inventions she listed in late 1979 did not include, though some of them undoubtedly built on, what she created during her research-and-development career with several British companies. In her 1978 interview with Gabrielle Taylor, she mentioned "about 30 ideas," which seems likely to be a modest estimate (based, perhaps, on considering the whole "378" car as a single "idea" rather than the multiple inventions that must have been involved, and probably *not* including the special tools she elsewhere says she invented). In any case, if we accept the figure of 30, then if she has continued to invent in the 1980s at the rate she did during 1973–9, she should have about 20 more completed inventions and 4–5 in development, for a total of about 55 (Gear, I-5/79; Kennedy-Way, L-11/14/79, *Kennedy Kinetics;* G. Taylor, 35).

Doreen Kennedy-Way is not only a highly versatile inventor, but a quintessential example of the specialist in machines.

## SUMMARY

The major conclusion to be drawn from this should already be obvious: contrary to the stereotype, women do invent machines. They also patent them, and in significant numbers. Moreover, women invent mechanical aids to a great many areas of human endeavor—agriculture, medicine, and manufacturing, for example, in addition to labor-saving machines and devices for production and other work in the home.

Beyond this, we would very much like to know 1) what percentage of women's inventions are mechanical, and 2) what percentage of all mechanical inventions are women's. As for the second question, since women's share of United States patents hovered around 1% in the 19th century, and even now has risen only to about 5%, the answer would scarcely justify the enormous effort needed to arrive at it. The first, however, is very much worth trying to answer, and the results can be used to suggest what women's share of mechanical inventions and patents would be if all else were equal.

Until now we have had just three major sources of statistical information on women inventors, all focusing exclusively on patents: LWP, WB/28, and Katherine Lee. Among the compilers of these sources, only Lee even attempts to arrive at a percentage of machines among her sample of women's patents, and her results, as already noted, are unsatisfactory.

The LWP compilers classified only the last brief installment of their work, covering patents granted between October 1, 1892, and March 1, 1895, just under two and a half years of the nearly ninety years covered by the work. And that classification—which I have elsewhere discussed as bizarre (Stanley, "Rose")[95]—not only contained no Machines category, but placed several mechanical inventions under headings that tend to diminish them (according to standard male-identified definitions of significance) or, if one reads no further than the heading, to confirm the stereotype of women preoccupied with "domestic" or "household" concerns. For example, the LWP compilers' class titled "Washing and Cleaning" contains seven washing machines and two mangles; a Mop-wringer, a Pressing frame, four Window cleaners or washers, a Steam-pressing attachment for sadirons, two Clothes dryers, a Swinging support for clothes lines, a Carpet-cleaning apparatus and two Carpet sweepers, to say nothing of a Folding handle for dustpans. This quick tally of obviously mechanical items—which does not even count the three Dish cleaners, the four Dish drainers, the Flatiron and the three Sadirons—gives a total of 23 out of the 52 patents listed in the category, or *nearly half.*

LWP's "Wearing Apparel" category contains a Tailor's drafting device and several other items for drafting and measuring, a Folding umbrella and several other umbrella-related patents, a Safety pin, Safety catches for brooches and hat pins, three Hook-and-eye inventions plus a Hook-and-eye safety pin, and a Combined shoe and glove buttoner, among other things. Most shocking—and most pertinent here—this category also conceals three industrial-scale machines: **Margaret Knight**'s Sole-cutting machine, her Feeding attachment for a sole-cutting machine, and her Lasting and sole-laying machine.

However, let us see what can be done with existing sources for the 19th century.

Among the 652 patents classified by LWP's compilers for the period from October 1, 1892, to March 1, 1895, machines or mechanical devices account for at least 89, or about 13.6% of the total. (This is my categorization; the LWP compilers, as noted, had no Machines category.) Among the 124 inventions listed by LWP's compilers for the single year of 1876, machines or mechanical devices account for 20 patents, or about 16.1% of the total. Combining the LWP classified group and my 1876 sample, we have 109 machines out of a total of 776, or nearly 14.5%.

Note that the percentage of machines seems to be higher in 1876 than in 1892–5. Even more interesting, in my 1876 study ("Conjurer"), when we look at patents omitted by the LWP compilers we find that *machines are the largest single category of omission.*[96] Thus

---

95. Since the published version ("Conjurer") omits the classification material, I should specify here that the LWP compilers followed neither the formal classes of the Patent Office at the time, nor the common-sense groupings later used by WB/28, but created ad hoc headings that reflected their stereotyped expectations of what women would invent.

96. Further research has modified some of the numbers in this paper, but this statement is still true.

perhaps if we had a truly complete LWP, the proportion of machines among women's 19th-century American patents might be significantly higher.

We might also have a different picture of women's scope within the machines category, for all of the omitted machines for 1876 are commercial, industrial, agricultural, or highly technical machines used in nondomestic environments and in endeavors incongruent with the 19th-century ideal of women's work: three manufacturing machines (two used in metal fabrication); dredging machinery; an agricultural machine; and an improvement in ship propulsion. Of the 20 machines listed by LWP for 1876, by contrast, 12 were either sewing or laundry machines, and of the two manufacturing machines listed, one was probably a candy-making machine whereas the other was used in making wigs. In fact, it seems that the stereotype separating women from machines blinded the compilers to certain of women's nontraditional inventions that were right under their noses ("Conjurer," App. A & C, Conclusion).

Before leaving the 19th century, however, we should note that even if LWP did represent the universe of women's inventions for the century, the ratio of women's machine patents to their patents in stereotypically female fields of invention, such as Wearing apparel, is unexpected: For 1876, for example, among the 124 listed inventions, machines numbered 20 and Wearing apparel numbered 29.

*20th century*

The Women's Bureau study, WB/28, provides us with a sample numerically comparable to LWP, over 5000 inventions patented by women. Notice, however, that these 20th-century women obtained a thousand more patents in ten years than their 19th-century sisters did in nearly ninety. WB/28 gives no names, but it has two great advantages over LWP: *all* of its patents are classified, not just a few; and its classification system makes sense. Moreover, WB/28 contains a few subcategories that are avowedly mechanical. Because of the way the information is presented, however, as discussed above, firm numbers can seldom be obtained unless the category or subcategory is exclusively mechanical (or unless the number of types of invention matches the numbers of patents listed, one for one), and the Bureau compilers, like the LWP compilers, do not break out an overall category of machines. Thus we are left once more with estimates—though with more fruitful estimates than previously available. What I want to do here is to look at the WB/28 compilation in a different way than in the preceding text, in order to arrive at some idea of the overall proportion of machines among these 5000-odd patents.

After recording the total number of patents in each category (and subcategory, where applicable), I calculated a minimum and an estimated/average number of *mechanical inventions* within each grouping, as follows: dividing the total number of patents by the number of *types* of invention listed, I arrived at an average number of patents per type, assuming an even distribution of the types. Then I counted the number of types of invention that could be classed as machines or mechanical in nature, and multiplied that by the figure for patents per type. This gave me my estimated or average number of mechanical inventions within any given category or subcategory.

As throughout the chapter, I am defining *mechanical inventions* rather broadly to include not only obvious machines, but devices that are collapsible, telescoping, folding, or extensible, or have such parts, or are attachments to machines. I also include mechanical toys and such devices as hooks and eyes, which might elsewhere be dismissed as "domestic" or "wearing apparel." Usually, on the other hand, I exclude electrical items, even though some of them may have mechanical aspects or parts, and rigid supports and frames, even though the Patent Office classifies the latter as Mechanical (i.e., not Chemical, not Electrical, and not a Design).

For the *minimum* number of mechanical patents I simply took the number of mechanical types in a category or subcategory and assumed that it contained only one patent. This would be the absolute minimum number of machines or mechanical devices that could be represented.

The rather striking discrepancy between the minimum and the estimated/average figure in many cases, and in the overall total, arises from the fact that some categories have many inventions of few types, so that each type represents, or could represent, as many as 15 inventions.

Then I added the minimum figures for all the categories to produce one total, and the estimated/average figures together to produce a second total. Using these two totals, I calculated a minimum and an estimated/average percentage of machine/mechanical patents among women's American patents 1905–21. The results are as follows:

| WB/28 Category & No. of Patents | | # Mechanical | | % Mechanical | |
|---|---|---|---|---|---|
| | | Min. | Est./Av. | Min. | Est./Av. |
| I. Agric./Forestry | 221 | 100 | 115.07 | 45.2 % | 52.06% |
| II. Mining/Quarrying | 14 | 4 | 5.09 | 28.57 | 35.35 |
| III. Manufacture | 223 | 123 | 141.471 | 55.1 | 63.4 |
| IV. Structural ... | 208 | 29 | 55.795 | 13.9 | 26.8 |
| V. Transportation | 345 | 169 | 240.7 | 48.98 | 69.76 |
| VI. Trade | 71 | 25 | 35.21 | 35.2 | 49.59 |
| VII. Hotel/Restaurant | 10 | 5 | 5.55 | 50 | 50.5 |
| VIII. Laundry/Drycl. | 6 | 4 | 4 | 66.6 | 66.6 |
| IX. Dressmak./Millin. | 116 | 7 | 51.6 | 5.9 | 43.7 |
| X. Office ... Equip. | 71 | 29 | 35.989 | 40.8 | 50.68 |
| XI. Fishing | 9 | 3 | 3.375 | 33.3 | 37.5 |
| XII. Household | 1385 | 97 | 540.075 | 7 | 38.99 |
| XIII. Ind'y/Agr./Home | 378 | 84 | 131.45 | 22.2 | 34.77 |
| XIV. Sci. Instr. ... | 76 | 34 | 47.85 | 44.7 | 62.9 |
| XV. Ordnance/Arms ... | 22 | 10 | 11.57 | 45.4 | 52.59 |
| XVI. Pers. Wear/Use | 1090* | 19 | 122.87 | 1.7 | 11.27 |
| XVII. Beauty/Barber ... | 46 | 6 | 15.331 | 13 | 33.3 |
| XVIII. Med./Surg./Dent. | 227 | 32 | 59.14 | 14 | 26 |
| XIX. Safety/Sanit. | 129 | 27 | 51.53 | 20.9 | 39.9 |
| XX. Education | 75 | 24 | 51.88 | 32 | 69.1 |
| XXI. Arts/Crafts | 67 | 29 | 34.02 | 43.28 | 50.77 |
| XXII. Amusement | 211 | 31 | 75.67 | 14.69 | 35.86 |
| XXIII. Misc. | 14 | 3 | 4.6 | 21.4 | 32.8 |
| TOTALS | 5016 | 894 | 1839.834 =1840 | 17.8%** | 36.68%** |

*This figure appears as 1000 in the master tabulation of WB/28 (15), but is obviously a typographical error, as the subcategory totals add up to 1090, and 1090 is necessary to reach the sample total of 5016.

**These percentages are not "totals" or averages of the columns above them, but represent the relationship between 894 and 1840, respectively, and the sample total of 5016.

Some interesting individual figures among the estimated averages are that 52% of women's agricultural patents, more than a third of their patents in mining, quarrying, and smelting, and nearly two-thirds of their patents in manufacturing are mechanical. More than two-thirds of women's patents in both education and transportation are mechanical, and even among their household inventions, nearly 40% are mechanical. Overall, as can be seen, women's mechanical patents amount to a minimum of 17.8% and a probably fairer average estimate of 36.68% of a sample of 5016 patents granted to American women in the early years of the 20th century.

Thus it would seem that American women still deserve to be called daughters of Athena, the Great Contriver.

# 5 Daughters of the Enchantress of Numbers and Grandma COBOL: Women Inventors and Innovators in Computers and Related Technology

I am much satisfied with this first child of mine. He is an uncommonly fine baby and will grow up to be a man of the first magnitude and power.
—Augusta Ada, Lady Lovelace

I believe that whoever has a controlling hand in the information revolution has a controlling hand on geopolitical influences.—Michael Dertouzos, 1982

Whatever you do, and whatever you build, never forget the relationship to the people who are going to use it.—Grace Hopper, 1978

A computer without software is like a car without a steering wheel.
—Sandra Kurtzig, 1987

## INTRODUCTION

Computers are among the most significant inventions of all time. They will transform human life as thoroughly as, and should be ranked with, such radical developments as language, the taming of fire, agriculture, writing, and dependable contraceptives. As Michael Dertouzos might have said, had he cared to speak English, the hand that rocks the computer rules the world.

That hand has seldom been pictured as female, though interestingly enough, at least one scholar traces the computer's origins to women's work. According to Theodore Wertime in the *Smithsonian Book of Invention* (1978: 240), "Like many of our other machines the computer goes back to the earliest Egyptian experiments with rotary grinders for grain. . . ." However, only two women are generally known as significant contributors to the invention and development of computers: **Ada, Countess of Lovelace,** and **Captain,** later **Rear Admiral, Grace Hopper.** And even for these women, most people could not say exactly what they did. Let us take a look, then, at these two pioneers before discussing the important but even lesser-known contributions of other women.

## PIONEERS

### Lady Lovelace—the Proto-Computer

The least interesting thing about **Augusta Ada, Lady Lovelace** (1815–52), for the purposes of this chapter, is that she was Lord Byron's child. She never saw her father after she was a month old, and probably inherited her mathematical ability from her mother, Lady Annabella Milbanke, Byron's Princess of Parallelograms. Like her mother, Ada studied geometry on her own, for fun—though a mathematical course of study was also imposed on her, to make her as different from her father as possible. Lady Milbanke described the eight-year-old Ada to Byron in a letter: "Not devoid of imagination, but it is chiefly exercised in connection with her mechanical ingenuity—the manufacture of ships, boats, etc." (Angluin, 7). Touring the industrial Midlands in her teens, Ada delighted in figuring out the workings of the machines.

At fourteen, Ada was tutored by the famous mathematician Augustus De Morgan. Tragically, she never knew that she might have done first-rate original work in mathematics. De Morgan, believing woman's constitution unfit for great mental exertion, did not tell her.

Instead he wrote confidentially to Ada's mother several years later, when Ada's only published work appeared, saying, "The tract about Babbage's machine is a pretty thing enough, but I think I could produce a series of extracts, out of Lady Lovelace's first queries upon new subjects, which would make a mathematician see that it was no criterion of what might be expected from her." She had, he said, the makings of "an original mathematical investigator, perhaps of first-rate eminence." Defending this concealment, he continued, "All women who have published mathematics hitherto have shown knowledge, and power of getting it, but no one (except possibly Maria Agnesi) has wrestled with the difficulties and shown a man's strength in getting over them." The reason was obvious to De Morgan: "The very great tension of mind which they require is beyond the strength of a woman's power of application. Lady L. has unquestionably as much power as would require all the strength of a man's constitution to bear the fatigue of thought to which it will unquestionably lead her." Thus De Morgan restricted his comments on her work to "very good" and "quite right."

In other words, *it was because she had such great power that he kept silence;* had she had less, he would have spoken out.

At seventeen she went to one of Charles Babbage's Saturday evening parties with Sophia De Morgan. A versatile genius, Babbage created innovations in the mail service, railroads, signalling devices, telescopic instruments, archaeological dating, locks, submarines, earthquake detectors, and even stage lighting. He had been inspired to improve calculating machines when he saw the many errors in navigational tables. His first attempt was the so-called Difference Engine, which used addition and subtraction to eliminate the labor of multiplication and division.

Babbage kept a small model of the Engine at home to show to guests. As Mrs. De Morgan wrote, "While other visitors gazed at the working of this beautiful instrument with the sort of expression, and I dare say the sort of feeling, that some savages are said to have shown on first seeing a looking glass or hearing a gun, . . . Miss Byron, young as she was, understood its workings, and saw the great beauty of the invention" (Angluin, 7).

In 1834 Ada heard Dr. Dionysius Lardner's lectures on the Difference Engine. She also continued to visit Babbage with Mrs. De Morgan or Mary Somerville, herself a famous mathematician. Meanwhile, at nineteen, Ada married William, eighth Baron King, later Lord Lovelace. Her marriage and children temporarily occupied her, but she would not long be content as socially fashionable wife and mother.

By 1842 Babbage had abandoned the Difference Engine for his new Analytical Engine. Directed by punched pasteboard cards (as developed for Jacquard looms about 1800), this machine was to have a collection of registers called the Store and a separate mechanism called the Mill, which could add, subtract, multiply, or divide two numbers to produce a third. An addressing mechanism would place that third number in any register of the Store. The English government, which had supported the Difference Engine, rejected this new idea.

The only interest in the Analytical Engine came from Italy, where Babbage's presentation inspired the mathematician L.F. Menabrea to write a paper on it (1842), and from Lady Lovelace, who continued to correspond with Babbage and often invited him to stay at her house. After the birth of her third child, she was impatient to resume her mathematical education. Gradually she developed the idea of helping Babbage with his work. She expressed herself very diffidently at first: "It strikes me that at some future time . . . my head may [prove] subservient to some of your purposes and plans . . . tho' I scarcely dare so exalt myself as hope, however humbly, that I can ever be intellectually worthy to attempt serving you. . . ." But she began at once to study the method of finite differences, which she knew was related to his work, reserving "a couple of hours daily" for her studies when in town. The difficulty she had in doing this can only be imagined. By June of 1843 she had translated Menabrea's paper and begun annotating it, extending his ideas and correcting some serious errors in Babbage's work (Gleiser, 40). Soon the Notes were longer than the translation. Impressed, Babbage urged Lady Lovelace to publish the Notes separately as an original paper, but she considered herself committed to the editor who had agreed to publish the translation. Later in 1843 the work appeared in *Taylor's Scientific Memoirs,* signed only "A.A.L."[1]

Soon she and Babbage were working together. At first she worked more as an apprentice, but eventually, according to some sources, as a collaborator (Gleiser, 40). She, too, became obsessed with the Analytical Engine. In the absence of government support, the two of them and Lord Lovelace worked out a scheme for betting on the horses. As with so many such schemes, they won a little at first, but then lost heavily. According to Dana Angluin of the Yale Computer Science Department, writing in the mid-1970s, the two men withdrew, but Ada continued, pawning her jewels repeatedly to pay her debts.[2]

Another Babbage-Lovelace brainchild was an automatic tic-tac-toe machine, which would be taken on tour to challenge all comers. They would charge admission to see the games, and thus make money for their project (Russ Adams, 81).

As it turns out, Babbage and Lady Lovelace were working on an impossible task. Even had money been unlimited, and Ada not died in her mid-30s, the Analytical Engine could not have been built. The precision milling and machining necessary for some of its parts simply did not yet exist. In fact, only recently has IBM been able to build the machine as originally conceived.

Thus we may never know the full extent of Ada Lovelace's contribution. Her 1843 paper is certainly a remarkable document. As Angluin comments, "Therein, a century before its time, is the concept of a general-purpose digital computer, developed to an amazing degree of sophistication." Various passages show that she or Menabrea or both had already grasped the idea of many modern concepts in computer design and programming: memory, central processor, switch register, program library, FOR-loops, indexing, comments, program trace, coder, keypuncher, analysis of algorithms. Menabrea had begun to write some programs, but Lady Lovelace carried the idea much further, and is often considered the world's first computer programmer (e.g., Burkitt & Williams). According to Russ Adams, writing in 1983, Lady Lovelace "was to be [Babbage's] assistant and be responsible for writing test programs for the machine" (78). Says Angluin, "I would guess that it was that exercise which led to the development of some of

1. For the text of this paper see Morrison & Morrison.
2. Many accounts of Lady Lovelace and the proto-computer focus on her gambling, her temperament, and her drug addiction, ignoring her achievement. Some have even portrayed her as a neglectful mother, though this was apparently not her children's view. No mention is made of the great cause the money would have served, or that the drugs were prescribed to dull the pain of advanced cancer. This pain, in fact, was so severe that when her dragon-mother decided she must break free of the drugs, mattresses had to be hung on the walls of her room so that she would not injure herself as she flung herself about in agony. My point is that all such reports are irrelevant to her contribution to computer history and technology, and it is a disservice to the reader as well as to Lady Lovelace to have most of the one paragraph she might receive in computer texts and histories devoted to her alleged character flaws (or her looks, or her famous father).

the more subtle and powerful concepts, e.g., looping and indexing" (Angluin, 6–7). She may have had a broader vision of the computer's possibilities than Babbage himself, for she predicted, among other things, that it would one day compose music.

She suggested to Babbage that he use binary storage, as in modern computers, instead of the decimal system he had first tried. She also invented conditional branching. This powerful programming tool, according to modern computer pioneer Dr. Grace Hopper, could have advanced the computing industry by fifteen years had it not been lost with the rest of Lady Lovelace's work (N. Martin, PC-1976).

For these innovations alone she deserves to be considered an important computer pioneer, though her contributions have not gone undisputed. In fact, the pendulum of professional opinion on Lady Lovelace's contribution to the proto-computer has swung twice, bringing her first from obscurity into perhaps too bright a limelight, then back toward discredit with the appearance of Dorothy Stein's book (1985), and finally to a possibly more balanced view with Joan Baum's 1987 biography. Whatever may be the final verdict, just as Babbage is considered the father of the modern computer, Ada spoke of herself as its mother: "I am much satisfied with this first child of mine. He is an uncommonly fine baby and will grow up to be a man of the first magnitude and power" (Adams, 78; Gleiser, 41).

An important computer language of the 1970s, considered for sole use by the Department of Defense, is called Ada.[3] By the 1980s Ada Lovelace was almost routinely mentioned in historical discussions of computers (see, e.g., Edwards & Broadwell; C. Evans; Burkitt & Williams, Feigenbaum & McCorduck; W.D. Gardner, 186–7), and not just as a gambling dilettante, but as a serious contributor to Babbage's work.

*Calculator Inventions*

Before moving on to the second major pioneer among women inventors in computer technology, let us take a quick look at some 19th- and early-20th-century women who worked, as Babbage began, with calculators. None of these women won lasting fame as inventors, though **Marie Pape-Carpantier** was a renowned French educator, **Emily Duncan** was mentioned in *Inventive Age* magazine, and **Edith Clarke** appears in *Notable American Women* (mod.). They illustrate that women worked in this area of technology before the computer:

**Marie Pape-Carpantier** (1815–78) of Paris invented a *boulier numérateur,* an educational calculator using colored balls to make the calculations. It was shown at the Paris Exhibition of 1878, in the Scholastic Department (Betham-Edwards, 154–5).

In 1882 **Mary E. Winter** of Galesburg, IL, with others, patented an Adding machine. In 1890 and 1895, respectively, **Christiana Newhaus** of Chicago and **Lydia D. Myers** of Philadelphia patented Abacuses. In 1903 **Christel Hamann** of Friedenau-Berlin, Germany, took out a U.S. patent on a Calculating machine.

**Emily C. Duncan** (1849–1934) received two patents for Calculators (1903, 1904) while living in Jennings, LA. She was the wife of another inventor, Eugene Duncan. Since according to the February 1905 issue of *Inventive Age,* her machines allowed one to

---

3. At least one computer scientist considers the naming highly inappropriate, and "a most dubious honor," especially since this is the only time a computer language has been named for a woman. Dr. Saj-Nicole Joni of Wellesley College, who refuses to work for the Department of Defense, feels that "this reinforces the whole myth of the woman-inventor/mathematician/woman-with-machines as evil. The stuff they want to write programs in Ada for is about death and destruction . . ." (L-2/28/84).

calculate the interest and remaining terms for loans of various amounts, I originally speculated that she may have worked in a bank, but this is evidently not so.

Emily Cornelius Forbes was born September 27, 1849, in Coral, IL, third child and third daughter of Anthony Forbes and Lydia Van Culert Forbes, both from New York State. A farm family, they had six children in all. Emily Forbes married her stepbrother, James Eugene Duncan, when she was only 14 and he only 17 or 18, home on furlough from the Civil War. She bore eight children, of whom five survived. Nothing is known of her schooling, though her granddaughter Emily Stothart, her namesake and biographer,[4] thinks she probably went to public elementary schools. Her husband's formal education ended at age 9, yet it is apparent from his poetry that he was "very knowledgeable about . . . literature, history, and current events," and his penmanship was so fine that people hired him to write invitations and documents for them (Stothart, *Home,* chap. 8).

Early in their marriage, the young couple lived in Wisconsin, first at Pine River, and later in Centralia. In the 1880s, however, they pioneered on the Dakota frontier, where they suffered from cold and loneliness, and where at least two of their children died for lack of medical attention. A heart-rending letter from Emily Duncan to her sister Lydia, written in 1892, says, in part, "Oh, Lydia, I hope you will never see the trouble we have seen in Dakota. . . . it was so hard to see poor little Arthur die and no doctor. Poor little darling, it was so hard to give him up."

And across the top of this same letter, she wrote, "You don't know how I miss you and wish you was here and could hold your hand but don't wish you had to live in such a cold country. I will send you a little piece of dried elk seeing we can have no more lunches" (*ibid.,* chap. 4). Emily Duncan must have been overjoyed when they moved back to Wisconsin.

Her first patent was for a Bevel square, interesting in that Eugene Duncan, a carpenter and cabinet-maker, had also patented "try" or bevel squares. This 1901 document gives her address as Centralia, WI. (Actually, the family had moved to Jennings, LA, in 1900, but presumably she had applied for this patent before leaving.) By this time Emily would have been 52. Only two of her surviving five children still lived at home, and one of them was already 25 and working as a schoolteacher. Life probably became easier, at least for a while. Though Eugene Duncan never profited greatly from any of his inventions (and the family has no record of Emily's receiving anything for hers), the 25-year-old daughter immediately got a teaching position. Eugene Duncan also evidently had a Civil War pension.

They lived in Jennings for six years, during which time Emily received her two calculator patents. Family records do not indicate how the Duncans became interested in calculators, as Eugene's livelihood continued to depend mainly on furniture-making, carpentry, and possibly a few oil paintings sold. In any case, he had patented a Computing scale before leaving Wisconsin, and, like Emily, patented a Calculator in 1904. As noted, her two calculators were designed specifically for uses involving credit or borrowing: her 1903 calculator makes it easy to calculate interest on any amount of money at 6–8%; and her 1904 invention allows one quickly to calculate the number of days a note has to run.

The invention work took place in the carpentry workshop behind the house. Emily

---

4. Mrs. Stothart, who spent Christmas, summer vacations, and other special occasions at the home of her Duncan grandparents, has researched their lives for a book entitled *Home at Last,* available from the author. She generously shared her information with me while her book was in progress (L-3/9, 3/20, and 5/5/84).

Stothart recalls at least one other invention—an envelope with a special opening string—but no one knows whether this was Emily's idea or Eugene's. Undoubtedly many inventions from these two fertile brains never reached the Patent Office.

The family may have underestimated the invention revenues, however. Sometime before 1909 Eugene had acquired an interest in a ladder-making factory, perhaps in connection with his patented convertible step/extension ladder (1909, 1912). He sold his factory interest in 1909 for enough to build a home in Elton, near Jennings, where he and Emily lived for the rest of their lives. Emily Stothart recalls a Steinway grand piano at the Elton house. Certainly Emily found it hard to manage after Eugene died, especially when joined by her widowed youngest daughter Marion and Marion's 13-year-old daughter, all living on Grandma's meager pension. However, she retained her spirit until the end. In the last summer of her life she climbed a fig tree to help the children pick figs, and just two months before she died, she hosted her own 85th birthday party with members of the Old Women's League (OWLS) of Jennings, and other neighbors and friends (Stothart, *Home, passim;* L-5/21/87).

**Edith Clarke** (1883–1959) was an electrical engineer who worked for many years for General Electric Company. She received the Society of Women Engineers' 1954 Achievement Award for her notable work with electric power systems and transmission. Of interest here is her first patent (1925), for a calculating device that allowed engineers to monitor or predict the performance of electrical transmission lines and systems without laboriously solving many complicated equations. Having spent several years as a human computer, she knew the need for the calculating device and charts that she created (Goff, 54–5, 58–9; *NAW*/mod.).

Clarke was by no means atypical in being employed to do calculations: women wishing to enter science were often hired as "computers," especially in astronomy. This background makes doubly appropriate the contributions of the women celebrated here.

## Grace Hopper—the Modern Computer

**Rear Admiral Grace Murray Hopper** (1906–    ), retired from the Navy in 1987, is one of the most important pioneers in recent computer history. Born Grace Brewster in New York City, but claiming Wolfeboro, NH, as home, she showed an early fascination with how things work. She was surrounded as a child by construction toys. "I built everything," she says, "ships, railroad cars, bridges, even furniture for my sister's dollhouse. Though not noted for her feminism (in the 1970s she returned all mail addressing her as "Ms."), she does realize how lucky she was that her civil engineer father gave his daughters the same training as his sons (Cimons, "Computer Language"). And she would have gone to Annapolis if she could have (Navy Dept. Press Release, Aug, 24, 1973: 2).

At age seven she took apart a clock. In order to put it back together, she took apart another one—and then another and another, till all seven of the family's clocks lay in pieces around her. Clocks seem to be an important motif in Admiral Hopper's life. As a Navy Captain she kept one running counterclockwise behind her Pentagon desk, to remind her that things need not forever be done the same way.

She majored in math, physics, and economics at Vassar, graduating Phi Beta Kappa in 1928. After getting her MA in math at Yale in 1930 she married Vincent Hopper (they divorced in 1945). While working on her PhD from Yale, granted in 1934, she taught at Vassar. In December 1943 she entered the Navy Reserve and attended the USNR Midshipman's School at Northampton, MA. This seems a surprising step for a promising young academic, but Dr. Hopper was the granddaughter of an admiral, Rear Admiral Alexander Wilson Russell. Admiral Russell, she remembers, took a dim view of women and cats on ships. When she went into the Navy, she put flowers on his grave and assured him she would not wreck his service.

It was the navy that introduced her to computers, ordering her as a lieutenant to the Bureau of Ordnance Computation Project at Harvard, where she worked on the Mark I and Mark II digital computers. Crusty (she once mused about shooting the next person who told her "But we've always done it this way") but soft-spoken, and completely original, Dr. Hopper has a way with words: "I walked into a room and the professor said, 'Lieutenant, that is a calculating machine.' I looked at a monstrous device stretching 51 feet across a large room. It was Mark I and I was stepping into history" (Cimons, "Computer Language").

In 1946, at war's end, Lt. Hopper resigned from her Vassar leave of absence and from active duty in the Naval Reserve (she had been denied transfer to the Regular Navy on grounds of age). She joined the Harvard faculty as a research fellow in engineering sciences and applied physics at the Computation Laboratory, where she continued to work on the Mark II and III computers.[5]

In 1949 she became a senior scientist for the Eckert-Mauchly Computer Corporation because they seemed closest of all the fledgling computer companies to having an operating computer—UNIVAC I. She stayed with Eckert-Mauchly, which became the Univac Division of Sperry-Rand, until 1967 when the Navy recalled her to active duty and Captain's rank to standardize its programming languages. She worked at the Pentagon for the next twenty years.

This employment history tells little of the brilliance of her achievement. At Harvard she created the first operating programs for the Mark I. At Univac she helped develop the UNIVAC I, the first commercially viable computer, and in 1952, as head of the programming group, she created the first *practical* compiler (Halcrow),[6] consisting of a group of error-free programs put on tape and assigned call numbers. Given any call number, a computer could search the tape, find the program, and follow it. Before this, computers had to be instructed, with long strings of numbers,[7] to do each separate addition, subtraction, or other arithmetic process. And no matter how many times the same calculation was done, the whole instruction had to be repeated. Essentially, computer operators had to start from scratch every time they approached the machine.

To put her achievement another way, *she had invented automatic programming,* a system enabling the computer to "write" its own program from key instructions. Herman Goldstine notes that as a result of her efforts Univac had "perhaps the first automatic coding systems in the United States . . . , called A-0 and A-1" (340). Soon she realized that even this intermediate step was unnecessary. Directed only by mathematical equations, the computer could call the pieces and put them together. When people were skeptical, Capt. Hopper decided to show them that she could make a computer do anything she could define. She set out to program a computer to do differential calculus. It took her a year.

5. While working on the Mark II she and her colleagues discovered the original computer bug—a huge moth stuck in a relay. The moth, taped to a logbook page to explain that day's problems, entered computer history, and the English language (Gilbert & Moore, 45).

6. Herman Goldstine credits Heinz Rutishauser of Zurich with "probably" inventing the first compiler; but whether because he died young, or because his ideas were ahead of their time, his work was not widely accepted or applied (337–8). Thus credit properly belongs, as here, with Grace Hopper.

7. At one point in her early programming career, she became so habituated to octal (base-8) arithmetic that she sometimes used it on her checkbook, with disastrous results (Gilbert & Moore, 45).

Ideas don't take so long to develop, she maintains. "It only takes one year to develop an idea; the other nine years are needed to get people to believe it" (J. Mason, 84).

Univac was delighted that she had a computer doing differential calculus, but reminded her that the customers really wanted data processing. That meant dealing with words. After writing 500 typical programs and identifying 30 verbs common to all of them, she created the first computer language consisting of words, FLOW-MATIC. Robert Bemer notes in his article on the history of COBOL (131) that FLOW-MATIC grew out of the A-series compilers and became operational for UNIVAC I and II in December 1956.[8] Then Dr. Hopper recommended that data processing be written in English. Once again she was told her idea was impossible—computers couldn't understand English. This time it took her three years to convince people.

From FLOW-MATIC and two other languages, one of them a modification of FLOW-MATIC, evolved COBOL (the *CO*mmon *B*usiness-*O*riented *L*anguage), used on thousands of government and business computers. Aside from its intrinsic virtues, COBOL has the added virtue of being usable on several different computers. Since Dr. Hopper not only invented FLOW-MATIC, but was instrumental in starting the COBOL project and served as one of the team's two technical advisors, she is usually credited with developing COBOL. Of course, as some have pointed out, her role was actually rather indirect by this stage, and it might be more accurate to call her the grandmother rather than the mother of COBOL.

But this is getting ahead of the story. With John Mauchly, Grace Hopper also did the coding for BINAC, which was essentially two computers linked together. This configuration gave BINAC the then-unique capacity to check itself for accuracy. It was the first stored-program computer to be completed in the United States, and nearly the first in the world (Shurkin, *Engines,* 230, 230–1n).[9]

Dr. Hopper was also one of the first, if not the first, to realize that software (programs and sets of programs) are an essential part of a computer system and should be included with any computer sold. In 1957 she wrote a memo to that effect, titled "Layette for a Computer."

Another of her important inventions is virtual storage. Virtual storage is a computer version of sleight of hand, an operating system allowing data and program segments to be swapped between peripheral and central storage in such a way that the *effective* storage capacity of a given computer is vastly increased. "Essentially," says Myles Walsh, virtual storage "allows programmers to assume that they have virtually unlimited computer storage at their disposal." For example, a computer with 2 million positions of computer storage can be made to look as if it had 16 million positions (69; Wexelblat, 16–7).

As if all this, and her more recent work in standardizing computer languages for all Navy operations, were not enough, Dr. Hopper became involved in the 1970s in a project to make computers faster than anyone imagined possible. The prevailing approach to more speed was to use very fast adders working simultaneously. If one of them drifts—and high-speed adders, she says, do drift—the results are useless. Dr. Hopper thought "we could build a better and faster computer with these slower micros if we'd just use the assembly-line concept. I could move my data along, split up and in parallel, letting

8. The *A* in the name of the A-series of compilers stood for "Algebraic," denoting compilers intended for algebraic and scientific applications. FLOW-MATIC was originally named B-O, the *B* standing for "Business." Says Bemer, "The concept of the compiler is largely due to Dr. Grace Murray Hopper, who was in charge of these projects" (Bemer, 131).
9. All Shurkin references in this chapter are to his *Engines of the Mind.*

each micro handle one operation. For certain problems, like the weather, it would work. Eventually, I could beat the velocity of light by having work done in parallel."[10]

Hyperbole or no, the value of parallel processing, which Dr. Hopper has been recommending at least since 1972,[11] has at last been recognized, and is being taken up everywhere. By late 1976 she had not persuaded anyone to try it, but she had already borrowed two computers and as soon as she got more was ready to try it out herself. Some ten years later, however, in the Sunday magazine section of a San Francisco Bay Area newspaper, William J. Broad said, "Instead of solving problems step by step—as most computers do throughout the world—the new machines are meant to break apart computational puzzles and solve their thousands or millions of separate parts all at once. The revolutionary approach goes by the unassuming name of parallel processing" (Broad, 11). If this sounds like an echo of Hopper, that is because it is.

The writer goes on to quote Dr. Kenneth G. Wilson, a Nobel laureate in physics: "There are problems we'd like to solve right now that require 100,000 times more computing power than is currently available. The only way to conceive of getting it is with parallel processing" (*ibid.*).

Like John Mauchly, a co-worker in the 1950s, Dr. Hopper became fascinated with the potential uses of computers in weather forecasting. Advance knowledge can be important in making natural disasters less disastrous, and in feeding the world. Better weather forecasting involves satellite photographs that often need to be enhanced before they can be interpreted. It takes one quadrillion computer operations to enhance one photograph—which in the 1970s could mean taking three days to enhance one group of photographs. By that time, the weather has happened (Fields; McPhilimy).

The list of Grace Hopper's honors and awards is very long: Naval Ordnance Development Award, 1946; Fellow, Institute of Electrical and Electronic Engineers, 1962; Society of Women Engineers Achievement Award, 1964; Data Processing Management Association's first Computer Sciences "Man of the Year" Award, 1969; American Federation of Information Processing Societies' Harry Goode Memorial Award, 1970; Honorary Doctor of Engineering degree, Newark School of Engineering, 1972; Wilbur Lucius Cross Medal, Yale University, 1972; Member, National Academy of Engineering, 1973; Legion of Merit (U.S. Navy), 1973; Distinguished Fellow, British Computer Society, 1973; Honorary Doctor of Laws degree, University of Pennsylvania, 1974; Honorary Doctor of Science Degree, Pratt Institute, 1976; W. Wallace McDowell Award, IEEE Computer Society, 1979, to name a few. As of the mid-1980s she was the only woman included in the tapes of computer pioneers held at the British Science Library in London.

She has published more than fifty papers and articles and had an award established in her honor, the Grace Murray Hopper Award, inaugurated by Univac in 1972, awarded annually by the Association for Computing Machinery to a young person for outstanding work in the computer field. Her promotion to Captain when over seventy years old and in the Naval Reserve required an act of Congress. By her final retirement in 1987, she had

10. Dr. Saj-Nicole Joni of Wellesley College notes, "This is simply not possible," and assumes the reference is to Dr. Hopper's work on "non-Von-Neumann architectures for supercomputers. In one sense, they are faster than the architecture, but they are not really faster than the speed of light . . ." (L-2/28/84).
11. Dr. Hopper's awareness of the possibilities of parallel processing may actually go back much further. She and John Mauchly did, after all, as Joel Shurkin points out (*Engines,* 230), do the coding for BINAC, an early double computer that worked in parallel. Astonishingly enough, this is Shurkin's only reference to Dr. Hopper in a book that purports to be a history of the computer.

reached her grandfather's rank, Rear Admiral. At that time she was featured on CBS Television's *60 Minutes* program. But she might be proudest of the praise of one of her many interviewers (J. Gilbert): "Grace Hopper is as much a part of the future as of the past."

These two pioneers are not alone. Women have contributed to the computer revolution in all areas, from theory and machine design to languages, and to various applications of computer technology in business, research, and daily lives. I will consider a sample of them here under four headings: hardware or machine design; software and language design; innovative applications; and artificial intelligence. Innovations in theory will sometimes be inherent in all but the third category.

## HARDWARE: MACHINE DESIGN

### ENIAC *Girls*

While Grace Hopper was meeting Mark I, the so-called ENIAC girls were programming the world's first general-purpose electronic computer, ENIAC, at the Moore School of Electrical Engineering in Philadelphia. Mostly math graduates from nearby colleges, they had themselves originally been hired as computers—to calculate shell and missile trajectories for the War Department.[12] At one time there were 200 young women doing firing tables on desk calculators (Shurkin, 126).

When ENIAC was ready to replace them, "The women were given a crash course in the machine's logic and other circuits, and were then expected to perform the 'clerical' task of designing and testing the ENIAC programs" (Simons, 63). Some of these women later went to Los Alamos to program the Manhattan Project computers. As soon as the computer designers and engineers realized that programming was not merely a clerical detail, but a creative and challenging endeavor, it was redefined as men's work. Meanwhile, these women had done important things. Programming would normally be discussed in connection with software, but original programming of this sort, accomplished by mechanical alterations and manipulations of dials and switches, helps define and refine the machine itself (cf. Shurkin, 164, 171).

Computer historian Nancy Stern lists the following women as important in the Moore School project: **Adele Goldstine, Kay Mauchly, Frances Bilas, Ruth Lichterman, Elizabeth Jennings, Marilyn Wescoff,** and **Betty Holberton.** Kay Mauchly was then Kathleen McNulty; Betty Holberton was then Elizabeth ("Betty") Snyder. Frances Bilas later married Homer Spence, one of the Army's technicians on the project. Betty Jennings married one of the Moore School engineers. Adele Goldstine was already married to Herman H. (then Lt.) Goldstine, the Army's liaison officer and project overseer. The others were "six of the best computers," chosen by Lt. Goldstine to learn to program ENIAC (Stern, L-4/12/ 84; H. Goldstine, 202, and L-5/30/84).

**Adele Goldstine** was born Adele Katz in New York City in 1920. A graduate of Hunter College and of the University of Michigan (MS), she came to the Moore School in 1942 when the Army detailed her husband to head the ENIAC project. A mathematician, Adele helped write ENIAC's documentation. She was ENIAC's first programmer and wrote the

---

12. Women's scientific work in computing has often been dismissed as routine—which it surely was in daily practice—and unimportant, which it just as surely was not. Indeed, an astronomy graduate student once quipped, more truly than he knew, that "American astronomy became preeminent over European astronomy because of two discoveries: Hale discovered money and Pickering discovered women"—as calculators, that is (V. Rubin, 60).

manual on its logical operation. She then taught the other women to program, either directly or through this manual. She also "wrote the first program to convert the ENIAC to a stored-program computer" (H. Goldstine, L-5/30/1984; cf. Shurkin, 126, 149, 206).

**Kathleen (Kay) McNulty** (1920– ) graduated as one of only three math majors from Chestnut Hill College near Philadelphia in 1942. She was 22 and looking for a job. She went with her friend **Fran Bilas** to an interview at the University of Pennsylvania for mathematicians to calculate bullet and missile trajectories. Both young women were hired. In 1948 Kay married one of the project engineers, John Mauchly, after his first wife Mary died in 1946.

As Kay Mauchly later described it, when ENIAC was nearing completion, "Somebody gave us a whole stack of blueprints, . . . the wiring diagrams for all the panels, . . . and they said, 'Here, figure out how the machine works and then figure out how to program it.' " As soon as someone explained one part of the machine, however, she and the others were off and running: "Well, once you knew how an accumulator worked, you could pretty well . . . trace the other circuits for yourself and figure this thing out" (Shurkin, 188). No sooner had they learned to program a trajectory on the machine than they heard ENIAC was to be tested with a problem from Los Alamos. It was ENIAC that told the Los Alamos scientists they were facing a solvable problem as they considered the trigger for the secret hydrogen bomb.

Later, when John von Neumann had conceived of rudimentary storage capacity for ENIAC, and Adele Goldstine had written the program and machine instructions for it, it was Kathleen McNulty and Richard Clippinger who designed the necessary new hardware and created the modifications to ENIAC, which had been delivered to the Army (Shurkin, 206).

Now the widow of John Mauchly, Kay Mauchly lives in the Philadelphia area. As part of what Joel Shurkin calls her "full and active life," she oversees the use and preservation of the abundant historical material left behind by her husband at his death.

**Frances Bilas** Spence lives or once lived in the Aberdeen, MD, area. **Elizabeth ("Betty") Jennings** is now apparently divorced (Goldstine, L-5/30/84; Shurkin, 126–8, 247, 326).[13]

### Other Hardware Pioneers

Another hardware pioneer is **Margaret K. Butler** (1924– ). As a staff mathematician at Argonne National Laboratories in the early 1950s, Margaret Butler helped develop one of the first digital computers for science. She participated in the evaluation and selection of the first commercial digital computer for scientific computation in 1956-7, and prepared and implemented programs for certain reactor equations, diffusion-theory equations, reactor kinetics, spherical harmonics problems, etc., both for UNIVAC and for AVIDAC. In addition, she worked on the logical design of Argonne's GEORGE computer. Both AVIDAC and GEORGE were descendants of the computer built at the Institute for Advanced Studies by the group of which **Thelma Estrin** (see below) was a part. Although Margaret Butler now works primarily in software, her research has encompassed such hardware or hardware-related areas as image-processing, information systems, computer-system performance, and scientific and engineering applications of computers.

---

13. I have no details on **Ruth Lichterman** or **Marilyn Wescoff,** and would welcome information from readers on their background, subsequent careers, and present whereabouts.

Butler, now a Senior Computer Scientist at Argonne, has well over fifty professional publications to her credit, and has been active in the Association for Computing Machinery, the American Nuclear Society, the Association for Women in Computing, and the IEEE Computer Society. She has been honored by mention in *Who's Who in Engineering, Who's Who of American Women,* and *Who's Who in America,* as well as *American Men and Women of Science* (/14,/15). In private life she is married to James W. Butler, and has a grown son. (*AMWS*/14,/15; Butler, c.v./1983; *WBWRA;* Zimmerman, L-6/19/84).

**Margaret Cavanaugh** worked for Margaret Butler in the early days at Argonne (mid-1950s) as an assistant mathematician in the Reactor Engineering Division. As elsewhere at that time, many of the Division's women were themselves computers, using hand calculators. However, the lab soon built its own scientific computer to handle precision calculations for the engineers—a room-sized device with a memory of about 2 kilobytes (comparable to the $29.95 Timex-Sinclair of the early 1980s). Cavanaugh was assigned to program it, using switches and punched paper tape and entering a hexadecimal number for each instruction. "I used to spend my time in the bathroom at night," she recalls, "figuring out how to save one byte!" That was her introduction to computers.

After taking time out for a family, she now designs systems and software for ITT, and lives in Encinitas, CA (Zimmerman, L-6/19/ 84, "Women in Computing," 87).

In 1979 **Thelma Estrin** (c. 1925–   ) had already been a member of the IEEE for thirty years, but was still one of only six women fellows. In 1984, accepting an IEEE Centennial Medal, Dr. Estrin reviewed her distinguished career. During her thirty-five-year member-ship, she said, she had "participated in and contributed to the emergence of: biomedical engineering, neuroscience, computer engineering, the information society, and the women's movement" ("Computers," 8.4.1).

Thelma Estrin has good reason to be interested in the women's movement. During World War II she and her husband Gerald, both history majors, were recruited for the war effort. She took an intensive engineering-assistant course at Stevens Institute of Tech-nology and was placed at Radio Receptor Company. Gerald enlisted in the Army and was sent to the Signal Corps. After the war they took graduate work in electronics at the University of Wisconsin, obtaining their BS, MS, and PhD degrees in electrical engineering "in record time." Gerald was then invited to join von Neumann's Princeton team to design and build the first "stored-program digital electronic computing machine." The couple moved to Princeton. Thelma applied to RCA, but, as she puts it "hiring a woman engineer just was not something they wanted to do at that time."

This became a turning point in her life. In later years she has said that, partly in reaction to Orwell's frightening vision of a technology-ridden society, she wanted to use her training and creativity humanely, and therefore chose to do research on the electrical activity of the brain. However, as she admits in her 1984 talk (8.4.1),

> I have related only half the story. The other half is that I had difficulty obtaining traditional employment as an electrical engineer, because my commitment to engineering research was viewed as transitory and not taken seriously. This attitude towards women's careers was very prevalent in the '50s.

A friend helped her to obtain a research position in the Electroencephalography (EEG) Department of the Neurological Institute in New York City. She endured a four-hour round-trip commute to the City from Princeton, compensated partly, as she now recalls, by being able to participate in the intellectual life at Princeton. She and her husband often dined with the von Neumanns, and "Johnny" was very much interested in her brain research.

Thelma Estrin's role in the earlier days of the Princeton group is difficult to document, and may have been informal. Dr. Herman Goldstine evidently intended to list her among the project engineers in his book on the history of the project, since the book's index places her on the page on which the list occurs; however, on the page itself (306), her name is unaccountably omitted.[14]

Goldstine's praise for these engineers is worth noting. Looking back from the 1970s, he called the project's engineering "the keystone in the arch built there" (306). Like Chief Engineer Julian Bigelow and his assistant and successor James Pomerene, "the other engineers were all excellent . . . all made important contributions and should be remembered." He then proceeds to forget one.

"Not only did the group develop and build a pioneering and successfully operating machine," says Goldstine, "they also founded a line of machines" that included as direct and collateral descendants ORDVAC, the first ILLIAC, JOHNNIAC, the Swedish BESK and SMIL, the German PERM, the Russian BESM, the Israeli WEIZAC, the Australian SILLIAC, and the Michigan State University MSUDC (306–7, 307n).

Among the group's specific engineering or hardware innovations were the asynchronous mode of operation; exploitation and improvement of F.C. Williams' storage tubes; and new chassis designs (308). Gerald Estrin is specifically credited with writing "what may well be the first engineering type of diagnostic code," a possible forerunner of 1970s diagnostic codes (309). Thelma was, by virtue of her equal training and her marriage to Gerald, ideally placed for at least indirect participation in the group, and may have contributed to some of these important developments.

Thelma Estrin herself notes that she worked for a summer (probably 1953) on the JOHNNIAC under Julian Bigelow, who had previously worked with Norbert Weiner. She does not specify her technical role in the JOHNNIAC project, but says that Bigelow encouraged her to think about the computer as an ideal tool for studying the stochastic properties of brain waves.

When Israeli scientists invited Gerald Estrin to come to the Weizmann Institute of Science and build a version of the JOHNNIAC there, she became "an early member of the small engineering group and made design changes on the machine." Thus she participated in building the first computer in the Middle East. The team arrived in Israel in December 1953. Eighteen months later the central processing unit of the WEIZAC was in operation. This Dr. Estrin calls "phenomenal" because no one in Israel knew anything about computers—and all vacuum tubes had to be imported.

Returning from Israel in 1955 to find the Princeton project complete, the Estrins moved to Los Angeles where Gerald had a professorship at UCLA. Excluded from the engineering school by nepotism rules, Thelma eventually became connected with the medical school. In 1960 she organized a conference on Computers in Brain Research for the new Brain Research Institute (BRI) there. About this time began her collaboration with Dr. Ross Adey, Professor of Anatomy, ham radio operator, and fellow of the IEEE. Together they designed and established, for BRI, *the first general-purpose computer facility for brain research*, which they called the Data Processing Laboratory (DPL).

14. When queried in 1984 about this omission, Dr. Goldstine replied that it was "inadvertent" (L-5/30/84). Supporting this reply is Dr. Estrin's presence, as noted, in the index with the sole occurrence of p. 306, indicating that her name was once—as late as page proofs—on the list of project engineers praised here (first Princeton paperback printing, 1980).

Formally organized in 1961, DPL was the first integrated computer laboratory expressly designed and established to develop computer technology for nervous system research. In 1962 NIH awarded them a major grant. That same year Dr. Estrin became a member of the Brain Research Institute. Over the next seventeen years, supported by NIH grants, DPL made pioneering contributions in the following areas: analog-to-digital conversion, data acquisition and processing, signal analysis, interactive graphics, modeling and simulation, distributed processing systems, and laboratory computer systems. In 1970 she became director of DPL and Principal Investigator on the NIH grants.

In 1980 she left BRI to join the Computer Science Department of the School of Engineering and Applied Science, nepotism rules having met a well-deserved demise. From the 1970s onward she has been deeply involved in improving women's position and opportunities in engineering. Two of her three daughters are engineers.

But this is getting ahead of the story. In her 1984 talk Dr. Estrin gives a chronological review of her most important technical achievements in electronics and computers:

While at the Neurological Institute in 1951, she redesigned an electronic frequency analyzer for bioelectric signals (Estrin & Hoefer). The frustrations of trying to record these signals in an analog mode with 1950s equipment, coupled with her WEIZAC experience in digital logic design, gave her a basis for introducing digital techniques for recording the impulse firing pattern of neurons in 1961 (Estrin, "Recording . . .").

The first NIH grant to the Brain Research Institute funded Dr. Estrin's proposal to design and implement an analog-to-digital conversion system for DPL. Her system was versatile enough to digitize both the amplitudes of electrical activity and the time intervals between neuronal discharges. Researchers could make data tapes in IBM format, which could then be analyzed at the Health Science Computing Facility at DPL (Estrin, "Conversion").

In 1963, on a Fulbright year in Israel, she worked on developing automated pattern-recognition techniques, with the idea of identifying EEG patterns preceding brain seizures and finding ways of using feedback to avert such seizures. This was too big a project for one year, but resulted in a paper given at a medical electronics conference in Belgium (Estrin, "Analog").

Working with Molly Brazier in her EEG laboratory after returning from Israel (1965–70), Thelma Estrin designed an analog-to-digital system that was "on line" for the computer at DPL. Says Dr. Estrin, "We could do simple processing in almost real time, and evaluate the data as we were recording" (Estrin, "Neurophysiological"). Also during the 1960s, with the help of Robert Uzgalis, she implemented her idea for presenting the EEG as a spatio-temporal pattern. Using new advances in computer graphics, she and Uzgalis presented the EEG as a series of spatial distributions which they then photographed under computer control. Their "EEG MOVIES" made possible new clinical insights (Estrin & Uzgalis).

During the 1970s Dr. Estrin studied novel uses of interactive graphics as a brain-research tool. She and colleagues designed a "comprehensive modeling program" called NEURAL to test hypotheses about the generation of electrical brain neuron signals called spike trains. They also developed a simulation program called BRAINMAPS to aid brain surgeons. BRAINMAPS allowed surgeons to choose or map out in advance the best approach to certain structures (Estrin, Sclabassi, & Buchness, 1974, 1975; Estrin, Wegner, & Bettinger; Harper, Sclabassi, & Estrin; Sclabassi, Buchness, & Estrin).

In the early 1980s Dr. Estrin worked with **Donna Hudson,** an engineering student at UCLA, to develop an expert system for microcomputers, following formalized decision rules. She credits Hudson with designing and implementing a system called EMERGE.

Written in PASCAL, EMERGE is "an interactive computer-based consultation program" originally designed as an "expert system" for emergency-room personnel treating patients with chest pain. The EMERGE software, however, can be used to create expert systems for "any specialized areas whose knowledge can be expressed as sets of inference rules" (Hudson & Estrin).

In 1984 she summed up the technical achievement of her career as follows:

> I started my career with the desire to be a scientist and study the sources of the electric fields on the surface of the head. In order to do this I had to develop computer-based instrumentation, and my significant achievements have been in engineering. . . . My contributions have been in innovating and transferring electronic and computer technology to neuroscience and clinical medicine.

Her predictions for the future application of her work to medicine (Chapter 2) show how actively and creatively her inventive mind is still working.

In later years Dr. Estrin has received recognition for her work in medical engineering, becoming, for example, the first woman member of the board of directors of the Biomedical Engineering Society. Thelma Estrin has a number of firsts to her credit, in fact. Among the more relevant here is becoming the first woman to chair the Signal-Processing and Information-Handling Technical Committee of IEEE. At this writing, she continues her research at UCLA, work that could form the basis for further developments in artificial intelligence (Goldstine, L-5/30/84, *Computer* . . . , 306ff, 307n; *WBWRA*, 192; Th. Estrin, L-8/18/87).

Another pioneer, entering the computer field in 1950, was **Dr. Ruth M. Davis** (1928– ). Though discussed below under Software, she deserves mention here since according to Geoff Simons, she programmed three of the first digital computers, SEAC, ORDVAC, and UNIVAC I, at a time when programming could still be classed at least partly as a hardware operation. (Presumably she was part of a team in these cases, as other names are connected with the development of these three computers.) She also pioneered the first robotics program in the Department of Defense, and sponsored the famous "Shakey" robot at SRI International (Simons, 31), perhaps while serving as Undersecretary for Research and Engineering for the Department of Defense (1977–9).

*Miniaturization*

When Grace Hopper entered computing in the 1940s, a computer was a monster over fifty feet long. Now an entire computer fits on a single silicon chip, and may soon fit inside a single cell. A crucial step in this thirty-year miniaturization revolution was the invention and development of printed circuits. At least three women have been important in this vital technical area. The award-winning work of **Elise Harmon** at Aerovox and **Edith Olson** with the United States Army in the 1950s have been discussed in the preceding chapter. Suffice it to say here that Olson compared the assignment with learning how to print all the books in the Library of Congress on a grain of rice.

**Elizabeth Zimmerman** is a research scientist in electronics. During the 1960s she worked at Philco's Advanced Technology Center in Blue Bell, PA. Her work "has contributed to high-frequency germanium devices, negative resistance diodes, parametric amplifiers, and active elements used in micro-electronic circuits." Dr. Zimmerman is mentioned in Irene Peden's important survey article on women in engineering careers (273).

**Helene E. Kulsrud** (1933– ) of Princeton, NJ, will be discussed at greater length below. Here we should note that as a member of the technical staff of RCA she received an RCA Achievement Award (1961) for work on Electron Guns, and an RCA Team Achievement Award (1966) for work on the Spectra computers. Since coming to the Institute for Defense Analyses in the late 1960s she has been involved in planning the IDA system for the CRAY-I computer, both hardware and software (Kulsrud, c.v./1984).

A pertinent patentable invention of the 1960s was **Cora L. Romero**'s Automatic symbol-reproducing device. Romero listed her residence as Havana, Cuba. Her 1962 patent was considered important enough to be mentioned by Patent Examiner Patricia Ives in her 1980 article on black and women inventors (125).

## Recent Achievements

### Business and personal computers

**Evelyn Berezin** (1925–  ) of Glen Cove, NY, has received considerable recognition in business (Powell), and today considers herself more entrepreneur than innovator (L-8/1/84). Nevertheless, her contributions to computer technology, and specifically to computer hardware, are noteworthy. Indeed, in 1972 she appeared among "Leaders in Electronics" in the *New York Times* Biographical Edition (L. Sloane).

Born into a poor family in the East Bronx, Evelyn Berezin was a business major in college when a sudden job offer turned her attention toward physics. Pursuing that interest after graduation (1946), she won an Atomic Energy Commission fellowship and did all the course work for a PhD in cosmic-ray physics at New York University. Because she found no physics job when her fellowship ran out, and because the new field of electronic data processing looked interesting, she went into it.

Physics' loss was computer technology's gain.

Evelyn Berezin has always innovated in hardware. When she first went to work as an engineer for "a little company in the Syrian area of Brooklyn" in 1951, they told her, "Design a computer," and she did. The fact that neither she nor almost anyone else had ever seen a computer in those days did not stop her, and twenty years later she called the task "a lot of fun—when I wasn't terrified" (L. Sloane, 257). This company was the Electronic Computer Corporation, and the computer was the Elecom 200. The computer as a whole was not patented, only some of its separate ideas (Berezin, L-8/1/84).

In 1953, for Underwood, *she designed the first office computer.* This "very significant" computer was never marketed, as Underwood sold out to Olivetti (Berezin, c.v.; L-8/1/84).

While heading the Logic Design Section at Teleregister Corporation in the late 1950s, Evelyn Berezin led a team that designed, among other things, on-line systems for three major banks. Her major achievement while at Teleregister, however, and the one of which she is proudest, was designing United Airlines' nationwide on-line computer-reservations system. "That," she says, "was the first large commercial (that is, non-military) on-line interactive system; and the fact that it provided fast response and worked reliably for about eleven years after its installation in the early 1960s was an achievement of which I am particularly proud" (L-7/6/84). The system, installed in sixty cities, was the forerunner of similar systems for other airlines.

In the 1960s Evelyn Berezin moved to Digitronics Corporation as manager of logic design. There she designed the first high-speed commercial digital communications terminal (150 characters a second; 1962) and a digital on-line race-track system, installed in 1965 at Roosevelt Raceway. A data communications terminal sends information from one place to another, like the Telex, only faster and without errors (or with fewer errors) (Berezin, L-8/1/84). The need for computers at race tracks had been evident at least since the near-riot of 1927 at Havre de Grace, MD,[15] and the market was potentially quite lucrative. By 1946, for example, American Totalizator was charging at least $150,000 a month in rental for each of its relay computers. Multiply this by more than fifty race tracks, and by the months of the racing season, and the potential profits become apparent (Shurkin, 234ff).

The story of Redactron and Data Secretary's successful challenge to IBM's editing typewriters has been told in Chapter 4 ("She Invading . . .").

Evelyn Berezin demurs at being called the mother of word-processing. She points out that the IBM machine, though crude, did do some word-processing functions, and that other precursors also existed. Furthermore, she was the President of Redactron and not its chief design engineer. However, she did have "a lot to say" about the design of the Data Secretary, and feels it is accurate to call her one of the pioneers of word-processing (L-7/6/84). Some surveys have shown that installing a word-processor can double an office's productivity (Burkitt & Williams, 46). Multiply that by tens of thousands of offices in the United States alone, and you get some idea of the significance of the improvement.

Berezin holds six patents that, as usual, are poor indicators of her inventive contribution. I have located the following five: Data processing system, assigned to the Underwood Corporation (1959); Arithmetic device (Elecom 50-Decimal Adder-Subtractor) (1960), with **Phyllis Hersh**; Control means with record-sensing for an electronic calculator (Elecom 50) (1961); Electronic data file processor (1962), with others; Information signal-generation apparatus, assigned to Digitronics Corporation (1972).

Selling Redactron to Burroughs in 1976, Evelyn Berezin moved from computer technology into investment management. President of Greenhouse Management Corporation of Glen Cove, NY, and a director of several other corporations, she is involved in such community service organizations as the American Woman's Economic Development Corporation. In 1975 she appeared on *Business Week*'s list of the top 100 businesswomen in the United States (Berezin, c.v., L-7/6, 8/1/84; Burkitt & Williams, 46–7; "She's Invading"; Powell, 88ff; L. Sloane, 257–8).

As computers advanced, they became steadily smaller, shrinking from room-size to fingernail-size or less in one human generation. Many good small machines now compete for a market once dominated by a few giants. Among these competitors in the early 1980s were the Systems 2600 and 3005, the Vector 4, and the 5005-E Multishare produced by Vector Graphic, Inc., of Thousand Oaks, CA. Vector's co-founders were **Lore Harp** (1944– ) and **Carole Ely** (1941– ) (Emerson, 206). Lore Harp's then-husband invented the company's original product, a memory board, and continued to specialize in technical development while Lore and Carole concentrated on marketing and management. As one interviewer put it, in the beginning (1976) Lore Harp didn't know a bit from a byte (Benner, 112). However, at least one outside source suggests a role for her in the design of one of Vector's excellent microcomputers, and she herself says that she played a role in their functionality and ergonomic design, as did Ms. Ely (Zimmerman, L-6/19/84; Harp, L-8/8/87). Moreover, it was Harp who decided that the company would design a small-business computer, involving both hardware and software innovations (Benner, 114); and she must have had rather specific ideas in mind for her husband, the technical expert, to realize.

From a beginning in a spare bedroom, with a shower stall for a warehouse, Harp and Ely

15. A horse named Cockney, listed at 9 to 1 at post time, won the race. But because last-minute betting had been so heavy, the payoff was delayed, and winners received much less than the posted odds. Needless to say, they were not pleased (Shurkin, 234).

built a multi-million-dollar business with 400 employees and over-the-counter sales of $36,000,000 in 1982 (Benner, 114; Gottschalk, 1; DATASOURCES/83). Such meteoric growth is almost commonplace in the computer industry. Recently more people are even beginning to do what Harp and Ely did all along: stress the human side of computers. What is still unusual, and thus significant for the question of whether woman-owned and -run businesses would really be different, is the child care and household help included automatically in the Vector Graphic benefits package. Says Sandra Emerson, "Their company stands as a landmark in the progress of women succeeding in and changing the world of technology." Thus in addition to their technical contribution to Vector's computers, Lore Harp—and Carole Ely—can be credited with significant innovations in the important socioeconomic area of employee-benefit packages (Emerson, 206).

Born in Düsseldorf, Germany, in 1944, the older of two children, Lore Harp visited the United States in 1966, just before entering college. The visit became a permanent stay, and she entered college here instead, taking a BA in anthropology. During her undergraduate years she married Bob Harp and bore two daughters. Later she took a year of law school and eventually got her MBA, from Pepperdine. According to Jean Deleage, former manager of a European venture-capital firm that invested $750,000 in Vector Graphic in July 1979, Lore Harp is "an absolutely exceptional manager. . . ." Deleage places her in a class with James Treybig and Robert Swanson, founders and presidents, respectively, of Tandem Computers and Genentech (a bio-engineering company near San Francisco). Continues Deleage (Benner, 112):

> She has enthusiasm. She has guts. She is precise. She knows every number. She fights for everything. She has leadership, and people like to work for her. I think she's going to be the first woman to start a company listed on the New York Stock Exchange.

Lore Harp has since left the computer field (Vector folded under a direct challenge from IBM in its major market), and founded Aplex Corporation in Northern California to market her invention, Le Funelle®, a disposable sanitary device enabling women to urinate standing up, and thus to deal with filthy public restrooms (Wartik). In 1988, Lore Harp, now married to a publisher, co-chaired the National Financial Committee for Michael Dukakis' presidential campaign (Benner; DATASOURCES/1983; Emerson, 206; Gottschalk, 1; Harp, PC-7/28/, 8/8/87 and Q-7/28/87; Palmer, 24; Wojahn, 45; Zimmerman, L-6/19/84).

Briefly mentioned by Gene Bylinsky (*Silicon,* 237) as president of Condesin, a Silicon Valley company, is **Lois Stone.** In 1986 Condesin was developing a new type of memory chip.

### Picture storage

In 1978 **Amaline Julianna Frank**'s new device for storing pictures and prints in a computer was featured in Stacy Jones' *New York Times* "Patents" column. Ms. Frank's technical achievement must have been considerable, for her hardware and method have cut in half the amount of information that must be coded, allowing much faster printing. Her employer, Bell Telephone, planned to use the method for printing the yellow pages (Jones, 7/29/78: 25).

### Display technology

Important parts of any computer's hardware are its display screen and the technology that makes its operations and results visible to users. **Anne Chiang** (1942–   ) of Cupertino, CA, holds patents in this highly technical field. Born in China, she was the middle child and second daughter of a biology professor and a secondary school teacher. Both parents encouraged her to achieve. Like so many successful women scientists, she was inspired by the biography of Marie Curie. Her elementary school not only taught her exciting math, but was a nurturing environment; her six years of secondary education in an all-girls' school were also inspiring, with encouragement from several teachers. Graduating third in a class of 500, she was excused from the entrance exams at the National Taiwan University.

At this point, Anne experienced the first overt barriers to her career goals: she wanted to study math, but was guided toward chemistry instead. She took a BS in chemistry at Taiwan, then a PhD at the University of Southern California, always gravitating toward subjects requiring thought rather than memorization. Neither at college nor in her work situations has she ever had a real mentor, though several friends have proved professionally helpful. Her graduate-school professors all took a wait-and-see attitude rather than offering strong encouragement or mentoring, even though she was doing well. Having a highly supportive spouse has in fact been the major sustaining factor through the years.

In 1972 Dr. Chiang joined Xerox's Palo Alto Research Center (PARC), working at first on elastomeric polymers for optical recording. Between 1977 and 1981 she worked on the company's electrophoretic display technology, becoming a project leader. This work led to two patents (1978, 1981) for Electrophoretic display compositions and devices. It also led to her being called "the leading electrophoretic scientist" of that period (Bartimo, 94).[16]

The 1981 patent deals with what as of 1984 she considered her most important invention. It is a composition patent for a colloid suspension that demanded understanding and exploiting the complex interrelations among four separate components, and their influence on device performance. Her new electrophoretic displays are stable, and much more pleasing in appearance than liquid-crystal displays, lending themselves well, for example, to the new lap computers. By 1984 Xerox had no product plans for this technology, though Dr. Chiang has had several inquiries from outside.

Partly in frustration at this lack of interest, she switched fields at PARC in 1982, to silicon thin-film devices. Specifically, she studies laser crystallization of silicon thin-films and the application of this technology to display, printing, and electro-optics. As Project Leader in Silicon Thin-Film Devices, she had four patents pending as of 1984, for laser crystallization, semiconductor processing, and photodetectors for document input scanners.

Dr. Chiang is a much more versatile and prolific scientist than her patents reveal. At Memorex Corporation, before joining Xerox, she worked on organic photoconductors, polymer rheology, dispersion of magnetic particles and their applications to microfilms and magnetic recording media. By 1984 she had already published 2 papers on polymer rheology, 3 papers on the photochemistry of coordination compounds, 8 papers on silicon thin-film devices, and 15 papers on electrophoretic displays. *She has also proposed more than 30 inventions in these varied fields:* microfilms, magnetic tapes, elastomeric light valves, optic disks, flat-panel displays, thin-film transistors, and photodetectors.

During her first ten years at PARC, because of her early inexperience, and the company's desire to keep some important inventions as trade secrets, only two patents actually resulted. Many ideas were released for publication. Now, she says, the tempo has picked up, and her work gets more fascinating every day (Chiang, Biog. and I-9/4/84; Bartimo; *AMWS*/15).

16. The writer of this article, however, did not mention Dr. Chiang by name.

Involved in research on digital image-processing is **Dr. Marilyn Noz** of New York University Medical Center. As of the early 1980s she was Associate Professor of Radiology and a physicist in the Division of Nuclear Medicine at Bellevue Hospital Center. Although she disclaims any inventions, certain titles in her 1982 bibliography of nearly sixty professional articles suggest that Dr. Noz may be defining "invention" too narrowly (see, e.g., Noz *et al.*). Moreover, her research may well have applications for the medical uses of computers (Noz, L-9/13/82, c.v./1982, PC-9/1/83).

*Bubble memory*

In the mid-1970s, magnetic bubbles were the great hope for economical, nonvolatile computer memory. Problems have arisen, however, including low data rates and a need for extra semiconductor chips for interfacing two computer systems. In 1985 came a breakthrough in higher densities and faster data rates for bubble-memory devices. **Subhadra "Dutta" Gupta,** then at the Westinghouse Research and Development Center, was a crucial member of the breakthrough team: using silicon-on-garnet technology, they deposited the first operational transistor on a magnetic substrate. Dutta Gupta is a specialist in laser annealing, the process used to heat and alter semiconductor properties. She is now at Materials Research Corporation.

This advance in bubble-memory devices was recognized as one of the year's top 100 innovations for 1985 by the editors of *Science Digest* (Howard, 50, 51).

**Lynn Conway** (1938–   ) spent her early career as an innovator in computer hardware. After studying physics at M.I.T. and taking her MS in electrical engineering from Columbia (1963), she joined IBM's research staff in 1964. At IBM she made "major contributions to the architecture of ultra-high-performance computing systems" (M. Marshall *et al.,* 104). Unfortunately, this was a non-360 system, just as IBM was standardizing its products under the 360 protocol. At Memorex (after 1969) Conway was processor architect and head of a team that developed a small business computer. At the working-prototype stage, Memorex cancelled the whole line. At Xerox PARC after 1973 she was initially charged with creating a combined optical character-recognition-facsimile system. Her design worked, though the machine was a behemoth. This project triggered her interest in Very Large Systems Integrated computers (VLSI). She eventually became known as an expert in the system design of VLSI chips. In 1975 she was working on a special-purpose architecture for image-processing. Bert Sutherland, who joined Xerox at that time, notes that it was an innovative architecture, even though he axed it as not something Xerox could use (Corcoran, 48).

However, Lynn Conway's major contribution to computer technology to date, and the one for which she received *Electronics* magazine's 1981 Achievement Award, lies in the area of automated design. It is a design *methodology* that is proving to be a stimulating and powerful solution to the excruciatingly complex design problems posed by VLSI systems. She evolved the method, in collaboration with Carver Mead of Cal Tech, who had long advocated similar ideas, and who served as a consultant on her Multi-chip project at Xerox (Wiener, L-10/1/87). They finished the draft of their book just in time for her prototype course at M.I.T. in fall 1978. The essence of the method is as simple as it is revolutionary: maintaining a certain regularity in design structure. Also involved is the fullest possible cooperation and communication among all parties concerned in a design problem.

Using the Defense Department's computer network, ARPANET, Conway sent designs and design critiques flying between M.I.T., computer experts and student designers all over the country. Hewlett-Packard Research in Palo Alto fabricated the chips, and students had actual chips in hand only six weeks after the end of the course. In 1980 the

Conway and Mead's textbook appeared, embodying the course, the method, and the results of their experience, their now-classic *Introduction to VLSI Systems.*[17]

The Conway-Mead method would not be patentable, and some would not call it an invention. In fact, it was a formulation and rationalization of the ideas of many others into a useful system of uniform and workable design rules (for example, they made sure their specifications could be met by most existing foundries), and rules for evaluating system performance, and the rapid dissemination of the system throughout the industry. But this uniformity was just what was needed at the time.[18] The method has already influenced the design of such commercial systems as the Motorola 68000 and the Intel iAPX-432, as well as a complex experimental floating-point processor designed by Digital Equipment Corporation and M.I.T.'s public-key encoding IC (40,000 transistors). Moreover, if the United States succeeds in meeting Japan's challenge on fifth-generation computers—the so-called smart machines, knowledge-based systems, or KIPS—it may be through this level of cooperation, now known as *collaboration technology.*

During the late 1970s, Conway maintained close contact with the developing VLSI community, sponsoring contests, giving talks, and helping Doug Fairbairn of her research team start a magazine (*Lambda,* which became *VLSI Design*). By the early 1980s, however, Xerox was losing interest, even though many researchers now relied on the chip-fabrication service. DARPA and PARC arranged for the Information Sciences Institute of the University of Southern California at Marina Del Rey to take over the service, now called MOSIS. Conway could now turn to other fields.

One such new interest was knowledge engineering,[19] designing and building expert systems. Recruiting Mark Stefik from Stanford, Conway now started her own knowledge-systems group at PARC. This group eventually created LOOPS (*LISP Object-Oriented Programming System*), an integrated collection of four programming paradigms, which she introduced by the route created for the chip-design method: short intensive courses and competitions like "Truckin'." This led to the Colab project, as Conway and Stefik conceived the idea of knowledge-engineering workstations.

PARC researchers became more involved with knowledge engineering, but even as this

---

17. Says Conway of the book, "In electronics, a new wave comes through in bits and pieces. Usually, after it has all evolved someone writes a book about it. What we decided to do was to write about it while it is still happening" (M. Marshall *et al.,* 105).

18. Of interest here are the comments of another inventor (and entrepreneur) in VLSI design, a former colleague at Xerox: Both the weakness and the strength of the Conway-Mead approach was that it did not take process variations or design-rule variations into account. As just noted, in order to develop the methodology and to prove experimentally its ability in the classroom, they used the design rules and standardized methods that many foundries—including Xerox—had on tap. Thus Xerox and others were able to process the designs and provide the turnaround times required for short courses.

The real impact is as follows:

*In academia,* the result of the multi-chip project was badly needed. VLSI design is not a theoretical question. Rather, it requires an experiential and prototyping capability to teach. Through the Multi-Chip Project a new type of engineering lab was constructed, one that was sorely needed.

*For industry,* that is for system houses, it indicated a way not to be dependent on chip (standard) or on semiconductor houses for everything. System prototypes could be generated with internal control and in reasonable time frames. "In truth," says Pat Wiener, "without this type of methodology as a guideline, no VLSI chip of reasonable yield could be designed. Chip yield then and now is the basis of the Japanese challenge. Chip yield relates back directly to cost. In short, this design approach sooner or later had to be recognized by [many more] designers. This book provided the sooner, and ensured that new graduates were exposed to this concept" (Wiener, L-10/1/87).

19. This term was coined by Stanford University computer scientist H. Penny Nii (Corcoran, 50).

happened, Lynn Conway was thinking about moving on. The VLSI success had its price. The fight to push through and defend that then-controversial project had brought Sutherland under fire and alienated Conway from PARC's top management. At about this time Robert Cooper, director of DARPA, offered her a position in Washington, DC, managing DARPA's still undefined supercomputer project, or strategic computing initiative. She accepted.[20]

Her task, and that of her successors, was and is to develop computers capable of symbolic reasoning at speeds a thousand times faster than those used in the military systems of the early 1980s (*AMWS*/15; Corcoran; Feigenbaum & McCorduck, 48–54f; M. Marshall *et al.;* Raloff, "U.S. Challenges").

Even with the DARPA job behind her, however, Lynn Conway found it difficult to get a suitable position in Silicon Valley. She has found a challenging niche in academia, instead, as Professor and Associate Dean of the College of Engineering at the University of Michigan in Ann Arbor. There, since 1985, she wants to be thought of as an "elder or mentor." In her spare time, she jogs, canoes, and goes river-rafting. Her family consists of two Siamese cats.

Let it not be thought, however, that she has given up either controversy or innovation. She continues to stir up interest and support for collaboration technology throughout the university, and she is one of three principal investigators for a project called Expres, which will enable researchers to collaborate and send a group proposal to NSF by electronic mail. She is also working with Richard Volz in the Robotics Laboratory at Michigan, developing "teleautomation," a combination of remote-control and automated devices to capitalize on the strengths of both people and machines (Corcoran).

A strong-willed and intensely private person, Lynn Conway is the daughter of a chemical engineer, who found that she excelled in science and math in high school. A revealing quotation for her personality is the one she stuck in a picture frame in her Ann Arbor office: "Play for more than you can afford to lose, and you will learn the game" (Corcoran, 50).

## Hardware and theory

A Massachusetts woman who may one day be numbered among a new wave of computer pioneers is **Susan Huhn Eustis** (1940–   ) of Groton and Lexington, CT. Daughter of a sales representative for voting machines, Susan went to public schools, including a tiny high school in Chautauqua, NY, and then on to Barnard, where she studied math and government. For the next fifteen years she sold and serviced mechanical voting machines, working with great success for the same company as her father. She first came to public attention in 1976 when, after teaching herself electronics and computer software and programming, she invented a compact electronic voting machine that represented a major advance in the field (see Applications, below).

Now married and the mother of two children, Susan Eustis divides her time between family and basic research in computer theory and design. She has become fascinated with what she calls psychic structure and has written a book about it. Her more recent computer inventions rely on this structure, in line with one of her major goals, which is to make computers more accessible to ordinary people. Unlike most inventors in this field, who tackle this problem through software, Eustis wants to modify the hardware. Computers

---

20. Another woman inventor among Xerox's employees at the time put it more bluntly: in her opinion, Conway left Xerox because she couldn't get promoted and, as a small-project leader, she found her projects never got anywhere (Wiener, PC-10/2/87).

began as linear in their processing. She visualizes a computer that works in three dimensions. As of 1982 she had applied for patents on two important inventions: a new central processing unit for computers (in collaboration with her husband), and chips for scrambling or encrypting data and programs. Both chips would be compatible with any computer. One patent, for a Compact data-processing system usable by relatively unskilled people, was granted in 1984, and appeared in Stacy Jones' column (Feb. 25: 30).

Susan Eustis has received more help from people outside than inside the computer industry. Dr. John Livingstone and other Boston physicians; Ellen Curtiss, Research Director at Arthur D. Little, and Alma Tryner, also of Arthur D. Little; her patent attorney; and John Tierney in public relations have been supportive and encouraging (J. Dietz; Eustis, I-6/22/82; "Non-Lazy Susan," 15).

## A Lady Edison

**Patricia ("Pat") Wiener** of La Honda, CA, might be called a Lady Edison of modern electronics. During several interviews in her home high above Silicon Valley in 1983, and in the Palo Alto office of her company, Page Automated Telecommunications Systems, Inc. (PATSI), in 1987 and 1988, we discussed some of her more than thirty electronics inventions. These inventions have led to three patents granted and a fourth pending, at least two of which are likely to be quite important. The others never reached patent stage because of company mergers and resultant chaos, or employers' keeping them as trade secrets. Her most recent work has produced inventions at or near the disclosure stage.

She considers herself basically a scientist, interested primarily in cancer and in matter and energy. Fascinated by the onset of disorder in nature as well as the causes of its wonderful symmetry, she looks for simplicity, whether in engineering or in physics. In technology and invention her overall goal is to create more meaningful capacity for specific tasks with less cost, less volume, and more performance. She wants computers to be user-friendly "in depth," and to bring to individuals the real performance and help they seek when they buy their machines.

The youngest child and only daughter of Russian immigrant parents in New York, Patricia Wiener seems to have taken after her brilliant father, who could do math problems in his head faster than she could do them on her slide rule. Insatiably curious, as a small child she frightened her mother by riding buses all over the city to see what was going on. She had her first business at the age of six, selling bottles of invisible ink made with her chemistry set, given to her by an older brother.

She is very much lefthanded,[21] and describes herself as visually oriented, block-oriented, and nonverbal in her thinking. She has demonstrated musical talent. This probably came from her mother's side: her great-uncle was flautist to the Czar before the Revolution. Her mother wanted her to be a pianist, and lessons began at age six with a prominent teacher. Patricia showed such talent that the teacher wanted her to devote herself to music, aiming for Carnegie Hall by age ten. She agreed, on condition that she could be a composer or a conductor. She had begun theory lessons at eight, and composed her first music before age ten. When the teacher said, "Women aren't conductors," she snapped back, "OK, in that case I guess I'll be a physicist." She knew women could be physicists because one of her heroes was Madame Curie.

---

21. When a teacher, aided by her mother, tried to change her handwriting from left to right, she became so angry that she deliberately broke her right arm in a skating fall.

This ambitious child had responsibility thrust on her early: her father died when she was thirteen, and she had to take care of the family fortunes because her mother had lost her voice in shock at her husband's death.

Pat went to an all-girls' high school, where her scientific ability soon became evident. When she built a Wilson cloud chamber, and started to teach herself calculus, a teacher took an interest in her and began tutoring her in the subject after school. In fact, three teachers were important influences during this period. In particular, the woman who headed the biology department understood her interdisciplinary approach to science, and encouraged her. The male heads of chemistry and physics also took an interest, and often stayed after school to help her with her self-education and building projects. At graduation she won the school's science prize. She also won Honorable Mention in the Westinghouse Science Talent Search. Her project had brought her consideration for top honors, and she had briefly dreamed of going to M.I.T. However, without the fellowship accompanying the top award, this was impossible.

Pat then entered Brooklyn College, where she worked her way—and helped support her mother—by dishwashing. She married at 18 and had her first child at 20. Following her husband to his changing jobs (he is now vice-president for engineering at a software-based Silicon Valley start-up company), she attended several colleges, and herself worked for several different companies. Her first job after high school (and after an eight-hour interview that ranged as far as her theories on matter and energy) was with Bell Labs, where she got hands-on trouble-shooting experience. She was one of the first women hired there in her field; indeed, one of the first hired that young in any professional engineering capacity.

Moving to California with her husband, she worked for several small California companies, and then for the Research Division of Librascope of Glendale. Some of her work there was pioneering work in magnetic-core memory for computers. This invention, which made possible the flexible and real-time operations of the Whirlwind computers, is usually—and properly—credited to Jay Forrester, head of the Whirlwind Project at IBM (Shurkin, 258, 278). However, Pat Wiener was pursuing this work independently at roughly the same period.[22]

Returning to school, she got her degree in physics from UCLA in 1958. Not wishing to work on atomic bombs or any offensive-war-related technology, and interested in electron transfer across membranes, she entered graduate school in biophysics with Dr. Jean Bath. Interrupted yet again by her husband's move to IBM in New York, she continued her graduate studies at NYU and Columbia while consulting to several electronics companies in research and development. During this period she developed a single-transistor, feedback, thin-film-based vumeter for a small instrumentation firm. (Such devices make sound levels visible for monitoring, as on stereo equipment and laboratory instruments, and measure distortion.)

During this period she also experienced discrimination so extreme that it now seems funny. At one company where she was working, the president walked through, saw her,

---

22. Magnetic-core memory is magnetic storage in which the data medium consists of magnetic cores. Each core is shaped like a miniature doughnut as small as 18/1000 of an inch in diameter (about the size of a grain of salt). The cores can be polarized in a binary manner, On-Off. Most computers made since 1975 use semiconductor chips as the memory medium instead (Edwards & Broadwell, 432–3). Pat Wiener was also involved in research in the ferromagnetic materials used in these devices. As an indication of her competence and reputation in this area, Dr. William Rubinstein, then setting up a new facility to make magnetic core memories, asked her to become general manager. Since the job would have taken her to Brooklyn and caused stress in her family, however, she turned it down.

and said, "I won't have women in my lab," whereupon she was fired, even though her supervisor said she was one of the best engineers he'd ever had. Columbia University refused her a fellowship because they said she would have children and leave. Apparently no one had noticed that her family of four children was already complete. They gave the fellowships to two 25-year-old single men who then did poorly; even so, when she asked again, the answer was no.

In 1965 she joined ASEA Electric, a Swedish firm. Her job was to bring their electronics technology up to state of the art. She did just that. After only two years, however, Sweden entered a recession, and all the electronic work was pulled back home. Then she joined Information Displays, Inc. (1967–9), where she invented entirely new techniques for generating vector-based graphics; in other words, a symbol generator. Hers was a high-speed, high-quality system still state of the art in the 1980s. Her inventions cut the size of the system from 14 boards to one, and the time required from 4 microseconds to 125 nanoseconds per stroke. It was also inexpensive to make—$150 to $180, compared to the $2500 price tag of comparable systems. Eight potential patents were involved here, but the company did not want to file for patents or even write papers on the nature of the system.

In 1969, with four others, including her husband, Pat Wiener founded a company called On-Line Computer. She became Vice-president for Research and Development and Engineering. The plan was to generate a computer on a silicon wafer. She invented a new memory system—which she did patent (1973). This was a string-based high-speed memory system. *It was one of the first systems on a single silicon chip.* She and her colleagues also put out one of the first 128-line asynchronous time-division multiplexers for the time-sharing industry (1970). That system was scheduled to be field-tested at Dartmouth (where time-sharing systems were integrated into the teaching) when the recession of 1970–1 hit.

In 1971 she formed her own company, Anarm, Inc. Its charter was to build vector-based graphic systems using the memory system she had developed at On-Line. Anarm produced a series of systems for the Federal Aviation Administration, and her invention was considered a third-generation display system. However, as the recession deepened, boding ill for start-up companies, she decided to sell the technology.

In 1973, as Senior Program Director at Otis Elevator Research & Development (a new organization), she developed a new electronic positioning system for elevators. Fifteen potential patents were involved. However, just as her system went into field testing, United Technology took over Otis, and in the resulting chaos she left—as did the patent attorneys. Her system, which was thoroughly documented, did nevertheless eventually reach the market in modified form.

In 1977 Pat Wiener became Manager of Advanced Technical Concepts at the corporate level for Xerox, Stamford, CT. She was responsible for advanced electronics, reporting to the group monitoring all the R&D being generated in the company. At Xerox she invented a new technique to allow multifunctional devices on a single large silicon substrate, and developed the concepts for the related products.

Pat Wiener has actively sought, through papers and other means, to move the industry towards viewing silicon wafers as a potential substrate, a substitute for printed circuit boards, and as a base for monolithic, integrated systems—a possible solution to many performance problems encountered since 1977. This idea is now being recognized as a potentially viable answer for speed, power, and interconnect questions, and as a significant asset in the development of parallel computing systems.

This work eventually gave rise to two patents: for an Integrated stylus array for printer operation (1982), and for an Image bit structuring apparatus and method (1983). In 1979 Xerox reorganized its research and development group. Because of the resulting chaos, and her husband's health, she left Xerox. She intended to develop this latest technology and form her own company, but decided the interacting technologies needed some further work. In 1980 she took a position as head of Memorex's new VLSI Center, where she was to develop some of the new interconnect and processing technology. Finding she would not be able to pursue this development, she left.

By 1981 she founded a small R&D company, returning to full-time consulting while writing her business plan and seeking venture capital to exploit the technology she had developed at Xerox. (A license agreement with Xerox was being negotiated.) The technology is still sufficiently different, even visionary, that competitors working in the same old direction should find it difficult to switch gears.

In 1983–5 Pat Wiener's consulting and research took her in new but related technology directions: large-area integration, wafer-scale integration, new modular systems. An important paper she presented at the IEEE's spring Computer Conference (Comp/con) in 1980, "New Technologies for Low-Cost Automated Telecommunications-Based Office System Products . . . ," reveals some of this new thinking. This work opens up whole different technological approaches to products to be developed—a three-dimensional or wafer-oriented approach, or what might be called a systems point of view rather than a transistor point of view. The approach calls for more flexibility in system-substrate size. Pat Wiener evaluates density (and merit) in VLSI systems not according to how many transistors are packed in per unit area, but according to how much *function* is gained per unit *volume*. In 1985 a patent related to that work was filed for by a client company.

In 1986, after multiple setbacks in her effort to raise major venture capital,[23] Pat Wiener tried a new tack: developing funding for her new technology through the Federal Small Business Innovation Research (SBIR) program established to meet the Japanese challenge. A multiple award-winner in this program, her company (PATSI) has developed new concepts in modeling techniques for the automation of complex systems (enterprises), using and expanding current AI methodologies.

Pat Wiener's latest inventive activities include automation of design, factories, and telecommunications systems; robotics; and the development of new devices using new materials to enhance density (as defined above), speed, and overall system throughput. Her current research includes super-conducting materials as well as overall systems design using these various technologies and new materials. She intends to "bootstrap" a major corporation using the avenue opened up by this SBIR program, developing multiple aspects of interactive technology, including both hardware and software.

Pat Wiener points out, in fact, that dividing computer technology into hardware and software, as in this chapter, is arbitrary and not always useful for her work, or for any of the best computer work. She usually works with entire systems, involving innovations in

---

23. Getting money from venture capitalists is difficult for anyone who is pushing the state of the art in high technology, says Wiener. Belying their name, venture people are not interested in ideas that will require extensive development, preferring the safer product that is nearly market-ready. Moreover, really new technology is the work of mavericks. Venture capitalists with enough vision, or enough technical understanding, to risk leaving the herd for such a maverick are rare indeed. Women seeking venture capital have a particularly difficult time. They face the double credibility gap of doubt that they can handle (*a*) high technology and (*b*) high finance.

both hardware and software. Nor, she points out, can software really be designed in a vacuum; it must always reflect, and its algorithms be effectively married to, the hardware on which it will run, or which it will cause to operate, if the system is to reach optimal function.

Pat Wiener has four successful children—two daughters (a choreographer who is artistic director of her own dance company, and a computer systems analyst) and two sons (an architect and a PhD candidate in biophysics). "Women don't strut about anything but life itself," she says. "Their creation is life itself, and any other creation beside that is almost irrelevant." Perhaps that is true, but society seems at present to take another view. In any case, she has as many of these "irrelevant" creations to her credit as all but a very few inventors in this book. She also has a positive attitude, enormous energy, broad interests, and, as the foregoing makes clear, many talents—science, invention, business, music. At one point, she even wanted to be a sculptor. Finally, she wants to help other women who would like to enter science and engineering, but she intends to help in what she believes the most effective way, by serving as a successful role model (Wiener, I-8/13/83 and 7/31, 8/6, 8/11, and 10/31/87; L-10/1/87; PC 10/2/87).

Patents that came to my attention as the book was going to press are two hardware inventions from 1979, both patented by European teams: **Henriette Magna** of Paris, France, and three male co-patentees (her name alphabetically last) received a U.S. patent for their Thermosensitive data-carrier designed for the recording of information and a Method of recording information on such a data-carrier. The patent is assigned to Thomson-Brandt of Paris. A team of Russian inventors protected their Semiconductor structure sensitive to pressure with an American patent. Of the nine co-inventors here, three are women: **Elena M. Kistova, Natalya I. Lukicheva,** and **Elena S. Jurova,** all of Moscow. The inventors assigned their patent to the "GIREDMET" Institute, USSR. These three women are also part of another Soviet team that patented a semiconductor invention earlier in 1979: Fabrication of a heterogeneous semiconductor structure with composition gradient utilizing a gas phase transfer process. This patent, assigned to the same institute, bears the names of 15 inventors, five of them women: in addition to Kistova, Lukicheva, and Jurova, they are **Alla N. Lupacheva** and **Marina A. Konstantinova,** both of Moscow.

## SOFTWARE

"For reasons no one has been able to fathom," says Joel Shurkin in *Engines of the Mind* (230n), "the best early programmers in computer history, going back to the **Countess of Lovelace,** were women."[24] Perhaps the reasons are less unfathomable than Shurkin thinks. Computer programming, after all, is essentially communicating with the computer.

### *Software Pioneers*

I have discussed most of the ENIAC programmers under Hardware, since at that stage programming was scarcely separable from machine design. However, at least one of these

---

24. *Early* is a key word here, for, as already noted, computing has had its own takeover, partly analogous to what happened in agriculture. As the importance of programming became clear, and as society became more dependent on computers—or realized that it would become so—men began to force women out of higher-level programming roles.

women went on to steadily more significant achievements in software. **Betty Snyder Holberton** collaborated with John Mauchly and others in the original programming of BINAC, the first stored-program computer in the United States and very nearly the first in the world (the British EDSAC was completed first, but BINAC ran its first program sooner; Shurkin, 239). She also worked with Grace Hopper in writing one of the first assembly languages—which Shurkin calls "a significant achievement" but relegates to a footnote (231n).

Grace Hopper credits Betty Holberton with a still more significant achievement: "I think the first step to tell us we could actually use a computer to write programs was Betty Holberton's 'Sort-Merge Generator,' " which she developed when the computer establishment was still saying that automatic programming would never happen. As we have seen, Dr. Hopper herself brought automatic programming to realization, presumably building on Holberton's work. Dr. Hopper points out that this Sort-Merge Generator also contained the first version of a virtual memory (Wexelblat, 9).

In 1959 Betty Holberton was Chief of the Programming Research Branch, Applied Mathematics Laboratory, at the David Taylor Model Basin in Washington, DC, where she helped to introduce Grace Hopper's A-2 compiler. She became a member of the Short-Range Committee that produced COBOL. Though not an official member of the final editing committee for the language, she was "present during most of the work in Boston, and contributed to the editing." For the ultimate refinements in COBOL demanded by CODASYL in early 1960, Betty Holberton co-chaired the editing committee and did all the actual work (Webstein was abroad). She also did the final editing to prepare the Short Range Committee report for the Government Printing Office (Sammet, L-9/10/84; Sammet/Wexelblat, 211–2).

As of 1978, when she served as a discussant for the COBOL session at the first conference on the history of programming languages, Betty Holberton was at the National Bureau of Standards (Bemer, 133; Shurkin, 230, 231n; Wexelblat, 9, 14, 202, 211–2).

Another pioneer credited by Grace Hopper is **Adele Mildred Coss.** In November of 1952 she created her Editing Generator, a program that would build an editing routine. Since editing was a besetting problem in the early days of programming, this innovation was a major advance. As then-Capt. Hopper described it, "Millie Coss went to work and in November of 1952 had operating a program which took in the format of the file, the format of the records, and the format of the output printed document, and produced the coding to get from one to the other. It also did a little bit of summing. . . ." According to Dr. Hopper this was the "ancestor of most RPGs [Report Generators] and those particular techniques" (Wexelblat, 12).

Both Coss and **Margaret Harper** (1919–   ) were members of the Programming Development Group created at Univac by **Grace Hopper** in the fall of 1951. Harper evidently joined in 1952 (*WWCF/*'63–'64: 93). Although Dr. Hopper deservedly receives major credit for the work of this group, which created a library of tested programs, looking toward the development of what she called "executive routines," compilers that are "general and intelligent," she credits the group as well: "This work is necessarily group research, and this account cannot be published without citing those members . . . primarily responsible for the achievement of these results" (Hopper, "Compiling Routines," 3; Hopper & Mauchly, 1252).

**Margaret Harper** worked for the Auerbach Corporation as a programming analyst in the 1960s, later taught at the University of Pennsylvania, and is now retired. Daughter of an auto dealer and a musician/homemaker, Margaret was educated in public elementary schools, a private high school, and the University of Pennsylvania (chemistry). She is a first child (older of two children). Both parents encouraged her to achieve, and she found role models in books. She is ambidextrous. However, she considers herself lacking in artistic talent, and does not really consider herself an inventor, saying, "I was best at development" (Q-6/15/84). Development, of course, is intrinsic to invention (as opposed to discovery) in this field as in others, and often requires innovation. Moreover, Grace Hopper does not give undeserved praise. Margaret Harper appears in *Who's Who in the Computer Field, 1963–'64*, and *Who's Who in Computers and Data Processing* for 1971 (vol. 1).

According to a Silicon Valley informant, **Ida Rhodes** did pioneering work in computers, but further research will be needed to establish the details.

Mentioned above as a hardware pioneer, **Margaret K. Butler** of Argonne Laboratories also did important early work in software. In the 1940s, as a Junior Mathematician in the Naval Reactor Division, she did some of the computational work underlying the *Nautilus* submarine prototype. She is a fellow of the American Nuclear Society. In the 1950s she wrote the initial AVIDAC floating-point package, plus "many utility and test routines, and several mathematical subroutines for the early Argonne-built machines" (Butler, c.v., 2,3). During those same years she developed computer programs to solve engineering problems and to aid in the design of nuclear reactors. As Associate Mathematician and Head of the Applications Programming Section of the Applied Mathematics Division at Argonne from April 1959 to November 1965, she (c.v., 2)

> directed the development of the AMD Program Library . . . and Argonne's first computer operating system. . . . In addition [she] actively participated in the preparation of an operating system for the IBM 704, the documentation of the CDC 3600 subroutine library, and the design and development of many computer programs, primarily in the reactor and high-energy physics areas—e.g., one-dimensional multigroup diffusion, Monte Carlo spark-chamber analysis, accelerator design studies.

During the late 1960s and early 1970s Butler did computer science research in the areas, among others, of image-processing and reactor physics computation. In the 1980s, as a Senior Computer Scientist, she carried out benchmark studies for evaluating laboratory computers, and did research in computing technology forecasting, the measurement and evaluation of computer performance, computer history, and the application of computers to scientific and engineering problems. She also worked on standards for computers and information processing.

Margaret Butler may be best remembered, however—as she would prefer—as creator and director of the National Energy Software Center (also called the Argonne Code Center). There she operates a clearinghouse for the worldwide exchange of computer programs for peaceful uses of nuclear energy, and develops world standards for computer technology (*AMWS*/14,/15; Butler, c.v./ 1983, L-5/25/84; *WBWRA*).

**Ruth M. Davis** (1928– ) is another pioneer. She is one of the six women in the computer field chosen for a profile by Rina Yarmish and Louise Grinstein in the early 1980s. Born in Sharpesville, PA, she received a PhD in math from the University of Maryland in 1955. Noteworthy in her varied employment history are positions as a mathematician at the National Bureau of Standards (1950–1), as head of the Operations Research Division at the David Taylor Model Basin (1955–61), and as Director of the Lister Hill National Center for Biomedical Communications (1968–70). During much of this time she held concurrent academic positions, at American University, the University of Maryland, and University of Pittsburgh. From 1972 to 1977 she directed the Institute for Computer Sciences and Technology at the National Bureau of Standards.

Ruth Davis has received many honors, notably for purposes of this chapter, the DPMA "Man of the Year" Award, which she became the second woman to receive (1979), and an honorary degree from Carnegie-Mellon University. She has also been chosen for high government positions, serving as Undersecretary for Research and Advanced Technology in the Department of Defense (1977–9) and Assistant Secretary for Resource Applications at the Department of Energy (1979–83).

Sources are uniformly admiring—and almost uniformly vague—about Davis's specific technical accomplishments. Carla Schanstra, writing in 1980, says that Dr. Davis has worked as a "computer programming innovator, university teacher, and government official" (73). Yarmish and Grinstein credit her with work in "systems design, information systems, and artificial intelligence," adding that modular programming is her area of interest in information-systems research (26,27). According to *WBWRA,* she worked for the Navy Department on nuclear reactor design in 1953 (180). Strangely, it is left to a British writer, Geoff L. Simons, to specify that "Ruth Davis programmed three of the first digital computers (SEAC, ORDVAC and UNIVAC I) . . ." (31). Fuller answers exist in her scientific and technical papers, numbering more than 100 by 1980 (Schanstra, 73; *WBWRA*)—and are still being created, for Ruth Davis is a computer pioneer who is still very much with us (Schanstra, 72, 73; Simons, 31; *WBWRA; WWA/1977–8; WWAW/* 1983–4; Yarmish & Grinstein, II: 26–7).

Two pioneers whose work predates modern digital computers, and who thus could scarcely have foreseen all the future applications of their work, are **Margaret K. Odell** and her collaborator, Robert C. Russell. In 1918 and 1922 they patented a surname-encoding process that captures sound-alike misspellings or other variants of a surname. This so-called Soundex System is used in data storage and retrieval for the U.S. Census. Not precisely software, but incorporated into software, it also has applications to sorting and searching functions of the computer (Knuth, 391–2).

## *Programming Languages*

### *Assembly language*

Before natural-language programming languages existed, as already noted, programmers had to talk to the computer in assembly language. **Elaine E. Boehme** was a member of the group, with A.L. Samuel and others, who first used the open addressing system in constructing address tables for an assembly program (Iverson, 153n).

### *FLOW-MATIC*

As already noted, **Grace Hopper** created FLOW-MATIC—which is to say that she initiated the effort and led the group that developed it—and is generally, if somewhat imprecisely, given credit for COBOL. This attribution may be imprecise, as her role in the actual development of the language was indirect, but is not without basis, since of the three languages giving rise to COBOL, one was FLOW-MATIC and another was a slight modification of it.

At least two other women receive major credit for computer languages: **Jean E. Sammet** of IBM for FORMAC, and **Adele Goldberg** for Smalltalk. But that is getting ahead of the story. Several women were involved in the creation of the many computer languages of the 1950s–'70s: FORTRAN, COBOL, LISP, BASIC, PL-1, and Ada, among others. Let us look briefly at these team inventors.

*FORTRAN*

**Lois M. Haibt** jointed the FORTRAN group in fall 1955, to plan and program Section 4 of the FORTRAN I Compiler. She is one of 13 co-authors of the paper documenting the FORTRAN language ("The FORTRAN Automatic Coding System," Backus *et al.,* 1957), and John Backus thanks her as one of the "principal movers in making FORTRAN a reality" (Wexelblat, 35, 40, 43–4, 57–8). Herman Goldstine lists her as one of Backus's colleagues (341n).

**Grace Elizabeth ("Libby") Mitchell** joined the FORTRAN group in 1957 to write the Programmer's Primer (IBM, 1957). This primer was so well written that scientists and engineers not trained in computers could easily use a machine. Later, aided by **Bernyce Brady** and Leroy May, Mitchell programmed the majority of the new code for FORTRAN II (1978). Backus considers Mitchell another of those "principal movers" in the development of FORTRAN (Wexelblat, 40–4). Jack Laffan notes several specific and noteworthy achievements: First, Mitchell developed a common statement in FORTRAN allowing programs to share data. Second, she developed methods for allowing subroutines in FORTRAN. Third, and most important, about 1955 she developed the first operating system for the IBM 704. *This was the original and forerunner of all future operating systems.* Later, she worked with Stan Poley on the Symbolic Optimal Assembly Program (SOAP) for the IBM 650 Computers. This was a program that took advantage of the geometry of the machine to speed up its operation.

Jack Laffan, who knew her well, supplies some personal details and impressions. She loved cats. She married a writer. "Everything she touched turned to gold," says Laffan—not that she became wealthy, but that her work was rich and useful. Mitchell worked at IBM from 1954 until mandatory retirement in the early 1970s, with a lucrative retirement package. At this writing she lives in Fishkill, NY (Laffan, I-5/6/85).

Another contributor to a version of FORTRAN, FORTRAN-H was **Bev Dales.** Coming to IBM with previous experience at Bell Labs, she worked with Jack Laffan in 1964–5. Dales did the arithmetic scan or parsing on FORTRAN-H. Laffan remembers her as one who could transform anything into a work of art and who drew beautiful pictures on blackboards with pastels. Now living in Hyde Park, NY, she is a full-time painter (Laffan, I-5/6/85).

*COBOL (1959)*

The roles of **Grace Hopper** and **Betty Holberton** in the creation of COBOL have already been described. Several other women played major roles in this team invention. On the Short Range Committee charged with creating the language was **Mary K. Hawes** of Burroughs Corporation, who had originally suggested the need for a common or universal business language (Bemer, 130). At the Committee's first meeting in June of 1959, Mrs. Hawes presented her ideas on data description, and was appointed to chair the task group on it. She made the main report on that aspect of the language development to the CODASYL Steering Committee (Backus/Wexelblat; Sammet/Wexelblat, 202, 204, 208–9).

**Jean Sammet,** then working at Sylvania Electric Products, was an important member of the Short Range Committee. She chaired the task group on procedural statements, and was one of the six people who produced the detailed language specifications for consideration by the Committee. This group worked for two solid weeks, sometimes around the clock, producing full specifications by November 7. The same six people became the editing committee, now chaired by Miss Sammet. The committee worked together in Boston, again often around the clock, and produced the final specifications in December 1959. (Sammet, DB, L-9/10/84; Sammet/Wexelblat, 209–11).

Four other women who contributed to COBOL were **Gertrude Tierney, Nora Taylor, Deborah Davidson,** and **Sue Knapp.** Tierney served with Jean Sammet on the Statement Task Group, and on the six-person group producing the language specifications and doing the editing. She was therefore one of the major contributors to the development of COBOL. Nora Taylor, though not officially a member of the editing committee, contributed to the final editing process in Boston. Davidson and Knapp worked with the Short-Range Committee (Sammet/Wexelblat, 209–11, 214).

## LISP (1960)

**Phyllis Fox** of M.I.T. wrote the LISP Programmer's Manual published by Research Laboratory of Electronics in 1960. This she was eminently qualified to do, since she had participated in the initial development of the language in the 1950s. As of the early 1980s, LISP was the language of choice for artificial intelligence applications of computers (Bernstein, 91; Wexelblat, 181, 183–4).

## JOVIAL (1962)

Several women were involved in developing the language that became known as JOVIAL. JOVIAL was a special-purpose system-programming language developed for a large Air Force system called SACCS. Since SACCS was to be developed from scratch, it would need not only new computers, but new programming techniques and operating systems, as well as a new language. In 1958 a project called CUSS (Compiler and Utility System for SACCS) was formed to create the needed language and compilers. The following women were project members: **Patricia Weaver, Donna Neeb,** and **Patricia Metz.** Only Jules Schwartz, the team's manager, had any previous experience with languages or compilers (Wexelblat, 371–2).

While CUSS was getting started in New Jersey, some California members of the initial language-investigation team started their own project (CLIP) to develop a SACCS language. The California project included at least one woman, **Ellen Clark.** Clark later co-authored the "CLIP Translator" (*Communications of the ACM* 4: 1[Jan. 1961]: 12–22). CLIP eventually stopped work, however, and left the field to CUSS (Wexelblat, 371, 372, 396).

## JOSS (early-1960s)

**Leola Cutler** and **Mary Lind** worked with J.C. ("Cliff") Shaw in producing nearly all the software for the language called JOSS (JOHNNIAC Open Shop System) in the early 1960s. One computer expert called JOSS "a truly remarkable achievement on the JOHNNIAC," from which "it wasn't hard to extrapolate what could be done with a modern machine." JOHNNIAC was a Rand Corporation computer (Wexelblat, 496, 509).

## PL-1 (mid-1970s)

**Bernice Weitzenhoffer** was an IBM employee in the early 1960s. With her extensive experience in FORTRAN and COBOL programming, as well as in compiler development, she was a natural choice to become one of IBM's representatives on the Advanced Language Development Committee formed in October of 1963 with members from SHARE, a user's group composed mainly of scientists. The language they created was PL-1 (Wexelblat, 553, 580).

*SNOBOL (1963)*

**Laura White Noll** implemented the symbol table and storage-management routines for the group creating the original SNOBOL language at Bell Labs in the early 1960s. The first usable implementation of SNOBOL took only three weeks, "a matter of pride within the group" (Wexelblat, 603). All told, Ms. Noll provided about four months of programming support to the development of this language (Wexelblat, 611).

She evidently stayed with the project as it expanded to create the later languages SNOBOL 2, 3, and 4—the last of which, according to Ralph Griswold, one of the project leaders, was not completed till 1968, and should really be considered a separate language. Noll may have been involved in the development of software for document-formatting with SNOBOL-4. Certainly she wrote the in-house technical report on it (Noll). Citing this report, Griswold notes that as SNOBOL-4 developed into a much larger project than any of the earlier SNOBOL languages, "Emphasis was placed on document-preparation systems, and we became leaders in introducing this technology to BTL. . . . We also developed a considerable amount of our own software . . . , including document-formatting languages with interfaces to line printers and photocomposition equipment" (Wexelblat, 611).

*SIMULA (early 1960s)*

Discussing the development of the SIMULA languages at a 1978 conference, Kristen Nygaard and Ole Johan Dahl do not mention any women involved in the original creation process. They do, however, credit a "very successful" implementation of SIMULA 67 on a DEC-10 computer by a Swedish team in 1973–4. An "important member" of that team was **Ingrid Wennerstrom.** In fact, Nygaard acknowledges that "SIMULA exists as a living language with an active user community [because of] a series of good implementations by some exceptionally competent teams" (Wexelblat, 477, 481).

*APL (mid-1960s), IDAL, DDT, etc.*

**Helene E. Kulsrud** (1933– ) has helped design several compilers and computer languages, debugging languages and tools. Taking her BA at Smith College in 1953 (*cum laude,* mathematics) and her MS at the University of Chicago in 1955 (astrophysics, thesis supervised by S. Chandrasekhar), she has worked in both academia and industry. She has been an assistant astronomer at Princeton University, head programmer for the Educational Testing Service, a member of the technical staff of RCA Laboratories, a research associate at Yale University, and a senior staff member of Aeronaut Research Associations Princeton, Inc. As of 1984 she was a member of the research staff and Deputy Head of the Institute for Defense Analyses in Princeton.

Since 1975 Kulsrud has been in charge of the design and implementation of a new language, IDAL, for system programming and mathematical applications. In the course of this work she produced a portable compiler that runs on the CEDC 6600 and 7600 and CRAY I computers, producing code for these machines and for the Data General ECLIPSE computer. IDAL is the first language to combine "true vector capabilities with recursive structures needed to manipulate bit fields." As of 1984 she was continuing to work on the IDAL compiler and to optimize the code.

Kulsrud considers IDAL her most important invention. However, she has also created a Meta-Compiler system that produced a compiler for the earlier APL, and she has designed an interactive debugging language for the Spectra system (at RCA in the 1950s–'60s), a graphic language (at Yale in the 1960s), and a special language and programming system for simulating electron optical devices (at RCA). While a Visiting Consultant at Lawrence

Livermore Labs in 1979, she worked on the LLL interactive debugging language called
DDT (Kulsrud, Q-1984, c.v./1984; *WWC&DP/* 1972–3; *AMWS/*15).

## Ada

One of the most-discussed computer languages of the 1970s and early '80s is Ada, named
after Ada, Countess of Lovelace, sometimes called the first computer programmer. Few
women are generally mentioned as contributing to this higher-order, standardized
language intended to be used by all the military services. However, Professor J.N. Buxton
of the University of Warwick in England notes that among the dozen or so people who
gave him advice and criticism on the draft of the STONEMAN report (one of a series of
reports on the requirements for Ada programming environments, and thus a contribution
to the development of the language) was **Patricia Santoni** of the Naval Ocean Systems
Center in San Diego, California (Buxton, L-11/11/80).

Dr. Santoni clarifies her contribution as follows (L-2/9/81):

> I was interested in and followed the work done to derive the requirements for the new DoD
> programming language and subsequently to develop the language itself, but I had no direct
> participation in those processes. What I assisted Professor Buxton with was the develop-
> ment of the requirements for a facility and a set of tools to support the use of the language.

In other words, she was involved in developing the *software-engineering environment* for
Ada.

Dr. Santoni is, in fact, one of the pioneers in the development of such environments, and
herein evidently lies her major contribution to the field of computing. As she explains,
people first instructed the computer in binary or octal (base-8) numbers. But as more and
more complex things were demanded of computers, we had to find better ways of
communicating with them. First came "more human-oriented mnemonics . . . which a
person could write, and another program (called an assembler) could translate into the
right numbers." The first effective breakthrough in this crucial area, as noted, was **Grace
Hopper**'s work.

But these mnemonics, or assembly languages, were still "too close to the way the
machine worked and too far below the level at which a person thinks," as Dr. Santoni puts
it. The next breakthrough was the so-called higher-order language (HOL), of which
FORTRAN was the first, in 1954. Among the early HOL's were ALGOL, PL-1, PASCAL, and
COBOL, each accompanied by its own compiler, or software to translate back into the
numbers demanded by the computer. The great value of COBOL—and thus of Grace
Hopper's contribution, notes Dr. Santoni in passing, was that, unlike most of the existing
HOL's, which were oriented to scientists thinking in terms of sine, cosine, and square
root, etc., COBOL was oriented toward the businessperson (and other ordinary users),
thinking in terms of files of data and formats for reports.

For several years these languages proliferated, until researchers realized that software
problems were becoming too complex to be solved by just another HOL. In 1968 a group
of people (including John Buxton) coined the term *software engineering* to express the
idea that software systems, like houses and bridges, need to be carefully engineered, their
requirements thoroughly understood, and the system architecture and design developed
according to sound principles. During the 1970s software development proceeded along
these lines, with the result that the newer languages are accompanied by all sorts of
programming or programming-development tools supporting all the ordered developmental phases. (Assemblers and compilers are examples of these tools.)

By the mid-1970s, many of these programming-development tools existed, but they were scattered around in universities and companies, and not generally accessible. In 1975 the Naval Ocean Systems Center in San Diego began gathering large numbers of these tools into a single facility. Dr. Santoni's work on this project involved investigating a great many methods and tools and researching how to make them all work together to support a complex and long-lived system life cycle. This kind of facility is now called a *software engineering environment.*

Dr. Santoni was invited to the original conference on the Ada support environment. Thereafter she met with and reviewed documents by both Buxton and his predecessors. By 1981 she was a member of the Air Force team charged with evaluating and choosing among three competitive designs for their Ada Programming Support Environment. This Air Force Ada environment was based on the work reported in STONEMAN, the document on which she worked most closely with Buxton.

In addition to her work for the Air Force—which qualifies as a contribution to the invention of a programming language as understood today—Dr. Santoni in 1981 was working on "building a more comprehensive support environment for two Navy activities."

Ada has "some very good features," according to Dr. Santoni, "especially concerning the careful writing and control of data structures and the ability to better write real-time programs." But Ada's real advantages, she says, have less to do with any features of the language itself than with the fact that it has the full support of the Department of Defense, industry, and the universities. Thus enough Ada compilers and other language-specific tools will be created, and the widespread benefits of this complete development will serve as an example for software development in general.

At least two other women participated in the development of Ada: **Jean Sammet** (see below) and **Dr. Edith McMahon** of Falls Church, VA. As of 1981, Dr. McMahon was head of the Computer Sciences Corporation design team, one of the three competitors being evaluated for the Air Force Ada contract (Santoni, L-2/9/81).

Ada's continuing importance is underscored in a 1986 *Fortune* magazine article called "The Growing Gap in Software." According to its author, John Paul Newport, the computer revolution is in trouble because software lags behind hardware. In the particularly troubled field of military software, Newport outlines three promising areas, of which the most important, at least in the near term, is broader use of Ada. Its advantages, both military and civilian, are multiple: according to a Lockheed expert, it can help companies achieve 50% gains in productivity; it prevents some common programming errors, and makes programs written in it easier to understand and to modify. Best of all, perhaps, Ada makes it easier to re-use programs (that is, to incorporate all or part of an existing program into a new one, either unchanged or only slightly modified) because it is compatible with so many different kinds of computers (Newport, 138).

## AXES

In the 1970s **Margaret Hamilton** of Cambridge, MA, published an AXES Syntax Description, which would suggest that she and her co-author, S. Zeldin, were the originators of the AXES language. They described AXES in an earlier publication as "A Specification Language Based on Completeness of Control" (see Hamilton & Zeldin refs.). And since Hamilton is first author, she was probably the major contributor to the project (Shirley, L-6/4/84). She seems to have founded Higher-Order Software of Cambridge, MA, to market her language and software innovations (see below).

*Smalltalk*

An important language invented in the 1970s and still being developed today is Smalltalk. Created by Alan Kay and **Adele Goldberg** (1945–  ) at Xerox's Palo Alto Research Center, this is a major advance in programming technology. Indeed, Smalltalk is a perfect example of the arbitrariness of my chapter divisions, for the name refers, not just to a language, but to an entire programming environment and a whole new dynamic mode of communication, including programming tools, re-usable frameworks,[25] and a graphical interface.

BASIC (*B*eginners' *A*ll-purpose *S*ymbolic *I*nstruction *C*ode), designed as a simple and powerful programming tool for students in all disciplines, was the first major attempt to make computers more accessible to ordinary users. LOGO was a powerful further step in that direction. Smalltalk is intended to be both simpler and more powerful than LOGO. From an artificial-intelligence perspective, Pat Wiener notes that Smalltalk reflects the way people think, allowing them to make deductions from incomplete information, and thus actually helps them to think as they program.

Kay originated the language in 1971, as an outgrowth of his dissertation work on the FLEX machine, which incorporated ideas from SIMULA and SKETCHPAD;[26] Adele Goldberg joined the project in 1973 and managed it from 1979 onward. She has worked with the Learning Research Group team to develop and polish the language and system to make it useful to professional programmers and ready it for public distribution. She created the 1980 version of the system called Smalltalk-80®, used in government, industrial R&D, and commercial software houses, as well as in some universities. Dr. Goldberg is now President of ParcPlace Systems, a Xerox spinoff that has become an independent international company selling software based on the Smalltalk object-oriented technology.[27]

Though occasionally used in school systems, and though some of its pedagogical materials were tested in junior high, Smalltalk is by no means a children's language. It is intended for people of all ages, and for high-level work. Indeed, it is already proving useful in artificial intelligence applications (see below).

In a 1981 article, Dr. Goldberg described Smalltalk as the software part of Alan Kay's

---

25. *Re-usable frameworks* are sets of program definitions that describe an application area, and therefore can be plugged in without change (re-used) for several tasks or problems in that area.

26. SIMULA was a programming language developed in 1967 by Norwegian researchers to provide the special support needed for computer simulations. SKETCHPAD was one of several significant research systems developed at M.I.T.'s Lincoln Laboratory in the 1960s. It was a computer-aided design system in which graphical objects could be directly manipulated. Alan Kay's FLEX machine was a design for a personal computer combining SIMULA-like simulation and SKETCHPAD graphical objects (A. Goldberg, L-6/8/88).

27. In this context an object is "a package of data and procedures that belong together. Specifically, all constants and the contents of all variables are objects." Kaehler and Patterson liken an *object* in Smalltalk to a *record* in PASCAL, "but much richer and more versatile" (148). Being object-oriented, then, is getting work done by "sending messages" to objects. Proponents of the language say this is not only the way the brain naturally works ("We perceive the world around us as made up of objects, and our brains arrange information into chunks"), but also good thinking ("When we put all the information associated with a question node into an instance of class Question, we didn't just clean up the Animal program. We allowed ourselves to think of that body of information and action as a single unit, namely, a question node"; Kaehler & Patterson, 148, 159).

Smalltalk-80 is a trademark of ParcPlace Systems.

personal computing vision, the Dynabook.[28] Dynabook is a notebook-size computer powerful enough to serve as reference library, architectural tool, musical instrument, and intelligent companion, all in one. It would take in and put out information in quantities approaching the capabilities of the human sensory system. Its visual output would surpass newsprint in quality and clarity, and its audio output would also be of high quality. There would be no appreciable lag between command and performance. In sum, the goals for the system were, and are, flexibility, resolution, and response.

In Dr. Goldberg's own more technical terms, it is a "hand-held, high-performance computer with a high-resolution display, input and output devices supporting visual and audio communication paths, and network connections to shared information resources" ("Introducing," 18). Dynabook has not yet been built, but as a San Francisco book reviewer noted in 1987, it "continues to seduce the dreamers of Silicon Valley." At the 1984 press conference introducing the Macintosh, reporters asked Steve Jobs what he wanted to do next. He said, "I want to built a Dynabook" (Markoff, D-1).

Innovative thinking is part of the job for Adele Goldberg. In addition to her work with Smalltalk, as a Xerox research manager she led projects to design other programming languages and environments, interface technology, and user applications. She has taught and held research positions at Stanford University and at the Universidade Pontificia Catolica in Rio de Janeiro. After taking a PhD in information sciences at the University of Chicago in 1973, she joined Xerox. She has lectured widely, including trips to Japan and China, and has published two books on Smalltalk. She has also edited a book on the history of personal workstations, based on a conference sponsored by the Association for Computing Machinery (ACM).

Dr. Goldberg has served as President (1984–6) and National Secretary of the ACM, as Editor-in-Chief of the journal *Computing Surveys,* and as a member of the Board of Directors of the American Federation of Information Processing Societies. She also has numerous article publications to her credit, and she edited a special issue of *Byte* magazine devoted entirely to Smalltalk (August 1981). In her introductory article to that issue, she recognizes other members of the Xerox Learning Research Group, among whom were three women: **Peggy Asprey, Laura Gould,** and **Diana Merry.**

Adele Goldberg has compiled an impressive innovative record at Xerox, first fully within the corporate fold, and more recently in a semi-independent subsidiary. At the time of her interview she had been in charge of her own independent company for just a few months; but hers will be a name to watch in the future of computer technology (Edwards & Broadwell, 430–1; A. Goldberg, biog., I-9/1987, L-5/24, 6/8/88, "Introducing," "Programmer . . ."; LRG, 3ff; Markoff; ParcPlace Systems press releases [1987]; Wiener, L-10/1/87, I-10/13/87).

*Real-time programming*

**Mary S. Pickett,** a staff research scientist at General Motors since 1971, has designed real-time programming languages for the Company ("Robot Abstraction").

*COBOL, FORMAC, Ada*

It seems fitting to close this discussion of women's contributions to programming languages with the work of **Jean E. Sammet** (1928–   ) of IBM. Not only was she a key

---

28. John Markoff of the San Francisco *Examiner,* in a 1987 review of three biographies of Steve Jobs, the Apple Computer pioneer, calls Dynabook "a vision . . . unfolded by Adele Goldberg and Alan Kay, Xerox Corp. computer scientists, in the early 1970s" (D-1).

member of the team that produced COBOL, as already noted, but more recently she has been involved with the Defense Department's important language Ada. Although she had no part in the language design, Miss Sammet served as one of the Defense Department's "Distinguished Reviewers" for Ada, is a member of the Ada Advisory Board (a Federal Advisory Committee organized by the Department), and organized and served for almost two years as the first chairperson of the Association for Computing Machinery's AdaTEC (now called SIGAda) Policy Subcommittee (Sammet, DB, L-8/28/87). From 1978 to 1984 she was the "divisional focal point on Ada" for the IBM Federal Systems Division, and the interface on Ada with the rest of the corporation. From 1981 to 1987 she had the corporate responsibility for IBM's activities in standardizing Ada.

When Rina Yarmish and Louise Grinstein chose to profile six noteworthy women in computer development, Jean Sammet was one of them.[29] They point out that she has influenced academia as well as industry, creating some of the earliest credit-bearing courses ever offered in programming. In addition to her early mathematics teaching while in graduate school at the University of Illinois, she has taught programming courses at Adelphi College, Northeastern University, UCLA, Mt. Holyoke, and IBM's Systems Research Institute. Her own academic background shows a BA from Mt. Holyoke in 1948, an MA from the University of Illinois in 1949 and an honorary ScD from Mt. Holyoke in 1978. In 1977 she was elected to membership in the National Academy of Engineering for "contributions to the development of higher-level programming languages and leadership in computer science and engineering" (Sammet, DB/1983).

Jean Sammet's name has become firmly associated with programming languages in general through her categorization and descriptions of and writings on programming-language history (see, e.g., her 1969 book, *Programming Languages: History and Fundamentals,* her numerous articles on high-level languages, and her various "Roster of Programming Languages" articles dating from 1967), and through her role as organizer, general chair, and program chair of the highly successful conference on the History of Programming Languages (HOPL) held by the ACM's Special Interest Group on Programming Languages (SIGPLAN) in June 1978 (see Wexelblat).

Miss Sammet's employment history, too, shows a focus on programming and programming languages. At Sperry Gyroscope (1955–8) she organized and supervised the company's first scientific programming group. While at Sylvania Electric (1958–61), she first headed the section on MOBIDIC Programming, then served as Staff Consultant for Programming Research (Sammet, DB/1983).

In 1961 she joined IBM, where she has stayed. At IBM she "started and directed work on language projects, including use of restricted English as a programming language, development of a natural language for mathematical problems, and concepts of user-defined language systems." As Programming Technology Planning Manager for the FSC Programming Laboratory of the Federal Systems Division (1968–74) she was "involved in software planning" as well as consulting and research on programming languages. As of 1983 she had returned to the position of Programming Language Technology Manager, in order to concentrate again on programming languages.

Jean Sammet prefers not to rank her several major contributions to computing. Perhaps her most noteworthy innovation for the purposes of this chapter, however, is her initial

---

29. Miss Sammet notes that the Yarmish & Grinstein discussion of her career contains some errors. Evidently she did *not,* as they report, design and develop the COBOL compiler for the MOBIDIC large-scale computer; nor, as of August 1987, has she made professional contributions to the theory of high-level programming languages (Sammet, L-6/28/87).

creation of the language called FORMAC at IBM in the early 1960s, and her management of the group that did the full development of the language. The key technical person in the group, who led much of the detailed work on FORMAC, was also a woman, **Elaine Bond.** Miss Sammet published "Introduction to FORMAC" in August 1964 (Sammet & Bond) and "FORMAC—An Experimental *FOR*mula *MA*nipulation *C*ompiler" in the same year (Bond *et al.*) In 1965 she received IBM's Outstanding Contribution Award for this achievement. The language has at least two firsts to its credit: it was the first practical language for manipulating formulas,[30] and it was the first example of adding formula-manipulation capabilities to an existing programming language (FORTRAN).

Both science and engineering rely heavily on the nonnumeric calculations usually taught in algebra and calculus. These problems are just as dull and time-consuming—and just as well suited to computer handling—as numerical computations. But early computer languages could not deal with this kind of work, variously called formula manipulation, symbolic mathematics, symbolic computation, nonnumerical mathematics, and nonnumerical computation. FORMAC's importance lies in its major contribution to the practical harnessing of computers for this whole new area of routine computing (Sammet, DB/1983).

Another highly significant achievement of Miss Sammet's is her founding of the ACM's Special Interest Committee on Symbolic and Algebraic Manipulation (SICSAM) in 1965. In so doing, she helped create a whole new branch of computer science.

Miss Sammet has received numerous honors recognizing these achievements. In addition to the honorary doctorate, the IBM award, and the Engineering Academy membership already mentioned, she has received Mt. Holyoke Alumnae Association's Centennial Award for "major contributions to the technological advance of programming science" (1972), was one of the first two honorary members of the national computer honorary society, Upsilon Pi Epsilon (1975), and was selected as honorary chair for the 2nd symposium on Symbolic and Algebraic Manipulation (1971). In 1985 she received the ACM Distinguished Service Award for "dedicated, tireless and dynamic leadership in service to ACM and the computing community; for advancing the art and science of computer programming languages and recording their history" (ACM Press release, Dec. 1985).

She has served as President of the ACM (1974–6), and from 1979 to 1987 was Editor-in-Chief of ACM *Computing Reviews* (DB) and *ACM Guide to Computing Literature.* She is also a member of the Board of the Computer Museum—where she herself, it is to be hoped, will find an honored place (Sammet, DB/1983, L-9/10/84 and 6/22/84, Q-6/22/84, Sammet entries, Bibliog.; Yarmish & Grinstein).

### Recent Software

As computers advance, software becomes more sophisticated. One of the most urgent needs in software is enabling different brands of computers to "talk" to each other. At least one of the many recent translator programs is the work of a female. In the early 1980s, at 17, **Viviane Baladi** of Switzerland won two important competitions with her system for making two different computers compatible. One was the Swiss competition "La Science

---

30. Before FORMAC, Miss Sammet points out, there were software programs usable for manipulating formulas, but otherwise there existed only one very small experimental attempt at a language (ALGY) with only a few commands, that evidently never got used outside the group that developed it (L-9/10/84).

Appelle les Jeunes"; the other was the European division of a Young Scientists and Inventors contest sponsored by the Philips Company (SWIA).

For some vital applications, two entirely different *types* of computer must be linked together. In 1982 **Harriett B. Rigas** of Washington State University received the Society of Women Engineers Achievement Award for contributions in electrical engineering and computer technology. Specifically, she "enabled the automatic patching system for an analog/hybrid computer." Rigas has a PhD in electrical engineering from the University of Kansas, and has chaired the Electrical and Computer Engineering Department at WSU. She is on the IEEE Board, and a member of Sigma Xi and Tau Beta Pi (SWE/1982).

Women have, in fact, been involved in virtually all areas of software development and innovation. Dianne Littwin, publisher of Wiley Professional Software, for example, noted late in 1983 that she receives unsolicited programs from women as well as from men. The women's submissions were less numerous, probably because unsolicited software has traditionally been in the engineering field (Russ Adams, 82).

By the 1970s **Margaret Hamilton** of Cambridge, MA, was creating what became known as higher-order software. Hamilton and her co-author S. Zeldin published "Higher Order Software—A Methodology for Defining Software," in 1976. Earlier, they originated AXES, "A Specification Language Based on Completeness and Control" (see Hamilton & Zeldin). **Steve Shirley,** founder of the English firm F International and herself an innovator, describes Hamilton's work as follows (L-6/4/84):

> The work in which she has been engaged is concerned with the provability of software and is of practical application, not merely relevant to small programs. The technique has been automated so it can be used by people who are neither mathematicians nor professional software people; the work also seems to extend to the high-level specification of systems (i.e. what-has-to-be-achieved) as well as down to the automatic generation of code (the intricate details of how-to-do-it).

Margaret Hamilton is evidently an entrepreneur as well as an inventor. In the early 1980s her firm, Higher-Order Software, relocated to Somerville, MA.

**Betty J. Mitchell** (1926–  ) of St. Albans, VA, is an excellent example of a woman innovating in the practical uses of computers in industry. From 1951 to her retirement in 1982 Mrs. Mitchell worked for one company, Union Carbide, but in an amazing variety of areas. Beginning at the Niagara Falls plant in emission spectrography and X-ray spectroscopy, she moved in 1963 to the Chemicals and Plastics Division in West Virginia, where she became an analytical supervisor, then Computer Center Supervisor, and finally Quality Assurance Laboratory Department Head. She was the first woman to hold a technical manager's position at any of the three Union Carbide Kanawha Valley locations. In 1978 she went to the Agricultural Products Division, where she became Technical Services Laboratory Department Head and, just before retirement, Plant Laboratory Department Head.

At the Chemicals and Plastics Division in 1963–75, she specialized in statistical approaches to solving technical problems in the laboratory and in the plant as a whole. As part of this approach—and what brings her to this chapter—she wrote "numerous computer programs of an innovative nature" (B. Mitchell, biog/1983).

Mrs. Mitchell has several technical publications to her credit, one of which won the Society for Applied Spectroscopy's Journal award for the best paper published in 1966 (B. Mitchell & Hopper). Professional-society memberships include the American Crystallographic Association and the Instrument Society of America (during pertinent employment), and the Society for Applied Spectroscopy and the American Society for Quality

Control (current). She appears in the 1969 edition of *Personalities of the South* and in the 1983–84 edition of *Who's Who in West Virginia,* and was honored as "Ms. Quality of 1972" by the Charleston section of the American Society for Quality Control.

Born Betty Jeanne Janiak in Niagara Falls, NY, she took her bachelor's degree in mathematics from the University of Michigan—and married—in 1948. Mrs. Mitchell has also studied at the Illinois Institute of Technology, the Rochester Institute of Technology, and the University of Texas, among others. Except for two years for childbearing—two daughters, now grown—she pursued a professional career. Retired from Union Carbide (1982), she lives in St. Albans, near Charleston, WV (B. Mitchell, biog/1983).

In 1984 **Joyce Wrenn** (c. 1936–  ) was Director of entry-system software publishing for IBM in Boca Raton, FL. In that position she had the final say about personal-computer software to be sold under IBM's label.

A winner since she won a high school scholarship at 13, Joyce Wrenn joined IBM in 1961 as a programmer in the Advanced Systems Development Division. Every project she worked on for the next twenty years, in various programming and management positions for the company, was successful. In 1970 she invented the "application customizer" (see Applications, below). She enjoys her work, considering her present directorship the most challenging, and the most fun, yet. She avoids retirement talk, and plans to live well into her nineties (Zientara, 57, 59–60). Further research into her programming years would probably reveal other noteworthy inventions.

Another software generalist is **Stephanie ("Steve") Shirley** (1933–  ), born Vera Stephanie Buchthal in Dortmund, Germany. Her father was a lawyer. Brought to England in 1939 to escape the Nazis, she was educated there, partly in private primary and grammar schools (all-girls' schools except for the initial primary years). Unable to afford university, she enrolled in evening classes at Sir John Cass College in London, where she got a BSc in mathematics. She worked first at the Dollis Hill Research Station of the Post Office (the British Post Office then managed telecommunications as well as the mails, and thus had a scientific and technical arm), where she did quite sophisticated work with calculators. Among her assignments were projecting traffic patterns on transatlantic telephone cables and designing futuristic electronic telephone exchanges. On the latter project she worked with T.H. Flowers of Bletchley fame.[31] Thus she gained both electronics experience and exposure to the possibilities of computers. She also continued to take classes, working toward a MSc in probability theory.

In 1959 she married Derek Shirley and soon thereafter left the Post Office for a new company called Computer Developments Limited (CDL). There, she helped to design the 1301 computer and became a skilled programmer and software engineer. By 1962, however, she realized how "irksome" it was to be an employee, and left to found Freelance Programmers. Her son was born the next year. She incorporated in 1964, becoming Freelance Programmers, Ltd. Part of her idea was to employ people like herself who wanted to work but did not fit into the ordinary industrial scheme of rigid work hours and days. Evolving from the systems division of Freelance Programmers was F2—a group which, among other things, designed systems for clients. Together FPL and F2 make up F I[nternational] Group/PLC, one of England's largest software houses. F.I. has

---

31. Bletchley Park was the site of Britain's secret World War II code-cracking operation, using the COLOSSUS computers. Led by the brilliant, doomed Alan Turing, this group cracked the Germans' *Enigma* code and thus played an important—perhaps crucial—role in the Allied victory. The COLOSSUS computers were also undeniably the world's first digital electronic computers—fed by paper tape and capable of absorbing data at the then-unheard-of speed of 5000 characters per second (Simons, 33–5).

had Danish, Dutch, and American subsidiaries, serving all of Scandinavia, the Benelux countries, and the United States.

Steve Shirley does not claim to be technically very innovative; yet by her own statement she has worked "as a senior member of teams involved in filter design using time series, statistical checking of random number sequences, real-time programs to control un-buffered peripheral equipment, simulation of traffic patterns, systems software for an ICL mainframe and some early work—very minor—on the design of electronic telephone exchanges." She would describe herself, she says, as a lateral thinker (L-5/22/84). Elsewhere she has said that her ability consists of being able to see *both* the forest and the trees. Geoff Simons notes that she designed software for her employer, CDL, in the early 1960s (29). Whatever may be history's verdict on her technical achievements, her business innovations and successes are indisputable. Alvin Toffler described F. I. as a "seed of the future" in his book *The Third Wave*. By 1981 her company could claim "a million-pound order, a large number of satisfied clients, and an annual turnover of around 3 million pounds" (Simons, 138). By 1983 F.I. had a net worth of nearly $2 million, with annual revenues approaching $10 million, a 66% return on capital employed, and sales growth of 26% a year ("National Academy . . . ," 1). By 1987 turnover had doubled yet again.

Steve Shirley is doubtless far too modest in evaluating her work, particularly her inventiveness. Not only did she devise a scheme for transferring corporate equity to freelancers, and pioneer new techniques for management of outworkers that "may well prove to be a pattern on which other companies, like Rank Xerox, will model their own schemes" (T. Lloyd, 68), but when a job assigned to one of her freelance panel goes badly wrong, as did an early job for an oil company, she can step in. Quickly learning FORTRAN, she got the work moving in a week and two weekends.

Public honors and professional and industry recognition have certainly come to her. Steve Shirley was a founder member of the British Computer Society, serving as Vice-President, 1979–82. In 1980 came her OBE (Officer, Order of the British Empire), for services to industry (Simons, 187). She is a member of the Department of Industry's Electronics and Avionics Requirements Board and chair of its Computing and Communications Committee, the first of a series of public appointments. She is ranked among the top 100 business people in the UK, and the top woman in business, and thus plays a major role as a model for younger women. In 1987 she became a Freeman of the City of London (T. Lloyd, chap. 4; "National Academy . . ."; Shirley, L-5/22/84; Simons, 5, 29–30, 136–7, 187).

Let us now look at some important types of special-purpose software and their women inventors.

### Inflight recorders

A striking example brought to my attention by Steve Shirley is the Concorde's black box, created in 1966–7 by a team of freelance programmers, all female (L-8/4/87). The project manager was **Ann Leach** (now **Moffatt**) (L-6/2/88).

### Timesharing

**Mary Elizabeth Kroening** of Computer Systems Design Group in San Diego, CA, with a male collaborator, designed and implemented an interactive timeshared operating system for an IMSAI 8080 microprocessor or personal computer. This was CHAOS II, and the year was 1979. CHAOS II differs from CHAOS I primarily in "the implementation of the directory system." Using standard, low-cost components, CHAOS II is intended to provide a "congenial, interactive, multi-programming environment," usability with several programming languages, a system of access to files and directories that allows file protection and sharing, and a logical, hierarchical system structure that is easy to implement and maintain. Developed on a Hewlett-Packard system, CHAOS II has a user interface modeled after Bell's UNIX system (taking into account the limitations of the IMSAI 8080 processor and floppy disks). It is written mainly in assembly language, and can be used with either a hard disk or a 16-bit microprocessor. Its primary programming language is BASIC, but as noted it fits with other languages as well. According to its developers, it is "a practical tool for the novice user, as well as a powerful tool for the systems programmer" (Kroening & Levinsky).

*Educational software*

One of the most exciting areas in recent programming history is educational software. **Dr. Ann Piestrup** (c. 1942– ), founder of the Learning Company in Menlo Park, CA, is a noteworthy innovator in this field. Dr. Piestrup's commitment to better education grew out of experience. Leaving a convent at 21, before her final vows, she taught in the black ghetto of Buffalo, NY. There she became frustrated with available teaching tools in an area where 20% of her fourth-grade students were functional illiterates.

To improve her credentials after arriving in Silicon Valley (San Francisco Peninsula, CA), Piestrup took a PhD in educational psychology from the University of California at Berkeley. She then sought more effective ways to teach children. She wanted to use electronic games—not the violent, space-invaders variety, but her own innovative games. Beginning with a small grant from Apple Computer, she opened the Learning Company in Portola Valley in 1979. Three of its early games, designed for the IBM-PCjr and for Peninsula schools, were JUGGLES BUTTERFLY, BUMBLES GAMES, and BUMBLE PLOT. The first demonstrates such basic concepts as right, left, above, and below, to three- to six-year-olds. The second upgrades math skills for children aged four to ten; and the third introduces graphics and enhanced math skills for students aged eight to 13.

The company incorporated in 1982 and moved to nearby Menlo Park in 1983. As of 1985 it was marketing 15 educational games for various computers. These games present basic concepts of math, algebra, spatial relations, logic, and computers in a warm tone, using whimsical stories, bright colors, cheery music, and humor. ROBOT ODYSSEY I, for example, takes players to Robotopolis, an underground city of robots. The object is to find a way out of the city. To progress through the treacherous maze, players must learn to build more and more complex robots. In doing so, they learn abstract concepts of logic such as *and, or,* and *not,* and begin to see how to design integrated circuits.

As of 1985 the Learning Company was a privately held corporation employing 28 people as researchers, technicians, educators, and marketing specialists. In fiscal 1982 it had sales of $1 million, and Gene Bylinsky in his book *Silicon Valley High Tech* called it "top of its field." Dr. Piestrup has come a long way from the convent (Buckland; Bylinsky *et al.,* 117; Perl, PC-1980).

Another contemporary educational innovator from the San Francisco Peninsula is **Joan Fischer Targ,** a junior high school educator in Palo Alto, CA. For her classes she wrote a computer program requiring students to teach others. This sure-fire educational technique has not often been applied to computer-aided instruction, since the idea has usually been for the computer to help the students. Targ reports that the girls using her program always say that the helping part is one of their favorite things about it, whereas the boys seldom say this (Zimmerman, ". . . Computing," 79f).

Considerably earlier is **Dorothea Pye Simon** (1913– ), mentioned by *Discover* magazine as part of a team that created computer programs to solve physics problems by "thinking them through" in different ways (June 1981: 41–4). Mrs. Simon herself downplays her contribution here, describing her role as that of "inventor's helper—running experiments, sometimes serving as naive subject, analyzing and summarizing data and protocols, editing manuscripts, etc. Ideas for the experiments and interesting interpretations of the data were largely my husband's" (D. Simon, L-7/10/84).

Without detracting in the least from the innovative work of Nobel laureate Dr. Herbert A. Simon, or of **Dr. Jill Larkin** (see below), I must note that in work as complex and collaborations as longstanding as this, it is hard to disentangle individual contributions. The resulting invention or innovation is truly a team effort (see Simon *et al.,* "Individual," "Tale," "Expert," "Models").

Dorothea Simon's independent innovation lies in the field of computer-aided instruction (CAI; Simon, L-7/10/84):

> The innovation that I myself developed . . . grew out of work I did as a research assistant at the Learning Research and Development Center of the University of Pittsburgh in the early 1970s. The project was to develop CAI lessons in language arts, specifically for spelling. The problem was to figure out how to teach spelling and test for competence on a computer with no audio capability.

Since in learning experiments subjects retained words and digits in short-term memory almost equally well whether the material was presented audially or visually, she and her colleagues wondered,

> Why could we not just show the children a word and ask them to spell it, expecting the same kind of performance as if the word had been pronounced? To find out, I ran a small pilot experiment which showed that visually presented words were spelled with enormously greater accuracy than orally presented words. How could that be?

She talked the problem over with her husband, who almost immediately saw that spelling information must be

> in the visual memory store. Shortly after that he sat down at the terminal and overnight had a program that would spell phonetically, although not very accurately. Later he wrote the paper, "Alternate Uses of Phonemic Information in Spelling," for which I provided all the data, analyses, references, etc.

She then did the programming and wrote the lessons for a CAI program based on that idea, which was used for a few years in a public elementary school. Both teachers and children liked it, but it was never formally evaluated as a teaching tool, or written up for publication.

Later, at Carnegie-Mellon University, she developed these ideas further and wrote her own paper called "Spelling—A Task Analysis." It is, she says, "still the best theory of spelling extant." As she notes, "This was surely innovative work, and the kind that deserves some sort of recognition."

Dorothea Simon was born in Chico, CA, the daughter of a bank accountant. Except for being a first child and having no brothers, she lacks most of my projected background characteristics for women inventors. Her mother was a homemaker; Dorothea attended public schools, she reports no mentors until she met Herbert Simon; she is righthanded, feels she has no artistic talent, and was educated outside the field of her research (AB and MA in political science from Berkeley and Chicago, respectively), although she did get an elementary teaching credential from Chatham College in Pittsburgh, PA, in 1969. To her knowledge, there have been no other inventors in her family. However, both her parents were "handy and moderately creative in all kinds of things about the house and garden," and were pleased to have a "top-of-the-class daughter," though they did not pressure her to achieve. She remembers having a fair amount of time to play, read, or walk by herself.

Unable to find a job in political science upon getting her MA in 1934, she became secretary to the head of the Political Science Department at the University of Chicago.[32] Here she met and married (1937) Herbert Simon, then a graduate student. The partnership, which produced three children, now grown, was intellectually fruitful as well: he became her mentor and role model. "From the beginning," she says, "we shared professional as well as other interests, and read and made suggestions on each other's theses, papers, and books. He has always taken seriously my suggestions, and in turn has been enormously helpful to me, both professionally and in my work with organizations" (Q-1984).

Now retired from Carnegie-Mellon, where she had been a Research Associate in the Psychology Department and her husband's collaborator since 1973, Dorothea Simon keeps her hand in by reviewing manuscripts that are sent her (D. Simon, Q-1984, L-7/10/84).

As director of the Carnegie-Mellon University Center for Design of Educational Computing, **Dr. Jill Larkin** (1943–   ) works in the forefront of educational software and systems. According to Dorothea Simon, Dr. Larkin "is making major contributions to the theory of problem-solving (writing innovative computer simulation programs as well as doing experiments with students), . . . and is taking leadership in the development of innovative instructional programs on campus" (Q-1984).

A first child whose father bought her a chemistry set, a microscope, and an erector set, she was encouraged to achieve by both parents. She also spent much time daydreaming. After public elementary and secondary schools she took her BA in math at Harvard, her MA in physics at Berkeley, and her PhD in science and mathematics education at Berkeley (1975). She taught at Milton Academy in Massachusetts, at the Tefari Mekonen School in Addis Ababa, Ethiopia, and at the University of California (Berkeley) before coming to Carnegie-Mellon as a Visiting Assistant Professor in the Psychology Department. In 1983 she became an Associate Professor and, as noted, Director of the Center for Design of Educational Computing.

Dr. Larkin describes her projects there in applying cognitive science to educational computing as follows:

1) designing human-computer interfaces that are flexible and easy to use for both frequent and infrequent users;

2) building computer-implemented instruction based on psychological theory, and using techniques from both modern computer science and the rich 20-year history of computer-assisted instruction. Some examples:

(*a*) building intelligent tutors in geometry, electrical engineering, and physics;

(*b*) designing self-explanatory computer-based tools to help students understand the logic of the task being done.

She also mentions a project to create models for learning and performance in two very different scientific disciplines, physics (highly structured) and organic chemistry (apparently unstructured). In both, Dr. Larkin and her colleagues are working on ways to represent knowledge, and then directly teach the modeled knowledge.

As can be seen, Dr. Larkin's work shades easily into artificial intelligence (AI). In 1984

32. Rosalyn Yalow also started as a secretary—at Columbia.

she considered her most important invention a model of thought processes supporting quantitative reasoning. Indeed, with her creation of FERMI,[33] an expert system for multi-domain problem-solving, she crosses the line. FERMI will therefore be discussed below in the AI section. Note, however, that FERMI would have its educational uses.

Dorothea Simon describes Dr. Larkin as "generous with her ideas, time, and expertise with colleagues and students." As of 1984 she had 41 professional papers to her credit, and was a member of several professional organizations, including the American Association of Physics Teachers, the American Educational Research Association, the American Psychological Association, and the National Association for Research in Science Teaching (D. Simon, c.v.-1984, "Expert"; Larkin, c.v.-1984, Q-1984 L-10/11/87; see also McCorduck, L-3/15/84).

*Personal software*

A large new area for computer software is personal computing, including programs for home management. An important pioneer here, who also works in educational software, is **Dr. Sandra Hutchins** (1946–  ). Dr. Hutchins is Technical Director, and a principal, of Emerson & Stern Associates, a software and systems design consulting firm in Del Mar, CA. Emerson & Stern produces and markets 2-Bit® Software, a line of educational programs, games, puzzles, and "software with a sense of humor." These twelve programs can be run by total novices on low-cost computers such as the early Timex/Sinclair models. Two of the twelve, WORKING MOTHER'S DILEMMA® and KEYSTROKE MANAGE-MENT® were chosen for a 1985 Smithsonian exhibit on American games.

Emerson & Stern advertises its ability to create user-oriented programs for any computer in any language. As of the mid-1980s, for instance, the company was producing college-entrance-exam-preparation programs for Simon & Schuster.

Far more fundamental than this potentially profitable software, however, is Dr. Hutchins' work on programs enabling computers to respond to human speech and to natural English. In computer terminology, she is creating a lexical analyzer. BIG EARS, her speech-recognition software package, was developed with support from the Department of Education's Small Business Innovation Research Program. Work on other modules of this system—the GOOF-PROOFER, the GRAMMATIC ALLY, and COMMON SENSE—was well under way at last report. Jan Zimmerman described Hutchins in 1983 as "a computer engineer who designs software for state-of-the-art voice recognition systems" (". . . Computing," 79).

An innovative and significant feature of all Emerson & Stern software is its awareness of social issues, particularly the special concerns of women, minorities, and the handicapped. The company has developed a humorous series of cartoon-based games previewing for eight- to twelve-year-olds some of the social situations they may encounter; a time-management package for women; a combined software/workbook package to help children protect themselves from physical assault and sexual abuse. Dr. Hutchins and her partner (physicist, writer, futurist, and entrepreneur Jan Zimmerman), have also been researching the best keyboard design for producing a written form of American Sign Language, and as of 1984 were working on special educational software for victims of head injuries (Emerson & Stern, 1984).

Sandra Hutchins brings an excellent academic and technical background to Emerson & Stern. Her undergraduate degree was in physics with a minor in linguistics (highest

33. An acronym for *F*lexible *E*xpert *R*easoner with *M*ultiple-domain *I*nterfacing, and also a tribute to Enrico Fermi, known for his reasoning powers.

honors). She has a PhD in information and computer science from the University of California at San Diego (1970), and twelve years' industry experience in the design and management of real-time software and hardware for communications, specifically voice-processing, message-switching, secure computing, modems, and personal computers. In 1981–2 she was Technical Director of Voice Processing at ITT; and she has served as Engineering Manager for two different corporations—Linkabit (1977–9) and ITT's Defense Communications Division. She holds at least two patents, one for digital compression of speech in the 1970s and one for computer recognition of speech in severe noise environments (1982). She is a member of the Institute of Electrical and Electronic Engineers, the Association for Computer Machinery, and two honorary organizations, Eta Kappa Nu and Sigma Xi. She taught at Purdue University in the 1970s (*AMWS*/14,/15; Hutchins, c.v., "Making Changes"; Emerson & Stern/1984).

Involved in personal-software development at a more elementary level—as of 1983 she worked in computer retail sales—is **Karenn Jagoda** of Washington, DC. Ms. Jagoda has an undergraduate degree in math and a master's degree in administrative science. Her interest in computers dates from childhood when her father bought her an analog computer kit. On an IBM personal computer provided by her employer, she has developed a home-management program to keep track of recipes, food, and cooking information. Ms. Jagoda calls this a "tough application," but thinks there is a lot of interest in it (Russ Adams, 82).

**Roberta Williams** of Oakhurst, CA, is best known for her innovations in computer games. However, in 1983 Sierra On-Line, the company she founded with her husband Ken, introduced HOMEWORD, an inexpensive and easy-to-use word-processing program for the IBM-PCjr. Roberta also designed such home-productivity software as HOMEWORD SPELLER, HOMEWORD FILER, HOMEWORD FINANCE, and HOMEWORD TAX (Hendra Assoc.'s, "Sierra . . . Software").

The ultimate in personal software to date is the computerized counterpart of **Frances GABe**'s Self-Cleaning House, the IntelliHome. Created by **Portia Isaacson** (1942–   ) and Egil Juliussen, this invention does more than bring a computer into the home; it virtually makes the home into a computer. Although IntelliHome, first built in Dallas, has features attractive to anyone, it is obviously a house that makes life easier for two-career couples. No one need stay home for repair people, for example. Both can go off to work while the computer admits the repair person (in response to the proper access code), and monitors the repair. The computer then dials either spouse at work and reports: "This is your house. The water heater repairman arrived at 10:33 A.M. He gained access using the temporary code provided and left at 11:51 A.M." If he had tried to enter any other rooms, the computer would have warned him through speakers: "You're being video-taped. Return to approved areas." If he failed to do so, the computer would then call both the homeowners and the police. Indeed, if an appliance breaks down while the owners are at work, the computer can summon a repair person on its own.

If the carpool arrives early, the computer can let them in (with proper access code, of course), where they will find their favorite news programs playing on the television.

Just before leaving work for home, the IntelliHome owner can call the computer and instruct it to draw bath water, specifying the desired temperature. As the owner's car pulls into the driveway (if it is dark), the outside lights automatically go on, and upon entering the front door, the arrival is announced by name in all occupied rooms.

IntelliHome is beautifully adapted for security. If an intruder appears outside, the computer's remote sensors trigger a stern voice on the outside speakers, "You are on private property. Please leave immediately!" Hidden video cameras will start photographing and after 30 seconds, sprinklers come on to soak the intruder thoroughly. If he stays any longer, the computer calls the police.

These are only a few of IntelliHome's many automatic capabilities—features that could make it the home of the 21st century. A particularly attractive one is that when the telephone rings, all television and radio sets are instantly muted (Capettini).

Portia Isaacson holds degrees in physics, mathematics, and computer science, including a PhD in computer science from Southern Methodist University (1974). Her technical contributions lie in personal-computer technology, computer architecture and operating systems, and—most outstandingly exemplified by IntelliHome—home automation. However, she is perhaps best known, at least before the IntelliHome, as an entrepreneur and futurist. She founded Future Computing, Inc., of Richardson, TX, in 1980. In only three years it boasted a staff of 125 and revenues of $3.8 million. In 1984, projecting revenues to triple, she sold the company to McGraw-Hill (Zientara, 60; P. Isaacson, c.v. 1988).

Dr. Isaacson has worked for Xerox as an Engineering Specialist (1974–5), for Recognition Equipment, Inc., as a Senior Systems Software Programmer, as a Scientific Programmer for Lockheed (1968), and as a Scientific Programmer-Analyst for Bell Helicopter (1967–8). She is currently, among other things, Vice-Chair and Director at Intellisys Corporation (1985– ). She has also taught at the University of Texas (Dallas), North Texas State University (1971–4) and Southern Methodist (Visiting Industrial Professor). She has received considerable professional recognition, including appointment to the IEEE Computer Society Governing Board (1980–2), as Managing Technical Editor for the journal IEEE *Computer* (1978–81), and to the Board of Directors of Microsoft Corporation.

Portia Isaacson has been called tough as nails; she calls herself "bull-headed," or single-focused. She was born into an Oklahoma dairy family where no one had ever gone to high school. The oldest child, she was competitive from the beginning, playing basketball in school and then getting involved in competitive public speaking. By age four she was already dreaming of being a scientist, playing with bottles and pretending to be a chemist, idolizing Marie Curie. Joining the Army for two years to get out of Oklahoma, she then married and had three children. When her husband left her with the three boys, she was only 21 and still in college. It took several student loans plus all available welfare programs, but she got her physics degree in three years. She remarried, enrolled in Southern Methodist for her PhD and, with her second husband, opened one of the first computer stores in the country.

She is still married to her second husband, co-designer on the IntelliHome, Egil Juliussen, formerly with Texas Instruments. She now divides her time between Dallas and Colorado Springs, CO, where her University of Life is located.

*Business software*

Software for business uses of computers may be a prime area for women inventors. Dianne Littwin of Wiley Professional Software noted in 1983 that the unsolicited software submissions she was seeing from women were business programs such as accounting templates (Russ Adams, 82). And the world's largest independent software company was founded by a woman to exploit business software that she had invented. That company is ASK Computer of Los Altos, California, and the woman is Sandra Kurtzig.

**Sandra Kurtzig** (1947– ) of Atherton, CA, started her company in 1972 with only $2,000. Though trained in aeronautical engineering, mathematics, and chemistry, she was then a 25-year-old homemaker looking to create a part-time job for herself while bringing up her two boys. Her first foray into software design was writing recipes for robot cooks.

However, she swiftly graduated to innovative programs for businesses, one of the earliest of which allowed a newspaper company to monitor its carriers. Later she created minicomputer programs and information systems that help manufacturers to optimize inventory, improve product quality, reduce operating expenses, and improve customer service. The ASK systems get information flowing from the various parts of a company to the department that needs it. By 1983, just a little more than ten years out, Kurtzig's company had 200 employees, $22 million a year in sales, and $2.3 million in profits. It also had a reputation for success.

Because she financed all of her company's growth from earned income rather than borrowing venture capital, Sandra Kurtzig has reaped a real bonanza, more than $65 million in ASK stock. Every ASK Computer employee also owns stock. ASK is one of the fastest-growing publicly held companies in the United States, with $76 million in sales in 1986. Kurtzig has recently been called one of the "heroes of Silicon Valley" (J. Adams, 42).

Little of the inner Sandra Kurtzig appears in most of the sources on this fabulously successful woman. The focus of these articles is, quite properly, her achievement as inventor and entrepreneur. Carolyn Caddes, however, in her *Portraits of Success,* a book of photographs of Silicon Valley pioneers, gives us the personal note (vii):

> Most of the subjects in the book treated me like an artist. A few treated me like a servant. Because of my Southern roots, I expect people to rise and shake hands when they are introduced. Not always so! But what a pleasure when John Wilson and Sandy Kurtzig walked out of their offices, down the halls, and into their lobbies to greet me, then helped lug my cumbersome equipment back to their offices.

Caddes also supplies some early background information. Sandra Kurtzig was born in Chicago, the daughter of a writer (mother) and a real-estate developer. The family moved to Los Angeles when she was eleven. She took a BA in chemistry and math at the University of California–Los Angeles; then a master's in aeronautical engineering at Stanford. She worked in technical and marketing positions with General Electric and TRW until she decided to leave full-time employment to care for her sons. No one, including Caddes, reveals anything about her marriage, except to say that the husband received a $20 million settlement in their divorce.

In 1985 Kurtzig left daily responsibilities at ASK. Thus freed, she did such things as cover women in business for *Good Morning, America,* and lecture in a Stanford University continuing education seminar on women taking charge of their lives (Taking Charge: A Stanford Focus on Professional Women, Mar. 14, 1987). However, as chairman of the ASK board she stayed involved in the company's long-range planning, and recently returned to an active role in the company. And she thinks big. She visualized half a billion dollars in yearly sales by 1990. After all, as Kurtzig puts it, a computer without software is like a car without a steering wheel (Bylinsky, *Silicon,* 117, 237; Caddes, vii, 80–1; Clurman; Gottschalk, 1f; Schanstra, 72ff; Wojahn, 45ff). For late information, see Kurtzig's own book, *CEO* (1991).

*Word-processing.* This is a vital aspect of business-oriented software. Writers and professors swear by their computers, and all but the smallest and most backward offices now have some sort of word-processing equipment. Thanks in part to her business success, **Evelyn Berezin** has received some recognition for her work in word-processing hardware. A word-processing software inventor with Wang Laboratories in Lowell, MA, **Jenny T.L. Chuang,** has not been so lucky. Although a co-creator of the Wang word-processing system, she is completely unknown.

Born Jenny Chen in 1946, she took her BSc in physics at the National Taiwan University (1967) and her MSc in physics at Brown University (1970). After teaching for a year at Mt. St. Vincent College in Yonkers, NY, she joined Wang Laboratories in Tewksbury, MA, in the early 1970s. There, as a system programmer, she was on the team that created the successful Wang 220 minicomputer family. She wrote the original 16-digit-accuracy floating-point math package for the Wang 2200 operating system. In the mid-1970s, as Assistant Manager of the System Department in a Wang subsidiary in Taiwan, she wrote several commercial turn-key packages in BASIC on the Wang 2200 Series computers.

In 1976–8 as a System Software Engineer, she was one of the original four programmers who created the highly successful Wang word-processing systems. She designed and developed the master programs for the operating system, the security procedures, the recovery programs, the indexer, and all the menu-driven user-interface programs for the Wang WP 20, 30, and 25 systems. In May of 1978 Dr. Wang chose Jenny Chuang to be in charge of all ideographic (as in Chinese or Japanese characters) product development. She and her team developed the first Chinese/Japanese Word Processing System on Wang OIS and VS product lines, and delivered it in 1979. In 1980 they began to develop an ideographic DP system on the Wang 220 MVP system, in which BASIC supported the input and output of the Chinese or Japanese characters. This product was successful enough to give rise to a joint venture between Wang and the People's Republic of China. By the time the company moved ideographic development to Taiwan it had become a $5 million operation.

Jenny Chuang feels that her physics background was valuable in her approach to the problems of the word-processing system, and that Wang's great success has been due in part to the simple and logical way she set things up. Yet when Harold Koplow of Wang was interviewed by *Computer World* and other publications in the field, as often happened during the ten years Ms. Chuang worked for him, he credited other team members, but never mentioned her.

Meanwhile, Ms. Chuang has continued to work for Wang, always holding positions that demanded innovation. From late 1981 to mid-1984 she was responsible for word-processing development on all of Wang's major product lines. She was Product Development Manager for WP Plus, which integrated text with graphs, images, voice, and database. From mid-1984 to early 1985 she was charged with designing and building a generic integrated document-processor for all Wang's current and future product lines. In April 1985 she became Director of Office Application Software Development, and head of a software development group, directing some 120 software engineers in the development of document-processing systems, decision-support systems data exchange development WITA-based products, and desktop systems application software (Chuang, PC-6/12/87, c.v./1987).

It will be interesting to see whether her solid record of achievement ever breaks into the public view. Perhaps, like Berezin and Kurtzig, she will have to start her own company first.

### Industrial software

*Robotic manufacturing.* A sophisticated form of software with applications to auto-mated manufacturing is the invention of **Mary S. Pickett** (c. 1946–  ), Staff Research Scientist in the Computer Science Department at General Motors, and colleagues. Featured in a two-page color ad in *High Technology* magazine in 1984 (with Pickett and co-inventor Robert Tilove), this is RoboTeach, the first computer system to integrate robotics, solid or 3-D modeling, and simulation. The problem is—or was—that robots had to be led through each new task by humans. This cumbersome method was fine for a few robots doing fairly simple tasks. Now, however, with GM and other manufacturers planning to install thousands of robots doing more varied and complex tasks, a more efficient and error-free robot-instruction method was needed.

To begin with, engineers needed to simulate the robot's work on computer. GMSOLID, an interactive geometric modeling system, is one component of the system. It provides geometrically complete (three-dimensional) representations of the robot's working environment and the spatial relations among its parts and between them and the robot. RoboTeach combines powerful programming languages to manipulate robots, GMSOLID to simulate the physical components of the robot and its environment, and a further advance in simulation that represents the actual process and motions of the robot during its task. Using this system, manufacturers can verify that the proposed motions will actually do the work desired, and various sequences of motions can be tried out. When the best one is found, it can be programmed into the robot's computer. By late 1985, a prototype of RoboTeach was in experimental use at GM. With the final model, engineers can foresee robots' collisions with other objects and modify the work space accordingly. They can automate entire manufacturing processes by computer, and cut design and training time from weeks to days.

Taking her BS in math from Iowa State University and her MS in computer science from Purdue, Mrs. Pickett joined GM in 1971. In addition to her work on GMSOLID and RoboTeach she has done, as mentioned above, research on the design of real-time programming languages.

Mary Pickett and her colleagues have brought robotic simulation a long way from stick-figure graphics. Not surprisingly, RoboTeach was chosen as one of the top 100 innovations for 1985 by the editors of *Science Digest* (Howard, 52; "The Robot Abstraction").

*Computer-aided design.* Industry—i.e., manufacturing—is one of the major users of computer-aided design. From 1963 to 1974 **Amy Lee Clark Spear** (1924–  ) of McLean, VA, worked for RCA's Aerospace Systems Division in Burlington, MA. As a Senior Engineering Scientist in Product Design she was primarily responsible for studies of the effectiveness of systems and served as Systems Engineer for the plant's design automation capability. In that position she did significant work in design automation, twice receiving RCA Team Awards for technical excellence, one in March of 1972. In 1971 the Design Automation System had bogged down in an effort to computerize completely the design of printed circuit boards. Mrs. Spear led a team charged with getting the project going again, and developing a fully operable Design Automation Printed Circuit Board System within a year. This they did. Under her direction, the program was reoriented around an interactive graphics system, which permitted the maximum human decision-making, with rapid response, but precluded human error. Trial boards were processed in December 1971.

The system accepts engineering drawings and generates completed artwork on printed circuit boards. The grid points for the layout of the printed circuit runs were fed directly into computer-controlled photo-etching process control and automated assembly equipment. This work ranges from relatively simple two-sided boards to the eight-layer boards typical of one kind of technical requirement. Costs were reduced 35% per board, creating an estimated first-quarter saving to the Company of $28,000 for two programs alone. Even larger (in terms of percentage) and equally significant time savings have also resulted from Mrs. Spear's work. Qualitative savings have resulted in other areas as well:

the system reduces duplication of effort, makes better use of drafting talent, and significantly reduces the need for engineering follow-up. Boards generated under Spear's design-automation system moved through fabrication and assembly with almost no quality-control rejections or rework requirements. Her system establishes a data base that allows a more consistent and more accurate final product, with such spin-offs as generating automatic test routines from the data base.

The system was ready in considerably less time than that taken by other divisions, and was offered for use elsewhere in the company. The RCA *Technical Excellence Newsletter* called Spear's invention "of the highest technical order" (RCA, *Technical . . .* No. 93). For details of the system and its operation, see Spear's description in the RCA *Chief Engineer's Newsletter* in 1974 (Spear, "ASD Design").

In the 1960s Mrs. Spear worked in the Reliability program for the Lunar Module Rendezvous Radar and Service Module Transponder. By 1966 she was project engineer for the Rendezvous radar, guiding the design effort from the critical prototype phase into production. In 1967 she became Manager of Product Assurance for the Lunar Module, charged with directing the reliability, quality, and configuration-management efforts for all the Lunar Module radars at a crucial period of changes in design. Just at that point the stringent size and weight limitations of Apollo's electronic packages resulted in densely packed modules that were difficult to assemble. Thus the modules were prone to certain potential failures not visible to existing quality-control inspections.

Amy Spear realized the need for, and devised, the necessary new testing—a minimum-time burn-in testing and temperature cycling to screen for those potential failures. She then put the new testing procedures into effect at the lowest assembly level that was economically feasible. Her approach was so successful in improving the reliability of the Rendezvous Radar that it was instituted for all Apollo electronic equipment.

Still earlier, at GE's Heavy Military Equipment Department (1959–62), Mrs. Spear worked in the Advanced Projects Development Section. There she determined what components would be needed for proposed new systems, and instituted programs to develop them if they did not already exist. She prepared the development specifications, monitored the development efforts, and coordinated the technology available in various GE laboratories to support the projects. By July 1962 she was a Systems Reliability Engineer in the Radio Guidance operation. There, she helped establish the rapid-response field-failure-reporting system used for GE's several guidance systems: the Mod III and Mark IV Guidance and/or Tracking radars for the *Atlas, Titan II,* and *Titan III* boosters used in the Air Force, Mercury, Ranger, and Gemini programs. She was also responsible for evaluating field and production failures and correcting them quickly.

Mrs. Spear currently works for the MITRE Corporation, a Federal Contract Research Center, a private, nonprofit corporation that does systems engineering studies for the Air Force. At the Omaha site, where she was located until 1988, MITRE specializes in Strategic Air Command systems. Her work, though significant—and recognized in 1984 by an IEEE Centennial Medal—became more administrative or educational than directly innovative (*BSBW*). At this writing she is Lead Engineer for Joint Tactical Command Control and Communications in the Joint C$^3$ Systems Department of MITRE's Washington, DC, operation, her division located in McLean, VA (Spear, L-7/11/88).

Amy Spear took her bachelor's degree in electrical engineering (BEE) at Cornell, her MSEE from Northeastern University in 1966, and her MS in Systems Engineering from Lowell Technological Institute in 1975. She is married and the mother of four daughters, two of whom are also Cornell engineers. She took out six years to spend exclusively with them. When she joined RCA she managed to get her achievement recognized in her salary despite this hiatus, and commended the company for its efforts to bring women's salaries into line with men's. She is a senior member of the IEEE, a fellow of the International Society of Logistics Engineers, and a fellow of the Society of Women Engineers (M. King, 69; Spear, L-7/1, 8/19/87, c.v.-6/87).[34]

### Miscellaneous Contributions to Software

According to Larry Druffel's 1979 article on Ada, the Defense Department's standardized high-order language, then still in development, **Marilyn Bohl** was "involved in software development for IBM." Unfortunately, Druffel does not specify her contribution, referring only to her "numerous" successful textbooks such as *Information Processing* (SRA, 1980), *A Guide for Programmers* (Prentice-Hall, 1978), *FORTRAN Programming and WATFIV* (SRA, 1973). However, since Bohl has been invited to comment on Druffel's analysis of Ada's effect on college courses, she is evidently important in the field (Druffel, 60–1).

## APPLICATIONS

The following women's inventions and innovations are most noteworthy for applying existing hardware and software to new uses. Obviously some modification of existing technology—sometimes entirely new software or hardware—may be necessary. Thus the classification is arbitrary, a matter of emphasis or original intent.

### Tailoring Applications to Users

In 1970 **Joyce Wrenn** of IBM in Boca Raton, FL, invented the "application customizer," one of her proudest achievements. IBM not only applied for a patent, but honored her at an awards dinner and with "a nice check" (Zientara, 59–60).

### Data Management

One of the most central and obvious applications of computers is data management. An interesting example of a woman innovating here is **Janet Marie Dearholt Esty** (1941–   ). As she told Lois Rich in her interview for *Millionairess* (136), when she worked for Univac she "built a new thing called data management, which was taking care of all the paper work and reporting requirements for government contracts." This would have been in the mid- to late 1960s (1965–8; Rich-McCoy, 127ff; *WWW*/17: 215).

### Computer Modeling of Complex Problems

In 1979 **Ann Corrigan** of Creative Software, Morristown, NJ, published an article describing her work in adapting the relatively simple Huntington II population models to run on "several popular microcomputers," i.e., personal computers. The particular use she suggests for these new models is education. As Ms. Corrigan correctly notes, "an understanding of the basic nature of population growth and decline is necessary for many

---

34. Amy Spear's mother was born a Blanchard. Spear's great-great-grandfather Blanchard was descended from Huguenots (L-8/19/87). Thus it is possible that Spear may be related in some way to Thomas and **Helen Blanchard** (Chap. 4).

of the critical problems facing humanity today," and yet students are not exposed to these ideas until college, if then. Her adaptation will enable even grade-schoolers to understand how population growth works (Corrigan).

## Medicine, Aids to the Handicapped, etc.

Among the most important applications of computers are medical. A recently fruitful area has been computerized aids to stroke or accident victims and others deprived of the use of limbs or senses. **Steve Shirley** of England has been involved in this area in an administrative and enabling way on a national level. In 1981 she organized an exhibition of micro-based aids for disabled people in the House of Commons. During the International Year for Disabled People she worked on several projects linking Information Technology to the needs of the disabled. Her interest is intensely personal as well as social (her only son is severely handicapped), but whether her involvement is technical and innovative as well as administrative is unclear (T. Lloyd, 64, 76). However, at about the same time, several American women became involved at the hands-on technical level.

A striking example here is the talking wheelchair devised by **Dr. Carol Simpson** and colleagues at Psycholinguistic Research Associates in Menlo Park, CA, in collaboration with an electronics engineer and a rehabilitation engineer from the Stanford Children's Hospital. A trial model was ready in 1980, using a computerized word processor and a Votrax phoneme synthesizer. Users construct sentences, which appear on a video screen, and then direct the synthesizer to speak them.

Interestingly enough, the original voice produced was male because, as Dr. Simpson told an interviewer, the necessary data on female voices did not exist. However, the chair's voice could be adjusted in pitch, rate, and volume; and Dr. Simpson and her colleagues were working to make the chair's voice sound more natural. Later models offer a female voice.

Perhaps the wheelchair's most useful feature is its allowance for a wide range of disabilities: trauma, stroke, cerebral palsy, and other conditions interfering with both speech and motor systems. The voice synthesizer will respond to information from a typewriter-like keyboard, a joystick, or clicks of a switch. Sentences once constructed can be stored for later use. With experience, users can learn to "speak" as fast as 30 words per minute ("Wheelchair Speaks . . ."). See, for example, Tira's story in *Ms.*, Sept. 1987. Such a chair enables the brilliant Stephen Hawking to communicate with the world, as seen in the 1992 film *A Brief History of Time.*

This wheelchair, under its technical name of Versatile Portable Speech Prosthesis— was one of the regional prizewinners in the Johns Hopkins First National Search for Applications of Personal Computing to Aid the Handicapped, held in 1981 (Hazan, L-1/22/ 82; IEEE, *Search*).

Other regional winners in this first Johns Hopkins Search were **Dr. Gwen Nugent** and **Casey Stone; Sandra J. Jackson; Dr. Rosamond Gianutsos** and **Carol Klitzner.** Gianutsos and Klitzner were national finalists as well, and Sandra Jackson's team, which included another woman, **Judy Maples Simmons,** was one of the top ten national winners.

At the time, **Dr. Gwen Curry Nugent** (1948– ) was Production Coordinator for the Educational Media Project for the Hearing-Impaired at the University of Nebraska in Lincoln. She had graduated from Texas Tech University in 1969 with a major in speech and communication, and received her PhD from the University of Nebraska in Educational Psychology and Measurement in 1979. Dr. Nugent and her collaborator **Priscilla**

**Garhart ("Casey") Stone** (1950–   ), also at the Educational Media Project, created "Think It Through," an interactive videodisk to teach independent thinking to hearing-impaired children.

As an example of the program, the first problem to be solved is an everyday mystery. Grocer Jones suddenly discovers he is out of food-coloring, and asks the students to help him figure out why. They must then work through the problem to solve the mystery. Unlike traditional computer-aided instruction or programmed learning, this program simulates a real-life problem-solving situation. Students explore various avenues of thought, changing paths at will. Paths not leading to a successful solution are not stigmatized as "wrong." The idea is to guide students toward systematic testing and elimination of incorrect hypotheses.

After an introductory sequence and a short explanation of defining a problem, the computer presents three possible definitions for Mr. Jones' problem. There is no single right answer; two of the three possible definitions presented will work and are thus acceptable. Likewise, Mr. Jones is not an authority figure. At one point in the thinking-through process, consulting him is an option, but usually he does not know the answer, and the students must continue alone. At each choice point, the computer branches to new options dictated by that choice.

The text and logic can easily be modified to meet individual instructional needs, just by rewriting parts of the computer program, while the visual material on the videodisk remains intact.

Dr. Nugent, who is married and has a young daughter, did college teaching and spent ten years in television before coming to the Nebraska project. As of 1982 she had 16 publications to her credit. **Priscilla ("Casey") Stone,** an oldest child who has other inventors in the family (brother and paternal great-grandfather), took her BA in philosophy, her MS in education, and further graduate work in educational psychology and computers. Set-breaking thinking was the rule rather than the exception in her home, where her mother (later an ordained minister) wrote children's books, and her father, a college professor and minister, played with language. Logic, too, was part of her training. She learned chess by age four. Later, she and her father played a game that involved devising ways of destroying New York City. This, says Stone, "required developing hypotheses and then logically following through from the first action to each successive and cumulative effect" (Hazan, L-1/22/82; IEEE, *Search*; Nugent, L-4/ 12, 4/28/82; Q-1982; C. Stone, L-5/28/82, Q-1982).

**Rosamond Gianutsos** (1945–   ) is Director of Cognitive Rehabilitation Services, Sunnyside, NY, Adjunct Associate Professor at the New York University Medical Center, and a clinical and research psychologist. Fourth daughter of a business administrator with mechanical engineering training, she was always the "smart one" in school. After private all-girls' secondary schools, she majored in psychology at Barnard (AB, 1966). Her family supported her through college, but balked at graduate school, supporting instead her lower-achieving brother. She nevertheless got her PhD at New York University in experimental psychology in 1970, aided by a federal traineeship. Dr. Gianutsos showed an early interest in the handicapped, serving in 1965 as an occupational-therapy trainee at Beth Abraham Hospital in the Bronx. In 1978–9, and thereafter on a part-time basis, she served as Senior Psychologist in Rehabilitation Medicine at the New York University Center, Bellevue Hospital, New York City. In 1981 she was a consultant to the Vocational Rehabilitation Department, State of Maine; in 1983, Associate Professor of Psychology at Adelphi University, Garden City, NY.

With **Carol Klitzner** of Computer Software Solutions in Forest Hills, NY, in the early

1980s, Dr. Gianutsos created a series of computer programs to help diagnose and treat brain-injured people. As already noted, these cognitive rehabilitation programs were among the regional winners and national finalists in the Johns Hopkins First National Search for Applications of Personal Computing to Aid the Handicapped (1981). Included in the package are two administrative programs, LOG and VISANGLE, providing respectively a continuous record of each patient's activity in the series, and a better understanding of the real-life implications of visual field test results. The seven therapeutic programs help to spot—and then provide exercises to treat—problems with visual perception and memory, the two areas of mental function most often affected by brain injury. Program SRWL (Speeded Reading of Word Lists), for example, can identify difficulty with any of four basic human information-processing functions: anchoring at the margin, horizontal scanning, identifying words within the perceptual span, and monitoring the periphery.

In the rehabilitative exercises, the computer presents word lists in three different formats—lists of nine words to be read left to right or right to left; words appearing one by one in the center of the screen; and single words appearing usually in the center, but sometimes unexpectedly off to the far left or right of the screen. Another program called SEARCH presents on the screen an empty box surrounded by an array of abstract shapes (8 rows, 8 columns). When a shape appears in the box, the patient tries to match it in the array. Upon finding the matching shape, the patient presses the space bar, and the therapist confirms or rejects the match. The computer stores both the search times to correct responses, and the full list of incorrect matches for display at the end of the test.

Four programs deal with memory, including one called SPAN, specifically designed to help restore short-term storage capacity. As of the early 1980s, the programming package, titled *Computer Programs for Cognitive Rehabilitation,* was available from Life Science Associates, Bayport, NY, to run on either a TRS-80 or an Apple II disk system.

Dr. Gianutsos does not consider this contribution especially creative. But in her next breath, telling what it is instead, she gives a perfect description of one kind of invention: "an exceptionally fruitful synthesis of my professional activities as a cognitive psychologist in rehabilitation with a computer background." Any programmer, she continues, could have written the programs. "But they would not have known which tasks would be useful and/or practical."

At Barnard Dr. Gianutsos participated in the National Science Foundation Program for Undergraduate Research (1965–6). In 1980 she was Principal Investigator on a grant for "Rehabilitation of Visual Field Defects and Hemi-imperception in the Brain-Injured." She belongs to the American Psychological Association, the American Congress of Rehabilitation, and the International Neuropsychological Association. She is a guest reviewer for three professional journals, including the *Journal of Neuropsychology,* and has at least twenty publications to her credit, including an in-house text, an edited volume of essays, and the computer programs just mentioned (Gianutsos, c.v.-1982, A-5/5/82, "Using Microcomputers"; "What Is . . ."; Hazan, L-1/22/82; IEEE, *Search*).

**Sandra Jackson** (1954–  ) of San Antonio, TX, with **Judy Maples Simmons** and a male colleague, developed "We Help More," a program that adapts computer-assisted instruction to learning- and emotionally-disabled children who are "mainstreamed," that is, integrated into ordinary classrooms. Teachers using the program report that these students work faster and with higher-than-usual levels of motivation and concentration when at the computer. Students are absent less and present fewer discipline problems on computer days. Since such children may number from 10 to 20% of all school-age children, the importance of "We Help More" and other such programs is indisputable.

Ms. Jackson, who has a master's degree in education, is a reading specialist in the San Antonio secondary schools. She taught herself computer programming on school computers, and has had no formal training in computer science. As a teacher, she has not experienced sex discrimination; but she was struck to find that she and her collaborator Judy Simmons were the only two women at the Johns Hopkins award banquet for top winners in the Search (Hazan, L-1/22/82; IEEE, *Search,* 59–61; Jackson, L-8/6/82, c.v./1982).

Other first-round winners were **Carolee Mountcastle, Jean M. Riley, Jamelle Loehr, Valarie Hanzelka, Mary G. Hatch, Margaret Barker, Peggy S. Eyrich** and **Margaret Maier.**[35]

### *Robots for the handicapped*

Mentioned also in the Health/Medicine chapter is the robot hand developed by students in a Stanford mechanical engineering class in 1985. Controlled by a microprocessor, the hand has mechanical fingers and an opposable thumb. It can understand anything sent in standard computer code (including foreign languages and data sent by telephone), and translate it into sign language. The user holds the hand and "reads" by touch what its motions say. It is intended primarily for sufferers of Usher's Syndrome, who first become deaf and then go blind; while still only deaf, they can learn sign language, so that when blindness strikes, they can still use the hand. The original idea for the hand came from the Southwest Research Institute of San Antonio, TX, which was looking for an alternative to Telebraille (converts electronic code into Braille's raised dots), was actually commissioned by the Smith-Kettlewell Institute for the Visually Impaired in San Francisco, and was funded by the National Institute for Handicapped Research. Among the four Stanford students working on the project was one young woman, **Patricia Lee** (Shurkin, "Robot").

Even more significant than a wheelchair that talks is one that listens: one, in fact, that can be controlled by the user's voice. Alsatian-born **Martine Kempf** (1958–  ) has significantly advanced the necessary technology for such a wheelchair by inventing a voice-recognition microcomputer called the Katalavox that measures only 1 by 10 by 4 inches, and weighs less than five pounds. Indeed, the prototype system was reduced to a single board weighing *less than one pound.* This she did at age 23, in 1982, while an astronomy student at Bonn University in Germany, learning the programming on her Apple PC. Her Katalavox is a breakthrough invention many times more efficient than comparable models from established computer firms. For example, it responds to a spoken command in .08 second, as opposed to $1/2$–2 seconds for her competitors.

Now, using this technology, she has created the world's first generally available electric wheelchair directed by a speech-recognition control system. This new chair, representing a quantum leap in wheelchair technology, was introduced at the 1983 convention of the International Society of Prosthetics and Orthetics in London.

It is noteworthy, considering some of the stereotypes about women and mechanical or electronic expertise, that Martine Kempf not only conceived the innovative software necessary for the Katalavox, but designed and built the hardware—designing the board and soldering the circuits herself.

In 1985 the French gave Martine the Prix Grand Siècle, which places her in the august company of Jacques Cousteau, Lord Mountbatten, and the great cellist Rostropovich. That same year, ABC television filmed the Katalavox story for the *Believe It or Not* show. In

35. There may have been a few more female winners: certain given names on the list were ambiguous, and as Paul Hazan notes, the winners' names were not categorized by sex (L-1/22/82).

1986 the Spinal Cord Society awarded her their annual silver medal for "special services to humanity, in recognition of her contribution to the life quality of paraplegics." In 1987, the town of Dossenheim near Strasbourg, France, named a street after her, Rue Martine Kempf, and the President of the European Parliament spoke at the ceremony.

Although the main or most lucrative application of Katalavox may lie in microsurgery (Health/Medicine), the wheelchair application may please Martine Kempf more, since the needs of the handicapped inspired her invention in the first place. In Germany she saw armless thalidomide victims unable to drive cars. Her father's company had adapted cars for foot-steering and foot-gear-shifting, but if a turn signal or a dimmer or wiper switch needed to be manipulated at the same time, the feet simply could not manage. Her father, himself wheelchair-bound by polio, had earlier designed a fully hand-operated automobile, and encouraged Martine and her brothers toward helping the handicapped. Seeing the difficulty with adapting cars for the armless, she got the idea of voice control, but the technology she visualized did not exist. Martine Kempf set out to create it, and did so. The first use of the Katalavox was in a car. In January 1984, the first driver's license was granted for a totally voice-controlled car. In April 1984, at an industrial exposition in Tokyo, the German Ministry of Science and Technology exhibited a Mercedes car with 50 secondary functions activated by the Katalavox. The car was a collaboration between Martine Kempf of KEMPF Strasbourg and Daimler-Benz of Stuttgart.

Martine is the youngest of three children (two brothers) of an entrepreneur-inventor and a medical technician (who has also taught, and now keeps the books for the family company). Though generally righthanded, she is partly ambidextrous and sometimes writes with her left hand. She has considerable musical talent, playing piano, recorder, bassoon, violin, and cello, in some cases with concert-level skill. She attended private co-ed schools, completing her secondary (baccalaureate) degree in Greece.

Martine Kempf now lives in Silicon Valley, U.S.A. (Sunnyvale, CA), where she continues her combined research, invention, and manufacturing operation in a facility that opened in 1987. The Katalavox is not patented, because she wishes to keep its technology secret. Her company is called simply "KEMPF." We may expect to hear a great deal more from this significant young woman inventor in the next few years. (Brower; Vare & Ptacek 166–8; Kempf, I-7/19/88; Q-7/19/88; KEMPF brochure/1988).

**K.G. Engelhardt** is the principal investigator at the Veterans Administration Hospital Robotics Laboratory. Since 1981 she and her Stanford colleagues have been working to adapt robots to help paralytics and oldsters live more independent lives. The project is doubly an application: applying computers to medicine, and adapting a multipurpose robot—originally built by Stanford students from off-the-shelf equipment—to serving the handicapped. In 1985 Englehardt and her crew announced to the press that the robot arm could now feed one of their spinal-injury patients.

Interestingly enough, the research team includes a dancer, **Tina Ebey** of Menlo Park, CA. She is the "motion consultant," helping programmer Dave Cannon identify coordinates that will move the robot gracefully (and thus presumably more efficiently).

Since the robot's potential users have unimpaired brains, the project seeks only to compensate for arms and legs. Thus the task is partly, as one project member put it, learning how to teach smart people to use smart machines. The robot recognizes only 58 words, simple commands like "right," "left," "up," "down," etc., and lacks the sensory capabilities to avoid an object in its path. Luckily, it recognizes "stop." Since even simple tasks require complex motions, directing the robot demands some skill, and therefore some training. People aged 5 to 90, with varied disabilities, have enrolled. After five sessions, all could direct the robot to give them a glass of water—the most common request.

Engelhardt's project seeks a marketable product. As of 1985, project members predicted that within a decade these robots will be available for about the cost of a new car—assuming that manufacturers decide to make them (Donnelly).

*Workstations for the handicapped*

At Boeing Computer Services' Artificial Intelligence Center in Bellevue, WA, **Caroline Fu** created a voice-controlled workstation for those who have lost the use of their arms. Ms. Fu's new "Hands-off" workstation features two voice-recognition units like those used at the VA, and two robotic arms made by Universal Machine Intelligence, Ltd, of England. The arms do such jobs as inserting floppy disks into the workstation's personal computer. Despite attempts to keep costs low, the workstation will probably cost $30,000–50,000. However, as Fu points out, the quadriplegic programmer testing the station at Boeing had a productivity level as high as or higher than that of ordinary employees. **K.G. Engelhardt** of the VA project just described was evaluator on the project. Using this voice-controlled station, says Dr. Engelhardt, a disabled programmer "can be completely vocationally independent . . ." (Donnelly, 17).

In 1988 Boeing decided to give the technology to a reliable robotic manufacturer, Prab Robots, Inc., of Kalamazoo, MI. Prab predicted that 250,000 of the workstations would be made in the next few years. A delighted Caroline Fu wrote, "My heart is bubbling with joy whenever I think of all the physically impaired people [who] can now be independent and working as equal contributors to the world's working force" (L-6/20/88).

Chinese-born, youngest of three daughters of a professor and scholar, Caroline Fu was first tutored at home, and then attended all-girls' schools, through the secondary level. She is righthanded, but shows artistic talent: drawing, dance, and music. Influenced by her father toward creativity and achievement, and encouraged by both parents to be independent and understanding when confronted by unfair system barriers, she took as her role model Jen-Hsung Wu,[36] a colleague of Nobelists Yang and Li in their prize-winning research.

Caroline Fu holds a BS in applied mathematics, electrical engineering, and physics (1970) and an MS in Computer Sciences (1972) from the University of Wisconsin. She has corporate technical experience at General Motors, Burroughs, Westinghouse, Ford, Abbott Laboratories, and Boeing (1983–   ). She counts her supervisors at Ford, Abbott, and Boeing as mentors. At Westinghouse she developed a software system for $CO_2$ laser fusion simulation, and a graphical comparative analysis of different techniques for extracting uranium from various sources. At Ford she developed an acquisition system to certify the pre-test conditions for car-crash tests, and designed and implemented an "all-purpose" engineering application system for analyzing data. At Abbott she was Project Manager and Chief Designer for voice-recognition and response-distributed processing systems for drug-safety research programs.

Caroline Fu serves on the Board of Directors of the National Service Robot Association, and is a member of the Acoustical Society of America, the American Voice

---

36. Alternate transliteration of **Chien-Shiung Wu** (1915–   ). Featured here in Chap. 4, Dr. Wu was a colleague of Drs. Dao Li (often transliterated "Lee") and Chen Ning Yang. Dr. Wu devised the famous experiments whereby Li and Yang overthrew the Parity Principle (and thus showed that a certain law of physics does not apply to weak interactions of subatomic particles, and thus is not universal). The two men won a Nobel Prize in 1957 for this proof.

Input/Output Society, the Society of Manufacturing Engineers (Senior Member, 1984–6), and the Association for Women in Computing. She has numerous conference presentations to her credit, and in 1988 organized and chaired the first session on "Artificial Intelligence in Acoustics" for the 1988 Conference of the Acoustical Society of America. She has written a monograph on her workstation. As of 1987 she is a Manager and Principal Scientist for a program in Speech Perception and Language Comprehension at Boeing Aerospace in Seattle, where she established and directs a research laboratory in speech and natural language processing. "I hope," she says, "in the very near future I can once again share news of Boeing's progress with the research world" (Fu, c.v., L-9/29/87, Q-9/29/87, "Hands-off," "Independent").

Such voice-controlled workstations could also be used by nonimpaired workers who simply have their hands otherwise occupied. The stations might prove especially useful, with some modifications already visualized, in hazardous environments such as space. Also already envisioned is a tele-operation capable of controlling multiple robots performing repair and service functions outside a space station (Fu, "Independent," 8).

### Other medical applications

The EMERGE system, a computerized aid to emergency-room personnel dealing with chest pain, developed by **Donna Hudson** under the direction of **Thelma Estrin,** should certainly be recalled here, as should Estrin's work with computerizing medical records. Because of her interest in brain research and her specialization in biomedical engineering, most of Estrin's computer inventions and innovations are actually medical or related applications of the computer. ·

In the early 1980s, under the title "Another Computer Triumph," researchers reported applying computers to ultrasonic visualization of tissues. The project began with the work of **Linda Joynt,** PhD, and her collaborator, an electrical engineer. The resulting technique adapts the tissue-characterization capabilities of ultrasound to analysis of tissue structure in abdominal organs. Traditionally, ultrasound scanning aims ultra-high-frequency sound waves at tissue. The waves scatter upon hitting the tissue, return to the transducer, and create an image. The new technique devised by Dr. Joynt and her colleague records ultrasound waves digitally on a computer so that the tissue structure can be inferred by analyzing the wave forms. More sensitive than either X-ray or conventional ultrasound, this new method can differentiate between normal and abnormal tissue, and can be applied to many organs. At Stanford in 1982 it was being used in studies of liver and spleen tissue. Eventually it may be used to detect abdominal tumors and breast cancer. The project was funded by the American Cancer Society ("Another Computer").

**Lorna E. Campbell** and **Jennifer P. Stone**, also of Stanford, should be mentioned as part of the team that developed a novel application of synchrotron radiation to obtain X-ray pictures of blood vessels. Successfully tested on animal hearts in 1981, the technique is a non-invasive alternative to catheterization in angiography, and could also be used in mass screening for coronary and cerebral artery disease. This collaboration between the Medical School and the Departments of Physics and Electrical Engineering at Stanford is mentioned here because a computer is involved: the two X-ray pictures taken after an organ is injected with iodine (using crystals to select the single X-ray frequency most sensitive to iodine) are at first as blurry as most X-ray images of soft tissue; but when the two images are fed into a computer, scientists can suppress information derived from tissues in general and enhance information coming from the iodine. The result is a sharp image despite the arteries' constant motion ("Scientists Develop Unique").

*Satellites, Rockets, Space Travel*

Modern artificial satellites and their rocket launchers depend heavily on computers for guidance, monitoring, control, and retrieval. **Dr. Alla G. Masevich**'s creation of the world's first satellite tracking system in six months has been discussed. She continued in charge of the Soviet satellite tracking program, and by the 1980s her hundred-station system was obviously computerized. Dr. Masevich's importance to the Soviet government can be gathered from her appointment by 1966 as Vice-chair of the USSR Academy of Sciences Astronomical Council.

**Yvonne Brill**'s work on satellite propulsion systems may involve computers, but this is not clear in my sources (Chap. 4).

Satellites have important uses in communications, astronomy, and weather observation and prediction. **Dr. Alva Matthews** was involved in the *Telstar* satellite project that linked the world in a live television network for the first time in 1962. Specifically, she was involved in the computer programming for *Telstar's* tracking antenna. The satellite, powered by solar cells, follows a low orbit. For 20 minutes of each orbit it links the USA and Europe so that television pictures—and up to 60 simultaneous telephone conversations—can be sent clearly.

Dr. Matthews took her pre-engineering education at Middlebury and Barnard Colleges, followed by a doctorate in engineering mechanics from Columbia. She has worked, in addition to the *Telstar* Project, as a senior research engineer with a private engineering consulting firm (Paul Weidlinger Associates in New York City), making significant contributions in many areas of engineering mechanics and structures, including the design of helicopter blades, wave propagation in soil and rock (with applications to nuclear-weapons impacts) and structural design of thin concrete shells. Many of her publications are recognized as pioneering, particularly in the area of pressure-wave propagation and vibrations in elastic media. She is a member of the American Society of Mechanical Engineers and the New York State Professional Engineers.

Her other computer-related work includes "the development and adaptation of large computer codes." As of the mid-1970s she was Associate Professor of Mechanical and Aerospace Science at the University of Rochester. She is married to A.R. Solomon (Harpur, 172; Malotke *et al.*, 29; SWE Award Booklet/1971; *WBWRA*, 189, 193–4).

In 1966 **Marjorie Rhodes Townsend** (1930– ) of Washington, DC, became Project Manager of NASA's Small Astronomy Satellite (SAS) Project, headquartered at the Goddard Space Flight Center, Greenbelt, MD. A 1973 *Science* article on Townsend reported, "She is responsible for the inception, design, construction, and testing of the satellites as well as the launches themselves" (Holden, 48). As early as 1959, when only in her late twenties, Marjorie Townsend directed for NASA the design of the ground data-processing system for the *Tyros II* weather satellite. She is the co-inventor of a digital telemetry system.

Marjorie Townsend is a superwoman of the **Rosalyn Yalow/ Betsy Ancker-Johnson** stamp. An only child, expected to achieve by an engineer-inventor father, she entered George Washington University at 15 and became the first woman to get an engineering degree there (1951). She got this degree despite the fact that she married during her junior year, took a job at the National Bureau of Standards so that her husband could finish medical school, and continued her own studies at night. During her eight years at the Bureau she rose to Senior Engineer in the Airborne Sonar Branch. She also bore four sons without taking extra leave.

A 1973 interviewer described "an open, somewhat girlish appearance and manner which does not detract from the obvious fact that she is a woman who knows what she's doing" (Holden, 48). Two of the three Small Astronomy Satellites scheduled in her project had been launched: SAS-A, or *Uhuru,* the world's first X-ray astronomy satellite (1970), and SAS-B, a gamma-ray satellite launched from Kenya in 1972. She was already planning for the third, scheduled for 1975. *Uhuru* has been credited with "open[ing] up the whole field of high-energy astronomy." The recent exciting work on black holes, for example, depends on X-ray and gamma-ray observations.

Among Townsend's honors are NASA's Exceptional Service Medal for her "outstanding technical and managerial leadership" of this Small Astronomy Satellite program, and the order of Knight of the Italian Republic, both in 1972. (The SAS project was a joint venture with Italy, which supplied the launch facilities and personnel at a platform built in the Indian Ocean off Kenya. One official told her she was the first NASA person who ever bothered to learn any Italian). In 1973 she won the Federal Women's Award for distinguished Government service.

Though trusting that barriers against professional women will disappear, Marjorie Townsend acknowledged that these barriers—and considerable stereotyping against female achievement—do exist. She thinks she escaped the stereotype that cripples so many girls at puberty partly because of her father's attitude, partly because as an only child she judged herself by adult rather than peer standards, and partly because she was so young in high school that the boys didn't really see her as a "girl." At the close of the interview, she was looking forward to extending the SAS project to an infrared satellite—another first (*AMWS*/15; Holden, 49; *WBWRA,* 90, 190).

**Dr. Mildred Mitchell,** head bionicist at the Air Force Avionics Laboratory, specializes in creating or adapting electronic devices to perform human functions, especially during space travel. Although she holds an advanced degree in mathematics, her first work in space research, in 1960, was in psychology (where she also has an advanced degree), leading the isolation studies on the Project Mercury astronauts. She has been a member of the Aerospace Medical Research Laboratory since 1958.

Specific inventions of hers are an artificial muscle, a nail-bender, and artificial biological clocks. The electronically controlled muscle, containing 130,000 separate fibers, is capable of lifting heavy equipment when a human muscle is incapacitated or otherwise occupied, as for instance in the high gravity of a rocket-propelled lift-off. The nail-bender is a small device that can be activated by a puff of air to bend an iron nail,[37] or presumably any desired piece of metal of similar strength. Dr. Mitchell and her team are also working to create means of communicating by touch, using small jets of air as stimuli.

These inventions will have implications for aids to the handicapped as well as for space travel.

## Computers and Science

Another important application of computers is scientific research. Recall that the first test of ENIAC was a massive computation designed to tell the Los Alamos scientists whether their problem with the hydrogen bomb trigger was solvable or not (Shurkin, 188–9). Special computers and entire new languages have been designed with science in mind.

---

37. According to Vare & Ptacek (170), the nail is actually bent by the air. All statements of these writers must be cautiously examined.

WordPerfect, a popular word-processing program of the 1980s and '90s, has a scientific version.

**Jean Sammet,** as we saw, organized Sperry Gyroscope's first scientific programming group in 1955. Still earlier, in the early 1950s, **Margaret Butler** was involved in developing both AVIDAC and GEORGE, two of the earliest digital computers for scientific purposes.

In the 1960s, **Suzette Harold** of Oxford, England, worked on information-retrieval projects for the Atlas computer of the Chiltern Atomic Energy Laboratory. Leaving work in 1966 to have a child, she later rejoined the computer industry through F International, Ltd, which enables computer professionals to work flexible hours from their homes. Through F International, in the late 1960s, she worked to create a critical path program for the Atomic Energy Research Establishment at Harwell, on 360/65 and /75 computers, and to automate a British standard.

With all of this she has still found time for politics (she stood for Parliament as a Conservative in 1974 and again in 1975), her family (she now has four children), and increased responsibility with F International. By 1976 she had become its Managing Director ("Calm View"; T. Lloyd, 72, 78; Simons, 138–9).

In 1975 **Dr. Nancy Martin** (c. 1942–   ) attended the Women's Idea Conference in Socorro, NM. An Assistant Professor of Computing and Information Science at the University of New Mexico, and a consultant to the Heuristic Programming Project at Stanford University, Dr. Martin was creating computer programs capable of designing genetics experiments.

After an undergraduate degree in math from Stanford, Nancy Martin took her PhD in Computing and Communication Science from the University of Michigan (1973). Her specialty was the analysis of algorithms and formal language theory. While at the University of New Mexico she served on the Women's Studies governing board and created a computerized bibliography on women in math, science, and engineering. In 1980 she became head of the Computer Science Department at the University of Michigan (Flint). In the early to mid-1980s she pursued independent research at the Wang Institute of Graduate Studies in Tynsboro, MA.

**Dr. Thelma Estrin**'s computerized laboratory for brain research is another important example here.

### Education and Statistics

In 1956–7 **Helene E. Kulsrud** was Head of Programming for the Educational Testing Service, Princeton, NJ. In that position she worked on statistical applications of computers, and pioneered the use of digital computers for reporting scores on the SAT exams (Kulsrud, c.v. /1984).

**Ann Corrigan**'s work in adapting computer models of population dynamics to personal computers, as noted above, was intended for educational uses (Corrigan).

### Business and Industry

**Carol Prisant** of Roslyn, NY, is only one of many women who have applied computers in a novel way to business. **Evelyn Berezin**'s pioneering office computer, improved editing typewriter, commercial airlines on-line reservation system, and on-line racetrack system leap to mind, for example, along with **Jenny Chuang**'s contributions to the Wang system. But Prisant's application is an ingenious one. In 1982, using a computerized catalogue of 50,000 artists' names, 600,000 prices, and five-year records of sales prices for these and other artists, she created a telephone art-appraisal service called Telepraisal.

Customers can dial Telepraisal and give the artist's name, the painting's date, and its dimensions. Prisant's computer acts on this information, spitting out a list of recent sales by that artist, and allowing Prisant and her son Barden to estimate the value of the painting in question. Data entered on the 50,000 artists may help the customer locate the artist, and computerized records of Telepraisal's own past sales and inquiries may even pinpoint a prospective buyer.

Carol Prisant's background is in antiques; her son's in art history. During their first year of operation they appraised some 4,000 paintings by telephone (Schultz, 25f).

**Betty J. Mitchell** (1926–   ), already mentioned as a programming innovator for Union Carbide Corporation, deserves further mention here. As head of the Quality Assurance Laboratory at the Institute, WV, plant after 1975, she both promoted and participated in the application of computers to reducing and calculating laboratory data in the highly complicated field of quality control (B. Mitchell, Biog/1983).

## Sports

A minor but interesting adaptation of computer technology to sports accessories is **Victoria H. Zevgolis**'s Tennisonic, a solar-powered liquid-crystal digital watch and tennis scorekeeper (1980). Zevgolis, a former employee of the U.S. Department of Health and Human Services, was inspired to invent her Tennisonic after a 1976 doubles match in which all four players forgot the score. Her collaborator, computer engineer Harry Kitchen of Gaithersburg, MD, is co-patentee. Zevgolis herself now lives in Silver Spring, MD, where the company she founded to exploit her invention, Sportsonic, is also located (Zevgolis, c.v., L-4/23 and 5/6/82, Sportsonic prospectus; Burgess, 8).

## Daily Life

The Rock-A-Bye Bear is a prime example of the kind of computer application that standard histories of technology might omit altogether or trivialize as a toy. As a contribution to child-care technology it counts rather differently:

No one can measure the human misery caused through the ages by babies screaming in the night—or much of the day, or both—but we can get some idea of its depth by the desperate means parents and other caretakers have used to put babies to sleep—whiskey, laudanum, and whiffs of oven gas among the more recent. Some cases of physical child abuse are triggered by uncontrollable and seemingly endless crying; and since the overwhelming majority of our prison population were abused as children, the chain of misery becomes not merely personal but social.

Thus it is both pleasant and significant to report that in the 1970s **Marie Shields** (c. 1928–   ) of Ft. Lauderdale, FL, invented a real aid to crying babies and their mothers. With the help of her partner Bob Bissett and an electronics engineer, Ms. Shields created the Rock-A-Bye Bear, a soft cuddly teddy bear containing a microcomputer that activates a tape of recorded human "womb sounds." It can be turned on and placed directly in the infant's crib. To get the crucial recording—the rhythmic "whooshing" of the pregnant woman's blood in the pelvic arteries—Shields and Bissett enlisted a local obstetrician and one of his near-term patients.

The recording got its first test amid an uproar of crying in the nursery of Holy Cross Hospital, Ft. Lauderdale. Eighteen screaming infants instantly quieted down, and some of the nurses even reportedly felt its effects. The Rock-A-Bye Bear works best on babies during the first three months of life. For such young babies, the familiar womb sounds are almost invariably soothing.[38]

By 1979 Shields and Bissett had sold about 1500 bears at $39.95 each, and were working on a model that would clip to the side of a crib or bassinet. They were planning to market the bears to airlines as part of in-flight equipment for dealing with inconsolable, screaming babies (Rieker, 72f).

The Rock-A-Bye Bear is also a prime example of the collaborative nature of much of modern invention. The invention is based on a Japanese doctor's observation reported in the United States in 1975. Shields and Bissett, who both work in construction, needed the help of an obstetrician, a pregnant woman, and an electronics engineer.

While engineering students at Stanford University in 1979, **Nancy Jackson** of Houston, TX, and three male collaborators designed a pocket-sized personal information system that was about five years ahead of its time. Index-card size and only an inch thick, the device contained most of the information needed for daily life, plus some for unusual situations such as travel. It could also be programmed to fit special individual needs. In order to learn what people really needed and wanted in a personal information system, they interviewed four people with various work- and lifestyles: an engineer, a home-maker, a student, and an administrator, spending at least eight hours with each.

The device incorporated a constant time and date display; television and radio receivers, each available at the touch of a keyboard button, and its own small liquid-crystal display (LCD). The keyboard design bore 14 other buttons (plus of course a calculator and alphabetic characters) that might be called information/decision buttons, including APPOINTMENTS, MAP, PHONE LIST, and TRANSLATIONS. Pressing APPOINTMENTS caused other buttons to change to become the branches of an appointments decision tree: year, month, day, hours. Then to view that day's appointments on the LCD, the user could simply press DAY. Or, to preview a future month's appointments s/he could press MONTH, at which point twelve of the other buttons would assume the names of the months. Pushing any one of them would then bring that month's appointments, if any, to the screen.

Input would enter the computer from its own keyboard or from magnetic strips. The Stanford professor who posed this design problem in Industrial Engineering 108 suggested that the automobile clubs might put their maps on suitable magnetic strips. Part of the MAP decision tree is a FREEWAYS branch. A print-out would be available when desired.

When Jackson and her fellow students created this design, the technology for most of the individual parts was available, but no one had combined them in this way. The professor praised their creativity, called their device "a reasonable extrapolation of where electronics technology is, and where it is going in, at most, five to ten years," and gave them each an A in the course (Press, "Students Design," 5).

---

38. The success of this invention prompts a speculation. Observers of so-called primitive societies have often remarked on their calm and quiet babies. Since women in such societies often carry their babies in a sling within easy reach of the breast, the infant's ear would also be very near the mother's heart and could probably hear or sense its beating. This practice might even give a clue to the prevalence of right-handedness among modern humans, in that it suggests a possible survival benefit of right-handedness: An infant carried just under the left breast, to free the right hand for gathering or other work, could hear the heart better, and thus remain calmer and be more positively viewed and treated by its mother, than the infant of a left-handed mother, who would presumably be carried just under the right breast.

## Libraries

As of the mid-1970s, **Henriette Avram** directed research and development for automating bibliographical services at the Library of Congress. In that position she has innovated in one of the most natural applications of computers, and one that is becoming increasingly important as the so-called information explosion continues: data storage and exchange for libraries. Specifically, Avram developed the standard format for machine-readable bibliographies. This format allows libraries to exchange bibliographic records freely without laborious conversions from one system to another. Ever since joining the Library of Congress in 1965, she has been one of the most influential figures in library automation worldwide (*WBWRA*, 179).

Unlike Avram, who is a Government employee operating at the national and world level, **Alicia Paige** (1921–   ), a black woman from Avon, MA, was for years a small-town librarian on Boston's South Shore, and her invention is directed toward the problems of local public libraries. However, the two inventions are actually very similar.

At the end of 1978 Paige's life looked bleak. She was 57 years old, her sister had just died, her divorce had just become final, and she was facing the sale of her house. Luckily, however, when she asked herself, "Where do I go from here?" she came up with an innovative answer. Remembering her frustration, and the frustration of library patrons, when interlibrary loans took six months, she thought, "Now, wouldn't it be interesting if I could put all of the South Shore library holdings on one database, put a CRT in each library, and make the librarians feel like they're big computer operators?" With the system she visualized in place, librarians could call for a book by author, title, or subject, and determine instantly whether any local library had it, and if so, which one.

Encouraged by her brothers, whom she consulted about the idea, Alicia Paige formed a company to exploit her innovation. Since she had used the money from her house for start-up capital, she was often terrified in the early days, for she had no other resources. She describes waking up in the night, sitting up in bed, and asking herself whether she was crazy. At the same time, however, she thought big. "I never thought of just a small business," she says. "I said, no, if I'm going to start a business, I might as well go big. Like IBM." By 1985 her company, Computer Engineering Associates, was doing about $6 million a year in sales. Alicia Paige says she would like to keep going until she has $1 billion in orders (C. Hartman, 47–50).

## Elections

A computer application that touches the lives of most people in developed countries, at least in election years, is the voting machine. Voting machines have been around for a long time, beginning with the purely mechanical ones invented in England in the late 1860s. Thomas Edison invented one, too, and voting machines were used in New York State elections as early as 1892.

In 1977 **Susan Huhn** of Groton, MA, patented a "Compact electronic voting machine" described by *People* magazine as "an electronic marvel that works like a sophisticated calculator, weighs only 35 pounds, and costs about the same as a mechanical voting machine." The invention received considerable publicity in advance of its patent date, with articles not only in *People,* but in the Boston *Globe* in 1976. The machine is both easier to operate and more efficient than earlier machines, instantly detecting voting errors. It can be carried in a lightweight aluminum case that quickly converts to a private polling booth. At its 1976 debut in a Boston union election, Huhn predicted that it would do to its 700-pound predecessors what the pocket calculator did to office adding machines.

The design is an electronic modification of Edison's, which was based on infinite resolution. In a 1982 telephone interview, Huhn (now Eustis) said that the design was still commercially viable and that many others had tried her approach but failed. Her background as the daughter of a voting-machine sales representative, and her own years spent selling and servicing such machines, gave her an edge. In the 1970s she set up her own elections consulting firm, Valid Vote, which also produced and marketed her machine. By 1982, however, she had licensed another company to do the manufacturing and, though retaining part ownership of Valid Vote, was devoting much of her time to the theoretical research mentioned above (CE, "Voting machines"; J. Dietz; *EB*/58, "Voting Machines"; S. Huhn, I-6/22/82; "A Non-Lazy Susan," 15).

## Law Enforcement

**Nancy Burson** used computers to create an "aging machine." Born into the family of a management consultant about 1948, Ms. Burson grew up in St. Louis. She showed artistic talent in early childhood and has worked for some years as an artist in New York City. While attending an art-and-technology exhibit in 1969 she got the idea for a computerized machine that would use a photograph of a person's face to create an image of the way that face would look ten or twenty years later. Computer scientists told her it couldn't be done. She persisted. When computer technology caught up with her thinking in the late 1970s, with the aid of M.I.T. programmers she devised her process and the necessary apparatus: with the original image frozen on a television screen, Burson matches it with its facial type, already stored in memory, then adds characteristics from a separate typed bank of aged faces. As she described the idea in a 1982 interview, "It's like taking a wrinkle mask off one face and putting it on another" (Calio, 94). The machine can also work in reverse to create a younger image, and can create composite faces.

The technology has obvious uses in medicine (plastic surgeons can show what patients might look like after surgery) and in film (when characters age during the course of a movie, directors rely on miracles of makeup, or find other actors to play the younger or older parts; with Burson's invention, the image of the original actor could itself be aged or "juvenilized," and this computer-transformed image could appear on the screen). Such a potentially lucrative use would be welcome to Nancy Burson, who spent some $35,000 to get her 1981 patent, with a male collaborator, for a Method and apparatus for producing an image of a person's face at a different age.[39]

But perhaps the most important use of her technology is in law enforcement. Police departments can "age" photographs of missing children or long-wanted criminals. The ability to create composite faces is useful where police have witnesses' descriptions of a suspect but no photographs. Burson's technology helped locate three missing children in 1988 ("Putting a Face").

As of 1982, when *People* magazine profiled Burson and her "aging machine," she planned to form her own company rather than sell her patent rights, despite several offers. As she put it, she came to New York "to make art—and become rich and famous" (Calio, 96). In 1986 she was featured in a *Smithsonian* magazine article on facial research. The

---

39. As this book goes to press, this potential became reality, though whether Burson's specific process was involved is unclear.

article gave no financial details, but did suggest that she retains control of her invention (McDermott, 113, 114, 122).

## Telecommunications

For years scientists have predicted that we would one day be able to see our telephone callers. Now this prediction has come true, at least experimentally. Funded by the Defense Advanced Research Projects Agency (DARPA), the Media Technology Laboratory at M.I.T. began a project called "Transmission of Presence." But precisely what is to be transmitted? Or, to put it another way, exactly how is "presence" to be transmitted?

Among the more acceptable solutions was devised by **Susan Brennan,** then a master's candidate at M.I.T. She proposed that caricatures be transmitted, and for her master's thesis developed the software to transform faces into caricatures. Starting with a realistic line drawing of the face, the software compares this drawing with an "average" face, consisting of a composite of all the faces stored in the computer. Then the operator exaggerates the differences between the drawing and the composite until a satisfactory caricature-likeness emerges. Automatically animated, the caricature then travels, driven by speech, over ordinary telephone lines. Though by 1986 Ms. Brennan's system had not been used commercially, it proved more successful than some other approaches, perhaps because it mimics the human brain's own strategy for recognizing faces.

Susan Brennan went from M.I.T. to become a researcher in the Hewlett-Packard Laboratories (McDermott, 120ff).

The contributions of **Yvonne Brill** and **Alva Matthews** are also pertinent here, as Brill's invention is used on RCA's communications satellites, and Dr. Matthews was involved in the development of the *Telstar* satellite for linking the United States and Europe for television transmissions.

## Games, Entertainment

**Roberta Williams** (1953– ) of Oakhurst, CA, has done something most of us only dream of: she makes a living from her fantasies. Williams designs microcomputer adventure games. She invented the first home-computer adventure game to use graphics, "Mystery House." Although Williams has worked as a computer operator and programmer, she did not become really interested in the computer world until she began inventing games, and it was her husband Kenneth who programmed "Mystery House" for the computer. Marketing the game through a computer magazine, the Williamses sold $8,000 worth in a single month. Soon they had their own company, Sierra On-Line, as of 1983 doing a $10 million business.

Roberta Williams' second game, "The Wizard and the Princess," continued her innovative track, being one of the first to be programmed on a disk rather than on a cassette. This change revolutionized the microcomputer game industry by making possible much longer games. Most games used six colors; "Wizard . . ." used 21.

Her most complex and interesting adventure game as of 1984 was called "Time Zone." This advanced game occupies both sides of six disks, making it ten times as long as most games, and usually requires months to complete. Programming it took a team of programmers two years. Involving time travel all the way back to the Stone Age and as far forward as AD 2082 to collect some necessary objects, this game attracts a different audience from that of the violent arcade video games. Whole families often play her games. This pleases Williams, herself the mother of two children. She told a 1983 interviewer that Sierra On-Line receives hundreds of letters from families asking for help with one of her games.

Another epic game she designed is the best-selling "Zork." A 12-disk epic,[40] "Zork" may challenge "Time Zone" for its time title (Krueger, 54).

A third-generation computer game designed by Roberta Williams is "King's Quest." This IBM-PC game offers animation, enhanced graphics and sound, and simulated 3-D screens, and is played in real time. The program accepts whole sentences of input from players, and there are two ways to solve each puzzle. The scoring gives more points for less obvious solutions, and bonuses for solving the game within a certain time.

Roberta Williams does plan to design arcade games, but if her career to date is any indication, these games will be unique. As mentioned above, she has also begun to design home-productivity software. She will undoubtedly continue to innovate in her home-computer games as well, experimenting with using "real time" and animation within her adventure games, for example—as she has already begun to do with "King's Quest." In the usual adventure game, the players determine the action. In this new type, the computer unexpectedly takes over at certain points, whereupon the players must figure out how to respond.

Asked about the origins of her inventiveness, Ms. Williams says that as a child she had a big imagination, loved to read, and "did not have that many friends or playmates." She was a born storyteller and loved solving problems. She liked books with female heroes, and was encouraged to achieve by her father, who was a county agricultural supervisor. However, she feels that she had no real mentors or role models. She was "too restless" to go to college, but had some technical training in computers. She is a first child, and reports being told she has some artistic talent, in drawing. She is left-handed.

Roberta Williams has made a striking success in a field whose commercial potential and impact on socialization make it highly significant (John Anderson; Bosveld; Delson; Dolan; Hendra, L-4/3/84; Krueger, 53, 54; Shapiro; Van Gelder, "Bucks"; R. Williams, Q-1984, L-3/13/84).

Several other women have recently entered the field of designing computer games. **Dona Bailey,** a former programmer for General Motors, joined the coin-operated-games division of Atari in 1980. The arcade game "Centipede" was her first project. She used different, lighter colors, and took great care to make the game visually arresting. Her track-ball for monitoring the player's position gives the players more direct kinesthetic feedback than in multi-button flying-target-type games, and they need not visualize everything in their heads. According to Bailey's partner on the project, "Centipede" was aimed at the women's market, and her viewpoint was essential to its success there. However, both men and women seem to like "Centipede," which has been described as "easy to learn, yet tough to master," and a combination of "personality, humor, simple controls, and attractive graphics." Says Anne Krueger, "Dona Bailey definitely knew what she was doing."

By 1984 Bailey had left Atari to join Videa, a design firm founded by three former Atari employees. She had been the only woman in coin-op engineering, and was suffering from what she called "fraternity burnout." She and her colleagues want to do "state-of-the-art games . . . that are funny without being sickeningly cute. Just really good entertainment" (Krueger, 52ff).

---

40. "Time Zone" is variously described as occupying 12 disks or occupying both sides of 6 disks. Thus we presumably have only a problem of computer-world semantics (many disks are called "double-sided" disks), and the two games are probably about the same length.

**Susan Forner** (1953–   ) is a free-lance game designer and graphics consultant for Dave Nutting Associates, the game-design subsidiary at Bally/Midway. A self-taught artist who studied computer and electronic visualization at the University of Illinois, Ms. Forner sees room for artists in the computer-games field because they can provide what engineers can't: "appealing graphics." As of 1984 she was working on a nonviolent, educational game. Neither a shoot-'em-up nor a strategy game, it is designed to appeal to novices at computer games, and particularly to women in their thirties (Krueger, 52).

**Linda Averett,** with her husband Ed, designs games for the Odyssey system. Since computer science was not offered at her college, she holds a degree in engineering physics. Ms. Averett estimated in 1983, however, that at least 50% of software graduates were women. Linda Averett usually does the programming for Averett & Associates, while Ed creates the graphics. Some of the games they have created are "K.C. Munchkin," "K.C.'s Crazy Chase," and "Pick Axe Pete" (*ibid.*).

**Janice Hendricks,** one of the first women to be hired full-time in video graphics and programming at Nutting Associates, designed the computer animation for "Joust," a popular arcade game marketed by Williams Electronics. Both Dona Bailey and Susan Forner have praised "Joust" for its sensitive graphics, ease of movement, and appeal to women. This is no surprise to Hendricks, who designed the "Joust" graphics to appeal to herself.

Originally a psychology major, with an art background, she was enchanted by the graphics at the Siggraph computer graphics show in 1978. She took a few computer courses to try to exorcise the enchantment, but the more she got involved, the more tempting computer graphics became. Eventually she decided to get her master's degree in engineering. Going to the 1979 Siggraph show, she introduced herself to people at the Nutting booth, and three months later she had a job with Nutting. Of that experience, she says, "It's a good place to get started. You're allowed to drift around and find out what you're good at. It's like school" (*ibid.*).

**Sue Currier** (1952–   ) came to the United States from Australia as a model. She began distributing computer games from her kitchen table in the lulls between modeling assignments. Her table-top enterprise grew into Softsync, a partnership with her husband that became one of the main software suppliers for the Timex-Sinclair 1000 (Krueger, 52f).

The comments of a male leader in the field on women's contribution to computer graphics and game design are stereotyped but interesting. Says Dave Nutting, "Women are better at creating the patterns, imagery, and atmosphere for games." And further, they "have more of a sense of feeling and color than men do. Games done by men work fine, but usually will look a bit stiff" (*ibid.*).

*Graphics*

A burgeoning new field of both art and business, with applications in science, printing, and publishing—not to mention computer games—is computer graphics. Museums have exhibited computer art, notably a small but exquisite show at the Science Museum in London in 1985. Books have appeared showcasing particularly beautiful examples of computer art, and classes in computer graphics have sprung up in community college and university extension programs. At the same time, computer graphics have their workaday applications for the charts and graphs of both business and science, and for desktop publishing. The necessary languages and other software may soon create a sideline industry in their own right.

Many examples of women's contributions in this field could doubtless be given. **Nancy Burson, Susan Brennan,** and **Roberta Williams** all have graphics inventions to their credit, as just noted, and at least one woman has designed a graphics language. **Helene Kulsrud** was a research associate in computer science at Yale in the mid-1960s. While there she studied computer graphics and designed a Graphic Language (Kulsrud, c.v. /1984, "General Purpose . . .").

## Military Applications

Some of the above-mentioned work on satellites, rockets, and space travel could equally well be considered here, inasmuch as satellites, their booster rockets, and much of the technology for space travel either originated in war- or defense-related research, or has warfare applications.

**Amy C. Spear,** mentioned above for innovations in computer-aided design, worked at General Electric in Syracuse, NY, on various projects in support of the Atlas and Titan missile guidance radars, as well as on the MOSAR radar and communication satellites. In 1963 she joined the Aerospace Systems Division of RCA in Burlington, MA, where she was a Senior Engineering Scientist in Product Design. While at RCA Spear received a Team Award for her work with six male colleagues on "the first LCSS (Land Combat Support System) Master File for the Army Missile Command." This system gave the Army a means of monitoring LCSS performance continuously in order to determine the operational readiness of weapons systems, locate problems, and start corrective action. The initial development phase of this computerized data bank (the Master File), which involved collecting, evaluating and organizing the data, programming the generation of the data bank and the output reports, and close coordination of all facets of the program with Missile Command, was completed in just 60 days. The successful results were then demonstrated on the Command computer.

In 1974 Amy Spear joined the MITRE Corporation, a Federal Contract Research Center (Bedford, MA), supporting the Air Force Systems Command. In 1977 she became a Group Leader on the Air Force Satellite Communications Program. Subsequently, she provided support for the Strategic Air Command in Nebraska. During these years she has worked at implementing computerized systems, but has not contributed to their design. (Spear, c.v./1987, L-7/1/87, 2/21/89).

## ARTIFICIAL INTELLIGENCE

An application of computers so important that it sometimes threatens to obscure all others, and thus deserves separate treatment here, is artificial intelligence (AI). Indeed, in post-industrial society, where knowledge replaces capital as the most important resource, the significance of computers in general, and AI in particular, can scarcely be overestimated. Artificial intelligence covers a broad spectrum of theoretical and practical work, ranging from "smart" copiers that can help their users use them, to projects aimed at "downloading" the contents of a human brain into a computer. The latter may sound like the stuff of fantasy or science fiction—one more attempt to achieve immortality—but the researchers involved are perfectly serious.

Since Mary Shelley's *Frankenstein* has been called the first exposition of artificial intelligence (Shurkin, 54), and since **Ada Lovelace** visualized the computer composing music, there is distinguished precedent for women's role in AI.

Several women already discussed are doing theoretical work pertinent to AI, such as

**Thelma Estrin,** formerly at the Brain Research Institute and now in the UCLA School of Engineering, and **Susan Huhn Eustis** on her own in New England. **Dr. Jill Larkin** is credited with "major contributions" to the theory of problem-solving (D. Simon, Q-1984).

Women have also contributed to AI in administrative and enabling ways: **Ruth Davis** is credited with unspecified contributions to artificial intelligence. By 1984, British computer innovator and entrepreneur **Steve Shirley** was reportedly preoccupied with AI, or as the British more often call it, knowledge engineering. One of thirty British industrialists invited by the Department of Industry to discuss the Japanese Fifth Generation Project in 1982, she worked to establish knowledge engineering as a goal for the British computer industry (T. Lloyd, 63, 76).

Not yet discussed, but featured by Yarmish and Grinstein as noteworthy for her work on automata theory, among other things, is **Sheila Adele Greibach** (1939– ) of Los Angeles. Graduating *summa cum laude* from Radcliffe with an AB in mathematics and applied linguistics in 1960, she took her PhD in applied mathematics at Harvard (1963), where she later taught. Since 1969 she has taught at UCLA, primarily in the Department of Computer Science. While at Harvard she worked on the Harvard Project in mathematical linguistics and automatic translation (1959–69). During the latter 1960s Dr. Greibach also consulted to System Development Corporation. By 1971 she had already appeared in *Who's Who in Computers and Data Processing.*

Dr. Greibach is the author of a book on the theory of program structures, published in 1975, and more than sixty professional papers. She is a member of the IEEE Computer Society, the American Mathematical Society, the Association for Computing Machinery, and the European Association for Theoretical Computer Science. She is Editor-in-Chief of the journal *Mathematical Systems Theory* and a referee for such journals as *Information and Control, IEEE Transactions on Computers, Proceedings of the American Mathematical Society,* and *Journal of the ACM.* She won the UCLA Woman of Science Award in 1983.

In private life she is Mrs. Jack Carlyle. Born in New York City, she is a first child, and daughter of an inventor. Her father, Dr. Emil Greibach, held more than thirty patents, including one for the first practical bone-conducting hearing aid.

Though her recent work is largely theoretical, it has implications for and applications to the efficiency of compilers, the relationship of structured design to structured programming techniques, and the use of top-down as opposed to bottom-up approaches to program design (Greibach, L-8/18/87; Yarmish & Grinstein II: 27).

No single chapter, much less a chapter section, can cover events in this exciting field, but let us take a sample of what women are doing, beginning with the more practical end of things.

### Hardware, System Design

Researchers realize that computer design is reaching the speed limit for linear processing, and that achieving the speeds necessary for expert systems, much less for true artificial intelligence, would demand—as **Grace Hopper** had long maintained—parallel processing. That is, instead of having the computer do one operation, store the result and then go on to the next operation, the problem would be divided into many operations to be done simultaneously by different units working in parallel, and the results combined as appropriate.

Beginning in 1986, as Head System Analyst and Deputy Head of the Computer Center at the Institute for Defense Analyses in Princeton, NJ, **Helene Kulsrud** has been studying the problems involved in using multiple large central processing units (CPUs) such as the CRAY II and S1 computers in this way (Kulsrud, c.v./1987).

### *"Smart"* Machines

In 1981 design executives at Xerox encountered a problem with their office-size copier Model 8200, which one writer called the Cuisinart of copiers. This giant (small-car-size) machine did everything—it reduced, enlarged, sorted, collated, and even bound copies automatically. But people had trouble operating it. In typical high-tech fashion, the Xerox people took their problem to a cognitive scientist, John Seely Brown, head of their own Intelligent Systems Laboratory at Palo Alto. When Brown and his group videotaped teams of two people using the machine, every user, from computer wizard to marketing vice-president became confused in the same way. If the copier was using people to supply its missing arms and legs the prostheses were a poor fit. The problem was to "make the copier adapt itself to the ability of the human beings using it."

The first specific problem was to make the computer a better "listener." Already it could sense both events (e.g., paper movements) and non-events (nothing happening for a certain interval). Unfortunately, if a user stopped for a break, to talk to someone else in the room or recheck something before copying, the computer might decide the job was done and clear its memory for the next job. (Kleiner, 6). Part of the team on this project was **Lucy Suchman,** a Stanford graduate in anthropology.

A still earlier practical project that began as a mechanical solution to a textile problem but developed into a computer application and eventually into a case for smart machines, was **Maria Luck-Szanto**'s tailor knitting machine (see Chap. 4).

### *Expert Systems*

According to Pamela McCorduck, author of *Machines Who Think,* and co-author of *The Fifth Generation,* as of 1984 only one woman could be considered a pioneer in AI: Japanese-born computer scientist **H. Penny Nii** of Stanford University. "Pioneer, connoisseur, and purist in knowledge engineering," Nii took her graduate work at Stanford, and by the mid-1980s had ten years' experience overseeing the construction of expert systems. Her method of operation, which involves learning from human experts in a given field so that she can train a computer to aid them, is the "essence of cooperation," and her work in expert systems, says McCorduck, is "unparalleled." Indeed, as noted elsewhere, she apparently coined the term *knowledge engineering.* Dr. Nii is married to Edward Feigenbaum, also at Stanford (L-3/15/84; *Fifth Generation,* 25, 79ff). Dr. Nii's reputation is international: Steve Shirley named her as one of two noteworthy women working in AI in the United States (L-4/8/87). The other was **Adele Goldberg.**

Ms. McCorduck speaks, however, of "a number of young scholars" in the field who are female, who have not yet made startling contributions but, she seems to imply, could well do so. Here, McCorduck names no names, though in *Machines Who Think,* she mentioned **Ruth Wagner** (noteworthy work in the early days at the Rand Corporation) and **Cynthia Solomon.**

McCorduck would not have known about **Donna Hudson**'s collaboration with **Dr. Thelma Estrin** on EMERGE, an expert system to aid emergency-room personnel in the handling of chest pain (Estrin, "Computers . . . ," 8.4.4; Hudson and Estrin). She may well have had in mind, among others, **Dr. Wendy Lehnert** (1950– ) who has done pioneering work in AI, and possibly even designed a schema for incorporating emotional states into AI systems.[41]

## Natural-Language Processing

In 1984 Wendy Lehnert won a Presidential Young Investigator Award for her work in artificial intelligence, trying to make computers think like human beings. She believes the key to this attempt is understanding how humans use natural language, how we convey and understand information. Traditionally the understanding of information has been gauged by such things as reading-comprehension tests, where the subject reads a paragraph and then answers questions about its content. Dr. Lehnert points out that such a test shows only how much we *remember,* not how we break the information down into pieces in order to understand and remember it. There is, after all, a storage problem involved for humans as well as for computers.

She believes a better key to this aspect of language understanding is how people would *summarize* the test paragraph, or any narrative they might read. Thus Dr. Lehnert is "teaching" her computer to summarize various types of stories, ranging from the very simple to the very complex. To do so, she breaks a narrative into its basic elements or "plot units," such as loss, retaliation, or competition, and maps the connections between units. Thus armed, PUGG, her computer's "plot unit graph generator," processes plot units into a graph, indicating which plot units overlap to make the best summary.

In technical terms, she has developed a process model for narrative text summarization. Another part of her work is developing natural-language interfaces for artificial-intelligence applications.

One goal in all this is to improve computers as aids to instruction (CAI). Currently available CAI software, says Dr. Lehnert, provides little more than drill-and-practice tutoring programs. Such programs may help with memorization and the solving of routine problems, but can never help students develop higher—or more flexible—reasoning strategies.

In addition to her university-based research, Dr. Lehnert has also served as a consultant to computer companies and other industries, either supervising contract research that could have had practical results or serving as a design consultant. For example, she was Vice-President for Research and later a design consultant on interactive advisory systems for Cognitive Systems, a company she co-founded in 1980. In recent years, she has "pulled back from the commercialization of AI technologies in order to better concentrate on more purely theoretical problems which may or may not result in fodder for the entrepreneurs."

Although computer programs cannot be patented, the innovative nature of Wendy Lehnert's work is clear. Professor of Computer and Information Science at the University of Massachusetts at Amherst, Dr. Lehnert comes to computer science from a background in mathematics, holding both bachelor's and master's degrees in math before taking her PhD in computer science at Yale in 1977. She spent one year in a private, all-girls' high school, and has at least one other inventor in her family tree, the creator of an unusual musical instrument. She credits a sixth-grade teacher for encouraging nonconformity, and mentor professors at both graduate and undergraduate levels; but found no role models either in books or among real women. Her parents showed no professional aspirations for

41. I am indebted to Dr. Saj-Nicole Joni of Wellesley for calling Dr. Lehnert to my attention (L-2/28/84).

her, though both encouraged her to do well in school, thus helping to give her an early love of learning.

By 1987 she had two books and 74 articles (including book chapters) to her credit. She is a member of the American Association for Artificial Intelligence, the Cognitive Science Society, and the Review Board of the *Journal of Artificial Intelligence,* among others, and she is an editor for a journal called *Cognitive Science.* She has a young son whom she calls "a source of great joy and inspiration to me at this (slightly advanced) stage of my life" (Joni, L-2/28/84; Lehnert, Q-1984, c.v.-1984, L-7/23/87; U. of MA news release, 4/23/84).

Prominent among British researchers working on natural-language processing is **Dr. Karen Sparck-Jones** (1935– ) of Cambridge University. An only child, talented in drawing, she attended an all-girls' secondary school and a women's college at Cambridge before taking her PhD in philosophy there in 1964. Her research project, conducted at the Cambridge Language Research Unit, was automatic natural-language processing and information retrieval. Her dissertation was published by the Edinburgh University Press under the title *Synonymy and Semantic Classification.*

Dr. Sparck-Jones emphasizes the "rather indefinite" and interactive or collaborative nature of much computer research, and the consequent difficulty of saying that X invented y. Her own work, she says, has been very much of this indefinite and interactive kind, so that she hesitates to claim "prime responsibility" for any given part of it, or even to classify it "simply" as invention. "The most that can be said," she insists, is that she has made a recognized contribution to certain "approaches to problems" in the application areas that interest her. She has indeed been recognized, and not merely in academia, for in 1984 Steve Shirley identified Dr. Sparck-Jones as one of the few British women doing innovative work in computing (L-5/22/84).

Dr. Sparck-Jones's interests fall under the general heading of linguistic information-processing. She has helped create computational systems capable of manipulating information expressed in natural language, "where that manipulation requires . . . what might be described as an understanding of linguistic expressions . . . as small as individual sentences [or] as extensive as a whole document text." Specifically, she has worked on document indexing and retrieval, devising means of choosing key concepts in documents and characterizing them in such a way that other people can gain access to them. This research has involved investigating the many different ways of expressing the same concept, and has led to the idea of clusters of related words in retrieval-indexing. "Any member of the cluster used by the potential reader can be matched up with any other member of the cluster" occurring in the same document. One of the things she is known for, then, is "semantic classification," finding bases in usage for grouping certain keywords together.

Relatedly, she has worked on weighting words occurring in a given document as more or less significant for indicating the content or usefulness of that document to a particular reader: words that occur frequently, for example, or words that have proved interesting to this same reader in the past.

Dr. Sparck-Jones has also worked on exploiting the idea that words fall into classes, "or . . . express common general concepts" for natural-language processing in general, and particularly for language-processing tasks requiring a "more refined and detailed interpretation of a piece of discourse . . . than the rather selective and schematic one . . . represented by document indexing." One specific area of her work here is enabling computers to interpret English-language questions to a database.

"Earlier on," she adds, she was concerned to undergird this research with some fundamental theory of semantic classification. Her work here, she notes, is original, if not especially well known, and her capsule summary of it thus worth quoting (L-9/9/87):

> . . . I have not been concerned with computers as pieces of hardware, or with things like programming languages. My concern has been with building computational systems to do something we do, namely process natural language material for the information it conveys; this is in the general area of artificial intelligence, but also has a wider relevance if one says that much of our information communication is done through natural language, and if one wishes computers to form a part of society's information processing world, as this depends on natural language, they will have to process language, too.

Besides her published dissertation, Dr. Sparck-Jones has written or edited five books, alone or in collaboration, and has numerous professional articles to her credit. She has been a Research Fellow of Newnham College, Cambridge, lectured at Cambridge, been a Visiting Research Associate of System Development Corporation in Santa Monica, CA, and subsequently a GEC Research Fellow and Senior Research Associate at the Computer Laboratory, Cambridge. She has also been an Official Fellow and Librarian of Darwin College, Cambridge, and served as Editor of the *Journal of Datamation* from 1977 to 1980. She has also been a member of the SERC and Alvey working groups on intelligent knowledge-based systems, and as of late 1987 was a member of the Alvey IKBS Advisory Group and Manager of the IKBS Natural Language Research Theme (Sparck-Jones, c.v./1987).

In private life she is married to Roger Needham, with whom she also sometimes collaborates in research. They have no children, and she notes that she has been "fortunate in not having to cope . . . with combining research and children."

Though she fears that, in the UK at least, computing is becoming more of a male-gender occupation, Dr. Sparck-Jones suggests that natural-language processing may be an area where women's contributions have been recognized. Specific women she considers worth mentioning are **Professor Barbara Grosz** of Harvard, **Dr. Candace Sidner** of Bolt, Beranek and Newman, and **Professor Bonnie Webber** of the University of Pennsylvania (L-9/9/87).

Also, according to **Gill Ringland** of International Computers Ltd in Brackwell, Berks., **Margaret Wellbank** of British Telecom has made "major contributions to knowledge acquisition technology" for expert systems (L-9/14/87). Ringland herself—a computer scientist, software developer and manager, business manager, and entrepreneur—has created noteworthy software inventions or innovations, among them a computer simulation revealing that rival theories on structural formulation of liquids could not be discriminated between. Known in the 1970s as "someone who could 'tame' new technical areas," Ringland devised a set of techniques still in use in 1987 for handling the design stage of the then-new transaction processing (on-line) systems. Her most difficult technical innovation, by her own estimate, was her work at Lawrence Berkeley Lab on data exchange between databases, now (with further development at the Lab after she left) an ANSI standard (Ringland, L-8/26/87).

**Dr. Jill Larkin** (1943–   ), Associate Professor of Psychology at Carnegie-Mellon University, has already been discussed in connection with the application of her research in cognitive science to educational computing. Here she deserves recognition for originating FERMI, an expert system for multi-domain problem-solving. She wrote the original proposal, and developed the system in collaboration with Jamie Carbonell, who now manages the project (Larkin *et al.;* Larkin, L-10/11/87; McCorduck, L-3/15/84).

Dr. Larkin describes FERMI (*F*lexible *E*xpert *R*easoner with *M*ultiple-domain *I*nferencing) as a prototype problem-solving machine. FERMI has already been used to solve problems in fluid statics and in two quite unrelated domains (DC circuits and computing centers of mass). Ultimately, Dr. Larkin and her colleagues feel FERMI should show how expert systems can be built to separate knowledge according to levels of generality, and even form a testable hypothesis as to how human experts separate knowledge to preserve generality. Finally, FERMI may be practically applied to testing how to communicate general knowledge so that human learners can learn to separate knowledge according to its level of generality (Larkin, c.v./1984; Larkin *et al.*).

So far, she considers her most important invention or innovation to have been a model of thought processes supporting quantitative reasoning. A glance at her c.v., however, shows numerous projects whose outcomes might be worth mentioning here.

As noted, **Dorothea Pye Simon,** a colleague until retiring, credits Dr. Larkin with "major contributions" to problem-solving theory (D. Simon, Q-1983; Larkin, c.v./1984, L-8/1/87).

Doing significant work at the University of California in Santa Barbara is **Susan Hackwood,** Director of the Center for Robotic Systems. Among her collaborators is **Evelyn L. Hu,** formerly with Bell Labs in Murray Hill, NJ, and living in Somerset, NJ. In 1979 Ms. Hu, with a male co-inventor, patented Superconducting junctions utilizing a binary semiconductor barrier. This 1979 invention is actually hardware, a physical component of computerized systems. The salient feature is its composition, containing both germanium and selenium (Wiener, PC-10/2, 11/6/87; U.S. Pat. Gazette).

**Pat Wiener**'s work in AI reflects her biophysics training. She is keenly interested in molecular electronics and neural networks and what they can teach us about developing viable high-speed computer systems. Her work also reflects her view of the history of AI, which she says began with an attempt to *duplicate* (and replace) human beings. The initial approach—which she considers off track—was to try to cram systems with all possible necessary facts so that decisions would be 100% accurate. Rather than trying to duplicate or replace human beings, she says, it makes better sense to provide them with high-quality *tools.* Almost anyone can get the right answer or decision with all possible and 100% accurate information; real intelligence consists of arriving at such answers with minimum information. In the context of *artificial* intelligence, then, the question is not how much information or knowledge you can store, but how you represent that knowledge, how you manage the flow of question and "thought" or choice processes, provide correction points, incorporate corrections or changes, etc. In short, what is necessary is a system not limited by facts, but deductively defined (Head & Wiener).

Pat Wiener's current AI work is applying appropriate AI concepts to the automation of such complex enterprises as manufacturing, design, testing, and robotics. Specifically, she has been involved in building the interfaces, both hardware and software, between test instruments and central analyzing computers, laying a viable physical basis for operating expert test systems. She and others dealt with the modeling aspects, and her company (PATSI) has introduced the concept of procedural entity-relation modeling, a methodology that takes time into account in the automation process, and simplifies the process of operating and defining a complex enterprise. It also clarifies the problem of the kinds of knowledge required for the expert or knowledge-based system being created, the means of representing that knowledge, and how it must flow.

In robotics Pat Wiener is working on physical techniques to increase the number of sensing centers in a given robot—to decentralize the sensory input as it were—to more closely reflect human multiple-input characteristics. As she so cogently argues, the real problem with robots, and with expert systems in general, is not so much the storage or even the processing of information, but the *input*. What we cannot duplicate is the human eye or ear, much less the human skin, where every cell sends information to the brain.[42] Along that line, she is designing systems and operative techniques to give mobile robotic devices a redundant recognition of spatial relationships (Wiener, I-8/11/87).

### The Tomorrow-Makers[43]

At the far end, in several senses, of AI research is the work of those who look beyond robots that relieve humans of drudge work, or replace lost limbs or senses, into realms where computers not only think, but educate themselves. Here, the most enthusiastic researchers admit to misgivings about machines turning on their makers; about makers with no particular loyalty to DNA, human or otherwise; about bacteria-size computers that could live on sunlight and eat rocks (thus destroying the planet as surely as any nuclear catastrophe); about "virus" programs that can be implanted to take over computers; and about computers that can't be unplugged or shut off. Here, too, researchers talk soberly of "downloading" the contents of the human brain into a machine brain that could be made self-reproducing, thus achieving the long-sought goal of immortality. As one chronicler of these tomorrow-makers, Grant Fjermedal, put it (122),

> I thought about how delightful it was to be human and live and how vibrant was this odd collection of individuals who somehow get along in a grouping called society. Just as I could sit for a long time and enjoy watching a rookery of gulls or puffins squawking and eating and taking off and landing. . . . Yet puffins and gulls haven't figured out a way to destroy their world, and humans have. . . . Even without artificially intelligent computers and robots entering the fray, it sometimes seemed there was precious little holding things together in a world so filled with the means of destruction.

In these far reaches of AI, Fjermedal profiles no women. The only woman mentioned is **Chris Peterson,** wife of the brilliant Eric Drexel, and he does not discuss her work. Drexel's particular passion is building almost unimaginably tiny robots and computers, so tiny that they could fit inside a single cell and change or repair the DNA. As Fjermedal explains it (167),

> Place such a robotic ship into each cell of your body and they might be able to push regenerative DNA buttons and repair damaged molecules to keep you forever young. You might also be able to change from one person to another. A unified pressing of different DNA keys could result in a change in height, . . . appearance, perhaps even . . . in sex. Depending upon just how much genetic information we carry from our evolutionary past, we may even be able to change from one species to another.

This, of course, is just one of the many scenarios for the possible roles of these microscopic computers. Peterson, an M.I.T. graduate in chemistry, evidently contributes in important if unspecified ways to her husband's work (Fjermedal, PC-spring 1987).

Fortunately, or unfortunately, depending on your viewpoint, women *are* working in these very advanced areas. If Fjermedal didn't find them, it was because he didn't really look. And the reason that he didn't really look was not malice, or even that he is a popular rather than a scholarly writer, but *that he expected not to find any women*. There may be a

---

42. For an illuminating, sometimes hilarious, view of AI researchers' attempts to give a robot even limited "sight," see Grant Fjermedal's *Tomorrow Makers,* 11ff.

43. My heading here is borrowed from the title of Grant Fjermedal's sobering book on the furthest reaches of AI in the United States and Japan (Macmillan, 1986).

difference in research focus or in intended principal use of the resulting technology among the female tomorrow-makers, and their projects will have less funding. But the advanced hardware and software and the necessary set-breaking thinking are women's province as well as men's.

## SUMMARY

In 1983 Ellen McClain wrote an optimistic article on women's participation in the computer industry for *Popular Computing*. Women make up 26% of all computer professionals in the work force, she notes, and according to a National Science Foundation study women receiving bachelor's degrees in computer science in 1978 and '79 are earning virtually equal salaries with their male counterparts. "It's going to take time (and a few retirements)," she admits, "for women to feel that their presence makes a difference in the computing field. Yet the industry is slowly evolving into one of the rare meritocracies in our society. Sex discrimination, already less of a problem than in many other businesses, seems to be receding quickly as a young industry gets younger" (69, 73–4).

Hopeful and persuasive words. A sobering note, however, is sounded by the membership statistics of the IEEE. As late as 1986, only 15,713, or just over 5.5%, of the 282,708 members were women. Worse, only 13 of the prestigious Fellows of the Institute were female out of a total of 3,678, only 128 of the senior members out of a total of 23,704, and only 8,610 of the full members out of 188,939.

On the hopeful side, women's membership in these various categories is increasing faster than the overall increase. From 1975 to 1986, for example, women Fellows more than doubled (increasing from 5 to 13), whereas the overall group of Fellows increased only from 2,589 to 3,678. However, the numbers for women are so small, both absolutely and as a percentage of the total, that this seems little cause for rejoicing. Female senior members also more than doubled their numbers, from only 59 in 1975 to the present 128, still, however, totalling less than 2% of all senior members. The greatest increase among voting members has been in the Member category, where women rose from only 1,494 in 1975 to 8,610 and now number nearly 4.6% of this category.

Among the student members, the increase in female participation is most striking of all, a rise of more than ten times from only 541 in 1975 to 5,821 in 1986. This is more than 12% of the overall student membership for that year (47,891).

Organizational statistics alone, of course, do not tell the tale. Nor has anyone yet written a history of computers and computing that does justice to women's role. This chapter is a first stab at that difficult task, hampered by what might be called the computernik personality: resolutely forward-looking, often too impatient to look back. My impression after this attempt is that in the early days, computing offered more or less equal opportunity—at least to those few women who, like the ENIAC girls and **Dr. Grace Hopper,** happened to be trained in math or were on the spot for one reason or another. But as soon as men realized the world-shaking possibilities of the computer, they began to professionalize the field, and to exclude women. Even where women did manage to enter, they have either chosen or been forced disproportionately into the software and ergonomic areas as opposed to hardware and theory.

Nevertheless, as should be abundantly apparent from the foregoing, women have made noteworthy contributions in *all* areas of computing, including basic materials research. In the late 1980s a team of physicists including **Dr. Angelica Stacy** of Berkeley created two new superconducting compounds that may make possible, among other things, smaller, faster, and more powerful computers than any known today. Dr. Stacy, a Presidential Young Investigator, and her colleagues at the Lawrence Berkeley Laboratory's Materials and Chemical Sciences Division, announced the new compounds in 1987. One of them, still secret, shows some superconducting behavior at temperatures as "warm" as minus 38 degrees Fahrenheit. This would make it usable in labs, factories, and even homes. Earlier, the team had created a compound of yttrium, barium, copper, and oxygen that became a superconductor at minus 280 degrees Fahrenheit. Even this was a tremendous advance over previously known compounds, which needed temperatures near absolute zero. The two newest compounds are being called "the most important technological development since the discovery of the transistor" ("Back to the Future," 13).

The computer revolution has taken place so quickly, moreover, that several of the women pioneers by sheer longevity have survived to a time when women's skills are becoming more desirable and necessary in the field. The era of raw competitive exuberance is gone, and cooperation—as shown by the work of **Lynn Conway** and **H. Penny Nii** and, perhaps most instructively of all, by the Japanese success story—bids fair to become the order of the day. Women are better equipped by their socialization for this new cooperative era of computers than are men. Unless and until, therefore, men quickly become experts in this vital skill, women should have a heyday in contributions to computer innovation in the last years of this century.

# Conclusion

Statistics are people with the tears washed off.—Ruth Sidel

As faces are to souls, so are patents to inventions. —N. Davenport

Women invent. Women have always invented. In fact, if we look back far enough, they may have been the first inventors, and were almost certainly the primary technologists of the species until well after sedentism, when agriculture became a full-time job instead of part of women's work. Women still invent. They invent significant things. They create breakthroughs and fundamental inventions as well as improvements and attachments. They even become professional inventors. And they do all this in the full range of human endeavor and technology, not just in the domestic sphere—which is, in any case, an umbrella for chemical, medical, mechanical, menstrual, child-socialization, and architectural technologies, among others. It is also, by some criteria, the most important sphere of all.

Women invent, but are not, until now, recognized as inventors.

This, to paraphrase Vladimir Nabokov's brilliant introduction to *Laughter in the Dark,* is the whole of the story and we might have left it at that had there not been profit and pleasure in the recapitulation.

The profit and pleasure here will lie in beginning for the first time to clarify—and to quantify—women's participation in invention.

## Statistics: Quantifying Women's Role

Patents, as a recent student of the British patent system remarked, are to inventions as faces to souls: an imperfect reflection, at best. Patents are an even less satisfactory record of female invention than of male invention, as this book repeatedly makes clear, and most of those few who have used patent statistics to study women inventors in the past have produced fragmentary and conflicting results—up to and including Patent Commissioners themselves. That is partly why any sort of statistical analysis would have been useless until now.

Realistically speaking, however, when it comes to overall numbers, patents are the major source of information, and some attempt to quantify my findings seems imperative.

## 19th Century: Patents

According to LWP, between May 5, 1809, when **Mary Kies** received the first U.S. patent granted to a woman, and the end of February 1895, women received 3975 U.S. patents, not counting design patents or reissues.[1] Since some 535,000 regular patents were granted, all told, during that period, women's share of patented invention during those

---

1. Unless otherwise specified, when I speak of total patents, I refer only to ordinary or "utility" patents, and do not include design or reissue patents. When plant patents and genetically engineered organism patents became available in the 20th century, they are likewise excluded unless otherwise noted.

years was less than 1%. It amounted, in fact, using LWP totals, to under three-quarters of 1%.[2]

LWP is not complete. My own research has uncovered many omissions, one of the more dramatic clusters occurring in the year of the Philadelphia Centennial: the LWP compilers omit at least 27 U.S. patents granted to women in 1876.[3] Since they list only 124 patents for that year, this means they missed approximately one woman's patent for every 4.5 they recorded. Compared with the tentatively corrected total of 151 (124 + 27) for that year, then, the omissions rate was nearly 18% (17.9%).

My correction of LWP is still in progress.[4] The 1890 list shows far fewer omissions— 18 so far—compared with 258 listed. Thus the compilers missed only about one patent for every 14 recorded, for an omissions rate of 6.99%. For 1882, however, they omitted at least 45 patents while listing only 74, for an omissions rate of nearly 39%. In 1867 they omitted at least 7 patents, listing 31; omissions rate, 18.4%. If 1876 represents an "average" accuracy for the LWP compilers, the overall omissions will be signifi-cant.

On the other hand, LWP includes a few males, such as Florence Veerkamp, Marie C.A. Ruffin, and Evelyn French.[5] However, the omissions far outnumber the false inclusions; and when an accurate count for the whole century is finally done, LWP's totals will be found significantly too low.

Even writers who use LWP (shown by the end dates of their statistical periods)[6] do not always agree with its figures—or with each other. The discrepancy may mean only that the writer includes design and reissue patents. Nineteenth-century journalist **Caroline Westcott Romney,** for example—herself a prolific inventor—counts 3458 patents granted to women through September 30, 1892, ("Women . . . ," 11) as opposed to LWP's total of 3323 through that date, the end date of Part II. Adding LWP's 134 designs and reissues to that date, however, gives 3457, almost precisely her total.

For others, there is no such easy answer. They seem to have miscounted LWP rather seriously, made their own count or estimate from patent records, or used another list.[7] Following are some of the more noteworthy conflicting reports of women's 19th-century patent activity, beginning with LWP's contemporaries and moving forward in time:

2. Rounding to 4000 (which, given LWP's omissions, is highly conservative), we get .747%

3. I originally suggested 33 omissions ("Conjurer"), but Florence Kroeber (two paents), Mildred Blakey (two patents), and Marie E.P. Audouin turn out to be male, and Lacy Simmons' gender is still ambiguous.

4. It will appear in connection with my projected biographical dictionary of 19th-century American women inventors, possibly available as a separate database.

5. Florence, especially with a Dutch surname, and before Florence Nightingale popularized the name for girls, must always be suspected as the anglicized form of Florens(z); and many French males are named Marie. Ruffin, for example, is Marie Charles Auguste Ruffin. Even today, an English Evelyn is likely to be male.

6. The three parts of LWP end on June 30 (July 1), 1888; September 30 (October 1), 1892; and February 28 (March 1), 1895. These seem arbitrary and unhelpful, not corresponding to decades, census years, or even year ends. However, the three compilations were evidently done at the instigation of feminists, notably Charlotte Smith, in connection with 1) the Patent Office centennial (1890); 2) the Columbian Exposition in Chicago, opening fall 1892; and 3) the Atlanta Exposition of 1895.

7. The specter of another major list of 19th-century women inventors has haunted my research, but I could never either authenticate or exorcise it. Charlotte Smith certainly planned such a list, but to my knowledge never managed to produce it. She acknowledged a list furnished her by Mr. Du Bois, Editor of *Inventive Age,* "of the women inventors that came under his observation," but gives no time span or indication of the numbers involved (*WI*/1: 2). Rev. Ada Bowles spent a dozen years researching women inventors and may have had similar plans, but all I have found is a report in the *Woman's Journal* (Mar. 18, 1899: 88) of a

References to LWP began to appear even before the compilation itself was published. Ida Tarbell's article "Women as Inventors," for example, appeared in the *Chautauquan* for March 1887. To indicate the importance of the forthcoming list, Tarbell cites Mary Lowe Dickinson,[8] in "a late issue of this magazine," as saying that only 334 patents have so far been issued to women, and that of 22,000 issued "last year" (1886: 22,508, incl. designs and reissues), only 90 were to women.[9] Mrs. Dickinson, says Tarbell, does not make careless statements, and doubtless used the best published sources for her figures. But "until Mr. Gill compiled his record, not even the half-truth was known" (355). Actually, Tarbell continues, *1935* U.S. patents have been granted to women through the end of 1886. But Tarbell's count is also low, even though she says the LWP manuscript is complete up to December 14, 1886: the published version of LWP lists 2019 women's patents by that date (2032 through the end of the year).

Tarbell's detailed counts for early years and decades are also at odds with LWP. She reports one patent for each of the years 1821, '22, '25, '28, '31, '34, and '41, whereas LWP has none for 1821—omitting **Sophia Welles'** important Wethersfield hats—but does have patents for 1815, '19, '23, and '26, '33 (two), '39 (two), and '40, and lists multiple patents rather than just one for 1828 (two), '31 (three) and '34 (three).

In the 1850s, Tarbell reports 13 patents, whereas LWP finds 23; in the 1860s, 216 *versus* LWP's 268.

For the single year 1876 she reports 136 U.S. patents granted to women. Here she exceeds LWP's published total of 124 by a dozen patents! Perhaps some of LWP's many omissions for that year were failures to transfer listings from a draft—which Tarbell saw—to a final manuscript. For 1886, up to December 14, she reports 169 patents. This, added to the 13 patents recorded by LWP for December 21 and 28, would equal 182, slightly exceeding LWP's 1886 total of 180. All in all, a puzzling series of discrepancies.

In April 1891 Charlotte Smith reports in the first issue of her *Woman Inventor* that women have 4000 patents to their credit, a very high figure considering LWP's total through March 1, 1891, of 2905. Later in that issue, by contrast, she says that women received 537 patents from July 1, 1888, to March 1, 1891. LWP lists 628. Since Smith's son worked in the Patent Office at this time, these discrepancies are doubly puzzling. Since she uses July 1, 1888 (the break date between Parts I and II of LWP), as her starting point, and since she was the main instigator of the original compilation, she was almost surely using LWP and not another published list.[10]

Indeed, a closer look at this passage makes unmistakable that her reference was to LWP, for she cites its early numbers in some detail. She correctly notes LWP's count of

---

speech she gave at the Massachusetts Woman Suffrage Association earlier that month, and an undocumented reference in Mozans to some of the material mentioned there. Mary Logan may have seen a list that extended into the early 20th century, but she gives no sources.

Several early lists of U.S. patentees exist (see Burke's *List;* Madison; Leggett; U.S. Patent Office, 1872), but they contain only a few women inventors (Leggett's list is by type of invention).

8. Educator, lecturer, philanthropist, and author, born 1830 (Logan, 713). Tarbell calls her "one of our noblest and best informed advocates for more work and better wages for women" (355).

9. LWP finds 180 women's patents for that year.

10. More interesting than the mistaken number here, whether the mistake is a typo or reliance on preliminary figures, is Smith's statement that, thanks to Patent Commissioner Mitchell, she can present a complete *classified* list of women's patents for these nearly three years. Did Smith herself propose to do the classification? Or was Part II of LWP originally intended to contain a classified listing like the one appended to Part III?

19 women's patents between **Mary Brush**'s corset patent of 1815 and July 10, 1840 (including the latter); 39 women's patents between 1840 and 1860 (including the latter); and in 1857 their largest annual number to date, six. With the 1860s, the discrepancies begin again, for she counts 261 "from the year 1860 to 1870" (which should, since she counted 1860 with her previous total, mean 1861 through '70), whereas LWP has 263 for 1861–9, and 60 for 1870, totalling 323. "From 1870 to 1880" (presumably 1871 through '80) she counts 897, whereas LWP has 891. Finally, as noted above, her count for July 1, 1888, through March 1, 1891, is nearly 100 patents too low.

Elsewhere in the same issue, Smith says that some 20,000 patents are granted each year in the United States, about 700 of them to women. Obviously, unless 1890 and '91 have shown dizzying increases over the 1880s, not reflected either in LWP or in my own count of 1890 (excluding designs), this not only conflicts sharply with her count of 537 women's patents granted between 1888 and '91 (c. 200/year), but is much higher than any other careful count. Indeed, by most counts, women did not reach 700 patents a year until the 1920s, and did not reach a 3.5% share of all patents granted until the 1980s!

In 1893, Mary Lockwood, charged with investigating women's patents for the Board of Lady Managers at the Columbian Exposition in Chicago, reported that "over 4000 patents have been issued to women from 1809 to 1893, 2000 of this number having issued since 1888 . . ." (Weimann, *Fair,* 428). Assuming that "to 1893" means "through 1892," and assuming that she was using LWP, as her statistical boundaries suggest, her figures are puzzling. LWP lists 3323 regular patents through September 1892 (up to October 1). Adding the 134 designs and reissues listed by LWP to that date brings the total, as noted above, to 3457.

Lockwood's further statement that half of women's 4000 patents were issued in the five years 1888 through 1892 is also puzzling. LWP's total for Part II, July 1, 1888, through the end of September (to October 1) 1892, is only 1026; even with the 75 designs and 1 reissue listed, it is still only 1102. Lockwood's counts diverge so far from LWP that it seems she may have used LWP as her base, retaining its periods, but supplemented the listing with her own research—or with another list.

Likewise, Lockwood's estimate that women averaged 229 patents a year in the decade preceding the Fair seems high.[11] According to LWP, women received 1913 patents in the decade 1883–92, for a yearly average of about 191. The 80 design and reissue patents recorded for the decade bring the total to 1993, for an LWP average of about 199 per year.

An 1899 article in *The Patent Record* ("Women as Inventors") maintains that before 1860 "there were scarcely half a dozen patents taken out by women," whereas LWP

Smith's instigation of LWP is recounted in this same issue of the *Woman Inventor* (*I*/2:3–4). She mentions calling upon Commissioners Marble (who first ordered lists of women inventors kept), Babson, Butterworth, Montgomery (under whom the compilation began), and Hall (who told her he lacked the necessary "clerical force"). She called upon Hall seventeen times, appealed to Congress, and appeared before the appropriations committees before finally getting the "necessary assistance." Under J.B. Marvin, Chief of Draftsman's Division, "the work of compiling 'Women Inventors' [i.e., LWP, Pt. I] occupied four clerks about ten days."

11. It may also conflict with her earlier statement that women received half (or 2000) of their overall total of 4000 patents between 1888 and 1892; that would be an average of 500 a year for those four years (1889–92, inclusive). Moreover, if women averaged 229 patents a year for the decade 1883–92, they would have received 2290 patents overall; if 2000 of them fall during the last four years, that leaves only 292 for the first six years, for an average of 48.6 per year, which is ridiculous.

records 55, and actually there were several more, including **Charity Long's** consumption remedy of 1812.[12]

After researching the topic for twelve years, Rev. Ada Bowles gave a talk on "Women as Inventors" to the Massachusetts Woman Suffrage Association in March 1899. As reported in the *Woman's Journal* (March 18, 1899: 88), she found 15 patents in the 25 years following **Mary Kies'** patent of 1808 (*sic*), i.e., presumably through 1833. LWP shows 14. For the next 25 years (through 1858), her total agrees with LWP: 35. For 1859–84 (through 1883?), however, Bowles finds 1503 women's patents, whereas LWP shows 1544. Then, according to the *WJ* report, for the twelve years 1884 through 1895 inclusive (presumably through February, since she says "the latest date to which the patent office reports have been published"), women have received 3905 patents. This last must be faulty reporting of her talk, perhaps compounded by a typographical error, for this is very close to LWP's overall total from 1809 through February 1895 (3975). An important item of further research will be to find Bowles' original work, or at least the paper as delivered to the W.S.A. However, since her figures are usually the same as or less than LWP's, she must not have made an independent count of women's patents.

Perhaps most surprising of the conflicting reports on women's patent share in the 19th century comes from the Patent Office itself. Commissioner Duell, in his report for 1900 (xi), says, "From 1790 to March 1, 1895, some 5,535 patents were granted to women." This figure is so high that I wondered whether he made his own count, not using LWP, whose total for regular patents, as mentioned, is 3975. Even including the 193 design and reissue patents tabulated by LWP raises LWP's overall total only to 4168. However, since Duell uses the cutoff date of Part III (March 1, 1895) as his endpoint—otherwise an awkward and unhelpful stopping place—I conclude that he used LWP. Was a line omitted from his text in which he extrapolated from LWP's totals to the end of the century? As will be seen below, if women continued to patent during the last half of the 1890s at the accelerating rate obtaining in 1890–4, they would have arrived at something close to Duell's total by December 31, 1899.

Yet in his next sentence Duell contradicts himself, saying, "It is a fair estimate that *out of every one thousand patents, one is granted to a woman*" (emphasis mine). That would mean only about 530 U.S. patents granted to women since the beginning, which is ten times fewer than he has just proposed. Substituting "thousand" for "hundred"—which would fit with his earlier statement—is difficult to blame on a typesetter.

H.J. Mozans, whose 1913 book *Woman in Science* was reprinted as late as the 1970s, and who devotes a chapter to women as inventors, is another who apparently uses LWP and yet often conflicts with it. His figures for the early decades of the 19th century differ only slightly from LWP's totals—though it is puzzling that they should differ at all in those scanty years which are so easy to count. By the 1860s the differences suddenly become striking. Between 1861 and 1871, says Mozans, 441 patents were awarded to women. LWP reports 323 from 1861 through 1870, inclusive. Even if Mozans includes designs and reissues, he could scarcely reach 441 unless he made his own count or supplemented LWP with another list.[13]

---

12. What the *Patent Record* writer probably meant, as a later writer stated more accurately from what sounds like the same source, is that before the Civil War women *averaged* fewer than six patents a year. According to LWP, women's total patents reached 6 in only one year (1857) before either 1860 or the Civil War.

13. Mozans refers to Ada C. Bowles, already mentioned as a researcher on women inventors, but the reference is vague and undocumented.

Another serious disagreement between Mozans and LWP appears in his figures for "the seven years following 1888"—essentially to the end of LWP (Parts II and III). Mozans says that women received 2526 patents during that time, which would bring his total for the entire LWP period up to 5130. LWP's total (regular patents only) is 3975. Again, even if Mozans counted or estimated to the end of 1895 (whereas LWP stops after February), and included designs and reissues, he could never reach such a large total by starting from LWP. What other sources Mozans may have used, and whether Mozans or LWP is closer to the correct total, remains to be determined by further research.

A 1940 *New York Times* article ("Women Gaining . . .") states correctly (according to LWP) that "Prior to the Civil War . . . , the number of patents granted to women averaged less than six a year." However, the writer goes on to say that "During the war this number jumped at times to over 100," whereas according to LWP, the most patents women received during any Civil War year was 25 (1863); and they never received 100 patents in any single year until 1876.

In 1975 Katherine Lee did a master's thesis at San Jose State University titled "Women and Patents: A Historical Investigation." Her work is noteworthy, despite its many errors, because she did not use LWP or any other secondary sources, but made her own count from the patent gazettes, the annual reports of the patent commissioners, and an early compilation of all patents granted 1790–1848 (presumably Burke's *List*). She examined only even-decade years and, after 1870, only every other week of the gazettes for those years. She often overcounts by assuming ambiguous names are female when actually they are male,[14] but she does find several women not listed by LWP. For example, she locates a remedy for consumption patented in 1812 by **Charity Long,** whereas LWP finds no women's patents between **Mary Kies** in 1809 and **Mary Brush** in 1815. Lee's 19th-century figures compared with LWP's are as follows (16):

| Lee | | LWP | |
|---|---|---|---|
| 1790–1848 | 19 | 1790 (1809)–1848 | 29 |
| 1850 | 3 | 1850 | 3 |
| 1860 | 10 | 1860 | 5 |
| 1870 | 60 | 1870 | 60 |
| 1880 | 44 X 2 = [ 88][15] | 1880 | 90 |
| 1890 | 108 X 2 =[216] | 1890 | 258 |

Lee's count of 10 for 1860 diverges further from LWP than the numbers alone show; for her text makes clear that she has located four patents not on LWP. The same can be said for the 1870 counts. Though the totals are identical, Lee finds at least four patents not listed by LWP—one of which appears to be a band saw or planing machine—and includes two design patents in her total as well. For 1880, where the totals again look very close, Lee finds at least five patents not listed by LWP, and includes three design patents.

14. The seriousness of this error can be seen in the 1940 figures, for example, where Lee counted Lavern Jordan as female (tellingly spelled "Laverne" in her text [37]) when actually he is male. Since Jordan received seven patents in 1940, and Lee found only 740 patents for women that year, all told, her total was off by nearly 1% by virtue of that mistaken assumption alone (38). Similarly, Lee read Aristid V. Grosse as *Astrid,* and therefore counted eight chemical patents as women's that actually belonged to men.
15. For 1880 and 1890, I double Lee's 26-week total to make her figures comparable with LWP's full-year totals.

In 1890, Lee finds two patents not on LWP, and includes one design patent in her total (48ff).

## LWP

Imperfect though it is, LWP is the only known systematic count of women's 19th-century U.S. patents with any pretensions to completeness, and since my comprehensive correction is itself incomplete, for present purposes I will rely mainly on LWP.

Three-quarters of 1%—or even 1%—of all U.S. patents may not seem an impressive figure. However, we must remember first of all that these patents are but the tip of an iceberg of women's *inventions* that were either never patented (mostly passed on freely to friends, neighbors, or children), or patented by men, particularly before passage of the Married Women's Property Acts. One potentially explorable part of the iceberg is the corpus of patents granted to men but assigned to women. It is not irrelevant, either, that the Patent Office and the patent system—including definitions of *invention* and *technology,* criteria for patentability, the kinds of things people were encouraged to invent, and what they received recognition for inventing—were set up by men, with reference to men's work.

Given all this, the rapid increase in women's share of patents over the century, as shown by the totals at the end of the three parts of LWP, is certainly impressive:

| | |
|---|---|
| May 5, 1809—June 26, 1888 | 2297 patents |
| July 1, 1888—Sept. 27, 1892 | 1026 patents |
| Oct. 1, 1892—Feb. 26, 1895 | 652 patents |
| TOTAL: | 3975 patents |

As can be seen, women received almost half as many patents between the summer of 1888 and the fall of 1892 as they did in the preceding nearly eight decades! To look at it another way, in the five decades between **Mary Kies'** patent and the end of the 1850s (through December 1859), according to LWP, women received only 55 patents, or 1.1 per decade. In the 1860s, however (January 1860–December 1869), under the powerful double impetus of the Civil War and the Married Women's Property Acts, women received 268 patents. This was almost five times as many as in the preceding five decades. Indeed, in the single year of 1869 they received more patents (56) than in those first five decades. And the curve continued to soar:

| | | |
|---|---|---|
| 1809–59 | 55 patents | 1.1/decade |
| 1860–69 | 268 | 268 |
| 1870–79 | 861 | 861 |
| 1880–89 | 1428 | 1428 |

2612 (TOTAL through 1880s)

From January 1890 through February 1895, women received nearly as many patents (1343) as they had in the whole decade of the 1880s, which was in turn more than half again (166%) as prolific as the 1870s. Patent activity in general was increasing over these decades, of course, but women's rate of increase overtook men's in the 1860s, and continued to accelerate faster throughout the century. The Women's Bureau of the U.S. Labor Department provides a useful tabulation, using LWP (WB/28: 12):

|  | PATENTS ISSUED TO | | | |
|  | Women | | Men | |
| Periods after 1790 | Average number/yr | % incr. over previous period | Average number/yr | % incr. over previous period |
|---|---|---|---|---|
| 45 yr. preceding 1836 | 0.4 | — | 200.9 | — |
| 9.5 yr. preceding 1845 | .7 | 85.0% | 456.9 | 106.8% |
| decade ending 1855 | 1.3 | 75.7 | 964.8 | 111.2 |
| decade ending 1865 | 10.1 | 676.9 | 3,767.4 | 290.5 |
| decade ending 1875 | 67.3 | 566.3 | 11,918.4 | 126.4 |
| decade ending 1885 | 106.0 | 57.5 | 16,079.3 | 34.9 |
| 9 yr. ending 1894 | 229.8 | 116.8 | 21,784.0 | 35.5 |

Since LWP stopped after February 1895, the Women's Bureau elected to omit that year; and data on women's patents for the last five years of the century must be dug out of the original records.

The Report of the Board of Commissioners representing New York State at the Cotton States and International Exposition held at Atlanta in 1895 says that "over five thousand American women had made inventions worthy of being recognized by the Government" (48). Even counting designs and reissues (193) this is significantly higher than LWP's total of 4168 (3975 + 193).[16]

George Walsh says in a *Patent Record* article of 1899 that 90% of women's U.S. patents were granted in the twenty years preceding his article. Women's patenting achievement in these two decades was indeed impressive—probably over 4000 regular patents. However, this two-decade share of their century's achievement, as will appear below, is probably nearer 80% than 90%.

Mozans, without naming his source, says that women received 3615 patents between 1895 and 1910 (346). This seems low. Since women received at least 328 patents in 1905, 400 in 1906, and 488 in 1910, the average for the decade 1901–10 should approach 400, and the total for that decade alone, 4000. This leaves fewer than zero for the five years in question, 1895–9. Of course, Mozans wrote before my source for these figures (WB/28) appeared, and may have been relying on partial samples or Patent Office reports. Even if we simply divide his 15-year total of 3615 by three to obtain an approximation for 1895–9, we get only 1205—which is too high (since these should be the lowest five years of the fifteen) yet significantly lower than the demonstrable output for 1890–4, and thus almost certainly wrong.

A possible calculation based on LWP is as follows: Women received 1308 patents for 1890–4 inclusive. A minimum or base estimate of their patent output for 1895–9 would therefore be 1308. Then, since women's output 1885–94 was more than double that for

---

16. *Patents* and *women* are not precisely comparable, since some women received several patents. On the other hand, some women had female co-patentees. As the former outnumbered the latter, the LWP number for women would actually be lower than 3975.

1875–84, and was still accelerating (as shown by figures from the early 20th century), we need an acceleration factor. Women's patents reached 328 a year by 1905 and 400 by 1906. Thus it seems reasonable to suggest that the average output for 1895–9 was higher than the 1890–4 average of 261+—say, perhaps 275–80. This would raise the 5-year total to 1375–1400, not including designs or reissues. Adding this to LWP's total of 3940 (through December 31, 1894) gives us an LWP-based estimate of 5315–5340 for women's total U.S. utility patents through 1899. Considering LWP's many omissions, this can safely be rounded to 5350.

### British women's patents

Far less has been written about British women inventors, if this can be imagined, than about U.S. women inventors. Judit Brody of the Science Museum Library finds 62 patents[17] registered in women's names from 1635 to 1852 approximately 0.39% of all British patents to that date ("Patterns," 4). As Brody explains, getting a British patent was both time-consuming and exceedingly costly: protection for England and Wales alone cost nearly 95 pounds; and extension to Scotland and Ireland cost a further 216 pounds—an enormous sum in those days, more than three years' income for most people.

A more extensive count for British women, through 1884, appears in a paper read by Helen Blackburn at the Patent and Trade-Mark Congress connected with the Columbian Exposition in Chicago in 1892–3 (cited by Romney, "Women," 11):[18]

| | |
|---|---|
| 17th century | 7(?) |
| 18th century | 10 |
| 19th century through 1852 | 40 |
| 19th century through 1884 | 178 |
| TOTAL | 235 |

American women, by contrast, had received 1707 patents by 1884 even by LWP's imperfect figures, and not including any design or reissue patents. Of course, overall totals for British patents are also much lower; thus the relative share for British women may be less discrepant than first appears.

### 19th Century: Other Indicators

Other indicators of women's share in 19th-century American invention are census data, the numbers of inventions women exhibited at the great expositions and fairs of the century, and the numbers of inventions they advertised in inventors' journals, seeking development partners, or offering a patent or an invention for sale. The latter two sources include unpatented as well as patented inventions, and select for those inventions being

17. This count exceeds the 57 obtainable from Blackburn's text through that date (see just below). However, both were based on Woodcroft's tabulation through 1852, as Brody's references and Blackburn's end dates make clear. As indicated by my question mark, Blackburn may not mention each patent for that early time.
18. Helen Blackburn (1842–1903) was a British suffragist, editor of *The Englishwoman's Review* (1881–90), and author of the standard work on woman suffrage in Britain. Despite queries directed to Girton College, Cambridge, where her papers are held, this paper has proved elusive. I would be grateful for a copy (or information leading me to a copy) of it.

actively marketed by their inventors. Presumably these were inventions with some commercial potential.

*Census data*

The census records are particularly interesting in that they document a very rare breed—women who classified themselves as inventors, i.e., as inventing for a living or as their main occupation. Professional inventors are rare in both sexes, of course, but have been thought virtually nonexistent among 19th-century women.

Before 1900 the occupation of inventor is not separately tabulated, but is combined with architects, designers, draftsmen, etc. During the late 19th century, the percentage of gainfully employed women reporting this group of occupations was actually rising, as were the total numbers, from 2,820 designers, draftsmen, and inventors in 1880 to 18,943 in 1900:[19]

| | |
|---|---|
| 1880 | 1.2% |
| 1890 | 1.9% |
| 1900 | 3.5% (1,041/29,524) |

When architects are separated out and designers, draftsmen, and inventors tabulated as a group for 1900, women's share rises to nearly 5% (4.97%: 942 out of 18,946). Unfortunately, there is probably no way either to separate out the number of women inventors or to link these numbers with names of known women patentees.

*Expositions*

A striking increase in women's inventive participation over the century can be seen between the Centennial Exposition of 1876 and the Columbian Exposition of 1892–3. At Philadelphia in 1876, women showed 79 inventions in the Women's Pavilion and won 14 Centennial awards (Warner/Trescott).

At Chicago in 1892–3, women showed at least 335 inventions and won at least 33 awards. This was the personal count of Mary Lockwood, chair of the Board of Lady Managers Patents Committee, who points out that gathering patents and models for the Exposition was severely hampered by lack of funds.[20] One woman alone—**Caroline Romney**—showed 13 or 14 inventions, only one of which, to my knowledge, was patented. This figure of 335 is not strictly comparable to the Philadelphia total, because it includes inventions exhibited in the State buildings (Weimann, *Fair*, 178), and other departments as well as (or instead of) the Woman's Building. However, it is a useful indicator.

The only published source I've found that focuses on the Woman's Building inventors at Chicago in useful detail (Romney, "Women . . .") is incomplete: a multi-part magazine article, only one part of which apparently ever appeared.[21] The issue of the *Journal of Industrial Education* carrying Part I may have been the magazine's last issue. This initial part, mostly occupied by introductory material, mentions 23 women exhibitors in the

---

19. In 1870, 352 persons classified themselves as inventors, but my source did not tabulate them by sex (*Occupations*/1900: xxxiii).

20. Chicago, Board of Lady Managers, Official Minutes, Report of Committee No. 28, Ethnology, Archaeology, etc.: 121. I am grateful to Jeanne Weimann, author of *The Fair Women*, for calling my attention to this reference.

21. It may be incomplete in other ways, too, for Deborah Warner says that **Margaret Colvin** exhibited her Triumph rotary washer at the Columbian Exposition as well as at the Centennial (112).

Inventions Room of the Woman's Building, covering surnames alphabetically through *H*. Romney comments that the Inventions Room was too small to accommodate all the women inventors who wished to exhibit, and that "the space allotted to women in the Manufactures and Liberal Arts Building is also similarly limited," so that women who were "at all dilatory" about making their applications were "left out in the cold." This, says Romney, accounts for the Exposition's "comparatively small display" of women's inventions, and is "certainly not a fair representation of what they have accomplished in this direction" (23). Romney evidently intended to name all or most of the women exhibitors she had observed, starting with the ones in the Woman's Building.[22]

*Inventors' journals*

The *Patent Record* (*PR*) was one of several inventors' journals started in the 19th century and continuing into the 20th. A count done of women's share of monthly *PR* lists—"Partners Wanted" and "Inventions and Patents for Sale"—for 1899 and 1900 shows markedly higher percentages for women's inventions than have ever been estimated for patents in these years, partly, of course, because some unpatented inventions are included.

In ten months of 1899, 313 inventions were listed whose inventors were seeking development partners. Of these, at least 17, or 5.43%, were women, and another 157 were listed by initials only or had ambiguous names.

In 1900 (whole year) the comparable figures were 847 inventions, 61 of them by women, and 382 names of undetermined gender. The minimum percentage of women here, even if all the ambiguous names and initials-only listings turned out to be male, was 7.2%.

Likewise, in the "Inventions . . . for Sale" listings, for the ten months of 1899, 39, or 4.9%, of the 795 inventions listed were women's, and nearly half (358) of the remainder were ambiguous names or initials-only listings. And of the 1506 inventions listed for 1900, 99, or more than 6.5%, were women and 596 uncertain.

Thus it appears that either women's inventions were somewhat more commercially viable than men's (about which, more below), women were somewhat less likely to patent their inventions than men, or other counts of women's patents for these years have been much too low—or all three. It should be noted, however, that the number of unpatented inventions included in the lists is quite small.

*The American Inventor* and *The Inventive Age,* both published, like *PR,* in Washington, DC, ran complete lists of all patents granted since the previous issue, and a "New Inventions" or "New Patents for Sale" column in each issue. The complete lists are of no more use than the Patent Office name lists, and the inventions/patents-for-sale lists use initials for first names so heavily that the research involved in obtaining a count of women's participation for these two periodicals would be monumental.[23] However, a check of the June 1, 1899, issue of the *American Inventor* shows that at least four out of 49 inventions for sale were women's (over 8%).

---

But Colvin does not appear in Romney's article, even though Romney carried the Woman's Building inventors through *H*. Of course, Colvin may have exhibited elsewhere in the Fair, even though at Philadelphia she exhibited in the Women's Pavilion.

22. Romney's article, "Women as Inventors and the Value of Their Inventions in Household Economics," appears in the September 1894 issue (vol. 9, no. 1) of *JIE*. I have been unable to locate the October or any subsequent issues, and would be grateful for any information on this publication, or on Romney's papers.

23. However, at least a sample study of *IA* on this point will appear in my projected 19th-century volume.

## 20th Century: Patents

After the turn of the century we have new benchmark sources for women's share of U.S. patents. The first of these is a count and compilation done by the Women's Bureau of the U.S. Department of Labor in the early 1920s, here called WB/28. This excellent source covers an artificial decade (ten select years) between 1905 and 1921: 1905 and '06, 1910 and '11, 1913 and '14, and 1918–21.

The WB/28 figures are as follows (13):

| YEAR | ALL PATENTS | PATENTS TO WOMEN | |
|------|-------------|------|------|
| | | # | % |
| 1905 | 29,734 | 328 | 1.1 |
| 1906 | 31,181 | 400 | 1.3 |
| 1910 | 35,168 | 488 | 1.4 |
| 1911 | 32,917 | 413 | 1.3 |
| 1913 | 33,941 | 501 | 1.5 |
| 1914 | 39,945 | 522 | 1.3 |
| 1918 | 38,569 | 666 | 1.7 |
| 1919 | 36,872 | 494 | 1.3 |
| 1920 | 37,164 | 638 | 1.7 |
| 1921 | 37,885 | 566 | 1.5 |
| TOTAL: | 353,426 | 5,016 | 1.4% |

Thus in the equivalent of a single decade, women received significantly more patents than in the more than 85 years tabulated by LWP (1809 through February 1895), and almost as many as in the entire 19th century, by present (LWP-based) estimates.

Other sources seem to disagree with WB/28 almost as often as with LWP. As already mentioned, if the progression suggested by WB/28's figures for 1905, '06, and '10 is indicative,[24] women should have received some 4000 patents during the first decade of the century. Thus Mozans' already-mentioned estimate of 3615 patents to women for the fifteen years 1895–1910 is almost certainly much too low, especially since he seems usually to include design and reissue patents. Since women were already averaging more than 261 patents a year for 1890–4 (LWP), and received 328 patents in 1905, 400 in 1906, and 488 in 1910, as noted, it seems more likely that for those 15 years 1896–1910 women averaged nearly 370 patents annually, for a total more like 5550, exclusive of designs and reissues. Since WB/28 omits 1895, however—and begins again in 1905, using only the selected years just mentioned, there is no precise way to compare Mozans' figures with WB/28.

Katherine Lee's counts for 26 weeks of 1910 and 1920 are 276 and 364, respectively; doubled, they are 552 and 728, both considerably higher than the WB/28 figures. However, the discrepancies may be accounted for in part by Lee's inclusion of design patents, which numbered 77 of her 1920 total, for example.[25]

---

24. It should be noted again here that WB/28's figures are always conservative, because the compilers omitted all patents whose patentees' names were ambiguous, and included only patents issued in unmistakably female names.

25. Doubling this to 154 would probably not be accurate, since one inventor, **Alice H. Flanders,** alone received 19 of those 77 design patents, a situation not likely to be repeated in the other 26 weeks of gazettes.

If women invented at roughly the same average rate in the eleven unselected years as in the select "decade," then they should have received *at least* 10,000 patents in the first 21 years of the 20th century (my estimate is 10,292, or roughly 10,300). This, added to the roughly 5350 estimated above for the 19th century (through December 1899), should produce a total of more than 15,000 patents for women by 1921.

However, this total is not acceded to women until a 1935 study, published in 1938 (Ives, "Patent," 113, citing J.N. Miller). If indeed women had received only 15,000 of the 2,100,000 U.S. patents by then issued, women's overall patent share would have been only about .7% (.714). This figure, not even as high as that suggested for the 19th century by LWP, is clearly much too low. How was it arrived at? Even if women showed no further acceleration in their patenting activity after 1920, and received only 5000 patents per decade, by 1935 they should have received nearly 23,000 patents, counting their 5350 from the 19th century, which would hold their patent share steady at just over 1%.

The same underestimates prevail into the 1940s. In 1940 a *New York Times* story ("Women Gaining") reported in connection with a meeting of the National Inventors' Congress that women had received a total of 20,000 patents. Once again, even if women were still getting only 5000 patents per decade, by 1940 they should have received at least another 2500 patents beyond their 1935 total, making a total of 25,500 patents, or more than 1% of the total. My own estimate is more than 28,000.

The same article, in unconscious conflict with its own previous figures, and showing awareness of WB/28, says that women received 666 patents in 1918. The writer also estimates that women received 600–650, or about 1.5%, of the 43,090 U.S. patents granted in 1939, and that 8–12 of the 700 patents being granted weekly at the time of writing were going to women (1.4–1.7%).

If this 1940 article is even close to correct, either in its very low overall estimate or in its yearly estimates (lower than those for 1905–21), then women's rate of patenting must have slowed—or men's increased—drastically in the 1930s. (The effects of the Great Depression, increasing industrialization and professionalization of engineering, chemistry, and design, and perhaps the beginning of the influx of foreign inventors—overwhelmingly male—that now swells the total of U.S. patents—will have to be evaluated. If corporate inventors were already beginning to outnumber individual inventors, women would be disproportionately affected.)

A first glance at Katherine Lee's results for her 26-week samples suggests, on the contrary, a continuing increase for women's patents:

1920: 364 × 2 = 728 patents to women

1930: 400 × 2 = 800 patents to women

1940: 740 × 2 = 1480 patents to women

In fact, her 1940 total represents women's highest annual share of patents in her entire study: 3.04%. Unfortunately, Lee's 1940 results contain two major errors, which together reduce her 26-week total by 15 patents.[26] Moreover, all of her totals include design patents.

---

WB/28 does not make clear whether design patents are included. I assume they are not, since the figures tabulated from LWP for purposes of comparison with the WB/28 results are those for regular patents only (12; Table I).

26. As already observed, Lee read Lavern Jordan as Laverne Jordan, and Aristid Grosse as Astrid Grosse (37), and counted their seven and eight patents, respectively, in her total.

Correcting the known error in 1940, and deleting design patents to make these figures comparable with foregoing estimates, we get:

1920: 364–77 designs = 287 × 2 = 574

1930: 400–95 designs = 305 × 2 = 610

1940: 725–458 designs = 267 × 2 = 534

If these figures are even in the ballpark, perhaps women did indeed receive only 5000–6000 patents per decade between 1920 and 1940. At this rate, as already mentioned, they should have had a minimum of more than 25,500 patents to their credit by 1940, and possibly more than 28,000. We know from WB/28, for example, that women received *at least* 638 patents in 1920, and the 1920s were boom years.

Lee's figures for 1950–70 are as follows:

1950: 573 [1146] 2.37%

1960: 512 [1024] 2.06%

1970: 817 [1634] 2.40%

Since I know of no major errors in Lee's figures for these years, we can merely correct for the design patents included:

1950: 573 − 168 designs = 405 × 2 = 810

1960: 512 − 67 designs = 445 × 2 = 890

1970: 817 − 88 designs = 729 × 2 = 1458

Taking the average of the two figures at the beginning and end of each decade, women should have received roughly the following numbers of patents:

1940s: 6,720

1950s: 8,500

1960s: 11,735

for an estimated total of 26,955. Using that very rough basis, and adding the minimum estimate of 25,500 patents received up to 1940, by the time the second wave of the women's rights movement hit in 1970, women should have had, at the very least, more than 50,000 patents to their credit (52,455), not counting design or plant patents. With my own estimate of more than 28,000, the total rises to more than 55,000.

However, until now there has been little more agreement in the 20th century than in the 19th about women's patent share.

No further comprehensive counts of women's contribution to U.S. patents seem to have been undertaken until the 1970s, when Lee's thesis was also published. James F. O'Bryon created *A Profile of Users of the U.S. Patent System: 1968 and 1973* as his MS thesis at M.I.T. (1975). In his sample of 1000 randomly selected patents, half granted in 1968 and half in 1973, O'Bryon found (43) that 98% were granted to men working alone or in groups; 1% to ambiguously named persons; and only 1% to women inventing alone or with men.

In 1984 the U.S. Office of Technology Assessment and Forecast (OTAF) started a study of women's patent share. Because of the difficulty of assigning gender to foreign names, the researchers confined their study to patents granted to U.S. residents. Even here, they had trouble with ambiguous names: their final given-name file contained 16,769 names, of which 13,202 were male, 3,389 female, and 178 ambiguous.

Results of the first installment of this unpublished study were as follows for patents with at least one female inventor and all patentees U.S. residents:

| | | |
|------|------|------|
| 1975 | 1022 | 2.2% |
| 1976 | 979 | 2.2% |
| 1977 | 1012 | 2.4% |
| 1978 | 1083 | 2.6% |
| 1979 | 813 | 2.7% |
| 1980 | 741 | 2.0% |
| 1981 | 1179 | 3.0% |
| 1982 | 907 | 2.7% |
| 1983 | 1098 | 3.2% |
| TOTAL | 8834 | 2.5% |

As can be seen, though there are fluctuations, the general trend for women is still upward, not only in absolute numbers but in percentages of patents granted to U.S. residents. Moreover, the OTAF researchers point out that because of the problem with ambiguous names, their 2.5% figure "is best described as a minimum number" ([3]).

In 1990, OTAF updated and published this study with adjusted results for 1977–86 and new results for 1987–8 (OTAF, 21):

**Number and percent of utility patents with at least one woman inventor**

| | | |
|------|--------|-------|
| 1977 | 968 | 2.33% |
| 1978 | 1068 | 2.59% |
| 1979 | 774 | 2.57% |
| 1980 | 696 | 1.86% |
| 1981 | 1140 | 2.91% |
| 1982 | 880 | 2.60% |
| 1983 | 1082 | 3.29% |
| 1984 | 1389 | 3.62% |
| 1985 | 1516 | 3.83% |
| 1986 | 1572 | 4.12% |
| 1987 | 1962 | 4.51% |
| 1988 | 2089 | 5.16% |
| TOTAL | 15,136 | 3.32% |

(For these figures, the patent's first-named inventor must be a U.S. resident on the patent date.)

Women's percentage of utility patents continues to grow, breaking 4% for the first time in 1986. Interestingly enough, when design and all other types of patents are added, women's percentages are slightly higher, rising from 2.6% in 1977 to 5.52% in 1988, for an overall average of 3.64% (OTAF, 1, 21).

Now for the first time we have some basis, albeit partial and conflicting, for estimating the numbers and percentages of women's share of United States patents for the 20th century, and also since 1809:

|  | 1790(1809)–12/31/1894 | 3940 |
|---|---|---|
| 19th Century: | 1895–99 | 1375–1400 |
|  |  | 5315–40 |
|  |  | (rounded up to) 5350 |
| 20th Century: | 1900–04 | 1765 |
|  | 1905–21 | 8527 |
|  | 1922–30 | 5400 |
|  | 1931–40 | 7000 |
|  | 1941–50 | 6720 |
|  | 1951–60 | 8500 |
|  | 1961–70 | 11,735 |
|  | 1971–4 | 3600 |
|  | 1975–6 | 2001 |
|  | 1977–88 | 15,136 |
|  |  | 70,384 |
|  |  | (rounded up to) 70,500 |
| Overall total through 1988: |  | 75,850[27] |

This would be nearly 2% of the 4,000,000+ patents granted by that date.

Two earlier writers have reached higher estimates, without revealing their calculations. In a 1980 article on black and women inventors, Patricia Carter Ives estimated that about 3% of all U.S. patents to that date had been issued to women (114). As some 4,200,000 regular patents had been issued by then, women would have about 126,000 U.S. patents to their credit. Since Ives herself was a patent examiner, her figures should reflect current information within the Patent Office. Even if she was including designs and plant patents as well as regular patents in her calculations, this percentage would represent a significantly higher number than our rough cumulative estimate, which would have reached only somewhat over 63,000 by the publication date of her article.

Using the OTAF methodology, with computer-matching of names and a sample limited to patents with at least one U.S. resident among the patentees, Fred Amram estimates women's yearly share of U.S. patents at 8% for 1985. This, says Amram, is up from 1.5% in 1951 (Lee reports 2.53% for 1950, and WB/28 had figures as high as this for the 19-teens and early '20s). Interestingly enough, Amram's estimate for women's overall share of U.S. patents (2%) is lower than Ives' ("Problem-solving," 1, 7).

---

27. This total is probably still conservative, for several reasons: First, the surveyors of patent records tend to miss more women's patents than they falsely include, especially where a woman's name does not come first on a patent, and in areas of endeavor that are rare for females to enter. Second, both WB/28 and OTAF omit all ambiguous names. And third, omitting patents of foreign origin means omitting all the Soviet patents, where women are relatively numerous—and where, moreover, gender is usually obvious from the feminine endings of the names. Figures are either taken directly or extrapolated from LWP, WB/28, K. Lee, and OTAF, in that order.

## *20th Century: Other Indicators*

Other sources for information on women's inventive role are census records, inventors' congresses and fairs, inventors' associations, invention-development organizations, invention consultants and patent attorneys, and the "Patents" column of the *New York Times*. As with the alternative indicators for the 19th century, some of these sources include unpatented as well as patented inventions.

### *Census records*

As noted for the 19th century, the inventor's occupation is rare for both sexes, only about 2300 so listing themselves each census in 1910, 1920, and 1930. In 1910, with 2347 inventors in a population of only 92.4 million, the percentage for this occupational category was .00254. Interestingly enough, as the total U.S. population rose from 92.4 million in 1910 to 123.1 million in 1930 (SAUS/1985: 6), the percentage classified as inventors was actually dropping rather dramatically, perhaps reflecting the much-touted decline of the individual inventor and the rise of the employed or corporate inventor, who would probably appear in the census as a chemist or an engineer (cf. O'Bryon, 46). Female inventors show this decline as well:

1910   44 inventors/2347 = 1.87%
1920   27 inventors/2376 = 1.11%
1930   21 inventors/2300 = 0.91%

Unfortunately, after 1930 inventors are no longer tabulated as a separate category.

### *Inventors' congresses*

Earlier in this century, the National Association of Manufacturers was known as the National Association of Manufacturers and Inventors, which in itself speaks volumes. In 1935 the National Inventors Congress was held in Chicago. According to the *Christian Science Monitor* (June 25), women were present, but no numbers are given.

In 1937 the Congress was held in New York City. The Providence *Journal* ("Women Inventors") reported, "More than 10% of the gadgets on display . . . were invented by women," and quoted a California woman inventor's estimate that "more than 11% of all new inventions are developed by women." Her name was not given.

In 1938 the association met in Chicago, with 200 tinkerers and inventors present, including "housewives." Seven women were mentioned by name ("Tinkerers End . . ."). Even if these were the only women present, which seems questionable, they would constitute 3.5% of the inventors at the congress.

### *Inventors' organizations*

These organizations differ vastly in size, sophistication, and activity, but are all organized by inventors themselves for their mutual benefit and aid.

One of the largest and best organized of these is a Southern California group variously known as the Inventors Workshop International (IWI) and the Inventors' Licensing and Marketing Association (ILMA). Two of the guiding lights of IWI/ILMA, Maggie Weisberg and Melvin Fuller, later operated the Inventors Workshop International Education Foundation of Tarzana, CA. As of 1988, ILMA and the Foundation had 250 women members, 10% of the total (Weisberg, L-2/10/88).

Dorothy Stephenson, then president of the Inventors Association of New England, stated in 1978 that IANE had women members, but none of them as yet had patents. IANE, it should be noted, was newly formed at that time.

Henry B. Ehrhardt of the Northwest Inventors Association, Everett, WA, reported in 1980 that "several women" attended their meetings from time to time, but "as I remember only three of them actually became members of the organization and received patents on their inventions" (L-n.d.[1980]).

The Confederacy of Mississippi Inventors had "only a few women" in their organization as of 1980, and none as yet with patents. However, Rudolph Paine, then president of CMI, noted that "one of the better two of my inventions was my wife's idea." Since that invention was not yet patented, he gave no details (Paine, L-11/19/80). Elsewhere in Mississippi, also in 1980, Charles Wright reported four women members for the Mississippi Society of Scientists & Inventors, but gave no total membership figures (L-10/22/80).

*Invention-development organizations*

In answer to a 1980 query, John Tilden of the Indianapolis Center for Advanced Research, noted that "we do have women inventors who occasionally submit their disclosures to us for evaluation, some of which are already patented," but gave no numbers or percentages.

The Center for Innovation (CFI) of the Montana Energy and MHD Research and Development Institute (MERDI), reported in 1980 that in three and one-half years of operation they had received inquiries from over 2000 inventors, about 5% of them women (inventors or small entrepreneurs). Most of the ideas in question were already successfully or unsuccessfully on the market, minor (non-functional) changes to something already marketed, or "non-licensable" because of high production, marketing, or advertising costs. At that date, CFI was not working with any women inventors on developing a new idea or business.

In 1976 the University of Oregon Center for Invention and Innovation estimated that women constituted 11% of their client files (H. Carroll, L-10/29/76).

As early as 1975, a for-profit invention-development firm, Lawrence Peska Associates of New York City, reported that 18–20% of its clients were women, and cited the Patent Office as estimating that about 20% of its more than 2,000 weekly patent applications came from women (Sorel, PC-1975).

*Inventions consultants/patent attorneys*

Richard A. Onanian, a private inventions consultant who was also connected with an invention development organization (Institute for Invention and Innovation of Arlington, MA), noted as early as the mid-1970s, "With regard to my professional consulting, I can state quite confidently that the numbers of female clients and inquiries from women have increased quite noticeably over the past few years" (L-9/8/76). He did not quantify this statement, but cited enrollment figures for his Boston and Cambridge adult-education classes in "Ideas, Inventions, and Patents." Female enrollment increased from 5 out of 66 in 1974 (nearly 8%) to 10 out of 39 (25%+) for (part of) 1976. Said Onanian, "Although it represents a very small sample, I believe it is indicative of the innovative contributions women are destined to make in the future."

Foreseeing the same increased contribution, but far less positive about it, was the chief patents counsel of a major American corporation who refused to cooperate with Fred Amram's study of women inventors, saying that women are receiving more patents than ever before, and adding, "In fact, at this rate we may be living in a matriarchy in fifteen years" ("Problem-solving," 4).

## Stacy Jones' "Patents" column

The late Stacy V. Jones compiled the *New York Times* "Patents" column for more than thirty years. The value of statistics based on his column is that he not only chose some of the most technologically significant patents from each week's Gazette, but also included certain patents with what might be called human interest, covering inventions significant by criteria similar to those suggested in this book. Most of these patents, it should be noted, would also have economic interest or significance, and Jones provided easily accessible information on patents and inventions being licensed, manufactured, and sold. In the decade 1978–87, Jones mentioned at least 2651 inventors of known gender in his weekly column.[28] Of those, 117, or 4.4%, were women.

The year-by-year figures show that in this decade there was a slight trend toward increasing recognition for women inventors, as measured by this highly visible column. A sociologist might say that women inventors were becoming more *salient* in American society. The average number of women mentioned each year for the second five years is only slightly higher than for the first (12 *vs.* 11.4). Note, though, that without the eight women mentioned in a special section of a 1979 column—just after **Barbara Askins** became the first sole woman inventor to be named Inventor of the Year—the difference would be greater: 12/yr. for 1983–7, *vs.* 9.8/yr. for 1978–82.

Even more noteworthy, perhaps, is that women's 4.4% of patents judged significant and/or interesting enough to be mentioned in Stacy Jones' column exceeds the usual estimates of women's patent share for these years. The OTAF estimates of that share for this decade ranged from 2.82% in 1978 to 4.9% in 1987 (including designs, plant patents, etc.).

*   *   *   *

To summarize a long and complex argument filled with conflicting data, no thorough and accurate count of numbers and percents of women's U.S. patents, with gender determinations for all ambiguous names, has ever yet been done. As a result, there is wide disagreement as to what these figures should be. Estimates for the 19th century range from a ridiculous one in a thousand (.1%), or somewhat over 600 patents total, to more than one in a hundred (over 1%), or more than 6000 patents (cf. Duell's estimated total through early 1895). The LWP compilers, the only ones who did an actual count, listed a total of just under 4000 (3975) regular patents with women patentees through the end of February 1895, or slightly less than .75 % (.747) of all numbered patents through that date. Extrapolating from LWP for the missing years 1895–9, we get a low first estimate for the whole 19th century of 5350–5375, or just under 1% (about .85%) of its 640,000-odd U.S. patents.

My latest research on LWP's omissions suggests that the overall omission rate will lie somewhere between the shocking 1882 figure of nearly 39% and the provisional 1890

---

28. For 1978–80 only females were counted; the number of males mentioned for those three years was estimated using an average of 4 patents per week for 52 weeks (minus, for 1979, the 13 weeks of the newspaper strike). Jones also mentioned 61 inventors with ambiguous names, some of whom may turn out to be female.

figure of nearly 7% (6.99%). Granting the compilers more years with low-rather than high-end omission rates, the average might fall, say, between 10 and 15%. If so, using 5350 as a base, women's share might come very close to 1% (5885–6181) of all regular numbered U.S. patents granted during the 19th century (640,166). However, omission rates for the four years more or less systematically studied (1867, 1876, 1882, 1890) average 23.1%, suggesting a corrected total for women of 6957, or *more than 1%* (1.09%) of regular numbered patents for the century.

Exposition figures show a dramatic increase in women's participation during the century, and gender tabulations of lists of inventions for sale or offering shares for development in inventors' periodicals indicate a higher percentage of women than the patent-list estimates.

Even in the 20th century, when some records are computerized, patent data are not recorded by sex, and the numbers are overwhelming for hand-eye counts. Estimates of women's overall share of U.S. patents held vary from a low of 1 or 2% to a high of 3%. This may not sound like a large variation, but as U.S. regular patents exceed 5 million, it represents a range between 50,000 and 150,000 patents.

Estimates of women's recent annual share of patents range from under 4% (OTAF) to 8% (Amram, "Problem-solving"). As in the 19th century, certain other measures suggest higher percentages (or higher significance) than the patent estimates. A sample of patents mentioned by Stacy Jones in the *New York Times* 1978–87, for example, contains 4.4% women, and several inventors' expositions, inventors' organizations, and inventions development groups report as high as 10% women members/clients/exhibitors. These last, of course, would include women who do not yet have patents, and unpatented as well as patented inventions.

In any case, there is no doubt that both women's numerical participation in invention and their salience or recognition in technology and invention are increasing. In 1977 the first woman—**Mary Ollidene Weaver**—was chosen (as part of a team) as Inventor of the Year, and in 1978 **Barbara Askins** was chosen alone as Inventor of the Year. Stacy Jones noted more women inventors for his *New York Times* "Patents" column in the last half of the decade 1978–87 than he did in the first half. In 1988 **Janine Jagger** became the first woman to be named Distinguished Inventor of the Year. In 1989, **Diane Pennica** was one of four Genentech scientists named Inventors of the Year for their invention of the drug t-PA, used to dissolve blood clots in heart-attack victims. In 1991 Nobel laureate **Gertrude Elion** became the first woman inducted into the National Inventors Hall of Fame, and other nominations are being prepared at this writing.

In short, women invent. And from this moment forward, it is to be hoped, they will be recognized as inventors.

# Appendix A-1:
# Detailed Argument of the
# Agricultural Takeover

The argument as presented here applies primarily to the Near East between c. 4000 and 2000 BC. However, broadly applicable patterns may emerge from this analysis.

As the women of a society perfected their horticultural skills and tools, they would naturally have increased the productivity of their gardens. At some point, the men must have begun to feel their hunting and fishing were no longer necessary, or certainly not so important as before. As gardens expanded with population, local big-game species either died out or were pushed far from the group's home base. (This may already have happened during the intensified gathering of the late prehorticultural phase.) Although Elise Boulding hesitates to speak of a role loss for men in her discussion of the period in *The Underside of History*, it seems to me a rather precise term for what was going on. The loss would have been particularly severe if, as at Çatal Hüyük, it was accompanied by the decline and disappearance of the hunters' male god, and the rise and dominance of a powerful goddess associated with (but not limited to) fertility, for this could have meant a status loss of a different order. Even without the religious factor, wherever hunting and fishing were men's major economic roles, the loss would have been traumatic.

Men in such a situation probably responded in several ways. They may have capitalized on the increasing scarcity of meat to make it a prestige item. Among the Mundurucu of Brazil, for example, meat was still the favored food of both sexes and hunting a central occupation in the culture when the Murphys studied them, even though the women's gardens and their farinha-making provided the largest single share of the diet (62). Men may have elaborated their hunting rituals and weaponcraft to occupy the lengthening intervals between hunts. On their ever-longer journeys after game, they may have sought something to bring home in lieu of meat; thus the beginnings of their specialization in trade. Eventually, however, they would surely have found new economic roles. The most prestigious economic role at this important transitional period was probably cultivation. Moreover, as population increased and gardens became fields, food production demanded more and more labor from the women. In societies where new fields were being cleared, men began at this stage to help women by taking over the heavy work of field-clearing.

Bemba men (Africa) have ritualized their field-clearing contribution, a specialized and dangerous form of tree-pollarding. Interestingly enough, although the women contribute heavily to field-clearing as well—piling and stacking the branches cut by the men, burning the piles, and spreading the ash on the fields—the men's pollarding ritual establishes them, in their own minds at least, as the owners of the fields.[1]

In much of the Near East, however, there were no new fields to clear; the frontiers lay in coaxing new productivity from existing fields. But cultivation was strongly entwined with Goddess worship. Before men could take over any significant or central part of the cultivative role, society itself had to change.

Specifically, 1) women must be weakened or overburdened in some way so as to want and need the help of men in cultivation; 2) when this need arose, men had to be ready with the spare time, the new vision, and new technology(ies) not available to women; 3) men had to develop new bases of power; while 4) women had to lose status and power; and 5) the religion must change.

I suggest that between a high point for women about 6000 BC, or whenever cultivation "triumphed" in a given area, and about 2400 BC when Urukagina's Law Code appeared in Sumer, all of these revolutionary developments did indeed occur in parts of the Near East. Many factors were involved, so tightly interwoven that only for purposes of discussion can we separate them out.

## Population

First and foremost among these factors is population. Traditionally we have assumed that a population explosion accompanies settled, cultivative life. More recently, this assumption has been attacked. Ruby

---

1. (J. Brown/Tiffany, 59–60). It is interesting to speculate that ritualization may have helped limit or ossify men's participation in cultivation at this rather low level. Moreover, it could be seen as a religious change.

Rohrlich prefers to downplay population as a factor in social and economic stratification ("State," 76). Mark Cohen and others have argued that a population explosion—and thus a food crisis—had already occurred before horticulture. This idea makes perfect sense wherever intensive seed-gathering and exploitation preceded settled gardening, for reasons that will become clear below. But whether the late-horticultural population advance is called an explosion or not, and whether it is the first or the second in human prehistory, it almost undeniably took place. Comparing community size in horticultural and agricultural societies in a standard cross-cultural sample, Martin and Voorhies find, for example (282), that villages (under 400 people) dominate among horticultural societies (79%), whereas towns (400 to 1000+) and cities (5000 to 50,000+) dominate among agricultural societies (total of 76%). Marija Gimbutas notes a "considerable increase in population" during the Karanovo III phase of the East Balkan Neolithic, about 5400-5300 BC (*Gods,* 30). For the Near East in general, Kent Flannery suggests a 60-fold increase in 6,000 years, from a Paleolithic density estimated at 0.1 per square kilometer, to between 1 and 2 people per square kilometer with early dry farming, to more than 6 per square kilometer after irrigation (93). Mellaart calculates that the population for Çatal Hüyük would have multiplied 528 times in only 800 years, and that the excess over observed size in later levels must have emigrated (*Neolithic,* 99).

For a root-crop system in the New Guinea Highlands, James Watson postulates a startling population explosion after the arrival of the sweet potato and the development of settled horticulture (302): "The Enga . . . could have reached their present population of 100,000 in 300 years from an initial population of 1,146 with an average annual rate increase of 1.5%."

Even if the population did not explode, the switch to agriculture could cause land pressures. As Flannery explains for Iran, for example, gatherer-hunters could use at least 35% of the land, whereas dry farmers find only 10% "arable," and *only 1% is irrigated* (Ucko & Dimbleby, 94; emphasis mine).

This rapid population growth is credited—or blamed—for a great many things. As Ester Boserup points out, when the population increased sufficiently, land could no longer be fallowed, but must be continuously cultivated, enormously increasing the workload (32). Disrupting, as it did, the traditional migratory patterns of local nomads, the change also had serious political and social consequences. Moreover, a hamlet of 40 people should find it far easier to operate on a subsistence economic basis and a consensual political basis than Lagash (Sumer) with its estimated 100,000 (Kramer, *Sumerians,* 88).

What is not usually considered, but what concerns us here, is the effect on women specifically. For a graphic example, compressed in time, of what happens to women on first exposure to settled life and an agricultural diet, consider the case of the foraging !Kung of Angola, Botswana, and Southwest Africa, recently settled into villages and shifted to a grain-based diet (Kolata, *passim*). Among the foraging !Kung the two sexes are roughly equal in status, with the women contributing 60-80% of the group's food by weight. Among the settled groups, however, women are losing their egalitarian status, and the *birth rate has risen 30% in only five years.*

Let us look more closely at this crucial example. The women seem to be losing status partly because they no longer contribute the major food supply (men leave the village to farm and care for Bantu cattle for wages), because they have lost mobility (they stay at home and take care of houses and children), and because their men are starting to imitate the male-dominated Bantu. Part of women's loss of mobility (which will aggravate their loss of status as their men learn the Bantu language and advance further into modernity) is due to their higher fertility. In the nomadic state, mostly because of less rich nutrition, puberty came late: 15.5 years. Women had their first child at about 19.5 years, and gave birth only about every four years thereafter (Kolata, 121)—partly because they nursed their babies for 3-4 years, and had too little body fat to ovulate during that time. There is also some evidence that nomadic !Kung women had seasonal or drought-induced periods of nonovulation. Now puberty comes earlier, and the interval between births has dropped 30 percent! Women stop nursing sooner and supplement their babies' diets with cow's milk and grains. Their own diets being richer, they ovulate regularly and have babies much oftener, with predictable effects on mobility[2] and division of labor.

The greater number of children has another, more subtle effect on the status of women. The nomadic !Kung had so few children that all ages and both sexes played together. This kind of play-grouping discourages the development of distinct games and roles for the two sexes (Kolata, 120). Now, with more

2. Relatively unimportant—arguably even nonexistent (see Dahlberg, ed.)—among fully nomadic gatherer-hunters, differential mobility has crippling economic and prestige or status effects on the less mobile members of a settled and modernizing cash/market economy.

children, and with boys going off with their fathers to herd cattle, this equalizing effect is lost (see also Draper, 89, *passim*).[3]

California anthropologist Shirley Lee observed analogous changes among the settled versus the nomadic Paiutes (MA thesis, Stanford, 1966).

Both !Kung and Paiute have skipped the early horticultural stages and been wrenched directly into farming with herding—a change imposed from outside rather than developing organically within their group—but many of the elements in our story are here.

What does not emerge so clearly without emphasis at this point, perhaps, is that *women's life quality and status can be undermined whether a population explosion takes place or not.* A *birth explosion* will suffice.

Excessive childbearing puts women at a new disadvantage in several ways: Very early and very late pregnancy and early lactation make the heaviest work tiring or even dangerous, and the more frequent the pregnancies—and the more the society depends on cultivation—the more of a disadvantage pregnancy becomes. Second, caring for several young children at once is not only physically and emotionally more exhausting but mentally more distracting than caring for just one. This alone could go far to explain why women did not rise to the challenge of the new problems of scale developing in late horticultural times in the Near East.

Third, densely settled, permanent populations fall prey to more infectious diseases than do nomadic groups. Some viruses, for example, need a certain population size and density in order to survive. Naturally, more children than adults contract these diseases, and their mothers usually care for them—as indeed women do for any sick person in the group. Thus yet another of women's traditional tasks and roles expands disproportionately with more births whether or not these children survive to adulthood; and their deaths are a further burden to women.

Fourth, several years of too-frequent pregnancies can undermine women's own general health and strength. A woman thus overburdened, finding cultivation more demanding every year and all her other tasks continuing unabated, is in no position to analyze the political consequences if a man offers to help her with soil-preparation.

An important related factor here is *diet.* A seed-based or seed-heavy diet contains more calories and more fat than a diet composed mainly of fruit and greens with occasional nuts and meat. The grain-based diet of sedentary true agriculturalists is both richer (in fat) and poorer (less varied, possibly poorer in protein and some vitamins) than the diet either of foragers or of horticulturalists who supplement their fare liberally with gathering, hunting, and fishing. Judith Brown's comparison of the Iroquois and Bemba (African millet-growers) diets is instructive here. She calls the Iroquois diet "varied, nutritious, and dependable"; the Bemba diet "monotonous, marginal," and "interrupted by near-famine periods" (Tiffany, ed., 62–3).

The larger grains of domesticated species of wheat contain much more fat than the seeds of their wild ancestors (Ucko & Dimbleby, 85). The precise relationship between dietary fat, body fat, and women's fertility is still controversial. However, the fact that fatty tissue makes estrogen, the nearly universal amenorrhea among true anorexics, the soaring birth rate just described for the settled !Kung women, and the earlier puberty of well-to-do girls in the Middle Ages and of Japanese girls after World War II, among other things, suggest that, up to a point, fat promotes fertility. In addition to supporting earlier puberty and more regular ovulation, it speeds the return of ovulation after pregnancy, undermining the traditional contraceptive technique of prolonged nursing (cf. Short).[4] Finally, it delays menopause (McDougall & McDougall, 186–7).

At the same time, domesticated wheat grain contains only half as much protein as that of wild species (Flannery, 80)—and cassava, the staple for hundreds of millions of people in Africa and South America, contains virtually no protein at all. Women with their greater need for iron and protein, could have become

---

3. Considering what boys allegedly learn about teamwork, competition, contact (and contracts) on the playing fields of Eton and elsewhere, the importance of this change can scarcely be overestimated. Unfortunately, since anthropologists have only recently seen gathering-hunting societies as egalitarian, no one even knew that a change had taken place (see Leacock, *Myths,* "Women").

4. Recent work in anthropology underscores the importance of fat in the diet (See Shipman, 68; David Perlman, "Facts"). The McDougalls are speaking mainly of animal fat, but of course domesticated animals are fatter than wild ones as well; and the changes in menarche for Japanese girls are striking: from 16.5 years in 1875 to 15.2 in 1950 and 13.9 in 1960!

weakened by anemia and/or more vulnerable to disease at certain periods of dietary adjustment during the transition from gathering/hunting to horticulture and from early to late horticulture with greater dependence on grain. Such an effect might be especially likely where the domesticated species were used only for milk, or where the men ate most of the meat. Mellaart reports evidence of chronic anemia (a characteristic bony lesion of the skull) in the skeletons at Çatal Hüyük—most of which were female— despite what he calls a varied diet there (*Neolithic*, 100; Allison, 77).[5]

One final note on the relation between diet and fertility. Since in at least one mammalian species breeding behavior depends entirely on the estrogen content of its food grasses ("E-vole-ution," 215), it would seem pertinent to ask whether domesticated wheat and other grains contain more or less estrogen than their wild-grass predecessors. Moreover, storing grain can evidently affect its estrogen content (Dr. Dean Edell, KGO radio, Feb. 1984). Dr. Judith Galaver of the University of Pittsburgh finds that "phyto-estrogens" in wheat, corn, rice, and hops have the chemical potential to cause female-hormone-like effects in rats, and may partly account (when these grains are used in alcoholic beverages) for the reproductive problems experienced by alcoholics of both sexes ("Organic Liquor"). Such possible relationships would seem an important area for further research.

## Domestication of Animals

The *domestication of animals* is relevant both to the birth explosion and to the changing economic and power relations between women and men. For one thing, the meat of domesticated animals contains more fat than game meat. For another, where the animals were milked, women probably took on another demanding job. Of the 48 societies practicing milking in Murdock & White's standard cross-cultural sample, only 15 assigned it exclusively to men (Murdock & Provost, 296). The animal milk supply undoubtedly allowed some motherless infants to survive. On the other hand, women who fed their babies supplementally on cow's or goat's milk probably stopped nursing sooner, leaving their infants less well nourished and less well protected against disease. It also meant that the women got pregnant again sooner, depriving the weanling of care and attention. Thus *a ready alternate milk supply, instead of reducing infant mortality, could actually have increased it, while at the same time increasing the birth rate*. Certainly, archaeologists have found large numbers of infants and young children buried under the floors of Neolithic houses, both in Old Europe and in the Near East (Gimbutas, *Macedonia*, 85; Mellaart, *Town*, 60). Many children may have survived just through the period when they cost women most time and energy, only to be replaced by another fragile life, a physically and emotionally draining cycle.

Some have suggested that domestication of animals helped undermine women's high status by revealing the male role in procreation. The ancients certainly thought so. In both Greek myth and Chinese myth the first king to learn about physical paternity is the one who imposes monogamy (Graves, *Myths* I: 97). Archaeologists maintain, however, that Paleolithic hunters knew this role as early as the time of the Altamira cave paintings, which show bisons mounting and other representations of the reproductive theme (Sherratt, 92).

If, on the other hand, it was not the hunters (i.e., presumably males in that society), but the reproducers, and arguably the butchers—the women—who had this knowledge, the question appears in a new light. We need not believe with Malinowski for a distant culture that women knew the secret and kept it from men—though this idea takes on new interest as well in light of recent suggestions that women did some of the cave art.[6] We need only realize the flexibility of *knowing* as a concept in communication between anthropologists and informants, and indeed in human psychology in general. Just as Mbuti women both know and do not know about the sacred flutes used in the group's central religious rite, because they are not supposed to know and the rite is now a male preserve, and just as many Germans did not "know" what went on in the death camps, late Paleolithic and early Neolithic men could have known/not known about

5. Malaria could also explain these lesions.
6. (E.g., Grahn/Spretnak, *Politics*, 267). This suggestion receives unexpected support from a recent analysis of the technology used at Lascaux Cave in France (about 17,000 years ago). The artists' calcium phosphate pigment, for instance, demanded heating animal bone to about 400 degrees Celsius, mixing it with calcite, and heating again to 1,000 degrees Celsius. Women's work is much more closely related to the technology necessary for achieving this than men's—as it would also seem to be with the stone lamps used for light in the caves, the tallow lamp fuel, the clays used in some pigments, the mortar and pestle used to grind other pigments, etc. (Leroi-Gourhan, 110, *passim*).

their procreative role.[7] In this scenario the importance of animal-domestication would lie in making this role *unmistakable* just as other factors spurred men to new action.

For the Near East, animals may have been domesticated as early or almost as early as plants, possibly by 9000-8000 BC (Flannery, 91; Mellaart, *Town,* 19). Since women's status, if Çatal Hüyük is indicative, reached an all-time high two or three millennia later, not seriously declining until around 3000 BC, animal-domestication must have had a much-delayed effect. The delay could be explained if women were—as myth, lactation capability, and previous specialization in infant foods would all argue—actually the first domesticators of animals, while men still hunted. At Zawi Chemi in the Near East, for example, though sheep were domesticated by 8900 BC, hunting continued to provide most of the meat (Mellaart, *Town,* 19). Only later, perhaps, as agriculture became a full-time job rather than part of women's overall subsistence role, hunting declined, and women were beginning to be overburdened, did men take over.

Another scenario for such a takeover involves the introduction by conquest or trade of a new domestic animal into a culture where the main domesticate is owned or controlled by women. The disastrous economic effects can be seen clearly in the simpler culture of the North American Plains Indians, Blackfoot, Hidatsa, and Cree. Pre-contact, their only domestic animal and beast of burden was the dog. Owned by the women, dogs were the measure of wealth in the group. When whites introduced horses—to the men—the culture was transformed, sometimes in a single generation. Women lost status as the buffalo, hunted from horseback, became the major source of both food and shelter. Women got new tasks, but not authority over them; and they lost not only their role as main food-providers, but their control over the group's wealth (AAUW, 11–2; Liberty).

It would be interesting to know, in any case, whether a male deity or principle, perhaps in the form of reverence for phallic images—or a husband for the Goddess—entered or reentered religion as a result of a new certainty of the male role in procreation. Gimbutas notes for Old Europe, for example, that *a male principle was separated from the previously androgynous Great Goddess about 6000 BC* (*Goddesses,* 196).

## Animal-Drawn Plough

The economic effects on women of animal domestication may have shown most severely in making possible animal-drawn ploughs. When the idea arose, about the 4th millennium BC in the Near East, of harnessing animals instead of people to the rope-traction ploughs of the day, men may have been effectively in charge, even if not yet considered the "owners," of the necessary animals. As already mentioned, they also had more time than the women. Thus, even though, as I argue elsewhere, the original inventor of the plough might well have been a woman, for some of the same reasons men took over herding (if they did), they later took over ploughing with animal-drawn ploughs.

In some places taboos grew up preventing women from working with cattle. Did the men create these taboos to protect their specialization—their new economic role—or to balance the women's religious taboos on planting? Did women create protective taboos of their own? A fascinating hint of an answer appears in the Riffian culture (North Africa) studied by Carleton Coon in the 1930s. Of the three crops "most deeply entwined with Riffian life"—rye, grapes, and olives—the oldest is rye. The Riffians prize rye as a richly nourishing and strength-producing food. Rye is never planted with domestic animals or with the plough, but is cultivated by hand (hoe) on unirrigated mountain slopes. Though all other grains are threshed by domestic animals on a threshing floor, the women thresh rye by hand with sticks. *It is taboo to thresh rye with cattle.* "The whole rye complex," observes Coon, "seems to antedate or at least to be negatively associated with the cattle and plough complex" (49).[8]

Coon is explicit, too, incidentally, about the sacred quality of the plough. Made of wood in several parts,

---

7. Also, of course, knowing is not the same as believing, and knowledge can be lost, just as women's ancient knowledge of herbal contraceptives and other anti-fertility technology was largely lost in Western culture. Moreover, as fundamentalist Christians today demonstrate, believing something is true for animals is not the same as believing it true for humans—who may be considered a different order of creation. It need not be an insult to primitive peoples to suggest that they believed this—assuming they knew that different is not the same as inferior. (See also Turnbull's comments on Mbuti beliefs about the impossibility of premarital pregnancy, in Dahlberg, ed.)

8. As a caution in interpreting the effects of men's herding on women's control of agriculture—and as an example of the extreme difficulty of constructing a paradigm for the takeover—note that Riffian women do

the plough is of an ancient type. The main body of the plough and the peg holding the plough pole to the yoke must be made of oak, the plough pole of oak or cedar. "Farmers usually make their own ploughs, though blacksmiths make the iron shares, and an especially gifted ploughmaker may make ploughs for his neighbors. Because of the plough's sacred quality, however, he cannot or does not" accept money for his work (Coon, 51).

The fate of the Sumerian Goddess Ninlil may reflect the impact of the new animal-drawn ploughs on women's role in agriculture. Credited with giving her people their knowledge of planting and harvesting methods, Ninlil was seen, when Enlil arrived in Sumer, as the daughter of the Goddess in Nippur. Enlil enters Nippur and rapes Ninlil (M. Stone, *When God*, 82; Kramer, *Sumerians*, 146; Gray, *Near Eastern*, 13),[9] or, as Merlin Stone puts it, "His introduction into the city of Nippur is associated mythologically with the rape of the daughter of the Goddess in Nippur." As in so many parallel Greek myths, Ninlil later appears as Enlil's wife. Enlil was known as the "lord of the plow" (Rohrlich, "State," 83), and Samuel Kramer says, "It was this same Enlil who fashioned the pickax and probably the plow as prototypes of the agricultural implements to be used by man" (*Mythology*, 42; *Sumerians*, 119; in the latter book, Kramer drops the "probably").

Suppose that, as in Africa in the 19th and 20th centuries AD, incomers with a more advanced agricultural technology entered Mesopotamia in the 4th millennium BC. Already male-oriented and/or worshipping a male deity, they introduced their new plough to the men rather than to the women, and consciously or unconsciously used their superior technology, as well as the power of the stylus, which they took over from the natives, to aggrandize their deity(ies) at the expense of the indigenous ones. Introduction of the plough—to men, by Quaker missionaries and others—was similarly disastrous for Seneca Indian women in North America, as Joan Jensen shows ("Native American"). Whatever may have happened, as already mentioned, the Sumerian words for "farmer," "plough," and "furrow" are borrowed words (M. Stone, *When God*, 84); and several representations on Sumerian cylinder seals show a deity carrying a plough and traveling in a boat (Kramer, *Sumerians*, 157-c).

## Large-Scale Irrigation

The animal-drawn plough might have affected women's cultivative role less drastically in the Near East had it not been linked, in a way seldom mentioned, with another new technology—*large-scale irrigation using water from great distances*. Axel Steensberg suggests that a heavy basalt ard-share from Bronze Age Syria was probably used to make small canals or water-furrows in an irrigated field, and that it could not have been pulled by a person, but would have required two oxen (30). Ploughing, then, must prepare the soil not just for sowing but also for watering. Moreover, new modes of ploughing and new modes of irrigation can both disrupt traditional landholding patterns (L. White, *Medieval Technology*, 44; Rohrlich, "State," 82). These disruptions could have contributed to women's loss of status: if the old patterns involved holding in common by matriclans, most changes might have been downhill for women.

---

the cattle-herding, and do it under astonishingly difficult conditions! "No house which contains a women," says Coon, "lacks a cow. . . ." In favorable weather the cattle graze in fields or on the "less rugged mountain slopes," tended by women or girls. During the winter, they stay in the house where the women feed them. "During the late winter, before the green shoots have appeared on the mountain sides, women often spend all their time in a search for weeds and grass to feed the family cow." Getting the cows to their mountain pastures "is often a difficult feat, on account of the narrowness and steepness of the paths; often four or more women, pulling on ropes, horns, and tail, are required to guide a cow over a dangerous place" (Coon, 40). Moreover, "curiously enough," the mating of the cattle is women's work. The owner of the bull and the owner of the cow "each sends his wife" to take the animals to a secluded place. "For any man to witness the mating of cattle would be a shameful experience, although women may look upon it with impunity" (*ibid.*).

Sheep-herding, on the other hand, is done by grown men, especially old men who cannot do much else (*ibid.*).

Interestingly enough, rye also yields more flour than other Riffian grains: 32 loaves of bread from one *mudd* of rye, as compared with 19–20 for barley and 24 for wheat.

9. Kramer calls this tale of rape and repeated deception of a prepubescent girl a "delightful myth" (*Mythology*, 43).

The region-wide irrigation system found in Mesopotamia by 3000 BC demanded an enormous investment of labor, not only initially, to dig or enlarge the canals and other works, but to keep them free of silt and in good repair. More than this, it demanded knowledge of distant sources of water, awareness of the best routes for bringing water from those sources, and a rather sophisticated level of engineering knowledge. Long-distance traders—or nomads—were likeliest to gain the geographical knowledge involved. The engineering knowledge may have arisen in one group and been transmitted to others, by immigration, conquest, or a subtle combination of the two. If, as I suggest, the bearers of this knowledge were already patriarchal in outlook, they would probably transmit their knowledge mainly to males. The males would meet them first if they entered with warlike intentions, would have a somewhat greater chance of meeting them in trade, and would have the mental freedom and time to see the possibilities and adapt the new technology to their own land.

The crucial role of large-scale irrigation technology in the agricultural takeover can be seen in Sumerian myths surrounding Enki, the water god. In earliest legends the Goddess Ninhursag seems to play a dominant role, her name preceding those of Enki and Enlil. Later, she appears as Enki's wife or sister, and has clearly lost what once was Hers. It seems significant that *Enki may not be of Sumerian origin* (Kramer, *Sumerians,* 79).

In the myth "Enki Establishes World Order" (written down after 2000 BC), Enki created the irrigation canals, "making the Tigris and Euphrates eat together," and he or his canal overseer "has carried off like fat the princely knee[10] from the palace." Just after this obscure line, Inanna, who has given up Her royal scepter, twice asks Enki where Her royal powers have gone. His consolation for Her "loss of powers as a result of the canal building" ends with a most illuminating sentence: "Inanna, you who do not know the distant wells, the fastening ropes, the inundation has come, the land is restored, the inundation of Enlil has come" (M. Stone, *When God,* 84).

Obviously, irrigation has taken on a new importance, both generally, in response to demands of population and trade, and in specific areas as drought or changes in rivercourses threatened a city's crop land. About 3500 BC, for example, thousands of acres reverted to desert in Sumer because of the loss of the eastern branch of the Euphrates (Schmandt-Besserat, *Legacy,* 2).

Note, too, that in the Cain-and-Abel-like dispute between Enten and Emesh, in which Emesh challenges his brother's claim to the position of "farmer of the gods," Enten responds, "Father Enlil, you have given me charge of the canals / I brought the water of abundance." And Enlil, deciding for Enten, says, "The life-producing waters of all the lands—Enten is in charge of them. . . . Emesh, my son, how do you compare yourself with [him]?" (Kramer, *Sumerians,* 219).

Graphic and linguistic evidence from Canaan may indicate a takeover within irrigation itself. Associations of the Goddess and water are ancient and widespread. Guardian spirits and deities of springs are usually female, and in one Egyptian myth rain is the breast milk of the overarching sky-goddess Nut. Ishtar was originally connected in a significant way with irrigation during the dry season. Indeed, says John Gray in his *Near Eastern Myths* (22), "Ishtar and Athtar may be etymologically connected with the Arabic verb *athara* meaning 'to irrigate.' " If so, it becomes crucial to know the age and origin of this verb, when Ishtar was transformed into the male Aththar, and when large-scale irrigation was developed or introduced in Canaan. Ishtar as dispenser of never-failing waters was presumably a pre-kingship theme, older than the representations of it reproduced by Gray from a Kassite temple at Uruk and in an Amorite sculpture from Mari (15th and 19th century BC). Thus the representation of King Gudea in precisely the same attitude as the Amorite Ishtar could reflect the transition in progress (22, 30, 63).

It may not be irrelevant, either, that according to one myth, irrigation was introduced by a war god, Ninurta (Chas. Singer I: 546).

## Professionalization

The combination of large-scale irrigation and complex ploughing technology—dictated partly by reliance on irrigation—transformed agriculture from one of women's economic activities—a part-time, seasonal

10.  Lap or throne? Cf. the concept of the Goddess as Throne, like Isis in Egypt (Stone, *When God,* 96) and the lap of the Goddess as the source of power and, later, kingship, symbolizing sexual union with Her or with Her priestess. "Lap" itself may be a euphemism in some translations for what might more accurately be rendered "cunt" (cf. the feminist critique of the Biblical translation of the Song of Songs in Shuttle & Redgrove, 19–23; also Grahn in Spretnak, *Politics,* 271; Kramer, *Sumerians,* 159).

job—to a full-time and year-round occupation; from something women taught to their daughters as part of their productive role, to the subject of numerous textbooks for farmers. One surviving example is the Sumerian Farmer's Almanac from Ur (and elsewhere; Kramer, *Sumerians,* 104ff, 340ff): when the fields are flooded in May and June the farmer must limit the waters to the proper depth. Once the waters depart, but while the field is still wet, shod oxen must be let in to stamp out weeds and help level the surface again, followed by a final leveling done with "small light axes," then by "men with pickaxes" going all around the field and smoothing it out, and finally by working over the whole field with a drag. While the field dries, the farmer readies all the necessary cultivating tools—including whips and goads to keep both laborers and beasts at their work.

When the field is dry enough, it must be thoroughly ploughed twice with two different deep-soil ploughs, then harrowed and raked thee times, and finally "pulverized with hammers." Only then can the actual planting begin, accomplished by yet another plough "with an attachment that carries the seed from a container through a narrow funnel down to the furrow." There must be just eight furrows to each strip (6–7 meters long); the seed must go exactly two "fingers" deep; and there are several different kinds of furrows. After planting, the field must be leveled yet again so that the barley will have the best chance to sprout. Birds must be scared off, and vermin charmed away. The growing barley must be watered at least three times—when it fills the bottom of the furrow, when it is about half grown, and when it has reached full height. If it shows no sign of the dread *samana* disease (probably a fungal rust) the farmer can water a fourth time to increase the yield by 10%.

All this, prior to the harvest, should give some idea why women lost out in agriculture.

## Patrilocal Residence

The support and status women receive in a society where brides bring their grooms home to mother, as among the Iroquois, or where the couple lives among the bride's kin, is too obvious to need retracing here. Equally obvious are the opposite effects of the opposite residence and descent patterns. Says Ruby Rohrlich, in fact, "The changeover to a patrilineal, patrilocal system led to the creation of the patriarchal family, probably the most important element in the creation of male supremacy" ("State," 84).

Moreover, patrilocality could have contributed to the exhausting effect on women of the birth explosion, as mothers would no longer have their own mothers, sisters, or other female relatives at hand to share the child-rearing burden.

A third effect, virtually never mentioned, is that this shift in descent and residence patterns, by interrupting the ancient mother-daughter continuity, may also have interrupted the traditional instruction in abortion and contraception techniques—women's control of their own fertility. The discontinuity would be particularly sharp where girls went to their future husbands' families before puberty. In the patrilineal and polygamous compounds of parts of Africa, a new bride gains status only by producing children. The loyalty of the mother-in-law—and of the midwife, the other traditional source of antifertility methods and information—is to the husband.

Whether this pattern developed organically or was introduced from outside, it would have drastic pernicious effects on women.

## Warfare

I said above that before men could take over agriculture, they must develop a new power base. This is exactly what happened. In addition to their specialization in distant trade and in animal husbandry, in ploughing and irrigation, they began to specialize in war. Near Eastern society, as it developed from scattered Neolithic agrovillages into clusters of small city-states, has been called a communal theocracy (Gray, 13, citing Frankfort), reminding us that the temple was the main center of power. Until about 4000 BC (and sometimes later), the Goddess was worshipped as supreme deity in these temples, and Her chief priestess held both religious and economic power. Although councils of elders or "assemblies" apparently existed, "Queen-priestess" is probably not a misnomer for this woman's role (cf. M. Stone, *When God,* 96; Spretnak, *Politics,* 23).

According to Ruby Rohrlich, who has written cogently on women's loss of status during this period, the people of the Neolithic towns and villages "drew together into larger territorial and political units" in the first place in response to raiding by nomadic herders ("State," 80–1). As the population of the region grew beyond the capacity of dry-farming or simple irrigation to support its agriculture, conflicts would have arisen over water. Conflicts would have arisen over land, too, as irrigation made some parcels far more valuable than others. Indeed, if Sumer was anything like Iran, only a tiny portion of the land could be irrigated. The builders of the far-flung canal system in place by 3000 BC would have felt a need to defend it. As rising population—and burgeoning trade—demanded ever-higher farm production, lands could no longer lie fallow (cf. Boserup, 32), which may have deprived nomads of customary winter-grazing rights. Their once-welcome arrival would now be met with resistance. In any case, by about the time the new canal system was finished, warfare became "chronic," as Rohrlich puts it, and "walled cities were a feature of the Sumerian landscape" ("State," 82). Myths suggest that women at first tried to solve conflicts peacefully, were nevertheless drawn into the wars, but eventually were excluded from military activities and decisions (*ibid.*, 90).

As the temporary (male) war-leaders appointed to deal with these conflicts refused to surrender their positions at war's end, they created a second power center in the society. Eventually these war-leaders set themselves up as "kings," usually seeking a legitimizing marriage with or investiture by the Queen-priestess or one of her daughters. If Enlil's treatment of Ninlil in Nippur is any reflection of this development, it was not the gradual and natural transition that has sometimes been suggested.[11] Says Rohrlich, "As military matters became paramount in economic and political decision-making, the successful generals eventually became the rulers, usually after a power struggle with the priesthood" (*ibid.*, 82).

## Religious Change: Patriarchal Deity

I have already hinted at the last factor I wish to discuss, *the intrusion of a patriarchal male deity (or deities) into a world where the Goddess had reigned supreme for thousands of years.* Western women have so long been deprived of a female deity, or even a feminine aspect to the godhead, that we can scarcely imagine what it would feel like to live in a Goddess-worshipping society. Mary, the mother of Jesus, has absorbed much of the unsatisfied longing for the divine feminine, at least among Catholics. But Mary, with her status totally derived from her motherhood, bears no resemblance to the mighty Queen of Heaven who existed before anything else was.[12]

11. Certainly all the warnings of the Hebrew prophet about what would happen to the people if they set up kings came true, especially for women.
12. To those accustomed to hearing Goddess-worship dismissed as a cult, the following opening of a prayer to Ishtar is illuminating (M. Stone, *Mirrors* I: 107):

Queen of Heaven, Goddess of the Universe,
the One who walked in terrible chaos
and brought life by the law of love
and out of chaos brought us harmony
and from chaos She has led us by the hand . . .

Consider, too, this invocation of the Sumerian Mother Goddess Nammu, perhaps the oldest recorded universal creator (*ibid.*, II: 37):

Ama Tu An Ki—
Mother who gave birth
to heaven and earth,
Primordial Creator of the Universe
who oversees the fashioning of life
and to each decrees their fate,
Oldest of the Old,
ancient, even among Sumerians,
Mother Primeval Sea left memories
in the land of the two rivers—
that it was She who had created all above,
that it was She who had created all below . . .

One contemporary American woman seems to have made the imaginative leap, however. Here is the Mormon heretic, feminist, and peace advocate Sonia Johnson reshaping her pantheon (118–19):

> Also soon after that fateful night, I discovered one day that men had made God in their own image to keep control of women. Why then could not women remake God in a way that would . . . empower everyone, make everyone whole?
>
> · · · · · · · · · · · · · · · · · · · · · · · · · · · · · · · · · · · · · · · · · · · · · · · · · · · · · · ·
>
> Thinking about this, I remembered Mother in Heaven, the divine being to whom Mormon doctrine attests (. . . put there, appropriately enough, by a female prophet). Remembered, sought her and found her. And loved her. Oh, how I loved and continue to love her! As I gradually reinstated her in her rightful position in my new heaven as equal in power and glory to Father—not in subordination . . . —I began to feel a wholeness and a personal power that transcended any happiness I had ever known. With Mother on her throne as a model for me in heaven, I felt wonderful. I felt wonderful knowing that femaleness is as divine, as desirable, as powerful as maleness. "No wonder men have felt so great for so long!" I marveled. . . . I felt marvelous to be able to identify with God. I felt like Boadicea, deeply and contentedly strong. Nothing seemed impossible.

I suggest that gradually—or perhaps in some cases rather suddenly—from the beginning of the 3rd millennium BC onward, women in the Middle East began to have Sonia Johnson's experience in reverse as male deities everywhere usurped the female—sons, lovers, or consorts becoming conquerors, rapists, and finally dominant husbands. Most students of the period, including Ruby Rohrlich ("State," 94) and Catherine Callaghan ("Wanderings," 25), seem to think this eclipse of the divine female *reflects* women's loss of status in the society. In the case of an indigenous development, like the earlier eclipse of the male hunting god at Çatal Hüyük, that might well be true. And we may be dealing here with a chicken-and-egg question. But since the male deity and the patriarchal virus in this case evidently arrived with nomadic invaders or immigrants from another culture, it may have been the other way around. The suggestion to males that God was, after all, male, may have emboldened them to take some of the steps they took—indeed, *may have made patriarchy thinkable*—at the same time as the repression and usurpation of female deities helped to undermine the confidence of women.

What is certain is that the religion did change, and the change is clearly visible in agricultural myth as well as elsewhere. Epitomizing this "transformation of Neolithic religious structures" between the late 4th and the 2nd millennia BC is the decline of Nammu. In early Sumerian myths, Nammu was the creator of all life and ancestress of all the gods, personification of the primeval sea; by the later pre-dynastic period her role was diluted so that creation was a joint effort between Nammu, Ninmah (another goddess), and the

---

The great Isis speaks for herself in Apuleius's *The Golden Ass* (M. Stone, *When God,* 22–3):

> I am Nature, the universal Mother,
> mistress of all elements, primordial
> child of time, sovereign of all
> things spiritual, queen of the dead,
> queen also of the immortals, the
> single manifestation of all gods and
> goddesses that are. My nod governs
> the shining heights of Heaven, the
> wholesome sea breezes, the lamentable
> silences of the world below. Though
> I am worshipped in many aspects, known
> by countless names, and propitiated
> with all manner of different rites,
> yet the whole round earth venerates me.

After listing some of the many names by which she is known in various lands, she concludes, "but both races of Aethiopians, whose lands the morning sun first shines upon, and the Egyptians who excel in ancient learning and worship me with ceremonies proper to my godhead, call me by my true name, namely Queen Isis."

water-god Enki. By about 3000 BC, the great Nammu is merely the consort of An—who had himself once been female (the sky seen as female; Rohrlich, "State," 85–6; cf. M. Stone, *When God,* 41).

In a parallel decline, in the late and literary Sumerian myths, instead of Ninlil giving cultivative knowledge and benefits directly to her people as she had originally done (M. Stone, *When God,* 3; *Mirrors II:* 39)—and as Isis, Ishtar, and Demeter had done elsewhere—two male deities, Enlil and Enki, send Ashnan the Grain Goddess to earth to make abundant its grain, equipping her with both plough and yoke; Enlil appoints Enten the farmer-god or farmer of the gods as his "steadfast and trustworthy field-worker"; and Enki becomes the progenitor of Uttu, Goddess of Plants (Kramer, *Mythology,* 42; *Sumerians* 218, 263). Indeed, Enlil, often called "Lord of the Plow" by Kramer, was by the advent of what is usually known as Sumerian civilization, about 2500 BC, one of the three or four chief deities in the pantheon (along with Ninhursag the ancient Earth/Creation Goddess; An the heaven god; and Enki the water god) and considered by some the supreme deity (*Sumerians,* 148, 118).[13]

Likewise in Greek myth, former Great Goddess and direct patroness of agriculture, Demeter, instructs Triptolemus in cultivation, then gives him seed-grain, a wooden plough, and a chariot drawn by serpents, and sends him out to teach all the world what he has learned. She also gives Phytalus a fig tree and teaches him how to cultivate it—undoubtedly including the ancient woman's secret of caprification, or pollinating the cultivated tree with wild-fig pollen. In yet another parallel, Arkas, first king of Arcadia, is credited with introducing agriculture, which he learned from Demeter (Graves, *Myths* I: 92, 95; Thomson, I: 131).

In summary, then, my scenario for the agricultural takeover is as follows: Preoccupied and possibly exhausted by more frequent childbearing, overworked in their many economic roles, deprived of matrilineal support networks and progressively deprived, too, of the psychological support provided by the Goddess religion, women may have become vulnerable to such a revolutionary change just at a time when men, relatively under-employed in the society, perhaps still suffering or remembering a role loss as hunting declined or disappeared, but gaining in power with their new specializations in animal husbandry, animal-powered ploughing, large-scale irrigation, long-distance trade[14]—and perhaps especially warfare—were both inspired and justified in acting to change their status by the new male-god religion beating on their temple doors. Certainly, wherever agriculture was so strongly integrated into the religion as in the Neolithic Near East, a religious change must precede—or immediately follow—a male takeover of cultivation. Certainly, too, it was males who would have had the greater leisure and freedom from distraction necessary to focus on and solve the problems of scale developing with the population explosion and the stratification of their society.

Earlier theories have accounted for the rise of private property and the stratification of society, but have not explained why so much of that property—and the higher strata of society—fell to males. Of all the factors I have examined, the three key factors seem to me to be *population* (without which none of the others, including institutionalized warfare, would apply); *patriarchy,* which surely received its impetus and model from the new one-male-god religion; and *professionalization.* Cultivation had been for millennia something women did in addition to all their other productive, reproductive, artistic, religious, and caring roles. In short, it was a part-time job. Unfortunately for them, as the Sumerian Farmer's Almanac shows, cultivation became not only a full-time job, but too complex to be learned casually from one's mother, at just about the same time that one of their other jobs (reproduction and child-care) increased its demands, and also just about the time that new technology and new religious ideas gave men the means, the backbone, and the justification to enter a previously forbidden field and eventually to dominate it. Thus women agriculturalists entered the shadows of a long dark age from which they have not yet emerged, at least in our culture.

13. Though An had once been supreme, says Kramer, available sources show Enlil as the leading deity by about this time (*Sumerians,* 118). Kramer repeatedly so calls him. (Cf., however, the caption to Plate XIV in Kramer, *Mythology:* "In the latter half of the third millennium the water-god Enki played a predominant role in Sumerian religion and myth." Perhaps this caption, like the accompanying illustration, is reproduced from Henri Frankfort's book on cylinder seals.) Indeed, Kramer seems at pains to improve Enlil's reputation, which he says suffered because early-excavated materials pictured Enlil as the gods' hatchet man (*ibid.,* 119). Some of Enlil's bad press, it occurs to me, may have been deserved, arising from the part taken by his followers in an ongoing agricultural, religious, and political takeover that was not entirely nonviolent.

14. At least by the Early Dynastic period, an actual class of merchants had arisen "that also acquired wealth, power and land under the aegis of the ruling class" (Rohrlich, "State," 81).

# Appendix A-2:
# Catherine Greene and the Cotton Gin

Since the story of Catherine Greene as sole or "real" inventor of the short-staple cotton gin seems to be alive and well despite denials in an authoritative biography of Whitney and in *Notable American Women*, it seems worth an extended discussion here. There are three basic positions in the controversy: First, that Catherine Greene was the real inventor and Whitney merely her model-maker; second, that Whitney was the inventor but Greene solved a crucial problem with the first model; and third, that the story of Greene's contribution is just a pleasant—or puerile—fantasy.

An early (1890) exponent of the first position is Matilda J. Gage (*/1: 3*):

Although the work on the model was done by the hands of Eli Whitney, yet Mrs. Greene originated the idea, and knowing Whitney to be a practical mechanic, she suggested his doing the work. This was during the winter of 1792-3, when he was a guest at her house near Savannah.

It was after one of the planters' regular evening laments on the horrendous difficulties of de-seeding the short-staple cotton, continues Gage, that Greene proposed her idea to Whitney, and he started to work (*ibid.; 4*):

The work was done in her house, and under her immediate supervision. The wooden teeth first tried did not do the trick well, and Whitney, despairing, was about to throw the work aside, when Mrs. Greene . . . proposed the substitution of wire. Whitney thereupon replaced the wooden by wire teeth, and within ten days from the first conception of Mrs. Greene's idea a small model was completed.

Regrettably, Gage does not reveal her source.[1] But in addition to her dramatic and uncompromising allegation that the machine was Greene's idea and built under her supervision, Gage advances two important specifics—the wire teeth, and the hope of profit—that lend credence to her report. Ann Forfreedom (PC-1982) and Julietta Arthur (315) have taken the royalties paid to Greene as evidence that Whitney acknowledged her technical contribution to the invention. Unfortunately for this argument, the same obligation could have arisen from a purely financial contribution.

Clearly she backed the invention. Jeannette Mirsky and Allan Nevins' biography of Whitney often gives the impression that the Miller-Whitney partnership to develop the gin was solely a contract between the two men (e.g., 66). However, Phineas Miller as plantation manager and the penniless Whitney could have done nothing on their own (see Mirsky and Nevins' discussion of a prenuptial agreement between Miller and Greene). Catherine Greene soon had to place all of her resources behind the partnership to save it; and the endless lawsuits over the gin forced her to sell her home plantation, Mulberry Grove, in 1800 (*NAW*). The partnership's finances eventually became so complicated that a court case with two referees was required to straighten it out.

More interesting to me is the first point about substituting wire teeth for wooden ones on the primary roller of the gin. Yet except for Martha Rayne (115), none of the other first-position advocates has taken it up (e.g., Armitage, 17; Blashfield, *Hellraisers*, 160-1; Russell Conwell, cited in Boulding, 686; Forfreedom, 207). Even Mozans, who seems otherwise to follow Gage's account, and Annie Nathan Meyer, who specifically cites Gage (280), do not mention it.

What reason do we have for believing Gage? We know little about Catherine Greene's childhood and schooling. According to *NAW*, she had "no pretensions to intellectual distinction." But her quick and intuitive intelligence—the very kind an inventor needs—is vividly described by Elizabeth Ellet (47-8):

She possessed, moreover, a marvelous quickness of perception, and the faculty of comprehending a subject with surprising readiness. Thus in conversation, she seemed to appreciate every thing said on almost any topic; and frequently would astonish others by the ease with which her mind took hold of the ideas presented. On one occasion . . . she looked over the books and collection of a Swedish botanist, making remarks from time to time which much interested him, and showed her an observer of no

---

1. I would be grateful to hear from scholars working on Matilda Gage, with information on these sources in particular, or the nature and reliability of her sources on invention in general.

common intelligence. This extraordinary activity of mind, and tact in seizing on points, so as to apprehend almost intuitively, distinguished her throughout life.

Both her childhood friend and first husband, Nathaniel Greene, and his father seem to have been blacksmiths, and Catherine was allegedly familiar with the operation of her father-in-law's anchor forge (Ellet, 49; Meyer, 280; *WI/1*). Some of the circulating 19th-century stories called Greene "a woman of great mechanical ingenuity" (Hammond, 93). Certainly it was she, and not Whitney, who had all the previous experience with cotton and the de-seeding problem. Certainly, too, she gave him room, board, workspace, and tools, and suggested that he try to build a short-staple cotton gin. If she did outline for him, even in the sketchiest form, what turned out to be a successful basic plan or approach, Whitney's speed in creating the first model—usually given as ten days—seems less astonishing.

The Shaker *Manifesto* for July 1890 reprinted a report of a public confession by Whitney that Mrs. General Greene gave him the idea of the gin (164). Though this wording could mean merely that she suggested he try to create the gin, the Shaker writer probably did not take it this way or it would never have gained space in the *Manifesto*. Moreover, the report seems to have originated in the periodical *Inventive Age*, whose editor ran a column on mechanical inventions. This column was rare outside the realm of women's magazines and feminist periodicals in regularly listing women's as well as men's patents.[2]

If in addition to "giving Whitney the idea" of the gin (however we interpret the phrase), Catherine Greene suggested using wire teeth rather than wooden ones, her contribution must be regarded as significant indeed. Clearly, had she been male, in that case, she would have been named co-inventor or assignee on the patent.

Equally clearly, of course, Eli Whitney was a mechanical genius. It is not inconceivable that one who made violins at age 12 could have solved such a frustrating problem in only ten days—and without ever having seen a cotton boll before coming south.

Modern scholarly sources generally take the other extreme position: that Catherine Greene made no meaningful *technical* contribution to the cotton gin whatsoever. Says *Notable American Women*, "There seems to be no basis for the legend which later credited her with contributing one of the basic principles of the design." Mirsky and Nevins, whose 1952 biography of Whitney is based on an examination of all the Whitney papers, say of the persistent brush-suggestion story, "If the story has an element of truth, it would be that she permitted the use of her hearth-brush when Whitney found it impossible to procure bristles to make the clearer for his model" (Author's Note, x).

On the side of this third position is the fact that the bitter battle for control of the gin made Whitney many enemies. Nothing could have suited them better than an important role for Catherine Greene in the original invention, and some of them might have stooped to inventing such a role. (Equally, however, it could explain why Greene's role was not announced.)

Another factor discussed by Mirsky and Nevins, though not in this context, is the idea that Whitney may have been more than half in love with the bewitching Mrs. Greene. Certain it is that he did not marry, though long economically able to do so, until after her death in 1814 (284):

> For 20 years, perhaps, Catherine's image so possessed him that even when he had established himself financially and could have married, he was unable to give himself to another woman. Only Catherine's death could end the spell.

Wives, lovers, sisters—female family members in general, according to the prevailing ideology—wish unselfishly to help their provider-males. They need or desire no recompense. The Beloved/colleague connection may have been the hardest of all such leaps for an 18th-century male to make. In other words, the more he loved Catherine, the less likely he would have been to credit her.

But the basic modern argument is simply a negative one: Eli Whitney did not credit Catherine Greene with technical help *in any surviving written source,* nor did Whitney's descendant and namesake Eli Whitney of New Haven do so when interviewed in the 1950s; therefore, she probably made no technical contribution. However, this negative evidence proves nothing. Are male inventors any more likely than other males to credit their female co-workers? Charles B. Hall, for example, gave his sister Julia no public credit for the aluminum-refining process that made him rich and famous, and spawned the corporate giant

2. The column, "Mechanical Inventions and Designs," appeared 1896–1911. I would be grateful for information on a list of women inventors the DuBois brothers (editor and publisher of *Inventive Age*) may have compiled c. 1913.

ALCOA. Yet her expert testimony—without which he would never have won his patent—indicates she must have been intimately involved in the invention process (Trescott, "Julia," *passim*).

We shall probably never know for sure. Between us and that far-off time at Mulberry Grove two curtains are drawn: the iron curtain of an inventor's necessary secrecy, and the silken curtain of a lady's prescribed modesty. The two curtains are equally opaque for us. Whitney wrote to no one during his original blaze of creation in Georgia; and his letters written to Greene during his Northern labors to make the gin patentable are missing. Were they burnt at his request as were some of Charles Hall's technical letters to Julia? Elizabeth Ellet gives us a striking instance of Greene's exquisite delicacy and unfailing adherence to the lady's code (54–5). And Matilda Gage answers the rhetorical question why Greene did not patent her invention (/1: 6):

> To have done so would have exposed her to the contumely and ridicule of her friends. Custom, that unwritten law, has for years frowned upon any attempt of woman to take such a step. If she has been gifted with inventive genius, she has either stifled its exercise . . . or, like Mrs. Greene, suffered some man to claim the award her due. She shrunk from the persecution that would have attended her claiming the patent, while by associating herself with Whitney as his partner in the manufacture of the machines, she hoped at least to share in the pecuniary advantages of the invention.

Even if Catherine Greene were otherwise inclined to claim credit for some role in the invention, she could scarcely have done so while her friend Eli was fighting for the very life of his patent. And to have done so years later, when victory came just as the patent was about to expire, might have seemed ridiculous indeed, not to say disloyal.

My own instincts incline me, despite specific denials by *NAW* and Mirsky and Nevins, to the second position—that Catherine Greene contributed something more important than tea and sympathy to Whitney's original model, most likely the use of bristles to keep lint from clogging the gin. Who, after all, had more experience than women at removing lint—with brushes?

This idea appeared in print as early as 1832 in a "Sketch of Eli Whitney" by William Scarborough in the *Southern Agriculturalist,* here paraphrased by M.B. Hammond (97):

> The inventor, it seems, had encountered a difficulty in the fact that the cotton lint . . . disengaged from the seed adhered to the teeth of the cylinder and impeded the work of the machine. He was greatly perplexed to know how to overcome this difficulty, when Mrs. Greene, who had been a witness to his fruitless efforts to disengage the cotton from the teeth of the cylinder, picked up the hearth-brush and laughingly remarked, "Why don't you use this?" Acting on this suggestion, Whitney returned to his work and added a second cylinder, studded with stiff bristles, revolving contiguous to, but in an opposite direction from, the other cylinder. This served to sweep the particles of cotton away as they were ginned and made the gin a practical success.

Hammond, author of an 1890s monograph on the United States cotton industry, calls the pro-Greene claims "puerile" (*ibid.*). Yet he seems to accept the brush-suggestion story even while denying its importance: "Valuable as this suggestion may have been to Whitney, it does not require a very discerning mind to distinguish between the importance of this suggestion and that of the invention itself" (*ibid.*). Granted the obvious distinction between the whole of an invention and any one of its parts, the distinction we really need to make here—and then discard for purposes of this argument—is between a technical understanding or description of the solution to the clogging problem and the *intuitive but no less valid understanding* that may have flashed into Greene's mind as she watched Whitney's struggle. Whether she "laughingly" held out the hearthbrush and said, "Why don't you use this?" or told him "I'd suggest some sort of continuous brushing action, using bristles about as stiff as these," the effect was the same. Had her hearer been less brilliant, she might have been less cryptic.

It must be noted, too, in reply to Hammond, that no invention becomes valuable unless it works. And Whitney's gin would never have worked without the brush-roller(s). The only real question is whether, left to himself, Whitney would have solved the problem—or whether schoolchildren a hundred years later would have learned some other name in connection with the cotton gin. But then, the whole course of American history might have been different.

# Appendix A-3:
## Selected 19th-Century Women's Dairy Patents

**Mary C. Atkinson,** Baltimore, MD—Milk-skimmer (1885)

**Anna E. Baldwin,** Newark, NJ—Improved milking machine (1879), plus three other patents: an Improvement in milk-separators and an Improvement in milk-coolers (1869); and an Improved process of treating milk to obtain useful products (1868); also a reissue of her Improved milk-cooler (1871)

**Ollie E. Belle,** Derby, KS—Churn (1891)

**Angelina G. (and A.F.) Bonham,** Seven-Mile Ford, VA—Hand and foot churn-power (1891)

**Mary J. Bridges,** Wimberly Hill, TX—Combined clothes-washer and churn (1883)

**Eliza Brough,** Greenville, MI—two patents: Improvement in reciprocating churns (1877); Improvement in churns (1877)

**Rebecca Sarah Brusie,** Sonora, CA—Improvement in churns (1877)

**Rebecca A. Campbell,** Quincy, MA—Milk pail (1883)

**Mary Couplin,** Thornberg, IA—Milk can elevator (1884)

**Charlotte Augusta Davenport** (later **Bacon**), Charlton Depot, MA—Detachable spout for milk pails (1885)

**Gabrielle De Nottbeck,** New York, NY—Reciprocating churn (1875)[1]

**Celinda A. Dodge,** Kinnikinnick, WI—Milk-strainer (1886)

**Frances A. (and I.A.) Dodge,** Brady, TX—Milk-cooler (1889)

**Alma A. Foster,** Pomeroy, OH—Butter-worker (1890)

**Mary E. Garnham** (and Fernando D. Hubbard), Carthage, NY—Improvement in milk-coolers or warmers (1876). Garnham also invented an improved hair-curling iron on her own (1876)

**Eliza German,** Fannin Co., TX—Churn (1883)

**Julia A. Graves,** Fair Haven, VT—Butter-worker (1888)

**Elvira C. Hale,** Steamburg, NY—Milk-vat (1872)

**Bettie Adela Malone** (and Alexander Malone Franklin), Simpson Co., KY—Churn dasher (1885)

**Mary E.J. Marr,** Jefferson LA—Improvement in churn dashers (1868)

**Alice E. Mendenhall,** Visalia, KY—Skimming device for milk cans (1887)

**Rebecca E. Miles,** Porterville, CA—Churn (1892)

**Bettie Mitchum,** Hammonville, KY—Churn (1889)

**Mamie M. Montgomery,** Gowensville, SC—Churn (1894)

**Maria A. Ober,** Chazy, NY—Improvement in churn dashers (1867)

**Susan M. Palmer,** Greene, NY—Improvement in churns (1872)

**Effie L. Prall,** Wellington, KS—Motor for churns (1887)

**Sarah E. Saul,** New York, NY—Improvement in churns (1865). Saul also patented a Life-preserving skirt (1870) and an "Improvement in metallic casings for lead pipes" (1876)

**Katharina Seifert,** Cincinnati, OH—two patents for Machine for degranulating cheese curd (1883, 1886)

**Maria L. Shade,** Fair Grove, MO—Milk vessel (1872)

**Mary A. Sheaffer,** Elizabethtown, PA—Process for manufacturing cheese (1869)

**Agnes W.H. Smith,** Beaver Dam, WI—two patents for Milk cans (1889, 1890)

**Lettie A. Smith,** Pineville, PA—Butter-worker (1853)

**Anna M.E. Stewart** (and Theron S. Stewart) Toledo, OH—Churn dasher (1888)

**Lydia W. Stiles,** Brooklyn, OH—Butter-worker (1860)

**Sarah L. Stuart,** Greenfield, IN—Improvement in churns (1877)

**Lulie F. Sullivan,** Holden, MO—Churn (1891)

**Mary Louise Traver,** Detroit, MI—Improvement in churn dashers (1876)

**Amanda E. Turvey,** Fredericktown, OH—Improvement in milk coolers (1879)

**Mattie A. Van Alstine,** Mountain Valley, AR—Milk-pail (1888)

---

1. Omitted from LWP.

**Hannah C. Warner,** Woodbury, NJ—Improvement in coolers for butter-workers (1879)

**Minerva A. Widger,** DeKalb, IL—Improvement in preparing rennet for making cheese, etc. (1874)

**Lizzie F. Wood,** W. Lebanon, NH—Aerator for milk and other liquids (1891)

**Emma J. Woodruff,** Chicago, IL—Process of treating butter (1885). This very interesting process, whose purpose may have been to sour the milk and/or make the butter come more quickly, consisted of adding to the milk white-wine rennet, sugar, salt, bicarbonate of soda, bicarbonate potassium, alum, and butter, in specified proportions.

# Appendix A-4:
## Animal Medicine

Veterinary medicine has proved even more of a male bastion than human medicine. Neither edition of *Notable American Women* has a single veterinarian in its occupational index, for example, and only now are the walls finally cracking.

Women have long "doctored" animals, of course, but predictably few of their remedies reached the patent office. **Margaret N. George**'s and **Dicie Croan**'s 19th-century specifics for hog and chicken cholera have already been mentioned. An Englishwoman who patented her veterinary invention in the United States in the 1880s also deserves to be noted here: **Georgina N. Laurence** or **Lawrence** of London patented an Apparatus for treating animals by electricity in 1889. Earlier that same year, Lawrence had received a British patent (#4960, May 25) for an Improved means of administering electricity to and otherwise treating horses and other animals. The apparatus consisted of a stock for confining the animal, a means of raising or lowering it as needed, and a means for administering the electric current, using a battery with zinc and platinum plates. On this patent she lists her address as 5 Nottingham Place, Regent's Park, Middlesex; her full name as Georgina McGuire Lawrance; and her occupation as "Medical electrician."[1]

Of the medical, surgical, and related inventions listed by the Women's Bureau study for the early 20th century, only one, a Veterinary medicine spoon, is specifically for the treatment of animals; however, some of the other items may have had their veterinary uses.

An example of a patented invention from more recent years is the work of **Eva Schoenberger** of Givatayim, Israel, who with male co-inventors received two U.S. patents for treating bovine mastitis in 1978.

Other innovative work by women in the veterinary field may or may not have been patentable. The first woman veterinary surgeon, **Dr. Aleen Cust,** who graduated from New College, Edinburgh, Scotland in 1900, did distinguished work in animal bacteriology for more than two decades, and served with the Royal Army Veterinary Corps during World War I. Another British woman, **Dr. Phyllis Croft,** pioneered in the use of EEGs to diagnose animal ailments. Here in the United States, **Dr. Jean R. Hagan** became well known for her work on poultry diseases, as did **Dr. Louise Lombard** for her work on animal cancers and tumors. **Dr. Jean Holzworth**'s award-winning work on cat diseases may be mentioned here, in view of cats' function in vermin-control on farms. **Dr. Phyllis Lose** opened the first privately owned equine clinic in the United States in Chester County, PA, in 1974. Designing and building the clinic herself, she undoubtedly incorporated her own innovations for adapting the structure to its highly specialized purpose (M. Lee, 24, 83).

A world-recognized authority in her field is **Dr. Lucy Lee** (1931–  ) of the Regional Poultry Research Laboratory in East Lansing, MI. There, as a research chemist, Dr. Lee is working on the control of Marek's disease in commercial poultry flocks. Born in China in 1931, she emigrated to the United States in the 1940s. After taking a bachelor's degree in biology and an MS in zoology, she got her PhD in biochemistry at Michigan State University in 1967. She is married and has two daughters (*WBWRA*, 32–3).

---

1. Interestingly enough, the London Post Office Directory for 1889 lists two Lawrences as dairymen and another as a Harley Street surgeon. Also, one H.N. Lawrence (none of these three) received two patents for administering electricity, but evidently to humans rather than animals. Further research could conceivably connect her with one or more of these families. It would also be interesting to know whether Georgina Lawrence knew the woman veterinarian who lived in N. Croyden at that time—Mrs. Emma Palmer of 2, Hope Cottages, Sydenham Road (Kelly's P.O. Directory, N. Suburbs, 1888).

# Appendix A-5:
## Food-Processing Machines, Devices

Following is a sample, taken from LWP, of 19th-century women's American patents for food-processing machines—raisin-and cherry-stoners, egg beaters, dough kneaders, and the like:

**Nancy M. Johnson,** Washington, DC—Ice Cream freezer (1843)
**Rosanna Carpenter,** Portsmouth, NH—Improved raisin-stoner (1864)
**Lucy Sawyer Nourse,** Templeton, MA—Flour sifter (1871)
**Evelyn L. Edwards,** Vineland, NY—Impr. in dough-kneading machines (1877)
**S. Jennie Renner,** Petersburg, PA—Impr. in apple corers and quarterers (1877)
**Eleanor W. Jones,** Cleveland, OH—Impr. in dough-mixers (1877)
**Sara Cooper,** Morrisonville, PA—Impr. in apple-quartering and coring machines (1877)
**Emma E. Orendorff,** Delavan, IL—Impr. in apple-corers (1878)
**Emily A. Stears,** Brooklyn, NY—Impr. in household presses for fruits, etc. (1878)
**Sarah A. Ulmer,** Portland, ME—Egg-beater (1880)
**Ellen B. Kendall,** Cincinnati, OH—Ice-shaver (1881)
**Mildred E. Ewell,** Winchester, VA—Dough-kneader (1881)
**Alice B. Wood,** Beaver Dam, WI—Corn popper (1881)
**Felicie F.N. Marais,** NY, NY—Fruit press (1881)
**Katherine A. Livingwood,** Womelsdorf, PA—Egg-beater (1883)
**Eliza M.C. Anderson,** NY, NY—Green corn cutter (1883)
**Hannah Zephyrene Gibson,** Oberlin, OH—Egg-beater (1884)
**Felice Molini,** San Francisco, CA—Pea-shelling machine (1884)
**Maria L. Mathews,** Newark, NJ—Fruit Press (1884)
**Jane Mikel,** Gillum, IL—Comb. squeezer and strainer (1886)
**Edith A. Marsh,** New Albany, IN—Egg-beater (1886)
**Frances V. Raymond,** Buffalo, NY—Beating & mixing machine (1886)
**Sarah E. Toucey,** NY, NY—Grape-seed extractor (1888)
**Mary F. Bishop,** Bridgeport, CT—Means for operating egg-beaters (1888)
**Nell L. Barron,** Barton, VT—Cake mixer (1889)
**Cecilia B. Darley,** Philadelphia, PA—Corn-cutting device (1890)
**Louise Stevenson,** Morristown, NJ—Ice-cream freezer (1890)
**Araliza Moncrief,** Brooklyn, NY—Raisin-seeder (1890)
**Fannie C. Davis,** Pana, IL—Dough-kneading, mixing device (1890)
**Louisa M. Thompson,** Atlanta, GA—Preparing green corn (1890)
**Lydia Stockwell,** Atchison, KS—Dough-kneader (1891)
**Louisa R. Alexander,** Elmira, NY—Rotary screen & sifter (1891)
**Annie Mangin,** Woodside, NY—Pastry fork (1892)
**Jenetta V. Bohannan,** Baltimore, MD—Meat-slicer (1892)
**Ettie E.B. Shleppy,** Crawfordsville, IN—Comb. beef-mangler (1893)
**Priscilla M. Burns,** St. Louis, MO—Flour-sifter with . . . cover (1893)
**Sara R. Thompson,** Henrietta, TX—Fruit press (1893)
**Jenny B. Fitch,** Utica, NY—Salad-mixing device (1893)
**Frances E. Paine,** Boston, MA—Egg-beater (1893)
**Jennie D. Harvey,** Wilkesbarre, PA—Mayonnaise mixer (1893)
**Jennie D. Harvey,** Wilkesbarre, PA—Egg-beater (1893)
**Rose B. Lavigne,** New Haven, CT—Meat tenderer (1893)
**Margaret A. Wilcox,** Chicago, IL—Dough mixer (1893)
**Arobine C. Mitchell,** Ennis, MT—Mixer or beater (1894)
Narcisse & **Theresia Heneault,** Chicago, IL—Vegetable slicer (1894)
**Annie C. Cook,** Council Bluffs, IA—Vegetable-slicing machine (1894)

# Appendix A-6: Stoves and Related Cooking Inventions Patented by Women in the 19th Century (U.S. Patents Listed on LWP)

**Emma Steinhauer,** Philadelphia—Adapting anthracite-burning stove for cooking (1831)

**Mary Van Vranken,** Washington, DC—Attachment for heating kettles, boilers by gas (1867)

**Mary E.A.W. Evard,** Leesburg, VA—Improvement in boiling apparatus (1868); Improvement in cooking stoves (1868); Toaster (1868)

**Sarah M. Clark,** Beaver Dam, WI—Reservoir cook-stove (1868)

**Margaret Armstrong,** W. Alexander, PA—Cooking stove (1869)

**Clarissa Preston,** Wheeling, WV—Oven (1869)

**Angelina Madison,** Cincinnati, OH—Charcoal cooking furnace (1869)

**Mary Ann Boughton,** Bridgeport, CT—Improvement in cooking-stoves (1871)

**Isabella C. Schramm,** Des Moines, IA—Boiler attachment for cooking & washing (1872)

**Mary Jones De Leon,** Baltimore, MD—Improvement in cooking apparatus (1873)

**Anna E. Neitz,** New York, NY—Improvement in cooking apparatus (1874)

**Minnie B. Lloyd,** New York, NY—Improvement in malt grain driers (1874)

**Mary T. Fitch,** Lockport, NY—Improvement in magazine cooking stoves (1874)

**Mary A. Scott,** Pakota, IN—Improvement in cooking apparatus (1874)

**Emma Miller,** Randolph, NY—Improvement for cooking apparatus (1877)

**Amanda E. Campbell,** Boulder, CO—Steam-cooking apparatus (1881)

**Cremora F. McCarty,** Boston, MA—Heating and cooking apparatus (1883)

**Emma G. Nunn,** Vassar, MI—Heat-indicator for oven (1884)

**Adele C. Philippi,** New Orleans, LA—Attachment for cooking stove (1886)

**Mary Ruggles,** Saratoga Springs, NY—Cooking Attachment for heating stoves (1886)

**Amelia A. Waterhouse,** San Francisco, CA—Reversible broiler or toaster (1889)

**Ada M. Throckmorton,** Chillicothe, OH—Broiler (1887)

**Julia A. Downey,** Oberlin, OH—Toaster (1887)

**Camilla Julier,** Hanging Rock, OH—Oven Thermometer (1889)

**Ida Smith,** W. Union, IA—Cooking, toasting, & shielding device for stoves (1891)

**Margaret E. Owen,** Lambourn, Eng.—Cooking range or stove (1891)

**Louise D. Fein,** Cincinnati, OH—Steam Cooker (1892)

**Virginia M. Cone,** Alameda, CA—Heat regulator for cooking stoves (1892)

**Margaret A. Wilcox,** Chicago, IL—Combined cooking and hot-water-heating stove (1892)

**Olive A. Wilcox,** St. Louis, MO—Bake-oven sliding drawer (1893)

**Drusilla F. Galer,** Palmyra, PA—Heating attachment for cookstoves (1893)

**Margaret E. Jehu,** Estherville, IA—Apparatus for cooking, broiling, baking (1894)

**Myra C. Farnum,** Port Jervis, NY—Oven (1894)

**Augusta R. Isaacs,** New York, NY—Firebox & grate for ranges, stoves & heaters (1894)

**Laura M. Stone,** Rochester, NY—Gas [cooking] stove (1894)

**Rebecca H. Hayes,** Galveston, TX—Cooking stove (1894)

**Mary A. Williams,** Kansas City, MO—Oven regulator (1894)

# Appendix A-7: Women's 19th-Century Canning and Preserving Inventions; Selected U.S. Patents[1]

**Rhoda Davis,** Brookhaven, NY—Impr. elastic cap for sealing cans and bottles (1857)

**Kate E. Ashley,** Williamsburg, NY—Impr. stand for preserve jars (1865)

**Nancy Patton,** Coles Co., IL—Impr. method of preserving eggs (1866)

**Nancy Patton,** Kansas, IL—Impr. composition for preserving eggs (1868)

**Catherine Bruner,** Marshall, MO—Preserving eggs . . . (1869)

**Mary E. Grigsby,** Putnamville, IN—Impr. fruit-straining apparatus (1870)

**Jane Carter,** Towanda, PA—Impr. in safety-cases for fruit jars (1870)

**Sarah B. Hertwig,** Cincinnati, OH—Funnel for canning fruit (1872)

**Eliza G. Haller,** Carlisle, PA—Fruit-jar (1873)

**Mary E. Lewis,** Mansfield, OH—Impr. in fruit-packers (1873)

**Abigail White,** Chunchula, AL—Composition for preserving eggs (1876; #172,677)

**Elizabeth E. Mefford** and **Susan Peebles,** Adrian, MI—Impr. in process for preserving fruit (1876)

**Elizabeth S. Hunt,** Cleveland, OH—Impr. in preserving packages for fruits, etc. (1876)

**Catherine Hastings,** Oswego, NY—Impr. in fruit-jars (1877)

**Catharine L. Darby,** Fort Motte, SC—Impr. in sealing devices for bottles, jars, etc. (1877)

**Lorinda Gallaher,** Wheeling, WV—Fruit-jar (1883)

**Mary A. Tipney,** Janesville, WI—Device for preserving fruits, etc. (1885)

**Frances M. Austen,** Dickinson, NY—Boiler for heating fruit in jars (1885)

**Elizabeth A. Taylor,** Baltimore, MD—Fruit jar (1885)

**L. Adelle Hapgood,** Randolph, NY—Apparatus for preserving eggs (1887)

**Helen L. Smith Hopper,** Hutchinston, KS—Apparatus for preparing canned food (1887)

**Mary J. Hamlin,** Bartramville, OH—Clamp for fruit-cans (1888)

**Ida Zander,** Milwaukee, WI—Jar-holder & kettle cover (1889)

**Mollie Ellis,** Bardstown, KY—Jar-holding funnel apparatus (1889)

**Martha A. Carter,** Amo, IN—Fruit canner (1889)

**Serine Unna,** Hastings, NE—Preserving fruit (1889)

**Rosalie Miller,** New York, NY—Preserve-jar lifter (1891)

**Belle S. Beam,** Strother, MO—Funnel for canning purposes (1891)

**Maggie Marshfield,** Chicago, IL—Preserving fruit (1892)

**Rebecca C. Sheeley,** Lyons, KS—Canning or preserving jar (1892)

**Artemisia F. Frost,** Shubert, NE—Holder for fruit jars (1892)

**Ina M. Fisher,** Burrows, MO—Jar-filler and -holder (1892)

**Mary E. Clark,** Corning, NY—Glass fruit jar protector (1893)

**Emma Feuerherd,** Coswig, Germany—Pot for preserving provisions on ice (1893)

**Ida L. McDermett,** Baird, TX—Preparing fruit for canning and preserving (1893)

**Electa Williams,** Pavilion, MI—Fruit jar (1894)

**Artemisia F. Frost,** Highland, KS—Rack or holder for fruit jars (1894)

**Helen A. Robinson,** Clymer, NY—Preserving jar (1894)

**Ellen M. Williston,** Wellsboro, PA—Apparatus for canning fruit (1894)

**Fannie Anker,** Jersey City, NJ—Preserving food compound (1894)

---

1. These come almost entirely from LWP; Abigail White's 1876 invention was omitted from LWP (see Stanley, "Conjurer").

# Appendix A-8:
# 19th-Century Women Inventors of
# Agricultural Machines, Attachments

This LWP-based list contains one British inventor who received an American patent for her work, and focuses on planting, cultivating, and harvesting machines and agricultural vehicles. An asterisk (*) indicates omission from LWP.

**Ollie Baughn,** Tobias, NE—Potato-planting attachment (1889)

**Zina A. Beecher,** Marysville, OH—Attachment for cultivators (1892)

**Harriet Bender,** Denver, CO—Garden seed drill (1893)

**Eliza J. Bentinck** (w/ J.A. Renner), Galveston, TX—Digging machine (1893)

**Harriet Carmont,** London, Eng.—Starter or brake for vehicles (1894)

**Annie H. Chilton,** Baltimore, MD—Horse detacher and brake, two patents (1891; 1894)

**Lucy A. Corning,** Rockford, IL—Baling press, two patents (both 1893)

**Marie Danos,** New Iberia, LA—Cultivator; Cotton scraper (both 1887). The cultivator is a big one with at least five blades, apparently to be pulled by a horse and guided at rear handles by a person. The "scraper" is also a cultivator of sorts, the improvement apparently involving the means of attaching the team to pull it.

**Elizabeth Dark** (w/ others), Quitman, MO—Combined harrow, marker, and cultivator (1880)

***Jessie C. Denson** (w/ Stephen B. Bell), Iamonia, FL—Plow (1882)

**Sarah Evans,** Pittsburgh, PA—Plow (1833). Since Sarah received this patent as executrix for G. Evans, her exact role in the invention is unclear.

**Lizzie B. Fleming,** Pierce City, MO—Wheel cleaner (1894)

**Miranda Fort,** Talbotton, GA—Improvement in plows (1870)

**Mary I. Goldsmith,** The Plains, VA—Combined plant-setter and fertilizer distributor (1881)

**Ethel O. Harvey,** Lubec, ME—Improvement in clevises (U-shaped metal device closed by pin, used to attach tongue of plow to draft chain, etc.; a safety hook or catch) (1875)

***Iris Hobson,** Ft. Scott, KS—Potato digger (1876)

**Martha Jones,** Amelia County, VA—Improvement in corn huskers and shellers (1868)

**Roseanna Mays,** Strasburg, OH—Fruit-picker (1892)

**Drusilla C. McConnell,** Williamstown, SC—Fender for plows (1891)

**Annie McFarlane,** San Bernardino, CA—Improvement in dumping wagons (1879)

**Emily V.D. Miller,** Vicksburg, MS—Hand cotton-picker (1883)

**Emma D. Mills,** Clinton, NY—Hop-stripping machine (1885); Hop-picking machine (1894)

**Elma M. (and John E.) Mitchell,** Salem, IA—Plow, five patents (1891–3) *(three of the five omitted from LWP).

**Amanda Morrison,** Wilmington, NC—Market truck (1883)

**Eliza H. Newcomb,** New York, NY—Improvement in fruit-gatherers (1863)

**Ella Pitts,** Palmer, KS—[Seed] Drill (1891)

**Ethel H. Porter,** Lincolnton, NC—Straw- and fodder-cutter (1834)

**Marie E. Ronat,** Rochelle, IL—Improvement in plows (1876)

***Philena Stanton,** Millbrook, MI—Harrow tooth (1883)

**Lena Strong** (w/ Alphonso B. Smith), San Francisco, CA—Fruit-picker (1886)

**Hope H. Tigner,** Stinson, GA—Harrow (1894)

**Sara W. Trabue,** Girard, IL—Baling press, three patents (Aug., Dec. 1887; 1888)

**Anna Trexler,** Sabin, MN—Combined plow and harrow (1888)

**Angeline Underwood,** Carrollton, IL—Land pulverizer (1875). Underwood, who also held two other patents, exhibited her machine at Philadelphia Centennial in 1876

**Mary C. Walling,** Brenham, TX—Cotton-picker (1885). This may have been a processing machine rather than one for detaching the bolls from the plants.

**Nell O. Westburg,** Sanborn, IA—Band-cutter and feeder for threshing machine (1893)
**Marian Whiles,** Woodville, MO—Hay rack (1891)
**Adeline Widmayer,** New York, NY—Dumping wagon (1895)
**Harriet D.T. Wilson,** New York, NY—One-wheeled vehicle (1892)
**Luna Wright** (w/ Samuel J. Miller), Economy, IN—Improvement in corn-plows (1868)

# Appendix A-9:
## Miscellaneous Agricultural Inventions

Considered in this section are areas of agricultural technology not studied extensively for this project, but too important to be omitted altogether. To avoid disputed rankings for significance, I present them in alphabetical order.

### Apiculture (Beekeeping)

In many primitive cultures—and in parts of Europe before the 18th century—honey was the only or major sweetener and an important preservative as well. Data on honey-gathering were available for only about a quarter of the 185 societies in Murdock and Provost's "representative sample" of the world's societies. Of these 48 societies, 39 assigned honey-collection to males, another 5 predominantly to males, 2 to both sexes, and only 2 to females (204, 207, 209).[1]

It is interesting to note, however, that in Neolithic Europe the Goddess was often visualized as a bee, and the Goddess Ker or Car was associated with beekeeping (Graves, *Myths* I: 33). Perhaps it was women who domesticated bees? Since honey had both religious and medicinal uses, and early artificial hives imitated the bees' own structures in woven straw and pottery (related to women's work; De Bono, 109), this seems a logical hypothesis. At the very least, women may have invented artificial beehives.

As late as Roman times, beekeeping was one of the three major branches of agriculture: horticulture, animal husbandry, and apiculture. The reason is obvious: apart from the bees' essential role in pollination, cane sugar had not yet reached Europe, and honey was still the major sweetener for all uses, the basis for an alcoholic beverage (mead), and an important preservative as well. Bees are still a vital part of the agricultural scheme of things, pollinating, for example, all our major fruit crops; but cane, corn, and beet sugars are now available, to say nothing of the artificial sweeteners; and mead is something we read about in medieval epics. Beekeeping was still more important in the 19th century than it is today, and several women received American patents for improved hives and other bee-related inventions:

**Thipena P. Hornbrook,** Wheeling, VA—Beehive (1861)
**Eliza Jane Donavan,** Indianapolis, IN—Improvement in beehives (1870)
**Elizabeth O'Connor,** Philadelphia, PA—Improvement in beehives (1871)
**Harriet A. Farnam,** South Bend, IN—Improvement in beehives (1871)
**Louisa C. Huff,** Louina, AL—Improvement in beekeepers' protective masks (1873)
**Emma Carter,** Folsom City, CA—Improvement in bee-feeding devices (1878)
**Sara E. (& Gordon) Finley,** Petersburg, TN—Bee-bucket (1880)

At least one of these women, **Harriet Farnam** (c. 1827– ), not merely patented her invention, but had it manufactured and sold. *Turner's South Bend Annual and Business Mirror* for 1872 advertises Mrs. H.A. Farnam's "Non-Swarmer." This improvement, attachable to any existing hive, allows the workers to come in or out, but queen and drones cannot escape. "Controlling the queen, we thereby control the swarm," is the central claim. Praising the device's "many points of merit," the ad continues:

> It prevents a superfluity of drones. It prevents robbing. It keeps swarms from going away by keeping the Queen in, compelling the swarm to return to the old stand when an empty hive may be placed and the box containing the Queen attached; thus the swarm virtually hive themselves.

Moreover, the Non-swarmer is affordable by all bee-keepers. Inquiries are to be addressed to the Italian Bee Company of Des Moines, Iowa. Presumably this company was the original manufacturer of the

---

1. Among the very primitive Tiwi of Melville Island, honey-gathering has traditionally been women's work, and thus the Tiwi—if Murdock and Provost follow Goodale, whom they cite—must be one of the two societies mentioned as assigning this work to women. However, when Goodale studied the Tiwi, men were beginning to gather honey (152).

Non-Swarmer; but by 1873, according to *Turner's Annual* for that year, the manufacturer was E.R. Farnam (Harriet's husband). Says the *Annual* writer, "The Non-Swarmer . . . has been used successfully by many of the most experienced bee-keepers in the country, all of whom speak of it in terms of high recommendation" (not surprisingly, this invention was born of an "intimate acquaintance with bee culture"; 11).

Harriet Adaline Farnam appears, even from the meager biographical information available, to have been a versatile inventor and a resourceful entrepreneur.

The Women's Bureau study lists only a Bee-feeder subclass among women's Agricultural inventions for the ten years covered. However, one **Frances A. Dunham** invented a valuable beehive improvement before 1910, for according to Minnie J. Reynolds (6–7), she showed it at the Suffrage Bazar held at the Martha Washington Hotel in New York City November 6–7, 1908. Dunham's invention, which Reynolds says was patented, was a comb foundation, which lessened bees' constructive labors, and also discouraged the development of drones, since its cells were all "worker-size." Reynolds called this "one of the most valuable agricultural inventions ever made." Evidently the first synthetic comb foundations had appeared just fifty years before, in 1843 (De Bono, 109).

## Pest and Vermin Control, Insecticides, Etc.

Women have contributed to pest-control technology from very ancient times, when they devised herbal, mechanical, or animal weapons against insects, worms, and crop-robbers. Otis T. Mason notes (*Share,* 18) that woman tamed the cat to protect her granaries, and Ralph Linton reports British Columbian Indian women putting up "various ingenious scarecrows" to keep the deer—and presumably also the crows—away (*Tree,* 90). Minoan houses contained tubes made expressly for snakes to crawl in and exterminate vermin (Rohrlich-Leavitt *et al.,* 39), and a basil plant still guards each house door in far southern Greece—a traditional fly repellent—or did at least when I visited Stoupa in the late 1970s. It is interesting to note how many herbal vermifuges (wormers) are also contraceptives or abortifacients. These 19th-century American patentees thus follow, all unknowing, in the footsteps of ancient foremothers:

**Harriet A. Farnam,** South Bend, IN—Improvement in flycatchers (1871)
**Lydia M. Gould,** Chicago, IL—Improvement in fly-traps (1872)
**Elizabeth Hooper,** Diamond Springs, CA—Compound for destroying bedbugs (1872)
**Elizabeth B. Johnston,** Germantown, KY—Improvement in devices for killing mosquitoes (1877)
**Kate M.** (and John Jay) **Strong,** Talladega, AL—Device for protecting tables, etc. from crawling insects (1880)
**Hannah Hey,** Philadelphia, PA—Vermin-trap (1881)
**Isabella S. Graves,** Fishkill-on-Hudson, NY—Insecticide (1884)
**Phebe R. Lamborn,** West Liberty, IA—Fly-screen-door attachment (1884)
**Rebecca McKee,** New York, NY—Insecticide (1886)
**Clara P. Brown,** New York, NY—Insect trap (1889)
**Jennie G.F. Johnson,** Mt. Vernon, NY—Insect trap (1889)
**Mary F. Sallade,** New York, NY—Insect trap (1889)
**Ann E. Hagstedt,** New York, NY—Compound for exterminating insects (1891)
**Mary P.C. Hooper,** New York, NY—Mosquito trap (1891)
**Mary A. Hawley,** Dixon, IL—Device for applying insecticides (1892)
**Mary A. Hawley,** Dixon, IL—Device for applying insecticides (1895)

Although to my knowledge none of these women gained lasting fame from their inventions, at least three were versatile multiple inventors; at least two were also entrepreneurs who manufactured (or had manufactured) and sold their inventions; one other caused a stir at the time by the court cases attending her application. **Harriet Farnam**'s Hoosier Fly-catcher was probably her best-known invention. This flytrap—which, incidentally, seems to work on the same basic principle as one I saw in use in rural California in 1981—was manufactured by Farnam's husband Edward in South Bend, IN, and advertised in *Turner's South Bend Directory* for 1871–2 under headlines reading "The flies like it" (156):

This invention was made after studying closely for years the habits of the pestiferous house fly, and is offered to the public with the perfect assurance that it will . . . catch and imprison in one day more flies than all the other known methods . . . combined . . . It is simple in construction, cheap, and will last a lifetime. Large numbers have already been sold, and in every instance have given unbounded satisfaction.

Since the flytrap-manufacturing business seems to have operated from the Farnams' residence; since Harriet, though formerly a postal clerk, listed herself as "keeping house" in the 1870 census, wherein Edward had other work (bookkeeping); and since the "infant" industry listed no hired employees in *Turner's South Bend Annual and Business Mirror* for 1872, the enterprise may actually have been Harriet's rather than his, or a family venture (one daughter was 18 in 1870), though pursued under Edward's name for the sake of propriety or ease in certain business dealings.

It is interesting to speculate that the Fly-catcher may have sold well enough to support the whole family, for Edward Farnam has no regular occupation listed between 1872 (clerk) and the time of the family's departure from South Bend. The latter is not conclusive, however, for the 1869–70 directory lists no occupation for him, either; and this—as well as much else—might be explained by intermittent poor health. Also, both Farnams owned real estate that might have brought them income.[2]

**Kate Strong,** with her husband John, invented an ingenious device to keep ants from climbing table legs: a concave flange fitted around the leg and coated underneath with chalk. Climbing ants, unable to cling to the chalk, would fall back to the floor. In 1875 they applied for a patent, but unfortunately at the same time one Savannah Cruikshank applied for a patent for the same kind of ant guard. In the resulting "interference" hearings, the first examiner ruled in favor of Cruikshank (a male), but the Strongs appealed and the Examiners-in-Chief found in their favor. This decision was appealed to the Commissioner of Patents, who ruled that neither party deserved a patent, as the principle was common knowledge. Cruikshank gave up at this point, but the Strongs persisted, taking their case to the courts. Although admitting the prior use of both flanges and chalk, the courts found in their favor on the basis of combined features, and, after nearly five years of struggle, the Strongs received their patent in 1880.

Doubtless a similar list could be compiled for Britain and several other nations, if time and funds allowed. At present, however, I have only the following significant example:

In the days before DDT and screen doors, flypaper was the state of the art for controlling flies. During the late 19th and early 20th centuries the sticky strips hanging from the ceiling were such a commonplace of summer life that no one would think to ask who invented flypaper. If anyone had asked, s/he might have been told it was Mrs. Beeton, author of the famed *Mrs. Beeton's Book of Household Management* (1861). And if Mrs. Beeton did not invent it, she did at least clearly outline what was surely some woman's invention—the use of treacle on strips of paper to catch flies. Edward De Bono classes this among "key devices" in his London *Times* inventions history (222). **Isabella M. Beeton** (1837–65) managed to sound like a housewife and domestic scientist of forty years' experience, born of a long line of same. Yet she was only 23 when she started the book that made her famous. Born in London, the eldest child of a large family, Isabella Mayson at age 19 married the publisher of the *Englishwoman's Domestic Magazine,* Sam Beeton. She worked and traveled with her husband, becoming assistant editor of the magazine after the birth of her first child. *Mrs. Beeton's* appeared serially at first. Gathered together and published in 1861, it undoubtedly contains a great many things that would have been classed as inventions had they been suggested by men. This book, like many of its kind, deserves further research. Isabella Beeton died of puerperal fever at age 28, shortly after the birth of her fourth child.

**Mary P. Carpenter,** later Hooper, of Buffalo, San Francisco, and New York, was one of the more prolific of known 19th-century American women inventors, with 13 patents spanning more than thirty years, from 1862 to 1894. Her sewing-machine and other inventions are discussed elsewhere (Chap. 4). The invention that concerns us here is her Mosquito trap, patented in 1891. Four years earlier (1887) she had patented a (mosquito-) Netting canopy for beds.

**Mary F. Sallade** of Philadelphia and New York City was also a prolific woman inventor by 19th-century standards, holding at least six patents for things as diverse as plaiting machines and bellows. Her last known patent was dated 1889, for an Insect trap. Unfortunately, I have no information on her except for what her patents reveal about her inventiveness—and about her residence: she lived in Philadelphia in the 1870s and in New York in the 1880s.

2. Information from Discovery Hall Museum, South Bend, IN.

### 20th century

The Women's Bureau study reveals very little inventive activity for women in pest control in the early 20th century. Under Planting, tilling and harvesting machinery and equipment, the study lists only one rather cryptic subcategory that is relevant: "Fertilizer, insecticide." Presumably this is a fertilizer-and-insecticide *spreader*. Under category F, Plant enemy exterminators, are listed two relevant subcategories (out of 8); Animal trap; Boll weevil exterminator. As always, neither names nor numbers are given, but four inventions are involved, at most.

**Dr. Margaret Shepard**'s and other above-mentioned inventions for preventing or curing plant diseases would also be relevant here. Quite obviously, the war against pests has moved into the laboratory. **Dr. Marjorie Hoy** of UC Berkeley is turning the weapons of genetics against the spider mites that cause $75 million a year in damage to California's almond, peach, and apple crops. Born about 1942, Dr. Hoy holds degrees from the University of Kansas and from UC in zoology, entomology, and biological sciences. As of this writing she is Professor of Entomological Sciences at Berkeley.

Artificial pest control for crop plants has traditionally taken one of three routes: finding or creating pesticidal chemicals; selecting or breeding plants for pest-resistance; and importing or aiding natural predators to control the pests. Indiscriminate use of the chemical approach has added a new urgency, and genetics a new potential, to the third approach; and both scientists and growers have learned from our chemical sorrows. The new ideal is an integrated management program (IMP) that incorporates all three traditional approaches, and seeks control—a new balance—rather than eradication. IMP pest-control plans emphasize biological over chemical controls, and call modern genetics to the aid of improving both the pest-resistance of the crop plants and the predatory efficiency or survival rate of the pest's predators. This is the context of Dr. Marjorie Hoy's work.

Under normal conditions, a predatory mite called *Metaseiulus occidentalis* controls the populations of all three spider mites infesting California's almond orchards—the Pacific, two-spotted, and European red mites. But chemicals such as Sevin (carbaryl) and Pounce/Ambush (permethrin) used against spider mites and against an even worse pest of almonds, the navel-orangeworm—disrupt this system of biological control, either by killing *M. occidentalis* and other predators, or by killing so many spider mites that their predators starve or leave the orchard. Dr. Hoy and her group have created populations of *M. occidentalis* that are resistant to some of these chemicals, notably Sevin.

This kind of thing, as Dr. Hoy herself points out, has been done before. What is new, and far more invention than discovery, is the practical application of this creation in the field: devising techniques for mass-producing the resistant mites, effectively dispersing them, and establishing them in the orchards, and for monitoring its population relative to its prey as well as its levels of pesticide resistance, so that chemicals can be cautiously applied to aid it, or new releases can augment its numbers, if needed. Results so far are highly promising, with resistant species successfully overwintering, and retaining their pesticide resistance for as long as three or four years.

This control program has promise not only for almonds, where it has been most extensively tested, but for apples, peaches, and even grapes. Everyone benefits if less pesticide is spread on our fields to percolate into the ground water. In addition, since the spider-mite pesticides in question are expensive, this IMP should reduce the cost of producing almonds and other affected crops (Hoy; Hoy *et al.*, "Managing" and "Aerial").

An earlier worker in the biological approach to pest control is **Dr. Odette Shotwell** of the U.S. Department of Agriculture in Illinois. Her long-term study of the chemistry of the Japanese beetle had a practical purpose: developing a bacterial disease that would attack and thus control this destructive pest of fruit trees, grape and berry vines, clover, and corn. Accidentally introduced into the United States about 1916, the beetle spread over several thousand square miles of the Atlantic seaboard, and as far west as eastern Ohio, causing some $10,000,000 in damage annually in the eastern states alone. Biological control has proved the best weapon against it, particularly the bacterial "milky disease" fatal to 90% of infested larvae (*EA*/83; *EB*/58; *WBWRA,* 32).

Dr. Shotwell's important work with antibiotics (Chap. 2) can be seen as another form of biological control, using fungi to attack bacteria.

### Windmills

Windmills were widely used in the 19th century to raise water for stock as well as for humans from the deep wells of the Midwest and West. They still dot the landscape in the more rural counties of California.

Three scattered examples of women's patents in this field—one from the United States, one from England (though a U.S. patent), and one from Canada:

**Mary Augusta Hill,** St. Catharine's, Ont.—Hill's Storm-Proof Wind Wheel (1871)
**Winfred (& C.) Silvester,** Windom, KS—Windmill regulator (1892)
**Sarah J. Rollason,** London, Eng.—Windmill or wind motor (1894)

Eldress **Emeline Hart** of the Shakers also invented an improved windmill, but seems not to have patented it.

# Appendix HM-1:
## LWP Medical Apparatus Patentees

**Carrie Aurelia Munro(e),** Salt Lake City, UT—Improvement in vapor baths (1874)

**Helen M. Rowley,** Van Wert, OH—Improvement in masks for medical purposes (1875)

**Anna L. Palmer,** Chariton, IA—Improvement in combined aspirators, concealed uterine cauterizers, and vaginal syringes (1879)

**Fanny C. Batcheller,** Boston, MA—Device for restoring facial symmetry (1881)

**Laura M. Adams,** New York, NY—Syringe (1881)

**Anna L. Olsen,** Brooklyn, NY—Respirator (1884)

**Margaret Hart,** New York, NY—Syringe (1885)

**Carrie Aurelia Munro(e),** Olive Branch, OH—Vapor bath (1886)

**Eliza Ellis Scott,** Hamilton, Ont.—Druggist's sieve (1886)

**Rachel S. Packson,** Emporia, KS—Speculum (1887)

**Eliza Jane Harding,** Denver, CO—Depurator [device that purifies, as in a pharmacy, by heating] (1887)

**Annie H.** (and M.G.) **Collins,** Camden, NJ—Syringe (1890)

**Lucy R. Meyer,** Chicago, IL—Syringe and syringe box (1892)

**Mary S. Woodward,** Chicago, IL—Inhaler (1892)

**Ida M. Hemsteger,** Chicago, IL—Protector for blister poultices (1894)

**Hilda K. Anderson,** Stockholm, Sweden—Disinfecting apparatus (1894)

# Appendix HM-2:
# Inventions in Medical Materials
# and Research Techniques

## Materials

Two minor 19th-century inventors belonging in this section are **Lucinda A. Bucknam** of Needham, MA, and **Jeanie E. Pierce** of Lewiston, ME, who patented processes for preparing absorbent cotton in 1880 and 1881, respectively.

Twentieth-century examples include **Christine Conklin** of Health Chem Corporation, New York City, who invented an award-winning bacteriostatic and flame-retardant safety fabric called Lectrolite, designed to replace carbon-filled, rubber-coated cotton fabrics. It is a polyvinyl chloride laminate reinforced with synthetic fabric. Since Lectrolite has so many medical and hospital uses—operating-room mattresses, pads, pillows; hospital sheeting, stretcher fabric, etc.—it deserves a brief mention here. Its 1976 price of $2.90 per yard at a 54-inch width seems commendably moderate as well ("Electrically . . .," 71).

Another anti-bacterial, flame-retardant fabric was patented by **Vidabelle O. Cirino** of Hereford, TX, and a male co-inventor in 1978. Cirino, a USDA employee, assigned her rights to the Department.

In 1970, **Eva T. Tuhos** received a patent for Electrophoresis films.[1] These films, used in such sophisticated chemical tests as comparisons between blood proteins, have medical applications, among others. And since Patricia Ives lists the invention in her 1980 article on black and women inventors (126), it may well be a significant one; Ives as usual gives no details.

Two Russian women received an American patent for a new material with surgical applications. **Nina Mikhailovna Tikhova** and **Vera F. Terekhova,** both of Moscow, protected a new magnesium-based alloy for use in bone surgery (1972).

## Research Techniques and Procedures

Since modern medicine advances through research, new research techniques and inventions contributing to medical research can be significant medical inventions. Women have contributed substantially to this field, beginning at least as early as the 19th century.

**Ellen Swallow Richards** has many scientific and technical accomplishments to her credit. Pertinent here is her invention of an artificial gastric juice composed of pepsin and hydrochloric acid, enabling digestion to be studied in the laboratory (R. Clarke; M. Stern, 134–5).

**Dr. Florence Seibert** (1897–1991) developed a method and apparatus for producing bacteria-free distilled water in a single process. Doctors had long wondered why distilled-water injections sometimes caused brief but severe fevers. Dr. Seibert found that distillation killed the bacteria in water, but did not destroy certain bacterial products, which rode through on water droplets and were injected into patients. She designed a special spray-catching trap for the still, and won Chicago University's Ricketts Prize for 1923. Her work made possible the widespread use of plasma injections during World War II. It also made safer, of course, the millions of routine injections given every day in hospitals around the world.

Since both Edna Yost (*AWS*, 177f) and Elizabeth O'Hern (57ff) have profiled Seibert, suffice it to say here that after polio destroyed her dream of being a doctor, she followed her passion for biochemistry into chemical research, taking her PhD at Yale in 1923. Her Ricketts Prize work at Chicago brought her to the attention of Dr. Esmond R. Long, who assigned her to isolate the specific chemical substance causing the positive skin reaction in the standard test for tuberculosis. After ten years (others had been trying for sixty),

---

1. Electrophoresis is a process in which a plate coated with agarose gel or another polymer is subjected to an electric field. Positively charged molecules gradually migrate toward the negative electrode; negatively charged, to the positive. The rate at which each molecule moves depends on the strength of its charge and on its own size or molecular weight. Thus, many charged biological components of a substance being tested will separate into easily recoverable fractions. Agarose, which is electrically neutral, does not interfere with the separation and, when formed into a membrane or film, can be used for sorting molecules by size.

she produced pure crystalline PPD (Purified Protein Derivative of the tubercle bacillus), designated PPD-S. This work, making tuberculosis tests significantly more reliable and bringing her worldwide recognition, involved inventing a second special filter. Seibert used gun-cotton membrane of a precise thickness over a porous clay filter, the combination calculated to admit only molecules of a certain size. She also made a special lot of PPD at the Mulford Biological Laboratories, to be set aside as the U.S. standard. Soon adopted by the World Health Organization, Dr. Seibert's PPD's became the international standard as well.

As early as the 1930s, honors began to come to her—a Guggenheim in 1937–8, the Trudeau Gold Medal in 1938, the Garvan Medal in 1942, and the John Scott Award. She received several honorary doctorates, including one in science from the University of Chicago; the first Achievement Award of the Association of University Women; and the National Achievement Award. In the 1940s Florence Seibert came to the Henry Phipps Institute of the University of Pennsylvania, where she eventually became Professor of Biochemistry, and continued her work on tuberculosis. Even after "retiring" to Florida, she continued her research at home, working now on cancer, until well into her eighties (*AMWS*/12; Barnard, 379ff; O'Hern, 57ff; Rossiter, 182, 274, 300, 307–10, 373 n76; *WBWRA;* Yost, *AWS,* 177–95).

**Dr. Florence Sabin** (1871–1953), one of the greatest U.S. scientists—and incidentally, a member of the board granting Florence Seibert's Guggenheim Fellowship—had three careers: medical research, medical teaching, and public health. Her major research triumphs were discovering the origin of the lymphatic vessels, direct observation of blood-vessel formation in a chick embryo, and studies in tuberculosis resistance.

Dr. Sabin appears here because she developed 1) the ingenious method she used to determine the embryological origins of the lymph vessels (Yost, *AWS,* 71), and 2) a technique for studying living cells by placing drops of blood, often drawn from her own arm, in a warm box, and noting their reactions to various stimuli. She is also usually credited with refining Leipzig Professor Spalteholz's technique for using a nontoxic tissue stain to study living cells—and thus inventing supra-vital staining, a technique adopted in laboratories the world over.[2]

Much honored, she has several "firsts" to her credit: she was the first woman on the full-time faculty of Johns Hopkins Medical School (1902), and later its first woman full professor (1917–25); first woman elected to the National Academy of Sciences (1925); first woman member of the Rockefeller Institute (1925–38), where she organized and headed the Department of Cellular Studies; first woman president of the American Association of Anatomists. Dr. Sabin received the National Achievement Award, the Trudeau Medal, the M. Carey Thomas Prize in science (1935), and honorary degrees from a dozen leading universities. When New York City honored Marie Curie in Carnegie Hall in 1921, it was Dr. Florence Sabin who represented American women of science. In 1952, just a year before she died, she won a Lasker Award. With all of this, she was, according to Edna Yost, "a warm, motherly type of woman whose handclasp meant helpfulness and whose whole being breathed the happy sense of completion that comes to those who have lived life and found it good" (Yost, *AWS,* 62–79; Stoddard, 361–71; *NAW*/mod.; *WBWRA*).

Pathologist **Maud Slye** (1869–1954) spent a lifetime investigating whether cancer-proneness is inherited. Although her first theory, that susceptibility to cancer resided in a single recessive gene, proved wrong, her carefully controlled work with mouse populations of known heredity contributed significantly to the heritability question. At one point she had 9,000 mice with pedigrees recorded for 100 generations. All told, she reared and kept pedigrees on over 150,000 mice. In fact, her innovation of using large populations of animals with fully known ancestry may have been her greatest scientific contribution. In popular terms, she "invented" genetically uniform mice as a research tool.

She also repeatedly advocated more autopsies so that the causes of patients' death could be known and recorded, and a central record bureau for human cancer statistics, so that all factors, both hereditary and environmental, in the onset of this dread disease could be evaluated.

Though she never received the giant grants of today, often paying for her research from her own pocket, Maud Slye's contributions to medical research were recognized and honored: a gold medal from the

---

2. However, according to *NAW*/mod., although Dr. Sabin used and helped popularize this important technique, one of her students, Herbert M. Evans, actually refined the method and introduced it into the United States. Whatever the truth of this particular case, Dr. Sabin did travel to Germany several summers to bring back ideas, and she did study in Leipzig for a time.

American Medical Association in 1914, the University of Chicago's Ricketts Prize in 1915, a gold medal from the American Radiological Society in 1922 and an honorary DSc from Brown University in 1937 (*CB*/1940: 743–5; McDowell & Umlauf, 368; *NAW*/mod.; Rossiter, 182, 348 n57; *WBWRA*).

Born in Bogotá to missionary parents, **Dr. Mary Caldwell** (1890–1972) returned to the United States for schooling, taking her AB from Western College for Women in Ohio, and her PhD from Columbia University. After teaching at Western, she became Columbia's first woman professor of chemistry and a "memorable" teacher, specializing in enzyme research. Many of today's standard principles of enzymology, especially regarding starch enzymes, originated in her laboratory. She invented several techniques for isolating enzymes, and won the 1960 Garvan Medal for one of them. Retiring in 1959 to a country home near Fishkill, NY, she continued to publish scientific papers until her death (*WBWRA*, 164, 165).

**Mary S. Shorb** (1907–  ) of Beltsville, MD, did work with important applications in the treatment of pernicious anemia, leading to a medical/pharmaceutical invention. In 1947 Shorb found that one of two factors necessary for the growth of *Lactobacillus lactis* Dorner was present in refined liver extracts in amounts almost directly proportional to their anti-anemic activity. Thus by testing for the so-called LLD factor, one could quickly determine the activity of a given batch of liver extract. In other words, Shorb had created a quick test for the effectiveness of the major therapy for pernicious anemia.

Her work also contributed to the discovery of one of the more important vitamins. When other researchers isolated a red crystalline compound from liver extracts determined clinically active by her assay, she showed that this red substance vigorously promoted the growth of *L. lactis* Dorner, and soon it proved to help pernicious anemia patients as well. The red crystalline compound—proved identical to the growth factor used in her assay—was Vitamin B-12. Author of 60 scientific papers, Dr. Shorb has been honored by, among other things, an honorary ScD degree from the University of Idaho, and the Shorb Lectureship established in her honor at the University of Maryland (Raper, 31; Shorb, 1987 Nomination to Maryland Women's Hall of Fame).

**Dr. Bernice Eddy** (1903–  ) of the U.S. Department of Biological Standards invented the early tissue-culture techniques for tumor viruses that made possible her landmark 1950s work in cancer virology. Dr. Eddy and her collaborator, **Dr. Sarah Stewart** (1906–76), not only discovered the polyoma virus—named SE Polyoma Virus in their honor—but showed that it causes many kinds of animal tumors—and is transmissible from animal to animal. Bernice Eddy was born in West Virginia. She took her PhD from the University of Cincinnati in 1927 and after various fellowships and some work for the Public Health Service, joined the National Institutes of Health in 1937. She married in 1938 and bore two children (O'Hern, 151–70; *WBWRA*, 220–1; *AMWS*/12).

In 1940 **Dr. Anna Goldfeder** (c. 1900–  ) became Director of the Cancer and Radiobiological Research Laboratory of New York City's Health and Hospitals Corporation. By then she had already worked in cancer research for ten years. Now after more than fifty years, she has many pioneering achievements to her credit, at least two of which seem pertinent here: She was among the first to use radioactive tracers in metabolic studies, but my source is unclear whether she originated or adapted this idea. She definitely developed an isogenic host-tumor system for evaluating new chemotherapeutic drugs, and a model for studying mouse mammary tumors with different growth rates. These are important research techniques or systems, both widely used in cancer research today (Moller & Friend).

Recent reports tell of cancer-research techniques—a precisely defined experimental system and clearer identification of the activating mechanism for a *ras* oncogene in that system. In 1984 **Saraswati Sukumar** and colleagues at the National Cancer Institute reported highly reproducible induction of mammary tumors in rats with nitroso-methyl-urea (NMU), and confirmed previous indications that the oncogene's malignant properties result from a single point mutation. This high degree of reproducibility "opens the door to a reexamination" of chemical causes of cancer. Dr. Sukumar's work should lead directly to "a model system for studying the malignant activation of an oncogene and the mechanism by which it contributes to the development of a tumor." Moreover, findings from such experiments should apply directly to human cancer ("Oncogenes," 73f).

Other important mid-century research was **Dr. Allene R. Jeanes'** dextrans work at the Department of Agriculture. As noted in the main text, dextrans carbohydrates, produced by bacteria, have important medical uses as blood-volume expanders, and thus in transfusions when whole blood is unavailable. Dr. Jeanes, who worked at the Peoria labs, developed assay methods for carbohydrates and their degradation products, established the B-512 strain of microorganism as a vigorous dextran producer, and developed a simple, rapid method for determining the structure of dextrans. At retirement in 1976, with more than sixty publications to her credit, she held ten patents in carbohydrate chemistry. In 1978 an issue of the journal

*Carbohydrate Chemistry* was dedicated to Dr. Jeanes and one other woman scientist (*WBWRA* 31, 88, 164; "Allene R. Jeanes"; Sandford).

A crucial step in killing any disease-causing organism is finding ways to culture it safely, cheaply, and in sufficient numbers for use in hundreds of experiments. Thus the inventor of practical culture techniques may deserve partial credit when the disease in question is ultimately cured or controlled. Such credit may belong to **Nina G. Kljeuva** and her colleagues **Rita V. Krzhevova, Rina V. Fradkina, Natalia N. Sukhareva,** who in 1968, with two male colleagues, patented a Process for cultivating trypanosomes. Trypanosomes cause, among other things, sleeping sickness. Dr. Kljeuva and her co-patentees, all of Moscow, assigned their American patent to the State Institute for Medical-Biological Preparations.

One of the most important new techniques for studying the human brain and for harnessing it to help heal the body is **Dr. Barbara Brown**'s pathbreaking development of biofeedback. A neurophysiologist at the Veterans Administration Hospital in Sepulveda, CA, Dr. Brown found that the brain emits at least four distinct kinds of waves, depending on its activity at the time—delta, the sleep pattern; theta, linked to creativity; beta, connected with mental concentration; and alpha, reflecting a relaxed state. If people could connect physical sensations with each emission, Dr. Brown hypothesized, perhaps they could learn to achieve the various states at will. The potential implications for medicine are limitless—from re-education of damaged muscles to slowing the heart and reducing blood pressure, controlling otherwise intractable pain, and managing stress.

The potential for learning about the brain, and about mind-body interaction, is also almost infinite. If, as Dr. Brown maintains, the brain can indeed control a single body cell, and people could learn to use that control, the research potential there alone is mind-boggling. Brown and others have continued working, with very interesting results, in the fertile field she created in the 1970s.

Barbara Brown belongs here not for her *discovery* of biofeedback, but for her innovative application of its findings to human health, and for her *invention* of two devices used to make alpha waves more vivid and memorable to patients and research subjects: the Alpha train and the Alpha wave racetrack. Concerned with the "relative abstractness" of the signals revealing brain or body activity, she reasoned that meaningful feedback signals should have "some symbolic meaning or interest" not achievable with ordinary tones and colors. Then came inspiration: "The most obvious toy of childhood, . . . the choo-choo train." Within hours she had adapted the apparatus so that the train would start when alpha waves appeared in a subject, and stop again when they disappeared.

Her invention proved entirely enchanting. For adults (who also loved the train), Dr. Brown invented another toy—a racing car set operated by brain waves. With this invention, two people could be wired up at once and race their cars against each other, competing for alpha-wave control (B. Brown, "Anatomy," 50, 51).

Much has been written about Barbara Brown and her achievements in both popular and scientific literature. Suffice it here to cite a few of Brown's works (see Bibliog.), and to quote Beatrice Levin, who opens her discussion of Dr. Brown by noting, "Women have been willing to pioneer in areas where men have hesitated to tread" (149).

**Dr. Marjorie Horning** of Houston is best and most deservedly known for her work with Reba Hill on movement of drugs across the placenta. As late as 1968, the placenta was considered a barrier keeping the fetus from harm. Dr. Horning, however, showed that virtually every drug taken by a pregnant woman reaches her unborn child, as taken or in the form of breakdown products. This revolutionary discovery underlies all recent work in preventing drug-induced birth defects. Dr. Horning also found that drugs taken by a nursing mother will reach her child through the breast milk.

Dr. Horning, as of the late 1970s Professor of Biochemistry at Baylor College of Medicine, 1977 winner of the Garvan Medal, had to overcome discrimination. After graduating from Goucher College in 1938 she worked as an unpaid aide in a University of Pennsylvania Laboratory because women could not be hired for lab work. She got her MS and PhD (1943) degrees from the University of Michigan, married, and in 1961 came to the Institute for Lipid Research at Baylor. Recently, with her husband, she won the Warner Lambert Lectureship—and her place in this book—for outstanding contributions in developing techniques for studying how drugs are broken down and used in the human body (Levin, 2, 171–3; *WBWRA,* 165, 221–2).

**Candace Pert** (c. 1947) is a name to watch in medical research. While still a graduate student at Johns Hopkins Medical School, she and her mentor Solomon Snyder discovered the brain's opiate receptors, the endorphins. Her name came first on the 1973 papers. This landmark discovery was an important step in the study of opiate addiction. Moreover, Pert and co-workers recently reported that the opiate receptors are involved in filtering sensory stimuli to the brain—and perhaps even in obesity ("Larger Role"; "Opiate Antagonist"). In fact, the endorphins may be the most important discovery in brain chemistry in the past decade. In a highly questionable move, frankly branded as sexism by one experienced science writer (Arehart-Treichel, "Winning"), Candace Pert was omitted from the 1978 Lasker Awards for the discovery of endorphins and enkephalins.

Now an independent investigator for the National Institute of Mental Health, and a wife and mother, Dr. Pert continues her work in brain chemistry (Grady, "Candace"; I. Kaye). She appears here because she and a male colleague at NIMH recently devised a simple method for visualizing drug and neurotransmitter receptors in the brain "Larger Role," 101, 103).

The importance of visualization—being able to see what is actually happening in living cells or cell preparations—can hardly be overestimated. Thus the importance of supra-vital staining, of techniques undisputedly attributed to Dr. Florence Sabin, and of Pert's method just mentioned. A related invention, in which biologist **Nina Allen** has played a role, is the videomicroscope, linking a powerful optical microscope and a television camera to allow live and moving cells to be studied for the first time at high contrast and high speeds. Eventually, an electron microscope can be linked to this system.

Allen's husband, Dartmouth biologist Robert Allen, has received major credit for the invention—which grew, like so many other inventions, out of an accident. But a recent article on the new system, which was announced at a meeting of the New York Microscopical Society in November 1980, mentions that he developed it "with the help of his wife, Nina, also a biologist." The Allens' system can "reveal in great detail a vivid world in which blood platelets simmer like geyser pools, and tiny bodies within the cells travel through fibers like traffic on a freeway" ("Live . . . ," 78f).

In the early 1980s, **Dr. Edna P. McCrane** and colleagues at the National Eye Institute reported a technique for locating the retinal cells that respond to blue light in primates. McCrane's technique consists of injecting Procion yellow dye into the eye, whereupon it is taken up selectively by the cells most sensitive to blue light—possibly the cells that carry the message from these receptors to the brain (McCrane).

These receptors interest both researchers and clinicians; researchers because the blue-sensitive cells seem specialized for color-determinations, and clinicians because they can indicate retinal disease. Researchers feel the technique may help them map the complex circuitry of the primate retina. Clinical applications and research overlap here as well: since the dye kills the blue-sensitive receptors, a dye injection creates an animal model of a human syndrome in which these color-determining cells deteriorate, producing an acquired color-blindness. Such a model allows highly precise research, possibly leading to a useful treatment.

Some possible uses of mouse-human "chimaeric" antibody and of chimaeric mice—and the six women involved in developing these dramatic research technologies—have been discussed in Chapter 2, but it is worth recalling here that Stanford's patent on this mouse is a first in patent history (Hofstadter, "Mice"; McCune *et al.*).

The last example here leads naturally to the next section, for it could not have happened without a special kind of medical record—that preserved in the frozen brains of victims of Alzheimer's and various other degenerative brain disease. Drs. Charles Marotta and **Elizabeth Sajdel-Sulkowska** reported in 1984 that Alzheimer brains lack a protein blocker of an enzyme that destroys RNA. The technique of using such tissue for molecular-chemical studies—in other words, of using frozen brains to study brain biochemistry—is itself an innovation (scientists had assumed such large molecules as DNA and RNA would be destroyed by post-mortem digestion processes). And the reported finding, as the news media noted, "raises the possibility of improved diagnosis and eventually a treatment" for a condition that costs the United States some $25 billion yearly. These Harvard researchers were supported by the National Institute of Aging ("Clue to Alzheimer's . . ."; Marotta & Sajdel-Sulkowska).

## *Medical Records*

It may seem strange to class medical record-keeping and centralized disease registries at research procedures, but a moment's thought reminds us that cigarette-smoking and lung cancer could never have been linked without statistical analysis of thousands of lung-cancer case records. In fact, much of what we know about cancer goes back to tumor registries or other meticulous records. Nor would we know the role of highly absorbent menstrual tampons in toxic shock syndrome without detailed case records and carefully gathered statistics.

Women have made important contributions to—perhaps actually originated—the idea of medical records and statistics, for their own sake and as a research tool. One of the earliest of these innovators, to my knowledge, was French obstetrician **Marie Louise Duges LaChapelle** (1769–1821). LaChapelle has other inventions and innovations to her credit, and is discussed at length elsewhere. Suffice it here to say that her statistical tables on 40,000 births settled such long-debated questions as the exact length of pregnancy, the average duration of labor, the frequency of difficult delivery caused by the shape of the pelvis, etc. (Hurd-Mead, *HWM,* 461, 490, 497–9, 500).

When **Dr. Marie Zakrzewska** and the Blackwell sisters established the New York Infirmary for Women and Children in New York City in 1857, Dr. Zakrzewska introduced true medical record-keeping. To the traditional name-and-address record she added age, sex, occupation, diagnosis, and treatment for each case.

**Dr. Charlotte Amanda Blake Brown** suggested the establishment of a tumor registry as early as 1887. **Dr. Eliza Mosher,** who worked into the early 20th century, pioneered systematic record-keeping for the student health services—themselves innovations—which she established at the University of Michigan and elsewhere (*NAW*). In this century, **Maud Slye** repeatedly advocated a central record bureau for human cancer statistics of the kind eventually established by the American Cancer Society.

As a kind of capstone to these early contributions comes the work of **Eleanor MacDonald.** After taking a degree in statistics (Radcliffe, 1928), she began the studies whereby she proved that complete records can contribute to controlling cancer. For five years she followed up all reported cancer deaths in Massachusetts with a house-to-house investigation. Her 1937 results, buttressed by two decades of autopsy records at a major Boston hospital, were immediately recognized as "the first accurate calibration of cancer mortality in the country." Not content with data-gathering, she took social action, urging communities to set up cancer-awareness centers, and doing a weekly radio show on public health. Noteworthy among her accomplishments is her establishment of the link between sun exposure and skin cancer, by showing that the incidence of melanoma increases with proximity to the equator. Though holding only a bachelor's degree, she became a full professor of epidemiology at the University of Texas. Retiring in the late 1970s, she wrote two books on environmental and genetic factors in cancer (*WBWRA,* 221).

**Dr. Helen Dyer,** noteworthy for her contribution to sulfa drugs, pioneered in another kind of cancer-related record-keeping, preparing the first comprehensive index of tumor chemotherapy. Her index was used by the Cancer Institute to guide its new chemotheraphy program (*WBWRA,* 220).

A farsighted woman whose record-keeping ideas in radiation exposure have regrettably not yet been accepted, is **Dr. Rosalie Bertell** (1929–     ). A PhD in mathematics, a Roman Catholic nun, and a former Senior Cancer Research Scientist at Roswell Park Memorial Institute in Buffalo, NY, Dr. Bertell studies the hazards of low-level radiation. As an indication of her perceived competence and international reputation, she has served on the Environmental Health Committee of the British Columbia Medical Association, the Environmental Pollutant Movement and Transformation Committee of the Environmental Protection Agency, the Energy Task Force of the National Council of Churches, and the Citizens' Advisory Committee to the President's Commission on Three Mile Island—among other consultancies.

Says Dr. Bertell, "This dream of a peaceful atom is empty . . . , for it leads either to quick death from the most lethal weapons [she contends that nuclear power serves to legitimize the operations necessary to maintain nuclear weapons] . . . or to a slow death from the ordinary pollution of this industry." The health risks do not stop with increased cancer rates. Radiation also increases susceptibility to infectious diseases and causes "premature aging, with earlier onset of heart disease, diabetes, arthritis, coronary-renal disease, and other debilitating chronic health problems." Today she would doubtless add AIDS to the list, as her theory supports the heterodox idea that AIDS is not "caused" by the HIV virus, but results from a sustained and multiple attack on the immune system.

Since the effects from radium or plutonium in drinking water and food are so subtle and so long delayed, Dr. Bertell contends that they can be detected only by proper statistical measurement of the population at risk from the hazard—not choosing so large a population as to dilute the effect, or so small a population as to exaggerate the time needed for damage to appear. Both such inappropriate populations, she contends, are now being used to convince people that low-level radiation is harmless. She proposes a sophisticated health-monitoring system focusing on more delicate health indicators—average age at diagnosis of chronic diseases; cancer occurrences grouped together for a common cause; and survival rates for immature infants. Dr. Bertell suggests using state-of-the-art techniques in epidemiology to apportion responsibility for harmful radiation effects. For example, a 20% increase in lung cancer in a given area might be attributable to careless handling of uranium mine tailings. She then proposes that "the persons or government agencies responsible for the carelessness be held responsible for 20% of the medical costs for all lung-cancer victims in that population." This, she says, would make individual lawsuits unnecessary and at the same time protect corporations from the "all-or-nothing cost bind."

Her system, called Health Watch International, would operate worldwide. It would use the latest technology to provide legal evidence of "subtle exploitation of one country by another either directly by the sale of harmful industry or technology, or indirectly by the polluting of the air and water systems which form the life-support system of another nation" (Bertell, 5–6). One can only wonder how many Chernobyls must cast their deadly pall before we adopt Dr. Bertell's innovation.

An appropriate closing example here is infant-mortality statistics. Death rates for children in their first year may be a more sensitive indicator of a nation's overall well-being than either gross national product or per capita income. Certainly, as the U.S. example shows, there is no one-to-one correspondence between a large GNP or a high per capita income and low infant mortality. In Washington, DC, the infant mortality rate is an outright disgrace, having jumped in a single year from 22.2 per thousand in 1979 to 24.6 per thousand in 1980. Japan's rate, by contrast, is nine per thousand.

In 1981 **Kathleen Newland** called attention to factors correlating directly with low infant mortality: breast-feeding for six months to a year; the mother's education; fewer and better-spaced births; and abstaining from both alcohol and tobacco ("Infant Mortality").

## Environmental Research Techniques

**Dorothy M. Settle** of the California Institute of Technology, with a male colleague, has adapted space-research techniques to the vital study of lead contamination in food. Since lead pollution (from gasoline) was so pervasive in the U.S., FDA studies on lead levels in food are suspect. Using laboratory techniques devised for moon-rock samples to achieve a lead-free test environment, Settle and her colleagues conclude that lead-soldered cans may account for half the lead in the American diet, and should probably be banned immediately. Canned tuna, for instance, contains a thousand times more lead than freshly caught fish ("Lead-Soldered").

Interestingly enough, an anonymous woman researcher (A.D.) warned against lead-soldered cans as early as 1870, and the controversy continues (Groth; K.A. Johnson; Austern & Mussman).

## Radioimmunoassay

One of the most powerful research techniques, and at the same time one of the most powerful diagnostic tools, of the twentieth century is radioimmunoassay (RIA). Discussed in the section on diagnostic tests, this path-breaking invention, which brought **Rosalyn Yalow** the 1977 Nobel Prize in Medicine, must be recalled here.

## Genetic Engineering

The techniques covered by the term *genetic engineering* are powerful research tools, beginning to be used in treatment. They have already begun to transform medicine, and will soon make it practically unrecognizable. They are discussed in the main text in a separate section.

# Appendix HM-3:
## Women's 19th-Century Nursing and Invalid-Care Patents, from LWP

**Martha P. Coleman,** Boston, MA—Improvement in drinking cups for the sick (1865)
**Hannah Conway,** Dayton, OH—Improved hospital bed (1868)
**Susan C. Currie,** New York, NY—Combined medicine spoon and bottle stopper (1869)
**Isabella Waller,** Cleveland, OH—Improved hospital bed (1872)
**Lydia Stewart,** San Francisco, CA—Medicated towel (1873)
**Maria Bradley,** New York, NY—Improvement in rubber drawers for invalids (1875)
**Phebe K. Stice** and William H. King, Swan Creek, IL—Improvement in invalid bedsteads (1879)
**Rosa O'Donnell,** San Francisco, CA—Improvement in attachments for invalids (1879)
**Jane Henry** and James R. Smith, Chicago, IL—Urinal for invalids (1880)
**Elizabeth J. Holcombe,** Syracuse, NY—Vaginal irrigator and urinal (1881)
**Annie S. Evans,** Kingston, Ont.,—Invalid bedstead (1882)
**Anna P. Fobes** (Forbes?), Syracuse, NY—Invalid bed (1883)
**Sophia Pelkey,** Eau Claire, WI—Medicine-spoon and vial holder (1883)
**Fannie Dickinson,** Chicago, IL—Bed-pan (1883)
**Carrie S. Murphy,** Dayton, OH—Flexible urinal (1883)
**Helen M. Burley,** Brooklyn, NY—Invalid bed (1884)
**Sarah Vincent Beal,** Ann Arbor, MI—Bed-pan (1885)
**Clara M. Kimball,** Milwaukee, WI—Compound for medicating bed quilts (1886)
**Margaret Hammond,** Port Madison, WI—Patient's elevator and perambulator (1886)
**Johanna Patten,** Crown Pt., IN—Comb. suspensory bandage and truss (1887)
**Elisabeth Gnion,** New York, NY—Medicine spoon (1887)
**Amanda DeVoe,** Brooklyn, NY—Invalid-chair (1887)
**Mary A. Hendricks,** Charleston, SC—Folding invalid-chair (1887)
**Kate Ryan,** Brooklyn, NY—Filling attachment for hot-water bags (1887)
**Kate Scanlon,** New York, NY—Poultice pan (1887)
**Hannah C. Bailey,** Utica, NY—Attachment for invalids' beds (1887)
**Elizabeth D. Staples** *et al.,* Goshen, NY—Hospital transfer-bed (1888)
**Annie Caller,** Albany, NY—Meat broiler (1889)
**Eleanor K. Howe,** Bridgeport, CT—Body brace (1889)
**Margaretta Riker,** New York, NY—Bedpan (1889)
**Lina M. Hoffman,** Philadelphia, PA—Invalid chair (1889)
**Agnes Hardie,** Leetonia, OH—Combination cabinet, commode, wardrobe, and invalid chair (1889)
**Julia C. McDowell,** Canton, OH—Cabinet stand for sickrooms (1889)
**Marie A. Campbell,** Chicago, IL—Adjustable pillow or body support (1890)
**Martha L. Abraham,** Chicago, IL—Bedpan (1890)
**Emma L. Tozer,** Canandaigua, NY—Invalid's undergarment (1890)
**Clara B. Dadisman,** Omaha, NE—Bunion protector (1890)
**Betsey A. Dugot,** Mallet Creek, OH—Bedpan (1890)
**Mary St. John,** New York, NY—Bedpan (1891)
**Mary A. Darrow,** New York, NY—Invalid chair (1891)
**Caroline S. Johnson** and M.E. Moore, Deering Center, ME—Invalid bed (1891)
**Mary A. Hawley,** Dixon, IL—Invalid's table (1891)
**Lizzie H. Fifield,** Manchester, NH—Invalid's robe (1891)
**Mary J. Cunningham,** Boston, MA—Bedpan (1892)
**Lydia Stockwell,** Atchison, KS—Bedpan (1892)
**Helen G. Monkhouse,** London, Eng.—Invalid lift (1892)
**Katherine A. Shaffer,** Harrisburg, PA—Pad for bedpan (1892)
**Queenie T. Lauderdale,** Dyersburg, TN—Invalid's hot tray (1893)
**Elizabeth Phillips,** London, Eng.—Bedpan or commode (1893)

# Appendix HM-4:
# Physical Medicine

One of the most important and controversial developments in physical medicine was the polio treatment devised by **Sister Elizabeth Kenny** (1886–1952). The summer polio epidemics that periodically terrorized the United States and other developed nations left thousands crippled, mostly children (thus the name infantile paralysis). In 1916 there were 27,363 reported cases in the United States alone, with 7,179 deaths (Hand, 55). The year Sister Kenny died, just before the vaccines came out, some 58,000 cases were reported.

Elizabeth Kenny was born in New South Wales, Australia, youngest daughter of an Irish veterinary surgeon and a woman who read poetry to her seven children. Elizabeth early showed the courage and initiative she would need in her long battle with the medical establishment. Encountering a death adder in the bush at the age of eight, she stunned it with a stick and brought the limp body home to show her mother (Kenny, 7–8).

At the suggestion of her good friend and mentor Dr. Aeneas McDonnell, Elizabeth entered nurse's training as preparation for her intended missionary work in India. But she soon decided that her mission was to the people of the Australian bush. Her fiancé did not understand her dedication. One day, her three years' training finished, she was leaving for a carefree day at the races with him when she was called to help a woman in premature labor. He wanted her to refuse. When she would not, he forced her to decide on the spot "whether you are to be married to me or to your vocation" (Kenny, 10). Luckily for thousands of polio victims, the vocation won.

Her first introduction to muscle ailments came at fourteen. Smaller and much weaker than the other children, her younger brother Bill tired so easily that the older ones often carried him piggyback to and from school. While recuperating at Dr. McDonnell's house from a badly broken wrist, Elizabeth studied a book on muscle structure to learn how to help her frail brother (Kenny, 12–3):

I had rigged up a mechanical wooden man with pulleys and strings to demonstrate the workings of the muscular system. Bill fell to and worked with a will. In a short time I was able to trace muscles from their origin and insertion and had gained a fair knowledge of their function. For his part, Bill developed surprisingly. . . . He could isolate the principal muscles of his body by voluntary contraction. His strength grew with the years, until he rejoiced in it as much as he had once resented his weakness.

Elizabeth Kenny faced her first polio case at age twenty-three. Riding up to the door of a ranch employee whose two-year-old daughter had become her special pet on an earlier visit, she found the child in agony. "One knee was drawn up toward the face and the foot was pointed downward. The little heel was twisted and turned outward. . . . One arm lay with flexed elbow across the chest. Any attempt to straighten a member caused the child extreme pain." Baffled, Sister Kenny rode out to telegraph Dr. McConnell. The reply came within hours: "Infantile paralysis. No known treatment. Do the best you can with the symptoms. . . ."

Luckily ignorant of the accepted treatment—which was to immobilize the patient and splint the affected limbs—Kenny did just as McDonnell suggested. She treated the painful spasm with heat. After trying dry heat and a linseed poultice without effect, she tried moist heat—a wool blanket torn into strips, wrung out of boiling water, and applied. After a few applications the child's crying stopped, and she fell asleep. When she woke again, she called at once for "them rags that wells my legs!" Thus, as Kenny wrote in her autobiography years later, "the little girl of the Australian bushland unknowingly spoke her approval of a treatment that was one day to become the subject of much heated debate among the learned members of the medical world" (Kenny, 24–5).

By 1913 Sister Kenny had her own cottage hospital in Clifton, South Queensland, where before leaving to nurse the wounded in World War I, she successfully treated several cases of polio. Even in those early days, before the medical opposition to her treatment was fully mobilized, the usual medical reaction to the recovery of one of her cases was that the child must not have had polio after all.

During her war service she was wounded by shrapnel in France. One knee was shot through and

**Octavia McPheters,** Denver, CO—Douche pan (1893)
**Annie Boden** *et al.,* San Francisco, CA—Invalid bed (1893)
**Hannah F. Harding,** Sydney, NSW, Australia—Invalid's mattress (1893)
**Charlotte H. Parliaman,** Palmyra, NY—Rest for invalids (1894)
**Leopoldina Harvey** and Jacob Amos, Los Angeles, CA—Shoulder- and back-bracing chair (1894)
**Sarah A. Potter,** Harper, KS—Invalid chair (1894)
**Regina** (and Rudolph) **Wangersheim,** Chicago—Invalid's bed-rest (1985)

fragments worked themselves out of her leg for years thereafter. She also developed a serious heart condition from overwork.[1] Nevertheless, she came home to resume nursing work in her native Australia, eventually opening a small clinic under an awning in the back yard of a house in Townsville, Queensland. There, using a sturdy table and a zinc bathtub, and aided only by the parents of her patients, she restored to useful lives the child victims of polio, cerebral diplegia, and birth palsy. In those days she never got patients in the acute stage of polio, but only those given up by the doctors, sometimes years after the initial infection.[2]

The story of the battle now joined between a determined nurse who knew she was on the right track, and a medical establishment unwilling to admit its mistake, has been told elsewhere. Suffice it to say here that its real victims were the children deprived of the Kenny treatment when 80–85% of Kenny's patients recovered use of their limbs (80–85% of the doctors' splinted patients were permanently paralyzed). The official attitude can be summed up by the reply of a senior orthopedic surgeon to Sister Kenny's request for a hospital bed for a little girl whose mother had brought her 800 miles on a train: "Doctors are not going to be taught by a nurse" (Kenny, 126).

There were a few exceptions to this obstructionist attitude in Australia and England, but it was the United States that gave her her real opportunity. She began in the Minneapolis–St. Paul area in 1938. Her demonstrations of rapid improvement in polio patients won her an invitation to lecture on physiotherapy at the University of Minnesota. The National Foundation for Infantile Paralysis supplied money to prolong her stay. Kenny's treatment of acute polio patients at Ancker Hospital was so effective that the medical observers and professors of orthopedic surgery and physical therapy at the University of Minnesota Medical School concluded, "We have no hesitation in saying that this method will form the basis of all future treatment."[3] When she left for Australia in January 1941, she was assured a place as a Special Guest Instructor at the school on her return. Her ward, Mary Kenny, was to stay and supervise the work in her absence.

From this point on, though there were still dissenters and obstacles, Elizabeth Kenny's treatment won acceptance, and she began to receive the recognition she had long deserved. In December 1941, the National Infantile Paralysis Foundation officially endorsed her treatment and her concept of the disease. This acceptance was announced over the air by Foundation President Basil O'Connor and Dr. Morris Fishbein, editor of the *Journal of the American Medical Association*. Almost immediately, she began her guest lectures at the University of Minnesota, to doctors, physiotherapists, and nurses interested in her treatment. Within a year after the Kenny method was established in Minneapolis, the incidence of residual paralysis in polio cases had dropped from 85% to 20%, and deformity was almost unknown.

1. Kenny is vague about her illness in her autobiography, *And They Shall Walk,* never naming it or telling any details of treatment. In fact, a quick reading of the passage in question could leave one with the impression that she cured herself through sheer force of will. This may in fact be near the truth, but she does also mention a specialist in Stuttgart and a pilgrimage to Lourdes. One thing is clear: though given four months to live in 1918 or '19, she lived an active, even strenuous, life until 1952. Her autobiography mentions only one later flare-up of the condition, and when she died in 1952 it was of Parkinsonism, not cardiac disease (Marlow, 265).

2. The treatment, when Kenny finally got access to acute-stage patients and could evolve it fully, consisted of moist hot packs during the acute stages, followed by early passive exercise and by active exercise as soon as possible. She also insisted that no muscle reflex tests be given during treatment because they threw the tortured muscles back into spasm. Her concept of polio's muscle symptoms was that muscles were in spasm and needed to be relaxed, kept from atrophying by passive exercise, and re-educated to their intended work. In this re-education, she used not only active exercise, but the patient's own will and concentration on the affected muscle or muscles. Just how fast the initial re-education could take place, at least in her hands, is shown in a remarkable incident that happened years later, in 1942, when a group of skeptical doctors came to investigate her work (Kenny, 255):

> After one of the patients was examined, the doctors, ten in all, pronounced the extensors of the leg completely paralyzed. I disagreed and stepped forward. Again, as I had done in my own wild out-back of Australia over thirty years before, I taught the muscle what to do, gave back to it its motor pattern, and then linked it up with the brain path. Full use of the muscle was restored in less than twenty seconds.

3. This statement, however, was mysteriously edited out of the report of this commission as it was published in the *Journal of the American Medical Association* (Kenny, 239).

A movie was made of Kenny's life, written by Mary McCarthy and starring Rosalind Russell. President Franklin Roosevelt, himself a polio victim, invited her to lunch at the White House. In September 1942, she became the first woman to win the American Congress of Physical Therapy's Distinguished Service Gold Key. That same fall she received the *Parents* Magazine Medal for "outstanding service to children." In December 1942 the Minneapolis Public Welfare Board dedicated the beautiful Elizabeth Kenny Institute clinic at Eighteenth Street and Chicago Avenue.

By 1943 Kenny-treatment centers were operating in Memphis, Little Rock, New Orleans, Louisville, St. Louis, Indianapolis, and Dallas. On May 2, 1943, she received one of her most gratifying honors, an honorary DSc from the University of Rochester in New York. The University's president added a personal note to his address that could fittingly have served as Sister Kenny's epitaph: "In the dark world of suffering you have lit a candle that will never be put out" (Kenny, 167). In 1952, the year Jonas Salk first tried his vaccine on human subjects (Hand, 55), Elizabeth Kenny's own candle went out.

In the brief decades since the Salk and Sabin vaccines, we have forgotten the terror of polio, and may need to recall, in order properly to evaluate Sister Kenny's invention of the only effective treatment, that more than half a million people, mostly children, once fell victim to the disease in a single generation. And where polio still rages, the Kenny treatment is still needed. As recently as 1986, 3.5 million children died from six infectious diseases, one of which was polio (WHO statistics, *Science News,* Apr. 5, 1986).

A fellow nurse wrote of Sister Kenny in the *American Nurses' Journal* (Kenny, A-1):

> But those hands are remarkable. They are full of healing. You wonder, as you watch . . . , if they are the result of thirty-two years of bringing healing to people, or if they were always strong and supple and magnetic. She moves them with a complete economy of motion but with a sureness and a grace that makes them beautiful. I watched her later as she sat on the stage before more than 3000 eager nurses. Her repose was more arresting than the march of a military band—and her hands were at peace in her lap. Yet in the films in which she showed her revolutionary methods of treatment, those busy hands were commanding. A single finger . . . raised in command, and the patient would make prodigious efforts to make his muscles comply.

Continued this observer, "The comments of the nurses after the meeting were interesting. Some had seen only the technique—and they marveled at its effectiveness. Others had seen the woman—and they were moved."

## Rolfing

The condition of her own young son led PhD biochemist **Ida Rolf** to study the rationales behind several major systems of medicine. Based on this study, she developed her system of "structural integration" (Rolf), usually called Rolfing. This system recognizes that both psychological and physical histories shape—sometimes deform—people's bodies, thickening connective tissue and tightening muscles in response to psychic as well as physical injury, revealing past tensions and unexpressed angers. These abnormal tightenings and thickenings interfere with the flow of fluids and sometimes even with the free passage of nerves and nerve impulses through the body. Rolf recommended a vigorous program of physical manipulations to break up these blocks, release the old angers and tensions, and restore the free flow of fluids, nerve impulses, and energy through the body. The structure, and particularly the alignment, of trouble areas would often be changed, and the whole person might look much different, standing, moving, and walking in a new way.

The treatment was and is controversial, not only because Ida Rolf had no medical degree and the treatment sessions can be quite painful, but because the treatment's rationale lies outside established medicine. Moreover, some of the fears and angers released at the sessions may be overwhelming for the patient without psychiatric help at hand; and what Dr. Rolf called unresolved disease from the body's past history might flare up in the course of treatment.

However, the Rolf system won many adherents and practitioners in the United States. Dr. Rolf founded the Rolf Institute in Boulder, CO, for further study of her method, and wrote a book, *The Integration of Human Structures* (1977).

## Dance Therapy

**Marian Chace** (1896–1970) must be mentioned as the creator of an entire new field of physical medicine.

She invented dance therapy as a formal variety of treatment for physical and, later, mental ills—and as a profession. Since she is profiled in *NAW,* suffice it here to note that she was the oldest of four children, and studied at Corcoran School of Art (1915) before she studied dance. She started dancing to release tension in her back muscles after an accident, but soon found that dance was her natural means of expression.

From Ruth St. Denis and Ted Shawn, in the 1920s, Chace learned their concepts of movement, and their theory of cultural influences on the way people move. As a dance teacher in the 1930s, she found her students treating dance as therapy. As doctors began to refer patients to her, she became better known for her ability to reach the mentally ill through movement and rhythm than for her art. She also created seminars for grade-school teachers on the use of movement in communication, and worked with troubled adolescents.

By 1942 Marian Chace's reputation as a dance therapist led Dr. Winifred Overholser to invite her to start a dance program at St. Elizabeth's Psychiatric Hospital in Washington—where, unaccountably, her official designation was Red Cross volunteer. There she developed totally new techniques and used methods never before tried with the mentally ill in a hospital setting. By 1944 she worked full-time at St. Elizabeth's. After studying psychotherapy there and with Frieda Fromm-Reichmann, she began to train other therapists, sometimes serving them as both teacher and therapist.

During the next 20 years, while she worked both at St. Elizabeth's and at Chestnut Lodge, a private psychiatric hospital in Maryland, with Frieda Fromm-Reichmann, Chace lectured at hospitals, universities, and clinics throughout the United States. In 1957 she started a summer dance-therapy-training workshop at the Turtle Bay Music School in New York City, offered every summer until her death.

Mental-health workers in Israel welcomed her 1964 visit there as the beginning of an important new direction in therapy in that country. The next year, at Chace's urging, a few dance therapists formed the American Dance Therapy Association, with Chace as first president. Other honors were the first Oveta Culp Hobby Award (1955) and an outstanding-service award from the Department of Health, Education, and Welfare (1956). Although she retired from St. Elizabeth's in 1966, she worked at Chestnut Lodge till her death in 1970. Her insight that the human body was the "most easily available . . . , freest means of healthful self-expression and emotional release" led her to develop techniques that have rescued many from mental illness (*NAW*/mod.; "Health Honors . . .").[4]

## Radiation Medicine

The earliest form of radiation therapy in modern medicine—irradiation with radium—became possible with **Marie Curie**'s discovery and purification of that element. Whether or not Mme Curie was also involved in the first medical application of her discovery, she seems to have realized, either independently or jointly with Pierre, that radium radiation kills cancer cells. That idea, called radiotherapy, opened up a whole new area of cancer treatment and, indeed, a whole new branch of medicine, but the Curies refused, as always, to profit from their genius. Such a highly significant achievement would go far to explain why Marie Curie is one of only two women profiled by John H. Talbott in his *Biographical History of Medicine* (1970).[5]

Mme Curie's landmark invention of the mobile X-ray unit, discussed under Medical Apparatus, is also relevant here.

Likewise, modern radio-isotope therapy would not have been possible without a discovery made by Marie's elder daughter **Irène Joliot-Curie** (1897–1956), and her husband Frédéric in the 1930s. In the course of an experiment designed to transmute aluminum into phosphorus by bombardment with alpha particles, the couple found they had created a new isotope of phosphorus. This substance continued to emit particles after the bombardment stopped—which meant, as they soon realized, that they had created artificial radioactivity. Within a year they had synthesized a whole group of radio-isotopes, bringing them the Nobel Prize for Chemistry in 1935. Their work belongs here because, once having discovered these isotopes, they deliberately set out to create more—presumably, given Irène's parents, not unmindful of the medical applications. In fact, as Eugene Hecht points out in *Physics in Perspective,* "The stage was set for

4. I am grateful to Laura Menard, then of the Wilbur Hot Springs (CA) staff (1981), for calling Marian Chace and the innovative nature of her work to my attention.
5. Talbott mentions a few other women, but these are largely royal patrons of medicine or research, or eminent patients. The other woman profiled is Dr. Florence R. Sabin.

our present-day medical, biological, and industrial uses of [radio-isotopes]" in this French laboratory more than half a century ago (343).

Much has been written—or is easily available—about the Curies. Yet Irène is far less known than her mother, seldom if ever mentioned in physics texts. Existing accounts deal more with politics than with science. With regard to my tentative woman-inventor's profile being postulated here, however, note that Irène was a firstborn child of achieving parents, a lover of the arts if not herself an artist, and privately tutored for much of her education. Marie and Irène are the only mother and daughter ever to win Nobel Prizes—and were long the only parent and child to do so. Despite her Nobel Prize, and later brilliant papers, Irène was excluded from the French Academy, though her husband was admitted. Like her mother, Irène died of the effects of radiation, driving herself through 10- to 12-hour days in the lab till the very end (E. Hecht, 343; Opfell, 165–82).

**Edith Quimby**'s invention of accurate ways to measure the actual tumor dose during radiation treatments and **Barbara Askins'** method of enhancing faint X-ray photos both represent important advances in radiotherapy technology. Both mean, among other things, greater safety for patients exposed to X-irradiation for diagnosis and treatment.

Edith Quimby (1891–1982) was born Edith Hinkley in Rockford, IL. Oldest of three children, she was blessed with an endlessly inquiring mind. Edith's father and a high school science teacher encouraged her curiosity. At Whitman College in Walla Walla, WA, she became the first woman to major in math and physics, graduating in 1912. She taught high school chemistry and physics to pay for graduate work. When her husband, whom she had met at UC–Berkeley, entered a PhD program in New York City, she became the assistant to Dr. Gioacchino Failla just as he began to develop a radiological research laboratory for the City's Memorial Hospital for Cancer and Allied Diseases. This was 1919. Only six years earlier, the first gram of radium produced in America sold for $120,000. Edith Quimby had entered the ground floor of a very new field.

Her first great contribution was to provide, after precise research, the tables and guidelines enabling radiologists to deliver an exact desired dose of radiation *to the tumor itself* in cancer therapy. Since the quality or strength of the radiation source, the field size, the dosage-delivery rate, the depth and tissue type of the tumor, the patient's body weight, and the distance of the radiation source from the tumor all vary, and all affect the amount of radiation reaching the tumor, finding the proper dosage was a headache for doctors and a nightmare for patients. Moreover, the dosage received by the skin and other intervening tissues could not be ignored. As Dr. Quimby herself put it,

> More than one radiologist, having established a routine pelvic cycle which seems satisfactory in the average case, has found himself confronted with disastrous bladder and intestinal damage after employing the same cycle on a woman weighing 100 pounds or less. If he had considered that the dose to the intestines in a woman 16 cm. in anterior-posterior diameter is of the order of twice as great as in a woman 24 cm. in diameter, he might not have fallen into this difficulty.

All this is to say nothing of the back-scatter problem and the exit dose. Into this chaos, Edith Quimby brought order. Her precise calculations, appearing in more than fifty papers spanning two decades, did not go unnoticed. She received an honorary ScD from her alma mater and, in 1940, the highest honor of the American Radium Society, its Janeway Medal (1940). In 1941 she won the Gold Medal of the Radiological Society of North America, an honor given to only one other woman before her (Marie Curie), and was appointed Assistant Professor of Radiology at Cornell Medical School. By 1954 she was Professor of Radiology at Columbia University's College of Physicians and Surgeons, and President of the American Radium Society.

Her later research dealt with radio-isotopes—methods of handling the isotopes themselves, best nursing practices for patients receiving such treatment, and precautions for avoiding radiation hazards. This latter work has implications both for public safety and for the environment.

Honors continued to come to Dr. Quimby, with an honorary ScD from Rutgers in 1956, and the Medal of the American Cancer Society in 1957. She sat on the Atomic Energy Commission's Committee for the Control and Distribution of Radioactive Isotopes and on the National Committee for Radiation Protection. She was also a long-time examiner for the American Board of Radiology, accrediting physicians as radiologists. She retired in 1960, becoming Professor Emerita of Radiology.

Edith Quimby and her husband, who was on the physics faculty at Columbia, lived for many years in Greenwich Village, with travel as their most noteworthy recreation. The couple had no children. Edith died in Manhattan at the age of 91 (*CB*/1949, 1983; *NYT* Obit., 10/13/82: A-28; Obit., *Physics Today,* Dec. 1982: 71; Quimby; Rossiter, . . . Struggles, 261–2; Yost, *WMS,* 94–107).

Following directly in Edith Quimby's footsteps is **Leela Krishnan,** a radio-oncologist at the University of Kansas Medical Center. Dr. Krishnan has been concerned about radiation delivered to the heart during lumpectomy-plus-radiation treatment of breast cancer. With traditional techniques, some areas of the heart could receive a dose of up to 4,000 rads in several weeks. To avoid possible damage or heart complications, she and her team have devised a combined breast-cancer treatment of X-rays and electron radiation that reduces the heart's radiation dose by about 60%. The treatment, reported in 1983, involves versatile linear accelerators ("Cutting Heart's Dose").

**Barbara S. Askins** (1939–  ), a physical chemist at the Marshall Space Flight Center in Alabama, became 1978 Inventor of the Year for her patented method of enhancing faint photographic negatives. She was only the second woman ever so honored by the Association for the Advancement of Invention and Innovation, headquartered in Arlington, VA, and the first honored alone, i.e., without collaborators.[6] Her invention, assigned to NASA, is a technically exciting means of getting useful prints from severely underexposed negatives. Undoubtedly one of the reasons for her honor was that the invention has applications in three fields: astronomy, space research, and medicine. As Ms. Askins herself noted, "Medical radiographs that are 80–90% underexposed can be increased to readable density and contrast by autoradiographic image intensification." This technique, she continued, consists of "combining the image silver of the radiograph with a radioactive compound, thiourea labeled with sulfur-35, and then making an autoradiograph from the activated negative." Her method works down to 96% underexposure (Askins, 684), which means that patients can be exposed to significantly less radiation from diagnostic X-rays, and physicians still obtain a useful photograph. The process can also be used for restoring faded family snapshots and photographs of historical interest.

Born Barbara Scott in Belfast, TN, Barbara Askins took her BS and MS degrees in chemistry at the University of Alabama, Huntsville. Author of several technical papers, she is a member of the American Chemical Society, Sigma Xi Honorary Research Society, the American Association for the Advancement of Science, and the World Future Society. In 1978, Ms. Askins was married to another chemist and living in Huntsville, AL, with her daughter and son (Askins; Stacy Jones, 7/22/78, 1/27/79, 2/24/79; Moynehan, 23; NASA PR, Jan. 1981).

Eminent cancer researcher **Dr. Anna Goldfeder** (1902–  ) has several innovations to her credit that pertain to radiology. She was among the first to study closely the effects of ionizing radiation on normal as well as cancerous tissues, differentiating for the first time between effects on growth and effects on metabolism. Her findings—that radiation interrupted growth more easily than metabolism and that effects differed with type of radiation—X-ray, radium, and radon—were *discoveries.* But the techniques she developed in her research, for measuring these effects on the growth of normal cells *in vitro,* were *inventions.* Moreover, her discoveries led to an improvement (increased thickness) in the lead shields used in radiotherapy.

Earlier, she had been the first to demonstrate (in mice) that radiation can totally destroy a malignant tumor, and the treated animal can then live out a normal lifespan. This she accomplished with lead shielding. With proper shielding, doses as large as 12,000 R did not even sterilize the mice. As a further result of her animal studies, Dr. Goldfeder suggested that radiologists expose multiple small areas to radiation in order to allow larger total dose without skin damage. This method proved effective.

In work similar to Edith Quimby's, Dr. Goldfeder determined the equivalent roentgen value of X-rays for gamma rays from radium. Here, too, her results had obvious therapeutic applications.

Using mice with bilateral mammary tumors, Dr. Goldfeder demonstrated that, contrary to prevailing medical opinion at the time, *all cancer nodules* in a given host or patient must be irradiated. She also developed an isogenic host-tumor system that proved useful in evaluating prospective radiation treatments. Her work on whole-body irradiation demonstrated the protective effect of vitamins and of isogenic bone marrow cells during such therapy; revealed variations in radiosensitivity among the various organs of the body; and established the release of enzymes through or from the cell membrane as a significant criterion of such sensitivity.

In 1976 Dr. Goldfeder looked back on a fifty-year career in cancer research and said she would do it all over again. The two contributions she herself emphasizes are her demonstration that properly shielded ionizing radiation can completely destroy a tumor—and her discovery that apparently identical tumors can

---

6. The first woman, **Ollidene Weaver,** received the 1977 award jointly with male collaborators.

respond differently to radiation, so that each tumor must be treated individually according to its biological characteristics (Goldfeder; Moller & Friend). The contributions of this distinguished Polish-born researcher to cancer therapy in general and to radiology in particular can scarcely be overestimated.

In 1981 Dr. E. Barrie Hughes and his research team at the Stanford Synchrotron Radiation Laboratory reported a non-invasive method, using synchrotron radiation, of X-raying the coronary arteries. The team included two women, **Lorna E. Campbell** and **Jennifer P. Stone.** Synchrotron radiation, obtained by whirling electrons in a ring at nearly the speed of light, is a very intense form of X-ray that can be finely tuned. It has been used in many kinds of scientific research into structures, but this is its first application to medical X-rays.

The 1981 announcement reported animal procedures but the researchers predicted success in human trials as well. The method provides an alternative to the expensive and painful catheterization procedures used in hospitals for obtaining pictures of coronary arteries and other important blood vessels. The patient would receive a simple injection of radioactive iodine in an arm vein, followed by two X-ray pictures, all on an out-patient basis. The synchrotron radiation works where ordinary X-rays could not, partly because it is delivered in a form 100,000 times more intense, and partly because the Stanford apparatus allows radiologists to select the single X-ray frequency most sensitive to iodine. With such intense radiation, exposure times can be very short, and dangerous scattering is virtually eliminated.

The final part of the invention is that the two X-ray images are computer-enhanced: information from the X-rays is fed into a computer and processed in such a way as to suppress information derived from intervening tissues and emphasize information derived from the iodine-filled structure. The result is an image that stays sharp despite motion in the arteries.

With such an inexpensive and low-risk procedure, mass screening programs for coronary and cerebral artery disease, similar to the programs long used for tuberculosis, become feasible. The developers predict that synchrotron X-ray diagnosis may be one of the most important advances since Roentgen discovered the X-ray tube in 1895 ("Scientists Develop").

In 1982 **Barbara Japp** and co-workers at the Ontario Cancer Institute in Toronto announced that they had devised an alternative treatment for large choroid malignant melanomas of the eye. The surgical treatment had become controversial, some fearing that it may actually stimulate the spread of the tumor. The new treatment, which shrank tumors in six patients, consists of implanting multiple concentric rings of iridium-192 wire into the sclera until shrinkage is achieved (Miller & Arehart-Treichel).

## Heat Therapy

One of the newest treatments for cancer—which may actually be an ancient treatment ("Hyperthermia's Hot Spot")—is heat. If heat therapy really is ancient, it may well have been a woman's remedy, since women were traditionally both healers and guardians of fire and hearth. In a section on Greek medical women before Hippocrates, for example, Kate Campbell Hurd-Mead reports a cure of epilepsy by malaria (*HWM,* 41). Beginning in the late 1880s, doctors deliberately infected syphilis patients with malaria, because the raging fever cured the venereal disease (*EB*/71, "Venereal Diseases"). Some of these experimenters noticed that the fever sometimes shrank tumors as well.

At least one woman doctor in modern times has worked in the area of heat therapy for cancer: **Dr. Joan Bull** of the University of Texas. As of 1981, Dr. Bull and her group were pursuing the systemic approach to hyperthermia, keeping anaesthetized patients at 107 degrees for two to four hours. Such drastic methods are used only where patients have widespread metastatic disease resistant to all other treatments. Although heat treatment is not yet considered a cure, it produces some striking remissions, and it seems to enhance other therapies. Hyperthermia reinforces chemotherapy, for example, by breaking cancer-cell membranes for drugs to enter. This can make a previously useless drug effective, and other drugs effective at lower dosages, thus minimizing toxic side effects. Heat therapy also enhances the effectiveness of radiation, attacking cells that are low in oxygen, while radiation attacks oxygen-rich cells (Grady, "Heat," 84, 85).

As of 1982, **Dr. Anna Goldfeder** was also doing research on hyperthermia in cancer therapy (Moller & Friend, 4).

Interestingly enough, Cuban-born inventor **Leonora Hafen** of Jackson Hole, WY, had the idea of using heat to destroy cancer cells when she was a little girl. The heat used in today's treatments is often produced either by ultrasound or by magnetic fields (Grady, "Heat," 84, 86). Ms. Hafen's idea was ultrasound (I-5/81).

More familiar are the everyday uses of heat to relieve the aches and pains of muscle strain, mild arthritis, etc. Although the modern soft hot water bottle probably had to await the invention of vulcanized rubber, early means of delivering heat to aching limbs were very likely women's work. Modern refinements of the hot water bottle—making it, for example, more attractive to a child in pain—are sometimes women's work as well, as in the case of **Helen** and **Julia Smit** of Evanston, IL, who have patented and made a prototype of a cuddly hot water bottle (Crossen).

In 1975 **Olga Arango** of Miami, FL, patented a Heat-and-cold applicator for supporting and supplying heat, or removing heat from certain portions of the body.

## *Music Therapy*

Another very ancient therapy exploits the healing effects of music; or, as the women I discuss would have it, music can open the way for people to heal themselves. This ancient technique was used by midwives in many cultures, and by priestesses in the ancient Near Eastern and pre-Hellenic Greek healing temples thousands of years ago. Indeed, according to Sophie Drinker (119), the Roman midwives' chief aid at childbirth was music. Some hint of the power of music survives in popular culture, as in Congreve's memorable lines, "Music has charms to soothe a savage breast, / To soften rocks, or bend a knotted oak." But its healing power, long ignored by the medical establishment, is only recently being rediscovered. I consider it here because music is sound, and sound is a vibration that acts in a physical way upon human bodies.

**Laurel Elizabeth Keyes** uses everyone's "personal key note or tone which, when discovered and sung, brings on a feeling of centeredness and well-being" to help people learn to heal themselves. She has written a book on her work (McGinnis, 14, n11). The large academic literature on music therapy deals almost entirely with what music can do in various medical (especially psychological) conditions, and not how or why (*ibid.*, n4). The "what," however, is impressive enough: music can slow heart rates and lower blood pressure, calm mental patients, help "socialize" angry youths and skid row alcoholics, help the multiply handicapped to learn, aid stroke patients in recovering or relearning language, banish headaches and backaches, etc.

Six other women working in this area as of 1980—a music therapist, a Shiatsu therapist, composers, and/or teachers—are **Ruth Anderson, Jeriann Hilderley, Annea Lockwood, Jean Mass, Ann McGinnis,** and **Pauline Oliveros. Ruth Anderson** used music to "find her way back to health" after a near-fatal illness. Unable to compose, she explored inner sound in meditation. Her first new work was an electronic composition called "Points," using only sine tones (as free from overtones as possible—like the sound of a wet glass being rubbed around its rim). When she played "Points" for her graduate composition class one night, the tired students reported feeling rested and energized. As this happened repeatedly, Anderson concluded that "the sine tones themselves acted like sonic acupressure."

Teacher, musician, and composer, **Jeriann Hilderley** uses chant as massage. **Ann Cain McGinnis** uses music as an adjunct to shiatsu.

Like Anderson, **Pauline Oliveros,** a well-known and innovative composer of electronic music, first explored the therapeutic effects of music for her own health. She noticed that when she sang long tones or played them on her accordion, she felt better, both physically and emotionally. Later she began exploring the idea with other women, who noticed the same effect. Oliveros maintains that taking care not to manipulate sound in a goal-oriented way—a point also emphasized by **Annea Lockwood**—but allowing subtle changes to occur, and relating in certain ways to the people around you who are also sounding, enhances the effect. Eventually, Oliveros incorporated this exploration into her music, which has departed radically from traditional forms and notation.

**Jean Mass** is a music therapist working, as of 1980, for the Lighthouse for the Blind in Manhattan and in the Therapy Training Program at Turtle Bay Music School. She considers music a basic need and accords it unequalled power to act on the human organism, but does not know how it acts, except that, as she puts it, music offers alternatives, or access to hidden and untouched resources. Most of the brain damage she sees is left-hemisphere damage in stroke victims, causing loss or partial loss of language. Mass finds that if these people can learn music, then by a technique called melodic intonation language can be retrieved from the left hemisphere and housed in the right. (Or perhaps the previously passive language already residing there can be unlocked.) The simple-sounding miracle works as follows: Mass sings a simple chant while at the same time touching the client's body and showing a picture of the thing or activity for which a word must be learned. Gradually, the client begins to sing the words back, and "eventually makes the necessary nervous system connections between the word and its lost meaning."

Still more interesting is the case of a three-year-old girl afflicted with brain seizures. She not only had no concept of language, but could not coordinate her brain signals well enough even to walk. Yet she learned to respond when Mass *sang* directions to her, and loved the drum Mass played for her so much that she learned to walk to get it (McGinnis, 12).

## Laser Therapy

A fitting high-tech closing to this section is the use of lasers in medicine. **Judith Walker** of the Walker Pain Institute, West Los Angeles, has used cold lasers in at least two innovative ways: to treat otherwise intractable pain; and to treat paralysis. Although laser analgesia apparently originated in Europe, Dr. Walker devised her own technique, tested it in a double-blind study, and reported the results to a seminar convened by the Food and Drug Administration's National Center for Devices and Radiological Health in Rockville, MD, in 1983.

Cold lasers, as opposed to the cutting and cauterizing lasers used in surgery, are low-power devices. They do not burn the skin, and indeed seem to have no harmful effects.

In Dr. Walker's study, an anaesthesiologist treated 26 patients with previously untreatable chronic pain. Their afflictions ranged from osteoarthritis to trigeminal neuralgia (intense facial pain). Her technique consisted of irradiating four sites on each side of the body, two near the wrist, one at the big toe and one at the heel—calculated to stimulate the radial and median nerves and the saphenous nerve, respectively. Dr. Walker chose these sites not only because of the areas these nerves control, but because so much of the brain cortex is occupied by signals from these sites. "By stimulating these three nerves," she says, "we're actually stimulating 90 to 95 percent of the cortical nerves." She used a one-milliwatt helium-neon laser with a wavelength of 632 micrometers, pulsing at 20 cycles per second for 40 seconds.

A control group received laser stimulation of nearby sites not reaching the designated nerves.

Patients evaluated the results after each treatment. The complete series was 30 treatments administered over ten weeks. Many of the subjects found slight, temporary relief after the first or second treatment, but significant and long-term or permanent relief usually required the full 30 sessions. Nineteen of the 26 reported relief at the end of the series, and 15 of the 19 were still pain-free after six months. Controls reported no such relief. However, 5 of 7 controls who later received proper stimulation then experienced relief from pain.

Although the patient series is small, and the findings must be replicated, Dave Lytle, coordinator of the Rockville seminar, calls Dr. Walker's technique "exciting" and probably very effective (Raloff, "Zapping Pain").

A popular source reported at about the same time (January 1983) that Dr. Walker's technique, called Subcutaneous Nerve Stimulation or SCNS, was a painless stimulation of nerves in the wrists and ankles that could relieve pain anywhere in the body. SCNS had a 98% or greater success rate with menstrual pain and the pain of endometriosis (Shelton, 15).

By October 1983, Dr. Walker had expanded her laser-analgesia research to rheumatoid arthritis, shingles, and experiments to restore motor function and control spasticity in paraplegics. In a double-blind clinical study funded by the American Paralysis Association, Dr. Walker used the same laser and stimulus areas as in the pain-relief study, but irradiated paraplegics for 30 seconds at a time, up to five times per day. Her preliminary conclusions were that the stimulation suppresses clonus (involuntary exaggeration of the knee-jerk), improves results on tests of spinal nerve transmission, and may help spasticity (Raloff, "Cold Lasers").

# Appendix SFA-1:
## Women's Auto-Erotic Technology

In Havelock Ellis's view, it was Japanese women who had brought mechanical auto-erotism to the highest art. Although these women also used phalluses (*engi*), made of paper or clay (H. Ellis, I: 168) or by attaching a stick to a *kabutogata* or hard condom (made of tortoiseshell, horn or leather and used for contraception and possibly to correct impotence), their supreme achievement was the *rin-no-tama,* also called *watama* or *ben-wa.* This sophisticated vaginal device consisted of two hollow brass balls about the size of pigeon's eggs, one empty and one containing mercury or a smaller ball of heavy metal. Women inserted the empty ball first, far enough to contact the cervix; then the second, until it touched the first, followed by a paper tampon to hold both in place. Then, "the slightest movement of the pelvis or thighs, or even spontaneous movement of the organs, causes the metal ball (or the quicksilver) to roll, and the resulting vibration produces a prolonged voluptuous titillation, a gentle shock as from a weak electric inductive apparatus."

The *rin-no-tama* spread to China (in fact, are sometimes known as "Chinese bells"), India, and Vietnam, and by the 18th century were known in France, as *pommes d'amour.* Still available in specialized shops and catalogues, now made of ivory or plastic, with one sphere empty and one containing mercury, or of steel "with a number of small metal tongues," *ben-wa* balls are described in contemporary sex guides and books on sexual technique (Comfort, 188; H. Ellis, I:167–8; S. Green, 96–7; Himes, 125; Reuben, "*ben-wa*"; Rosenbaum, 93). According to Shirley Green, Japanese women sometimes used little brass balls with bells inside "to produce a musical accompaniment" to intercourse (97).

*Ben-wa* balls are by no means an isolated phenomenon. Ranging from such ancient techniques as swinging in swings—in the ancient Near East and perhaps elsewhere a sacred act bound up with women's rituals for Goddess-worship—to such 20th-century techniques as leaning against a spinning washer or dryer—women have found ways to sexual pleasure. Not surprisingly, Japanese women using *rin-no-tama* often swung in hammocks or rocked in rocking chairs to heighten their internal sensations (H. Ellis, I:168; Hawkes, *World* II: 106–8; Rosenbaum, 92–3).

Swings were also known from ancient Greece (470 BC) and Rome to Mexico (AD 300–600), 17th-century Russia, and 19th-century France, at times with echoes of their former religious role. The Athenians, for example, celebrated a swinging festival. American Indian women swung by their hands from tree branches; and modern Greek and Cretan girls still swing in the religious festivals for Easter and St. George's Day. Today, the erotic function of swings is seldom mentioned. Says Bernard Rudofsky (86),

> To be sure, our common house-and-garden swing is but a pale copy of a one-time bold device for generating violent motion and emotion. Neither in scale nor in operation does it compare to its counterpart of the remote past, nor does it so much as hint at its erstwhile delights. . . . [A]ncient sources are more explicit. Greek vases show women on swings assisted by those reliable purveyors of orgiastic pleasures, satyrs, maenads and winged *erotes.* Clearly, swinging then had therapeutic functions that today are entrusted to the psychiatrist.

Rocking chairs and hobby horses, too, have been seats of pleasure (Rudofsky, 90–1; H. Ellis, *ibid.,* 174). And although by the 19th century in Europe and America the makers of such items were usually male, women may well have originated them long ago, perhaps by analogy with their infants' rocking cradles. Later, women would adapt horseback-riding, bicycles,[1] treadle sewing machines, and, as already mentioned, washing machines or dryers, to their pleasure-technology purposes (Rudofsky, 92; Rosenbaum, 93; H. Ellis, 175–8). Indeed, some 19th-century women whose work obliged them to operate a double-treadle sewing machine for ten or twelve hours a day complained bitterly about the unavoidable sexual stimulation (Offen).

1. The possible erotic effects of bicycles may underlie reformer Charlotte Smith's violent opposition to the bicycle craze of the late 19th century. Smith seemed to have a penchant for unpopular and losing causes, and sometimes her voice seems a lone one, but she was far from alone in being horrified by young women riding bicycles in great numbers. Many felt that "when cycling leads to sexual excitement the fault lies more with the woman than with the machine" (H. Ellis, I: 178).

It would scarcely be fair to leave this section without a counter example, from the camp of those who saw masturbation not as a natural expression of sexuality, but as a sin or perversion. **Ellen E. Perkins,** a nurse from Beaver Bay, MN, invented and patented a preventive suit in 1908. Presumably intended mainly for inmates of institutions, this item covered the whole torso, with a lock for the crotch, whose key was to be retained by an attendant. Perkins claimed that it caused little if any discomfort (Blashfield, *Hellraisers,* 167).

# Appendix SFA-2:
# Aphrodisiacs

Conditions as common as high blood pressure can lead to male impotence or flagging desire. Thus male aphrodisiacs have been a significant area of invention for a very long time. The most ancient of these preparations may have been women's inventions, partly because fertility in all its aspects was originally woman's province, and partly because so many of the preparations were also foods or condiments—or otherwise part of a technological cluster with women's work (medicines, perfumes, cosmetics, dyes, mortuary preparations, ritual drugs). Robert Graves mentions myrrh, for example, used both in perfumes and in burial preparations, as a "well-known aphrodisiac." It is also a traditional emmenagogue, or menses-promoter (*Myths* I; 72–3; Nelson, 4).

Aphrodisiacs traditionally fall into three types: ingested, smoked or inhaled, and used in a bath or massaged into the skin with oil, such as jasmine essence. Where any specialization exists in the culture, midwives were the likely inventors, guardians, and dispensers of aphrodisiac technology. Athenian midwives evidently prescribed them, for example, and the Roman *sagae,* though they specialized in abortion, had aphrodisiacs as well (Rich, *Born,* 121; Hurd-Mead, *HWM,* 73).

Jeanne Rose, in *Herbs and Things,* devotes an entire chapter to aphrodisiacs and lists 54 aphrodisiac herbs or substances in her alphabetical section (5), including alcohol in small quantities, anemone, cannabis (marijuana), cantharides, celery, clove, ergot, fo-ti-tieng (*Hydrocotyle asiatica minor*), ginseng, hashish, jasmine, kava kava, maidenhair, male fern, mandrake, meat diet, musk, navel wort, parsley, periwinkle, valerian, wild poppy, and yohimbe. The reputed aphrodisiac qualities of asparagus (J. Rose, 40) may account for some of the mystique surrounding this vegetable. Rose particularly favors a combined extract of damiana and saw palmetto berries, taken three times a day before meals (56, 141). Interestingly enough, one of damiana's species names is *aphrodisiaca* (*Turneria aphrodisiaca;* 56).

Modern medicine usually dismisses such preparations as useless. However, recent experiments have confirmed aphrodisiac effects for at least one of these traditional preparations, yohimbe, from the *Corynanthe yohimbe* tree of Africa. It contains a crystalline alkaloid called yohimbine ($C_{21}H_{22}N_2O_3$), which stimulates breeding behavior in male rats. Stanford researchers led by Dr. Julian Davidson and including one woman, **Dr. Erla Smith,** reported in 1984 that yohimbine "increases arousal in sexually experienced male rats, facilitates copulatory behavior (including ejaculation) in sexually naive males, and induces sexual activity in males that were previously sexually inactive." The authors conclude, "These data suggest that yohimbine may be a true aphrodisiac . . ." (Hofstadter, "Folk"; J. Clark *et al.*).

Tattooing and other physical aphrodisiacs, or changes induced in the human body itself, should be mentioned in passing here. As tattooing is often intended to increase sexual attractiveness, it may form part of a puberty rite (Delaney *et al.,* 27); and as it is (or once was) often done by women, we can hypothesize that women invented this nonchemical aphrodisiac—and art form—in several cultures (cf. J. Rose, 84). Considerable technology is involved in achieving permanence, avoiding infection, etc.

Among the Kung-San of Africa, women massage their infant daughters' genitals so that they will dangle enticingly by the time the girl reaches puberty (H. Fisher, *Contract,* 23).

A striking example to close this section is the Marquesan "vagina medicine" and its related technology. Women of many cultures have used alum and other herbal astringents to increase sexual pleasure for their lovers, and thus directly or indirectly for themselves—and to increase their own desirability—by shrinking the vagina. The Marquesan example is particularly noteworthy, in that it involves a system of treatments beginning when girl infants are only a few weeks old and continuing until puberty. Ideally, several different astringents are used in a particular sequence, but some of the necessary plants grow in isolated and remote areas and are not always obtainable, in which case women adapt the treatment to available ingredients. Robert Suggs suggests that one of the plants used on Nuku Hiva was *tu'ei'au* (*Phyllanthus pacificus* var. *uahukensis*). The Marquesans thought this treatment tightened the vagina, gave it better muscle tone, and suppressed vaginal secretions and odor.

As late as the 1960s, according to Suggs, "The use of vaginal astringents is by no means a tottering survival. Most young couples use one or more of the astringents on their female children. Marquesans residing in Tahiti have requested that [astringent] be prepared and sent to them. . . ." Suggs notes that although the men often fetched the ingredients for the "vagina medicine," "it was the women who seemed to know most about the preparations" (39– 41).

Nor is the Marquesan example entirely isolated. Among the Baganda and the Shaheli of Africa, older women teach girls to enlarge their labia before they reach puberty. Dahomeyan pubescent girls learn to massage the vagina to cause thickening and muscular development of the labia. These techniques are intended to increase the woman's sexual pleasure as well as the man's (Delaney *et al.,* 27).

A modern report that deserves further research is the 1950s experience of an American woman with a silver cervical cap. This cap, prescribed for her by an Austrian Jewish woman physician who had fled the Nazis, not only worked perfectly for birth control, but "enhanced intercourse" to the point where she did not need clitoral stimulation to reach orgasm. "I'm sure you know," she told Barbara Seaman in the 1970s, "that when Freud first described the 'vaginal orgasm,' most of his women patients were wearing some sort of pessary or cap." This woman, not named in my source, but described as a professor, said further that if she could find her old silver cap, she would start wearing it again, dents and all (Seaman & Seaman, 235–6).

# *Appendix SFA-3:*
# *Women's Technology for Counterfeiting Virginity; Sex as Business*

The same astringents that shrink the vagina for pleasure can be used to simulate a lost virginity. In cultures where virginity is prized, the technology of counterfeiting virginity can be raised to a high art.

The following "remedies for damaged maidenheads" are attributed to **Trotula** of Salerno, which dates them to the 11th century AD (compiled c. 13th century; cited in O'Faolain & Martines, 143):

1) Soaking a soft cloth in a mixture of ground sugar, white of egg mixed into rainwater in which alum, fleabane, and the dry wood of "a vine" have been boiled down with "other similar herbs," and thoroughly washing and bathing her vagina with this solution.
2) Making a suppository soaked in rainwater mixed with well-ground fresh oak bark, and inserting the suppository shortly before intercourse.
3) Using plantain, sumac . . . , oak galls, large black bryony, and alum cooked in rainwater and "fomenting" the private parts with this mixture.
4) Carefully applying a leech to the labia, so that blood trickles out and forms a crust on the orifice; "the flux of blood will tighten the passage. . . ."

Trotula herself most likely did not invent these preparations and techniques, but it seems virtually certain that they were women's inventions.

Sigrid Undset, a Nobel laureate and meticulous researcher, reveals the existence in 14th-century Norway of herbal preparations and techniques for simulating or artificially restoring virginity. Here is a drunken priest accusing a local wise woman and healer at a banquet in *Kristin Lavransdatter* (I:52–3):

> "Be still, Lady Aashild," cried Sira Sigurd: "do you not talk of the long tail evil drags after it. You sit here as though you were mistress in the house, and not Ragnfrid. But 'tis strange you could not help her child—have you no more of that strong water you dealt in once, which could make whole the sheep already boiling in the pot, and turn women to maids in the bridal-bed? Think you I know not of the wedding in this very parish where you made a bath for the bride that was no maid . . .?"

The use of pigeon's blood or other nonhuman blood for the stained sheets displayed outside the bridal chamber after the wedding night is famous from literature as well as from reality. Even if the idea itself were not considered an invention, some of the techniques used to contain the blood and dispense it at the right instant without alerting the bridegroom—like inserting blood-soaked sponges or fish bladders into the vagina in advance—were ingenious indeed.

Others, like Trotula's leech device, which seems to have been prevalent, and the use of small pieces of broken glass, were both painful and dangerous, testifying to the urgency felt by the women—or to the greed of their procurers. A 16th-century Belgian physician reported that certain Spanish women "greedy for money, are in the custom of selling girls several times over as virgins by falsifying the hymen with trickles of blood and other materials." Such greed, of course, was not limited to Spain or to any particular century. A character out of 15th-century English fiction, an enterprising bawd named Celestina, for example, used needle and thread to pass off young women "three severall times for a virgin" (O'Faolain & Martines, 143; Bloch, 166–8; S. Green, 66).

"Alum," says Shirley Green in her *Curious History of Contraception,* "is an astringent that tightens the vagina and manages to give a fair imitation of virginity" (65–6). Its uses for this purpose were common knowledge on one level, as the 17th-century English play *Sodom* unmistakably reveals:

> The already Cuckold getts a Maidenhead,
> Which (is) a toyle, made of restringent aide.
> Cunt wash't with Allom makes a Whore a Maid.

According to Ivan Bloch (167), writing specifically of England, a sure method of restoring lost virginity was published in the *Pall Mall Gazette* in 1885. This method was customary in the Middle Ages and up to the 17th century, but then became "neglected and rare" until revived in a veritable "defloration mania"—i.e., a desire for virgins as sexual partners—during the 18th century. He lists astringent plants and leeches in addition to the broken glass and sponges and fish bladders already mentioned. He cites two specific astringent recipes or formulas: one a vapor produced by putting a piece of red-hot iron or brick in vinegar, and the other a decoction of such by-now familiar-sounding herbs or plant products as acorns, sloes, myrrh, Provençal roses, cypress nuts, *unguentum adstringeus feirnelli,* water distilled from myrrh.

Virginity could be simulated even after childbirth, as a Chinese observer reports from Cambodia in the 13th century AD. A poultice of hot rice and salt applied to the perineum for 24 hours after birth made the woman ready for immediate intercourse (as was the custom; nonresponsive husbands could be discarded), and also "causing an astringency which seems to renew the young mother's virginity" (Goldsmith, 78).

As to who were the guardians, inventors, and improvers of these techniques, it seems likely that most were women, although after the male takeover or professionalization of medicine, which happened far earlier in the Near East than in Europe, male surgeons and physicians may have entered the field. Bloch mentions "scholarly monks and Arabian doctors of the Middle Ages" who brought this technique to a high art, but also cites a French source as saying that "all the great ladies and gentlemen of the 17th century knew recipes for the restoration of virginity," and that barbers and bathkeepers—often women—were particularly skilled, as were prostitutes and procuresses. In the fashionable brothels of 18th-century London, such practices were "general." Two names here are **Charlotte Hayes,** who had a "reputation in this art," and who claimed that a girl could "lose her virginity five hundred times and still pass as a virgin," and that "great dealer in human beings" **Mme Jeffries** of Chelsea. Midwives also sometimes had a sideline in patching up virgins (Bloch, 157, 166–8).

"Granny-women" in the American Ozarks played a crucial sexual-counselor's role well into this century, instructing experienced brides in counterfeiting virginity by means of concealed noise-making devices as well as by canny use of astringents. One such auditory technique gone hilariously awry is recounted by Vance Randolph in "Fireworks Under the Bed" (41–2). Randolph points out that this tale is part of a "substantial body of tales and jokes concerned with deception on the wedding night to cover up the loss of the bride's virginity."

Lest we laugh too much at earlier ages, however, we should note that surgical operations to reattach a torn hymen, or create one by plastic surgery where one never existed, still occur in the U.S.; and that Lady Diana Spencer, before she could become the current Princess of Wales had to prove that she was *virgo intacta.*

Before leaving this section I wish to underscore the pertinence of alum and its many uses to my cluster-of-technologies arguments for women as likely inventors in various areas: Alum turns up repeatedly in women's work, in food-preservation (especially pickling), in dyeing (as mordant), in medicine, and in contraception, in addition to the uses just noted. And these are just the more familiar uses. Alum also turns up in women's recipes or formulas for a long-lasting red ink, for clarifying tallow, for preserving grasses for dried arrangements, for a paste used to repair cracks in concrete floors, and for a waterproofing compound for tweeds (Juster, 20, 23, 58, 107, 123). Such multiple associations of a substance or a technology with one sex or gender in the division of labor can buttress the argument for connecting that sex with the technological advances in question.

## Sex as Business

Prostitution is by no means the oldest profession. Women were potters, weavers, cultivators, musicians, mourners, morticians, cosmeticians, scribes, traders, midwives, healers, priestesses, and many other things before they were ever prostitutes. And it is long past time to put to rest the Judaeo-Christian slander that the women who celebrated the Goddess in Her temples with sexual rites were prostitutes. It is also time for scholars as well as preachers to stop calling these consecrated women "temple harlots" or "sacred prostitutes" and give them their actual titles according to which group of priestesses they belonged to—or simply call them *sexual celebrants.*

Prostitution does have a long history, during part of which it was a major repository for our culture's information on both sex-pleasure and anti-fertility technique. With the progressive subjugation of the wife from the end of the Bronze Age to the 19th century, and the holocaust of witches that blighted early modern Europe, prostitutes, along with the vanishing midwives and herbal healers—and, it now seems, nuns—were left to carry on certain vital forms of knowledge.

Commercial sex has its own technology—seldom classed or studied as such. One example from a subspecialty of the trade must suffice here: flagellation. Whatever we may think of this practice (usually blamed on a childhood that links love with pain), flagellation seems to have spawned its own minor technology, with switches chosen from specific plants and carefully graded for suppleness; means of maintaining that suppleness, means of reviving lost sensation (and lost senses), apprenticeships for practitioners, etc. Women, as brothel owners, seem likely to have been the inventors and improvers as well as the major purveyors of this art and technology.

Indeed, in Victorian London certain madams specialized in flagellation,[1] among them the woman who mainly concerns us here, **Theresa Berkley** (d. 1836). Henry Spencer Ashbee, Victorian connoisseur and bibliographer of pornography, who considered flagellation "[t]he secret sexual propensity which the English most cherish," called her the queen of her profession. Ashbee, the major source on Berkley (147ff), calls her *Mrs.* Berkley (though whether this reflected actual marital status is not clear), locates her at 28 Charlotte Street, Portland Place (actually the Charlotte Street near Marylebone, now Hallam Street), and calls her a perceptive student of her clients' lightest caprices and deepest desires, perfectly willing to gratify them "if her avarice was rewarded in return." Her "instruments of torture," Ashbee continues, were more numerous than those of any other "governess":

> Her supply of birch was extensive, and kept in water, so that it was always green and pliant; she had shafts with a dozen whip thongs on each of them; a dozen different sizes of cat-o'-nine-tails, some with needle points worked into them; various kinds of thin bending canes; leather straps like coat traces; battledoors, made of thick sole-leather, with inch nails run through to docket and curry comb tough hides rendered callous by many years' flagellation. Holly brushes, furze brushes; a prickly evergreen called butcher's bush; and during the summer . . . China vases filled with a constant supply of green nettles, with which she often restored the dead to life.

Many of these were undoubtedly ancient technologies, and it is impossible at this remove to identify her own refinements. What brings Theresa Berkley to this chapter is her invention in the spring of 1828 of "a notorious machine . . . to flog gentlemen upon." Ashbee (149) says it was invented *for* her, but indicates that she had the original idea and hired someone to manufacture the piece according to her directions and requirements. Bloch (296) and others have simply credited her with the invention.

Strictly speaking, it was not so much machine as apparatus, an articulated and adjustable padded framework for supporting her "incorrigibles"[2] at the desired angle for their whippings in her establishment. Bloch has the clearest description (296): "This apparatus for the flogging of men . . . was really an adjustable ladder which could be extended to a considerable length, on which the victim was securely strapped, openings being left for head and genitals." This so-called "Berkley horse" or Chevalet was featured in an engraving from Theresa's Memoirs, reproduced by Ashbee.[3] Though notorious in its own day, it was thus not quite unique, nor was it short-lived. It was apparently still in use in Paris at the time of the 1900 Exhibition, and "greatly appreciated by men of the world"; it also appears in modern flagellation pornography (Bloch, 296 nl). Steven Marcus, bringing Berkley's device to light again for today's readers,

---

1. Says Ashbee (147): "At the early part of this century, very sumptuously fitted-up establishments, exclusively devoted to the administration of the birch, were not uncommon in London. . . . It would be easy to form a very lengthy list of these female flagellants. . . . Mrs. Collett was a noted whipper, and George the IV is known to have visited her. . . ."
2. The fiction that the madam was a governess correcting her bad-boy students was apparently almost universal between flagellators and their customers; and repeated punishment was necessary because the clients were "incorrigible." Bloch reports a letter Theresa Berkley received from a would-be client, addressing her as "Honoured lady," and saying he had visited a certain Mme Wilson in Marylebone as well as several other governesses, but they had made no impression on him. Finally he had prevailed upon "your close friend, Count G." to introduce him to her, at which he jumped for joy, "because I have been told of your famous apparatus, the Chevalet, which should succeed in punishing sufficiently undisciplined young men like myself" (Bloch, 297–8). Whether this extended euphemism was in part a code to protect certain correspondence from prying eyes, or a psychological self-deception the victims found necessary to support them against the condemnation of their own society, or a reflection of the public-school origins of the preference—or all three—is difficult to say.
3. Sometimes described as enabling many gentlemen to be flogged at once, the device as depicted in this engraving accommodates only a single devotee, and those who imagined mass-production floggings may have confused Berkley's device with the brainchild of an 18th-century *bon vivant* and fellow Londoner, Chace Price, which would have accommodated 40 men at once (Bloch, 284)!

compared it to a large football blocking-dummy, and called it "perversity's contribution to the Industrial Revolution" (68).[4]

The Berkley horse may not be Theresa Berkley's only contribution to what we might call high-tech flagellation, however, since, according to Ashbee (150),

> Mrs. Berkley had also, in her second floor, a hook and pulley attached to the ceiling, by which she could draw a man up by his hands.

Bloch also mentions her "compound pulley" allowing her to suspend a client and flog him in that position (297).

Ashbee hopes that Theresa Berkley's Memoirs—which he has evidently seen, for he reproduces one illustration from the work and mentions another (this pulley)—will "soon" be published. Her executor, Dr. Vance of Cork Street, had persuaded the publisher to suspend publication, but Vance had recently (1870s?; cf. date of Ashbee's first ed.) died. I have been unable to find these Memoirs, either published or unpublished, in the British Museum; and this same Dr. Vance *destroyed all seven boxes of Berkley's correspondence,* presumably because some of her clients came from the noblest families in Britain. Thus the Queen of Flagellants remains a shadowy figure.[5] However, she must indeed have been, as Ashbee noted, a "thorough woman of business." In only eight years of operation, she amassed what was for those times an immense fortune, over £10,000—which she left to her missionary brother. He self-righteously rejected the ill-gotten gains, which then went to Dr. Vance, who "refused to administer," and the estate evidently went to the Crown (one original source included by Ashbee in his 1877 edition says the total was £12,500, besides other property; Ashbee, 148, 150; Ashbee/1877, interp. after xlvi, p. 3 of letter; Wallechinsky & Wallace, *Almanac,* 964).

Theresa Berkley may not have been alone as an inventor of flagellation aids. According to Ashbee, **Sarah L. Potter,** alias **Stewart** (d.1873), mistress of Stewart's Academy, and another London specialist in flagellation, had among her equipment a folding ladder with straps, birch rods, furze brooms, and "secret implements," all of which came to light because of a court case of July 1863 (Ashbee, 148; Ashbee/1877, II: 30).

4. According to Ashbee (150), "The original horse is among the models of the Society of Arts at the Adelphi, and was presented by Dr. Vance, her executor." However, Ashbee wrote in the 1880s. In the intervening century, the Adelphi's patent models have been scattered, and neither mail inquiries nor personal visits to London have been successful in locating this model.

5. Indeed, one might almost suspect a deliberate conspiracy of the fates as well as among ordinary mortal guardians of morality to blot out her memory: She died *just one year too soon* for the detailed death records that would have given me vital information about her. The year I went to the British Museum specifically to locate her memoirs, all of the *B* drawers of the manuscript catalogue had been sent off to the typist—apparently without copies of the cards retained! Some of these priceless drawers fell off the back of the truck on which they were being transported, and no one can be absolutely certain that all cards were recovered. At that time, I searched the Ashbee Collection on my own without success.

According to London researcher Mike Pentelow, the building where Berkley reigned as Queen of Flagellants no longer exists (posh apartments stand there today), but judging by business cards found in phone booths in the area in 1994, flaggellation is alive and well. I have tried to trace Theresa's missionary brother, but none of the major denominations sending missionaries to Australia during the early 19th century lists a Berkley (or variant spellings—which may mean that Berkley was indeed her married name). I would be grateful for information leading to the brother (and thus perhaps to Sarah Theresa Berkley's origins and maiden name) and/or to her autobiography.

# Appendix SFA-4:
## Selected 19th-Century Women's U.S. Patents in Menstrual Technology (LWP)

**Gertrude Campbell,** Chicago, IL—Monthly protector (1883)
**Louise Lange,** New York, NY—Catamenial sack (1883)
**Nannette Amia,** Brooklyn, NY—Catamenial sack (1884)
**Johanna Patten,** Crown Pt., IN—Combined suspensory bandage and truss (1887)
**Anrie Valon Robinson,** Fort Ann, NY—Catamenial sack (1887)
**Emma J. Gooch,** Washington, DC—Napkin supporter (1887)
**Pamelia M. Boleman,** New York, NY—Catamenial sack (1888)
**Emma A. Wiley,** Los Angeles, CA—Catamenial sack (1889)
**Cornelia Steinmetz,** Baltimore, MD—Catamenial sack (1889)
**Ellen M. Preston,** Providence, RI—Napkin belt (1890)
**Georgiana Fuller,** Gloversville, NY—Catamenial sack (1890)
**Annie J. Miller,** Dansville, NY—Catamenial sack (1891)
**Annie Willoughby,** Philadelphia, PA—Catamenial sack support (1891)
**Emma H. Carpenter,** Springfield, VT—Catamenial sack or bandage (1892)
**Roza I. Odell** and **Clara S. Howell,** Baltimore, MD—Catamenial sack (1892)
**Eliza Kirwin,** Indianapolis, IN—Catamenial sack (1894)
**Emma J. Pike,** London, Eng.—Suspensory belt (1894)

# Appendix SFA-5:
# Abortion

A section closely parallel to the main-text contraception section could be written on women's role in abortion technology, beginning with its significance in the modern world. With hunger rampant and contraception imperfect even where available; with consciousness of overpopulation and the cost of rearing a child in the developed world both soaring, that significance can scarcely be overstated. In the early 1980s an estimated one out of four of all pregnancies ended in abortion, for a total of 40 million abortions each year, worldwide. In the United States it was about one in three; in the Soviet Union, more than one in two! Eight thousand women a month depend on Medi-Cal abortions in the State of California alone, and these are just the reported, legal abortions ("Abortion Statistics"; ACLU *News* XLVI: 6(Aug./Sept. 1981): 1; Billington, L-1981; Isaacson, 21).

The section on abortion technology would precede the contraception section, for many herbs found to produce abortions were probably refined into contraceptives, and physical abortion techniques almost certainly led to IUD technology. However, since no one doubts women's role in early abortion technology, as admirably documented by George Devereux[1] and in modern nonestablishment abortion technology, I will limit my discussion to four particularly noteworthy inventions, two "primitive" technologies that are also quite modern and two modern ones, including a dual-purpose device patented in the 1970s.

## Eskimo Women's Concealed Probe

As mentioned in the contraceptive section, primitive and not-so-primitive women have for millennia caused abortion by inserting long, thin instruments into the uterus—such forerunners of the metal coathanger as sharpened sticks, the midrib of a leaf, a wooden stiletto whose handle represents Kupo, the god of abortions (Hawaii), a pointed root smeared with musk, and a walrus bone (Devereux, 36)—with the intention of breaking the membranes, killing the fetus, or otherwise bringing on premature labor. Many of the operators showed great skill.

Some Eskimo women had developed their walrus-bone probe into a highly sophisticated instrument. They shielded the bone's sharp point with a fur cover until the instrument was safely inside the uterus, thus avoiding injury to vagina and cervix. When the bone was in place, they could pull back the fur shield with a string of reindeer sinew and free the point to do its work. A second string sheathed the blade again for withdrawal (Devereux, 221).

## Cervix-Dilating "Tents"

American Indian women used slippery elm sticks inserted into the cervix (Devereux, 302). Apparently the bark contains the active ingredient, for sometimes (perhaps most often) a roll of the bark alone is used. In the late 1970s an American woman reported that in the 1940s she had used this method, handed down to her from her mother, whose grandmother was Cherokee (Soma, 94). Likewise, in an ancient technique, Japanese women inserted a bundle of dried *Laminaria* seaweed into the cervix. Expanding as it absorbs moisture, the seaweed gradually dilates the cervix so that labor begins. So effective and gentle (relative to instant, instrumental dilation) is this preparation that modern medicine has adopted it for legal, clinic abortions (Hale & Pion; Newton; Seaman & Seaman, 292; J. Strauss *et al.*). The *Laminaria* bundle is called a "tent" in medical terms: any substance inserted into a small opening in order to stretch it or keep it open (Newton, 446). Doctors find fewer harmful effects on later child-bearing ability with this form of

---

1. Devereux' book does, however, contain some startling information. In many groups the men frown on abortion (or claim to white male anthropologists and missionaries that they do), and the women use their technology in secret; in others, women may be forced to abort by husbands, lovers, male relatives, or medicine men (Devereux 12–3, 16–7, 20, 174, 220). Indeed, 50 of the 370 tribes Devereux deals with had some form of forced abortion (364–71)!

For an update of Devereux see, e.g., Farnsworth *et al.;* De Laszlo & Henshaw; S. Gage; *SN,* Oct. 27, 1979: 293.

dilatation (Harlap). Women who have experienced it, however, find it by no means painless (Van Gelder, ". . . Seaweed," 113).

Note that *Laminaria* seaweed has long been used for food in Japan, and has even been cultivated when natural beds became depleted. Some of the common names of this brown alga are tang, tangle, sea tangle, redware, sea wand, and sea girdle (Chapman & Chapman, 95ff; Hale & Pion, 833).

## 20th-Century Contributions

Contemporary women have made at least two noteworthy advances in abortion technology: Vitamin C abortions and the Del-Em Menstrual Extractor. Although the Vitamin C method is still experimental, women report success with massive doses of the vitamin taken at the first suspicion of pregnancy, either just before or at the time of the next expected period. Most of the women have used 6 grams a day for five days; others have taken the entire dose at once, and still others have taken half the daily dose twice a day. Although the dosage and its timing can vary, success seems to demand very large doses of the vitamin, taken close to an expected period and quite early in the pregnancy, probably no later than six weeks from the last normal period (S. Gage, 1, 40; Seaman & Seaman, 292; cf. also Farnsworth *et al.*, 583).

Women are also investigating the possibility that Vitamin C may be an effective contraceptive or back-up method of contraception. Taken around ovulation time, it could affect the "hospitable" cervical mucus believed to aid sperm in reaching the uterus (S. Gage, 40), whether by making it more acidic and thus hostile to sperm, by changing its consistency, or by some as yet unknown mechanism. Some women report using Vitamin C in conjunction with a diaphragm.

Although the mechanism by which Vitamin C might prevent or interrupt pregnancy is not yet clear,[2] women are continuing to experiment on themselves and share their results with other women through the medical self-help movement. Suzann Gage contrasts this process favorably with "male-controlled research . . . that uses women as uninformed experimental subjects" (40).

## The Del-Em

**Lorraine Rothman,** a leader in the women's self-help movement of the late 1960s, invented the Del-Em Menstrual Extractor in 1971 and patented it in 1974. Not billed as an abortion apparatus, the Del-Em was used in advanced Self-Help Clinics in a women's research project, in which laywomen extracted each other's periods in a single procedure instead of allowing the usual several days of bleeding. However, the Del-Em's usefulness for an early suction abortion, up to about eight weeks after the last normal menstrual period, is obvious, and many women have safely so used it.

The device consists essentially of a collection jar connected by flexible tubing to a cannula for insertion into the uterus, and also connected to a 50-cc. syringe to provide suction. Ms. Rothman has purposely designed the Del-Em so that any woman can make one for herself with materials mostly available either in her own home or at a hardware store. For instance, an ordinary Mason jar can serve as the collection jar, and a cannula can be made from the soft plastic tubing in pump spray cans and bottles. A large cork can serve for the two-hole (rubber) stopper that would otherwise have to come from a laboratory supply house. The syringe can be a kitchen syringe of a kind available from gourmet cookware shops, etc. Suzann Gage's book tells where to buy any parts that cannot easily be improvised, such as the *necessary safety valve* between syringe and jar, to prevent air from being pumped into the uterus (Fishel, 56–7; S. Gage, 17ff; Rothman & Punnett).

For later-stage abortions, the same general apparatus will work but needs somewhat stronger suction. Women have been inventive in devising sources of this necessary suction, using vacuum cleaners, milking machines, bicycle pumps, and hand pumps. Bicycle pumps, of course must be converted to suck air in rather than pump air out, by attaching the plunger of the pump to its rod backwards and then putting it back into the pump barrel. Suzann Gage cautions that the suction must not be too strong, or it may damage the uterus, and suggests using about sixty pounds of pressure, or about the amount produced by putting a finger on one end of a plastic straw and sucking as hard as possible on the other. She also suggests two variations on the syringe as sources of suction: the Mighty Vac hand pump, and burning an alcohol-soaked

2. If Vitamin C is indeed an effective weapon against cancer as Dr. Linus Pauling maintains, it may act directly on the fetus, which is, like a cancer, a mass of rapidly dividing cells.

rag in a second jar connected (instead of the syringe) to the second tube. The burning alcohol exhausts all the air in the jar and produces a vacuum.

Clearly, women have been no less ingenious in their contributions to abortion technology than in any of the other areas thus far discussed. On the contrary, if necessity is indeed the mother of invention, womens' inventions here should have a certain edge. As this book goes to press, there is a concerted attack in progress on U.S. women's right to safe, legal abortion. Both women's ancient herbal remedies and devices like the Del-Em will be increasingly important until the current wave of fanaticism subsides (cf. Tumulty articles).

# Appendix M-1:
# 19th-Century U.S. Sewing Machine Patents Granted to Women

Abbreviations:
NYC = New York City
SM = Sewing machine
* = Not on LWP

**Hannah D. Conrad,** Dayton, OH—Improvement in SMs (1857)
**Abbey S. Smith,** Lockport, NY—Improved starting mechanism (1863)
**Abby H. Price,** NYC—Improvement in machine-made ruffles (1863)
**Mary J. West,** Watertown, NY—Improvement in marking attachments (1866)
**Cornelia F. Ingraham,** Indianapolis, IN—Improved bobbin (1866)
**Mary A. Duffy,** NYC—Improvement in SM gages for tucking . . . etc. (1866)
**Anna Weissenborn,** NYC—Improved tuck-maker/creaser (1867)
**Mary T. Fitch,** Lockport, NY—Improved ruffling attachment (1867)
**Sarah F. Browne,** Savannah, GA—Improvement in SM markers (1867)
**Mary A. Duffy,** NYC—Improvement in marking gages for SMs (1868)
**Anna P. Rogers,** Quincy, IL—Improved tuck-creasing attachment (1869)
**Anna Weissenborn,** NYC—Improvement in treadles for SMs (1869)
**Sara Tutton,** Tunkhannock, PA—Improved presser-foot (1869)
**Caroline Garcin,** Colmar, France—Drive mechanism (1869)
**Anna P. Rogers,** Quincy, IL—Guide for SMs (1869)
**Anna Hancock,** NYC—Improvement in SMs (1869)
**Mary Dewey,** New Albany, IN—Soap-holding attachment (1869)
**Mary W. Welty,** NYC—Shield for SMs (1870)
**Jennie L. Lake,** Brooklyn, NY—Improvement in SMs (1870)
**Sarah A. Sexton,** NYC—Improvement in SM seats (1870)
**Celia P. Clark,** Lock Haven, PA—Improvement in needle-sharpening attachments for SMs (1871)
**Eleanor C. Sproul,** Green Pt., NY—Improvement in braid-guiding attachments for SMs (1871)
**Mary E. Antrim,** Philadelphia, PA—Attachment for SMs
**Sarah O. Matteson,** Chicago, IL—Improved attachment for SMs (1871)
**Elizabeth P. Smith,** Chicago, IL—Improved tuck-creaser for SMs (1872)
**Augusta Hoover,** Bethlehem, IN—Quilting attachment (1872)
**Sarah Perry,** NYC—Guide for hemmers (1872)
**Mary A. Williams,** Chicago, IL—Improved needle-threader for SMs (1873)
**Mary Jane Pitrat,** Gallipolis, OH—Improvement in thread-cutters (1873)
**Mary Dewey,** Chicago, IL—Improved quilting attachment for SMs (1874)
**Eliza Maria Jones,** Brockville, Canada—Improvement in tuck-creasing devices (1874)
**Elizabeth F. Shaw,** Brooklyn, NY—Improvement in dress-protectors for SMs (1875)
**Lillian Roosevelt,** Hempstead, TX—Improved SM thread guard (1875)
**Elizabeth Sloan,** NYC—Improved plaiting attachment for SMs (1876)
**Georgiana L. Townsend,** Philadelphia, PA—Improvement in device for operating SMs (1876)
**Julia E. Snapp,** Georgetown, IL—Improvement in spinning attachments of sewing machines (1876)
**Mary Duff,** Benton, IL—Improvement in hemmers for SMs (1876)
**Hanna G. Crawford,** Peabody, KS—Improved extension for SM table (1877)
**Eliza Ann Vance,** Gallipolis, OH—Improvement in tuckers (1877)
**Susan DeLamater Granger,** Ann Arbor, MI—Improved SM chair (1878)
**Charlotte A. Van Cort** and **Sarah A. Shepherd,** Washington, DC—Improved SM treadle (1879)
**Rebecca M. Hicks,** Minneapolis, MN—SM receptacle and extension-table leaf (1880)
**Medora D. Barringer,** Silver Creek, MI—Thread-cutter (1880)
**Caroline Clement,** Penhook, VA—Tension device for SMs (1881)

Mary G. Price, St. Louis, MO—Tuck-marker (1881)

Emmeline A. Winaus, NYC—Tucking and plaiting scale for SMs (1881)

Alice J. Winslow, Kalamazoo, MI—Thread-holder and cutter (1881)

*De Lana Sheplie, Boston, MA—SM attachment (1882)

*Mary E. Davis, White Cloud, KS—Thread-cutter (1882)

Anna L. Smith, Edmore, MI—Ruffling attachment for SMs (1882)

*Martha S. Moss, Indianapolis, IN—Perforating attachment for SMs (1882)

Mary E. Smith, Southbury, CT—Lamp-supporting bracket for SMs (1883)

Elizabeth Chavers, Siddon, MI—SM shuttle (1883)

Ella J. Moore, Brimfield, IL—Device for holding material to be sewn (1883)

Cornelia T. Freeman, Elizabeth, NJ—Guide-setter for SMs (1885)

Emma J. Swartout, Danbury, CT—Machine for sewing hat-tips (1885)

Elizabeth Cheshire, Cincinnati, OH—Book-SM (1885)

Elizabeth Cheshire, Cincinnati, OH—[Book-sewing] Needle (1885)

Charlotte Lenz, Cleveland, OH—SM (1885)

Sarah A. Shepherd, Washington, DC—SM treadle (1885)

Lizzie McKeogh, Pittsburgh, PA—Footpad for SM treadles (1886)

Elizabeth Laufer, Cincinnati, OH—SM table (1886)

Cornelia H. Owen, DeLand, FL—Method of hemstitching (1888)

Minna G. Blake, Vicksburg, MI—SM table (1888)

Emma M. Ackerman, Quincy, MA—SM (1888)

Emma F. Briggs, San Marcos, TX—Motor for SM (1888)

Mary E. Hunter, Osborn, OH—Guide for SMs (1888)

Alice M. Perkins, La Crosse, WI—Looping attachment for SMs (1888)

Annie Lewis, Galveston, TX—Bobbin-winder (1888)

Alice A. Crawford, Austin, TX—Tuck-creasing and -marking attachment for SMs (1888)

Mary F. West, Binghamton, NY—Thread-holder and cutter (1888)

Lucy J. Persall, Fort Edward, NY—Tuck-holder (1889)

Alice LaG. Mayo, Great Falls, MT—Hemming attachment (1889)

Alice LaG. Mayo, Great Falls, MT—Guide for SMs (1890)

Louisa A. Harrison, Boston, MA—Presser foot (1890)

Jennie Horvath, NYC—Overseaming SM (1891)

Maria M. Bostik, Washington, DC—Combined chair and SM top (1891)

Jessie L. Gardner, St. Joseph, MO—Hand appliance for operating SMs (1891)

Mary L. Birdsong, Arcola, MI—SM operating attachment (1892)

Elizabeth Calm, NYC—SM (1892)

Catharina Booss, NYC—Guiding attachment for SMs (1892)

Minnetie R. Thurston, Kalamazoo, MI—Cylindrical SM needle (1892)

Anna H. Clayton, Louisville, KY—SM motor (1892)

Eda F. Hustis, Milwaukee, WI—SM brake (1892)

Rachel A. Shellard, Virginia City, NV—SM attachment (1893)

Mary E. Hall, Boston, MA—Hemstitching attachment (1893)

Louisa E. Webber, Buckhannon, WV—Hemming attachment (1893)

Elizabeth B. Look, Boston, MA—Thread-cutter or holder (1893)

Julia E. Snapp, Danville, IL—Thread-cutter for SMs (1893)

Mary Tobener, Gold Hill, NV—Fabric-folding attachment (1894)

Mary Riker, Yonkers, NY—Binder attachment for SMs (1894)

Therese R. Fischer, Baltimore, MD—Threader for SM needles (1894)

[Wm. G. Cummins and] Emma A. Monro, McMinnville, TN—Trimming attachment for SMs (1894)

# *Appendix M-2:*
# *Women Patentees of Dishwashing Tools and Machines (U.S.), 1870–Early 1895*

Abbreviations:
DW = Dish-washer
DWM = Dish-washing machine
DCM = Dish-cleaning machine
* = Not on LWP

**Mary Hobson,** Williamsburg, MO—Improved DW (1870)
**\*Glory Ann Wells,** Luzerne, NY—DW (1870)
**Harriet C. Robertson,** Saginaw, MI—Improvement in DWs (1870)
**Mary E. Lewis,** Mansfield, OH—Improvement in DWs (1871)
**Catharine Woodruff,** Antioch, CA—Improvement in DWMs (1872)
**Charlotte H. Sterling,** Gambier, OH—DW (1872)
**Maria E. Whiteside,** Polo, IL—Improvement in DWs (1875)
**Julia J. Hoffman,** Garnett, KS—Improvement in DWs (1877)
**Elizabeth W. Letts,** Ripon, WI—Improvement in DCMs (1877)
**Emma S. Hampton,** Detroit, MI—Improvement in DWs (1878)
**Althea F. Whitney,** Cleveland, OH—Improvement in DWs (1879)
**Lucinda Warren,** Providence, RI—Improvement in DWs (1879)
**Margaret A. Brass,** Juniata, NE—DW (1879)
**Sallie J. Anderson,** Bartlett, TN—DWM (1882)
**\*Sally M. Fenton,** Salinas, CA—DW (1885)
**Phebe E. Cox,** Readington, NJ—DWM (1887)
**Lina Sloan,** Wassau, WI—DW (1887)
**Florence S. Miller,** Hoopeston, IL—DW (1889)
**Virginia C. Caradine,** Sherman, TX—Apparatus for washing and draining dishes (1890)
**Catharine M. Bryan,** Sharon, IA—Dish-washing and -straining machine (1890)
**Margaret A. Wilcox,** Chicago, IL—Comb. clothes-and dish-washer (1890)
**Harriet E. Davis,** Topeka, KS—DW (1890)
**Grace M. Hawes,** Birchardville, TX—DW (1890)
**Addie D. Miller,** Portland, OR—DWM (1890)
**Mahala J. Parmley,** Footeville, WI—DW (1891)
**Florence S. Miller,** Sibley, IL—DW (1891)
**Huldah A. Shepard,** Nelsonville, OH—Comb. steam-cooker and DW (1891)
**Eliza A.H. Wood** and **Minnie W. Gordon,** Bloomfield, FL—DW (1892)
**Mary A. Crosby,** Durand, WI—DW (1892)
**Mary A. Crosby,** Durand, WI—Dish-cleaner (1894)
**Belle Yancy,** Carlinville, IL—Dish-cleaner (1895)

# Appendix M-3:
# Women Patentees of Washing Machines and Related Inventions (U.S.), 1862–94

Abbreviations:
WM = Washing machine
CW = Clothes-washer
CD = Clothes-drainer
CS = Clothes-strainer
W = Wringer
NYC = New York City
* = Not on LWP

**Ellen B. Boyce,** Green Point, NY—WM (1862)
**Ariadna B. Mercier,** Providence, RI—Improved CW (1866)
***Charity Pendleton,** Iowa City, IA—WM (1867)
**Adelia Waldron,** San Jose, CA—Improved WM (1867)
**Caroline F. Fleming,** Belleville, IL—Improved WM (1868)
**Eliza D. Hunt,** NYC—Improved WM (1868)
**Anna M. Smith,** Pittsburgh, PA—Improved WM (1870)
**Sarah J. Clark,** Richmond, IN—Improvement in WMs (1871)
**Jessie H. Murray,** Kirkwood, NY—Improvement in WMs and Ws (1871)
**Eliza A. Turnbull,** Springfield, OH—Improvement in WMs (1872)
**Elizabeth B. Hull,** Clifton, IL—Improvement in WMs (1872)
**Martha E. Binford,** Jackson, TN—WM (1872)
**Semele Short,** Cincinnati, OH—Improvement in WMs (1873)
**Mary A. Barnes,** Olympia, WA Terr.—Improvement in boiler WMs (1874)
**Mary Ann Johnson,** Washington, DC—Improvement in Ws (1875)
**Sara Lindsley,** Collins, NY—Improvement in WMs (1876)
**Rosa Heilmann,** NYC—Improvement in WMs (1876)
**Julia C. Smith,** Ashton, IL—Improvement in boiler WMs (1879)
**Emily A. Clark,** Alexandria, NE—WM (1880)
**Charlotte C. Howell,** Kent's Station, IN—WM (1881)
**Maggie Bigham,** Pontotoc Co., MI—WM (1881)
**Anna M. Dennison,** Chicago, IL—CS (1882)
**Eliza J. Whitlow,** Mexico, MO—WM (1882)
**Rachel P. Smith,** Chicago, IL—WM (1882)
**Mary J. Bridges,** Wimberly Mill, TX—Comb. CW and churn (1883)
**Sophia Steelman,** Somers Point, NJ—WM (1883)
**Matilda G. [and Andrew J.] Guffin,** Rushville, IN—WM (1885)
**Sarah Sewell,** Mark Center, OH—Comb. WM and teeter or seesaw (1885)
**Sabina W. Cook,** Dayton, WA Terr.—WM (1886)
**Lucinda N. Senter,** Red Bluff, CA—WM (1887)
**Louisa Reinhardt,** Louisville, KY—WM (1888)
**Rosa Joel,** NYC—WM (1889)
**Johanna Starck,** Englewood, IL—WM (1890)
**Margaret A. Wilcox,** Chicago, IL—Comb. Dish-washer/WM (1890)
**Johanna Anderson,** Braceville, IL—WM (1890)
**Pamelia Downing,** Bay City, MI—WM (1890)
**Mary Lumley,** Versailles, MO—WM (1891)
**Annie L. Eversmeyer,** Laporte City, IA—CD (1891)
**Frances S. Dowell,** Eureka Springs, AR—WM (1891)
**Mary A. Jack,** Braceville, IL—WM (1892)

Sallie F. Gibson, Princess, KY—WM (1892)
Marian R. Gibbs, Cherry Creek, NY—WM (1892)
Lucinda J. Wilson, Hardy, TX—WM (1892)
Marcia McKeever, Promise City, MO—WM (1893)
Mary A. Marks, Toledo, OH—WM (1893)
Margaret E. Churchill, St. Joseph, MO—WM (1893)
Louise Kelly, Rosebank, NY—WM (1894)
Eliza C. Burt, Detroit, MI—WM (1894)
Mildred M. Lord, Milwaukee, WI—WM (1894)

# Appendix M-4:
# Electrical and Related Inventions, 1882–99

**Elizabeth Morey,** NYC—Comb. electric lamp and generator (1882)
**Ida C. Himmer,** NYC—Electric battery (1884)
**Ida C. Himmer,** NYC—Secondary electric clock (1884)
**Mary E. Shaffer,** NYC—Noninductive electric cable (1884)
**Mary McMullin,** Middlesex, Eng.—Electric brush (1884)
**Ella N. Gaillard,** NYC—Electric illuminating apparatus (1886)
**Ida C. Himmer,** NYC—Electric winding device for clocks (1886)
**Nina F. Whitney,** Columbus, OH—Electric alarm clock (1894)
**Maria R. Hirsch,** Milwaukee, WI—Commutator brush (1895)
**Eliza Pettet,** Farmer City, IL—Carbon contact (1899)

# Appendix M-5:
# Other 20th-Century Machines:
# Military Machines

WB/28 has a class, XV, for Ordnance, firearms, and ammunition. Women have been recruited as munitions workers in several wars since the factory system arose, and women's participation in war has been underreported in Western history at least since the Crusades. A more balanced view of women and war, long overdue, will probably soon begin to emerge from current research. However, if American women's patents 1905–21 are any indication, war will not be shown to be a major preoccupation of women. The following 22 inventions (17 types) constitute just .4% of the total of women's inventions for the ten chosen years, even though the study period encompassed World War I:[1]

| | |
|---|---|
| Automatic pistol | Primer |
| Bomb-launching apparatus | Railway torpedo |
| Cane-gun | Rear sight for guns |
| Cartridge tube filler | Resilient missile |
| Flashlight attachment, firearms | Single trigger mechanism |
| Front sight for firearms | Submarine mine |
| Incendiary ball | Top for powder cans |
| Ordnance | Woven cartridge carrier |
| Percussion and ignition fuse | |

I have not located any of these inventors. One Lea W. Rockwell of Washington, DC patented a Percussion and ignition fuse in 1920, but Lea turns out to be male.

During World War II, Hollywood actress **Hedy Lamarr** patented a radio guidance system for torpedoes (1942 ) with composer George Antheil. Those who would discount this achievement should recall that Hedy was once married to Fritz Mandl, owner of the largest munitions works in Austria, and had been present when he and various experts discussed new devices (Antheil, 330). The patent was granted to Miss Lamarr under her real name, Hedy Kiesler Markey, for a "Secret communication system." The patent was classified as secret because the device had actually been built, and the Navy was testing it with the intent of using it.

Also during World War II, an unsung black inventor named **Henrietta Bradberry** invented and patented a Torpedo discharge means in 1945. In 1943 she had patented a Bed rack (*not* for a torpedo bed, but one of the human-bedroom variety). This diverse pair of patents suggests a versatile mind whose owner we would like to know better. Patricia Ives lists Bradberry among Black Patentees (119) in her 1980 article on black and women patentees. Presence on the Ives listings often indicates an invention of some importance, but Ives provides no details. Bradberry lived in Chicago.

**Gina Piediscalzzi** of Burbank, CA, was a childhood "inventor" of military machines. At age eight she read in one of her father's magazines that the Army was seeking new weapons designs. She drew her idea for a military vehicle, vaguely resembling a bulldozer. The officers who received her letter, realizing the sketch was the work of a child, responded with a warm letter, saying they would keep her idea on file. Frustrated in the mid-1970s by her lack of technical know-how, Ms. Piediscalzzi had joined Inventors Workshop International and was studying electronics and plastics technology. She has several inventions to her credit (Weisberg, L-8/16/ 76).

Belonging here as well are the several women scientists and engineers mentioned in the section on electrical/mechanical technology, whose inventions were intended or in any case used for modern guided-missile technology, such as the Soviet and American satellite-tracking specialists **Alla Masevich** and **Marjorie Rhodes Townsend,** and the American miniaturization specialist **Dr. Edith Olson.**

---

1. Fred Amram dates all these inventions to "during World War I alone" (e.g. "Woman's Work . . . ," 14–5; cf. also "Innovative . . . ," 12). Actually, though the WB/28 wording is confusing just here, and of course many of them were inspired by the War, these inventions come from the same period, 1905–21 inclusive, as all the other inventions discussed in the report.

During World War II, **Mabel Rockwell** worked with the Navy. Her work involved research and development on special instrumentation for underwater propulsion systems and submarine guidance (SWE, . . . *Technical Contributions,* 28). Dr. Rockwell later worked on the Polaris missile as well, but eventually left the project as she became opposed to war. The inventions she must inevitably have created in these two periods of work were probably never patented because they would have been secret.

An outstanding example not previously mentioned is **Dr. Irmgard Flügge-Lotz** (1903–74). Born and educated in Germany, where she was encouraged in her interest in technical subjects both by her mother, whose family had worked in construction for generations, and by her father, a mathematician. From watching zeppelins being tested as a child, she graduated to a specialization in aeronautics, entering the Technische Hochschule of Hanover in 1923. She was often the only woman in her classes. Taking her doctorate in engineering in 1929, she began almost at once to do distinguished work with one of the founders of modern aeronautics. As early as 1931 she developed the Lotz Method of calculating the distribution of lift in aircraft wings of diverse shape, which became adopted throughout the world.

Later she became interested in automatic flight controls for aircraft, and made important contributions to discontinuous or "on-off" control systems. Such controls have the advantage of simplicity: having only two or three input settings, they are simple and inexpensive to manufacture, and sturdy and reliable in service. Although Dr. Flügge-Lotz's work was mainly theoretical, as John Spreiter of Stanford reminds us, "a theory for their performance had to be developed before design could be undertaken with confidence" (2).

By 1938 she had risen to be Head of the Department of Theoretical Aerodynamics at Göttingen University. After the Nazis rose to power, however, her husband Wilhelm Flügge was denied promotion because of his anti-Nazi views; she was doubly discriminated against as anti-Nazi and female. When she and her husband came to the United States in 1948, he was immediately made a professor of engineering, whereas she was offered only a lectureship. She had to wait until 1960 before becoming the first woman full professor in the Engineering Department at Stanford University (Spreiter, 2):

> By the middle 1950s, it seemed evident to almost everyone at Stanford that Dr. Flügge-Lotz was carrying on all the duties of a full professor, but without official recognition. In fact, it was hard for students to understand why she was a Lecturer rather than a Professor. . . . The same question arose on the international scene in the summer of 1960 when she was the only woman delegate from the United States at the First Congress of International Federation of Automatic Control in Moscow. By then, the disparity had become apparent to all. Before school opened for autumn quarter, she was appointed to the rank of full Professor in both Engineering Mechanics and Aeronautics and Astronautics.

Dr. Flügge-Lotz published two books, both on automatic controls, and more than fifty scientific papers. In 1970 she was honored by the Society of Women Engineers with its Achievement Award. She became a Fellow of the American Institute of Aeronautics and Astronautics (second woman so honored) and a senior member of the Institute of Electrical and Electronic Engineers. Some of her work was applicable to satellite control, which of course includes rockets and missiles, and she has been called the female Werner von Braun (Hunt, PC/1980; *NAW*/mod.; Malotke, 21; Peden, 271–2; Spreiter; SWE Award Booklet, 1970;

A contemporary example is **Merri Wood** (c. 1946–   ) of Los Alamos, featured in a 1985 article on the "Bomb Makers." The article, published on the 40th anniversary of the bombing of Hiroshima, called Wood a "thermonuclear weapons designer." Naturally, no details of her work are available, as the projects are classified, but it is worth noting that she appears on the cover of the magazine that ran the article (*Discover,* August 1985), which would suggest some importance for her role at Los Alamos (Mohs, cover, 29–30).

# Appendix M-6:
# A Sample of Women's Horse Transport and Related Inventions, U.S. Patents, 1861–94

\* = Not on LWP

**Sarah J. Wheeler,** New Britain, CT—Curry-comb (1861)
**Clara A. Bartelett,** Ferndale, CA—Improvement in side-saddles (1864)
**Clara A. Bartelett,** Ferndale, CA—Improvement in side-saddles (1866)
**Sarah Ruth,** Philadelphia, PA—Improvement in sunshades for horses (1868)
**Anna M. Bardwell,** Amherst, MA—Carriage-curtain fastening (1872)
**Helen Macker,** Boston—Improvement in amalgams for coating harness trimmings (1872)
**Sarah Ruth,** Philadelphia, PA—Sunshade for horse (1873)
**Ethel O. Harvey,** Lubec, ME—Improvement in clevises (1875)
**Lottie W. Timmens**—Fayetteville, NY—Improvement in snap-hooks (1876)
**\*Jennie H. Spofford,** Philadelphia, PA—Riding saddle (1876)
**Louisa F. Sleeper,** Philadelphia, PA—Improvement in detaching horses (1877)
**\*Louisa F. Sleeper,** Philadelphia, PA—Detaching apparatus for horses (1877)
**Ellen Crebs,** Huntsville, PA—Improvement in flynets for horses (1877)
**Ivy J. Hart,** Chandler, IN—Wagon jack (1878)
**Emma J. Osborne,** Easley, SC—Baggage attachment for vehicles (1880)
**Hannah H. Bowker,** Boston—Unhitching device for stable-stalls (1880)
**Isabella Hillen,** Cincinnati, OH—Stirrup (1881)
**Juliaette W.B. Carpenter,** Medford, MA—Halter (1883)
**Sarah H. Stewart,** Concord, NH—Checkrein-hook (1884)
**Sue E. Smither** (w/ Wm.), Keene, KY—Trace attachment (1884)
**Lydia A. Blood,** Chicago—Tug for harness (1884)
**Leila C. Harrison,** New Haven, CT—Hitching device (1884)
**Lydia A. Blood,** Chicago, IL—Tug for harness (1885)
**Myra L. Eckles,** Northfield, MN—Riding saddle (1885)
**Annah E. Bussell,** Nashua, NH—Rein-holder (1886)
**Sarah J. Hull,** Stella, NE—Rein-muff (1886)
**Sarah M. Hardenbergh,** Le Fever Falls, NY—Carriage-seat (1886)
**Martha B. Hancock,** Syracuse, NY—Wheel attachment for sleighs (1887)
**Julia B. Wood,** Manchester, NH—Draft-hook for vehicles (1888)
**Mary A. Forward,** Kalamazoo, MI—Rein-holder (1888)
**Letitia V. Luce,** New Orleans, LA—Sunshade for vehicles (1888)
**Mattie M. Marsh,** Moscow, ID—Folding step for vehicles (1889)
**Nellie M. Cahoone,** Newark, NJ—Harness pad (1889)
**Mary Ingersoll,** Erie, PA—Check hook for harness (1889)
**Emma Allen,** Freeport, IL—Whip socket (1889)
**Florence J. Conner,** Cincinnati, OH—Trace support (1890)
**Carrie A. Grant,** Cambridgeport, MA—Vehicle axle (1890)
**Eliza A. Nichols,** Stormville, TX—Wagon step (1890)
**Missouri E. Wroe,** Brenham, TX—Trace and trace-chain supporter (1890)
**Clara Turney,** Sausalito, CA—Horse-controlling device (1890)
**\*Matilda M. Busby,** Salt Lake City, UT Terr. [w/ 2 males] Brake (for carriage) (1890)
**Jane C. Havens,** Philadelphia, PA—Rein guide (1890)
**Mary J. McArthur,** Cleveland, OH—Axle lubricator (1890)
**Annie H. Chilton,** Baltimore, MD—Horse detacher and brake (1891)
**Kitina M. Alvord,** Huntington, MA—Draft attachment for sleighs (1891)
**Elizabeth V. Stryker,** Plainfield, IL—Fly net for horses (1891)
**Annie R. Chittenden,** Osceola, IA—Road cart (1891)

**Letitia V. Luce,** New Orleans, LA—Sunshade for vehicles (1892)

**Etta Post,** Oakland, NJ—Carriage step (1892)

**Kate M. Dean,** Memphis, TN—Rein support (1892)

**Fannie L. Gee,** Burnsville, AL—Sidesaddle (1892)

**Hattie L. Philips,** Wisner, MI—Top prop for carriages (1893)

**Sophia Giesecke,** St. Louis, MO—Horse brush (1893)

**Bertha A.J. Block,** Racine, WI—Wood horse collar and manufacture of same (1893)

**Harriet Carmont,** London, Eng.—Starter or brake for vehicles (1894)

**Annie H. Chilton,** Baltimore, MD—Combined horse detacher and brake (1894)

**Mary E. Poupard,** London, Eng.—Sectional horseshoe (1894)

**Mary E. Poupard,** London, Eng.—Horseshoe (1894)

**Mary E. Poupard,** London, Eng.—Segmental horseshoe (1894)

**Lizzie B. Fleming,** Pierce City, MO—Wheel cleaner (1894)

# Appendix M-7:
# Women's 19th-Century Railroad and Related Inventions, U.S. Patents, 1864–94

Abbreviations:
NYC = New York City
* = Not on LWP

**Mary Jane Montgomery,** NYC—Improvement in locomotive wheels (1864)
**Eliza Dexter Murfey,** NYC—16 patents dealing with packings and bearings for railway axles, journals, etc. (1870)
  Material for packing journals and bearings (May 10)
  Material for packings and bearings (July 12)
  *3 patents for a Journal-box (all Oct. 4)
  Impregnating fibrous material for packing, etc. (Oct. 11)
  Saturating fibrous materials with powdered substances for bearings and packings (Oct. 11)
  Material for packing and bearings (Oct. 11)
  *Material for journals and bearings (Oct. 25)
  *Composition for bearings (Nov. 1)
  *Spindle step (Nov. 15)
  *Material for bearings and journals (Nov. 15)
  Mode of preventing the heating of axles or journals (Nov. 15)
  *Material for bearings and journals (Dec. 27)
  *Bearing and packing material (Dec. 27)
  *Manufacture of material for journals, bearings, and packings (Dec. 27)
**Augusta M. Rodgers,** Brooklyn, NY—Improvement in railroad car heaters (1871)
**Carrie R. Laman,** Painted Post, NY—Improvement in lubricating railway journals (1871)
**Augusta M. Rodgers,** Brooklyn, NY—Improvement in conveyors of smoke and cinders for locomotives (Apr. 4, 1871)
**Augusta M. Rodgers,** Brooklyn, NY—Improvement in conveyors of smoke and cinders for locomotives (May 9, 1871)
**Eliza D. Murfey,** NYC—4 patents relating to railway journal packings and bearings (all 1871)
  Improvement in bearings and packings (Aug. 1)
  Improvement in journal-bearings (Dec. 12)
  Improvement in materials for packings and bearings (Dec. 12)
  Improvement in packings and bearings for journals, etc. (Dec.   12)
**Katharine E. Holmes.** Cambridgeport, MA—Improvement in railway-car safety apparatus (1871)
**Betsey Ann Worden,** Scranton, PA—Improvement in car couplings (1872)
**Eliza D. Murfey,** NYC—Improvement in packings for pistons, etc. (1872)
**Augusta Rodgers,** Brooklyn, NY—Improvement in spark-arrestors (1872)
**Sarah Mahan,** Cleveland, OH—Improvement in steps for berths (1872)
**Mary G. Briggs,** Boston, MA—Spring-seat for furniture or cars (1872)
**Claudia B. Turnbull,** Washington, DC—Improvement in street-car awnings (1873)
**Maria L. Ghirardini,** Providence, RI—Improvement in rails for street-railways (1873)
**Harriet B. Devlan,** Jersey City, NJ—Improvement in packings for railroad journal-boxes (1874)
**Clara T. Crane** (w/ Wellsly), Auburn, NY—Improvement in car-axle boxes (1874)
***Eliza D. Murfey,** NYC—Bearing surface for spindles (1875)
***Eliza D. Murfey,** NYC—Preparation for bearing surface for spindles (1875)
**Leonora E. Yates,** Washington, DC—Railway tie (1876)
**Amanda L. Waggoner** (w/ John), Bridgeport, NJ—Improvement in station indicators (1876)
**Maria L. Ghirardini,** Pawtucket, RI—Improvement in chimneys (1876)
**Theursa D. Baldwin,** Albion, MI—Improvement in car-couplings (1877)
**Elizabeth Holt,** Pittsburgh, PA—Improvement in packings for pistons, rods, etc. (1877)

**Laura Jane Gott,** La Grange, OH—Improvement in car-coupling pins (1878)
**Louisa B. Simpson,** Lawrence, KS—Apparatus for destroying vegetation on railroads (1878)
**Sarah A. Haydock,** Ostrander, OH—Improvement in car-couplings (1878)
**Sarah A. Haydock,** Ostrander, OH—Improvement in car-couplings (1879)
**Louisa B. Simpson,** Lawrence, KS—Apparatus for destroying vegetation on railroads (1879)
**Margaret E. Skerritt,** Albany, NY—Improvement in car-steps (1879)
**Mary E. Walton,** NYC—Improvement in locomotive and other chimneys (1879)
**Eleanor A. McMann,** Cleveland, OH—Guard for sleeping-car berths (1880)
**Marie A. Vosburgh,** Kalamazoo, MI—Danger signal (1880)
**Eleanor A. McMann,** Cleveland, OH—Safe-guard for sleeping-berths (1880)
**Eleanor A. McMann,** Cleveland, OH—Guard for sleeping-car berths (1880)
**Christina Beggs,** Allegheny, PA—Car-seat (1880)
**Mary E. Walton,** NYC—Elevated railway (1881)
**Florella L. Kinsman,** Magog, Quebec, CAN—Car-heater (1881)
**Christena Beggs,** Allegheny City, PA—Limb-support for car-seats (1881)
**Nancy P. Wilkerson,** Terre Haute, IN—Cattle-car (1881)
**Annie K. Pentz,** Clearfield, PA—Stock-car (1882)
**Mary Jeffries Whaley,** Lincoln, NE—Covering the slot of cable roads (1882)
**Maggie A. Moran,** Ft. Wayne, IN—Car-coupling (1882)
**Susan L. Sinclair,** Allegheny, PA—Method of filling the recesses in the tread of car-wheels (1885)
**Betsey A. Maxey,** Knoxville, IL—Car-coupling (1885)
**\*Geneva Armstrong,** Elmira, NY—Stock-car (1885)
**Catharine A. Williamson** (w/ Budd Rawlings), St. Louis, MO—Street-car (1886)
**Estella Case,** NYC—Rubber sound-deadener and packing band (1886)
**Cornelia C. Wood,** Sibley, IA—Bunk for railway cars (1887)
**Susan L. Sinclair,** Allegheny, PA—Car-wheel (1887)
**Mary J. Watson,** Sacramento, CA—Street and station indicator (1888)
**Ellen E.L. Woodward,** Chicago, IL—Safety-car (1888)
**Sarah E. Davidson,** Glasgow, KY—Car-coupling (1888)
**Catherine A. Bond,** Seneca, NY—Car-coupling (1888)
**Mary E. Farnham,** New Orleans—Car starter (1889)
**Emily S. Hutchinson,** Peekskill, NY—Car-window screen (1889)
**Harriett B. Devlan,** Jersey City, NJ—Packing for journals (1889)
**Isabella L. Harry** (w/ C.F.R.), West Salem, OH—Car-coupling (1889)
**Mary I. Riggin,** Newark, NJ—Railway crossing gate (1890)
**Alice G. Middleton,** Philadelphia, PA—Transfer apparatus for traction cable cars (1890)
**Catherine L. Gibbon,** Albany, NY—Construction of side-bearing railway tracks (1890)
**Catherine L. Gibbon,** Albany, NY—Construction of railway tracks (1890)
**Marian M. Green,** County Line, TN—Combined railway chair and crosstile (1890)
**Lucy Gaddis,** Gold Hill, NM—Car-coupling (1891)
**Dora Hirsh,** Lancaster, PA—Car-coupling (1891)
**Sarah A. Weathersby,** Killeen, TX—Car-coupling (1891)
**Irene N. Soly,** Montreal, CAN—Station-indicator (1891)
**Catherine L. Gibbon** (w/ T.H.), NYC—Construction of railroad tracks (1891)
**Catherine L. Gibbon,** NYC—Compound rail for railway (1891)
**Ernestine S.W. Rosenthal,** Philadelphia, PA—Draft guard for cars (1891)
**Alice A. Whipple,** Providence, RI—Apparatus for sanding railway tracks (1892)
**Mary E. Cook,** Amity, OR—Railway car stove (1893)
**Mame Lester,** Logansport, IN—Attachment for unloading boxcars (1893)
**Margaret A. Wilcox,** Chicago, IL—Car heater (1893)
**Helen K. Ingram,** Jacksonville, FL—Railroad car (1894)
**Marguerite Maidhof** (w/ Victor E.), NYC—Car fender (1894)
**Minnie McPhail** (w/ Frank Kopicke), Taunton, MN—Car-coupling (1894)

# Appendix M-8:
## Elena Popescu (1887–1944)

*The following sketch was written by her son, Commandant Constantin Copaciu, of Monaco, in 1981. It is based partly on the diary she kept from age nine, much of which was regrettably lost during World War II.*

The exterior life of my mother was very simple [she married an Army lieutenant and had one child]: her inner life, however, is more difficult to describe. She was aloof, implacable and isolated in her own world. She was disapproved of by her parents and many other people, who considered her too liberal and too intellectual for the pragmatic society in which she lived.

She loved music from the time when, as a child, she got a "phonograph"—one of the first in the world—as a gift from her eldest brother-in-law. "She doesn't like to play the piano, but she spends all day long with her horrible mechanical music," said her mother. However, the uncertainty of this new musical gadget led her to feel the ambiguity which lies at the core of music, blurring the line between sentiment and real feelings; between the ironies of comedy and the pathos of drama.

Many years later, she taught her niece about "Mozart's divertimento, Wagnerian complexity, the strength of Beethoven and the delicate touch of Chopin and others such as Haydn and Handel." This I remember from a letter my wife sent me during the Second World War. In another letter my wife wrote, "Last night I found Aunt Elena doing her embroidery—she was very skilful in lace-making—by moonlight and the light of the anti-aircraft fire." This was during the blackout when Bucharest was bombed.

"In a peaceful environment, surrounded by sisters and brothers, under the protection of my parents in a lovely patriarchal country, my youth was full of romanticism and beauty, but beyond the gates of my wonderful house, every evening at sunset, I saw the crowds of blackened people going home," she wrote. This contact with industrial ugliness impressed her deeply in contrast with the fairy-tale world in which she lived. This could be the origin of her early rebellion against the importance which her parents and relatives attached to conservatism and tradition. "I never understood their inability to perceive moving things, which are the sense of life," she wrote in her diary.

In her innocence of life she could not understand why her mother was so horrified by "this child which is no girl and not even a boy" (her mother's words). "I did not appreciate today my mother's remarks about my visits to the Gypsies," she noted in her diary. She was full of courage and loved tight-rope risks and defying conventions. The marvelous and interesting pages about Gypsy life she wrote in her diary gave a lot of information about these fascinating people who, at that time, wandered throughout Europe: their folklore and mythology, which goes back centuries; beautiful women wearing colorful costumes; their dancing; predicting the future and preparing love potions she found fascinating.

At the age of 19 she evolved a distinctive appearance and life style which were very clearly reflected in her journal. Her appearance was the reverse of what her parents thought a young lady ought to look like. One way to find emancipation and liberty and "maybe happiness" was to conform to the rules by marrying whilst at the same time avoiding the system, which would have obliged her to marry the man of her parents' choice. She was of course supported by her younger brother. She had to fight, but in the end won, and married the man she wanted.

"Nicolas [Copaciu, her husband] sometimes astonishes people with his lack of warmth, but to me he is very different," she wrote somewhere; and later on, "We of course have rows, but there is never any threat to our marriage because it never grows stale, each of us being tolerant and admiring of the other's audacity." Her husband was a daring soldier and a great soul. It would be necessary to read his letters from the trenches of the First World War to understand how deeply their love was rooted. They were different from each other, but they formed a single being.

During the withdrawal of the Army in the battle for Rumania in 1916, her husband, who was in charge of an ammunition train, considered his position too comfortable for a soldier and left the command of the train to his wife, to go and fight in the front lines.

That was the beginning of a saga. A young woman commander of a unit behind the front lines, knowing nothing of Army affairs, was not common. She wrote in her diary, "Today was a very exciting and funny day. I was asked to go before the Russian Army Staff, five kilometers from Iv——, a small Rumanian village. Yesterday we saw brave Cossacks—'Look, Mama, like my toy soldiers,' cried my child—with their splendid band music, marching towards the front, only several kilometers away. We admired this proud and beautiful Army marching in formation on the snowbound road, filling us with hope that the Rumanian Army and our country would be saved. . . ."

And farther on: "I took Tic [her son] with me and, conducted by a peasant in his little coach, I went to the Staff. I was received by the mocking and amused glances of the young officers who introduced me and my son to the General and two officers, who very politely invited me to sit down and offered me tea, cakes, and French cognac, and to Tic a box of candy and a beautiful little Orthodox icon. I felt as if I were at a ladies' party in Bucharest.

"Abruptly the general said that he had been informed that there was a Rumanian ammunition train not far from his Army and had asked to see the person in charge—presumably an officer—and was therefore very surprised to be confronted by a woman! I was sure that he knew the situation perfectly well. He asked me about my rank in the Army and whether there are many Rumanian women who fight in the front lines. I replied that my husband asked me to replace him whilst he fought in the front lines, and that I know that by his side there now fights a very young girl who volunteered at the beginning of the war. And I added that I am very proud of her because she is heroic and I believe she will be the Jeanne d'Arc of Rumania! The General silently studied me for a moment and, laughing, he said, 'Are you not jealous?' 'Why?' I said. 'I am sure she will soon be my best friend.'

"The General then said he wanted to know whether I could supply his Army with ammunition if necessary. I replied, 'Of course, but only with the approval of the Rumanian Army Staff. Even my husband could not obtain ammunition from "my" train,' I added, without approval.' An echo of laughter passed through the room."

From the beginning of the war until the end, the narrative spins along, alive with descriptions of war-time exploits. It is an original, sentimental, serious, and often very funny picture.

In fact, it is in the area of war, heroism and absurdities—an area of exceptions—that she formed her inner conviction and developed her character. I often felt her to be a mixture of Scarlett O'Hara and the mother of Pearl Buck. She was so much a law unto herself, so honest, so innocent, so idealistic, and so courageous that she tended to be taken for granted. So it comes as no surprise that she settled with a disturbing blandness for telling the whole truth, as far as she could see it, and nothing but the truth. Her journal was essential for her existence and self-understanding.

Her social energy, like her dedication to the tasks in the front lines as well as squalid work in hospitals during the war, and her attitude and emotion towards the hardship of workers and peasants between 1910 and 1920—the passion of the protest meetings and the behavior of the Government, the poverty of the children and the parents' fierceness yet loyalty in the fight against the invaders—alarmed her family and enchanted many people, generally poor and unknown.

Her family, which never understood the little girl, failed again to understand the woman of the 1930s. [Nevertheless] respecting and loving them, she continued to be the least conservative of them. Only her little brother ever knew that she got typhus in the war hospital and was saved only by the miracle of a Russian doctor. She was in a coma for many days, and another of my boyhood [memories] is that when she woke from her long sleep she looked at me and her beloved brother, . . . realizing from whence she returned, [and] I saw for the first and only time the light of tears in her eyes.

That illness weakened her for the rest of her life. However, her social [concern and work] continued as before, with discretion, simplicity, and charitableness. We were astonished after her death to receive a huge number of letters from families and individuals who had benefited from her help. Her aid was very diversified; she helped poor workers, peasant families, and old people, and pleaded with the Army Staff during the Second World War for young recruits and only sons of poor widows.

When one of her older sisters died, she considered it her duty to bring up her pretty, orphaned niece. She did not listen to family opinion but thought only of the girl, who, with grace and innocence, expressed her willingness to go with the Aunt of her dreams. Torn between her devotion to her sick husband—to die before long—and [anxiety] for her son always on the seas or on foreign missions, she made of this girl a fighting little woman, a friend and confidante. And later on, when her son fell in love with the girl, she fought once more against the family's disapproval and found herself at last alone, her beloved brother having died a year before.

When she died after much illness at the age of 57 on 5 September 1944, my wife alone (at the express wish of my mother) attended her funeral in . . . Vienna, not far from the tomb of Beethoven and the Memorial of Mozart. Her last wish for "no sad songs and no roses on my head" was respected.

She had loved flowers. I remember how on her deathbed she looked at a bunch of roses and whispered the splendid verses of Paul Valéry: "le beau en ce qui passe, attestant ce qui dure, donne à 'ma vie même' un sceau de l'éternité." I was surprised that she substituted "à la fleur même" with "à ma vie même," but I did not dare to correct her. Was she thinking of flowers, or perhaps of her own life? I don't know.

# Bibliography

## ABBREVIATIONS USED IN TEXT REFERENCES AND BIBLIOGRAPHY

| | |
|---|---|
| AAUW. | See American Association of University Women |
| AJPH: | *American Journal of Public Health* |
| AMWS: | *American Men and Women of Science;* early editions *AMS* (*American Men of Science*) |
| AW. | See *American Women* titles |
| AWG: | *American Woman's Gazetteer;* see Sherr & Kazickas |
| AWIS: | American Women in Science |
| AWS: | *American Women of Science;* see Edna Yost |
| CAB: | *Cyclopedia of American Biography* |
| CB: | *Current Biography;* numbers refer to edition year |
| CE: | *Columbia Encyclopedia;* numbers specify editions |
| c.v./: | curriculum vitae (of inventors discussed); year after slash refers to date of version sent me |
| DAB: | *Dictionary of American Biography* |
| DHEW/ | |
| CHS: | Dept. of Health, Education, and Welfare, Center for Health Statistics |
| DNB: | *Dictionary of National Biography* (British) |
| DSB: | *Dictionary of Scientific Biography* |
| EA: | *Encyclopedia Americana* |
| EB: | *Encyclopaedia Britannica* |
| I-: | Interview; date following dash is date of interview |
| IEEE: | Institute of Electrical and Electronic Engineers |
| L-: | Letter(s); date(s) or year range(s) given |
| LD/00: | "Women as Inventors," *Literary Digest* |
| LRG: | Learning Research Group |
| LSS: | Letter to the Secretary of State (early U.S. patent reports) |
| LWP: | List of Women Patentees; see United States Patent and Trademark Office, *Women Inventors to Whom . . .* |
| MPH: | Master's in Public Health |
| NASA: | National Aeronautics and Space Administration |
| NAW: | *Notable American Women, 1607–1950* |
| NAW/ | |
| mod.: | *Notable American Women: The Modern Period* |
| NCAB: | *National Cyclopedia of American Biography* |
| NCW: | *New Century for Women* |
| NOW: | National Organization for Women |
| NPR: | National Public Radio |
| NSF: | National Science Foundation |
| NUC/ | |
| Pre- | |
| 1956: | *National Union Catalog of Pre-1956 Imprints* |
| NYT: | *New York Times* |
| OED: | *Oxford English Dictionary* (13-vol. ed.) |
| OTAF: | Office of Technology Assessment and Forecast |
| Q-: | Inventor's questionnaire completed and sent me by inventor; date refers to year completed |
| PC-: | Personal communication, including phone calls, postcards, letters, conversations, mainly from inventors; date is date of sending or occurrence |
| PDR: | *Physicians' Desk Reference* |

PRI:            Psoriasis Research Institute
RCA:            Radio Corporation of America
SF:             San Francisco
*SN:*           *Science News*
SWE:            Society of Women Engineers
SWIA:           Swedish Women Inventors' Association
USDA:           United States Department of Agriculture
VOIR.           See *Vsesoyuznoye . . .*
*WA:*           *Women of Achievement;* see Field *et al.,* eds.
*WABI:*         *World Almanac Book of Inventions;* see Giscard d'Estaing
WB/28,
/43, / 75.      See United States Department of Labor, Women's Bureau
*WBWRA:*        *Women's Book of World Records and Achievements;* see Lois Decker O'Neill, ed.
*WEMS:*         *Woman's Encyclopedia of Myths and Secrets;* see Barbara Walker, ed.
WI/1 and
/2:             *The Woman Inventor;* see Charlotte Smith
WIPO:           World Intellectual Property Organization; see Swedish Women Inventors' Association
*WMS:*          *Women of Modern Science;* see Edna Yost
*WW:*           *The Working Woman*
*WWA:*          *Who's Who in America;* number refers to year
*WWCDP:*        *Who's Who in Computers and Data Processing*
*WWCF:*         *Who's Who in the Computer Field*
*WWW:*          *Who's Who in the West*
*WWWA:*         *Who Was Who in America*
WWWNA
AUTHORS:        *World Who's Who of North American Authors*
*WWWS:*         *World Who's Who in Science . . .*

A.D., Letter to the Editors *re* Julia Colman, "Are Tin Fruit Cans a Source of Metallic Poisoning?" *Scientific American,* Aug. 20, 1870; cited in Shenton, 116.

Abbot, Walter S., and William E. Harrison, comps, *The First One Hundred Years of McKeesport.* McKeesport, PA: The Centennial Historical Committee, 1894.

Abbott, Willis J., and John C. Winston, *Notable Women in History.* London: Greening & Co., 1913.

"Abortion Statistics," *Science News* 119: 15 (Apr. 11, 1981): 236.

Adamczyk, Alice (Schomburg Center for Research in Black Culture, New York, NY), L-5/21, 6/22, 9/17/81.

Adams, James L., "Big Kids' Toys," *Stanford Magazine* 16: 2 (Summer 1988): 40–4.

Adams, Richard E.W., "Rio Azul," *National Geographic* 169: 4 (Apr. 1986): 420–65.

Adams, Russ, "Women in Computing," *Interface Age* 8: 12 (Dec. 1983): 78ff.

Adamson, Colin, "The Penetration of Invention into Society," *The Inventor* (London), Apr. 1971: 3–11.

Adelson, Suzanne, "Kids Aren't So Lonely in San Diego . . . ," *People* 14: July 14, 1980: 106f.

"Airborne Recipes," *Time,* June 7, 1948: 61–2.

Airy, Emery, "Business Memories," in Opal Eckert and Leola Sweeney, eds, *Tales of Nodaway County.* Maryville, MO: Nodaway County Historical Society, 1977.

Albaum, Gerald, *The Independent Inventor,* Explorations in Invention and Innovation Series III. Eugene, OR: Experimental Center for the Advancement of Invention and Innovation, University of Oregon, 1976.

Aldige, Julie, "Jazzercise: Dancing to Stay Fit," Portola Valley, CA, *Country Almanac,* June 27, 1984: 18.

Aldington, Richard, and Delano Ames, rev., *Larousse Mythologie Générale,* Felix Guirand, ed. London: Hamlyn, 1959.

Alexander, Marianne E., "Anna Eugenia Emma Schneider, 1889–, Flour Miller," in Helmes.

Alic, Margaret, *Hypatia's Heritage: A History of Women in Science from Antiquity to the Late Nineteenth Century.* London: Women's Press, 1986.

Alic, Margaret, "Women and Technology in Ancient Alexandria: Maria and Hypatia," *Women's Studies International Quarterly* 4: 3 (Fall 1981): 305–12.

Allen, Henry, "The Patent Truth of a Bygone Age," *Washington Post,* May 11, 1990: D1, 4.

"Allene R. Jeanes: Garvan Medal," *Chemical and Engineering News,* Apr. 23, 1956: 1984.

Allison, Marvin J., "Paleopathology in Ancient Peru," *Natural History* 88: 2 (Feb. 1979): 74–82.

"Alpha Wave of the Future," *Time,* Jan. 4, 1974: 22.

Alphandery, Marie Fernande, *Dictionnaire des inventeurs français.* Paris: Editions Seghers, 1962.

Alva, Walter, et al., "Splendors of the Moche . . . ," *National Geographic* 177: 6 (June 1990): 2–47.

Amariani, John F., *Dictionary of American Food and Drink.* New Haven, CT: Ticknor & Fields, 1983.

American Association of University Women, *Taking Hold of Technology: Topic Guide for 1981–83.* Washington, DC: AAUW, 1981.

*American Biographies,* Wheeler Preston, ed. New York: Harper & Brothers, 1940; repr., Detroit: Gale, 1974.

*American Inventor* magazine (Cupertino, CA), 1977–.

*American Men and Women of Science,* ed. Jaques Cattell. New York: Bowker, 1906–.

*American Women: The Standard Biographical Dictionary of Notable Women,* 3 vols. (1935–6, 1937–8, 1939–40). Los Angeles: American Publications, 1936–9.

*American Women, 1935–1940: A Composite Biographical Dictionary.* Detroit: Gale, 1981.

"America's First Straw Bonnet," Norwood, MA, *Messenger,* Sept. 29, 1927.

Ammons, Virgie M. (Inventor, Eglon, WV), L-1/4/82; PC, Spring 1983.

Amram, Fred, "The Innovative Woman," unpublished paper sent by author, 1981.

Amram, Fred, "The Innovative Woman," *New Scientist,* May 24, 1984: 10–2.

Amram, Fred, "Invention as Problem-Solving: Special Contributions of Female Inventors," paper presented at the National Women's Studies Association Conference, Madison, WI, 1988.

Amram, Fred, "Woman's Work Includes Invention," *The Woman Engineer* (New York) 1: 4 (Spring 1981): 29–33ff.

Amram, Fred, *et al.,* "Her Works Praise Her: An Exhibition of Inventions by Women," Goldstein Gallery, University of Minnesota, Minneapolis, June 12–Dec. 16, 1988.

"An Ancient People's Antibiotic," San Francisco *Chronicle,* Sept. 22, 1980: 11.

Andersen, Robert N. (Researcher, Rensselaer Co. Historical Soc., Troy, NY), L-10/26/81, 4/7, 4/30/87.

Anderson, Bonnie S., and Judith P. Zinsser, *A History of Their Own: Women in Europe from Prehistory to the Present.* New York: Harper & Row, 1988.

Anderson, John, "The Dark Crystal," *Creative Computing,* Mar. 1983: 168–74.

Anderson, Mary, ed., "Women's Contributions in the Field of Invention," U.S. Dept. of Labor, Women's Bureau, Bulletin No. 28, Washington, DC, 1923.

Andrassy, Stella, "Solar Water Heaters," United Nations Conference on New Sources of Energy, Rome, Aug. 21–31, 1961. *Proceedings* 5 (Solar Energy, II). New York: United Nations, 1964.

Andrewes, Christopher H., *The Natural History of Viruses.* New York: Norton, 1967.

Angluin, Dana, "Lady Lovelace and the Analytical Engine," *Newsletter of the Association for Women in Mathematics* 6: 1 (1976): 5–10, and 6: 2 (1976): 6–8.

Anon. ["An American Physician"], *Reproductive Control; Or, A Rational Guide to Matrimonial Happiness* . . . Cincinnati, 1855; repr., 1974, in Rosenberg & Rosenberg.

"Another Computer Triumph: Ultrasound Tissue Characterization," *Stanford University Friends of Radiology Newsletter,* Winter 1982.

Antheil, George, *Bad Boy of Music.* Garden City, NY: Doubleday, 1945.

"Antibodies Tie Up Malaria Parasite," *Science News,* Feb. 9, 1959: 106.

*Appleton's Cyclopaedia of American Biography,* 6 vols. New York: D. Appleton & Co., 1888–9; repr., Detroit: Gale, 1968.

Aptekar, Herbert H., *Anjea: Infanticide, Abortion, and Contraception in Savage Society.* New York: William Godwin, 1931.

Archambault, JoAllyn, "Women Inventing the Wheel," in Zimmerman, ed., *Future* . . .

Ardrey, Robert L., *American Agricultural Implements: A Review of Invention and Development.* . . . Chicago: Published by the author, 1894.

Arehart-Treichel, Joan, "Tumor-Associated Antigens: Attacking Lung Cancer," *Science News* 118 (July 12, 1980): 26–8.

Arehart-Treichel, Joan, "Vitamin C for the Cervix," *Science News* 123: 2 (Jan. 8, 1983): 23.

Arehart-Treichel, Joan, "Winning and Losing: The Medical Awards Game," *Science News,* Feb. 24, 1979: 120f.

Aristophanes, *Lysistrata,* in C.A. Robinson, Jr., ed., *An Anthology of Greek Drama.* New York: Holt, Rinehart, 1965 (1949).

Armitage, Sue, "History of Women and Technology," in Judy Smith, ed., 17–9.

Arthur, Julietta K., "Noted Wives Remind Us . . . ," *Independent Woman* 18: 10 (Oct. 1939): 314–5ff.

Arthur, Marylin, " 'Liberated' Women: The Classical Era," in Bridenthal & Koonz, eds.

Asbury, Herbert, *Gem of the Prairie: An Informal History of the Chicago Underworld.* New York: Knopf, 1940.

Ashbee, Henry Spencer, *Index Librorum Prohibitorum,* London: Privately printed, 1877.

Ashton, James, *The Book of Nature.* . . . New York: Brother Jonathan Office, 1865; Arno repr., 1974.

Askins, Barbara S., "Autoradiographic Image Intensification: Applied in Medical Radiography," *Science,* Feb. 10, 1978: 684–5.

"Athens Woman Pleased with Popularity of Her Invention," Athens, OH, *Messenger,* Fri., Apr. 2, 1965: 6.

"Atlantic Flight to the West . . . A Woman Passenger," *The Times* (London), Sept. 1, 1927.

Auel, Jean (Author, Portland, OR), PC-1981.

Auel, Jean, *Clan of the Cave Bear.* New York: Crown, 1980.

Austern, H. Thomas, and Harry Mussman, "Container-Contributed Lead as a Part of Environmental Exposure to Lead," *Agriculture and Human Values* 3 (Winter-Spring 1986): 157–70.

Australian Broadcasting System series, "The Inventors," 1978–9.

[Averell, Georgene H.], Obit., *New York Times,* Aug. 28, 1963.

Avery, Catherine, ed., *The New Century Classical Handbook.* New York: Appleton-Century-Crofts, 1962.

Ayensu, Edward, "The Beginnings of Agriculture," in Doster *et al.,* eds.

Ayrton, Hertha, *The Electric Arc.* London: "The Electrician" Printing and Publishing Co., 1902.

Ayrton, Hertha, "The Origin and Growth of Ripple Mark," *Proceedings of the Royal Society* A, vol. 84, 1910.

Azzi, Robert, "Damascus: Syria's Uneasy Eden" *National Geographic* 145: 4 (Apr. 1974): 512–35.

Babbage, Charles, *Passages from the Life of a Philosopher.* London: Longman, Green, 1864; repr., Farnsborough, Eng.: Gregg International Publishers, 1969.

"Back to the Future: Big Science at Cal," *California Monthly* 97: 5 (Apr. 1987): 13.

Backus, John, "The History of FORTRAN I, II, and III," in Wexelblat, ed.

Bailey, L.H., ed., *Cyclopedia of American Agriculture.* New York: Macmillan, 1910.

Baker, Ernest J., "Betsey Metcalf and the First American Straw Bonnet," *The Hat Industry,* June 1923: 59–62.

Baker, R., *New and Improved: Inventors and Inventions That Have Changed the Modern World.* London: British Museum Publications, 1976.

Baldwin, Catherine, "Taking the Sting Out of Allergies," *Stanford Magazine,* Centennial ed., Sept. 1991: 178.

Baldwin, Richard S., *The Fungus Fighters: Two Women Scientists and Their Discovery.* Ithaca, NY: Cornell University Press, 1981.

Baldwin, Robert, PhD (Biochemistry Dept., Stanford University), PC- 1980.

Ballard, Philip, and Roberta A. Ballard, "Glucocorticoids in Prevention of Respiratory Distress Syndrome," *Hospital Practice,* Sept. 1980: 81–7.

Ballard, Roberta A., MD (Chief, Dept. of Pediatrics, and Dir., Maternal/ Child Health Services, Mt. Zion Hospital and Medical Center, San Francisco, CA), L-12/9/86; c.v.

Ballard, Roberta A., *et al.,* "The Hospital Alternative Birth Center: Is It Safe? Experience in 1000 Cases from 1976 to 1980," *Journal of Perinatology* 5 (1980): 3:61–4.

Ballou, Adin, *History of the Hopedale Community.* Lowell, MA: Thompson & Hill, 1897.

Bancroft, Hubert Howe, *The Book of the Fair: An Historical and Descriptive Presentation of the World's Science, Art, and Industry, as Viewed through the Columbian Exposition at Chicago in 1893.* Chicago: Bancroft, 1893.

Banner, Josephine [Josephina de Vasconcellos] (Artist, inventor; Ambleside, Cumbria, Eng.), L:-12/12/ 79, 2/20/80.

Bannerman, Jean, *Leading Ladies: Canada.* Belleville, Ontario: Mika Publishing Co., 1977.

Baring-Gould, Sabine, *Curious Myths of the Middle Ages,* Boston: Roberts, 1889.

Barnard, Eunice F., "Women Microbe Hunters," *Independent Woman,* Dec. 1936: 379ff.

Barnard, Kathy (Ref. Librarian, Indiana State Library, Indianapolis), L-8/8/83.

Barry, Tina (Dir., Corporate Communications, Kimberly-Clark Corp., Dallas, TX), L-1986–87.

Bartimo, Jim, "A 'Real' Computer in Your Lap?" *Info-World,* May 7, 1984: 91–4.

Bass, Ellen, "Tampons," *Chrysalis,* No. 9 (Fall 1979): 47–9.

Bassetti, Ann (Researcher, Arlington, VA), L-11/30/80.

Batchelder, Samuel, *Introduction and Early Progress of the Cotton Manu- facture in the United States.* Clifton, NJ: Augustus M. Kelley, 1972 (1863).

Bates, Daisy, *The Passing of the Aborigines.* London: John Murray, 1952 (1938).

Baum, Joan, *The Calculating Passion of Ada Bryon.* Hamden, CT: Archon, 1987.

Baxandall, Rosalyn, Linda Gordon, and Susan Reverby, eds, *America's Working Women: A Documentary History—1600 to the Present.* New York: Vintage, 1976.

Bay Laurel, Alicia, *Living on the Earth.* New York: Vintage, 1971 (1970).

Bayliss-Smith, Tim, "Constraints on Population Growth: The Case of the Polynesian Outlier Atolls in the Precontact Period," *Human Ecology* 2: 4(1974): 259–95.

Beall, Otho, Jr., and Richard H. Shyrock, *Cotton Mather: First Significant Figure in American Medicine.* Baltimore: Johns Hopkins, 1954.

Beauchamp, Virginia (English Dept., University of Maryland), PC-1988–present.

Becker-Colonna, Andreina L., *Ancient Italic Civilizations: Part One, Sardinia.* San Francisco: San Francisco State University, 1963.

Beecher, Catharine, and Harriet Beecher Stowe, *The American Woman's Home.* New York: Arno, 1971 (1869).

Begley, Sharon, "Lost Treasures of the Maya," *Newsweek,* June 4, 1984: 73.

Behrend, B.E., "The Achievements of Benjamin G. Lammé," *Electric Journal,* June 1919, n.p.; repr. Westinghouse Electric & Manufacturing Company Sales Letter No. 313 (June 9): 3–6.

Behrman, Daniel, *Solar Energy: The Awakening Science.* Boston: Little, Brown, 1976.

Belanger, Jane (Coord., Pediatric/Perinatal Admin., Mt. Zion Hospital, San Francisco, CA), "Summary of Professional Activities, Roberta Ballard, M.D.," n.d.; sent by author, 1986.

Bell, Diane, *Daughters of the Dreaming.* London: Allen & Unwin, 1984.

Bell, Susan G., *Women: From the Greeks to the French Revolution.* Belmont, CA: Wadsworth, 1973.

Bemer, Robert, "A View of the History of COBOL," *Honeywell Computer Journal* 5: 3 (1971): 130–5.

Bender, Barbara, *Farming in Prehistory: From Hunter-Gatherer to Food-Producer.* New York: St. Martin's, 1975.

Benedict, R.G., F.H. Stodola, O[dette] Shotwell, *et al.,* "A New Streptomycin," *Science* 112 (1950): 77–8.

Benerito, Ruth R., H.Z. Jung, and T.L. Ward, "Use of Cold Plasma in the Modification of Cotton," preprint of the 16th Textile Chemistry and Processing Conference, Mar. 17–19, 1976.

Benerito, Ruth R., J.B. McKelvey, and R.J. Berni, "Built-in Lubrication for Cottons, *Journal of the American Oil Chemistry Council* 48 (1971): 284–90.

Benerito, Ruth R., T.L. Ward, and O. Hinojosa, "Instrumental Analyses of Modified Celluloses," in J.C. Arthur, Jr., ed., *ACS Symposium Series,* U.S.A., No. 260. Washington, DC: American Chemical Society, 1984.

"Benjamin C. Lammé," typescript, Westinghouse Corporation, Public Relations files, n.d. (sent me by Guenter Holzer).

Benjamin, Marcus, "Woman's Achievement in Invention and Science," in King, comp.

Benner, Susan, "Cashing in on Computers," *Working Woman* 7 (Sept. 1982): 112–4.

Berezin, Evelyn (Pres., Greenhouse Management Corp., Glen Cove, NY), c.v./1984; L-7/6, 8/1/84.

Berg, Marie (Researcher, Minneapolis, MN), PC-6/25/88 *et seq.*

Bergsvik, Maj-Britt R. (Inventor, Linköping, Sweden), L-9/5/86.

Bernard, Hollin C., "Invents New Multiple Aspirator," New York *Amsterdam News,* Dec. 23, 1972: B-10.

Berney, Esther S., *A Collector's Guide to Pressing Irons and Trivets.* New York: Crown, 1977.

Bernstein, Jeremy, *The Analytical Engine: Computers—Past, Present, and Future.* New York: Morrow, 1981.

Berry, Carolyn, "Quill Art," *Heresies,* Winter 1978: 112–3.

Bertell, Rosalie, "New Structures for Growth," paper presented at World Future Studies Conference ("Science and Technology and the Future"), May 8–10, 1979.

"Bertha Lammé Feicht," *Ohio State University Monthly* 7: 9 (May 1916) 35.

Bessette, Monica, PhD (History of Art and Architecture, Brown Univ., Providence, RI), L-10/1/91.

Betham-Edwards, Matilda, *Six Life Studies of Famous Women, 1880.* Freeport, NY: Books for Libraries Press, repr. 1972.

Bethel Township Schools, Clark County, OH, Course of Study and Regulations, 1910; class lists; grade reports of Bertha Lammé, 1884–8. Lammé Family papers, privately held.

"Beulah Henry" Profile, *Coronet,* July 1939: 159.

Bever, Marilynn A., "The Women of M.I.T., 1871–1941: Who They Were, What They Achieved." M.I.T. undergraduate thesis, 1976.

Bickel, Lennard, *Rise Up to Life: A Biography of H.W. Florey.* New York: Scribner's, 1972.

Billings, Evelyn L., John J. Billings, and Maurice Catarinich, *Atlas of Ovulation Method.* Collegeville, MN: The Liturgical Press, 1974.

Billings, John R., "American Inventions and Discoveries in Medicine, Surgery, and Practical Sanitation," *Scientific American,* July 4, 1891: 10.

Billington, James H. (Dir., Woodrow Wilson International Center for Scholars, Smithsonian Institution, Washington, DC), L- [1981].

Bindocci, Cynthia (West Virginia University, Morgantown), L-1982-6; PC-3/2/84.

Bindocci, Cynthia, "Identification of Technology Content for Science and Technology Centers," doctoral dissertation, West Virginia University (Morgantown), 1983.

Bindocci, Cynthia, *Women and Technology: An Annotated Bibliography.* New York: Garland, 1993.

Binford, Sally, "Myths and Matriarchies," *Human Behavior* 8 (May 1979): 62–6.

Bingham, Stephen D., comp., *Early History of Michigan with Biographies of State Officers.* Lansing, MI: Thorp & Godfrey, 1888.

Bird, Caroline, *Enterprising Women.* New York: Norton, 1976.

Bishop, J. Leander, *A History of American Manufactures from 1608 to 1860.* 2 vols. Philadelphia: Edward Young & Co., 1864.

Bishop, Jerry, "Successes in Lung-Cancer Vaccine Trials . . . ," *Wall Street Journal,* Sept. 16, 1987: 31.

Blackburn, Helen, "Notes on Preparations for Chicago Exhibition," *Englishwoman's Review,* Jan. 15, 1893: 64–5.

Blackburn, John, "Nurse's Inventions Are a Real Lifesaver . . . ," *National Enquirer,* Nov. 19, 1991: 35.

Blair, Homer P. (Vice-Pres. for Patents and Licenses, Itek Corporation), PC-1979 to Dorothy Stephenson; L-1980–83.

Blair, Seabury, "You Never Saw a Purple Cow?" Bremerton, WA, *Sun* "Weekender" sec., May 30, 1981: 33.

Blakeslee, Sandy, Mary Lou Allen, and Marilyn Yalom, *Women in the Sciences and Engineering.* Stanford, CA: Center for Research on Women, 1978.

[Blanchard, Miss Helen Augusta], Obit., Portland, ME, *Evening Express and Advertiser,* Jan. 13, 1922: 36.

Blandford, Percy, *Old Farm Tools: An Illustrated History.* Ft. Lauderdale, FL: Gale Research Co., 1976.

Blashfield, Jean, "The First Nude Film Sensation and Other Offbeat Tales About Hollywood's Women," *National Enquirer,* July 21, 1981: 24.

Blashfield, Jean, *Hellraisers, Heroines, and Holy Women.* New York: St. Martin's Press, 1980.

Blau, Kathy, "Grace Murray Hopper," student paper, University of New Mexico (Albuquerque), submitted to Nancy Martin, Dept. of Computing and Information Science, 1978–9.

Bleek, Wolf, *Sexual Relationships and Birth Control in Ghana: A Case Study of a Rural Town.* Amsterdam: University of Amsterdam 1976.

Bloch, Ivan, *Sexual Life in England,* tr. William Fostern. London: Transworld, 1965.

Blodgett, Harriet, *Centuries of Female Days: Englishwomen's Private Diaries.* New Brunswick, NJ: Rutgers University Press, 1988.

"Blood Sugar Control for Healthy Babies," *Science News* 121: 9 (Feb. 27, 1982): 134.

Board of Lady Managers, *Approved Official Minutes of the World's Columbian Commission,* Sixth Session. Chicago: Rand McNally, 1894.

Boccaccio, Giovanni, *Of Famous Women,* tr. Guido A. Guerino. New Brunswick, NJ: Rutgers University Press, 1963.

"Body Worship," *Newsweek,* Jan. 28, 1974: 67.

Boesch, Christophe, and Hedwig Boesch, "Sex Bias Between Brawn and Brain in Chimps," *New Scientist,* Jan. 14, 1982: 81.

Boesch, Christophe, and Hedwig Boesch, "Sex Differences in the Use of Natural Hammers by Wild Chimpanzees: A Preliminary Report," *Journal of Human Evolution* 10 (1981): 585–93.

Bolemon, Jay (Physics, University of Central Florida, Orlando), L-1/30/82.

Bond, E[laine] R., *et al.*, "FORMAC—An Experimental FORmula MAnipulation Compiler," *Proceedings* of the 19th National Conference of the Association for Computing Machinery. New York: ACM, 1964.

Bordley, James III, and A. McGehee Harvey, *Two Centuries of American Medicine, 1776–1976.* Philadelphia: Saunders, 1976.

Bose, Christine E., and Philip L. Bereano, "Household Technologies: Burden or Blessing?" in Zimmerman, ed., *Technological.*

Boserup, Ester, *Woman's Role in Economic Development.* New York: St. Martin's, 1970.

Bossen, Laurel, "Women in Modernizing Societies," in Sharon W. Tiffany, ed.

The Boston Women's Health Book Collective, *Our Bodies, Ourselves,* 2nd ed. New York: Simon & Schuster, 1984.

Bosveld, Jane, "A Wizard of Computer Fantasy," *Ms.,* Aug. 1983: 20.

Boulding, Elise, *The Underside of History.* Boulder, CO: Westview Press, 1976.

Bourgeois, Jean-Louis, "GEO Exploration: Splendors in the Mud," *Geo* 5: Aug. 1983: 18–25.

Bouton, Katherine, "The Nobel Pair: Mavericks of Medical Research in a Shared Quest," *New York Times Magazine,* Jan. 29, 1989: 28–9.

Bowden, B., *Faster Than Thought: A Symposium on Digital Computing Machines.* London: Pitman & Sons, 1953.

Bowen, Roger, "Betsey Baker Was a Bonnet Maker," *Yankee Magazine,* Mar. 1967: 71–3ff.

[Bowles, Ada C.], "Women as Inventors," *The Woman's Journal,* Mar. 18: 1899: 88.

Boyer, Mrs. Robert L. (Daughter of Bertha Lammé, Pittsburgh, PA), L-1983.

Branca, Patricia, *The Silent Sisterhood.* Pittsburgh: Carnegie-Mellon University Press, 1975.

Branca, Patricia, *Women in Europe Since 1750.* London: Croom Helm, 1978.

Bray, Francesca (East Asian History of Science Library, Cambridge, Eng.), L-1979-80.

Bray, Francesca, *Agriculture,* vol. 6, pt. II, in Joseph Needham, ed., *Science and Civilisation in China.* Cambridge: Cambridge University Press, 1984.

Breasted, James H., *A History of Egypt from the Earliest Times to the Persian Conquest.* New York: Scribner's Sons, 1909.

Brenner, Edward J. (Executive Director, American Association for the Advancement of Invention and Innovation, Alexandria, VA), PC-3/3/80.

Bridenthal, Renate, and Claudia Koonz, eds, *Becoming Visible: Women in European History.* Boston: Houghton Mifflin, 1977.

Bridenthal, Renate, Claudia Koonz, and Susan Stuard, eds, *Becoming Visible: Women in European History,* 2nd ed. Boston: Houghton Mifflin, 1987.

Brinks, Jim S. (Prof. of Animal Breeding and Genetics, Colorado State University, Ft. Collins, CO), PC-3/28/85.

Britzman, Agnita (Inventor, Long Beach, CA), Q-3/7/83; L-7/12/83).

Broad, William J., "Computers Learn to Think Faster," San Francisco *Examiner,* Nov. 4, 1984: 11–2.

Brody, Jane, "Study Links a Pre-Cancer Condition and Users of Pill," *New York Times,* Sept. 29, 1970: 84: 1.

Brody, Judit (British Science Library, London, Eng.), L-1984- present.

Brody, Judit, "Patterns of Patents: Early British Inventions by Women," a paper presented at the XVIIth International History of Science Congress, Berkeley, CA, July 31 – Aug. 8, 1985.

Brookfield, H. C., "The Ecology of Highland Settlement: Some Suggestions," *American Anthropologist,* Special Publication, Part 2, 66: 4 (Aug. 1964): 20–38.

Brookfield, H. C., "The Highland Peoples of New Guinea," *Geographical Journal* 127: 4 (Dec. 1961): 437ff.

Brose, David, *et al., Ancient Art of the American Woodland Indians.* New York: Abrams, 1985.

Brower, Montgomery, "Thanks to Martine Kempf's Little Black Box, When People Talk, Machines Listen—and Obey," *People* 25: 21 (May 26, 1986): 73ff.

Brown, Barbara B., "The Anatomy of a Phenomenon: Me and BFT," *Psychology Today,* Aug. 1974: 48–57ff.

Brown, Barbara B., "Biofeedback: An Exercise in 'Self-Control,' " *Saturday Review,* Feb. 22, 1975: 22–6.

Brown, Barbara B., *New Mind, New Body; Bio-Feedback: New Directions for the Mind.* New York: Harper & Row, 1974.

Brown, Barbara B., *Stress and Art of Biofeedback.* New York: Harper & Row, 1977.

Brown, Judith K., "Economic Organization and the Position of Women Among the Iroquois," in Tiffany, ed.

Brown, Judith K., "A Note on the Division of Labor by Sex," *American Anthropologist* 72 (1970): 1073–8.

Brown, Kenneth, *Inventors at Work: Interviews with Sixteen Notable American Inventors.* Redmond, WA: Microsoft Press, 1988.

Brown, Patricia Leigh, "Smokeless Future for the Ashtray?" *New York Times,* May 7, 1987: 26.

Brown, Victoria, *Uncommon Lives of Common Women: The Missing Half of Wisconsin History.* Madison, WI: Wisconsin Feminists Project Fund, 1975.

Brown, Whitney Cartwright (Granddaughter of Jessie Cartwright, Girdwood, AK), PC-4/29/83.

Brownlee, W. Elliot, and Mary M. Brownlee, *Women in the American Economy: A Documentary History, 1675 to 1929.* New Haven, CT: Yale University Press, 1976.

Bruemmer, Fred, "Eskimos Are Warm People," *Natural History* 90: 10 (Oct. 1981): 42–9.

Brun, Hertha Otis, "Household Inventions," *The American Kitchen Magazine,* January 1896; repr. (abr.), *Scientific American* 74 (Feb. 15, 1896): 99.

Bryant, Alice G., "An Auricle Cap," *Laryngoscope,* Jan. 1906: 79.

[Bryant, Alice G.] "Dr. Alice Bryant Dies . . . ," *New York Times,* July 27, 1942.

[Bryant, Alice G.] "Dr. Alice G. Bryant, a Boston Physician," *Boston Globe,* July 26, AP.

Bryant, Alice G., "Ear Nose and Throat Instrument Box," *Laryngoscope,* Sept. 1917: 703–7.

Bryant, Alice G., "An Ear, Nose and Throat Treatment Table," *American Academy of Ophthalmology and Otolaryngology* (1915): 311.

Bryant, Alice G., "An Electric Foot-Switch," *Laryngoscope,* Dec. 1916: 1383.

[Bryant, Alice G.], "A Lady and a Doctor," *Technology Review,* May 1, 1939.

Bryant, Alice G., "Lip-Reading of Vital Value to the Hard of Hearing," *The Medical Woman's Journal,* June 1931: 141–4.

Bryant, Alice G., "Lip Reading Shall Not Fail," *The Medical Woman's Journal,* Feb. 1930: 34–6.

Bryant, Alice G., "A New Tongue Depressor," *Laryngoscope,* Feb. 1905: 116.

Bryant, Alice G., "A New Tonsil Tenaculum," *Laryngoscope,* Dec. 1905: 947.

Bryant, Alice G., "The Use of Cold Wire Snare in the Removal of Hypertrophied Tonsils," *The British Medical Journal* (Section of Laryngology and Otology), Nov. 17, 1906: 1390–5.

[Bryant, Alice G.], Vassar College Alumnae Association (information provided me from Bryant's file).

[Bryant, Alice G.], *Vassar College Bulletin* 29: 2 (1939): 51.

[Bryant, Alice G.], *Vassar Quarterly* 17:1 (Feb. 1932): 69.

Bryant, Alice G., "What I Think of the Woman's Party," *Equal Rights* 11: 24 (July 26, 1924): 189.

[Bryant, Alice G.], "Woman Pioneer Specialist Dies," Charlotte, NC *Observer,* July 27, 1942.

Bryenton, Elisabeth (Inventor, Lakewood, OH), Q-3/7/85.

Buckland, Bruce, "Learning Company Designs PCjr. Software," Portola Valley, CA, *Country Almanac,* Nov. 9, 1983: 32.

Budoff, Penny Wise, MD, *No More Hot Flashes and Other Good News.* New York: Putnam, 1983.

Budoff, Penny Wise, MD, *No More Menstrual Cramps and Other Good News.* New York: Putnam, 1980.

Bullough, Vern L., "Female Physiology, Technology, and Women's Liberation," in Trescott, ed.

Bullough, Vern L., *The Subordinate Sex: A History of Attitudes Towards Women.* Urbana: University of Illinois Press, 1973.

Burgess, George, "A Watch Whose Time Has Come," *The Hopewell (VA) News,* Mar. 30, 1982: 8.

Burke, Edmund, comp., *List of Patents for Inventions and Designs, Issued by the United States, from 1790 to 1847. . . .* Washington, DC: U.S. Patent Office, 1847.

Burkitt, Alan, and Elaine Williams, *The Silicon Civilisation.* London: W.H. Allen, 1980.

Burleigh, D.W. (Technical & Production Dir., Britax, Excelsior Ltd., Weybridge, Surrey, England), L-9/11/85.

Burnet, Sir Macfarlane, *The Natural History of Infectious Diseases.* Cambridge: Cambridge University Press, 1972.

Burnham, Phyllis (MS Curator, Regional History Collections, Western Michigan University, Kalamazoo) L-3/2, 4/22/82.

Burpee Company, Warminster, PA, news releases on Alice Vonk's $10,000 white marigold, Aug. 28, 1975.

Burpee, David, "In Search of the White Marigold," *Gardening,* Mar./Apr. 1979: 8–9.

Burt, McKinley, Jr., *Black Inventors of America.* Portland, OR: National Book Co., 1969.

Burton, Jean, *Lydia Pinkham Is Her Name.* New York: Farrar, Straus, 1949.

Bush, Corlann G., "The Barn Is His, the House Is Mine: Agricultural Technology and Sex Roles," in George Daniels and Mark Rose, eds, *Energy and Transport.* Beverly Hills, CA: Sage, 1982.

Bush, Corlann G., "Women and the Assessment of Technology: To Think, to Be; to Unthink, to Free," in Joan Rothschild, ed., *Machina.*

Bush, Corlann G., ed., *Taking Hold of Technology.* . . . Washington, DC: American Association of University Women, 1981.

*Business Week* eds, "A Lady's Hand Guides Fight on Heart Disease," *Business Week,* Nov. 20, 1965: 130ff.

Butler, Margaret (Argonne National Laboratory, Argonne, IL), c.v./1983; L-5/25/84.

Buxton, J.N. (Computer Science, University of Warwick, Coventry, Eng.), L-11/11/80.

Bylinsky, Gene, "Alza Finally Finds a Cure for Losses," *Fortune,* Apr. 28, 1986: 80.

Bylinsky, Gene, *et al., Silicon Valley High Tech.* Hong Kong: International, 1985.

Byrn, Edward W., "The Progress of Invention During the Past Fifty Years," *Scientific American* 75 (July 25, 1896): 82–3.

Byrn, Edward W., *The Progress of Invention in the Nineteenth Century.* New York: Munn, 1900.

C.W.C. and S.D., "The Hospice: A Decent Way to Die," *Ms.,* Nov. 1981: 96.

Caddes, Carolyn, *Portraits of Success: Impressions of Silicon Valley Pioneers.* Palo Alto, CA: Tioga Press, 1986.

Cahill, Colleen (Ref. Librarian, Indiana Historical Society Library, Indianapolis), L-8/16/83.

Calio, Jim, "Artist Nancy Burson and Her Age Machine Are Making People Old Before Their Time," *People* 17: 22 (June 7, 1982): 93–4f.

Callaghan, Catherine A., "Patridominance and Proto-Utian Words for 'Man,' 'Woman,' and 'Person,' " unpublished paper sent by author (Linguistics Dept., Ohio State University, Columbus), 1985.

Callaghan, Catherine A., "The Wanderings of the Goddess: Language and Myth in Western Culture," Symposium on Women, Language, and Culture, Georgetown University, Mar. 16, 1977. Pub. in *Phoenix: New Directions in the Study of Man* 3:2 (Fall/Winter 1979): 25–37; repr. in Linda Clark et al, eds, *Image-Breaking/Image-Building.* New York: Pilgrim, 1981.

"Calling the Shots," Kaiser-Permanente Health Plan (Oakland, CA), *Planning for Health,* Fall 1984: 9.

"Calm View of Management," *Datamation,* Aug. 1976: 13–4.

Cameron, John R. (Dept. of Medical Physics, University of Wisconsin, Madison), L-2/16, 4/20/82.

Campbell, Dugald, *On the Trail of the Veiled Tuareg: An Account of These Mysterious Nomadic Warriors Whose Home Is the Trackless Desert and Whose History Fades into the Far Past.* Philadelphia: Lippincott, 1928[?].

Canby, Thomas Y., "The Anasazi," *National Geographic* 162: 5 (Nov. 1982): 554–92.

"Cancer Vaccines in the Works," *Science News* 115 (Apr. 14, 1979): 248.

Capettini, Roger, "Home, Smart Home," *National Enquirer,* Jan. 12, 1988: 31.

"The Car Coupler," *Inventive Age,* Feb. 10, 1891: 2.

Carmichael, Bill, *Incredible Collectors, Weird Antiques, and Odd Hobbies.* Englewood Cliffs, NJ: William E. Carmichael, 1971.

"Carpenter Self-Threading and Self-Setting Needle for Sewing Machines," *Scientific American,* Sept. 10, 1870: 164.

Carroll, Bernice A., ed., *Liberating Women's History: Theoretical and Critical Essays.* Champaign/ Urbana: University of Illinois Press, 1976.

Carroll, Helen (University of Oregon Center for Invention and Innovation, College of Business Administration, Eugene, OR), L-10/29/76.

Carter, Art, "Jerseyan Gives Invention to French Government," *Afro- American* Magazine, Nov. 10, 1951: 6.

Carter, E.F., *Dictionary of Inventions and Discoveries.* New York: Philosophical Library, 1967.

[Cartwright, Jessie W.], Obits.: Chicago *Daily News, Sun-Times, Tribune,* June 17, 1977.

"Cashing In on Cookie Craze," *New York Times,* Jan. 31, 1983: Y-1.

Cavendish, Richard, ed., *Man, Myth, and Magic: An Illustrated Encyclopedia of the Supernatural.* New York: Marshall Cavendish, 1970.

Cercek, Lea, and Boris Cercek, "Apparent Tumour Specificity with the SCM Test," *British Journal of Cancer* 31 (1975): 252–3.

Cercek, Lea, and Boris Cercek, "Application of the Phenomenon of Changes in the Structuredness of Cytoplasmic Matrix (SCM) in the Diagnosis of Malignant Disorders: A Review," *European Journal of Cancer* 13 (Sept. 1977): 903–15.

Cercek, Lea, and Boris Cercek, "Changes in the Structuredness of Cytoplasmic Matrix (SCM) in Human Lymphocytes Induced by PHA and Cancer Basic Protein as Measured in Single Cells," *British Journal of Cancer* 33 (May 1976): 539–43.

Chaff, Sandra L., *et al. Women in Medicine: A Bibiliography of the Literature on Women Physicians.* Metuchen, NJ: Scarecrow Press, 1977.

Chamney, Anne R. (Engineer, inventor, London, Eng.), L-1/18/84; Q-1/18/ 84.

Chamney, Anne R. "Bioengineering," *Woman Engineer* (London) 10: 9 (1968): 4–6.

Chapman, V.J., and D.J. Chapman, *Seaweeds and Their Uses.* London: Methuen, 1980.

Chapple, Eliot D., and Carleton S. Coon, *Principles of Anthropology.* New York: Holt, 1942.

Cheke, Val, *The Story of Cheese-Making in Britain.* London: Routledge & Kegan Paul, 1959.

Chen, A., "Marijuana and the Reproductive Cycle," *Science News* 123: 13 (Mar. 26, 1983): 197.

Chiang, Anne (Researcher, inventor, Xerox Corporation, Palo Alto, CA), I-9/4/84; Q-9/4/84; Biog./1984; L-9/24/84.

Chicago, Judy, *The Dinner Party: A Symbol of Our Heritage.* Garden City, NY: Doubleday/Anchor, 1979.

Chicago, World's Columbian Exposition, 1893, Official Catalogue. Chicago: W.B. Conkey, 1893.

"Chicken Pox Vaccine—Good Results," San Francisco *Chronicle,* May 31, 1984: 1f.

Childe, Gordon, *What Happened in History.* Harmondsworth, Eng.: Penguin, 1943.

Chilton, Mary-Dell, "Genetic Engineering—Prospects for Use in Crop Management," *British Crop Protection Conference—Pests and Diseases—1984.* Croydon, Eng.: British Crop Protection Council, 1984.

Chilton, Mary-Dell, "A Vector for Introducing New Genes into Plants," *Scientific American* 248: 6 (June 1983): 50–9.

Chilton, Mary-Dell, *et al.,* "Stable Incorporation of Plasmid DNA into Higher Plant Cells: The Molecular Basis of Crown Gall Tumorigenesis," *Cell* 11 (1977): 263–71.

Chuang, Jenny (Computer scientist, inventor, Wang Labs, Lowell, MA), PC-6/12/87; c.v./1987.

Ciarlante, Marjorie (Archivist, Scientific, Economic, and Natural Resources Br., Civil Archives Div., National Archives, Washington, DC), L-11/24/86.

Ciarlante, Marjorie, "A Statistical Profile of Eminent American Inventors, 1790–1860: Social Origins and Role," PhD dissertation, Northwestern University, 1978.

Cimons, Marlene, "Computer Language Invented by a Woman," Long Island *Press,* Oct. 30, 1973: 11.

Cimons, Marlene, "Entrepreneurs: Turning Cottage Industry into Big Business and Other Tips for Inventive Women," *Ms.,* Nov. 1981: 83–4.

Cimons, Marlene, "Navy's Computer Expert: Call Her Captain, Not Ms.," Los Angeles *Times,* Oct. 21, 1973, "View" sec.: 1, 6.

Clark, Alice, *The Working Lives of Women in the Seventeenth Century.* London: Routledge & Kegan Paul, 1981 (1919).

Clark, John T., Erla R. Smith, and Julian M. Davidson, "Enhancement of Sexual Motivation in Male Rats by Yohimbine," *Science* 225 (Aug. 24, 1984): 847–8.

Clarke, Charles, ed., "Journal Kept While Crossing the Plains," Part III, *Historical Society of Southern California Quarterly* 59 (Fall 1977): 264.

Clarke, Dennis J. (Meriden Public Library, Meriden, CT), L-3/13, 5/4/ 1987.

Clarke, Robert, *Ellen Swallow: The Woman Who Founded Ecology.* Chicago: Follett, 1973.

Cleary, Donald P., *Great American Grands: The Success Formulas That Made Them Famous.* New York: Fairchild, 1981.

"Clever Women Inventors," *The Patent Record,* Dec. 1899: 26.

"Cloth Is Coated to Shed Stains," *Business Week,* Oct. 6, 1956: 197–8.

"Clue to Alzheimer's Disease," San Jose *Mercury News,* Aug. 24, 1984: 1.

Clurman, Carol, "Three Wise Women," *USA Weekend* (San Francisco Peninsula *Times-Tribune* Magazine), Mar. 13–5, 1987: 7.

Clymer, Eleanor, and Lillian Erlich, *Modern American Career Women.* New York: Dodd, Mead, 1939.

Cohen, Mark Nathan, *The Food Crisis in Prehistory: Overpopulation and the Origins of Agriculture.* New Haven, CT: Yale University Press, 1977.

Cohen, Stanley N., "Manipulation of Genes," *Scientific American* 233: 1 (July 1975): 24–33.

Cohen, Stanley N., Annie C.Y. Chang, Herbert W. Boyer, and Robert B. Helling, "Construction of Biologically Functional Bacterial Plasmids in Vitro," *Proceedings of the National Academy of Sciences* 70: 11 (Nov. 1973): 3240–4.

Cohn, Victor, *Sister Kenny: The Woman Who Challenged the Doctors.* Minneapolis: University of Minnesota Press, 1975.

Cole, Doris, *From Tipi to Skyscraper: A History of Women in Architecture.* Boston: i Press, 1973.

Cole, Sonia, *The Prehistory of East Africa.* New York: Macmillan, 1963.

Colegrave, Suki, *The Spirit of the Valley: Masculine and Feminine in the Human Psyche.* Los Angeles: Tarcher, 1979.

Colman, Julia, "Are Tin Fruit Cans a Source of Metallic Poisoning?" *Scientific American,* June 18, 1970; cited in James Shenton, ed., 116.

Combellick, John (Grandson of inventor, Marion, IA), PC-3/5/91; L-5/21/ 91.

Comfort, Alex, *The Joy of Sex.* New York: Crown, 1972.

Conard-Pyle Co. (West Grove, PA), "Blue Princess, Blue Angel: Two Beautiful New Holly Girls," press release, n.d. [1973?].

Conard-Pyle Co., "Three New Blue Hollies," press release, n.d. [1975?].

Conard-Pyle Co., "The Unique Marvelous Blue Hollies," press release, n.d. [1975?].

Connecticut, Report of the Commissioners from Connecticut of the Columbian Exposition of 1893, at Chicago, *Connecticut at the World's Fair;* also *Report of the Work of the Board of Lady Managers of Connecticut,* Hartford, CT: Case, Lockwood, & Brainard, 1898.

Connolly, Lisa, "Ida Rolf," *Human Behavior* 6(May 1977): 17–23.

Convis, Orville P., "Why Trains Have Red Taillights," *Christian Science Monitor,* Aug. 4, 1986.

Conwell, Esther (Physicist, inventor, Xerox Corp., Rochester, NY), Q-1983, c.v./1983, L-6/2/83.

Conwell, Jennie Hayden, Inaugural editorial, Ladies' Department, *Minneapolis Daily Chronicle,* weekly ed., *Conwell's Star of the North,* c. 1865, in Agnes Rush Burr, *Russell H. Conwell and His Work.* Philadelphia: J.C. Winston, 1926, 141–4.

Conwell, Russell H., *Acres of Diamonds,* 1877; repr. in William R. Webb, ed. Kansas City, MO: Hallmark, 1968. Also repr. in Mary Cohart, ed., *Unsung Champions of Women.* Albuquerque: University of New Mexico Press, 1975, 250–65.

Coon, Carleton S., *Tribes of the Rif,* Harvard African Studies IX. Cambridge, MA: Peabody Museum, 1931.

Coontz, Stephanie, and Peta Henderson, eds. *Women's Work, Men's Property: The Origins of Gender and Class.* London: Verso, 1986.

Cooper, Carolyn C. (History of Science, Yale University, New Haven, CT), L-8/30/85.

Cooper, Grace Rogers, *The Sewing Machine: Its Invention and Development.* Washington DC: Smithsonian Institution Press, 1976.

Copaciu, Cdt. Constantin (Monte Carlo, Monaco), L-8/18, 11/17/81.

Corcoran, Elizabeth, "She Incites Chip Revolutions," *IEEE Spectrum* 24: 12 (Dec. 1987): 46–51.

Corn, Joseph J. (VTSS Program, Stanford University, Stanford, CA), PC-1985 - present.

Corn, Joseph J., "Making Flying 'Thinkable': Women Pilots and the Selling of Aviation, 1927–1940," *American Quarterly,* 31 (Fall 1979): 557–71.

Corrigan, Ann, "Computer Modeling of Simple Population Dynamics for Education," in Lucas & Adams, eds, *Personal Computing Proceedings.* New York: AIFPS, 1979, 179–86.

Coston, Martha C., *A Signal Success: The Work and Travels of Mrs. Martha J. Coston.* Philadelphia: Lippincott, 1886.

Coulter, Harris L., and Barbara L. Fisher, *DPT: A Shot in the Dark.* New York: Harcourt Brace, 1985.

Coville, F.V., "The Panamint Indians of California," *American Anthropologist* 5 (Oct. 1892): 351ff.

Cowan, Ruth, "From Virginia Dare to Virginia Slims: Women and Technology in American Life." *Technology and Culture* 20: 1 (Jan. 1979): 51–63; repr. in Trescott, ed.

Cowan, Ruth, "Less Work for Mother?" *American Heritage of Invention and Technology* 2: 3 (Spring 1985): 57–63.

Cowan, Ruth, *More Work for Mother: The Ironies of Household Technology from the Open Hearth to the Microwave.* New York: Basic Books, 1983.

Cox, Jerry, "These 'Birds' Help Kids See," The Sacramento *Union,* Feb. 5, 1981, extra: 1, 9.

Cox, June (Acad. Dir., Gifted Students Inst., Arlington, TX), PC-1979.

Cramp, Arthur, *Nostrums and Quackery.* Chicago: A.M.A. Press, 1911.

Cranmer, J.H., Giuliana C. Tesoro, and D.R. Uhlmann, "Chemical Modification of Carbon Fiber Surfaces with Organic Polymer Coatings," *Independent Engineer in Chemical Products, Research and Development* 21 (1982): 185–90.

Creger, William P., Roy Maffley, and Rose Payne, "Memorial Resolution: Judith Graham Pool, 1919–1975," Stanford University *Campus Report,* Sept. 24, 1975.

Crittenden, Ann, "Closing the Muscle Gap," in Twin, ed.

Crossen, Cynthia, "It Took Years to Get People to Stop Wearing Hats . . .," *Wall Street Journal,* Dec. 2, 1985.

*Crystal Palace, Official Catalogue of the Exhibition.* London, 1851.

Culkin, John, "Qwerty and Beyond." *New York Times,* Mar. 15, 1978: 27: 1.

"The Cup," Tassaway® ad, *McCall's,* Aug. 1970: R-5.

Curie, Eve, *Madame Curie,* tr. Vincent Sheean. New York: Garden City Publishing, 1940.

"Curling Iron Inventor," *Hue,* March 1958: 26–7.

*Current Biography,* Charles Moritz, *et al.,* eds. New York: Wilson, 1940–.

Curry, Mrs. Pat (Genealogical researcher, Blanchard descendant, Richmond, CA), PC-1/30/86; L-2/2/86, 3/30/87; Blanchard genealogy.

Curwen, E. Cecil, "Querns," *Antiquity* 11: 42 (June 1937): 133–51.

"Cutting Heart's Dose in Breast Therapy," *Science News* 124: 24 (Dec. 10, 1983): 376.

*The Cyclopaedia of American Biography.* New York: Press Association Compilers, 1918.

Dahlberg, Frances, ed., *Woman the Gatherer.* New Haven, CT: Yale University Press, 1981.

Dal Molin, Armando (Music Typographers' Union, Oyster Bay, NY), PC-3/14/80.

Dall, Caroline, *The College, the Market, and the Court; or, Woman's Relation to Education, Labor, and Law.* Concord, NH: Comford Press, 1867.

Dall, Caroline, *Historical Pictures Retouched.* Boston: Walker Wise, 1860.

Dallas, E.S., *Kettner's Book of the Table.* London: Dulau & Co., 1877.

Dallman, Peter R., and Judith Graham Pool, "Treatment of Hemophilia with Factor VIII Concentrates," *New England Journal of Medicine* 278: 4 (Jan. 25, 1968): 199–202.

Daly, Mary, *Gyn/Ecology: The Metaethics of Radical Feminism.* Boston: Beacon, 1978.

Damour, Felix, *Louise Bourgeois: Her Life, Her Work.* Paris: Jouve & Boyer, 1900.

Daniel, Neil (English Dept., Texas Christian University, Fort Worth, TX) PC, 1979–80.

Daniels, Farrington, and John A. Duffie, eds., *Solar Energy Research.* Madison, WI: University of Wisconsin Press, 1955.

Dann, C. Marshall, "Patents: Another Way to Publish." *Physics Today,* March 1978: 23ff.

Dasent, George W., ed., *Norwegian Folk Tales.* New York: Appleton, 1859.

Dash, Bhagwan, and R.N. Basu, "Methods for Sterilization and Contraception in Ancient and Medieval India," *Indian Journal of History of Science* 3 (May 1968): 9–24.

Dash, Joan, *A Life of One's Own: Three Gifted Women and the Men They Married.* New York: Harper & Row, 1973.

Davenport, Walter Rice, *Biography of Thomas Davenport, the "Brandon Blacksmith," Inventor of the Electric Motor.* Montpelier: The Vermont Historical Society, 1929.

Davidson, Caroline, *A Woman's Work Is Never Done.* London: Chatto & Windus, 1982.

Davis, Adelaide (Alumni Dir., Queens College, Charlotte, NC), PC-1987-9.

Davis, Delia T., "Book-Making in the West," *The Critic* 37: 3 (Sept. 1900): 232–41.

Davis, Elizabeth Gould, *The First Sex.* New York: Putnam's, 1971; Penguin paperback ed., 1973.

Davis, Hayden E. (Vice-Pres., Weight Watchers International, Manhasset, NY), L-6/21/85.

Davis, Hugh J., *Intrauterine Devices for Contraception: The IUD.* Baltimore: Williams and Wilkins, 1971.

Davis, J.G., *Cheese.* London: Churchill Livingstone, 1976.

Davison, Jean, "What Makes a Woman?" Paper on Kikuyu Women presented at the Center for Research on Women, Stanford University, Apr. 2, 1986.

"Debbi Fields," *People,* Mar. 23, 1981: 102.

De Bono, Edward, *Eureka: An Illustrated History of Inventions from the Wheel to the Computer.* New York: Holt, Rinehart, 1974.

De Camp, L. Sprague, *The Heroic Age of American Invention: Thirty-Two Men Who Made the Modern American Era.* Garden City, NY: Doubleday, 1961.

Deevey, Edward S., Jr., "The Human Population," *Scientific American* 203: 3 (Sept. 1960): 195–204.

Degler, Carl, *At Odds: Women and the Family in America from the Revolution to the Present.* New York: Oxford, 1980.

De Kruif, Paul, *Microbe Hunters.* New York: Harcourt, 1966.

Delaney, Janice, Mary Jane Lupton, and Emily Toth, *The Curse: A Cultural History of Menstruation.* New York: Dutton, 1976.

Delatiner, Barbara, "An Amateur's Luck Leads to New Hybrid Hollies." *New York Times,* Dec. 17, 1978, "Leisure" sec.: 48f.

De Laszlo, Henry, and Paul S. Henshaw, "Plant Materials Used by Primitive Peoples to Affect Fertility," *Science* 119 (May 7, 1954): 626–31.

Delson, James, "A Young Girl's Fantasy Turns to Fortune," *Family Computing,* Dec. 1983: 66–9.

Deming, Edward, and Faith Andrews, *Work and Worship: The Economic Order of the Shakers.* Greenwich, CT: New York Graphic Society, 1974.

Dennell, Robin W., "Archaeobotany and Early Farming in Europe," *Archeology,* Jan. 1978: 8ff.

Denney, Mike, "A Hospital for Patients," *Image* (San Francisco *Chronicle and Examiner* Magazine), Aug. 31, 1986: 5.

Devereux, George, *A Study of Abortion in Primitive Societies,* New York: Julian Press, 1955.

DeWitt, Robert, "Wizard and the Princess: Computer Fantasy Comes True," *ANTIC: The Atari Resource,* Nov. 1983: 23–5.

"Diagnosis by DNA," *Discover,* June 1981: 11–2.

Diamonstein, Barbaralee, *Open Secrets: Ninety-four Women in Touch with Our Time.* New York: Viking, 1970.

*Dictionary of American Biography,* American Council of Learned Societies, eds, 20 vols, 6 suppls. New York: Scribner's, 1928–36, 1944, 1958, 1973, 1974, 1977, 1980.

*Dictionary of National Biography* (British), ed. Leslie Stephen. London: Smith, Tide & Co., 1886– .

*Dictionary of Scientific Biography.* Charles C. Gillispie, ed., for the American Council of Learned Societies, 8 vols. New York: Scribner's, 1980 (1970 – ).

"The Diet-Behavior Connection," *Science News* 124: 8 (Aug. 20, 1983): 125.

"Diet Chain Looks to Spray Pounds Away," *USA Today,* June 10, 1986: 2B.

"Diet Pizzas Could Solve Egg Surplus," *Star,* July 6, 1982: 16.

Dietz, Fred, comp., *A Leaf from the Past.* New York: Rowland & Ives, 1914.

Dietz, Jean, "She's Helping to Miniaturize the Vote." Boston *Globe,* May 21, 1976: 41.

Diggs, Irene, *Black Inventors.* Chicago: Institute of Positive Education, 1975.

Dingwall, Eric John, *The Girdle of Chastity: A Fascinating History of Chastity Belts.* New York: Clarion, 1959 (1931).

Diolé, Phillippe, *The Forgotten People of the Pacific.* Woodbury, NY: Barrona, 1976.

"Discovery . . . August 2," *Ms.,* Jan. 1985: 23.

Djerassi, Carl (Chemistry, Stanford University) PC-10/27, 11/9/81.

Djerassi, Carl, *The Politics of Contraception.* New York: W.W. Norton, 1979.

Djerassi, Carl, "Steroid Oral Contraceptives," *Science* 151 (1966): 1055–61.

"Dr. Aletta Jacobs," *The Medical Woman's Journal* 45: 6 (June 1938): 175.

"Dr. Isabella Herb, Noted Anesthetist, Retires after Long Career," *The Presbyterian Hospital of the City of Chicago Bulletin* 33:5 (June–July 1941): 1f.

"Dr. Kendrick Dies," National NOW *Times,* Dec./Jan. 1980–81: 12.

"Dr. Wu Wins Award," *Journal of the American Association of University Women,* Oct. 1959: 15.

Dodds, Tom, "Safety Firsts," *Family Safety,* Summer 1974: 12.

Dodge, Norton T., *Women in the Soviet Economy: Their Role in Economic, Scientific, and Technical Development.* Baltimore: Johns Hopkins University Press, 1966.

Doetsch, R.N., and T.M. Cook, *Introduction to Bacteria and Their Ecobiology.* Lancaster, Eng.: Medical and Technical Publishing Co., 1973.

Dolan, Carrie, "A Bit of Old-Style Imagination Leads to a High-Tech Success," *The Wall Street Journal,* Feb. 7, 1983.

Donnelly, Kathleen, "Robotics," *Palo Alto Weekly,* Jul. 31, 1985: 15–17.

Doster, Alexis, III, *et al.,* eds, *The Smithsonian Book of Invention.* Washington, DC: Smithsonian Exposition Books, 1978.

Doughan, David (Assistant, Fawcett Library, London, Eng.). PC-1979 to present.

Doughan, David, and Denise Sanchez, *Feminist Periodicals, 1855–1984: An Annotated Critical Bibliography of British, Irish, and International Titles.* Brighton, Eng.: Harvester, 1987.

Douglas, Emily Taft, *Remember the Ladies.* New York: Putnam's, 1966.

Downing, Marymay, "Prehistoric Goddesses: The Cretan Challenge," *Journal of Feminist Studies in Religion* 1:1: 7–22.

Drake, Francis S., *Dictionary of American Biography Including Men of the Time.* Boston: James R. Osgood & Co. 1872; repr., Detroit: Gale, 1974.

Draper, Patricia, "!Kung Women: Contrasts in Sexual Egalitarianism in Foraging and Sedentary Contexts," in Reiter, ed.

The Draper Co., *Textile Texts for Cotton Manufacturers. . . .* Hopedale, MA: Draper, 1903.

The Draper Co., *Labor-Saving Looms.* Hopedale, MA: Draper, 1904.

Drinker, Sophie L., *Music and Women.* New York: Coward, McCann, 1948.

Druffel, Larry E., "Ada: How Will It Affect College Course Offerings?" *Interface* 3 (Fall 1979): 58–61.

Duberman, Martin B., Fred Eggan, and Richard O. Clemmer, "Documents in Hopi Indian Sexuality: Imperialism, Culture, and Resistance," *Radical History Review* 30 (Spring/Summer 1979): 99–130.

Du Bois, Cora, *The People of Alor: A Social-Psychological Study of an East Indian Island.* New York: Harper, 1961 (1944).

Dudley, Rosemary J., "She Who Bleeds but Does Not Die," *Heresies* 5 (Spring 1978): 112–6.

Duffy, John, *The Healers: The Rise of the Medical Establishment.* New York: McGraw-Hill, 1976.

Dufty, William, *Sugar Blues.* Radnor, PA: Chilton, 1975.

Dunkling, Leslie, *First Names First.* London: Dent, 1977.

Dunkling, Leslie, and William Gosling. *The Facts on File Dictionary of First Names.* New York: Facts on File, 1983.

Dunmire, John R. (Sr. Ed., Garden Dept., *Sunset* Magazine, Menlo Park, CA 94025), L-11/8/85.

Eagan, Andrea Boroff, "The Contraceptive Sponge: Easy—But Is It Safe?" *Ms.* 12: 7 (Jan. 1984): 94–5.

Eagan, Andrea Boroff, "The Selling of Pre-Menstrual Syndrome," *Ms.* 12:4 (Oct. 1983): 26ff.

Eagle, Mary K., ed., *The Congress of Women.* 2 vols., Chicago: Conkey, 1894.

Eakins, Pamela, "The Rise of the Free-standing Birth Center: Principles and Practice," *Women and Health* 9:4 (Winter 1984): 49–64.

Eakins, Pamela, and Gary A. Richwald, "Free-Standing Birth Centers in California: Structure, Cost, Medical Outcome and Issues," Final Report of MCH Contract #85-86865. Sacramento: State of California Dept. of Health and Human Services, Div. of Maternal and Child Health, 1986.

Earle, Alice M., *Colonial Dames and Goodwives.* New York: Macmillan, 1924.

Edell, Dr. Dean (Medical expert, KGO-radio, -TV, San Francisco).

Eder, James F., "The Caloric Returns to Food Collecting: Disruption and Change Among the Batak of the Philippine Tropical Forest," *Human Ecology* 6:1 (1978): 55–69.

Edwards, Diane, "Nerve Transplant: Proceed with Caution," *Science News* 131:9 (Feb. 28, 1987): 135.

Edwards, D[iane], "Thalidomide: Is There a Silver Lining?" *Science News* 131: 13 (Mar. 28, 1987): 198.

Edwards, Linden F., "Dr. Mary Edwards Walker (1832–1919): Charlatan or Martyr?" *Ohio State Medical Journal* 54: Part I (Sept. 1958): 1160–62; Part II (Oct. 1958): 1296–98.

Edwards, Perry, and Bruce Broadwell, *Data Processing: Computers in Action.* Belmont, CA: Wadsworth, 1979.

Ehle, John, *The Cheeses and Wines of England and France.* New York: Harper & Row, 1972.

Ehrenreich, Barbara, and Deirdre English, *Witches, Midwives, and Nurses: A History of Women Healers.* Old Westbury, NY: Feminist Press, 1973.

Ehrhardt, Henry B. (Northwest Inventors Association, P.O. Box 5303, Everett, WA 98206), L-n.d. [1980].

Eikorf-Stork, N., *The World Atlas of Cheese.* London: Paddington Press, 1976.

"Electrically Conductive Fabric," *Industrial Research,* Oct. 1976: 71.

*Electronic Engineering Times,* "Engineers in the News." June 6, 1973.

"Electronic Translation of Fabric and Toile Designs Suggested," *The Hosiery Trade Journal,* May 1963: 91–4.

"Eleventh International Exhibition of Inventions," *The Times* (London), Oct. 3, 1935: 9.

Ellet, Elizabeth F., *The Eminent and Heroic Women of America.* New York: Arno Press, 1974.

Ellis, Franklin, *History of Berrien and Van Buren Counties.* Philadelphia: D.W. Ensign & Co., 1880.

Ellis, Havelock, *Studies in the Psychology of Sex,* 4 vols. New York: Random House, 1936 (1897–1928).

Ellis, William S., "Romania: Maverick on a Tightrope," *National Geographic* 148: 5 (Nov. 1975): 688–709.

Ellis-Simons, Pamela, "Judi Jazzercise," *Continental,* June 1987: 30ff.

Emerson, Sandra, "Bambi Meets Godzilla: Life in the Corporate Jungle," in Zimmerman, ed., *Technological.*

Emerson, & Stern Associates, Computer Software (San Diego, CA), brochure, [1986].

Englemore, Ellie A. (Exec. Assist. to Edward A. Feigenbaum, Computer Science Dept., Stanford University, Stanford, CA), L-2/27/84.

Estioko-Griffin, Agnes, and P. Bion Griffin, "Woman the Hunter: The Agta," in Frances Dahlberg, ed.

Estrin, Thelma (Computer scientist, brain researcher, inventor, UCLA), L-8/18/87.

Estrin, Thelma A., "Analog to Digital Conversion of Physiological Data," *Medical Electronics,* proceedings of the Fifth International Conference, Liège, 1963. Desoer, Belgium, 1965.

Estrin, Thelma A., "Computers, Neuroscience, and Women," *Proceedings, Sixth Annual Conference, IEEE Engineering in Medicine and Biology Society,* Sept. 15–17, 1984.

Estrin, Thelma A., "A Conversion System for Neuroelectric Data," *Electroencephalography and Clinical Neurophysiology* 14 (1962): 414–6.

Estrin, Thelma A., "Neurophysiological Research Using a Remote Time- Shared Computer," *1966 Rochester Conference on Data Acquisition and Processing in Biology and Medicine.* London: Pergamon, 1968.

Estrin, Thelma A., "On-Line Electroencephalographic Digital Computing System," *Electroencephalography and Clinical Neurophysiology* 18 (1965): 524–6.

Estrin, Thelma A., "Recording the Impulse Firing Pattern of Neurons Utilizing Digital Techniques," *Digest of the 1961 International Conference on Medical Electronics,* 1961.

Estrin, Thelma A., and P.F.A. Hoefer, "Revised Frequency Analyzer for Bioelectric Potentials in the Sub-Audio Range." *Review of Scientific Instruments* 25 (1954): 840–1.

Estrin, Thelma A., and R.C. Uzgalis, "Computerized Display of Spatio-Temporal EEG Patterns," *IEEE Transactions on Bio-Medical Engineering,* BME-16: 3 (1969): 192–6.

Estrin, Thelma A., R.J. Sclabassi, and R. Buchness, "Applications of Computer Networking in Neuroscience," *Proceedings of 5th Annual Meeting of the Society for Neuroscience,* New York, Nov. 2–6, 1975.

Estrin, Thelma A., R. Sclabassi, and R. Buchness, "Computer Graphics Applications to Neurosurgery," *Proceedings, First World Conference on Medical Information,* Stockholm, Sweden, August 1974: 831–6.

Estrin, Thelma A., J.V. Wegner, and R. Bettinger, "Computer-Generated Brain Maps," *Proceedings, San Diego Biomedical Symposium,* San Diego, CA, Feb. 5–7, 1975: 369–74.

"Eureka!" London *Sunday Times* Colour Supplement, Apr. 1971.

Eustis, Susan Huhn (Inventor, entrepreneur, computer scientist, Groton, MA), I-6/22/82; PC-6/22/82.

Evans, Christopher, *The Micro Millennium.* New York: Washington Square Press, 1981 (1979).

Evans, George Ewart, *Ask the Fellows Who Cut the Hay.* London: Faber, 1965.

Evans, George Ewart, *The Farm and the Village.* London: Faber, 1969.

Evans-Pritchard, Sir Edward, ed., *Peoples of the Earth,* 20 vols. Oakland, ME: Danbury House, 1972–3.

"E-vole-ution: Spring Is in the Grass," *Science News* 120: 14 (Oct. 3, 1981): 215.

"Ex-Directors Acquitted: Mr. and Mrs. Miles to Design Planes Again," London *Daily Telegraph,* June 3, 1950.

"Exercise and Late Menstruation," *Science News,* Oct. 17, 1981: 249.

Fabricant, Florence, "Ice Cream," American Heritage Society, *Americana* 3: 3 (July 1975): 19.

"Facts and Figures About Dutch Women in Medicine," *Journal of the American Medical Women's Association* 13: 6 (June 1958): 251–3.

Faegre, Torvald, *Tents: Architecture of Nomads.* Garden City, NY: Doubleday/Anchor, 1979.

Fagan, Brian M., and *Scientific American* eds, *Prehistoric Times.* San Francisco: Freeman, 1983.

Fairfax, Beatrice. See Marie Manning.

Faragher, John Mack, "The Midwestern Farming Family, 1859," in Kerber and Mathews, eds.

Farmer, Lydia H. *What America Owes to Women: The National Exposition Souvenir.* Buffalo, NY: Moulton, 1893.

Farnsworth, Norman R., Audrey S. Bingel, Geoffrey A. Cordell, Frank A. Crane, and Harry S. Fong, "Potential Value of Plants as Anti-Fertility Agents." *Journal of Pharmaceutical Science,* 64: 4 (Apr. 1975): 535–98; 64: 5 (May 1975): 717–54.

Feigenbaum, Edward. See Englemore, Ellie.

Feigenbaum, Edward, and Pamela McCorduck, *The Fifth Generation: Artificial Intelligence and Japan's Computer Challenge to the World.* Reading, MA: Addison-Wesley, 1983.

Feldman, Susan, ed., *African Myths and Tales.* New York: Dell, 1963.

Felton, Bruce, and Mark Fowler, *Famous Americans You Never Knew Existed.* New York: Stein & Day, 1979.

"Female Inventive Talent," *Scientific American* 23: (Sept. 17, 1870): 184.

Fenton, Alexander, *The Northern Isles: Orkney and Shetland.* Edinburgh: John Donald, 1978.

Fera, Darla, comp., *Women in American Agriculture: A Select Bibliography.* Washington, DC: U.S. Department of Agriculture, 1977.

Field, Laura, *et al.,* eds, *Women of Achievement.* New York: House of Field, 1940.

Fields, Sidney, "Only Human: Grandma of Navy's Computers," New York *Daily News,* Sept. 23, 1975.

Fink, Deborah, and Dorothy Schwieder, "Iowa Farm Women in the 1930s—A Reassessment," unpublished paper sent by Fink, 1984.

Fink, Virginia S., "What Work Is Real? Changing Roles of Farm and Ranch Wives in Southeastern Ohio," *Journal of Rural Studies* 7: 1/2(1991): 17–22.

Finlay, A.C., G[ladys] L. Hobby, *et al.,* "Terramycin, a New Antibiotic," *Science* 111 (1950): 85.

"Fire Destroys Home of Famous Cookie," San Francisco *Chronicle,* Jan. 2, 1985: 3.

"The First Class of '82," *Cleveland Magazine,* Jan. 1982: 43.

"First Woman in Inventors Hall of Fame," *Washington Post,* Mar. 2, 1991: D4.

Fischer, Mary A., "A California Scientist Remodels a Perpetual Problem's Ancient Solution: The Contraceptive Sponge," *People,* May 2, 1983: 57–8.

Fishel, Elizabeth. "Women's Self-Help Movement," *Ramparts,* Nov. 1973: 29–31ff.

Fisher, Elizabeth, *Woman's Creation: Sexual Evolution and the Shaping of Society.* Garden City, NY: Doubleday, 1979.

Fisher, Helen E., "Prehistoric Sex," *Science Digest* 90: 5 (May 1982): 34f.

Fisher, Helen E., *The Sex Contract: The Evolution of Human Behavior.* New York: Morrow, 1982.

Fisher, Marc, "Teenage Scientist: Discovery Could Revolutionize Agriculture," *Princeton Alumni Weekly;* repr., *Ms.,* July 1980: 23.

Fisher, Marc, and Carol Kovach, "Teenager's Find May Revolutionize Agriculture," *American Horticulturalist,* Nov. 1980: 1–2.

Fitch, Steve, "Freshman's Research in Botany Could Save Oil," *The Daily Princetonian,* Feb. 21, 1980: 1, 5.

"Five-Year Checkup," *Scientific American* 244:1 (Jan. 1981): 82.

Fjermedal, Grant (Author, Seattle, WA), PC-Spring 1987.

Fjermedal, Grant, *Magic Bullets.* New York: Macmillan, 1984.

Fjermedal, Grant, *The Tomorrow Makers.* New York: Macmillan, 1986.

Flannery, Kent V., "Origins and Ecological Effects of Early Domestication in Iran and the Near East," in Ucko & Dimbleby.

Fletcher, Susan, "Raggedy Ann and Andy Help Kids Report Crime," *Ms.,* Jan. 1981: 20.

"Flexitime Cited as Aid to Women." *Women's Agenda* (Women's Action Alliance, New York), 4:2 (Mar./Apr. 1979).

Foote, E[dward], Jr., MD, *The Radical Remedy in Social Science; or, Borning Better Babies Through Regulating Reproduction and Controlling Conception.* New York: Murray Hill, 1886. Arno repr. 1974, in Rosenberg & Rosenberg, eds.

Forbes, R.J., *Studies in Ancient Technology.* Leiden: E.J. Brill, 1964.

Forché, Carolyn, "Upheaval in El Salvador: Stories of Three Women," *Ms.,* Jan. 1980: 91ff.

*Foremost Women in Communications,* Barbara Love, ed. New York: Bowker, 1970.

Forfreedom, Ann (Feminist scholar, Sacramento and Oakland, CA), L-1979–83.

Forfreedom, Ann, ed., *Women Out of History: A Herstory Anthology.* Sacramento, CA: Forfreedom, 1972.

Formanek-Brunell, Miriam, "The Factory and the Nursery: Business Women in the American Doll Industry, 1870–1930," a paper presented at the Seventh Berkshire Conference on the History of Women, Wellesley College, June 1987.

Forster, Margaret, *Significant Sisters*. New York: Knopf, 1985.

"Fortunes in Small Inventions," *Scientific American* 63 (Oct. 11, 1890): 225.

Fowler, W[illiam] A., G[eoffrey] R. Burbidge, and E. Margaret Burbidge, "Stellar Evolution and the Synthesis of Elements, *Astrophysics Journal* 122 (Sept. 1955): 271–85.

"Frances Gabe's Self Cleaning House . . . ," *People* 17: 12 (Mar. 29, 1982): 38, 41.

Frank, Elizabeth Pope, "Hello . . . Hello . . . I Need Help!" *Good Housekeeping,* June 1980: 123ff.

Frank, Ruth F., *Something New Under the Sun: Building Connecticut's First Solar House.* Andover, MA: Brick House Publishing Co., 1980.

Franklin, Deborah, "Microbubbles Enhance Diagnostic Value of Ultrasound," Stanford University *Campus Report,* July 14, 1982: 10.

Frantz, Virginia, "An Evaluation of Radioactive Iodine Therapy in Metastatic Thyroid Cancer," *Journal of Clinical Endocrinology,* Sept. 1950: 1084–91.

Fraser, Antonia, *The Weaker Vessel.* New York: Vintage, 1982.

Frazer, Sir James G., *The Golden Bough* (abr.) New York: Macmillan, 1947.

Frazer, Sir James G., *Myths on the Origin of Fire.* London: Macmillan, 1930.

Frederick, Arthur L. (General Patent Counsel, Curtiss-Wright Corp., Wood-Ridge, NJ 07075), L-8/1/84.

Freke, Elizabeth. *Mrs. Elizabeth Freke, Her Diary: 1671 to 1714.* Mary Carbery, ed., Cork: Guy, 1913.

"French Officials Accept Gift of N.J. Inventor," *Afro-American,* Sept. 6, 1952: 18.

Frese, Anne (Dir. Niles Community Library, Niles, MI), L-12/10/86 and 1/15/87.

Friberg, Jöran, "Numbers and Measures in the Earliest Written Records," *Scientific American* 250: 2 (Feb. 1984): 110–8.

Friedman, Nancy, "TSS and the Sponge," *New West,* Mar. 1981: 60, 62.

[Friend, Dr. Charlotte], Obit., San Francisco *Chronicle,* Jan. 17, 1987: 15.

Frigo, Victoria, "How to Patent Your Invention." *Ms.,* Feb. 1976: 14–5.

*Fromages de France.* Paris: Société Civile d'Information et d'Edition des Services Agricole, 1953.

Froslid, Kenneth, "Helen Taussig, MD—Savior of Blue Babies." *Today's Health,* Aug. 1968: 48ff.

Fryer, Peter, *The Birth Controllers.* London: Secker & Warburg, 1965.

Fu, Caroline (Mgr., Speech Perception/Language Understanding/AI/ Intelligent Robotics, Boeing Aerospace, Seattle), c.v./1987, Q-9/29/87, L-1987–88.

Fu, Caroline, "A Hands-off Workstation," in Gavriel Salvendy, ed., *Advances in Human Factors/ Ergonomics,* vol. 10B: *Cognitive Engineering in the Design of Human-Computer Interaction and Expert Systems.* Amsterdam: Elsevier, 1987.

Fu, Caroline, "An Independent Workstation for a Quadriplegic," in World Rehabilitation Fund, *International Exchange of Experts and Information on Rehabilitation,* Monograph No. 37, *Interactive Robot Aids,* 1986.

Fukuda, Hitoshi, *Irrigation in the World: Comparative Developments.* Tokyo: University of Tokyo Press, 1976.

Fuller, Edmund, *Tinkers and Genius.* New York: Hastings House, 1955.

Fuller, Melvin, "Frances Gabe and Her Self-Cleaning House, *The Lightbulb* 10: 2 (Feb. 1979): 1, 11ff.

Furst, Alan, "The Golden Gage of Goo: How Debbi and Randy Fields Unleashed a Cookie Monster," *Esquire,* Dec. 1984: 324–30.

Fussell, G.E., *Farming Technique from Prehistoric to Modern Times.* New York: Pergamon, 1965.

GABe, Frances (Newberg, OR), Q-5/19/85; L-6/13/85.

GABe, Frances, "The GABe Self-Cleaning House," in Zimmerman, ed., *Technological.*

Gage, Matilda Joslyn, *Woman as Inventor.* Fayetteville: New York State Woman Suffrage Association, 1870 [Gage/1]; repr. (rev.) in *North American Review* 136, May 1883 [Gage/2].

Gage, Matilda J., *Woman, Church, and State.* Salem, NH: Ayer, 1893.

Gage, Suzann, *When Birth Control Fails.* Hollywood, CA: Speculum Press, 1979.

Galdone, Paul, *The Little Red Hen.* New York: Seabury, 1973.

Garbutt, Joan (Acquisitions editor, Philadelphia, PA), L-2/5, 3/22/83.

Gardner, W. David, "Computer Software and Services: Information Solutions for Growth Companies," *Inc.,* Dec. 1983: 173–87.

Garmon, Linda, "To Make Bones About It," *Science News* 119: 20 (May 16, 1981): 317–8.

Garrison, Vivian, "Rural Ethnic Support Systems Focus of NIMH Study," *ADHHA News* 7 (2/6/81): 2ff.

Garrison, Webb, "Many Major Inventions Were the Work of Women," *National Enquirer,* June 10, 1986: 13.

Gear, Edward (Institute of Patentees and Inventors, London, Eng.), I-5/79.

Geeyer, Robert F., Lois B. Wilson, Ferez S. Nallaseth, Paul F. Milner, Michael Bittner, and John T. Wilson, "Direct Identification of Sickle-Cell Anemia (SS) by Blot Hybridization," *Proceedings of the National Academy of Sciences* 78: 8 (Aug. 1981): 5081–5.

"Genetic Licensing Signs Eleven Firms," Stanford University *Observer,* May 1985: 9.

"Genetic Research Results in Breakthrough Against Hepatitis," Colusa, CA, *Sun-Herald,* Aug. 5, 1981: C-4.

Gerrier, Arthur J. (Maine Historical Society, Portland, ME), L-3/1, 26/82.

Gianutsos, Rosamond (Psychology, Adelphi University, Garden City, NY), L-5/5/82, c.v./1982, Q-5/5/82.

Gianutsos, Rosamond, "Using Microcomputers for Cognitive Rehabilitation," a paper presented to the American Psychological Association, Los Angeles, CA, Aug. 1981.

Gianutsos, Rosamond, "What Is Cognitive Rehabilitation?" *Journal of Rehabilitation,* July–Sept. 1980: 36–40.

Gibbon, John H., Jr., "Application of a Mechanical Heart and Lung Apparatus to Cardiac Surgery," *Minnesota Medicine* 37: 1 (Jan. 1954): 171–7.

Gibbon, John H., Jr., "The Development of the Heart-Lung Apparatus," *Review of Surgery* 27: 4 (July–Aug. 1970): 231–44.

[Gibbon, John H., Jr.], Obit., *New York Times,* Feb. 6, 1973: 40.

Gibbs, Angelica, "With Palette Knife and Skillet," *New Yorker,* May 28, 1949: 34ff.

Giedion, Siegfried, *Mechanization Takes Command: A Contribution to Anonymous History.* Oxford: Oxford University Press, 1948.

Gies, Joseph, and Frances Gies, *The Ingenious Yankees.* New York: Crowell, 1976.

Gilbert, Jean K., "Pioneer in Computers Combines Past with Future." *The Arizona Republic,* Women's Forum, Feb. 18, 1972: 45.

Gilbert, K.R., *Sewing Machines.* London: H.M. Stationery Office, 1970.

Gilbert, Lynn, and Gaylen Moore, *Particular Passions.* New York: Clarkson N. Potter, 1981.

Gilbert, Susan, "The Minds Behind the Top One Hundred," *Science Digest* 93: 12 (Dec. 1985): 27–63.

Gilgamesh. See N.K. Sanders.

Gill, Barrie, "Lily's Machine Types Music," London *Daily Herald,* Jan. 24, 1961.

Gilligan, Carol, *In a Different Voice: Psychological Theory and Women's Development.* Cambridge, MA: Harvard University Press, 1982.

Gimbutas, Marija, *The Goddesses and Gods of Old Europe, 6500 to 3500 BC: Myths and Cult Images.* London: Thames & Hudson, 1982.

Gimbutas, Marija, *The Gods and Goddeses of Old Europe, 7000 to 3500 BC: Myths, Legends, and Cult Images.* Berkeley: University of California Press, 1974.

Gimbutas, Marija, ed., *Neolithic Macedonia.* Los Angeles: University of Southern California Press, 1976.

Giscard d'Estaing, Valérie-Anne, *The World Almanac Book of Inventions.* New York: Ballantine, 1985.

Gleason, Venetia Taft (Granddaughter of Amy Johns Taft, Palo Alto, CA), PC-Jan. 1979.

Gleiser, Molly, "Lady Lovelace and the Difference Engine," *Computer Decisions,* May 1975: 38–41.

GMG Publishing, *Black Women: Achievements Against the Odds.* New York: GMG Publishing, 1981.

Godin, Jean Baptiste André, *Social Solutions,* tr. Marie Howland. New York: Lovell, 1886.

Godson, G. Nigel, "Molecular Approaches to Malaria Vaccines," *Scientific American,* May 1985: 52–9.

Godson, G. Nigel, et al., "Identification and Chemical Synthesis of a Tandemly Related Immunogenic Region of *Plasmodium knowlesi* Circumsporozoite Protein," *Nature* 305: 5929 (Sept. 1, 1983): 29–33.

Goff, Alice C., *Women Can Be Engineers.* Ann Arbor, MI: Edwards, 1946.

"Gold Drink May Beat Arthritis," *Star,* Aug. 28, 1984: 6.

Goldberg, Adele (Pres., ParcPlace Systems, Palo Alto, CA), Biog.-9/87; I-9/87; L-5/24/88.

Goldberg, Adele, "Introducing the Smalltalk-80 System," *Byte,* Aug. 1981: 14–26.

Goldberg, Adele, "Programmer as Reading," *IEEE Software,* Sept. 1987: 62–70.

Goldberg, Adele, *Smalltalk-80: The Interactive Programming Environment.* Reading, MA: Addison-Wesley, 1984.

Goldberg, Adele, and David Robson, *Smalltalk-80: The Language and Its Implementation.* Reading, MA: Addison-Wesley, 1983.

Goldberg, Leslie, "Staying King of the Hill in Toyland," San Francisco *Examiner,* Nov. 3, 1985: S-1-2.

Goldfeder, Anna, "An Overview of Fifty Years of Cancer Research: Autobiographical Essay," *Cancer Research* 36 (Jan. 1976): 1–9.

Goldsmith, Judith, *Childbirth Wisdom from the World's Oldest Societies.* New York: Congdon & Weed, 1984.

Goldstine, Herman (American Philosophical Society, Philadelphia), L-5/ 30, 8/18/84.

Goldstine, Herman, *The Computer from Pascal to Von Neumann.* Princeton, NJ: Princeton University Press, 1972.

Gomme, Arthur Allan, *Patents of Invention: Origin and Growth of the Patent System in Britain.* London: Longmans, Green, 1946.

Good, Edwin M., *Giraffes, Black Dragons, and Other Pianos.* Stanford, CA: Stanford University Press, 1982.

Goodale, Jane C., *Tiwi Wives: A Study of the Women of Melville Island, North Australia.* Seattle: University of Washington Press, 1971.

Goodall, Jane, "The Behavior of Free-Living Chimpanzees in the Gombe Stream Reserve," *Animal Behavior Monographs* I, 1968: 165–311.

Goodall, Jane, "Tool-Using and Aimed Throwing in a Community of Free-Living Chimpanzees," *Nature* 201 (Mar. 28, 1964): 1264–6.

Goodall, Jane, "Tool-Using in Primates and Other Vertebrates," *Advances in Studies of Behavior,* New York: Academy Press, 1970, vol. 3: 195–250.

Goodall, Jane, and David A. Hamburg, "Chimpanzee Behavior as a Model for the Behavior of Early Man: New Evidence on Possible Origins of Human Behavior. *American Handbook of Psychiatry* 6 (1974): 14–43.

Goodman, Mary Elena (Researcher, Menlo Park, CA), PC-1979.

Goodwater, Leanna, *Women in Antiquity: An Annotated Bibliography.* Metuchen, NJ: Scarecrow Press, 1975.

Gordon, Beverly, *Shaker Textile Arts.* Hanover, NH: University Press of New England, 1980.

Gordon, James Edward, *Structures; or, Why Things Don't Fall Down.* New York: Plenum, 1978.

Gore, Rick, "An Age-Old Challenge Grows," *National Geographic* 156: 5 (Nov. 1979): 594–639.

Gornick, Vivian, "Women in Science: A Passion for Discovery," *Ms.* 12: 4 (Oct. 1983): 50–2ff.

Gottschalk, Earl C., Jr., "Distaff Owners: Women Start Up Their Own Businesses . . . ," *The Wall Street Journal,* May 17, 1983: 1f.

Grady, Denise, "Candace Pert: Addicted to Research," *Discover,* Dec. 1981: 54–60.

Grady, Denise, "Heat Therapy for Cancer," *Discover,* June 1981: 84–6.

Grahn, Judy, "From Sacred Blood to the Curse and Beyond," in Spretnak, ed.

Grahn, Judy, *The Queen of Wands.* Trumansburg, NY: Crossing Press, 1982.

Grant, Michael, *Myths of Greeks and Romans.* Cleveland: World Publishing, 1962.

Graves, Robert, *The Greek Myths,* 2 vols. Baltimore: Penguin, 1955.

Gray, John, *Near Eastern Mythology.* London: Hamlyn, 1969.

Green, Maureen, "A French Family of Rose Hybridizers and the Best-Selling Magic It Works." *Smithsonian,* April 1969: 45ff.

Green, Shirley, *The Curious History of Contraception.* New York: St. Martin's, 1971.

Gregor, Thomas, "Privacy and Extra-Marital Affairs in a Tropical Forest Community," in Daniel Gross, ed., *Peoples and Cultures of Native South America.* New York: Natural History Press, 1973, 242–60.

Gregory, Annadora F., *Pioneer Days in Crete, Nebraska.* Lincoln, NE: State Journal Printing Co., c. 1937.

Greibach, Sheila Adele (Computer scientist, teacher, inventor, Computer Science Dept., UCLA), L-8/18/87.

Gribble, Katherine J., "Searching for Women Who Lost Their Names," paper presented at National Women's Studies Conference, Columbus, OH, June 26–30, 1983.

Grigson, Geoffrey, and Charles H. Gibbs-Smith, eds, *Things: A Volume of Objects Devised by Man's Genius Which Are the Measure of His Civilization.* London: Grosvenor, 1954.

Gross, Martin L., *The Doctors.* New York: Random House, 1966.

Grossworth, Marvin, "Don't Just Stand There—Invent!" *Science Digest,* Jan. 1976: 50ff.

Groth, Edward, III, "Lead in Canned Food," *Agriculture and Human Values* 3 (Winter–Spring 1986): 91–145.

Gruson, Sydney, "San Juan Talks Open on Birth Control; Theme Held Key to Caribbean Problems," *New York Times,* May 13, 1955; 8.

Gunby, Olive F., "Women Inventors," *Scientific American* Supplement No. 1325 (May 25, 1901): 21241–2.

Hafen, Leonora (Inventor, Jackson Hole, WY), I-5/81.

Hafter, Daryl (History, Eastern Michigan University, Ypsilanti, MI), L-5/12/80.

Hafter, Daryl, "Gender Formation from a Working Class Viewpoint: Guildwomen in Eighteenth-Century Rouen," *Proceedings of the Western Society for French History* 16 (1989).

Hafter, Daryl, "The Programmed Brocade Loom and the 'Decline of the Drawgirl,' " in Trescott, ed.

Hafter, Daryl, "[Report on the] International Conference on the Role of Women in the History of Science, Technology, and Medicine in the Nineteenth and Twentieth Centuries—Veszprém, Hungary, Aug. 15–9, 1983," *Technology and Culture* 16 (1985): 262–7.

Halcrow, Allan R., "The First Lady of Computing," *Interface Age,* Dec. 1983: 86.

Hale, Ralph W., and Ronald J. Pion, eds, "Laminaria: An Underutilized Clinical Adjunct," *Clinical Obstetrics and Gynecology* 15 (1972): 829–50.

Hall, Stephen S., "Eureka! Inventors, and How They Got Their Wheels Turning," *Smithsonian* 15 (Aug. 1984): 111–12ff.

Halsbury, Earl of, "Invention and Technological Progress" (4th Annual Spooner Lecture), *The Inventor* (London), June 1971: 10–34.

Hamamsy, Laila S., "The Role of Women in a Changing Navaho Society," in Tiffany.

Hambleton, Ronald, *The Branding of America.* Dublin, NH: Yankee Books, 1987.

Hamilton, Alice, *Exploring the Dangerous Trades: The Autobiography of Alice Hamilton.* Boston: Little, Brown, 1943.

Hamilton, Margaret, and S. Zeldin, *AXES Syntax Description,* TR-4. Cambridge, MA: Higher Order Software, Dec. 1976.

Hamilton, Margaret, and S. Zeldin, *The Foundations of AXES: A Specification Language Based on Completeness of Control,* Doc. R-964. Cambridge, MA: Charles Stark Draper Laboratory, Mar. 1976.

Hamilton, Margaret, and S. Zeldin, *Integrated Software Development System/ Higher Order Software Conceptual Description,* TR-3. Cambridge, MA: Higher Order Software, Nov. 1976.

Hamilton, Margaret, and S. Zeldin, "Higher Order Software—A Methodology for Defining Software," *IEEE Transactions on Software Engineering,* vol. SE-2, no. 1, 1976.

Hamilton, Margaret, and S. Zeldin, "Reliability in Terms of Predictability," Compsac '78, *Proceedings,* Chicago, IEEE Computer Society cat. no. 78CH1338-3C, Nov. 1978.

Hamilton, Mildred, "A Nursing Alternative," San Francisco *Examiner,* Dec. 28, 1983: E-7.

Hammerton, J.A., ed., *Concise Universal Biography.* London: Educational Book Co., 1934–5; 845.

[Hammond, M.B.] "Correspondence of Eli Whitney Relative to the Invention of the Cotton Gin," *American Historical Review* 3 (Oct.–Jul. 1898): 90–127.

Hanaford, Phebe A., *Daughters of America.* Augusta, ME: True & Co., 1883.

Hand, Douglas, "1955: The Making of the Polio Vaccine," *American Heritage of Invention and Technology* 1: 1 (Summer 1985): 54–7.

Handlin, David P., *The American Home: Architecture and Society, 1815–1915.* Boston: Little, Brown, 1979.

Handy, M.P., ed., *World's Columbian Exposition, 1893, Official Catalogue.* Part 1. Chicago: Conkey, 1893.

Hansson, Vidar, Martin Ritzen, Kenneth Purvis, and Frank S. French, eds, *Endocrine Approach to Male Contraception.* Copenhagen: Scriptor, 1978.

Hard, William, "Adventures in Industrial Citizenship," *Reader's Digest,* Mar. 1940: 35ff.

Harlap, Susan, *et al.,* "Prospective Study of Spontaneous Fetal Losses After Induced Abortions," *New England Journal of Medicine* 301 (Sept. 27, 1979): 677–81.

Harness, Arminta (Past Pres., Society of Women Engineers, Richland, WA), PC-1979.

Harp, Lore (Aplex Corp., San Mateo, CA), PC-7/28, 8/28/87; Q-1987.

Harper, Margaret H. (Computer programmer, analyst, inventor, Philadelphia, PA), Q-6/15/84.

Harper, R.M., R.S. Sclabassi, and Thelma Estrin, "Time Series Analysis and Sleep Research," *IEEE Transactions on Automatic Control,* vol. AC-19, no. 6, Dec. 1974: 932–43.

Harpur, Patrick, ed., *The Timetable of Technology.* New York: Hearst Books, 1982.

Harris, Anne Sutherland, and Linda Nochlin, *Women Artists, 1550–1950.* Los Angeles: Museum Associates, 1976.

Harris, Barbara J., *Beyond Her Sphere: Women and the Professions in American History.* Westport, CT: Greenwood, 1978.

Harris, Linda (Public Relations, Burpee Co., Warminster, PA), L-8/18/82.

Harris, Marvin, *Culture, People, Nature: An Introduction to General Anthropology,* 2nd ed. New York: Crowell, 1975.

Harris, Rivkah, "The *Naditu* Woman," in R.D. Biggs and J.A. Brinkman, eds, *Studies Presented to A. Leo Oppenheim, June 7, 1964.* Chicago: The Oriental Institute of the University of Chicago, 1964.

Harrison, Jane E., *Myths of Greece and Rome.* London: E. Benn, 1928.

Harrison, Michelle, *Self-Help for Pre-Menstrual Syndrome.* New York: Random House, 1985.

Hart, Emeline, "Colloquy Between Procrastination and Alacrity," Shaker *Manifesto* 19: 8 (Aug. 1889): 182–3.

Hartman, Curtis, "The Spirit of Independence," *Inc.,* July 1985: 47–50.

Hartman, Mary S., and Lois Banner, *Clio's Consciousness Raised: New Perspectives on the History of Women.* New York: Harper & Row, 1974.

Hartung, John, "Light, Puberty and Aggression: A Proximal Mechanism Hypothesis," *Human Ecology* 6: 3 (1978): 273–97.

"Harvard Astronomer Wins 1957 Achievement Award," *Journal of the American Association of University Women,* Oct. 1957, 15–6.

Hawkes, Jacquetta, *The Atlas of Early Man.* New York: St. Martin's, 1976.

Hawkes, Jacquetta, ed., *The World of the Past.* New York, Simon & Schuster, 1963.

Hawkes, Jacquetta, and Sir Leonard Woolley, *Prehistory and the Beginning of Civilization.* New York: Harper & Row, 1963.

Hawkins, Mary Hannah, "Nineteenth Century Household Patents by Women," in Betsy Warrior and Lisa Leghorn, eds, *Houseworker's Handbook,* 3rd ed. Cambridge, MA, 1975, 88.

Hayden, Brian, "Population Control Among Hunter/Gatherers," *World Archaeology* 4:1 (June 1972): 205–21.

Hayden, Dolores, *The Grand Domestic Revolution: A History of Feminist Designs for American Homes, Neighborhoods, and Cities.* Cambridge, MA: M.I.T. Press, 1981.

Hayden, Dolores, "Making Housing Work for People," *Ms.* 12: 7 (Jan. 1984): 69–71.

Hayden, Dolores, "Redesigning the Domestic Workplace," *Chrysalis,* No. 1 (1977): 19–29.

Hayden, Joan S. (Genealogist, Maine Historical Society, Portland, ME), L-3/8, 4/2/82.

Hayne, Arthur P., "Olives," in *A Report of Work of the Agricultural Experiment Stations of the University of California, 1894–95.* Sacramento, CA.: State Printing Office, 1896: 195ff.

Hazan, Paul L. (Program Dir., The Johns Hopkins University Applied Physics Laboratory, Laurel, MD), L-1/22/82.

Head, Sabin, and Patricia Wiener, "Procedural Entity-Relation Modeling for the Enterprise of Multi-processor Digital Systems Design," Final Project Report, U.S. Dept. of Defense Small Business Innovation Research Program, Feb. 15, 1987.

"Health Honors Given: Hospital Aide Receives First Oveta Culp Hobby Award," *New York Times,* Apr. 13, 1956: 27: 1.

"Heart-Lung Device Success in Surgery," *New York Times,* May 30, 1953.

Hecht, Eugene, *Physics in Perspective.* Reading, MA: Addison-Wesley, 1980.

Hecht, Jeff, "Pulsed Lasers Clear 'Port Wine' Birthmarks," *New Scientist,* Apr. 24, 1986: 32.

Heggie, Barbara, "Ice Woman," *New Yorker,* Sept. 6, 1941: 23ff.

Hellerstein, Erna O., Leslie Parker Hume, and Karen M. Offen, eds., *Victorian Women: A Documentary Account of Women's Lives in Nineteenth-Century England, France, and the United States.* Brighton, Eng.: Harvester, 1981.

Helmes, Winifred G., ed., *Notable Maryland Women.* Cambridge, MD: Tidewater, 1977.

Hendra, Barbara J., Associates, "Sierra On-Line Brings Software into the Home," News Release, n.d. [1984].

Hendra, Barbara J. (Barbara J. Hendra Associates, New York) L-4/3/84.

Henrotin, Ellen, "An Outsider's View of the Woman's Exhibit," *Cosmopolitan* 15 (Sept. 1893): 560–6.

"Hepatitis Vaccine," *Discover,* Dec. 1980: 12.

Herizon Women's Collective [Jackson, OH], "Women to Be Proud Of—Historical Portraits of Notable Jackson County Women," Exhibit at Jackson, OH, City Library, Nov. 1982.

Herlihy, David, "The Natural History of Medieval Women," *Natural History* 87: 3 (Mar. 1978): 56–67.

"Her Master's Voice," *The Observer* (London), Apr. 4, 1963.

Hern, Warren, "Knowledge and Use of Herbal Contraceptives in a Peruvian Amazon Village," *Human Organization,* 35: 1 (Spring 1976): 9–19.

Hernandez, Connie Sutherland (Researcher, Springfield, MO), PC-3/26/81.

Herodas, *The Mimes and Fragments,* ed. A.D. Knox. Cambridge: Cambridge University Press, 1922.

Herodotus, *History,* tr. George Rawlinson. New York: Tudor, 1928.

Herr, E.M., "Benjamin G. Lammé," editorial, *Electric Journal,* June 1919; repr. Westinghouse Electric and Manufacturing Company Sales Letter No. 313, June [1919]: 1–2.

Her Say, "Mothers of Invention," *Matrix* 6: 11 (Mar. 1982).

Herter, George, and Bertha Herter, *Bull Cook and Authentic Historical Recipes and Practices,* 3 vols. Waseca, MN: Herter's, 1960–74.

Heschong, Lisa, "Thermal Necessity," *Natural History* 90: 10 (Oct. 1981): 32–7.

Hewat, Mathew L., MD, *Bantu Folk Lore.* Westport, CT: Negro Universities Press, 1906 (1970).

"Hey, Did You Know That . . . ," *In Site* (Trade Annual of the National Association of Confectioners), 1982.

Heyn, Ernest V., *The Fire of Genius: Inventions of the Past Century.* New York: Doubleday/Anchor, 1976.

Heyn, Ernest V., "Ten Better Mousetraps," *Saturday Review* 5 (July 22, 1978): 20–1.

Heywood, William S., *History of Hopedale Community,* Lowell, MA: Thompson & Hill, 1897.

Hicks, Jim, *The Empire Builders.* Alexandria, VA: Time-Life Books: 1974.

Higham, C.F.W., "Prehistoric Rice Cultivation in Southeast Asia, *Scientific American* 250: 4 (Apr. 1984): 100–7.

Hillinger, Charles, "Inventor Building 'Self-Cleaning' House," Los Angeles *Times,* Nov. 1, 1981, Part I-B: 3.

Himes, Norman E., *Medical History of Contraception.* New York: Schocken, 1970 (1936).

Hinchcliffe, Frances, *Knit One, Purl One: Historic and Contemporary Knitting From the V & A's Collection.* London: Victoria and Albert Museum, [1985].

Hinding, Andrea, ed., *Women's History Sources: A Guide to Archives and Manuscript Collections in the United States.* New York: Bowker, 1979.

Hinkley, Edith Quimby, "Some Practical Considerations Regarding the Employment of Various Qualities of Roentgen Rays in Therapy," *Radiology* 38: 3 (Mar. 1942): 261–73.

*History of Dearborn and Ohio Counties, Indiana.* Chicago: Weakley, 1885.

Hitchens, Arthur P., and Morris C. Leikind, "Introduction of Agar-agar into Bacteriology," *Journal of Bacteriology* 37 (1939): 485–93.

Hitchings, Catherine F., *Universalist and Unitarian Women Ministers.* Boston: Universalist Historical Society, 1975.

Hobday, Stephen (Engineer, inventor, Farnsborough, Hants., Eng.), L-6/79.

Hobson, Laura Z., "Follies Girl to Scientist," *Independent Woman* 20: 10 (Oct. 1941): 297–8f.

Hofstadter, Laura, "Allergy Shots May Avert Asthma Effects . . . ," Stanford University *Campus Report,* Sept. 16, 1987: 9.

Hofstadter, Laura, "Chemical Selectively Kills Tumors," Stanford University *Campus Report,* May 17, 1987: 13.

Hofstadter, Laura, "Combination Antibody May Fight MS," Stanford University *Campus Report,* Jan. 23, 1985: 17.

Hofstadter, Laura, "Experimental Drug May Fight Cancer with More Power," Stanford University *Campus Report,* July 3, 1985: 18.

Hofstadter, Laura, "Folk Aphrodisiac Works in Male Rats," Stanford University *Campus Report,* Sept. 5, 1984: 15.

Hofstadter, Laura, "Medical School Students Design Synthetic Helix," Stanford University *Campus Report,* Feb. 17, 1988: 12.

Hofstadter, Laura, "Mice with Human Immune Cells Offer New Approach to Studying Disease," Stanford University *Campus Report,* Sept. 21, 1988: 1, 21.

Hofstadter, Laura, "Monoclonal Antibodies Offer Anti-Rejection Promise," Stanford University *Campus Report,* July 29, 1987: 7.

Hofstadter, Laura, "New Techniques Improve Safety of Vaccinations," Stanford University *Campus Report,* Oct. 23, 1985: 12, 15.

Hofstadter, Laura, "Patent Covers Lab-Constructed DNA Molecules," Stanford University *Campus Report,* Sept. 5, 1984: 1f.

Hofstadter, Laura, " 'Piggyback' Vaccines May Offer Enhanced Protection," Stanford University *Campus Report,* Dec. 18, 1985: 11.

Hofstadter, Laura, "Protein Substance May Be Useful in Treating AIDS," Stanford University *Campus Report,* Mar. 14, 1984: 18.

Holden, Constance, "NASA Satellite Project: The Boss Is a Woman," *Science* 179: 4068 (Jan. 5, 1973): 48-9.

Holmberg, Allan R., *Nomads of the Long Bow: The Siriono of Eastern Bolivia.* Garden City, NY: 1969 (1950).

[Holmes, Verena, Obit.], *The Chartered Mechanical Engineer,* 1964.

[Holmes, Verena], *The Woman Engineer* (London) 5: 5 (Dec. 1940): 70.

Holzer, Guenter, "Bertha Lammé: First Woman Electrical Engineer," unpublished paper (1983), sent me by the author.

Holzer, Guenter (Retired engineer, St. Cloud, MN), PC-1982-3.

Hopkins, Arthur John, *Alchemy: Child of Greek Philosophy.* New York: AMS Press, 1967.

Hopkins, Keith, "Contraception in the Roman Empire," *Comparative Studies in Society and History* 8 (Oct. 1965): 124-51.

Hopkins, Timothy, *The Kelloggs in the Old World and the New,* vol. 2. San Francisco: Sunset Press, 1903.

Hopper, Grace Murray (Pioneer computer inventor), c.v./1979.

"Hopper, Grace Murray," *Navy Times,* Nov. 21, 1973.

Hopper, Grace Murray, "Compiling Routines," *Computers and Automation* 2: 4 (May 1953).

Hopper, Grace Murray, "Dispersal of Computing Power," *Proceedings,* 1973 Conference, Data Processing Institute. Ottawa, 74-80.

Hopper, Grace Murray, "The Education of a Computer," Paper presented at the meetings of the Association for Computing Machinery, May 2-3, 1952.

Hopper, Grace Murray, "Possible Futures and Present Actions." *Proceedings,* Fifth Australian Computer Conference, Brisbane, 1972, 272-3.

Hopper, Grace Murray, "Possible Futures—Hardware, Software, and People," Commencement Address, Technological Institute, Northwestern University, Evanston, IL, June 1978.

Hopper, Grace Murray, "Preliminary Definitions, Data-Processing Compiler," Univac Division internal document, Jan. 31, 1955 (15pp.), sent me by (then-Capt.) Hopper.

Hopper, Grace Murray, "Standardization and the Future of Computers," *Data Processing* 14: 329-35.

Hopper, Grace Murray, "Technology: Future Directions." *Proceedings,* 1974 Conference, Data Processing Institute, Ottawa, 251-6.

Hopper, Grace M., and John Mauchly, "Influence of Programming Techniques on the Design of Computers," *Proceedings of the Institute of Radio Engineering* 41: 10 (Oct. 1953): 1250-4.

Hopping, Mary E., "Incidents of Pioneer Life, As I Remember and As I Have Been Told," typescript, California State Library, Sacramento.

Hopson, Janet, "The Freeze-drying Technique Makes for Movable Feasts," *Smithsonian* 14: 4 (July 1983): 90-7.

Horgan, Edward R., *The Shaker Holy Land.* Harvard, MA: Harvard Common Press, 1982.

Hornsby, Jeremy, *The Story of Inventions.* New York: Crescent, 1977.

Hornung, Clarence, *The Way It Was: New York, 1850-1890.* New York: Schocken, 1977.

"The Hospice: A Decent Way to Die," *Ms.,* Nov. 1981: 96.

Hossie, Linda, "Society a Hurdle for Women Inventors," *Toronto Globe and Mail,* Mar. 6, 1986: A3.

Houlihan, Sherida, and John Wotiz, "Women in Chemistry Before 1900." *Journal of Chemical Education* 52: 6 (June 1975): 362-4.

"Household Inventions," *Scientific American* 74 (Feb. 15, 1896): 99 (condensed from an article by Hannah Otis Brun in *American Kitchen Magazine,* Boston, Jan. 1896).

[Howard, Helen, *et al.*], "The Year's Top One Hundred Innovations and the Men and Women Behind Them," *Science Digest* 93: 12 (Dec. 1985): 27-63.

Howes, Durwood, ed., *American Women, 1935-40: A Composite Biographical Dictionary.* Detroit: Gale, 1981.

"How to Knit a Yacht," *Time,* Mar. 15, 1963: 77.

Hoy, Marjorie, "Genetic Improvement of *Metaseiulus occidentalis:* Implementing Pesticide-Resistant Strains—Progress and Problems," in P.J. Cameron *et al.,* eds, *Proceedings,* Australasian Workshop on Development and Implementation of IPM, Mt. Albert Research Centre, Auckland, NZ, July 20–22, 1982. Auckland: Government Printer, 1982.

Hoy, Marjorie, with William W. Barnett, Lonnie C. Hendricks, Darryl Castro, Daniel Cahn, Walter J. Bentley. "Managing Spider Mites with Pesticide-Resistant Predators," *California Agriculture* 38: 7, 8 (July/ Aug., 1984): 18.

Hoy, Marjorie, with Hugo E. van de Baan, J.J. Rob Groot, and Ross P. Field, "Aerial Movements of Mites in Almonds: Implications for Pest Management," *California Agriculture* 38: 9 (Sept. 1984): 20.

Høyrup, Else, *Women and Mathematics, Science, and Engineering: A Partially Annotated Bibliography with Emphasis on Mathematics.* . . . Roskilde, Denmark: University Library, 1978.

Hrdy, Sarah Blaffer, "Heat Loss," *Science/83,* Oct. 1983: 73–8.

Hrdy, Sarah Blaffer, *The Woman That Never Evolved.* Cambridge, MA: Harvard University Press, 1981.

Hudson, Donna L., and Thelma Estrin, "EMERGE—A Data-Driven medical Decision-Making Aid," *IEEE Transactions on Pattern Analysis and Machine Intelligence,* vol. PAMI-6, no. 1, Jan. 1984: 87–91.

Hufnagel, V., *No More Hysterectomies.* New York: New American Library, 1988.

Hughes, Muriel Joy, *Women Healers in Medieval Life and Literature.* New York: King's Crown Press, 1943.

Huhn, Charles C. (Dir., Public Relations, Bissell, Grand Rapids, MI), L-4/25/88.

Huhn, S. See Eustis, Susan Huhn.

Humphreys, Mary Gay, "The Ways of Women Ranchers," *Century,* 83: 4 (Feb. 1912): 525–34.

"A Hundred Pockets for This Old Lady," *New York Times,* Nov. 7, 1908: 14:5.

Hunt, Margaret (Researcher, Cambridge, MA), L-3/78, PC-1980.

Hurd-Mead, Kate Campbell. See Mead, Kate C. Hurd-

Hurt, Douglas R., "The First Farmers in the Ohio Country," *Agriculture and Human Values* 2: 3 (Summer 1985): 5.

Hutchens, Alma R., *et al., Indian Herbalogy of North America.* Toronto: Merco, 1973 (1969).

Hutchins, Sandra E. (Computer software designer, engineer, inventor, Emerson & Stern Associates, Del Mar, CA), c.v./1984.

Hutchins, Sandra E., "Making Changes: Women in Technical Policy-Making," in Jan Zimmerman, ed., *Future.*

Hyams, Edward, *A History of Gardens and Gardening.* New York: Praeger, 1971.

Hyde, Ida, "A Micro-electrode and Unicellular Stimulation," *Biological Bulletin* 40 (1921): 130–3.

Hymowitz, Carol, and Michaele Weissman, *A History of Women in America.* New York: Bantam, 1978.

*Hypatia's Sisters: Biographies of Women Scientists, Past and Present.* Seattle: Feminists Northwest, 1976.

"Hyperthermia's Hot Spot Is Engineering," *Science News* 119 (May 2, 1981): 277–8.

"Idea Awards Given to Eight Workers," *New York Times,* Mar. 4, 1944: 21:1.

"Ideas That Made Money." See J.M.

IEEE *Search.* See Institute of Electrical and Electronic Engineers.

"If Your Maxi Doesn't Look Like This . . . ," ad for "Always Plus" Maxi-pads, *People,* June 30, 1986: 40–1.

"Infant Mortality: What Trends Show," *Science News* 129: 8 (Feb. 22, 1986): 123.

"Information Useful to Patentees," *Scientific American* 5 (Dec. 7, 1861): 363.

Ingels, Margaret, "Petticoats and Slide Rules." *Midwest Engineer,* Aug. 1952.

"The Innovators: Wonder Fabric Wizards," *Textile World* (McGraw-Hill's International Textile Magazine), Dec. 1972.

Institute of Electrical and Electronic Engineers, Annual Reports of the Secretary, 1977, 1980, 1983, 1986.

Institute of Electrical and Electronic Engineers, *Proceedings of the Johns Hopkins First National Search for Applications of Personal Computing to Aid the Handicapped.* Silver Spring, MD: IEEE, 1981.

International Exhibition of Inventions, Eleventh, London: *The Times,* Oct. 3, 1935: 9.

*Inventive Age,* ed. and pub. James T. DuBois and Robert G. DuBois. Washington, DC, 1889–1914.

*The Inventor* (London): Institute of Patentees and Inventors, Staple Inn Bldgs, London, 1928– .

"Inventors Hall of Fame Honors Gertrude B. Elion," *Wall Street Journal,* Feb. 27, 1991:B3.

Ireland, Norma O., *Index to Women of the World.* Westwood, MA: Faxon, 1970; suppl. ed., Metuchen, NJ: Scarecrow Press, 1988.

Irvin, Helen Deiss (Scholar, lawyer, Lexington, KY, and Arlington, MA), L-1979–82.

Irvin, Helen Deiss, "The Machine in Utopia: Shaker Women and Technology," *Women's Studies International Quarterly* 4:3 (Fall 1981): 313–9.

Irvin, Helen Deiss, *Women in Kentucky.* Lexington: University Press of Kentucky, 1979.

Irwin, Inez Haynes, *Angels and Amazons.* Garden City, NY: Doubleday, 1933.

Isaacson, Portia (Computer scientist, inventor, Colorado Springs, CO), c.v./1988.

Isaacson, Walter, "The Battle over Abortion," *Time* 117:14 (Apr. 6, 1981): 20.

Israel, Fred L., ed., *1897 Sears Roebuck Catalogue.* New York: Chelsea House, 1968.

Israel, Paul, *From the Machine Shop to the Industrial Laboratory: Telegraphy and the Changing Context of American Invention, 1830–1920.* Baltimore: Johns Hopkins, 1992.

Itani, Jun'ichiro, "The Society of Japanese Monkeys," *Japan Quarterly* 8 (1961): 421.

Iverson, Kenneth E., *A Programming Language.* New York: Wiley, 1962.

Ives, Patricia Carter, "Creativity and Inventions: Black Women with Inventive Genius," National Society of Black Engineers *News,* Oct./Nov. 1984: 47–9.

Ives, Patricia Carter, "Patent and Trademark Innovations of Black Americans and Women." *Journal of the Patent Office Society* 62:2 (Feb. 1980): 108–26.

J.M., "Ideas That Made Money," *Christian Science Monitor,* Jan. 10, 1942: 15.

Jackson, Crowell B. "Jackie" (Inventor, New York, NY), Q-1978; L-1978–85.

Jackson, Donald D., "If Women Needed a Quick Pick-me-up, Lydia Provided One," *Smithsonian* 15: 4 (July 1984): 107–19.

Jackson, Sandra (Teacher, software inventor, San Antonio, TX), c.v./1982, L-8/1/82.

Jackson, Sandra, "We Help More," IEEE *Search.*

Jacobs, Aletta H., MD, "Holland's Pioneer Woman Doctor," *The Medical Woman's Journal* 35: 9 (Sept. 1928): 257–9.

Jacobsen, Thorkild, "Notes on Nintur," in Giorgio Buccellati, ed., *Approaches to the Study of the Ancient Near East: A Volume of Studies Offered to Ignace Jay Gelb.* Rome: Biblical Institute Press, 1972, 274–98.

Jacobsen, Thorkild, and Robert M. Adams, "Salt and Silt in Ancient Mesopotamian Agriculture," *Science* 128: 3334 (Nov. 21, 1958): 1251–8.

James, Edward T., Janet Wilson James, and Paul S. Boyer, eds, *Notable American Women, 1607–1950: A Biographical Dictionary,* 3 vols. Cambridge, MA: Harvard University Press, 1971.

Janeway, Elizabeth, ed., *Women: Their Changing Roles.* New York: Arno Press, 1978.

Jarman, M.R., G.N. Bailey, and H.N. Jarman, eds, *Early European Agriculture: Its Foundations and Development,* Cambridge: Cambridge University Press, 1982.

Jayne, Walter Addison, *The Healing Gods of Ancient Civilizations.* New Haven, CT: Yale University Press, 1925.

Jeffrey, Julie Roy, *Frontier Women: The Trans-Mississippi West, 1840–1880.* New York: Hill & Wang, 1979.

Jeffries, John P., "The New Alchemy of Photovoltaics," *Sierra,* Nov./Dec. 1983: 45–9.

Jensen, Joan M., "Churns and Butter Making in the Mid-Atlantic Farm Economy, 1750–1850," in Porter and Mulligan.

Jensen, Joan M., "Cloth, Butter, and Boarders: Women's Household Production for the Market," *Review of Radical Economics* 12:2 (Summer 1980).

Jensen, Joan M., "Native American Women and Agriculture: A Seneca Case Study," *Sex Roles* 3 (1977): 5: 423–46.

Jensen, Joan M., *With These Hands: Women Working on the Land.* Old Westbury, NY: Feminist Press, 1981.

Jensen, Oliver, *The Revolt of American Women.* New York: Harcourt Brace, 1971 (1952).

"Jessie Cartwright: Our Name in the Hall of Fame," Grant Hospital of Chicago, *Grant Gazette* 11: 2 (Apr./May 1972): 1.

Jett, Marva N. (USDA, SEA, Washington, DC), L-5/29/80.

Jobin, Judith, "How To Turn Your Idea into a Money-Maker," *Woman's Day,* Sept. 1, 1978: 60ff.

Jobin, Judith (Writer), PC-3/3/80.

Johnson, Kirk A., "Perspective Paper: Lead in Canned Food," *Agriculture and Human Values* 3 (Winter-Spring 1986): 146–56.

Johnson, Rebecca, *How to Hatch, Brood, Feed, and Prevent Chicks from Dying in the Shell.* Maxwell, IA: L.R. Shepherd, 1906.

Johnson, Rebecca, "A Woman's Remarkable Success with Poultry," *Tenth Annual Iowa Yearbook of Agriculture (1909).* Des Moines: Iowa Dept. of Agriculture, 1910.

Johnson, Sonia, *From Housewife to Heretic.* Garden City, NY: Doubleday, 1983.

Johnson, William Weber, "Two New Exhibitions Explore the Dark Mysteries of the Maya," *Smithsonian* 17: 2 (May 1986): 38–46+.

Johnston, Terry, "Dvorak *vs.* Qwertyuiop = Speed," Stanford University *Observer,* Apr. 1985: 4.

Jones, Glyn, *Welsh Legends and Folk-Tales.* London: Oxford University Press, 1955.

Jones, Stacy V. (*New York Times* "Patents" columnist, Arlington, VA), PC-1979–80.

Jones, Stacy V., "Patents" column, *New York Times,* 1952–88. Cited by date and page in text.

Jones, Stacy V., "Patents: The Woman's Side of Patent Activity," *New York Times,* Feb. 24, 1979.

Joni, Saj-Nicole (Computer Science, Wellesley College, Wellesley, MA), L-2/28/84.

Juster, Norton, *So Sweet to Labor: Rural Women in America, 1865–1895.* New York: Viking, 1979.

Kaehler, Ted, and Dave Patterson, "A Small Taste of Smalltalk," *Byte,* Aug. 1986: 145.

Kaempffert, Waldemar, ed., *A Popular History of American Invention.* New York: Scribner's, 1924.

Kalat, James, *Biological Psychology.* Belmont, CA: Wadsworth, 1980, 1988.

Kane, Joseph N., *Famous First Facts . . . in American History,* 4th ed. New York: H.W. Wilson, 1981.

Karsten, Rafael, "A Totalitarian State of the Past: The Civilization of the Inca Empire in Ancient Peru," *Commentationes Humanarum, Litterarum* (Elsinore, Denmark) 16: 1 (1949): 90ff.

Kary, Lee, "She Started Something," *Mountain Life and Work* 40:4 (Winter 1965): 8–11.

Kash, Sara, "Birth Control Survey: Sterilization Tops List in U.S.," *Ms.,* Jan. 1984: 17.

Kawatzky, Vivian, "What Did a Woman Ever Invent?" Milwaukee *Sentinel,* Nov. 29, 1961: H-3.

Kaye, Ivan, "Eight Americans Chart the Mysteries of the Ultimate Frontier: The Human Brain," *People* 11:5 (Feb. 5, 1979): 24–9.

Kaye, Marvin, *The Story of Monopoly, Silly Putty, Bingo, Twister, Frisbee, Scrabble, et cetera.* New York: Stein & Day, 1977 (1973).

Keenan, Katherine, "Lilian Vaughan Morgan (1870–1952): Her Life and Work," *American Zoologist* 23 (1983): 867–76.

Keenan, Katherine, and Harriet Laine, "A Biographical Survey of Women in Biology," a paper presented at the National Women's Studies Association Convention, Humboldt State University, Arcata, CA, June 1982.

Keithler, Susan N. (Public Relations, Personal Products Corp., Milltown, NJ), L-1/5/83.

Kellogg, Clara Louise, *Memoirs of an American Prima Donna.* New York: G.P. Putnam's Sons, 1913.

Kempf, Martine (Inventor, entrepreneur, Sunnyvale, CA), I-7/18/88, Q-7/18/88, L-7/18/88.

Kempf, Martine, "History of the Katalovox," Sunnyvale, CA: KEMPF, 1987.

Kendrick, Pearl L., "Use of Alum-Treated Pertussis Vaccine, and of Alum-Precipitated Combined Pertussis Vaccine and Diphtheria Toxoid, for Active Immunization," *American Journal of Public Health* 32 (June 1943): 615–26.

Kennedy, David M., *Birth Control in America: The Career of Margaret Sanger.* New Haven, CT: Yale University Press, 1970.

Kennedy, Rankin, *The Book of the Motor Car.* London: Caxton, [191?].

Kennedy-Way, Doreen, *Kennedy Kinetics* (brochure). Dudley, West Midlands, Eng.: [1979].

Kennedy-Way, Doreen (Dudley, West Midlands, Eng.), L-11/14/79.

Kenny, Sister Elizabeth (with Martha Ostenso), *And They Shall Walk.* New York: Dodd, Mead, 1943.

Kerber, Linda K., and Jane DeHart Mathews, eds, *Women's America: Refocusing the Past.* New York: Oxford University Press, 1982.

Kerenyi, Carl, *Eleusis: Archetypal Image of Mother and Daughter.* Princeton, NJ: Princeton University Press, 1967.

*Kettner's Book of the Table.* See E.S. Dallas.

"Killing Bacteria Effectively and Safely," *Science News* 130:18 (Nov. 11, 1986): 280.

Kilmer, Anne Draffkorn, "The Mesopotamian Concept of Overpopulation and Its Solution as Reflected in the Mythology," *Orientalia* 41 (n.s., fasc. 2), 1972: 160–77.

Kimberly-Clark Corporation (Neenah, WI), *Product Fact Sheet, Kotex Feminine Napkins* [1979].

Kimberly-Clark Corporation, Neenah, WI, *Supplemental Data Booklet, Kotex Products* (1978).

King, Mary Sarah, "Two Engineers, Two Approaches," Boston *Globe*, July 20, 1972: 64f.

King, William C., comp., *Woman: Her Position, Influence, and Achievement Throughout the Civilized World*. Springfield, MA: King-Richardson, 1902.

Kippley, Sheila, *Breast-Feeding and Natural Child Spacing: The Ecology of Natural Mothering*. New York: Harper & Row, 1974.

Kirkbride, Diana, "Beidha: An Early Neolithic Village in Jordan," *Archaeology* 19:3 (June 1966): 199–207.

Kirschner, Heidi, *Fireless Cookery*. Seattle: Madrona, 1981.

Kitchen, Clarissa (Palo Alto, CA), PC-8/24/83.

Kitzinger, Sheila, *Women as Mothers*. New York: Random House, 1978.

Klein, Aaron E., and Cynthia Klein, *The Better Mousetrap: A Miscellany of Gadgets, Labor-Saving Devices, and Inventions That Intrigue*. New York: Beaufort, 1982.

Kleinberg, Susan J., "Technology and Women's Work: The Lives of Working Class Women in Pittsburgh, 1870–1900," in Trescott, ed.

Kleinegger, Christine, "Labor-Saving and Money-Spending: Farm Women's Strategies for Acquiring Labor-Saving Devices in the Early Twentieth Century," a paper presented at the University of Connecticut Conference on Women, Work and Technology, Storrs, Oct. 1984.

Knight, Kate Brannon, *The History of the Work of Connecticut Women at the World's Columbian Exposition, Chicago, 1893*. Hartford, CT: Hartford Press, 1898.

Knowles, Melita, "Persevering Briton Invents Typewriter for Music," *Christian Science Monitor*, Aug. 14, 1962.

Knuth, Donald E., *The Art of Computer Programming*. Reading, MA: Addison-Wesley 1973 (1968).

Kobes, Marjorie (Librarian, Dvoracek Memorial Library, Wilber, NE), L-5/11/87.

Kolata, Gina B., "!Kung Hunter-Gatherers: Feminism, Diet, and Birth Control," in Michael H. Logan and Edward E. Hunt, Jr., *Health and the Human Condition*. North Scituate, MA: Duxbury Press, 1978, 118–22.

Kolbjørnsen, Astrid, and Oddleiv Apneseth, "Eneste kvinne på oppfinnermessen [Only Woman at Inventors' Fair]," Bergen *Tidende* [*Tidings*], Apr. 4, 1986: Trade/Industry page.

Koral, April, "A Young Inventor Sheds Light on a Tricky Problem," *Seventeen*, Apr. 1977: 104.

Kosikowski, Frank V. (Food Science, Cornell University, Ithaca, NY), L-5/31, 6/6, 7/1/85.

Kosikowski, Frank V., "Cheese," *Scientific American*, May 1985: 88–92ff.

Kosikowski, Frank V., *Cheese and Fermented Milk Foods*, 2nd ed. Ann Arbor, MI: Edwards Bros., 1977.

Kotulak, Ronald, "Sorry, Mary, You're No Hero After All," Chicago *Tribune*, Oct. 1, 1976: 1, 4ff.

Kovach, Carol, "Bryenton Makes Strides," Fairview Park [Cleveland], OH, *Sun Herald*, July 24, 1980: 25.

Kraditor, Aileen S., ed., *Up from the Pedestal*. Chicago: Quadrangle Books, 1968.

Kramer, Samuel N., *Sumerian Mythology*. Philadelphia: University of Pennsylvania Press, 1972.

Kramer, Samuel N., *The Sumerians: Their History, Culture, and Character*. Chicago: University of Chicago Press, 1963.

Kranzberg, Melvin, and Carroll W. Pursell, Jr., *Technology in Western Civilization*, 2 vols. New York: Oxford University Press, 1967.

Kroeber, Theodora, *Ishi in Two Worlds*. Berkeley: University of California Press, 1961.

Kroeber, Theodora, *Ishi: Last of His Tribe*. Berkeley, CA: Parnassus Press, 1964.

Kroening, Mary E., and Jeff L. Levinsky, "CHAOS II: Design and Implementation of a Timeshared Microcomputer System," in Lucas and Adams, eds, *Personal Computing Proceedings*. New York: AIFPS, 1979.

Krueger, Ann, "Welcome to the Club," *Video Games*, 1983: 51–3ff.

Kuehnl, Eileen K., "Maude Adams, an American Idol . . . ," PhD dissertation, University of Wisconsin, Madison, 1984.

Kulsrud, Helene E. (Computer scientist, inventor, Princeton, NJ), Q-8/22/ 84, c.v./1984.

Kulsrud, Helene E., "A General Purpose Graphic Language," *Communications of the Association for Computing Machinery* 11 (1968): 4.

Kulsrud, Helene E., R.D. Schultz, and R.S. Knight, *IDAL Reference Manual*, July 1979.

Kundsin, Ruth, ed., *Women and Success: The Anatomy of Achievement*. New York: Morrow, 1974.

Kunitz, Stanley J., and Howard Haycraft, eds, *American Authors, 1600–1900: A Biographical Dictionary of American Literature.* New York: H.W. Wilson, Co., 1938.

Kupry, Natalia, *Profiles in Labor: Essays About Heroes of Socialist Labour.* Moscow: Progress Publishers, 1980.

Kurtzig, Sandra L. [with Tom Parker], *CEO: Building a $400 Million Company from the Ground Up.* New York: W.W. Norton, 1991.

*LD/00: Literary Digest* article on recent progress of U.S. women inventors. See "Women as Inventors . . ."

Lacey, Louise, *Lunaception: A Feminine Odyssey into Fertility and Contraception.* New York: Warner, 1976 (1974).

*Ladies' Home Journal* eds, "Women in the Eighties," *Ladies' Home Journal,* Nov. 1979: 187–94.

*The Ladies' Who's Who.* London: International Art and Publishing Co., 1919.

"Lady Edison . . . ," *Literary Digest,* May 8, 1937: 20–1.

Laffan, Jack (Ret. IBM computer scientist, Upper Falls, MD), I-5/6/85.

Lambert, Eleanor, *World of Fashion: People, Places, Resources.* New York: Bowker, 1976.

Lammé, Benjamin, *Benjamin Garver Lammé: An Autobiography.* New York: Putnam's Sons, 1926.

Lampe, David, "Give Me a Home, Where the Snow Monkeys Roam," *Discover* 9: 7 (July 1988): 36–43.

Lamperti, Claudia (Printer, publisher, Norwich, VT), L-1/28/84.

Lancaster, C.S., "Women, Horticulture, and Society in Sub-Saharan Africa," *American Anthropologist* 78 (1976): 539–64.

Lancaster, Jane, "On the Evolution of Tool-Using Behavior," *American Anthropologist* 70 (1968): 56–66.

Lancaster, Jane, *Primate Behavior and the Emergence of Human Culture.* New York: Holt, Rinehart, 1975.

Lancaster, Paul, "Crazy about Invention," *American Heritage of Invention and Technology* 1:1 (Summer 1985): 42–53.

Landes, Ruth, *The Ojibwa Woman.* New York: Norton, 1971 [1969].

Langer, William L., "Checks on Population Growth: 1750–1850," *Scientific American* 226:2 (Feb. 1972), 92–9.

Langone, John, "The Quest for the Male Pill," *Discover,* Oct. 1982: 26–30.

Langone, John, "Riddle of the Tampon," *Discover,* Dec. 1980: 26–9.

LaPerriere, Dr. Robert (Dermatology, Kaiser Medical Center, Sacramento, CA), PC-5/5/86.

"A Larger Role for Opiate Receptors," *Science News* 119: 7.

Larkin, Jill (Psychology, Carnegie-Mellon University, Pittsburgh, PA), Q-8/17/84; c.v./1984; L-8/1, 10/11/87.

Larkin, Jill, *et al.,* "FERMI: A Flexible Expert Reasoner with Multi-Domain Inferencing," *Cognitive Science* 12: 1 (Jan.–Mar. 1988): 101–38.

Larned, Ellen D., *History of Windham County, Connecticut,* vol. II, 1760–1880. Chester, CT: Pequot, 1976 (1881).

"La Roe, Dr. Else K.," *Medical Woman's Journal* 56:2 (Feb. 1949): 59.

La Roe, Else K., *Woman Surgeon.* New York: Popular Library, 1961 (1957).

*Larousse Gastronomique* (Prosper Montagne, ed.). New York: Crown, 1961.

"Lasker Awards," *Science News* 122:21 (Nov. 20, 1982): 326.

Laverick, Elizabeth (The Woman's Engineering Society, London), Q-12/ 14/83; L-1/12, 4/7, 12/14/83.

Lawrence, Virginia, "Contraception," *Undercurrents* 29 (Aug.–Sept. 1978): 26.

Laws, Judith L., *The Second X: Sex Role and Social Role.* New York: Elsevier, 1979.

Lea, Tom, *The King Ranch.* Boston: Little, Brown, 1957.

Leacock, Eleanor B., *Myths of Male Dominance: Collected Articles on Women Cross-Culturally.* New York: Monthly Review Press, 1981.

Leacock, Eleanor B., "Women in Egalitarian Society," in Bridenthal, Koonz, and Stuard, eds.

"Lead-soldered Cans: A Serious Hazard?" *Science News* 117 (Mar. 22, 1980): 180–1.

"Learning from Local Folk Healers," *Science News* 119 (Feb. 21, 1981): 118.

Learning Research Group, *Personal Dynamic Media.* Palo Alto, CA: Xerox Corp., 1976.

Lecler, René, "Hommage à Fromage," *Saturday Review,* June 24, 1972: 77.

Lee, Katherine E., "Women and Patents: A Historical Investigation." MA thesis in Librarianship, San Jose State University, 1975.

Lee, Mary Price, *Ms. Veterinarian.* Philadelphia: Westminister Press, 1976.

Lee, Richard B., and Irven DeVore, eds, *Man the Hunter.* Chicago: Aldine, 1968.

Lee, Shirley Wilson, "A Survey of Acculturation in the Intermontane Area of the United States." MA thesis, Anthropology, Stanford University, 1966.

Lefkowitz, Mary R. "Princess Ida, the Amazons, and a Women's College Curriculum," *Times Literary Supplement,* Nov. 27, 1981: 1399–1401.

Leggett, M.D., comp., *Patents for Inventions . . . 1790–1873,* 3 vols. Washington, DC: Government Printing Office, 1874.

Lehnert, Wendy (Computer scientist, inventor, Dept. of Computer and Information Science, University of Massachusetts, Amherst), Q-9/19/ 84, c.v./1984, L-7/23/87.

Leibowitz, Lila, *Females, Males, Families: A Biosocial Approach.* North Scituate, MA: Duxbury, 1978.

Leigh-Smith, Alice, and Walter Minder, "Experimental Evidence of the Existence of Element 85 in the Thorium Family," *Nature,* Dec. 26, 1942: 767–8.

Leitner, Susan T., "Cotton Candy," *The Rotarian,* Mar. 1983: 35.

Lempert, Ted, "Algae Application May Supplant Oil," New Brunswick, NJ, *Home News,* Mar. 2, 1980.

Lentin, Judith, "Gillian Loban—Artist and Inventor," Women Inventors Project *Focus* 2: 3 (Oct. 1988): 2.

Leonard, Jonathan, "The Ghost of Yellow Jack," *Harvard Magazine,* Mar.–Apr. 1981, 20–7.

Leonard, Jonathan, and the Time/Life eds, *The First Farmers.* New York: Time/Life, 1973.

Leonard, Oliver B., "Pioneers of Plainfield," Plainfield, NJ, *Courier News,* Oct. 10, 1916–Jan. 25, 1918.

Lerner, Gerda, *The Creation of Patriarchy.* Oxford: Oxford University Press, 1986.

Lerner, Gerda, *The Majority Finds Its Past.* Oxford: Oxford University Press, 1979.

Leroi-Gourhan, Arlette, "The Archaeology of Lascaux Cave," *Scientific American* 246: 6 (June 1982): 104–12.

Lessa, William A., and Evon Z. Vogt, *Reader in Comparative Religion: An Anthropological Approach.* New York: Harper & Row, 1972 (1958).

Levin, Beatrice S., *Women and Medicine.* Metuchen, NJ: Scarecrow Press, 1980.

Levoy, Gregg, "Fighting Fire with Care," *Image* (San Francisco *Chronicle and Examiner*), Oct. 19, 1986: 9–10.

Lewallen, Eleanor, "The Saga of Sally Simmons Sponges," *The Mendocino Grapevine,* No. 333 (Nov. 5, 1980): 16.

Lewin, Tamar, "A New Test for Cancers," *New York Times,* May 12, 1987: 30

[Lewis, Amelia], "The People's Stove and Cooking Utensils," *The National Food and Fuel Reformer,* Dec. 5, 1874: 80.

Lewy, Hildegard, "Nitokris-Naqi'a," *Journal of Near Eastern Studies* 11 (1952): 264–86.

Liberty, Margot, "Hell Came with Horses: Plains Indian Women in the Equestrian Era," *The Magazine of Western History* 32: 3 (1982): 10–9.

*Liberty's Women,* Robert McHenry, ed. Springfield, MA: Merriam, 1980.

"Licensing Plan Inaugurated for Genetic Engineering," Stanford University *Observer,* Oct. 1981: 1f.

Lichty, Erle, "Demons and Population Control," *Expedition* 13:2 (Winter 1971): 22–6.

*Lightbulb* (magazine of Inventors Workshop International, Tarzana, CA), 1972–87.

Lindheim, Burton, "Dione Lucas, TV Cooking Teacher, 62." *New York Times,* Dec. 19, 1971: 60.

Lindholm, Cherry, and Charles Lindholm, "Taboos of the Dugum Dani," *Science Digest,* Jan./Feb. 1981: 82–7ff.

Lindsay, Bertha, and Lillian Phelps, *Industries and Inventions of the Shakers.* Canterbury, NH: Canterbury Shakers.

Lindsay, Jack, *The Origins of Alchemy in Graeco-Roman Egypt.* London: Frederick Muller, 1970.

Linton, Ralph, "The Marquesas," in Abram Kardiner, ed., *The Individual and His Society.* New York: Columbia University Press, 1939.

Linton, Ralph, *The Tree of Culture.* New York: Knopf, 1957.

"The Little Red Hen." See Galdone, Paul.

"Live from Dartmouth," *Discover,* Jan. 1981, 78, 81.

Livingstone, David, *Missionary Travels and Researches in South Africa.* London: John Murray, 1857.

Lloyd, Tom, *Dinosaur and Co.: Studies in Corporate Evolution.* London: Routledge, 1984.

Loban, Gillian (Artist, teacher, inventor, Alberta, CAN), L-11/29/88; FREEHAND® brochure, [1988].

Lockwood, Mary S., *Yesterdays in Washington,* 2 vols. Rosslyn, VA: Commonwealth Co., [1915].

Logan, Mary S., *The Part Taken by Women in American History.* Wilmington, DE: Perry-Nalle, 1912; Arno repr., 1972.

London, Crystal Palace Exhibition, Official Catalogue, London, 1851.

Longstreet, Stephen, *The Queen Bees: The Women Who Shaped America.* Indianapolis: Bobbs-Merrill, 1979.

Lönnrot, Elias, comp., *The Kalevala,* tr. Francis Peabody Magoun, Jr. Cambridge, MA: Harvard University Press, 1963.

[Los Angeles, CA, Chamber of Commerce,] "The Work of One Woman," *Southern California Business,* Nov. 1926: 26f.

Loughead, Flora Haines, *Dictionary of Given Names with Origins and Meanings,* 2nd ed. Glendale, CA: Clark, 1958.

Lovell, Sir Bernard, "The Erudite Shepherdess of Home-Made Moons," *Saturday Review,* Sept. 7, 1963: 44–6.

Lowe, Jeannette (Staff Horticulturist and Public Relations, W. Atlee Burpee Co., Warminster, PA), L-7/21/83.

Lucas, Jay P., and Russell E. Adams, eds, *Personal Computing Proceedings* (Proceedings of the National Computer Conference Personal Computing Festival, June 4–7) New York, 1979.

Luck-Szanto, Maria (Designer, inventor, London, Eng.), I-4/25/84, Q-5/15/ 84, L-1984–8.

Luckett, Margie H., *Maryland Women.* Baltimore: King Bros., 1937.

Lurd, Sharon B., and Carolyn Patton-Crowder, *Appalachian Women: A Learning/Teaching Guide.*

Lynch, Lisa, "Scientists Postpone DNA-Corn Planting," Stanford *Daily,* July 3, 1984: 7.

Maas, Martha (Dept. of Music, Ohio State University, Columbus), L-9/25/ 83.

Maccoby, Eleanor, and Carol Jacklin, eds, *The Psychology of Sex Differences.* Stanford, CA: Stanford University Press, 1974.

Macdonald, Anne L., *Feminine Ingenuity: Women and Invention in America.* New York: Ballantine, 1992.

MacFadyen, J. Tevere, "Educated Monkeys Help the Disabled to Help Themselves," *Smithsonian,* Oct. 1986: 125–6.

Mackenzie, Donald, *Egyptian Myth and Legend.* London: Gresham, [1913].

Mackin, James A. (NASA, Lewis Research Center, Cleveland, OH), L-6/9/83 to Diane Swendeman.

Macksey, Joan, and Kenneth Macksey, *The Book of Women's Achievements.* New York: Stein & Day, 1975.

*Macmillan Dictionary of Women's Biography.* See Uglow, ed.

MacQueen, Kenneth G., "Cybernetics and Haute Couture," *New Scientist,* Feb. 1963: 450–2.

Maddox, Brenda, "Byronic, Despite Everything" (review of Joan Baum, *The Calculating Passion of Ada Byron*), *New York Times Book Review,* Oct. 5, 1986, VII, 44: 2.

*Mademoiselle* Magazine, Achievement Award issues (Jan.) 1947–59; Merit Award issues (Jan.) 1943–6.

Madison, James, comp., *List of Names and Patents [1790–1805].* Washington, DC: U.S. Patent Office, [1806].

Magatti, Anita (Patent Counsel, Schering-Plough Corp., Madison, NJ), L-9/5/86.

Mahaffy, Anne R. (Assist. Product Dir., Johnson & Johnson Health Care Div., New Brunswick, NJ), L-12/2/76.

Mainiero, Linda, ed., *American Women Writers: A Critical Reference Guide from Colonial Times to the Present,* 4 vols. New York: Frederick Unger, 1979–80.

Maitland, H.B., and M.C. Maitland, "Cultivation of Vaccinia Virus without Tissue Culture," *Lancet* 2 (1928): 596–7.

"Malaria and the Green Revolution," *Science News* 120: 14 (Oct. 3, 1981): 213.

Malinowski, Bronislav, *The Sexual Life of Savages in Northwestern Melanesia. . . .* London: Routledge, 1929.

Malkinson, Frederick D., "The Heat's On," *Archives of Dermatology* 116 (1980): Aug. 885–7.

Malone, Dumas, ed., *Dictionary of American Biography,* 26 vols. New York: Scribner's, 1935–6.

Malotke, Mary E., *et al.,* comps., "Women's Technical Contributions: A Brief History of 89 Technical Women's Accomplishments in Science and Engineering," a paper presented at the 1977 National Convention of the Society of Women Engineers; 2nd ed., Barbara C. Faust, for the 1980 SWE National Convention.

Malt, Lillian (Teacher, inventor, Lillian Malt Training Services, Godalming, Surrey, Eng.), L-8/6, 8/27/80.

Malt, Lillian "Keyboard Design in the Electronic Era," Unpublished paper presented at the Printing Industry Research Association Eurotype- Forum, London, Sept. 14–5, 1977.

Manabe, Yukio, "Laminaria Tent for Gradual and Safe Cervical Dilatation." *American Journal of Obstetrics and Gynecology,* 110:5 (July 1971): 743–5.

Mandel, William M. (Soviet expert, Berkeley, CA), PC-7/8, 9/5/1979.

Mander, Jerry, *Four Arguments for the Elimination of Television.* New York: Morrow, 1978.

Manning, Marie, *Ladies Now and Then.* New York: E.P. Dutton, 1944.

Mano, D. Keith, "Inventors Expo 1978," *National Review,* Aug. 18, 1978: 1035–6.

Marcus, Steven, *The Other Victorians.* New York: Basic Books, 1966.

Marden, Orison S., "Women as Inventors," in *The Consolidated Encyclopedic Library.* New York: Emerson Press, 1903, 4086–9.

Margolin, Malcolm, *The Ohlone Way: Indian Life in the San Francisco–Monterey Bay Area.* Berkeley, CA: Heyday, 1978.

Mark, Norman, *Mayors, Madams, and Madmen.* Chicago: Chicago Review Press, 1979.

Markoff, John, "Authors Byte into Core of Apple's History," San Francisco *Examiner,* Sept. 20, 1987: D-1f.

Marks, Geoffrey, and William K. Beatty, *Women in White,* New York: Scribner's, 1972.

Marlow, Joan, *The Great Women.* New York: A & W Publishers, 1979.

Marmon, Lucretia, "Barbie Doll Developer Ruth Handler Offers a New Look to Mastectomy Victims," *People,* Apr. 11, 1977: 80–1.

Marotta, Charles, and Elizabeth Sajdel-Sulkowska, "Alzheimer's Disease Brains: Alterations in RNA Levels . . . ," *Science* 225 (Aug. 31, 1984): 947–9.

Marriott, Alice, "The First Tipi," in E. Adamson Hoebel, Jesse D. Jennings, Elmer R. Smith, eds, *Readings in Anthropology.* New York: McGraw-Hill, 1955.

Marsden, Kate, "The Leper," in Mary K. Eagle, ed.

Marsden, Kate, *On Sledge and Horseback to Outcast Siberian Lepers.* London: Record Press, [1893].

Marshall, Agnes B., *Victorian Ices and Ice Cream,* London: A.B. Marshall, 1885; repr. Scribner's, 1976.

Marshall, Martin, Larry Waller, and Howard Wolff, "The 1981 Achievement Award," *Electronics,* Oct. 20, 1981: 102–5.

Martin, M. Kay, and Barbara Voorhies, *Female of the Species.* New York: Columbia University Press, 1975.

Martin, Nancy, "Computerized Bibliography of Women in Math, Science, and Engineering" (Unpublished, sent by author).

Martin, Nancy, "In the Beginning We Were There: An Historical Overview of Women in Engineering/ Science," Society of Women Engineers, *Proceedings,* 1979: 36–52.

Marvingt, Marie, "Comme J'ai ConÇu le Premier Avion Sanitaire," *Revue de la Société Scientifique et Historique de Documentation Aérienne,* Jan.–Feb. 1940.

"Mary Hopkinson Weds Dr. Gibbon," *New York Times,* Jan. 28, 1931.

Mason, John F., "Grand Lady of Software." *Electronic Design* 22 (Oct. 25, 1976): 82–6.

Mason, Otis T., *Cradles of the North American Indians.* Seattle, WA: Shorey, 1889.

Mason, Otis T., *The Origins of Invention.* London: W. Scott, 1902.

Mason, Otis T., *Woman's Share in Primitive Culture.* New York: Appleton, 1894.

Mason-Hohl, Elizabeth, *The Diseases of Women: A Translation of [Trotula's] "Passionibus mulierum curandum."* Los Angeles: Ward Ritchie, 1940.

Masterson, Pat, "She Won't Give Up," *The Patriot,* June 1976.

Matthaei, Julie A., *An Economic History of Women in America: Women's Work, the Sexual Division of Labor, and the Development of Capitalism.* New York: Schocken, 1982.

Matthews, Alva, "Some Pioneers," talk given to the Society of Women Engineers, c. 1978.

Matthiessen, Peter, *The Snow Leopard.* New York: Bantam, 1979 (1978).

Maurer, Allan, "Living Batteries," *Omni,* Dec. 1982: 44–5.

" 'Maverick' Woman Joins Ranks of Men in Inventors Hall of Fame," San Jose *Mercury-News,* Mar. 10, 1991: 2E.

Maxwell, Nicole, *Witch Doctor's Apprentice.* Boston: Houghton Mifflin, 1961.

Mayer, Marlene (daughter of inventor Rose B. Null, Des Plaines, IL), I-11/15/89.

Mayne, Ethel Colburn, *The Life and Letters of Lady Noel Byron.* London: R. West, 1973 (1929).

McAdoo, Maisie, "Birth Control," *Ms.,* May 1983: 113–4.

McCarroll, June A.W., "White Line First in Highway Safety," *The Clubwoman,* Oct. 1938, 10ff.

McClain, Ellen Jaffe, "Do Women Resist Computers?" *Popular Computing* 2:3 (Jan. 1983): 66–78.

McCorduck, Pamela (Writer, New York), L-3/15/84.

McCorduck, Pamela, *Machines Who Think.* San Francisco: W.H. Freeman, 1979.

McCormack, Patricia, "Much Skepticism on Natural Birth Control," Los Angeles *Times,* June 27, 1980: pt. V, 8–9.

McCowen, Jennie, MD, "Women in Iowa," *Annals of Iowa* 3 (Oct. 1884): 97–113.

McCrane, Edna P., "Staining of Blue-Sensitive Cones of the Macaque Retina by a Fluorescent Dye," *Science* 213 (Sept. 11, 1981): 1278–81.

McCullough, Joan, *First of All: Significant "Firsts" by American Women.* New York: Holt, Rinehart, 1980.

McCune, J.M., *et al.,* "The SCID-hu Mouse . . . ," *Science* 241 (Sept. 23, 1988): 1632–9.

McDaniel, Susan A., Helene Cummins, and Rachelle S. Beauchamp, "Mothers of Invention? Meshing the Roles of Inventor, Mother, and Worker," *Women's Studies International Forum* 11:1 (1988): 1–12.

McDaniels, David K., *The Sun: Our Future Energy Source.* New York: Wiley, 1979.

McDermott, Jeanne, "Face to Face, It's the Expression That Bears the Message," *Smithsonian* 16:12 (Mar. 1986): 113–22.

McDougall, John A., and Mary McDougall, *The McDougall Plan.* Piscataway, NJ: New City Publishers, 1983.

McDowell, Barbara, and Hana Umlauf, eds., *The Good Housekeeping Woman's Almanac.* New York: Newspaper Enterprise Association, 1977.

McGinnis, Ann Cain, "Music and Healing," *Heresies,* No. 10: 3:2 (1980): 11–14.

*The McGraw-Hill Encyclopedia of World Biography.* New York: McGraw- Hill, 1973.

McGrew, W.C., "The Female Chimpanzee as a Human Evolutionary Prototype," in Dahlberg, ed.

McHenry, Robert, ed., *Liberty's Women.* Springfield, MA: Merriam, 1980.

McIntyre, Loren, "The Lost Empire of the Incas," *National Geographic,* 144:6 (Dec. 1973): 729–87.

McLaughlin, Terence, "Ada Byron, the Romantic Computer Programmer." *British Science News,* 1977, 6135–6.

McMurran, Kristin, "Frances Gabe's Self-Cleaning House Could Mean New Rights of Spring for Housewives," *People* 17:12 (Mar. 19, 1982): 38, 41.

McPhilimy, Glennis, "First Lady of Computers Predicts a Bright Future." Boulder, CO, *Sunday Camera,* Feb. 4, 1979: 33.

McVeigh, J.C., *Sun Power: An Introduction to the Applications of Solar Power.* New York: Pergamon, 1977.

Mead, Kate Campbell Hurd-, *A History of Women in Medicine.* Haddam, CT: Haddam Press, 1938; repr. Boston: Milford House, 1973.

Mead, Kate Campbell Hurd-, "Trotula," *Isis* 14 (1930): 349–67.

Mead, Margaret, *Growing Up in New Guinea.* New York: Morrow, 1975 (1930).

"Mechanical Inventions and Designs . . . ," *Inventive Age,* June 1, 1911: 6.

"Medical Women of Today," *The Medical Woman's Journal,* 40 (Jan. 1933): 18.

Medicine, Beatrice (Anthropologist, Dakota Sioux), PC-5/1980.

Mednick, Martha, *et al., Women and Achievement: Social and Motivational Analyses.* Washington, DC: Hemisphere, 1975.

Meinel, Aden B., and Marjorie P. Meinel, *Applied Solar Energy: An Introduction.* Reading, MA: Addison-Wesley, 1976.

Meinel, Aden B., and Marjorie P. Meinel, "Physics Looks at Solar Energy," *Physics Today,* Feb. 1972: 44–50.

Melcher, Marguerite Fellows, *The Shaker Adventure.* Princeton, NJ: Princeton University Press, 1941.

Mellaart, James, "Çatal Hüyük: A Neolithic Town in Anatolia," *Proceedings of the British Royal Academy* 51 (1965): 201ff.

Mellaart, James, *Çatal Hüyük: A Neolithic Town in Anatolia.* New York: McGraw-Hill, 1967.

Mellaart, James, *The Neolithic of the Near East.* New York: Scribner's, 1973.

"A Memorial to Dr. Aletta Jacobs," *The Medical Woman's Journal* 38: 9 (Sept. 1931): 235.

Meyer, Annie Nathan, *Woman's Work in America.* New York: Holt, 1891; Arno repr., 1972.

Michaud, Sabrina, and Roland Michaud, "Trek to Lofty Hunza and Beyond," *National Geographic* 148: 5 (Nov. 1975): 644–69.

Miller, G. Tyler, *Living in the Environment,* 2nd ed. Belmont, CA: Wadsworth, 1979.

Miller, Howard E. (Westwood Public Library, Westwood, MA), L-1/16, 2/6, 3/17/1980.

Miller, Howard E., "Betsy Metcalf Baker," unpublished paper sent by author, 1980, 2 pp.

Miller, James Nevin, "All Inventors Are Not Male," Washington *Star,* July 17, 1938.

Miller, Julie Ann, "Cell-e-vision," *Science News* 119: 15 (Apr. 11, 1981); 234–5f.

Miller, Julie Ann, "Colorful Views of Vision," *Science News* 120: 14 (Oct. 3, 1981): 220–1.

Miller, Julie Ann, "Crystals Worth Millions of Sneezes," *Science News* 123: 11 (Mar. 12, 1983): 165.

Miller, Julie Ann, "Microbial Antifreeze: Gene Splicing Takes to the Field," *Science News* 124: 9 (Aug. 27, 1983): 132.

Miller, Julie Ann, "New Biotech Tool: Recombinant RNA," *Science News* 125: 2 (Jan. 14, 1984): 23.

Miller, Julie Ann, "Pulling Together a Cluster Theory of Immune Response," *Science News* 121: 5 (Feb. 6, 1982): 85.

Miller, Julie Ann, and Joan Arehart-Treichel, "Radioactive Rings Shrink Eye Tumor," *Science News* 122 (Nov. 6, 1982): 300.

Mirsky, Jeannette, and Allan Nevins, *The World of Eli Whitney.* New York: Macmillan, 1952.

"Miss Anne Chamney," *The Woman Engineer* (London), 10: 9 (1968): 7.

"Miss Henry, Charlotte Woman, 'Lady Edison,' " Charlotte, NC, *News,* Oct. 21, 1930.

"Miss Hosmer's Discoveries," New York *Evening Post,* Oct. 26, 1878.

"Miss Szanto's New Fabric, Strong as Tweed, Soft as Wool," *Textile Bulletin,* Feb. 22, 1946.

"Miss Verena Holmes," *The Woman Engineer* (London) 9: 15 (1965): 2, 14.

Mitchell, Betty (Inventor, St. Albans, VA), PC/Biog-1983.

Mitchell, Betty, "Applications of Computerized Statistical Techniques in Quantitative X-Ray Analysis," in J.B. Newkirk and G.R. Mallett, eds, *Advances in X-Ray Analysis* II. New York: Plenum, 1983.

Mitchell, Betty, and F.N. Hopper, ''Digital Computer Calculation and Correction of Matrix Effects in X-Ray Spectroscopy," *Applied Spectroscopy* 20 (May–June, 1966): 172ff.

Mitchell, Martha, "History of Women in Mining from the Middle Ages to the Enactment of Protective Legislation," Society of Women Engineers, *Proceedings,* San Francisco, 1979: 33–5.

Moehrle, Doris (Inventor, Las Vegas, NV), L-10/19/79, PC-7/17/83.

Mohr, James, C., "Abortion in America," in Kerber and Mathews.

Mohs, Mayo, "America's Fiftieth Biggest Business," *Discover* 6: 8 (Aug. 1985): 24–35.

Moller, Mary, and Charlotte Friend, "Dr. Anna Goldfeder Honored," AWIS (Association for Women in Science) *Newsletter* 11: 2 (Apr./May, 1982): 4.

Molnos, Angela, ed., *Culture Sources for Population Planning in East Africa.* Munich: Weltforum, 1973.

Monaghan, Patricia, *Women in Myth and Legend.* London: Junction Books, 1981. (U.S.: *The Book of Goddesses and Heroines.* New York: Dutton, 1981.)

"Monkeys Provide Hands for Disabled," San Francisco *Chronicle,* Dec. 19, 1982: E-9.

Montgomery, James, *Cotton Manufactures of the United States . . . and . . . Great Britain.* New York: Burt Franklin, 1840.

Montmasson, Joseph-Marie, *Invention and the Unconscious.* New York: Harcourt, Brace, 1932.

Moolman, Valerie, *Women Aloft.* Alexandria, VA.: Time/Life Books, 1981.

Moore, Doris Langley, *Ada, Countess of Lovelace, Byron's Legitimate Daughter:* London: John Murray, 1977.

Morell, Parker, *Lillian Russell: The Era of Plush.* New York: Random House, 1940.

Morgan, Jane, and Elisabeth Roden (Informants, Palo Alto, CA), PC-5/16/ 86.

Morris, Helen (Engineer, informant, Sunnyvale, CA), L-11/9/88, PC-10/15/88.

Morrison, Philip, and Emily Morrison, eds, *Charles Babbage and His Calculating Engine.* New York: Dover, 1961.

Morrow, John F., Stanley N. Cohen, Annie C.Y. Chang, *et al.,* "Replication and Transcription of Eukaryotic DNA in *Escherichia coli,*" *Proceedings of the National Academy of Sciences* 71: 5 (May 1974): 1743–7.

Mort, Cynda, "Career Built on Child Safety," Fort Wayne, IN, *News-Sentinel,* Mar. 20, 1976.

Moseley, Mabeth, *Irascible Genius: The Life of Charles Babbage, Inventor.* London: Hutchinson Co., 1964.

Mosher, Clelia Duel, "Normal Menstruation . . . ," *Johns Hopkins Hospital Bulletin* 12 (1901): 178–9.

Mosher, Clelia Duel, *Woman's Physical Freedom.* New York: The Woman's Press, 1923.

Moskowitz, Milt, "The 'Success' Story of Rely Tampons," San Francisco *Chronicle,* Oct. 18, 1980: 29.

Mossinghof, Gerald, and Barbara Luxembourg, comps., *Profile of Female Inventors.* Washington, DC: Office of Technology Assessment and Forecast, 1984.

Mott, Luther, *A History of American Magazines, 1865–1885,* 5 vols. Cambridge, MA: Harvard University Press, 1938–68.

Moussa, Farag, *Les Femmes Inventeurs Existent: Je Les Ai Rencontrées*. Geneva: Editions Farag Moussa, 1986.

Moynehan, Barbara, "Gazette: Three Mothers of Invention," *Ms.,* Apr. 1979: 23.

Mozans, H.J., *Woman in Science*. New York: Appleton & Co., 1913; repr. M.I.T. Press, 1976.

"Mrs. Potter Palmer," Pittsburgh *Leader,* Mar. 18, 1892.

"Mrs. Vonk's Victory," *Time* 106: 10 (Sept. 8, 1975): 6.

Mumford, Lewis, *The Myth of the Machine*. New York: Harcourt, Brace, 1966.

Mumford, Lewis, *Technics and Civilization*. New York: Harcourt, Brace, 1962.

Munson, Richard, and Barrett Stambler, "Competing for the Sun," *Technology Review,* Nov./Dec. 1982: 12–13.

Munzer, Martha E., *Unusual Careers*. New York: Knopf, 1962.

Murdock, George P., and Caterina Provost, "Factors in the Division of Labor by Sex," *Ethnology* 12: 2 (Apr. 1973): 203–25.

Murdock, George P., and Douglas R. White, "Standard Cross-Cultural Sample," *Ethnology* 8: 4 (Oct. 1969): 329–69.

Murphy, Jim, "Weird and Wacky Inventions," *Flights of Color,* Lexington, MA: Ginn & Co., 1982 (1978).

Murphy, Yolanda, and Robert F. Murphy, *Women of the Forest*. New York: Columbia University Press, 1974.

Murray, Geoffrey, "Japanese Women Poke Holes in Men's Claims to Inventors' Fame," *Christian Science Monitor,* Feb. 18, 1982: 1, 9.

"Musical Typewriter," *Manchester Guardian,* Jan. 24, 1961.

"Musigraph Brings Fame but No Fortune," *The Times* (London), Jan. 14, 1967.

Myres, Sandra J. (History, University of Texas–Arlington), L-1980-7.

Myres, Sandra J., *Westering Women and the Frontier Experience, 1800–1915*. Albuquerque: University of New Mexico Press, 1982.

Nader, Ralph, "Toll-Free Hotlines," *Ladies Home Journal* 95 (May 1978): 42.

Napoli, Mary Ann, "Fire-and-Ice Medicine," *Self,* Apr. 1984: 107–10.

Nathan, Maud, *The Story of an Epoch-Making Movement*. Garden City, NY: Doubleday, Page, 1926.

"N[ational] A[cademy of] S[ciences] Awards," *Physics Today* 17 (June 1964): 50–1.

*National Cyclopaedia of American Biography,* 57 vols. New York and Clifton, NJ: James T. White and Co., 1892–1977; reprint, vols. 1–50, Ann Arbor: University Microfilms, 1967–71.

National Public Radio, "51%," July 4, 1991: Interview with Gertrude Elion.

Neal, Julia, *The Kentucky Shakers*. Lexington, KY.: University of Kentucky Press, 1977.

Neary, John, "The Search for a White Marigold," *Horticulture* 54: 2 (Feb. 1976): 20–8.

Nelson, Chris, *Self-Help Home Remedies,* rev. ed., Chico, CA: Feminist Women's Health Center (Chico, CA), 1976.

Nerlove, Sara B., "Women's Workload and Infant-Feeding Practices: A Relationship with Demographic Implications," *Ethnology* 13: 2 (Apr. 1974): 207–14.

Netting, Robert M., "Marital Relations in the Jos Plateau of Nigeria," *American Anthropologist* 71 (1969): 1037–46.

Neushul, Peter, "Marie Stopes and the Development of Birth Control Technology," a paper presented at the 1989 Conference of the Society for the History of Technology, Sacramento, CA.

"A New Alternative," *Time* 132: 7 (Aug. 15, 1988): 65.

"New Cancer Detection Test Announced," *Science,* Jan. 17, 1981: 37.

*New Century for Women,* published by the women of the Centennial Exposition, Philadelphia, 1876.

"A New Old Remedy," *Discover,* Aug. 1985: 11.

"New Products," *Business Week,* Feb. 23, 1957: 110f.

"New Weapon in War Against Breast Cancer," San Francisco *Chronicle,* Nov. 12, 1981: 1, 22.

Newport, John Paul, Jr., "The Growing Gap in Software," *Fortune,* Apr. 28, 1986: 132–42.

Newton, Burritt W., "Laminaria Tent: Relic of the Past or Modern Medical Device?" *American Journal of Obstetrics and Gynecology* 13: 4 (June 15, 1972): 442–8.

"Nine Patents Issued to Women in Week," *New York Times,* Mar. 9, 1941: 4-D.

"The 1974 I-R 100 Award Winners," *Industrial Research,* Oct. 1974: 23–32.

"The Nobel Pair," *New York Times* Magazine, Jan. 29, 1989: 28ff.

Nofziger, Margaret, *A Cooperative Method of Natural Birth Control,* 2nd ed. Summertown, TN: The Book Publishing Co., 1978.

Noll, Laura White, "A Text Formatting Program for Phototypesetting Documents," unpublished technical report, Holmdel, NJ: Bell Labs, Apr. 15, 1971.

"A Non-Lazy Susan Devises a Voting Machine," *People,* Oct. 4, 1976: 15.

Noonan, John T., Jr., *Contraception: A History of Its Treatment by the Catholic Theologians and Canonists.* Cambridge, MA: Harvard University Press, 1965.

Nordhoff, Charles, *The Communistic Societies of the United States.* New York: Harper & Bros., 1875.

Norman, Bruce, *The Inventing of America.* New York: Taplinger, 1976.

Norton, Mary Beth, *Liberty's Daughters: The Revolutionary Experience of American Women, 1750–1800.* Boston: Little, Brown, 1980.

Nostbakker, Janis, and Jack Humphrey, *The Canadian Inventions Book.* Toronto: Greay de Pencier, 1976.

*Notable American Women, 1607–1950: A Biographical Dictionary,* 3 vols., Edward T. James, Janet Wilson James, and Paul S. Boyer, eds, Cambridge, MA: Harvard University Press, 1971.

*Notable American Women: The Modern Period—A Biographical Dictionary.* Barbara Sicherman and Carol Hurd Green, eds. Cambridge, MA: Harvard University Press, 1980.

Novotny, Ann, *Alice's World—The Life and Photography of an American Original: Alice Austen, 1866–1952.* Old Greenwich, CT: Chatham Press, 1976.

Nowak, Mariette, *Eve's Rib: A Revolutionary New View of the Female.* New York: St. Martin's, 1980.

Noz, Marilyn (Assoc. Prof. of Radiology, Bellevue Hospital, New York University Med. Ctr. New York), L-9/13/81, c.v./1982, PC-9/1/83.

Noz, Marilyn, *et al.,* "A Modular Computer System for the Nuclear Medicine/Ultrasound Laboratory," *Radiology* 124 (Sept. 1977): 759–62.

Null, Fred (Marin Co., CA, son of inventor Rose B. Null), I-10/28/89.

Nussenzweig, Ruth S., "Progress in Malaria Vaccine Development: Characterization of Protective Antigens," *Scandinavian Journal of Infectious Diseases,* Suppl. 36, 1982: 40–5.

Oakley, Ann, *Sex, Gender and Society.* New York: Harper, 1972.

O'Bryon, James Frederick, "A Profile of Users of the U.S. Patent System: 1968 and 1973." MS thesis, M.I.T., 1975.

O'Faolain, Julia, and Lauro Martines, eds, *Not in God's Image.* New York: Harper & Row, 1973.

Offen, Karen, "Women, Technology, and Public Hygiene: An Episode in the Early Social History of the Sewing Machine," unpublished paper sent by author, Oct. 1985; published as " 'Powered by a Woman's Foot': A Documentary Introduction to the Sexual Politics of the Sewing Machine in Nineteenth Century France," *Women's Studies International Forum* 11:2 (1988): 93–101.

O'Hern, Elizabeth M., *Profiles of Pioneer Women Scientists.* Washington, DC: Acropolis, 1986.

Ohio State University, Columbus, Alumni Records, Russell S. Feicht; Bertha A. Lammé.

Ohio State University, Columbus, Office of the Registrar, official transcripts, Bertha Avanelle Lammé, Class of 1893.

Ohio State University, Columbus, Sixteenth Annual Commencement Program, June 14, 1893.

Ohnuki-Tierney, Emiko, "The Shamanism of the Ainu . . . ," *Ethnology* 12:1 (1973): 15ff.

*Old Home Week, July 3–9, 1910.* McKeesport, PA, 1910.

"Olney Medalist," *Chemical and Engineering News,* June 10, 1963: 70.

O'Mahoney, B.M.E., "Henrietta Vansittart—Britain's First Woman Engineer?" *The Woman Engineer* (London) 13: 4 (Spring 1983): 1–2f.

Onanian, Richard A. (Inst. for Invention and Innovation, Arlington, MA), L-9/8/76.

"Oncogenes in Vivo," *Scientific American* 250:2 (Feb. 1984): 73–6.

"One Half Our History" (An exhibit at Widener Library, Harvard University, Sept. 10–Oct. 7, 1982), *Schlesinger Library Newsletter,* Fall 1982.

"One Theme but Many Variations," *Business Week,* Apr. 13, 1963: 148f.

"£1,000 to Aid Production of Music Typewriter," *The Times* (London), Jan. 24, 1961.

O'Neill, Lois Decker, ed., *The Women's Book of World Records and Achievements.* Garden City, NY: Anchor Books, 1979.

Opfell, Olga S., *The Lady Laureates: Women Who Have Won the Nobel Prize.* Metuchen, NJ: Scarecrow, 1978.

"Opiate Antagonist Counters Obesity," *Science News* 119: 7 (Feb. 14, 1981): 103.

Oram, Charlotte (Woman Health International, Washington, DC), L-12/13/1982.

Orcutt, Samuel, and Ambrose Beardsley, MD, *The History of the Old Town of Derby, Connecticut, 1642–1880.* Springfield, MA: Press of Springfield Printing Co., 1880.

Oregon, University of, Center for Invention and Innovation (College of Business Administration, Eugene, OR). See Carroll, Helen.

Orel, Jeannette V., "Give the Marlboro Man His Own Range," Los Angeles *Times*, Dec. 16, 1975, II: 7.

Orel, J[eannette] V., "Smoke-Sorbing Device," United States Dept. of Health *Smoking and Health Bulletin*, Apr. 1980: 148.

Orel, Jeannette V. (Smoke-Trapper®, Los Angeles, CA), L-5/26, 8/5, 8/19/87; Q-8/5/87.

Orenberg, Elaine K. (Pharmaceutical consultant, Los Altos, CA), L-3/15,23/88; PC-1988.

Orenberg, Elaine K., D.G. Deneau, and E.M. Farber, "Response of Chronic Psoriatic Plaques to Localized Heating Induced by Ultrasound," *Archives of Dermatology* 116 (1980): 893–7.

Orente, Rose, "Where Are All the Mothers of Invention?" New York *Times*, Jan. 9, 1980: C16.

"Organic Liquor Ingredients Thought to Cause Female Hormonal Effects," San Francisco *Peninsula Times Tribune*, July 11, 1984: A-3.

Ormsbee, Mary Whitton, *These Were the Women.* New York: Hastings House, 1954.

Ortiz-Molina, M.G., T.Y. Toong, and Giuliana C. Tesoro, "Smoldering Combustion of Flexible Polyurethane Foams and Its Transition to Flaming or Extinguishment," *Proceedings of the 17th International Symposium of the Combustion Institute* (1979). Pittsburgh: Combustion Institute, 1979.

Ortner, Sherry, and Harriet Whitehead, eds, *Sexual Meanings: The Cultural Construction of Gender and Sexuality.* Cambridge: Cambridge University Press, 1981.

Orton, Christine, "A Queen of Music," *The Times* (London), Feb. 28, 1966.

O'Shea, John, and Marek Zvelebil, "Oleneostrovski Mogilnik: Reconstructing the Social and Economic Organization of Prehistoric Foragers in Northern Russia," *Journal of Anthropological Archaeology* 3 (1984): 1–40.

OTAF (Office of . . . Technology Assessment and Forecast . . . ), U.S. Patent and Trademark Office, Dept. of Commerce, *Buttons to Biotech: U.S. Patenting by Women, 1977 to 1988.* Washington, DC, 1990.

Owen, May *et al.*, "Three Generations Look at Their Profession," *Texas Medicine* 71 (Jan. 1975): 91–100.

*The Oxford Companion to American History.* New York: Oxford University Press, 1966.

"P-74 Inventors' Awards," *The Times* (London), Feb. 12, 1975.

Paige, Karen, and Jeffrey Paige, *The Politics of Reproductive Ritual.* Berkeley: University of California Press, 1981.

Paine, Rudolph (Confederacy of Mississippi Inventors, Vicksburg, MS), L-2/1989.

Palmer, Gladys P., "Lifestyles of the Rich and Democratic," *Image* (San Francisco *Chronicle and Examiner)* June 5, 1988: 18–25.

Panabaker, Janet, *Inventing Women: Profiles of Women Inventors.* Waterloo, Ont.: Women Inventors Project, 1991.

Pandharipande, Rajeshwari, "Fertility and Power: The Image of Women in the Early Civilizations of India," a paper presented at the National Women's Studies Association Convention, Urbana, IL, 1986.

Parmiter, Charles, "Simple New Blood Test Detects Cancer Early and Pinpoints Location," *National Enquirer*, Dec. 28, 1976: 20.

Parrinder, Geoffrey, *African Mythology.* London: Hamlyn, 1967.

Parsons, Claudia, "Verena Through the Eyes of Her Friends," *Woman Engineer* (London) 9: 13 (1964): 2–6.

Parton, James, *et al.*, *Eminent Women of the Age; Being Narratives of the Lives and Deeds of the Most Prominent Women of the Present Generation.* Hartford, CT: Betts & Co., 1868.

Partridge, Michael, *Farm Tools Through the Ages.* Reading, Eng.: Osprey, 1973.

"Patent for Gene Splicing, Cloning, Awarded Stanford," Stanford University *Observer*, Jan. 1981: 1f.

Pauley, Gay, "She Wants to Correct Computer's Image." Long Island *Press*, Sept. 24, 1975.

Paulme, Denise, ed., *Women of Tropical Africa*, tr. H.M. Wright. Berkeley: University of California Press, 1963.

Paulsen, Twila (now Okken; San Bruno, CA), L=, Q=8/30/1988.

Pauwels, Louis, and Jacques Bergier, *The Morning of the Magicians*, tr. Rollo Myers. London: Granada, 1971.

Pearson, Elmer R., and Julia Neal, *The Shaker Image.* New York, New York Graphic Society, 1974.

Peden, Irene Carswell, "Women in Engineering Careers," in John K. Whinnery, ed., *The World of Engineering.* New York: McGraw-Hill, 1965.

"Peg Sherlock: She's Where the Action Is," *People*, Fall 1971: 1–4.

Pelling, Margaret (Wellcome Unit for the History of Medicine, Oxford, Eng.), L-12/23/86.

Pelling, Margaret, "Healing the Sick Poor: Social Policy and Disability in Norwich, 1550–1640," *Medical History* 29 (1985): 115–37.

Pelling, Margaret, "Occupational Diversity: Barber Surgeons and the Trades of Norwich, 1550–1640, " *Bulletin of the History of Medicine* 56 (1982): 484.

Pelling, Margaret, "Tradition and Diversity: Medical Practice in Norwich, 1550–1640," in Leo S. Olschki, ed., *Scienze, Credenze Occulte, Livelli di Cultura,* Proceedings of the Convegno Interrazinoli di Studi, Florence, June 26–30, 1980.

Pelling, Margaret, and Charles Webster, "Medical Practitioners," in Charles Webster, ed., *Health, Medicine and Mortality in the Sixteenth Century.* Cambridge: Cambridge University Press. 1979.

Pennington, Mary E., *et al., Eggs.* Chicago: Progress Publications, 1933.

"Perfected by a Woman's Doctor," ad for MEDS tampons, *Good Housekeeping,* June 1943: 100.

Perl, Teri (Mathematician, author, Palo Alto, CA), PC-1980.

Perl, Teri, *Math Equals.* Reading, MA: Addison-Wesley, 1978.

Perles, Catherine, "Hearth and Home in the Old Stone Age," *Natural History,* Oct. 1981: 38–41.

Perlman, Alice, and Polly Perlman, "Women's Power in the Ancient World," *Women's Caucus: Religious Studies* 3 (Summer 1975): 4–6.

Perlman, David, "Facts from Hyenas' Mouths," San Francisco *Chronicle,* Jul. 6, 1985: 10.

Perlman, David, "S.F. Researcher Finds New Test for Sickle Cell," San Francisco *Chronicle,* Jan. 6, 1982: 2.

Perrier, Dorothy M., and Ruth R. Benerito, "Catalysis of the Cellulose-cyclic Urea Reaction by Built-in Acid Groups," *Textile Research Journal* 51 (1971): 680–5.

Perrier, Dorothy M., and Ruth R. Benerito, "Electron Donor Properties of Tertiary Amines in Cellulose Anion Exchanges," in J.C. Arthur, ed., *ACS Symposium Series, U.S.A.* No. 48. Chicago: American Chemical Society, 1977.

"Personal Glimpses," *Literary Digest,* Mar. 24, 1928: 58.

Personal Products Co. (Milltown, NJ), "About Personal Products . . . and the Development of Feminine Protection . . . ," [1983], 3 pp.

"Personals About Inventors," *Patent Record,* May 1900: 4.

Peska, Lawrence, Associates (New York, NY), "Women Inventors on the Rise," Press release, 1976.

Peters, Tom, and Bob Waterman, *A Year of Excellence, 1985: The Peters and Waterman MBWA Diary.* New York, NY: Not Just Another/Random House, 1984.

Peterson, Virginia K. (Harshe-Rotman, Inc. [Public relations firm for Norge], Chicago, IL), "Biography of Mrs. Jessie Cartwright," 2 pp., n.d. (University of Illinois/Chicago Circle Library, Cartwright Papers).

Petit, Charles, "Bay Area Scientists See Cheap Hepatitis Vaccine," San Francisco *Chronicle,* Nov. 19, 1981: 32.

Pfaff, Timothy, "A Conversation with Ed Roberts," *California Monthly,* Jan.–Feb. 1985: 9.

Pfeiffer, John E., "Cro-Magnon Hunters Were Really Us, Working Out Strategies for Survival," *Smithsonian* 17:7 (Oct. 1986): 75–84.

Pfeiffer, John E., "The Mysterious Rise and Decline of Monte Alban," *Smithsonian,* Feb. 1980: 62–75.

PG&E (Pacific Gas & Electric Co.) *Progress* (monthly newsletter to customers).

Phelan, Mary Kay, *Probing the Unknown: The Story of Dr. Florence Sabin.* New York: Crowell, 1969.

Phelps, Ethel J., *Tatterhood and Other Tales.* Old Westbury, NY: Feminist Press, 1978.

Phelps, Robert H., ed., *Men in the News–1958: Personality Sketches from "The New York Times."* Philadelphia: Lippincott, 1959.

Phillips, Ralph W., "Cattle," *Scientific American,* June 1958: 51–9.

*Physicians' Desk Reference.* Oradell, NJ: Medical Economics Co., 1946–  .

"Pill Box Sounds Reminder," *Popular Science,* March 1975: 20.

Pinckney, Callan (with Sallie Bateson), *Callanetics: Ten Years Younger in Ten Hours.* New York: Morrow, 1984.

Pinckney, Eliza, Letterbook. South Carolina Historical Society, Charleston.

Pinnell, Evelyn M., "Fabrics Designed by Hungarian Boon to British Home Workers," *Christian Science Monitor,* May 23, 1947: 8A.

"Pioneers Called to Great Beyond," *The Payette* (ID) *Independent* 44: 47 (Feb. 14, 1935): 1.

Pisan, Christine de, *The Book of the City of Ladies,* tr. Earl Jeffrey Richards. New York: Persea Books, 1982.

"Plants Bearing Imported Genes," *Science News* 123: 12 (Mar. 19, 1983): 184.

Plaster, W. Emory, Jr. (Loudon County Historical Society, Leesburg, VA), L-7/12/82.

Plastic Applications (Boca Raton, FL), Business plan, n.d. (sent me by J. Waldron, Pres., 1983): 11.

Plume, Christian, *Le Livre de Fromage*. Paris; Editions des Deux Coqs d'Or, 1968.

Podesta, Victoria, "Toxic Shock Syndrome May Be Due to Super Fibers," *National NOW Times*, Dec./Jan., 1980-1.

Podo, F., *et al.* [Letter research report], to Bernard A. Shapiro, *NMR Newsletter*, No. 328, Jan. 1986.

"Poliovirus Read from Start to Finish," *Science News* 120:1 (July 4, 1981): 6.

Pomeroy, Sarah, *Goddesses, Whores, Wives and Slaves: Women in Classical Antiquity*. New York: Schocken Books, 1975.

Pomo de Ayala, Felipe Huaman, *Letter to a King: A Peruvian Chief's Account of Life Under the Incas and Under Spanish Rule*. New York: Dutton, 1978.

Pool, Judith G., and John F. Bunker, "Women in Medicine," *Hospital Practice* 7:8 (Aug. 1972): 109-16.

Pool, Judith G., and J. Robinson, "Assay of Plasma Antihaemophilic Globulin (AHG)," *British Journal of Haematology* 5 (1959): 17-23.

Pool, Judith G., and Angela E. Shannon, "Production of High-Potency Concentrates of Antihemophilic Globulin in a Closed-Bag System," *New England Journal of Medicine* 273:27 (Dec. 30, 1965): 1443-7.

Porter, Glenn and William H. Mulligan, Jr., eds, *Working Papers from the Regional Economic History Research Center* 5: 2, 3. Wilmington, DE: Eleutherian Mills-Hagley Foundation, 1982.

Post, Robert C., ed., *1876: A Centennial Exhibition*. Washington, DC: Smithsonian Institution, 1976.

Pound, Arthur, *Transportation Progress; The History of Self-Propelled Vehicles from Earliest Times Down to the Modern Motor Car*. Garden City, NY: Doubleday, 1934.

Powdermaker, Hortense, *Life in Lesu: The Study of a Melanesian Society in New Ireland*. New York: Norton, 1971 (1933).

Powell, Jim, "Cashing In," *Savvy*, Apr. 1983: 88ff.

Poyner, Edna (National Home Economics Association, Washington, DC), PC-5/15, 18/83.

"Presidents, Actors, Millionaires Try Their Hands at Invention," *Literary Digest*, Mar. 24, 1928: 53-8.

Press, Harry, "Cutting Edge of Research (Past, Present): Is Obsidian Blade Sharpest Surgical Tool?" Stanford University *Observer*, June 1985: 8.

Press, Harry, "Students Design Pocket-Sized Information System," Stanford University *Observer*, Oct. 1979: 5.

Preston, Richard M., "The Blue-Green Wonder: Will It Help Feed the World?" *Country Journal*, Mar. 1982: 88-94.

Preston, Wheeler, ed., *American Biographies*. New York: Harper & Bros, 1940; repr., Detroit: Gale, 1974.

"Preventing Hydrocephalus in Infants," *Science News*, Sept. 19, 1981: 181.

Pritikin, Nathan, *The Pritikin Permanent Weight-Loss Manual*. New York: Bantam, 1982 (1981).

Progress of Woman's Rights," *Scientific American* 29 (1873): 7.

Psoriasis Research Institute (Palo Alto, CA), *Newsletter* 7:3 (1988): 2.

Pugh, Mary Jo (Bentley Historical Library, University of Michigan, Ann Arbor), L-5/4, 6/4/82.

Pursell, Carroll, "Women Inventors in America," *Technology and Culture* 22: 3 (July 1981): 545-9.

"Putting a Face on the Future," *USA Today*, Oct. 29, 1987: 2A.

"Queen of Physics," *Newsweek*, May 20, 1963, 92f.

"Question: Why Is the Male Ego Like a California Olive?" *Ms.* Gazette, Oct. 1979: 30.

Quimby, Edith H., "Some Practical Considerations Regarding the Employment of Various Qualities of Roentgen Rays in Therapy," *Radiology* 38: 3 (Mar. 1942): 261-73.

Quinn, J.M., *History of the State of California, Oakland, and Environs*, 2 vols. Los Angeles: Historic Record Co., 1907.

"QWERTYUIOP? Lillian Malt Has Discovered a Better Way," *People*, Mar. 28, 1977: 36.

Radetsky, Peter, "The Rise and (Maybe Not the) Fall of Toxic Shock Syndrome," *Science/85*, Feb. 1985, 73-9.

Radford, Claire (Pio Pico Museum, Whittier, CA), L-2-3/80.

Raloff, Janet, "Antiviral Drug Could Cut Chicken Pox Short," *Science News* 140: 22 (Nov. 30, 1991): 358.

Raloff, Janet, "Cold Lasers Employed to Treat Paralysis," *Science News* 124:16 (Oct. 15, 1983): 245.

Raloff, Janet, "Making Blood Share More of Its Oxygen," *Science News* 129: 17 (Apr. 26, 1986): 260.

Raloff, Janet, "U.S. Challenges Japan in Planned 'New-Generation' Computers," *Science News* 123: 25 (June 18, 1983): 390.

Raloff, Janet, "Zapping Pain: Hot Prospects for the Cold Laser," *Science News* 123: 7 (Feb. 12, 1983): 100.

Rambo, Ralph, *Lady of Mystery: Sarah Winchester.* Pioneer Series No. 1. San Jose, CA: The Rosicrucian Press, 1967.

[Ramey, Estelle,] "An Interview with Dr. Estelle Ramey," *Georgetown Medical Bulletin* 24: 1 (Aug. 1970): 4–11.

Rand, Peter, *The Private Rich.* New York: Crown, 1984.

Randell, Brian, ed., *The Origins of Digital Computers,* 3rd ed. Berlin: Springer-Verlag, 1982 (1973).

Randolph, Vance, ed., *Pissing in the Snow and Other Ozark Folktales.* New York: Avon, 1976.

Ransom, Helen (Connecticut Historical Society, Hartford, CT), L-1981–2.

Raper, Kenneth B., "A Decade of Antibiotics in America," *Mycologia,* 45: 1 (Jan.–Feb. 1952): 1–59.

Raper, Kenneth B., and Dorothy I. Fennell, *The Genus* Aspergillus. Baltimore: Williams & Wilkins, 1965.

"Rapid Diagnosis of Legionellosis," *Science News,* 119 (June 6, 1981): 338.

Rattray, Robert, *Religion and Art in Ashanti.* London: Oxford University Press, 1959.

Rau, Hans, *Solar Energy.* New York: Macmillan, 1964 (1958).

Ravenel, Harriott H., *Eliza Lucas Pinckney.* New York: Scribner's Sons, 1896.

Rayne, Martha Louise, *What Can A Woman Do, or, Her Position in the Business and Literary World.* Petersburgh, NY: Eagle, 1893.

Rayner, George H., "Non-Inventive Woman." London *Daily Telegraph,* Dec. 28, 1934.

Rayner, George H., "Women's Inventions." London *Daily Telegraph,* Dec. 30, 1933.

RCA *Technical Excellence Newsletter,* No. 77 (Dec. 21, 1970); No. 93 (Apr. 14, 1972).

Read, K.E., "Cultures of the Central Highlands, New Guinea," *Southwestern Journal of Anthropology* 10: 1 (Spring 1956).

["Recollections of a Schoolmate"], a fragment, Lammé Family papers, privately held.

Reed, Elizabeth W. (Geneticist, author, St. Paul, MN), PC-1979 to present.

Reed, Evelyn, *Woman's Evolution: From Matriarchal Clan to Patriarchal Family.* New York: Pathfinder, 1975.

Rees, Abraham, *et al.,* eds, *The Cyclopaedia or Universal Dictionary of Arts, Sciences, and Literature,* 39 vols. London: Longman, Hurst, 1819.

Reiter, Rayna R., ed., *Toward an Anthropology of Women.* New York: Monthly Review Press, 1975.

Rennie, John, "Proteins 2, Malaria 0: Malaria-Free Mice Offer Clues for Developing a Human Vaccine," *Scientific American* 265: 1 (July 1991): 24–5.

Reuben, David. *Everything You Ever Wanted to Know About Sex but Were Afraid to Ask.* New York: D. McKay Co., 1969.

Reynolds, Francis (Reynolds Engineering, Redmond, WA), L-1980–2.

Reynolds, Minnie J., "Women as Inventors," Inter-urban Woman Suffrage Series No. 6, 1908; repr. in part from New York *Sun,* Oct. 25, 1908.

Ribando, Curtis P. (Patent Advisor, Agricultural Research, USDA, N. Central Region, Northern Regional Research Center, Peoria, IL), L-3/26/80; PC-4/1980.

Rich, Adrienne, *Of Woman Born.* New York: Norton, 1976.

Rich-McCoy, Lois, *Millionairess.* New York: Harper & Row, 1978.

Richman, Tom, "One Man's Family," *Inc.,* Nov. 1983: 151–6.

Richman, Tom, "A Tale of Two Companies," *Inc.,* July 1984: cover story.

Richmond, Mary L., ed., *Shaker Literature: A Bibliography.* Hanover, NH: University Press of New England, 1976.

Richter, Derek, ed., *Women Scientists: The Road to Liberation.* London: Macmillan, 1982.

Riegel, Robert, *The American Feminists.* Lawrence: University of Kansas Press, 1963.

Rieker, Jane, "How Do You Handle a Baby Having a Wail of a Time? Try Slosky the Bear," *People,* Dec. 17, 1979: 72–5.

Rimmer, B.M. (Industrial Property Section, British Library Science Reference Library, London, Eng.), L-9/15/81.

Ringland, Gillian (Computer scientist, software developer/manager, business manager, entrepreneur, inventor, ICL Office Systems, Bracknell, Berks., UK), c.v./1987, L-8/26, 9/14/87; Q-8/25/87.

Ritzenthaler, Mary Lynn (Assist. MS Librarian, University of Illinois/Chicago Circle, Chicago, IL), L-1/3, 7/31/79.

"RN Awarded $520,000 for Her Invention," *American Journal of Nursing* 87:7 (July 1987): 982.

Robelen, Horace B. (Corp. Secretary, Larus & Bro., Richmond, VA), L-11/7/83.

Robertson, Patrick, ed., *Shell Book of Firsts.* London: Ebury Press, 1974.

Robin, Dr. Eugene, "A Risk/Benefit Look at Childhood Vaccines and Routine X-Rays," *Stanford Magazine,* Summer 1986: 8.

Robins, Elizabeth, "Ancilla's Share," in Mary Cohart, ed., *Unsung Champions of Women.* Albuquerque: University of New Mexico Press, 1975, 250–65.

Robinson, Donald, "America's Seventy-five Most Important Women, *Ladies' Home Journal* 87: 1 (Jan. 1971): 71–3.

"The Robot Abstraction," *High Technology* 4: 7 (July 1984): 13–5.

*Rodale's Encyclopedia of Organic Gardening.* Emmaus, PA: Rodale, 1978.

Rodgers, Dorothy, *A Personal Book.* New York: Harper & Row, 1977.

Rohrlich, Ruby (Anthropologist, New York, NY), PC-1980–present.

Rohrlich, Ruby, "State Formation in Sumer and the Subjugation of Women," *Feminist Studies* 6: 1 (Spring 1980): 76–102.

Rohrlich-Leavitt, Ruby, "Women in Transition: Crete and Sumer," in Bridenthal and Koonz, *Becoming Visible: Women in European History.* Boston: Houghton Mifflin, 1977.

Rohrlich-Leavitt, Ruby, Barbara Sykes, and Elizabeth Weatherford, "Aboriginal Woman: Male and Female Anthropological Perspectives," in Gerit Huizer and Bruce Mannheim, eds, *Politics of Anthropology.* Hawthorne, NY: Mouton, 1979.

Rolf, Ida P., "Structural Integration," *Confina Psychiatrica* 16 (1973): 69–79.

Romney, Caroline Westcott, "Four Months in Old Mexico," in Mary K. Eagle, ed.

Romney, Caroline Westcott, "Women as Inventors, and the Value of Their Inventions in Household Economics," *Journal of Industrial Education* 9:1 (Sept. 1894): 1–11ff (Pt. I).

Ronen, Simcha, *Flexible Working Hours: An Innovation in the Quality of Work Life.* New York: McGraw-Hill, 1981.

Ronhaar, J.H., *Woman in Primitive Mother-right Societies.* The Hague: J.B. Wolters, 1931.

Roosevelt's Aid Asked to Free Mrs. Slater . . . , *New York Times,* June 21, 1931: 6: 2.

Roper, Tim, "A Question of Taste," *New Scientist,* Mar. 29, 1984: 30–3.

Rose, Deborah Lee, "Genetics vs. Genocide," *California Monthly,* Feb. 1987: 12–5.

Rose, Jeanne, *Herbs and Things: Jeanne Rose's Herbal.* New York: Grosset and Dunlap, 1972.

Rosenbaum, Maj-Britt, MD, "Vibrators: Turning On to Pleasure," *Mademoiselle,* Jan. 1981: 92ff.

Rosenberg, Charles, and Carroll Smith-Rosenberg, eds, *Birth Control and Family Planning in Nineteenth-Century America.* New York: Arno, 1974.

Rosenthal, Evelyn, ed., *Women, Aging and Ageism.* Binghamton, NY: Haworth, 1990.

Rosner, Aria C., "How We Do It," *Journal of School Health* 46:8 (Oct. 1976): 480–1.

Ross, H. Ling, *Natives of Sarawak and British North Borneo.* New York: Truslove & Comba, 1896.

Ross, Ishbel, *Crusades and Crinolines: The Life and Times of Ellen Curtis Demorest and William Jennings Demorest.* New York: Harper & Row, 1963.

Rosser, Roland (National Research and Development Council, London). L-1980–present.

Rossiter, Margaret, *Women Scientists in America: Struggles and Strategies to 1940.* Baltimore: Johns Hopkins University Press, 1982.

Rossman, Joseph, "Women Inventors," *Journal of the Patent Office Society* 10: (Sept. 1927): 18ff.

Roth, Laura M., and Nancy M. O'Fallon, *Women in Physics.* New York: American Physical Society, 1983.

Roth, Shirley H. (Chemist, patentee, Cities Service Co., Cranbury, NJ), PC-1973, 981.

Roth, Walter E., "An Inquiry into the Animism and Folklore of the Guiana Indians," 30th Annual Report, Bureau of American Ethnology, Washington, DC: Smithsonian, 1908–9.

Rothman, Lorraine, and Laura Punnett, "Menstrual Extraction: Procedures, Politics," *Quest* 4: 3 (Summer 1978): 44–60.

Rothschild, Joan, ed., *Machina Ex Dea: Feminist Perspectives on Technology.* New York: Pergamon, 1983.

Rothschild, Joan, ed., *Women, Technology, and Innovation.* New York: Pergamon, 1982.

Rubin, Lillian, *Women of a Certain Age.* New York: Harper, 1979.

Rubin, Vera, "Women's Work," Science/86, July/Aug. 1986: 58–65.

Rubinstein, Helena, *My Life for Beauty.* New York: Simon & Schuster, 1964.

Ruddle, Kenneth, Dennis Johnson, Patricia K. Townsend, and John D. Rees, *Palm Sago: A Tropical Starch from Marginal Lands.* Honolulu: University Press of Hawaii, 1978.

Rudofsky, Bernard, *Now I Lay Me Down to Eat.* Garden City, NY: Doubleday, 1980.

"Running Woe," *Time,* Jan. 25, 1982: 50.

"Runs a Business on Outworkers," People of Vision sec., *Business* (London) 81: 3 (Mar. 1951): 57.

Ruys, A. Charlotte, "Pioneer Medical Women in the Netherlands," *Journal of the American Medical Women's Association* 7: 3 (Mar. 1952): 99–101.

Sabin, Florence, *Bryn Mawr College Fiftieth Anniversary.* Bryn Mawr, PA: Bryn Mawr College, 1935.

St. Clair, Ethelwyn Powell (granddaughter of inventor, Broaddus, TX), L-5/1/84 to Emily Stothart.

Saltus, Richard, "Brain Cell Transplants May Help Senility Victims of the Future," San Francisco *Examiner,* June 29, 1982: A4.

Salvatore, Diane, "A Very Special Love Story," *Ladies' Home Journal,* May 1986: 54–61.

Sammet, Jean E. (Mgr., Programming Language Technology, IBM Federal Systems Division, Bethesda, MD), Detailed Bibliography (DB)/1981; L-1984-7.

Sammet, Jean E., "The Early History of COBOL," in Wexelblat.

Sammet, Jean E., "History of IBM's Technical Contributions to High Level Programming Languages," *IBM Journal of Research and Development* 25: 5 (Sept. 1981): 520–34.

Sammet, Jean E., "An Overview of High-Level Languages," in Marshall Yovits, ed., *Advances in Computers,* vol. 20. New York: Academic Press, 1981.

Sammet, Jean E., *Programming Languages: History and Fundamentals.* Englewood Cliffs, NJ: Prentice-Hall, 1969.

Sammet, Jean E., "The Use of English as a Programming Language," *Communications of the ACM* 9: 3 (Mar. 1966): 228–30.

Sammet, Jean E., and Elaine R. Bond, "Introduction to FORMAC," *IEEE Transactions on Electronic Computers,* vol. EC-13, no. 4 (Aug. 1964): 386–94.

Sanday, Peggy R., "Toward a Theory of the Status of Women," *American Anthropologist* 75 (1973): 1682–1700.

Sanday, Peggy R., *Female Power and Male Dominance: On the Origins of Sexual Inequality.* Cambridge: Cambridge University Press, 1981.

Sanders, N.K., *The Epic of Gilgamesh.* Harmondsworth, Eng.: Penguin, 1979 (1960).

Sandford, Paula A., "Allene R. Jeanes," *Carbohydrate Research* 66 (1978): 3–5.

Sandilands, G.S., "Josephina de Vasconcellos," *The Studio,* Jan. 1956: 8–11.

Sandoz, Mari, *Cheyenne Autumn.* New York: Hastings, 1953.

San Francisco Mechanic's Institute, *Report of the Eighth Industrial Exhibition,* 1871: 122.

Sanger, Margaret, *Woman and the New Race.* New York: Truth, 1920.

Santoni, Patricia (Naval Ocean Systems Center, San Diego, CA), L-2/9/81.

Sarton, George, *A History of Science,* 2 vols. Cambridge, MA: Harvard University Press, 1952, 1959.

Schanstra, Carla, "Women in Infosystems: Climbing the Corporate Ladder," *Infosystems,* May 1980: 72ff.

Schieber, Linda, "Philatelic Fallacy," *Savvy,* Nov. 1982: 91.

Schlissel, Lillian, *Women's Diaries of the Westward Journey.* New York: Schocken, 1982.

Schmandt-Besserat, Denise, "The Envelopes That Bear the First Writing." *Technology and Culture* 21: 3 (July 1980): 357–85.

Schmandt-Besserat, Denise, ed., *Early Technologies,* Invited Lectures on the Middle East at the University of Texas at Austin, vol. 3. Malibu, CA: Undena, 1979.

Schmandt-Besserat, Denise, ed., *The Legacy of Summer.* Malibu, CA: Undena, 1976.

Schmidt, Dolores Barraccano, with Earl Robert Schmidt, "The Invisible Woman: The Historian as Professional Magician," in Forfreedom.

Schmidt, Hubert G., *Agriculture in New Jersey: A Three-Hundred-Year History.* New Brunswick, NJ: Rutgers University Press, 1973.

Schmookler, Jacob, "Inventors Past and Present." *The Review of Economics and Statistics* 39: 3 (Aug. 1957): 321–33.

Schneider, David M., "Abortion and Depopulation on a Pacific Island," in Benjamin D. Paul, ed., *Health, Culture, and Community.* New York: Russell Sage, 1955.

Schneider, David M., and Kathleen Gough, eds, *Matrilineal Kinship.* Berkeley: University of California Press, 1961.

Schultz, Leslie M., "Dial-a-de Kooning," *Inc.,* Nov. 1983: 25f.

Schumacher, Dorin, "Hidden Death: The Sexual Effects of Hysterectomy," in Evelyn Rosenthal, ed., *Women, Aging, and Ageism.* Binghamton, NY: Haworth, 1990.

Schwarz, Joel, "A Touch of Glass," *Science/81,* Sept. 1981: 79–80.

Schweich, Cindy, "Making Bright Ideas Pay," *McCall's,* Oct. 1980: 73.

*Science Digest* Editors, "America's Top One Hundred Young Scientists," *Science Digest,* Dec. 1984: 40–71.

Scientists Develop Unique Way to Take X-Ray Pictures," Stanford *Observer,* Oct. 1981: 2.

Sclabassi, R.J., R. Buchness, and Thelma Estrin, "Interactive Graphics in the Analysis of Neuronal Spike Train Data," *Computers in Biology and Medicine* 6 (1976): 163–78.

"Scotchgard: A Stain Repellent," *Consumer Reports* 26 (Jan. 1961): 5–6.

Scott, Anne Firor, *Making the Invisible Women Visible.* Urbana: University of Illinois Press, 1984.

Scott, Charles F., "Benjamin Garver Lammé," Ohio State University *Monthly,* Oct. 1924: 1–8.

Seaman, Barbara, and Gideon Seaman, *Women and the Crisis in Sex Hormones.* New York: Bantam, 1978.

*Sears Roebuck Catalogue,* 1897, ed. Fred L. Israel. New York: Chelsea House, 1968.

Seidenbaum, Art, "No Room to Fume," Los Angeles *Times,* Apr. 1, 1975, CC.

"Serious Monkey Business," *Planning for Health,* Kaiser Foundation Health Plan Newsletter, Spring 1984: 9.

Sessions, Gordon M., *et al., Traffic Devices: Historical Aspects Thereof.* Washington, DC: Institute of Traffic Engineers, 1971, 101–7.

1750 Arch Records, *New Music for Electronic and Recorded Media,* No. 1765. Berkeley, CA, 1977?

Sewall, May Wright, ed., *The World's Congress of Representative Women.* Chicago: Rand McNally, 1894.

Seymour, Morris, *et al., Connecticut at the World's Fair: Report of the Commissioners from Connecticut of the Columbian Exhibition of 1893 at Chicago; also, Report of the Work of the Board of Lady Managers of Connecticut.* Hartford, CT: Case, Lockwood & Brainard, 1898.

Shaker Almanac: *New and Favorite Receipts of the Shakers and Illustrated Almanac for 1883.* New York: A.J. White, 1882. (Other issues referred to by year.)

Shapiro, Neil, "The World's Longest Game," *Popular Mechanics,* Jul. 1982: 54–5.

Sharnoff, Sylvia Duran, "Lowly Lichens Offer Beauty—and Food, Drugs, and Perfume," *Smithsonian* 15: 1 (Apr. 1984): 135–43.

Sharp, Evelyn, *Hertha Ayrton: A Memoir.* London: Edward Arnold & Co., 1926.

Sharp, Lauriston, "Steel Axes for Stone Age Australians," in Edward H. Spicer, ed., *Human Problems in Technological Change.* New York: Wiley, 1967 (1952).

Sharpe, Samuel, *The History of Egypt from the Earliest Times Till the Conquest by the Arabs AD 640.,* 6th ed., London: Bell, 1876.

Sharrer, G. Terry. (Assoc. Curator, Div. of Extractive Industries, Smithsonian Institution), L-3/5/1980.

Shaw, Dr. Jane E. (Pres. ALZA Research, ALZA Corp., Palo Alto, CA), L-12/29/81, 7/30/86.

Shaw, Jane E., "Drug Delivery Systems," in *Annual Reports in Medicinal Chemistry* 15. New York: Academic Press, 1980: 302–14.

Shaw, Jane E., and S.K. Chandrasekaran, "Controlled Topical Delivery of Drugs for Systemic Action," *Drug Metabolism Reviews* 8 (1978): 2: 223–33.

Shaw, Jane E., and John Urquhart, "Programmed, Systemic Drug Delivery by the Transdermal Route," *TIPS,* April 1980: 208–11.

Shea, John G., *The American Shakers and Their Furniture.* Cincinnati, OH: Litton Educational Publishing, 1971.

"She Had a Better Idea," *National Enquirer,* May 20, 1973: 33.

"She Wants Fog," London *Daily Herald,* Oct. 5, 1953.

Sheldon, May French, "An African Expedition," in Mary K. Eagle, ed.

Shell, Ellen Ruppel, "Chemists Whip Up a Tasty Mess of Artificial Flavors," *Smithsonian* 17: 2 (May 1986).

"Shell-Shocked," *Family Weekly,* Jan. 9, 1983: 34.

Shelton, Linda K. "Relief from Chronic Pain: Alternative Treatments," *Ms.,* Jan. 1983: 15.

Shenton, James, ed., *Free Enterprise Forever: "Scientific American" in the 19th Century.* New York: Images Graphiques, 1977.

Shephard, Dr. M. Claire (Imperial Chemical Industries, Plant Protection Div., Jealott's Hill Research Station, Bracknell, Berks., Eng.), L-12/10/ 81, 3/16/85.

Shephard, Margaret C. (M. Claire), "Factors Which Influence the Biological Performance of Pesticides," *Proceedings of the 1981 British Crop Protection Conference—Pests and Diseases:* 711–21.

Shephard, Margaret C. (M. Claire), "PP969: A Broad-Spectrum Systemic Fungicide for Injection or Soil Application," Imperial Chemical Industries (Bracknell) (unpublished paper sent by author).

Shephard, Margaret C. (M. Claire), *et al.,* "Sensitivity to Ethirimol of Powdery Mildew from UK Barley Crops," *Proceedings of the 8th British Insecticide and Fungicide Conference (1975)* I: 639ff.

Sheraton, Mimi, "More Thoughts on Smoking and Food," *New York Times,* "Style," Nov. 13, 1982.

Sheridan, Mary, "Young Women Leaders in China," *Signs* 2: 1 (Autumn 1976): 59–88.

Sherlock, Margaret H. (Chemist, patentee, Bloomfield, NJ), L-1986-7; Q-9/24/86; c.v./1979.

Sherman, Patsy O. (Chemist, inventor, Chemical Resources Div., Minnesota Mining & Manufacturing Co., St. Paul, MN), Q-1983.

Sherr, Lynn, and Jurate Kazickas, *American Woman's Gazetteer.* New York: Bantam, 1976.

Sherr, Lynn, and Jurate Kazickas, "New Voyager: Landmarks of Our Own," *Ms.,* Aug. 1976: 59–62.

Sherr, Lynn, and Jurate Kazickas, eds, *The Woman's Calendar.* New York: Universe Books, 1971–9.

Sherratt, Andrew, ed., *The Cambridge Encyclopedia of Archeology.* New York: Crown/Cambridge, 1980.

Sherrod, Pamela, "Inventor's Grandchild Turns Mother of Invention," Chicago *Tribune,* Mar. 9, 1987, Sec. 4: 11.

"She's Invading IBM's Stronghold," *Business Week,* Dec. 2, 1972: 52.

Sheth, Madhuri, PhD (National Institute for Training in Industrial Engineering, Viharhake, Bombay, India) PC-7/85.

Shipman, Pat, "Silent Bones, Broken Stones," *Discover* 6: 8 (Aug. 1985): 66–9.

Shirley, Stephanie ("Steve") (Entrepreneur, inventor, F International Group Ltd, Chesham, Bucks., Eng.), L-1984-8.

Shirley, Steve, "The Distributed Office," *Journal of the Royal Society of Arts,* June 1987: 1–12.

Shizuru, Judith, *et al.,* "Islet Allograft Survival After a Single Course of Treatment . . . with Antibody to L3Tv," *Science* 237: 4812 (July 17, 1987): 278–80.

Shorb, Mary S., "Unidentified Growth Factors for *Lactobacillus lactis* in Refined Liver Extracts," *Journal of Biological Chemistry* 169 (1947): 455–6.

Short, R.V., "Breast Feeding," *Scientific American* 250: 4 (Apr. 1984): 23–9.

Shostak, Marjorie, "!Kung Women," paper presented at Stanford University, Institute for Research on Women and Gender, Dec. 2, 1986.

Shostak, Marjorie, "Life Before Horticulture," *Horticulture,* 55: 2 (Feb. 1977): 38ff.

Shostak, Marjorie, *Nisa: The Life and Words of a Kung Woman.* New York: Random House, 1982.

Shotwell, Walt, "Ads Aim to Give Iowa a Fruitful Tomorrow . . . ," Des Moines *Register,* Sept. 24, 1990: 12B.

Shuck, Chris, "Student Develops a New Fertilizer," Trenton, NJ, *The Sunday Trentonian,* Mar. 2, 1980: 20.

Shulder, Diane, and Florynce Kennedy, *Abortion Rap,* New York: McGraw-Hill, 1971.

Shurcliff, William A., *New Inventions in Low Cost Solar Heating: One Hundred Daring Schemes Tried and Untried.* Andover, MA: Brick House Publishing Co., 1979.

Shurkin, Joel, *Engines of the Mind,* New York: Norton, 1984.

Shurkin, Joel, "Robot Hand Brings Sign Language to the Deaf-Blind," Stanford University *Campus Report,* July 31, 1985: 7.

Shurkin, Joel, "Walbot, Corn Geneticist, Does Pioneer Research on the Farm," Stanford University *Campus Report,* Dec. 2, 1981: 5, 13.

Shuttle, Penelope, and Peter Redgrove, *The Wise Wound: Eve's Curse and Everywoman.* New York: Marek, 1978.

Sicherman, Barbara, and Carol Hurd Green, eds, *Notable American Women: The Modern Period—A Biographical Dictionary.* Cambridge, MA: Harvard University Press, 1980.

Sicherman, Barbara (*Notable American Women* Project, Radcliffe College, Cambridge, MA), L-1/9/80.

Siegel, Micki, "Mothers of Invention," *Good Housekeeping,* Sept. 1984: 72–80.

Sikic Branimir I., Nina F. Friend, *et al.,* "Dissociation of Antitumor Potency from Anthracycline Cardiotoxicity in a Doxorubican Analog," *Science* 228: (June 28, 1985): 1544–6.

Silberner, Joanne, "86 Laskers: AIDS, VD, Growth Work," *Science News,* Sept. 27, 1986: 197.

Silberner, Joanne, "AIDS Studies Suggest New Directions, Therapies," *Science News* 130:25, 26 (Dec. 20, 27, 1986): 388–9.

Silberner, Joanne, "Collaborators Cohen, Levi-Montalcini Win Medical Nobel," *Science News* 130 (Oct. 18, 1986): 244.

Silberner, Joanne, "Enzyme Ruse Successful," *Science News,* Mar. 7, 1987: 153.

Silberner, Joanne, "Sneaking n a Therapeutic Enzyme," *Science News* 129: 18 (May 3, 1986): 277.

Silvanov, Beverley (Inventor, consultant, London, Eng.), L-11/14/79, 7/10/86.

Silvanov, Beverley, "Medical Invention," *The Inventor* (London), Sept. 1976: 16–17.

Simon, Dorothea P. (Ret. education researcher, inventor, Pittsburgh, PA), c.v./1984; Q-6/14/84; L-7/10, 10/20/84.

Simon Dorothea P., "Spelling—A Task Analysis," *Instructional Science* 5 (1976): 277–302.

Simon, Dorothea P., and H.A. Simon, "Alternate Uses of Phonemic Information in Spelling," *Review of Educational Research* 43 (1973): 115–37.

Simon, Dorothea P., and H.A. Simon, "Individual Differences in Solving Physics Problems," in R.S. Siegler, ed., *Children's Thinking: What Develops?* Hillsdale, NJ: Erlbaum, 1978.

Simon, Dorothea P., and H.A. Simon, "A Tale of Two Protocols," in J. Lochhead and J. Clement, eds, *Cognitive Process Instruction.* Philadelphia: Franklin Institute, 1979.

Simon, Dorothea P., *et al.,* "Expert and Novice Performance in Solving Physics Problems," *Science* 208 (1980): 1335–42.

Simon, Dorothea P., *et al.,* "Models of Competence in Solving Physics Problems," *Cognitive Science* 4 (1980): 4: 317–43.

Simons, Geoff L., *Women in Computing,* Manchester, Eng.: National Computing Centre, 1981.

Singer, Charles, *et al., A History of Technology,* 4 vols. + suppls. Oxford: Oxford University Press, 1954.

Singer, Connie, "For 200,000 Fatsos, Carol Kiebala's Scolding Fridge Keeps Them Watching Their Weight," *People* 12: 23 (Dec. 3, 1979):81f.

Sjöö, Monica, and Barbara Mor, *The Ancient Religion of the Great Cosmic Mother of All.* Trondheim, Norway: Rainbow, 1981; rev. ed., New York: Harper & Row, 1984 (1981).

Skelton, Nancy, "No More Ifs, Ands, Butts—It's Law," Los Angeles *Times,* Apr. 14, 1985: 1ff.

Sklar, Kathryn K. *Catharine Beecher: A Study in American Domesticity.* New Haven, CT: Yale University Press, 1973.

Sloan, Jan Butin (Kansas City Art Institute, Kansas City, MO), L-9/25/86.

Sloane, Bonnie K., "In Brief," National NOW *Times,* Jan./Feb. 1982: 4.

Sloane, Leonard, "Doing It Herself," *New York Times Biographical Edition,* February 1972: 257–8.

Smith, Anne D. (Librarian, Marshallville Public Library, Marshallville, GA), L-4/1/85.

Smith, Charlotte, ed. and pub., *The Woman Inventor,* April, June 1891.

Smith, Denis, "Lessons from the History of Technology," *The Inventor* (London), Jan. 1978: 6ff.

Smith, Judy, *Something Old, Something New, Something Borrowed, Something Due: Women and Appropriate Technology.* Butte, MT: National Center for Appropriate Technology, 1978.

Smith, Judy, ed., *Women and Technology: Deciding What's Appropriate.* Proceedings of a conference on Women and Technology, Missoula, MT, Apr. 27-9, 1979. Butte, MT: National Center for Appropriate Technology, 1979.

Smith, Naomi, "How Can I Improve It?" *Flights of Color.* Lexington, MA: Ginn & Co., 1982.

Smith, William, "Pioneer in Computers," *New York Times,* Sept. 5, 1971.

Smither, Robert K., (Physicist, inventor: Argonne National Laboratory, Argonne, IL), PC-1978–present.

Smither, R[obert] K., "Crystal Diffraction Lenses for Imaging Gamma-Ray Telescope," invited paper for the 13th Texas Symposium on Relativistic Astrophysics, Chicago, IL, Dec. 15–18, 1986.

Smither, R[obert] K., "New Method for Focusing X-Rays and Gamma Rays," *Review of Scientific Instruments* 53: 2 (Feb. 1982): 131–41.

Society of Women Engineers, *Women's Technical Contributions* (Achievement Award booklets). New York: SWE, 1952 – .

Sokolov, Raymond, "A Root Awakening." *Natural History* 87: 9 (Nov. 1978): 34ff.

Solecki, Ralph S., *Shanidar: The First Flower People.* New York: Knopf, 1971.

Soma, Rose, "Stories from Yesterday . . . and Tomorrow?" *Ms.,* May 1979: 93–4.

"Something for Everyone at the Women's Exhibit," New Orleans *Times-Picayune,* June 5, 1983, "Vivant" sec.: 1.

Sorel, Kitty (Lawrence Peska Assoc., New York), PC-1979–80.

Soule, Ethel V., "Fireplace and Oven Utensils for Cooking in the Early American Kitchen," *The North Jersey Highlander* (North Jersey Highlands Historical Society, Newfoundland, NJ) 7: 4 (1971): 21–8.

Soule, Dr. J., *Science of Reproduction and Reproductive Control.* New York: U.S. District Court, 1856, in Rosenberg & Rosenberg.

Southam, Anna, "Historical Review of Intrauterine Devices," *Proceedings of the 2nd International Conference on Intra-Uterine Contraception,* New York, 1964: 3–5. Amsterdam: Excerpta Medica, 1965.

Sparck-Jones, Karen (Assist. Dir. of Research, Cambridge University Computer Laboratory, Cambridge, Eng.), Q-1987; c.v./1987; L-9/9/87, 8/24/88.

Sparck-Jones, Karen, "A Statistical Interpretation of Term Specificity and Its Application in Retrieval," *Journal of Documentation* 28 (1971): 11–21.

Sparck-Jones, Karen, *Synonymy and Semantic Classification.* Edinburgh: Edinburgh University Press, 1986.

Sparck-Jones, Karen, and S.E. Robertson, "Relevance Weighting of Search Terms," *Journal of the American Society for Information Sciences* 27 (1976): 129–46.

Spear, Amy C. (Engineer, inventor, MITRE Washington, McLean, VA), L-1977–8, c.v./1987.

Spear, Amy C., "ASD Design Automation System," *RCA Chief Engineer's Newsletter,* No. 141, Feb. 11, 1974.

Spitzer, Dan, "Putting Food First," *California Living* magazine (San Francisco *Chronicle* and *Examiner*), Sept. 20, 1981: 8–11.

Spreiter, John R. (Chair, Dept. of Engineering, Stanford University), *et al.,* "Memorial Resolution, Irmgard Flügge-Lotz, 1903–1974."

Spretnak, Charlene, *Lost Goddesses of Ancient Greece: A Collection of Pre-Hellenic Mythology.* Berkeley, CA: Moon Books, 1978.

Spretnak, Charlene, ed., *Politics of Women's Spirituality.* Garden City, NY: Doubleday/Anchor, 1982.

Sprigg, June, *By Shaker Hands.* New York: Knopf, 1975.

"Sprouts Recruited in Cancer Fight," *Science News* 113–14 (Dec. 9, 1978): 409.

Stage, Sarah, *Female Complaints: Lydia Pinkham and the Business of Women's Medicine.* New York: Norton, 1979.

Stanford, Anne, *The Countess of Fork.* Canoga Park, CA: Orirana Press, 1985.

Stanley, Autumn, "The Champion of Women Inventors," *American Heritage of Invention and Technology,* Summer 1992: 22–6.

Stanley, Autumn, "The Creation of Patriarchy Revisited," *Mentalities/Mentalités* 7:2, (1992).

Stanley, Autumn, "The Feminist Debate in History of Technology: Do Mothers Invent?" in Cheris Kramarae and Dale Spender, eds, *The Knowledge Explosion.* New York: Teachers College Press, 1992.

Stanley, Autumn, "From Africa to America: Black Women Inventors," in Zimmerman, ed., *Technological.*

Stanley, Autumn, "Inventing the Wheel: Women as Automotive Inventors in the Early Twentieth Century," a paper presented at the Organization of American Historians Conference, New York, NY, April 1986.

Stanley, Autumn, "Invention as Utopian Thinking: Feminist Transformations in Technology," a paper presented at the National Women's Studies Association Conference, Towson, MD, June 1989.

Stanley, Autumn, "Invention Begins at Forty: Older Women of the Nineteenth Century as Inventors," in Rosenthal, ed.

Stanley, Autumn, "Mothers of Invention: A Chronicle," *The Lightbulb* (Journal of the Inventors Workshop International, Tarzana, CA) 12: 2 (May 1982): 26–8.

Stanley, Autumn, "Mothers of Invention: A Preliminary Research Report," *Proceedings,* Society of Women Engineers Conference, San Francisco, 1979.

Stanley, Autumn, "Once and Future Power: Women as Inventors," *Women's Studies International Forum.* 15: 2 (1992).

Stanley, Autumn, "The Patent Office Clerk as Conjurer: The Vanishing Lady Trick in a Nineteenth-Century Historical Source," in Barbara Wright, ed.

Stanley, Autumn, "Professional Women Inventors of the Nineteenth Century," a paper presented at the Society for the History of Technology Conference, Sacramento, CA 1989.

Stanley, Autumn, "A Rose by Any Other Name: Omissions and Dysclassifications in a List of Nineteenth-Century American Women Patentees, 1876 and 1890," a paper presented at the Conference on Women, Work, and Technology, University of Connecticut, Storrs, October 1984.

Stanley, Autumn, "Women Hold up Two-Thirds of the Sky: Notes for a Revised History of Technology," in Joan Rothschild, ed., *Machina.*

Stanley, Donna B. (Informant, Jackson, OH), PC-1979 – present.

Stark, John, "Sarah Winchester's Ghostly Penance," San Francisco *Examiner,* Aug. 29, 1976, "Scene" sec.: 1, 4.

Stark, Rachael M., "A Year and a Day: The Celtic Concepton of the Otherworld and the Otherworldly," Unpublished senior honors thesis, University of Massachusetts–Amherst, 1981.

Starr, Michael, "She Did Not Lead a Movement," *American History Illustrated* 15: 5 (Aug. 1980): 44–7.

"Start When You Please," *Time* 119: (Jan. 10, 1977): 51–2.

Steensberg, Axel, *Stone Shares of Ploughing Instruments from the Bronze Age of Syria.* Det Kongelige Danske Videnskabernes Selskab, Historisk-filosofiske Meddelelser, 47, 6. Copenhagen, 1977.

Stein, Dorothy, *Ada: A Life and a Legacy.* Cambridge: M.I.T. Press, 1985.

Stellman, Jeanne Mager, *Women's Work, Women's Health: Myths and Realities.* New York: Pantheon, 1977.

Stephenson, Dorothy (Pres., Inventors Association of New England, Cambridge, MA), L-1978–81.

Sterling, Dorothy, *Black Foremothers: Three Lives.* Old Westbury, NY: Feminist Press, 1979.

Stern, Ava, *The Self-Made Woman* (Unpublished MS; chap. on Bette Graham sent me by Graham, 1981).

Stern, Madeline B., *We the Women: Career Firsts of Nineteenth-Century America.* New York: Schulte, 1963.

Stern, Nancy (Computer Science, Hofstra University, Hempstead, NY), L-4/12, 9/4/84.

Stern, Nancy, *From ENIAC to UNIVAC: An Appraisal of the Eckert-Manchly Computers.* Bedford, MA: Digital Press, 1981.

Stevenson, Jeanne Hackley, "Eleanor Albert Bliss, 1899–, Sulfa Compounds Researcher," in Helmes.

Steward, John F., *The Reaper.* New York: Greenberg, 1931.

Stewart, Benning, "They Achieve," Indianapolis *Star,* May 4, 1947.

Stoddard, Hope, *Famous American Women.* New York: Crowell, 1970.

Stodola, F.H., O.L. Shotwell, *et al.,* "Hydroxy-Streptomycin, a New Antibiotic from *Streptomycetes griseocarneus,*" *Journal of the American Chemical Society* 73 (1951): 2290–3.

Stokes, Donald, "Engineer, Surgeon Join Forces to Seek Aid for Deaf People," Stanford University *Campus Report,* July 31, 1985: 5.

Stokes, Donald, "Hilgard Fights Children's Pain with Hypnotherapy," Stanford University *Campus Report,* Dec. 19, 1984: 5–6.

Stone, Casey G. (Educational Media Production Project for the Hearing- Impaired, University of Nebraska–Lincoln), L-5/28/82.

Stone, Casey G. and Gwen C. Nugent, "Think It Through: An Interactive Videodisc for the Hearing-Impaired," in *Proceedings of the Johns Hopkins First National Search for Applications of Personal Computing to Aid the Handicapped:* 49ff. Los Angeles, IEEE Computer Society, 1981.

Stone, Elizabeth, "A Madame Curie from the Bronx," *The New York Times Magazine,* April 9, 1978: 29–34f.

Stone, Judith, "Cells for Sale," *Discover* 9: 8 (Aug. 1988): 32–9.

Stone, Merlin, *Ancient Mirrors of Womanhood,* 2 vols. New York: Sibylline Press, 1979.

Stone, Merlin, *When God Was a Woman.* New York: Harcourt Brace, 1976.

Storch, Dr. Marcie, *How to Relieve Cramps and Other Menstrual Problems.* New York: Workman, 1982.

Stothart, Emily (Granddaughter and biographer of inventor Emily Duncan, Coushatta, LA), L-1984–8.

Stothart, Emily, *Home at Last: Eugene and Emily J. Duncan.* Bossier City, LA: Everett Publishing Co., 1987.

Stowe, Harriet Beecher, ed., *Our Famous Women: An Authorized Record of the Lives and Deeds of Distinguished American Women of our Times.* Hartford, CT: Worthington, 1884.

Stowe, Harriet Beecher, *The Pearl of Orr's Island: A Story of the Coast of Maine.* Boston: Houghton Mifflin, 1862.

Strathern, Marilyn, *Women in Between: Female Roles in a Male World: Mount Hagen, New Guinea.* London: Seminar Press, 1972.

Straus, Ralph, *Carriages and Coaches.* London: M. Secker, 1912.

Strauss, John H., *et al.,* "Laminaria Use in Midtrimester Abortions . . . ," *American Journal of Obstetrics and Gynecology.* 134: 3 (1979): 260–4.

Stroup, Sue (Mrs. Jacob), "Reminiscences of Mrs. Jacob Stroup and Mrs. G.W. Brinno, Pioneers of Washoe, Fayette County, in the Early Seventies." Typescript, Idaho Historical Society (Boise), n.d.

Strouse, Jean, "Sweet Temptation," *Newsweek,* Oct. 6, 1980: 93Aff.

Stuard, Susan Mosher, "Dame Trot," *Signs* 1: 2 (Winter 1975): 537–42.

Stuard, Susan Mosher, ed., *Women in Medieval Society.* Philadelphia: University of Pennsylvania Press, 1976.

Stuart, Sandra Lee, comp., *Who Won What When.* Secaucus, NJ: Lyle Stuart, 1977.

Stubblefield, Phillip G., "Laminaria Tents and Legal Abortion: The Experience of the 1970's," in G.S. Berger *et al.,* eds, *Second Trimester Abortion: Perspectives after a Decade of Experience.* Littletown, MA: PSG Publishing Co., 1980.

"Students Develop New Cable-Car Grip," Stanford University *Observer,* April 1979: 5.

"A Study in Feminine Invention," *Scientific American,* Oct. 1924: 260.

Stutley, Margaret, and James Stutley, eds, *Harper's Dictionary of Hinduism.* New York: Harper & Row, 1977.

Suggs, Robert C., *Marquesan Sexual Behavior.* New York: Harcourt, Brace, 1966.

Suid, Murray, and Ron Harris, *Made in America: Eight Great All-American Creations.* Reading, MA: Addison-Wesley, 1978.

Sutherland, Connie, "Kohe Malamalama," unpublished MPH thesis, University of Hawaii School of Public Health, 1977.

Swedish Women Inventors' Association, "The Woman Inventor," WIPO Exhibit, Geneva, Mar. 1985.

Sweet, Ellen, "A Failed Revolution," *Ms.* 16: 9 (Mar. 1988): 75–9.

"Sweet Success: Mrs. Fields Cookies Goes East," *Fortune,* Feb. 7, 1983: 9.

Tangley, L., "Plant Compounds Control Moth Morphology," *Science News* 121 (Mar. 6, 1982): 150.

Tangri, Sandra Schwartz, "Determinants of Occupational Role Innovation Among College Women," in Mednick *et al.*

Tannahill, Reay, *Food in History.* New York: Stein & Day, 1973.

Tanner, James J., "Population Limitation Today and in Ancient Polynesia," *Bioscience* 25: (Aug. 1975): 513–6.

Tanner, Nancy M., *On Becoming Human: A Model of the Transition from Ape to Human and the Reconstruction of Early Human Social Life.* Cambridge: Cambridge University Press, 1981.

Tanner, Nancy M., and Adrienne Zihlman, "Women in Evolution, Part I: Innovation and Selection in Human Origins." *Signs* 1: 3 (Spring 1976): 585–608.

Tarbell, Ida C., "Women as Inventors." *The Chautauquan* 7: 6 (Mar. 1887): 355–7.

Taussig, F.W., *Inventors and Money-Makers; Lectures on Some Relations Between Economics and Psychology. . . .* New York: Macmillan, 1915.

Taussig, Helen, "Dangerous Tranquility" (editorial), *Science,* May 25, 1962: 136.

Taylor, Erma, "She Painted the White Line on Our Highway." *Pageant,* Sept. 1946: 53–5.

Taylor, Frank Sherwood, *The Alchemists: Founders of Modern Chemistry.* New York: Schuman, 1949.

Taylor, Gabrielle, "How Could We Do Without Them?" *Women's Realm* (England), Dec. 16, 1978: 32–5.

Taylor, Norman, *Plant Drugs That Changed the World.* New York: Dodd, Mead, 1965.

Tee, G.J., "Doris Langley Moore, *Ada, Countess of Lovelace, Byron's Legitimate Daughter"* (review), *Annals of Science* 36: 1 (Jan./Feb. 1979): 89–91.

Teleki, Geza, "The Omnivorous Chimpanzee," *Scientific American,* January 1973: 32–42.

Telkes, Maria, "Latent Heat Storage." *Proceedings,* Symposium on Energy from the Sun, Chicago, IL, Apr. 3-7, 1978: 325–37.

Telkes, Maria, "Solar Cooking Ovens." *Solar Energy: A Journal of Solar Energy Science and Engineering,* 3: 1 (Jan. 1959): 1ff.

"The $10,000 Marigold," *Science Digest* 78 (Nov. 1975): 19–20.

*Tennessee Centennial and International Exposition, May 1-Oct. 31, 1897; Catalogue [of the] Women's Department,* Nashville: Burch, Hinton, and Co., 1897.

Terman, David S., *et al.,* "Preliminary Observations of the Effects on Breast Adenocarcinoma of Plasma Perfused over Immobilized Protein A," *New England Journal of Medicine* 305: 20 (Nov. 1981): 1195–1200.

Tesoro, Giuliana C., "An Effective New Antistatic Finish," *Modern Textiles Magazine* 38:1 (1957): 47–8.

Tesoro, Giuliana C. See also Valko, E.I.,; Ortiz-Molina, M.G.; and Cranmer, J.H.

Tesoro, Giuliana C., and A. Oroslan, "Reactions of Cellulose with Unsymmetrical Sulfones," *Textile Research Journal* 33 (1963): 2: 93–107.

Testorff, Ken, "The Lady Is a Captain: She Teaches Computers How to Talk." *All Hands,* No. 700, May 1975: 32–5.

" 'Test Tube' Calf: Food and Fertility," *Science News* 120: 1 (July 4, 1981): 7.

Teubal, Savina, *Sarah the Priestess: The First Matriarch of Genesis.* Athens, OH: Swallow, 1984.

"Thalidomide Makes a Comeback," *Discover,* Sept. 1988: 6.

Tharp, Louise Hall, "Bonnet Girls," *New England Galaxy,* Winter 1960: 3–10.

Thayn, Florian H. (Head, Art and Reference Div., Architect of the Capitol, Washington, DC), L-10/31/86 to Portia James.

Theiler, M[ax], "Studies on the Action of Yellow-fever Virus in Mice." *Annals of Tropical Medicine and Parasitology* 24 (1930): 249–72.

Thom, C., and W. Fisk, *The Book of Cheese.* New York: Macmillan, 1918.

Thomas, Martha J.B. (Chemist, corporate research director, inventor, Danvers, MA), Q-7/11/83.

Thompson (Mary Hopkinson Gibbon) Obit, Boston *Globe,* Apr. 11,1986.

[Thomsen, Dietrick E.], "Do Veterinarians Need Ethylene Oxide?" *Science News* 130: 6 (Aug. 9, 1986): 89.

Thomsen, Dietrick E., "The Shiny Gray Flexible Electric Sandwich Wrap," *Science News* 121: 5 (Feb. 6, 1982): 90–1.

Thomson, George, *Studies in Ancient Greek Society,* 2 vols. New York: International Publishers, 1949, 1955.

Thorpe, Susan, "The Cure for Cramps: It Took a Woman Doctor." *Ms.,* Nov. 1980: 36.

"Three Researchers Win Nobel Prize in Medicine," Stanford *Daily,* Oct. 18, 1988: 3.

"Three Share Nobel for Medicine," San Francisco *Examiner,* Oct. 17,1988: 1.

Thune, Halldis A. (Inventor, former nurse, Hønefoss, Norway), L-4/24/86 to present; Q-1989.

Thune, Halldis A., "Women as Inventors," unpublished paper sent by author, 1987.

Tide-Mark Press, *Women Working Calendar.* Hartford, CT, 1983.

Tierney, John, "Fanisi's Choice," *Science,* Jan./Feb. 1986: 26–42.

Tiffany, Sharon W., ed., *Women and Society: An Anthropological Reader.* St. Albans, VT: Eden Press, 1979.

Tifft, Susan, "Debate on the Boundary of Life," *Time,* Apr. 11, 1983: 68–9.

Tilden, John H. (Innovation Consultant, Indianapolis Center for Adv. Research, Indianapolis), L-10/14/ 80).

"Tinkerers End 'Picnic Slide,' " *Christian Science Monitor,* Apr. 11, 1938: 1, 16.

"Tiny Pilgrims Who Seek Medical Miracles," San Francisco *Chronicle,* Mon., Dec. 14, 1981: 11.

Tobin, Andy (Herbert J. Rozoff Associates, Chicago), L-4/26, 5/7/79.

"Too Busy for Anniversaries," *Medical Woman's Journal* 33: 11 (Nov. 1926): 321–2.

Torre, Susana, ed., *Women in American Architecture: A Historic and Contemporary Perspective.* New York: Whitney Library of Design(Watson-Guptill), 1977.

Townley, Calvert, "The Edison Medal," *Electric Journal,* June 1919.

Townsend, Patricia K., "Sago Production in a New Guinea Economy," *Human Ecology* 2:3 (1974): 217–36.

Trescott, Martha M. (Scholar, author, Urbana, IL, and Dallas, TX), PC-1979 – present.

Trescott, Martha M., "A History of Women Engineers in the United States, 1850–1975: A Progress Report," Society of Women Engineers National Conference, Jan. 1979, *Proceedings.* San Francisco: Society of Women Engineers 1979: 2–14.

Trescott, Martha M., "Julia B. Hall and Aluminum," *Journal of Chemical Education* 54: 1 (Jan. 1977): 24–5.

Trescott, Martha M., "Lillian Moller Gilbreth and the Founding of Modern Industrial Engineering," in Rothschild, ed., *Machina. . . .*

Trescott, Martha M., *New Images, New Paths: Women Engineers in American History in Their Own Words.* In preparation.

Trescott, Martha M., *The Rise of the American Electrochemicals Industry, 1880–1910: Studies in the American Technological Environment.* Westport, CT: Greenwood, 1981.

Trescott, Martha M., ed., *Dynamos and Virgins Revisited: Womenand Technological Change in History.* Metuchen, NJ: Scarecrow Press,1979.

Trinkaus, Erik, and William W. Howells "The Neanderthals," *Scientific American,* 241: 6 (Dec. 1979): 118–33.

Triplett, Glover B., Jr., and David M. Van Doren, Jr., "Agriculture Without Tillage," *Scientific American,* Jan. 1977: 28ff.

Trotter, A.P., "Mrs. Ayrton's Work on the Electric Arc," *Nature* 113: 2828 (Jan. 12, 1924): 48.

Trotula. See Elizabeth Mason-Hohl.

Truxal, Carol, "The Woman Engineer," *IEEE Spectrum,* Apr. 1983: 58–62.

Tucker, Jonathan B., "Biochips: Can Molecules Compute?" *High Technology* 4: 2 (Feb. 1984): 36–47.

Tucker, Linda S., "Amelia's Bloomers." *Ms.,* Dec. 1979: 115–18.

Tucker, Theresa (Herizon Collective, Jackson, OH), L-8/20, 8/25, 9/1, 9/4/84.

Tul, Viviana (Researcher, Palo Alto, CA), PC-1983–4.

Tulk, Waldo (The Historical Society of Pennsylvania, Philadelphia), L-9/3/82.

Tumulty, Karen, "Abortion: Few Doctors at the Front," Los Angeles *Times,* Aug. 14, 1989: 1ff.

Tumulty, Karen, "Feminists Teaching Home Procedure," Los Angeles *Times.* Aug. 14, 1989: 15.

Turnbull, Colin, *The Forest People: A Study of the Pygmies of the Congo.* New York: Simon & Schuster, 1961.

Turnbull, Colin, "Mbuti Womanhood," in Dahlberg, ed.

Tuska, C.D., *Independent Inventors and the Patent System.* Subcommittee on Patents, Trademarks, and Copyrights; Committee on the Judiciary, U.S. Senate, Study No. 28. Washington, DC: Government Printing Office, 1961.

Tutelian, Louise, "Mothers of Invention," *Glamour,* June 1982: 135.

Twin, Stephanie L., ed., *Out of the Bleachers: Writings on Women and Sport.* Old Westbury, NY: The Feminist Press, 1979.

"Two Studies of Thyroid Function: Nuclear Medicine," Stanford University Medical Center, *Friends of Radiology Newsletter,* Winter 1982.

Ucko, Peter J., and G.W. Dimbleday, eds, *The Domestication and Exploitation of Plants and Animals.* Chicago, IL: Aldine, 1969.

Uglow, Jennifer S., ed., *The Macmillan Dictionary of Women's Biography.* London: Macmillan, 1982.

Undset, Sigrid, *Kristin Lavransdatter,* tr. Charles Archer and J.S. Scott. New York: Knopf, 1946 (1930).

Ungerer, Miriam, "A Few Cold Facts About Ice Cream," *New York Times,* Aug. 27, 1980: C1f.

United States Congress, House, Select Committee on Population, *Hearings on Fertility and Contraception in America,* 95th Congress, vol. III, 1978: 131.

United States Dept. of Agriculture, *Cheese Varieties and Descriptions, Agriculture Handbook* 54. Washington, DC: Government Printing Office, 1969.

United States Dept. of Agriculture, Nomination of Ollidene Weaver for AAII Inventor-of-the Year Award, 1976.

United States Dept. of Labor, Women's Bureau, Bulletin No. 28, *Women's Contributions in the Field of Invention.* Washington, DC, 1923.

United States Dept. of Labor, Women's Bureau, Bulletin No. 43. *Women as Inventors,* Washington, DC, 1941.

United States Dept. of Labor, Women's Bureau, "Women Inventors," 1975 (unpublished paper sent by the Director, Aug. 1985), 4 pp.

United States, Navy Dept., Press Release, "Grace Murray Hopper Promoted," Aug. 24, 1973.

United States Patent and Trademark Office, *Official Gazette.* Washington, DC, 1872– .

United States Patent and Trademark Office, *Report,* 1790–1836. New York: Gales & Seaton.

United States Patent and Trademark Office, *Revolutionary Ideas: Patents and Progress in America.* Washington, DC, [1976].

United States Patent and Trademark Office, *Women Inventors to Whom Patents Have Been Granted by the United States Government, 1790 to July 1, 1888,* with suppls. Washington, DC: Government Printing Office, 1888, 1892, 1895.

Univac *News,* "Commander Grace Hopper, United States Navy, Retired. . . . To Be Honored by Computer Organization," Press Release, Sept. 11, 1970.

"Update: DNA Controversy," *Mother Jones,* Apr. 1977: 7.

"Update on Contraceptives: What's Safe? Effective? Convenient?" *Changing Times* 37: 10 (Oct. 1983): 72–6.

Uselding, Paul (Economics, University of Illinois, Champaign-Urbana), PC-1982.

Valko, E.I., and Giuliana C. Tesoro, "New and Durable Antistatic Finishes," *Modern Textile Magazine* 38: 7 (1958): 62–70.

Van Allen, Judith, " 'Sitting on a Man': Colonialism and the Lost Political Institutions of Igbo Women," in Tiffany.

Van Arsdale, Peter W., "Population Dynamics Among Asmat Hunter-Gatherers of New Guinea: Data, Methods, Comparisons," *Human Ecology* 6:4 (1978): 435–67.

Vanak, Joann, "Time Spent in Housework," *Scientific American* 231 (Nov. 1974): 116–20.

Van Gelder, Lindsy, "What Does Seaweed Have to Do with Abortion?" *Ms.,* Nov. 1980: 112ff.

Van Gelder, Lindsy, "Where the Bucks Are," *Mademoiselle,* Jan. 1984: 104–5ff.

Van Leeuwen, Jeannette, "Have You Heard About Scotchgard?" *Good Housekeeping* 145 (Oct. 1957): 164.

Van Loon, Maurits, "Mureybat: An Early Village in Inland Syria," *Archaeology* 19: 3 (June 1966): 215–6.

Vann, Richard, "Women in Preindustrial Capitalism," in Bridenthal and Koonz.

Vare, Ethlie, and Greg Ptacek, *Mothers of Invention from the Bra to the Bomb . . .* New York: Morrow, 1988.

Vaughn, Agnes Carr, *The House of the Double Axe: The Palace at Knossos.* Garden City, NY: Doubleday, 1959.

Vavasour, Lady Anne, *My Last Tour and First Work; or, A Visit to the Baths of Wildbad and Rippoldsau.* London: Hugh Cunningham, 1842.

Vetter, Betty M., "Women in the Natural Sciences." *Signs* 1:3 (1976): 713–20.

Vetter, Betty M., "Working Women Scientists and Engineers," *Science* 207 (Jan. 4, 1980): 28–34.

Viacrucis, Kiyo Sato-(School nurse, inventor; Sacramento, CA), I-8/25/81; L-9/8/81.

Vicinus, Martha, *Suffer and Be Still: Women in the Victorian Age.* Bloomington: Indiana University Press,, 1972.

Vietmeyer, Noel (Author), L-12/27/81.

Vietmeyer, Noel, "The Greening of the Future." *Quest,* Sept. 1979: 25ff.

"Vionnet, Couturier, Dies at 98; Innovator Created the Bias Cut," *New York Times,* Mar. 5, 1975: 42.

[Vionnet, Madeliene], "Transition" [Obit.], *Newsweek,* Mar. 17, 1975: 55.

"A Visit to the Kitchen of Mrs. Fields," *Woman's Day,* June 11, 1985: 108–12.

"Vitamin E for Premature Infants," *Science News* 120 (Dec. 19, 26, 1981): 393.

Vogel, Virgil J., *American Indian Medicine.* Norman: University of Oklahoma Press, 1970.

Voltaire, *Dictionnaire Philosophique.* Paris: Garnier Bros., 1764.

Vonk, Alice (Developer of white marigold, Sully, IA), Q-7/1983.

*Vsesoyuznoye Obschchestvo Izobretatelei I Ratsionalizatorov Mezhdu IV i V s'ezdami* [The All-Union Society of Inventors and Rationalizers Between Its Fourth and Fifth Congresses]. Moscow: Profizdat, 1978.

Wade, Nicholas, "Tinkering with the Genes," in Doster *et al.*

Wajcman, Judy, *Feminism Confronts Technology.* University Park: Pennsylvania State University Press, 1991.

Waldor, Matthew K., Leonore Herzenberg, Mae Lim, *et al.,* "Reversal of Experimental Encephalomyelitis with Monoclonal Antibody to a T-Cell Subset Marker," *Science,* Jan 25, 1985: 415–7.

Waldron, John (Pres., Plastic Applications, Boca Raton, FL), L-1/10/83.

Walker, Alice, "The Strangest Dinner Party I Ever Went To," *Ms.,* July/Aug. 1982: 58–9.

Walker, Barbara, "Motherhood and Power," *Womanspirit* 8: 31 (Spring 1982): 52–5.

Walker, Barbara, ed., *The Woman's Encyclopedia of Myths and Secrets.* San Francisco: Harper & Row, 1983.

Walker, Benjamin, ed., *The Hindu World,* 2 vols. New York: Praeger, 1968.

Walker, D.E. (Science Reference Library, British Library, London), L-1981-3.

Walker, Geraldine M. (Hon. Sec'y, Riding for the Disabled, Medstead, Hants., Eng.), L-11/5/81.

Wallas, Ada (Mrs. Graham), *Before the Bluestockings.* London: Allen & Unwin, 1929.

Wallechinsky David, and Irving Wallace, *The People's Almanac.* Garden City, NY: Doubleday, 1975.

Walsh, George E., "Women Inventors," *The Patent Record,* Mar. 1899: 9.

Walsh, Myles E., *Understanding Computers.* New York: Wiley, 1982.

Wamsley, James S., with Anne M. Cooper, *Idols, Victims, Pioneers; Virginia's Women from 1607.* Richmond: Virginia State Chamber of Commerce, 1976.

Ward, Earlene, "Sierra On-Line Has Staff Dealing with Technical Computer Software," *Sierra Star* (Oakhurst, CA), Aug. 3, 1983.

Ward, Maria E., *The Common Sense of Bicycling: Bicycling for Ladies.* New York: Brentano's, 1896.

Ward, T.L., and Ruth R. Benerito, "Coatings Formed at the Interface of Glass and Plumbite-Treated Cotton," *Thin Solid Films* 53 (1978): 73–9.

Wärn, Eva, and Lennart Nilsson (Workshop Dirs., Swedish Inventors' Association, Stockholm, Sweden), L-11/6/85 to Halldis Thune.

Warner, Deborah J., "Fashion, Emancipation, Reform, and the Rational Undergarment," *Dress* 4 (1978): 24–9.

Warner, Deborah, J. "The Women's Pavilion," In Post.

Warner, Deborah, J., "Women Inventors at the Centennial," in Martha Trescott, ed.

Warner, Hal, "First Woman Patent Holder Resident of Early Killingly," Hartford (CT) *Times* May 5, 1962: 3.

Warrior, Betsy, "Necessity Is the Mother of Invention," in Betsy Warrior and Lisa Leghorn, eds, *Houseworker's Handbook,* 3rd ed. Cambridge, MA: Women's Center, 1975.

Wartik, Nancy, "Le Funelle: The Equalizer," *Ms.,* Dec. 1986: 25.

Washburn, Robert C., *The Life and Times of Lydia Pinkham.* New York: Putnam's, 1931.

Washington, Sharon (Animation painter, inventor, Pacoima, CA), PC-Fall 1981.

Wasson, Albert H., *The Road to Eleusis: The Unveiling of the Mysteries.* New York: Harcourt, 1978.

Waterhouse, J.A.H., *Cancer Handbook of Epidemiology and Prognosis.* London: Churchill Livingstone, 1974.

Waterson, A.P., and Lise Wilkinson, *Introduction to the History of Virology.* Cambridge: Cambridge University Press, 1978.

Watson, James B., "From Hunting to Horticulture in the New Guinea Highlands." *Ethnology* 4: 3 (July 1965): 295–309.

Way, Doreen Kennedy-. See Kennedy-Way, Doreen.

Weibal, Robert E., *et al.,* "Live Attenuated Varicella Virus Vaccine: Efficacy Trial in Healthy Children," *New England Journal of Medicine* 310: 22 (May 31, 1984): 1409–15.

Weimann, Jeanne M. (Independent scholar, author; Monroe, CT), PC-1980–present.

Weimann, Jeanne M., *The Fair Women: The Story of the Woman's Building, World's Columbian Exposition, Chicago, 1893.* Chicago: Academy Chicago, 1981.

Weiner, Annette B., *Women of Value, Men of Renown: New Perspectives in Trobriand Exchange.* Austin: University of Texas Press, 1976.

Weiner, Michael A., *Earth Medicine—Earth Foods: Plant Remedies, Drugs, and Natural Foods of the North American Indians.* New York: Macmillan, 1972.

Weintraub, Pamela, "Live from Dartmouth" (Invention feature), *Discover,* Jan. 1981: 78f.

Weintraub, Pamela, "Computerized Contraception" (Invention feature), *Discover,* Apr. 1981: 58.

Weisberg, Maggie (Inventors' Workshop International; IWI Education Foundation, Tarzana, CA), L-1976–present.

"We're Keeping the Feathers Dry," Minnesota Mining & Manufacturing Co. ad, *Smithsonian,* Feb. 1983: 39.

"We're Protecting Your Food with All Our Mites," ad, UC–Berkeley Foundation, *California Monthly,* Mar./Apr. 1985 (inside front cover).

Wernick, Robert, "An Unknown Lady of Vix and Her Buried Treasures," *Smithsonian* 16: 12 (Mar. 1986): 140–60.

Wertheimer, Barbara, *We Were There: The Story of Working Women in America.* New York: Pantheon, 1977.

Wertime, Theodore, "Tradition and Progress," in Doster *et al.*

Westendorf, Wolfhart, *Painting, Sculpture, and Architecture of Ancient Egypt.* New York: Abrams, 1968.

Westin, Jeane, *Making Do: How Women Survived the '30's.* Chicago: Follett, 1976.

Westinghouse Electric Company, Pittsburgh, "Benjamin G. Lammé"; "Russell S. Feicht," sections from a typescript, n.p., n.d. (Sent me by Guenter Holzer, 1983).

Wexelblat, Richard, ed., *History of Programming Languages.* New York: Academic Press, 1981.

Wheat, Valerie, "The Red Rains: A Period Piece." *Chrysalis* 1: (1977): 91–3.

"Wheelchair Speaks for Those Who Can't," Stanford University *Campus Report,* Mar. 19, 1980: 10.

White, A[nna] J., "The Mystery Explained," *Shaker Almanac for 1886.* New York: A.J. White, 6ff.

White, Anna J., and Leila S. Taylor, *Shakerism: Its Meaning and Message.* Columbus, OH: Fred J. Heer, 1904.

White, George S., *A Memoir of Samuel Slater, Connected with a History of the Rise and Progress of the Cotton Manufacture in England and America.* Philadelphia, 1836.

White, Lynn, Jr., *Dynamo and Virgin Reconsidered: Essays in the Dynamism of Western Culture.* Cambridge: M.I.T. Press, 1968 (1971).

White, Lynn, Jr., *Medieval Religion and Technology*. Berkeley: University of California Press, 1978.

White, Lynn, Jr., *Medieval Technology and Social Change*. Oxford: Oxford University Press, 1962.

Whitehead, Harriet, "The Bow and the Burden Strap: A New Look at Institutionalized Homosexuality in Native North America," in Ortner & Whitehead, eds.

Whitman, Mary (Granddaughter of inventor, informant, Springfield, VA), PC-1979–89.

Whitman, Trudy, "On the Polio Trail," *Independent Woman* 27, Jan. 1948: 11–2.

Whitworth, Brad, "Close Up," *Measure* (Hewlett-Packard Corp. magazine, Palo Alto, CA), Sept.-Oct., 1980: 17.

*Who's Who in America*. Chicago: Marquis Who's Who, var. eds.

*Who's Who in the Computer Field*. Newtonville, MA: Berkeley Enterprises, 1954–64.

*Who's Who in Computers and Data Processing 1971*, 5th ed. New York: Quadrangle, 1971.

*Who's Who in Government*, Chicago: Marquis Who's Who, 1971-2–.

*Who's Who in the West*, 17th ed. Chicago: Marquis, 1980–1.

*Who's Who of American Women*. Chicago: Marquis, 1962 – .

*Who Was Who Among North American Authors, 1921–1939*, Detroit: Gale Research, 1976.

*Who Was Who in America*, Chicago: Marquis, 1942.

"Whose Cells Are They?" *Science/85, Feb. 1985: 6.*

Wiener, Pat (Inventor, scientist, entrepreneur; Page Automated Telecommunications Systems, La Honda and Palo Alto, CA), I-1983; PC-1983–present.

Wiener, Pat, "New Technologies for Low-Cost Automated Telecommunications-Based Office Systems Products: LSI↔VLSI↔LAI," *Proceedings* of the IEEE Spring Computer Conference (Comp/Con). Silver Spring, MD: IEEE, 1980, 16ff.

Wik, Reynold, *Henry Ford and Grass-roots America*. Ann Arbor: University of Michigan Press, 1972.

Willard, Frances E., and Mary A. Livermore, eds, *A Woman of the Century: 1470 Biographical Sketches Accompanied by Portraits of Leading American Women in All Walks of Life*. Buffalo: Moulton, 1893; repr., Gale Research, Detroit, 1967.

Willard, Mary J. (MD, innovator; Helping Hands: Simian Aides for the Disabled, New Rochelle, NY), L-7/1/86.

Willard, Mary J., Judi Zazula, *et al.*, "Training a Capuchin *(Cebus apella)* to Perform as an Aide for a Quadriplegic," *Primates* 23: 4 (Oct. 1982): 520–32.

Willard, Mary J., *et al.*, "The Psycho-Social Impact of Simian Aides on Quadriplegics," *Einstein Quarterly Journal of Biological Medicine* 3 (1985): 104–8.

Williams, Ernest E., "Industrial Fort Wayne, No. 17: The Toidey Company, Inc.," Fort Wayne, IN, *News-Sentinel*.

Williams, Greer, *Virus Hunters*. New York: Knopf, 1959.

Williams, Roberta (Inventor, entrepreneur: Sierra On-Line, Oakhurst, CA), L-3/13/84; Q-2/84.

Williams, Selma R., *Demeter's Daughters: The Women Who Founded America, 1587–1787*. New York: Atheneum, 1976.

Williams, Sharlotte Neely, "The Argument Against the Physiological Determination of Female Roles: A Reply to Pierre L. van den Berghe's Rejoinder to Williams' Extension of Brown's Article," *American Anthropologist* 75: 5 (Oct. 1973): 1725–8.

Williams, Sharlotte Neely, "The Limitations of the Male/Female Activity Distinction Among Primates: An Extension of Judith K. Brown's 'A Note on the Division of Labor by Sex,' " *American Anthropologist* 73: 3 (June 1971): 805–6.

Willis, Mary-Sherman, "Cervical Caps: Old and Yet Too New," *Science News* 116: 25, 26 (Dec. 22, 29, 1979): 431f.

Wilson, Joan Hoff, "Dancing Dogs of the Colonial Period," *Early American Literature* 7: 3 (Winter 1973): 225ff.

Wilson, Lois B. (Inventor; Dept. of Cell and Molecular Biology, Medical College of Georgia, Augusta), c.v./1981; L-9/8/81.

"Wins Invention Medal," *New York Times*, Oct. 11, 1934.

Wiser, Vivian (Historian, Agricultural History Branch, USDA, Washington, DC), PC-10/25/79; L-1/21/80, 2/8/85.

Wiser, Vivian, "Women in American Agriculture," paper presented to the Economic Research Service, USDA, Freer Gallery Auditorium, Washington, DC, Nov. 18, 1975 (sent me by the author).

Wiser, Vivian, "Women in the U.S.D.A.," *Agricultural History* 50: 1 (Jan. 1976).

Wiswell, Phil, "Roberta Williams, Time-Traveling Adventuress," *Electronic Fun with Computers and Games,* Mar. 1984: 30–3.

Wojahn, Ellen, "Why There Aren't More Women in This Magazine," *Inc.,* July 1986: 45.

"Woman Discovers Better Steel Process," *Popular Science Monthly,* Feb. 1928: 51.

"Woman Engineer Honored," *New York Times,* June 8, 1958: 79.

"Woman Fights for Invention," Sacramento *Bee,* Mar. 21, 1980: B2.

"Woman Has New Aeroplane," *New York Times,* July 31, 1908: 12: 1.

*The Woman Inventor,* ed. and pub. by Charlotte Smith, Washington, DC, Apr. and June 1891.

"Woman Inventor Found Dead in Blazing Flat," *The Times* (London), May 5, 1959.

"Woman Perfects Torpedo," *New York Times,* Dec. 17, 1899: 8: 1.

"A Woman's Remarkable Invention," *The Patent Record,* Sept. 1900: 7.

*Woman's Who's Who of America.* New York: American Commonwealth Co., 1914–5.

"Women as Inventors," *The Christian Science Monitor,* June 25, 1935.

"Women as Inventors," *Literary Digest,* Jan. 6, 1900: 14–5.

"Women as Inventors," *The Patent Record,* May 1899: 5.

"Women as Inventors," *The Times* (London), Oct. 3, 1935: 9.

"Women as Inventors," *The Woman's Journal* Mar, 18, 1899: 88 (Report of a talk by Ada C. Bowles).

"Women Gaining in Patent Rolls," *New York Times,* Aug. 11, 1940: D3.

"Women in Electricity," *New York Sun,* Nov. 26, 1899, n.p.

"Women in Electricity," *Woman's Journal* (Boston, MA), Dec. 2, 1899.

"Women Inventors," Providence, RI, *Journal,* June 16, 1937.

"Women Inventors," *Scientific American,* Aug. 19, 1899: 123 (Report on Ada C. Bowles lecture).

"Women Inventors," *Scientific American,* Apr. 4, 1903: 247.

"Women Inventors Profile: Dr. Mary Anne White," Women Inventors Project *Focus* 2: 2 (June 1988): 2–3.

Women Inventors Project (Waterloo, Ont.), *Focus,* 1987– .

*Women of Achievement.* New York: House of Field, 1940.

"Women to Show Inventions," *New York Sun,* Sun., Oct. 25, 1908: 6.

"Women Who Are Inventors," *New York Times,* Oct. 19, 1913, cited in Janeway, 55–6.

Women's Idea Conference: The Positive Approach. New Mexico Institute of Mining and Technology, Socorro, NM, 1975 (Report of the conference).

Wonder Woman Foundation (New York), 1983 awards announcement, n.d. [Dec. 1983].

Wood, Clive, *Intrauterine Devices.* London: Butterworths, 1971.

Wood, Clive, and Beryl Sutters, *The Fight for Acceptance: A History of Contraception.* Aylesbury, Eng.: Medical and Technical Publishing, 1970.

Woodcroft, Bennet, *Alphabetical Index of Patentees, 1617–1852.* London: Evelyn, Adams & MacKay, 1969.

Woodward, Helen Beal, *The Bold Women.* New York: Farrar, Straus, 1953.

Wooldridge, Clifton R., *Hands Up! In the World of Crime, or Twelve Years a Detective.* Chicago: Thomas & Thomas, 1901.

"The Work of One Woman," *Southern California Business,* Nov. 1926: 26f.

*World Who's Who in Science: A Biographical Dictionary of Notable Scientists from Antiquity to the Present.* Chicago, Marquis, 1968.

Wright, Barbara D., ed., *Women, Work and Technology: Transformations.* Ann Arbor: University of Michigan Press, 1987.

Wright, Charles H. (Mississippi Soc. of Scientists and Inventors, Sandhill, MS), L-10/22/80.

Wright, David (Science, Technology, and Society Program, Michigan State University, E. Lansing, MI), PC-3/15/88.

Wright, Jessie, MD, "The Respir-Aid Rocking Bed in Poliomyelitis," *American Journal of Nursing* 47 (July 1947): 454–5.

Wu Huai-Yin (Dir., Dept. of Physical Chemistry, First Medical College, Shanghai), L-4/5/82.

"Wu Receives Wolf Prize in Physics," *Physics Today,* May 1978: 81.

Yalow, Rosalyn, Speech delivered at the American Physical Society meetings in New York City, January 1979.

Yancopoulos, G.D., and F.W. Alt, "Reconstruction of an Immune System," *Science* 241: Sept. 23, 1988: 1581–4.

Yarbrough, Mrs. J.A., "Carolina Girl in New York Wins Title of Lady Edison," undated newsp. clipping, Beulah Henry file, Carolina Room, Charlotte, NC, Public Library, c. 1930.

Yarmish, Rina J., and Louise S. Grinstein, "Brief Notes on Six Women in Computer Development," *Journal of Computers in Mathematics and Science Teaching,* Winter 1982: 38–9 (Part I), and Spring 1983: 26–7 (Part II).

Yeager, Anne, Obit., San Francisco *Examiner,* Nov. 4, 1984: B9.

"Yeast Engineered to Produce Hepatitis B Coat," *Science News* 120: 6(Aug 8, 1981): 84.

"Yeasts Receive Nitrogen Fixation Genes," *Science News,* Feb. 21, 1981: 119.

Yonge, Charlotte M., *History of Christian Names,* rev. ed. London: Macmillan, 1884.

Yost, Edna, *American Women of Science.* Philadelphia: Lippincott, 1943.

Yost, Edna (with Lillian Gilbreth), *Normal Lives for the Disabled.* New York: Macmillan, 1944.

Yost, Edna, *Women of Modern Science.* New York: Dodd, 1959.

"You Have Ruined My Volaille," ad in *Motorland,* Apr. 1974: 21.

"Young Ladies Locating Lands" (from The Milwaukee *Wisconsin), New York Times,* Sept. 8, 1882: 5: 2.

Yulsman, Tom, "The One Hundred: Who They Are and What They Think," *Science Digest,* Dec. 1984: 78–9ff.

Zaffaroni, Alejandro, "Delivering Drugs," *Chemtech,* Feb. 1980: 82–8.

Zahm, John Augustine, pseud., see Mozans, H.J.

Zehring, Robin, "New Ways to Step Out," *California Living* magazine, San Francisco *Chronicle & Examiner,* May 9, 1976: 30ff.

Zevgolis, Victoria H. (Inventor, entrepreneur), c.v./1982; L-4/23, 5/6/82; Sportsonic prospectus, 1981.

Zientara, Marguerite, "Five Powerful Women," *Info World,* May 21, 1984: 57–61.

Zihlman, Adrienne, "Women and Evolution, Part II: Subsistence and Social Organization Among Early Hominids," *Signs* 4: 1 (Autumn 1978): 4–20f. (For Part I, see Nancy Tanner.)

Zihlman, Adrienne L., "Women as Shapers of the Human Adaptation," in Dahlberg, ed.

Zimmerman, Jan (Consultant, author; San Diego, CA), L-1982–7.

Zimmerman, Jan, ed., *Future, Technology, and Women,* Proceedings of the Conference held at San Diego State University, Mar. 6–8, 1981. San Diego, CA: Women's Studies Department, SDSU, 1981.

Zimmerman, Jan, ed., *The Technological Woman: Interfacing with Tomorrow.* New York: Praeger, 1983.

Zimmerman, Jan, "Women in Computing: Meeting the Challenges in an Automated Industry (II)," *Interface Age* 8: 12 (Dec. 1983): 79ff.

Zobel, Louise Purwin, and Jerome Fremont, MD, "Medical Miracles" *Peninsula Magazine,* June 1986: 46–70.

Zvelebil, Marek, "Postglacial Foraging in the Forests of Europe," *Scientific American* 254: 5 (May 1986): 104–15.

Zyda, Joan, "Smokeless Ashtray Is Hot New Invention," Chicago *Tribune,* Aug. 1, 1978, sect. 2.

# Index

Names and subjects are combined in a single index, with the names of women inventors in bold type for instant identification.

menstrual technology
(*continued*)
sponges, 218
takeover, 213
tampons, 212f, 215, 218–20
undergarments, 218
mental health, culture and,
200
mental retardation, 189
prevention, 185, 188f
mentors (for women inventors),
455
brother as, 350–1
colleagues as, 491
husband as, 477
lack of, 393, 451, 501
parents as, 34
teachers as, 34, 456, 506
**Mercer, Sophia L.,** 315
**Mercier, Ariadna B.,** 594
mercury bichloride contraceptive, 259
Merino sheep, 43
**Merry, Diana,** 469
**Meserve, Kathleen,** 38
Mesolithic, xxiv
**Messalina,** 97
metabolism
defect-diagnosis, 185, 189,
203
iodine, 186
monitoring, 186, 563, 565
protein, 189
metal goods, manufacturing,
284–5, 326–7, 342, 360, 378,
401–2, 407
metallurgy, 147, 284–5, 360,
378, 407, 413, 599
computer aids, 491
metal-punching machine, 326,
342
metal-smelting, xxxii, 58, 432
metal-working machines, 326–
7
metate and muller, 8, 284, 287
**Metcalf, Bets(e)y,** 320–1. *See
also* **Baker, Betsy**
**Metz, Patricia,** 464
**Mey, Amelia,** 66
**Meyenberg, Julia,** 305
**Meyer, Lucy R.,** 273, 560
MIBO inventions
Mini-, trolley, 182
Postbox, 182–3
Shower Cabin, 182
Transport, 182
**Michejda, Maria,** 153
microelectric circuits, 386
microfilm technology, 451
microwave oven, 57, 63, 364
microwave/related inventions,
389, 393
**Middleton, Alice,** 333, 602

midwives, 87, 89n5, 90, 101,
194ff, 210, 226ff, 232ff,
239n, 255, 257, 274n71, 282,
577, 581
ancient Hebrew, 92
and contraception, 254–5,
273, 274n71
instruments, 92–3, 239n, 255
inventions by, 273, 581
male, 233, 283
Norwich, England, 101n13
Roman, 239n, 577
and technology transmission,
235, 581, 584
training, 104, 108, 233ff
U.S. ban on, 276n
migration, as population control, 243n33
**Mikel, Jane,** 550
**Mikhalina, Alexandra M.,**
406
**Mikhlina, Eva E.,** 141
Milbanke, Lady Annabella,
433ff
mildew control, 40f
**Miles, Rebecca E.,** 547
military inventions. *See* warfare
milk
aerator, 45
bottles, paper, 408–9
condensed, 330n35
coolers, 45f, 47–8, 67f
pasteurized, 330n35
separator, 45, 370
sterilizing apparatus, 107,
330n
yields, 48
milking, gender and, 536
milking machine, 45f, 370
women's invention, 46
milking techniques, 48
milking-trainer, 48
**Millar, Pamela,** 48
**Miller, Addie D.,** 593
**Miller, Anna L.,** 37
**Miller, Annie J.,** 587
**Miller, Elizabeth,** 169
**Miller, Emily V. D.,** 553
**Miller, Emma,** 551
**Miller, Florence S.,** 593
Miller, G. Tyler, 84n62
**Miller, Lisa,** 136
**Miller, Maria E. B.,** 325
**Miller, Rosalie,** 552
millet, 17
millinery inventions, 327, 431
milling balls (flint), 8
**Mills, Emma,** xxix, 81, 303,
366, 553
mills, ochre, 8
mills, rolling, 407
**Milson, Hannah,** 107–8, 310
miners, women as first, 285

**Minerva,** 2, 13, 91, 285, 287,
291, 305
/Medica, 89
miniaturization (electronics),
387, 447–8, 449
mining, 285, 431f
mining machine, 340
**Minnick, Maggie E.,** 301
Minoan women, 223
**Miracle, Virginia S.,** 328
miscarriage (preventive), 211.
224–5
**Mishel, Madeline.** *See*
**Hauptman, Madeline
Mishel**
**Missett, Judi S.,** 174
missile/related inventions, 382,
387, 597
mistletoe, 89, 262
*misy,* 261, 264
**Mitchell, Arobine C.,** 550
**Mitchell, Betty J.,** 472, 496
**Mitchell, Elma M.,** 81, 553
**Mitchell, Grace,** 463
**Mitchell, Dr. Mildred,** 183,
494
**Mitchum, Bettie,** 547
**Miti-Miti,** 285
mixers. *See under* food-process-
ing technology
mobile X-ray units, 108, 201,
573
mobility
female, xxxv, 27–8, 534–5
male, 27
**Mobley, Carolyn R.,** 217
Moche people (Peru), xxi, 91
Modoc women, 226–7
**Moe, Bertha,** 364
**Moehrle, Doris,** 214, 218, 277–
8, 283
**Mofort, Marie,** 70
MOGO-1, 182
*mokh,* 251f
molasses-related process, 54
molds, 122. *See also* fungi;
yeast
as early antibiotics, 122–3
"Moldy Mary," 124n
**Molini, Felice,** 550
Molnos, Angela, 241
monarchy, xxvii
**Moncrief, Araliza,** 550
monitors, medical, 113, 195f
monkeys
as aides to disabled, xx,
xxiv
inventions of, 4–5
snow, 4–5, 87n, 244
**Monkhouse, Helen G.,** 568
monoclonal antibodies, 136,
138, 150, 204
chimaeric, 138, 204

product, 216ff, 276, 364, 378, 423, 428
transport, 329f, 379, 382, 418f, 599f
workplace, 356–7, 412, 414
safety pin, 316, 330, 429
sagae, 581. *See also* abortion-ists; midwives
sage, 222f, 231, 262
sago as food/medicine, 88
sago-processing, xxi, xxxii, 18, 20–1
as first engineering, xxi, xxxii, 20
technological change in, xxiii, 7f
women's invention, 21
St. Denis, Ruth, 573
**St. John, Mary,** 568
salad-mixer, 52
**Salamon, Karlene W.,** 141
**Salina,** 41–2
salinization, 14
**Sallade, Mary,** 312, 556f
salves, 88, 100, 115, 126, 231
**Sammet, Jean E.,** 462, 463–4, 467, 469–7, 495
**Sammuramat,** 15, 288–90. *See* **Semiramis**
Samothea, 96
Sanchez, Denise, xiii, 59n43
Sanday, Peggy, xxiv
**Sanderson, Martha H.,** 301
sandwich, new, 71
Sanger, Margaret, 265, 274–5ff, 283, 349
Sanio-Hiowe, 21n15
"sanitary devices," 216
sanitary equipment, 206f
sanitary napkin, 213–18, 330n
absorbency, 216f
anti-distortion, 217
belt, adhesive, 216, 218
beltless, 216, 218
casing, 216
comfort, 217
disposable, 215, 217
disposal means, 217
light-days, 216
maxi-, 216ff
mini-, 216
**Santoni, Patricia,** 466–7
Sarah, 284n44
**Sara(s)vati,** 285
satellites, 391–2, 419–20. See also *satellites by name*
data-processing, 392
design, 392, 493
guidance, 484, 493, 598
infrared (astronomy), 494
propulsion, 419, 493
testing, 493

tracking, 391f, 484, 493, 597
**Sato-Viacrucis, Kiyo,** 175, 192–3
Satrup ard, 13
**Saul, Sarah,** 45, 547
**Saunders, Dr. Cicely,** 196
Saw palmetto, 231, 581
saws, 286, 318f
band-, 319, 341
circular, 286, 341
-sharpener, 395–6
scaffolds, window-washing, 315
scaldhead, 101
scales, 416
scalpels (midwives'), 92
**Scanlon, Kate,** 568
scarecrows, 556
scarlatina cure, 115
scarlet fever, xxxiii, 129, 156–7, 187–8
mortality, 157
**Schaefer, Amalie,** 307, 315
**Schaefer, Louise,** 324
**Schaefer, Marie,** 37
Schiebinger, Londa, xii
**Schiffman, Dr. Susan S.,** 173
schistosomiasis, 162, 204
Schmidt, Hubert, 78
**Schmidt, Dr. Ingeborg,** 193
**Schneider, Anna E. E.,** 55, 72, 404
**Schneider, Mary,** 301
**Schoenberger, Eva,** 549
**Schoental, Miss,** 126
schooling. *See* education and in-vention
**Schott, Mathilde,** 144
**Schrama-de Pauw, Agnes,** 391
**Schramm, Isabella C.,** 551
**Schroeder, Becky,** v, ix
**Schultze, Augusta,** 179
**Schultze, Maria,** 319
Schwieder, Dorothy, 50
sciatica remedy, 97f
SCID-hu mouse, 165–6
scissors/shears, 80, 82, 315f, 371, 395
pinking, 316
scolding fridge, 173–4
scopolamine (TTS), 142
Scotchgard®, xii
**Scott, Eliza E.,** 560
**Scott, Mary A.,** 551
screening tests (medical), 153, 185–7
screw
-driver, 316
in eyeless needle, 347
gimlet-pointed, 314n
propeller, 338
woman's invention, 285–6

sculpture method, new, 181, 343n, 352
scythe manufacture, 82
**Seago, Lydia A.,** 326
**Seaman, Barbara,** 275–6n, 279
Sears
catalogue, 272
women and vacuum clean-ers, 397
seaweed, uses, 16, 35, 227, 251, 258, 588–9. *See also* algae; *gums by name; Laminaria; other species by name*
secondhand-smoke preventive, 167
secret societies, women's, 88
security system, home, 390
sedentism, xixf, 513
and fertility/population, 242–3, 282, 533–6
as watershed for women, xix, 29, 243, 533–6
seed-grading machine, 371, 376
seedling beds, 17
seeds
gathering, 7
processing, 8
treatment (pre-planting), 16, 33–4, 40f
**Seibert, Dr. Florence,** 561–2
**Seidler, Neola,** 216
**Seifert, Katharina,** 46, 547
**Seigel, Mother,** 87, 116–17
Seigel's Syrup, 116ff
analysis, 117
**Seitz, Mary A.,** 327
seizure-prevention, 446
self-cleaning fish tank, 391
Self-Cleaning House (SCH), 63, 363, 366, 374–5, 399
self-help movement, 589
contraception, 276, 283
medicine, 200–1, 221, 589
self-image (of women inven-tors). *See under* inventive achievement
self-pleasure technology, 579–80
self-waiting table, 317–18
**Selligue, Félicité R.,** 338
semaphore, 378
**Semiramis,** 15, 30, 253, 284, 288–90
**Semple, A. C.,** 326
Seneca shaman-healers, 93
**Senter, Lucinda,** 594
Serjdetour, 385
serum treatments, 128f
set-breaking thinking, xix
**Settle, Dorothy,** 567
**Sewell, Sarah,** 302, 594
sewing aids, 315–16, 324

# About the Author

AUTUMN STANLEY was originally a literature major, taking graduate (Stanford University) and undergraduate (Transylvania College) degrees and teaching briefly at the college level in this field. However, science was another strong interest, and she minored informally in biology. While working as a developmental editor for a college textbook publisher, she undertook to recognize more women's achievements in science texts. As this program expanded to embrace other fields she was struck by the total lack of information on women as inventors, and decided to remedy it. The result, after more than a dozen years of research and writing, is this book.

Along the way, she refocused her career on gender roles in society, presenting papers on her work at national and international conferences, and publishing articles in professional journals and in essay collections on women and technology and other areas of gender studies. The most recent appears in Dale Spender and Cheris Kramarae, eds., *Knowledge Explosion: Generations of Feminist Scholarship,* Teachers College Press, Columbia University, 1992.

From 1984 to 1988 she was an Affiliated Scholar at the Institute for Research on Women and Gender at Stanford. Presently she is an independent scholar, with membership in the Institute for Historical Study (Berkeley, CA). Current projects include a biography of economic feminist and reformer Charlotte Smith (1840–1917).

ADIRONDACK COMMUNITY COLLEGE
LIBRARY
BAY ROAD